The UK Pesticide Guide 2011

Editor: M. A. Lainsbury, BSc(Hons)

BCPC

www.cabi.org

© 2011 CAB International and BCPC (British Crop Production Council). All rights reserved. No part of this publication may be reproduced in any form or by any means, electronically, mechanically, by photocopying, recording or otherwise, without the prior permission of the copyright owners.

CABI is one of the world's foremost publishers of databases, books and compendia in agriculture and applied life sciences. It has a worldwide reputation for producing high quality, value-added information, drawing on its links with the scientific community. CABI produces CAB Abstracts – the leading agricultural A&I database; together with the Crop Protection Compendium which provides solutions for identifying, preventing and solving crop health problems. CABI is a not-for-profit organization. For further information, please contact CABI, Nosworthy Way, Wallingford, Oxon OX10 8DE, UK.

Telephone: (01491) 832111
Fax: (01491) 833508
e-mail: enquiries@cabi.org
Web: www.cabi.org

BCPC (British Crop Production Council) promotes the use of good science and technology in understanding and applying effective and sustainable crop production. Its key objectives are to identify developing issues in the science and practice of crop protection and production and provide informed, independent analysis; to publish definitive information for growers, advisors and other stakeholders; to organise conferences and symposia; and to stimulate interest and learning. BCPC is a Registered Charity and a Company limited by Guarantee. Further details available from BCPC, 7 Omni Business Centre, Omega Park, Alton, Hampshire GU34 2QD, UK.

Telephone: (01420) 593200
Fax: (01420) 593209
e-mail: md@bcpc.org
Web: www.bcpc.org

ISBN 978-1-84593-830-7

Contents

Disclaimer	v
Editor's Note	vi
Changes Since 2010 Edition	vii
Introduction	xii
Notes on Contents and Layout	xiv

SECTION 1 CROP/PEST GUIDE	**1**
Crop/Pest Guide Index	3
Crop/Pest Guide	5
SECTION 2 PESTICIDE PROFILES	**83**
SECTION 3 PRODUCTS ALSO REGISTERED	**591**
SECTION 4 ADJUVANTS	**633**
SECTION 5 USEFUL INFORMATION	**669**
Pesticide Legislation	671
Regulation 1107/2009 - The Replacement for EU 91/414	671
The Food and Environment Protection Act 1985 (FEPA) and Control of Pesticides Regulations 1986 (COPR)	671
Plant Protection Products Regulation	672
Dangerous Preparations Directive	672
The Review Programme	672
Control of Substances Hazardous to Health Regulations 1988 (COSHH)	673
Certificates of Competence – the roles of BASIS and the NPTC	673
Maximum Residue Levels	674
Approval (On-label and Off-label)	676
Statutory Conditions of Use	676
Types of Approval	676
Withdrawal of Approval	676
Off-label Extension of Use	677
Using Crop Protection Chemicals	678
Use of Herbicides In or Near Water	678
Use of Pesticides in Forestry	679
Pesticides Used as Seed Treatments	681
Aerial Application of Pesticides	686
Resistance Management	690
Preparation in advance	690
Using crop protection products	690
International Action Committees	691
Poisons and Poisoning	692
Chemicals Subject to the Poison Law	692
Part I Poisons	692
Part II Poisons	692
Occupational Exposure Limits	692
First Aid Measures	693
General measures	693
Reporting of pesticide poisoning	693
Additional information	693

Environmental Protection 695
   Protection of Bees 695
      Honey bees 695
      Wild bees 695
   The Campaign Against Illegal Poisoning of Wildlife 695
   Water Quality 696
      Where to get information 696
      Protecting surface waters 696
      Local Environmental Risk Assessments for Pesticides 697
      Groundwater regulations 698
      Integrated Farm Management (IFM) 698

SECTION 6 APPENDICES   701

Appendix 1 Suppliers of Pesticides and Adjuvants   703
Appendix 2 Useful Contacts   711
Appendix 3 Keys to Crop and Weed Growth Stages   714
Appendix 4 Key to Hazard Classifications and Safety Precautions   722
Appendix 5 Key to Abbreviations and Acronyms   729
Appendix 6 Definitions   732
Appendix 7 References   734

INDEX OF PROPRIETARY NAMES OF PRODUCTS   737

Disclaimer

Every effort has been made to ensure that the information in this book is correct at the time of going to press, but the Editor and the publishers do not accept liability for any error or omission in the content, or for any loss, damage or other accident arising from the use of the products listed herein. Omission of a product does not necessarily mean that it is not approved or that it is unavailable.

It is essential to follow the instructions on the approved label when handling, storing or using any crop protection product. Approved 'off-label' extensions of use are undertaken entirely at the risk of the user.

Editor's Note

This is the 24th edition of this annual publication and, with the impending changes to the approval process, it becomes increasingly important to know the approval status and expiry date of plant protection products. The turmoil caused by the discussions on what should replace EU directive 91/414 is largely over – its replacement, 1107/2009, came into force on 14 December 2009 and will be applied to new applications for approval from 14 June 2011. On that date, all products holding a current approval under 91/414 will be deemed to be authorised under 1107/2009 but, when reviewed at the due date, will be required to meet the new criteria (see section 5). Now that the workload from this exercise is lessening, the Chemicals Regulation Directorate (CRD) has been able to grant some new approvals. All approvals granted under COPR were due to expire by 31/12/2013 but, to allow more time for active ingredients to be listed on Annex 1, this 'back-stop' date has now been extended by CRD to 31/12/2021. This edition of *The UK Pesticide Guide* contains a significant number of new mixtures, and two significant new active ingredients listed for the first time are bixafen and isopyrazam – both cereal fungicides heralding a new class of actives.

While the framework of 1107/2009 is now in place, some of its terms, such as 'endocrine disruptor', have yet to be precisely defined. Where the thresholds are set will have a big influence on which actives are lost and which are retained but, under current thinking, it is likely that the days of vinclozolin approvals are numbered. The new Water Directive is also likely to have an impact on pesticide approvals when it is introduced in 2012, but current data show encouraging signs of improved water quality as pesticide users take greater care to keep pesticides out of water. As approvals are lost, for whatever reason, this publication provides a very useful guide to what alternatives are available from the pesticide armoury – see section 1.

This book is an annual publication and provides a snapshot of what is approved at the time of publication. If you need more up-to-date information, the content is also available as a web-based subscription database, which not only is updated more frequently, but also contains much more information about each product than we can possibly squeeze into the book. With the increase in mobile internet access, this is likely to become an increasingly popular alternative way of accessing this information. For the first time in 2011, it also includes access to a pest and disease identification program (*IdentiPest*). This is part of our programme of continuous improvement, and will help to ensure the right products are selected for the right targets.

Each active ingredient entry carries a classification on its mode of action, and this is important for managing the risk of resistance. By ensuring that pests and diseases are subjected to a range of different modes of action, both within crops and across years, the risk of reduced pesticide activity is lessened. For people who transport these products but are not especially concerned about their use, addition of information on the UN Numbers, Transport Code and Packaging Group has been completed after its introduction last year. This information can also be of value to the emergency services when faced with a pesticide emergency.

Many of the new products listed in this edition are parallel or generic approvals based on existing products, but please bear in mind that the sources of these products do not generaly contribute to the development of new active ingredients. As the thresholds for approval become increasingly difficult to achieve, the cost of gaining new approvals is increasing. Parallel or generic products can offer cost savings over the originals, but it is important to strike a balance between maintaining a fair price and continuing to fund further research. We express our gratitude to all those people who have provided the information for the compilation of this guide. As always, criticisms and suggestions for improvement are always welcome and, in particular, if you notice any errors or omissions in the entries, please let me know via the email address below.

M. Lainsbury
Editor
ukpg@bcpc.org

The information in this publication has been selected from official sources, from the suppliers' product labels and from the product manuals of the pesticides approved for use in the UK under the Control of Pesticides Regulations 1986 (COPR) or the Plant Protection Products Regulations 1995 (PPPR). The content is based on information received by the Editor up to October 2010.

Changes Since 2010 Edition

Pesticides and adjuvants added or deleted since the 2010 edition are listed below. Every effort has been made to ensure the accuracy of this list, but late notifications to the Editor are not included.

Products Added

Products new to this edition are shown here. In addition, products that were listed in the previous edition whose PSD/HSE registration number and/or supplier have changed are included.

Product	Reg. No.	Supplier	Product	Reg. No.	Supplier
Acanto Prima	14971	DuPont	Clayton Oust SC	14871	Clayton
Acclaim	14905	European Ag	Clayton Smelter	14937	Clayton
Agriguard Chlorothalonil	14839	AgriGuard	Clayton Tebucon EW	14824	Clayton
			Clenecrop Super	14628	Nufarm UK
Agritox	14894	Nufarm UK	Clipper	14820	Makhteshim
Agritox 50	14814	Nufarm UK	Clomaz 36 CS	14938	Goldengrass
Ally Max SX	14835	DuPont	Cohort	15035	Makhteshim
Alpha Bromolin 225 EC	14864	Makhteshim	Concept	14642	ChemSource
			Conga	15011	Nufarm UK
Aquarius	14712	Makhteshim	Coragen	14930	DuPont
Aria	14771	AgriGuard	County Mark	A0689	Greenaway
Atom	15038	Nufarm UK	Credo	14577	DuPont
Aviator 235 Xpro	15026	Bayer CropScience	Crown	14830	AgriGuard
			CS Azoxy	14954	ChemSource
Barclay D-Quat	14833	Barclay	CS Chlormequat	14946	ChemSource
Becki	14677	ChemSource	CTL 500	14812	Goldengrass
Benta 480 SL	14940	Sharda	Curator	14955	Syngenta
Betanal MaxxPro	15086	Bayer CropScience	Cutaway	14445	Syngenta
			Dartagnan WG	14963	ChemSource
BiPlay SX	14836	DuPont	Datura	14915	AgriChem BV
Biscaya	15014	Bayer CropScience	Decabane	14985	BASF
Blazer M	15084	Headland	Dessicash 200	14848	Sharda
Bonser	A0715	Nufarm UK	Diqua	14032	Sharda
Bontima	14899	Syngenta	Diquash	14849	Agrichem
Boogie	15061	Bayer CropScience	Dow Shield 400	14984	Dow
			Dragoon Gold	14973	Belcrop
Bravo 500	14548	Syngenta	EA Cloporam	14828	European Ag
Broadway Sunrise	14960	Dow	EA Difcon	14965	European Ag
Calibre SX	15032	DuPont	EA Epoxi	14832	European Ag
Careca	14948	AgriChem BV	EA Tebuzole	15000	European Ag
CCC 720	14891	European Ag	Echo	14982	AgriGuard
Champion	14599	AgriGuard	Eco-flex	A0696	Greenaway
Chikita	14675	ChemSource	Eradicoat	13724	Certis
Clayton Bestow	14996	Clayton	Firefly 155	14818	Bayer CropScience
Clayton Bramble	14856	Clayton			
Clayton Cajole	14995	Clayton	Flagon 400 EC	14921	Makhteshim
Clayton Orleans	14827	Clayton	Flanker	14760	DuPont

Product	Reg. No.	Supplier
Folio Gold	14368	Syngenta
Fubol Gold WG	14605	Syngenta
Fungazil 100 SL	14999	Certis
Galivor	14986	BASF
Glyfo TDI	14743	Q-Chem
Grainstore	14932	Pan Agriculture
Harness	14803	ChemSource
Headland Spruce	15089	Headland
Headway	14396	Syngenta
Heritage Maxx	14787	Syngenta
Hithane	14852	ChemSource
Howitzer	14942	ChemSource
HY-MCPA	14927	Agrichem
Intracrop Boost	A0709	Intracrop
Intracrop Cogent	A0707	Intracrop
Intracrop Inca	A0701	Intracrop
Intracrop Incite	A0702	Intracrop
Intracrop Retainer NF	A0711	Intracrop
Intracrop Tonto	A0708	Intracrop
Kayak	14847	Syngenta
K-Obiol EC 25	13573	Lodi UK
K-Obiol ULV 6	13572	Lodi UK
Larke	14914	Nufarm UK
Lexi	14890	AgChem Access
Liberate	14872	ChemSource
Lupo	14931	Nufarm UK
Majestik	14831	Certis
Mascot Micronised Lawn Sand	14763	Rigby Taylor
MCPA 25%	14893	Nufarm UK
Megaflo	15034	Dow
Meridian	14676	AgChem Access
Movon	14784	Bayer CropScience
Mutiny	14805	ChemSource
Name TBC	15027	Bayer CropScience
Nappa	14881	ChemSource
Nautile DG	14692	United Phosphorus
Nautilus	14961	ChemSource
Neosorexa Bait Blocks	H6817	BASF
Neosorexa Gold Ratpacks	H8302	BASF
Neosorexa Pasta Bait	H7758	BASF
Nightjar	14656	AgChem Access
Nufarm MCPA 750	14892	Nufarm UK
Nufosate Ace	14959	Nufarm UK
Oklar SX	15037	DuPont
Orca	14970	ChemSource
Orian	14798	ChemSource

Product	Reg. No.	Supplier
Pan Proteb	14896	Pan Agriculture
Penncozeb WDG	14719	United Phosphorus
Ponder	14693	ChemSource
potassium bicarbonate	00000	various
Primo Maxx	14780	Syngenta
Proline 275	14790	Bayer CropScience
Propel Flo	15044	Clayton
Proper Flo	15102	Globachem
Propi 25 EC	14939	Sharda
Proshield	14767	Scotts
Prothio	14945	ChemSource
Pure Glyphosate 360	14834	Pure Amenity
Pure Max	14952	Pure Amenity
Pure Turf	14851	Pure Amenity
Quad-Glob 200 SL	14758	Q-Chem
Raza	14800	ChemSource
Reaper	14924	AgChem Access
Relva	14873	Belcrop
Resplend	14975	BASF
Rock	14928	Goldengrass
Roller	A0694	Agrovista
Rumo	14883	DuPont
Rustler	14951	ChemSource
Scout	14779	AgriGuard
Setanta 50 WP	14943	AgriGuard
Setanta Flo	14678	AgriGuard
Shyfo	15040	Sharda
Sienna	14991	BASF
Siltra Xpro	15082	Bayer CropScience
Skyway 285 Xpro	15028	Bayer CropScience
Solitaire 50 WP	14962	AgriGuard
Sorexa D	H8242	BASF
Sphere ASBO	15064	Sphere
Stamen	14895	AgriGuard
Standon Santa Fe 50 WP	14966	Standon
Standon Santa Fe Flo	14727	Standon
Standon Yorker	14751	Standon
Standon Zing PQF	14949	Standon
Statis	13079	AgriGuard
Stiletto	14729	BASF
Supreme	14838	AgriGuard
Sven	14859	Interfarm
Sword	15099	Makhteshim
Symbol	14769	United Phosphorus
Tanker	15016	Nufarm UK

Product	Reg. No.	Supplier	Product	Reg. No.	Supplier
Teamforce	14923	AgriChem BV	UPL Diquat	14944	United Phosphorus
Tebucon 250	14789	Goldengrass	Valbon	14868	Certis
Temper Turf	14819	AgriGuard	Valiant	14456	Pan Amenity
Tempt	14227	Chiltern	Velocity	A0697	Agrovista
Thrash	15067	Pan Agriculture	Video DG	14850	United Phosphorus
Timpani	14651	Nufarm UK	Vigon	14785	Bayer CropScience
Tonik	14761	AgriGuard
Trafalgar	14888	Pan Amenity
Traton SX	14837	DuPont	Waila	14976	ChemSource
Trotter	14236	Agrovista	Whistle	14757	BASF
Trounce	14222	Chiltern	Woomera	14972	Belcrop
Twister	14885	ChemSource	Zampro DM	15013	BASF
Unicur	14776	Bayer CropScience	Zimbrail	14735	DuPont
			Zip	14806	AgriChem BV
Unix	14846	Syngenta
UPL B Zone	14808	United Phosphorus

Products Deleted

The appearance of a product name in the following list may not necessarily mean that it is no longer available. In cases of doubt, refer to Section 3 or the supplier.

Product	Reg. No.	Supplier	Product	Reg. No.	Supplier
Addit	A0693	Koppert	Betosip Combi	13254	Sipcam
Agrotech Chlorthalonil 500 SC	13176	Agrotech Trading	Blagden Professional Sodium Chlorate Weedkiller	12624	Blagden
Akolin 330 EC	14282	Aako
Alpha Briotril	13580	Makhteshim	Blazer	14126	Headland
Alpha Briotril Plus 19/19	13579	Makhteshim	Blizzard	14363	AgriGuard
			B-Nine SG	12734	Certis
Alpha Bromolin 225 EC	13431	Makhteshim	Brasson	10560	Monsanto
			Burex 430 SC	12619	United Phosphorus
Alpha Bromotril P	13326	Makhteshim
Alpha Propachlor 50 SC	13585	Makhteshim	Casoron G	07926	Miracle
			Casoron G	09022	Chemtura
Alpha Propachlor 65 WP	04807	Makhteshim	Casoron G	09023	Nomix Enviro
			Casoron G	10709	Scotts
Ashlade CP	06481	Nufarm UK	Casoron G	11632	Certis
Atom	15038	Nufarm UK	Casoron G4	07927	Miracle
Banlene Super	10053	Bayer CropScience	Casoron G4	09215	Chemtura
			Casoron G4	10708	Scotts
Barclay Clinger	A0198	Barclay	Casoron G-SR	07925	Miracle
Barleyquat B	13965	Taminco	CDP Minis	12955	De Sangosse
BASF Dimethoate 40	00199	BASF	Celmitron 70WDG	13630	SD Agchem
Basta	13820	Bayer CropScience	Certis Dichlobenil Granules	13705	Certis
Beetaweed	12260	AgriGuard	Cheer 500	14418	Goldengrass
Beetup Flo	11916	United Phosphorus	Clayton Florin	14351	Clayton

Product	Reg. No.	Supplier
Clayton Lenacil 80 W	09488	Clayton
Clayton Oxen G	12770	Clayton
Cleancrop Mandrake	12500	United Agri
Clenecorn Super	14628	Nufarm UK
Condor 5	12484	Doff Portland
Cooke's Professional Sodium Chlorate Weedkiller with Fire Depressant	06796	Cooke's Chemicals
CTL 500	14290	Goldengrass
Curzate 60 WG	13904	Headland
Daconil Turf	09265	Scotts
Dacthal W-75	10289	AMVAC
Dacthal W-75	11323	Certis
Deadfast Sodium Chlorate Weed Killer - Professional Use	12510	Growing Success
Decimate	13419	Certis
Deosan Chlorate Weedkiller (Fire Suppressed)	08521	Johnson Diversey
Dican	13685	Agform
Dichlobenil Granules	11871	Certis
Dicore	12542	AgriChem BV
Dixie 6	11790	Greencrop
Doff Sodium Chlorate Weedkiller	06049	Doff Portland
Duo 400 SC	12561	AgriGuard
Duo 400 SC	14232	AgriGuard
Embargo G	12690	Barclay
Exit Wetex	11276	Bayer CropScience
Ferramol	12274	Certis
Flagon 400 EC	13470	Makhteshim
Fungaflor 100EC	12978	Certis
Gamit 36 CS	12598	Belchim
Gem Sodium Chlorate Weedkiller	04276	Metcalf
Graduate	10776	Bayer CropScience
Greencrop Vegex	12439	Greencrop
Growing Success Sodium Chlorate Weedkiller	10787	Growing Success
Guardsman	05494	Chiltern
Headland Cedar	12180	Headland
Headland Guard 2000	A0652	Headland
Headland Tolerate	10774	Headland
Holster	A0682	Greenaway
Impetus	11021	Headland
Jazz	13941	Interfarm

Product	Reg. No.	Supplier
Landgold Amidosulfuron	09021	Landgold
Landgold Amidosulfuron	12110	Teliton
Landgold Azzox	14003	Goldengrass
Landgold Bentazone SL	13418	Teliton
Landgold Chlorothalonil 50	13473	Teliton
Landgold Chlorothalonil 50	14011	Goldengrass
Landgold Clodinafop	13329	Teliton
Landgold Cyper 100	14105	Goldengrass
Landgold Deltaland	12114	Teliton
Landgold Deputy	11975	Landgold
Landgold Deputy	12115	Teliton
Landgold Difenoconazole	09964	Landgold
Landgold Epoxi	14062	Makhteshim
Landgold Epoxiconazole	09821	Landgold
Landgold Epoxiconazole	12117	Teliton
Landgold Epoxiconazole	14025	Goldengrass
Landgold Epoxiconazole FM	08806	Landgold
Landgold Fenpropidin 750	08973	Landgold
Landgold Fenpropimorph 750	10472	Landgold
Landgold Fenpropimorph 750	12118	Teliton
Landgold Fluazinam	08060	Landgold
Landgold Fluazinam	12119	Teliton
Landgold Fluroxypyr	13356	Teliton
Landgold Glyphosate 360	13977	Goldengrass
Landgold Metamitron	06287	Landgold
Landgold Metamitron	12123	Teliton
Landgold Metazachlor 50	09726	Landgold
Landgold Metazachlor 50	12124	Teliton
Landgold Metazachlor 50 SC	12133	Teliton
Landgold Metribuzin WG	12468	Teliton
Landgold Metsulfuron	13421	Teliton
Landgold Nicosulfuron	13464	Teliton
Landgold Nicosulfuron	13997	Goldengrass
Landgold Piccant	13770	Goldengrass
Landgold Pirimicarb 50	14022	Goldengrass

Product	Reg. No.	Supplier
Landgold PQF 100	10763	Landgold
Landgold PQF 100	12127	Teliton
Landgold Strobilurin 250	12128	Teliton
Landgold Strobilurin KE	09908	Landgold
Landgold Strobilurin KF	09196	Landgold
Landgold Tacet	13036	Teliton
Landgold Tepraloxydim	13379	Teliton
Landgold Tepraloxydim	14026	Goldengrass
Landgold TFS 50	08941	Landgold
Landgold Tralkoxydim	08604	Landgold
Landgold Tralkoxydim	12130	Teliton
Lenazar Flowable	11068	Hermoo
Luxan Dichlobenil Granules	09250	Luxan
Lynx S	13276	De Sangosse
Majestik	-	Certis
Malvi	12862	AgChem Access
Markate 50	13529	Agrovista
Midstream GSR	11674	Scotts
Nicotine 40% Shreds	05725	Dow
Nufarm Cropoil Gold	A0446	Nufarm UK
Optimol	12997	De Sangosse
Optimol XL	12937	De Sangosse
Osorno	12486	Nufarm UK
Panache 40	10632	DAPT
Permesso	14108	AgChem Access
Pesta M	13275	De Sangosse
Portman Weedmaster	06018	Agform
Proline	12084	Bayer CropScience
Proshield	14767	Scotts
Quail	14506	Syngenta
Questor	A0495	United Phosphorus
Ramrod 20 Granular	10313	Monsanto
Ramrod Flowable	10314	Monsanto
Redigo Twin	12314	Bayer CropScience
Repulse	11328	Certis
Resimate	13789	Certis
Rizolex	09673	Certis
Sakarat Ready-to-Use (Whole Wheat)	H6808	Killgerm
Sakarat X	H6809	Killgerm
Scotts Dichlo G Macro	12011	Scotts

Product	Reg. No.	Supplier
Scotts Dichlo G Micro	11857	Scotts
Sewercide Cut Wheat Rat Bait	H6805	Killgerm
Sewercide Whole Wheat Rat Bait	H6806	Killgerm
Sierraron G	09263	Scotts
Sierraron G	09675	Scotts
Sierraron G4	10491	Scotts
Sipcam Solo-D	12617	Sipcam
Sodium Chlorate	06294	Marlow Chemical
Sorexa Checkatube	H6702	Sorex
Standon Dichlobenil 6 G	08874	Standon
Standon Epoxifen	08972	Standon
Standon Kresoxim FM	08922	Standon
Standon Kresoxim Super	09794	Standon
Standon Metazachlor 50	05581	Standon
Standon Metazachlor-Q	09676	Standon
Standon Tralkoxydim	09579	Standon
Strathclyde Sodium Chlorate Weedclear	12015	Dremm
Strength AC	H7469	Certis
Strobiplus 250 EC	15094	Agform
Talunex	13798	Certis
Tasker 75	10544	Headland
Titan 480	12596	AgriGuard
Torch Extra	12140	Bayer CropScience
Tripart Sentinel	03250	Tripart
Tripart Sentinel 2	05140	Tripart
Turfclear	07506	Scotts
Valbon	12603	Certis
Vermigon Bait	H7925	Certis
Vermigon Blocks	H7926	Certis
Vermigon Pasta	H8191	Certis
Vermigon Pellets	H7941	Certis
Vermigon Sewer Blocks	H7927	Certis
Vertalec	04781	Koppert
Viking Granules	11859	Nomix Enviro
Viking Granules	13883	Certis
Warefog 25	06776	Whyte Agrochemicals
Weedmaster SC	08793	Agform
XL-All Nicotine 95%	07402	Vitax

Introduction

Purpose

The primary aim of this book is to provide a practical handbook of pesticides, plant growth regulators and adjuvants that the farmer or grower can realistically and legally obtain in the UK and to identify the purposes for which they can be used. It is designed to help in the identification of products appropriate to a particular problem. In addition to uses recommended on product labels, details are provided of those uses which do not appear on product labels but which have been granted specific off-label approval (SOLAs). As well as identifying the products available, the book provides guidance on how to use them safely and effectively but without giving details of doses, volumes, spray schedules or approved tank mixtures. Sections 5 and 6 provide essential background information on a wide range of pesticide-related issues including legislation, codes of practice, poisons and treatment of poisoning, products for use in special situations and weed and crop growth stage keys.

While we have tried to cover all other important factors, this book does not provide a full statement of product recommendations. Before using any pesticide product it is essential that the user should read the label carefully and comply strictly with the instructions it contains. **This *Guide* does not substitute for the product label.**

Scope

The *Guide* is confined to pesticides registered in the UK for use in arable agriculture, horticulture (including amenity use), forestry and areas in or near water. Within these fields of use there are about 2500 products in the UK with current approval. The *Guide* is a reference for virtually all of them given in two sections. Section 2 gives full details of the products notified to the editor as available on the market. Products are included in this section only if requested by the supplier and supported by evidence of approval. Section 3 gives brief details of all other registered products with extant approval.

Section 2 lists some 450 individual active ingredients and mixtures. For each entry a list is shown of the available approved products together with the name of the supplier from whom each may be obtained. All types of pesticide covered by Control of Pesticide Regulations or Plant Protection Products Regulations are included. This embraces acaricides, algicides, herbicides, fungicides, insecticides, lumbricides, molluscicides, nematacides and rodenticides together with plant growth regulators used as straw shorteners, sprout inhibitors and for various horticultural purposes. The total number of products included in section 2 (i.e. available in the market in 2010) is about 1460.

The *Guide* also gives information (in section 4) on around 130 authorised adjuvants which, although not active pesticides themselves, may be added to pesticide products to improve their effectiveness. The *Guide* does not include products solely approved for amateur, home and garden, domestic, food storage, public or animal health uses.

Sources of Information

The information in this edition has been drawn from these authoritative sources:

- approved labels and product manuals received from suppliers of pesticides up to October 2010.
- websites of the Chemicals Regulation Directorate (CRD, www.pesticides.gov.uk) and the Health and Safety Executive (HSE, www.hse.gov.uk).

Introduction

Criteria for Inclusion

To be included, a pesticide must meet the following conditions:
- it must have extant approval under UK pesticides legislation
- information on the approved uses must have been provided by the supplier
- it must be expected to be on the UK market during the currency of this edition.

When a company changes its name, whether by merger, takeover or joint venture, it is obliged to re-register its products in the name of the new company, and new MAPP numbers are normally assigned. Where stocks of the previously registered product remain legally available, both old and new identities are included in the *Guide* and remain until approval of the former lapses or stocks are notified as exhausted.

Products that have been withdrawn from the market and whose approval will finally lapse during 2010 are identified in each profile. After the indicated date, sale, storage or use of the product bearing that approval number becomes illegal. Where there is a direct replacement product, this is indicated.

The Voluntary Initiative

The Voluntary Initiative (VI) is a programme of measures, agreed by the crop protection industry and related organisations with Government, to minimise the environmental impacts of crop protection products. The programme has provided a framework of practices and principles to help Government achieve its objective of protection of water quality and enhancement of farmland biodiversity. Many of the environmental protection schemes launched under the VI represent current best practice.

The first five-year phase of the programme formally concluded in March 2006, but it is continuing with a rolling two-year review of proposals and targets; current proposals include a new online crop protection management plan and greater involvement in catchment-sensitive farming (see below).

A key element of the VI has been the provision of environmental information on crop protection products. Members of the Crop Protection Association (CPA) have committed to do this by producing Environmental Information Sheets (EISs) for all their marketed professional products.

EISs reinforce and supplement information on a product label by giving specific environmental impact information in a standardised format. They highlight any situations where risk management is essential to ensure environmental protection. Their purpose is to provide user-friendly information to advisers (including those in the amenity sector), farmers and growers on the environmental impact of crop protection products, to allow planning with a better understanding of the practical implications.

Links are provided to give users rapid access to EISs on the VI website: www.voluntaryinitiative.org.uk.

Catchment-sensitive Farming

Catchment-sensitive Farming (CSF) is the government's response to climate change and the new European Water Directive. Rainfall events are likely to become heavier, with a greater risk of soil, nutrients and pesticides ending up in the waterways. The CSF programme is investigating the effects of specific targeted advice on 20 catchment areas perceived as being at risk, and new entries in this *Guide* can help identify the pesticides that pose the greatest risk of water contamination (see Environmental Safety).

Notes on Contents and Layout

The book consists of six main sections:

1 Crop/Pest Guide
2 Pesticide Profiles
3 Products Also Registered
4 Adjuvants
5 Useful Information
6 Appendices

1 Crop/Pest Guide

This section enables the user to identify which active ingredients are approved for a particular crop/pest combination. The crops are grouped as shown in the Crop Index. For convenience, some crops and pests have been grouped into generic units (for example, 'cereals' are covered as a group, not individually). Indications from the Crop/Pest Guide must always be checked against the specific entry in the pesticide profiles section because a product may not be approved on all crops in the group. Chemicals indicated as having uses in cereals, for example, may be approved for use only on winter wheat and winter barley, not for other cereals. Because of the difference in wording of product labels, it may sometimes be necessary to refer to broader categories of organism in addition to specific organisms (e.g. annual grasses and annual weeds in addition to annual meadowgrass).

2 Pesticide Profiles

Each active ingredient and mixture of ingredients has a separate numbered profile entry. The entries are arranged in alphabetical order using the common names approved by the British Standards Institution and used in *The Pesticide Manual* (15th Edition; BCPC, 2009). Where an active ingredient is available only in mixtures, this is stated. The ingredients of the mixtures are themselves ordered alphabetically, and the entries appear at the correct point for the first named active ingredient.

Below the profile title, a brief phrase describes the main use of the active ingredient(s). This is followed where appropriate by the mode of action code(s) as published by the Fungicide, Herbicide and Insecticide Resistance Action Committees.

Within each profile entry, a table lists the products approved and available on the market in the following style:

Product name	Main supplier	Active ingredient contents	Formulation type	Registration number
1 Broadsword	United Phosphorus	200:85:65 g/l	EC	09140
2 Greengard	SumiAgro Amenity	200:85:65 g/l	EC	11715
3 Nu-Shot	Nufarm UK	200:85:65 g/l	EC	11148

Many of the **product names** are registered trademarks, but no special indication of this is given. Individual products are indexed under their entry number in the Index of Proprietary Names at the back of the book. The **main supplier** indicates the marketing outlet through which the product may be purchased. Full addresses and contact details of suppliers are listed in Appendix 1. For mixtures, the **active ingredient contents** are given in the same order as they appear in the profile heading. The **formulation types** are listed in full in the Key to Abbreviations and Acronyms (Appendix 5). The **registration number** normally refers to the registration with the Chemicals Regulation Directorate (CRD). In cases where the products registered with the Health and Safety Executive are included, HSE numbers are quoted, e.g. H0462.

Below the product table, a **Uses** section lists all approved uses notified by suppliers to the Editor by October 2010, giving both the principal target organisms and the recommended crops or situations (identified in ***bold italics***). Where there is an important condition of use, or where the approval is off-label, this is shown in parentheses as (*off-label*). Numbers in square brackets refer to the numbered products in the table above. Thus, in the example shown above, a use approved for Nu-Shot (product 3) but not for Broadsword or Greengard (products 1 and 2) appears as:

Annual dicotyledons in **rotational grassland** [3]

Any **Specific Off-Label Approvals (SOLAs)** for products in the profile are detailed below the list of approved uses. Each SOLA has a separate entry and shows the crops to which it applies, the Notice of Approval number, the expiry date (if due to fall during the current edition), and the product reference number in square brackets. SOLAs do not appear on the product label and are undertaken entirely at the risk of the user.

Below the SOLA paragraph, **Notes** are listed under the following headings. Unless otherwise stated, any reference to dose is made in terms of product rather than active ingredient. Where a note refers to a particular product rather than to the entry generally, this is indicated by numbers in square brackets as described above.

Approval information	Information of a general nature about the approval status of the active ingredient or products in the profile is given here. Notes on approval for aerial or ULV application are given.
	Inclusion of the active ingredient in Annex I under EC Directive 1107/2009 is shown, as well as any acceptance from the British Beer and Pub Association (BBPA) for its use on barley or hops for brewing or malting.
	Where a product approval will finally expire in 2011, the expiry date is shown here.
Efficacy	This entry identifies factors important in making the most effective use of the treatment. Certain factors, such as the need to apply chemicals uniformly, the need to use appropriate volume rates, and the need to prevent settling out of the active ingredient in the spray tank, are assumed to be important for all treatments and are not emphasised in individual profiles.
Restrictions	Notes are included in this section where products are subject to the Poisons Law, and where the label warns about organophosphorus and/or anticholinesterase compounds. Factors important in optimising performance and minimising the risk of crop damage, including statutory conditions relating to the maximum permitted number of applications, are listed. Any restrictions on crop varieties that may not be sprayed are mentioned here.
Environmental safety	Where the label specifies an environmental hazard classification, it is noted together with the associated risk phrases. Any other special operator precautions and any conditions concerning withholding livestock from treated areas are specified. Where any of the products in the profile are subject to Category A, Category B or broadcast air-assisted buffer zone restrictions under the LERAP scheme, the relevant classification is shown.
	Other environmental hazards are also noted here, including potential dangers to livestock, game, wildlife, bees and fish. The need to avoid drift onto neighbouring crops and the need to wash out equipment thoroughly after use are important with all pesticide treatments, but may receive special mention here if of particular significance.
Crop-specific information	Instructions about the timing of application or cultivations that are specific to a particular crop, rather than generally applicable, are mentioned here. The latest permitted use and harvest intervals (if any) are shown for individual crops.
Following crops guidance	Any specific label instructions about what may be grown after a treated crop, whether harvested normally or after failure, are shown here,

Notes on Contents and Layout

Hazard classification and safety precautions	The label hazard classifications and precautions are shown using a series of letter and number codes which are explained in Appendix 4. Hazard warnings are now given in full, with the other precautions listed under the following subheadings:

- Risk phrases
- Operator protection
- Environmental protection
- Consumer protection
- Storage and disposal
- Treated seed
- Vertebrate/rodent control products
- Medical advice

This section is given for information only and should not be used for the purpose of making a COSHH assessment without reference to the actual label of the product to be used.

3 Products also Registered

Products with an extant approval for all or part of the year of the edition in which they appear are listed in this section if they have not been requested by their supplier or manufacturer for inclusion in Section 2. Details shown are the same (apart from formulation) as those in the product tables of Section 2 and, where relevant, an approval expiry date is shown. Not all products listed here will be available in the market, but this list, together with the products in Section 2, comprises a comprehensive listing of all approved products in the UK for uses within the scope of the *Guide* for the year in question.

4 Adjuvants

Adjuvants are listed in a table ordered alphabetically by product name. For each product, details are shown of the main supplier, the authorisation number and the mode of action (e.g. wetter, sticker, etc.) as shown on the label. A brief statement of the uses of the adjuvant is given. Protective clothing requirements and label precautions are listed using the codes from Appendix 4.

5 Useful Information

This section summarises legislation covering approval, storage, sale and use of pesticides in the UK. There are brief articles on the broader issues concerning the use of crop protection chemicals, including resistance management. Lists are provided of products approved for use in or near water, in forestry, as seed treatments and for aerial application. Chemicals subject to the Poisons Laws are listed, and there is a summary of first aid measures if pesticide poisoning should be suspected. Finally, this section provides guidance on environmental protection issues and covers the protection of bees and the use of pesticides in or near water.

6 Appendices

Appendix 1	Gives names, addresses, telephone and fax numbers of all companies listed as main suppliers in the pesticide profiles. E-mail and website addresses are also listed where available.
	Where a supplier is no longer trading under that name (usually following a merger or takeover), but products under that name are still listed in the *Guide* because they are still available in the supply chain, a cross-reference indicates the new 'parent' company from which technical or commercial information can be obtained.
Appendix 2	Gives names, addresses, telephone and fax numbers of useful contacts, including the National Poisons Information Service. Website addresses are included where available.

Notes on Contents and Layout

Appendix 3	Gives details of the keys to crop and weed growth stages, including the publication reference for each. The numeric codes are used in the descriptive sections of the pesticide profiles (Section 2).
Appendix 4	Shows the full text for code letters and numbers used in the pesticide profiles (Section 2) to indicate personal protective equipment requirements and label precautions.
Appendix 5	Shows the full text for the formulation abbreviations used in the pesticide profiles (Section 2). Other abbreviations and acronyms used in the *Guide* are also explained here.
Appendix 6	Provides full definitions of officially agreed descriptive phrases for crops or situations used in the pesticide profiles (Section 2) where misunderstandings can occur.
Appendix 7	Shows a list of useful reference publications, which amplify the summarised information provided in this section.

SECTION 1
CROP/PEST GUIDE

Crop/Pest Guide Index

Important Note: The Crop/Pest Guide Index refers to pages on which the subject can be located.

Arable and vegetable crops
Agricultural herbage
    Grass 5
    Herbage legumes 5
    Herbage seed 6
Brassicas
    Brassica seed crops 7
    Brassicas, general 7
    Fodder brassicas 7
    Leaf and flowerhead brassicas 8
    Mustard 9
    Root brassicas 10
    Salad greens 11
Cereals
    Barley 12
    Cereals, general 18
    Maize/sweetcorn 18
    Oats 19
    Rye/triticale 21
    Undersown cereals 25
    Wheat 25
Edible fungi
    Mushrooms 32
Fruiting vegetables
    Aubergines 32
    Cucurbits 32
    Peppers 33
    Tomatoes 33
Herb crops
    Herbs 33
Leafy vegetables
    Endives 34
    Lettuce 35
    Spinach 35
    Watercress 36
Legumes
    Beans (Phaseolus) 36
    Beans (Vicia) 37
    Lupins 38
    Peas 38
Miscellaneous arable
    Miscellaneous arable industrial crops 39
    Miscellaneous arable crops 40
    Miscellaneous arable situations 41
Miscellaneous field vegetables
    All vegetables 41
Oilseed crops
    Linseed/flax 41
    Miscellaneous oilseeds 42
    Oilseed rape 43
    Soya 45
    Sunflowers 45
Root and tuber crops
    Beet crops 45
    Carrots/parsnips 47
    Miscellaneous root crops 48
    Potatoes 49
Stem and bulb vegetables
    Asparagus 50
    Celery/chicory 51
    Globe artichokes/cardoons 51
    Onions/leeks/garlic 52
    Vegetables 53

Flowers and ornamentals
Flowers
    Bedding plants, general 53
    Bulbs/corms 54
    Miscellaneous flowers 54
    Pot plants 55
    Protected bulbs/corms 55
    Protected flowers 55
Miscellaneous flowers and ornamentals
    Miscellaneous uses 55
    Soils 56
Ornamentals
    Nursery stock 56
    Trees and shrubs 58
    Woody ornamentals 58

Forestry
Forest nurseries, general
    Forest nurseries 59
Forestry plantations, general
    Forestry plantations 59
Miscellaneous forestry situations
    Cut logs/timber 60
Woodland on farms
    Woodland 60

Fruit and hops
All Bush fruit
    All currants 61
    All protected bush fruit 62
    All vines 62
    Bilberries/blueberries/cranberries 62
    Bush fruit, general 63
    Gooseberries 63
    Miscellaneous bush fruit 63
Cane fruit
    All outdoor cane fruit 64
    All protected cane fruit 65
Hops, general
    Hops 65
Miscellaneous fruit situations
    Fruit crops, general 66
    Fruit nursery stock 67
Other fruit
    All outdoor strawberries 67

Crop/Pest Guide Index

    Protected miscellaneous fruit 68
    Rhubarb 68
Tree fruit
    All nuts 69
    All pome fruit 69
    All stone fruit 71
    Miscellaneous top fruit 72

Grain/crop store uses
Stored seed
    Stored grain/rape/linseed 72

Grass
Turf/amenity grass
    Amenity grassland 72
    Managed amenity turf 73

Non-crop pest control
Farm buildings/yards
    Farm buildings 74
    Farmyards 74
Farmland pest control
    Farmland situations 74
Miscellaneous non-crop pest control
    Manure/rubbish 74
    Miscellaneous pest control
        situations 75

Protected salad and vegetable crops
Protected brassicas
    Protected brassica vegetables 75
    Protected salad brassicas 75

Protected crops, general
    All protected crops 76
    Glasshouses 76
    Protected vegetables, general 76
    Soils and compost 76
Protected fruiting vegetables
    Protected aubergines 77
    Protected cucurbits 77
    Protected tomatoes 78
Protected herb crops
    Protected herbs 79
Protected leafy vegetables
    Mustard and cress 79
    Protected leafy vegetables 80
    Protected spinach 80
Protected legumes
    Protected peas and beans 80
Protected root and tuber vegetables
    Protected carrots/parsnips/celeriac .. 81
    Protected root brassicas 81
Protected stem and bulb vegetables
    Protected celery/chicory 81
    Protected onions/leeks/garlic 81

Total vegetation control
Aquatic situations, general
    Aquatic situations 82
Non-crop areas, general
    Miscellaneous non-crop situations ... 82
    Non-crop farm areas 82
    Paths/roads etc 82

Crop/Pest Guide

Important note: For convenience, some crops and pests or targets have been brought together into generic groups in this guide, e.g. 'cereals', 'annual grasses'. It is essential to check the profile entry in Section 2 *and* the label to ensure that a product is approved for a specific crop/pest combination, e.g. winter wheat/blackgrass.

Arable and vegetable crops

Agricultural herbage - Grass

Crop control	Desiccation	glyphosate
Diseases	Crown rust	propiconazole
	Drechslera leaf spot	propiconazole
	Powdery mildew	propiconazole
	Rhynchosporium	propiconazole
Pests	Aphids	esfenvalerate
	Birds/mammals	aluminium ammonium sulphate
	Flies	chlorpyrifos, cypermethrin, esfenvalerate
	Leatherjackets	chlorpyrifos, methiocarb (*seed admixture*)
	Slugs/snails	methiocarb (*seed admixture*)
Plant growth regulation	Quality/yield control	sulphur
Weeds	Broad-leaved weeds	2,4-D, 2,4-D + dicamba, 2,4-D + dicamba + dichlorprop-P, 2,4-D + dicamba + triclopyr, 2,4-D + MCPA, 2,4-DB, 2,4-DB + linuron + MCPA, 2,4-DB + MCPA, amidosulfuron, aminopyralid + fluroxypyr, aminopyralid + triclopyr, asulam, citronella oil, clopyralid, clopyralid (*off-label*), clopyralid (*off-label - spot treatment*), clopyralid (*spot treatment*), clopyralid + fluroxypyr + triclopyr, clopyralid + triclopyr, clopyralid + triclopyr (*off-label - via weed wiper*), dicamba + MCPA + mecoprop-P, dicamba + mecoprop-P, dichlorprop-P, fluroxypyr, fluroxypyr + triclopyr, glyphosate, glyphosate (*wiper application*), MCPA, MCPA + MCPB, MCPB, mecoprop-P, metsulfuron-methyl, thifensulfuron-methyl
	Crops as weeds	fluroxypyr
	Grass weeds	glyphosate
	Weeds, miscellaneous	aminopyralid + triclopyr, fluroxypyr, glufosinate-ammonium, glyphosate
	Woody weeds/scrub	asulam

Agricultural herbage - Herbage legumes

Crop control	Desiccation	diquat
Weeds	Broad-leaved weeds	2,4-DB, carbetamide, diquat (*off-label*), isoxaben (*off-label - grown for game cover*), MCPA + MCPB, propyzamide
	Crops as weeds	carbetamide
	Grass weeds	carbetamide, diquat (*seed crop*), fluazifop-P-butyl (*off-label*), propyzamide, tri-allate

Crop/Pest Guide - Arable and vegetable crops

| | Weeds, miscellaneous | diquat (*seed crop*) |

Agricultural herbage - Herbage seed

Crop control	Desiccation	diquat
Diseases	Crown rust	propiconazole
	Damping off	thiram (*seed treatment*)
	Disease control/foliar feed	chlorothalonil + mancozeb (*off-label*), difenoconazole (*off-label*), mancozeb (*off-label*), maneb (*off-label*), prochloraz (*off-label*), prochloraz + propiconazole (*off-label*), spiroxamine + tebuconazole (*off-label*), tebuconazole (*off-label*)
	Drechslera leaf spot	propiconazole
	Foliar diseases	azoxystrobin (*off-label*), azoxystrobin + cyproconazole (*off-label*), azoxystrobin + fenpropimorph (*off-label*), chlorothalonil (*off-label*), chlorothalonil + flutriafol (*off-label*), cyproconazole (*off-label*), cyproconazole + cyprodinil (*off-label*), cyproconazole + quinoxyfen (*off-label*), cyproconazole + trifloxystrobin (*off-label*), difenoconazole (*off-label*), dimoxystrobin + epoxiconazole (*off-label*), epoxiconazole (*off-label*), epoxiconazole + fenpropimorph (*off-label*), epoxiconazole + fenpropimorph + kresoxim-methyl (*off-label*), epoxiconazole + kresoxim-methyl (*off-label*), epoxiconazole + kresoxim-methyl + pyraclostrobin (*off-label*), epoxiconazole + pyraclostrobin (*off-label*), famoxadone + flusilazole (*off-label*), fenpropimorph (*off-label*), fenpropimorph + flusilazole (*off-label*), fenpropimorph + pyraclostrobin (*off-label*), fluquinconazole (*off-label*), fluquinconazole + prochloraz (*off-label*), metconazole (*off-label*), prothioconazole + tebuconazole (*off-label*), pyraclostrobin (*off-label*), tebuconazole (*off-label*), tebuconazole + triadimenol (*off-label*), trifloxystrobin (*off-label*)
	Powdery mildew	cyflufenamid (*off-label*), epoxiconazole (*off-label*), fenpropimorph + quinoxyfen (*off-label*), propiconazole, sulphur (*off-label*)
	Rhynchosporium	propiconazole
Pests	Aphids	deltamethrin (*off-label*), dimethoate, esfenvalerate (*off-label*), lambda-cyhalothrin (*off-label*), lambda-cyhalothrin + pirimicarb (*off-label*), tau-fluvalinate (*off-label*), zeta-cypermethrin (*off-label*)
	Beetles	deltamethrin (*off-label*)
	Caterpillars	deltamethrin (*off-label*)
	Flies	esfenvalerate
	Leafhoppers	deltamethrin (*off-label*)
	Pests, miscellaneous	alpha-cypermethrin (*off-label*), lambda-cyhalothrin (*off-label*), tau-fluvalinate (*off-label*), zeta-cypermethrin (*off-label*)
	Thrips	deltamethrin (*off-label*)
Plant growth regulation	Growth control	trinexapac-ethyl
Weeds	Broad-leaved weeds	amidosulfuron + iodosulfuron-methyl-sodium (*off-label*), asulam (*off-label*), bifenox (*off-label*), bromoxynil (*off-label*), bromoxynil + diflufenican + ioxynil (*off-label*), carfentrazone-ethyl (*off-label*), carfentrazone-ethyl +

Crop/Pest Guide - Arable and vegetable crops

		flupyrsulfuron-methyl (*off-label*), clopyralid (*off-label*), dicamba + MCPA + mecoprop-P, dichlorprop-P + ioxynil (*off-label*), dichlorprop-P + MCPA + mecoprop-P (*off-label*), diflufenican (*off-label*), diflufenican + flufenacet (*off-label*), diflufenican + flurtamone (*off-label*), ethofumesate, florasulam (*off-label*), florasulam + fluroxypyr (*off-label*), iodosulfuron-methyl-sodium (*off-label*), MCPA, MCPB, mecoprop-P, pendimethalin (*off-label*), propyzamide, sulfosulfuron (*off-label*)
	Crops as weeds	clopyralid (*off-label*), iodosulfuron-methyl-sodium (*off-label - from seed*)
	Grass weeds	clodinafop-propargyl (*off-label*), diflufenican + flufenacet (*off-label*), diflufenican + flurtamone (*off-label*), ethofumesate, fenoxaprop-P-ethyl (*off-label*), pendimethalin (*off-label*), propyzamide, sulfosulfuron (*off-label*)

Brassicas - Brassica seed crops

Diseases	Alternaria	iprodione (*pre-storage only*)
Weeds	Broad-leaved weeds	carbetamide, propyzamide
	Crops as weeds	carbetamide
	Grass weeds	carbetamide, propyzamide

Brassicas - Brassicas, general

Diseases	Botrytis	Bacillus subtilis (*off-label*)
	Downy mildew	chlorothalonil, chlorothalonil (*moderate control*), copper oxychloride (*off-label*)
Pests	Aphids	deltamethrin (*off-label*), pymetrozine (*off-label*), spirotetramat (*off-label*)
	Beetles	deltamethrin (*off-label*)
	Caterpillars	deltamethrin (*off-label*)
	Pests, miscellaneous	deltamethrin (*off-label*)
	Whiteflies	spirotetramat (*off-label*)
Plant growth regulation	Quality/yield control	sulphur
Weeds	Broad-leaved weeds	pyridate (*off-label*)
	Grass weeds	fluazifop-P-butyl (*off-label*)

Brassicas - Fodder brassicas

Diseases	Alternaria	azoxystrobin, fludioxonil + metalaxyl-M + thiamethoxam (*moderate control*), iprodione (*off-label*)
	Bacterial blight	copper oxychloride (*off-label*)
	Black rot	copper oxychloride (*off-label*)
	Bottom rot	copper oxychloride (*off-label*)
	Disease control/foliar feed	difenoconazole (*off-label*)
	Downy mildew	fludioxonil + metalaxyl-M + thiamethoxam, fosetyl-aluminium (*off-label*)
	Phytophthora	copper oxychloride (*off-label*)
	Powdery mildew	azoxystrobin + difenoconazole
	Pythium	fludioxonil + metalaxyl-M + thiamethoxam (*reduction*)
	Ring spot	azoxystrobin, difenoconazole (*off-label*)
	Seed-borne diseases	fludioxonil + metalaxyl-M + thiamethoxam

Crop/Pest Guide - Arable and vegetable crops

	Spear rot	copper oxychloride (*off-label*)
	Stem canker	copper oxychloride (*off-label*)
	White blister	azoxystrobin, azoxystrobin + difenoconazole
Pests	Aphids	cypermethrin, esfenvalerate, fatty acids (*off-label*), pirimicarb, pymetrozine (*off-label*)
	Beetles	alpha-cypermethrin, deltamethrin, fludioxonil + metalaxyl-M + thiamethoxam (*reduction of damage*)
	Caterpillars	alpha-cypermethrin, Bacillus thuringiensis (*off-label*), cypermethrin, deltamethrin, esfenvalerate
	Flies	chlorpyrifos (*off-label*)
	Pests, miscellaneous	dimethoate (*off-label*)
	Thrips	fatty acids (*off-label*)
Weeds	Broad-leaved weeds	clomazone (*off-label*), clopyralid, metazachlor (*off-label*), napropamide, pyridate (*off-label*)
	Crops as weeds	fluazifop-P-butyl (*stockfeed only*)
	Grass weeds	fluazifop-P-butyl (*stockfeed only*), metazachlor (*off-label*), napropamide

Brassicas - Leaf and flowerhead brassicas

Diseases	Alternaria	azoxystrobin, boscalid + pyraclostrobin, chlorothalonil + metalaxyl-M (*moderate control*), difenoconazole, iprodione, iprodione (*off-label*), prothioconazole, tebuconazole, tebuconazole + trifloxystrobin
	Bacterial blight	copper oxychloride (*off-label*)
	Black rot	copper oxychloride (*off-label*)
	Botrytis	chlorothalonil, chlorothalonil (*moderate control*), chlorothalonil (*off-label*), iprodione (*off-label*)
	Bottom rot	copper oxychloride (*off-label*)
	Damping off	thiram (*seed treatment*), tolclofos-methyl
	Disease control/foliar feed	fluopicolide + propamocarb hydrochloride (*off-label*)
	Downy mildew	chlorothalonil, chlorothalonil (*moderate control*), chlorothalonil (*off-label*), chlorothalonil + metalaxyl-M, fosetyl-aluminium (*off-label*), propamocarb hydrochloride
	Foliar diseases	flusilazole (*off-label*)
	Light leaf spot	prothioconazole, tebuconazole, tebuconazole + trifloxystrobin
	Phoma leaf spot	tebuconazole + trifloxystrobin
	Phytophthora	copper oxychloride (*off-label*), propamocarb hydrochloride
	Powdery mildew	azoxystrobin + difenoconazole, prothioconazole, tebuconazole, tebuconazole + trifloxystrobin
	Pythium	metalaxyl-M, propamocarb hydrochloride
	Rhizoctonia	tolclofos-methyl (*off-label*)
	Ring spot	azoxystrobin, boscalid + pyraclostrobin, chlorothalonil, chlorothalonil + metalaxyl-M (*reduction*), difenoconazole, difenoconazole (*off-label*), prothioconazole, tebuconazole, tebuconazole (*off-label*), tebuconazole + trifloxystrobin
	Spear rot	copper oxychloride (*off-label*)
	Stem canker	copper oxychloride (*off-label*), flusilazole (*off-label*), prothioconazole

Crop/Pest Guide - Arable and vegetable crops

	Storage rots	metalaxyl-M (*off-label*)
	White blister	azoxystrobin, azoxystrobin + difenoconazole, boscalid + pyraclostrobin, boscalid + pyraclostrobin (*qualified minor use*), chlorothalonil + metalaxyl-M, mancozeb + metalaxyl-M (*off-label*), tebuconazole + trifloxystrobin
Pests	Aphids	acetamiprid (*off-label*), bifenthrin, chlorpyrifos, cypermethrin, deltamethrin (*off-label*), dimethoate, dimethoate (*off-label*), esfenvalerate, fatty acids, fatty acids (*off-label*), lambda-cyhalothrin + pirimicarb, pirimicarb, pymetrozine, pymetrozine (*off-label*), pymetrozine (*useful levels of control*), pyrethrins, spirotetramat, thiacloprid
	Beetles	alpha-cypermethrin, deltamethrin, deltamethrin (*off-label*), lambda-cyhalothrin + pirimicarb, pyrethrins
	Birds/mammals	aluminium ammonium sulphate
	Caterpillars	alpha-cypermethrin, Bacillus thuringiensis, Bacillus thuringiensis (*off-label*), bifenthrin, chlorpyrifos, cypermethrin, deltamethrin, deltamethrin (*off-label*), diflubenzuron, esfenvalerate, indoxacarb, indoxacarb (*off-label*), lambda-cyhalothrin, lambda-cyhalothrin + pirimicarb, pyrethrins, spinosad, teflubenzuron (*off-label*)
	Cutworms	chlorpyrifos
	Flies	chlorpyrifos, chlorpyrifos (*off-label*), teflubenzuron (*off-label*)
	Leatherjackets	chlorpyrifos
	Mealybugs	fatty acids
	Pests, miscellaneous	alpha-cypermethrin (*off-label*), deltamethrin (*off-label*), dimethoate, dimethoate (*off-label*)
	Scale insects	fatty acids
	Slugs/snails	metaldehyde, methiocarb
	Spider mites	fatty acids
	Thrips	fatty acids (*off-label*)
	Whiteflies	acetamiprid (*off-label*), bifenthrin, chlorpyrifos, cypermethrin, fatty acids, lambda-cyhalothrin, lambda-cyhalothrin + pirimicarb, pyrethrins, spirotetramat
Weeds	Broad-leaved weeds	carbetamide, clomazone (*off-label*), clopyralid, clopyralid (*off-label*), metazachlor, metazachlor (*off-label*), napropamide, napropamide (*off-label*), pendimethalin, pendimethalin (*off-label*), propyzamide (*off-label*), pyridate, pyridate (*off-label*)
	Crops as weeds	carbetamide, cycloxydim, pendimethalin, tepraloxydim
	Grass weeds	carbetamide, cycloxydim, fluazifop-P-butyl (*off-label*), metazachlor, metazachlor (*off-label*), napropamide, napropamide (*off-label*), pendimethalin, pendimethalin (*off-label*), propyzamide (*off-label*), tepraloxydim

Brassicas - Mustard

Crop control	Desiccation	diquat (*off-label*), glyphosate
Diseases	Alternaria	fludioxonil + metalaxyl-M + thiamethoxam (*moderate control*)
	Disease control/foliar feed	boscalid (*off-label*), difenoconazole (*off-label*), tebuconazole (*off-label*)
	Downy mildew	fludioxonil + metalaxyl-M + thiamethoxam

Crop/Pest Guide - Arable and vegetable crops

	Foliar diseases	cyproconazole (*off-label*), difenoconazole (*off-label*), tebuconazole (*off-label*)
	Pythium	fludioxonil + metalaxyl-M + thiamethoxam (*reduction*)
	Seed-borne diseases	fludioxonil + metalaxyl-M + thiamethoxam
Pests	Aphids	lambda-cyhalothrin (*off-label*), lambda-cyhalothrin + pirimicarb (*off-label*), pirimicarb (*off-label*), spirotetramat (*off-label*), tau-fluvalinate (*off-label*)
	Beetles	beta-cyfluthrin + imidacloprid (*off-label - seed treatment*), deltamethrin, fludioxonil + metalaxyl-M + thiamethoxam (*reduction of damage*), thiacloprid
	Caterpillars	deltamethrin
	Flies	tefluthrin (*off-label - seed treatment*)
	Pests, miscellaneous	lambda-cyhalothrin (*off-label*), tau-fluvalinate (*off-label*)
	Weevils	deltamethrin
	Whiteflies	lambda-cyhalothrin (*off-label*), spirotetramat (*off-label*)
Weeds	Broad-leaved weeds	bifenox (*off-label*), carbetamide (*off-label*), clopyralid (*off-label*), clopyralid + picloram (*off-label*), glyphosate, metazachlor (*off-label*), metazachlor + quinmerac (*off-label*), napropamide (*off-label*)
	Crops as weeds	glyphosate, propaquizafop (*off-label*)
	Grass weeds	carbetamide (*off-label*), cycloxydim (*off-label*), fluazifop-P-butyl (*off-label*), glyphosate, metazachlor (*off-label*), metazachlor + quinmerac (*off-label*), napropamide (*off-label*), propaquizafop (*off-label*), tepraloxydim (*off-label*)
	Weeds, miscellaneous	glyphosate

Brassicas - Root brassicas

Diseases	Alternaria	azoxystrobin (*off-label*), azoxystrobin + difenoconazole (*off-label*), cyprodinil + fludioxonil (*off-label*), fenpropimorph (*off-label*), iprodione + thiophanate-methyl (*off-label*), tebuconazole, tebuconazole (*off-label*)
	Botrytis	propamocarb hydrochloride (*off-label*)
	Crown rot	fenpropimorph (*off-label*), iprodione + thiophanate-methyl (*off-label*)
	Damping off	propamocarb hydrochloride (*off-label*), thiram (*off-label - seed treatment*), thiram (*seed treatment*), tolclofos-methyl (*off-label*)
	Disease control/foliar feed	tebuconazole (*off-label*)
	Downy mildew	metalaxyl-M (*off-label*), propamocarb hydrochloride (*off-label*)
	Foliar diseases	azoxystrobin (*off-label*), flusilazole (*off-label*), tebuconazole (*off-label*), tebuconazole + trifloxystrobin (*off-label*)
	Fungus diseases	tebuconazole (*off-label*)
	Phytophthora	propamocarb hydrochloride (*off-label*)
	Powdery mildew	azoxystrobin (*off-label*), azoxystrobin + difenoconazole (*off-label*), fenpropimorph (*off-label*), sulphur, sulphur (*off-label*), tebuconazole, tebuconazole (*off-label*)
	Pythium	metalaxyl-M, metalaxyl-M (*off-label*)
	Rhizoctonia	azoxystrobin (*off-label*), tolclofos-methyl (*off-label*), tolclofos-methyl (*with fleece or mesh covers*), tolclofos-methyl (*without covers*)

	Ring spot	tebuconazole (*off-label*)
	Sclerotinia	boscalid + pyraclostrobin (*off-label*), cyprodinil + fludioxonil (*off-label*), tebuconazole (*off-label*)
	Seed-borne diseases	thiram (*off-label - seed treatment*)
	Stem canker	flusilazole (*off-label*)
	White blister	metalaxyl-M (*off-label*), propamocarb hydrochloride (*off-label*)
Pests	Aphids	deltamethrin (*off-label*), esfenvalerate, fatty acids (*off-label*), lambda-cyhalothrin + pirimicarb (*off-label*), pirimicarb, pirimicarb (*off-label*)
	Beetles	deltamethrin, deltamethrin (*off-label*)
	Caterpillars	deltamethrin, deltamethrin (*off-label*), esfenvalerate, lambda-cyhalothrin (*off-label*)
	Cutworms	Bacillus thuringiensis (*off-label*), cypermethrin (*off-label*), lambda-cyhalothrin (*off-label*), lambda-cyhalothrin + pirimicarb (*off-label*)
	Flies	chlorpyrifos (*off-label*), lambda-cyhalothrin (*off-label*)
	Leafhoppers	deltamethrin (*off-label*)
	Pests, miscellaneous	deltamethrin (*off-label*), dimethoate (*off-label*), lambda-cyhalothrin (*off-label*)
	Thrips	deltamethrin (*off-label*), fatty acids (*off-label*)
	Weevils	lambda-cyhalothrin (*off-label*)
Plant growth regulation	Quality/yield control	sulphur
Weeds	Broad-leaved weeds	clomazone (*off-label*), clopyralid, dimethenamid-p + metazachlor (*off-label*), glyphosate, metamitron (*off-label*), metazachlor, pendimethalin (*off-label*), prosulfocarb (*off-label*)
	Crops as weeds	cycloxydim, fluazifop-P-butyl (*stockfeed only*), glyphosate
	Grass weeds	cycloxydim, cycloxydim (*off-label*), fluazifop-P-butyl (*stockfeed only*), glyphosate, metamitron (*off-label*), metazachlor, pendimethalin (*off-label*), propaquizafop, propaquizafop (*off-label*), prosulfocarb (*off-label*), tepraloxydim (*off-label*)
	Weeds, miscellaneous	glyphosate, glyphosate (*off-label*), metamitron (*off-label*)

Brassicas - Salad greens

Diseases	Alternaria	azoxystrobin (*off-label - for baby leaf production*), iprodione (*off-label*)
	Aphanomyces	fatty acids (*off-label*)
	Bacterial blight	copper oxychloride (*off-label*)
	Black rot	copper oxychloride (*off-label*)
	Botrytis	fenhexamid (*off-label - baby leaf production*), propamocarb hydrochloride (*off-label*)
	Bottom rot	copper oxychloride (*off-label*)
	Damping off	cymoxanil + fludioxonil + metalaxyl-M (*off-label*), fosetyl-aluminium + propamocarb hydrochloride (*off-label - baby leaf production*), propamocarb hydrochloride (*off-label*), tolclofos-methyl
	Downy mildew	azoxystrobin (*off-label - for baby leaf production*), copper oxychloride (*off-label*), dimethomorph + mancozeb (*off-label*), dimethomorph + mancozeb (*off-label - for baby*

Crop/Pest Guide - Arable and vegetable crops

		leaf production), fosetyl-aluminium (*off-label*), fosetyl-aluminium (*off-label - for baby leaf production*), fosetyl-aluminium + propamocarb hydrochloride (*off-label - baby leaf production*), propamocarb hydrochloride, propamocarb hydrochloride (*off-label*)
	Foliar diseases	fosetyl-aluminium (*off-label*)
	Phytophthora	copper oxychloride (*off-label*), propamocarb hydrochloride, propamocarb hydrochloride (*off-label*)
	Pythium	metalaxyl-M, propamocarb hydrochloride
	Rhizoctonia	tolclofos-methyl (*off-label*), tolclofos-methyl (*off-label - baby leaf production only*)
	Ring spot	difenoconazole (*off-label*), tebuconazole (*off-label*)
	Spear rot	copper oxychloride (*off-label*)
	Stem canker	copper oxychloride (*off-label*)
	White blister	boscalid + pyraclostrobin (*off-label*)
Pests	Aphids	acetamiprid (*off-label*), chlorpyrifos, cypermethrin (*off-label*), cypermethrin (*off-label - outdoor and protected crops*), deltamethrin (*off-label*), esfenvalerate, fatty acids (*off-label*), lambda-cyhalothrin (*off-label*), pirimicarb, pirimicarb (*for baby leaf production*), pirimicarb (*off-label*), pirimicarb (*off-label - baby leaf production only*), pirimicarb (*off-label - for baby leaf production*), pymetrozine (*off-label*), pymetrozine (*off-label - for baby leaf production*), spirotetramat (*off-label*), thiacloprid (*off-label*), thiacloprid (*off-label - baby leaf production*)
	Beetles	deltamethrin (*off-label*), deltamethrin (*off-label - baby leaf production*)
	Birds/mammals	aluminium ammonium sulphate
	Caterpillars	alpha-cypermethrin (*off-label - for baby leaf production*), Bacillus thuringiensis (*off-label*), chlorpyrifos, cypermethrin (*off-label - outdoor and protected crops*), deltamethrin (*off-label*), deltamethrin (*off-label - baby leaf production*), diflubenzuron (*off-label - for baby leaf production*), esfenvalerate, teflubenzuron (*off-label*)
	Cutworms	chlorpyrifos
	Flies	chlorpyrifos, chlorpyrifos (*off-label*), teflubenzuron (*off-label*), tefluthrin (*off-label - seed treatment*)
	Leafhoppers	deltamethrin (*off-label*)
	Leatherjackets	chlorpyrifos
	Pests, miscellaneous	alpha-cypermethrin (*off-label*), deltamethrin (*off-label*), dimethoate (*off-label*), lambda-cyhalothrin (*off-label*)
	Thrips	deltamethrin (*off-label*), fatty acids (*off-label*)
	Whiteflies	chlorpyrifos, spirotetramat (*off-label*)
Weeds	Broad-leaved weeds	clomazone (*off-label*), metazachlor (*off-label*), napropamide (*off-label*), pendimethalin (*off-label*), pendimethalin (*off-label - for baby leaf production*), propyzamide (*off-label*)
	Grass weeds	fluazifop-P-butyl (*off-label*), metazachlor (*off-label*), napropamide (*off-label*), pendimethalin (*off-label*), propyzamide (*off-label*)

Cereals - Barley

Crop control	Desiccation	diquat, diquat (*stockfeed only*), glyphosate
Diseases	Covered smut	carboxin + thiram (*seed treatment*), clothianidin + prothioconazole (*seed treatment*), clothianidin +

Crop/Pest Guide - Arable and vegetable crops

	prothioconazole + tebuconazole + triazoxide (*seed treatment*), fludioxonil (*seed treatment*), fludioxonil + flutriafol (*seed treatment*), fludioxonil + tefluthrin, fluquinconazole + prochloraz (*seed treatment*), fuberidazole + imidacloprid + triadimenol (*seed treatment*), fuberidazole + triadimenol (*seed treatment*), guazatine + imazalil (*seed treatment*), prochloraz + triticonazole (*seed treatment*), prothioconazole (*seed treatment*), prothioconazole + tebuconazole + triazoxide (*seed treatment*)
Ear blight	prothioconazole + tebuconazole
Eyespot	azoxystrobin + cyproconazole (*reduction*), bixafen + prothioconazole (*reduction of incidence and severity*), boscalid + epoxiconazole (*moderate*), boscalid + epoxiconazole (*moderate control*), carbendazim + flusilazole, chlorothalonil + cyproconazole, cyproconazole (*useful reduction*), cyproconazole + cyprodinil, cyprodinil, cyprodinil + picoxystrobin, epoxiconazole (*reduction*), epoxiconazole + fenpropimorph (*reduction*), epoxiconazole + fenpropimorph + kresoxim-methyl (*reduction*), epoxiconazole + kresoxim-methyl (*reduction*), epoxiconazole + metconazole (*reduction*), epoxiconazole + prochloraz, fluoxastrobin + prothioconazole (*reduction*), fluoxastrobin + prothioconazole + trifloxystrobin (*reduction*), flusilazole, prochloraz, prochloraz + propiconazole, prochloraz + proquinazid + tebuconazole, prochloraz + tebuconazole, prothioconazole, prothioconazole + spiroxamine, prothioconazole + spiroxamine + tebuconazole (*reduction*), prothioconazole + tebuconazole, prothioconazole + tebuconazole (*reduction*), prothioconazole + trifloxystrobin (*reduction*), prothioconazole + trifloxystrobin (*reduction in severity*)
Foot rot	carboxin + thiram (*seed treatment*), clothianidin + prothioconazole (*seed treatment*), clothianidin + prothioconazole + tebuconazole + triazoxide (*seed treatment*), fludioxonil (*seed treatment*), fludioxonil + flutriafol (*seed treatment*), fludioxonil + tefluthrin, fluquinconazole + prochloraz (*seed treatment*), fuberidazole + imidacloprid + triadimenol (*seed treatment*), fuberidazole + imidacloprid + triadimenol (*seed treatment - reduction*), fuberidazole + triadimenol (*seed treatment*), guazatine (*seed treatment*), guazatine (*seed treatment - reduction*), guazatine + imazalil (*seed treatment*), guazatine + imazalil (*seed treatment - moderate control*), prochloraz + triticonazole (*seed treatment*), prothioconazole (*seed treatment*), prothioconazole + tebuconazole + triazoxide (*seed treatment*)
Late ear diseases	bixafen + prothioconazole, fluoxastrobin + prothioconazole, fluoxastrobin + prothioconazole + trifloxystrobin, prothioconazole, prothioconazole + spiroxamine + tebuconazole, prothioconazole + tebuconazole
Leaf stripe	carboxin + thiram (*seed treatment*), carboxin + thiram (*seed treatment - reduction*), clothianidin + prothioconazole (*seed treatment*), clothianidin + prothioconazole + tebuconazole + triazoxide (*seed treatment*), fludioxonil (*seed treatment - reduction*),

Crop/Pest Guide - Arable and vegetable crops

	fludioxonil + flutriafol (*seed treatment*), fludioxonil + tefluthrin (*partial control*), fuberidazole + triadimenol (*seed treatment*), guazatine (*seed treatment*), guazatine + imazalil (*seed treatment*), imazalil, prochloraz + triticonazole (*seed treatment*), prothioconazole (*seed treatment*), prothioconazole + tebuconazole + triazoxide (*seed treatment*), tebuconazole
Loose smut	carboxin + thiram (*seed treatment - reduction*), clothianidin + prothioconazole (*seed treatment*), clothianidin + prothioconazole + tebuconazole + triazoxide (*seed treatment*), fludioxonil + flutriafol (*seed treatment*), fluquinconazole + prochloraz (*seed treatment*), fuberidazole + imidacloprid + triadimenol (*seed treatment*), fuberidazole + triadimenol (*seed treatment*), prochloraz + triticonazole (*seed treatment*), prothioconazole (*seed treatment*), prothioconazole + tebuconazole + triazoxide (*seed treatment*), tebuconazole
Net blotch	azoxystrobin, azoxystrobin + chlorothalonil, azoxystrobin + cyproconazole, azoxystrobin + fenpropimorph, bixafen + prothioconazole, boscalid + epoxiconazole, carboxin + thiram (*seed treatment*), chlorothalonil + cyproconazole, chlorothalonil + cyproconazole + propiconazole (*reduction*), chlorothalonil + picoxystrobin, clothianidin + prothioconazole + tebuconazole + triazoxide (*seed treatment*), cyproconazole (*useful reduction*), cyproconazole + cyprodinil, cyproconazole + picoxystrobin (*moderate control*), cyproconazole + propiconazole, cyproconazole + trifloxystrobin, cyprodinil, cyprodinil + isopyrazam, cyprodinil + picoxystrobin, epoxiconazole, epoxiconazole + fenpropimorph, epoxiconazole + fenpropimorph + kresoxim-methyl, epoxiconazole + fenpropimorph + metrafenone, epoxiconazole + fenpropimorph + pyraclostrobin, epoxiconazole + kresoxim-methyl, epoxiconazole + kresoxim-methyl + pyraclostrobin, epoxiconazole + metconazole, epoxiconazole + metrafenone, epoxiconazole + prochloraz, epoxiconazole + pyraclostrobin, famoxadone + flusilazole, fenpropimorph + flusilazole, fenpropimorph + pyraclostrobin, fluoxastrobin + prothioconazole, fluoxastrobin + prothioconazole + trifloxystrobin, flusilazole, fuberidazole + imidacloprid + triadimenol (*seed treatment - seed-borne only*), guazatine (*seed treatment*), guazatine + imazalil (*seed treatment - moderate control*), imazalil, mancozeb, maneb, metconazole (*reduction*), picoxystrobin, prochloraz, prochloraz + proquinazid + tebuconazole (*moderate control*), prochloraz + tebuconazole, propiconazole + tebuconazole, prothioconazole, prothioconazole + spiroxamine, prothioconazole + spiroxamine + tebuconazole, prothioconazole + tebuconazole, prothioconazole + tebuconazole + triazoxide (*seed treatment*), prothioconazole + trifloxystrobin, pyraclostrobin, spiroxamine + tebuconazole, tebuconazole, tebuconazole + triadimenol, trifloxystrobin
Powdery mildew	azoxystrobin, azoxystrobin + cyproconazole, azoxystrobin + fenpropimorph, bixafen + prothioconazole, boscalid + epoxiconazole,

Crop/Pest Guide - Arable and vegetable crops

	carbendazim + flusilazole, chlorothalonil + cyproconazole, chlorothalonil + cyproconazole + propiconazole, cyflufenamid, cyproconazole, cyproconazole + cyprodinil, cyproconazole + propiconazole, cyproconazole + trifloxystrobin, cyprodinil, cyprodinil + isopyrazam, cyprodinil + picoxystrobin, epoxiconazole, epoxiconazole + fenpropimorph, epoxiconazole + fenpropimorph + kresoxim-methyl, epoxiconazole + fenpropimorph + metrafenone, epoxiconazole + fenpropimorph + pyraclostrobin, epoxiconazole + kresoxim-methyl, epoxiconazole + metconazole (*moderate control*), epoxiconazole + metrafenone, epoxiconazole + prochloraz, epoxiconazole + prochloraz (*moderate control*), fenpropidin, fenpropimorph, fenpropimorph + flusilazole, fenpropimorph + kresoxim-methyl, fenpropimorph + pyraclostrobin, fenpropimorph + quinoxyfen, fluoxastrobin + prothioconazole, fluoxastrobin + prothioconazole + trifloxystrobin, fluquinconazole + prochloraz, flusilazole, flusilazole (*moderate control*), flutriafol, fuberidazole + imidacloprid + triadimenol (*seed treatment*), metconazole, metconazole (*moderate control*), metrafenone, picoxystrobin, prochloraz, prochloraz + proquinazid + tebuconazole, prochloraz + tebuconazole, propiconazole, propiconazole + tebuconazole, proquinazid, prothioconazole, prothioconazole + spiroxamine, prothioconazole + spiroxamine + tebuconazole, prothioconazole + tebuconazole, prothioconazole + trifloxystrobin, quinoxyfen, spiroxamine, spiroxamine + tebuconazole, sulphur, tebuconazole, tebuconazole + triadimenol
Ramularia leaf spots	bixafen + prothioconazole, cyprodinil + isopyrazam, epoxiconazole + metconazole, epoxiconazole + metconazole (*moderate control*)
Rhynchosporium	azoxystrobin, azoxystrobin + chlorothalonil, azoxystrobin + chlorothalonil (*moderate control*), azoxystrobin + cyproconazole (*moderate control*), azoxystrobin + fenpropimorph, bixafen + prothioconazole, boscalid + epoxiconazole (*moderate*), boscalid + epoxiconazole (*moderate control*), chlorothalonil, chlorothalonil (*moderate control*), chlorothalonil + cyproconazole, chlorothalonil + cyproconazole + propiconazole (*reduction*), chlorothalonil + flusilazole, chlorothalonil + picoxystrobin, chlorothalonil + propiconazole, cyproconazole, cyproconazole + cyprodinil, cyproconazole + picoxystrobin (*moderate control*), cyproconazole + propiconazole, cyproconazole + trifloxystrobin, cyprodinil, cyprodinil (*moderate control*), cyprodinil + isopyrazam, cyprodinil + picoxystrobin, epoxiconazole, epoxiconazole + fenpropimorph, epoxiconazole + fenpropimorph + kresoxim-methyl, epoxiconazole + fenpropimorph + metrafenone, epoxiconazole + fenpropimorph + pyraclostrobin, epoxiconazole + kresoxim-methyl, epoxiconazole + kresoxim-methyl + pyraclostrobin, epoxiconazole + metconazole, epoxiconazole + metrafenone, epoxiconazole + prochloraz, epoxiconazole + pyraclostrobin, famoxadone + flusilazole, fenpropidin (*moderate control*), fenpropimorph, fenpropimorph +

Crop/Pest Guide - Arable and vegetable crops

	flusilazole, fenpropimorph + kresoxim-methyl, fenpropimorph + pyraclostrobin, fluoxastrobin + prothioconazole, fluoxastrobin + prothioconazole + trifloxystrobin, fluquinconazole + prochloraz, flusilazole, flusilazole (*moderate control*), flutriafol, mancozeb, maneb, metconazole, picoxystrobin, prochloraz, prochloraz + propiconazole, prochloraz + proquinazid + tebuconazole (*moderate control*), prochloraz + tebuconazole, propiconazole, propiconazole + tebuconazole, prothioconazole, prothioconazole + spiroxamine, prothioconazole + spiroxamine + tebuconazole, prothioconazole + tebuconazole, prothioconazole + trifloxystrobin, pyraclostrobin (*moderate*), spiroxamine (*reduction*), spiroxamine + tebuconazole, tebuconazole, tebuconazole + triadimenol, trifloxystrobin
Rust	azoxystrobin, azoxystrobin + chlorothalonil, azoxystrobin + cyproconazole, azoxystrobin + fenpropimorph, bixafen + prothioconazole, boscalid + epoxiconazole, carbendazim + flusilazole, chlorothalonil + cyproconazole, chlorothalonil + cyproconazole + propiconazole, chlorothalonil + picoxystrobin, chlorothalonil + propiconazole, cyproconazole, cyproconazole + cyprodinil, cyproconazole + picoxystrobin, cyproconazole + propiconazole, cyproconazole + trifloxystrobin, cyprodinil + isopyrazam, cyprodinil + picoxystrobin, epoxiconazole, epoxiconazole + fenpropimorph, epoxiconazole + fenpropimorph + kresoxim-methyl, epoxiconazole + fenpropimorph + metrafenone, epoxiconazole + fenpropimorph + pyraclostrobin, epoxiconazole + kresoxim-methyl, epoxiconazole + kresoxim-methyl + pyraclostrobin, epoxiconazole + metconazole, epoxiconazole + metrafenone, epoxiconazole + prochloraz, epoxiconazole + pyraclostrobin, famoxadone + flusilazole, fenpropidin, fenpropidin (*moderate control*), fenpropimorph, fenpropimorph + flusilazole, fenpropimorph + pyraclostrobin, fluoxastrobin + prothioconazole, fluoxastrobin + prothioconazole + trifloxystrobin, fluquinconazole + prochloraz, fluquinconazole + prochloraz (*seed treatment*), flusilazole, flutriafol, fuberidazole + imidacloprid + triadimenol (*seed treatment*), mancozeb, maneb, metconazole, picoxystrobin, prochloraz + proquinazid + tebuconazole, prochloraz + tebuconazole, propiconazole, propiconazole + tebuconazole, prothioconazole, prothioconazole + spiroxamine, prothioconazole + spiroxamine + tebuconazole, prothioconazole + tebuconazole, prothioconazole + trifloxystrobin, pyraclostrobin, spiroxamine, spiroxamine + tebuconazole, tebuconazole, tebuconazole + triadimenol, trifloxystrobin
Seed-borne diseases	fluquinconazole, ipconazole
Septoria	flusilazole
Snow mould	fludioxonil (*seed treatment*), fludioxonil + flutriafol (*seed treatment*), fludioxonil + tefluthrin, fuberidazole + imidacloprid + triadimenol (*seed treatment - reduction*), guazatine + imazalil (*seed treatment*)
Soil-borne diseases	ipconazole
Sooty moulds	mancozeb, maneb, prothioconazole + tebuconazole

Crop/Pest Guide - Arable and vegetable crops

	Take-all	azoxystrobin (*reduction*), azoxystrobin + chlorothalonil (*reduction*), azoxystrobin + chlorothalonil (*reduction only*), azoxystrobin + cyproconazole (*reduction*), azoxystrobin + fenpropimorph (*reduction*), fluoxastrobin + prothioconazole (*reduction*), fluquinconazole + prochloraz (*seed treatment - reduction*), silthiofam (*seed treatment*)
Pests	Aphids	alpha-cypermethrin, bifenthrin, chlorpyrifos, clothianidin (*seed treatment*), clothianidin + prothioconazole (*seed treatment*), clothianidin + prothioconazole + tebuconazole + triazoxide (*seed treatment*), cypermethrin, cypermethrin (*autumn sown*), cypermethrin (*moderate control*), deltamethrin, esfenvalerate, fuberidazole + imidacloprid + triadimenol (*seed treatment*), lambda-cyhalothrin, lambda-cyhalothrin + pirimicarb, pirimicarb, tau-fluvalinate, zeta-cypermethrin
	Birds/mammals	aluminium ammonium sulphate
	Flies	alpha-cypermethrin, alpha-cypermethrin (*some control if present at application*), chlorpyrifos, cypermethrin, cypermethrin (*autumn sown*), fludioxonil + tefluthrin
	Leafhoppers	clothianidin (*seed treatment*), clothianidin + prothioconazole (*seed treatment*), clothianidin + prothioconazole + tebuconazole + triazoxide (*seed treatment*)
	Leatherjackets	chlorpyrifos
	Midges	chlorpyrifos
	Slugs/snails	clothianidin (*seed treatment*), clothianidin + prothioconazole (*seed treatment*), clothianidin + prothioconazole + tebuconazole + triazoxide (*seed treatment - reduction*)
	Wireworms	clothianidin (*seed treatment*), clothianidin + prothioconazole (*seed treatment*), clothianidin + prothioconazole + tebuconazole + triazoxide (*seed treatment - reduction*), fludioxonil + tefluthrin, fuberidazole + imidacloprid + triadimenol (*reduction of damage*)
Plant growth regulation	Growth control	2-chloroethylphosphonic acid, 2-chloroethylphosphonic acid + mepiquat chloride, chlormequat, chlormequat + 2-chloroethylphosphonic acid, chlormequat + 2-chloroethylphosphonic acid + mepiquat chloride, mepiquat chloride + prohexadione-calcium, trinexapac-ethyl
	Quality/yield control	2-chloroethylphosphonic acid + mepiquat chloride (*low lodging situations*), chlormequat, mepiquat chloride + prohexadione-calcium, sulphur
Weeds	Broad-leaved weeds	2,4-D, 2,4-D + MCPA, 2,4-DB, amidosulfuron, amidosulfuron + iodosulfuron-methyl-sodium, bifenox, bromoxynil, bromoxynil + diflufenican + ioxynil, bromoxynil + ioxynil, carfentrazone-ethyl, carfentrazone-ethyl + mecoprop-P, carfentrazone-ethyl + metsulfuron-methyl, chlorotoluron, chlorotoluron (*off-label*), chlorotoluron + diflufenican, clopyralid, clopyralid + florasulam + fluroxypyr, dicamba + MCPA + mecoprop-P, dicamba + mecoprop-P, dichlorprop-P, dichlorprop-P + ioxynil, dichlorprop-P + MCPA + mecoprop-P, diflufenican, diflufenican + flufenacet, diflufenican + flufenacet + flurtamone, diflufenican + flurtamone,

Crop/Pest Guide - Arable and vegetable crops

		diflufenican + mecoprop-P, diflufenican + pendimethalin, florasulam, florasulam + fluroxypyr, flufenacet + pendimethalin, flupyrsulfuron-methyl, fluroxypyr, fluroxypyr + thifensulfuron-methyl + tribenuron-methyl, glyphosate, iodosulfuron-methyl-sodium, MCPA, mecoprop-P, metsulfuron-methyl, metsulfuron-methyl + thifensulfuron-methyl, metsulfuron-methyl + tribenuron-methyl, pendimethalin, pendimethalin + picolinafen, picolinafen, prosulfocarb, thifensulfuron-methyl + tribenuron-methyl, tribenuron-methyl
	Crops as weeds	amidosulfuron + iodosulfuron-methyl-sodium, diflufenican, diflufenican + flufenacet, diflufenican + flurtamone, florasulam, fluroxypyr, glyphosate, iodosulfuron-methyl-sodium, metsulfuron-methyl + tribenuron-methyl, pendimethalin
	Grass weeds	chlorotoluron, chlorotoluron (*off-label*), chlorotoluron + diflufenican, chlorotoluron + diflufenican (*off-label*), diflufenican, diflufenican + flufenacet, diflufenican + flufenacet + flurtamone, diflufenican + flurtamone, diflufenican + pendimethalin, diquat (*animal feed*), fenoxaprop-P-ethyl, flufenacet + pendimethalin, flupyrsulfuron-methyl, glyphosate, pendimethalin, pendimethalin + picolinafen, pinoxaden, prosulfocarb, prosulfocarb (*off-label*), tralkoxydim, tri-allate
	Weeds, miscellaneous	diquat (*animal feed only*), fluroxypyr, glyphosate

Cereals - Cereals, general

Crop control	Desiccation	diquat, glyphosate (*off-label*), glyphosate (*off-label - for wild bird seed production*)
Diseases	Powdery mildew	cyflufenamid (*off-label*)
Pests	Flies	chlorpyrifos
	Leatherjackets	methiocarb (*reduction*)
	Pests, miscellaneous	lambda-cyhalothrin (*off-label*)
	Slugs/snails	methiocarb
Plant growth regulation	Growth control	chlormequat (*off-label*)
Weeds	Broad-leaved weeds	2,4-DB + MCPA, bromoxynil + prosulfuron (*off-label*), dicamba + MCPA + mecoprop-P (*off-label*), diflufenican (*off-label*), fluroxypyr (*off-label - grown for wild bird seed production*), isoxaben (*off-label - grown for game cover*), MCPA, MCPB (*off-label*), mecoprop-P (*off-label*), metsulfuron-methyl + tribenuron-methyl (*off-label*)
	Grass weeds	chlorotoluron (*off-label*)
	Weeds, miscellaneous	clopyralid (*off-label - game cover*), diquat, glyphosate (*off-label*), glyphosate (*off-label - grown for wild bird seed production*)

Cereals - Maize/sweetcorn

Diseases	Damping off	fludioxonil + metalaxyl M, thiram (*seed treatment*)
	Eyespot	flusilazole (*off-label*)
	Foliar diseases	flusilazole (*off-label*), pyraclostrobin (*off-label*)
	Fusarium	fludioxonil + metalaxyl M
	Pythium	fludioxonil + metalaxyl M

Crop/Pest Guide - Arable and vegetable crops

Pests	Aphids	pirimicarb, pirimicarb (*off-label*), pymetrozine (*off-label*)
	Caterpillars	Bacillus thuringiensis (*off-label*), indoxacarb (*off-label*)
	Flies	chlorpyrifos, clothianidin, lambda-cyhalothrin (*off-label*)
	Pests, miscellaneous	lambda-cyhalothrin (*off-label*)
	Slugs/snails	methiocarb
	Symphylids	clothianidin (*reduction*)
	Wireworms	clothianidin
Weeds	Broad-leaved weeds	bromoxynil, bromoxynil + prosulfuron, bromoxynil + terbuthylazine, bromoxynil + terbuthylazine (*off-label*), clopyralid, flufenacet + isoxaflutole, flufenacet + isoxaflutole (*off-label*), fluroxypyr, fluroxypyr (*off-label*), isoxaben (*off-label - grown for game cover*), mesotrione, mesotrione (*off-label*), mesotrione + terbuthylazine, mesotrione + terbuthylazine (*off-label*), nicosulfuron, nicosulfuron (*off-label*), pendimethalin, pendimethalin (*off-label*), pendimethalin (*off-label - under covers*), pendimethalin + terbuthylazine, rimsulfuron, S-metolachlor, S-metolachlor (*moderately susceptible*)
	Crops as weeds	fluroxypyr, fluroxypyr (*off-label*), mesotrione, nicosulfuron, pendimethalin, rimsulfuron
	Grass weeds	bromoxynil + terbuthylazine (*off-label*), flufenacet + isoxaflutole, flufenacet + isoxaflutole (*off-label*), mesotrione, mesotrione (*off-label*), mesotrione + terbuthylazine, mesotrione + terbuthylazine (*off-label*), nicosulfuron, nicosulfuron (*off-label*), pendimethalin, pendimethalin (*off-label*), pendimethalin + terbuthylazine, S-metolachlor
	Weeds, miscellaneous	fluroxypyr

Cereals - Oats

Crop control	Desiccation	diquat, diquat (*stockfeed only*), glyphosate
Diseases	Bunt	fluoxastrobin + prothioconazole
	Covered smut	carboxin + thiram (*seed treatment*), fluoxastrobin + prothioconazole, prochloraz + triticonazole (*seed treatment*)
	Crown rust	azoxystrobin, azoxystrobin + cyproconazole, azoxystrobin + fenpropimorph, bixafen + prothioconazole, boscalid + epoxiconazole, cyproconazole, cyproconazole + picoxystrobin, epoxiconazole + fenpropimorph + metrafenone, epoxiconazole + fenpropimorph + pyraclostrobin, epoxiconazole + kresoxim-methyl + pyraclostrobin, epoxiconazole + metrafenone, epoxiconazole + pyraclostrobin, epoxiconazole + pyraclostrobin (*qualified minor use*), fenpropimorph + pyraclostrobin, fluoxastrobin + prothioconazole, picoxystrobin, prochloraz + proquinazid + tebuconazole (*qualified minor use recommendation*), propiconazole, prothioconazole, prothioconazole + spiroxamine, prothioconazole + spiroxamine + tebuconazole, prothioconazole + tebuconazole, pyraclostrobin, tebuconazole, tebuconazole (*reduction*), tebuconazole + triadimenol
	Eyespot	bixafen + prothioconazole (*reduction of incidence and severity*), boscalid + epoxiconazole (*moderate control*),

Crop/Pest Guide - Arable and vegetable crops

		epoxiconazole + fenpropimorph + kresoxim-methyl (*reduction*), epoxiconazole + kresoxim-methyl (*reduction*), fluoxastrobin + prothioconazole, fluoxastrobin + prothioconazole (*Reduction of incidence and severity*), prothioconazole, prothioconazole + spiroxamine, prothioconazole + spiroxamine + tebuconazole, prothioconazole + tebuconazole
	Foot rot	carboxin + thiram (*seed treatment*), clothianidin + prothioconazole (*seed treatment*), difenoconazole + fludioxonil, fludioxonil (*seed treatment*), fludioxonil + flutriafol (*seed treatment*), fludioxonil + tefluthrin, fluoxastrobin + prothioconazole, fuberidazole + imidacloprid + triadimenol (*seed treatment - reduction*), fuberidazole + triadimenol (*seed treatment*), guazatine (*seed treatment - reduction*), prochloraz + triticonazole (*seed treatment*), prothioconazole (*seed treatment*)
	Leaf spot	difenoconazole + fludioxonil, fludioxonil (*seed treatment*), fludioxonil + flutriafol (*seed treatment*), fludioxonil + tefluthrin, fuberidazole + imidacloprid + triadimenol (*seed treatment*), fuberidazole + triadimenol (*seed treatment*), guazatine (*seed treatment*), guazatine + imazalil (*seed treatment*)
	Loose smut	carboxin + thiram (*seed treatment - reduction*), clothianidin + prothioconazole (*seed treatment*), fluoxastrobin + prothioconazole, fuberidazole + imidacloprid + triadimenol (*seed treatment*), fuberidazole + triadimenol (*seed treatment*), prochloraz + triticonazole (*seed treatment*), prothioconazole (*seed treatment*)
	Powdery mildew	azoxystrobin, azoxystrobin + cyproconazole, azoxystrobin + fenpropimorph, bixafen + prothioconazole, boscalid + epoxiconazole, cyflufenamid, cyproconazole, epoxiconazole, epoxiconazole + fenpropimorph, epoxiconazole + fenpropimorph + kresoxim-methyl, epoxiconazole + fenpropimorph + metrafenone, epoxiconazole + fenpropimorph + pyraclostrobin, epoxiconazole + kresoxim-methyl, epoxiconazole + metrafenone, fenpropimorph, fenpropimorph + kresoxim-methyl, fenpropimorph + pyraclostrobin, fenpropimorph + quinoxyfen, fluoxastrobin + prothioconazole, fuberidazole + imidacloprid + triadimenol (*seed treatment*), metrafenone (*evidence of mildew control on oats is limited*), picoxystrobin, prochloraz + proquinazid + tebuconazole, propiconazole, proquinazid, prothioconazole, prothioconazole + spiroxamine, prothioconazole + spiroxamine + tebuconazole, prothioconazole + tebuconazole, quinoxyfen, sulphur, tebuconazole, tebuconazole + triadimenol
	Seed-borne diseases	difenoconazole + fludioxonil
	Snow mould	fludioxonil (*seed treatment*)
Pests	Aphids	bifenthrin, clothianidin (*seed treatment*), clothianidin + prothioconazole (*seed treatment*), cypermethrin, cypermethrin (*moderate control*), deltamethrin, fuberidazole + imidacloprid + triadimenol (*seed treatment*), lambda-cyhalothrin, lambda-cyhalothrin + pirimicarb, pirimicarb, zeta-cypermethrin
	Birds/mammals	aluminium ammonium sulphate
	Flies	chlorpyrifos, cypermethrin

Crop/Pest Guide - Arable and vegetable crops

	Leafhoppers	clothianidin (*seed treatment*), clothianidin + prothioconazole (*seed treatment*)
	Leatherjackets	chlorpyrifos
	Midges	chlorpyrifos
	Slugs/snails	clothianidin (*seed treatment*), clothianidin + prothioconazole (*seed treatment*)
	Thrips	chlorpyrifos
	Wireworms	clothianidin (*seed treatment*), clothianidin + prothioconazole (*seed treatment*), fludioxonil + tefluthrin, fuberidazole + imidacloprid + triadimenol (*reduction of damage*)
Plant growth regulation	Growth control	chlormequat, trinexapac-ethyl
	Quality/yield control	chlormequat, sulphur
Weeds	Broad-leaved weeds	2,4-D, 2,4-D + MCPA, 2,4-DB, amidosulfuron, bromoxynil, bromoxynil + ioxynil, carfentrazone-ethyl, carfentrazone-ethyl + flupyrsulfuron-methyl, carfentrazone-ethyl + mecoprop-P, carfentrazone-ethyl + metsulfuron-methyl, clopyralid, clopyralid + florasulam + fluroxypyr, dicamba + MCPA + mecoprop-P, dicamba + mecoprop-P, dichlorprop-P, dichlorprop-P + ioxynil, dichlorprop-P + MCPA + mecoprop-P, diflufenican (*off-label*), diflufenican + flupyrsulfuron-methyl (*off-label*), diflufenican + flurtamone (*off-label*), florasulam, florasulam + fluroxypyr, flupyrsulfuron-methyl, flupyrsulfuron-methyl + thifensulfuron-methyl, fluroxypyr, glyphosate, MCPA, mecoprop-P, metsulfuron-methyl, metsulfuron-methyl + thifensulfuron-methyl, metsulfuron-methyl + tribenuron-methyl, thifensulfuron-methyl + tribenuron-methyl, tribenuron-methyl
	Crops as weeds	florasulam, flumioxazine (*off-label*), fluroxypyr, glyphosate, metsulfuron-methyl + tribenuron-methyl
	Grass weeds	carfentrazone-ethyl + flupyrsulfuron-methyl, diflufenican + flupyrsulfuron-methyl (*off-label*), diflufenican + flurtamone (*off-label*), diquat (*animal feed*), flumioxazine (*off-label*), flupyrsulfuron-methyl + thifensulfuron-methyl, glyphosate
	Weeds, miscellaneous	diquat (*animal feed only*), fluroxypyr, glyphosate

Cereals - Rye/triticale

Diseases	Blue mould	fuberidazole + triadimenol (*seed treatment*)
	Bunt	clothianidin + prothioconazole (*seed treatment*), fluoxastrobin + prothioconazole, prothioconazole (*seed treatment*)
	Disease control/foliar feed	chlorothalonil (*off-label*), chlorothalonil + mancozeb (*off-label*), difenoconazole (*off-label*), mancozeb (*off-label*), tebuconazole (*off-label*)
	Ear blight	epoxiconazole + metconazole (*good reduction*), epoxiconazole + prochloraz (*good reduction*), tebuconazole (*off-label*), thiophanate-methyl (*reduction only*)
	Eyespot	bixafen + prothioconazole (*reduction of incidence and severity*), bixafen + prothioconazole + spiroxamine, bixafen + prothioconazole + tebuconazole (*reduction of*

	incidence and severity), epoxiconazole + fenpropimorph + kresoxim-methyl (*reduction*), epoxiconazole + kresoxim-methyl (*reduction*), epoxiconazole + metconazole (*reduction*), epoxiconazole + prochloraz, fluoxastrobin + prothioconazole (*reduction*), prochloraz, prochloraz + tebuconazole, prothioconazole, prothioconazole + spiroxamine, prothioconazole + spiroxamine + tebuconazole (*reduction*), prothioconazole + tebuconazole, prothioconazole + tebuconazole (*reduction*)
Foliar diseases	azoxystrobin + chlorothalonil (*off-label*), azoxystrobin + cyproconazole (*off-label*), chlorothalonil (*off-label*), chlorothalonil + flutriafol (*off-label*), cyproconazole (*off-label*), cyproconazole + cyprodinil (*off-label*), cyproconazole + propiconazole (*off-label*), cyproconazole + trifloxystrobin (*off-label*), difenoconazole (*off-label*), dimoxystrobin + epoxiconazole (*off-label*), epoxiconazole + kresoxim-methyl + pyraclostrobin (*off-label*), epoxiconazole + pyraclostrobin (*off-label*), famoxadone + flusilazole (*off-label*), fenpropidin (*off-label*), fenpropimorph + flusilazole (*off-label*), fenpropimorph + pyraclostrobin (*off-label*), fluquinconazole (*off-label*), fluquinconazole + prochloraz (*off-label*), prochloraz (*off-label*), pyraclostrobin (*off-label*), tebuconazole (*off-label*), trifloxystrobin (*off-label*)
Foot rot	carboxin + thiram (*seed treatment*), clothianidin + prothioconazole (*seed treatment*), fluoxastrobin + prothioconazole, fuberidazole + triadimenol (*seed treatment*), prochloraz + triticonazole (*seed treatment*), prothioconazole (*seed treatment*)
Fusarium	thiophanate-methyl
Late ear diseases	bixafen + prothioconazole, bixafen + prothioconazole + spiroxamine, bixafen + prothioconazole + tebuconazole, prothioconazole + spiroxamine
Loose smut	fluoxastrobin + prothioconazole
Net blotch	prothioconazole + tebuconazole
Powdery mildew	azoxystrobin, azoxystrobin + cyproconazole, azoxystrobin + fenpropimorph, bixafen + prothioconazole, bixafen + prothioconazole + spiroxamine, bixafen + prothioconazole + tebuconazole, cyflufenamid, cyproconazole, epoxiconazole, epoxiconazole + fenpropimorph, epoxiconazole + fenpropimorph + kresoxim-methyl, epoxiconazole + fenpropimorph + metrafenone, epoxiconazole + kresoxim-methyl, epoxiconazole + metconazole (*moderate control*), epoxiconazole + metconazole (*moderate reduction*), epoxiconazole + metrafenone, epoxiconazole + prochloraz (*moderate control*), fenpropimorph, fenpropimorph + kresoxim-methyl, fenpropimorph + quinoxyfen, fluoxastrobin + prothioconazole, prochloraz, prochloraz + tebuconazole, propiconazole, proquinazid, prothioconazole, prothioconazole + spiroxamine, prothioconazole + spiroxamine + tebuconazole, prothioconazole + tebuconazole, quinoxyfen, spiroxamine, spiroxamine + tebuconazole, sulphur, sulphur (*off-label*), tebuconazole, tebuconazole (*off-label*), tebuconazole + triadimenol
Rhynchosporium	azoxystrobin, azoxystrobin + cyproconazole (*moderate control*), azoxystrobin + fenpropimorph, azoxystrobin +

		fenpropimorph (*reduction*), bixafen + prothioconazole, bixafen + prothioconazole + spiroxamine, bixafen + prothioconazole + tebuconazole, epoxiconazole, epoxiconazole + fenpropimorph, epoxiconazole + fenpropimorph + kresoxim-methyl, epoxiconazole + fenpropimorph + metrafenone, epoxiconazole + kresoxim-methyl, epoxiconazole + metrafenone, epoxiconazole + prochloraz, fenpropimorph + kresoxim-methyl, fluoxastrobin + prothioconazole, prochloraz, prochloraz + tebuconazole, propiconazole, prothioconazole, prothioconazole + spiroxamine, prothioconazole + spiroxamine + tebuconazole, prothioconazole + tebuconazole, spiroxamine (*reduction*), tebuconazole
	Rust	azoxystrobin, azoxystrobin + cyproconazole, azoxystrobin + fenpropimorph, bixafen + prothioconazole, bixafen + prothioconazole + spiroxamine, bixafen + prothioconazole + tebuconazole, cyproconazole, epoxiconazole, epoxiconazole + fenpropimorph, epoxiconazole + fenpropimorph + kresoxim-methyl, epoxiconazole + fenpropimorph + metrafenone, epoxiconazole + kresoxim-methyl, epoxiconazole + metconazole, epoxiconazole + metrafenone, epoxiconazole + prochloraz, fenpropimorph, fluoxastrobin + prothioconazole, prochloraz + tebuconazole, propiconazole, prothioconazole, prothioconazole + spiroxamine, prothioconazole + spiroxamine + tebuconazole, prothioconazole + tebuconazole, spiroxamine, spiroxamine + tebuconazole, tebuconazole, tebuconazole (*off-label*), tebuconazole + triadimenol
	Seed-borne diseases	fludioxonil (*off-label*), fludioxonil + tefluthrin (*off-label*), fluquinconazole (*off-label*), fluquinconazole + prochloraz (*off-label*), silthiofam (*off-label*)
	Septoria	bixafen + prothioconazole, bixafen + prothioconazole + spiroxamine, bixafen + prothioconazole + tebuconazole, epoxiconazole, epoxiconazole + fenpropimorph, epoxiconazole + fenpropimorph + kresoxim-methyl, epoxiconazole + fenpropimorph + metrafenone, epoxiconazole + kresoxim-methyl, epoxiconazole + metconazole, epoxiconazole + metrafenone, epoxiconazole + prochloraz, fenpropimorph + kresoxim-methyl (*reduction*), prochloraz, propiconazole, tebuconazole (*off-label*)
	Stripe smut	difenoconazole + fludioxonil
	Take-all	azoxystrobin + fenpropimorph (*reduction*)
	Tan spot	bixafen + prothioconazole, bixafen + prothioconazole + spiroxamine, bixafen + prothioconazole + tebuconazole
Pests	Aphids	clothianidin (*seed treatment*), clothianidin + prothioconazole (*seed treatment*), cypermethrin, cypermethrin (*moderate control*), deltamethrin (*off-label*), dimethoate, lambda-cyhalothrin (*off-label*), lambda-cyhalothrin + pirimicarb, pirimicarb, tau-fluvalinate (*off-label*), zeta-cypermethrin (*off-label*)
	Flies	chlorpyrifos, chlorpyrifos (*off-label*), cypermethrin
	Leafhoppers	clothianidin (*seed treatment*), clothianidin + prothioconazole (*seed treatment*)
	Leatherjackets	chlorpyrifos, chlorpyrifos (*off-label*)
	Midges	chlorpyrifos, chlorpyrifos (*off-label*)

Crop/Pest Guide - Arable and vegetable crops

	Pests, miscellaneous	alpha-cypermethrin (*off-label*), chlorpyrifos, chlorpyrifos (*off-label*), deltamethrin (*off-label*), lambda-cyhalothrin (*off-label*), zeta-cypermethrin (*off-label*)
	Slugs/snails	clothianidin (*seed treatment*), clothianidin + prothioconazole (*seed treatment*)
	Wireworms	clothianidin (*seed treatment*), clothianidin + prothioconazole (*seed treatment*)
Plant growth regulation	Growth control	2-chloroethylphosphonic acid, 2-chloroethylphosphonic acid + mepiquat chloride, chlormequat, chlormequat (*off-label*), chlormequat + 2-chloroethylphosphonic acid (*off-label*), chlormequat + 2-chloroethylphosphonic acid + mepiquat chloride (*off-label*), chlormequat + imazaquin (*off-label*), mepiquat chloride + prohexadione-calcium, trinexapac-ethyl
	Quality/yield control	mepiquat chloride + prohexadione-calcium
Weeds	Broad-leaved weeds	2,4-D, amidosulfuron, amidosulfuron + iodosulfuron-methyl-sodium, bifenox, bromoxynil (*off-label*), bromoxynil + diflufenican + ioxynil, bromoxynil + ioxynil, carfentrazone-ethyl, carfentrazone-ethyl (*off-label*), carfentrazone-ethyl + flupyrsulfuron-methyl, carfentrazone-ethyl + metsulfuron-methyl, chlorotoluron, chlorotoluron (*off-label*), chlorotoluron + diflufenican, clopyralid (*off-label*), dicamba + MCPA + mecoprop-P, dicamba + MCPA + mecoprop-P (*off-label*), dicamba + mecoprop-P, dicamba + mecoprop-P (*off-label*), dichlorprop-P + ioxynil, diflufenican, diflufenican + flufenacet (*off-label*), diflufenican + flurtamone (*off-label*), diflufenican + pendimethalin, florasulam (*off-label*), florasulam + fluroxypyr (*off-label*), florasulam + pyroxsulam, flupyrsulfuron-methyl, fluroxypyr, iodosulfuron-methyl-sodium, iodosulfuron-methyl-sodium + mesosulfuron-methyl (*off-label*), MCPA, MCPA + MCPB (*off-label*), MCPB (*off-label*), mecoprop-P (*off-label*), metsulfuron-methyl, metsulfuron-methyl + tribenuron-methyl, metsulfuron-methyl + tribenuron-methyl (*off-label*), pendimethalin, pendimethalin + pyroxsulam, sulfosulfuron (*off-label*), thifensulfuron-methyl + tribenuron-methyl, tribenuron-methyl
	Crops as weeds	amidosulfuron + iodosulfuron-methyl-sodium, diflufenican, florasulam + pyroxsulam, fluroxypyr, metsulfuron-methyl + tribenuron-methyl, pendimethalin, pendimethalin + pyroxsulam (*from seed*)
	Grass weeds	carfentrazone-ethyl + flupyrsulfuron-methyl, chlorotoluron, chlorotoluron (*off-label*), chlorotoluron + diflufenican, clodinafop-propargyl, diflufenican, diflufenican + flufenacet (*off-label*), diflufenican + flurtamone (*off-label*), diflufenican + pendimethalin, fenoxaprop-P-ethyl (*off-label*), florasulam + pyroxsulam, iodosulfuron-methyl-sodium (*from seed*), iodosulfuron-methyl-sodium + mesosulfuron-methyl (*off-label*), pendimethalin, pendimethalin + pyroxsulam, pendimethalin + pyroxsulam (*MS only*), prosulfocarb (*off-label*), sulfosulfuron (*off-label*), tralkoxydim, tri-allate
	Weeds, miscellaneous	fluroxypyr

Cereals - Undersown cereals

Weeds	Broad-leaved weeds	2,4-D, 2,4-DB + linuron + MCPA (*undersown with clover*), 2,4-DB + MCPA, dicamba + MCPA + mecoprop-P, dicamba + MCPA + mecoprop-P (*grass only*), dicamba + mecoprop-P, dichlorprop-P, MCPA, MCPA (*red clover or grass*), MCPA + MCPB, MCPB

Cereals - Wheat

Crop control	Desiccation	glyphosate
Diseases	Blue mould	fuberidazole + imidacloprid + triadimenol (*seed treatment*), fuberidazole + triadimenol (*seed treatment*)
	Bunt	carboxin + thiram (*seed treatment*), clothianidin + prothioconazole (*seed treatment*), difenoconazole + fludioxonil, fludioxonil (*seed treatment*), fludioxonil + flutriafol (*seed treatment*), fludioxonil + tefluthrin, fluoxastrobin + prothioconazole, fluquinconazole (*seed treatment*), fluquinconazole + prochloraz (*seed treatment*), fuberidazole + imidacloprid + triadimenol (*seed treatment*), fuberidazole + triadimenol (*seed treatment*), prochloraz + triticonazole (*off-label - seed treatment*), prochloraz + triticonazole (*seed treatment*), prothioconazole (*seed treatment*)
	Disease control/foliar feed	chlorothalonil (*off-label*), chlorothalonil + mancozeb (*off-label*), difenoconazole (*off-label*), mancozeb (*off-label*)
	Drechslera leaf spot	fenpropimorph + pyraclostrobin
	Ear blight	boscalid + epoxiconazole (*good reduction*), dimoxystrobin + epoxiconazole, epoxiconazole, epoxiconazole (*reduction*), epoxiconazole + fenpropimorph, epoxiconazole + fenpropimorph + kresoxim-methyl (*reduction*), epoxiconazole + fenpropimorph + metrafenone (*reduction*), epoxiconazole + fenpropimorph + pyraclostrobin (*reduction*), epoxiconazole + kresoxim-methyl (*reduction*), epoxiconazole + metconazole (*good reduction*), epoxiconazole + metrafenone (*reduction*), epoxiconazole + prochloraz (*good reduction*), epoxiconazole + pyraclostrobin (*apply during flowering*), epoxiconazole + pyraclostrobin (*reduction*), metconazole (*reduction*), prothioconazole + tebuconazole, prothioconazole + tebuconazole (*moderate control*), spiroxamine + tebuconazole, tebuconazole, tebuconazole + triadimenol, thiophanate-methyl (*reduction only*)
	Eyespot	azoxystrobin + cyproconazole (*reduction*), bixafen + prothioconazole (*reduction of incidence and severity*), bixafen + prothioconazole + spiroxamine, bixafen + prothioconazole + tebuconazole (*reduction of incidence and severity*), boscalid + epoxiconazole (*moderate*), boscalid + epoxiconazole (*moderate control*), carbendazim + flusilazole, chlorothalonil + cyproconazole, cyproconazole (*useful reduction*), cyproconazole + cyprodinil, cyproconazole + quinoxyfen (*reduction*), cyprodinil, cyprodinil + picoxystrobin (*moderate control*), epoxiconazole (*Reduction*), epoxiconazole + fenpropimorph (*reduction*), epoxiconazole + fenpropimorph + kresoxim-methyl (*reduction*), epoxiconazole + kresoxim-methyl (*reduction*), epoxiconazole + metconazole (*reduction*),

Crop/Pest Guide - Arable and vegetable crops

	epoxiconazole + prochloraz, fluoxastrobin + prothioconazole, fluoxastrobin + prothioconazole (*reduction*), fluoxastrobin + prothioconazole + trifloxystrobin (*reduction*), flusilazole, metrafenone (*reduction*), picoxystrobin (*reduction*), prochloraz, prochloraz + propiconazole, prochloraz + proquinazid + tebuconazole, prochloraz + tebuconazole, prothioconazole, prothioconazole + spiroxamine, prothioconazole + spiroxamine + tebuconazole (*reduction*), prothioconazole + tebuconazole, prothioconazole + tebuconazole (*reduction*), prothioconazole + trifloxystrobin
Foliar diseases	azoxystrobin (*off-label*), azoxystrobin + chlorothalonil (*off-label*), azoxystrobin + cyproconazole (*off-label*), azoxystrobin + fenpropimorph (*off-label*), boscalid + epoxiconazole (*off-label*), chlorothalonil (*off-label*), chlorothalonil + flutriafol (*off-label*), cyproconazole (*off-label*), cyproconazole + cyprodinil (*off-label*), cyproconazole + propiconazole (*off-label*), cyproconazole + trifloxystrobin (*off-label*), difenoconazole (*off-label*), dimoxystrobin + epoxiconazole (*off-label*), epoxiconazole (*off-label*), epoxiconazole + fenpropimorph (*off-label*), epoxiconazole + fenpropimorph + kresoxim-methyl (*off-label*), epoxiconazole + kresoxim-methyl (*off-label*), epoxiconazole + kresoxim-methyl + pyraclostrobin (*off-label*), epoxiconazole + pyraclostrobin (*off-label*), famoxadone + flusilazole (*off-label*), fenpropidin (*off-label*), fenpropimorph (*off-label*), fenpropimorph + flusilazole (*off-label*), fenpropimorph + kresoxim-methyl (*off-label*), fluquinconazole + prochloraz (*off-label*), pyraclostrobin (*off-label*), trifloxystrobin (*off-label*)
Foot rot	carboxin + thiram (*seed treatment*), clothianidin + prothioconazole (*seed treatment*), difenoconazole + fludioxonil, fludioxonil (*seed treatment*), fludioxonil + flutriafol (*seed treatment*), fludioxonil + tefluthrin, fluoxastrobin + prothioconazole, fluoxastrobin + prothioconazole (*reduction*), fluquinconazole + prochloraz (*seed treatment*), fuberidazole + imidacloprid + triadimenol (*seed treatment - reduction*), fuberidazole + triadimenol (*seed treatment*), guazatine (*seed treatment - reduction*), prochloraz + triticonazole (*off-label - seed treatment*), prochloraz + triticonazole (*seed treatment*), prothioconazole (*seed treatment*)
Fusarium	thiophanate-methyl
Late ear diseases	azoxystrobin, azoxystrobin + fenpropimorph (*moderate control*), bixafen + prothioconazole, bixafen + prothioconazole + spiroxamine, bixafen + prothioconazole + tebuconazole, chlorothalonil + flutriafol, cyproconazole, cyproconazole + propiconazole, fenpropidin + tebuconazole, fluoxastrobin + prothioconazole, fluoxastrobin + prothioconazole + trifloxystrobin, picoxystrobin, prochloraz + proquinazid + tebuconazole (*reduction only*), prochloraz + tebuconazole, prothioconazole, prothioconazole + spiroxamine, prothioconazole + spiroxamine + tebuconazole, prothioconazole + tebuconazole, prothioconazole + trifloxystrobin
Loose smut	clothianidin + prothioconazole (*seed treatment*), fludioxonil + flutriafol (*seed treatment*), fluoxastrobin +

Crop/Pest Guide - Arable and vegetable crops

	prothioconazole, fuberidazole + imidacloprid + triadimenol (*seed treatment*), fuberidazole + triadimenol (*seed treatment*), prochloraz + triticonazole (*off-label - seed treatment*), prochloraz + triticonazole (*seed treatment*), prothioconazole (*seed treatment*)
Powdery mildew	azoxystrobin, azoxystrobin + cyproconazole, bixafen + prothioconazole, bixafen + prothioconazole + spiroxamine, bixafen + prothioconazole + tebuconazole, boscalid + epoxiconazole, carbendazim + flusilazole, chlorothalonil + cyproconazole, chlorothalonil + cyproconazole + propiconazole (*moderate*), chlorothalonil + flutriafol, cyflufenamid, cyproconazole, cyproconazole + cyprodinil, cyproconazole + propiconazole, cyproconazole + quinoxyfen, cyproconazole + trifloxystrobin, cyprodinil (*moderate control*), epoxiconazole, epoxiconazole (*off-label*), epoxiconazole + fenpropimorph, epoxiconazole + fenpropimorph + metrafenone, epoxiconazole + kresoxim-methyl, epoxiconazole + metconazole (*moderate control*), epoxiconazole + metrafenone, epoxiconazole + prochloraz (*moderate control*), fenpropidin, fenpropimorph, fenpropimorph + flusilazole, fenpropimorph + quinoxyfen, fluoxastrobin + prothioconazole, fluoxastrobin + prothioconazole + trifloxystrobin, fluquinconazole, fluquinconazole + prochloraz (*moderate control*), flusilazole, flutriafol, fuberidazole + imidacloprid + triadimenol (*seed treatment*), metconazole (*moderate control*), metrafenone, prochloraz, prochloraz + proquinazid + tebuconazole, prochloraz + tebuconazole, propiconazole, propiconazole + tebuconazole, proquinazid, proquinazid (*off-label*), prothioconazole, prothioconazole + spiroxamine, prothioconazole + spiroxamine + tebuconazole, prothioconazole + tebuconazole, prothioconazole + trifloxystrobin, quinoxyfen, spiroxamine, spiroxamine + tebuconazole, sulphur, tebuconazole, tebuconazole + triadimenol
Rust	azoxystrobin, azoxystrobin + chlorothalonil, azoxystrobin + cyproconazole, azoxystrobin + fenpropimorph, bixafen + prothioconazole, bixafen + prothioconazole + spiroxamine, bixafen + prothioconazole + tebuconazole, boscalid + epoxiconazole, carbendazim + flusilazole, chlorothalonil + cyproconazole, chlorothalonil + cyproconazole + propiconazole, chlorothalonil + flutriafol, chlorothalonil + picoxystrobin, chlorothalonil + propiconazole, cyproconazole, cyproconazole + cyprodinil, cyproconazole + picoxystrobin, cyproconazole + propiconazole, cyproconazole + quinoxyfen, cyproconazole + trifloxystrobin, cyprodinil + picoxystrobin, difenoconazole, dimoxystrobin + epoxiconazole, epoxiconazole, epoxiconazole + fenpropimorph, epoxiconazole + fenpropimorph + kresoxim-methyl, epoxiconazole + fenpropimorph + metrafenone, epoxiconazole + fenpropimorph + pyraclostrobin, epoxiconazole + kresoxim-methyl, epoxiconazole + kresoxim-methyl + pyraclostrobin, epoxiconazole + metconazole, epoxiconazole + metrafenone, epoxiconazole + prochloraz, epoxiconazole + pyraclostrobin, famoxadone + flusilazole, fenpropidin (*moderate control*),

Crop/Pest Guide - Arable and vegetable crops

	fenpropimorph, fenpropimorph + flusilazole, fenpropimorph + pyraclostrobin, fluoxastrobin + prothioconazole, fluoxastrobin + prothioconazole + trifloxystrobin, fluquinconazole, fluquinconazole (*seed treatment*), fluquinconazole + prochloraz, fluquinconazole + prochloraz (*seed treatment*), fluquinconazole + prochloraz (*seed treatment - moderate control*), flusilazole, flutriafol, fuberidazole + imidacloprid + triadimenol (*seed treatment*), mancozeb, mancozeb (*useful control*), maneb, metconazole, picoxystrobin, prochloraz + proquinazid + tebuconazole, prochloraz + tebuconazole, propiconazole, propiconazole + tebuconazole, prothioconazole, prothioconazole + spiroxamine, prothioconazole + spiroxamine + tebuconazole, prothioconazole + tebuconazole, prothioconazole + trifloxystrobin, pyraclostrobin, spiroxamine, spiroxamine + tebuconazole, tebuconazole, tebuconazole + triadimenol, trifloxystrobin
Seed-borne diseases	difenoconazole + fludioxonil, fludioxonil (*off-label*), fluquinconazole, fluquinconazole (*off-label*), fluquinconazole + prochloraz (*off-label*), ipconazole, silthiofam (*off-label*)
Septoria	azoxystrobin, azoxystrobin + chlorothalonil, azoxystrobin + cyproconazole, azoxystrobin + fenpropimorph, bixafen + prothioconazole, bixafen + prothioconazole + spiroxamine, bixafen + prothioconazole + tebuconazole, boscalid + epoxiconazole, carbendazim + flusilazole, carboxin + thiram (*seed treatment*), chlorothalonil, chlorothalonil + cyproconazole, chlorothalonil + cyproconazole + propiconazole, chlorothalonil + flusilazole, chlorothalonil + flutriafol, chlorothalonil + mancozeb, chlorothalonil + picoxystrobin, chlorothalonil + propiconazole, chlorothalonil + tebuconazole, cyproconazole, cyproconazole + cyprodinil, cyproconazole + picoxystrobin, cyproconazole + propiconazole, cyproconazole + quinoxyfen, cyproconazole + trifloxystrobin, cyprodinil + picoxystrobin, difenoconazole, difenoconazole + fludioxonil, dimoxystrobin + epoxiconazole, epoxiconazole, epoxiconazole + fenpropimorph, epoxiconazole + fenpropimorph + kresoxim-methyl, epoxiconazole + fenpropimorph + metrafenone, epoxiconazole + fenpropimorph + pyraclostrobin, epoxiconazole + kresoxim-methyl, epoxiconazole + kresoxim-methyl + pyraclostrobin, epoxiconazole + metconazole, epoxiconazole + metrafenone, epoxiconazole + prochloraz, epoxiconazole + pyraclostrobin, famoxadone + flusilazole, fenpropidin + tebuconazole, fenpropimorph + flusilazole, fenpropimorph + kresoxim-methyl (*reduction*), fenpropimorph + pyraclostrobin, fludioxonil (*seed treatment*), fludioxonil + flutriafol (*seed treatment*), fludioxonil + tefluthrin, fluoxastrobin + prothioconazole, fluoxastrobin + prothioconazole (*reduction*), fluoxastrobin + prothioconazole + trifloxystrobin, fluquinconazole, fluquinconazole (*seed treatment*), fluquinconazole + prochloraz, fluquinconazole + prochloraz (*seed treatment*), fluquinconazole + prochloraz (*seed treatment - moderate control*), flusilazole, flutriafol, fuberidazole +

Crop/Pest Guide - Arable and vegetable crops

		imidacloprid + triadimenol (*seed treatment*), fuberidazole + imidacloprid + triadimenol (*seed treatment - reduction*), guazatine (*seed treatment*), guazatine (*seed treatment - partial control*), mancozeb, mancozeb (*reduction*), maneb, metconazole, picoxystrobin, prochloraz, prochloraz + propiconazole, prochloraz + proquinazid + tebuconazole, prochloraz + proquinazid + tebuconazole (*moderate control*), prochloraz + tebuconazole, prochloraz + triticonazole (*off-label - seed treatment*), prochloraz + triticonazole (*seed treatment*), propiconazole, propiconazole + tebuconazole, prothioconazole, prothioconazole + spiroxamine, prothioconazole + spiroxamine + tebuconazole, prothioconazole + tebuconazole, prothioconazole + trifloxystrobin, pyraclostrobin, spiroxamine + tebuconazole, tebuconazole, tebuconazole + triadimenol, trifloxystrobin
	Sharp eyespot	fluoxastrobin + prothioconazole, fluoxastrobin + prothioconazole (*reduction*)
	Snow mould	difenoconazole + fludioxonil, fludioxonil (*seed treatment*), fludioxonil + flutriafol (*seed treatment*), fludioxonil + tefluthrin
	Soil-borne diseases	ipconazole
	Sooty moulds	boscalid + epoxiconazole, epoxiconazole (*reduction*), epoxiconazole + fenpropimorph (*reduction*), epoxiconazole + fenpropimorph + kresoxim-methyl (*reduction*), epoxiconazole + fenpropimorph (*reduction*), epoxiconazole + fenpropimorph + metrafenone (*reduction*), epoxiconazole + kresoxim-methyl (*reduction*), epoxiconazole + metconazole (*good reduction*), epoxiconazole + metrafenone (*reduction*), epoxiconazole + prochloraz (*good reduction*), fluoxastrobin + prothioconazole (*reduction*), mancozeb, maneb, propiconazole, prothioconazole + tebuconazole, prothioconazole + tebuconazole (*reduction*), spiroxamine + tebuconazole, tebuconazole, tebuconazole + triadimenol
	Take-all	azoxystrobin (*reduction*), azoxystrobin + chlorothalonil (*reduction*), azoxystrobin + chlorothalonil (*reduction only*), azoxystrobin + cyproconazole (*reduction*), azoxystrobin + fenpropimorph (*reduction*), fluoxastrobin + prothioconazole (*reduction*), fluquinconazole (*seed treatment - reduction*), fluquinconazole + prochloraz (*seed treatment - reduction*), silthiofam (*seed treatment*)
	Tan spot	bixafen + prothioconazole, bixafen + prothioconazole + spiroxamine, bixafen + prothioconazole + tebuconazole, boscalid + epoxiconazole (*reduction only*), chlorothalonil + picoxystrobin, dimoxystrobin + epoxiconazole, epoxiconazole + metconazole (*moderate control*), epoxiconazole + pyraclostrobin, fluoxastrobin + prothioconazole, fluoxastrobin + prothioconazole + trifloxystrobin, prothioconazole, prothioconazole + spiroxamine, prothioconazole + tebuconazole
Pests	Aphids	alpha-cypermethrin, bifenthrin, chlorpyrifos, clothianidin (*seed treatment*), clothianidin + prothioconazole (*seed treatment*), cypermethrin, cypermethrin (*autumn sown*), cypermethrin (*moderate control*), deltamethrin, deltamethrin (*off-label*), dimethoate, esfenvalerate, flonicamid, fuberidazole + imidacloprid + triadimenol (*seed treatment*), lambda-cyhalothrin, lambda-cyhalothrin + pirimicarb, pirimicarb, tau-fluvalinate, tau-

		fluvalinate (*off-label*), zeta-cypermethrin, zeta-cypermethrin (*off-label*)
	Beetles	lambda-cyhalothrin
	Birds/mammals	aluminium ammonium sulphate
	Caterpillars	lambda-cyhalothrin
	Flies	alpha-cypermethrin, alpha-cypermethrin (*some control if present at application*), chlorpyrifos, chlorpyrifos (*off-label*), cypermethrin, cypermethrin (*autumn sown*), dimethoate, fludioxonil + tefluthrin, lambda-cyhalothrin
	Leafhoppers	clothianidin (*seed treatment*), clothianidin + prothioconazole (*seed treatment*)
	Leatherjackets	chlorpyrifos, chlorpyrifos (*off-label*)
	Midges	chlorpyrifos, chlorpyrifos (*off-label*), lambda-cyhalothrin, thiacloprid
	Pests, miscellaneous	alpha-cypermethrin (*off-label*), chlorpyrifos, chlorpyrifos (*off-label*), deltamethrin (*off-label*), zeta-cypermethrin (*off-label*)
	Slugs/snails	clothianidin (*seed treatment*), clothianidin + prothioconazole (*seed treatment*)
	Suckers	lambda-cyhalothrin
	Thrips	chlorpyrifos
	Weevils	lambda-cyhalothrin
	Wireworms	clothianidin (*seed treatment*), clothianidin + prothioconazole (*seed treatment*), fludioxonil + tefluthrin, fuberidazole + imidacloprid + triadimenol (*reduction of damage*)
Plant growth regulation	Growth control	2-chloroethylphosphonic acid, 2-chloroethylphosphonic acid (*off-label*), 2-chloroethylphosphonic acid + mepiquat chloride, 2-chloroethylphosphonic acid + mepiquat chloride (*off-label*), chlormequat, chlormequat (*off-label*), chlormequat + 2-chloroethylphosphonic acid, chlormequat + 2-chloroethylphosphonic acid (*off-label*), chlormequat + 2-chloroethylphosphonic acid + mepiquat chloride, chlormequat + 2-chloroethylphosphonic acid + mepiquat chloride (*off-label*), chlormequat + imazaquin, chlormequat + mepiquat chloride, mepiquat chloride + prohexadione-calcium, mepiquat chloride + prohexadione-calcium (*off-label*), trinexapac-ethyl, trinexapac-ethyl (*off-label - cv A C Barrie only*)
	Quality/yield control	chlormequat + imazaquin, mepiquat chloride + prohexadione-calcium, mepiquat chloride + prohexadione-calcium (*off-label*), sulphur
Weeds	Broad-leaved weeds	2,4-D, 2,4-D + MCPA, 2,4-DB, 2,4-DB + MCPA, amidosulfuron, amidosulfuron (*off-label*), amidosulfuron + iodosulfuron-methyl-sodium, amidosulfuron + iodosulfuron-methyl-sodium (*off-label*), bifenox, bifenox (*off-label*), bromoxynil, bromoxynil (*off-label*), bromoxynil + diflufenican + ioxynil, bromoxynil + ioxynil, carfentrazone-ethyl, carfentrazone-ethyl + flupyrsulfuron-methyl, carfentrazone-ethyl + flupyrsulfuron-methyl (*off-label*), carfentrazone-ethyl + mecoprop-P, carfentrazone-ethyl + metsulfuron-methyl, chlorotoluron, chlorotoluron (*off-label*), chlorotoluron + diflufenican, clodinafop-propargyl + prosulfocarb, clopyralid, clopyralid (*off-label*), clopyralid + florasulam + fluroxypyr, dicamba + MCPA + mecoprop-P, dicamba + MCPA + mecoprop-P (*off-label*), dicamba + mecoprop-

P, dicamba + mecoprop-P (*off-label*), dichlorprop-P, dichlorprop-P + ioxynil, dichlorprop-P + MCPA + mecoprop-P, diflufenican, diflufenican + flufenacet, diflufenican + flufenacet (*off-label*), diflufenican + flufenacet + flurtamone, diflufenican + flupyrsulfuron-methyl, diflufenican + flurtamone, diflufenican + iodosulfuron-methyl-sodium + mesosulfuron-methyl, diflufenican + mecoprop-P, diflufenican + pendimethalin, florasulam, florasulam + fluroxypyr, florasulam + fluroxypyr (*off-label*), florasulam + pyroxsulam, flufenacet + pendimethalin, flumioxazine, flupyrsulfuron-methyl, flupyrsulfuron-methyl + thifensulfuron-methyl, fluroxypyr, fluroxypyr + thifensulfuron-methyl + tribenuron-methyl, glyphosate, iodosulfuron-methyl-sodium, iodosulfuron-methyl-sodium (*off-label*), iodosulfuron-methyl-sodium + mesosulfuron-methyl, iodosulfuron-methyl-sodium + mesosulfuron-methyl (*off-label*), iodosulfuron-methyl-sodium + mesosulfuron-methyl (*post-em up to 8 true leaves*), MCPA, MCPB (*off-label*), mecoprop-P, mecoprop-P (*off-label*), metsulfuron-methyl, metsulfuron-methyl + thifensulfuron-methyl, metsulfuron-methyl + thifensulfuron-methyl (*off-label*), metsulfuron-methyl + tribenuron-methyl, metsulfuron-methyl + tribenuron-methyl (*off-label*), pendimethalin, pendimethalin (*off-label*), pendimethalin + picolinafen, pendimethalin + pyroxsulam, picolinafen, prosulfocarb, prosulfocarb (*off-label*), sulfosulfuron, thifensulfuron-methyl + tribenuron-methyl, tribenuron-methyl

Crops as weeds
amidosulfuron + iodosulfuron-methyl-sodium, diflufenican, diflufenican + flufenacet, diflufenican + flurtamone, diflufenican + iodosulfuron-methyl-sodium + mesosulfuron-methyl, florasulam, florasulam + pyroxsulam, flumioxazine, fluroxypyr, glyphosate, iodosulfuron-methyl-sodium (*off-label - from seed*), metsulfuron-methyl + tribenuron-methyl, pendimethalin, pendimethalin + pyroxsulam (*from seed*)

Grass weeds
carfentrazone-ethyl + flupyrsulfuron-methyl, chlorotoluron, chlorotoluron (*off-label*), chlorotoluron + diflufenican, chlorotoluron + diflufenican (*off-label*), clodinafop-propargyl, clodinafop-propargyl + pinoxaden, clodinafop-propargyl + pinoxaden (*from seed*), clodinafop-propargyl + prosulfocarb, diflufenican, diflufenican + flufenacet, diflufenican + flufenacet (*off-label*), diflufenican + flufenacet + flurtamone, diflufenican + flufenacet + flurtamone (*moderately susceptible*), diflufenican + flupyrsulfuron-methyl, diflufenican + flurtamone, diflufenican + iodosulfuron-methyl-sodium + mesosulfuron-methyl, diflufenican + pendimethalin, fenoxaprop-P-ethyl, florasulam + pyroxsulam, flufenacet + pendimethalin, flumioxazine, flupyrsulfuron-methyl, flupyrsulfuron-methyl + thifensulfuron-methyl, glyphosate, iodosulfuron-methyl-sodium (*from seed*), iodosulfuron-methyl-sodium + mesosulfuron-methyl, iodosulfuron-methyl-sodium + mesosulfuron-methyl (*off-label*), iodosulfuron-methyl-sodium + mesosulfuron-methyl (*post em up to GS29*), iodosulfuron-methyl-sodium + mesosulfuron-methyl (*post em up to GS31*), iodosulfuron-methyl-sodium + mesosulfuron-methyl (*post em up to GS39 (EMR resistant up to GS29)*), iodosulfuron-methyl-sodium +

		mesosulfuron-methyl (*post-em up to GS30*), iodosulfuron-methyl-sodium + mesosulfuron-methyl (*post-em up to GS31*), pendimethalin, pendimethalin (*off-label*), pendimethalin + picolinafen, pendimethalin + pyroxsulam, pendimethalin + pyroxsulam (*MS only*), pinoxaden, propoxycarbazone-sodium, prosulfocarb, prosulfocarb (*off-label*), sulfosulfuron, sulfosulfuron (*moderate control of barren brome*), sulfosulfuron (*moderate control only*), tralkoxydim, tri-allate, tri-allate (*off-label*)
	Weeds, miscellaneous	fluroxypyr, glyphosate

Edible fungi - Mushrooms

Diseases	Bacterial blotch	sodium hypochlorite (commodity substance)
Pests	Flies	deltamethrin (*off-label*), diflubenzuron (*off-label - other than mushrooms*)

Fruiting vegetables - Aubergines

Diseases	Botrytis	azoxystrobin (*off-label*), pyrimethanil (*off-label*)
	Didymella stem rot	azoxystrobin (*off-label*)
	Phytophthora	azoxystrobin (*off-label*), propamocarb hydrochloride
	Powdery mildew	azoxystrobin (*off-label*), sulphur (*off-label*)
	Pythium	propamocarb hydrochloride
Pests	Aphids	fatty acids (*off-label*), pirimicarb (*off-label*)
	Leaf miners	oxamyl (*off-label*)
	Thrips	fatty acids (*off-label*)
	Whiteflies	Lecanicillium lecanii

Fruiting vegetables - Cucurbits

Diseases	Botrytis	mepanipyrim (*off-label*)
	Disease control/foliar feed	mancozeb
	Phytophthora	propamocarb hydrochloride
	Powdery mildew	azoxystrobin (*off-label*), bupirimate (*off-label*), bupirimate (*outdoor only*), myclobutanil (*off-label*)
	Pythium	propamocarb hydrochloride
Pests	Aphids	fatty acids, fatty acids (*off-label*), pirimicarb, pirimicarb (*off-label*)
	Mealybugs	fatty acids
	Pests, miscellaneous	thiacloprid (*off-label*)
	Scale insects	fatty acids
	Spider mites	fatty acids
	Thrips	fatty acids (*off-label*)
	Whiteflies	fatty acids
Weeds	Broad-leaved weeds	isoxaben (*off-label*), propyzamide (*off-label*)
	Grass weeds	propyzamide (*off-label*)

Fruiting vegetables - Peppers

Diseases	Damping off	propamocarb hydrochloride (*off-label*)
	Phytophthora	propamocarb hydrochloride
	Pythium	propamocarb hydrochloride
Pests	Aphids	fatty acids, pirimicarb
	Mealybugs	fatty acids
	Scale insects	fatty acids
	Spider mites	fatty acids
	Whiteflies	fatty acids

Fruiting vegetables - Tomatoes

Diseases	Botrytis	fenhexamid (*off-label*), pyrimethanil (*off-label*)
	Damping off	copper oxychloride
	Foot rot	copper oxychloride
	Phytophthora	copper oxychloride, propamocarb hydrochloride
	Pythium	propamocarb hydrochloride
Pests	Aphids	fatty acids, pirimicarb, pyrethrins
	Beetles	pyrethrins
	Caterpillars	pyrethrins
	Mealybugs	fatty acids
	Scale insects	fatty acids
	Spider mites	fatty acids
	Whiteflies	fatty acids, pyrethrins

Herb crops - Herbs

Crop control	Desiccation	diquat (*off-label*)
Diseases	Alternaria	cyprodinil + fludioxonil (*off-label*)
	Botrytis	Bacillus subtilis (*off-label*), cyprodinil + fludioxonil (*off-label*), fenhexamid (*off-label*), propamocarb hydrochloride (*off-label*)
	Damping off	cymoxanil + fludioxonil + metalaxyl-M (*off-label*), fosetyl-aluminium + propamocarb hydrochloride (*off-label*), propamocarb hydrochloride (*off-label*)
	Disease control/foliar feed	mancozeb + metalaxyl-M (*off-label*), prochloraz (*off-label*)
	Downy mildew	copper oxychloride (*off-label*), dimethomorph + mancozeb (*off-label*), fosetyl-aluminium (*off-label*), fosetyl-aluminium + propamocarb hydrochloride (*off-label*), metalaxyl-M (*off-label*), propamocarb hydrochloride (*off-label*)
	Foliar diseases	difenoconazole (*off-label*), fosetyl-aluminium (*off-label*), tebuconazole (*off-label*)
	Phytophthora	propamocarb hydrochloride (*off-label*)
	Powdery mildew	azoxystrobin (*off-label*), tebuconazole (*off-label*)
	Pythium	metalaxyl-M (*off-label*)
	Ring spot	azoxystrobin (*off-label*)
	Rust	azoxystrobin (*off-label*), tebuconazole (*off-label*)
	Sclerotinia	cyprodinil + fludioxonil (*off-label*)
	Seed-borne diseases	thiram (*seed soak*)
	White blister	boscalid + pyraclostrobin (*off-label*)

Crop/Pest Guide - Arable and vegetable crops

Pests	Aphids	acetamiprid (*off-label*), cypermethrin (*off-label*), deltamethrin (*off-label*), fatty acids (*off-label*), lambda-cyhalothrin (*off-label*), lambda-cyhalothrin + pirimicarb (*off-label*), pirimicarb (*off-label*), spirotetramat (*off-label*)
	Beetles	deltamethrin (*off-label*), deltamethrin (*off-label - baby leaf production*)
	Caterpillars	Bacillus thuringiensis (*off-label*), deltamethrin (*off-label*), deltamethrin (*off-label - baby leaf production*), diflubenzuron (*off-label*)
	Cutworms	lambda-cyhalothrin (*off-label*)
	Flies	lambda-cyhalothrin (*off-label*), tefluthrin (*off-label - seed treatment*)
	Leafhoppers	deltamethrin (*off-label*)
	Pests, miscellaneous	deltamethrin (*off-label*), lambda-cyhalothrin (*off-label*), spinosad
	Thrips	deltamethrin (*off-label*), fatty acids (*off-label*), pirimicarb (*off-label*)
	Whiteflies	spirotetramat (*off-label*)
Weeds	Broad-leaved weeds	asulam (*off-label*), bentazone (*off-label*), chloridazon (*off-label*), chloridazon + quinmerac (*off-label*), clomazone (*off-label*), clopyralid (*off-label*), lenacil (*off-label*), linuron, linuron (*off-label*), metamitron (*off-label*), metazachlor (*off-label*), metazachlor + quinmerac (*off-label*), pendimethalin, pendimethalin (*off-label*), phenmedipham (*off-label*), propyzamide (*off-label*), prosulfocarb (*off-label*)
	Crops as weeds	pendimethalin, propaquizafop (*off-label*)
	Grass weeds	asulam (*off-label*), chloridazon (*off-label*), fluazifop-P-butyl (*off-label*), lenacil (*off-label*), linuron, linuron (*off-label*), metamitron (*off-label*), metazachlor (*off-label*), metazachlor + quinmerac (*off-label*), pendimethalin, pendimethalin (*off-label*), propaquizafop (*off-label*), propyzamide (*off-label*), prosulfocarb (*off-label*), tepraloxydim (*off-label*)
	Weeds, miscellaneous	diquat (*off-label*), glyphosate (*off-label*)

Leafy vegetables - Endives

Diseases	Botrytis	cyprodinil + fludioxonil (*off-label*), fenhexamid (*off-label*), propamocarb hydrochloride (*off-label*)
	Damping off	propamocarb hydrochloride (*off-label*), thiram (*off-label*)
	Downy mildew	boscalid + pyraclostrobin (*off-label*), copper oxychloride (*off-label*), dimethomorph + mancozeb (*off-label*), fosetyl-aluminium (*off-label*), fosetyl-aluminium + propamocarb hydrochloride (*off-label*), propamocarb hydrochloride (*off-label*)
	Foliar diseases	fosetyl-aluminium (*off-label*)
	Phytophthora	propamocarb hydrochloride (*off-label*)
Pests	Aphids	acetamiprid (*off-label*), cypermethrin (*off-label*), deltamethrin (*off-label*), fatty acids (*off-label*), lambda-cyhalothrin (*off-label*), pirimicarb (*off-label*), pymetrozine (*off-label*), spirotetramat (*off-label*)
	Caterpillars	Bacillus thuringiensis (*off-label*), diflubenzuron (*off-label*)
	Pests, miscellaneous	deltamethrin (*off-label*), lambda-cyhalothrin (*off-label*), spinosad

	Thrips	fatty acids (off-label), pirimicarb (off-label)
	Whiteflies	spirotetramat (off-label)
Weeds	Broad-leaved weeds	pendimethalin (off-label), propyzamide (off-label), S-metolachlor (off-label)
	Grass weeds	pendimethalin (off-label), propyzamide (off-label), S-metolachlor (off-label)

Leafy vegetables - Lettuce

Diseases	Botrytis	boscalid + pyraclostrobin, cyprodinil + fludioxonil (off-label), fenhexamid (off-label), iprodione, propamocarb hydrochloride
	Bottom rot	boscalid + pyraclostrobin
	Damping off	propamocarb hydrochloride, propamocarb hydrochloride (off-label), thiram (off-label)
	Disease control/foliar feed	mancozeb + metalaxyl-M (off-label)
	Downy mildew	copper oxychloride (off-label), dimethomorph + mancozeb (off-label), fosetyl-aluminium (off-label), mancozeb (off-label), propamocarb hydrochloride, propamocarb hydrochloride (off-label)
	Foliar diseases	fosetyl-aluminium (off-label)
	Sclerotinia	azoxystrobin (off-label)
	Soft rot	boscalid + pyraclostrobin
Pests	Aphids	acetamiprid (off-label), cypermethrin, cypermethrin (off-label), deltamethrin, deltamethrin (off-label), fatty acids, fatty acids (off-label), lambda-cyhalothrin (off-label), lambda-cyhalothrin + pirimicarb, pirimicarb, pirimicarb (off-label), pirimicarb (outdoor crops), pymetrozine (off-label), pyrethrins, spirotetramat, spirotetramat (off-label)
	Beetles	deltamethrin
	Caterpillars	Bacillus thuringiensis (off-label), cypermethrin, deltamethrin, diflubenzuron (off-label), pyrethrins
	Cutworms	cypermethrin, deltamethrin, lambda-cyhalothrin, lambda-cyhalothrin + pirimicarb
	Leafhoppers	deltamethrin
	Mealybugs	fatty acids
	Pests, miscellaneous	deltamethrin (off-label), lambda-cyhalothrin (off-label), spinosad
	Scale insects	fatty acids
	Slugs/snails	methiocarb
	Spider mites	fatty acids
	Thrips	deltamethrin, fatty acids (off-label)
	Whiteflies	fatty acids, spirotetramat (off-label)
Weeds	Broad-leaved weeds	pendimethalin (off-label), propyzamide, propyzamide (off-label), propyzamide (outdoor crops)
	Grass weeds	pendimethalin (off-label), propyzamide, propyzamide (off-label), propyzamide (outdoor crops)

Leafy vegetables - Spinach

Diseases	Damping off	cymoxanil + fludioxonil + metalaxyl-M (off-label), fosetyl-aluminium + propamocarb hydrochloride (off-label)

Crop/Pest Guide - Arable and vegetable crops

	Downy mildew	boscalid + pyraclostrobin (*off-label*), copper oxychloride (*off-label*), fosetyl-aluminium (*off-label*), fosetyl-aluminium + propamocarb hydrochloride (*off-label*), metalaxyl-M (*off-label*)
	Foliar diseases	fosetyl-aluminium (*off-label*)
	White tip	spirotetramat (*off-label*)
Pests	Aphids	cypermethrin (*off-label*), fatty acids (*off-label*), pirimicarb (*off-label*), spirotetramat (*off-label*)
	Caterpillars	Bacillus thuringiensis (*off-label*), cypermethrin (*off-label - outdoor and protected crops*), diflubenzuron (*off-label*)
	Flies	tefluthrin (*off-label - seed treatment*)
	Slugs/snails	methiocarb
	Thrips	fatty acids (*off-label*)
Weeds	Broad-leaved weeds	chloridazon (*off-label*), chloridazon + quinmerac (*off-label*), clopyralid (*off-label*), lenacil (*off-label*)
	Grass weeds	chloridazon (*off-label*), lenacil (*off-label*)

Leafy vegetables - Watercress

Diseases	Bacterial blight	copper oxychloride (*off-label*)
	Bottom rot	copper oxychloride (*off-label*)
	Damping off	metalaxyl-M (*off-label*), propamocarb hydrochloride (*off-label*)
	Downy mildew	fosetyl-aluminium (*off-label*), fosetyl-aluminium (*off-label - during propagation*), metalaxyl-M (*off-label*), propamocarb hydrochloride (*off-label - under protection*)
	Phytophthora	copper oxychloride (*off-label*), fosetyl-aluminium (*off-label - during propagation*), propamocarb hydrochloride (*off-label*), propamocarb hydrochloride (*off-label - under protection*)
	Pythium	copper oxychloride (*off-label - during propagation*), fosetyl-aluminium (*off-label - during propagation*), propamocarb hydrochloride (*off-label - under protection*)
	Rhizoctonia	copper oxychloride (*off-label - during propagation*)
	Spear rot	copper oxychloride (*off-label*)
	Stem canker	copper oxychloride (*off-label*)
Pests	Caterpillars	Bacillus thuringiensis (*off-label*)

Legumes - Beans (Phaseolus)

Crop control	Desiccation	diquat (*off-label*)
Diseases	Alternaria	iprodione (*off-label*)
	Botrytis	azoxystrobin (*off-label*), cyprodinil + fludioxonil (*moderate control*), iprodione (*off-label*)
	Damping off	thiram (*seed treatment*)
	Rust	tebuconazole (*off-label*)
	Sclerotinia	cyprodinil + fludioxonil
Pests	Aphids	fatty acids, pirimicarb
	Caterpillars	Bacillus thuringiensis (*off-label*), lambda-cyhalothrin (*off-label*)
	Flies	chlorpyrifos (*off-label*), chlorpyrifos (*off-label - harvested as a dry pulse*), chlorpyrifos (*off-label - harvested dry as a pulse*)
	Mealybugs	fatty acids

Crop/Pest Guide - Arable and vegetable crops

	Pests, miscellaneous	lambda-cyhalothrin (*off-label*)
	Scale insects	fatty acids
	Spider mites	fatty acids
	Whiteflies	fatty acids
Weeds	Broad-leaved weeds	bentazone, linuron, linuron (*off-label*), pendimethalin, pendimethalin (*off-label*), S-metolachlor (*off-label*)
	Crops as weeds	cycloxydim
	Grass weeds	cycloxydim, cycloxydim (*off-label*), fluazifop-P-butyl (*off-label*), linuron, linuron (*off-label*), pendimethalin (*off-label*), S-metolachlor (*off-label*)

Legumes - Beans (Vicia)

Crop control	Desiccation	diquat, glufosinate-ammonium, glyphosate
Diseases	Ascochyta	azoxystrobin (*off-label*)
	Botrytis	azoxystrobin (*off-label*), cyprodinil + fludioxonil (*moderate control*)
	Chocolate spot	boscalid + pyraclostrobin (*moderate*), chlorothalonil, chlorothalonil (*moderate control*), chlorothalonil + cyproconazole, chlorothalonil + pyrimethanil, cyproconazole, iprodione + thiophanate-methyl, tebuconazole
	Damping off	thiram (*seed treatment*)
	Downy mildew	chlorothalonil + metalaxyl-M, cymoxanil + fludioxonil + metalaxyl-M (*off-label - seed treatment*), fosetyl-aluminium
	Rust	azoxystrobin, boscalid + pyraclostrobin, chlorothalonil + cyproconazole, cyproconazole, metconazole, metconazole (*off-label*), tebuconazole, tebuconazole (*off-label*)
	Sclerotinia	cyprodinil + fludioxonil
Pests	Aphids	cypermethrin, cypermethrin (*off-label*), esfenvalerate, fatty acids, fatty acids (*off-label*), lambda-cyhalothrin, lambda-cyhalothrin + pirimicarb, pirimicarb, zeta-cypermethrin (*off-label*)
	Beetles	lambda-cyhalothrin, lambda-cyhalothrin (*off-label*)
	Birds/mammals	aluminium ammonium sulphate
	Caterpillars	Bacillus thuringiensis (*off-label*), lambda-cyhalothrin
	Mealybugs	fatty acids
	Pests, miscellaneous	lambda-cyhalothrin (*off-label*), zeta-cypermethrin (*off-label*)
	Scale insects	fatty acids
	Spider mites	fatty acids
	Thrips	fatty acids (*off-label*), lambda-cyhalothrin (*off-label*)
	Weevils	alpha-cypermethrin, cypermethrin, deltamethrin, esfenvalerate, lambda-cyhalothrin, lambda-cyhalothrin + pirimicarb, zeta-cypermethrin
	Whiteflies	fatty acids
Weeds	Broad-leaved weeds	bentazone, carbetamide, clomazone, clomazone (*off-label*), clomazone + linuron, glyphosate, imazamox + pendimethalin, imazamox + pendimethalin (*off-label*), isoxaben + terbuthylazine, isoxaben + terbuthylazine (*off-label*), linuron, pendimethalin (*off-label*), propyzamide, prosulfocarb (*off-label*)

Crop/Pest Guide - Arable and vegetable crops

	Crops as weeds	carbetamide, cycloxydim, fluazifop-P-butyl, glyphosate, pendimethalin (*off-label*), propyzamide, quizalofop-P-ethyl, quizalofop-P-tefuryl, tepraloxydim
	Grass weeds	carbetamide, clomazone, clomazone + linuron, cycloxydim, cycloxydim (*off-label*), diquat, fluazifop-P-butyl, fluazifop-P-butyl (*off-label*), glyphosate, isoxaben + terbuthylazine (*off-label*), linuron, pendimethalin (*off-label*), propaquizafop, propyzamide, prosulfocarb (*off-label*), quizalofop-P-ethyl, quizalofop-P-tefuryl, tepraloxydim, tepraloxydim (*off-label*), tri-allate, tri-allate (*off-label*)
	Weeds, miscellaneous	diquat, glyphosate

Legumes - Lupins

Crop control	Desiccation	diquat (*off-label*), glyphosate (*off-label*)
Diseases	Ascochyta	azoxystrobin (*off-label*), metconazole (*qualified minor use*)
	Botrytis	cyprodinil + fludioxonil (*off-label*), metconazole (*qualified minor use*)
	Damping off	cymoxanil + fludioxonil + metalaxyl-M (*off-label*), thiram (*off-label - seed treatment*)
	Downy mildew	cymoxanil + fludioxonil + metalaxyl-M (*off-label*), fosetyl-aluminium (*off-label*)
	Foliar diseases	chlorothalonil (*off-label*)
	Pythium	cymoxanil + fludioxonil + metalaxyl-M (*off-label*)
	Rust	azoxystrobin (*off-label*), metconazole (*qualified minor use*)
	Seed-borne diseases	fosetyl-aluminium (*off-label*), thiram (*off-label - seed treatment*)
Pests	Aphids	deltamethrin (*off-label*), fatty acids (*off-label*), lambda-cyhalothrin (*off-label*), lambda-cyhalothrin + pirimicarb (*off-label*), zeta-cypermethrin (*off-label*)
	Pests, miscellaneous	alpha-cypermethrin (*off-label*), deltamethrin (*off-label*), lambda-cyhalothrin (*off-label*), zeta-cypermethrin (*off-label*)
	Thrips	fatty acids (*off-label*)
Weeds	Broad-leaved weeds	carbetamide (*off-label*), clomazone (*off-label*), isoxaben + terbuthylazine (*off-label*), pendimethalin (*off-label*), pyridate (*off-label*)
	Crops as weeds	propaquizafop (*off-label*)
	Grass weeds	carbetamide (*off-label*), cycloxydim (*off-label*), fluazifop-P-butyl (*off-label*), pendimethalin (*off-label*), propaquizafop (*off-label*), tepraloxydim (*off-label*), tri-allate (*off-label*)
	Weeds, miscellaneous	glyphosate (*off-label*)

Legumes - Peas

Crop control	Desiccation	diquat, glufosinate-ammonium, glufosinate-ammonium (*not for seed*), glyphosate
Diseases	Alternaria	iprodione (*off-label*)
	Ascochyta	azoxystrobin, boscalid + pyraclostrobin (*moderate control only*), chlorothalonil, chlorothalonil +

Crop/Pest Guide - Arable and vegetable crops

		pyrimethanil, cymoxanil + fludioxonil + metalaxyl-M (*seed treatment*), cyprodinil + fludioxonil, metconazole (*reduction*)
	Botrytis	chlorothalonil, chlorothalonil (*moderate control*), chlorothalonil + cyproconazole, chlorothalonil + pyrimethanil, cyprodinil + fludioxonil (*moderate control*), iprodione (*off-label*), metconazole (*reduction*)
	Damping off	cymoxanil + fludioxonil + metalaxyl-M (*seed treatment*), thiram (*seed treatment*)
	Downy mildew	cymoxanil + fludioxonil + metalaxyl-M (*seed treatment*), fosetyl-aluminium (*off-label - seed treatment*)
	Mycosphaerella	chlorothalonil, chlorothalonil (*moderate control*), chlorothalonil + pyrimethanil, cyprodinil + fludioxonil, metconazole (*reduction*)
	Powdery mildew	sulphur (*off-label*)
	Pythium	cymoxanil + fludioxonil + metalaxyl-M (*seed treatment*)
	Rust	metconazole
	Sclerotinia	cyprodinil + fludioxonil
	Seed-borne diseases	fosetyl-aluminium (*off-label*)
Pests	Aphids	alpha-cypermethrin, alpha-cypermethrin (*reduction*), bifenthrin, cypermethrin, esfenvalerate, fatty acids, lambda-cyhalothrin, lambda-cyhalothrin + pirimicarb, pirimicarb, thiacloprid, zeta-cypermethrin
	Beetles	lambda-cyhalothrin
	Birds/mammals	aluminium ammonium sulphate
	Caterpillars	alpha-cypermethrin, cypermethrin, deltamethrin, lambda-cyhalothrin, lambda-cyhalothrin + pirimicarb, zeta-cypermethrin
	Mealybugs	fatty acids
	Midges	deltamethrin, lambda-cyhalothrin, lambda-cyhalothrin + pirimicarb, thiacloprid
	Scale insects	fatty acids
	Spider mites	fatty acids
	Weevils	alpha-cypermethrin, cypermethrin, deltamethrin, esfenvalerate, lambda-cyhalothrin, lambda-cyhalothrin + pirimicarb, zeta-cypermethrin
	Whiteflies	fatty acids
Weeds	Broad-leaved weeds	bentazone, clomazone, clomazone + linuron, flumioxazine (*off-label*), glyphosate, imazamox + pendimethalin, isoxaben + terbuthylazine, linuron, MCPB, pendimethalin, pendimethalin (*off-label*)
	Crops as weeds	cycloxydim, fluazifop-P-butyl, flumioxazine (*off-label*), glyphosate, pendimethalin, quizalofop-P-ethyl, quizalofop-P-tefuryl, tepraloxydim
	Grass weeds	clomazone, clomazone + linuron, cycloxydim, cycloxydim (*off-label*), diquat (*dry harvested*), fluazifop-P-butyl, glyphosate, linuron, pendimethalin, pendimethalin (*off-label*), propaquizafop, quizalofop-P-ethyl, quizalofop-P-tefuryl, tepraloxydim, tri-allate
	Weeds, miscellaneous	diquat (*dry harvested only*), glyphosate

Miscellaneous arable - Industrial crops

Crop control	Desiccation	glyphosate (*off-label*), glyphosate (*off-label - for wild bird seed production*)

Crop/Pest Guide - Arable and vegetable crops

Pests	Aphids	lambda-cyhalothrin (*off-label*)
	Pests, miscellaneous	lambda-cyhalothrin (*off-label*)
	Thrips	lambda-cyhalothrin (*off-label*)
	Whiteflies	lambda-cyhalothrin (*off-label*)
Weeds	Broad-leaved weeds	2,4-D (*off-label*), flufenacet + isoxaflutole (*off-label*), fluroxypyr (*off-label*), metsulfuron-methyl (*off-label*), metsulfuron-methyl + thifensulfuron-methyl (*off-label*), pendimethalin (*off-label*)
	Grass weeds	flufenacet + isoxaflutole (*off-label*), pendimethalin (*off-label*), propoxycarbazone-sodium (*off-label*)
	Weeds, miscellaneous	glyphosate (*off-label*), glyphosate (*off-label - grown for wild bird seed production*)

Miscellaneous arable - Miscellaneous arable crops

Crop control	Desiccation	glyphosate (*off-label*), glyphosate (*off-label - for wild bird seed production*)
Diseases	Botrytis	Bacillus subtilis (*off-label*), iprodione + thiophanate-methyl (*off-label*)
	Disease control/foliar feed	azoxystrobin (*off-label*), boscalid (*off-label*), boscalid + pyraclostrobin (*off-label*), potassium bicarbonate (commodity substance)
	Downy mildew	boscalid + pyraclostrobin (*off-label*)
	Powdery mildew	kresoxim-methyl (*off-label*)
	Sclerotinia	Coniothyrium minitans
Pests	Aphids	lambda-cyhalothrin (*off-label*), lambda-cyhalothrin + pirimicarb (*off-label*), maltodextrin, pyrethrins
	Birds/mammals	aluminium ammonium sulphate
	Caterpillars	indoxacarb (*off-label*), pyrethrins
	Pests, miscellaneous	indoxacarb (*off-label*), lambda-cyhalothrin (*off-label*), lambda-cyhalothrin + pirimicarb (*off-label*)
	Slugs/snails	ferric phosphate, metaldehyde, metaldehyde (*around*), metaldehyde (*except potatoes and cauliflowers*), metaldehyde (*excluding potato and cauliflower*)
	Spider mites	maltodextrin, pyrethrins
	Thrips	lambda-cyhalothrin (*off-label*), pyrethrins
	Whiteflies	lambda-cyhalothrin (*off-label*), maltodextrin
Plant growth regulation	Growth control	trinexapac-ethyl (*off-label - for wild bird seed production*)
Weeds	Broad-leaved weeds	2,4-D (*off-label*), bentazone (*off-label*), bromoxynil (*off-label*), bromoxynil + prosulfuron (*off-label*), carfentrazone-ethyl (*before planting*), clomazone (*off-label*), desmedipham + ethofumesate + phenmedipham (*off-label*), dicamba + mecoprop-P (*off-label*), diquat, diquat (*around*), ethofumesate (*off-label*), florasulam (*off-label*), florasulam + fluroxypyr (*off-label*), flufenacet + isoxaflutole (*off-label*), flufenacet + pendimethalin (*off-label*), fluroxypyr (*off-label*), isoxaben (*off-label - grown for game cover*), linuron (*off-label*), mesotrione (*off-label*), metsulfuron-methyl (*off-label*), pendimethalin (*off-label*), pendimethalin + picolinafen (*off-label*), propyzamide (*off-label*), pyridate (*off-label*), sulfosulfuron (*off-label*), thifensulfuron-methyl + tribenuron-methyl (*off-label*)

Crop/Pest Guide - Arable and vegetable crops

	Crops as weeds	carfentrazone-ethyl (*before planting*), quizalofop-P-tefuryl
	Grass weeds	diquat (*around only*), ethofumesate (*off-label*), flufenacet + isoxaflutole (*off-label*), flufenacet + pendimethalin (*off-label*), linuron (*off-label*), mesotrione (*off-label*), pendimethalin (*off-label*), pendimethalin + picolinafen (*off-label*), pinoxaden (*off-label*), propyzamide (*off-label*), quizalofop-P-tefuryl, sulfosulfuron (*off-label*), tepraloxydim (*off-label*), tri-allate (*off-label - for wild bird seed production*)
	Weeds, miscellaneous	clopyralid (*off-label - game cover*), diquat (*around only*), glyphosate, glyphosate (*before planting*), glyphosate (*off-label*), glyphosate (*off-label - grown for wild bird seed production*), glyphosate (*pre-sowing/planting*)

Miscellaneous arable - Miscellaneous arable situations

Diseases	Soil-borne diseases	dazomet (*soil fumigation*)
Pests	Free-living nematodes	dazomet (*soil fumigation*)
	Slugs/snails	metaldehyde
	Soil pests	dazomet (*soil fumigation*)
Weeds	Broad-leaved weeds	amitrole, citronella oil, glufosinate-ammonium, glufosinate-ammonium (*uncropped*), glyphosate
	Crops as weeds	amitrole (*barley stubble*), fluazifop-P-butyl, glyphosate, tepraloxydim
	Grass weeds	amitrole, fluazifop-P-butyl, glufosinate-ammonium, glufosinate-ammonium (*uncropped*), glyphosate, tepraloxydim
	Weeds, miscellaneous	2,4-D + dicamba + triclopyr, cycloxydim, dazomet (*soil fumigation*), fluazifop-P-butyl, glufosinate-ammonium, glyphosate, glyphosate (*wiper application*), metsulfuron-methyl, tepraloxydim, thifensulfuron-methyl

Miscellaneous field vegetables - All vegetables

Diseases	Soil-borne diseases	dazomet (*soil fumigation*)
Pests	Free-living nematodes	dazomet (*soil fumigation*)
	Soil pests	dazomet (*soil fumigation*)
Plant growth regulation	Quality/yield control	sulphur
Weeds	Broad-leaved weeds	glufosinate-ammonium
	Grass weeds	glufosinate-ammonium
	Weeds, miscellaneous	dazomet (*soil fumigation*), glufosinate-ammonium

Oilseed crops - Linseed/flax

Crop control	Desiccation	diquat, glufosinate-ammonium, glyphosate
Diseases	Botrytis	metconazole (*off-label*), tebuconazole (*reduction*)
	Disease control/foliar feed	boscalid (*off-label*), difenoconazole (*off-label*), metconazole (*off-label*), prochloraz (*off-label*)
	Foliar diseases	cyproconazole (*off-label*), difenoconazole (*off-label*)
	Powdery mildew	tebuconazole
	Seed-borne diseases	prochloraz (*seed treatment*)

Crop/Pest Guide - Arable and vegetable crops

Pests	Aphids	cypermethrin (*off-label*), lambda-cyhalothrin (*off-label*), lambda-cyhalothrin + pirimicarb (*off-label*), pirimicarb (*off-label*), tau-fluvalinate (*off-label*)
	Beetles	beta-cyfluthrin + imidacloprid (*off-label - seed treatment*), bifenthrin, zeta-cypermethrin
	Pests, miscellaneous	alpha-cypermethrin (*off-label*), lambda-cyhalothrin (*off-label*)
	Whiteflies	lambda-cyhalothrin (*off-label*)
Weeds	Broad-leaved weeds	amidosulfuron, amidosulfuron + iodosulfuron-methyl-sodium (*off-label*), bentazone, bifenox (*off-label*), bromoxynil (*off-label*), carbetamide (*off-label*), clopyralid, clopyralid (*off-label*), clopyralid + picloram (*off-label*), flupyrsulfuron-methyl (*off-label*), glyphosate, mesotrione (*off-label*), metazachlor (*off-label*), metazachlor + quinmerac (*off-label*), metsulfuron-methyl, napropamide (*off-label*), prosulfocarb (*off-label*)
	Crops as weeds	cycloxydim, fluazifop-P-butyl, glyphosate, mesotrione (*off-label*), propaquizafop (*off-label*), quizalofop-P-ethyl, quizalofop-P-tefuryl, tepraloxydim
	Grass weeds	carbetamide (*off-label*), cycloxydim, diquat, fluazifop-P-butyl, flupyrsulfuron-methyl (*off-label*), glyphosate, mesotrione (*off-label*), metazachlor (*off-label*), metazachlor + quinmerac (*off-label*), napropamide (*off-label*), propaquizafop, propaquizafop (*off-label*), prosulfocarb (*off-label*), quizalofop-P-ethyl, quizalofop-P-tefuryl, tepraloxydim, tri-allate (*off-label*)
	Weeds, miscellaneous	diquat, glyphosate

Oilseed crops - Miscellaneous oilseeds

Crop control	Desiccation	diquat (*off-label*), glyphosate (*off-label*)
Diseases	Alternaria	boscalid + pyraclostrobin (*off-label*)
	Botrytis	boscalid (*off-label*), boscalid + pyraclostrobin (*off-label*)
	Damping off	cymoxanil + fludioxonil + metalaxyl-M (*off-label - seed treatment*), thiram (*off-label - seed treatment*)
	Disease control/foliar feed	azoxystrobin + chlorothalonil (*off-label*), boscalid (*off-label*), chlorothalonil + metalaxyl-M (*off-label*), difenoconazole (*off-label*), propiconazole (*off-label*), tebuconazole (*off-label*)
	Downy mildew	chlorothalonil (*off-label*), cymoxanil + fludioxonil + metalaxyl-M (*off-label - seed treatment*), dimethomorph + mancozeb (*off-label*), mancozeb (*off-label*)
	Fire	boscalid + pyraclostrobin (*off-label*)
	Foliar diseases	azoxystrobin (*off-label*), cyproconazole (*off-label*), difenoconazole (*off-label*), tebuconazole (*off-label*)
	Powdery mildew	azoxystrobin (*off-label*), sulphur (*off-label*)
	Pythium	cymoxanil + fludioxonil + metalaxyl-M (*off-label - seed treatment*), fludioxonil + metalaxyl-M + thiamethoxam (*off-label*), thiram (*off-label - seed treatment*)
	Sclerotinia	boscalid (*off-label*)
	Seed-borne diseases	prochloraz (*off-label - seed treatment*)
Pests	Aphids	deltamethrin (*off-label*), lambda-cyhalothrin (*off-label*), lambda-cyhalothrin + pirimicarb (*off-label*), pirimicarb (*off-label*), tau-fluvalinate (*off-label*)

Crop/Pest Guide - Arable and vegetable crops

	Beetles	beta-cyfluthrin + clothianidin (*off-label*), beta-cyfluthrin + imidacloprid (*off-label - seed treatment*), deltamethrin (*off-label*), lambda-cyhalothrin (*off-label*)
	Caterpillars	deltamethrin (*off-label*)
	Leafhoppers	deltamethrin (*off-label*)
	Pests, miscellaneous	alpha-cypermethrin (*off-label*), lambda-cyhalothrin (*off-label*)
	Thrips	deltamethrin (*off-label*)
	Whiteflies	lambda-cyhalothrin (*off-label*)
Weeds	Broad-leaved weeds	asulam (*off-label*), bifenox (*off-label*), carbetamide (*off-label*), clomazone (*off-label*), clopyralid (*off-label*), clopyralid + picloram (*off-label*), ethofumesate (*off-label*), fluroxypyr (*off-label*), metazachlor (*off-label*), metazachlor + quinmerac (*off-label*), napropamide (*off-label*), pendimethalin (*off-label*), prosulfocarb (*off-label*)
	Crops as weeds	fluroxypyr (*off-label*), propaquizafop (*off-label*)
	Grass weeds	asulam (*off-label*), carbetamide (*off-label*), cycloxydim (*off-label*), ethofumesate (*off-label*), fluazifop-P-butyl (*off-label*), metazachlor (*off-label*), metazachlor + quinmerac (*off-label*), napropamide (*off-label*), pendimethalin (*off-label*), propaquizafop (*off-label*), prosulfocarb (*off-label*), tepraloxydim (*off-label*)
	Weeds, miscellaneous	diquat (*off-label*)

Oilseed crops - Oilseed rape

Crop control	Desiccation	diquat, glufosinate-ammonium, glyphosate
Diseases	Alternaria	azoxystrobin, azoxystrobin + cyproconazole, boscalid, difenoconazole, fludioxonil + metalaxyl-M + thiamethoxam (*moderate control*), iprodione, iprodione + thiophanate-methyl, metconazole, prochloraz, tebuconazole
	Black scurf and stem canker	difenoconazole, iprodione + thiophanate-methyl, prothioconazole + tebuconazole, tebuconazole
	Botrytis	chlorothalonil, chlorothalonil (*moderate control*), iprodione, iprodione + thiophanate-methyl, prochloraz
	Damping off	thiram (*seed treatment*)
	Downy mildew	chlorothalonil, chlorothalonil (*moderate control*), fludioxonil + metalaxyl-M + thiamethoxam, mancozeb
	Ear blight	thiophanate-methyl (*reduction only*)
	Fusarium	thiophanate-methyl
	Light leaf spot	carbendazim + flusilazole, cyproconazole, difenoconazole, famoxadone + flusilazole, flusilazole, iprodione + thiophanate-methyl, metconazole, prochloraz, prochloraz + propiconazole, propiconazole (*reduction*), prothioconazole + tebuconazole, prothioconazole + tebuconazole (*moderate control*), prothioconazole + tebuconazole (*moderate control only*), tebuconazole
	Phoma leaf spot	carbendazim + flusilazole, cyproconazole, prothioconazole + tebuconazole
	Powdery mildew	sulphur
	Pythium	fludioxonil + metalaxyl-M + thiamethoxam (*reduction*)
	Ring spot	tebuconazole (*reduction*)

Crop/Pest Guide - Arable and vegetable crops

	Sclerotinia	azoxystrobin, azoxystrobin + cyproconazole, boscalid, iprodione + thiophanate-methyl, picoxystrobin, prochloraz, prochloraz + tebuconazole, prothioconazole, prothioconazole + tebuconazole, tebuconazole
	Seed-borne diseases	fludioxonil + metalaxyl-M + thiamethoxam
	Stem canker	carbendazim + flusilazole (*reduction*), famoxadone + flusilazole (*suppression*), metconazole (*reduction*), prochloraz, prochloraz + propiconazole, prochloraz + thiram (*seed treatment*), prothioconazole + tebuconazole, tebuconazole
	White leaf spot	prochloraz
Pests	Aphids	acetamiprid, alpha-cypermethrin, beta-cyfluthrin + clothianidin, cypermethrin, deltamethrin, lambda-cyhalothrin, lambda-cyhalothrin + pirimicarb, pirimicarb, tau-fluvalinate
	Beetles	alpha-cypermethrin, beta-cyfluthrin + clothianidin, beta-cyfluthrin + imidacloprid (*seed treatment*), bifenthrin, cypermethrin, deltamethrin, fludioxonil + metalaxyl-M + thiamethoxam (*reduction of damage*), indoxacarb, lambda-cyhalothrin, lambda-cyhalothrin + pirimicarb, tau-fluvalinate, thiacloprid, zeta-cypermethrin
	Birds/mammals	aluminium ammonium sulphate
	Caterpillars	lambda-cyhalothrin
	Flies	beta-cyfluthrin + clothianidin, beta-cyfluthrin + clothianidin (*reduction of feeding activity only*)
	Midges	alpha-cypermethrin, bifenthrin, cypermethrin, cypermethrin (*some coincidental control only*), lambda-cyhalothrin, lambda-cyhalothrin + pirimicarb, zeta-cypermethrin
	Slugs/snails	methiocarb
	Weevils	alpha-cypermethrin, bifenthrin, cypermethrin, deltamethrin, lambda-cyhalothrin, lambda-cyhalothrin + pirimicarb, zeta-cypermethrin
Plant growth regulation	Growth control	tebuconazole
	Quality/yield control	sulphur
Weeds	Broad-leaved weeds	bifenox (*off-label*), carbetamide, clomazone, clomazone + metazachlor, clopyralid, clopyralid + picloram, clopyralid + picloram (*off-label*), dimethenamid-p + metazachlor, dimethenamid-p + metazachlor + quinmerac, glyphosate, metazachlor, metazachlor + quinmerac, napropamide, propyzamide, pyridate (*off-label*)
	Crops as weeds	carbetamide, cycloxydim, fluazifop-P-butyl, glyphosate, propyzamide, quizalofop-P-ethyl, quizalofop-P-tefuryl, tepraloxydim
	Grass weeds	carbetamide, clomazone, clomazone + metazachlor, cycloxydim, dimethenamid-p + metazachlor + quinmerac, diquat, fluazifop-P-butyl, glyphosate, metazachlor, metazachlor + quinmerac, napropamide, propaquizafop, propyzamide, quizalofop-P-ethyl, quizalofop-P-tefuryl, tepraloxydim, tri-allate (*off-label*)
	Weeds, miscellaneous	diquat, glyphosate

Oilseed crops - Soya

Diseases	Damping off	thiram (*seed treatment - qualified minor use*), thiram (*seed treatment - qualified minor use*)
Weeds	Broad-leaved weeds	bentazone (*off-label*)

Oilseed crops - Sunflowers

Crop control	Desiccation	diquat (*off-label*)
Pests	Slugs/snails	methiocarb
Plant growth regulation	Growth control	daminozide
Weeds	Broad-leaved weeds	isoxaben (*off-label - grown for game cover*), pendimethalin
	Crops as weeds	pendimethalin
	Grass weeds	pendimethalin

Root and tuber crops - Beet crops

Diseases	Alternaria	cyprodinil + fludioxonil (*off-label*)
	Aphanomyces	hymexazol (*off-label - seed treatment*)
	Black leg	hymexazol (*seed treatment*)
	Botrytis	iprodione (*off-label*)
	Cercospora leaf spot	azoxystrobin + cyproconazole, cyproconazole + trifloxystrobin, epoxiconazole + pyraclostrobin
	Damping off	cymoxanil + fludioxonil + metalaxyl-M (*off-label - seed treatment*)
	Disease control/foliar feed	flusilazole (*off-label*)
	Downy mildew	boscalid + pyraclostrobin (*off-label*), copper oxychloride (*off-label*), fosetyl-aluminium (*off-label*), fosetyl-aluminium + propamocarb hydrochloride, metalaxyl-M (*off-label*)
	Foliar diseases	cyproconazole (*off-label*), epoxiconazole + pyraclostrobin (*off-label*), flusilazole (*off-label*), fosetyl-aluminium (*off-label*)
	Powdery mildew	azoxystrobin + cyproconazole, carbendazim + flusilazole, cyproconazole, cyproconazole + trifloxystrobin, difenoconazole + fenpropidin, epoxiconazole + pyraclostrobin, flusilazole, quinoxyfen, sulphur, sulphur (*off-label*)
	Pythium	metalaxyl-M, metalaxyl-M (*off-label*)
	Ramularia leaf spots	azoxystrobin + cyproconazole, cyproconazole, cyproconazole + trifloxystrobin, difenoconazole + fenpropidin, epoxiconazole + pyraclostrobin, propiconazole (*reduction*)
	Rhizoctonia	azoxystrobin (*off-label*)
	Root malformation disorder	azoxystrobin (*off-label*), metalaxyl-M (*off-label*)
	Rust	azoxystrobin + cyproconazole, carbendazim + flusilazole, cyproconazole + trifloxystrobin, difenoconazole + fenpropidin, epoxiconazole + pyraclostrobin, fenpropimorph (*off-label*), flusilazole, propiconazole
	Sclerotinia	cyprodinil + fludioxonil (*off-label*)
	Seed-borne diseases	thiram (*seed soak*)

Crop/Pest Guide - Arable and vegetable crops

Pests	Aphids	acetamiprid (*off-label*), beta-cyfluthrin + clothianidin (*seed treatment*), cypermethrin, cypermethrin (*off-label*), cypermethrin (*off-label - outdoor and protected crops*), deltamethrin (*off-label*), dimethoate, dimethoate (*excluding Myzus persicae*), fatty acids (*off-label*), imidacloprid, lambda-cyhalothrin, lambda-cyhalothrin (*off-label*), lambda-cyhalothrin + pirimicarb, lambda-cyhalothrin + pirimicarb (*off-label*), oxamyl, oxamyl (*off-label*), pirimicarb, pirimicarb (*off-label*), spirotetramat (*off-label*), zeta-cypermethrin (*off-label*)
	Beetles	chlorpyrifos, chlorpyrifos (*off-label*), deltamethrin, imidacloprid, lambda-cyhalothrin, lambda-cyhalothrin (*off-label*), lambda-cyhalothrin + pirimicarb, lambda-cyhalothrin + pirimicarb (*off-label*), oxamyl, oxamyl (*off-label*), tefluthrin (*seed treatment*), thiamethoxam (*seed treatment*)
	Birds/mammals	aluminium ammonium sulphate
	Caterpillars	Bacillus thuringiensis (*off-label*), cypermethrin, cypermethrin (*off-label - outdoor and protected crops*), lambda-cyhalothrin, lambda-cyhalothrin (*off-label*)
	Cutworms	Bacillus thuringiensis (*off-label*), cypermethrin, lambda-cyhalothrin, lambda-cyhalothrin (*off-label*), lambda-cyhalothrin + pirimicarb, lambda-cyhalothrin + pirimicarb (*off-label*), methiocarb (*reduction*), zeta-cypermethrin
	Flies	lambda-cyhalothrin, oxamyl, oxamyl (*off-label*)
	Free-living nematodes	oxamyl, oxamyl (*off-label*)
	Leaf miners	dimethoate, imidacloprid, lambda-cyhalothrin, lambda-cyhalothrin (*off-label*), lambda-cyhalothrin + pirimicarb, lambda-cyhalothrin + pirimicarb (*off-label*), thiamethoxam (*seed treatment*)
	Leatherjackets	chlorpyrifos, chlorpyrifos (*off-label*), methiocarb (*reduction*)
	Millipedes	imidacloprid, methiocarb (*reduction*), oxamyl, oxamyl (*off-label*), tefluthrin (*seed treatment*), thiamethoxam (*seed treatment*)
	Pests, miscellaneous	chlorpyrifos (*off-label*), deltamethrin (*off-label*), lambda-cyhalothrin (*off-label*), zeta-cypermethrin (*off-label*)
	Slugs/snails	methiocarb
	Springtails	imidacloprid, tefluthrin (*seed treatment*), thiamethoxam (*seed treatment*)
	Symphylids	imidacloprid, tefluthrin (*seed treatment*), thiamethoxam (*seed treatment*)
	Thrips	fatty acids (*off-label*)
	Weevils	lambda-cyhalothrin
	Whiteflies	spirotetramat (*off-label*)
	Wireworms	thiamethoxam (*seed treatment - reduction*)
Plant growth regulation	Quality/yield control	sulphur
Weeds	Broad-leaved weeds	carbetamide, chloridazon, chloridazon (*off-label*), chloridazon + ethofumesate, chloridazon + metamitron, chloridazon + metamitron (*off-label*), chloridazon + quinmerac, chloridazon + quinmerac (*off-label*), chlorpropham + metamitron, clopyralid, clopyralid (*off-label*), desmedipham + ethofumesate + lenacil + phenmedipham, desmedipham + ethofumesate + metamitron + phenmedipham, desmedipham +

Crop/Pest Guide - Arable and vegetable crops

	ethofumesate + phenmedipham, desmedipham + phenmedipham, diquat, ethofumesate, ethofumesate + metamitron, ethofumesate + metamitron + phenmedipham, ethofumesate + phenmedipham, glufosinate-ammonium, glyphosate, lenacil, lenacil (*off-label*), lenacil + triflusulfuron-methyl, lenacil + triflusulfuron-methyl (*off-label*), metamitron, phenmedipham, propyzamide, triflusulfuron-methyl, triflusulfuron-methyl (*off-label*)
Crops as weeds	carbetamide, cycloxydim, fluazifop-P-butyl, glyphosate, glyphosate (*wiper application*), lenacil + triflusulfuron-methyl, propaquizafop (*off-label*), propyzamide, quizalofop-P-ethyl, quizalofop-P-tefuryl, tepraloxydim
Grass weeds	carbetamide, chloridazon, chloridazon (*off-label*), chloridazon + ethofumesate, chloridazon + metamitron, chloridazon + metamitron (*off-label*), chloridazon + quinmerac, chlorpropham + metamitron, cycloxydim, desmedipham + ethofumesate + lenacil + phenmedipham, desmedipham + ethofumesate + metamitron + phenmedipham, desmedipham + ethofumesate + phenmedipham, diquat, ethofumesate, ethofumesate + metamitron, ethofumesate + metamitron + phenmedipham, ethofumesate + phenmedipham, fluazifop-P-butyl, fluazifop-P-butyl (*off-label*), glufosinate-ammonium, glyphosate, lenacil, lenacil (*off-label*), metamitron, propaquizafop, propaquizafop (*off-label*), propyzamide, quizalofop-P-ethyl, quizalofop-P-tefuryl, tepraloxydim, tepraloxydim (*off-label*), tri-allate
Weeds, miscellaneous	diquat, glufosinate-ammonium, glyphosate, glyphosate (*off-label*)

Root and tuber crops - Carrots/parsnips

Diseases	Alternaria	azoxystrobin, azoxystrobin (*off-label*), azoxystrobin + difenoconazole, azoxystrobin + difenoconazole (*off-label*), boscalid + pyraclostrobin (*moderate*), cyprodinil + fludioxonil (*off-label*), fenpropimorph (*off-label*), iprodione + thiophanate-methyl (*off-label*), mancozeb, tebuconazole, tebuconazole + trifloxystrobin
	Cavity spot	metalaxyl-M, metalaxyl-M (*off-label*), metalaxyl-M (*reduction only*)
	Crown rot	fenpropimorph (*off-label*), iprodione + thiophanate-methyl (*off-label*)
	Disease control/foliar feed	mancozeb
	Foliar diseases	tebuconazole + trifloxystrobin (*off-label*)
	Powdery mildew	azoxystrobin, azoxystrobin (*off-label*), azoxystrobin + difenoconazole, azoxystrobin + difenoconazole (*off-label*), boscalid + pyraclostrobin, fenpropimorph (*off-label*), sulphur (*off-label*), tebuconazole, tebuconazole + trifloxystrobin
	Pythium	cymoxanil + fludioxonil + metalaxyl-M (*off-label - seed treatment*)
	Sclerotinia	boscalid + pyraclostrobin (*moderate*), cyprodinil + fludioxonil (*off-label*), tebuconazole, tebuconazole + trifloxystrobin
	Seed-borne diseases	thiram (*seed soak*)

Crop/Pest Guide - Arable and vegetable crops

Pests	Aphids	deltamethrin (*off-label*), fatty acids (*off-label*), lambda-cyhalothrin + pirimicarb, lambda-cyhalothrin + pirimicarb (*off-label*), pirimicarb, thiacloprid
	Beetles	deltamethrin (*off-label*)
	Birds/mammals	aluminium ammonium sulphate
	Caterpillars	deltamethrin (*off-label*)
	Cutworms	Bacillus thuringiensis (*off-label*), cypermethrin (*off-label*), lambda-cyhalothrin, lambda-cyhalothrin + pirimicarb, lambda-cyhalothrin + pirimicarb (*off-label*)
	Flies	lambda-cyhalothrin, lambda-cyhalothrin (*off-label*), tefluthrin (*off-label - seed treatment*)
	Leafhoppers	deltamethrin (*off-label*)
	Pests, miscellaneous	deltamethrin (*off-label*), lambda-cyhalothrin (*off-label*)
	Stem nematodes	oxamyl
	Thrips	deltamethrin (*off-label*), fatty acids (*off-label*)
Plant growth regulation	Growth control	maleic hydrazide (*off-label*)
Weeds	Broad-leaved weeds	clomazone, flumioxazine (*off-label*), isoxaben (*off-label*), linuron, metamitron (*off-label*), metribuzin (*off-label*), pendimethalin, pendimethalin (*off-label*), prosulfocarb (*off-label*)
	Crops as weeds	cycloxydim, fluazifop-P-butyl, flumioxazine (*off-label*), pendimethalin, tepraloxydim
	Grass weeds	cycloxydim, fluazifop-P-butyl, fluazifop-P-butyl (*off-label*), linuron, metamitron (*off-label*), pendimethalin, pendimethalin (*off-label*), propaquizafop, prosulfocarb (*off-label*), tepraloxydim, tepraloxydim (*off-label*)
	Weeds, miscellaneous	glyphosate (*off-label*), metamitron (*off-label*)

Root and tuber crops - Miscellaneous root crops

Crop control	Desiccation	diquat (*off-label*)
Diseases	Alternaria	azoxystrobin (*off-label*), azoxystrobin + difenoconazole (*off-label*), cyprodinil + fludioxonil (*off-label*), fenpropimorph (*off-label*)
	Crown rot	fenpropimorph (*off-label*)
	Disease control/foliar feed	tebuconazole (*off-label*)
	Foliar diseases	tebuconazole (*off-label*), tebuconazole + trifloxystrobin (*off-label*)
	Leaf spot	chlorothalonil (*off-label*)
	Phytophthora	difenoconazole (*off-label*)
	Powdery mildew	azoxystrobin (*off-label*), azoxystrobin + difenoconazole (*off-label*), fenpropimorph (*off-label*)
	Rust	azoxystrobin (*off-label*)
	Sclerotinia	azoxystrobin (*off-label*), cyprodinil + fludioxonil (*off-label*)
	Septoria	difenoconazole (*off-label*)
Pests	Aphids	deltamethrin (*off-label*), fatty acids (*off-label*), pirimicarb (*off-label*), pymetrozine (*off-label*)
	Cutworms	Bacillus thuringiensis (*off-label*), cypermethrin (*off-label*), lambda-cyhalothrin (*off-label*)
	Flies	lambda-cyhalothrin (*off-label*)

Crop/Pest Guide - Arable and vegetable crops

	Pests, miscellaneous	deltamethrin (*off-label*), dimethoate (*off-label*), lambda-cyhalothrin (*off-label*)
	Thrips	fatty acids (*off-label*)
	Weevils	lambda-cyhalothrin (*off-label*)
Weeds	Broad-leaved weeds	asulam (*off-label*), clomazone (*off-label*), linuron, linuron (*off-label*), metribuzin (*off-label*), pendimethalin (*off-label*), propyzamide (*off-label*), prosulfocarb (*off-label*), S-metolachlor (*off-label*)
	Grass weeds	asulam (*off-label*), fluazifop-P-butyl (*off-label*), linuron (*off-label*), pendimethalin (*off-label*), propyzamide (*off-label*), prosulfocarb (*off-label*), S-metolachlor (*off-label*)
	Weeds, miscellaneous	glyphosate (*off-label*), metribuzin (*off-label*)

Root and tuber crops - Potatoes

Crop control	Desiccation	carfentrazone-ethyl, diquat, glufosinate-ammonium
Diseases	Alternaria	azoxystrobin + chlorothalonil (*off-label*)
	Black dot	azoxystrobin
	Black scurf and stem canker	azoxystrobin, flutolanil (*tuber treatment*), imazalil + pencycuron (*tuber treatment*), imazalil + pencycuron (*tuber treatment - reduction*), pencycuron (*tuber treatment*), tolclofos-methyl (*off-label - tuber treatment only to be used with automatic planters*), tolclofos-methyl (*tuber treatment*), tolclofos-methyl (*tuber treatment only to be used with automatic planters*)
	Dry rot	imazalil, thiabendazole, thiabendazole (*tuber treatment - post-harvest*)
	Fusarium	imazalil
	Gangrene	imazalil, thiabendazole, thiabendazole (*tuber treatment - post-harvest*)
	Phytophthora	ametoctradin + dimethomorph, ametoctradin + mancozeb, amisulbrom, benthiavalicarb-isopropyl + mancozeb, boscalid + pyraclostrobin (*off-label*), chlorothalonil, chlorothalonil + mancozeb, chlorothalonil + propamocarb hydrochloride, copper oxychloride, cyazofamid, cymoxanil, cymoxanil + famoxadone, cymoxanil + mancozeb, cymoxanil + propamocarb, dimethomorph + mancozeb, fenamidone + propamocarb hydrochloride, fluazinam, fluazinam (*off-label*), fluazinam + metalaxyl-M, fluopicolide + propamocarb hydrochloride, mancozeb, mancozeb + metalaxyl-M, mancozeb + zoxamide, mandipropamid, maneb
	Powdery scab	fluazinam (*off-label*)
	Rhizoctonia	tolclofos-methyl (*chitted seed treatment*), tolclofos-methyl (*off-label*), tolclofos-methyl (*off-label - chitted seed treatment*)
	Scab	fluazinam (*off-label*)
	Seed-borne diseases	iprodione
	Silver scurf	imazalil, imazalil + pencycuron (*tuber treatment - reduction*), thiabendazole, thiabendazole (*tuber treatment - post-harvest*)
	Skin spot	imazalil, thiabendazole, thiabendazole (*tuber treatment - post-harvest*)

Crop/Pest Guide - Arable and vegetable crops

Pests	Aphids	acetamiprid, cypermethrin, esfenvalerate, flonicamid, lambda-cyhalothrin, lambda-cyhalothrin + pirimicarb, oxamyl, pirimicarb, pymetrozine, thiacloprid, thiamethoxam
	Beetles	chlorantraniliprole, lambda-cyhalothrin
	Caterpillars	cypermethrin, lambda-cyhalothrin
	Cutworms	chlorpyrifos, cypermethrin, lambda-cyhalothrin (*If present at time of application*), lambda-cyhalothrin + pirimicarb, zeta-cypermethrin
	Cyst nematodes	ethoprophos, fosthiazate, oxamyl
	Free-living nematodes	fosthiazate (*reduction*), oxamyl
	Leatherjackets	methiocarb (*reduction*)
	Slugs/snails	metaldehyde, methiocarb
	Weevils	lambda-cyhalothrin
	Wireworms	ethoprophos, fosthiazate (*reduction*)
Plant growth regulation	Growth control	chlorpropham, chlorpropham (*thermal fog*), ethylene, ethylene (*in store*), maleic hydrazide
	Quality/yield control	sulphur
Weeds	Broad-leaved weeds	bentazone, carfentrazone-ethyl, clomazone, clomazone + linuron, clomazone + metribuzin, diquat, flufenacet + metribuzin, glufosinate-ammonium, linuron, metribuzin, pendimethalin, prosulfocarb, rimsulfuron
	Crops as weeds	carfentrazone-ethyl, cycloxydim, metribuzin, pendimethalin, quizalofop-P-tefuryl, rimsulfuron
	Grass weeds	clomazone + linuron, clomazone + metribuzin, cycloxydim, diquat (*weed control and dessication*), flufenacet + metribuzin, glufosinate-ammonium, linuron, metribuzin, pendimethalin, propaquizafop, prosulfocarb, quizalofop-P-tefuryl
	Weeds, miscellaneous	diquat, diquat (*weed control or dessication*), glufosinate-ammonium (*not for seed*)

Stem and bulb vegetables - Asparagus

Diseases	Botrytis	azoxystrobin + chlorothalonil (*qualified minor use recommendation*), boscalid + pyraclostrobin (*off-label*), cyprodinil + fludioxonil (*off-label*)
	Downy mildew	metalaxyl-M (*off-label*)
	Rust	azoxystrobin, azoxystrobin + chlorothalonil (*moderate control only*), azoxystrobin + difenoconazole (*off-label*), boscalid + pyraclostrobin (*off-label*), difenoconazole (*off-label*)
	Stemphylium	azoxystrobin, azoxystrobin + chlorothalonil (*qualified minor use recommendation*), azoxystrobin + difenoconazole (*off-label*), cyprodinil + fludioxonil (*off-label*)
Pests	Aphids	fatty acids (*off-label*)
	Beetles	cypermethrin (*off-label*), thiacloprid (*off-label*)
	Thrips	fatty acids (*off-label*)
Weeds	Broad-leaved weeds	clomazone (*off-label*), clopyralid (*off-label*), isoxaben (*off-label*), linuron, metamitron (*off-label*), metribuzin (*off-label*), pendimethalin (*off-label*), pyridate (*off-label*), triclopyr (*off-label - directed treatment*)

	Grass weeds	fluazifop-P-butyl (*off-label*), metamitron (*off-label*), pendimethalin (*off-label*), triclopyr (*off-label - directed treatment*)
	Weeds, miscellaneous	glyphosate, glyphosate (*off-label*), metamitron (*off-label*), metribuzin (*off-label*)

Stem and bulb vegetables - Celery/chicory

Diseases	Alternaria	iprodione (*off-label*)
	Botrytis	azoxystrobin (*off-label*), cyprodinil + fludioxonil (*off-label*), iprodione (*off-label*)
	Damping off	thiram (*off-label*)
	Downy mildew	copper oxychloride (*off-label*), dimethomorph + mancozeb (*off-label*), fosetyl-aluminium (*off-label*)
	Foliar diseases	fosetyl-aluminium (*off-label*)
	Phytophthora	difenoconazole (*off-label*), fosetyl-aluminium (*off-label - for forcing*)
	Rhizoctonia	azoxystrobin (*off-label*)
	Sclerotinia	azoxystrobin (*off-label*)
	Seed-borne diseases	thiram (*seed soak*)
	Septoria	chlorothalonil (*qualified minor use*), copper oxychloride
Pests	Aphids	acetamiprid (*off-label*), cypermethrin (*off-label*), deltamethrin (*off-label*), fatty acids (*off-label*), pirimicarb (*off-label*), pirimicarb (*off-label - for forcing*), pymetrozine (*off-label*)
	Beetles	deltamethrin (*off-label*)
	Caterpillars	Bacillus thuringiensis (*off-label*), deltamethrin (*off-label*), diflubenzuron (*off-label*), lambda-cyhalothrin (*off-label*)
	Cutworms	lambda-cyhalothrin (*off-label*)
	Flies	lambda-cyhalothrin (*off-label*)
	Leafhoppers	deltamethrin (*off-label*)
	Pests, miscellaneous	deltamethrin (*off-label*), lambda-cyhalothrin (*off-label*), spinosad
	Thrips	deltamethrin (*off-label*), fatty acids (*off-label*)
Plant growth regulation	Quality/yield control	gibberellins
Weeds	Broad-leaved weeds	asulam (*off-label*), linuron (*off-label*), metamitron (*off-label*), pendimethalin (*off-label*), propyzamide (*off-label*), prosulfocarb (*off-label*), triflusulfuron-methyl (*off-label*)
	Crops as weeds	propaquizafop (*off-label*)
	Grass weeds	asulam (*off-label*), fluazifop-P-butyl (*off-label*), linuron (*off-label*), metamitron (*off-label*), pendimethalin (*off-label*), propaquizafop (*off-label*), propyzamide (*off-label*), prosulfocarb (*off-label*)

Stem and bulb vegetables - Globe artichokes/cardoons

Diseases	Downy mildew	azoxystrobin (*off-label*)
	Powdery mildew	azoxystrobin (*off-label*), myclobutanil (*off-label*)
Weeds	Broad-leaved weeds	metamitron (*off-label*)
	Grass weeds	fluazifop-P-butyl (*off-label*), metamitron (*off-label*)

Crop/Pest Guide - Arable and vegetable crops

Stem and bulb vegetables - Onions/leeks/garlic

Diseases	Alternaria	iprodione (*off-label*)
	Bacterial blight	copper oxychloride (*off-label*)
	Blight	fluoxastrobin + prothioconazole (*control*)
	Botrytis	azoxystrobin + chlorothalonil, chlorothalonil, chlorothalonil (*moderate control*), cyprodinil + fludioxonil (*off-label*), fluoxastrobin + prothioconazole (*useful reduction*), iprodione, iprodione (*off-label*)
	Bottom rot	copper oxychloride (*off-label*)
	Cladosporium	fluoxastrobin + prothioconazole (*useful reduction*)
	Collar rot	iprodione, iprodione (*off-label*)
	Damping off	thiram (*off-label - seed treatment*), thiram (*seed treatment*)
	Disease control/foliar feed	mancozeb
	Downy mildew	azoxystrobin, azoxystrobin (*off-label*), azoxystrobin + chlorothalonil, benthiavalicarb-isopropyl + mancozeb (*off-label*), dimethomorph + mancozeb (*off-label*), fluoxastrobin + prothioconazole (*control*), fosetyl-aluminium (*off-label*), mancozeb, mancozeb (*off-label*), mancozeb + metalaxyl-M (*off-label*), mancozeb + metalaxyl-M (*useful control*), metalaxyl-M (*off-label*), propiconazole (*off-label*)
	Foliar diseases	chlorothalonil (*off-label*), fosetyl-aluminium (*off-label*)
	Fusarium	thiophanate-methyl (*off-label*)
	Phytophthora	copper oxychloride (*off-label*), mancozeb (*off-label*), propamocarb hydrochloride
	Purple blotch	azoxystrobin, azoxystrobin + difenoconazole (*moderate control*), prothioconazole, tebuconazole + trifloxystrobin
	Pythium	metalaxyl-M, propamocarb hydrochloride
	Rust	azoxystrobin, azoxystrobin + difenoconazole, fenpropimorph, fluoxastrobin + prothioconazole (*useful reduction*), mancozeb, prothioconazole, tebuconazole, tebuconazole + trifloxystrobin
	Seed-borne diseases	thiram (*off-label - seed treatment*)
	Spear rot	copper oxychloride (*off-label*)
	Stem canker	copper oxychloride (*off-label*)
	Storage rots	copper oxychloride (*off-label*)
	White rot	boscalid + pyraclostrobin (*off-label*), tebuconazole (*off-label*)
	White tip	azoxystrobin + difenoconazole (*qualified minor use*), boscalid + pyraclostrobin (*off-label*), dimethomorph + mancozeb (*off-label*), mancozeb (*off-label*), tebuconazole + trifloxystrobin
Pests	Aphids	deltamethrin (*off-label*), fatty acids (*off-label*), lambda-cyhalothrin (*off-label*)
	Beetles	deltamethrin (*off-label*)
	Caterpillars	deltamethrin (*off-label*)
	Cutworms	Bacillus thuringiensis (*off-label*), chlorpyrifos
	Flies	tefluthrin (*off-label - seed treatment*)
	Free-living nematodes	oxamyl (*off-label*)
	Leafhoppers	deltamethrin (*off-label*)

Crop/Pest Guide - Flowers and ornamentals

	Pests, miscellaneous	abamectin (*off-label*), chlorpyrifos (*off-label*), deltamethrin (*off-label*), lambda-cyhalothrin (*off-label*), oxamyl (*off-label*)
	Stem nematodes	oxamyl (*off-label*)
	Thrips	abamectin (*off-label*), chlorpyrifos (*off-label*), deltamethrin (*off-label*), fatty acids (*off-label*), lambda-cyhalothrin (*off-label*), spinosad
Plant growth regulation	Growth control	maleic hydrazide, maleic hydrazide (*off-label*)
Weeds	Broad-leaved weeds	bentazone (*off-label*), chloridazon (*off-label*), clopyralid, clopyralid (*off-label*), flumioxazine (*off-label*), fluroxypyr (*off-label*), glyphosate, ioxynil, linuron (*off-label*), metazachlor (*off-label*), pendimethalin, pendimethalin (*off-label*), pendimethalin (*pre + post emergence treatment*), prosulfocarb (*off-label*), pyridate, pyridate (*off-label*)
	Crops as weeds	cycloxydim, fluazifop-P-butyl, flumioxazine (*off-label*), fluroxypyr (*off-label*), glyphosate, pendimethalin, propaquizafop (*off-label*), prosulfocarb (*off-label*), tepraloxydim
	Grass weeds	chloridazon (*off-label*), cycloxydim, cycloxydim (*off-label*), fluazifop-P-butyl, fluazifop-P-butyl (*off-label*), glyphosate, linuron (*off-label*), metazachlor (*off-label*), pendimethalin, pendimethalin (*off-label*), propaquizafop, propaquizafop (*off-label*), prosulfocarb (*off-label*), tepraloxydim, tepraloxydim (*off-label*)
	Weeds, miscellaneous	glyphosate, glyphosate (*off-label*)

Stem and bulb vegetables - Vegetables

Diseases	Botrytis	Bacillus subtilis (*off-label*)
Pests	Aphids	fatty acids (*off-label*)
	Thrips	fatty acids (*off-label*)

Flowers and ornamentals

Flowers - Bedding plants, general

Diseases	Botrytis	thiram
	Phytophthora	propamocarb hydrochloride
	Powdery mildew	bupirimate
	Pythium	propamocarb hydrochloride
	Rust	azoxystrobin (*off-label - in pots*), thiram
Pests	Aphids	imidacloprid, thiacloprid
	Beetles	thiacloprid
	Flies	imidacloprid
	Weevils	imidacloprid, thiacloprid
	Whiteflies	imidacloprid, thiacloprid
Plant growth regulation	Flowering control	paclobutrazol
	Growth control	chlormequat, daminozide, paclobutrazol

Flowers - Bulbs/corms

Diseases	Basal stem rot	chlorothalonil (*off-label*), thiabendazole (*off-label*)
	Botrytis	Bacillus subtilis (*off-label*), chlorothalonil (*off-label*), tebuconazole (*off-label*), tebuconazole (*off-label - for galanthamine production*), thiabendazole (*off-label*), thiram
	Crown rot	metalaxyl-M (*off-label*)
	Downy mildew	metalaxyl-M (*off-label*)
	Fusarium	thiabendazole, thiabendazole (*post-lifting or pre-planting dip*), thiabendazole (*post-lifting spray*)
	Ink disease	chlorothalonil (*qualified minor use*)
	Phytophthora	propamocarb hydrochloride
	Powdery mildew	bupirimate
	Pythium	propamocarb hydrochloride
Pests	Flies	chlorpyrifos (*off-label*)
Plant growth regulation	Flowering control	paclobutrazol
	Growth control	chlormequat, paclobutrazol
Weeds	Broad-leaved weeds	bentazone, carfentrazone-ethyl (*off-label*), pendimethalin (*off-label*)
	Crops as weeds	cycloxydim, pendimethalin (*off-label*)
	Grass weeds	cycloxydim, pendimethalin (*off-label*), S-metolachlor (*off-label*)

Flowers - Miscellaneous flowers

Diseases	Black spot	captan, kresoxim-methyl, myclobutanil
	Disease control/foliar feed	tebuconazole (*off-label*)
	Foliar diseases	tebuconazole (*off-label*)
	Powdery mildew	bupirimate, kresoxim-methyl, myclobutanil
	Rust	myclobutanil, penconazole
Pests	Aphids	cypermethrin (*off-label*), deltamethrin (*off-label*), imidacloprid, lambda-cyhalothrin + pirimicarb (*off-label*), pirimicarb (*off-label*)
	Beetles	deltamethrin (*off-label*)
	Caterpillars	deltamethrin (*off-label*)
	Cutworms	lambda-cyhalothrin + pirimicarb (*off-label*)
	Flies	imidacloprid, lambda-cyhalothrin (*off-label*)
	Leafhoppers	deltamethrin (*off-label*)
	Pests, miscellaneous	deltamethrin (*off-label*), lambda-cyhalothrin (*off-label*)
	Thrips	deltamethrin (*off-label*)
	Weevils	imidacloprid
	Whiteflies	imidacloprid
Plant growth regulation	Flowering control	paclobutrazol
	Growth control	paclobutrazol
Weeds	Broad-leaved weeds	linuron (*off-label*), metribuzin (*off-label*), pendimethalin (*off-label*), propyzamide
	Crops as weeds	propaquizafop (*off-label*)

| | Grass weeds | cycloxydim (off-label), fluazifop-P-butyl (off-label), linuron (off-label), pendimethalin (off-label), propaquizafop (off-label), propyzamide, tepraloxydim (off-label) |

Flowers - Pot plants

Diseases	Phytophthora	fosetyl-aluminium, propamocarb hydrochloride
	Pythium	propamocarb hydrochloride
	Root rot	fosetyl-aluminium
Pests	Aphids	deltamethrin, imidacloprid, thiacloprid
	Beetles	thiacloprid
	Caterpillars	deltamethrin
	Flies	imidacloprid
	Mealybugs	deltamethrin, petroleum oil
	Scale insects	deltamethrin, petroleum oil
	Spider mites	petroleum oil
	Weevils	imidacloprid, thiacloprid
	Whiteflies	deltamethrin, imidacloprid, thiacloprid
Plant growth regulation	Flowering control	paclobutrazol
	Fruiting control	paclobutrazol
	Growth control	chlormequat, daminozide, paclobutrazol

Flowers - Protected bulbs/corms

Pests	Aphids	pirimicarb

Flowers - Protected flowers

Diseases	Black spot	kresoxim-methyl
	Crown rot	metalaxyl-M (off-label)
	Disease control/foliar feed	difenoconazole (off-label)
	Downy mildew	metalaxyl-M (off-label)
	Powdery mildew	kresoxim-methyl
	Ring spot	difenoconazole (off-label - ground grown (or pot grown - 20050159)), difenoconazole (off-label - ground grown (or pot grown - SOLA 20050159))
	Rust	azoxystrobin (off-label), difenoconazole (off-label), difenoconazole (off-label - ground grown (or pot grown - 20050159)), difenoconazole (off-label - ground grown (or pot grown - SOLA 20050159))
Pests	Aphids	pirimicarb
	Spider mites	tebufenpyrad

Miscellaneous flowers and ornamentals - Miscellaneous uses

Crop control	Miscellaneous non-selective situations	benzoic acid
Diseases	Disease control/foliar feed	boscalid + pyraclostrobin (off-label), potassium bicarbonate (commodity substance)
	Powdery mildew	physical pest control
	Sclerotinia	Coniothyrium minitans
Pests	Aphids	physical pest control, pyrethrins

Crop/Pest Guide - Flowers and ornamentals

	Birds/mammals	aluminium ammonium sulphate
	Caterpillars	diflubenzuron, pyrethrins
	Mealybugs	physical pest control
	Scale insects	physical pest control
	Slugs/snails	ferric phosphate, metaldehyde, metaldehyde (*around*), metaldehyde (*do not apply to cauliflowers*), metaldehyde (*e*), methiocarb (*outdoor only*)
	Spider mites	physical pest control, pyrethrins
	Suckers	physical pest control
	Thrips	pyrethrins
	Whiteflies	physical pest control
Weeds	Broad-leaved weeds	carfentrazone-ethyl (*before planting*), diquat, diquat (*around*), propyzamide
	Crops as weeds	carfentrazone-ethyl (*before planting*)
	Grass weeds	diquat (*around only*), glyphosate, propyzamide
	Weeds, miscellaneous	diquat (*around only*), glyphosate, glyphosate (*before planting*), glyphosate (*pre-sowing/planting*)

Miscellaneous flowers and ornamentals - Soils

Diseases	Soil-borne diseases	metam-sodium
Pests	Free-living nematodes	metam-sodium
	Soil pests	metam-sodium
Weeds	Weeds, miscellaneous	metam-sodium

Ornamentals - Nursery stock

Diseases	Black spot	myclobutanil
	Botrytis	Bacillus subtilis (*off-label*), boscalid + pyraclostrobin (*off-label*), chlorothalonil, chlorothalonil (*moderate control*), cyprodinil + fludioxonil, iprodione, thiram, thiram (*except Hydrangea*)
	Damping off	fosetyl-aluminium + propamocarb hydrochloride (*off-label*), tolclofos-methyl
	Disease control/foliar feed	azoxystrobin (*off-label*), mancozeb, mancozeb + metalaxyl-M (*off-label*), mandipropamid (*off-label*), metrafenone (*off-label*)
	Downy mildew	benthiavalicarb-isopropyl + mancozeb (*off-label*)
	Foliar diseases	picoxystrobin (*off-label*), pyraclostrobin (*off-label*), thiamethoxam (*off-label*), trifloxystrobin (*off-label*)
	Foot rot	thiophanate-methyl (*off-label*), tolclofos-methyl
	Fungus diseases	prochloraz
	Leaf curl	thiophanate-methyl (*off-label*)
	Phytophthora	fosetyl-aluminium, metalaxyl-M, propamocarb hydrochloride
	Powdery mildew	cyflufenamid (*off-label*), metrafenone, myclobutanil, physical pest control, quinoxyfen (*off-label*)
	Pythium	metalaxyl-M, metalaxyl-M (*off-label*), propamocarb hydrochloride
	Root rot	tolclofos-methyl
	Rust	myclobutanil, propiconazole (*off-label*)

Crop/Pest Guide - Flowers and ornamentals

	Seed-borne diseases	iprodione
	Stem canker	thiophanate-methyl (*off-label*)
Pests	Aphids	acetamiprid, Beauveria bassiana, cypermethrin, deltamethrin, dimethoate, esfenvalerate, flonicamid (*off-label*), imidacloprid, imidacloprid (*container grown*), lambda-cyhalothrin (*off-label*), physical pest control, pirimicarb, pymetrozine, pymetrozine (*off-label*), pyrethrins, thiacloprid
	Beetles	Beauveria bassiana, thiacloprid
	Capsid bugs	cypermethrin, deltamethrin
	Caterpillars	Bacillus thuringiensis, cypermethrin, deltamethrin, diflubenzuron, indoxacarb, indoxacarb (*off-label*), pyrethrins, spinosad (*off-label*), teflubenzuron, teflubenzuron (*off-label*)
	Cutworms	cypermethrin
	Flies	imidacloprid, imidacloprid (*container grown*)
	Free-living nematodes	fosthiazate (*off-label*)
	Leaf miners	dimethoate, oxamyl (*off-label*), thiacloprid (*off-label*)
	Mealybugs	deltamethrin, petroleum oil, physical pest control
	Pests, miscellaneous	esfenvalerate, indoxacarb (*off-label*), lambda-cyhalothrin (*off-label*), methoxyfenozide (*off-label*), spinosad (*off-label*), teflubenzuron (*off-label*), thiacloprid (*off-label*)
	Scale insects	deltamethrin, petroleum oil, physical pest control
	Slugs/snails	metaldehyde
	Spider mites	abamectin, bifenthrin, dimethoate, etoxazole (*off-label*), petroleum oil, physical pest control, spirodiclofen (*off-label*), spiromesifen (*off-label*)
	Suckers	physical pest control
	Thrips	abamectin, cypermethrin, deltamethrin, deltamethrin (*off-label*), spinosad, teflubenzuron (*off-label*), thiacloprid (*off-label*), thiamethoxam (*off-label*)
	Weevils	chlorpyrifos, imidacloprid, imidacloprid (*container grown*), thiacloprid
	Whiteflies	acetamiprid, Beauveria bassiana, cypermethrin, deltamethrin, flonicamid (*off-label*), imidacloprid, imidacloprid (*container grown*), Lecanicillium lecanii, physical pest control, pymetrozine (*off-label*), spiromesifen (*off-label*), teflubenzuron, thiacloprid, thiacloprid (*off-label*)
Plant growth regulation	Growth control	1-naphthylacetic acid, 2-chloroethylphosphonic acid (*off-label*), 4-indol-3-ylbutyric acid, chlormequat, daminozide, gibberellins (*off-label*), indol-3-ylacetic acid, prohexadione-calcium (*off-label*)
Weeds	Broad-leaved weeds	2,4-D (*off-label*), bentazone (*off-label*), clopyralid, desmedipham + ethofumesate + phenmedipham (*off-label*), diquat, ethofumesate (*off-label*), florasulam (*off-label*), florasulam + fluroxypyr (*off-label*), flufenacet + isoxaflutole (*off-label*), flumioxazine (*off-label*), fluroxypyr (*off-label*), glufosinate-ammonium, imazamox + pendimethalin (*off-label*), isoxaben, isoxaben (*off-label*), linuron, linuron (*off-label*), metsulfuron-methyl (*off-label*), napropamide, oxadiazon, pendimethalin (*off-label*), propyzamide
	Crops as weeds	flumioxazine (*off-label*)

Crop/Pest Guide - Flowers and ornamentals

	Grass weeds	diquat, ethofumesate (*off-label*), fluazifop-P-butyl (*off-label*), flufenacet + isoxaflutole (*off-label*), glufosinate-ammonium, linuron (*off-label*), maleic hydrazide (*off-label*), napropamide, oxadiazon, pendimethalin (*off-label*), propyzamide, quizalofop-P-tefuryl (*off-label*), tepraloxydim (*off-label*)
	Mosses	maleic hydrazide + pelargonic acid
	Weeds, miscellaneous	carfentrazone-ethyl (*off-label*), diquat, glyphosate, glyphosate (*off-label*), maleic hydrazide + pelargonic acid

Ornamentals - Trees and shrubs

Diseases	Disease control/foliar feed	mancozeb
	Downy mildew	benthiavalicarb-isopropyl + mancozeb (*off-label*)
	Fungus diseases	prochloraz
	Powdery mildew	cyflufenamid (*off-label*), penconazole
	Rhizoctonia	chloropicrin (*soil fumigation*)
	Scab	penconazole
Pests	Aphids	deltamethrin, imidacloprid (*off-label*), lambda-cyhalothrin (*off-label*)
	Birds/mammals	aluminium ammonium sulphate, warfarin (*nuts*)
	Capsid bugs	deltamethrin
	Caterpillars	Bacillus thuringiensis, deltamethrin, diflubenzuron, teflubenzuron (*off-label*)
	Mealybugs	deltamethrin
	Pests, miscellaneous	lambda-cyhalothrin (*off-label*), teflubenzuron (*off-label*)
	Scale insects	deltamethrin
	Slugs/snails	ferric phosphate
	Thrips	deltamethrin, teflubenzuron (*off-label*)
	Whiteflies	deltamethrin
Weeds	Broad-leaved weeds	asulam, clopyralid (*off-label*), isoxaben, metazachlor, propyzamide
	Grass weeds	cycloxydim (*off-label*), metazachlor, propyzamide, quizalofop-P-tefuryl (*off-label*)
	Mosses	maleic hydrazide + pelargonic acid
	Weeds, miscellaneous	glyphosate, glyphosate (*wiper application*), glyphosate + sulfosulfuron, maleic hydrazide + pelargonic acid, propyzamide
	Woody weeds/scrub	asulam, glyphosate

Ornamentals - Woody ornamentals

Diseases	Fungus diseases	prochloraz
Pests	Aphids	fatty acids
	Mealybugs	fatty acids
	Scale insects	fatty acids
	Spider mites	fatty acids
	Whiteflies	fatty acids
Weeds	Bindweeds	oxadiazon
	Broad-leaved weeds	glufosinate-ammonium, napropamide, oxadiazon, propyzamide

Crop/Pest Guide - Forestry

	Grass weeds	glufosinate-ammonium, napropamide, oxadiazon, propyzamide
	Weeds, miscellaneous	glyphosate

Forestry

Forest nurseries, general - Forest nurseries

Crop control	Chemical stripping/thinning	glyphosate
Diseases	Botrytis	boscalid + pyraclostrobin (*off-label*), cyprodinil + fludioxonil, fenhexamid (*off-label*), iprodione (*off-label*), mepanipyrim (*off-label*)
	Disease control/foliar feed	azoxystrobin (*off-label*), mancozeb + metalaxyl-M (*off-label*)
	Foliar diseases	fluoxastrobin + prothioconazole (*off-label*)
	Powdery mildew	cyflufenamid (*off-label*), metrafenone (*off-label*), proquinazid (*off-label*), quinoxyfen (*off-label*)
Pests	Aphids	lambda-cyhalothrin (*off-label*), pirimicarb, pymetrozine (*off-label*)
	Birds/mammals	aluminium ammonium sulphate
	Caterpillars	indoxacarb (*off-label*), teflubenzuron (*off-label*)
	Pests, miscellaneous	indoxacarb (*off-label*), lambda-cyhalothrin (*off-label*), teflubenzuron (*off-label*), thiacloprid (*off-label*)
	Spider mites	bifenazate (*off-label*)
	Thrips	teflubenzuron (*off-label*)
	Weevils	alpha-cypermethrin (*off-label*)
	Whiteflies	pymetrozine (*off-label*)
Weeds	Broad-leaved weeds	2,4-D (*off-label*), carfentrazone-ethyl (*off-label*), ethofumesate (*off-label*), florasulam (*off-label*), flumioxazine (*off-label*), fluroxypyr (*off-label*), linuron, linuron (*off-label*), metsulfuron-methyl (*off-label*), napropamide, pendimethalin (*off-label*), pendimethalin + picolinafen (*off-label*), propyzamide, propyzamide (*off-label*)
	Crops as weeds	flumioxazine (*off-label*)
	Grass weeds	ethofumesate (*off-label*), linuron (*off-label*), napropamide, pendimethalin (*off-label*), pendimethalin + picolinafen (*off-label*), propaquizafop, propyzamide, propyzamide (*off-label*), tepraloxydim (*off-label*)
	Weeds, miscellaneous	diquat (*off-label*), glyphosate
	Woody weeds/scrub	glyphosate

Forestry plantations, general - Forestry plantations

Crop control	Chemical stripping/thinning	glyphosate, glyphosate (*stump treatment*)
Pests	Birds/mammals	aluminium ammonium sulphate, warfarin
	Caterpillars	diflubenzuron
	Weevils	alpha-cypermethrin (*off-label*), cypermethrin
Plant growth regulation	Growth control	gibberellins (*off-label*), glyphosate

Crop/Pest Guide - Forestry

Weeds	Aquatic weeds	glyphosate, propyzamide
	Broad-leaved weeds	2,4-D + dicamba + triclopyr, clopyralid (off-label), glufosinate-ammonium, glyphosate, glyphosate (wiper application), isoxaben, metamitron (off-label), metazachlor, napropamide (off-label), pendimethalin (off-label), propyzamide, triclopyr
	Grass weeds	cycloxydim, glufosinate-ammonium, glyphosate, metamitron (off-label), metazachlor, napropamide (off-label), propaquizafop, propyzamide
	Weeds, miscellaneous	carfentrazone-ethyl (off-label), glufosinate-ammonium, glyphosate, metamitron (off-label), propyzamide
	Woody weeds/scrub	2,4-D + dicamba + triclopyr, asulam, asulam (off-label), glyphosate, triclopyr

Miscellaneous forestry situations - Cut logs/timber

Pests	Beetles	chlorpyrifos, cypermethrin
Weeds	Grass weeds	propaquizafop

Woodland on farms - Woodland

Crop control	Chemical stripping/ thinning	glyphosate
Pests	Aphids	lambda-cyhalothrin (off-label)
	Beetles	lambda-cyhalothrin (off-label)
	Pests, miscellaneous	lambda-cyhalothrin (off-label)
	Sawflies	lambda-cyhalothrin (off-label)
	Thrips	lambda-cyhalothrin (off-label)
	Whiteflies	lambda-cyhalothrin (off-label)
Plant growth regulation	Growth control	gibberellins (do not exceed 2.5 g per 5 litres when used as a seed soak), gibberellins (off-label)
	Plant growth regulation, miscellaneous	gibberellins (qualified minor use)
Weeds	Broad-leaved weeds	2,4-D (off-label), 2,4-D + dicamba + triclopyr, florasulam (off-label), flumioxazine (off-label), fluroxypyr (off-label), lenacil (off-label), metamitron (off-label), metazachlor, metsulfuron-methyl (off-label), napropamide (off-label), pendimethalin (off-label), propyzamide
	Crops as weeds	fluazifop-P-butyl, flumioxazine (off-label)
	Grass weeds	cycloxydim, fluazifop-P-butyl, lenacil (off-label), metamitron (off-label), metazachlor, napropamide (off-label), pendimethalin (off-label), propaquizafop, propyzamide
	Weeds, miscellaneous	amitrole (off-label), glyphosate, glyphosate (off-label)
	Woody weeds/scrub	2,4-D + dicamba + triclopyr, glyphosate

Fruit and hops

All bush fruit - All currants

Diseases	Botrytis	Bacillus subtilis (*off-label*), boscalid + pyraclostrobin, boscalid + pyraclostrobin (*off-label*), cyprodinil + fludioxonil (*qualified minor use recommendation*), fenhexamid, pyrimethanil (*off-label*)
	Disease control/foliar feed	sulphur (*off-label*)
	Downy mildew	copper oxychloride (*off-label*)
	Leaf spot	boscalid + pyraclostrobin (*moderate control only*), boscalid + pyraclostrobin (*off-label*), chlorothalonil, chlorothalonil (*qualified minor use*), dodine, dodine (*off-label*)
	Powdery mildew	boscalid + pyraclostrobin (*moderate control only*), boscalid + pyraclostrobin (*off-label*), bupirimate, bupirimate (*off-label*), fenpropimorph (*off-label*), kresoxim-methyl, kresoxim-methyl (*off-label*), myclobutanil, myclobutanil (*off-label*), penconazole, penconazole (*off-label*)
	Rust	copper oxychloride
	Verticillium wilt	chloropicrin (*off-label*)
Pests	Aphids	chlorpyrifos, fatty acids (*off-label*), pirimicarb, pymetrozine (*off-label*)
	Capsid bugs	chlorpyrifos
	Caterpillars	Bacillus thuringiensis (*off-label*), chlorpyrifos, diflubenzuron, diflubenzuron (*off-label*)
	Gall, rust and leaf & bud mites	sulphur, tebufenpyrad (*off-label*)
	Midges	lambda-cyhalothrin (*off-label*)
	Pests, miscellaneous	lambda-cyhalothrin (*off-label*), thiacloprid (*off-label*)
	Sawflies	lambda-cyhalothrin (*off-label*)
	Spider mites	bifenthrin (*off-label*), chlorpyrifos, chlorpyrifos (*non OP resistant strains only*), clofentezine (*off-label*)
	Thrips	fatty acids (*off-label*)
Weeds	Bindweeds	oxadiazon, oxadiazon (*off-label*)
	Broad-leaved weeds	asulam, asulam (*off-label*), flufenacet + metribuzin (*off-label*), glufosinate-ammonium, isoxaben, isoxaben (*off-label*), lenacil (*off-label*), MCPB, MCPB (*off-label*), napropamide, napropamide (*off-label*), oxadiazon, oxadiazon (*off-label*), pendimethalin, pendimethalin (*off-label*), propyzamide
	Crops as weeds	fluazifop-P-butyl, pendimethalin
	Grass weeds	fluazifop-P-butyl, fluazifop-P-butyl (*off-label*), flufenacet + metribuzin (*off-label*), glufosinate-ammonium, lenacil (*off-label*), napropamide, napropamide (*off-label*), oxadiazon, oxadiazon (*off-label*), pendimethalin, pendimethalin (*off-label*), propyzamide
	Weeds, miscellaneous	glufosinate-ammonium, glyphosate (*off-label*)

All bush fruit - All protected bush fruit

Pests	Aphids	pymetrozine (*off-label*), pyrethrins
	Caterpillars	pyrethrins, thiacloprid (*off-label*)
	Midges	thiacloprid (*off-label*)
	Pests, miscellaneous	thiacloprid (*off-label*)
	Whiteflies	Lecanicillium lecanii (*off-label*)

All bush fruit - All vines

Diseases	Botrytis	Bacillus subtilis (*off-label*), fenhexamid, pyrimethanil (*off-label*)
	Disease control/foliar feed	mancozeb
	Downy mildew	benthiavalicarb-isopropyl + mancozeb (*off-label*), copper oxychloride, fosetyl-aluminium (*off-label*)
	Powdery mildew	fenbuconazole (*off-label*), kresoxim-methyl (*off-label*), meptyldinocap, myclobutanil (*off-label*), sulphur, tebuconazole + trifloxystrobin (*off-label*)
Pests	Mealybugs	petroleum oil
	Scale insects	petroleum oil
	Spider mites	petroleum oil
	Whiteflies	Lecanicillium lecanii (*off-label*)
Weeds	Bindweeds	oxadiazon
	Broad-leaved weeds	glufosinate-ammonium, isoxaben, oxadiazon, propyzamide (*off-label*)
	Grass weeds	glufosinate-ammonium, oxadiazon, propyzamide (*off-label*)
	Weeds, miscellaneous	glufosinate-ammonium, glyphosate (*off-label*)

All bush fruit - Bilberries/blueberries/cranberries

Diseases	Botrytis	Bacillus subtilis (*off-label*), cyprodinil + fludioxonil (*qualified minor use recommendation*), fenhexamid (*off-label*), pyrimethanil (*off-label*)
	Disease control/foliar feed	sulphur (*off-label*)
	Downy mildew	copper oxychloride (*off-label*)
	Leaf spot	dodine (*off-label*)
	Powdery mildew	fenpropimorph (*off-label*), penconazole (*off-label*)
Pests	Aphids	fatty acids (*off-label*), pirimicarb (*off-label*), pymetrozine (*off-label*)
	Caterpillars	Bacillus thuringiensis (*off-label*), diflubenzuron (*off-label*)
	Gall, rust and leaf & bud mites	tebufenpyrad (*off-label*)
	Pests, miscellaneous	thiacloprid (*off-label*)
Weeds	Bindweeds	oxadiazon (*off-label*)
	Broad-leaved weeds	asulam (*off-label*), flufenacet + metribuzin (*off-label*), isoxaben (*off-label*), lenacil (*off-label*), MCPB (*off-label*), napropamide (*off-label*), oxadiazon (*off-label*), pendimethalin (*off-label*), propyzamide (*off-label*)
	Grass weeds	fluazifop-P-butyl (*off-label*), flufenacet + metribuzin (*off-label*), lenacil (*off-label*), napropamide (*off-label*), oxadiazon (*off-label*), pendimethalin (*off-label*), propyzamide (*off-label*)

Crop/Pest Guide - Fruit and hops

	Weeds, miscellaneous	glufosinate-ammonium, glyphosate (*off-label*)

All bush fruit - Bush fruit, general

Diseases	Phytophthora	fluazinam (*off-label*)
Pests	Aphids	pyrethrins
	Birds/mammals	aluminium ammonium sulphate
	Caterpillars	pyrethrins
	Mealybugs	petroleum oil
	Scale insects	petroleum oil
	Spider mites	petroleum oil
Plant growth regulation	Quality/yield control	sulphur

All bush fruit - Gooseberries

Diseases	Botrytis	Bacillus subtilis (*off-label*), chlorothalonil, cyprodinil + fludioxonil (*qualified minor use recommendation*), fenhexamid, pyrimethanil (*off-label*)
	Downy mildew	copper oxychloride (*off-label*)
	Leaf spot	chlorothalonil (*qualified minor use*), dodine (*off-label*)
	Powdery mildew	bupirimate, chlorothalonil, fenpropimorph (*off-label*), kresoxim-methyl (*off-label*), myclobutanil, penconazole (*off-label*), sulphur
Pests	Aphids	chlorpyrifos, fatty acids (*off-label*), pirimicarb, pymetrozine (*off-label*)
	Capsid bugs	chlorpyrifos
	Caterpillars	Bacillus thuringiensis (*off-label*), chlorpyrifos, diflubenzuron (*off-label*)
	Gall, rust and leaf & bud mites	tebufenpyrad (*off-label*)
	Midges	lambda-cyhalothrin (*off-label*)
	Pests, miscellaneous	lambda-cyhalothrin (*off-label*), thiacloprid (*off-label*)
	Sawflies	lambda-cyhalothrin (*off-label*)
	Spider mites	chlorpyrifos, chlorpyrifos (*non OP resistant strains only*)
	Thrips	fatty acids (*off-label*)
Weeds	Bindweeds	oxadiazon
	Broad-leaved weeds	asulam (*off-label*), flufenacet + metribuzin (*off-label*), isoxaben, lenacil (*off-label*), MCPB (*off-label*), napropamide, oxadiazon, pendimethalin, propyzamide
	Crops as weeds	fluazifop-P-butyl, pendimethalin
	Grass weeds	fluazifop-P-butyl, flufenacet + metribuzin (*off-label*), lenacil (*off-label*), napropamide, oxadiazon, pendimethalin, propyzamide
	Weeds, miscellaneous	glyphosate (*off-label*)

All bush fruit - Miscellaneous bush fruit

Diseases	Alternaria	iprodione (*off-label*)
	Botrytis	cyprodinil + fludioxonil (*off-label*), fenhexamid, iprodione (*off-label*)
	Downy mildew	metalaxyl-M (*off-label*)

Crop/Pest Guide - Fruit and hops

	Foliar diseases	fosetyl-aluminium (*off-label*)
	Powdery mildew	sulphur
Pests	Capsid bugs	lambda-cyhalothrin (*off-label*)
	Caterpillars	indoxacarb (*off-label*)
	Pests, miscellaneous	lambda-cyhalothrin (*off-label*)
	Wasps	lambda-cyhalothrin (*off-label*)
Plant growth regulation	Growth control	prohexadione-calcium (*off-label*)
Weeds	Grass weeds	fluazifop-P-butyl (*off-label*)
	Weeds, miscellaneous	glyphosate (*off-label*)

Cane fruit - All outdoor cane fruit

Crop control	Desiccation	carfentrazone-ethyl (*off-label*)
Diseases	Botrytis	Bacillus subtilis (*off-label*), fenhexamid, iprodione, pyrimethanil (*off-label*)
	Botrytis fruit rot	cyprodinil + fludioxonil
	Cane blight	boscalid + pyraclostrobin (*off-label*), tebuconazole (*off-label*)
	Cane spot	copper oxychloride, copper oxychloride (*off-label*)
	Downy mildew	metalaxyl-M (*off-label*)
	Phytophthora	fluazinam (*off-label*), tebuconazole (*off-label*)
	Powdery mildew	azoxystrobin (*off-label*), bupirimate (*outdoor only*), fenpropimorph (*off-label*), myclobutanil
	Purple blotch	boscalid + pyraclostrobin (*off-label*), copper oxychloride
	Root rot	fluazinam (*off-label*)
	Verticillium wilt	chloropicrin (*off-label*)
Pests	Aphids	chlorpyrifos, chlorpyrifos (*off-label*), deltamethrin (*off-label*), fatty acids (*off-label*), pirimicarb, pirimicarb (*off-label*), pymetrozine (*off-label*), pyrethrins
	Beetles	chlorpyrifos, deltamethrin
	Birds/mammals	aluminium ammonium sulphate
	Capsid bugs	lambda-cyhalothrin (*off-label*)
	Caterpillars	Bacillus thuringiensis, Bacillus thuringiensis (*off-label*), diflubenzuron (*off-label*), pyrethrins
	Gall, rust and leaf & bud mites	tebufenpyrad (*off-label*)
	Mealybugs	petroleum oil
	Midges	chlorpyrifos
	Pests, miscellaneous	chlorpyrifos (*off-label*), deltamethrin (*off-label*), lambda-cyhalothrin (*off-label*), spinosad (*off-label*), thiacloprid (*off-label*)
	Scale insects	petroleum oil
	Slugs/snails	metaldehyde
	Spider mites	bifenthrin (*off-label*), chlorpyrifos, chlorpyrifos (*non OP resistant strains only*), chlorpyrifos (*off-label*), clofentezine (*off-label*), petroleum oil, tebufenpyrad (*off-label*)
	Thrips	fatty acids (*off-label*), spinosad (*off-label*)
	Weevils	lambda-cyhalothrin (*off-label*)

Plant growth regulation	Quality/yield control	sulphur
Weeds	Bindweeds	oxadiazon, oxadiazon (*off-label*)
	Broad-leaved weeds	asulam (*off-label*), glufosinate-ammonium, isoxaben, isoxaben (*off-label*), lenacil (*off-label*), napropamide, napropamide (*off-label*), oxadiazon, oxadiazon (*off-label*), pendimethalin, propyzamide, propyzamide (*England only*)
	Crops as weeds	fluazifop-P-butyl, pendimethalin
	Grass weeds	fluazifop-P-butyl, fluazifop-P-butyl (*off-label*), glufosinate-ammonium, napropamide, napropamide (*off-label*), oxadiazon, oxadiazon (*off-label*), pendimethalin, propyzamide, propyzamide (*England only*)
	Weeds, miscellaneous	glufosinate-ammonium

Cane fruit - All protected cane fruit

Crop control	Desiccation	carfentrazone-ethyl (*off-label*)
Diseases	Botrytis	pyrimethanil (*off-label*)
	Cane blight	boscalid + pyraclostrobin (*off-label*)
	Phytophthora	fluazinam (*off-label*)
	Powdery mildew	azoxystrobin (*off-label*), myclobutanil (*off-label*)
	Purple blotch	boscalid + pyraclostrobin (*off-label*)
	Root rot	fluazinam (*off-label*)
	Rust	boscalid + pyraclostrobin (*off-label*)
	Verticillium wilt	chloropicrin (*off-label*)
Pests	Aphids	pymetrozine (*off-label*), pyrethrins
	Beetles	thiacloprid (*off-label*)
	Capsid bugs	thiacloprid (*off-label*)
	Caterpillars	pyrethrins
	Gall, rust and leaf & bud mites	abamectin (*off-label*), tebufenpyrad (*off-label*)
	Mites	abamectin (*off-label*)
	Pests, miscellaneous	thiacloprid (*off-label*)
	Spider mites	abamectin (*off-label*), clofentezine (*off-label*), tebufenpyrad (*off-label*)
	Whiteflies	Lecanicillium lecanii (*off-label*)

Hops, general - Hops

Crop control	Chemical stripping/ thinning	diquat
	Desiccation	diquat (*off-label*)
Diseases	Botrytis	Bacillus subtilis (*off-label*), fenhexamid (*off-label*), iprodione (*off-label*)
	Disease control/foliar feed	boscalid + pyraclostrobin (*off-label*), cyazofamid (*off-label*), mancozeb + metalaxyl-M (*off-label*), thiophanate-methyl (*off-label - do not harvest for human or animal consumption (including idling) within 12 months of treatment.*)
	Downy mildew	chlorothalonil, copper oxychloride, cymoxanil (*off-label*), fosetyl-aluminium, metalaxyl-M (*off-label*)

Crop/Pest Guide - Fruit and hops

	Powdery mildew	bupirimate, cyflufenamid (*off-label*), fenpropimorph (*off-label*), kresoxim-methyl (*off-label*), myclobutanil (*off-label*), penconazole, quinoxyfen (*off-label*), sulphur
Pests	Aphids	acetamiprid (*off-label*), bifenthrin, cypermethrin, deltamethrin, flonicamid (*off-label*), imidacloprid, lambda-cyhalothrin (*off-label*), pymetrozine (*off-label*), spirotetramat (*off-label*), tebufenpyrad, thiamethoxam (*off-label*)
	Caterpillars	indoxacarb (*off-label*), spinosad (*off-label*)
	Free-living nematodes	fosthiazate (*off-label*)
	Mealybugs	petroleum oil
	Pests, miscellaneous	indoxacarb (*off-label*), lambda-cyhalothrin (*off-label*), methoxyfenozide (*off-label*), spinosad (*off-label*), thiacloprid (*off-label*)
	Scale insects	petroleum oil
	Spider mites	abamectin (*off-label*), bifenthrin, petroleum oil, spirodiclofen (*off-label*), tebufenpyrad
Plant growth regulation	Growth control	diquat, prohexadione-calcium (*off-label*)
Weeds	Bindweeds	oxadiazon
	Broad-leaved weeds	asulam, bentazone (*off-label*), flufenacet + isoxaflutole (*off-label*), flumioxazine (*off-label*), imazamox + pendimethalin (*off-label*), isoxaben, oxadiazon, pendimethalin, pendimethalin (*off-label*), propyzamide (*off-label*)
	Crops as weeds	fluazifop-P-butyl, flumioxazine (*off-label*), pendimethalin
	Grass weeds	diquat, fluazifop-P-butyl, flufenacet + isoxaflutole (*off-label*), maleic hydrazide (*off-label*), oxadiazon, pendimethalin, pendimethalin (*off-label*), propyzamide (*off-label*), tepraloxydim (*off-label*)
	Weeds, miscellaneous	diquat, glyphosate (*off-label*)

Miscellaneous fruit situations - Fruit crops, general

Diseases	Botrytis	Bacillus subtilis (*off-label*), fenhexamid (*off-label*), iprodione (*off-label*)
	Disease control/foliar feed	azoystrobin (*off-label*), boscalid + pyraclostrobin (*off-label*), mancozeb + metalaxyl-M (*off-label*), thiophanate-methyl (*off-label* - any fruit harvested within 12 months of treatment must be destroyed.)
	Downy mildew	metalaxyl-M (*off-label*)
	Powdery mildew	cyflufenamid (*off-label*), kresoxim-methyl (*off-label*), proquinazid (*off-label*)
Pests	Aphids	acetamiprid (*off-label*), fatty acids, lambda-cyhalothrin (*off-label*), pymetrozine (*off-label*), thiamethoxam (*off-label*)
	Caterpillars	indoxacarb (*off-label*), spinosad (*off-label*)
	Free-living nematodes	fosthiazate (*off-label*)
	Mealybugs	fatty acids, petroleum oil
	Pests, miscellaneous	lambda-cyhalothrin (*off-label*), methoxyfenozide (*off-label*), spinosad (*off-label*), thiacloprid (*off-label*)
	Scale insects	fatty acids, petroleum oil
	Spider mites	fatty acids, petroleum oil, spirodiclofen (*off-label*)
	Whiteflies	fatty acids, pymetrozine (*off-label*)

Crop/Pest Guide - Fruit and hops

Plant growth regulation	Fruiting control	ethylene (*in store*)
	Growth control	2-chloroethylphosphonic acid (*off-label*), prohexadione-calcium (*off-label*)
Weeds	Broad-leaved weeds	carfentrazone-ethyl (*off-label*), flufenacet + isoxaflutole (*off-label*), flumioxazine (*off-label*), imazamox + pendimethalin (*off-label*), pendimethalin (*off-label*), propyzamide (*off-label*)
	Crops as weeds	flumioxazine (*off-label*)
	Grass weeds	fluazifop-P-butyl (*off-label*), flufenacet + isoxaflutole (*off-label*), maleic hydrazide (*off-label*), pendimethalin (*off-label*), propyzamide (*off-label*), tepraloxydim (*off-label*)
	Weeds, miscellaneous	glyphosate (*off-label*)

Miscellaneous fruit situations - Fruit nursery stock

Weeds	Broad-leaved weeds	metazachlor
	Grass weeds	metazachlor

Other fruit - All outdoor strawberries

Diseases	Black spot	azoxystrobin (*off-label*), boscalid + pyraclostrobin, cyprodinil + fludioxonil (*qualified minor use*)
	Botrytis	Bacillus subtilis (*off-label*), boscalid + pyraclostrobin, captan, chlorothalonil, chlorothalonil (*moderate control*), chlorothalonil (*off-label*), cyprodinil + fludioxonil, fenhexamid, iprodione, mepanipyrim, pyrimethanil, thiram
	Crown rot	chloropicrin (*soil fumigation*), fosetyl-aluminium, fosetyl-aluminium (*off-label*)
	Foliar diseases	fosetyl-aluminium
	Leaf spot	chlorothalonil (*off-label*)
	Powdery mildew	boscalid + pyraclostrobin, bupirimate, fenpropimorph (*off-label*), kresoxim-methyl, myclobutanil, penconazole (*off-label*), quinoxyfen (*off-label*), sulphur
	Red core	chloropicrin (*soil fumigation*), fosetyl-aluminium, fosetyl-aluminium (*off-label*)
	Verticillium wilt	chloropicrin (*soil fumigation*), thiophanate-methyl (*off-label*)
Pests	Aphids	chlorpyrifos, fatty acids (*off-label*), pirimicarb, pymetrozine (*off-label*)
	Beetles	methiocarb
	Birds/mammals	aluminium ammonium sulphate
	Capsid bugs	thiacloprid (*off-label*)
	Caterpillars	Bacillus thuringiensis, chlorpyrifos
	Free-living nematodes	chloropicrin (*soil fumigation*)
	Slugs/snails	methiocarb
	Spider mites	abamectin (*off-label*), bifenthrin, chlorpyrifos, chlorpyrifos (*non OP resistant strains only*), clofentezine (*off-label*), fenpyroximate (*off-label*), spirodiclofen (*off-label*), tebufenpyrad
	Tarsonemid mites	abamectin (*off-label*), abamectin (*off-label - in propagation*)
	Weevils	chlorpyrifos, chlorpyrifos (*non-protected crops only*)

Crop/Pest Guide - Fruit and hops

Weeds	Broad-leaved weeds	asulam (*off-label*), clopyralid, glufosinate-ammonium, isoxaben, lenacil (*off-label*), metamitron (*off-label*), napropamide, pendimethalin, phenmedipham, propyzamide
	Crops as weeds	cycloxydim, fluazifop-P-butyl, pendimethalin
	Grass weeds	cycloxydim, fluazifop-P-butyl, glufosinate-ammonium, metamitron (*off-label*), napropamide, pendimethalin, propyzamide, S-metolachlor (*off-label*)
	Weeds, miscellaneous	glufosinate-ammonium

Other fruit - Protected miscellaneous fruit

Diseases	Black spot	azoxystrobin (*off-label*), boscalid + pyraclostrobin, cyprodinil + fludioxonil (*qualified minor use*)
	Botrytis	Bacillus subtilis (*reduction of damage to fruit*), boscalid + pyraclostrobin, cyprodinil + fludioxonil, iprodione, iprodione (*off-label*), mepanipyrim, thiram
	Botrytis fruit rot	Bacillus subtilis (*reduction of damage to fruit*)
	Crown rot	fosetyl-aluminium (*off-label*)
	Disease control/foliar feed	boscalid + pyraclostrobin (*off-label*), mancozeb + metalaxyl-M (*off-label*)
	Foliar diseases	fosetyl-aluminium (*off-label*)
	Powdery mildew	boscalid + pyraclostrobin, bupirimate, kresoxim-methyl, penconazole (*off-label*), quinoxyfen (*off-label*)
	Red core	fosetyl-aluminium (*off-label*)
Pests	Aphids	acetamiprid (*off-label*), deltamethrin (*off-label*), pymetrozine (*off-label*)
	Beetles	deltamethrin (*off-label*)
	Capsid bugs	thiacloprid (*off-label*)
	Caterpillars	Bacillus thuringiensis (*off-label*), deltamethrin (*off-label*), indoxacarb (*off-label*)
	Leafhoppers	deltamethrin (*off-label*)
	Pests, miscellaneous	deltamethrin (*off-label*), indoxacarb (*off-label*), thiacloprid (*off-label*)
	Spider mites	abamectin (*off-label*), bifenazate, bifenazate (*off-label*), clofentezine (*off-label*), etoxazole (*off-label*), fenpyroximate (*off-label*), spirodiclofen (*off-label*), spiromesifen (*off-label*)
	Tarsonemid mites	abamectin (*off-label*)
	Thrips	deltamethrin (*off-label*)
	Whiteflies	Lecanicillium lecanii (*off-label*), pymetrozine (*off-label*), spiromesifen (*off-label*)
Plant growth regulation	Quality/yield control	gibberellins

Other fruit - Rhubarb

Diseases	Bacterial blight	copper oxychloride (*off-label*)
	Damping off	thiram (*off-label - seed treatment*)
	Downy mildew	mancozeb + metalaxyl-M (*off-label*)
	Foliar diseases	copper oxychloride (*off-label*)
	Phytophthora	difenoconazole (*off-label*)

Pests	Aphids	deltamethrin (*off-label*), fatty acids (*off-label*), pirimicarb (*off-label*)
	Beetles	deltamethrin (*off-label*)
	Caterpillars	Bacillus thuringiensis (*off-label*), deltamethrin (*off-label*)
	Leafhoppers	deltamethrin (*off-label*)
	Pests, miscellaneous	deltamethrin (*off-label*)
	Thrips	deltamethrin (*off-label*), fatty acids (*off-label*)
Plant growth regulation	Quality/yield control	gibberellins
Weeds	Broad-leaved weeds	metamitron (*off-label*), pendimethalin (*off-label*), propyzamide, propyzamide (*outdoor*)
	Crops as weeds	propaquizafop (*off-label*)
	Grass weeds	metamitron (*off-label*), pendimethalin (*off-label*), propaquizafop (*off-label*), propyzamide, propyzamide (*outdoor*)
	Weeds, miscellaneous	glyphosate (*off-label*)

Tree fruit - All nuts

Diseases	Bacterial blight	copper oxychloride (*off-label*)
	Bacterial canker	copper oxychloride (*off-label*)
	Bottom rot	copper oxychloride (*off-label*)
	Phytophthora	copper oxychloride (*off-label*)
	Spear rot	copper oxychloride (*off-label*)
	Stem canker	copper oxychloride (*off-label*), tebuconazole (*off-label*)
Pests	Aphids	lambda-cyhalothrin (*off-label*)
	Caterpillars	diflubenzuron (*off-label*)
	Mites	tebufenpyrad (*off-label*), thiacloprid (*off-label*)
	Pests, miscellaneous	lambda-cyhalothrin (*off-label*)
Weeds	Bindweeds	oxadiazon (*off-label*)
	Broad-leaved weeds	2,4-D + dichlorprop-P + MCPA + mecoprop-P (*off-label*), asulam (*off-label*), dicamba + MCPA + mecoprop-P (*off-label*), dicamba + MCPA + mecoprop-P (*off-label - undersown with rotational grass*), fluroxypyr (*off-label*), glufosinate-ammonium, isoxaben (*off-label*), oxadiazon (*off-label*), pendimethalin (*off-label*), propyzamide (*off-label*)
	Grass weeds	fluazifop-P-butyl (*off-label*), glufosinate-ammonium, oxadiazon (*off-label*), pendimethalin (*off-label*), propyzamide (*off-label*)
	Weeds, miscellaneous	glufosinate-ammonium, glyphosate (*off-label*)

Tree fruit - All pome fruit

Crop control	Sucker/shoot control	glyphosate
Diseases	Alternaria	cyprodinil + fludioxonil
	Botrytis	iprodione (*off-label*), pyrimethanil (*off-label*), thiram
	Botrytis fruit rot	cyprodinil + fludioxonil, thiram
	Collar rot	copper oxychloride, copper oxychloride (*off-label*), fosetyl-aluminium
	Crown rot	fosetyl-aluminium
	Fusarium	cyprodinil + fludioxonil

Crop/Pest Guide - Fruit and hops

	Leaf curl	copper oxychloride (*off-label*)
	Penicillium rot	cyprodinil + fludioxonil
	Phytophthora	mancozeb + metalaxyl-M (*off-label*)
	Powdery mildew	boscalid + pyraclostrobin, bupirimate, dithianon + pyraclostrobin, fenbuconazole (*reduction*), kresoxim-methyl (*off-label*), kresoxim-methyl (*reduction*), meptyldinocap (*off-label*), myclobutanil, myclobutanil (*off-label*), penconazole, sulphur, sulphur (*off-label*)
	Scab	boscalid + pyraclostrobin, captan, captan (*off-label*), dithianon, dithianon (*off-label*), dithianon + pyraclostrobin, dodine, dodine (*off-label*), fenbuconazole, kresoxim-methyl, mancozeb, maneb, myclobutanil, pyrimethanil, pyrimethanil (*off-label*), sulphur, sulphur (*off-label*), thiram
	Stem canker	copper oxychloride, copper oxychloride (*off-label*), tebuconazole (*off-label*), thiophanate-methyl (*off-label*)
	Storage rots	captan, cyprodinil + fludioxonil, thiophanate-methyl (*off-label*), thiram
Pests	Aphids	acetamiprid, chlorpyrifos, chlorpyrifos (*off-label*), cypermethrin, deltamethrin, flonicamid, pirimicarb, thiacloprid, thiamethoxam
	Capsid bugs	chlorpyrifos, chlorpyrifos (*off-label*), cypermethrin, deltamethrin
	Caterpillars	Bacillus thuringiensis (*off-label*), chlorantraniliprole, chlorpyrifos, chlorpyrifos (*off-label*), Cydia pomonella GV, cypermethrin, deltamethrin, diflubenzuron, diflubenzuron (*off-label*), fenoxycarb, indoxacarb, methoxyfenozide, spinosad
	Gall, rust and leaf & bud mites	diflubenzuron, spirodiclofen
	Midges	thiacloprid (*off-label*)
	Pests, miscellaneous	chlorpyrifos (*off-label*), thiacloprid (*off-label*)
	Sawflies	chlorpyrifos, cypermethrin, deltamethrin
	Scale insects	spirodiclofen
	Spider mites	bifenthrin, chlorpyrifos, chlorpyrifos (*non OP resistant strains only*), clofentezine, fenpyroximate, spirodiclofen, tebufenpyrad
	Suckers	abamectin (*effective against larval stages but little activity against adults*), chlorpyrifos, cypermethrin, deltamethrin, diflubenzuron, lambda-cyhalothrin, spirodiclofen, thiamethoxam
	Weevils	chlorpyrifos
	Whiteflies	acetamiprid
Plant growth regulation	Fruiting control	1-methylcyclopropene (*off-label - post harvest only*), 1-methylcyclopropene (*off-label - post harvest use only*), 1-methylcyclopropene (*post-harvest use*), gibberellins, paclobutrazol, paclobutrazol (*off-label*)
	Growth control	2-chloroethylphosphonic acid (*off-label*), gibberellins, gibberellins (*off-label*), paclobutrazol, paclobutrazol (*off-label*), prohexadione-calcium
	Quality/yield control	gibberellins, gibberellins (*off-label*)
Weeds	Bindweeds	oxadiazon
	Broad-leaved weeds	2,4-D, 2,4-D + dichlorprop-P + MCPA + mecoprop-P, amitrole, asulam, asulam (*off-label*), clopyralid (*off-label*), dicamba + MCPA + mecoprop-P, fluroxypyr (*off-label*),

Crop/Pest Guide - Fruit and hops

		glufosinate-ammonium, isoxaben, oxadiazon, pendimethalin, pendimethalin (*off-label*), propyzamide, propyzamide (*off-label*)
	Crops as weeds	pendimethalin
	Grass weeds	amitrole, fluazifop-P-butyl (*off-label*), glufosinate-ammonium, glyphosate, oxadiazon, pendimethalin, pendimethalin (*off-label*), propyzamide, propyzamide (*off-label*)
	Weeds, miscellaneous	amitrole, amitrole (*off-label*), glufosinate-ammonium, glyphosate, glyphosate (*off-label*)

Tree fruit - All stone fruit

Crop control	Sucker/shoot control	glyphosate
Diseases	Bacterial canker	copper oxychloride
	Blossom wilt	boscalid + pyraclostrobin (*off-label*), cyprodinil + fludioxonil (*off-label*), fenbuconazole (*off-label*), myclobutanil (*off-label*)
	Botrytis	fenhexamid (*off-label*)
	Leaf curl	copper oxychloride, copper oxychloride (*off-label*)
	Powdery mildew	myclobutanil (*off-label*)
	Rust	myclobutanil (*off-label*)
	Sclerotinia	fenbuconazole (*off-label*), myclobutanil (*off-label*)
Pests	Aphids	acetamiprid, chlorpyrifos, chlorpyrifos (*Non OP resistant strains only*), cypermethrin, deltamethrin, pirimicarb, pirimicarb (*off-label*), tebufenpyrad (*off-label*), thiacloprid (*off-label*)
	Capsid bugs	chlorpyrifos
	Caterpillars	Bacillus thuringiensis (*off-label*), chlorpyrifos, cypermethrin, deltamethrin, diflubenzuron, indoxacarb (*off-label*)
	Gall, rust and leaf & bud mites	diflubenzuron
	Pests, miscellaneous	thiacloprid (*off-label*), thiacloprid (*off-label - under temporary protective rain covers*)
	Sawflies	deltamethrin
	Spider mites	chlorpyrifos, chlorpyrifos (*non OP resistant strains only*), clofentezine, fenpyroximate (*off-label*)
	Suckers	chlorpyrifos
	Weevils	chlorpyrifos
	Whiteflies	acetamiprid
Plant growth regulation	Growth control	paclobutrazol (*off-label*)
	Quality/yield control	gibberellins (*off-label*)
Weeds	Broad-leaved weeds	asulam, asulam (*off-label*), glufosinate-ammonium, isoxaben, pendimethalin, propyzamide, propyzamide (*off-label*)
	Crops as weeds	pendimethalin
	Grass weeds	fluazifop-P-butyl (*off-label*), glufosinate-ammonium, pendimethalin, propyzamide, propyzamide (*off-label*)
	Weeds, miscellaneous	amitrole (*off-label*), glufosinate-ammonium, glyphosate, glyphosate (*off-label*)

Crop/Pest Guide - Grain/crop store uses

Tree fruit - Miscellaneous top fruit

Diseases	Rhizoctonia	chloropicrin (*soil fumigation*)
Pests	Birds/mammals	aluminium ammonium sulphate
Plant growth regulation	Quality/yield control	sulphur

Grain/crop store uses

Stored seed - Stored grain/rape/linseed

Pests	Birds/mammals	aluminium ammonium sulphate
	Food/grain storage pests	chlorpyrifos-methyl, d-phenothrin + tetramethrin, pirimiphos-methyl
	Mites	physical pest control
	Pests, miscellaneous	aluminium phosphide, deltamethrin, deltamethrin (*off-label*), magnesium phosphide

Grass

Turf/amenity grass - Amenity grassland

Diseases	Anthracnose	azoxystrobin, azoxystrobin + propiconazole (*moderate control*), chlorothalonil + fludioxonil + propiconazole (*reduction*), propiconazole (*qualified minor use*), tebuconazole + trifloxystrobin
	Brown patch	azoxystrobin, chlorothalonil + fludioxonil + propiconazole, iprodione, propiconazole (*qualified minor use*)
	Crown rust	azoxystrobin
	Dollar spot	azoxystrobin + propiconazole, chlorothalonil + fludioxonil + propiconazole, iprodione, propiconazole, tebuconazole + trifloxystrobin
	Fairy rings	azoxystrobin
	Foliar diseases	fosetyl-aluminium (*off-label*)
	Fusarium	azoxystrobin, azoxystrobin + propiconazole, chlorothalonil + fludioxonil + propiconazole, iprodione, propiconazole, tebuconazole + trifloxystrobin, trifloxystrobin
	Melting out	azoxystrobin, iprodione
	Red thread	iprodione, tebuconazole + trifloxystrobin, trifloxystrobin
	Seed-borne diseases	azoxystrobin + propiconazole
	Snow mould	iprodione
	Take-all patch	azoxystrobin
Pests	Birds/mammals	aluminium ammonium sulphate, aluminium ammonium sulphate (*anti-fouling*)
	Flies	chlorpyrifos
	Leatherjackets	chlorpyrifos
	Slugs/snails	metaldehyde
Plant growth regulation	Growth control	trinexapac-ethyl
Weeds	Broad-leaved weeds	2,4-D, 2,4-D + dicamba, 2,4-D + dicamba + dichlorprop-P, aminopyralid + fluroxypyr, asulam, asulam (*not fine turf*), clopyralid + florasulam + fluroxypyr, clopyralid +

Crop/Pest Guide - Grass

		triclopyr, dicamba + MCPA + mecoprop-P, ethofumesate, florasulam + fluroxypyr, mecoprop-P
	Crops as weeds	florasulam + fluroxypyr, pinoxaden (*reduction only*)
	Grass weeds	ethofumesate, maleic hydrazide (*off-label*), pinoxaden (*off-label*)
	Mosses	carfentrazone-ethyl + mecoprop-P, ferrous sulphate
	Weeds, miscellaneous	carfentrazone-ethyl + mecoprop-P, florasulam + fluroxypyr, glyphosate, glyphosate (*wiper application*)
	Woody weeds/scrub	clopyralid + triclopyr

Turf/amenity grass - Managed amenity turf

Crop control	Miscellaneous non-selective situations	glufosinate-ammonium
Diseases	Anthracnose	azoxystrobin, azoxystrobin + propiconazole (*moderate control*), chlorothalonil, chlorothalonil + fludioxonil + propiconazole (*reduction*), propiconazole (*qualified minor use*), tebuconazole + trifloxystrobin
	Brown patch	azoxystrobin, chlorothalonil + fludioxonil + propiconazole, iprodione, propiconazole (*qualified minor use*)
	Crown rust	azoxystrobin
	Disease control/foliar feed	carbendazim, iprodione
	Dollar spot	azoxystrobin + propiconazole, chlorothalonil, chlorothalonil + fludioxonil + propiconazole, iprodione, propiconazole, pyraclostrobin (*useful reduction*), tebuconazole + trifloxystrobin
	Fairy rings	azoxystrobin
	Foliar diseases	fosetyl-aluminium (*off-label*)
	Fusarium	azoxystrobin, azoxystrobin + propiconazole, chlorothalonil, chlorothalonil + fludioxonil + propiconazole, iprodione, myclobutanil, prochloraz + tebuconazole, propiconazole, pyraclostrobin (*moderate control*), pyraclostrobin (*moderate control only*), tebuconazole + trifloxystrobin, trifloxystrobin
	Melting out	azoxystrobin, iprodione
	Pythium	fosetyl-aluminium (*off-label*)
	Red thread	chlorothalonil, iprodione, pyraclostrobin, tebuconazole + trifloxystrobin, trifloxystrobin
	Seed-borne diseases	azoxystrobin + propiconazole
	Snow mould	iprodione
	Take-all patch	azoxystrobin
Pests	Aphids	esfenvalerate
	Birds/mammals	aluminium ammonium sulphate, aluminium ammonium sulphate (*anti-fouling*)
	Chafers	imidacloprid
	Earthworms	carbendazim
	Flies	chlorpyrifos, esfenvalerate
	Leatherjackets	chlorpyrifos, imidacloprid
	Pests, miscellaneous	imidacloprid
	Slugs/snails	metaldehyde

Plant growth regulation	Growth control	trinexapac-ethyl
	Quality/yield control	sulphur
Weeds	Broad-leaved weeds	2,4-D, 2,4-D + dicamba, 2,4-D + dicamba + fluroxypyr, 2,4-D + florasulam, clopyralid + 2,4-D + MCPA, clopyralid + diflufenican + MCPA, clopyralid + florasulam + fluroxypyr, clopyralid + fluroxypyr + MCPA, dicamba + dichlorprop-P + ferrous sulphate + MCPA, dicamba + dichlorprop-P + MCPA, dicamba + MCPA + mecoprop-P, dichlorprop-P + ferrous sulphate + MCPA, dichlorprop-P + MCPA, ethofumesate (*off-label*), ferrous sulphate + MCPA + mecoprop-P, florasulam + fluroxypyr, MCPA + mecoprop-P, mecoprop-P
	Crops as weeds	2,4-D + dicamba + fluroxypyr, 2,4-D + florasulam, dichlorprop-P + MCPA, florasulam + fluroxypyr, pinoxaden (*reduction only*)
	Grass weeds	amitrole (*off-label - on amitrole resistant turf*), ethofumesate (*off-label*), pinoxaden (*off-label*), tepraloxydim (*off-label*)
	Mosses	carfentrazone-ethyl + mecoprop-P, dicamba + dichlorprop-P + ferrous sulphate + MCPA, dichlorprop-P + ferrous sulphate + MCPA, ferrous sulphate, ferrous sulphate + MCPA + mecoprop-P
	Weeds, miscellaneous	2,4-D + dicamba, carfentrazone-ethyl + mecoprop-P, florasulam + fluroxypyr, glyphosate (*pre-establishment*), glyphosate (*pre-establishment only*)

Non-crop pest control

Farm buildings/yards - Farm buildings

Pests	Birds/mammals	aluminium ammonium sulphate, bromadiolone, chlorophacinone, difenacoum, warfarin
	Flies	diflubenzuron, d-phenothrin + tetramethrin, tetramethrin
	Food/grain storage pests	alpha-cypermethrin
	Pests, miscellaneous	alpha-cypermethrin, cypermethrin, pyrethrins
	Wasps	d-phenothrin + tetramethrin
Weeds	Weeds, miscellaneous	glyphosate

Farm buildings/yards - Farmyards

Pests	Birds/mammals	bromadiolone, difenacoum

Farmland pest control - Farmland situations

Pests		aluminium phosphide

Miscellaneous non-crop pest control - Manure/rubbish

Pests	Flies	diflubenzuron

Crop/Pest Guide - Protected salad and vegetable crops

Miscellaneous non-crop pest control - Miscellaneous pest control situations

Pests   Birds/mammals   carbon dioxide (commodity substance), paraffin oil (commodity substance) (*egg treatment*)

Protected salad and vegetable crops

Protected brassicas - Protected brassica vegetables

Diseases	Bacterial blight	copper oxychloride (*off-label*)
	Black rot	copper oxychloride (*off-label*)
	Bottom rot	copper oxychloride (*off-label*)
	Damping off	fosetyl-aluminium + propamocarb hydrochloride (*off-label*)
	Downy mildew	fosetyl-aluminium (*off-label*)
	Phytophthora	copper oxychloride (*off-label*)
	Spear rot	copper oxychloride (*off-label*)
	Stem canker	copper oxychloride (*off-label*)
	White blister	boscalid + pyraclostrobin (*off-label*)
Pests	Aphids	pirimicarb (*off-label*), pyrethrins, spirotetramat (*off-label*)
	Beetles	deltamethrin (*off-label*), pyrethrins
	Caterpillars	Bacillus thuringiensis (*off-label*), deltamethrin (*off-label*), pyrethrins
	Mites	abamectin (*off-label*)
	Pests, miscellaneous	dimethoate (*off-label*)
	Whiteflies	pyrethrins, spirotetramat (*off-label*)
Weeds	Grass weeds	pendimethalin (*off-label*)

Protected brassicas - Protected salad brassicas

Diseases	Bacterial blight	copper oxychloride (*off-label*)
	Bottom rot	copper oxychloride (*off-label*)
	Damping off	fosetyl-aluminium + propamocarb hydrochloride (*off-label - baby leaf production*)
	Downy mildew	copper oxychloride (*off-label*), fosetyl-aluminium (*off-label*), fosetyl-aluminium (*off-label - for baby leaf production*), fosetyl-aluminium + propamocarb hydrochloride (*off-label - baby leaf production*)
	Foliar diseases	fosetyl-aluminium (*off-label*)
	Phytophthora	copper oxychloride (*off-label*)
	Ring spot	difenoconazole (*off-label*)
	Spear rot	copper oxychloride (*off-label*)
	Stem canker	copper oxychloride (*off-label*)
Pests	Aphids	acetamiprid (*off-label*), deltamethrin (*off-label*), lambda-cyhalothrin (*off-label*), lambda-cyhalothrin + pirimicarb (*off-label - for baby leaf production*), pymetrozine (*off-label*), pymetrozine (*off-label - for baby leaf production*)
	Beetles	deltamethrin (*off-label*), lambda-cyhalothrin (*off-label*)
	Caterpillars	Bacillus thuringiensis (*off-label*), deltamethrin (*off-label*), lambda-cyhalothrin (*off-label*)
	Leafhoppers	deltamethrin (*off-label*)

Crop/Pest Guide - Protected salad and vegetable crops

	Pests, miscellaneous	deltamethrin (*off-label*), dimethoate (*off-label*), lambda-cyhalothrin (*off-label*)
	Thrips	deltamethrin (*off-label*)
	Whiteflies	Lecanicillium lecanii (*off-label*)
Weeds	Broad-leaved weeds	pendimethalin (*off-label*)
	Grass weeds	pendimethalin (*off-label*)

Protected crops, general - All protected crops

Diseases	Sclerotinia	Coniothyrium minitans
	Soil-borne diseases	dazomet (*soil fumigation*)
Pests	Aphids	Beauveria bassiana, pyrethrins
	Beetles	Beauveria bassiana, pyrethrins
	Caterpillars	pyrethrins
	Free-living nematodes	dazomet (*soil fumigation*)
	Mealybugs	pyrethrins
	Scale insects	pyrethrins
	Slugs/snails	ferric phosphate, metaldehyde
	Soil pests	dazomet (*soil fumigation*)
	Spider mites	pyrethrins
	Thrips	pyrethrins
	Whiteflies	Beauveria bassiana, pyrethrins
Weeds	Weeds, miscellaneous	dazomet (*soil fumigation*)

Protected crops, general - Glasshouses

Diseases	Phytophthora	fosetyl-aluminium
	Root rot	fosetyl-aluminium
	Soil-borne diseases	metam-sodium
Pests	Free-living nematodes	metam-sodium
	Pests, miscellaneous	pyrethrins
	Soil pests	metam-sodium
Weeds	Weeds, miscellaneous	metam-sodium

Protected crops, general - Protected vegetables, general

Diseases	Botrytis	propamocarb hydrochloride (*off-label*)
	Damping off	propamocarb hydrochloride (*off-label*)
	Downy mildew	propamocarb hydrochloride (*off-label*)
	Phytophthora	propamocarb hydrochloride (*off-label*)
	White blister	propamocarb hydrochloride (*off-label*)
Pests	Aphids	pirimicarb (*off-label*), pymetrozine (*off-label*)
	Whiteflies	Lecanicillium lecanii (*off-label*), pymetrozine (*off-label*)

Protected crops, general - Soils and compost

Diseases	Root rot	metam-sodium
	Sclerotinia	metam-sodium
	Soil-borne diseases	metam-sodium
Pests	Cyst nematodes	metam-sodium

Crop/Pest Guide - Protected salad and vegetable crops

	Free-living nematodes	metam-sodium
	Millipedes	metam-sodium
	Pests, miscellaneous	dimethoate (*off-label* - for watercress propagation)
	Root-knot nematodes	metam-sodium
	Soil pests	metam-sodium
	Symphylids	metam-sodium
	Wireworms	metam-sodium
Weeds	Weeds, miscellaneous	metam-sodium

Protected fruiting vegetables - Protected aubergines

Diseases	Botrytis	cyprodinil + fludioxonil (*off-label*), fenhexamid (*off-label*), iprodione (*off-label*), thiophanate-methyl (*off-label*)
	Damping off	fosetyl-aluminium + propamocarb hydrochloride (*off-label*)
	Powdery mildew	myclobutanil (*off-label*), sulphur (*off-label*)
	Sclerotinia	thiophanate-methyl (*off-label*)
Pests	Aphids	acetamiprid, deltamethrin (*off-label*), pirimicarb (*off-label*), pymetrozine (*off-label*)
	Caterpillars	Bacillus thuringiensis (*off-label*), indoxacarb, teflubenzuron (*off-label*)
	Leaf miners	thiacloprid (*off-label*)
	Mites	etoxazole
	Pests, miscellaneous	deltamethrin (*off-label*), teflubenzuron (*off-label*)
	Spider mites	abamectin (*off-label*), spiromesifen (*off-label*)
	Thrips	abamectin (*off-label*), deltamethrin (*off-label*), spinosad, teflubenzuron (*off-label*), thiacloprid (*off-label*)
	Whiteflies	acetamiprid, pymetrozine (*off-label*), spiromesifen (*off-label*), teflubenzuron (*off-label*), thiacloprid (*off-label*)

Protected fruiting vegetables - Protected cucurbits

Diseases	Botrytis	chlorothalonil, cyprodinil + fludioxonil (*off-label*), fenhexamid (*off-label*), thiophanate-methyl (*off-label*)
	Damping off	fosetyl-aluminium + propamocarb hydrochloride (*off-label*), propamocarb hydrochloride (*off-label*)
	Downy mildew	azoxystrobin (*off-label*), metalaxyl-M (*off-label*)
	Fusarium	cyprodinil + fludioxonil (*off-label*)
	Mycosphaerella	cyprodinil + fludioxonil (*off-label*)
	Phytophthora	propamocarb hydrochloride, propamocarb hydrochloride (*off-label*)
	Powdery mildew	azoxystrobin (*off-label*), bupirimate, imazalil, myclobutanil (*off-label*), sulphur (*off-label*)
	Pythium	fosetyl-aluminium + propamocarb hydrochloride, propamocarb hydrochloride
	Sclerotinia	thiophanate-methyl (*off-label*)
Pests	Aphids	acetamiprid, deltamethrin, deltamethrin (*off-label*), pirimicarb (*off-label*), pymetrozine, pymetrozine (*off-label*), pyrethrins (*off-label*)
	Caterpillars	Bacillus thuringiensis, Bacillus thuringiensis (*off-label*), deltamethrin, indoxacarb, teflubenzuron (*off-label*)
	Leaf miners	oxamyl (*off-label*), thiacloprid (*off-label*)

Crop/Pest Guide - Protected salad and vegetable crops

	Mealybugs	deltamethrin, petroleum oil
	Pests, miscellaneous	deltamethrin (*off-label*), teflubenzuron (*off-label*)
	Scale insects	deltamethrin, petroleum oil
	Spider mites	abamectin, abamectin (*off-label*), petroleum oil, spiromesifen (*off-label*)
	Thrips	abamectin, deltamethrin, deltamethrin (*off-label*), spinosad, teflubenzuron (*off-label*), thiacloprid (*off-label*)
	Whiteflies	acetamiprid, cypermethrin, deltamethrin, Lecanicillium lecanii, Lecanicillium lecanii (*off-label*), pymetrozine (*off-label*), spiromesifen (*off-label*), teflubenzuron (*off-label*), thiacloprid (*off-label*)
Weeds	Broad-leaved weeds	isoxaben (*off-label*)

Protected fruiting vegetables - Protected tomatoes

Diseases	Botrytis	azoxystrobin (*off-label*), chlorothalonil, cyprodinil + fludioxonil (*off-label*), fenhexamid (*off-label*), iprodione, pyrimethanil (*off-label*)
	Damping off	copper oxychloride, propamocarb hydrochloride, propamocarb hydrochloride (*off-label*)
	Didymella stem rot	azoxystrobin (*off-label*)
	Disease control/foliar feed	maneb
	Foot rot	copper oxychloride
	Leaf mould	chlorothalonil
	Phytophthora	azoxystrobin (*off-label*), chlorothalonil, copper oxychloride, propamocarb hydrochloride
	Powdery mildew	azoxystrobin (*off-label*), bupirimate (*off-label*), myclobutanil (*off-label*), sulphur (*off-label*)
	Pythium	fosetyl-aluminium + propamocarb hydrochloride, propamocarb hydrochloride, propamocarb hydrochloride (*off-label*)
	Verticillium wilt	thiophanate-methyl (*off-label*)
Pests	Aphids	acetamiprid, deltamethrin, fatty acids, pirimicarb, pymetrozine (*off-label*), pyrethrins
	Beetles	pyrethrins
	Caterpillars	Bacillus thuringiensis, deltamethrin, indoxacarb, pyrethrins, teflubenzuron (*off-label*)
	Leaf miners	abamectin, abamectin (*off-label*), oxamyl (*off-label*), thiacloprid (*off-label*)
	Mealybugs	deltamethrin, fatty acids, petroleum oil, pyrethrins, pyrethrins (*off-label*)
	Mirid bugs	pyrethrins, pyrethrins (*off-label*)
	Mites	etoxazole
	Pests, miscellaneous	teflubenzuron (*off-label*)
	Scale insects	deltamethrin, fatty acids, petroleum oil
	Spider mites	abamectin, fatty acids, petroleum oil
	Thrips	deltamethrin, spinosad, teflubenzuron (*off-label*), thiacloprid (*off-label*)
	Whiteflies	acetamiprid, deltamethrin, fatty acids, Lecanicillium lecanii, pymetrozine (*off-label*), pyrethrins, spiromesifen, teflubenzuron (*off-label*), thiacloprid (*off-label*)
Plant growth regulation	Growth control	2-chloroethylphosphonic acid (*off-label*)

Crop/Pest Guide - Protected salad and vegetable crops

Protected herb crops - Protected herbs

Diseases	Alternaria	iprodione (*off-label*)
	Botrytis	cyprodinil + fludioxonil (*off-label*), fenhexamid (*off-label*), iprodione (*off-label*), propamocarb hydrochloride (*off-label*), pyrimethanil (*off-label*)
	Damping off	fosetyl-aluminium + propamocarb hydrochloride (*off-label*), propamocarb hydrochloride (*off-label*)
	Disease control/foliar feed	prochloraz (*off-label*)
	Downy mildew	copper oxychloride (*off-label*), fosetyl-aluminium (*off-label*), fosetyl-aluminium + propamocarb hydrochloride (*off-label*), metalaxyl-M (*off-label*), propamocarb hydrochloride (*off-label*)
	Powdery mildew	sulphur (*off-label*)
	Rhizoctonia	azoxystrobin (*off-label*)
	Seed-borne diseases	fosetyl-aluminium, fosetyl-aluminium (*off-label*)
	White blister	boscalid + pyraclostrobin (*off-label*)
Pests	Aphids	acetamiprid (*off-label*), deltamethrin (*off-label*), fatty acids (*off-label*), lambda-cyhalothrin + pirimicarb (*off-label*), pirimicarb (*off-label*), pymetrozine (*off-label*), pyrethrins (*off-label*), spirotetramat (*off-label*), thiacloprid (*off-label*)
	Beetles	deltamethrin (*off-label*)
	Caterpillars	Bacillus thuringiensis (*off-label*), deltamethrin (*off-label*)
	Leafhoppers	deltamethrin (*off-label*)
	Mites	abamectin (*off-label*)
	Pests, miscellaneous	abamectin (*off-label*), deltamethrin (*off-label*), spinosad (*off-label*)
	Thrips	deltamethrin (*off-label*), fatty acids (*off-label*), pirimicarb (*off-label*)
	Whiteflies	Lecanicillium lecanii (*off-label*), pymetrozine (*off-label*), spirotetramat (*off-label*)
Weeds	Broad-leaved weeds	pendimethalin (*off-label*), propyzamide (*off-label*)
	Grass weeds	pendimethalin (*off-label*), propyzamide (*off-label*)

Protected leafy vegetables - Mustard and cress

Diseases	Botrytis	cyprodinil + fludioxonil (*off-label*), propamocarb hydrochloride (*off-label*)
	Damping off	thiram (*off-label*)
	Downy mildew	copper oxychloride (*off-label*), dimethomorph + mancozeb (*off-label*), propamocarb hydrochloride (*off-label*)
Pests	Aphids	acetamiprid (*off-label*), fatty acids (*off-label*), spirotetramat (*off-label*)
	Pests, miscellaneous	spinosad
	Thrips	fatty acids (*off-label*)
	Whiteflies	spirotetramat (*off-label*)
Weeds	Broad-leaved weeds	pendimethalin (*off-label*), propyzamide (*off-label*)
	Grass weeds	pendimethalin (*off-label*), propyzamide (*off-label*)

Crop/Pest Guide - Protected salad and vegetable crops

Protected leafy vegetables - Protected leafy vegetables

Diseases	Botrytis	boscalid + pyraclostrobin, cyprodinil + fludioxonil (*off-label*), iprodione, propamocarb hydrochloride
	Bottom rot	boscalid + pyraclostrobin, tolclofos-methyl
	Damping off	propamocarb hydrochloride, propamocarb hydrochloride (*off-label*)
	Downy mildew	boscalid + pyraclostrobin (*off-label*), copper oxychloride (*off-label*), fosetyl-aluminium, fosetyl-aluminium (*off-label*), fosetyl-aluminium + propamocarb hydrochloride, propamocarb hydrochloride
	Foliar diseases	fosetyl-aluminium (*off-label*)
	Pythium	fosetyl-aluminium + propamocarb hydrochloride
	Rhizoctonia	azoxystrobin (*off-label*)
	Ring spot	difenoconazole (*off-label*)
	Soft rot	boscalid + pyraclostrobin
Pests	Aphids	acetamiprid (*off-label*), deltamethrin (*off-label*), lambda-cyhalothrin (*off-label*), lambda-cyhalothrin + pirimicarb (*off-label*), pirimicarb, pirimicarb (*off-label*), pymetrozine (*off-label*), pyrethrins, spirotetramat (*off-label*), thiacloprid (*off-label*)
	Beetles	deltamethrin (*off-label*), lambda-cyhalothrin (*off-label*)
	Caterpillars	Bacillus thuringiensis (*off-label*), deltamethrin (*off-label*), lambda-cyhalothrin (*off-label*), pyrethrins
	Leafhoppers	deltamethrin (*off-label*)
	Mites	abamectin (*off-label*)
	Pests, miscellaneous	abamectin (*off-label*), lambda-cyhalothrin (*off-label*)
	Thrips	deltamethrin (*off-label*), pirimicarb (*off-label*)
	Whiteflies	cypermethrin, Lecanicillium lecanii, spirotetramat (*off-label*)
Weeds	Broad-leaved weeds	pendimethalin (*off-label*), propyzamide (*off-label*)
	Grass weeds	pendimethalin (*off-label*), propyzamide (*off-label*)

Protected leafy vegetables - Protected spinach

Diseases	Downy mildew	copper oxychloride (*off-label*), fosetyl-aluminium (*off-label*), metalaxyl-M (*off-label*)
	Foliar diseases	fosetyl-aluminium (*off-label*)
Pests	Aphids	acetamiprid (*off-label*), deltamethrin (*off-label*), pirimicarb (*off-label*), spirotetramat (*off-label*)
	Beetles	deltamethrin (*off-label*)
	Caterpillars	Bacillus thuringiensis (*off-label*), deltamethrin (*off-label*)
	Leafhoppers	deltamethrin (*off-label*)
	Pests, miscellaneous	dimethoate (*off-label*)
	Thrips	deltamethrin (*off-label*)
	Whiteflies	spirotetramat (*off-label*)

Protected legumes - Protected peas and beans

Pests	Leaf miners	oxamyl (*off-label*)
	Whiteflies	Lecanicillium lecanii, Lecanicillium lecanii (*off-label*)

Protected root and tuber vegetables - Protected carrots/parsnips/celeriac

Pests	Pests, miscellaneous	dimethoate (off-label)
Weeds	Broad-leaved weeds	isoxaben (off-label - temporary protection), metribuzin (off-label)

Protected root and tuber vegetables - Protected root brassicas

Diseases	Botrytis	propamocarb hydrochloride (off-label)
	Damping off	propamocarb hydrochloride (off-label)
	Downy mildew	propamocarb hydrochloride (off-label)
	Phytophthora	propamocarb hydrochloride (off-label)
	Rhizoctonia	tolclofos-methyl (off-label)
	White blister	propamocarb hydrochloride (off-label)
Pests	Aphids	pirimicarb (off-label)

Protected stem and bulb vegetables - Protected celery/chicory

Diseases	Botrytis	azoxystrobin (off-label), cyprodinil + fludioxonil (off-label)
	Downy mildew	copper oxychloride (off-label), fosetyl-aluminium (off-label), fosetyl-aluminium + propamocarb hydrochloride (off-label)
	Foliar diseases	fosetyl-aluminium (off-label)
	Phytophthora	azoxystrobin (off-label - for forcing)
	Rhizoctonia	azoxystrobin (off-label), tolclofos-methyl (off-label)
	Sclerotinia	azoxystrobin (off-label)
Pests	Aphids	acetamiprid (off-label), deltamethrin (off-label), lambda-cyhalothrin + pirimicarb (off-label), pirimicarb (off-label), pymetrozine (off-label)
	Beetles	deltamethrin (off-label)
	Caterpillars	Bacillus thuringiensis (off-label), deltamethrin (off-label)
	Leafhoppers	deltamethrin (off-label)
	Mites	abamectin (off-label)
	Pests, miscellaneous	abamectin (off-label), deltamethrin (off-label)
	Thrips	deltamethrin (off-label)
Weeds	Grass weeds	pendimethalin (off-label)

Protected stem and bulb vegetables - Protected onions/leeks/garlic

Pests	Aphids	deltamethrin (off-label)
	Beetles	deltamethrin (off-label)
	Caterpillars	deltamethrin (off-label)
	Leafhoppers	deltamethrin (off-label)
	Pests, miscellaneous	deltamethrin (off-label)
	Stem nematodes	oxamyl (off-label)
	Thrips	deltamethrin (off-label)

Total vegetation control

Aquatic situations, general - Aquatic situations

Plant growth regulation	Growth control	glyphosate
Weeds	Aquatic weeds	2,4-D, glyphosate
	Grass weeds	glyphosate
	Weeds, miscellaneous	glyphosate

Non-crop areas, general - Miscellaneous non-crop situations

Weeds	Broad-leaved weeds	asulam, triclopyr
	Weeds, miscellaneous	flazasulfuron, glyphosate
	Woody weeds/scrub	glyphosate, triclopyr

Non-crop areas, general - Non-crop farm areas

Weeds	Broad-leaved weeds	2,4-D + dicamba + triclopyr, dicamba + MCPA + mecoprop-P, glufosinate-ammonium, metsulfuron-methyl, picloram, triclopyr
	Grass weeds	glufosinate-ammonium, glyphosate
	Weeds, miscellaneous	glufosinate-ammonium, glyphosate, picloram
	Woody weeds/scrub	glyphosate, picloram, triclopyr

Non-crop areas, general - Paths/roads etc

Pests	Birds/mammals	aluminium ammonium sulphate, aluminium ammonium sulphate (*anti-fouling*)
	Slugs/snails	metaldehyde
Weeds	Broad-leaved weeds	2,4-D + dicamba + triclopyr
	Grass weeds	glyphosate, quizalofop-P-tefuryl (*off-label*)
	Mosses	maleic hydrazide + pelargonic acid
	Weeds, miscellaneous	acetic acid, acetic acid (*Foliage kill only. Regrowth may occur from roots.*), diflufenican + glyphosate, flazasulfuron, glyphosate, glyphosate + sulfosulfuron, maleic hydrazide + pelargonic acid
	Woody weeds/scrub	2,4-D + dicamba + triclopyr, glyphosate

SECTION 2
PESTICIDE PROFILES

abamectin

1 abamectin

A selective acaricide and insecticide for use in ornamentals and other protected crops
IRAC mode of action code: 6

Products

1	Agrimec	Syngenta	18 g/l	EC	14491
2	Clayton Abba	Clayton	18 g/l	EC	13808
3	Dynamec	Syngenta Bioline	18 g/l	EC	13331

Uses
- Insect control in **protected chives** *(off-label)*, **protected cress** *(off-label)*, **protected frise** *(off-label)*, **protected herbs (see appendix 6)** *(off-label)*, **protected lamb's lettuce** *(off-label)*, **protected lettuce** *(off-label)*, **protected parsley** *(off-label)*, **protected radicchio** *(off-label)*, **protected scarole** *(off-label)* [3]
- Insect pests in **leeks** *(off-label)* [1]
- Leaf and bud mite in **protected blackberries** *(off-label)*, **protected raspberries** *(off-label)* [3]
- Leaf miner in **protected cherry tomatoes** *(off-label)* [3]; **protected tomatoes** [2, 3]
- Mites in **protected blackberries** *(off-label)*, **protected cress** *(off-label)*, **protected frise** *(off-label)*, **protected herbs (see appendix 6)** *(off-label)*, **protected lamb's lettuce** *(off-label)*, **protected leaf brassicas** *(off-label)*, **protected lettuce** *(off-label)*, **protected radicchio** *(off-label)*, **protected raspberries** *(off-label)*, **protected scarole** *(off-label)* [2]
- Pear sucker in **pears** *(effective against larval stages but little activity against adults)* [1]
- Red spider mites in **hops** *(off-label)* [1]; **protected blackberries** *(off-label)*, **protected cayenne peppers** *(off-label)*, **protected peppers** *(off-label)*, **protected raspberries** *(off-label)* [3]; **protected strawberries** *(off-label)* [2, 3]; **strawberries** *(off-label)* [2]
- Spider mites in **hops** *(off-label)* [1]
- Tarsonemid mites in **protected strawberries** *(off-label)*, **strawberries** *(off-label)* [2]; **strawberries** *(off-label - in propagation)* [3]
- Thrips in **leeks** *(off-label)* [1]
- Two-spotted spider mite in **ornamental plant production, protected cucumbers, protected ornamentals, protected tomatoes** [2, 3]; **protected aubergines** *(off-label)* [3]
- Western flower thrips in **ornamental plant production, protected cucumbers, protected ornamentals** [2, 3]; **protected aubergines** *(off-label)* [3]

Specific Off-Label Approvals (SOLAs)
- **hops** *20091581* [1]
- **leeks** *20101466* [1]
- **protected aubergines** *20070421* [3]
- **protected blackberries** *20101940* [2], *20072290* [3]
- **protected cayenne peppers** *20070422* [3]
- **protected cherry tomatoes** *20070422* [3]
- **protected chives** *20070430* [3]
- **protected cress** *20101939* [2], *20070430* [3]
- **protected frise** *20101939* [2], *20070430* [3]
- **protected herbs (see appendix 6)** *20101939* [2], *20070430* [3]
- **protected lamb's lettuce** *20101939* [2], *20070430* [3]
- **protected leaf brassicas** *20101939* [2]
- **protected lettuce** *20101939* [2], *20070430* [3]
- **protected parsley** *20070430* [3]
- **protected peppers** *20070422* [3]
- **protected radicchio** *20101939* [2], *20070430* [3]
- **protected raspberries** *20101940* [2], *20072290* [3]
- **protected scarole** *20101939* [2], *20070430* [3]
- **protected strawberries** *20090342* [2], *20070423* [3]
- **strawberries** *20090342* [2], *(in propagation) 20070423* [3]

Approval information
- Abamectin included in Annex I under EC Directive 91/414

SEE SECTION 3 FOR PRODUCTS ALSO REGISTERED

Pesticide Profiles

Efficacy guidance
- Abamectin controls adults and immature stages of the two-spotted spider mite, the larval stages of leaf miners and the nymphs of Western Flower Thrips, plus a useful reduction in adults
- Treat at first sign of infestation. Repeat sprays may be required
- For effective control total cover of all plant surfaces is essential, but avoid run-off
- Target pests quickly become immobilised but 3-5 d may be required for maximum mortality
- Indoor applications should be made through a hydraulic nozzle applicator or a knapsack applicator. Outdoors suitable high volume hydraulic nozzle applicators should be used
- Limited data shows abamectin only slightly harmful to Anthocorid bugs and so compatible with biological control systems in which Anthocorid bugs are important [1]

Restrictions
- Number of treatments 6 on protected tomatoes and cucumbers (only 4 of which can be made when flowers or fruit present); not restricted on flowers but rotation with other products advised
- Maximum concentration must not exceed 50 ml per 100 l water
- Do not mix with wetters, stickers or other adjuvants
- Do not use on ferns (*Adiantum* spp) or Shasta daisies
- Do not treat protected crops which are in flower or have set fruit between 1 Nov and 28 Feb
- For use on all varieties of pears [1]
- Do not treat cherry tomatoes (but protected cherry tomatoes may be treated off-label)
- Consult manufacturer for list of plant varieties tested for safety
- There is insufficient evidence to support product compatibility with integrated and biological pest control programmes
- Unprotected persons must be kept out of treated areas until the spray has dried

Crop-specific information
- HI 3 d for protected edible crops
- On tomato or cucumber crops that are in flower, or have started to set fruit, treat only between 1 Mar and 31 Oct. Seedling tomatoes or cucumbers that have not started to flower or set fruit may be treated at any time
- Some spotting or staining may occur on carnation, kalanchoe and begonia foliage

Environmental safety
- Dangerous for the environment
- Very toxic to aquatic organisms
- High risk to bees. Do not apply to crops in flower or to those in which bees are actively foraging. Do not apply when flowering weeds are present
- Keep in original container, tightly closed, in a safe place, under lock and key
- Where bumble bees are used in tomatoes as pollinators, keep them out for 24 h after treatment
- Broadcast air-assisted LERAP [1] (5 m)

Hazard classification and safety precautions
Hazard Harmful, Dangerous for the environment
Transport code 9
Packaging group III
UN number 3082
Risk phrases R22a, R50, R53a
Operator protection A, C, D, H, K, M; U02a, U05a [1-3]; U20a [2, 3]; U20c [1]
Environmental protection E02a [2, 3] (until spray has dried); E12a, E15b, E34, E38 [1-3]; E12e [2, 3]; E12f, E22c [1]; E17b [1] (5 m)
Storage and disposal D01, D02, D05, D09b, D10c, D12a
Medical advice M04a

2 acetamiprid

A neonicotinoid insecticide for use in top fruit and horticulture
IRAC mode of action code: 4A

Products

1	Gazelle SG	Certis	20% w/w	SG	13725
2	Insyst	Certis	20% w/w	SP	13414

FOR FULL CONDITIONS OF USE ALWAYS READ THE PRODUCT LABEL

acetamiprid

Uses
- Aphids in **apples**, **cherries**, **cress** *(off-label)*, **frise** *(off-label)*, **herbs (see appendix 6)** *(off-label)*, **hops** *(off-label)*, **lamb's lettuce** *(off-label)*, **lettuce** *(off-label)*, **ornamental plant production**, **parsley** *(off-label)*, **pears**, **plums**, **protected aubergines**, **protected cress** *(off-label)*, **protected frise** *(off-label)*, **protected herbs (see appendix 6)** *(off-label)*, **protected hops** *(off-label)*, **protected lamb's lettuce** *(off-label)*, **protected lettuce** *(off-label)*, **protected ornamentals**, **protected parsley** *(off-label)*, **protected peppers**, **protected radicchio** *(off-label)*, **protected rocket** *(off-label)*, **protected salad brassicas** *(off-label)*, **protected scarole** *(off-label)*, **protected soft fruit** *(off-label)*, **protected spinach** *(off-label)*, **protected tomatoes**, **radicchio** *(off-label)*, **rocket** *(off-label)*, **salad brassicas** *(off-label)*, **scarole** *(off-label)*, **soft fruit** *(off-label)*, **spinach** *(off-label)*, **top fruit** *(off-label)* [1]; **brussels sprouts** *(off-label)*, **seed potatoes**, **spring oilseed rape**, **ware potatoes**, **winter oilseed rape** [2]
- Whitefly in **apples**, **cherries**, **ornamental plant production**, **pears**, **plums**, **protected aubergines**, **protected ornamentals**, **protected peppers**, **protected tomatoes** [1]; **brussels sprouts** *(off-label)* [2]

Specific Off-Label Approvals (SOLAs)
- *brussels sprouts* 20072866 [2]
- *cress* 20101101 [1]
- *frise* 20082234 [1]
- *herbs (see appendix 6)* 20101101 [1]
- *hops* 20082857 [1]
- *lamb's lettuce* 20082234 [1]
- *lettuce* 20082234 [1]
- *parsley* 20082234 [1]
- *protected cress* 20101101 [1]
- *protected frise* 20082234 [1]
- *protected herbs (see appendix 6)* 20101101 [1]
- *protected hops* 20082857 [1]
- *protected lamb's lettuce* 20082234 [1]
- *protected lettuce* 20082234 [1]
- *protected parsley* 20082234 [1]
- *protected radicchio* 20082234 [1]
- *protected rocket* 20082234 [1]
- *protected salad brassicas* 20082234 [1]
- *protected scarole* 20082234 [1]
- *protected soft fruit* 20082857 [1]
- *protected spinach* 20101101 [1]
- *radicchio* 20082234 [1]
- *rocket* 20082234 [1]
- *salad brassicas* 20082234 [1]
- *scarole* 20082234 [1]
- *soft fruit* 20082857 [1]
- *spinach* 20101101 [1]
- *top fruit* 20082857 [1]

Approval information
- Acetamiprid included in Annex I under EC Directive 91/414

Efficacy guidance
- Best results obtained from application at the first sign of pest attack or when appropriate thresholds are reached
- Thorough coverage of foliage is essential to ensure best control. Acetamiprid has contact, systemic and translaminar activity

Restrictions
- Maximum number of treatments 1 per yr for cherries, ware potatoes; 2 per yr or crop for apples, pears, plums, ornamental plant production, protected crops, seed potatoes
- Do not use more than two applications of any neonicotinoid insecticide (e.g. acetamiprid, clothianidin, imidacloprid, thiacloprid) on any crop. Previous soil or seed treatment with a neonicotinoid counts as one such treatment
- 2 treatments are permitted on seed potatoes but must not be used consecutively [2]

SEE SECTION 3 FOR PRODUCTS ALSO REGISTERED

Pesticide Profiles

Crop-specific information
- HI 3 d for protected crops; 14 d for apples, cherries, pears, plums, potatoes

Environmental safety
- Harmful to aquatic organisms
- Acetamiprid is slightly toxic to predatory mites and generally slightly toxic to other beneficials
- Broadcast air-assisted LERAP [1] (18 m); LERAP Category B [1, 2]

Hazard classification and safety precautions
Hazard Dangerous for the environment
Risk phrases R22a [1]; R52, R53a [1, 2]
Operator protection A [1, 2]; H [1]; U02a, U19a [1, 2]; U20a [2]; U20b [1]
Environmental protection E15b, E16a, E16b [1, 2]; E17b [1] (18 m); E38 [2]
Storage and disposal D05 [1]; D09a, D10b [1, 2]; D12a [2]

3 acetic acid

A non-selective herbicide for non-crop situations

Products

1	Acetum	Unicrop	240 g/l	SL	13881
2	Natural Weed & Moss Spray No 1	Headland Amenity	240 g/l	SL	14509

Uses
- Annual and perennial weeds in **hard surfaces** *(Foliage kill only. Regrowth may occur from roots.)*, **natural surfaces not intended to bear vegetation** *(Foliage kill only. Regrowth may occur from roots.)*, **permeable surfaces overlying soil** *(Foliage kill only. Regrowth may occur from roots.)* [1]
- General weed control in **hard surfaces**, **natural surfaces not intended to bear vegetation**, **permeable surfaces overlying soil** [2]

Efficacy guidance
- Best results obtained from treatment of young tender weeds less than 10 cm high
- Treat in spring and repeat as necessary throughout the growing season
- Treat survivors as soon as fresh growth is seen
- Ensure complete coverage of foliage to the point of run-off
- Rainfall after treatment may reduce efficacy

Restrictions
- No restriction on number of treatments

Following crops guidance
- There is no residual activity in the soil and sowing or planting may take place as soon as treated weeds have died

Environmental safety
- Harmful to aquatic organisms
- High risk to bees
- Do not apply to crops in flower or to those in which bees are actively foraging. Do not apply when flowering weeds are present
- Keep people and animals off treated dense weed patches until spray has dried. This is not necessary for areas with occasional low growing prostrate weeds such as on pathways

Hazard classification and safety precautions
Hazard Irritant
UN number N/C
Risk phrases R36, R37, R38, R52
Operator protection A, C, H [1, 2]; D [2]; U05a, U09a, U19a [1, 2]; U20b [2]; U20c [1]
Environmental protection E12a, E12e, E15a
Storage and disposal D01, D02, D09a, D12a

FOR FULL CONDITIONS OF USE ALWAYS READ THE PRODUCT LABEL

4   alpha-cypermethrin

A contact and ingested pyrethroid insecticide for use in arable crops and agricultural buildings
IRAC mode of action code: 3

Products

1	A-Cyper 100EC	Goldengrass	100 g/l	EC	14175
2	Antec Durakil 1.5 SC	Antec Biosentry	15 g/l	SC	H7560
3	Antec Durakil 6SC	Antec Biosentry	60 g/l	SC	H7559
4	Contest	BASF	15% w/w	WG	10216

Uses

- Aphids in *spring barley, spring oilseed rape, spring wheat, winter barley, winter oilseed rape, winter wheat* [1]
- Brassica pod midge in *spring oilseed rape* [1]; *winter oilseed rape* [1, 4]
- Cabbage seed weevil in *spring oilseed rape, winter oilseed rape* [1, 4]
- Cabbage stem flea beetle in *spring oilseed rape* [1]; *winter oilseed rape* [1, 4]
- Cabbage stem weevil in *spring oilseed rape, winter oilseed rape* [1]
- Cabbage white butterfly in *salad brassicas* (off-label - for baby leaf production) [4]
- Caterpillars in *broccoli, brussels sprouts, cabbages, calabrese, cauliflowers, kale* [1, 4]
- Cereal aphid in *spring barley, spring wheat, winter barley, winter wheat* [4]
- Diamond-back moth in *salad brassicas* (off-label - for baby leaf production) [4]
- Flea beetle in *broccoli, brussels sprouts, cabbages, calabrese, cauliflowers, kale* [1, 4]
- Insect pests in *borage for oilseed production* (off-label), *chinese cabbage* (off-label), *choi sum* (off-label), *collards* (off-label), *durum wheat* (off-label), *grass seed crops* (off-label), *linseed* (off-label), *lupins* (off-label), *pak choi* (off-label), *spring rye* (off-label), *triticale* (off-label), *winter rye* (off-label) [4]
- Lesser mealworm in *poultry houses* [2, 3]
- Pea and bean weevil in *broad beans, combining peas, spring field beans, vining peas, winter field beans* [1, 4]
- Pea aphid in *combining peas, vining peas* [1]; *combining peas* (reduction), *vining peas* (reduction) [4]
- Pea moth in *combining peas, vining peas* [1, 4]
- Pine weevil in *forest* (off-label), *forest nurseries* (off-label) [4]
- Pollen beetle in *spring oilseed rape, winter oilseed rape* [1, 4]
- Poultry red mite in *poultry houses* [2, 3]
- Rape winter stem weevil in *winter oilseed rape* [4]
- Small white butterfly in *salad brassicas* (off-label - for baby leaf production) [4]
- Yellow cereal fly in *spring barley* (some control if present at application), *spring wheat* (some control if present at application), *winter barley* (some control if present at application), *winter wheat* (some control if present at application) [1]; *winter barley, winter wheat* [4]

Specific Off-Label Approvals (SOLAs)
- *borage for oilseed production* 20052268 [4]
- *chinese cabbage* 20052265 [4]
- *choi sum* 20052265 [4]
- *collards* 20052265 [4]
- *durum wheat* 20052269 [4]
- *forest* 20062391 [4]
- *forest nurseries* 20062391 [4]
- *grass seed crops* 20052269 [4]
- *linseed* 20052268 [4]
- *lupins* 20052267 [4]
- *pak choi* 20052265 [4]
- *salad brassicas* (for baby leaf production) 20011221 [4]
- *spring rye* 20052269 [4]
- *triticale* 20052269 [4]
- *winter rye* 20052269 [4]

Approval information
- Alpha-cypermethrin included in Annex I under EC Directive 91/414
- Accepted by BBPA for use on malting barley

SEE SECTION 3 FOR PRODUCTS ALSO REGISTERED

Pesticide Profiles

Efficacy guidance
- For cabbage stem flea beetle control spray oilseed rape when adult or larval damage first seen and about 1 mth later [4]
- For flowering pests on oilseed rape apply at any time during flowering, against pollen beetle best results achieved at green to yellow bud stage (GS 3.3-3.7), against seed weevil between 20 pods set and 80% petal fall (GS 4.7-5.8) [4]
- Spray cereals in autumn for control of cereal aphids, in spring/summer for grain aphids. (See label for details) [4]
- For flea beetle, caterpillar and cabbage aphid control on brassicas apply when the pest or damage first seen or as a preventive spray. Repeat if necessary [4]
- For pea and bean weevil control in peas and beans apply when pest attack first seen and repeat as necessary [4]
- For lesser mealworm control in poultry houses apply a coarse, low-pressure spray as routine treatment after clean-out and before each new crop. Spray vertical surfaces and ensure an overlap onto ceilings. It is not necessary to treat the floor [2, 3]
- Use the highest recommended concentration in poultry houses where extreme residual action is required or surfaces are dirty or highly absorbent [2, 3]

Restrictions
- Maximum number of treatments 4 per crop on edible brassicas, 3 per crop on winter oilseed rape, peas, 2 on beans, spring oilseed rape (only 1 after yellow bud stage - GS 3,7) [4]
- Maximum number of applications in animal husbandry use 2 when used in premises that are occupied by poultry [2, 3]
- Apply up to 2 sprays on cereals in autumn and spring, 1 in summer between 1 Apr and 31 Aug. See label for details of rates and maximum total dose [4]
- Only 1 aphicide treatment may be applied in cereals between 1 Apr and 31 Aug and spray volume must not be reduced in this period [4]
- Do not apply to a cereal crop if any product containing a pyrethroid or dimethoate has been applied after the start of ear emergence (GS 51) [4]

Crop-specific information
- Latest use: before the end of flowering for oilseed rape; before 31 Aug for cereals [4]
- HI vining peas 1 d; brassicas 7 d; combining peas, broad beans, field beans 11 d [4]
- For summer cereal application do not spray within 6 m from edge of crop and do not reduce volume when used after 31 Mar [4]

Environmental safety
- Dangerous for the environment [4]
- Very toxic to aquatic organisms [4]
- Extremely dangerous to fish or other aquatic life. Do not contaminate surface waters or ditches with chemical or used container [2, 3]
- Dangerous to bees [2, 3]
- Where possible spray oilseed rape crops in the late evening or early morning or in dull weather [4]
- Do not spray within 6 m of the edge of a cereal crop after 31 Mar in yr of harvest [4]
- Do not apply directly to poultry; collect eggs before application [2, 3]
- LERAP Category A [1, 4]

Hazard classification and safety precautions
Hazard Harmful, Dangerous for the environment [1, 4]; Flammable [1]
Transport code 3 [1]; 9 [2-4]
Packaging group III
UN number 1993 [1]; 3077 [4]; 3082 [2, 3]
Risk phrases R20, R22b, R66 [1]; R22a, R48, R50, R53a [1, 4]
Operator protection A, H [1-4]; C [1]; U02a, U19a [1-3]; U02b, U09a [2, 3]; U04a, U11, U14, U20a [1]; U05a, U10 [1, 4]; U20b [2-4]
Environmental protection E02a [2, 3] (until dry); E05a, E12c, E13a [2, 3]; E15a [1-4]; E16c, E16d [1, 4]; E34 [1]; E38 [4]
Consumer protection C05, C06, C07, C09, C11 [2, 3]
Storage and disposal D01, D02, D05, D10b [1, 4]; D09a, D12a [1-4]; D11a [2, 3]
Medical advice M04a, M05b [1]; M05a [4]

FOR FULL CONDITIONS OF USE ALWAYS READ THE PRODUCT LABEL

5 aluminium ammonium sulphate

An inorganic bird and animal repellent

Products

1	Curb Crop Spray Powder	Sphere	88% w/w	WP	02480
2	Liquid Curb Crop Spray	Sphere	83 g/l	SC	03164
3	Rezist	Sphere	88% w/w	WP	08576
4	Sphere ASBO	Sphere	83.3 g/l	SC	15064

Uses
- Animal repellent in *agricultural premises, all top fruit, broad beans, bush fruit, cane fruit, carrots, flowerhead brassicas, forest nursery beds, forestry plantations, grain stores, leaf brassicas, peas, permanent grassland, spring barley, spring field beans, spring oats, spring oilseed rape, spring wheat, strawberries, sugar beet, winter barley, winter field beans, winter oats, winter oilseed rape, winter wheat* [1, 2]; *all edible crops (outdoor), all non-edible crops (outdoor), amenity grassland, forest, hard surfaces, managed amenity turf* [4]; *amenity vegetation* [1]
- Bird repellent in *agricultural premises, all top fruit, broad beans, bush fruit, cane fruit, carrots, flowerhead brassicas, forest nursery beds, forestry plantations, grain stores, leaf brassicas, peas, permanent grassland, spring barley, spring field beans, spring oats, spring oilseed rape, spring wheat, strawberries, sugar beet, winter barley, winter field beans, winter oats, winter oilseed rape, winter wheat* [1, 2]; *all edible crops (outdoor), all non-edible crops (outdoor), amenity grassland, forest, hard surfaces, managed amenity turf* [4]; *amenity vegetation* [1]
- Dogs in *amenity grassland* (anti-fouling), *hard surfaces* (anti-fouling), *managed amenity turf* (anti-fouling) [3]
- Moles in *amenity grassland, hard surfaces, managed amenity turf* [3]

Efficacy guidance
- Apply as overall spray to growing crops before damage starts or mix powder with seed depending on type of protection required
- Spray deposit protects growth present at spraying but gives little protection to new growth
- Product must be sprayed onto dry foliage to be effective and must dry completely before dew or frost forms. In winter this may require some wind

Crop-specific information
- Latest use: no restriction

Hazard classification and safety precautions
  Operator protection U05a [1, 2, 4]; U20a [1-4]
  Environmental protection E15a [1-4]; E19b [1, 2, 4]
  Storage and disposal D01, D02, D05, D12a [1, 2, 4]; D09a [3]; D11a [1-4]
  Medical advice M03 [1, 2, 4]

6 aluminium phosphide

A phosphine generating compound used against vertebrates and grain store pests
IRAC mode of action code: 24A

Products

1	Degesch Fumigation Tablets	Rentokil	56% w/w	GE	09313
2	Phostoxin	Rentokil	56% w/w	GE	09315
3	Talunex	Certis	56% w/w	GE	14608

Uses
- Insect pests in *stored grain* [1]
- Moles in *farmland* [2, 3]
- Rabbits in *farmland* [2, 3]
- Rats in *farmland* [2, 3]

Approval information
- Accepted by BBPA for use in stores for malting barley

SEE SECTION 3 FOR PRODUCTS ALSO REGISTERED

Efficacy guidance
- Product releases poisonous hydrogen phosphide gas in contact with moisture
- Place fumigation tablets in grain stores as directed [1]
- Place pellets in burrows or runs and seal hole by heeling in or covering with turf. Do not cover pellets with soil. Inspect daily and treat any new or re-opened holes [2]

Restrictions
- Aluminium phosphide is subject to the Poisons Rules 1982 and the Poisons Act 1972. See Section 5 for more information
- Only to be used by operators instructed or trained in the use of aluminium phosphide and familiar with the precautionary measures to be taken. See label and HSE Guidance Notes for full precautions
- Only open container outdoors [2] and for immediate use. Keep away from liquid or water as this causes immediate release of gas. Do not use in wet weather
- Do not use within 3 m of human or animal habitation. Before application ensure that no humans or domestic animals are in adjacent buildings or structures. Allow a minimum airing-off period of 4 h before re-admission

Environmental safety
- Product liberates very toxic, highly flammable gas
- Dangerous to fish or other aquatic life. Do not contaminate surface waters or ditches with chemical or used container [1, 2]
- Prevent access by livestock, pets and other non-target mammals and birds to buildings under fumigation and ventilation [1]
- Pellets must never be placed or allowed to remain on ground surface
- Do not use adjacent to watercourses
- Take particular care to avoid gassing non-target animals, especially those protected under the Wildlife and Countryside Act (e.g. badgers, polecat, reptiles, natterjack toads, most birds). Do not use in burrows where there is evidence of badger or fox activity, or when burrows might be occupied by birds
- Dust remaining after decomposition is harmless and of no environmental hazard
- Keep in original container, tightly closed, in a safe place, under lock and key
- Dispose of empty containers as directed on label

Hazard classification and safety precautions
Hazard Very toxic, Highly flammable [1-3]; Harmful [2, 3]
Transport code 4.1 6.1 [3]; 4.3 [1, 2]
Packaging group I
UN number 1397
Risk phrases R21, R26, R28
Operator protection A, D, H; U01, U07, U13, U19a, U20a [1-3]; U05a [2, 3]; U05b, U18 [1]
Environmental protection E02a [1] (4 h min); E02a [2, 3] (4 h); E02b [1]; E13b, E34 [1-3]
Storage and disposal D01, D02, D07, D09b, D11b
Vertebrate/rodent control products V04a [2, 3]
Medical advice M04a

7   ametoctradin

A potato blight fungicide available only in mixtures
FRAC mode of action code: 45

See also ametoctradin + dimethomorph
ametoctradin + mancozeb

FOR FULL CONDITIONS OF USE ALWAYS READ THE PRODUCT LABEL

8 ametoctradin + dimethomorph

A systemic and protectant fungicide for potato blight control
FRAC mode of action code: 40 + 45

Products
1	Resplend	BASF	300:225 g/l	SC	14975
2	Zampro DM	BASF	300:225 g/l	SC	15013

Uses
- Blight in *potatoes*

Approval information
- Ametoctradin not yet included in Annex 1 under EC Directive 91/414. Dimethomorph listed in Annex 1

Efficacy guidance
- Application to dry foliage is rainfast within 1 hour of drying on the leaf.
- Will give effective control of phenylamide-resistant strains of blight

Crop-specific information
- HI 7 days for potatoes

Hazard classification and safety precautions
Hazard Harmful, Dangerous for the environment
UN number N/C
Risk phrases R22a, R52, R53a
Operator protection U05a
Environmental protection E15b, E34, E38
Storage and disposal D01, D02, D09a, D10c
Medical advice M03, M05a

9 ametoctradin + mancozeb

A protectant fungicide mixture for potato blight control
FRAC mode of action code: 45 + M3

Products
Decabane	BASF	80:480 g/l	WG	14985

Uses
- Blight in *potatoes*

Approval information
- Ametoctradin not yet included in Annex 1 under EC Directive 91/414. Mancozeb listed in Annex 1

Efficacy guidance
- Application to dry foliage is rainfast within 1 hour of drying on the leaf.
- Will control phenylamide-resistant strains of late blight

Crop-specific information
- HI for potatoes 7 days

Environmental safety
- LERAP Category B

Hazard classification and safety precautions
Hazard Harmful, Dangerous for the environment
Transport code 9
Packaging group III
UN number 3077
Risk phrases R50, R53a, R63
Operator protection A, D, H; U02a, U05a
Environmental protection E15b, E16a, E16b, E34, E38
Storage and disposal D01, D02, D09a, D10c, D12a
Medical advice M05a

SEE SECTION 3 FOR PRODUCTS ALSO REGISTERED

10   amidosulfuron

A post-emergence sulfonylurea herbicide for cleavers and other broad-leaved weed control in cereals
HRAC mode of action code: B

Products

1	Eagle	Bayer CropScience	75% w/w	WG	07318
2	Squire Ultra	Interfarm	75% w/w	WG	14401

Uses
- Annual dicotyledons in **durum wheat**, **durum wheat** (off-label), **linseed**, **spring barley**, **spring oats**, **spring rye**, **spring wheat**, **triticale**, **winter barley**, **winter oats**, **winter rye**, **winter wheat** [1]; **grassland** [2]
- Charlock in **durum wheat** [1]
- Cleavers in **durum wheat**, **durum wheat** (off-label), **linseed**, **spring barley**, **spring oats**, **spring rye**, **spring wheat**, **triticale**, **winter barley**, **winter oats**, **winter rye**, **winter wheat** [1]
- Docks in **grassland** [2]

Specific Off-Label Approvals (SOLAs)
- **durum wheat** 20061297 [1]

Approval information
- Accepted by BBPA for use on malting barley
- Amidosulfuron included in Annex I under EC Directive 91/414

Efficacy guidance
- For best results apply in spring (from 1 Feb) in warm weather when soil moist and weeds growing actively. When used in grassland following cutting or grazing, docks should be allowed to regrow before treatment
- Weed kill is slow, especially under cool, dry conditions. Weeds may sometimes only be stunted but will have little or no competitive effect on crop
- May be used on all soil types unless certain sequences are used on linseed. See label
- Spray is rainfast after 1 h
- Cleavers controlled from emergence to flower bud stage. If present at application charlock (up to flower bud), shepherds purse (up to flower bud) and field forget-me-not (up to 6 leaves) will also be controlled
- Amidosulfuron is a member of the ALS-inhibitor group of herbicides and products should be used in a planned Resistance Management strategy. See Section 5 for more information

Restrictions
- Maximum number of treatments 1 per crop [1]
- Use after 1 Feb and do not apply to rotational grass after 30 Jun, or to permanent grassland after 15 Oct [2]
- Do not apply to crops undersown or due to be undersown with clover or alfalfa [1]
- Do not spray crops under stress, suffering drought, waterlogged, grazed, lacking nutrients or if soil compacted
- Do not spray if frost expected
- Do not roll or harrow within 1 wk of spraying
- Specific restrictions apply to use in sequence or tank mixture with other sulfonylurea or ALS-inhibiting herbicides. See label for details. There are no recommendations for mixtures with metsulfuron-methyl products on linseed
- Certain mixtures with fungicides are expressly forbidden. See label for details

Crop-specific information
- Latest use: before first spikelets just visible (GS 51) for cereals; before flower buds visible for linseed; 15 Oct for grassland
- Broadcast cereal crops should be sprayed post-emergence after plants have a well established root system

Following crops guidance
- If a treated crop fails cereals may be sown after 15 d and thorough cultivation

FOR FULL CONDITIONS OF USE ALWAYS READ THE PRODUCT LABEL

amidosulfuron + iodosulfuron-methyl-sodium

- After normal harvest of a treated crop only cereals, winter oilseed rape, mustard, turnips, winter field beans or vetches may be sown in the same year as treatment and these must be preceded by ploughing or thorough cultivation
- Only cereals may be sown within 12 mth of application to grassland [2]
- Cereals or potatoes must be sown as the following crop after use of permitted mixtures or sequences with other sulfonylurea herbicides in cereals. Only cereals may be sown after the use of such sequences in linseed

Environmental safety
- Take care to wash out sprayers thoroughly. See label for details
- Avoid drift onto neighbouring broad-leaved plants or onto surface waters or ditches

Hazard classification and safety precautions
Hazard Dangerous for the environment [2]
Transport code 9
Packaging group III
UN number 3077
Risk phrases R50, R53a [2]
Operator protection U20a [1]; U20b [2]
Environmental protection E07b [2] (1 week); E15a [1, 2]; E38, E41 [2]
Storage and disposal D10a [1, 2]; D12a [2]

11 amidosulfuron + iodosulfuron-methyl-sodium

A post-emergence sulfonylurea herbicide mixture for cereals
HRAC mode of action code: B + B

See also iodosulfuron-methyl-sodium

Products
1	Chekker	Bayer CropScience	12.5:1.25% w/w	WG	10955
2	Sekator	Interfarm	5:1.25% w/w	WG	14746

Uses
- Annual dicotyledons in **durum wheat** *(off-label)*, **grass seed crops** *(off-label)*, **linseed** *(off-label)* [1]; **spring barley, spring rye, spring wheat, triticale, winter barley, winter rye, winter wheat** [1, 2]
- Chickweed in **spring barley, spring rye, spring wheat, triticale, winter barley, winter rye, winter wheat** [1, 2]
- Cleavers in **spring barley, spring rye, spring wheat, triticale, winter barley, winter rye, winter wheat** [1, 2]
- Mayweeds in **spring barley, spring rye, spring wheat, triticale, winter barley, winter rye, winter wheat** [1, 2]
- Volunteer oilseed rape in **spring barley, spring rye, spring wheat, triticale, winter barley, winter rye, winter wheat** [1, 2]

Specific Off-Label Approvals (SOLAs)
- *durum wheat 20060858* [1]
- *grass seed crops 20060858* [1]
- *linseed 20101125* [1]

Approval information
- Amidosulfuron and iodosulfuron-methyl-sodium included in Annex I under EC Directive 91/414
- Accepted by BBPA for use on malting barley

Efficacy guidance
- Best results obtained from treatment in warm weather when soil is moist and the weeds are growing actively
- Weeds must be present at application to be controlled
- Dry conditions resulting in moisture stress may reduce effectiveness
- Weed control is slow especially under cool dry conditions
- Occasionally weeds may only be stunted but they will normally have little or no competitive effect on the crop

SEE SECTION 3 FOR PRODUCTS ALSO REGISTERED

- Amidosulfuron and iodosulfuron are members of the ALS-inhibitor group of herbicides and products should be used in a planned Resistance Management strategy. See Section 5 for more information

Restrictions
- Maximum number of treatments 1 per crop
- Must only be applied between 1 Feb in yr of harvest and specified latest time of application
- Do not apply to crops undersown or to be undersown with grass, clover or alfalfa
- Do not roll or harrow within 1 wk of spraying
- Do not spray crops under stress from any cause or if the soil is compacted
- Do not spray if rain or frost expected
- Do not apply in mixture or in sequence with any other ALS inhibitor

Crop-specific information
- Latest use: before first spikelet of inflorescence just visible (GS 51)
- Treat drilled crops after the 2-leaf stage; treat broadcast crops after the plants have a well-established root system
- Applications to spring barley may cause transient crop yellowing

Following crops guidance
- Cereals, winter oilseed rape and winter field beans may be sown in the same yr as treatment provided they are preceded by ploughing or thorough cultivation. Any crop may be sown in the spring of the yr following treatment
- A minimum of 3 mth must elapse between treatment and sowing winter oilseed rape

Environmental safety
- Dangerous for the environment
- Toxic to aquatic organisms
- Take extreme care to avoid damage by drift onto broad-leaved plants outside the target area or onto ponds, waterways and ditches
- Observe carefully label instructions for sprayer cleaning
- LERAP Category B

Hazard classification and safety precautions
Hazard Irritant, Dangerous for the environment
Transport code 9
Packaging group III
UN number 3077
Risk phrases R36, R51, R53a
Operator protection A, C, H; U05a, U08, U11, U14, U15, U20b
Environmental protection E15a, E16a, E16b, E38
Storage and disposal D01, D02, D10a, D12a

12 aminopyralid

A pyridine carboxylic acid herbicide available only in mixtures, approvals suspended pending an investigation into compost contamination
HRAC mode of action code: O

13 aminopyralid + fluroxypyr

A foliar acting herbicide mixture for use in grassland
HRAC mode of action code: O + O

See also fluroxypyr

Products
1	Forefront	Dow	30:100 g/l	EO	14701
2	Mileway	Dow	30:100 g/l	EW	14702
3	Synero	Dow	30:100 g/l	EW	14708

Uses
- Buttercups in *amenity grassland* [2, 3]; *permanent grassland, rotational grass* [1]

FOR FULL CONDITIONS OF USE ALWAYS READ THE PRODUCT LABEL

aminopyralid + fluroxypyr

- Chickweed in **amenity grassland** [2, 3]; **permanent grassland, rotational grass** [1]
- Dandelions in **amenity grassland** [2, 3]; **permanent grassland, rotational grass** [1]
- Docks in **amenity grassland** [2, 3]; **permanent grassland, rotational grass** [1]
- Stinging nettle in **amenity grassland** [2, 3]; **permanent grassland, rotational grass** [1]
- Thistles in **amenity grassland** [2, 3]; **permanent grassland, rotational grass** [1]

Approval information
- Fluroxypyr included in Annex I under EC Directive 91/414 while aminopyralid has not yet been included.
- Approval expiry 29 Jul 2011 [1-3]

Efficacy guidance
- For best results and to avoid crop check, grass and weeds must be growing actively
- Allow 2-3 wk after cutting for hay or silage for sufficient regrowth to occur before spraying and leave 7 d afterwards to allow maximum translocation
- Where there is a high reservoir of weed seed or a historically high weed population a programmed approach may be needed involving a second treatment in the following yr
- Control may be reduced if rain falls within 1 h of spraying

Restrictions
- Maximum number of treatments 1 per yr.
- Do not apply to leys less than 1 year old.
- Do not apply by hand-held equipment.
- Do not use on grassland that will be used for animal feed, bedding, composting or mulching within 1 calender year of application.
- Do not use on grassland that will be grazed by animals other than cattle or sheep.
- Use of an antifoam is compulsory
- Do not use any treated plant material, or manure from animals fed on treated crops, for composting or mulching
- Do not use on crops grown for seed
- Manure from animals fed on pasture or silage treated with [1] should not leave the farm

Crop-specific information
- Treatment will kill clover
- Treatment may occasionally cause transient yellowing of the sward which is quickly outgrown
- Late treatments may lead to a slight transient leaning of grass that does not affect yield

Following crops guidance
- Do not drill clover or other legumes within 4 mth of treatment, or potatoes in the spring following treatment in the previous autumn
- In the event of failure of newly seeded treated grassland, grass may be re-seeded immediately, or wheat may be sown provided 4 mth have elapsed since application
- Residues in plant tissues, including manure, may affect succeeding susceptible crops of peas, beans, other legumes, carrots, other Umbelliferae, potatoes, tomatoes, lettuce and other Compositae. These crops should not be sown within 3 mth of ploughing up treated grassland

Environmental safety
- Dangerous for the environment
- Toxic to aquatic organisms
- Keep livestock out of treated areas for at least 7 d following treatment and until poisonous weeds, such as ragwort, have died down and become unpalatable
- To protect groundwater do not apply between 1 Sept and 28 Feb
- Avoid damage by drift onto susceptible crops, non-target plants or waterways

Hazard classification and safety precautions
Hazard Irritant, Dangerous for the environment
Transport code 9
Packaging group III
UN number 3082
Risk phrases R38, R41, R51, R53a, R67
Operator protection A, C; U05a, U08, U11, U14, U23a
Environmental protection E07a, E15b, E38
Consumer protection C01
Storage and disposal D01, D02, D05, D09a, D10b, D12a

SEE SECTION 3 FOR PRODUCTS ALSO REGISTERED

Pesticide Profiles

14 aminopyralid + triclopyr

A foliar acting herbicide mixture for broad-leaved weed control in grassland
HRAC mode of action code: O + O

See also triclopyr

Products
  Pharaoh   Dow   30:240g/l   EW   14731

Uses
- Buttercups in **grassland**
- Common nettle in **grassland**
- Dandelions in **grassland**
- Docks in **grassland**
- Thistles in **grassland**

Approval information
- Triclopyr included in Annex I under EC Directive 91/414 while aminopyralid has not yet been included.
- Approval expiry 29 Jul 2011

Efficacy guidance
- For best results and to avoid crop check, grass and weeds must be growing actively
- Use adeqate water volume to ensure good weed coverage. Increase water volume if necessary where the weed population is high and where the grass is dense
- Allow 2-3 wk after cutting for hay or silage for sufficient regrowth to occur before spraying and leave 7 d afterwards to allow maximum translocation
- Where there is a high reservoir of weed seed or a historically high weed population a programmed approach may be needed involving a second treatment in the following yr
- Control may be reduced if rain falls within 1 h of spraying

Restrictions
- Maximum number of treatments 1 per yr.
- Do not apply to leys less than 1 year old.
- Do not apply by hand-held equipment.
- Do not use on grassland that will be used for animal feed, bedding, composting or mulching within 1 calender year of application.
- Do not use on grassland that will be grazed by animals other than cattle or sheep.
- Use of an antifoam is compulsory
- Do not use any treated plant material, or manure from animals fed on treated crops, for composting or mulching
- Do not use on crops grown for seed
- Manure from animals fed on pasture or silage from treated crops should not leave the farm

Crop-specific information
- Latest use: 7 d before grazing or harvest for grassland
- Late applications may lead to transient leaning of grass which does not affect final yield

Following crops guidance
- Ensure that all plant remains of a treated crop have completely decayed before planting susceptible crops such as peas, beans and other legumes, sugar beet, carrots and other Umbelliferae, potatoes and tomatoes, lettuce and other Compositae
- Do not plant potatoes, sugar beet, vegetables, beans or other leguminous crops in the calendar yr following application

Environmental safety
- Dangerous for the environment
- Toxic to aquatic organisms
- Keep livestock out of treated areas for at least 7 d after treatment or until foliage of any poisonous weeds such as ragwort has died and become unpalatable
- To protect groundwater do not apply to grass leys less than 1 yr old

FOR FULL CONDITIONS OF USE ALWAYS READ THE PRODUCT LABEL

- Take extreme care to avoid drift onto susceptible crops, non-target plants or waterways. All conifers, especially pine and larch, are very sensitive and may be damaged by vapour drift in hot conditions
- LERAP Category B

Hazard classification and safety precautions
Hazard Irritant, Dangerous for the environment
Transport code 9
Packaging group III
UN number 3082
Risk phrases R43, R51, R53a
Operator protection A, H; U05a, U08, U14, U23a
Environmental protection E06a (7 days); E15b, E15c, E16a, E38
Consumer protection C01
Storage and disposal D01, D02, D09a, D10b, D12a

15 amisulbrom

A fungicide for use in potatoes
FRAC mode of action code: C4

Products
Shinkon   Nissan   200 g/l   SC   13722

Uses
- Late blight in **potatoes**

Environmental safety
- LERAP Category B

Hazard classification and safety precautions
Hazard Dangerous for the environment
Risk phrases R50, R53a
Operator protection A, C; U05a, U15, U20b
Environmental protection E13b, E15a, E16a, E34, E38
Storage and disposal D01, D02, D05, D09a, D10b, D12a

16 amitrole

A translocated, foliar-acting, non-selective triazole herbicide
HRAC mode of action code: F3

Products
Weedazol-TL   Nufarm UK   225 g/l   SL   11968

Uses
- Annual and perennial weeds in **apricots** (off-label), **cherries** (off-label), **peaches** (off-label), **plums** (off-label), **quinces** (off-label)
- Annual dicotyledons in **fallows**, **headlands**, **stubbles**
- Annual grasses in **fallows**, **headlands**, **stubbles**
- Barren brome in **apple orchards**, **pear orchards**
- Couch in **apple orchards**, **fallows**, **headlands**, **pear orchards**, **stubbles**
- Creeping thistle in **fallows**, **stubbles**
- Docks in **fallows**, **headlands**, **stubbles**
- General weed control in **apple orchards**, **farm woodland** (off-label), **pear orchards**
- Grass weeds in **managed amenity turf** (off-label - on amitrole resistant turf)
- Perennial dicotyledons in **apple orchards**, **fallows**, **headlands**, **pear orchards**
- Perennial grasses in **apple orchards**, **fallows**, **headlands**, **pear orchards**
- Volunteer potatoes in **stubbles** (barley stubble)

Specific Off-Label Approvals (SOLAs)
- *apricots* 20051981
- *cherries* 20051981

SEE SECTION 3 FOR PRODUCTS ALSO REGISTERED

Pesticide Profiles

- **farm woodland** *20051980*
- **managed amenity turf** *(on amitrole resistant turf) 20051979*
- **peaches** *20051981*
- **plums** *20051981*
- **quinces** *20051981*

Approval information
- Amitrole included in Annex I under EC Directive 91/414

Efficacy guidance
- In non-crop land may be applied at any time from Apr to Oct. Best results achieved in spring or early summer when weeds growing actively. For coltsfoot, hogweed and horsetail summer and autumn applications are preferred
- Uptake is via foliage and heavy rain immediately after application will reduce efficacy. Amitrole is less affected by drought than some residual herbicides and remains effective for up to 2 mth
- Applications made in summer may not give complete control of couch if past the shooting stage or not actively growing
- Effective crop competition and efficient ploughing are essential for good couch control

Restrictions
- Maximum number of treatments 1 per yr
- Keep off suckers or foliage of desirable trees or shrubs
- Do not spray areas into which the roots of adjacent trees or shrubs extend
- Application to land intended for spring barley should be in the preceding autumn, not the spring
- Do not spray on sloping ground when rain imminent and run-off may occur
- Do not spray if foliage is wet or rain imminent
- Do not use in low temperatures or in drought
- Do not mix product with acids

Crop-specific information
- Latest use: before end Jun or after harvest for apple and pear orchards; end Oct for headlands; end Oct and at least 2 wk before cultivation and drilling for stubbles and fallows

Following crops guidance
- Amitrole breaks down fairly quickly in medium and heavy soils and 3 wk should be allowed between application and sowing or planting. On sandy soils the interval should be 6 wk

Environmental safety
- Harmful to aquatic organisms
- Keep livestock out of treated areas for at least two weeks following treatment and until poisonous weeds, such as ragwort, have died down and become unpalatable
- Harmful to fish or other aquatic life. Do not contaminate surface waters or ditches with chemical or used container

Hazard classification and safety precautions
Hazard Harmful
Risk phrases R48, R52, R53a, R63
Operator protection A, C; U05a, U08, U19a, U20b
Environmental protection E07a, E15a
Storage and disposal D01, D02, D09a, D10b

17   asulam

A translocated carbamate herbicide for control of docks and bracken
HRAC mode of action code: I

Products

1	Agrotech Asulam	AgChem Access	400 g/l	SL	12243
2	Asulox	United Phosphorus	400 g/l	SL	13175
3	Brack-N	AgriGuard	400 g/l	SL	13138
4	Formule 1	AgriChem BV	400 g/l	SL	13255
5	Greencrop Frond	Greencrop	400 g/l	SL	11912

FOR FULL CONDITIONS OF USE ALWAYS READ THE PRODUCT LABEL

asulam

Uses
- Annual grasses in **celeriac** *(off-label)*, **chicory** *(off-label)*, **chicory root** *(off-label)*, **herbs (see appendix 6)** *(off-label)* [2]
- Annual meadow grass in **celeriac** *(off-label)*, **chicory** *(off-label)*, **chicory root** *(off-label)*, **herbs (see appendix 6)** *(off-label)* [2]
- Bracken in **amenity vegetation** [4, 5]; **forest** [1-5]; **forest** *(off-label)* [2]; **grassland** [4]; **permanent grassland** [1-3, 5]
- Brome grasses in **poppies for morphine production** *(off-label)* [2]
- Docks in **almonds** *(off-label)*, **chestnuts** *(off-label)*, **hazel nuts** *(off-label)*, **walnuts** *(off-label)* [2, 5]; **amenity grassland**, **apples**, **grassland**, **pears** [4]; **amenity grassland** *(not fine turf)*, **apple orchards**, **pear orchards**, **permanent grassland** [1-3, 5]; **amenity vegetation** [1-4]; **blackberries** *(off-label)*, **blueberries** *(off-label)*, **celeriac** *(off-label)*, **chicory** *(off-label)*, **chicory root** *(off-label)*, **clover seed crops** *(off-label)*, **cranberries** *(off-label)*, **damsons** *(off-label)*, **gooseberries** *(off-label)*, **herbs (see appendix 6)** *(off-label)*, **loganberries** *(off-label)*, **mint** *(off-label)*, **nectarines** *(off-label)*, **parsley** *(off-label)*, **poppies for morphine production** *(off-label)*, **quinces** *(off-label)*, **raspberries** *(off-label)*, **redcurrants** *(off-label)*, **strawberries** *(off-label)*, **tarragon** *(off-label)*, **whitecurrants** *(off-label)* [2]; **blackcurrants**, **cherries**, **hops**, **plums** [1-5]; **road verges**, **rotational grass**, **waste ground** [1, 2]
- Meadow grasses in **poppies for morphine production** *(off-label)* [2]

Specific Off-Label Approvals (SOLAs)
- **almonds** 20070816 [2], 20060866 [5]
- **blackberries** 20070799 [2]
- **blueberries** 20070799 [2]
- **celeriac** 20082738 expires 11 Feb 2011 [2]
- **chestnuts** 20070816 [2], 20060866 [5]
- **chicory** 20082739 expires 11 Feb 2011 [2]
- **chicory root** 20082739 expires 11 Feb 2011 [2]
- **clover seed crops** 20070825 [2]
- **cranberries** 20070799 [2]
- **damsons** 20070799 [2]
- **forest** 20070823 [2]
- **gooseberries** 20070799 [2]
- **hazel nuts** 20070816 [2], 20060866 [5]
- **herbs (see appendix 6)** 20082740 expires 11 Feb 2011 [2]
- **loganberries** 20070799 [2]
- **mint** 20070822 [2]
- **nectarines** 20070799 [2]
- **parsley** 20070822 [2]
- **poppies for morphine production** 20070826 [2]
- **quinces** 20070799 [2]
- **raspberries** 20070799 [2]
- **redcurrants** 20070799 [2]
- **strawberries** 20070824 [2]
- **tarragon** 20070822 [2]
- **walnuts** 20070816 [2], 20060866 [5]
- **whitecurrants** 20070799 [2]

Approval information
- May be applied through CDA equipment
- Approved for aerial application on bracken in agricultural grassland, amenity grassland, forestry and rough upland intended for grazing [2]. See Section 5 for more information
- Approved for use near surface waters. See Section 5 for more information
- Accepted by BBPA for use on hops

Efficacy guidance
- Spray bracken when fronds fully expanded but not senescent, usually Jul-Aug; docks in full leaf before flower stem emergence
- Bracken fronds must not be damaged by stock, frost or cutting before treatment
- Uptake and reliability of bracken control may be improved by use of specified additives - see label. Additives not recommended on forestry land

SEE SECTION 3 FOR PRODUCTS ALSO REGISTERED

Pesticide Profiles

- To allow adequate translocation do not cut or admit stock for 14 d after spraying bracken or 7 d after spraying docks. Preferably leave undisturbed until late autumn
- Complete bracken control rarely achieved by one treatment. Survivors should be sprayed when they recover to full green frond, which may be in the ensuing year but more likely in the second year following initial application

Restrictions
- Maximum number of treatments 1 per crop or 1 per yr
- Do not apply in drought or hot, dry conditions
- Do not use in pasture before mowing for hay

Crop-specific information
- Latest use: Aug-Sep for all situations. See label for details
- In forestry areas some young trees may be checked if sprayed directly (see label)
- In fruit crops apply as a directed spray
- Do not treat blackcurrant cuttings, hop sets or weak hills
- Some grasses and herbs will be damaged by full dose. Most sensitive are cocksfoot, Yorkshire fog, timothy, bents, annual meadow-grass, daisies, docks, plantains, saxifrage
- Apply as spot treatment in parsley, mint and tarragon, not directly to crop

Following crops guidance
- Allow at least 6 wk between spraying and planting any crop

Environmental safety
- Dangerous for the environment
- Very toxic to aquatic organisms
- Keep livestock out of treated areas for at least two weeks following treatment and until poisonous weeds, such as ragwort, have died down and become unpalatable
- The use of asulam near surface waters has been considered by CRD. Whilst every care should be taken to avoid contamination, any that does occur during the normal course of spraying should offer no harm to operators, to users and consumers of the water, to domestic and farm animals and to wildlife. Before spraying such areas the appropriate regulatory authority should be notified

Hazard classification and safety precautions
Hazard Irritant, Dangerous for the environment [1-3, 5]
Transport code 9 [1-3, 5]
Packaging group III [1-3, 5]
UN number 3082 [1-3, 5]; N/C [4]
Risk phrases R43, R50 [1-3, 5]; R52 [4]; R53a [5]
Operator protection A, C, D, H [1-3, 5]; M [1, 2]; M [3] (for ULV application); M [5]; U05a [4]; U08 [5]; U14 [1-3, 5]; U19a, U20b [1-3] (ULV use); U19a, U20b [4, 5]; U20c [1-3]
Environmental protection E07a [1-3, 5]; E07d [4]; E15a [1-5]; E38 [1-7]
Storage and disposal D01, D02 [4]; D05 [5]; D09a [1-5]; D10a [3-5]; D10c [1, 2]; D12a [1-4]

18   azoxystrobin

A systemic translaminar and protectant strobilurin fungicide for a wide range of crops
FRAC mode of action code: 11

Products

1	Amicron	AgriGuard	250 g/l	SC	13336
2	Amistar	Syngenta	250 g/l	SC	10443
3	Aubrac	AgChem Access	250 g/l	SC	13483
4	Azzox	Goldengrass	250 g/l	SC	14296
5	Clayton Belfry	Clayton	250 g/l	SC	12886
6	CS Azoxy	ChemSource	250 g/l	SC	14954
7	Harness	ChemSource	50% w/w	WG	14803
8	Heritage	Syngenta	50% w/w	WG	13536
9	Heritage Maxx	Syngenta	95 g/l	DC	14787
10	Panama	Pan Agriculture	50% w/w	SG	13950
11	Standon Azoxystrobin	Standon	250 g/l	SC	09515

FOR FULL CONDITIONS OF USE ALWAYS READ THE PRODUCT LABEL

azoxystrobin

Uses
- Alternaria in **broccoli, brussels sprouts, cabbages, calabrese, carrots, cauliflowers, collards, kale, spring oilseed rape, winter oilseed rape** [1-6]; **chicory root** *(off-label)*, **horseradish** *(off-label)*, **parsnips** *(off-label)*, **salad brassicas** *(off-label - for baby leaf production)* [2]
- Anthracnose in **amenity grassland, managed amenity turf** [7-10]
- Ascochyta in **broad beans** *(off-label)*, **lupins** *(off-label)* [2]; **combining peas, vining peas** [1-6]
- Black dot in **potatoes** [1-6]
- Black scurf and stem canker in **potatoes** [1-6]
- Black spot in **protected strawberries** *(off-label)*, **strawberries** *(off-label)* [2]
- Botrytis in **broad beans** *(off-label)*, **celery (outdoor)** *(off-label)*, **dwarf beans** *(off-label)*, **navy beans** *(off-label)*, **protected celery** *(off-label)*, **runner beans** *(off-label)* [2]
- Brown patch in **amenity grassland, managed amenity turf** [7-10]
- Brown rust in **spring barley, spring wheat, winter barley, winter wheat** [1-6, 11]; **spring rye, triticale, winter rye** [1-6]
- Crown rust in **amenity grassland, managed amenity turf** [7-10]; **spring oats, winter oats** [1-6]
- Didymella in **inert substrate aubergines** *(off-label)*, **inert substrate tomatoes** *(off-label)* [2]
- Didymella stem rot in **aubergines** *(off-label)*, **protected tomatoes** *(off-label)* [2]
- Disease control in **all edible seed crops grown outdoors** *(off-label)*, **all non-edible seed crops grown outdoors** *(off-label)*, **forest nurseries** *(off-label)*, **ornamental plant production** *(off-label)*, **protected forest nurseries** *(off-label)*, **protected ornamentals** *(off-label)*, **protected soft fruit** *(off-label)*, **soft fruit** *(off-label)* [2, 3]
- Downy mildew in **artichokes** *(off-label)*, **garlic** *(off-label)*, **inert substrate courgettes** *(off-label)*, **inert substrate cucumbers** *(off-label)*, **inert substrate gherkins** *(off-label)*, **protected courgettes** *(off-label)*, **protected cucumbers** *(off-label)*, **protected gherkins** *(off-label)*, **protected marrows** *(off-label)*, **protected melons** *(off-label)*, **protected pumpkins** *(off-label)*, **protected squashes** *(off-label)*, **salad brassicas** *(off-label - for baby leaf production)*, **salad onions** *(off-label)*, **shallots** *(off-label)* [2]; **bulb onions** [1-6]
- Fairy rings in **amenity grassland, managed amenity turf** [7-10]
- Foliar disease control in **crambe** *(off-label)*, **durum wheat** *(off-label)*, **radishes** *(off-label)* [2]; **grass seed crops** *(off-label)* [2, 11]
- Fusarium patch in **amenity grassland, managed amenity turf** [7-10]
- Glume blotch in **spring wheat, winter wheat** [1-6, 11]
- Grey mould in **aubergines** *(off-label)*, **inert substrate aubergines** *(off-label)*, **inert substrate tomatoes** *(off-label)*, **protected tomatoes** *(off-label)* [2]
- Late blight in **aubergines** *(off-label)*, **inert substrate aubergines** *(off-label)*, **inert substrate tomatoes** *(off-label)*, **protected tomatoes** *(off-label)* [2]
- Late ear diseases in **spring wheat, winter wheat** [1-6, 11]
- Melting out in **amenity grassland, managed amenity turf** [7-10]
- Net blotch in **spring barley, winter barley** [1-6, 11]
- Phytophthora in **protected chicory** *(off-label - for forcing)* [2]
- Powdery mildew in **artichokes** *(off-label)*, **aubergines** *(off-label)*, **blackberries** *(off-label)*, **chicory root** *(off-label)*, **chives** *(off-label)*, **courgettes** *(off-label)*, **herbs (see appendix 6)** *(off-label)*, **horseradish** *(off-label)*, **inert substrate aubergines** *(off-label)*, **inert substrate courgettes** *(off-label)*, **inert substrate cucumbers** *(off-label)*, **inert substrate gherkins** *(off-label)*, **inert substrate tomatoes** *(off-label)*, **parsley** *(off-label)*, **parsnips** *(off-label)*, **poppies for morphine production** *(off-label)*, **protected blackberries** *(off-label)*, **protected cayenne peppers** *(off-label)*, **protected courgettes** *(off-label)*, **protected cucumbers** *(off-label)*, **protected gherkins** *(off-label)*, **protected marrows** *(off-label)*, **protected melons** *(off-label)*, **protected peppers** *(off-label)*, **protected pumpkins** *(off-label)*, **protected raspberries** *(off-label)*, **protected squashes** *(off-label)*, **protected tomatoes** *(off-label)*, **raspberries** *(off-label)* [2]; **carrots, spring oats, triticale, winter oats, winter rye** [1-6]; **spring barley, winter barley** [1-6, 11]; **spring wheat, winter wheat** [11]
- Purple blotch in **leeks** [1-6]
- Rhizoctonia in **celery (outdoor)** *(off-label)*, **protected celery** *(off-label)*, **protected chives** *(off-label)*, **protected herbs (see appendix 6)** *(off-label)*, **protected lettuce** *(off-label)*, **protected parsley** *(off-label)* [2]; **red beet** *(off-label)* [3]; **swedes** *(off-label)*, **turnips** *(off-label)* [2, 3]
- Rhynchosporium in **spring barley, winter barley** [1-6, 11]; **spring rye, triticale, winter rye** [1-6]
- Ring spot in **broccoli, brussels sprouts, cabbages, calabrese, cauliflowers, collards, kale** [1-6]; **chives** *(off-label)*, **herbs (see appendix 6)** *(off-label)*, **parsley** *(off-label)* [2]

SEE SECTION 3 FOR PRODUCTS ALSO REGISTERED

- Root malformation disorder in **red beet** *(off-label)* [2]
- Rust in **asparagus**, **leeks**, **spring field beans**, **winter field beans** [1-6]; **chicory root** *(off-label)*, **chives** *(off-label)*, **herbs (see appendix 6)** *(off-label)*, **lupins** *(off-label)*, **parsley** *(off-label)* [2]
- Sclerotinia in **celeriac** *(off-label)*, **celery (outdoor)** *(off-label)*, **lettuce** *(off-label)*, **protected celery** *(off-label)* [2]
- Sclerotinia stem rot in **spring oilseed rape**, **winter oilseed rape** [1-6]
- Septoria leaf blotch in **spring wheat**, **winter wheat** [1-6, 11]
- Stemphylium in **asparagus** [1-6]
- Take-all in **spring barley** *(reduction)*, **spring wheat** *(reduction)*, **winter barley** *(reduction)*, **winter wheat** *(reduction)* [1-6]
- Take-all patch in **amenity grassland**, **managed amenity turf** [7-10]
- White blister in **broccoli**, **brussels sprouts**, **cabbages**, **calabrese**, **cauliflowers**, **collards**, **kale** [1-6]
- White rust in **chrysanthemums** *(off-label - in pots)*, **protected chrysanthemums** *(off-label)* [2]
- Yellow rust in **spring wheat**, **winter wheat** [1-6, 11]

Specific Off-Label Approvals (SOLAs)
- *all edible seed crops grown outdoors* 20090443 expires 31 Dec 2011 [2], 20100372 expires 31 Dec 2011 [3]
- *all non-edible seed crops grown outdoors* 20090443 expires 31 Dec 2011 [2], 20100372 expires 31 Dec 2011 [3]
- *artichokes* 20051814 expires 31 Dec 2011 [2]
- *aubergines* 20021533 expires 31 Dec 2011 [2]
- *blackberries* 20030365 expires 31 Dec 2011 [2]
- *broad beans* 20032311 expires 31 Dec 2011 [2]
- *celeriac* 20031862 expires 31 Dec 2011 [2]
- *celery (outdoor)* 20011041 expires 31 Dec 2011 [2]
- *chicory root* 20051813 expires 31 Dec 2011 [2]
- *chives* 20021293 expires 31 Dec 2011 [2]
- *chrysanthemums (in pots)* 20011684 expires 31 Dec 2011 [2]
- *courgettes* 20051985 expires 31 Dec 2011 [2]
- *crambe* 20081341 expires 31 Dec 2011 [2]
- *durum wheat* 20061722 expires 31 Dec 2011 [2]
- *dwarf beans* 20032311 expires 31 Dec 2011 [2]
- *forest nurseries* 20090443 expires 31 Dec 2011 [2], 20100372 expires 31 Dec 2011 [3]
- *garlic* 20061724 expires 31 Dec 2011 [2]
- *grass seed crops* 20061722 expires 31 Dec 2011 [2], 20061095 expires 31 Dec 2011 [11]
- *herbs (see appendix 6)* 20021293 expires 31 Dec 2011 [2]
- *horseradish* 20061721 expires 31 Dec 2011 [2]
- *inert substrate aubergines* 20011685 expires 31 Dec 2011 [2]
- *inert substrate courgettes* 20011685 expires 31 Dec 2011 [2]
- *inert substrate cucumbers* 20011685 expires 31 Dec 2011 [2]
- *inert substrate gherkins* 20011685 expires 31 Dec 2011 [2]
- *inert substrate tomatoes* 20011685 expires 31 Dec 2011 [2]
- *lettuce* 20011465 expires 31 Dec 2011 [2]
- *lupins* 20061723 expires 31 Dec 2011 [2]
- *navy beans* 20032311 expires 31 Dec 2011 [2]
- *ornamental plant production* 20090443 expires 31 Dec 2011 [2], 20100372 expires 31 Dec 2011 [3]
- *parsley* 20021293 expires 31 Dec 2011 [2]
- *parsnips* 20061721 expires 31 Dec 2011 [2]
- *poppies for morphine production* 20031137 expires 31 Dec 2011 [2]
- *protected blackberries* 20051194 expires 31 Dec 2011 [2]
- *protected cayenne peppers* 20021295 expires 31 Dec 2011 [2]
- *protected celery* 20011041 expires 31 Dec 2011 [2]
- *protected chicory (for forcing)* 20051813 expires 31 Dec 2011 [2]
- *protected chives* 20030659 expires 31 Dec 2011 [2]
- *protected chrysanthemums* 20011684 expires 31 Dec 2011 [2]
- *protected courgettes* 20021533 expires 31 Dec 2011 [2]
- *protected cucumbers* 20021533 expires 31 Dec 2011 [2]

FOR FULL CONDITIONS OF USE ALWAYS READ THE PRODUCT LABEL

azoxystrobin

- **protected forest nurseries** *20090443 expires 31 Dec 2011* [2], *20100372 expires 31 Dec 2011* [3]
- **protected gherkins** *20021533 expires 31 Dec 2011* [2]
- **protected herbs (see appendix 6)** *20030659 expires 31 Dec 2011* [2]
- **protected lettuce** *20030659 expires 31 Dec 2011* [2]
- **protected marrows** *20070371 expires 31 Dec 2011* [2]
- **protected melons** *20070371 expires 31 Dec 2011* [2]
- **protected ornamentals** *20090443 expires 31 Dec 2011* [2], *20100372 expires 31 Dec 2011* [3]
- **protected parsley** *20030659 expires 31 Dec 2011* [2]
- **protected peppers** *20021295 expires 31 Dec 2011* [2]
- **protected pumpkins** *20070371 expires 31 Dec 2011* [2]
- **protected raspberries** *20051194 expires 31 Dec 2011* [2]
- **protected soft fruit** *20090443 expires 31 Dec 2011* [2], *20100372 expires 31 Dec 2011* [3]
- **protected squashes** *20070371 expires 31 Dec 2011* [2]
- **protected strawberries** *20021294 expires 31 Dec 2011* [2]
- **protected tomatoes** *20021533 expires 31 Dec 2011* [2]
- **radishes** *20081448 expires 31 Dec 2011* [2]
- **raspberries** *20030365 expires 31 Dec 2011* [2]
- **red beet** *20040614 expires 31 Dec 2011* [2], *20100371 expires 31 Dec 2011* [3]
- **runner beans** *20032311 expires 31 Dec 2011* [2]
- **salad brassicas** *(for baby leaf production) 20011465 expires 31 Dec 2011* [2]
- **salad onions** *20021687 expires 31 Dec 2011* [2]
- **shallots** *20061724 expires 31 Dec 2011* [2]
- **soft fruit** *20090443 expires 31 Dec 2011* [2], *20100372 expires 31 Dec 2011* [3]
- **strawberries** *20021294 expires 31 Dec 2011* [2]
- **swedes** *20040614 expires 31 Dec 2011* [2], *20100371 expires 31 Dec 2011* [3]
- **turnips** *20040614 expires 31 Dec 2011* [2], *20100371 expires 31 Dec 2011* [3]

Approval information
- Azoxystrobin included in Annex I under EC Directive 91/414
- Accepted by BBPA for use on malting barley
- Approval expiry 31 Dec 2011 [1-11]

Efficacy guidance
- Best results obtained from use as a protectant or during early stages of disease establishment or when a predictive assessment indicates a risk of disease development
- Azoxystrobin inhibits fungal respiration and should always be used in mixture with fungicides with other modes of action
- Treatment under poor growing conditions may give less reliable results
- For good control of *Fusarium* patch in amenity turf and grass repeat treatment at minimum intervals of 2 wk
- Azoxystrobin is a member of the QoI cross resistance group. Product should be used preventatively and not relied on for its curative potential
- Use product in cereals as part of an Integrated Crop Management strategy incorporating other methods of control, including where appropriate other fungicides with a different mode of action. Do not apply more than two foliar applications of QoI containing products to any cereal crop
- There is a significant risk of widespread resistance occurring in *Septoria tritici* populations in UK. Failure to follow resistance management action may result in reduced levels of disease control
- On cereal crops product must always be used in mixture with another product, recommended for control of the same target disease, that contains a fungicide from a different cross resistance group and is applied at a dose that will give robust control
- Strains of barley powdery mildew resistant to QoIs are common in the UK

Restrictions
- Maximum number of treatments 1 per crop for potatoes; 2 per crop for brassicas, peas, cereals, oilseed rape; 4 per crop or yr for onions, carrots, leeks, amenity turf [1, 2, 5]
- Maximum total dose ranges from 2-4 times the single full dose depending on crop and product. See labels for details
- On turf the maximum number of treatments is 4 per yr but they must not exceed one third of the total number of fungicide treatments applied
- Do not use where there is risk of spray drift onto neighbouring apple crops
- The same spray equipment should not be used to treat apples

SEE SECTION 3 FOR PRODUCTS ALSO REGISTERED

Pesticide Profiles

Crop-specific information
- Latest use: at planting for potatoes; grain watery ripe (GS 71) for cereals; before senescence for asparagus
- HI: 10 d for carrots;14 d for broccoli, Brussels sprouts, bulb onions, cabbages, calabrese, cauliflowers, collards, kale, vining peas; 21 d for leeks, spring oilseed rape, winter oilseed rape; 36 d for combining peas, 35 d for field beans [1, 2, 5]
- In cereals control of established infections can be improved by appropriate tank mixtures or application as part of a programme. Always use in mixture with another product from a different cross-resistance group
- In turf use product at full dose rate in a disease control programme, alternating with fungicides of different modes of action
- In potatoes when used as incorporated treatment apply overall to the entire area to be planted, incorporate to 15 cm and plant on the same day. In-furrow spray should be directed at the furrow and not the seed tubers
- Applications to brassica crops must only be made to a developed leaf canopy and not before growth stages specified on the label
- Heavy disease pressure in brassicae and oilseed rape may require a second treatment
- All crops should be treated when not under stress. Check leaf wax on peas if necessary
- Consult processor before treating any crops for processing
- Treat asparagus after the harvest season. Where a new bed is established do not treat within 3 wk of transplanting out the crowns
- Do not apply to turf when ground is frozen or during drought

Environmental safety
- Dangerous for the environment
- Very toxic to aquatic organisms
- Avoid spray drift onto surrounding areas or crops, especially apples, plums or privet
- LERAP Category B [1-6] (potatoes only)

Hazard classification and safety precautions
Hazard Dangerous for the environment [1-9, 11]
Transport code 9
Packaging group III
UN number 3077 [7, 8, 10]; 3082 [1-6, 9, 11]
Risk phrases R50, R53a [1-9, 11]
Operator protection A [1-4]; A [5] (during treatment of potatoes); A [6, 9]; U05a [1-9]; U09a, U19a [1-6, 11]; U20b [1-9, 11]
Environmental protection E15a [1-6, 11]; E15b [7-9]; E16a [1-6] (potatoes only); E16b [2-6] (potatoes only); E38 [1-9]
Storage and disposal D01, D02 [2-9]; D03 [7-9]; D05, D09a [1-9, 11]; D07 [1]; D10b [5, 7-9, 11]; D10c [1-4, 6]; D12a [1-9]
Medical advice M05a [7-9]

19   azoxystrobin + chlorothalonil

A preventative and systemic fungicide mixture for cereals
FRAC mode of action code: 11 + M5

See also chlorothalonil

Products

1	Amistar Opti	Syngenta	100:500 g/l	SC	14582
2	Curator	Syngenta	80:400 g/l	SC	14955
3	Olympus	Syngenta	80:400 g/l	SC	13797

Uses
- Alternaria in **potatoes** *(off-label)* [3]
- Botrytis in **asparagus** *(qualified minor use recommendation)*, **bulb onions**, **garlic**, **shallots** [3]
- Brown rust in **spring barley**, **spring wheat**, **winter barley**, **winter wheat** [1, 2]
- Disease control in **poppies for morphine production** *(off-label)* [1]
- Downy mildew in **bulb onions**, **garlic**, **shallots** [3]

FOR FULL CONDITIONS OF USE ALWAYS READ THE PRODUCT LABEL

azoxystrobin + chlorothalonil

- Foliar disease control in **durum wheat** *(off-label)*, **spring rye** *(off-label)*, **triticale** *(off-label)*, **winter rye** *(off-label)* [1]
- Glume blotch in **spring wheat**, **winter wheat** [1, 2]
- Net blotch in **spring barley**, **winter barley** [1, 2]
- Rhynchosporium in **spring barley**, **winter barley** [2]; **spring barley** *(moderate control)*, **winter barley** *(moderate control)* [1]
- Rust in **asparagus** *(moderate control only)* [3]
- Septoria leaf blotch in **spring wheat**, **winter wheat** [1, 2]
- Stemphylium in **asparagus** *(qualified minor use recommendation)* [3]
- Take-all in **spring barley** *(reduction)*, **spring wheat** *(reduction)*, **winter barley** *(reduction)*, **winter wheat** *(reduction)* [1]; **spring barley** *(reduction only)*, **spring wheat** *(reduction only)*, **winter barley** *(reduction only)*, **winter wheat** *(reduction only)* [2]
- Yellow rust in **spring barley**, **winter barley** [2]; **spring wheat**, **winter wheat** [1, 2]

Specific Off-Label Approvals (SOLAs)
- **durum wheat** *20091987 expires 31 Dec 2011* [1]
- **poppies for morphine production** *20101246 expires 31 Dec 2011* [1]
- **potatoes** *20090711 expires 31 Dec 2011* [3]
- **spring rye** *20091987 expires 31 Dec 2011* [1]
- **triticale** *20091987 expires 31 Dec 2011* [1]
- **winter rye** *20091987 expires 31 Dec 2011* [1]

Approval information
- Azoxystrobin and chlorothalonil included in Annex I under EC Directive 91/414
- Accepted by BBPA for use on malting barley
- Approval expiry 31 Dec 2011 [1-3]

Efficacy guidance
- Best results obtained from applications made as a protectant treatment or in earliest stages of disease development. Further applications may be needed if disease attack is prolonged
- Reduction of barley spotting occurs when used as part of a programme with other fungicides
- Control of *Septoria* and rust diseases may be improved by mixture with a triazole fungicide
- Azoxystrobin is a member of the QoI cross resistance group. Product should be used preventatively and not relied on for its curative potential
- Use product in cereals as part of an Integrated Crop Management strategy incorporating other methods of control, including where appropriate other fungicides with a different mode of action. Do not apply more than two foliar applications of QoI containing products to any cereal crop
- There is a significant risk of widespread resistance occurring in *Septoria tritici* populations in UK. Failure to follow resistance management action may result in reduced levels of disease control

Restrictions
- Maximum number of treatments 2 per crop
- Maximum total dose on barley equivalent to one full dose treatment
- Do not use where there is risk of spray drift onto neighbouring apple crops
- The same spray equipment should not be used to treat apples

Crop-specific information
- Latest use: before beginning of heading (GS 51) for barley; before caryopsis watery ripe (GS 71) for wheat

Environmental safety
- Dangerous for the environment
- Very toxic to aquatic organisms
- LERAP Category B

Hazard classification and safety precautions
  Hazard Toxic [1]; Harmful [2, 3]; Dangerous for the environment [1-3]
  Transport code 9
  Packaging group III
  UN number 3082
  Risk phrases R20 [2, 3]; R22a, R23 [1]; R37, R40, R41, R43, R50, R53a [1-3]
  Operator protection A, C, H; U02a, U05a, U09a, U11, U14, U15, U19a [1-3]; U20a [2]; U20b [1, 3]
  Environmental protection E15a [1, 3]; E15b [2]; E16a, E34, E38 [1-3]

SEE SECTION 3 FOR PRODUCTS ALSO REGISTERED

Pesticide Profiles

Storage and disposal D01, D02, D09a, D10c, D12a [1-3]; D05 [1, 3]
Medical advice M04a [1, 3]

20   azoxystrobin + cyproconazole

A contact and systemic broad spectrum fungicide mixture for cereals and oilseed rape
FRAC mode of action code: 11 + 3

See also cyproconazole

Products
Priori Xtra   Syngenta   200:80 g/l   SC   11518

Uses
- Alternaria in **spring oilseed rape**, **winter oilseed rape**
- Brown rust in **spring barley**, **spring rye**, **spring wheat**, **winter barley**, **winter rye**, **winter wheat**
- Cercospora leaf spot in **fodder beet**, **sugar beet**
- Crown rust in **spring oats**, **winter oats**
- Eyespot in **spring barley** *(reduction)*, **spring wheat** *(reduction)*, **winter barley** *(reduction)*, **winter wheat** *(reduction)*
- Foliar disease control in **durum wheat** *(off-label)*, **grass seed crops** *(off-label)*, **triticale** *(off-label)*
- Glume blotch in **spring wheat**, **winter wheat**
- Net blotch in **spring barley**, **winter barley**
- Powdery mildew in **fodder beet**, **spring barley**, **spring oats**, **spring rye**, **spring wheat**, **sugar beet**, **winter barley**, **winter oats**, **winter rye**, **winter wheat**
- Ramularia leaf spots in **fodder beet**, **sugar beet**
- Rhynchosporium in **spring barley** *(moderate control)*, **spring rye** *(moderate control)*, **winter barley** *(moderate control)*, **winter rye** *(moderate control)*
- Rust in **fodder beet**, **sugar beet**
- Sclerotinia stem rot in **spring oilseed rape**, **winter oilseed rape**
- Septoria leaf blotch in **spring wheat**, **winter wheat**
- Take-all in **spring barley** *(reduction)*, **spring wheat** *(reduction)*, **winter barley** *(reduction)*, **winter wheat** *(reduction)*
- Yellow rust in **spring wheat**, **winter wheat**

Specific Off-Label Approvals (SOLAs)
- *durum wheat* 20052891
- *grass seed crops* 20052891
- *triticale* 20052891

Approval information
- Azoxystrobin included in Annex I under EC Directive 91/414
- Accepted by BBPA for use on malting barley

Efficacy guidance
- Best results obtained from treatment during the early stages of disease development
- A second application may be needed if disease attack is prolonged
- Azoxystrobin is a member of the QoI cross resistance group. Product should be used preventatively and not relied on for its curative potential
- Use product as part of an Integrated Crop Management strategy incorporating other methods of control, including where appropriate other fungicides with a different mode of action. Do not apply more than two foliar applications of QoI containing products to any cereal crop
- There is a significant risk of widespread resistance occurring in *Septoria tritici* populations in UK. Failure to follow resistance management action may result in reduced levels of disease control
- Strains of wheat and barley powdery mildew resistant to QoIs are common in the UK. Control of wheat powdery mildew can only be relied upon from the triazole component
- Where specific control of wheat mildew is required this should be achieved through a programme of measures including products recommended for the control of mildew that contain a fungicide from a different cross-resistance group and applied at a dose that will give robust control

FOR FULL CONDITIONS OF USE ALWAYS READ THE PRODUCT LABEL

azoxystrobin + difenoconazole

- Cyproconazole is a DMI fungicide. Resistance to some DMI fungicides has been identified in Septoria leaf blotch which may seriously affect performance of some products. For further advice contact a specialist advisor and visit the Fungicide Resistance Action Group (FRAG)-UK website

Restrictions
- Maximum total dose equivalent to two full dose treatments
- Do not use where there is risk of spray drift onto neighbouring apple crops
- The same spray equipment should not be used to treat apples

Crop-specific information
- Latest use: up to and including anthesis complete (GS 69) for rye and wheat; up to and including emergence of ear complete (GS 59) for barley and oats; BBCH79 (nearly all pods at final size) or 30 days before harvest, whichever is sooner for oilseed rape

Environmental safety
- Dangerous for the environment
- Very toxic to aquatic organisms

Hazard classification and safety precautions
Hazard Harmful, Dangerous for the environment
Transport code 9
Packaging group III
UN number 3082
Risk phrases R22a, R50, R53a, R63
Operator protection A; U05a, U09a, U19a, U20b
Environmental protection E15b, E34, E38
Storage and disposal D01, D02, D05, D09a, D10c, D12a
Medical advice M03

21 azoxystrobin + difenoconazole

A broad spectrum fungicide mixture for field crops
FRAC mode of action code: 11 + 3

See also difenoconazole

Products
Amistar Top	Syngenta	200:125 g/l	SC	12761

Uses
- Alternaria blight in **carrots**, **horseradish** *(off-label)*, **parsley root** *(off-label)*, **parsnips** *(off-label)*, **salsify** *(off-label)*
- Powdery mildew in **broccoli**, **brussels sprouts**, **cabbages**, **calabrese**, **carrots**, **cauliflowers**, **collards**, **horseradish** *(off-label)*, **kale**, **parsley root** *(off-label)*, **parsnips** *(off-label)*, **salsify** *(off-label)*
- Purple blotch in **leeks** *(moderate control)*
- Rust in **asparagus** *(off-label)*, **leeks**
- Stemphylium in **asparagus** *(off-label)*
- White blister in **broccoli**, **brussels sprouts**, **cabbages**, **calabrese**, **cauliflowers**, **collards**, **kale**
- White tip in **leeks** *(qualified minor use)*

Specific Off-Label Approvals (SOLAs)
- *asparagus* 20070831
- *horseradish* 20061476
- *parsley root* 20061476
- *parsnips* 20061476
- *salsify* 20061476

Approval information
- Azoxystrobin and difenoconazole included in Annex I under EC Directive 91/414

Efficacy guidance
- Best results obtained from applications made in the earliest stages of disease development or as a protectant treatment following a disease risk assessment
- Ensure the crop is free from any stress caused by environmental or agronomic effects

SEE SECTION 3 FOR PRODUCTS ALSO REGISTERED

Pesticide Profiles

- Azoxystrobin is a member of the QoI cross resistance group. Product should be used preventatively and not relied on for its curative potential
- Use as part of an Integrated Crop Management strategy incorporating other methods of control, including where appropriate other fungicides with a different mode of action. Do not apply more than two foliar applications of QoI containing products

Restrictions
- Maximum number of treatments 2 per crop
- Do not apply where there is a risk of spray drift onto neighbouring apple crops
- Consult processors before treating a crop destined for processing

Crop-specific information
- HI 14 d for carrots; 21 d for brassicas, leeks
- Minimum spray interval of 14 d must be observed on brassicas

Environmental safety
- Dangerous for the environment
- Very toxic to aquatic organisms

Hazard classification and safety precautions
Hazard Irritant, Dangerous for the environment
Transport code 9
Packaging group III
UN number 3082
Risk phrases R20, R43, R50, R53a
Operator protection A, H; U05a, U09a, U19a, U20b
Environmental protection E15b, E38
Storage and disposal D01, D02, D05, D09a, D10c, D12a

22 azoxystrobin + fenpropimorph

A protectant and eradicant fungicide mixture for cereals
FRAC mode of action code: 11 + 5

See also fenpropimorph

Products
1	Amistar Pro	Syngenta	100:280 g/l	SE	10513
2	Aspect	Syngenta	100:280 g/l	SE	10516

Uses
- Brown rust in *spring barley, spring rye, spring wheat, triticale, winter barley, winter rye, winter wheat* [1, 2]
- Crown rust in *spring oats, winter oats* [1, 2]
- Foliar disease control in *durum wheat* (off-label), *grass seed crops* (off-label) [1]
- Glume blotch in *spring wheat, winter wheat* [1, 2]
- Late ear diseases in *spring wheat* (moderate control), *winter wheat* (moderate control) [1, 2]
- Net blotch in *spring barley, winter barley* [1, 2]
- Powdery mildew in *spring barley, spring oats, spring rye, triticale, winter barley, winter oats, winter rye* [1, 2]
- Rhynchosporium in *spring barley, spring rye, triticale* (reduction), *winter barley, winter rye* [1, 2]
- Septoria leaf blotch in *spring wheat, winter wheat* [1, 2]
- Take-all in *spring barley* (reduction), *spring rye* (reduction), *spring wheat* (reduction), *triticale* (reduction), *winter barley* (reduction), *winter rye* (reduction), *winter wheat* (reduction) [1, 2]
- Yellow rust in *spring wheat, winter wheat* [1, 2]

Specific Off-Label Approvals (SOLAs)
- *durum wheat* 20060962 [1]
- *grass seed crops* 20060962 [1]

Approval information
- Azoxystrobin and fenpropimorph included in Annex I under EC Directive 91/414
- Accepted by BBPA for use on malting barley

FOR FULL CONDITIONS OF USE ALWAYS READ THE PRODUCT LABEL

azoxystrobin + propiconazole

Efficacy guidance
- Best results obtained from application before infection following a disease risk assessment, or when disease first seen in crop
- Results may be less reliable when used on crops under stress
- Treatments for protection against ear disease should be made at ear emergence
- Azoxystrobin is a member of the QoI cross resistance group. Product should be used preventatively and not relied on for its curative potential
- Use product as part of an Integrated Crop Management strategy incorporating other methods of control, including where appropriate other fungicides with a different mode of action. Do not apply more than two foliar applications of QoI containing products to any cereal crop
- There is a significant risk of widespread resistance occurring in *Septoria tritici* populations in UK. Failure to follow resistance management action may result in reduced levels of disease control
- Strains of barley powdery mildew resistant to QoIs are common in the UK
- In wheat product must always be used in mixture with another product, recommended for control of the same target disease, that contains a fungicide from a different cross resistance group and is applied at a dose that will give robust control

Restrictions
- Maximum total dose equivalent to 2 full dose treatments
- Do not use where there is risk of spray drift onto neighbouring apple crops
- The same spray equipment should not be used to treat apples

Crop-specific information
- Latest use: before early milk stage (GS 73) for wheat, barley; up to and including grain watery ripe stage (GS 71) for oats, rye, triticale
- HI 5 wk

Environmental safety
- Dangerous for the environment
- Very toxic to aquatic organisms

Hazard classification and safety precautions
Hazard Harmful, Dangerous for the environment
Transport code 9
Packaging group III
UN number 3082
Risk phrases R20, R38, R43, R50, R53a, R63
Operator protection A, H; U02a, U05a, U09a, U14, U15, U19a, U20b
Environmental protection E15a, E38
Storage and disposal D01, D02, D05, D09a, D10c, D12a
Medical advice M03

23   azoxystrobin + propiconazole

A contact and systemic broad spectrum fungicide mixture for use on grass
FRAC mode of action code: 11 + 3

Products
Headway	Syngenta	62.5:104 g/l	EC	14396

Uses
- Anthracnose in **amenity grassland** *(moderate control)*, **managed amenity turf** *(moderate control)*
- Dollar spot in **amenity grassland**, **managed amenity turf**
- Fusarium diseases in **amenity grassland**, **managed amenity turf**
- Microdochium nivale in **amenity grassland**, **managed amenity turf**

Approval information
- Azoxystrobin and propiconazole included in Annex I under EC Directive 91/414

Environmental safety
- LERAP Category B

Hazard classification and safety precautions
Hazard Dangerous for the environment

SEE SECTION 3 FOR PRODUCTS ALSO REGISTERED

Pesticide Profiles

Transport code 9
Packaging group III
UN number 3082
Risk phrases R50, R53a
Operator protection A; U05a, U09a, U19a, U20a
Environmental protection E15b, E16a, E16b, E34, E38
Storage and disposal D01, D02, D05, D09a, D10c, D12a

24   Bacillus subtilis

A bacterial fungicide for the control of Botrytis cinerea
FRAC mode of action code: F6

Products

Serenade ASO	Fargro	13.96 g/l	SC	14318

Uses
- Botrytis in **bilberries** (off-label), **blackcurrants** (off-label), **blueberries** (off-label), **bulb vegetables** (off-label), **canary grass** (off-label), **cane fruit** (off-label), **cranberries** (off-label), **figs** (off-label), **fruiting vegetables** (off-label), **gooseberries** (off-label), **herbs (see appendix 6)** (off-label), **hops** (off-label), **leafy vegetables** (off-label), **legumes** (off-label), **ornamental plant production** (off-label), **redcurrants** (off-label), **ribes hybrids** (off-label), **root & tuber crops** (off-label), **stem vegetables** (off-label), **strawberries** (off-label), **table grapes** (off-label), **top fruit** (off-label), **vegetable brassicas** (off-label), **wine grapes** (off-label)
- Botrytis fruit rot in **protected strawberries** (reduction of damage to fruit)
- Grey mould in **protected strawberries** (reduction of damage to fruit)

Specific Off-Label Approvals (SOLAs)
- *bilberries* 20090246
- *blackcurrants* 20090246
- *blueberries* 20090246
- *bulb vegetables* 20090246
- *canary grass* 20090246
- *cane fruit* 20090246
- *cranberries* 20090246
- *figs* 20090246
- *fruiting vegetables* 20090246
- *gooseberries* 20090246
- *herbs (see appendix 6)* 20090246
- *hops* 20090246
- *leafy vegetables* 20090246
- *legumes* 20090246
- *ornamental plant production* 20090246
- *redcurrants* 20090246
- *ribes hybrids* 20090246
- *root & tuber crops* 20090246
- *stem vegetables* 20090246
- *strawberries* 20090246
- *table grapes* 20090246
- *top fruit* 20090246
- *vegetable brassicas* 20090246
- *wine grapes* 20090246

Approval information
- Bacillus subtilis included in Annex I under EC Directive 91/414

Efficacy guidance
- Alternating applications with fungicides using a different mode of action is recommended for resistance management.
- Do not apply using irrigation equipment.
- For maximum effectiveness, start applications before disease development.
- Apply in a minimum water volume of 400 l/ha.

FOR FULL CONDITIONS OF USE ALWAYS READ THE PRODUCT LABEL

Bacillus thuringiensis

Restrictions
- Consult processor before using on crops grown for processing

Hazard classification and safety precautions
Hazard Irritant
Risk phrases R43
Operator protection A, H; U05a, U11, U14, U20b
Environmental protection E15b
Storage and disposal D01, D02, D05, D10a, D16

25 Bacillus thuringiensis

A bacterial insecticide for control of caterpillars
IRAC mode of action code: 11

Products
Dipel DF   Interfarm   32000 IU/mg   WG   14119

Uses
- Caterpillars in **amenity vegetation**, **apples** (off-label), **blackberries** (off-label), **borage** (off-label), **broad beans** (off-label), **broccoli**, **brussels sprouts**, **cabbages**, **calabrese** (off-label), **cauliflowers**, **celery (outdoor)** (off-label), **cherries** (off-label), **chinese cabbage** (off-label), **chives** (off-label), **choi sum** (off-label), **collards** (off-label), **crab apples** (off-label), **dwarf beans** (off-label), **french beans** (off-label), **frise** (off-label), **herbs (see appendix 6)** (off-label), **kale** (off-label), **leaf spinach** (off-label), **lettuce** (off-label), **ornamental plant production**, **pak choi** (off-label), **parsley** (off-label), **pears** (off-label), **protected aubergines** (off-label), **protected broccoli** (off-label), **protected brussels sprouts** (off-label), **protected cabbages** (off-label), **protected calabrese** (off-label), **protected cauliflowers** (off-label), **protected cayenne peppers** (off-label), **protected celery** (off-label), **protected chinese cabbage** (off-label), **protected chives** (off-label), **protected choi sum** (off-label), **protected collards** (off-label), **protected cucumbers**, **protected frise** (off-label), **protected herbs (see appendix 6)** (off-label), **protected kale** (off-label), **protected lettuce** (off-label), **protected ornamentals**, **protected parsley** (off-label), **protected peppers**, **protected rhubarb** (off-label), **protected salad brassicas** (off-label), **protected scarole** (off-label), **protected spinach** (off-label), **protected tomatoes**, **protected watercress** (off-label), **radicchio** (off-label), **raspberries**, **rhubarb** (off-label), **rubus hybrids** (off-label), **runner beans** (off-label), **salad brassicas** (off-label), **scarole** (off-label), **spinach** (off-label), **spring cabbage** (off-label), **strawberries**, **sweetcorn** (off-label), **watercress** (off-label)
- Cutworms in **bulb onions** (off-label), **carrots** (off-label), **garlic** (off-label), **horseradish** (off-label), **leeks** (off-label), **parsley root** (off-label), **parsnips** (off-label), **red beet** (off-label), **salsify** (off-label), **shallots** (off-label)
- Silver Y moth in **broad beans** (off-label), **dwarf beans** (off-label), **french beans** (off-label), **runner beans** (off-label), **sweetcorn** (off-label)
- Winter moth in **bilberries** (off-label), **blackcurrants** (off-label), **blueberries** (off-label), **cranberries** (off-label), **gooseberries** (off-label), **redcurrants** (off-label), **vaccinium spp.** (off-label), **whitecurrants** (off-label)

Specific Off-Label Approvals (SOLAs)
- *apples* 20091070
- *bilberries* 20091069
- *blackberries* 20091058
- *blackcurrants* 20091069
- *blueberries* 20091069
- *borage* 20091070
- *broad beans* 20091067
- *bulb onions* 20091065
- *calabrese* 20091064
- *carrots* 20091068
- *celery (outdoor)* 20091064
- *cherries* 20091070
- *chinese cabbage* 20091064
- *chives* 20091070

SEE SECTION 3 FOR PRODUCTS ALSO REGISTERED

Pesticide Profiles

- *choi sum* 20091064
- *collards* 20091064
- *crab apples* 20091070
- *cranberries* 20091069
- *dwarf beans* 20091067
- *french beans* 20091067
- *frise* 20091070
- *garlic* 20091065
- *gooseberries* 20091069
- *herbs (see appendix 6)* 20091070
- *horseradish* 20091068
- *kale* 20091064
- *leaf spinach* 20091070
- *leeks* 20091062
- *lettuce* 20091070
- *pak choi* 20091064
- *parsley* 20091070
- *parsley root* 20091068
- *parsnips* 20091068
- *pears* 20091070
- *protected aubergines* 20091059
- *protected broccoli* 20091064
- *protected brussels sprouts* 20091064
- *protected cabbages* 20091064
- *protected calabrese* 20091064
- *protected cauliflowers* 20091064
- *protected cayenne peppers* 20091060
- *protected celery* 20091064
- *protected chinese cabbage* 20091064
- *protected chives* 20091070
- *protected choi sum* 20091064
- *protected collards* 20091064
- *protected frise* 20091070
- *protected herbs (see appendix 6)* 20091070
- *protected kale* 20091064
- *protected lettuce* 20091070
- *protected parsley* 20091070
- *protected rhubarb* 20091064
- *protected salad brassicas* 20091070
- *protected scarole* 20091070
- *protected spinach* 20091070
- *protected watercress* 20091063
- *radicchio* 20091070
- *red beet* 20091061
- *redcurrants* 20091069
- *rhubarb* 20091064
- *rubus hybrids* 20091058
- *runner beans* 20091067
- *salad brassicas* 20091070
- *salsify* 20091068
- *scarole* 20091070
- *shallots* 20091065
- *spinach* 20091070
- *spring cabbage* 20091064
- *sweetcorn* 20091057
- *vaccinium spp.* 20091069
- *watercress* 20091063
- *whitecurrants* 20091069

FOR FULL CONDITIONS OF USE ALWAYS READ THE PRODUCT LABEL

Approval information
- Bacillus thuringiensis included in Annex I under EC Directive 91/414

Efficacy guidance
- Pest control achieved by ingestion by caterpillars of the treated plant vegetation. Caterpillars cease feeding and die in 1-3 d
- Apply as soon as larvae appear on crop and repeat every 7-10 d until the end of the hatching period
- Good coverage is essential, especially of undersides of leaves. Spray onto dry foliage and do not apply if rain expected within 6 h

Restrictions
- No restriction on number of treatments on edible crops
- Apply spray mixture as soon as possible after preparation

Crop-specific information
- HI zero

Environmental safety
- Store out of direct sunlight

Hazard classification and safety precautions
UN number N/C
Operator protection A, C, D; U05a, U15, U19a, U20c
Environmental protection E15a
Storage and disposal D01, D02, D05, D09a, D11a

26 Beauveria bassiana

It is an entomopathogenic fungus causing white muscardine disease. It can be used as a biological insecticide to control a number of pests such as termites, whitefly and some beetles

Products
	Naturalis-L	Belchim	7.16% w/w	OD	14655

Uses
- Aphids in *protected edible crops, protected ornamentals*
- Beetles in *protected edible crops, protected ornamentals*
- Whitefly in *protected edible crops, protected ornamentals*

Hazard classification and safety precautions
UN number N/C
Operator protection A, H

27 benalaxyl

A phenylamide (acylalanine) fungicide available only in mixtures
FRAC mode of action code: 4

28 bentazone

A post-emergence contact benzothiadiazinone herbicide
HRAC mode of action code: C3

Products
1	Basagran SG	BASF	87% w/w	SG	08360
2	Benta 480 SL	Sharda	480 g/l	SL	14940
3	Bentazone 480	Goldengrass	480 g/l	SL	14423
4	Mac-Bentazone 480 SL	AgChem Access	480 g/l	SL	13598
5	Troy 480	AgriChem BV	480 g/l	SL	12341
6	UPL B Zone	United Phosphorus	480 g/l	SL	14808
7	Zone 48	AgriGuard	480 g/l	SL	13165

Uses
- Annual dicotyledons in **broad beans, linseed, narcissi, potatoes, runner beans, spring field beans, winter field beans** [1-7]; **bulb onions** *(off-label)*, **garlic** *(off-label)*, **leeks** *(off-label)*,

SEE SECTION 3 FOR PRODUCTS ALSO REGISTERED

Pesticide Profiles

shallots *(off-label)*, **soya beans** *(off-label)* [1]; **combining peas, dwarf beans, vining peas** [5]; **french beans** [1-4, 6, 7]; **navy beans** [1, 3-7]; **peas** [1-4, 7]
- Chickweed in **chives** *(off-label)*, **game cover** *(off-label)*, **hops** *(off-label)*, **ornamental plant production** *(off-label)*, **salad onions** *(off-label)* [1]
- Cleavers in **chives** *(off-label)*, **game cover** *(off-label)*, **hops** *(off-label)*, **ornamental plant production** *(off-label)*, **salad onions** *(off-label)* [1]
- Common storksbill in **leeks** *(off-label)* [1]
- Fool's parsley in **leeks** *(off-label)* [1]
- Groundsel in **chives** *(off-label)*, **game cover** *(off-label)*, **hops** *(off-label)*, **ornamental plant production** *(off-label)*, **salad onions** *(off-label)* [1]
- Mayweeds in **bulb onions** *(off-label)*, **chives** *(off-label)*, **game cover** *(off-label)*, **garlic** *(off-label)*, **hops** *(off-label)*, **ornamental plant production** *(off-label)*, **salad onions** *(off-label)*, **shallots** *(off-label)* [1]

Specific Off-Label Approvals (SOLAs)
- **bulb onions** *20061631 expires 31 Jul 2011* [1]
- **chives** *20090793* [1]
- **game cover** *20082819 expires 31 Jul 2011* [1]
- **garlic** *20061631 expires 31 Jul 2011* [1]
- **hops** *20082819 expires 31 Jul 2011* [1]
- **leeks** *20061630 expires 31 Jul 2011* [1]
- **ornamental plant production** *20082819 expires 31 Jul 2011* [1]
- **salad onions** *20090793* [1]
- **shallots** *20061631 expires 31 Jul 2011* [1]
- **soya beans** *20061629 expires 31 Jul 2011* [1]

Approval information
- Bentazone included in Annex I under EC Directive 91/414
- Approval expiry 31 Jul 2011 [1-7]

Efficacy guidance
- Most effective control obtained when weeds are growing actively and less than 5 cm high or across. Good spray cover is essential
- The addition of specified adjuvant oils is recommended for use on some crops to improve fat hen control. Do not use under hot or humid conditions. See label for details
- Split dose application may be made in all recommended crops except peas and generally gives better weed control. See label for details

Restrictions
- Maximum number of treatments normally 2 per crop but check label
- Crops must be treated at correct stage of growth to avoid danger of scorch. See label for details
- Not all varieties of recommended crops are fully tolerant. Use only on tolerant varieties named in label. Do not use on forage or mange-tout varieties of peas
- Do not use on crops which have been affected by drought, waterlogging, frost or other stress conditions
- Do not apply insecticides within 7 d of treatment
- Leave 14 d after using a post-emergence grass herbicide and carry out a leaf wax test where relevant or 7 d where treatment precedes the grass herbicide
- Do not spray at temperatures above 21 °C. Delay spraying until evening if necessary
- Do not apply if rain or frost expected, if unseasonably cold, if foliage wet or in drought
- A minimum of 6 h (preferably 12 h) free from rain is required after application
- May be used on selected varieties of maincrop and second early potatoes (see label for details), not on seed crops or first earlies

Crop-specific information
- Latest use: before shoots exceed 15 cm high for potatoes and spring field beans (or 6-7 leaf pairs); 4 leaf pairs (6 pairs or 15 cm high with split dose) for broad beans; before flower buds visible for French, navy, runner and winter field beans, and linseed; before flower buds can be found enclosed in terminal shoot for peas
- Best results in narcissi obtained by using a suitable pre-emergence herbicide first
- Consult processor before using on crops for processing
- A satisfactory wax test must be carried out before use on peas
- Do not treat narcissi during flower bud initiation

FOR FULL CONDITIONS OF USE ALWAYS READ THE PRODUCT LABEL

benthiavalicarb-isopropyl

Environmental safety
- Dangerous for the environment
- Harmful to aquatic organisms

Hazard classification and safety precautions
Hazard Harmful [1-4, 6, 7]; Irritant [2, 5]; Dangerous for the environment [7]
Packaging group III [1]
UN number 2588 [1]; N/C [2-7]
Risk phrases R22a [1-4, 6, 7]; R36 [5, 7]; R41 [1, 4]; R43 [1-7]; R52, R53a [1, 3, 4, 6]
Operator protection A [1-7]; C [1, 3-5, 7]; H [5, 6]; U05a, U08, U19a [1-7]; U11 [1, 3-6]; U14 [1, 3, 4, 6]; U20a [2]; U20b [1, 3-7]
Environmental protection E15a [2]; E15b, E34 [1, 3-7]
Storage and disposal D01, D02, D09a [1-7]; D05, D07 [7]; D10b [1, 3, 4, 6]; D10c [2, 5, 7]
Medical advice M03 [1, 3-7]; M05a [1, 3, 4, 6]

29 benthiavalicarb-isopropyl

An amino acid amide carbamate fungicide available only in mixtures
FRAC mode of action code: 40

30 benthiavalicarb-isopropyl + mancozeb

A fungicide mixture for potatoes
FRAC mode of action code: 40 + M3

See also mancozeb

Products
Valbon	Certis	1.75:70% w/w	WG	14868

Uses
- Blight in **potatoes**
- Downy mildew in **amenity vegetation** *(off-label)*, **bulb onions** *(off-label)*, **garlic** *(off-label)*, **ornamental plant production** *(off-label)*, **shallots** *(off-label)*, **table grapes** *(off-label)*, **wine grapes** *(off-label)*

Specific Off-Label Approvals (SOLAs)
- *amenity vegetation 20101513*
- *bulb onions 20101518*
- *garlic 20101518*
- *ornamental plant production 20101513*
- *shallots 20101518*
- *table grapes 20101973*
- *wine grapes 20101973*

Approval information
- Benthiavalicarb-isopropyl and mancozeb included in Annex I under EC Directive 91/414

Efficacy guidance
- Apply as a protectant spray commencing before blight enters the crop irrespective of growth stage following a blight warning or where there is a local source of infection
- In the absence of a warning commence the spray programme before the foliage meets in the rows
- Repeat treatment every 7-10 d depending on disease pressure
- Increase water volume as necessary to ensure thorough coverage of the plant
- Use of air assisted sprayers or drop legs may help to improve coverage

Restrictions
- Maximum number of treatments 6 per crop
- Do not use more than three consecutive treatments. Include products with a different mode of action in the programme

Crop-specific information
- HI 7 d for potatoes

SEE SECTION 3 FOR PRODUCTS ALSO REGISTERED

Environmental safety
- Very toxic to aquatic organisms
- Risk to non-target insects or other arthropods. Avoid spraying within 6 m of field boundary
- LERAP Category B

Hazard classification and safety precautions
Hazard Irritant
Transport code 9
Packaging group III
UN number 3077
Risk phrases R40, R43, R50, R53a
Operator protection A, D, H; U14
Environmental protection E15a, E16a, E22c
Storage and disposal D01, D02, D05, D09a, D11a, D12b

31 benzoic acid

An organic horticultural disinfectant

Products

Menno Florades	Fargro	90 g/kg	SL	13985

Uses
- Bacteria in *all non-edible crops (outdoor)*
- Fungal spores in *all non-edible crops (outdoor)*
- Viroids in *all non-edible crops (outdoor)*
- Virus in *all non-edible crops (outdoor)*

Hazard classification and safety precautions
Hazard Irritant, Flammable
Transport code 3
Packaging group III
UN number 1987
Risk phrases R41, R67
Operator protection A, C, H; U05a, U11, U13, U14, U15
Environmental protection E15a
Storage and disposal D01, D02, D09a, D10a
Medical advice M05a

32 beta-cyfluthrin

A non-systemic pyrethroid insecticide available only in mixtures
IRAC mode of action code: 3

33 beta-cyfluthrin + clothianidin

An insecticidal seed treatment mixture for beet
IRAC mode of action code: 3 + 4A

See also clothianidin

Products

1	Modesto	Bayer CropScience	80:400 g/l	FS	14029
2	Poncho Beta	Bayer CropScience	53:400 g/l	FS	12076

Uses
- Aphids in *winter oilseed rape* [1]
- Beet virus yellows vectors in *fodder beet* (seed treatment), *sugar beet* (seed treatment) [2]
- Cabbage root fly in *winter oilseed rape* (reduction of feeding activity only) [1]
- Cabbage stem flea beetle in *winter oilseed rape* [1]
- Flea beetle in *poppies for morphine production* (off-label) [1]
- Turnip sawfly larvae in *winter oilseed rape* [1]

FOR FULL CONDITIONS OF USE ALWAYS READ THE PRODUCT LABEL

beta-cyfluthrin + imidacloprid

Specific Off-Label Approvals (SOLAs)
- *poppies for morphine production* 20090479 [1]

Approval information
- Beta-cyfluthrin and clothianidin included in Annex I under EC Directive 91/414

Efficacy guidance
- In addition to control of aphid virus vectors, product improves crop establishment by reducing damage caused by symphylids, springtails, millipedes, wireworms and leatherjackets
- Additional control measures should be taken where very high populations of soil pests are present
- Product reduces direct feeding damage by foliar pests such as pygmy mangold beetle, beet flea beetle, mangold fly
- Product is not active against nematodes
- Treatment does not alter physical characteristics of pelleted seed and no change to standard drill settings should be necessary

Restrictions
- Maximum number of treatments: 1 per seed batch
- Product must be co-applied with a colouring dye
- Product must only be applied as a coating to pelleted seed using special treatment machinery

Crop-specific information
- Latest use: pre-drilling

Environmental safety
- Extremely dangerous to fish or other aquatic life. Do not contaminate surface waters or ditches with chemical or used container

Hazard classification and safety precautions
    Hazard Harmful [1, 2]; Dangerous for the environment [1]
    Transport code 9
    Packaging group III
    UN number 3082
    Risk phrases R22a [1, 2]; R50, R53a [1]
    Operator protection A, H; U04a, U05a, U07, U13, U14 [1, 2]; U20b, U24 [2]; U20c [1]
    Environmental protection E03, E15a, E38 [1]; E13a [2]; E34, E36a [1, 2]
    Storage and disposal D01, D02, D09a, D14 [1, 2]; D05 [2]; D12a [1]
    Treated seed S02, S04b, S07 [1, 2]; S03, S04a, S05, S06a, S06b [2]
    Medical advice M03

34 beta-cyfluthrin + imidacloprid

An insecticide mixture for seed treatment of oilseed rape
IRAC mode of action code: 3 + 4A

See also imidacloprid

Products

1	Chinook Blue	Bayer CropScience	100:100 g/l	LS	11262
2	Chinook Colourless	Bayer CropScience	100:100 g/l	LS	11206

Uses
- Cabbage stem flea beetle in **evening primrose** (off-label - seed treatment), **honesty** (off-label - seed treatment), **linseed** (off-label - seed treatment), **mustard** (off-label - seed treatment), **winter oilseed rape** (seed treatment) [1, 2]; **spring oilseed rape** (seed treatment) [1]
- Flea beetle in **evening primrose** (off-label - seed treatment), **honesty** (off-label - seed treatment), **linseed** (off-label - seed treatment), **mustard** (off-label - seed treatment) [1, 2]; **spring oilseed rape** (seed treatment), **winter oilseed rape** (seed treatment) [1]

Specific Off-Label Approvals (SOLAs)
- *evening primrose* (seed treatment) 20052262 [1], (seed treatment) 20052263 [2]
- *honesty* (seed treatment) 20052262 [1], (seed treatment) 20052263 [2]
- *linseed* (seed treatment) 20052262 [1], (seed treatment) 20052263 [2]
- *mustard* (seed treatment) 20052262 [1], (seed treatment) 20052263 [2]

SEE SECTION 3 FOR PRODUCTS ALSO REGISTERED

Pesticide Profiles

Approval information
- Beta-cyfluthrin and imidacloprid included in Annex I under EC Directive 91/414

Efficacy guidance
- Drill treated seed as soon as possible after treatment
- Product will reduce damage by early attacks of flea beetle and may also reduce subsequent larval damage. However a follow-up spray treatment may be required if pest activity is heavy and prolonged

Restrictions
- Maximum number of treatments 1 per batch of seed
- Do not use on seed with more than 9% moisture content, on sprouted seed or on cracked, split or otherwise damaged seed
- Product must be co-applied with a colouring agent [2]
- Must be applied using seed treatment machine recommended by manufacturer
- Allow treated seed to dry before packaging. Drying can be assisted by subsequent application of talc
- Product can cause a transient tingling or numbing sensation to exposed skin. Avoid skin contact with the product, treated seed or dust throughout all operations in the treatment plant and during drilling

Crop-specific information
- Latest use: before drilling

Environmental safety
- Dangerous for the environment
- Toxic to aquatic organisms
- Extremely dangerous to fish or other aquatic life. Do not contaminate surface waters or ditches with chemical or used container
- Do not use treated seed as food or feed
- Dangerous to birds, game and other wildlife. Treated seed should not be left on the soil surface. Bury spillages
- Do not broadcast treated seed

Hazard classification and safety precautions
    Hazard Harmful, Dangerous for the environment
    Transport code 6.1
    Packaging group III
    UN number 3352
    Risk phrases R22a, R51, R53a
    Operator protection A, H; U04a, U05a, U07, U20b
    Environmental protection E13a, E34, E36a, E38
    Storage and disposal D01, D02, D05, D09a, D12a, D14
    Treated seed S01, S02, S03, S04c, S05, S06a, S06b, S07, S08
    Medical advice M03

35 bifenazate

A bifenazate acaricide for use on protected strawberries
IRAC mode of action code: 25

Products
    Floramite 240 SC   Certis   240 g/l   SC   13686

Uses
- Spider mites in *protected forest nurseries* (off-label), *protected hops* (off-label)
- Two-spotted spider mite in *protected strawberries*

Specific Off-Label Approvals (SOLAs)
- *protected forest nurseries* 20082851
- *protected hops* 20082851

Approval information
- Bifenazate included in Annex I under EC Directive 91/414

FOR FULL CONDITIONS OF USE ALWAYS READ THE PRODUCT LABEL

Efficacy guidance
- Bifenazate acts by contact knockdown after approximately 4 d and a subsequent period of residual control
- All mobile stages of mites are controlled with occasionally some ovicidal activity
- Best results obtained from a programme of two treatments started as soon as first spider mites are seen
- To minimise possible development of resistance use in a planned Resistance Management strategy. Alternate at least two products with different modes of action between treatment programmes of bifenazate. See Section 5 for more information

Restrictions
- Maximum number of treatments 2 per yr for protected strawberries
- After use in protected environments avoid re-entry for at least 2 h while full ventilation is carried out allowing spray to dry on leaves
- Personnel working among treated crops should wear suitable long-sleeved garments and gloves for 14 d after treatment
- Do not apply as a low volume spray

Crop-specific information
- HI: 7 d for protected strawberries
- Carry out small scale tolerance test on the strawberry cultivar before large scale use

Environmental safety
- Dangerous for the environment
- Toxic to aquatic organisms
- Harmful to non-target predatory mites. Avoid spray drift onto field margins, hedges, ditches, surface water and neighbouring crops

Hazard classification and safety precautions
Hazard Irritant, Dangerous for the environment
Transport code 9
Packaging group III
UN number 3082
Risk phrases R43, R51, R53a
Operator protection A, H; U02a, U04a, U05a, U10, U14, U15, U19a, U20b
Environmental protection E15b, E38
Storage and disposal D01, D02, D05, D09a, D10b, D12b
Medical advice M03

36 bifenox

A diphenyl ether herbicide for use in cereals and (off-label) in oilseed rape.
HRAC mode of action code: E

Products

1	Aspire	AgriGuard	480 g/l	SC	14303
2	Clayton Belstone	Clayton	480 g/l	SC	14033
3	Fox	Makhteshim	480 g/l	SC	11981
4	Oram	Makhteshim	480 g/l	SC	13410
5	Sabine	Makhteshim	480 g/l	SC	13439

Uses
- Annual dicotyledons in **canary flower (echium spp.)** (off-label), **evening primrose** (off-label), **honesty** (off-label), **linseed** (off-label), **mustard** (off-label), **oilseed rape** (off-label) [2, 4, 5]; **durum wheat** (off-label), **grass seed crops** (off-label) [2-5]; **triticale**, **winter barley**, **winter rye**, **winter wheat** [1, 2]
- Cleavers in **triticale, winter barley, winter rye, winter wheat** [3-5]
- Field pansy in **triticale, winter barley, winter rye, winter wheat** [3-5]
- Field speedwell in **triticale, winter barley, winter rye, winter wheat** [3-5]
- Forget-me-not in **triticale, winter barley, winter rye, winter wheat** [3-5]
- Geranium species in **spring oilseed rape** (off-label), **winter oilseed rape** (off-label) [3]
- Ivy-leaved speedwell in **triticale, winter barley, winter rye, winter wheat** [3-5]

SEE SECTION 3 FOR PRODUCTS ALSO REGISTERED

Pesticide Profiles

- Poppies in *triticale, winter barley, winter rye, winter wheat* [3-5]
- Red dead-nettle in *triticale, winter barley, winter rye, winter wheat* [3-5]

Specific Off-Label Approvals (SOLAs)
- *canary flower (echium spp.)* *20081839* [2], *20073141* [4], *20073143* [5]
- *durum wheat* *20081840* [2], *20072364* [3], *20073142* [4], *20073144* [5]
- *evening primrose* *20081839* [2], *20073141* [4], *20073143* [5]
- *grass seed crops* *20081840* [2], *20072364* [3], *20073142* [4], *20073144* [5]
- *honesty* *20081839* [2], *20073141* [4], *20073143* [5]
- *linseed* *20081839* [2], *20073141* [4], *20073143* [5]
- *mustard* *20081839* [2], *20073141* [4], *20073143* [5]
- *oilseed rape* *20081839* [2], *20073141* [4], *20073143* [5]
- *spring oilseed rape* *20060321* [3]
- *winter oilseed rape* *20060321* [3]

Approval information
- Accepted by BBPA for use on malting barley
- Bifenox included in Annex I under EC Directive 91/414

Efficacy guidance
- Bifenox is absorbed by foliage and emerging roots of susceptible species
- Best results obtained when weeds are growing actively with adequate soil moisture

Restrictions
- Maximum number of treatments: one per crop
- Do not apply to crops suffering from stress from whatever cause
- Do not apply if the crop is wet or if rain or frost is expected
- Avoid drift onto broad-leaved plants outside the target area

Crop-specific information
- Latest use: before 2nd node detectable (GS 32) for all crops

Environmental safety
- Dangerous for the environment
- Very toxic to aquatic organisms
- Do not empty into drains

Hazard classification and safety precautions
Hazard Dangerous for the environment
Transport code 9
Packaging group III
UN number 3082
Risk phrases R50, R53a
Operator protection A; U05a, U08, U20b
Environmental protection E15a, E38 [1-5]; E19b, E34 [2-5]
Storage and disposal D01, D02, D09a, D12a
Medical advice M03

37 bifenthrin

A contact and residual pyrethroid acaricide/insecticide for use in agricultural and horticultural crops
IRAC mode of action code: 3

Products

1	Brigade 80 SC	Belchim	80 g/l	SC	12853
2	Starion Flo	Belchim	80 g/l	SC	12455
3	Talstar 80 Flo	Belchim	80 g/l	SC	12352

Uses
- Aphids in *broccoli, brussels sprouts, cabbages, calabrese, cauliflowers* [1-3]; *combining peas, spring barley, spring oats, spring wheat, vining peas, winter barley, winter oats, winter wheat* [1, 2]
- Cabbage seed weevil in *spring oilseed rape, winter oilseed rape* [1, 2]
- Cabbage stem flea beetle in *winter oilseed rape* [1, 2]
- Caterpillars in *broccoli, brussels sprouts, cabbages, calabrese, cauliflowers* [1-3]

FOR FULL CONDITIONS OF USE ALWAYS READ THE PRODUCT LABEL

- Damson-hop aphid in *hops* [1-3]
- Flea beetle in *linseed, spring oilseed rape, winter oilseed rape* [1, 2]
- Fruit tree red spider mite in *apples, pears* [1-3]
- Pod midge in *spring oilseed rape, winter oilseed rape* [1, 2]
- Pollen beetle in *spring oilseed rape, winter oilseed rape* [1, 2]
- Rape winter stem weevil in *winter oilseed rape* [1, 2]
- Spider mites in *blackcurrants* (off-label) [2, 3]
- Two-spotted spider mite in *blackberries* (off-label), *hops, ornamental plant production, raspberries* (off-label), *strawberries* [1-3]
- Virus vectors in *winter barley, winter oats, winter wheat* [1, 2]
- Whitefly in *broccoli, brussels sprouts, cabbages, calabrese, cauliflowers* [1-3]

Specific Off-Label Approvals (SOLAs)
- *blackberries* 20093378 expires 30 May 2011 [1], 20093379 expires 30 May 2011 [2], 20093382 expires 30 May 2011 [3]
- *blackcurrants* 20093380 expires 30 May 2011 [2], 20093381 expires 30 May 2011 [3]
- *raspberries* 20093378 expires 30 May 2011 [1], 20093379 expires 30 May 2011 [2], 20093382 expires 30 May 2011 [3]

Approval information
- Accepted by BBPA for use on malting barley and hops. Bifenthrin has not been included in Annex 1.
- Approval expiry 30 May 2011 [1-3]

Efficacy guidance
- Timing of application varies with crop and pest. See label for details
- Good spray cover of upper and lower plant surfaces essential to achieve effective pest control. Spray volumes should be increased where crop is dense

Restrictions
- Maximum number of treatments 5 per yr for hops; 4 per crop on brassicas; 2 per yr for apples, pears, strawberries. Other crops either 2 per yr or a maximum total dose is stipulated. See label
- Do not apply to a cereal crop if any product containing a pyrethroid insecticide or dimethoate has been applied after the start of ear emergence (GS 51)
- Consult processor before use on crops for processing
- Do not treat ornamental specimens in blossom or under stress

Crop-specific information
- Latest use: before 31 Mar in yr of harvest for cereals; before late milk stage (GS 77) for wheat, before early milk stage (GS 73) for barley, oats [1, 2];
- HI 2 d for brassicas, 3 d for peas; 8 wk for linseed [1, 2], oilseed rape [1, 2]

Environmental safety
- Dangerous for the environment
- Very toxic to aquatic organisms
- High risk to bees. Do not apply to crops in flower, or to those in which bees are actively foraging, except as directed on named crops. Do not apply when flowering weeds are present [1-3]
- Extremely dangerous to bees. Do not apply to crops in flower or to those in which bees are actively foraging. Do not apply when flowering weeds are present
- Extremely dangerous to fish or other aquatic life. Do not contaminate surface waters or ditches with chemical or used container
- High risk to non-target insects or other arthropods. Not compatible with IPM systems
- Do not spray cereal crops within 6 m of the field boundary or after 31 Mar in the yr of harvest
- Keep in original container, tightly closed, in a safe place, under lock and key [1-3]
- Broadcast air-assisted LERAP (30 m); LERAP Category A

Hazard classification and safety precautions
Hazard Harmful, Dangerous for the environment
Transport code 9
Packaging group III
UN number 3082
Risk phrases R20, R22a, R43, R50, R53a
Operator protection A, H; U02a, U04a, U05a, U08, U19a, U20b

SEE SECTION 3 FOR PRODUCTS ALSO REGISTERED

Pesticide Profiles

Environmental protection E12a, E13a, E15b, E16c, E16d, E34, E38; E12f (named crops - see label); E17b (30 m); E22a (or after 31 Mar)
Storage and disposal D01, D02, D05, D09b, D10b, D12a
Medical advice M03, M05b

38   bixafen

A succinate dehydrogenase inhibitor (SDHI) fungicide available only in mixtures
FRAC mode of action code: 3 + 7

39   bixafen + prothioconazole

Succinate dehydrogenase inhibitor (SDHI) + triazole fungicide mixture for disease control in cereals
FRAC mode of action code: 3 + 7

Products
1	Aviator 235 Xpro	Bayer CropScience	75:160 g/l	EC	15026
2	Siltra Xpro	Bayer CropScience	60:200 g/l	EC	15082

Uses
- Brown rust in *spring barley, winter barley* [2]; *spring wheat, triticale, winter rye, winter wheat* [1]
- Crown rust in *spring oats, winter oats* [2]
- Ear diseases in *spring barley, winter barley* [2]; *spring wheat, triticale, winter wheat* [1]
- Eyespot in *spring barley* (reduction of incidence and severity), *spring oats* (reduction of incidence and severity), *winter barley* (reduction of incidence and severity), *winter oats* (reduction of incidence and severity) [2]; *spring wheat* (reduction of incidence and severity), *triticale* (reduction of incidence and severity), *winter wheat* (reduction of incidence and severity) [1]
- Glume blotch in *spring wheat, triticale, winter wheat* [1]
- Net blotch in *spring barley, winter barley* [2]
- Powdery mildew in *spring barley, spring oats, winter barley, winter oats* [2]; *spring wheat, triticale, winter rye, winter wheat* [1]
- Ramularia leaf spots in *spring barley, winter barley* [2]
- Rhynchosporium in *spring barley, winter barley* [2]; *winter rye* [1]
- Septoria leaf blotch in *spring wheat, triticale, winter wheat* [1]
- Tan spot in *spring wheat, triticale, winter wheat* [1]
- Yellow rust in *spring barley, winter barley* [2]; *spring wheat, triticale, winter wheat* [1]

Approval information
- Prothioconazole included in Annex I under EC Directive 91/414 but bixafen not yet included
- Accepted by BBPA for use on malting barley

Efficacy guidance
- Resistance to some DMI fungicides has been identified in Septoria leaf blotch (Mycosphaerella graminicola) which may seriously affect the performance of some products. For further advice on resistance management in DMIs contact your agronomist or specialist advisor, and visit the FRAG-UK website.

Environmental safety
- LERAP Category B

Hazard classification and safety precautions
Hazard Harmful, Dangerous for the environment
Transport code 9
Packaging group III
UN number 3082
Risk phrases R22a, R43, R51, R53a, R63
Operator protection U05a, U09a, U20b
Environmental protection E15b, E16a, E16b, E34, E38
Storage and disposal D01, D02, D09a, D10b, D12a

FOR FULL CONDITIONS OF USE ALWAYS READ THE PRODUCT LABEL

40 bixafen + prothioconazole + spiroxamine

A broad-spectrum fungicide mixture for disease control in cereals
FRAC mode of action code: 3 + 5 + 7

Products
Boogie   Bayer CropScience   50:100:250 g/l   EC   15061

Uses
- Brown rust in *spring wheat, triticale, winter rye, winter wheat*
- Ear diseases in *spring wheat, triticale, winter rye, winter wheat*
- Eyespot in *spring wheat, triticale, winter rye, winter wheat*
- Glume blotch in *spring wheat, triticale, winter rye, winter wheat*
- Powdery mildew in *spring wheat, triticale, winter rye, winter wheat*
- Rhynchosporium in *triticale, winter rye*
- Septoria leaf blotch in *spring wheat, triticale, winter rye, winter wheat*
- Tan spot in *spring wheat, triticale, winter rye, winter wheat*
- Yellow rust in *spring wheat, triticale, winter rye, winter wheat*

Approval information
- Prothioconazole and spiroxamine included in Annex I under EC Directive 91/414 but bixafen not yet included

Efficacy guidance
- Resistance to some DMI fungicides has been identified in Septoria leaf blotch (Mycosphaerella graminicola) which may seriously affect the performance of some products. For further advice on resistance management in DMIs contact your agronomist or specialist advisor, and visit the FRAG-UK website.

Restrictions
- No more than two applications of SDH inhibitors must be applied to the same cereal crop.

Environmental safety
- LERAP Category B

Hazard classification and safety precautions
Hazard Harmful, Dangerous for the environment
Risk phrases R20, R22a, R41, R50, R53a
Operator protection A, C, H; U05a, U09b, U11, U20a, U26
Environmental protection E15b, E16a, E16b, E34, E38
Storage and disposal D01, D02, D05, D09a, D10b, D12a
Medical advice M03

41 bixafen + prothioconazole + tebuconazole

A succinate dehydrogenase inhibitor (SDHI) + triazole fungicide mixture for disease control in cereals
FRAC mode of action code: 3 + 7

Products
1 Name TBC   Bayer CropScience   75:110:90 g/l   EC   15027
2 Skyway 285 Xpro   Bayer CropScience   75:110:100 g/l   EC   15028

Uses
- Brown rust in *spring wheat, triticale, winter rye, winter wheat*
- Ear diseases in *spring wheat, triticale, winter wheat*
- Eyespot in *spring wheat* (reduction of incidence and severity), *triticale* (reduction of incidence and severity), *winter wheat* (reduction of incidence and severity)
- Glume blotch in *spring wheat, triticale, winter wheat*
- Powdery mildew in *spring wheat, triticale, winter rye, winter wheat*
- Rhynchosporium in *winter rye*
- Septoria leaf blotch in *spring wheat, triticale, winter wheat*
- Tan spot in *spring wheat, triticale, winter wheat*
- Yellow rust in *spring wheat, triticale, winter wheat*

SEE SECTION 3 FOR PRODUCTS ALSO REGISTERED

Approval information
- Prothioconazole and tebuconazole included in Annex I under EC Directive 91/414 but bixafen not yet included

Efficacy guidance
- Resistance to some DMI fungicides has been identified in Septoria leaf blotch (Mycosphaerella graminicola) which may seriously affect the performance of some products. For further advice on resistance management in DMIs contact your agronomist or specialist advisor, and visit the FRAG-UK website.

Environmental safety
- LERAP Category B

Hazard classification and safety precautions
Hazard Harmful, Dangerous for the environment
Transport code 9
Packaging group III
UN number 3082
Risk phrases R22a, R43, R51, R53a, R63
Operator protection A, H; U05a, U09a, U20b
Environmental protection E15b, E16a, E16b, E34, E38
Storage and disposal D01, D02, D09a, D10b, D12a

42 boscalid

A translocated and translaminar anilide fungicide
FRAC mode of action code: 7

Products
1	Bosco WG	Goldengrass	50% w/w	WG	14455
2	Filan	BASF	50% w/w	WG	11449
3	Fulmar	AgChem Access	50% w/w	WG	14073

Uses
- Alternaria in **spring oilseed rape, winter oilseed rape** [1-3]
- Botrytis in **poppies for morphine production** *(off-label)* [2]
- Disease control in **all edible seed crops grown outdoors** *(off-label)*, **all non-edible seed crops grown outdoors** *(off-label)*, **borage for oilseed production** *(off-label)*, **canary flower (echium spp.)** *(off-label)*, **evening primrose** *(off-label)*, **honesty** *(off-label)*, **linseed** *(off-label)*, **mustard** *(off-label)* [2]
- Sclerotinia in **poppies for morphine production** *(off-label)* [2]
- Sclerotinia stem rot in **spring oilseed rape, winter oilseed rape** [1-3]

Specific Off-Label Approvals (SOLAs)
- *all edible seed crops grown outdoors* 20093231 [2]
- *all non-edible seed crops grown outdoors* 20093231 [2]
- *borage for oilseed production* 20093229 [2]
- *canary flower (echium spp.)* 20093229 [2]
- *evening primrose* 20093229 [2]
- *honesty* 20093229 [2]
- *linseed* 20093229 [2]
- *mustard* 20093229 [2]
- *poppies for morphine production* 20093222 [2]

Approval information
- Boscalid included in Annex I under EC Directive 91/414
- Accepted by BBPA on malting barley before ear emergence only
- Approval expiry 31 Jul 2011 [1, 3]

Efficacy guidance
- For best results apply as a protectant spray before symptoms are visible
- Applications against *Sclerotinia* should be made in high disease risk situations at early to full flower
- *Sclerotinia* control may be reduced when high risk conditions occur after flowering, leading to secondary disease spread

FOR FULL CONDITIONS OF USE ALWAYS READ THE PRODUCT LABEL

boscalid + epoxiconazole

- Applications against *Alternaria* should be made at full flowering. Later applications may result in reduced levels of control
- Ensure adequate spray penetration and good coverage

Restrictions
- Maximum total dose equivalent to two full dose treatments
- Do no treat crops to be used for seed production later than full flowering stage
- Avoid spray drift onto neighbouring crops

Crop-specific information
- Latest use: up to and including when 50% of pods have reached final size (GS 75) for oilseed rape

Environmental safety
- Dangerous for the environment
- Toxic to aquatic organisms
- Product represents minimal hazard to bees when used as directed. However local bee-keepers should be notified if crops are to be sprayed when in flower

Hazard classification and safety precautions
    Hazard Dangerous for the environment
    Transport code 9
    Packaging group III
    UN number 3077
    Risk phrases R51, R53a
    Operator protection U05a, U20c
    Environmental protection E15a, E38
    Storage and disposal D01, D02, D05, D09a, D10c, D12a

43 boscalid + epoxiconazole

A broad spectrum fungicide mixture for cereals
FRAC mode of action code: 7 + 3

See also epoxiconazole

Products

1	Chord	BASF	210:75 g/l	SC	13928
2	Deuce	BASF	233:67 g/l	SC	13240
3	Injun	Goldengrass	233:67 g/l	SC	14305
4	Tracker	BASF	233:67 g/l	SC	12295
5	Whistle	BASF	210:75 g/l	SC	14757

Uses
- Brown rust in **spring barley**, **spring wheat**, **winter barley**, **winter wheat** [1-5]
- Crown rust in **spring oats**, **winter oats** [1-5]
- Eyespot in **spring barley** *(moderate)*, **spring wheat** *(moderate)*, **winter barley** *(moderate)*, **winter wheat** *(moderate)* [2-4]; **spring barley** *(moderate control)*, **spring oats** *(moderate control)*, **spring wheat** *(moderate control)*, **winter barley** *(moderate control)*, **winter oats** *(moderate control)*, **winter wheat** *(moderate control)* [1, 5]
- Foliar disease control in **durum wheat** *(off-label)* [4]
- Fusarium ear blight in **spring wheat** *(good reduction)*, **winter wheat** *(good reduction)* [1, 5]
- Glume blotch in **spring wheat**, **winter wheat** [1, 5]
- Net blotch in **spring barley**, **winter barley** [1-5]
- Powdery mildew in **spring barley**, **spring oats**, **spring wheat**, **winter barley**, **winter oats**, **winter wheat** [1, 5]
- Rhynchosporium in **spring barley** *(moderate)*, **winter barley** *(moderate)* [2-4]; **spring barley** *(moderate control)*, **winter barley** *(moderate control)* [1, 5]
- Septoria leaf blotch in **spring wheat**, **winter wheat** [1-5]
- Sooty moulds in **spring wheat**, **winter wheat** [1, 5]
- Tan spot in **spring wheat** *(reduction only)*, **winter wheat** *(reduction only)* [1, 5]
- Yellow rust in **spring barley**, **spring wheat**, **winter barley**, **winter wheat** [1, 5]

Specific Off-Label Approvals (SOLAs)
- *durum wheat* 20070462 [4]

SEE SECTION 3 FOR PRODUCTS ALSO REGISTERED

Approval information
- Boscalid and epoxiconazole included in Annex I under EC Directive 91/414
- Accepted by BBPA for use on malting barley (before ear emergence only)

Efficacy guidance
- Apply at the start of foliar or stem based disease attack
- Optimum effect against eyespot achieved by spraying between leaf-sheath erect and second node detectable stages (GS 30-32)
- Epoxiconazole is a DMI fungicide. Resistance to some DMI fungicides has been identified in Septoria leaf blotch which may seriously affect performance of some products. For further advice contact a specialist advisor and visit the Fungicide Resistance Action Group (FRAG)-UK website

Restrictions
- Maximum total dose equivalent to two full dose treatments [2, 4]

Crop-specific information
- Latest use: before cereal ear emergence

Environmental safety
- Dangerous for the environment
- Toxic to aquatic organisms
- Avoid drift onto neighbouring crops. May cause damage to broad-leaved plant species
- LERAP Category B

Hazard classification and safety precautions
Hazard Harmful, Dangerous for the environment
Transport code 9 [2-4]
Packaging group III [2-4]
UN number 3082 [2-4]; N/C [1, 5]
Risk phrases R36, R40, R51, R53a, R62, R63
Operator protection A, C [1-5]; H [1, 2, 4, 5]; U05a, U20b
Environmental protection E15a, E16a, E38 [1-5]; E34 [1, 5]
Storage and disposal D01, D02, D08, D10c, D12a [1-5]; D09a [1, 5]
Medical advice M03

44 boscalid + pyraclostrobin

A protectant and systemic fungicide mixture
FRAC mode of action code: 7 + 11

See also pyraclostrobin

Products

1	Bellis	BASF	25.2:12.8% w/w	WG	12522
2	Signum	BASF	26.7:6.7% w/w	WG	11450

Uses
- Alternaria in **cabbages**, **calabrese**, **carrots** *(moderate)*, **cauliflowers**, **poppies for morphine production** *(off-label)* [2]
- American gooseberry mildew in **blackcurrants** *(moderate control only)* [2]
- Black spot in **protected strawberries**, **strawberries** [2]
- Blight in **potatoes** *(off-label)* [2]
- Blossom wilt in **cherries** *(off-label)*, **mirabelles** *(off-label)*, **plums** *(off-label)*, **protected cherries** *(off-label)*, **protected mirabelles** *(off-label)*, **protected plums** *(off-label)* [2]
- Botrytis in **asparagus** *(off-label)*, **forest nurseries** *(off-label)*, **ornamental plant production** *(off-label)*, **poppies for morphine production** *(off-label)*, **protected forest nurseries** *(off-label)*, **protected ornamentals** *(off-label)*, **redcurrants** *(off-label)*, **whitecurrants** *(off-label)* [2]
- Bottom rot in **lettuce**, **protected lettuce** [2]
- Cane blight in **blackberries** *(off-label)*, **protected blackberries** *(off-label)*, **raspberries** *(off-label)* [2]
- Chocolate spot in **spring field beans** *(moderate)*, **winter field beans** *(moderate)* [2]
- Disease control in **all edible crops (outdoor and protected)** *(off-label)*, **all non-edible crops (outdoor)** *(off-label)*, **all protected non-edible crops** *(off-label)*, **protected hops** *(off-label)*,

FOR FULL CONDITIONS OF USE ALWAYS READ THE PRODUCT LABEL

boscalid + pyraclostrobin

  protected soft fruit (off-label), *protected top fruit* (off-label), *soft fruit* (off-label), *top fruit* (off-label) [2]; *hops* (off-label) [1, 2]
- Downy mildew in *bulls blood* (off-label), *chard* (off-label), *endives* (off-label), *protected endives* (off-label), *spinach* (off-label), *spinach beet* (off-label) [2]
- Grey mould in *blackcurrants*, *lettuce*, *protected lettuce*, *protected strawberries*, *strawberries* [2]
- Late blight in *potatoes* (off-label) [2]
- Leaf and pod spot in *combining peas* (moderate control only), *vining peas* (moderate control only) [2]
- Leaf spot in *blackcurrants* (moderate control only), *redcurrants* (off-label), *whitecurrants* (off-label) [2]
- Poppy fire in *poppies for morphine production* (off-label) [2]
- Powdery mildew in *apples* [1]; *carrots*, *protected strawberries*, *redcurrants* (off-label), *strawberries*, *whitecurrants* (off-label) [2]
- Purple blotch in *blackberries* (off-label), *protected blackberries* (off-label), *raspberries* (off-label) [2]
- Ring spot in *broccoli*, *brussels sprouts*, *cabbages*, *calabrese*, *cauliflowers* [2]
- Rust in *asparagus* (off-label), *protected blackberries* (off-label), *spring field beans*, *winter field beans* [2]
- Scab in *apples*, *pears* [1]
- Sclerotinia in *carrots* (moderate), *horseradish* (off-label) [2]
- Soft rot in *lettuce*, *protected lettuce* [2]
- White blister in *broccoli*, *brussels sprouts*, *cabbages* (qualified minor use), *calabrese* (qualified minor use), *herbs (see appendix 6)* (off-label), *leaf brassicas* (off-label), *protected herbs (see appendix 6)* (off-label), *protected leaf brassicas* (off-label) [2]
- White rot in *bulb onions* (off-label), *garlic* (off-label), *salad onions* (off-label), *shallots* (off-label) [2]
- White tip in *leeks* (off-label) [2]

Specific Off-Label Approvals (SOLAs)
- *all edible crops (outdoor and protected)* 20102111 [2]
- *all non-edible crops (outdoor)* 20102111 [2]
- *all protected non-edible crops* 20102111 [2]
- *asparagus* 20102105 [2]
- *blackberries* 20102110 [2]
- *bulb onions* 20102108 [2]
- *bulls blood* 20102136 [2]
- *chard* 20102136 [2]
- *cherries* 20102109 [2]
- *endives* 20102136 [2]
- *forest nurseries* 20102119 [2]
- *garlic* 20102108 [2]
- *herbs (see appendix 6)* 20102115 [2]
- *hops* 20082822 [1], 20102111 [2]
- *horseradish* 20093375 [2]
- *leaf brassicas* 20102115 [2]
- *leeks* 20102134 [2]
- *mirabelles* 20102109 [2]
- *ornamental plant production* 20091842 [2]
- *plums* 20102109 [2]
- *poppies for morphine production* 20101233 [2]
- *potatoes* 20101942 [2]
- *protected blackberries* 20102102 [2]
- *protected cherries* 20102109 [2]
- *protected endives* 20102136 [2]
- *protected forest nurseries* 20102119 [2]
- *protected herbs (see appendix 6)* 20102115 [2]
- *protected hops* 20102111 [2]
- *protected leaf brassicas* 20102115 [2]
- *protected mirabelles* 20102109 [2]

SEE SECTION 3 FOR PRODUCTS ALSO REGISTERED

- ***protected ornamentals** 20091842* [2]
- ***protected plums** 20102109* [2]
- ***protected soft fruit** 20102111* [2]
- ***protected top fruit** 20102111* [2]
- ***raspberries** 20102110* [2]
- ***redcurrants** 20102114* [2]
- ***salad onions** 20102107* [2]
- ***shallots** 20102108* [2]
- ***soft fruit** 20102111* [2]
- ***spinach** 20102136* [2]
- ***spinach beet** 20102136* [2]
- ***top fruit** 20102111* [2]
- ***whitecurrants** 20102114* [2]

Approval information
- Boscalid and pyraclostrobin included in Annex I under EC Directive 91/414
- Accepted by BBPA for use on malting barley (before ear emergence only)

Efficacy guidance
- On brassicas apply as a protectant spray or at the first sign of disease and repeat at 3-4 wk intervals depending on disease pressure [2]
- Ensure adequate spray penetration and coverage by increasing water volume in dense crops [2]
- For best results on strawberries apply as a protectant spray at the white bud stage. Applications should be made in sequence with other products as part of a fungicide spray programme during flowering at 7-10 day intervals [2]
- On carrots and field beans apply as a protectant spray or at the first sign of disease with a repeat treatment if needed, as directed on the label [2]
- On lettuce apply as a protectant spray 1-2 wk after planting [2]
- Optimum results on apples and pears obtained from a protectant treatment from bud burst [1]
- Application as the final 2 sprays on apples and pears gives reduction in storage rots [1]
- Pyraclostrobin is a member of the QoI cross-resistance group of fungicides and should be used in programmes with fungicides with a different mode of action

Restrictions
- Maximum total dose equivalent to three full dose treatments on brassicas; two full dose treatments on all other field crops [2]
- Maximum number of treatments (including other QoI treatments) on apples and pears 4 per yr if total number of applications is 12 or more, or 3 per yr if the total number is fewer than 12 [1]
- Do not use more than 2 consecutive treatments on apples and pears, and these must be separated by a minimum of 2 applications of a fungicide with a different mode of action [1]
- Use a maximum of three applications per yr on brassicas and no more than two per yr on other field crops [2]
- Do not use consecutive treatments; apply in alternation with fungicides from a different cross resistance group and effective against the target diseases [2]
- Do not apply more than 6 kg/ha to the same area of land per yr [2]
- Consult processor before use on crops for processing
- Applications to lettuce and protected salad crops may only be made between 1 Apr and 31 Oct [2]

Crop-specific information
- HI 21 d for field beans; 14 d for brassica crops, carrots, lettuce; 7 d for apples, pears; 3 d for strawberries

Environmental safety
- Dangerous for the environment
- Very toxic to aquatic organisms
- Broadcast air-assisted LERAP [1] (40 m); LERAP Category B [2]

Hazard classification and safety precautions
    Hazard Harmful, Dangerous for the environment
    Transport code 9
    Packaging group III
    UN number 3077
    Risk phrases R22a, R50, R53a
    Operator protection A; U05a, U20c [1, 2]; U13 [1]

FOR FULL CONDITIONS OF USE ALWAYS READ THE PRODUCT LABEL

bromadiolone

Environmental protection E15a, E38 [1, 2]; E16a [2]; E17b [1] (40 m); E34 [1]
Storage and disposal D01, D02, D10c, D12a [1, 2]; D08 [1]; D09a [2]
Medical advice M03 [1]; M05a [1, 2]

45 bromadiolone

An anti-coagulant coumarin-derivative rodenticide

Products

1	Bromag	Killgerm	0.005% w/w	RB	H8285
2	Bromag Fresh Bait	Killgerm	0.005% w/w	RB	H8557
3	Endorats Premium Rat Killer	Irish Drugs	0.005% w/w	GB	H6725
4	Sakarat Bromabait	Killgerm	0.005% w/w	RB	H7902
5	Tomcat 2	Antec Biosentry	0.005% w/w	PT	H6736
6	Tomcat 2 Blox	Antec Biosentry	0.005% w/w	BB	H6731

Uses
- Mice in **farm buildings** [4-6]; **farm buildings/yards** [1, 2]; **farmyards** [4]
- Rats in **farm buildings** [3-6]; **farm buildings/yards** [1, 2]; **farmyards** [4]

Efficacy guidance
- Ready-to-use baits are formulated on a mould-resistant, whole-wheat base
- Use in baiting programme. Place baits in protected situations, sufficient for continuous feeding between treatments
- Chemical is effective against warfarin- and coumatetralyl-resistant rats and mice and does not induce bait shyness
- Use bait bags where loose baiting inconvenient (eg behind ricks, silage clamps etc)
- The resistance status of the rodent population should be assessed when considering the choice of product to use

Restrictions
- For use only by professional operators

Environmental safety
- Access to baits by children, birds and animals, particularly cats, dogs, pigs and poultry, must be prevented
- Baits must not be placed where food, feed or water could become contaminated
- Remains of bait and bait containers must be removed after treatment and burned or buried
- Rodent bodies must be searched for and burned or buried. They must not be placed in refuse bins or on rubbish tips
- Take extreme care to prevent domestic animals having access to the bait

Hazard classification and safety precautions
Operator protection A [1]; U13 [1-6]; U20b [1, 2, 4-6]
Environmental protection E15a [5, 6]
Storage and disposal D05, D07 [1, 2, 4]; D09a [1-6]; D11a [1, 2, 4-6]
Treated seed S06a [3]
Vertebrate/rodent control products V01a, V03a, V04a [3, 5, 6]; V01b, V03b, V04b [1, 2, 4]; V02 [1-6]
Medical advice M03 [1, 2, 4-6]

46 bromoxynil

A contact acting HBN herbicide
HRAC mode of action code: C3

Products

1	Alpha Bromolin 225 EC	Makhteshim	225 g/l	EC	14864
2	Butryflow	Nufarm UK	401.58 g/l	SC	14056
3	Flagon 400 EC	Makhteshim	400 g/l	EC	14921

Uses
- Annual dicotyledons in **durum wheat** *(off-label)*, **grass seed crops** *(off-label)*, **spring barley**, **spring oats**, **spring rye** *(off-label)*, **spring wheat**, **triticale** *(off-label)*, **winter barley**, **winter oats**,

SEE SECTION 3 FOR PRODUCTS ALSO REGISTERED

Pesticide Profiles

winter rye (off-label), *winter wheat* [1, 3]; *forage maize*, *grain maize*, *linseed* (off-label), *sweetcorn* [2]; *game cover* (off-label) [3]

Specific Off-Label Approvals (SOLAs)
- *durum wheat* 20102118 [1], 20102116 [3]
- *game cover* 20102116 [3]
- *grass seed crops* 20102118 [1], 20102116 [3]
- *linseed* 20101232 [2]
- *spring rye* 20102118 [1], 20102116 [3]
- *triticale* 20102118 [1], 20102116 [3]
- *winter rye* 20102118 [1], 20102116 [3]

Approval information
- Bromoxynil included in Annex I under EC Directive 91/414
- Accepted by BBPA for use on malting barley

Efficacy guidance
- Spray when main weed flush has germinated and the largest are at the 4 leaf stage
- Weed control can be enhanced by using a split treatment spraying each application when the weeds are seedling to 2 true leaves.

Restrictions
- Maximum number of treatments 1 per crop or yr or maximum total dose equivalent to one full dose treatment
- Do not apply with oils or other adjuvants
- Do not apply using hand-held equipment or at concentrations higher than those recommended
- Do not apply during frosty weather, drought, when soil is waterlogged, when rain expected within 4 h or to crops under any stress
- Take particular care to avoid drift onto neighbouring susceptible crops or open water surfaces

Crop-specific information
- Latest use: before 10 fully expanded leaf stage of crop for maize, sweetcorn; before crop 20 cm tall and before flower buds visible for linseed; before 2nd node detectable (GS 32) for cereals
- Foliar scorch, which rapidly disappears without affecting growth, will occur if treatment made in hot weather or during rapid growth

Environmental safety
- Very toxic to aquatic organisms
- Keep livestock out of treated areas for at least 6 wk after treatment
- High risk to bees. Do not apply to crops in flower or to those in which bees are actively foraging
- Do not contaminate surface waters or ditches with chemical or used container
- LERAP Category B

Hazard classification and safety precautions
Hazard Harmful, Dangerous for the environment [1-3]; Flammable [1]
Transport code 3 [1, 3]; 9 [2]
Packaging group III
UN number 1993 [1, 3]; 3082 [2]
Risk phrases R20, R22a, R43, R50, R53a, R63 [1-3]; R22b, R36, R37, R38 [1, 3]
Operator protection A, H [1-3]; C [1, 3]; K [2]; U02a, U03, U08, U10, U11, U13, U19a, U20a [1, 3]; U05a, U14, U15, U23a [1-3]
Environmental protection E06a [1, 3] (6 wk); E12d [1, 3]; E15a, E16a, E34, E38 [1-3]
Consumer protection C02a [1, 3] (6 wk)
Storage and disposal D01, D02, D05, D09a, D12a [1-3]; D10b [1, 3]; D10c [2]
Medical advice M03 [2]; M04a, M05b [1, 3]

FOR FULL CONDITIONS OF USE ALWAYS READ THE PRODUCT LABEL

bromoxynil + diflufenican + ioxynil

47 bromoxynil + diflufenican + ioxynil

A selective contact and translocated herbicide for cereals
HRAC mode of action code: C3 + F1 + C3

See also diflufenican
ioxynil

Products
Capture   Nufarm UK   300:50:200 g/l   SC   12514

Uses
- Annual dicotyledons in **grass seed crops** *(off-label)*, **spring barley, spring wheat, triticale, winter barley, winter rye, winter wheat**
- Chickweed in **spring barley, spring wheat, triticale, winter barley, winter rye, winter wheat**
- Field speedwell in **spring barley, spring wheat, triticale, winter barley, winter rye, winter wheat**
- Ivy-leaved speedwell in **spring barley, spring wheat, triticale, winter barley, winter rye, winter wheat**
- Knotgrass in **spring barley, spring wheat, triticale, winter barley, winter rye, winter wheat**
- Mayweeds in **spring barley, spring wheat, triticale, winter barley, winter rye, winter wheat**

Specific Off-Label Approvals (SOLAs)
- **grass seed crops** 20052240

Approval information
- Bromoxynil, diflufenican and ioxynil included in Annex I under EC Directive 91/414
- Accepted by BBPA for use on malting barley

Efficacy guidance
- Best results obtained on small weeds and when competition removed early
- Good spray coverage essential for good activity

Restrictions
- Maximum number of treatments 1 per crop
- Do not treat crops undersown or to be undersown
- Do not treat frosted crops or those that are under stress from any cause
- Do not apply by hand-held equipment or at concentrations higher than those recommended

Crop-specific information
- Latest use: before 2nd node detectable (GS 32)

Following crops guidance
- In the event of crop failure winter wheat may be drilled immediately after normal cultivations. Winter barley may be re-drilled after ploughing
- Land must be ploughed and an interval of 12 wk elapse after treatment before planting spring crops of wheat, barley, oilseed rape, peas, field beans, sugar beet, potatoes, carrots, edible brassicas or onions
- Successive treatments with any products containing diflufenican can lead to soil build-up and inversion ploughing to 150 mm must precede sowing any following non-cereal crop. Even where ploughing occurs some crops may be damaged

Environmental safety
- Dangerous for the environment
- Toxic to aquatic organisms
- Keep livestock out of treated areas for at least 2 wk after treatment
- Harmful to bees. Do not apply to crops in flower or to those in which bees are actively foraging
- LERAP Category B

Hazard classification and safety precautions
Hazard Harmful, Dangerous for the environment
Transport code 9
Packaging group III
UN number 3082
Risk phrases R22a, R50, R53a, R63, R69
Operator protection A, C, H; U05a, U08, U13, U19a, U20b

SEE SECTION 3 FOR PRODUCTS ALSO REGISTERED

Pesticide Profiles

Environmental protection E06a (2 wk); E12d, E12e, E15a, E16a, E34, E38
Consumer protection C02a (2 wk)
Storage and disposal D01, D02, D05, D09a, D10b, D12a
Medical advice M03

48 bromoxynil + ioxynil

A contact acting post-emergence HBN herbicide for cereals
HRAC mode of action code: C3 + C3

See also ioxynil

Products

1	Mextrol-Biox	Nufarm UK	200:200 g/l	EC	14697
2	Oxytril CM	Bayer CropScience	200:200 g/l	EC	14511
3	Stellox	Nufarm UK	200:200 g/l	EC	14696

Uses
- Annual dicotyledons in **spring barley, spring oats, spring rye, spring wheat, triticale, winter barley, winter oats, winter rye, winter wheat**

Approval information
- Bromoxynil and ioxynil included in Annex I under EC Directive 91/414
- Accepted by BBPA for use on malting barley

Efficacy guidance
- Best results achieved on young weeds growing actively in a highly competitive crop
- Do not apply during periods of drought or when rain imminent (some labels say 'if likely within 4 or 6 h')
- Recommended for tank mixture with hormone herbicides to extend weed spectrum. See labels for details

Restrictions
- Maximum number of treatments 1 per crop
- Do not spray crops stressed by drought, waterlogging or other factors
- Do not roll or harrow for several days before or after spraying. Number of days specified varies with product. See label for details
- Do not apply by hand-held equipment or at concentrations higher than those recommended

Crop-specific information
- Latest use: before 2nd node detectable stage (GS 31)
- HI (animal consumption) 6 wk
- Apply to winter or spring cereals from 1-2 fully expanded leaf stage, but before second node detectable (GS 32)
- Spray oats in spring when danger of frost past. Do not spray winter oats in autumn
- Apply to undersown cereals pre-sowing or pre-emergence of legume provided cover crop is at correct stage. Only spray trefoil pre-sowing
- On crops undersown with grasses alone or on direct re-seeds apply from 2-leaf stage of grass

Environmental safety
- Dangerous for the environment
- Very toxic to aquatic organisms
- Keep livestock out of treated areas for at least 6 wk after treatment and until poisonous weeds, such as ragwort, have died down and become unpalatable
- LERAP Category B

Hazard classification and safety precautions
  Hazard Harmful, Dangerous for the environment
  Transport code 9
  Packaging group III
  UN number 3082
  Risk phrases R22a, R22b, R43, R50, R53a, R63, R66, R67
  Operator protection A, C [1-3]; H [1, 2]; U05a, U08, U13, U19a, U20b, U23a
  Environmental protection E07a, E12d, E12e, E13b, E16a, E34, E38
  Consumer protection C02a (14 d)

FOR FULL CONDITIONS OF USE ALWAYS READ THE PRODUCT LABEL

Storage and disposal D01, D02, D09a, D11a, D12a
Medical advice M03, M05b

49 bromoxynil + prosulfuron

A contact and residual herbicide mixture for maize
HRAC mode of action code: C3 + B

See also prosulfuron

Products

Jester   Syngenta   60:3% w/w   WG   13500

Uses
- Annual dicotyledons in **forage maize**, **game cover** *(off-label)*, **grain maize**, **millet** *(off-label)*
- Black bindweed in **forage maize**, **grain maize**
- Chickweed in **forage maize**, **grain maize**
- Fat hen in **millet** *(off-label)*
- Hemp-nettle in **forage maize**, **grain maize**
- Knotgrass in **forage maize**, **grain maize**
- Mayweeds in **forage maize**, **grain maize**

Specific Off-Label Approvals (SOLAs)
- **game cover** *20100322 expires 30 Jun 2011*
- **millet** *20101803 expires 30 Jun 2011*

Approval information
- Bromoxynil and prosulfuron included in Annex I under EC Directive 91/414
- Approval expiry 30 Jun 2011

Efficacy guidance
- Prosulfuron is a member of the ALS-inhibitor group of herbicides and products should be used in a planned Resistance Management strategy. See Section 5 for more information

Restrictions
- Maximum total dose equivalent to one full dose treatment
- Do not treat crops grown for seed production
- Consult before use on crops intended for processing
- Do not use in frosty weather or on crops under stress
- Do not tank mix with organophosphate insecticides or apply in tank mix or sequence with any other sulfonylurea
- Do not apply by knapsack sprayer or in volumes less than those recommended
- Take care to wash out sprayers thoroughly. See label for details

Crop-specific information
- Latest use: before 7 crop leaves unfolded for maize
- Apply post-emergence up to when the crop has six unfolded leaves
- Product must be used with a non-ionic wetter that is not an organosilicone
- Apply in cool conditions, or during the evening, to avoid scorch

Following crops guidance
- Winter or spring cereals, winter or spring beans, spring sown peas or oilseed rape may be sown following normal harvest of a treated crop. Mould-board ploughing to 20 cm is recommended in some cases

Environmental safety
- Dangerous for the environment
- Very toxic to aquatic organisms
- Take special care to avoid drift outside the target area

Hazard classification and safety precautions
Hazard Toxic, Dangerous for the environment
Transport code 6.1
Packaging group III
UN number 2588
Risk phrases R22a, R23, R36, R43, R50, R53a, R63

SEE SECTION 3 FOR PRODUCTS ALSO REGISTERED

Pesticide Profiles

Operator protection A, D, H; U05a, U11, U15, U19a, U20c, U23a
Environmental protection E15b, E34, E38
Storage and disposal D01, D02, D05, D09a, D10c, D12a
Medical advice M04a

50 bromoxynil + terbuthylazine

A post-emergence contact and residual herbicide for maize
HRAC mode of action code: C3 + C1

See also terbuthylazine

Products
Templar   Makhteshim   200:300 g/l   SC   10254

Uses
- Annual dicotyledons in **forage maize**, **forage maize (under plastic mulches)** *(off-label)*, **grain maize under plastic mulches** *(off-label)*, **sweetcorn** *(off-label)*, **sweetcorn under plastic mulches** *(off-label)*
- Annual grasses in **forage maize (under plastic mulches)** *(off-label)*, **grain maize under plastic mulches** *(off-label)*, **sweetcorn under plastic mulches** *(off-label)*

Specific Off-Label Approvals (SOLAs)
- *forage maize (under plastic mulches)* 20091844
- *grain maize under plastic mulches* 20091844
- *sweetcorn* 20072179
- *sweetcorn under plastic mulches* 20091844

Approval information
- Bromoxynil included in Annex I under EC Directive 91/414
- Terbuthylazine has not been included in Annex 1 but the date of withdrawal has yet to be set.

Efficacy guidance
- Best results obtained from treatment when main flush of weeds has germinated and largest at 4 leaf stage
- Control can be enhanced by use of split dose treatment but second spray must be applied before crop canopy covers ground
- Soils should be moist at time of treatment and should not be disturbed afterwards
- Residual activity may be reduced on soils with high OM content, or where high amounts of slurry or manure have been applied

Restrictions
- Maximum total dose equivalent to one full dose treatment
- Foliar scorch may occur after treatment in hot conditions or when growth particularly rapid. In such conditions spray in the evening
- Do not apply during frosty weather, drought, when soil is waterlogged or when rain expected within 12 hr
- Do not treat crops under any stress whatever
- Product should be used only where damaging populations of competitive weeds have emerged. Yield may be reduced when used in absence of significant weed populations
- Succeeding crops may not be planted until the spring following application. Mould board plough to 150 mm before doing so

Crop-specific information
- Apply before 9 fully expanded leaf stage

Following crops guidance
- Keep livestock out of treated areas for at least 6 wk

Environmental safety
- Dangerous for the environment
- Very toxic to aquatic organisms
- Keep livestock out of treated areas for at least 6 wk after treatment

FOR FULL CONDITIONS OF USE ALWAYS READ THE PRODUCT LABEL

bupirimate

- Harmful to bees. Do not apply to crops in flower or to those in which bees are actively foraging. Do not apply when flowering weeds are present
- Do not apply by hand-held equipment or at concentrations higher than those recommended

Hazard classification and safety precautions
Hazard Harmful, Dangerous for the environment
Transport code 9
Packaging group III
UN number 3082
Risk phrases R20, R22a, R50, R53a
Operator protection A, C, H; U02a, U05a, U08, U13, U14, U15, U19a, U20a, U23a
Environmental protection E06a (6 wk); E12d, E12e, E13b, E38
Storage and disposal D01, D02, D12a
Medical advice M04a

51 bupirimate

A systemic aminopyrimidinol fungicide active against powdery mildew
FRAC mode of action code: 8

Products
Nimrod   Makhteshim   250 g/l   EC   13046

Uses
- Powdery mildew in **apples**, **begonias**, **blackcurrants**, **chrysanthemums**, **courgettes** *(outdoor only)*, **gooseberries**, **hops**, **marrows** *(outdoor only)*, **pears**, **protected cucumbers**, **protected strawberries**, **protected tomatoes** *(off-label)*, **pumpkins** *(off-label)*, **raspberries** *(outdoor only)*, **redcurrants** *(off-label)*, **roses**, **squashes** *(off-label)*, **strawberries**, **whitecurrants** *(off-label)*

Specific Off-Label Approvals (SOLAs)
- *protected tomatoes* 20070997
- *pumpkins* 20070994
- *redcurrants* 20082082
- *squashes* 20070994
- *whitecurrants* 20082082

Approval information
- Accepted by BBPA for use on hops

Efficacy guidance
- On apples during periods that favour disease development lower doses applied weekly give better results than higher rates fortnightly
- Not effective in protected crops against strains of mildew resistant to bupirimate

Restrictions
- Maximum number of treatments or maximum total dose depends on crop and dose (see label for details)

Crop-specific information
- HI: 1 d for apples, pears, strawberries; 2 d for cucurbits; 7 d for blackcurrants; 8 d for raspberries; 14 d for gooseberries, hops
- Apply before or at first signs of disease and repeat at 7-14 d intervals. Timing and maximum dose vary with crop. See label for details
- With apples, hops and ornamentals cultivars may vary in sensitivity to spray. See label for details
- If necessary to spray cucurbits in winter or early spring spray a few plants 10-14 d before spraying whole crop to test for likelihood of leaf spotting problem
- On roses some leaf puckering may occur on young soft growth in early spring or under low light intensity. Avoid use of high rates or wetter on such growth
- Never spray flowering begonias (or buds showing colour) as this can scorch petals
- Do not mix with other chemicals for application to begonias, cucumbers or gerberas

Environmental safety
- Dangerous for the environment
- Toxic to aquatic organisms
- Flammable

SEE SECTION 3 FOR PRODUCTS ALSO REGISTERED

Pesticide Profiles

- Product has negligible effect on *Phytoseiulus* and *Encarsia* and may be used in conjunction with biological control of red spider mite

Hazard classification and safety precautions
Hazard Irritant, Flammable, Dangerous for the environment
Transport code 3
Packaging group III
UN number 1993
Risk phrases R22b, R38, R51, R53a
Operator protection A, C; U05a, U20b
Environmental protection E15a
Storage and disposal D01, D02, D09a, D10c
Medical advice M05b

52 captan

A protectant phthalimide fungicide with horticultural uses
FRAC mode of action code: M4

Products
1	Alpha Captan 80 WDG	Makhteshim	80% w/w	WG	07096
2	Alpha Captan 83 WP	Makhteshim	83% w/w	WP	04806

Uses
- Black spot in **roses** [1, 2]
- Botrytis in **strawberries** [1, 2]
- Gloeosporium rot in **apples** [1, 2]
- Scab in **apples**, **pears** [1, 2]; **quinces** *(off-label)* [1]

Specific Off-Label Approvals (SOLAs)
- *quinces* 20091388 [1]

Approval information
- Captan included in Annex I under EC Directive 91/414

Restrictions
- Maximum number of treatments 12 per yr on apples and pears as pre-harvest sprays
- Product must not be used as dip or drench on apples or pears [2]
- Do not use on apple cultivars Bramley, Monarch, Winston, King Edward, Spartan, Kidd's Orange or Red Delicious or on pear cultivar D'Anjou
- Do not mix with alkaline materials or oils
- Do not use on fruit for processing
- Powered visor respirator with hood and neck cape must be used when handling concentrate

Crop-specific information
- HI apples, pears 14 d; strawberries 7 d
- For control of scab apply at bud burst and repeat at 10-14 d intervals until danger of scab infection ceased
- For suppression of fruit storage rots apply from late Jul and repeat at 2-3 wk intervals
- For black spot control in roses apply after pruning with 3 further applications at 14 d intervals or spray when spots appear and repeat at 7-10 d intervals
- For grey mould in strawberries spray at first open flower and repeat every 7-10 d [1]
- Do not leave diluted material for more than 2 h. Agitate well before and during spraying

Environmental safety
- Dangerous for the environment
- Very toxic to aquatic organisms

Hazard classification and safety precautions
Hazard Toxic [2]; Harmful [1]; Dangerous for the environment [1, 2]
Transport code 6.1 [2]; 9 [1]
Packaging group III
UN number 2588 [2]; 3077 [1]
Risk phrases R23, R37, R38 [2]; R36, R40, R43, R50, R53a [1, 2]; R41 [1]

FOR FULL CONDITIONS OF USE ALWAYS READ THE PRODUCT LABEL

carbendazim

Operator protection A [1, 2]; D, E, H [1]; U05a, U11, U19a [1, 2]; U09a, U20c [1]; U14, U15, U20b [2]
Environmental protection E15a
Storage and disposal D01, D02, D09a, D11a, D12a

53 carbendazim

A systemic benzimidazole fungicide with curative and protectant activity
FRAC mode of action code: 1

Products

1 Mascot Systemic	Rigby Taylor	500 g/l	SC	14488
2 Ringer	Barclay	500 g/l	SC	13289

Uses
- Disease control in *managed amenity turf* [1]
- Wormcast formation in *managed amenity turf* [2]

Approval information
- Carbendazim has been included in Annex I under Directive 91/414.
- Accepted by BBPA for use on malting barley

Efficacy guidance
- Products vary in the diseases listed as controlled for several crops. Labels must be consulted for full details and for rates and timings
- Mostly applied as spray or drench. Spray treatments normally applied at first sign of disease and repeated after 1 mth if required
- Apply as drench rather than spray where red spider mite predators are being used
- Do not apply during drought conditions. Rain or irrigation after treatment may improve control
- Leave 4 mth intervals where used only for worm cast control [2]
- To delay appearance of resistant strains alternate treatment with non-MBC fungicide. Eyespot in cereals and *Botrytis cinerea* in many crops are now widely resistant. To avoid the development of resistance, a maximum of 2 applications of any MBC product (thiophanate-methyl or carbendazim) are allowed in any one crop. Avoid using MBC fungicides alone.

Restrictions
- Maximum number of treatments (including applications of any product containing benomyl, carbendazim or thiophanate-methyl) varies with crop treated and product used - see labels for details
- Do not treat crops or turf suffering from drought or other physical or chemical stress
- Consult processors before using on crops for processing
- Not compatible with alkaline products such as lime sulphur

Crop-specific information
- Latest use: before 2nd node detectable (GS32) for cereals
- HI 2 d for inert substrate cucumbers, protected tomatoes; 7-14 d (depends on dose) for apples, pears; 14 d for mushrooms, protected celery; 21 d for oilseed rape, dwarf beans, lettuce, 6 mth for cane fruit
- On turf apply as preventative treatment in spring or autumn during periods of high disease risk [2]
- Apply as a drench to control soil-borne diseases in cucumbers and tomatoes and as a pre-planting dip treatment for bulbs

Environmental safety
- Dangerous for the environment
- Very toxic to aquatic organisms
- After use dipping suspension must not be discharged directly into ditches or drains

Hazard classification and safety precautions
  Hazard Toxic, Dangerous for the environment [2]
  Transport code 9 [2]
  Packaging group III [2]
  UN number 3082 [2]
  Risk phrases R46, R50, R53a, R60, R61, R68 [2]

SEE SECTION 3 FOR PRODUCTS ALSO REGISTERED

Pesticide Profiles

Operator protection A, C, H, M [2]; U14, U19a, U20b [2]
Environmental protection E14a, E34, E38 [2]
Storage and disposal D05, D08, D09a, D10a, D12a [2]
Medical advice M04a [2]

54   carbendazim + flusilazole

A broad-spectrum systemic and protectant fungicide for cereals, oilseed rape and sugar beet
FRAC mode of action code: 1 + 3

See also flusilazole

Products

1	Contrast	DuPont	125:250 g/l	SC	13107
2	Harvesan	DuPont	125:250 g/l	SC	13106
3	Punch C	DuPont	125:250 g/l	SC	13023

Uses
- Brown rust in *spring barley, spring wheat, winter barley, winter wheat*
- Canker in *spring oilseed rape (reduction), winter oilseed rape (reduction)*
- Eyespot in *spring barley, spring wheat, winter barley, winter wheat*
- Glume blotch in *spring wheat, winter wheat*
- Light leaf spot in *spring oilseed rape, winter oilseed rape*
- Phoma leaf spot in *spring oilseed rape, winter oilseed rape*
- Powdery mildew in *spring barley, spring wheat, sugar beet, winter barley, winter wheat*
- Rust in *sugar beet*
- Septoria leaf blotch in *spring wheat, winter wheat*
- Yellow rust in *spring barley, spring wheat, winter barley, winter wheat*

Approval information
- Carbendazim included in Annex I under Directive 91/414 and flusilazole has been reinstated in Annex 1 under Directive 91/414 following an appeal
- Accepted by BBPA for use on malting barley

Efficacy guidance
- Apply at early stage of disease development or in routine preventive programme
- Most effective timing of treatment on cereals varies with disease. See label for details
- Higher rate active against both MBC-sensitive and MBC-resistant eyespot
- Rain occurring within 2 h of spraying may reduce effectiveness
- To prevent build-up of resistant strains of cereal mildew tank mix with approved morpholine fungicide
- Flusilazole is a DMI fungicide. Resistance to some DMI fungicides has been identified in Septoria leaf blotch which may seriously affect performance of some products. For further advice contact a specialist advisor and visit the Fungicide Resistance Action Group (FRAG)-UK website

Restrictions
- Maximum number of treatments 1 per crop on sugar beet; 2 per crop on cereals and oilseed rape
- Do not apply to crops under stress or during frosty weather

Crop-specific information
- Latest use on cereals varies with crop and dose used - see label for details; before first flower opened stage for oilseed rape
- HI 7 wk for sugar beet
- Treat oilseed rape in autumn when leaf lesions first appear and spring from the start of stem extension when disease appears

Environmental safety
- Dangerous for the environment
- Toxic to aquatic organisms
- Dangerous to fish or other aquatic life. Do not contaminate surface waters or ditches with chemical or used container

Hazard classification and safety precautions
Hazard Toxic, Dangerous for the environment
Transport code 9

FOR FULL CONDITIONS OF USE ALWAYS READ THE PRODUCT LABEL

carbetamide

Packaging group III
UN number 3082
Risk phrases R22a, R40, R41, R43, R46, R51, R53a, R60, R61
Operator protection A, C, H, M; U05a, U11, U19a, U20b
Environmental protection E13b, E34, E38
Storage and disposal D01, D02, D05, D09a, D10b, D12a
Medical advice M04a

55   carbetamide

A residual pre- and post-emergence carbamate herbicide for a range of field crops
HRAC mode of action code: K2

Products

1	Crawler	Makhteshim	60% w/w	WG	11840
2	Riot	Makhteshim	60% w/w	WG	13380
3	Scrum	Makhteshim	60% w/w	WG	13386

Uses
- Annual dicotyledons in *cabbage seed crops, collards, fodder rape seed crops, kale seed crops, lucerne, red clover, sainfoin, spring cabbage, sugar beet seed crops, swede seed crops, turnip seed crops, white clover, winter field beans, winter oilseed rape* [1-3]; *evening primrose* (off-label), *honesty* (off-label), *linseed* (off-label), *lupins* (off-label), *mustard* (off-label) [1]
- Annual grasses in *cabbage seed crops, collards, fodder rape seed crops, kale seed crops, lucerne, red clover, sainfoin, spring cabbage, sugar beet seed crops, swede seed crops, turnip seed crops, white clover, winter field beans, winter oilseed rape* [1-3]; *evening primrose* (off-label), *honesty* (off-label), *linseed* (off-label), *lupins* (off-label), *mustard* (off-label) [1]
- Volunteer cereals in *cabbage seed crops, collards, fodder rape seed crops, kale seed crops, lucerne, red clover, sainfoin, spring cabbage, sugar beet seed crops, swede seed crops, turnip seed crops, white clover, winter field beans, winter oilseed rape* [1-3]

Specific Off-Label Approvals (SOLAs)
- *evening primrose* 20061236 [1]
- *honesty* 20061236 [1]
- *linseed* 20061236 [1]
- *lupins* 20061237 [1]
- *mustard* 20061236 [1]

Efficacy guidance
- Best results pre- or early post-emergence of weeds under cool, moist conditions. Adequate soil moisture is essential
- Dicotyledons controlled include chickweed, cleavers and speedwell
- Weed growth stops rapidly after treatment but full effects may take 6-8 wk to develop
- Various tank mixes are recommended to broaden the weed spectrum. See label for details
- Always follow WRAG guidelines for preventing and managing herbicide resistant weeds. See Section 5 for more information

Restrictions
- Maximum number of treatments 1 per crop for all crops
- Do not treat any crop on waterlogged soil
- Do not use on soils with more than 10% organic matter as residual activity is impaired
- Do not apply during prolonged periods of cold weather when weeds are dormant

Crop-specific information
- HI 6 wk for all crops
- Apply to brassicas from mid-Oct to end-Feb provided crop has at least 4 true leaves (spring cabbage, spring greens), 3-4 true leaves (seed crops, oilseed rape)
- Apply to established lucerne or sainfoin from Nov to end-Feb
- Apply to established red or white clover from Feb to mid-Mar

SEE SECTION 3 FOR PRODUCTS ALSO REGISTERED

Pesticide Profiles

Following crops guidance
- Succeeding crops may be sown 2 wk after treatment for brassicas, field beans, 8 wk after treatment for peas, runner beans, 16 wk after treatment for cereals, maize
- Ploughing is not necessary before sowing subsequent crops

Environmental safety
- Do not graze crops for at least 6 wk after treatment
- Some pesticides pose a greater threat of contamination of water than others and carbetamide is one of these pesticides. Take special care when applying carbetamide near water and do not apply if heavy rain is forecast

Hazard classification and safety precautions
  UN number n/a
  Operator protection U20c
  Environmental protection E15a
  Storage and disposal D09a, D11a

56   carbon dioxide (commodity substance)

A gas for the control of trapped rodents and other vertebrates

Products
  carbon dioxide   various   99.9%   GA   -

Uses
- Birds in *traps*
- Mice in *traps*
- Rats in *traps*

Approval information
- Approval for the use of carbon dioxide as a commodity substance was granted on 8 October 1993 by Ministers under regulation 5 of the Control of Pesticides Regulations 1986
- Only to be used where a licence has been issued in accordance with Section 16(1) of the Wildlife and Countryside Act 1981
- Carbon dioxide is being supported for review in the fourth stage of the EC Review Programme under Directive 91/414. Whether or not it is included in Annex I, the existing commodity chemical approval will have to be revoked because substances listed in Annex I must be approved for marketing and use under the EC regime. If carbon dioxide is included in Annex I, products containing it will need to gain approval in the normal way if they carry label claims for pesticidal activity

Efficacy guidance
- Use to destroy trapped rodent pests
- Use to control birds covered by general licences issued by the Agriculture and Environment Departments under Section 16(1) of the Wildlife and Countryside Act (1981) for the control of opportunistic bird species, where birds have been trapped or stupefied with alphachloralose/seconal

Restrictions
- Operators must wear self-contained breathing apparatus when carbon dioxide levels are greater than 0.5% v/v
- Operators must be suitably trained and competent

Environmental safety
- Unprotected persons and non-target animals must be excluded from the treatment enclosures and surrounding areas unless the carbon dioxide levels are below 0.5% v/v

Hazard classification and safety precautions
  Operator protection G

57   carboxin

A carboxamide fungicide available only in mixtures
FRAC mode of action code: 7

FOR FULL CONDITIONS OF USE ALWAYS READ THE PRODUCT LABEL

58 carboxin + thiram

A fungicide seed dressing for cereals
FRAC mode of action code: 7 + M3

See also thiram

Products
    Anchor   Chemtura   200:200 g/l   FS   08684

Uses
- Bunt in **winter wheat** *(seed treatment)*
- Covered smut in **spring barley** *(seed treatment)*, **spring oats** *(seed treatment)*, **winter barley** *(seed treatment)*, **winter oats** *(seed treatment)*
- Fusarium foot rot and seedling blight in **spring barley** *(seed treatment)*, **spring oats** *(seed treatment)*, **spring rye** *(seed treatment)*, **spring wheat** *(seed treatment)*, **triticale** *(seed treatment)*, **winter barley** *(seed treatment)*, **winter oats** *(seed treatment)*, **winter rye** *(seed treatment)*, **winter wheat** *(seed treatment)*
- Leaf stripe in **spring barley** *(seed treatment)*, **winter barley** *(seed treatment - reduction)*
- Loose smut in **spring barley** *(seed treatment - reduction)*, **spring oats** *(seed treatment - reduction)*, **winter barley** *(seed treatment - reduction)*, **winter oats** *(seed treatment - reduction)*
- Net blotch in **spring barley** *(seed treatment)*
- Septoria seedling blight in **spring wheat** *(seed treatment)*, **winter wheat** *(seed treatment)*
- Stinking smut in **spring wheat** *(seed treatment)*

Approval information
- Thiram included in Annex I under EC Directive 91/414
- Accepted by BBPA for use on malting barley

Efficacy guidance
- Apply through suitable liquid flowable seed treating equipment of the batch treatment or continuous flow type where a secondary mixing auger is fitted
- Drill flow may be affected by treatment. Always re-calibrate seed drill before use

Restrictions
- Maximum number of treatments 1 per batch of seed
- Do not treat seed with moisture content above 16%
- Do not apply to cracked, split or sprouted seed
- Do not store treated seed from one season to the next

Crop-specific information
- Latest use: pre-drilling

Environmental safety
- Dangerous for the environment
- Very toxic to aquatic organisms
- Do not use treated seed as food or feed
- Treated seed harmful to game and wildlife

Hazard classification and safety precautions
    Hazard Harmful, Dangerous for the environment
    Transport code 9
    Packaging group III
    UN number 3082
    Risk phrases R22a, R48, R50, R53a
    Operator protection A, D, H; U05a, U09a, U20c
    Environmental protection E15a, E34, E38
    Storage and disposal D01, D02, D05, D06a, D09a, D10a
    Treated seed S01, S02, S04b, S05, S06a, S07
    Medical advice M03

SEE SECTION 3 FOR PRODUCTS ALSO REGISTERED

59   carfentrazone-ethyl

A triazolinone contact herbicide
HRAC mode of action code: E

Products

1	Aurora 50 WG	Belchim	50% w/w	WG	11613
2	Carfen	AgriGuard	50% w/w	WG	13005
3	Headlite	AgChem Access	60 g/l	ME	13624
4	Shark	Belchim	60 g/l	ME	12762
5	Spotlight Plus	Belchim	60 g/l	ME	12436

Uses

- Annual and perennial weeds in *forest* (off-label), **ornamental plant production** (off-label), *protected forest* (off-label), **protected ornamentals** (off-label) [4]
- Annual dicotyledons in **all edible crops (outdoor)** (before planting), **all non-edible crops (outdoor)** (before planting), **forest nurseries** (off-label), *potatoes*, **protected forest nurseries** (off-label), **protected soft fruit** (off-label), **soft fruit** (off-label) [4]; **grass seed crops** (off-label), **spring rye** (off-label), **winter rye** (off-label) [1]; *narcissi* (off-label) [5]
- Black bindweed in *potatoes* [4]
- Cleavers in **durum wheat**, **spring barley**, **spring oats**, **spring wheat**, **triticale**, **winter barley**, **winter oats**, **winter wheat** [1, 2]; *potatoes* [4]
- Desiccation in **blackberries** (off-label), **protected blackberries** (off-label), **protected raspberries** (off-label), **protected rubus hybrids** (off-label), *raspberries* (off-label), **rubus hybrids** (off-label) [4]
- Fat hen in *potatoes* [4]
- Haulm destruction in **seed potatoes**, **ware potatoes** [3, 5]
- Ivy-leaved speedwell in **durum wheat**, **spring barley**, **spring oats**, **spring wheat**, **triticale**, **winter barley**, **winter oats**, **winter wheat** [1, 2]; *potatoes* [4]
- Knotgrass in *potatoes* [4]
- Redshank in *potatoes* [4]
- Volunteer oilseed rape in **all edible crops (outdoor)** (before planting), **all non-edible crops (outdoor)** (before planting), *potatoes* [4]

Specific Off-Label Approvals (SOLAs)

- *blackberries* 20080551 [4]
- *forest* 20080552 [4]
- *forest nurseries* 20082922 [4]
- *grass seed crops* 20060954 [1]
- *narcissi* 20091003 [5]
- *ornamental plant production* 20080552 [4]
- *protected blackberries* 20080551 [4]
- *protected forest* 20080552 [4]
- *protected forest nurseries* 20082922 [4]
- *protected ornamentals* 20080552 [4]
- *protected raspberries* 20080551 [4]
- *protected rubus hybrids* 20080551 [4]
- *protected soft fruit* 20082922 [4]
- *raspberries* 20080551 [4]
- *rubus hybrids* 20080551 [4]
- *soft fruit* 20082922 [4]
- *spring rye* 20060954 [1]
- *winter rye* 20060954 [1]

Approval information

- Carfentrazone-ethyl included in Annex I under EC Directive 91/414
- Accepted by BBPA for use on malting barley and hops

Efficacy guidance

- Best weed control results achieved from good spray cover applied to small actively growing weeds [1, 2, 4]
- Carfentrazone-ethyl acts by contact only; see label for optimum timing on specified weeds. Weeds emerging after application will not be controlled [1, 2, 4]

FOR FULL CONDITIONS OF USE ALWAYS READ THE PRODUCT LABEL

carfentrazone-ethyl

- For weed control in cereals use as two spray programme with one application in autumn and one in spring [1, 2]
- Efficacy of haulm destruction will be reduced where flailed haulm covers the stems at application [5]
- For potato crops with very dense vigorous haulm or where regrowth occurs following a single application a second application may be necessary to achieve satisfactory desiccation. A minimum interval between applications of 7 d should be observed to achieve optimum performance [5]

Restrictions
- Maximum number of treatments 2 per crop for cereals (1 in Autumn and 1 in Spring) [1, 2]; 1 per crop for weed control in potatoes [4]; 1 per yr for pre-planting treatments [4]
- Maximum total dose for potato haulm destruction equivalent to 1.6 full dose treatments [5]
- Do not treat cereal crops under stress from drought, waterlogging, cold, pests, diseases, nutrient or lime deficiency or any factors reducing plant growth [1, 2]
- Do not treat cereals undersown with clover or other legumes [1, 2]
- Allow at least 2-3 wk between application and lifting potatoes to allow skins to set if potatoes are to be stored [5]
- Follow label instructions for sprayer cleaning
- Do not apply through knapsack sprayers
- Contact processor before using a split dose on potatoes for processing

Crop-specific information
- Latest use: 1 mth before planting edible and non-edible crops; before 3rd node detectable (GS 33) on cereals [1, 2]
- HI 7 d for potatoes [5]
- For weed control in cereals treat from 2 leaf stage [1, 2]
- When used as a treatment prior to planting a subsequent crop apply before weeds exceed maximum sizes indicated in the label [4]
- If treated potato tubers are to be stored then allow at least 2-3 wk between the final application and lifting to allow skins to set [5]

Following crops guidance
- No restrictions apply on the planting of succeeding crops 1 mth after application to potatoes for haulm destruction or as a pre-planting treatment, or 3 mth after application to cereals for weed control
- In the event of failure of a treated cereal crop, all cereals, ryegrass, maize, oilseed rape, peas, sunflowers, *Phacelia*, vetches, carrots or onions may be planted within 1 mth of treatment

Environmental safety
- Dangerous for the environment
- Very toxic to aquatic organisms
- Some non-target crops are sensitive. Avoid drift onto broad-leaved plants outside the treated area, or onto ponds waterways or ditches

Hazard classification and safety precautions
    Hazard Irritant, Dangerous for the environment
    Transport code 9
    Packaging group III
    UN number 3077 [1, 2]; 3082 [3-5]
    Risk phrases R38 [3, 5]; R43, R50, R53a [1-5]
    Operator protection A, H; U05a, U14 [1-5]; U08, U13 [1, 2]; U20a [3-5]
    Environmental protection E15a [1, 4]; E15b [2, 3, 5]; E34 [1, 2]; E38 [1-5]
    Storage and disposal D01, D02, D09a, D12a [1-5]; D10b [1, 2]; D10c [3-5]
    Medical advice M05a [1]

SEE SECTION 3 FOR PRODUCTS ALSO REGISTERED

60 carfentrazone-ethyl + flupyrsulfuron-methyl

A foliar and residual acting herbicide for cereals
HRAC mode of action code: E + B

See also flupyrsulfuron-methyl

Products
Lexus Class   DuPont   33.3:16.7% w/w   WG   10809

Uses
- Annual dicotyledons in **durum wheat** *(off-label)*, **grass seed crops** *(off-label)*, **triticale, winter oats, winter rye, winter wheat**
- Blackgrass in **triticale, winter oats, winter rye, winter wheat**

Specific Off-Label Approvals (SOLAs)
- *durum wheat* 20080895
- *grass seed crops* 20080895

Approval information
- Carfentrazone-ethyl and flupyrsulfuron-methyl included in Annex I under EC Directive 91/414

Efficacy guidance
- Best results obtained when applied to small actively growing weeds
- Good spray cover of weeds must be obtained
- Increased degradation of active ingredient in high soil temperatures reduces residual activity
- Weed control may be reduced in dry soil conditions but susceptible weeds germinating soon after treatment will be controlled if adequate soil moisture present
- Symptoms may not be apparent on some weeds for up to 4 wk, depending on weather conditions
- Blackgrass should be treated from 1 leaf but before first node
- Flupyrsulfuron-methyl is a member of the ALS-inhibitor group of herbicides. To avoid the build up of resistance do not use any product containing an ALS-inhibitor herbicide with claims for control of grass weeds more than once on any crop
- Use these products as part of a resistance management strategy that includes cultural methods of control and does not use ALS inhibitors as the sole chemical method of grass weed control

Restrictions
- Maximum number of treatments 1 per crop
- Do not use on barley, on crops undersown with grasses or legumes, or any other broad-leaved crop
- Do not apply within 7 d of rolling
- Do not treat any crop suffering from drought, waterlogging, pest or disease attack, nutrient deficiency, or any other stress factors
- Specific restrictions apply to use in sequence or tank mixture with other sulfonylurea or ALS-inhibiting herbicides. See label for details

Crop-specific information
- Latest use: before 31 Dec in yr of sowing for winter oats, winter rye, triticale; before 1st node detectable (GS 31) for winter wheat
- Slight chlorosis and stunting may occur in certain conditions. Recovery is rapid and yield not affected

Following crops guidance
- Only cereals, oilseed rape, field beans, clover or grass may be sown in the yr of harvest of a treated crop
- In the event of crop failure only winter or spring wheat may be sown 1-3 mth after treatment. Land should be ploughed and cultivated to 15 cm minimum before resowing

Environmental safety
- Dangerous for the environment
- Very toxic to aquatic organisms
- Take extreme care to avoid damage by drift outside the target area or onto surface waters or ditches

FOR FULL CONDITIONS OF USE ALWAYS READ THE PRODUCT LABEL

carfentrazone-ethyl + mecoprop-P

- Spraying equipment should not be drained or flushed onto land planted, or to be planted, with trees or crops other than cereals and should be thoroughly cleansed after use - see label for instructions

Hazard classification and safety precautions
Hazard Irritant, Dangerous for the environment
Transport code 9
Packaging group III
UN number 3077
Risk phrases R43, R50, R53a
Operator protection A, H; U05a, U08, U14, U19a, U20b
Environmental protection E15b, E38
Storage and disposal D01, D02, D09a, D10b, D11a, D12a

61 carfentrazone-ethyl + mecoprop-P

A foliar applied herbicide for cereals
HRAC mode of action code: E + O

See also mecoprop-P

Products

1	Jewel	Scotts	1.5:60% w/w	WG	14327
2	Pan Glory	Pan Amenity	1.5:60% w/w	WG	14487
3	Platform S	Belchim	1.5:60% w/w	WG	14380

Uses
- Annual and perennial weeds in **amenity grassland, managed amenity turf** [1, 2]
- Charlock in **spring barley, spring oats, spring wheat, winter barley, winter oats, winter wheat** [3]
- Chickweed in **spring barley, spring oats, spring wheat, winter barley, winter oats, winter wheat** [3]
- Cleavers in **spring barley, spring oats, spring wheat, winter barley, winter oats, winter wheat** [3]
- Field speedwell in **spring barley, spring oats, spring wheat, winter barley, winter oats, winter wheat** [3]
- Ivy-leaved speedwell in **spring barley, spring oats, spring wheat, winter barley, winter oats, winter wheat** [3]
- Moss in **amenity grassland, managed amenity turf** [1, 2]
- Red dead-nettle in **spring barley, spring oats, spring wheat, winter barley, winter oats, winter wheat** [3]

Approval information
- Carfentrazone-ethyl and mecoprop-P included in Annex I under EC Directive 91/414
- Accepted by BBPA for use on malting barley

Efficacy guidance
- Best results obtained when weeds have germinated and growing vigorously in warm moist conditions
- Treatment of large weeds and poor spray coverage may result in reduced weed control

Restrictions
- Maximum number of treatments 2 per crop. The total amount of mecoprop-P applied in a single yr must not exceed the maximum total dose approved for any single product for the crop/situation
- Do not treat crops suffering from stress from any cause
- Do not treat crops undersown or to be undersown

Crop-specific information
- Latest use: before 3rd node detectable (GS 33)
- Can be used on all varieties of wheat and barley in autumn or spring from the beginning of tillering
- Early sown crops may be prone to damage if treated after period of rapid growth in autumn

Following crops guidance
- In the event of crop failure, any cereal, maize, oilseed rape, peas, vetches or sunflowers may be sown 1 mth after a spring treatment. Any crop may be planted 3 mth after treatment

SEE SECTION 3 FOR PRODUCTS ALSO REGISTERED

Pesticide Profiles

Environmental safety
- Dangerous for the environment
- Very toxic to aquatic organisms
- Keep livestock out of treated areas for at least two weeks following treatment and until poisonous weeds, such as ragwort, have died down and become unpalatable
- Some pesticides pose a greater threat of contamination of water than others and mecoprop-P is one of these pesticides. Take special care when applying mecoprop-P near water and do not apply if heavy rain is forecast

Hazard classification and safety precautions
Hazard Harmful, Dangerous for the environment
Transport code 9
Packaging group III
UN number 3077
Risk phrases R22a, R41, R43, R50, R53a
Operator protection A, C, H [1-3]; M [3]; U05a, U11, U13, U14, U20b [1-3]; U08 [3]; U09a [1, 2]
Environmental protection E07a, E34, E38 [1-3]; E15a [3]; E15b [1, 2]
Storage and disposal D01, D02, D09a, D10b [1-3]; D12a [3]
Medical advice M05a

62   carfentrazone-ethyl + metsulfuron-methyl

A foliar applied herbicide mixture for cereals
HRAC mode of action code: E + B

See also metsulfuron-methyl

Products
Asset Express   AgriGuard   40:10% w/w   WG   14378

Uses
- Annual dicotyledons in **durum wheat**, **spring barley**, **spring oats**, **spring wheat**, **triticale**, **winter barley**, **winter oats**, **winter wheat**

Approval information
- Carfentrazone-ethyl and metsulfuron-methyl included in Annex I under EC Directive 91/414
- Accepted by BBPA for use on malting barley
- Approval expiry 30 Jun 2011 [1]

Efficacy guidance
- Best results achieved from applications made in good growing conditions
- Good spray cover of weeds must be obtained
- Growth of weeds is inhibited within hours of treatment but the time taken for visible colour changes to appear will vary according to species and weather
- Product has short residual life in soil. Under normal moisture conditions susceptible weeds germinating soon after treatment will be controlled
- Metsulfuron-methyl is a member of the ALS-inhibitor group of herbicides and products should be used in a planned Resistance Management strategy. See Section 5 for more information

Restrictions
- Maximum number of treatments 1 per crop
- Product must only be used after 1 Feb
- Do not use on any crop suffering stress from drought, waterlogging, cold, pest or disease attack, or nutrient deficiency
- Do not use on crops undersown with grass or legumes or on any broad-leaved crop
- Do not apply within 7 d of rolling
- Specific restrictions apply to use in sequence or tank mixture with other sulfonylurea or ALS-inhibiting herbicides. See label for details
- Do not tank mix with products containing chlorpyrifos, diclofop-methyl, difenzoquat, flamprop-methyl, flutriafol, propiconazole or tralkoxydim

Crop-specific information
- Latest use: before 3rd node detectable (GS 33)
- Can be used on all soil types

FOR FULL CONDITIONS OF USE ALWAYS READ THE PRODUCT LABEL

- Apply in the spring from the 3-leaf stage on all crops
- Slight necrotic spotting of crops can occur under certain crop and soil conditions. Recovery is rapid and there is no effect on grain yield or quality

Following crops guidance
- Only cereals, oilseed rape, field beans or grass may be sown as a following crop in the same calendar yr. Any crop may follow in the next spring
- In the event of failure of a treated crop, only winter wheat may be sown 1-3 mth later after ploughing and cultivating to at least 15 cm

Environmental safety
- Dangerous for the environment
- Very toxic to aquatic organisms
- Take extreme care to avoid drift onto broad-leaved plants outside the target area or onto ponds, waterways or ditches, or onto land intended for cropping
- Spraying equipment should not be drained or flushed onto land planted, or to be planted, with trees or crops other than cereals and should be thoroughly cleansed after use - see label for instructions

Hazard classification and safety precautions
Hazard Irritant, Dangerous for the environment
Transport code 9
Packaging group III
UN number 3077
Risk phrases R43, R50, R53a
Operator protection A, H; U05a, U08, U14, U19a, U20b
Environmental protection E15b, E38
Storage and disposal D01, D02, D09a, D11a, D12a

63   chlorantraniliprole

An ingested and contact insecticide for insect pest control in apples and pears and, in Ireland only, for colorado beetle control in potatoes.
IRAC mode of action code: 28

Products
Coragen   DuPont   200 g/l   SC   14930

Uses
- Codling moth in *apples*, *pears*
- Colorado beetle in *potatoes*

Efficacy guidance
- Can be used as part of an Integrated Pest Management programme.
- For best fruit protection in apples and pears, apply before egg hatch.

Environmental safety
- Broadcast air-assisted LERAP (10 m)

Hazard classification and safety precautions
Hazard Dangerous for the environment
Transport code 9
Packaging group III
UN number 3082
Risk phrases R50, R53a
Environmental protection E15b, E22b, E38; E17b (10 m)
Storage and disposal D01, D02, D05, D09a, D12a

SEE SECTION 3 FOR PRODUCTS ALSO REGISTERED

Pesticide Profiles

64 chloridazon

A residual pyridazinone herbicide for beet crops
HRAC mode of action code: C1

Products

1	Better DF	Sipcam	65% w/w	SG	06250
2	Better Flowable	Sipcam	430 g/l	SC	04924
3	Parador	United Phosphorus	430 g/l	SC	12372
4	Pyramin DF	BASF	65% w/w	WG	03438
5	Takron	BASF	430 g/l	SC	11627
6	Tempest	AgriGuard	430 g/l	SC	12688

Uses

- Annual dicotyledons in **beet leaves** *(off-label)*, **chard** *(off-label)*, **herbs (see appendix 6)** *(off-label)*, **spinach** *(off-label)*, **spinach beet** *(off-label)* [4]; **bulb onions** *(off-label)*, **leeks** *(off-label)*, **salad onions** *(off-label)* [1, 4]; **fodder beet**, **mangels**, **sugar beet** [1-6]; **garlic** *(off-label)*, **shallots** *(off-label)* [1]
- Annual meadow grass in **beet leaves** *(off-label)*, **bulb onions** *(off-label)*, **chard** *(off-label)*, **herbs (see appendix 6)** *(off-label)*, **leeks** *(off-label)*, **salad onions** *(off-label)*, **spinach** *(off-label)*, **spinach beet** *(off-label)* [4]; **fodder beet**, **mangels**, **sugar beet** [1-6]

Specific Off-Label Approvals (SOLAs)

- **beet leaves** *20101103* [4]
- **bulb onions** *20101088* [1], *20101102* [4]
- **chard** *20101103* [4]
- **garlic** *20101088* [1]
- **herbs (see appendix 6)** *20101103* [4]
- **leeks** *20101088* [1], *20101102* [4]
- **salad onions** *20101088* [1], *20101102* [4]
- **shallots** *20101088* [1]
- **spinach** *20101103* [4]
- **spinach beet** *20101103* [4]

Approval information

- Chloridazon included in Annex I under EC Directive 91/414

Efficacy guidance

- Absorbed by roots of germinating weeds and best results achieved pre-emergence of weeds or crop when soil moist and adequate rain falls after application
- Application rate for some products depends on soil type. Check label for details

Restrictions

- Maximum number of treatments generally 1 per crop (pre-emergence) for fodder beet and mangels; 1 (pre-emergence) + 3 (post-emergence) per crop for sugar beet, but labels vary slightly. Check labels for maximum total dose for the crop to be treated
- Maximum total dose for onions and leeks equivalent to one full dose recommended for these crops [4]
- Do not use on Coarse Sands, Sands or Fine Sands or where organic matter exceeds 5%
- A maximum total dose of 2.6 kg/ha chloridazon may only be applied every third year on the same field

Crop-specific information

- Latest use: pre-emergence for fodder beet and mangels; normally before leaves of crop meet in row for sugar beet, but labels vary slightly; up to and including second true leaf stage for onions and leeks [4]
- Where used pre-emergence spray as soon as possible after drilling in mid-Mar to mid-Apr on fine, firm, clod-free seedbed
- Where crop drilled after mid-Apr or soil dry apply pre-drilling and incorporate to 2.5 cm immediately afterwards
- Various tank mixes recommended on sugar beet for pre- and post-emergence use and as repeated low dose treatments. See label for details
- Crop vigour may be reduced by treatment of crops growing under unfavourable conditions including poor tilth, drilling at incorrect depth, soil capping, physical damage, pest or disease

FOR FULL CONDITIONS OF USE ALWAYS READ THE PRODUCT LABEL

damage, excess seed dressing, trace-element deficiency or a sudden rise in temperature after a cold spell

Following crops guidance
- Winter cereals or any spring sown crop may follow a treated beet crop harvested at the normal time and after ploughing to at least 150 mm
- In the event of failure of a treated crop only a beet crop or maize may be drilled, after cultivation

Environmental safety
- Dangerous for the environment
- Very toxic to aquatic organisms

Hazard classification and safety precautions
Hazard Harmful [1, 4]; Irritant [2, 3, 5, 6]; Dangerous for the environment [1-6]
Transport code 9 [3-6]
Packaging group III [3-6]
UN number 3077 [4]; 3082 [3, 5, 6]; N/C [1, 2]
Risk phrases R22a [1, 4]; R36, R37, R38 [6]; R43 [2, 3, 5, 6]; R50, R53a [1-6]
Operator protection A [3, 5, 6]; U05a, U08, U20b [1-6]; U14, U19a [2, 3, 5, 6]; U15 [6]
Environmental protection E13b [3]; E15a [1-6]; E38 [1-5]
Storage and disposal D01, D02, D09a [1-6]; D10b [2, 3, 5, 6]; D10c [3]; D11a [1, 4]; D12a [1-5]
Medical advice M05a

65 chloridazon + ethofumesate

A contact and residual herbicide for beet crops
HRAC mode of action code: C1 + N

See also ethofumesate

Products
Magnum	BASF	275:170 g/l	SC	11727

Uses
- Annual dicotyledons in **fodder beet, sugar beet**
- Annual meadow grass in **fodder beet, sugar beet**

Approval information
- Chloridazon and ethofumesate included in Annex I under EC Directive 91/414

Efficacy guidance
- Best results achieved on a fine firm clod-free seedbed when soil moist and adequate rain falls after spraying. Efficacy and crop safety may be reduced if heavy rain falls just after incorporation
- Effectiveness may be reduced under conditions of low pH
- May be applied by conventional or repeat low dose method. See label for details

Restrictions
- Maximum number of treatments 1 pre-emergence for fodder beet; 1 pre-emergence plus 3 post-emergence for sugar beet
- May be used on soil classes Loamy Sand - Silty Clay Loam. Additional restrictions apply for some tank mixtures
- Do not treat beet post-emergence with recommended tank mixtures when temperature is, or is likely to be, above 21 °C on day of spraying or under conditions of high light intensity
- If a mixture with phenmedipham is applied to a crop previously treated with a pre-emergence herbicide and the crop is suffering from stress from whatever cause, no further such post-emergence applications may be made
- A maximum total dose of 2.6 kg/ha chloridazon may only be applied every third year on the same field.
- The maximum total dose must not exceed 1.0 kg ethofumesate per hectare in any three year period

Crop-specific information
- Latest use: pre-crop emergence for fodder beet; before crop leaves meet between rows for sugar beet
- Apply up to cotyledon stage of weeds

SEE SECTION 3 FOR PRODUCTS ALSO REGISTERED

Pesticide Profiles

- Crop vigour may be reduced by treatment of crops growing under unfavourable conditions including poor tilth, drilling at incorrect depth, soil capping, physical damage, excess nitrogen, excess seed dressing, trace element deficiency or a sudden rise in temperature after a cold spell. Frost after pre-emergence treatment may check crop growth

Following crops guidance
- In the event of crop failure only sugar beet, fodder beet or mangels may be re-drilled
- Any crop may be sown 3 mth after spraying following ploughing to 15 cm

Environmental safety
- Dangerous for the environment
- Very toxic to aquatic organisms

Hazard classification and safety precautions
Hazard Harmful, Dangerous for the environment
Transport code 9
Packaging group III
UN number 3082
Risk phrases R22a, R50, R53a
Operator protection A, H; U05a, U08, U19a, U20a
Environmental protection E15a, E38
Storage and disposal D01, D02, D09a, D10b, D12a
Medical advice M05a

66 chloridazon + metamitron

A contact and residual herbicide mixture for sugar beet
HRAC mode of action code: C1 + C1

See also metamitron

Products
    Volcan Combi   Sipcam   300:280 g/l   SC   10256

Uses
- Annual dicotyledons in **fodder beet** *(off-label)*, **sugar beet**
- Annual meadow grass in **fodder beet** *(off-label)*, **sugar beet**

Specific Off-Label Approvals (SOLAs)
- *fodder beet 20101104*

Approval information
- Chloridazon included in Annex I under EC Directive 91/414

Efficacy guidance
- Best results achieved from a sequential programmme of sprays
- Pre-emergence use improves efficacy of post-emergence programme. Best results obtained from application to a moist seed bed
- First post-emergence treatment should be made when weeds at early cotyledon stage and subsequent applications made when new weed flushes reach this stage
- Weeds surviving an earlier treatment should be treated again after 7-10 d even if no new weeds have appeared

Restrictions
- Maximum number of treatments 1 pre-emergence followed by 3 post-emergence
- Take advice if light soils have a high proportion of stones

Crop-specific information
- Latest use: when leaves of crop meet in rows
- Tolerance of crops growing under stress from any cause may be reduced
- Crops treated pre-emergence and subsequently subjected to frost may be checked and recovery may not be complete

Following crops guidance
- After the last application only sugar beet or mangels may be sown within 4 mth; cereals may be sown after 16 wk. Land should be mouldboard ploughed to 15 cm and thoroughly cultivated before any succeeding crop

FOR FULL CONDITIONS OF USE ALWAYS READ THE PRODUCT LABEL

chloridazon + quinmerac

Environmental safety
- Harmful to fish or other aquatic life. Do not contaminate surface waters or ditches with chemical or used container

Hazard classification and safety precautions
Hazard Irritant
UN number N/C
Risk phrases R36, R38
Operator protection U08, U14, U15, U20a
Environmental protection E13c, E34
Storage and disposal D09a, D10b

67   chloridazon + quinmerac

A herbicide mixture for use in beet crops
HRAC mode of action code: C1 + O

See also quinmerac

Products
Fiesta T   BASF   360:60 g/l   SC   11734

Uses
- Annual dicotyledons in **beet leaves** *(off-label)*, **chard** *(off-label)*, **fodder beet**, **herbs (see appendix 6)** *(off-label)*, **mangels**, **spinach** *(off-label)*, **spinach beet** *(off-label)*, **sugar beet**
- Annual meadow grass in **fodder beet**, **mangels**, **sugar beet**
- Chickweed in **fodder beet**, **mangels**, **sugar beet**
- Cleavers in **beet leaves** *(off-label)*, **chard** *(off-label)*, **fodder beet**, **herbs (see appendix 6)** *(off-label)*, **mangels**, **spinach** *(off-label)*, **spinach beet** *(off-label)*, **sugar beet**
- Field speedwell in **fodder beet**, **mangels**, **sugar beet**
- Ivy-leaved speedwell in **fodder beet**, **mangels**, **sugar beet**
- Mayweeds in **fodder beet**, **mangels**, **sugar beet**
- Poppies in **fodder beet**, **mangels**, **sugar beet**

Specific Off-Label Approvals (SOLAs)
- *beet leaves* 20101086
- *chard* 20101086
- *herbs (see appendix 6)* 20101086
- *spinach* 20101086
- *spinach beet* 20101086

Approval information
- Chloridazon and quinmerac included in Annex I under EC Directive 91/414.

Efficacy guidance
- Best results obtained when adequate soil moisture is present at application and afterwards to form an active herbicidal layer in the soil
- A programme of pre-emergence treatment followed by post-emergence application(s) in mixture with a contact herbicide optimises weed control and is essential for some difficult weed species such as cleavers
- Treatment pre-emergence only may not provide sufficient residual activity to give season-long weed control
- Effectiveness may be reduced under conditions of low pH
- Always follow WRAG guidelines for preventing and managing herbicide resistant weeds. See Section 5 for more information

Restrictions
- Maximum total dose on sugar beet equivalent to one treatment pre-emergence plus two treatments post-emergence, all at maximum individual dose; maximum total dose on fodder beet and mangels equivalent to one full dose pre-emergence treatment
- Heavy rain shortly after treatment may check crop growth particularly when it leaves water standing in surface depressions
- Treatment of stressed crops or those growing in unfavourable conditions may depress crop vigour and possibly reduce stand

SEE SECTION 3 FOR PRODUCTS ALSO REGISTERED

Pesticide Profiles

- Do not use on Sands or soils of high organic matter content
- Where rates of nitrogen higher than those generally recommended are considered necessary, apply at least 3 wk before drilling
- A maximum total dose of 2.6 kg/ha chloridazon may only be applied every third year on the same field

Crop-specific information
- Latest use: before plants meet between the rows for sugar beet; pre-emergence for fodder beet, mangels

Following crops guidance
- In the spring following normal harvest of a treated crop sow only cereals, beet or mangel crops, potatoes or field beans, after ploughing
- In the event of failure of a treated crop only a beet crop may be drilled, after cultivation

Environmental safety
- Dangerous for the environment
- Toxic to aquatic organisms
- To reduce movement to groundwater do not apply to dry soil or when heavy rain is forecast

Hazard classification and safety precautions
Hazard Irritant, Dangerous for the environment
Transport code 9
Packaging group III
UN number 3082
Risk phrases R43, R51, R53a
Operator protection A; U05a, U08, U14, U19a, U20b
Environmental protection E15a, E38
Storage and disposal D01, D02, D08, D09a, D10c, D12a
Medical advice M03, M05a

68 chlormequat

A plant-growth regulator for reducing stem growth and lodging

Products

1	3C Chlormequat 720	BASF	720 g/l	SL	13973
2	Adjust	Taminco	620 g/l	SL	13961
3	Agriguard Chlormequat 700	AgriGuard	700 g/l	SL	09782
4	Agriguard Chlormequat 720	AgriGuard	720 g/l	SL	09919
5	Agrovista 3 See 750	Agrovista	750 g/l	SL	14797
6	Barclay Holdup	Barclay	700 g/l	SL	11365
7	Barleyquat B	Taminco	620 g/l	SL	13962
8	Belcocel	Taminco	720 g/l	SL	11881
9	Bettaquat B	Taminco	620 g/l	SL	13965
10	CCC 720	European Ag	720 g/l	SL	14891
11	Clayton Coldstream	Clayton	400 g/l	SL	12724
12	Clayton Standup	Clayton	700 g/l	SL	11760
13	CS Chlormequat	ChemSource	700 g/l	SL	14946
14	Fargro Chlormequat	Fargro	460 g/l	SL	02600
15	Greencrop Carna	Greencrop	600 g/l	SL	09403
16	Hive	Nufarm UK	730 g/l	SL	11392
17	K2	Taminco	620 g/l	SL	13964
18	Mandops chlormequat	Taminco	700 g/l	SL	13970
19	Manipulator	Taminco	620 g/l	SL	13963
20	Mirquat	Nufarm UK	730 g/l	SL	11406
21	New 5C Cycocel	BASF	645 g/l	SL	01482
22	New 5C Quintacel	Nufarm UK	645 g/l	SL	12074
23	Selon	Taminco	620 g/l	SL	13966
24	Sigma PCT	Nufarm UK	460 g/l	SL	11209
25	Stabilan 700	Nufarm UK	700 g/l	SL	11393
26	Standon Girder 720	Standon	720 g/l	SL	14251
27	Terbine	Nufarm UK	730 g/l	SL	11407

Uses
- Growth regulation in **durum wheat** *(off-label)* [5, 8, 9, 16, 17, 21, 25]; **ornamental plant production** [5]; **spring barley** [2, 7, 19]; **spring durum wheat** *(off-label)*, **spring triticale**

FOR FULL CONDITIONS OF USE ALWAYS READ THE PRODUCT LABEL

chlormequat

  (off-label) [17]; **spring oats**, **spring wheat**, **winter barley**, **winter oats**, **winter wheat** [19, 26]; **spring rye** (off-label) [8, 9, 17, 18]; **triticale** [13, 16, 25, 27]; **triticale** (off-label) [9, 17]; **winter rye** [26]; **winter rye** (off-label) [5, 8, 9, 17, 18]
- Growth retardation in **spring barley** [17]
- Increasing yield in **spring barley** [2, 7, 18, 23]; **spring oats**, **winter oats** [2, 7, 23]; **winter barley** [1-3, 6, 7, 10, 15, 18, 21-23]
- Lodging control in **durum wheat** (off-label) [2, 9, 19]; **spring barley** [2, 7, 17-19, 23]; **spring oats**, **winter oats** [1-8, 10-13, 15-23, 25, 27]; **spring rye** [1, 5, 10, 21, 22]; **spring rye** (off-label), **winter rye** (off-label) [2, 7, 9, 18, 19]; **spring wheat**, **winter wheat** [1-6, 8-13, 15-25, 27]; **triticale** [1, 5, 10, 13, 16, 20-22, 25, 27]; **triticale** (off-label) [2, 7, 9, 19]; **winter barley** [1-8, 10-13, 16-20, 23-25, 27]; **winter rye** [1, 3-5, 10-13, 16, 20-22, 25, 27]
- Stem shortening in **bedding plants**, **camellias**, **hibiscus trionum**, **lilies** [12-14, 16, 20, 25, 27]; **geraniums**, **poinsettias** [1, 10, 12-14, 16, 20-22, 25, 27]

Specific Off-Label Approvals (SOLAs)
- **durum wheat** 20081166 [2], 20101081 [5], 20060952 [8], 20081173 [9], 20063619 [16], 20081175 [17], 20081179 [19], 20061298 [21], 20063487 [25]
- **spring durum wheat** 20081175 [17]
- **spring rye** 20081166 [2], 20081170 [7], 20060952 [8], 20081173 [9], 20081175 [17], 20081177 [18], 20081179 [19]
- **spring triticale** 20081175 [17]
- **triticale** 20081166 [2], 20081170 [7], 20081173 [9], 20081175 [17], 20081179 [19]
- **winter rye** 20081166 [2], 20101081 [5], 20081170 [7], 20060952 [8], 20081173 [9], 20081175 [17], 20081177 [18], 20081179 [19]

Approval information
- Chlormequat included in Annex 1 under EC Directive 91/414
- Approved for aerial application on wheat and oats [3, 15, 16, 20, 24]; on winter barley [15]; on triticale. See Section 5 for more information
- Accepted by BBPA for use on malting barley
- Approval expiry 30 Nov 2011 [3, 4, 12, 15]

Efficacy guidance
- Most effective results on cereals normally achieved from application from Apr onwards, on wheat and rye from leaf sheath erect to first node detectable (GS 30-31), on oats at second node detectable (GS 32), on winter barley from mid-tillering to leaf sheath erect (GS 25-30). However, recommendations vary with product. See label for details (n.b. [14] only approved on flowers)
- Influence on growth varies with crop and growth stage. Risk of lodging reduced by application at early stem extension. Root development and yield can be improved by earlier treatment
- Results on barley can be variable
- In tank mixes with other pesticides on cereals optimum timing for herbicide action may differ from that for growth reduction. See label for details of tank mix recommendations
- Most products recommended for use on oats require addition of approved non-ionic wetter. Check label
- Some products are formulated with trace elements to help compensate for increased demand during rapid growth

Restrictions
- Maximum number of treatments or maximum total dose varies with crop and product and whether split dose treatments are recommended. Check labels
- Do not use on very late sown spring wheat or oats or on crops under stress (n.b. [14] only approved on flowers)
- Mixtures with liquid nitrogen fertilizers may cause scorch and are specifically excluded on some labels
- Do not use on soils of low fertility unless such crops regularly receive adequate dressings of nitrogen
- At least 6 h, preferably 24 h, required before rain for maximum effectiveness. Do not apply to wet crops
- Check labels for tank mixtures known to be incompatible
- Not to be used on food crops [14]

Crop-specific information
- Latest use varies with crop and product. See label for details

SEE SECTION 3 FOR PRODUCTS ALSO REGISTERED

Pesticide Profiles

- May be used on cereals undersown with grass or clovers (n.b. [14] only approved on flowers)
- Ornamentals to be treated must be well established and growing vigorously. Do not treat in strong sunlight or when temperatures are likely to fall below 10 °C
- Temporary yellow spotting may occur on poinsettias. It can be minimised by use of a non-ionic wetting agent - see label

Environmental safety
- Harmful to aquatic organisms [12, 14, 16, 20, 25]
- Wash equipment thoroughly with water and wetting agent immediately after use and spray out. Spray out again before storing or using for another product. Traces can cause harm to susceptible crops sprayed later
- Do not use straw from treated cereals as horticultural growth medium or mulch [21]

Hazard classification and safety precautions
Hazard Harmful [1-5, 7-18, 20-27]; Irritant [6]
Transport code 8 [1, 3-6, 8-15, 25-27]
Packaging group III [1, 3-6, 8-15, 25-27]
UN number 1760 [1, 3-6, 8-15, 25-27]
Risk phrases R21 [3-5, 11, 26]; R22a [1-18, 20-27]; R36 [18]; R52 [2, 7, 12-14, 16, 18, 20, 23, 25, 26]; R53a [2, 7, 12-14, 16, 20, 23, 25, 26]
Operator protection A [1-18, 20-27]; C [18]; U05a, U19a [1-18, 20-27]; U08 [1-12, 14, 15, 17, 18, 21, 23, 24, 27]; U09a [13, 16, 20, 22, 25, 26]; U13 [6, 14]; U20a [6, 22]; U20b [1-5, 7-18, 20, 21, 23-27]
Environmental protection E15a [1, 2, 4-18, 20-27]; E34 [1-18, 20-27]
Consumer protection C01 [6]
Storage and disposal D01, D02, D09a [1-18, 20-27]; D05 [3-6, 11-16, 20, 22, 25]; D10a [6, 12, 13, 25, 26]; D10b [2-5, 7-9, 11, 14-18, 20, 23, 24, 27]; D10c [1, 10, 21, 22]; D12a [11]
Medical advice M03 [1-18, 20-27]; M05a [1, 10, 12-14, 16, 20-22, 25]

69   chlormequat + 2-chloroethylphosphonic acid

A plant growth regulator for use in cereals

See also 2-chloroethylphosphonic acid

Products
1	Greencrop Tycoon	Greencrop	305:155 g/l	SL 09571
2	Strate	Bayer CropScience	360:180 g/l	SL 10020
3	Upgrade	Bayer CropScience	360:180 g/l	SL 10029

Uses
- Growth regulation in **durum wheat** *(off-label)*, **spring rye** *(off-label)*, **triticale** *(off-label)*, **winter rye** *(off-label)* [2, 3]
- Lodging control in **spring barley**, **winter barley**, **winter wheat** [1-3]

Specific Off-Label Approvals (SOLAs)
- *durum wheat* 20061541 [2], 20063100 [3]
- *spring rye* 20061541 [2], 20063100 [3]
- *triticale* 20061541 [2], 20063100 [3]
- *winter rye* 20061541 [2], 20063100 [3]

Approval information
- Chlormequat and 2-chloroethylphosphonic acid (ethephon) included in Annex I under EC Directive 91/414
- Accepted by BBPA for use on malting barley
- In 2006 CRD required that all products containing 2-chloroethylphosphonic acid should carry the following warning in the main area of the container label: "2-chloroethylphosphonic acid is an anticholinesterase organophosphate. Handle with care"

Efficacy guidance
- Best results obtained when crops growing vigorously
- Recommended dose varies with growth stage. See labels for details and recommendations for use of sequential treatments

FOR FULL CONDITIONS OF USE ALWAYS READ THE PRODUCT LABEL

chlormequat + 2-chloroethylphosphonic acid + mepiquat chloride

Restrictions
- 2-chloroethylphosphonic acid is an anticholinesterase organophosphorus compound. Do not use if under medical advice not to work with such compounds
- Maximum number of treatments 1 per crop; maximum total dose equivalent to one full dose treatment
- Product must always be used with specified non-ionic wetter - see labels
- Do not use on any crop in sequence with any other product containing 2-chloroethylphosphonic acid
- Do not spray when crop wet or rain imminent
- Do not spray during cold weather or periods of night frost, when soil is very dry, when crop diseased or suffering pest damage, nutrient deficiency or herbicide stress
- If used on seed crops grown for certification inform seed merchant beforehand
- Do not use on wheat variety Moulin or on any winter varieties sown in spring [1]
- Do not use on spring barley variety Triumph [1]
- Do not treat barley on soils with more than 10% organic matter [1]
- Do not use in programme with any other product containing 2-chloroethylphosphonic acid [1]
- Only crops growing under conditions of high fertility should be treated

Crop-specific information
- Latest use: before flag leaf ligule/collar just visible (GS 39) or 1st spikelet visible (GS 51) for wheat or barley at top dose; or before flag leaf sheath opening (GS 47) for winter wheat at reduced dose
- Apply before lodging has started

Environmental safety
- Dangerous for the environment [2, 3]
- Harmful to aquatic organisms [1]
- Harmful to fish or other aquatic life. Do not contaminate surface waters or ditches with chemical or used container [2, 3]
- Do not use straw from treated cereals as a horticultural growth medium or as a mulch [1]

Hazard classification and safety precautions
Hazard Harmful [1-3]; Dangerous for the environment [2, 3]
Transport code 8
Packaging group III
UN number 3265
Risk phrases R22a, R37 [1-3]; R41 [2, 3]; R52 [1]
Operator protection A [1-3]; C [2, 3]; U05a, U08, U19a, U20b [1-3]; U11, U13 [2, 3]
Environmental protection E13c [2, 3]; E15a [1]; E34 [1-3]
Storage and disposal D01, D02, D09a, D10b [1-3]; D05 [2, 3]
Medical advice M01, M03 [1-3]; M05a [1]

70 chlormequat + 2-chloroethylphosphonic acid + mepiquat chloride

A plant growth regulator for reducing lodging in cereals

See also 2-chloroethylphosphonic acid
mepiquat chloride

Products
Cyclade   BASF   230:155:75 g/l   SL   08958

Uses
- Growth regulation in **durum wheat** (off-label), **spring rye** (off-label), **triticale** (off-label), **winter rye** (off-label)
- Lodging control in **spring barley**, **winter barley**, **winter wheat**

Specific Off-Label Approvals (SOLAs)
- *durum wheat* 20063098
- *spring rye* 20063098
- *triticale* 20063098
- *winter rye* 20063098

SEE SECTION 3 FOR PRODUCTS ALSO REGISTERED

Approval information
- Chlormequat and 2-chloroethylphosphonic acid (ethephon) and mepiquat chloride included in Annex I under EC Directive 91/414
- Accepted by BBPA for use on malting barley
- In 2006 CRD required that all products containing 2-chloroethylphosphonic acid should carry the following warning in the main area of the container label: "2-chloroethylphosphonic acid is an anticholinesterase organophosphate. Handle with care"

Efficacy guidance
- Best results achieved in a vigorous, actively growing crop with adequate fertility and moisture
- Optimum timing on all crops is from second node detectable stage (GS 32)
- Recommended for use as part of an intensive growing system which includes provision for optimum fertilizer treatment and disease control

Restrictions
- 2-chloroethylphosphonic acid is an anticholinesterase organophosphorus compound. Do not use if under medical advice not to work with such compounds
- Maximum number of treatments 1 per crop
- Maximum total dose depends on spraying regime adopted. See label
- Must be used with a non-ionic wetting agent
- Do not apply to stressed crops or those on soils of low fertility unless receiving adequate dressings of fertilizer
- Do not apply in temperatures above 21 °C or if crop is wet or if rain expected
- Do not treat variety Moulin nor any winter varieties sown in spring
- Do not use in a programme with any other product containing 2-chloroethylphosphonic acid
- Do not apply to barley on soils with more than 10% organic matter (winter wheat may be treated)
- Notify seed merchant in advance if use on a seed crop is proposed

Crop-specific information
- Latest use: before first spikelet of ear visible (GS 51) using reduced dose on winter barley; before flag leaf sheath opening (GS 47) using reduced dose on winter wheat; before flag leaf just visible on spring barley
- May be applied to crops undersown with grasses or clovers
- Treatment may cause some delay in ear emergence

Environmental safety
- Harmful to aquatic organisms
- Do not use straw from treated crops as a horticultural growth medium

Hazard classification and safety precautions
Hazard Harmful
Transport code 8
Packaging group III
UN number 3265
Risk phrases R22a, R37, R52
Operator protection A; U05a, U08, U19a, U20b
Environmental protection E15a, E34
Storage and disposal D01, D02, D09a, D10c
Medical advice M01, M03, M05a

71 chlormequat + imazaquin

A plant growth regulator mixture for winter wheat

See also imazaquin

Products
1	Meteor	BASF	368:0.8 g/l	SL	10403
2	Standon Imazaquin 5C	Standon	368:0.8 g/l	SL	08813
3	Upright	BASF	368:0.8 g/l	SL	10404

Uses
- Growth regulation in **spring rye** *(off-label)*, **triticale** *(off-label)*, **winter rye** *(off-label)* [1]
- Increasing yield in **winter wheat** [1-3]
- Lodging control in **winter wheat** [1-3]

FOR FULL CONDITIONS OF USE ALWAYS READ THE PRODUCT LABEL

chlormequat + mepiquat chloride

Specific Off-Label Approvals (SOLAs)
- ***spring rye*** *20063099* [1]
- ***triticale*** *20063099* [1]
- ***winter rye*** *20063099* [1]

Approval information
- Imazaquin included in Annex I under EC Directive 91/414

Efficacy guidance
- Apply to crops during good growing conditions or to those at risk from lodging
- On soils of low fertility, best results obtained where adequate nitrogen fertilizer used

Restrictions
- Maximum number of applications 1 per crop (2 per crop at split dose)
- Do not treat durum wheat
- Do not apply to undersown crops
- Do not apply when crop wet or rain imminent

Crop-specific information
- Latest use: before second node detectable (GS 31)
- Apply as single dose from leaf sheath lengthening up to and including 1st node detectable or as split dose, the first from tillers formed to leaf sheath lengthening, the second from leaf sheath erect up to and including 1st node detectable

Environmental safety
- Dangerous for the environment
- Toxic to aquatic organisms
- Do not use straw from treated cereals as horticultural growth medium or mulch

Hazard classification and safety precautions
Hazard Harmful [1-3]; Dangerous for the environment [2]
Transport code 8
Packaging group III
UN number 1760
Risk phrases R20, R52 [1, 3]; R22a, R36, R53a [1-3]; R51 [2]
Operator protection A, C; U05a, U08, U13, U19a, U20b [1-3]; U11, U15 [1, 3]
Environmental protection E15a, E34
Storage and disposal D01, D02, D05, D09a, D10b [1-3]; D06c, D12a [1, 3]
Medical advice M03 [1-3]; M05a [1, 3]

72   chlormequat + mepiquat chloride

A plant growth regulator for reducing lodging in wheat

See also mepiquat chloride

Products

Stronghold   BASF   345:115 g/l   SL   09134

Uses
- Lodging control in **winter wheat**

Approval information
- Chlormequat and mepiquat chloride included in Annex I under EC Directive 91/414

Efficacy guidance
- Optimum timing is when leaf sheaths erect (GS 30)
- Benefit will vary according to crop and stage of growth at application

Restrictions
- Maximum total dose equivalent to one full dose treatment
- Do not apply to stressed crops or those on soils of low fertility unless receiving adequate dressings of fertilizer
- Do not treat crops where significant foot diseases, especially take-all, are expected
- Do not treat crops on soils of low fertility
- Do not apply in temperatures above 21 °C or if crop is wet or if rain expected

SEE SECTION 3 FOR PRODUCTS ALSO REGISTERED

Pesticide Profiles

- Do not treat any winter varieties sown in spring
- Notify seed merchant in advance if use on a seed crop is proposed

Crop-specific information
- Latest use: before 3rd node detectable (GS 33)
- Apply during good growing conditions at the correct timings - see label
- May be applied to crops undersown with grasses or clovers
- Treatment may cause some delay in ear emergence
- Mixtures with liquid fertilizers may cause scorching in some circumstances

Environmental safety
- Do not use straw from treated crops as a horticultural growth medium or mulch

Hazard classification and safety precautions
Hazard Harmful
Transport code 8
Packaging group III
UN number 1760
Risk phrases R22a
Operator protection A; U05a, U08, U19a, U20b
Environmental protection E15a, E34
Storage and disposal D01, D02, D09a, D10c
Medical advice M03, M05a

73   2-chloroethylphosphonic acid

A plant growth regulator for cereals and various horticultural crops

See also chlormequat + 2-chloroethylphosphonic acid
chlormequat + 2-chloroethylphosphonic acid + imazaquin
chlormequat + 2-chloroethylphosphonic acid + mepiquat chloride

Products

1	Agriguard Cerusite	AgriGuard	480 g/l	SL	11494
2	Agrotech Ethephon	AgChem Access	480 g/l	SL	12170
3	Becki	ChemSource	480 g/l	SL	14677
4	Cerone	Bayer CropScience	480 g/l	SL	09985
5	Ethefon 480	Goldengrass	480 g/l	SL	14392
6	Pan Ethephon	Pan Agriculture	480 g/l	SL	13957
7	Pan Stiffen	Pan Agriculture	480 g/l	SL	13956

Uses
- Growth regulation in **apple for cider making** (off-label), **apples** (off-label), **durum wheat** (off-label), **protected ornamentals** (off-label), **protected tomatoes** (off-label) [4]
- Lodging control in **spring barley** [1-5]; **triticale, winter barley, winter rye, winter wheat** [1-7]

Specific Off-Label Approvals (SOLAs)
- *apple for cider making* 20091381 [4]
- *apples* 20102501 [4]
- *durum wheat* 20061195 [4]
- *protected ornamentals* 20091382 [4]
- *protected tomatoes* 20091383 [4]

Approval information
- 2-chloroethylphosphonic acid (ethephon) included in Annex I under EC Directive 91/414
- Approved for aerial application on winter barley [1, 4]. See Section 5 for more information
- In 2006 CRD required that all products containing this active ingredient should carry the following warning in the main area of the container label: "2-chloroethylphosphonic acid is an anticholinesterase organophosphate. Handle with care"
- Accepted by BBPA for use on malting barley

Efficacy guidance
- Best results achieved on crops growing vigorously under conditions of high fertility
- Optimum timing varies between crops and products. See labels for details
- Do not spray crops when wet or if rain imminent

FOR FULL CONDITIONS OF USE ALWAYS READ THE PRODUCT LABEL

Restrictions
- 2-chloroethylphosphonic acid is an anticholinesterase organophosphorus compound. Do not use if under medical advice not to work with such compounds
- Maximum number of treatments 1 per crop or yr
- Do not spray crops suffering from stress caused by any factor, during cold weather or period of night frost nor when soil very dry
- Do not apply to cereals within 10 d of herbicide or liquid fertilizer application
- Do not spray wheat or triticale where the leaf sheaths have split and the ear is visible

Crop-specific information
- Latest use: before 1st spikelet visible (GS 51) for spring barley, winter barley, winter rye; before flag leaf sheath opening (GS 47) for triticale, winter wheat
- HI cider apples, tomatoes 5 d

Environmental safety
- Dangerous for the environment [1, 6, 7]
- Harmful to aquatic organisms
- Avoid accidental deposits on painted objects such as cars, trucks, aircraft

Hazard classification and safety precautions
Hazard Irritant [1-7]; Dangerous for the environment [1, 6, 7]
Transport code 8
Packaging group III
UN number 3265
Risk phrases R36 [1]; R37, R41, R52, R53a [2-7]; R38 [1-7]
Operator protection A, C; U05a, U08, U13, U20b [1-7]; U11 [2-5]
Environmental protection E13c, E38 [2-5]; E15a [1, 6, 7]; E34 [1]
Storage and disposal D01, D02, D09a, D10b [1-7]; D12a [2-5]
Medical advice M01 [2-7]

74 2-chloroethylphosphonic acid + mepiquat chloride

A plant growth regulator for reducing lodging in cereals

See also mepiquat chloride

Products
1	Clayton Mepiquat	Clayton	155:305 g/l	SL	12360
2	Standon Mepiquat Plus	Standon	155:305 g/l	SL	09373
3	Terpal	BASF	155:305 g/l	SL	02103

Uses
- Growth regulation in **durum wheat** (off-label) [3]
- Increasing yield in **winter barley** (low lodging situations) [1, 3]
- Lodging control in **spring barley** [1, 3]; **triticale, winter barley, winter rye, winter wheat** [1-3]

Specific Off-Label Approvals (SOLAs)
- *durum wheat* 20063104 [3]

Approval information
- 2-chloroethylphosphonic acid (ethephon) and mepiquat chloride included in Annex I under EC Directive 91/414
- In 2006 CRD required that all products containing 2-chloroethylphosphonic acid should carry the following warning in the main area of the container label: "2-chloroethylphosphonic acid is an anticholinesterase organophosphate. Handle with care"
- Accepted by BBPA for use on malting barley

Efficacy guidance
- Best results achieved on crops growing vigorously under conditions of high fertility
- Recommended dose and timing vary with crop, cultivar, growing conditions, previous treatment and desired degree of lodging control. See label for details
- May be applied to crops undersown with grass or clovers
- Do not apply to crops if wet or rain expected as efficacy will be impaired

SEE SECTION 3 FOR PRODUCTS ALSO REGISTERED

Pesticide Profiles

Restrictions
- 2-chloroethylphosphonic acid is an anticholinesterase organophosphorus compound. Do not use if under medical advice not to work with such compounds
- Maximum number of treatments 2 per crop
- Add an authorised non-ionic wetter to spray solution. See label for recommended product and rate
- Do not treat crops damaged by herbicides or stressed by drought, waterlogging etc
- Do not treat crops on soils of low fertility unless adequately fertilized
- Do not use in a programme with any other product containing 2-chloroethylphosphonic acid
- Do not apply to winter cultivars sown in spring or treat winter barley, triticale or winter rye on soils with more than 10% organic matter (winter wheat may be treated)
- Do not apply at temperatures above 21 °C

Crop-specific information
- Latest use: before ear visible (GS 49) for winter barley, spring barley, winter wheat and triticale; flag leaf just visible (GS 37) for winter rye
- Late tillering may be increased with crops subject to moisture stress and may reduce quality of malting barley

Environmental safety
- Do not use straw from treated cereals as a mulch or growing medium

Hazard classification and safety precautions
Hazard Harmful
Transport code 8
Packaging group III
UN number 3265
Risk phrases R22a, R37, R53a [1-3]; R36 [2, 3]; R41, R53b [1]
Operator protection A, C [3]; U05a, U11 [1]; U20b [1-3]
Environmental protection E15a
Storage and disposal D01, D02, D09a [1-3]; D05, D10b, D12a [1]; D08, D10c [2, 3]
Medical advice M01 [1-3]; M05a [2, 3]

75 chlorophacinone

An indandione anticoagulant rodenticide

Products
1	Drat Rat Bait	B H & B	0.005% w/w	RB	H6743
2	Endorats	Irish Drugs	0.005% w/w	RB	H6744

Uses
- Rats in *farm buildings* [1, 2]
- Voles in *farm buildings* [1]

Efficacy guidance
- Chemical formulated with oil, thus improving weather resistance of bait
- Use in baiting programme
- Bait stations should be sited where rats active, by rat holes, along runs or in harbourages. Place bait in suitable containers
- Replenish baits every few days and remove unused bait when take ceases or after 7-10 d

Restrictions
- For use only by professional operators
- Resistance status of target population should be taken into account when considering choice of rodenticide
- Bait stations may be sited conveniently but bait should be inaccessible to non-target animals and protected from prevailing weather

Environmental safety
- Prevent access to baits by children, domestic animals and birds; see label for other precautions required
- Harmful to game, wild birds and animals

Hazard classification and safety precautions
Operator protection U13 [1, 2]; U20a [2]; U20b [1]

FOR FULL CONDITIONS OF USE ALWAYS READ THE PRODUCT LABEL

Storage and disposal D09a [1, 2]; D10a [1]; D11a [2]
Vertebrate/rodent control products V01a, V02, V03a, V04a

76   chloropicrin

A highly toxic horticultural soil fumigant
IRAC mode of action code: 8B

Products

K & S Chlorofume   K & S Fumigation   99.3% w/w   VP   08722

Uses
- Crown rot in **strawberries** (soil fumigation)
- Nematodes in **strawberries** (soil fumigation)
- Red core in **strawberries** (soil fumigation)
- Replant disease in **all top fruit** (soil fumigation), **hardy ornamentals** (soil fumigation)
- Verticillium wilt in **blackcurrants** (off-label), **protected blackberries** (off-label), **protected raspberries** (off-label), **protected rubus hybrids** (off-label), **raspberries** (off-label), **rubus hybrids** (off-label), **strawberries** (soil fumigation)

Specific Off-Label Approvals (SOLAs)
- **blackcurrants** 20072373
- **protected blackberries** 20072373
- **protected raspberries** 20072373
- **protected rubus hybrids** 20072373
- **raspberries** 20072373
- **rubus hybrids** 20072373

Efficacy guidance
- Treat pre-planting
- Apply with specialised injection equipment
- For treating small areas or re-planting a single tree, a hand-operated injector may be used. Mark the area to be treated and inject to 22 cm at intervals of 22 cm
- Polythene sheeting (150 gauge) should be progressively laid over soil as treatment proceeds. The margin of the sheeting around the treated area must be embedded or covered with treated soil. Remove progressively after at least 4 d provided good air movement conditions prevail. Aerate soil for 15 d before planting
- Carry out a cress test before planting

Restrictions
- Chloropicrin is subject to the Poisons Rules 1982 and the Poisons Act 1972. See Section 5 for more information
- Before use, consult the code of practice for the fumigation of soil with chloropicrin. 2 fumigators must be present
- Maximum number of treatments 1 per crop
- Remove contaminated gloves, boots or other clothing immediately and ventilate them in the open air until all odour is eliminated
- Keep unprotected persons out of treated areas and adjacent premises until advised otherwise by professional operator
- Keep treated areas covered with sheets of low gas permeability for at least 4 d after treatment
- Avoid releasing vapour into the air when persistent still air conditions prevail

Crop-specific information
- Latest use: at least 20 d before planting

Following crops guidance
- Carry out a cress test before replanting treated soil

Environmental safety
- Avoid treatment or vapour release when persistent still air conditions prevail
- Dangerous to game, wild birds and animals
- Dangerous to bees
- Dangerous to fish or other aquatic life. Do not contaminate surface waters or ditches with chemical or used container

SEE SECTION 3 FOR PRODUCTS ALSO REGISTERED

Pesticide Profiles

- Dangerous to livestock. Keep all livestock out of treated areas until advised otherwise by fumigator
- Keep in original container, tightly closed, in a safe place, under lock and key

Hazard classification and safety precautions
Hazard Very toxic
Transport code 6.1
Packaging group I
UN number 1580
Risk phrases R26, R27, R28, R34, R37
Operator protection A, G, H, K, M; U01, U04a, U05a, U10, U19a, U19b, U20a
Environmental protection E02a, E06b (until advised); E15a, E34
Storage and disposal D01, D02, D09b, D11a, D14
Medical advice M04a

77 chlorothalonil

A protectant chlorophenyl fungicide for use in many crops and turf
FRAC mode of action code: M5

See also azoxystrobin + chlorothalonil
carbendazim + chlorothalonil

Products

1	Acclaim	European Ag	500 g/l	SC	14905
2	Agriguard Chlorothalonil	AgriGuard	500 g/l	SC	14839
3	Bravo 500	Syngenta	500 g/l	SC	14548
4	Busa	AgChem Access	500 g/l	SC	13452
5	CTL 500	Goldengrass	500 g/l	SC	14812
6	Daconil Weatherstik	Syngenta	500 g/l	SC	12636
7	Greencrop Orchid 2	Greencrop	500 g/l	SC	12769
8	Greencrop Orchid B	Greencrop	500 g/l	SC	12251
9	Juliet	ChemSource	500 g/l	SC	14557
10	Mascot Contact	Rigby Taylor	500 g/l	SC	13718
11	Rezacur	Sherriff Amenity	500 g/l	SC	13717
12	Sanspor	Syngenta	500 g/l	SC	13313
13	Sonar	Syngenta	720 g/l	SC	13083
14	Supreme	AgriGuard	500 g/l	SC	14838

Uses

- American gooseberry mildew in **gooseberries** [4, 9, 12]
- Anthracnose in **managed amenity turf** [6, 10, 11]
- Ascochyta in **combining peas** [4, 7-9, 12, 13]
- Basal stem rot in **narcissi** *(off-label)* [3]
- Blight in **potatoes, protected tomatoes** [4, 7-9, 12, 13]
- Botrytis in **broccoli, brussels sprouts, cabbages, cauliflowers, protected ornamentals, spring oilseed rape, strawberries, winter oilseed rape** [7, 8, 12, 13]; **broccoli** *(moderate control)*, **brussels sprouts** *(moderate control)*, **bulb onions** *(moderate control)*, **cabbages** *(moderate control)*, **cauliflowers** *(moderate control)*, **combining peas** *(moderate control)*, **protected ornamentals** *(moderate control)*, **spring oilseed rape** *(moderate control)*, **strawberries** *(moderate control)*, **winter oilseed rape** *(moderate control)* [4, 9]; **bulb onions, combining peas** [12]; **calabrese** *(off-label)* [13]; **gooseberries** [4, 9, 12]; **protected cucumbers, protected tomatoes** [4, 7-9, 12, 13]; **strawberries** *(off-label)* [3]
- Celery leaf spot in **celery (outdoor)** *(qualified minor use)* [4, 7-9, 12, 13]
- Chocolate spot in **spring field beans, winter field beans** [7, 8, 12, 13]; **spring field beans** *(moderate control)*, **winter field beans** *(moderate control)* [4, 9]
- Currant leaf spot in **blackcurrants, gooseberries** *(qualified minor use)*, **redcurrants** *(qualified minor use)* [4, 7-9, 12, 13]
- Disease control in **durum wheat** *(off-label)*, **spring rye** *(off-label)*, **triticale** *(off-label)*, **winter rye** *(off-label)* [3]
- Dollar spot in **managed amenity turf** [6, 10, 11]
- Downy mildew in **brassica seed beds, broccoli, brussels sprouts, cabbages, cauliflowers, spring oilseed rape, winter oilseed rape** [7, 8, 12, 13]; **brassica seed beds** *(moderate control)*,

FOR FULL CONDITIONS OF USE ALWAYS READ THE PRODUCT LABEL

chlorothalonil

- broccoli *(moderate control)*, **brussels sprouts** *(moderate control)*, **cabbages** *(moderate control)*, **cauliflowers** *(moderate control)*, **spring oilseed rape** *(moderate control)*, **winter oilseed rape** *(moderate control)* [4, 9]; **calabrese** *(off-label)* [13]; **hops** [4, 7-9, 12, 13]; **poppies for morphine production** *(off-label)* [3]
- Foliar disease control in **durum wheat** *(off-label)*, **garlic** *(off-label)*, **grass seed crops** *(off-label)*, **lupins** *(off-label)*, **shallots** *(off-label)*, **spring rye** *(off-label)*, **winter rye** *(off-label)* [13]
- Fusarium patch in **managed amenity turf** [6, 10, 11]
- Glume blotch in **spring wheat** [1-5, 9, 12, 14]; **winter wheat** [1-5, 7-9, 12-14]
- Ink disease in **irises** *(qualified minor use)* [4, 7-9, 12, 13]
- Leaf mould in **protected tomatoes** [4, 9]
- Leaf rot in **bulb onions** [4, 7-9, 12, 13]
- Leaf spot in **celeriac** *(off-label)*, **strawberries** *(off-label)* [3]
- Mycosphaerella in **combining peas** [7, 8, 12, 13]; **combining peas** *(moderate control)* [4, 9]
- Neck rot in **bulb onions** [4, 7-9, 12, 13]; **narcissi** *(off-label)* [3]
- Red thread in **managed amenity turf** [6, 10, 11]
- Rhynchosporium in **spring barley**, **winter barley** [7, 8, 12]; **spring barley** *(moderate control)*, **winter barley** *(moderate control)* [1-5, 9, 14]
- Ring spot in **broccoli**, **brussels sprouts**, **cabbages**, **cauliflowers** [7, 8, 13]
- Septoria leaf blotch in **spring wheat** [1-5, 9, 12, 14]; **winter wheat** [1-5, 7-9, 12-14]

Specific Off-Label Approvals (SOLAs)
- *calabrese* 20072053 [13]
- *celeriac* 20091508 [3]
- *durum wheat* 20091506 [3], 20072050 [13]
- *garlic* 20072052 [13]
- *grass seed crops* 20072050 [13]
- *lupins* 20072051 [13]
- *narcissi* 20091824 [3]
- *poppies for morphine production* 20100170 [3]
- *shallots* 20072052 [13]
- *spring rye* 20091506 [3], 20072050 [13]
- *strawberries* 20091509 [3]
- *triticale* 20091506 [3]
- *winter rye* 20091506 [3], 20072050 [13]

Approval information
- Chlorothalonil included in Annex I under EC Directive 91/414. However a Standing Committee decision in July 2007 to reduce to the limit of determination the MRL set for chlorothalonil in blackberries and raspberries led to revocation of approvals for use on these crops
- Approval for aerial spraying on potatoes [7, 8, 12, 13]. See Section 5 for more information
- Accepted by BBPA for use on malting barley and hops
- Approval expiry 31 Aug 2011 [4, 6-13]

Efficacy guidance
- For some crops products differ in diseases listed as controlled. See label for details and for application rates, timing and number of sprays
- Apply as protective spray or as soon as disease appears and repeat as directed
- On cereals activity against Septoria may be reduced where serious mildew or rust present. In such conditions mix with suitable mildew or rust fungicide
- May be used for preventive and curative treatment of turf but treatment at a late stage of disease development may not be so successful and can leave bare patches of soil requiring renovation. Products are not recommended for curative control of Fusarium patch or Anthracnose [6, 10, 11]
- Optimum results on managed amenity turf obtained from treatment in spring or autumn when conditions most favourable for disease development [6, 10, 11]

Restrictions
- Maximum number of treatments and maximum total doses vary with crop and product - see labels for details
- Operators of vehicle mounted equipment must use a vehicle fitted with a cab and a forced air filtration unit with a pesticide filter complying with HSE Guidance Note PM74 or to an equivalent or higher standard when making broadcast or air-assisted applications
- Must only be used on established turf [6, 10, 11]

SEE SECTION 3 FOR PRODUCTS ALSO REGISTERED

Pesticide Profiles

- On managed amenity turf must only be applied by a pedestrian controlled sprayer or vehicle mounted/drawn equipment [6, 10, 11]
- Do not mow or water turf for 24 h after treatment. Do not add surfactant or mix with liquid fertilizer [6, 10, 11]

Crop-specific information
- Latest time of application to cereals varies with product. Consult label
- Latest use for winter oilseed rape before flowering; end Aug in yr of harvest for blackcurrants, gooseberries, redcurrants;
- HI 8 wk for field beans; 6 wk for combining peas; 28 d or before 31 Aug for post-harvest treatment for blackcurrants, gooseberries, redcurrants; 14 d for onions, strawberries; 10 d for hops; 7-14 d for broccoli, Brussels sprouts, cabbages, cauliflowers, celery, onions, potatoes; 3 d for cane fruit; 12-48 h for protected cucumbers, protected tomatoes
- For Botrytis control in strawberries important to start spraying early in flowering period and repeat at least 3 times at 10 d intervals
- On strawberries some scorching of calyx may occur with protected crops

Environmental safety
- Dangerous for the environment
- Very toxic to aquatic organisms
- Do not spray from the air within 250 m horizontal distance of surface waters or ditches
- Broadcast air-assisted LERAP (18 m) [7, 8, 12, 13]
- Broadcast air-assisted LERAP [1-5, 7-9, 12-14] (18 m); LERAP Category B [1-14]

Hazard classification and safety precautions
Hazard Harmful, Dangerous for the environment
Transport code 9 [1-3, 5-8, 10-14]
Packaging group III [1-3, 5-8, 10-14]
UN number 3082 [1-3, 5-8, 10-14]
Risk phrases R20, R36, R40, R43, R50, R53a [1-14]; R22a [13]; R37 [1-7, 9-14]; R38 [8]
Operator protection A, C, H [1-14]; J [1, 3, 4, 7-9, 12, 13]; M [1, 3, 4, 6-13]; U02a, U09a, U19a [1-9, 12-14]; U05a, U20a [1-14]
Environmental protection E15a, E16a [1-14]; E16b [1-5, 7-14]; E17b [1-5, 7-9, 12-14] (18 m); E18 [1-5, 7-9, 12-14]; E34 [6, 7, 10, 11]; E38 [1-6, 8, 9, 12-14]
Storage and disposal D01, D02, D05, D09a [1-14]; D10a [10, 11]; D10b [7]; D10c [1-6, 8, 9, 12-14]; D12a [1-9, 12-14]
Medical advice M05a [7]

78 chlorothalonil + cyproconazole

A systemic protectant and curative fungicide mixture for cereals
FRAC mode of action code: M5 + 3

See also cyproconazole

Products

1	Alto Elite	Syngenta	375:40 g/l	SC	08467
2	Octolan	Syngenta	375:40 g/l	SC	11675
3	SAN 703	Syngenta	375:40 g/l	SC	11676

Uses
- Brown rust in **spring barley, winter barley, winter wheat** [1-3]
- Chocolate spot in **spring field beans, winter field beans** [1-3]
- Eyespot in **winter barley, winter wheat** [1-3]
- Glume blotch in **winter wheat** [1]
- Grey mould in **combining peas** [1-3]
- Net blotch in **spring barley, winter barley** [1-3]
- Powdery mildew in **spring barley, winter barley, winter wheat** [1-3]
- Rhynchosporium in **spring barley, winter barley** [1-3]
- Rust in **spring field beans, winter field beans** [1]
- Septoria leaf blotch in **winter wheat** [1-3]
- Yellow rust in **spring barley, winter barley, winter wheat** [1-3]

FOR FULL CONDITIONS OF USE ALWAYS READ THE PRODUCT LABEL

chlorothalonil + cyproconazole + propiconazole

Approval information
- Chlorothalonil included in Annex I under EC Directive 91/414
- Accepted by BBPA for use on malting barley

Efficacy guidance
- Apply at first signs of infection or as soon as disease becomes active
- A repeat application may be made if re-infection occurs
- For established mildew tank-mix with an approved mildewicide
- When applied prior to third node detectable (GS 33) a useful reduction of eyespot will be obtained
- Cyproconazole is a DMI fungicide. Resistance to some DMI fungicides has been identified in Septoria leaf blotch which may seriously affect performance of some products. For further advice contact a specialist advisor and visit the Fungicide Resistance Action Group (FRAG)-UK website

Restrictions
- Maximum total dose equivalent to 2 full dose treatments
- Do not apply at concentrations higher than recommended

Crop-specific information
- Latest use: before beginning of anthesis (GS 60) for barley; before caryopsis watery ripe (GS 71) for wheat
- HI peas, field beans 6 wk
- If applied to winter wheat in spring at GS 30-33 straw shortening may occur but yield is not reduced

Environmental safety
- Dangerous for the environment
- Very toxic to aquatic organisms
- LERAP Category B

Hazard classification and safety precautions
Hazard Harmful, Dangerous for the environment
Transport code 9
Packaging group III
UN number 3082
Risk phrases R20, R37, R40, R41, R43, R50, R53a
Operator protection A, C, H, M; U05a, U19a [1-3]; U11, U20a [1]; U12, U20b [2, 3]
Environmental protection E15a, E16b [1]; E15b [2, 3]; E16a, E38 [1-3]
Storage and disposal D01, D02, D09a, D10c, D12a [1-3]; D05 [1]
Medical advice M04a

79 chlorothalonil + cyproconazole + propiconazole

A broad-spectrum fungicide mixture for cereals
FRAC mode of action code: M5 + 3 + 3

See also cyproconazole
propiconazole

Products
Cherokee   Syngenta   375:50:62.5 g/l   SE   13251

Uses
- Brown rust in **spring barley**, **winter barley**, **winter wheat**
- Net blotch in **spring barley** *(reduction)*, **winter barley** *(reduction)*
- Powdery mildew in **spring barley**, **winter barley**, **winter wheat** *(moderate)*
- Rhynchosporium in **spring barley** *(reduction)*, **winter barley** *(reduction)*
- Septoria leaf blotch in **winter wheat**
- Yellow rust in **winter wheat**

Approval information
- Chlorothalonil and propiconazole included in Annex I under EC Directive 91/414

Efficacy guidance
- Best results obtained from treatment at the early stages of disease development

SEE SECTION 3 FOR PRODUCTS ALSO REGISTERED

Pesticide Profiles

- Cyproconazole and propiconazole are DMI fungicides. Resistance to some DMI fungicides has been identified in Septoria leaf blotch which may seriously affect performance of some products. For further advice contact a specialist advisor and visit the Fungicide Resistance Action Group (FRAG)-UK website
- Product should be used as part of an Integrated Crop Management strategy incorporating other methods of control and including, where appropriate, other fungicides with a different mode of action
- Product should be used preventatively and not relied on for its curative potential
- Always follow FRAG guidelines for resistance management. See Section 5 for more information

Restrictions
- Maximum total dose equivalent to two full dose treatments on wheat and barley
- Do not treat crops under stress
- Spray solution must be used as soon as possible after mixing

Crop-specific information
- Latest use: before caryopsis watery ripe (GS 71) for winter wheat; before first spikelet just visible (GS 51) for barley
- When applied to winter wheat in the spring straw shortening may occur which causes no loss of yield

Environmental safety
- Dangerous for the environment
- Very toxic to aquatic organisms
- LERAP Category B

Hazard classification and safety precautions
Hazard Harmful, Dangerous for the environment
Transport code 9
Packaging group III
UN number 3082
Risk phrases R20, R36, R37, R40, R43, R50, R53a
Operator protection A, C, H; U02a, U05a, U09a, U10, U11, U14, U15, U19a, U20b
Environmental protection E15b, E16a, E16b, E34, E38
Storage and disposal D01, D02, D05, D07, D09a, D10c, D12a
Medical advice M04a

80 chlorothalonil + fludioxonil + propiconazole

A fungicide treatment for amenity grassland
FRAC mode of action code: M5 + 12 + 3

*See also fludioxonil
propiconazole*

Products
Instrata   Syngenta   362:14.5:56.9 g/l   SE   14154

Uses
- Anthracnose in **amenity grassland** *(reduction)*, **managed amenity turf** *(reduction)*
- Brown patch in **amenity grassland**, **managed amenity turf**
- Dollar spot in **amenity grassland**, **managed amenity turf**
- Fusarium patch in **amenity grassland**, **managed amenity turf**

Approval information
- Chlorothalonil, fludioxonil and propiconazole included in Annex I under EC Directive 91/414.

Environmental safety
- LERAP Category B

Hazard classification and safety precautions
Hazard Toxic, Dangerous for the environment
Transport code 9
Packaging group III
UN number 3082

FOR FULL CONDITIONS OF USE ALWAYS READ THE PRODUCT LABEL

Risk phrases R23, R36, R40, R43, R50, R53a
Operator protection A, C, H; U02a, U05a, U09a, U10, U11, U19a, U19c, U20b
Environmental protection E15b, E16a, E34, E38
Storage and disposal D01, D02, D09a, D12a
Medical advice M04a

81   chlorothalonil + flusilazole

A fungicide mixture for wheat and barley
FRAC mode of action code: M5 + 3

See also flusilazole

Products

1	Midas	DuPont	200:80 g/l	SC	13285
2	Scout	AgriGuard	200:80 g/l	SC	14779

Uses
- Glume blotch in **spring wheat**, **winter wheat**
- Rhynchosporium in **spring barley**, **winter barley**
- Septoria leaf blotch in **spring wheat**, **winter wheat**

Approval information
- Chlorothalonil included in Annex I under EC Directive 91/414.
- Flusilazole has been reinstated in Annex 1 under Directive 91/414 following an appeal

Efficacy guidance
- Best results achieved from treatment when diseases are at low levels and before they spread to new growth
- Under heavy disease pressure a second application may be made 3-4 wk later
- Flusilazole is a DMI fungicide. Resistance to some DMI fungicides has been identified in Septoria leaf blotch which may seriously affect performance of some products. For further advice contact a specialist advisor and visit the Fungicide Resistance Action Group (FRAG)-UK website

Restrictions
- Maximum number of treatments 2 per crop
- Do not apply to crops under stress
- Do not apply during frosty weather

Crop-specific information
- Latest use: end of flowering (GS 69) for wheat; end of heading (GS 59) for barley

Environmental safety
- Dangerous for the environment
- Very toxic to aquatic organisms
- LERAP Category B

Hazard classification and safety precautions
   Hazard Toxic, Dangerous for the environment
   Transport code 8 [2]; 9 [1]
   Packaging group III
   UN number 1760 [2]; 3082 [1]
   Risk phrases R37, R40, R41, R43, R50, R53a, R61
   Operator protection A, C, H, M; U05a, U11, U20b, U23a
   Environmental protection E15a, E16a, E34, E38
   Storage and disposal D01, D02, D05, D09a, D10b, D12a
   Medical advice M04a

SEE SECTION 3 FOR PRODUCTS ALSO REGISTERED

82 chlorothalonil + flutriafol

A systemic eradicant and protectant fungicide for winter wheat
FRAC mode of action code: M5 + 3

See also flutriafol

Products

1	Impact Excel	Headland	300:47 g/l	SC	11547
2	Prospa	Headland	300:47 g/l	SC	11548

Uses
- Brown rust in **winter wheat**
- Foliar disease control in **durum wheat** (off-label), **grass seed crops** (off-label), **spring rye** (off-label), **triticale** (off-label), **winter rye** (off-label)
- Late ear diseases in **winter wheat**
- Powdery mildew in **winter wheat**
- Septoria leaf blotch in **winter wheat**
- Yellow rust in **winter wheat**

Specific Off-Label Approvals (SOLAs)
- *durum wheat* 20052890 [1], 20052889 [2]
- *grass seed crops* 20052890 [1], 20052889 [2]
- *spring rye* 20052890 [1], 20052889 [2]
- *triticale* 20052890 [1], 20052889 [2]
- *winter rye* 20052890 [1], 20052889 [2]

Approval information
- Chlorothalonil included in Annex I under EC Directive 91/414

Efficacy guidance
- Generally disease control and yield benefit will be optimised when application is made at an early stage of disease development
- Apply as soon as disease is seen establishing in the crop

Restrictions
- Maximum number of treatments 2 per crop on winter wheat

Crop-specific information
- Latest use: before early milk stage (GS 73)
- On certain cultivars with erect leaves high transpiration can result in flag leaf tip scorch. This may be increased by treatment but does not affect yield

Environmental safety
- Dangerous for the environment [1, 2]
- Very toxic to aquatic organisms [1, 2]
- Dangerous to fish or other aquatic life. Do not contaminate surface waters or ditches with chemical or used container [2]
- LERAP Category B

Hazard classification and safety precautions
Hazard Toxic, Dangerous for the environment
Transport code 6.1
Packaging group II
UN number 2902
Risk phrases R23, R40, R41, R43, R50, R53a
Operator protection A, C, H; U02a, U05a, U09a, U11, U19a, U20b
Environmental protection E13b [2]; E15a [1]; E16a, E38 [1, 2]
Storage and disposal D01, D02, D09a, D10b [1, 2]; D05 [2]

FOR FULL CONDITIONS OF USE ALWAYS READ THE PRODUCT LABEL

83 chlorothalonil + mancozeb

A multi-site protectant fungicide mixture
FRAC mode of action code: M5 + M3

See also mancozeb

Products

1	Adagio	Interfarm	201:274 g/l	SC	10796
2	Guru	Interfarm	286:194 g/l	SC	10801

Uses
- Blight in **potatoes** [1]
- Disease control in **durum wheat** *(off-label)*, **grass seed crops** *(off-label)*, **spring rye** *(off-label)*, **winter rye** *(off-label)* [2]
- Glume blotch in **winter wheat** [2]
- Septoria leaf blotch in **winter wheat** [2]

Specific Off-Label Approvals (SOLAs)
- **durum wheat** 20100049 expires 31 Dec 2011 [2]
- **grass seed crops** 20100049 expires 31 Dec 2011 [2]
- **spring rye** 20100049 expires 31 Dec 2011 [2]
- **winter rye** 20100049 expires 31 Dec 2011 [2]

Approval information
- Chlorothalonil and mancozeb included in Annex I under EC Directive 91/414
- Approval expiry 31 Dec 2011 [1, 2]

Efficacy guidance
- Start spray treatments on potatoes immediately after a blight warning or just before the haulm meets in the row [1]
- It is essential to start the blight spray programme before the disease appears in the crop [1]
- Repeat treatments for blight at 7, 10 or 14 d intervals depending on blight risk and continue until haulm is to be destroyed [1]
- Best results on winter wheat achieved from protective applications to the flag leaf. If disease already present on lower leaves treat as soon as flag leaf is just visible (GS 37) [2]
- Activity on Septoria may be reduced in presence of severe mildew infection [2]

Restrictions
- Maximum number of treatments 5 per crop for potatoes [1]
- Maximum total dose on winter wheat equivalent to 1.5 full dose treatments [2]
- Do not treat crops under stress from frost, drought, waterlogging, trace element deficiency or pest attack [2]
- Broadcast air assisted applications must only be made by equipment fitted with a cab with a forced air filtration unit plus a pesticide filter complying with HSE Guidance Note PM 74 or an equivalent or higher standard

Crop-specific information
- Latest use: before grain watery ripe (GS 71) for winter wheat
- HI potatoes 7 d
- Irrigated potato crops should be sprayed immediately after irrigation [1]

Environmental safety
- Dangerous for the environment
- Very toxic to aquatic organisms
- LERAP Category B

Hazard classification and safety precautions
Hazard Irritant, Dangerous for the environment
Transport code 9
Packaging group III
UN number 3082
Risk phrases R37, R43, R50, R53a [1, 2]; R40 [2]
Operator protection A, C, H, M; U02a, U05a, U08, U10, U11, U13, U14, U19a [1, 2]; U20a [1]; U20b [2]

SEE SECTION 3 FOR PRODUCTS ALSO REGISTERED

Environmental protection E15a [1]; E16a, E16b, E38 [1, 2]
Storage and disposal D01, D02, D05, D09a, D10b, D12a
Medical advice M04a [1]

84 chlorothalonil + metalaxyl-M

A systemic and protectant fungicide for various crops
FRAC mode of action code: M5 + 4

See also metalaxyl-M

Products
  Folio Gold   Syngenta   500:37.5 g/l   SC   14368

Uses
- Alternaria in **brussels sprouts** *(moderate control)*, **cauliflowers** *(moderate control)*
- Disease control in **poppies for morphine production** *(off-label)*
- Downy mildew in **broad beans**, **brussels sprouts**, **cauliflowers**, **spring field beans**, **winter field beans**
- Ring spot in **brussels sprouts** *(reduction)*, **cauliflowers** *(reduction)*
- White blister in **brussels sprouts**, **cauliflowers**

Specific Off-Label Approvals (SOLAs)
- *poppies for morphine production* 20090251

Approval information
- Chlorothalonil and metalaxyl-M included in Annex I under EC Directive 91/414

Efficacy guidance
- Apply at first signs of disease or when weather conditions favourable for disease pressure
- Best results obtained when used in a full and well-timed programme. Repeat treatment at 14-21 d intervals if necessary
- Treatment of established disease will be less effective
- Evidence of effectiveness in bulb onions, shallots and leeks is limited

Restrictions
- Maximum total dose equivalent to 2 full doses on broad beans, field beans, cauliflowers, calabrese; 3 full doses on Brussels sprouts, leeks, onions and shallots

Crop-specific information
- HI 14 d for all crops

Environmental safety
- Dangerous for the environment
- Very toxic to aquatic organisms
- LERAP Category B

Hazard classification and safety precautions
  Hazard Harmful, Dangerous for the environment
  Transport code 9
  Packaging group III
  UN number 3082
  Risk phrases R20, R36, R37, R38, R40, R43, R50, R53a
  Operator protection A, C, H; U05a, U09a, U15, U20a
  Environmental protection E15a, E16a, E38
  Consumer protection C02a (14 d)
  Storage and disposal D01, D02, D05, D09a, D10c, D12a

FOR FULL CONDITIONS OF USE ALWAYS READ THE PRODUCT LABEL

85 chlorothalonil + picoxystrobin

A broad spectrum fungicide mixture for cereals
FRAC mode of action code: M5 + 11

See also picoxystrobin

Products

1	Credo	DuPont	500:100 g/l	SC	14577
2	Zimbrail	DuPont	500:100 g/l	SC	14735

Uses
- Brown rust in **spring barley**, **spring wheat**, **winter barley**, **winter wheat**
- Glume blotch in **spring wheat**, **winter wheat**
- Net blotch in **spring barley**, **winter barley**
- Rhynchosporium in **spring barley**, **winter barley**
- Septoria leaf blotch in **spring wheat**, **winter wheat**
- Tan spot in **spring wheat**, **winter wheat**
- Yellow rust in **spring wheat**, **winter wheat**

Approval information
- Chlorothalonil and picoxystrobin included in Annex I under EC Directive 91/414
- Accepted by BBPA for use on malting barley

Efficacy guidance
- Best results obtained as a protectant treatment, or when disease first seen in crop, applied in good growing conditions with adequate soil moisture
- Disease control for 4-6 wk normally achieved during stem elongation
- Results may be less reliable when used on crops under stress
- Treatments for protection against ear disease should be made at ear emergence
- Picoxystrobin is a member of the QoI cross resistance group. Product should be used preventatively and not relied on for its curative potential
- Use product as part of an Integrated Crop Management strategy incorporating other methods of control, including where appropriate other fungicides with a different mode of action. Do not apply more than two foliar applications of QoI containing products to any cereal crop
- Control of many diseases may be improved by use in mixture with an appropriate triazole
- There is a significant risk of widespread resistance occurring in *Septoria tritici* populations in UK. Failure to follow resistance management action may result in reduced levels of disease control
- Strains of barley powdery mildew resistant to QoIs are common in the UK

Restrictions
- Maximum number of treatments 2 per crop of wheat or barley

Crop-specific information
- Latest use: before first spikelet just visible (GS 51) for barley; before grain watery ripe (GS 71) for wheat

Environmental safety
- Dangerous for the environment
- Very toxic to aquatic organisms
- LERAP Category B

Hazard classification and safety precautions
   Hazard Harmful, Dangerous for the environment
   Transport code 9
   Packaging group III
   UN number 3082
   Risk phrases R20, R37, R40, R43, R50, R53a
   Operator protection A, H; U05a, U09a, U14, U15, U19a, U20a
   Environmental protection E15a, E16a, E16b, E38
   Storage and disposal D01, D02, D05, D09a, D10c, D12a

SEE SECTION 3 FOR PRODUCTS ALSO REGISTERED

86 chlorothalonil + propamocarb hydrochloride

A contact and systemic fungicide mixture for blight control in potatoes
FRAC mode of action code: M5 + 28

See also propamocarb hydrochloride

Products
Pan Magician	Pan Agriculture	375:375 g/l	SC	11992

Uses
- Blight in *potatoes*

Approval information
- Chlorothalonil and propamocarb hydrochloride included in Annex I under EC Directive 91/414

Efficacy guidance
- Commence treatment early in the season as soon as there is risk of infection
- In the absence of a blight warning treatment should start just before potatoes meet along the row
- Use only as a protectant. Stop use when blight readily visible (1% leaf area destroyed)
- Repeat sprays at 10-14 d intervals depending on blight infection risk. See label for details
- Complete blight spray programme after end Aug up to haulm destruction with protectant fungicides

Restrictions
- Maximum total dose equivalent to 6 full dose treatments on potatoes
- Apply to dry foliage. Do not apply if rainfall or irrigation imminent

Crop-specific information
- HI 7 d for potatoes

Environmental safety
- Dangerous for the environment
- Very toxic to aquatic organisms
- LERAP Category B

Hazard classification and safety precautions
Hazard Harmful, Dangerous for the environment
Transport code 9
Packaging group III
UN number 3082
Risk phrases R20, R40, R41, R43, R50, R53a
Operator protection A, C, H; U05a, U08, U11, U14, U19a, U20a
Environmental protection E15a, E16a, E16b, E34, E38
Storage and disposal D01, D02, D05, D09a, D10b, D12a

87 chlorothalonil + propiconazole

A systemic and protectant fungicide for winter wheat and barley
FRAC mode of action code: M5 + 3

See also propiconazole

Products
Prairie	Syngenta	250:62.5 g/l	SC	13994

Uses
- Brown rust in *spring barley*, *spring wheat*, *winter barley*, *winter wheat*
- Glume blotch in *spring wheat*, *winter wheat*
- Rhynchosporium in *spring barley*, *winter barley*
- Septoria leaf blotch in *spring wheat*, *winter wheat*
- Yellow rust in *spring wheat*, *winter wheat*

Approval information
- Chlorothalonil and propiconazole included in Annex I under EC Directive 91/414
- Accepted by BBPA for use on malting barley

FOR FULL CONDITIONS OF USE ALWAYS READ THE PRODUCT LABEL

chlorothalonil + pyrimethanil

Efficacy guidance
- On wheat apply from start of flag leaf emergence up and including when ears just fully emerged (GS 37-59), on barley at any time to ears fully emerged (GS 59)
- Best results achieved from early treatment, especially if weather wet, or as soon as disease develops

Restrictions
- Maximum number of treatments 2 per crop or 1 per crop if other propiconazole based fungicide used in programme

Crop-specific information
- Latest use: up to and including emergence of ear just complete (GS 59).
- HI 42 d

Environmental safety
- Irritating to eyes, skin and respiratory system
- Risk of serious damage to eyes
- Dangerous to fish or other aquatic life. Do not contaminate surface waters or ditches with chemical or used container
- LERAP Category B

Hazard classification and safety precautions
    Hazard Harmful, Dangerous for the environment
    Transport code 9
    Packaging group III
    UN number 3082
    Risk phrases R20, R36, R37, R40, R43, R50, R53a
    Operator protection A, C, H; U02a, U05a, U11, U20b
    Environmental protection E15b, E16a, E16b, E38
    Storage and disposal D01, D02, D09a, D10c, D12a

88   chlorothalonil + pyrimethanil

A fungicide mixture for use in combining peas
FRAC mode of action code: M5 + 9

See also pyrimethanil

Products

Walabi	BASF	375:150 g/l	SC	12265

Uses
- Ascochyta in **combining peas**
- Botrytis in **combining peas**
- Chocolate spot in **spring field beans, winter field beans**
- Mycosphaerella in **combining peas**

Approval information
- Chlorothalonil and pyrimethanil included in Annex I under EC Directive 91/414.

Efficacy guidance
- Use only as a protectant spray in a two spray programme
- For best results ensure good foliar cover

Restrictions
- Maximum total dose equivalent to two full dose treatments
- Consult processors before use on crops for processing

Crop-specific information
- HI 6 wk for combining peas; 8 wk for field beans

Environmental safety
- Dangerous for the environment
- Very toxic to aquatic organisms
- LERAP Category B

SEE SECTION 3 FOR PRODUCTS ALSO REGISTERED

Pesticide Profiles

Hazard classification and safety precautions
   Hazard Harmful, Dangerous for the environment
   Transport code 9
   Packaging group III
   UN number 3082
   Risk phrases R36, R40, R43, R50, R53a
   Operator protection A, C, H; U02a, U11
   Environmental protection E15a, E16a
   Storage and disposal D01, D02, D05, D09a, D10c, D12a
   Medical advice M03

89 chlorothalonil + tebuconazole

A protectant and systemic fungicide mixture for winter wheat
FRAC mode of action code: M5 + 3

See also tebuconazole

Products
1	Pentangle	Nufarm UK	500:180 g/l	SC	13746
2	Timpani	Nufarm UK	250:90 g/l	SC	14651

Uses
- Septoria leaf blotch in **winter wheat**

Approval information
- Chlorothalonil and tebuconazole included in Annex I under EC Directive 91/414

Efficacy guidance
- Latest time of application before anthesis (GS61)
- Maximum individual dose 1.0 l/ha. Where disease pressure is severe, a second application may be required but applications after flag leaf visible (GS39) may only result in a reduction of disease severity

Environmental safety
- LERAP Category B

Hazard classification and safety precautions
   Hazard Very toxic, Dangerous for the environment
   Transport code 9
   Packaging group III
   UN number 3082
   Risk phrases R26, R37, R40, R41, R43, R50, R53a, R63
   Operator protection A, C, H; U02a, U05a, U09a, U10, U11, U13, U14, U15, U19a, U19b
   Environmental protection E15a, E16a, E16b, E34, E38
   Storage and disposal D01, D02, D09a, D10b, D12a
   Medical advice M04a

90 chlorotoluron

A contact and residual urea herbicide for cereals
HRAC mode of action code: C2

Products
1	Alpha Chlorotoluron 500	Makhteshim	500 g/l	SC	04848
2	Chute	Makhteshim	90% w/w	WG	13520
3	Dicurane	Makhteshim	500 g/l	SC	14276
4	Dicurane 70SC	Makhteshim	700 g/l	SC	14277
5	Dicurane 90WDG	Makhteshim	90% w/w	WG	14273
6	Rotor	Makhteshim	700 g/l	SC	13511
7	Tolugan 700	Makhteshim	700 g/l	SC	08064
8	Tolurex 90 WDG	Makhteshim	90% w/w	WG	11403

FOR FULL CONDITIONS OF USE ALWAYS READ THE PRODUCT LABEL

chlorotoluron

Uses
- Annual dicotyledons in ***durum wheat*** [2-5, 8]; ***spring barley*** *(off-label)* [5, 6]; ***spring rye*** *(off-label)*, ***winter rye*** *(off-label)* [1, 5]; ***spring wheat*** [6]; ***spring wheat*** *(off-label)* [5]; ***triticale, winter barley, winter wheat*** [1-8]
- Annual grasses in ***durum wheat*** [2, 3, 5, 8]; ***rye*** *(off-label)* [8]; ***spring barley*** *(off-label)*, ***spring wheat*** *(off-label)* [2-5, 7, 8]; ***spring rye*** *(off-label)*, ***winter rye*** *(off-label)* [1, 3-5, 7]; ***spring wheat*** [6]; ***triticale, winter barley, winter wheat*** [1-3, 5-8]
- Annual meadow grass in ***durum wheat, triticale, winter barley, winter wheat*** [4, 5]; ***rye*** *(off-label)* [8]; ***spring barley*** *(off-label)* [2-8]; ***spring rye*** *(off-label)*, ***winter rye*** *(off-label)* [3-5, 7]; ***spring wheat*** *(off-label)* [2-5, 7, 8]
- Blackgrass in ***durum wheat*** [2, 4, 8]; ***spring barley*** *(off-label)*, ***spring wheat*** [6]; ***triticale, winter barley, winter wheat*** [1, 2, 4, 6-8]
- Italian ryegrass in ***durum wheat, triticale, winter barley, winter wheat*** [4]
- Perennial ryegrass in ***durum wheat, triticale, winter barley, winter wheat*** [4]
- Rough meadow grass in ***durum wheat*** [2, 4, 8]; ***spring barley*** *(off-label)*, ***spring wheat*** [6]; ***triticale, winter barley, winter wheat*** [1, 2, 4, 6-8]
- Wild oats in ***durum wheat*** [2, 4, 8]; ***spring barley*** *(off-label)*, ***spring wheat*** [6]; ***triticale, winter barley, winter wheat*** [1, 2, 4, 6-8]

Specific Off-Label Approvals (SOLAs)
- ***rye*** *20100350 expires 31 Aug 2011* [8]
- ***spring barley*** *20090215 expires 31 Aug 2011* [2], *20090290 expires 31 Aug 2011* [3], *20090292 expires 31 Aug 2011* [4], *20090291 expires 31 Aug 2011* [5], *20090218 expires 31 Aug 2011* [6], *20090207 expires 31 Aug 2011* [7], *20100350 expires 31 Aug 2011* [8]
- ***spring rye*** *20092465 expires 31 Aug 2011* [1], *20090290 expires 31 Aug 2011* [3], *20090292 expires 31 Aug 2011* [4], *20090291 expires 31 Aug 2011* [5], *20092467 expires 31 Aug 2011* [7]
- ***spring wheat*** *20090215 expires 31 Aug 2011* [2], *20090290 expires 31 Aug 2011* [3], *20090292 expires 31 Aug 2011* [4], *20090291 expires 31 Aug 2011* [5], *20090207 expires 31 Aug 2011* [7], *20100350 expires 31 Aug 2011* [8]
- ***winter rye*** *20092465 expires 31 Aug 2011* [1], *20090290 expires 31 Aug 2011* [3], *20090292 expires 31 Aug 2011* [4], *20090291 expires 31 Aug 2011* [5], *20092467 expires 31 Aug 2011* [7]

Approval information
- Chlorotoluron included in Annex I under EC Directive 91/414
- Accepted by BBPA for use on malting barley
- Approval expiry 31 Aug 2011 [1-8]

Efficacy guidance
- Best results achieved by application soon after drilling. Application in autumn controls most weeds germinating in early spring
- For wild oat control apply within 1 wk of drilling, not after 2-leaf stage. Blackgrass and meadow grasses controlled to 5 leaf, ryegrasses to 3 leaf stage
- Any trash or burnt straw should be buried and dispersed during seedbed preparation
- Control may be reduced if prolonged dry conditions follow application
- Harrowing after treatment may reduce weed control
- Always follow WRAG guidelines for preventing and managing herbicide resistant weeds. See Section 5 for more information

Restrictions
- Maximum number of treatments 1 per crop
- Use only on listed crop varieties. See label. Ensure seed well covered at drilling
- Apply only as pre-emergence spray in durum wheat, pre- or post-emergence in wheat, barley or triticale
- Do not apply pre-emergence to crops sown after 30 Nov
- Do not apply to crops severely checked by waterlogging, pests, frost or other factors
- Do not use on undersown crops or those due to be undersown
- Do not apply post-emergence in mixture with liquid fertilizers
- Do not roll for 7 d before or after application to an emerged crop
- Do not use on soils with more than 10% organic matter
- Crops on stony or gravelly soils may be damaged, especially after heavy rain

SEE SECTION 3 FOR PRODUCTS ALSO REGISTERED

Pesticide Profiles

Crop-specific information
- Latest use: pre-emergence for durum wheat. Post emergence timings on other crops vary - see labels
- Early sown crops may be damaged if applied before or during a period of rapid autumn growth
- Autumn treated crops must be drilled by 1 Dec and spring treated crops must be drilled by 1 Feb

Environmental safety
- Dangerous for the environment
- Very toxic to aquatic organisms. May cause long-term adverse effects in the aquatic environment
- Some pesticides pose a greater threat of contamination of water than others and chlorotoluron is one of these pesticides. Take special care when applying chlorotoluron near water and do not apply if heavy rain is forecast

Hazard classification and safety precautions
Hazard Harmful [1, 3, 4, 6]; Irritant [1]; Dangerous for the environment [1-8]
Transport code 9
Packaging group III
UN number 3077 [2, 5, 8]; 3082 [1, 3, 4, 6, 7]
Risk phrases R22a, R36, R38 [1]; R40, R63 [3, 4, 6]; R50, R53a [1-8]
Operator protection A [1-8]; C [1-3, 5-8]; H [4, 6]; U05a [1-3, 5-8]; U08 [1-5, 7, 8]; U09a [6]; U19a [1-8]; U20a [7]; U20b [1-6, 8]
Environmental protection E15a, E34 [1-8]; E38 [3, 4, 6]
Storage and disposal D01, D02, D09a, D10b [1-8]; D05 [3]; D12a [3, 4, 6]; D12b [1, 2, 5, 7, 8]
Medical advice M03 [1-8]; M04a [3, 4, 6]

91   chlorotoluron + diflufenican

A soil and foliar acting herbicide mixture for winter cereals
HRAC mode of action code: C2 + F1

See also diflufenican

Products

1 Agena	Nufarm UK	600:25 g/l	SC	14051
2 Buckler	Nufarm UK	600:25 g/l	SC	13476
3 Dicurane Surpass	Makhteshim	400:25 g/l	SC	14544
4 Dicurane XL	Makhteshim	450:15 g/l	SC	14569
5 Gloster	Makhteshim	400:25 g/l	SC	14037
6 Hekla	Makhteshim	450:15 g/l	SC	14298
7 Kula	Makhteshim	450:10 g/l	SC	14299
8 Steel	Nufarm UK	620:22.5 g/l	SC	14111
9 Tawny	Makhteshim	400:25 g/l	SC	14038

Uses
- Annual dicotyledons in **durum wheat**, **triticale** [3-7, 9]; **winter barley**, **winter wheat** [1-9]
- Annual grasses in **durum wheat**, **triticale**, **winter barley**, **winter wheat** [4, 6, 7, 9]; **spring barley** (off-label), **spring wheat** (off-label) [7]
- Annual meadow grass in **durum wheat**, **triticale** [4, 6, 7]; **winter barley**, **winter wheat** [1, 2, 4, 6-8]

Specific Off-Label Approvals (SOLAs)
- **spring barley** 20102059 [7]
- **spring wheat** 20102059 [7]

Approval information
- Chlorotoluron and diflufenican included in Annex I under EC Directive 91/414

Efficacy guidance
- Best results obtained in crops sown in a fine firm seedbed with clods no greater than fist size and adequate moisture present at spraying. Good even spray coverage of the soil is essential
- Good weed control depends on burial and dispersal of any trash or straw before drilling
- Weed control may be reduced in prolonged periods of below average rainfall
- In direct drilled crops soil surface should be cultivated and the drill slots closed before spraying
- Spring germinating weeds will not be controlled

FOR FULL CONDITIONS OF USE ALWAYS READ THE PRODUCT LABEL

chlorpropham

- Fluffy seedbeds should be rolled before treatment
- Harrowing a treated crop may reduce the level of weed control

Restrictions
- Maximum number of treatments 1 per crop for wheat, barley
- Do not apply if heavy rain is expected within 4 h or if crop is stressed
- Do not roll for 7 d before or after application. Do not roll autumn crops until spring
- Do not treat undersown crops, or those to be undersown
- Do not apply on soils with more than 10% organic matter
- Treat only listed varieties of wheat and barley

Crop-specific information
- Latest use: before end Feb for winter barley, winter wheat
- Transient leaf discolouration occasionally occurs after treatment of wheat or barley
- Treatment of wheat or barley on stony or gravelly soils may cause damage especially if heavy rain falls after application
- Early sown crops of wheat or barley may be damaged if treated during a period of rapid autumn growth

Following crops guidance
- Any crop may be drilled or planted following normal harvest of a treated crop provided the soil is ploughed to 150 mm
- In the event of failure of a treated crop only winter wheat or winter barley may be drilled immediately after ploughing. A period of 12 wk must elapse after ploughing before spring sowing of wheat, barley, oilseed rape, peas, beans, sugar beet, potatoes, carrots, edible brassicas or onions
- Successive treatments with any products containing diflufenican can lead to soil build-up and inversion ploughing to 150 mm must precede sowing any following non-cereal crop. Even where ploughing occurs some crops may be damaged

Environmental safety
- Dangerous for the environment
- Very toxic to aquatic organisms
- Do not apply to dry, cracked or waterlogged soils
- Some pesticides pose a greater threat of contamination of water than others and chlorotoluron is one of these pesticides. Take special care when applying chlorotoluron near water and do not apply if heavy rain is forecast
- LERAP Category B

Hazard classification and safety precautions
Hazard Harmful, Dangerous for the environment
Transport code 9
Packaging group III
UN number 3082
Risk phrases R40 [1-7, 9]; R43 [1, 2, 8]; R50, R53a, R63 [1-9]
Operator protection A, H [1-9]; C [9]; U05a [1-9]; U08, U13, U19a [1, 2, 4, 6-8]; U14, U15 [3, 5, 9]; U20a [4, 6, 7]; U20b [1, 2, 8]
Environmental protection E15a [1, 2, 4, 6-8]; E16a, E38 [1-9]; E16b [3, 5, 9]; E34 [4, 6, 7]
Storage and disposal D01, D02, D12a [1-9]; D05 [1, 2, 4, 6, 7]; D09a [1, 2, 4, 6-8]; D10b [4, 6-8]; D10c [1, 2]
Medical advice M04a [3, 5, 9]

92 chlorpropham

A residual carbamate herbicide and potato sprout suppressant
HRAC mode of action code: K2

See also chloridazon + chlorpropham + metamitron

Products

1	BL 500	United Phosphorus	500 g/l	HN	14387
2	Gro-Stop 100	Certis	300 g/l	HN	14182
3	Gro-Stop Fog	Certis	300 g/l	HN	14183
4	Gro-Stop Solid	Certis	100% w/w	BR	14103

SEE SECTION 3 FOR PRODUCTS ALSO REGISTERED

Pesticide Profiles

Products – continued

5	MSS CIPC 50 M	United Phosphorus	500 g/l	HN 14388
6	MSS Sprout Nip	Whyte Agrochemicals	100% w/w	HN 14184
7	Pro-Long	United Phosphorus	500 g/l	HN 14389

Uses
- Sprout suppression in **ware potatoes** [2, 5, 6]; **ware potatoes** *(thermal fog)* [1, 3, 4, 7]

Approval information
- Chlorpropham included in Annex I under EC Directive 91/414
- Some products are formulated for application by thermal fogging. See labels for details

Efficacy guidance
- Apply weed control sprays to freshly cultivated soil. Adequate rainfall must occur after spraying. Activity is greater in cold, wet than warm, dry conditions
- For sprout suppression apply with suitable fogging or rotary atomiser equipment or sprinkle granules over dry tubers before sprouting commences. Repeat applications may be needed. See labels for details
- Best results on potatoes obtained in purpose-built box stores with suitable forced draft ventilation. Potatoes in bulk stores should not be stacked more than 3 m high
- Blockage of air spaces between tubers prevents circulation of vapour and consequent loss of efficacy
- It is important to treat potatoes before the eyes open to obtain best results
- Effectiveness of fogging reduced in non-dedicated stores without proper insulation and temperature controls. Best results obtained at 5-10 °C and 75-80% humidity

Restrictions
- Maximum number of treatments normally 1 per batch for sprout suppression of potatoes; 1 per crop or yr for other named crops or situations
- Excess rainfall after application may result in crop damage
- Do not use on Sands, Very Light soils or soils low in organic matter
- Poor conditions at drilling or planting, soil compaction, surface capping, waterlogging or attack by pests may result in crop damage
- On crops under glass high temperatures and poor ventilation may cause crop damage
- Only clean, mature, disease-free potatoes should be treated for sprout suppression. Use of chlorpropham can inhibit tuber wound healing and the severity of skin spot infection in store may be increased if damaged tubers are treated
- Do not fog potatoes with a high level of skin spot
- Do not use on potatoes for seed. Do not handle, dry or store seed potatoes or any other seed or bulbs in boxes or buildings in which potatoes are being or have been treated
- Do not remove treated potatoes from store for sale or processing for at least 21 d after application

Crop-specific information
- Latest use: 3 d before drilling carrots; 7d before planting out celery; 21 d before removal for sale or processing for potatoes; before tulip leaves unfurl or flower bulbs 5 cm high
- Cure potatoes according to label instructions before treatment and allow 3 wk between completion of loading into store and first treatment
- Apply weed control sprays to seeded crops pre-emergence of crop or weeds, to onions as soon as first crop seedlings visible, to planted crops a few days before planting, to bulbs immediately after planting, to fruit crops in late autumn-early winter. See label for further details

Following crops guidance
- There is a risk of damage to seed potatoes which are handled or stored in boxes or buildings previously treated with chlorpropham

Environmental safety
- Dangerous for the environment
- Toxic to aquatic organisms
- Keep unprotected persons out of treated stores for at least 24 h after application
- Keep in original container, tightly closed, in a safe place, under lock and key

Hazard classification and safety precautions
  Hazard Toxic, Highly flammable [1, 5-7]; Harmful [2, 3]; Dangerous for the environment [1-7]
  Transport code 6.1 [1-3, 5, 7]; 9 [4, 6]
  Packaging group III

FOR FULL CONDITIONS OF USE ALWAYS READ THE PRODUCT LABEL

chlorpropham with cetrimide

UN number 2902 [1-3, 5, 7]; 3077 [4, 6]
Risk phrases R23, R24, R36, R39 [1, 5-7]; R25 [1, 7]; R40, R42, R43 [2, 3]; R51, R53a [1-5, 7]
Operator protection A, D, E, H, M [1-7]; C [1, 5-7]; J [1, 4-7]; U01, U19b [4]; U02a [2-4]; U04a, U14, U15, U20a [2, 3]; U05a, U08, U19a [1-7]; U20b [1, 4-7]
Environmental protection E02a [1, 5-7] (24 h); E15a [1-3, 5-7]; E15b, E38 [4]; E34 [1-7]
Consumer protection C02b [2, 3, 6]; C02c [1, 5, 7]; C12 [2, 3]
Storage and disposal D01, D02, D05 [1-7]; D06b [1, 5, 6]; D07, D11a, D12a [2-4]; D09b [1-3, 5-7]; D10a, D12b [1, 5-7]
Medical advice M04a

93   chlorpropham with cetrimide

A soil-acting herbicide for lettuce under cold glass - Last product withdrawn prior to approval expiry 31 July 2010
HRAC mode of action code: K2

94   chlorpropham + metamitron

A contact and residual herbicide mixture for beet crops
HRAC mode of action code: K2 + C1

See also metamitron

Products
    Newgold   Whyte Agrochemicals   24.5:280 g/l   SE   13937

Uses
- Annual dicotyledons in *fodder beet, mangels, sugar beet*
- Grass weeds in *fodder beet, mangels, sugar beet*

Approval information
- Chlorpropham included in Annex I under EC Directive 91/414

Efficacy guidance
- May be used pre-emergence alone or post-emergence in tank mixture or with an authorised adjuvant oil
- Use as part of a spray programme. For optimum efficacy a full programme of pre- and post-emergence sprays is recommended
- Ideally one pre-emergence application should be followed by a repeat overall low dose programme on mineral soils. On organic soils a programme of up to five post-emergence sprays is likely to be most effective
- Product combines residual and contact activity. Adequate soil moisture and a fine firm seedbed are essential for optimum residual activity

Restrictions
- Maximum number of treatments one pre-emergence plus three post-emergence or, where no pre-emergence treatment has been applied, up to five post-emergence treatments with an adjuvant oil

Crop-specific information
- Latest use: before crop foliage meets across the rows

Following crops guidance
- Only sugar beet, fodder beet or mangels should be sown as following crops within 4 mth of treatment. After normal harvest of a treated crop land should be mouldboard ploughed to 15 cm after which any spring crops may be sown or planted

Environmental safety
- Dangerous for the environment
- Very toxic to aquatic organisms

Hazard classification and safety precautions
    Hazard Dangerous for the environment
    Transport code 9

SEE SECTION 3 FOR PRODUCTS ALSO REGISTERED

Pesticide Profiles

Packaging group III
UN number 3082
Risk phrases R22a, R36, R38, R43, R50, R53a
Operator protection A, C; U05a, U08, U11, U14, U15, U19a, U20b
Environmental protection E15a, E34
Storage and disposal D01, D02, D09a, D10c
Medical advice M03

95 chlorpyrifos

A contact and ingested organophosphorus insecticide and acaricide
IRAC mode of action code: 1B

Products

1	Agriguard Chlorpyrifos	AgriGuard	480 g/l	EC	13298
2	Ballad	Headland	480 g/l	EC	11659
3	Chlobber	AgChem Access	480 g/l	EC	13723
4	Clayton Pontoon 480EC	Clayton	480 g/l	EC	14555
5	Crossfire 480	Dow	480 g/l	EC	12516
6	Cyren	Headland	480 g/l	EC	11028
7	Dursban WG	Dow	75% w/w	WG	09153
8	Equity	Dow	480 g/l	EC	12465
9	Govern	Dow	75% w/w	WG	13223
10	Parapet	Dow	75 % w/w	WG	12773
11	Pyrinex 48 EC	Makhteshim	480 g/l	EC	13534
12	Suscon Green Soil Insecticide	Fargro	10% w/w	CG	06312

Uses

- Ambrosia beetle in *cut logs* [1-4, 6-11].
- Aphids in *apples, gooseberries, pears, plums, raspberries, strawberries* [1-4, 6, 8-11]; *blackberries* (off-label) [8, 10]; *blackcurrants* [1, 3, 4, 9-11]; *broccoli, cabbages, calabrese, cauliflowers, chinese cabbage* [1-4, 6-11]; *brussels sprouts* [2, 3, 6, 11]; *currants* [2, 6, 8]; *quinces* (off-label) [8]; *redcurrants, whitecurrants* [1, 9, 10]
- Apple blossom weevil in *apples* [1-4, 6, 8-11]; *pears, plums* [1]
- Apple sucker in *apples* [1, 4, 9, 10]; *pears, plums* [1]
- Bean seed fly in *green beans* (off-label) [9]; *green beans* (off-label - harvested as a dry pulse) [6, 7]
- Cabbage root fly in *broccoli, brussels sprouts, cabbages, calabrese, cauliflowers* [1-4, 6-11]; *chinese cabbage* [1-3, 6-11]; *chinese cabbage* (off-label), *green beans harvested dry as a pulse* (off-label) [11]; *choi sum* (off-label), *collards* (off-label), *kale* (off-label), *kohlrabi* (off-label), *mooli* (off-label), *pak choi* (off-label), *radishes* [6, 7, 9-11]; *green beans* (off-label - harvested dry as a pulse) [10]; *tatsoi* (off-label) [6, 11]
- Capsids in *apples, gooseberries, pears* [1-4, 6, 8-11]; *blackcurrants* [1, 3, 4, 9-11]; *currants* [2, 6, 8]; *plums* [1]; *quinces* (off-label) [8]; *redcurrants, whitecurrants* [1, 9, 10]
- Caterpillars in *apples, pears* [2, 3, 6, 8, 11]; *blackcurrants* [1, 3, 4, 9-11]; *broccoli, cabbages, calabrese, cauliflowers, chinese cabbage* [1-4, 6-11]; *brussels sprouts* [2, 3, 6, 11]; *currants* [2, 6, 8]; *gooseberries* [1-4, 6, 8-11]; *quinces* (off-label) [8]; *redcurrants, whitecurrants* [1, 9, 10]
- Cereal aphid in *spring barley, spring wheat, winter barley, winter wheat* [4]
- Codling moth in *apples, pears* [2-4, 6, 8-11]
- Cutworms in *broccoli, cabbages, calabrese, cauliflowers, chinese cabbage* [1, 3, 4, 7-11]; *brussels sprouts* [3, 11]; *bulb onions* [2-4, 6-11]; *onions* [1]; *potatoes grown for seed* [4]; *seed potatoes* [2, 3, 7-11]
- Damson-hop aphid in *plums* [3, 9-11]; *plums* (Non OP resistant strains only) [4]
- Elm bark beetle in *cut logs* [2, 6]
- Frit fly in *amenity grassland* [1, 2, 5, 6]; *durum wheat, spring rye, triticale, winter rye* [7]; *durum wheat* (off-label), *spring rye* (off-label), *triticale* (off-label), *winter rye* (off-label) [8, 10]; *forage maize* [2-4, 6-11]; *grassland* [1, 4]; *maize* [1]; *managed amenity turf* [2-6, 11]; *permanent grassland* [2, 3, 6-11]; *rotational grass* [2, 3, 6, 7, 9-11]; *spring barley, winter barley* [1, 4, 8]; *spring oats, spring wheat, winter oats, winter wheat* [1-4, 6-11]
- Insect pests in *blackberries* (off-label) [6, 9, 11]; *durum wheat, leeks* (off-label), *spring rye, triticale, winter rye* [7]; *durum wheat* (off-label), *spring rye* (off-label), *triticale* (off-label), *winter rye* (off-label) [9, 11]; *fodder beet* (off-label) [7, 9-11]; *quinces* (off-label) [8]

FOR FULL CONDITIONS OF USE ALWAYS READ THE PRODUCT LABEL

chlorpyrifos

- Larch shoot beetle in *cut logs* [1-4, 6-11]
- Leatherjackets in *amenity grassland* [1, 2, 5, 6]; *broccoli, cabbages, calabrese, cauliflowers, chinese cabbage* [1, 2, 6-10]; *brussels sprouts* [2, 6]; *durum wheat, spring rye, triticale, winter rye* [7]; *durum wheat* (off-label), *spring rye* (off-label), *triticale* (off-label), *winter rye* (off-label) [8, 10]; *fodder beet* (off-label) [8]; *grassland* [1, 4]; *managed amenity turf* [2-6, 11]; *permanent grassland* [2, 3, 6-11]; *rotational grass* [2, 3, 6, 7, 9-11]; *spring barley, spring oats, spring wheat, sugar beet, winter barley, winter oats, winter wheat* [1-4, 6-11]
- Mealy aphid in *plums* [3, 11]
- Mealy plum aphid in *plums* [4]
- Midges in *raspberries* [4]
- Narcissus fly in *bulbs/corms* (off-label) [11]; *flower bulbs* (off-label) [6]
- Opomyza in *spring barley, spring oats, spring wheat, winter barley, winter oats, winter wheat* [1]
- Pear sucker in *pears* [4, 9, 10]
- Pine shoot beetle in *cut logs* [1-4, 6-11]
- Pygmy mangold beetle in *fodder beet* (off-label) [8]; *sugar beet* [1, 2, 6-10]
- Raspberry beetle in *raspberries* [1-4, 6, 8-11]
- Raspberry cane midge in *raspberries* [1-3, 6, 8-11]
- Red spider mites in *apples, pears, plums, raspberries* [2, 3, 6, 8-11]; *apples* (non OP resistant strains only), *blackcurrants* (non OP resistant strains only), *gooseberries* (non OP resistant strains only), *pears* (non OP resistant strains only), *plums* (non OP resistant strains only), *raspberries* (non OP resistant strains only), *strawberries* (non OP resistant strains only) [4]; *blackberries* (off-label) [8]; *blackcurrants* [1, 3, 9-11]; *currants* [2, 6, 8]; *gooseberries, strawberries* [1-3, 6, 8-11]; *redcurrants, whitecurrants* [1, 9, 10]
- Sawflies in *apples* [2-4, 6, 8-11]; *pears* [8]
- Strawberry blossom weevil in *strawberries* [1, 8]
- Suckers in *apples, pears* [2, 3, 6, 8, 11]
- Summer-fruit tortrix moth in *apples* [4, 9, 10]
- Thrips in *leeks* (off-label) [11]; *spring oats, spring wheat, winter oats, winter wheat* [2, 6]
- Tortrix moths in *apples, pears, plums, strawberries* [1-4, 6, 8-11]
- Vine weevil in *ornamental plant production* [12]; *strawberries* [2, 3, 6, 8-11]; *strawberries* (non-protected crops only) [4]
- Wheat bulb fly in *durum wheat, spring rye, triticale, winter rye* [7]; *durum wheat* (off-label), *spring rye* (off-label), *triticale* (off-label), *winter rye* (off-label) [8, 10]; *spring barley, winter barley* [1, 4, 7-10]; *spring oats, spring wheat, winter oats, winter wheat* [1-4, 6-11]
- Wheat-blossom midge in *durum wheat, spring rye, triticale, winter rye* [7]; *durum wheat* (off-label), *spring rye* (off-label), *triticale* (off-label), *winter rye* (off-label) [8, 10]; *spring barley, winter barley* [1, 2, 6]; *spring oats, winter oats* [1, 2, 6-11]; *spring wheat, winter wheat* [1, 2, 4, 6-11]
- Whitefly in *broccoli, cabbages, calabrese, cauliflowers, chinese cabbage* [1, 2, 6-10]; *brussels sprouts* [2, 6]
- Winter moth in *apples, pears, plums* [1-4, 6, 8-11]
- Woolly aphid in *apples* [2-4, 6, 8-11]; *pears* [2, 6, 8]

Specific Off-Label Approvals (SOLAs)
- *blackberries* 20052977 [6], 20052964 [8], 20100034 [9], 20061585 [10], 20100044 [11]
- *bulbs/corms* 20100043 [11]
- *chinese cabbage* 20100038 [11]
- *choi sum* 20050237 [6], 20031390 [7], 20100036 [9], 20061587 [10], 20100038 [11]
- *collards* 20050237 [6], 20031390 [7], 20100036 [9], 20061587 [10], 20100038 [11]
- *durum wheat* 20052965 [8], 20100037 [9], 20061583 [10], 20100042 [11]
- *flower bulbs* 20050236 [6]
- *fodder beet* 20091120 [7], 20052965 [8], 20091121 [9], 20100035 [9], 20091122 [10], 20100045 [11]
- *green beans* (harvested as a dry pulse) 20050235 [6], (harvested as a dry pulse) 20031391 [7], 20100033 [9], (harvested dry as a pulse) 20061586 [10]
- *green beans harvested dry as a pulse* 20100040 [11]
- *kale* 20050237 [6], 20031390 [7], 20100036 [9], 20061587 [10], 20100038 [11]
- *kohlrabi* 20050235 [6], 20031391 [7], 20100033 [9], 20061586 [10], 20100040 [11]
- *leeks* 20081469 [7], 20100039 [11]

SEE SECTION 3 FOR PRODUCTS ALSO REGISTERED

Pesticide Profiles

- **mooli** *20050235* [6], *20031391* [7], *20100033* [9], *20061586* [10], *20100040* [11]
- **pak choi** *20050237* [6], *20031390* [7], *20100036* [9], *20061587* [10], *20100038* [11]
- **quinces** *20052974* [8]
- **radishes** *20050235* [6], *20031391* [7], *20100033* [9], *20061586* [10], *20100040* [11]
- **spring rye** *20052965* [8], *20100037* [9], *20061583* [10], *20100042* [11]
- **tatsoi** *20050237* [6], *20100038* [11]
- **triticale** *20052965* [8], *20100037* [9], *20061583* [10], *20100042* [11]
- **winter rye** *20052965* [8], *20100037* [9], *20061583* [10], *20100042* [11]

Approval information
- Chlorpyrifos included in Annex I under EC Directive 91/414
- Accepted by BBPA for use on malting barley
- Following the environmental review of chlorpyrifos uses on ware potatoes, carrots, conifers (drench), forestry trees (dip) and cereals (for aphids and yellow cereal fly) were revoked in September 2005. In addition the maximum permitted number of treatments, or the maximum total dose, on several other crops was reduced.
- In 2006 CRD required that all products containing this active ingredient should carry the following warning in the main area of the container label: "Chlorpyrifos is an anticholinesterase organophosphate. Handle with care"
- Approval expiry 30 Jun 2011 [1, 3]
- Approval expiry 31 Dec 2011 [12]

Efficacy guidance
- Brassicas raised in plant-raising beds may require retreatment at transplanting
- Activity on field crops may be reduced when soil temperature below 5 °C or on organic soils
- In dry conditions the effect of granules applied as a surface band may be reduced
- Where pear suckers resistant to chlorpyrifos occur control is unlikely to be satisfactory
- Efficacy against frit fly and leatherjackets in grass may be reduced if applied during periods of frost
- Thorough incorporation by hand or mixing machine is essential when used in potting media for ornamental plant production
- When used for control of vine weevil in ornamental plant production the recently emerged and young larvae are controlled by incorporation into the growing media. A single application will control successive generations for the growing cycle of plants kept in the same containers for 2 yr, with partial control for a third yr. Where plants are potted-on into larger containers the fresh growing media must also be treated to obtain control [12]

Restrictions
- Contains an anticholinesterase organophosphate compound. Do not use if under medical advice not to work with such compounds
- Maximum number of treatments and timing vary with crop, product and pest. See label for details
- Do not apply to sugar beet under stress or within 4 d of applying a herbicide
- Once incorporated, treated growing media must be used within 30 d
- Do not use product in growing media used for aquatic or semi-aquatic plants, and not for propagation of any edible crop
- Do not graze lactating cows on treated pasture for 14 d after treatment

Crop-specific information
- Latest use 4d after transplanting edible brassica crops; end Jul for sugar beet; varies for other crops. See individual labels
- HI range from 7 d to 6 wk depending on crop. See labels
- For vine weevil control in ornamental nursery stock incorporate in growing medium when plants first potted from rooted cutting stage. Treat the fresh growing medium when plants are potted on into larger containers [12]
- For turf pests apply from Nov where high larval populations detected or damage seen
- On lettuce apply only to strong well developed plants when damage first seen, or on professional advice
- In apples use pre-blossom up to pink/white bud and post-blossom after petal fall
- Test tolerance of glasshouse ornamentals before widescale use for propagating unrooted cuttings, or potting unusual plants and new species, and when using any media with a high content of non-peat ingredients

Environmental safety
- Dangerous for the environment

FOR FULL CONDITIONS OF USE ALWAYS READ THE PRODUCT LABEL

- Very toxic to aquatic organisms
- Flammable [5, 8]
- Keep livestock out of treated areas for at least 14 d after treatment [2, 5-8]
- High risk to bees. Do not apply to crops in flower or to those in which bees are actively foraging. Do not apply when flowering weeds are present [2, 5-10]
- Chlorpyrifos is dangerous to some beneficial arthropods, especially parasitoid wasps, ground beetles, rove beetles and hoverflies. Environmental protection advice is not to spray summer applications within 12 m of the edge of the growing crop
- Broadcast air-assisted LERAP [1, 3-5, 7-11] (18 m); LERAP Category A [1-11]

Hazard classification and safety precautions
Hazard Harmful, Dangerous for the environment [1-12]; Flammable [1, 4, 5, 8, 11]
Transport code 3 [4, 11]; 6.1 [1, 2, 5, 6, 8]; 9 [7, 9, 10, 12]
Packaging group III [1, 2, 4-12]
UN number 1993 [4, 11]; 3017 [1, 5, 8]; 3018 [2, 6]; 3077 [7, 9, 10, 12]
Risk phrases R20 [2-6, 8, 11]; R22a, R53a [1-12]; R22b, R38 [1-6, 8, 11]; R36 [3-5, 8, 11]; R37 [1, 4, 5, 8]; R40 [11]; R41, R43 [3, 11]; R50 [1-11]; R51 [12]; R67 [1]
Operator protection A [1-12]; C [2-4, 6, 11]; H [1-4, 6, 11, 12]; P [8]; U02a [3, 11, 12]; U05a [1-6, 8, 11]; U08 [2-11]; U09b [1]; U11 [4, 5, 8]; U14 [2, 3, 6, 11]; U15 [11]; U19a [1-11]; U20b [1-12]
Environmental protection E06a [1-8, 11] (14 d); E12a [1, 2, 6, 9, 10]; E12c [3-5, 7, 8, 11]; E12e, E16c, E16d [1-11]; E13a [1-3, 6, 11]; E15a [1, 4, 5, 7-10, 12]; E17a [2, 6] (18 m); E17b [1, 3-5, 7-11] (18 m); E22a [1]; E34 [2-4, 6-11]; E38 [3-5, 7-10, 12]
Consumer protection C01 [12]
Storage and disposal D01 [2-12]; D02 [3-5, 7, 8, 11]; D05 [2, 3, 6, 12]; D09a [2-8, 11, 12]; D09c [9, 10]; D10b [2-11]; D11a [12]; D12a [3-5, 7-10, 12]; D12b [2, 6]
Medical advice M01 [1, 4, 5, 7-12]; M03 [1, 2, 4-11]; M04a [12]; M05b [1-6, 8, 11]

96 chlorpyrifos-methyl

An organophosphorus insecticide and acaricide for grain store use
IRAC mode of action code: 1B

Products

| | Reldan 22 | Dow | 225 g/l | EC | 12404 |

Uses
- Grain storage pests in **stored grain**
- Pre-harvest hygiene in **grain stores**

Approval information
- Chlorpyrifos-methyl included in Annex I under EC Directive 91/414
- Accepted by BBPA for use in stores for malting barley
- In 2006 CRD required that all products containing this active ingredient should carry the following warning in the main area of the container label: "Chlorpyrifos-methyl is an anticholinesterase organophosphate. Handle with care"

Efficacy guidance
- Apply to grain after drying to moisture content below 14%, cooling and cleaning
- Insecticide may become depleted at grain surface if grain is being cooled by continuous extraction of air from the base leading to reduced control of grain store pests especially mites
- Resistance to organophosphorus compounds sometimes occurs in insect and mite pests of stored products

Restrictions
- Contains an anticholinesterase organophosphorus compound. Do not use if under medical advice not to work with such compounds
- Maximum number of treatments 1 per batch or 1 per store, prior to storage
- Only treat grain in good condition
- Do not treat grain intended for sowing
- Minimum of 90 d must elapse between treatment of grain and removal from store for consumption or processing

SEE SECTION 3 FOR PRODUCTS ALSO REGISTERED

Pesticide Profiles

Crop-specific information
- May be applied pre-harvest to surfaces of empty store and grain handling machinery and as admixture with grain
- May be used in wheat, barley, oats, rye or triticale stores

Environmental safety
- Dangerous for the environment
- Very toxic to aquatic organisms

Hazard classification and safety precautions
    Hazard Harmful, Dangerous for the environment
    Transport code M6
    Packaging group III
    UN number 3082
    Risk phrases R22b, R41, R50, R53a
    Operator protection A, C, H, M; U05a, U08, U19a, U20b
    Environmental protection E15a, E34
    Storage and disposal D01, D02, D05, D09a, D10b
    Medical advice M01, M03, M05b

97 citronella oil

A natural plant extract herbicide

Products

	Barrier H	Barrier	22.9% w/w	EW	10136

Uses
- Ragwort in **land temporarily removed from production, permanent grassland**

Efficacy guidance
- Best results obtained from spot treatment of ragwort in the rosette stage, during dry still conditions
- Aerial growth of ragwort is rapidly destroyed. Longer term control depends on overall management strategy
- Check for regrowth after 28 d and re-apply as necessary

Crop-specific information
- Contact with grasses will result in transient scorch which is outgrown in good growing conditions

Environmental safety
- Apply away from bees
- Harmful to fish or other aquatic life. Do not contaminate surface waters or ditches with chemical or used container
- Keep livestock out of treated areas for at least 2 wk and until foliage of any poisonous weeds such as ragwort has died and become unpalatable

Hazard classification and safety precautions
    Hazard Irritant
    Risk phrases R36, R38
    Operator protection A, C; U05a, U20c
    Environmental protection E07b (2 wk); E12g, E13c
    Storage and disposal D01, D02, D05, D09a, D11a

98 clodinafop-propargyl

A foliar acting herbicide for annual grass weed control in wheat, triticale and rye
HRAC mode of action code: A

Products

1	Clodinafop 240	Goldengrass	240 g/l	EC	14417
2	Sword	Makhteshim	240 g/l	EC	15099
3	Topik	Syngenta	240 g/l	EC	12333
4	Tuli	AgChem Access	240 g/l	EC	13302
5	Viscount	Syngenta	240 g/l	EC	12642

FOR FULL CONDITIONS OF USE ALWAYS READ THE PRODUCT LABEL

clodinafop-propargyl

Uses
- Blackgrass in **durum wheat**, **spring rye**, **spring wheat**, **triticale**, **winter rye**, **winter wheat** [1-5]; **grass seed crops** (off-label) [3]
- Rough meadow grass in **durum wheat**, **spring rye**, **spring wheat**, **triticale**, **winter rye**, **winter wheat** [1-5]; **grass seed crops** (off-label) [3]
- Wild oats in **durum wheat**, **spring rye**, **spring wheat**, **triticale**, **winter rye**, **winter wheat** [1-5]; **grass seed crops** (off-label) [3]

Specific Off-Label Approvals (SOLAs)
- **grass seed crops** 20090210 [3]

Approval information
- Clodinafop-propargyl included in Annex I under EC Directive 91/414

Efficacy guidance
- Spray when majority of weeds have germinated but before competition reduces yield
- Products contain a herbicide safener (cloquintocet-mexyl) that improves crop tolerance to clodinafop-propargyl
- Optimum control achieved when all grass weeds emerged. Wait for delayed germination on dry or cloddy seedbed
- A mineral oil additive is recommended to give more consistent control of very high blackgrass populations or for late season treatments. See label for details
- Weed control not affected by soil type, organic matter or straw residues
- Control may be reduced if rain falls within 1 h of treatment
- Clodinafop-propargyl is an ACCase inhibitor herbicide. To avoid the build up of resistance do not apply products containing an ACCase inhibitor herbicide more than twice to any crop. In addition do not use any product containing clodinafop-propargyl in mixture or sequence with any other product containing the same ingredient
- Use these products as part of a resistance management strategy that includes cultural methods of control and does not use ACCase inhibitors as the sole chemical method of grass weed control
- Applying a second product containing an ACCase inhibitor to a crop will increase the risk of resistance development; only use a second ACCase inhibitor to control different weeds at a different timing
- Always follow WRAG guidelines for preventing and managing herbicide resistant weeds. See Section 5 for more information

Restrictions
- Maximum number of treatments 1 per crop
- Do not use on barley or oats
- Do not treat crops under stress or suffering from waterlogging, pest attack, disease or frost
- Do not treat crops undersown with grass mixtures
- Do not mix with products containing MCPA, mecoprop-P, 2,4-D or 2,4-DB
- MCPA, mecoprop, 2,4-D or 2,4-DB should not be applied within 21 d before, or 7 d after, treatment

Crop-specific information
- Latest use: before second node detectable stage (GS 32) for durum wheat, triticale, rye; before flag leaf sheath extending (GS 41) for wheat
- Spray in autumn, winter or spring from 1 true leaf stage (GS 11) to before second node detectable (GS 32) on durum, rye, triticale; before flag leaf sheath extends (GS 41) on wheat

Following crops guidance
- Any broad leaved crop or cereal (except oats) may be sown after failure of a treated crop provided that at least 3 wk have elapsed between application and drilling a cereal
- After normal harvest of a treated crop any broad leaved crop or wheat, durum wheat, rye, triticale or barley should be sown. Oats and grass should not be sown until the following spring

Environmental safety
- Dangerous for the environment
- Very toxic to aquatic organisms

Hazard classification and safety precautions
Hazard Dangerous for the environment
Transport code 9
Packaging group III

SEE SECTION 3 FOR PRODUCTS ALSO REGISTERED

Pesticide Profiles

UN number 3082
Risk phrases R50, R53a
Operator protection A, C, H, K; U02a, U05a, U09a, U20a
Environmental protection E15a, E38
Storage and disposal D01, D02, D05, D09a, D10c, D12a

99 clodinafop-propargyl + pinoxaden

A foliar acting herbicide mixture for winter wheat
HRAC mode of action code: A + A

See also pinoxaden

Products
Traxos   Syngenta   100:100 g/l   EC   12742

Uses
- Italian ryegrass in **winter wheat**
- Perennial ryegrass in **winter wheat** *(from seed)*
- Wild oats in **winter wheat**

Approval information
- Clodinafop-propargyl included in Annex I under EC Directive 91/414
- Approved for use on crops for brewing by BBPA

Efficacy guidance
- Products contain a herbicide safener (cloquintocet-mexyl) that improves crop tolerance to clodinafop-propargyl
- Best results obtained from treatment when all grass weeds have emerged. There is no residual activity
- Broad-leaved weeds are not controlled
- Treat before emerged weed competition reduces yield
- For ryegrass control use as part of a programme including other products with activity against ryegrass
- For blackgrass control must be used as part of an integrated control strategy
- Grass weed control may be reduced if rain falls within 1 hr of application
- Clodinafop-propargyl and pinoxaden are ACCase inhibitor herbicides. To avoid the build up of resistance do not apply products containing an ACCase inhibitor herbicide more than twice to any crop. In addition do not use any product containing clodinafop-propargyl or pinoxaden in mixture or sequence with any other product containing the same ingredient
- Use as part of a resistance management strategy that includes cultural methods of control and does not use ACCase inhibitors as the sole chemical method of grass weed control
- Applying a second product containing an ACCase inhibitor to a crop will increase the risk of resistance development; only use a second ACCase inhibitor to control different weeds at a different timing
- Always follow WRAG guidelines for preventing and managing herbicide resistant weeds. See Section 5 for more information

Restrictions
- Maximum number of treatments 1 per crop
- Product must always be used with specified adjuvant. See label
- Do not use on other cereals
- Do not treat crops under stress from any cause
- Do not treat crops undersown with grass or grass mixtures
- Avoid use of hormone-containing herbicides in mixture or in sequence. See label for restrictions on tank mixes

Crop-specific information
- Latest use: before flag leaf sheath extended (before GS 41) for winter wheat
- All varieties of winter wheat may be treated

Following crops guidance
- There are no restrictions on succeeding crops in a normal rotation

FOR FULL CONDITIONS OF USE ALWAYS READ THE PRODUCT LABEL

clodinafop-propargyl + prosulfocarb

- In the event of failure of a treated crop ryegrass, maize, oats or any broad-leaved crop may be planted after a minimum interval of 4 wk from application

Environmental safety
- Dangerous for the environment
- Toxic to aquatic organisms

Hazard classification and safety precautions
    Hazard Irritant, Dangerous for the environment
    Transport code 9
    Packaging group III
    UN number 3082
    Risk phrases R36, R38, R43, R51, R53a
    Operator protection A, C, H, K; U05a, U09a, U20b
    Environmental protection E15b, E38
    Storage and disposal D01, D02, D05, D09a, D10c, D11a, D12a

100   clodinafop-propargyl + prosulfocarb

A foliar acting herbicide mixture for grass and broad-leaved weed control in winter wheat
HRAC mode of action code: A + N

See also prosulfocarb

Products
    Auxiliary   BASF   10:800 g/l   EC   14576

Uses
- Annual dicotyledons in **winter wheat**
- Annual grasses in **winter wheat**

Approval information
- Clodinafop-propargyl and prosulfocarb included in Annex I under EC Directive 91/414

Efficacy guidance
- To avoid the build up of resistance do not apply products containing an ACCase inhibitor herbicide more than twice to any crop. In addition, do not use this product in mixture or sequence with any other product containing clodinafop-propargyl.

Restrictions
- Do not use on barley or oats
- Do not spray crops undersown with grass mixtures
- When [1] is applied first, leave 7 days before applying hormone herbicides. If mecoprop-p or 2,4-DB containing products are applied first, leave 14 days before [1] is applied. If MCPA or 2,4-D containing products are applied first, leave 21 days before [1] is applied.
- The cereal seed must be covered by 3 cm of soil and for best results apply to a firm, moist seedbed free of clods.

Crop-specific information
- Only one application per crop
- Latest use before 5 tiller stage (GS25)

Environmental safety
- LERAP Category B

Hazard classification and safety precautions
    Hazard Harmful, Dangerous for the environment
    Transport code 9
    Packaging group III
    UN number 3082
    Risk phrases R22b, R51, R53a
    Operator protection A, H; U02a, U05a, U08, U20c
    Environmental protection E15b, E16a, E16b, E38
    Storage and disposal D01, D02, D05, D09a, D10c, D12a
    Medical advice M05b

SEE SECTION 3 FOR PRODUCTS ALSO REGISTERED

Pesticide Profiles

101 clofentezine

A selective ovicidal tetrazine acaricide for use in top fruit
IRAC mode of action code: 10A

Products
Apollo 50 SC   Makhteshim   500 g/l   SC   10590

Uses
- Red spider mites in **apples**, **cherries**, **pears**, **plums**
- Spider mites in **blackcurrants** *(off-label)*, **protected blackberries** *(off-label)*, **protected raspberries** *(off-label)*, **protected strawberries** *(off-label)*, **raspberries** *(off-label)*, **strawberries** *(off-label)*

Specific Off-Label Approvals (SOLAs)
- **blackcurrants** 20012269
- **protected blackberries** 20012268
- **protected raspberries** 20012268
- **protected strawberries** 20012271
- **raspberries** 20012269
- **strawberries** 20012269

Approval information
- Clofentezine included in Annex I under EC Directive 91/414

Efficacy guidance
- Acts on eggs and early motile stages of mites. For effective control total cover of plants is essential, particular care being needed to cover undersides of leaves

Restrictions
- Maximum number of treatments 1 per yr for apples, pears, cherries, plums

Crop-specific information
- HI apples, pears 28 d; cherries, plums 8 wk
- For red spider mite control spray apples and pears between bud burst and pink bud, plums and cherries between white bud and first flower. Rust mite is also suppressed
- On established infestations apply in conjunction with an adult acaricide

Environmental safety
- Harmful to aquatic organisms
- Product safe on predatory mites, bees and other predatory insects

Hazard classification and safety precautions
UN number N/C
Risk phrases R52
Operator protection U05a, U08, U20b
Environmental protection E15a
Storage and disposal D01, D02, D05, D09a, D11a

102 clomazone

An isoxazolidinone residual herbicide for oilseed rape, field beans, combining and vining peas
HRAC mode of action code: F3

Products

1	Centium 360 CS	Belchim	360 g/l	CS	13846
2	Cirrus CS	Belchim	371 g/l	CS	13860
3	Clomaz 36 CS	Goldengrass	360 g/l	CS	14938
4	Concept	ChemSource	360 g/l	CS	14642
5	Echo	AgriGuard	360 g/l	CS	14982
6	Fiddle	Goldengrass	360 g/l	CS	14451
7	Gadwall	AgChem Access	360 g/l	CS	14522
8	Gamit 36 CS	Belchim	360 g/l	CS	13861

FOR FULL CONDITIONS OF USE ALWAYS READ THE PRODUCT LABEL

clomazone

Uses
- Annual dicotyledons in **asparagus** *(off-label)*, **broad beans** *(off-label)*, **broccoli** *(off-label)*, **brussels sprouts** *(off-label)*, **cabbages** *(off-label)*, **calabrese** *(off-label)*, **cauliflowers** *(off-label)*, **chinese cabbage** *(off-label)*, **choi sum** *(off-label)*, **collards** *(off-label)*, **game cover** *(off-label)*, **herbs (see appendix 6)** *(off-label)*, **kale** *(off-label)*, **lupins** *(off-label)*, **pak choi** *(off-label)*, **spring cabbage** *(off-label)*, **swedes** *(off-label)*, **sweet potato** *(off-label)* [8]; **combining peas, spring field beans, spring oilseed rape, vining peas, winter field beans, winter oilseed rape** [5]; **poppies for morphine production** *(off-label)* [1, 8]
- Annual meadow grass in **combining peas, spring field beans, spring oilseed rape, vining peas, winter field beans, winter oilseed rape** [5]
- Chickweed in **carrots, potatoes** [3, 8]; **combining peas, spring field beans, spring oilseed rape, vining peas, winter field beans, winter oilseed rape** [1, 2, 4, 6, 7]
- Cleavers in **carrots, potatoes** [3, 8]; **combining peas, spring field beans, spring oilseed rape, vining peas, winter field beans, winter oilseed rape** [1, 2, 4, 6, 7]
- Fool's parsley in **carrots, potatoes** [3, 8]; **combining peas, spring field beans, spring oilseed rape, vining peas** [1, 2, 4, 6, 7]
- Red dead-nettle in **carrots, potatoes** [3, 8]; **combining peas, spring field beans, spring oilseed rape, vining peas, winter field beans, winter oilseed rape** [1, 2, 4, 6, 7]
- Shepherd's purse in **carrots, potatoes** [3, 8]; **combining peas, spring field beans, spring oilseed rape, vining peas, winter field beans, winter oilseed rape** [1, 2, 4, 6, 7]

Specific Off-Label Approvals (SOLAs)
- **asparagus** 20091276 [8]
- **broad beans** 20091281 [8]
- **broccoli** 20091275 [8]
- **brussels sprouts** 20091275 [8]
- **cabbages** 20091275 [8]
- **calabrese** 20091275 [8]
- **cauliflowers** 20091275 [8]
- **chinese cabbage** 20091275 [8]
- **choi sum** 20091275 [8]
- **collards** 20091275 [8]
- **game cover** 20091280 [8]
- **herbs (see appendix 6)** 20100130 [8]
- **kale** 20091275 [8]
- **lupins** 20091274 [8]
- **pak choi** 20091275 [8]
- **poppies for morphine production** 20080593 [1], 20091283 [8]
- **spring cabbage** 20091275 [8]
- **swedes** 20091282 [8]
- **sweet potato** 20091279 [8]

Approval information
- Clomazone included in Annex I under EC Directive 91/414

Efficacy guidance
- Best results obtained from application as soon as possible after sowing crop and before emergence of crop or weeds
- Uptake is via roots and shoots. Seedbeds should be firm, level and free from clods. Loose puffy seedbeds should be consolidated before spraying
- Efficacy is reduced on organic soils, on dry cloddy seedbeds and if prolonged dry weather follows application
- Clomazone acts by inhibiting synthesis of chlorophyll pigments. Susceptible weeds emerge but are chlorotic and die shortly afterwards
- Season-long control of weeds may not be achieved
- Always follow WRAG guidelines for preventing and managing herbicide resistant weeds. See Section 5 for more information

Restrictions
- Maximum number of treatments one per crop
- Crops must be covered by a minimum of 20 mm settled soil. Do not apply to broadcast crops. Direct-drilled crops should be harrowed across the slits to cover seed before spraying

SEE SECTION 3 FOR PRODUCTS ALSO REGISTERED

Pesticide Profiles

- Do not use on compacted soils or soils of poor structure that may be liable to waterlogging
- Do not use on Sands or Very Light soils or those with more than 10% organic matter
- Do not treat two consecutive crops of carrots with clomazone in one calendar yr
- Consult manufacturer or your advisor before use on potato seed crops
- Do not overlap spray swaths. Crop plants emerged at time of treatment may be severely damaged

Crop-specific information
- Latest use: pre-emergence of crop
- Severe, but normally transient, crop damage may occur in overlaps on field beans
- Some transient crop bleaching may occur under certain climatic conditions and can be severe where heavy rain follows application. This is normally rapidly outgrown and has no effect on final crop yield. Overlapping spray swaths may cause severe damage to field beans

Following crops guidance
- Following normal harvest of a spring or autumn treated crop, cereals, oilseed rape, field beans, combining peas, potatoes, maize, turnips, linseed or sugar beet may be sown
- In the event of failure of an autumn treated crop, winter cereals or winter beans may be sown in the autumn if 6 wk have elapsed since treatment. In the spring following crop failure combining peas, field beans or potatoes may be sown if 6 wk have elapsed since treatment, and spring cereals, maize, turnips, onions, carrots or linseed may be sown if 7 mth have elapsed since treatment
- In the event of a failure of a spring treated crop a wide range of crops may be sown provided intervals of 6-9 wk have elapsed since treatment. See label for details
- Prior to resowing any listed replacement crop the soil should be ploughed and cultivated to 15 cm

Environmental safety
- Take extreme care to avoid drift outside the target area, or on to ponds, waterways or ditches as considerable damage may occur. Apply using a coarse quality spray

Hazard classification and safety precautions
Hazard Irritant [5]
UN number N/C
Risk phrases R43 [5]
Operator protection A; U05a, U14, U20b
Environmental protection E15b
Storage and disposal D01, D02, D09a, D10b

103 clomazone + linuron

An isoxazolidinone and urea residual herbicide for peas, field beans and potatoes
HRAC mode of action code: F3 + C2

See also linuron

Products
1	Lingo	Belchim	45:250 g/l	CC	14075
2	Linzone	Belchim	45:250g/l	CC	14688

Uses
- Annual dicotyledons in **combining peas**, **potatoes**, **spring field beans**, **vining peas**, **winter field beans**
- Annual meadow grass in **combining peas**, **potatoes**, **spring field beans**, **vining peas**, **winter field beans**

Approval information
- Clomazone and linuron included in Annex I under EC Directive 91/414

Restrictions
- Do not apply to broadcast crops
- Do not apply on light soils if heavy rain is expected within 24 hours of application
- Transient crop bleaching may occur under some conditions. This is normally outgrown with no effect on final yield

FOR FULL CONDITIONS OF USE ALWAYS READ THE PRODUCT LABEL

clomazone + metazachlor

Following crops guidance
- Following normal harvest of an autumn treated crop, cereals, oilseed rape, field beans, combining peas, potatoes, maize, turnip, linseed and sugar beet may be sown. Prior to planting, soil should be ploughed and cultivated to a minimum depth of 15 cm.
- Following normal harvest of a spring treated crop, cereals, oilseed rape, field beans, combining peas, potatoes, maize, linseed, sugar beet and turnip may be sown. Prior to planting, soil should be ploughed and cultivated to a minimum depth of 15 cm. Do not sow or plant lettuce in the same season as an application of [1]

Hazard classification and safety precautions
Hazard Toxic, Dangerous for the environment
Transport code 9
Packaging group III
UN number 3082
Risk phrases R22a, R40, R48, R50, R53a, R61, R62
Operator protection A, H; U02a, U05a, U13, U14, U15, U19a
Environmental protection E15a, E23, E34, E38
Storage and disposal D01, D02, D12a
Medical advice M04a

104 clomazone + metazachlor

A mixture of isoxazolidinone and residual anilide herbicides for weed control in oilseed rape
HRAC mode of action code: F3 + K3

See also metazachlor

Products
Nimbus CS   BASF   33.3:250   SC   14092

Uses
- Annual dicotyledons in **winter oilseed rape**
- Annual grasses in **winter oilseed rape**
- Cleavers in **winter oilseed rape**

Approval information
- Clomazone and metazachlor included in Annex I under EC Directive 91/414

Restrictions
- A maximum total dose of 1000 g.ai/ha may only be applied every third year to the same field

Environmental safety
- LERAP Category B

Hazard classification and safety precautions
Hazard Irritant, Dangerous for the environment
Transport code 9
Packaging group III
UN number 3082
Risk phrases R43, R50, R53a
Operator protection A, H; U05a, U08, U14, U19a, U20b
Environmental protection E07a, E15b, E16a, E38
Storage and disposal D01, D02, D08, D09a, D10c, D12a
Medical advice M05a

105 clomazone + metribuzin

A mixture of isoxazolidinone and triazinone residual herbicides for use in potatoes
HRAC mode of action code: F3 + C1

See also metribuzin

Products
Metric   Belchim   60:233 g/l   CC   14214

SEE SECTION 3 FOR PRODUCTS ALSO REGISTERED

Pesticide Profiles

Uses
- Annual dicotyledons in **potatoes**
- Annual meadow grass in **potatoes**

Approval information
- Clomazone and metribuzin included in Annex I under EC Directive 91/414

Restrictions
- For use on specified varieties of potato only
- Safety to daughter tubers has not been tested; Consult manufacturer before treating seed crops

Environmental safety
- LERAP Category B

Hazard classification and safety precautions
Hazard Dangerous for the environment
Transport code 9
Packaging group III
UN number 3082
Risk phrases R50, R53a
Operator protection A, H; U05a, U19a, U20b
Environmental protection E15b, E16a, E16b, E34, E38
Storage and disposal D01, D02, D09a, D10b, D12a
Medical advice M05a

106 clopyralid

A foliar translocated picolinic herbicide for a wide range of crops
HRAC mode of action code: O

See also bromoxynil + clopyralid

Products

1	Dow Shield	Dow	200 g/l	SL	10988
2	Dow Shield 400	Dow	400 g/l	SL	14984
3	Glopyr 200 SL	Globachem	200 g/l	SL	10979
4	Lontrel 200	Dow	18.02% w/w	SL	11558
5	Vivendi 200	AgriChem BV	200 g/l	SL	12782

Uses
- Annual dicotyledons in **broccoli, brussels sprouts, bulb onions, cabbages, calabrese, fodder rape, forage maize, kale, linseed, salad onions, strawberries, sweetcorn** [1, 3-5]; **cauliflowers** [1, 4, 5]; **conifers and broadleaved trees** *(off-label)*, **permanent grassland** *(off-label)*, **rotational grass** [4]; **fodder beet, mangels, ornamental plant production, red beet, spring barley, spring oats, spring oilseed rape, spring wheat, sugar beet, swedes, turnips, winter barley, winter oats, winter oilseed rape, winter wheat** [1-5]; **grassland** [2, 5]; **permanent grassland** [1, 3, 4]
- Clovers in **grass seed crops** *(off-label)* [2]
- Corn marigold in **broccoli, brussels sprouts, bulb onions, cabbages, calabrese, cauliflowers, fodder rape, forage maize, kale, linseed, permanent grassland, salad onions, strawberries, sweetcorn** [1, 3, 4]; **fodder beet, mangels, ornamental plant production, red beet, spring barley, spring oats, spring oilseed rape, spring wheat, sugar beet, swedes, turnips, winter barley, winter oats, winter oilseed rape, winter wheat** [1-4]; **grassland** [2]; **permanent grassland** *(off-label)*, **rotational grass** [4]
- Creeping thistle in **broccoli, brussels sprouts, bulb onions, cabbages, calabrese, cauliflowers, fodder rape, forage maize, kale, linseed, permanent grassland, salad onions, strawberries, sweetcorn** [1, 3, 4]; **fodder beet, mangels, ornamental plant production, red beet, spring barley, spring oats, spring oilseed rape, spring wheat, sugar beet, swedes, turnips, winter barley, winter oats, winter oilseed rape, winter wheat** [1-4]; **grassland** [2]; **permanent grassland** *(off-label)*, **rotational grass** [4]
- General weed control in **canary grass** *(off-label - game cover)*, **quinoa** *(off-label - game cover)*, **tanka millet** *(off-label - game cover)*, **white millet** *(off-label - game cover)* [4]
- Groundsel in **chard** *(off-label)*, **spinach** *(off-label)* [2]; **chives** *(off-label)*, **herbs (see appendix 6)** *(off-label)*, **leaf spinach** *(off-label)*, **parsley** *(off-label)* [4]; **spinach beet** *(off-label)* [2, 4]

FOR FULL CONDITIONS OF USE ALWAYS READ THE PRODUCT LABEL

clopyralid

- Mayweeds in **apples** *(off-label)*, **borage for oilseed production** *(off-label)*, **chard** *(off-label)*, **crab apples** *(off-label)*, **durum wheat** *(off-label)*, **pears** *(off-label)*, **spinach** *(off-label)*, **spring rye** *(off-label)*, **winter rye** *(off-label)* [2]; **broccoli, brussels sprouts, bulb onions, cabbages, calabrese, cauliflowers, fodder rape, forage maize, kale, linseed, salad onions, strawberries, sweetcorn** [1, 3-5]; **canary flower (echium spp.)** *(off-label)*, **evening primrose** *(off-label)*, **poppies for morphine production** *(off-label)*, **spinach beet** *(off-label)* [2, 4]; **chives** *(off-label)*, **herbs (see appendix 6)** *(off-label)*, **leaf spinach** *(off-label)*, **parsley** *(off-label)*, **permanent grassland** *(off-label)*, **rotational grass, spring linseed** *(off-label)* [4]; **collards** *(off-label)*, **garlic** *(off-label)*, **shallots** *(off-label)* [3, 4]; **fodder beet, mangels, ornamental plant production, red beet, spring barley, spring oats, spring oilseed rape, spring wheat, sugar beet, swedes, turnips, winter barley, winter oats, winter oilseed rape, winter wheat** [1-5]; **grassland** [2, 5]; **honesty** *(off-label)*, **mustard** *(off-label)* [2-4]; **permanent grassland** [1, 3, 4]; **triticale** *(off-label)* [2, 3]
- Sowthistle in **poppies for morphine production** *(off-label)* [2]
- Thistles in **apple orchards** *(off-label)*, **grass seed crops** *(off-label)*, **pear orchards** *(off-label)*, **permanent grassland** *(off-label - spot treatment)*, **spring linseed** *(off-label)* [4]; **apples** *(off-label)*, **borage for oilseed production** *(off-label)*, **crab apples** *(off-label)*, **durum wheat** *(off-label)*, **pears** *(off-label)*, **spring rye** *(off-label)*, **trees** *(off-label)*, **triticale** *(off-label)*, **winter rye** *(off-label)* [2]; **asparagus** *(off-label)*, **canary flower (echium spp.)** *(off-label)*, **evening primrose** *(off-label)*, **poppies for morphine production** *(off-label)* [2, 4]; **broccoli, brussels sprouts, bulb onions, cabbages, calabrese, cauliflowers, fodder beet, fodder rape, forage maize, grassland, kale, linseed, mangels, ornamental plant production, red beet, salad onions, spring barley, spring oats, spring oilseed rape, spring wheat, strawberries, sugar beet, swedes, sweetcorn, turnips, winter barley, winter oats, winter oilseed rape, winter wheat** [5]; **collards** *(off-label)*, **garlic** *(off-label)*, **shallots** *(off-label)* [3, 4]; **honesty** *(off-label)*, **mustard** *(off-label)* [2-4]; **permanent grassland** *(spot treatment)* [1, 4]

Specific Off-Label Approvals (SOLAs)
- *apple orchards* 20062444 [4]
- *apples* 20102080 [2]
- *asparagus* 20102079 [2], 20062445 [4]
- *borage for oilseed production* 20102086 [2]
- *canary flower (echium spp.)* 20102086 [2], 20061933 [4]
- *canary grass* (game cover) 20062450 [4]
- *chard* 20102081 [2]
- *chives* 20062442 [4]
- *collards* 20102689 [3], 20061932 [4]
- *conifers and broadleaved trees* 20062449 [4]
- *crab apples* 20102080 [2]
- *durum wheat* 20102085 [2]
- *evening primrose* 20102086 [2], 20061933 [4]
- *garlic* 20102686 [3], 20061936 [4]
- *grass seed crops* 20102084 [2], 20062448 [4]
- *herbs (see appendix 6)* 20062442 [4]
- *honesty* 20102086 [2], 20102685 [3], 20062446 [4]
- *leaf spinach* 20062443 [4]
- *mustard* 20102086 [2], 20102685 [3], 20061933 [4]
- *parsley* 20062442 [4]
- *pear orchards* 20062444 [4]
- *pears* 20102080 [2]
- *permanent grassland* (spot treatment) 20062448 [4], 20062448 [4]
- *poppies for morphine production* 20102082 [2], 20062447 [4]
- *quinoa* (game cover) 20062450 [4]
- *shallots* 20102686 [3], 20061936 [4]
- *spinach* 20102081 [2]
- *spinach beet* 20102081 [2], 20062443 [4]
- *spring linseed* 20061933 [4]
- *spring rye* 20102085 [2]
- *tanka millet* (game cover) 20062450 [4]
- *trees* 20102083 [2]

SEE SECTION 3 FOR PRODUCTS ALSO REGISTERED

- ***triticale*** *20102085* [2], *20102688* [3]
- ***white millet*** *(game cover) 20062450* [4]
- ***winter rye*** *20102085* [2]

Approval information
- Clopyralid included in Annex I under EC Directive 91/414
- Accepted by BBPA for use on malting barley

Efficacy guidance
- Best results achieved by application to young actively growing weed seedlings. Treat creeping thistle at rosette stage and repeat 3-4 wk later as directed
- High activity on weeds of Compositae family. For most crops recommended for use in tank mixes. See label for details

Restrictions
- Maximum total dose varies between the equivalent of one and two full dose treatments, depending on the crop treated. See labels for details
- Do not apply to cereals later than the second node detectable stage (GS 32)
- Do not apply when crop damp or when rain expected within 6 h
- Do not use straw from treated cereals in compost or any other form for glasshouse crops. Straw may be used for strawing down strawberries
- Straw from treated grass seed crops or linseed should be baled and carted away. If incorporated do not plant winter beans in same year
- Do not use on onions at temperatures above 20 °C or when under stress
- Do not treat maiden strawberries or runner beds or apply to early leaf growth during blossom period or within 4 wk of picking. Aug or early Sep sprays may reduce yield

Crop-specific information
- Latest use: 7 d before cutting grass for hay or silage; before 3rd node detectable (GS 33) for cereals; before flower buds visible from above for oilseed rape, linseed
- HI grassland 7 d; apples, pears, strawberries 4 wk; maize, sweetcorn, onions, Brussels sprouts, broccoli, cabbage, cauliflowers, calabrese, kale, fodder rape, oilseed rape, swedes, turnips, sugar beet, red beet, fodder beet, mangels, sage, honesty 6 wk
- Timing of application varies with weed problem, crop and other ingredients of tank mixes. See labels for details
- Apply as directed spray in woody ornamentals, avoiding leaves, buds and green stems. Do not apply in root zone of families Compositae or Leguminosae

Following crops guidance
- Do not plant susceptible autumn-sown crops in same year as treatment. Do not apply later than Jul where susceptible crops are to be planted in spring. See label for details

Environmental safety
- Dangerous for the environment [3]
- Harmful to aquatic organisms [1, 3, 4]
- Wash spray equipment thoroughly with water and detergent immediately after use. Traces of product can damage susceptible plants sprayed later
- Keep livestock out of treated areas for at least 7 d and until foliage of any poisonous weeds such as ragwort has died and become unpalatable
- Some pesticides pose a greater threat of contamination of water than others and clopyralid is one of these pesticides. Take special care when applying clopyralid near water and do not apply if heavy rain is forecast

Hazard classification and safety precautions
Hazard Dangerous for the environment [3]
UN number N/C
Risk phrases R50 [3]; R52 [1, 2, 4]; R53a [1-4]
Operator protection A, C; U05a [5]; U08, U19a, U20b [1-5]
Environmental protection E07a [1-4]; E15a [1-5]; E34 [5]
Storage and disposal D01 [1, 2, 4, 5]; D02 [5]; D05, D09a, D10b [1-5]; D12a [1, 2, 4]

FOR FULL CONDITIONS OF USE ALWAYS READ THE PRODUCT LABEL

107 clopyralid + 2,4-D + MCPA

A translocated herbicide mixture for managed amenity turf
HRAC mode of action code: O + O + O

See also 2,4-D
MCPA

Products
Esteem   Vitax   35:150:175 g/l   SL   12555

Uses
- Annual dicotyledons in **managed amenity turf**
- Perennial dicotyledons in **managed amenity turf**

Approval information
- Clopyralid, 2,4-D and MCPA included in Annex I under EC Directive 91/414

Efficacy guidance
- Apply when soil is moist and weeds growing actively, normally between Apr and Oct
- Ensure sufficient leaf area for uptake and select appropriate water volume to achieve good spray coverage
- To allow maximum translocation do not cut grass for 3 d after treatment

Restrictions
- Maximum number of treatments on managed amenity turf 1 per yr
- Do not spray in periods of drought unless irrigation is applied
- Do not apply if night temrperatures are low, if ground frost is imminent or in periods of prolonged cold weather
- Do not treat turf less than 1 yr old
- Do not use any treated plant materials for composting or mulching
- Do not mow for 3 d before or after treatment

Crop-specific information
- Consult manufacturer or carry out small scale safety test on turf grass cultivars not previously treated

Following crops guidance
- If reseeding of treated turf is required allow at least 6 wk after treatment before drilling grasses into the sward
- Where treated land is subsequently to be sown with broad-leaved plants allow an interval of at least 6 mth

Environmental safety
- Dangerous for the environment
- Harmful to aquatic organisms
- Wash spray equipment thoroughly with water and detergent immediately after use. Traces of product can damage susceptible plants sprayed later
- Do not spray outside the target area and avoid drift onto non-target plants
- Some pesticides pose a greater threat of contamination of water than others and clopyralid is one of these pesticides. Take special care when applying clopyralid near water and do not apply if heavy rain is forecast

Hazard classification and safety precautions
Hazard Harmful, Dangerous for the environment
Transport code M6
Packaging group III
UN number 3082
Risk phrases R22a, R37, R41, R52, R53a
Operator protection A, C, H; U05a, U11, U15
Environmental protection E15a, E34
Storage and disposal D01, D02, D12a

SEE SECTION 3 FOR PRODUCTS ALSO REGISTERED

Pesticide Profiles

108 clopyralid + diflufenican + MCPA

A selective herbicide for use in managed amenity turf
HRAC mode of action code: O + F1 + O

See also diflufenican
MCPA

Products
Spearhead   Bayer Environ.   20:15:300 g/l   SL   09941

Uses
- Annual dicotyledons in **managed amenity turf**
- Perennial dicotyledons in **managed amenity turf**

Approval information
- Clopyralid, diflufenican and MCPA included in Annex I under EC Directive 91/414

Efficacy guidance
- Best results achieved by application when grass and weeds are actively growing
- Treatment during early part of the season is recommended, but not during drought

Restrictions
- Maximum number of treatments 1 per yr
- Only use when sward is satisfactorily established and regular mowing has begun
- Turf sown in spring or early summer may be ready for treatment after 2 mth. Later sown turf should not be sprayed until growth is resumed in the following spring
- Avoid mowing within 3-4 d before or after treatment
- Do not use cuttings from treated area as a mulch for any crop

Environmental safety
- Extremely dangerous to fish or other aquatic life. Do not contaminate surface waters or ditches with chemical or used container
- Keep livestock out of treated areas/away from treated water for at least 1 wk and until foliage of any poisonous weeds, such as ragwort, has died and become unpalatable
- Avoid drift. Small amounts of spray can cause serious injury to herbaceous plants, vegetables, fruit and glasshouse crops
- Keep livestock out of treated areas
- Some pesticides pose a greater threat of contamination of water than others and clopyralid is one of these pesticides. Take special care when applying clopyralid near water and do not apply if heavy rain is forecast
- LERAP Category B

Hazard classification and safety precautions
Hazard Harmful, Irritant
Transport code 9
Packaging group III
UN number 3082
Risk phrases R20, R21, R22a, R36, R38
Operator protection A, C; U02a, U05a, U08, U19a, U20b
Environmental protection E06a (unstipulated period); E13a, E16a, E16b, E34
Storage and disposal D01, D02, D05, D09a, D10b
Medical advice M03

FOR FULL CONDITIONS OF USE ALWAYS READ THE PRODUCT LABEL

109 clopyralid + florasulam + fluroxypyr

A translocated herbicide mixture for cereals
HRAC mode of action code: O + B + O

See also florasulam
fluroxypyr

Products

1	Galaxy	Dow	80:2.5:100 g/l	EC	14085
2	Praxys	Dow	80:2.5;100 g/l	EC	13912

Uses
- Annual dicotyledons in **amenity grassland, lawns, managed amenity turf** [2]; **spring barley, spring oats, spring wheat, winter barley, winter oats, winter wheat** [1]
- Chickweed in **spring barley, spring oats, spring wheat, winter barley, winter oats, winter wheat** [1]
- Cleavers in **spring barley, spring oats, spring wheat, winter barley, winter oats, winter wheat** [1]

Approval information
- Clopyralid, florasulam and fluroxypyr are all included in Annex I under Directive 91/414
- Approval expiry 31 Dec 2011 [1, 2]

Efficacy guidance
- Best results obtained when weeds are small and actively growing
- Effectiveness may be reduced when soil is very dry
- Use adequate water volume to achieve complete spray coverage of the weeds
- Florasulam is a member of the ALS-inhibitor group of herbicides

Restrictions
- Maximum number of treatments 1 per yr for all crops
- Do not spray when crops are under stress from any cause
- Do not apply through CDA applicators
- Do not roll or harrow for 7 d before or after application
- Do not use any treated plant material for composting or mulching
- Do not use manure from animals fed on treated crops for composting
- Specific restrictions apply to use in sequence or tank mixture with other sulfonylurea or ALS-inhibiting herbicides. See label for details

Crop-specific information
- Latest use: before second node detectable for oats; before third node detectable for spring barley and spring wheat; before flag leaf detectable in winter barley and winter wheat.

Following crops guidance
- Where residues of a treated crop have not completely decayed by the time of planting a succeeding crop, avoid planting peas, beans and other legumes, carrots and other Umbelliferae, potatoes, lettuce and other Compositae, glasshouse and protected crops
- Where the product has been used in mixture with certain named products (see label) only cereals or grass may be sown in the autumn following harvest. Otherwise cereals, oilseed rape, grass or vegetable brassicas as transplants may be sown as a following crop in the same calendar yr as treatment. Oilseed rape may show some temporary reduction of vigour after a dry summer, but yields are not affected
- In addition to the above, field beans, linseed, peas, sugar beet, potatoes, maize, clover (for use in grass/clover mixtures) or carrots may be sown in the calendar yr following treatment
- In the event of failure of a treated crop in spring, only spring wheat, spring barley, spring oats, maize or ryegrass may be sown

Environmental safety
- Dangerous for the environment
- Very toxic to aquatic organisms
- Take extreme care to avoid drift outside the target area
- Some pesticides pose a greater threat of contamination of water than others and clopyralid is one of these pesticides. Take special care when applying clopyralid near water and do not apply if heavy rain is forecast

SEE SECTION 3 FOR PRODUCTS ALSO REGISTERED

Pesticide Profiles

Hazard classification and safety precautions
   Hazard Harmful, Dangerous for the environment
   Transport code 9
   Packaging group III
   UN number 3082
   Risk phrases R20, R36, R38, R50, R53a
   Operator protection A, H [1, 2]; C [1]; U05a, U11, U19a, U20a
   Environmental protection E15b, E34, E38
   Storage and disposal D01, D02, D09a, D10b, D12a [1, 2]; D05 [2]
   Medical advice M03 [1, 2]; M05a [1]

110   clopyralid + fluroxypyr + MCPA

A translocated herbicide mixture for use in sports and amenity turf
HRAC mode of action code: O + O + O

See also fluroxypyr
MCPA

Products
   Greenor   Rigby Taylor   20:40:200 g/l   ME   10909

Uses
 - Annual dicotyledons in **managed amenity turf**

Approval information
 - Clopyralid, fluroxypyr and MCPA included in Annex I under EC Directive 91/414

Efficacy guidance
 - Best results achieved when weeds actively growing and turf grass competitive
 - Treatment should normally be between Apr-Sep when the soil is moist
 - Do not apply during drought unless irrigation is applied
 - Allow 3 d before or after mowing established turf to ensure sufficient weed leaf surface present to allow uptake and movement

Restrictions
 - Maximum number of treatments 2 per yr
 - Do not treat grass under stress from frost, drought, waterlogging, trace element deficiency, disease or pest attack
 - Do not treat if night temperatures are low, when frost is imminent or during prolonged cold weather

Crop-specific information
 - Treat young turf only in spring when at least 2 mth have elapsed since sowing
 - Allow 5 d after mowing young turf before treatment
 - Product selective on a number of turf grass species (see label) but consultation or testing recommended before treatment of any cultivar

Environmental safety
 - Dangerous for the environment
 - Very toxic to aquatic organisms
 - Wash spray equipment thoroughly with water and detergent immediately after use. Traces of product can damage susceptible plants sprayed later
 - Some pesticides pose a greater threat of contamination of water than others and clopyralid is one of these pesticides. Take special care when applying clopyralid near water and do not apply if heavy rain is forecast

Hazard classification and safety precautions
   Hazard Irritant, Dangerous for the environment
   Transport code 9
   Packaging group III
   UN number 3082
   Risk phrases R36, R43, R50, R53a
   Operator protection A, C; U05a, U08, U14, U19a, U20b

FOR FULL CONDITIONS OF USE ALWAYS READ THE PRODUCT LABEL

Environmental protection E15a, E38
Storage and disposal D01, D02, D05, D09a, D10b, D12a

111   clopyralid + fluroxypyr + triclopyr

A foliar acting herbicide mixture for grassland
HRAC mode of action code: O + O + O

See also fluroxypyr
triclopyr

Products
  Pastor   Dow   50:75:100 g/l   EC   11168

Uses
- Annual dicotyledons in **rotational grass**
- Docks in **permanent grassland, rotational grass**
- Stinging nettle in **permanent grassland**
- Thistles in **permanent grassland**

Approval information
- Clopyralid, fluroxypyr and triclopyr included in Annex I under EC Directive 91/414

Efficacy guidance
- Apply in spring or autumn depending on weeds present or as a split treatment in spring followed by autumn
- Treatment must be made when weeds and grass are actively growing
- It is important to ensure sufficient leaf area is present for uptake especially on established docks and thistles
- On large well established docks and where there is a large soil seed reservoir further treatment in the following year may be needed
- To allow maximum translocation in the weeds do not cut grass for 4 wk after treatment

Restrictions
- Maximum number of treatments 1 per yr at full dose or 2 per yr at half dose
- Do not spray in drought, very hot or very cold weather
- Do not treat sports or amenity turf
- Product kills or severely checks clover and should not be used where clover is an important constituent of the sward
- Do not roll or harrow 10 d before or 7 d after treatment
- Do not allow spray or drift to reach other crops, amenity plantings, gardens, ponds, lakes or water courses

Crop-specific information
- HI 7 d
- Application during active growth ensures minimal check to grass. Newly sown grass may be treated from the third leaf visible stage
- Product may be used in established grassland which is under non-rotational setaside arrangements
- Very occasionally some yellowing of the sward may occur after treatment but is quickly outgrown

Following crops guidance
- Residues in incompletely decayed plant tissue may affect succeeding susceptible crops such as peas, beans and other legumes, carrots and related crops, potatoes, tomatoes, lettuce
- Do not plant susceptible autumn-sown crops in the same yr as treatment with product. Spring sown crops may follow if treatment was before end July in the previous yr

Environmental safety
- Dangerous for the environment
- Toxic to aquatic organisms
- Keep livestock out of treated areas for at least 7 d following treatment and until foliage of poisonous weeds such as ragwort has died and become unpalatable
- Wash spray equipment thoroughly with water and detergent immediately after use. Traces of product can damage susceptible plants sprayed later

SEE SECTION 3 FOR PRODUCTS ALSO REGISTERED

Pesticide Profiles

- Some pesticides pose a greater threat of contamination of water than others and clopyralid is one of these pesticides. Take special care when applying clopyralid near water and do not apply if heavy rain is forecast

Hazard classification and safety precautions
Hazard Harmful, Flammable, Dangerous for the environment
Transport code 3
Packaging group III
UN number 1993
Risk phrases R22b, R37, R38, R41, R43, R51, R53a, R67
Operator protection A, C; U02a, U05a, U08, U11, U14, U19a, U20b
Environmental protection E07a, E15a, E34, E38
Consumer protection C01
Storage and disposal D01, D02, D05, D09a, D10b, D12a
Medical advice M05b

112   clopyralid + picloram

A post-emergence herbicide mixture for oilseed rape
HRAC mode of action code: O + O

See also picloram

Products

1	EA Cloporam	European Ag	267:67 g/l	SL	14828
2	Galera	Dow	267:67 g/l	SL	11961
3	Legara	AgChem Access	267:67 g/l	SL	13888
4	Nugget	AgriGuard	267:67 g/l	SL	13121
5	Piccant	Goldengrass	267:67 g/l	SL	14294
6	Renegade	ChemSource	267:67 g/l	SL	14507

Uses
- Cleavers in **evening primrose** *(off-label)*, **honesty** *(off-label)*, **linseed** *(off-label)*, **mustard** *(off-label)*, **spring oilseed rape** *(off-label)* [2]; **winter oilseed rape** [1-6]
- Mayweeds in **evening primrose** *(off-label)*, **honesty** *(off-label)*, **linseed** *(off-label)*, **mustard** *(off-label)*, **spring oilseed rape** *(off-label)* [2]; **winter oilseed rape** [1-6]

Specific Off-Label Approvals (SOLAs)
- *evening primrose* 20060960 [2]
- *honesty* 20060960 [2]
- *linseed* 20060960 [2]
- *mustard* 20060960 [2]
- *spring oilseed rape* 20061437 [2]

Approval information
- Clopyralid and picloram included in Annex I under EC Directive 91/414

Efficacy guidance
- Best results obtained from treatment when weeds are small and actively growing
- Cleavers that germinate after treatment will not be controlled

Restrictions
- Maximum total dose equivalent to one full dose treatment
- Do not treat crops under stress from cold, drought, pest damage, nutrient deficiency or any other cause
- Do not roll or harrow for 7 d before or after spraying
- Do not apply through CDA applicators
- Do not use any treated plant material for composting or mulching
- Do not use manure from animals fed on treated crops for composting
- Chop and incorporate all treated plant remains in early autumn, or as soon as possible after harvest, to release any residues into the soil. Ensure that all treated plant remains have completely decayed before planting susceptible crops

Crop-specific information
- Latest Use: before flower buds visible above crop canopy for winter oilseed rape

FOR FULL CONDITIONS OF USE ALWAYS READ THE PRODUCT LABEL

clopyralid + triclopyr

Following crops guidance
- Any crop may be sown in the calendar yr following treatment
- Ploughing or thorough cultivation should be carried out before planting leguminous crops
- Do not attempt to plant peas, beans, other legumes, carrots, other umbelliferous crops, potatoes, lettuce, other Compositae, or any glasshouse or protected crops if treated crop remains have not fully decayed by the time of planting
- In the event of failure of an autumn treated crop only oilseed rape, wheat, barley, oats, maize or ryegrass may be sown in the spring and only after ploughing or thorough cultivation

Environmental safety
- Dangerous for the environment
- Toxic to aquatic organisms
- Take extreme care to avoid drift onto crops and non-target plants outside the target area
- Some pesticides pose a greater threat of contamination of water than others and clopyralid is one of these pesticides. Take special care when applying clopyralid near water and do not apply if heavy rain is forecast

Hazard classification and safety precautions
 Hazard Dangerous for the environment
 UN number N/C
 Risk phrases R51, R53a
 Operator protection A, C; U05a
 Environmental protection E15a, E34
 Storage and disposal D01, D02, D05, D07, D09a

113 clopyralid + triclopyr

A perennial and woody weed herbicide for use in grassland
HRAC mode of action code: O + O

See also triclopyr

Products

1	Blaster	Headland Amenity	60:240 g/l	EC	13267
2	Grazon 90	Dow	60:240 g/l	EC	13117
3	Thistlex	Dow	200:200 g/l	EC	11533

Uses
- Brambles in *amenity grassland* [1]
- Broom in *amenity grassland* [1]
- Creeping thistle in *permanent grassland, rotational grass* [3]
- Docks in *amenity grassland* [1]
- Gorse in *amenity grassland* [1]
- Perennial dicotyledons in *amenity grassland* [1]; *permanent grassland* (off-label - via weed wiper) [2]
- Stinging nettle in *amenity grassland* [1]
- Thistles in *amenity grassland* [1]

Specific Off-Label Approvals (SOLAs)
- *permanent grassland* (via weed wiper) 20063161 [2]

Approval information
- Clopyralid and triclopyr included in Annex I under EC Directive 91/414

Efficacy guidance
- Must be applied to actively growing weeds
- Correct timing crucial for good control. Spray stinging nettle before flowering, docks in rosette stage in spring, creeping thistle before flower stems 15-20 cm high, brambles, broom and gorse in Jun-Aug
- Allow 2-3 wk regrowth after grazing or mowing before spraying perennial weeds
- Where there is a large reservoir of weed seed in the soil further treatment in the following yr may be needed

Restrictions
- Maximum number of treatments 1 per yr

SEE SECTION 3 FOR PRODUCTS ALSO REGISTERED

Pesticide Profiles

- Only use on permanent pasture or rotational grassland established for at least 1 yr
- Do not apply where clover is an important constituent of sward
- Do not roll or harrow within 10 d before or 7 d after spraying
- Do not cut grass for 21 d before or 28 d after spraying
- Do not use any treated plant material for composting or mulching, and do not use manure for composting from animals fed on treated crops
- Do not apply by hand-held rotary atomiser equipment
- Do not allow drift onto other crops, amenity plantings or gardens, ponds, lakes or water courses. All conifers, especially pine and larch, are very sensitive

Crop-specific information
- Latest use: 7 d before grazing or cutting grass
- Some transient yellowing of treated swards may occur but is quickly outgrown

Following crops guidance
- Residues in plant tissues which have not completely decayed may affect succeeding susceptible crops such as peas, beans, other legumes, carrots, parsnips, potatoes, tomatoes, lettuce, glasshouse and protected crops
- Do not plant susceptible autumn-sown crops (eg winter beans) in same year as treatment and allow at least 9 mth from treatment before planting a susceptible crop in the following yr
- Do not direct drill kale, swedes, turnips, grass or grass mixtures within 6 wk of spraying
- Do not spray after end Jul where susceptible crops are to be planted in the next spring

Environmental safety
- Dangerous for the environment
- Very toxic to aquatic organisms
- Keep livestock out of treated areas for at least 7 d after spraying and until foliage of any poisonous weeds such as ragwort or buttercup has died down and become unpalatable
- Some pesticides pose a greater threat of contamination of water than others and clopyralid is one of these pesticides. Take special care when applying clopyralid near water and do not apply if heavy rain is forecast

Hazard classification and safety precautions
Hazard Harmful [1, 2]; Irritant [3]; Dangerous for the environment [1-3]
Transport code 3 [1, 2]
Packaging group III [1, 2]
UN number 1993 [1, 2]; N/C [3]
Risk phrases R22a, R38, R43, R50 [1, 2]; R41, R53a [1-3]; R52 [3]
Operator protection A, C [1-3]; H, M [1, 2]; U02a, U05a, U11, U20b [1-3]; U08, U14, U19a, U23b [1, 2]; U15, U23a [3]
Environmental protection E07a, E15a [1-3]; E23, E34, E38 [1, 2]
Consumer protection C01 [1, 2]
Storage and disposal D01, D02, D05, D09a, D10b, D12a
Medical advice M03 [3]

114 clothianidin

A nitromethylene neonicotinoid insecticide
IRAC mode of action code: 4A

See also beta-cyfluthrin + clothianidin

Products
1	Deter	Bayer CropScience	250 g/l	FS	12411
2	Poncho	Bayer CropScience	600 g/l	SC	13910

Uses
- Frit fly in **forage maize**, **grain maize**, **sweetcorn** [2]
- Leafhoppers in **durum wheat** *(seed treatment)*, **triticale** *(seed treatment)*, **winter barley** *(seed treatment)*, **winter oats** *(seed treatment)*, **winter rye** *(seed treatment)*, **winter wheat** *(seed treatment)* [1]

FOR FULL CONDITIONS OF USE ALWAYS READ THE PRODUCT LABEL

clothianidin + prothioconazole

- Slugs in **durum wheat** *(seed treatment)*, **triticale** *(seed treatment)*, **winter barley** *(seed treatment)*, **winter oats** *(seed treatment)*, **winter rye** *(seed treatment)*, **winter wheat** *(seed treatment)* [1]
- Symphylids in **forage maize** *(reduction)*, **grain maize** *(reduction)*, **sweetcorn** *(reduction)* [2]
- Virus vectors in **durum wheat** *(seed treatment)*, **triticale** *(seed treatment)*, **winter barley** *(seed treatment)*, **winter oats** *(seed treatment)*, **winter rye** *(seed treatment)*, **winter wheat** *(seed treatment)* [1]
- Wireworm in **durum wheat** *(seed treatment)*, **triticale** *(seed treatment)*, **winter barley** *(seed treatment)*, **winter oats** *(seed treatment)*, **winter rye** *(seed treatment)*, **winter wheat** *(seed treatment)* [1]; **forage maize**, **grain maize**, **sweetcorn** [2]

Approval information
- Clothianidin included in Annex I under EC Directive 91/414
- Accepted by BBPA for use on malting barley

Efficacy guidance
- May only be used in conjunction with manufacturer's approved seed treatment application equipment or by following procedures given in the operating instructions
- Treated seed should preferably be drilled in the same season
- Evenness of seed cover improved by simultaneous application with a small volume (1.5-3 litres/tonne) of water
- Calibrate drill for treated seed and drill at 2.5-4 cm into firm, well prepared seedbed
- Use minimum 125 kg treated seed per ha
- Seed should be drilled to a depth of 40 mm into a well prepared and firm seedbed. If seed is present on the surface, or if spills have occurred, the field should be harrowed and rolled if conditions are appropriate
- When aphid activity is unusually late, or is heavy and prolonged in areas of high risk, and mild weather predominates, follow-up foliar aphicide may be required
- Incidental suppression of leafhoppers in early spring (Mar/Apr) may be achieved but if a specific attack develops an additional foliar insecticide may be required

Restrictions
- Maximum number of seed treatments one per batch
- Product must be fully re-dispersed and homogeneous before use
- Do not use on seed with more than 16% moisture content, or on sprouted, cracked or skinned seed
- All seed batches should be tested to ensure they are suitable for treatment

Crop-specific information
- Latest use: pre-drilling

Environmental safety
- Harmful to game and wildlife. Treated seed should not be left on the surface. Bury spillages.

Hazard classification and safety precautions
   Hazard Harmful, Dangerous for the environment [2]; Irritant [1]
   UN number N/C
   Risk phrases R22a, R50, R53a [2]; R43 [1]
   Operator protection A, H; U04a, U05a, U13 [1, 2]; U07, U14, U20b [1]; U20c [2]
   Environmental protection E03, E15b, E34 [1, 2]; E36a [1]; E38 [2]
   Storage and disposal D01, D02, D09a, D14 [1, 2]; D05 [1]; D10d, D12a [2]
   Treated seed S01, S02, S03, S04b, S05, S06b, S07, S08 [1]; S04d [2]; S06a [1, 2]
   Medical advice M03

115   clothianidin + prothioconazole

A combined fungicide and insecticide seed dressing for cereals
IRAC mode of action code: 4A

See also prothioconazole

Products
Redigo Deter   Bayer CropScience   250:50 g/l   FS   12423

SEE SECTION 3 FOR PRODUCTS ALSO REGISTERED

Uses
- Bunt in **durum wheat** *(seed treatment)*, **triticale** *(seed treatment)*, **winter rye** *(seed treatment)*, **winter wheat** *(seed treatment)*
- Covered smut in **winter barley** *(seed treatment)*
- Fusarium foot rot and seedling blight in **durum wheat** *(seed treatment)*, **triticale** *(seed treatment)*, **winter barley** *(seed treatment)*, **winter oats** *(seed treatment)*, **winter rye** *(seed treatment)*, **winter wheat** *(seed treatment)*
- Leaf stripe in **winter barley** *(seed treatment)*
- Leafhoppers in **durum wheat** *(seed treatment)*, **triticale** *(seed treatment)*, **winter barley** *(seed treatment)*, **winter oats** *(seed treatment)*, **winter rye** *(seed treatment)*, **winter wheat** *(seed treatment)*
- Loose smut in **durum wheat** *(seed treatment)*, **winter barley** *(seed treatment)*, **winter oats** *(seed treatment)*, **winter wheat** *(seed treatment)*
- Slugs in **durum wheat** *(seed treatment)*, **triticale** *(seed treatment)*, **winter barley** *(seed treatment)*, **winter oats** *(seed treatment)*, **winter rye** *(seed treatment)*, **winter wheat** *(seed treatment)*
- Virus vectors in **durum wheat** *(seed treatment)*, **triticale** *(seed treatment)*, **winter barley** *(seed treatment)*, **winter oats** *(seed treatment)*, **winter rye** *(seed treatment)*, **winter wheat** *(seed treatment)*
- Wireworm in **durum wheat** *(seed treatment)*, **triticale** *(seed treatment)*, **winter barley** *(seed treatment)*, **winter oats** *(seed treatment)*, **winter rye** *(seed treatment)*, **winter wheat** *(seed treatment)*

Approval information
- Clothianidin and prothioconazole included in Annex I under EC Directive 91/414
- Accepted by BBPA for use on malting barley

Efficacy guidance
- May only be used in conjunction with manufacturer's approved seed treatment application equipment or by following procedures given in the operating instructions
- Treated wheat seed should preferably be drilled in the same season. Treated seed of other cereals must be used in the same season
- Evenness of seed cover improved by simultaneous application with a small volume (1.5-3 litres/tonne) of water
- Calibrate drill for treated seed and drill at 2.5-4 cm into firm, well prepared seedbed
- Use minimum 125 kg treated seed per ha
- When aphid activity unusually late, or is heavy and prolonged in areas of high risk, and mild weather predominates, follow-up foliar aphicide may be required
- Incidental suppression of leafhoppers in early spring (Mar/Apr) may be achieved but if a specific attack develops an additional foliar insecticide may be required
- Protection against slugs only applies to germinating seeds and not to aerial parts after emergence
- Where very high populations of aphids, wireworms or slugs occur additional specific control measures may be needed

Restrictions
- Maximum number of seed treatments one per batch for all cereals
- Product must be fully re-dispersed and homogeneous before use
- Do not use on seed with more than 16% moisture content, or on sprouted, cracked or skinned seed
- All seed batches should be tested to ensure they are suitable for treatment

Crop-specific information
- Latest use: pre-drilling of all cereals

Environmental safety
- Harmful to aquatic organisms

Hazard classification and safety precautions
Hazard Irritant
UN number N/C
Risk phrases R43, R52, R53a
Operator protection A, H; U14, U19a, U20b
Environmental protection E15a, E34, E38

FOR FULL CONDITIONS OF USE ALWAYS READ THE PRODUCT LABEL

clothianidin + prothioconazole + tebuconazole + triazoxide

Storage and disposal D01, D02, D09a, D11a, D12a
Treated seed S01, S02, S03, S04b, S05, S06a, S07, S08

116   clothianidin + prothioconazole + tebuconazole + triazoxide

A fungicide and insecticide seed treatment mixture for winter barley
IRAC mode of action code: 4A

See also prothioconazole
tebuconazole
triazoxide

Products

Raxil Deter	Bayer CropScience	333.3:33.3:20:13.3 g/l	FS	13030

Uses
- Barley yellow dwarf virus vectors in **winter barley** *(seed treatment)*
- Covered smut in **winter barley** *(seed treatment)*
- Fusarium foot rot and seedling blight in **winter barley** *(seed treatment)*
- Leaf stripe in **winter barley** *(seed treatment)*
- Leafhoppers in **winter barley** *(seed treatment)*
- Loose smut in **winter barley** *(seed treatment)*
- Net blotch in **winter barley** *(seed treatment)*
- Slugs in **winter barley** *(seed treatment - reduction)*
- Wireworm in **winter barley** *(seed treatment - reduction)*

Approval information
- Clothianidin and prothioconazole included in Annex I under EC Directive 91/414
- Accepted by BBPA for use on malting barley
- Approval expiry 30 May 2011

Efficacy guidance
- May only be used in conjunction with manufacturer's approved seed treatment application equipment or by following procedures given in the operating instructions
- Treated barley seed must be used in the same season
- Evenness of seed cover improved by simultaneous application with a small volume (1.5-3 litres/tonne) of water
- Calibrate drill for treated seed and drill at 4 cm into firm, well prepared seedbed
- Use minimum 125 kg treated seed per ha
- When aphid activity unusually late, or is heavy and prolonged in areas of high risk, and mild weather predominates, follow-up foliar aphicide may be required
- Leafhoppers can be suppressed in early spring (Mar/Apr) but if a specific attack develops an additional foliar insecticide may be required
- Protection against slugs only applies to germinating seeds and not to aerial parts after emergence
- Where very high populations of aphids, wireworms or slugs occur additional specific control measures may be needed

Restrictions
- Maximum number of seed treatments one per batch
- Product must be fully re-dispersed and homogeneous before use
- Do not use on seed with more than 16% moisture content, or on sprouted, cracked or skinned seed
- All seed batches should be tested to ensure they are suitable for treatment

Crop-specific information
- Latest use: pre-drilling of winter barley

Environmental safety
- Harmful to aquatic organisms

Hazard classification and safety precautions
  Hazard Irritant
  Transport code 9
  Packaging group III

SEE SECTION 3 FOR PRODUCTS ALSO REGISTERED

UN number 3082
Risk phrases R43, R52, R53a
Operator protection A, D, H; U07, U14, U20a
Environmental protection E15a, E34, E36a, E38
Storage and disposal D01, D02, D09a, D12a, D14
Treated seed S01, S02, S03, S04b, S05, S07, S08

117 Coniothyrium minitans

A fungal parasite of sclerotia in soil

Products
   Contans WG   Belchim   1000 IU/mg   WG   12616

Uses
- Sclerotinia in **all edible crops (outdoor), all non-edible crops (outdoor), protected crops**

Efficacy guidance
- *C.minitans* is a soil acting biological fungicide with specific action against the resting bodies (sclerotia) of *Sclerotinia sclerotiorum* and *S.minor*
- Treat 3 mth before disease protection is required to allow time for the infective sclerotia in the soil to be reduced
- A post-harvest treatment of soil and debris prevents further contamination of the soil with sclerotia produced by the previous crop
- For best results the soil should be moist and the temperature between 12-20 °C
- Application should be followed by soil incorporation into the surface layer to 10 cm with a rotovator or rotary harrow. Thorough spray coverage of the soil is essential to ensure uniform distribution
- If soil temperature drops below 0 °C or rises above 27 °C fungicidal activity is suspended but restarts when soil temperature returns within this range
- In glasshouses untreated areas at the margins should be covered by a film to avoid spread of spores from untreated sclerotia

Restrictions
- Maximum number of treatments 1 per crop or situation
- Do not plough or cultivate after treatment
- Do not mix with other pesticides, acids, alkalines or any product that attacks organic material
- Store product under cool dry conditions away from direct sunlight and away from heat sources

Crop-specific information
- Latest use: pre-planting
- HI: zero

Following crops guidance
- There are no restrictions on following crops and no waiting interval is specified

Hazard classification and safety precautions
   Hazard Harmful
   UN number N/C
   Risk phrases R42
   Operator protection A, C, D; U05a, U19a
   Environmental protection E15b, E34
   Storage and disposal D01, D02, D12a
   Medical advice M03

FOR FULL CONDITIONS OF USE ALWAYS READ THE PRODUCT LABEL

copper oxychloride

118 copper oxychloride

A protectant copper fungicide and bactericide
FRAC mode of action code: M1

Products

1	Cuprokylt	Unicrop	50% w/w (copper)	WP	00604
2	Cuprokylt FL	Unicrop	270 g/l (copper)	SC	08299
3	Headland Inorganic Liquid Copper	Headland	256 g/l (copper)	SC	13009

Uses

- Bacterial canker in **cherries** [1, 3]; **cob nuts** *(off-label)*, **hazel nuts** *(off-label)*, **walnuts** *(off-label)* [2]; **plums** [1, 2]
- Bacterial rot in **broccoli** *(off-label)*, **brussels sprouts** *(off-label)*, **cabbages** *(off-label)*, **calabrese** *(off-label)*, **cauliflowers** *(off-label)*, **chinese cabbage** *(off-label)*, **choi sum** *(off-label)*, **cob nuts** *(off-label)*, **collards** *(off-label)*, **hazel nuts** *(off-label)*, **kale** *(off-label)*, **pak choi** *(off-label)*, **protected broccoli** *(off-label)*, **protected brussels sprouts** *(off-label)*, **protected cabbages** *(off-label)*, **protected calabrese** *(off-label)*, **protected cauliflowers** *(off-label)*, **protected chinese cabbage** *(off-label)*, **protected choi sum** *(off-label)*, **protected collards** *(off-label)*, **tatsoi** *(off-label)*, **walnuts** *(off-label)*, **watercress** *(off-label)* [3]; **bulb onions** *(off-label)*, **garlic** *(off-label)*, **leeks** *(off-label)*, **salad onions** *(off-label)*, **shallots** *(off-label)* [1, 3]; **rhubarb** *(off-label)* [1, 2]
- Black rot in **broccoli** *(off-label)*, **brussels sprouts** *(off-label)*, **cabbages** *(off-label)*, **calabrese** *(off-label)*, **cauliflowers** *(off-label)*, **chinese cabbage** *(off-label)*, **collards** *(off-label)*, **kale** *(off-label)* [1]; **protected brassica seedlings** *(off-label)* [2]
- Blight in **broccoli** *(off-label)*, **brussels sprouts** *(off-label)*, **bulb onions** *(off-label)*, **cabbages** *(off-label)*, **calabrese** *(off-label)*, **cauliflowers** *(off-label)*, **chinese cabbage** *(off-label)*, **choi sum** *(off-label)*, **collards** *(off-label)*, **garlic** *(off-label)*, **kale** *(off-label)*, **leeks** *(off-label)*, **pak choi** *(off-label)*, **protected broccoli** *(off-label)*, **protected brussels sprouts** *(off-label)*, **protected cabbages** *(off-label)*, **protected calabrese** *(off-label)*, **protected cauliflowers** *(off-label)*, **protected chinese cabbage** *(off-label)*, **protected choi sum** *(off-label)*, **protected collards** *(off-label)*, **salad onions** *(off-label)*, **shallots** *(off-label)*, **tatsoi** *(off-label)*, **watercress** *(off-label)* [3]; **cob nuts** *(off-label)*, **hazel nuts** *(off-label)*, **walnuts** *(off-label)* [2, 3]; **potatoes**, **tomatoes (outdoor)** [1-3]; **protected tomatoes** [1, 2]
- Buck-eye rot in **protected tomatoes**, **tomatoes (outdoor)** [1-3]
- Cane spot in **blackberries** *(off-label)*, **loganberries**, **raspberries**, **rubus hybrids** *(off-label)* [1, 2]
- Canker in **apples**, **pears** [1-3]; **apples** *(off-label)* [2]; **broccoli** *(off-label)*, **brussels sprouts** *(off-label)*, **bulb onions** *(off-label)*, **cabbages** *(off-label)*, **calabrese** *(off-label)*, **cauliflowers** *(off-label)*, **chinese cabbage** *(off-label)*, **choi sum** *(off-label)*, **cob nuts** *(off-label)*, **collards** *(off-label)*, **garlic** *(off-label)*, **hazel nuts** *(off-label)*, **kale** *(off-label)*, **leeks** *(off-label)*, **pak choi** *(off-label)*, **protected broccoli** *(off-label)*, **protected brussels sprouts** *(off-label)*, **protected cabbages** *(off-label)*, **protected calabrese** *(off-label)*, **protected cauliflowers** *(off-label)*, **protected chinese cabbage** *(off-label)*, **protected choi sum** *(off-label)*, **protected collards** *(off-label)*, **salad onions** *(off-label)*, **shallots** *(off-label)*, **tatsoi** *(off-label)*, **walnuts** *(off-label)*, **watercress** *(off-label)* [3]
- Celery leaf spot in **celery (outdoor)** [1-3]
- Collar rot in **apples**, **apples** *(off-label)* [2]
- Damping off in **protected tomatoes**, **tomatoes (outdoor)** [1, 2]
- Downy mildew in **bilberries** *(off-label)*, **blueberries** *(off-label)*, **cranberries** *(off-label)*, **gooseberries** *(off-label)*, **redcurrants** *(off-label)*, **whitecurrants** *(off-label)* [1, 2]; **chard** *(off-label)*, **cress** *(off-label)*, **frise** *(off-label)*, **herbs (see appendix 6)** *(off-label)*, **lamb's lettuce** *(off-label)*, **protected chard** *(off-label)*, **protected cress** *(off-label)*, **protected frise** *(off-label)*, **protected herbs (see appendix 6)** *(off-label)*, **protected lamb's lettuce** *(off-label)*, **protected radicchio** *(off-label)*, **protected salad brassicas** *(off-label)*, **protected scarole** *(off-label)*, **protected spinach** *(off-label)*, **protected spinach beet** *(off-label)*, **radicchio** *(off-label)*, **salad brassicas** *(off-label)*, **scarole** *(off-label)*, **spinach** *(off-label)*, **spinach beet** *(off-label)*, **wine grapes** [3]; **hops** [1-3]
- Foliar disease control in **rhubarb** *(off-label)* [3]
- Foot rot in **protected tomatoes**, **tomatoes (outdoor)** [1, 2]
- Leaf curl in **apricots** *(off-label)*, **peaches** [1, 2]; **quinces** *(off-label)* [1]

SEE SECTION 3 FOR PRODUCTS ALSO REGISTERED

Pesticide Profiles

- Pseudomonas storage rots in **bulb onions** *(off-label)*, **garlic** *(off-label)*, **leeks** *(off-label)*, **salad onions** *(off-label)*, **shallots** *(off-label)* [1]
- Purple blotch in **blackberries** [3]
- Pythium in **watercress** *(off-label - during propagation)* [2]
- Rhizoctonia in **watercress** *(off-label - during propagation)* [2]
- Rhizoctonia rot in **broccoli** *(off-label)*, **brussels sprouts** *(off-label)*, **bulb onions** *(off-label)*, **cabbages** *(off-label)*, **calabrese** *(off-label)*, **cauliflowers** *(off-label)*, **chinese cabbage** *(off-label)*, **choi sum** *(off-label)*, **cob nuts** *(off-label)*, **collards** *(off-label)*, **garlic** *(off-label)*, **hazel nuts** *(off-label)*, **kale** *(off-label)*, **leeks** *(off-label)*, **pak choi** *(off-label)*, **protected broccoli** *(off-label)*, **protected brussels sprouts** *(off-label)*, **protected cabbages** *(off-label)*, **protected calabrese** *(off-label)*, **protected cauliflowers** *(off-label)*, **protected chinese cabbage** *(off-label)*, **protected choi sum** *(off-label)*, **protected collards** *(off-label)*, **salad onions** *(off-label)*, **shallots** *(off-label)*, **tatsoi** *(off-label)*, **walnuts** *(off-label)*, **watercress** *(off-label)* [3]
- Rust in **blackcurrants** [1-3]
- Spear rot in **broccoli** *(off-label)*, **brussels sprouts** *(off-label)*, **cabbages** *(off-label)*, **calabrese** *(off-label)*, **cauliflowers** *(off-label)*, **chinese cabbage** *(off-label)*, **collards** *(off-label)*, **kale** *(off-label)* [1, 3]; **bulb onions** *(off-label)*, **choi sum** *(off-label)*, **cob nuts** *(off-label)*, **garlic** *(off-label)*, **hazel nuts** *(off-label)*, **leeks** *(off-label)*, **pak choi** *(off-label)*, **protected broccoli** *(off-label)*, **protected brussels sprouts** *(off-label)*, **protected cabbages** *(off-label)*, **protected calabrese** *(off-label)*, **protected cauliflowers** *(off-label)*, **protected chinese cabbage** *(off-label)*, **protected choi sum** *(off-label)*, **protected collards** *(off-label)*, **salad onions** *(off-label)*, **shallots** *(off-label)*, **tatsoi** *(off-label)*, **walnuts** *(off-label)*, **watercress** *(off-label)* [3]; **protected brassica seedlings** *(off-label)* [2]

Specific Off-Label Approvals (SOLAs)
- *apples* 982336 [2]
- *apricots* 20063134 [1], 20063140 [2]
- *bilberries* 20063131 [1], 20063137 [2]
- *blackberries* 20063132 [1], 20063139 [2]
- *blueberries* 20063131 [1], 20063137 [2]
- *broccoli* 20010115 [1], 20080156 [3]
- *brussels sprouts* 20010115 [1], 20080156 [3]
- *bulb onions* 991127 [1], 20080156 [3]
- *cabbages* 20010115 [1], 20080156 [3]
- *calabrese* 20010115 [1], 20080156 [3]
- *cauliflowers* 20010115 [1], 20080156 [3]
- *chard* 20080157 [3]
- *chinese cabbage* 20010115 [1], 20080156 [3]
- *choi sum* 20080156 [3]
- *cob nuts* 990385 [2], 20080156 [3]
- *collards* 20010115 [1], 20080156 [3]
- *cranberries* 20063131 [1], 20063137 [2]
- *cress* 20080157 [3]
- *frise* 20080157 [3]
- *garlic* 991127 [1], 20080156 [3]
- *gooseberries* 20063131 [1], 20063137 [2]
- *hazel nuts* 990385 [2], 20080156 [3]
- *herbs (see appendix 6)* 20080157 [3]
- *kale* 20010115 [1], 20080156 [3]
- *lamb's lettuce* 20080157 [3]
- *leeks* 991127 [1], 20080156 [3]
- *pak choi* 20080156 [3]
- *protected brassica seedlings* 20010117 [2]
- *protected broccoli* 20080156 [3]
- *protected brussels sprouts* 20080156 [3]
- *protected cabbages* 20080156 [3]
- *protected calabrese* 20080156 [3]
- *protected cauliflowers* 20080156 [3]
- *protected chard* 20080157 [3]
- *protected chinese cabbage* 20080156 [3]

FOR FULL CONDITIONS OF USE ALWAYS READ THE PRODUCT LABEL

copper oxychloride

- **protected choi sum** *20080156* [3]
- **protected collards** *20080156* [3]
- **protected cress** *20080157* [3]
- **protected frise** *20080157* [3]
- **protected herbs (see appendix 6)** *20080157* [3]
- **protected lamb's lettuce** *20080157* [3]
- **protected radicchio** *20080157* [3]
- **protected salad brassicas** *20080157* [3]
- **protected scarole** *20080157* [3]
- **protected spinach** *20080157* [3]
- **protected spinach beet** *20080157* [3]
- **quinces** *20063133* [1]
- **radicchio** *20080157* [3]
- **redcurrants** *20063131* [1], *20063137* [2]
- **rhubarb** *20063135* [1], *20063138* [2], *20080158* [3]
- **rubus hybrids** *20063132* [1], *20063139* [2]
- **salad brassicas** *20080157* [3]
- **salad onions** *991127* [1], *20080156* [3]
- **scarole** *20080157* [3]
- **shallots** *991127* [1], *20080156* [3]
- **spinach** *20080157* [3]
- **spinach beet** *20080157* [3]
- **tatsoi** *20080156* [3]
- **walnuts** *990385* [2], *20080156* [3]
- **watercress** *(during propagation) 20001538* [2], *20080156* [3]
- **whitecurrants** *20063131* [1], *20063137* [2]

Approval information
- Approved for aerial application on potatoes [1]. See Section 5 for more information
- Accepted by BBPA for use on hops

Efficacy guidance
- Spray crops at high volume when foliage dry but avoid run off. Do not spray if rain expected soon
- Spray interval commonly 10-14 d but varies with crop, see label for details
- If buck-eye rot occurs, spray soil surface and lower parts of tomato plants to protect unaffected fruit [1, 2]
- A follow-up spray in the following spring should be made to top fruit severely infected with bacterial canker

Restrictions
- Maximum number of treatments 3 per crop for apples, blackberries, grapevines, pears; 2 per crop for watercress. Not specified for other crops
- Some peach cultivars are sensitive to copper. Treat non-sensitive varieties only [1, 2]

Crop-specific information
- HI hops 7 d
- Slight damage may occur to leaves of cherries and plums

Environmental safety
- Dangerous for the environment
- Very toxic to aquatic organisms
- Keep all livestock out of treated areas for at least 3 wk

Hazard classification and safety precautions
Hazard Harmful [2]; Dangerous for the environment [1-3]
Transport code 6.1 [2]; 9 [1, 3]
Packaging group III
UN number 3010 [2]; 3077 [1]; 3082 [3]
Risk phrases R22a, R50 [2]; R51 [1, 3]; R53a [1-3]
Operator protection A [3]; U20a [1, 3]; U20c [2]
Environmental protection E06a [1-3] (3 wk); E13c, E34, E38 [1, 3]; E15a [2]
Storage and disposal D01, D09a [1-3]; D05, D10c, D12a [2]; D10b [1, 3]

SEE SECTION 3 FOR PRODUCTS ALSO REGISTERED

119 cyazofamid

A cyanoimidazole sulfonamide protectant fungicide for potatoes
FRAC mode of action code: 21

Products

1	Linford	AgChem Access	400 g/l	SC	13824
2	Ranman Twinpack	Belchim	400 g/l	KL	11851
3	Swallow	AgChem Access	400 g/l	SC	14078

Uses
- Blight in *potatoes* [1-3]
- Disease control in *hops* (off-label) [2]

Specific Off-Label Approvals (SOLAs)
- *hops* 20082918 [2]

Approval information
- Cyazofamid included in Annex I under EC Directive 91/414

Efficacy guidance
- Apply as a protectant treatment before blight enters the crop and repeat every 7-10 d depending on severity of disease pressure
- Commence spray programme immediately the risk of blight in the locality occurs, usually when the crop meets along the rows
- Product must always be used with organosilicone adjuvant provided in the twin pack
- To minimise the chance of development of resistance no more than three applications should be made consecutively (out of a permissible total of six) in the blight control programme. For more information on Resistance Management see Section 5

Restrictions
- Maximum number of treatments 6 per crop (no more than three of which should be consecutive)
- Mixed product must not be allowed to stand overnight
- Consult processor before using on crops intended for processing

Crop-specific information
- HI 7 d

Environmental safety
- Dangerous for the environment
- Very toxic to aquatic organisms
- Do not empty into drains

Hazard classification and safety precautions
 Hazard Harmful (adjuvant); Dangerous for the environment
 Transport code 9 [2]
 Packaging group III [2]
 UN number 3082 [2]
 Risk phrases R20, R36, R48 (adjuvant); R41, R50, R53a
 Operator protection A, C; U02a, U05a, U11, U15, U20a; U19a (adjuvant)
 Environmental protection E15a, E19b, E34, E38
 Storage and disposal D01, D02, D09a, D10c, D12a
 Medical advice M03

120 cycloxydim

A translocated post-emergence cyclohexanedione oxime herbicide for grass weed control
HRAC mode of action code: A

Products

	Laser	BASF	200 g/l	EC	12930

Uses
- Annual grasses in *amenity vegetation* (off-label), *borage for oilseed production* (off-label), *broad beans* (off-label), *brussels sprouts*, *bulb onions*, *cabbages*, *canary flower (echium spp.)* (off-label), *carrots*, *cauliflowers*, *combining peas*, *dwarf beans*, *early potatoes*, *edible*

FOR FULL CONDITIONS OF USE ALWAYS READ THE PRODUCT LABEL

cycloxydim

  podded peas *(off-label),* ***evening primrose*** *(off-label),* ***farm forestry, flower bulbs, fodder beet, forest, garlic*** *(off-label),* ***honesty*** *(off-label),* ***horseradish*** *(off-label),* ***leeks, linseed, lupins*** *(off-label),* ***maincrop potatoes, mallow (althaea spp.)*** *(off-label),* ***mangels, mustard*** *(off-label), **parsnips, runner beans*** *(off-label),* ***salad onions, shallots*** *(off-label),* ***spring field beans, spring oilseed rape, strawberries, sugar beet, swedes, turnips*** *(off-label),* ***vining peas, winter field beans, winter oilseed rape***
- Black bent in ***brussels sprouts, bulb onions, cabbages, carrots, cauliflowers, combining peas, dwarf beans, early potatoes, flower bulbs, fodder beet, leeks, linseed, maincrop potatoes, mangels, parsnips, salad onions, spring field beans, spring oilseed rape, strawberries, sugar beet, swedes, vining peas, winter field beans, winter oilseed rape***
- Blackgrass in ***brussels sprouts, bulb onions, cabbages, carrots, cauliflowers, combining peas, dwarf beans, early potatoes, flower bulbs, fodder beet, leeks, linseed, maincrop potatoes, mangels, parsnips, salad onions, spring field beans, spring oilseed rape, strawberries, sugar beet, swedes, vining peas, winter field beans, winter oilseed rape***
- Couch in ***brussels sprouts, bulb onions, cabbages, carrots, cauliflowers, combining peas, dwarf beans, early potatoes, flower bulbs, fodder beet, leeks, linseed, maincrop potatoes, mangels, parsnips, salad onions, spring field beans, spring oilseed rape, strawberries, sugar beet, swedes, vining peas, winter field beans, winter oilseed rape***
- Creeping bent in ***brussels sprouts, bulb onions, cabbages, carrots, cauliflowers, combining peas, dwarf beans, early potatoes, flower bulbs, fodder beet, leeks, linseed, maincrop potatoes, mangels, parsnips, salad onions, spring field beans, spring oilseed rape, strawberries, sugar beet, swedes, vining peas, winter field beans, winter oilseed rape***
- Green cover in ***land temporarily removed from production***
- Onion couch in ***brussels sprouts, bulb onions, cabbages, carrots, cauliflowers, combining peas, dwarf beans, early potatoes, flower bulbs, fodder beet, leeks, linseed, maincrop potatoes, mangels, parsnips, salad onions, spring field beans, spring oilseed rape, strawberries, sugar beet, swedes, vining peas, winter field beans, winter oilseed rape***
- Perennial grasses in ***amenity vegetation*** *(off-label),* ***borage for oilseed production*** *(off-label),* ***broad beans*** *(off-label),* ***canary flower (echium spp.)*** *(off-label),* ***edible podded peas*** *(off-label), **evening primrose*** *(off-label),* ***farm forestry, forest, garlic*** *(off-label),* ***honesty*** *(off-label),* ***horseradish*** *(off-label),* ***lupins*** *(off-label),* ***mallow (althaea spp.)*** *(off-label),* ***mustard*** *(off-label),* ***runner beans*** *(off-label),* ***shallots*** *(off-label),* ***turnips*** *(off-label)*
- Volunteer cereals in ***brussels sprouts, bulb onions, cabbages, carrots, cauliflowers, combining peas, dwarf beans, early potatoes, flower bulbs, fodder beet, leeks, linseed, maincrop potatoes, mangels, parsnips, salad onions, spring field beans, spring oilseed rape, strawberries, sugar beet, swedes, vining peas, winter field beans, winter oilseed rape***
- Wild oats in ***brussels sprouts, bulb onions, cabbages, carrots, cauliflowers, combining peas, dwarf beans, early potatoes, flower bulbs, fodder beet, leeks, linseed, maincrop potatoes, mangels, parsnips, salad onions, spring field beans, spring oilseed rape, strawberries, sugar beet, swedes, vining peas, winter field beans, winter oilseed rape***

Specific Off-Label Approvals (SOLAs)
- *amenity vegetation 20081092*
- *borage for oilseed production 20081094*
- *broad beans 20081136*
- *canary flower (echium spp.) 20081094*
- *edible podded peas 20081096*
- *evening primrose 20081094*
- *garlic 20081097*
- *honesty 20081094*
- *horseradish 20081098*
- *lupins 20081093*
- *mallow (althaea spp.) 20081098*
- *mustard 20081094*
- *runner beans 20081096*
- *shallots 20081097*
- *turnips 20081095*

Efficacy guidance
- Best results achieved when weeds small and have not begun to compete with crop. Effectiveness reduced by drought, cool conditions or stress. Weeds emerging after application are not controlled

SEE SECTION 3 FOR PRODUCTS ALSO REGISTERED

- Foliage death usually complete after 3-4 wk but longer under cool conditions, especially late treatments to winter oilseed rape
- Perennial grasses should have sufficient foliage to absorb spray and should not be cultivated for at least 14 d after treatment
- On established couch pre-planting cultivation recommended to fragment rhizomes and encourage uniform emergence
- Split applications to volunteer wheat and barley at GS 12-14 will often give adequate control in winter oilseed rape. See label for details
- Apply to dry foliage when rain not expected for at least 2 h
- Cycloxydim is an ACCase inhibitor herbicide. To avoid the build up of resistance do not apply products containing an ACCase inhibitor herbicide more than twice to any crop. In addition do not use any product containing cycloxydim in mixture or sequence with any other product containing the same ingredient
- Use these products as part of a resistance management strategy that includes cultural methods of control and does not use ACCase inhibitors as the sole chemical method of grass weed control
- Applying a second product containing an ACCase inhibitor to a crop will increase the risk of resistance development; only use a second ACCase inhibitor to control different weeds at a different timing
- Always follow WRAG guidelines for preventing and managing herbicide resistant weeds. See Section 5 for more information

Restrictions
- Maximum number of treatments 1 per crop or yr in most situations. See label
- Must be used with authorised adjuvant oil. See label
- Do not apply to crops damaged or stressed by adverse weather, pest or disease attack or other pesticide treatment
- Prevent drift onto other crops, especially cereals and grass

Crop-specific information
- HI cabbage, cauliflower, calabrese, salad onions 4 wk; peas, dwarf beans 5 wk; bulb onions, carrots, parsnips, strawberries 6 wk; sugar and fodder beet, leeks, mangels, potatoes, field beans, swedes, Brussels sprouts, winter field beans 8 wk; oilseed rape, soya beans, linseed 12 wk
- Recommended time of application varies with crop. See label for details
- On peas a crystal violet wax test should be done if leaf wax likely to have been affected by weather conditions or other chemical treatment. The wax test is essential if other products are to be sprayed before or after treatment
- May be used on ornamental bulbs when crop 5-10 cm tall. Product has been used on tulips, narcissi, hyacinths and irises but some subjects may be more sensitive and growers advised to check tolerance on small number of plants before treating the rest of the crop
- May be applied to land temporarily removed from production where the green cover is made up predominantly of tolerant crops listed on label. Use on industrial crops of linseed and oilseed rape on land temporarily removed from production also permitted.

Following crops guidance
- Guideline intervals for sowing succeeding crops after failed treated crop: field beans, peas, sugar beet, rape, kale, swedes, radish, white clover, lucerne 1 wk; dwarf French beans 4 wk; wheat, barley, maize 8 wk
- Oats should not be sown after failure of a treated crop

Environmental safety
- Dangerous for the environment
- Toxic to aquatic organisms
- Harmful to fish or other aquatic life. Do not contaminate surface waters or ditches with chemical or used container

Hazard classification and safety precautions
    Hazard Harmful, Dangerous for the environment
    Transport code 9
    Packaging group III
    UN number 3082
    Risk phrases R22b, R38, R51, R53a
    Operator protection A, C; U05a, U08, U20b
    Environmental protection E15a, E38

FOR FULL CONDITIONS OF USE ALWAYS READ THE PRODUCT LABEL

Storage and disposal D01, D02, D09a, D10c, D12a
Medical advice M05b

121 Cydia pomonella GV

An baculovirus insecticide for codling moth control

Products
Cyd-X   Certis   10 g/l granulovirus   SC   13535

Uses
- Codling moth in **apples**, **pears**

Approval information
- Cydia pomonella included in Annex I under EC Directive 91/414

Efficacy guidance
- Optimum results require a precise spraying schedule commencing soon after egg laying and before first hatch
- First generation of codling moth normally occurs in first 2 wk of Jun in UK
- Apply in sufficient water to achieve good coverage of whole tree
- Normally 3 applications at intervals of 8 sunny days are needed for each codling moth generation
- Fruit damage is reduced significantly in the first yr and will reduce the population of codling moth in the following yr

Restrictions
- Maximum number of treatments 3 per season
- Do not tank mix with copper or any products with a pH lower than 5 or higher than 8
- Do not use as the exclusive measure for codling moth control. Adopt an anti-resistance strategy. See Section 5 for more information
- Consult processor before use on crops for processing

Crop-specific information
- HI: 14 d for apples, pears
- Evidence of efficacy and crop safety on pears is limited

Environmental safety
- Product has no adverse effect on predatory insects and is fully compatible with IPM programmes

Hazard classification and safety precautions
  Hazard Harmful
  Risk phrases R42, R43
  Operator protection A, D, H; U14, U19a
  Environmental protection E15a
  Storage and disposal D12a

122 cyflufenamid

An amidoxime fungicide for cereals
FRAC mode of action code: U6

Products
Cyflamid   Certis   50 g/l   EW   12403

Uses
- Powdery mildew in **amenity vegetation** (off-label), **durum wheat**, **forest nurseries** (off-label), **grass seed crops** (off-label), **hops** (off-label), **ornamental plant production** (off-label), **rye** (off-label), **soft fruit** (off-label), **spring barley**, **spring oats**, **spring wheat**, **triticale**, **winter barley**, **winter oats**, **winter rye**, **winter wheat**

Specific Off-Label Approvals (SOLAs)
- *amenity vegetation* 20070512
- *forest nurseries* 20082915
- *grass seed crops* 20060602
- *hops* 20082915

SEE SECTION 3 FOR PRODUCTS ALSO REGISTERED

Pesticide Profiles

- *ornamental plant production* 20070512
- *rye* 20060602
- *soft fruit* 20082915

Approval information
- Cyflufenamid included in Annex 1 under EC Directive 91/414
- Accepted by BBPA on malting barley

Efficacy guidance
- Best results obtained from treatment at the first visible signs of infection by powdery mildew
- Sustained disease pressure may require a second treatment
- Disease spectrum may be broadened by appropriate tank mixtures. See label
- Product is rainfast within 1 h
- Must be used as part of an Integrated Crop Management programme that includes alternating use, or mixture, with fungicides with a different mode of action effective against powdery mildew

Restrictions
- Maximum number of treatments 2 per crop on all recommended cereals
- Apply only in the spring
- Do not apply to crops under stress from drought, waterlogging, cold, pests or diseases, lime or nutrient deficiency or other factors affecting crop growth

Crop-specific information
- Latest use: before start of flowering of cereal crops

Following crops guidance
- No restrictions

Environmental safety
- Dangerous for the environment
- Very toxic to aquatic organisms
- Avoid drift onto ponds, waterways or ditches

Hazard classification and safety precautions
Hazard Dangerous for the environment
Transport code 9
Packaging group III
UN number 3082
Risk phrases R50, R53a
Operator protection A; U05a
Environmental protection E15b, E34
Storage and disposal D01, D02, D05, D10c, D12b

123 cymoxanil

A cyanoacetamide oxime fungicide for potatoes
FRAC mode of action code: 27

*See also chlorothalonil + cymoxanil
             cyazofamid + cymoxanil*

Products

1	Cymbal	Belchim	45% w/w	WG	13312
2	Drum	Belchim	45% w/w	WG	13853
3	Harpoon	Headland	60% w/w	WG	14278
4	Option	DuPont	60% w/w	WG	11834
5	Sipcam C 50	Sipcam	50% w/w	WP	10610
6	Sipcam C50 WG	Sipcam	50% w/w	WG	14650

Uses
- Blight in **potatoes** [1-6]
- Downy mildew in **hops** *(off-label)* [4]

Specific Off-Label Approvals (SOLAs)
- *hops* 20102250 [4]

FOR FULL CONDITIONS OF USE ALWAYS READ THE PRODUCT LABEL

cymoxanil + famoxadone

Efficacy guidance
- Product to be used in mixture with specified mixture partners in order to combine systemic and protective activity. See labels for details
- Commence spray programme as soon as weather conditions favourable for disease development occur and before infection appears. At latest the first treatment should be made as the foliage meets along the rows
- Repeat treatments at 7-14 day intervals according to disease incidence and weather conditions
- Treat irrigated crops as soon as possible after irrigation and repeat at 10 d intervals

Restrictions
- Maximum total dose equivalent to four full dose treatments [5]. Not specified for other products
- Minimum spray interval 7 d
- Do not allow packs to become wet during storage and do not keep opened packs from one season to another
- Cymoxanil must be used in tank mixture with other fungicides

Crop-specific information
- HI for potatoes 7 d [5]; 14 d [1]
- Consult processor before treating potatoes grown for processing

Environmental safety
- Dangerous for the environment
- Very toxic to aquatic organisms

Hazard classification and safety precautions
Hazard Harmful [3-6]; Irritant [1, 2]; Dangerous for the environment [1-4]
Transport code 9 [1-4]
Packaging group III [1-4]
UN number 3077 [1-4]; N/C [5, 6]
Risk phrases R22a [3-6]; R43 [1, 2]; R50, R53a [1-4]
Operator protection A [1, 3-6]; H [1]; U05a [1-6]; U08 [3-6]; U09a, U14 [1, 2]; U19a [3, 4]; U20a [1]; U20b [2-6]
Environmental protection E13a, E34 [3, 4]; E13c [5, 6]; E15a [1, 2]; E38 [1, 2, 4]
Storage and disposal D01, D02, D09a, D11a [1-6]; D12a [1, 2, 4]
Medical advice M03 [3-6]

124   cymoxanil + famoxadone

A preventative and curative fungicide mixture for potatoes
FRAC mode of action code: 27 + 11

See also famoxadone

Products
Tanos   DuPont   25:25% w/w   WG   10677

Uses
- Blight in **potatoes**

Approval information
- Famoxadone included in Annex I under EC Directive 91/414

Efficacy guidance
- Commence spray programme before infection appears as soon as weather conditions favourable for disease development to occur. At latest the first treatment should be made as the foliage meets along the rows
- Repeat treatments at 7-10 day intervals according to disease incidence and weather conditions
- Reduce spray interval if conditions are conducive to the spread of blight
- Spray as soon as possible after irrigation
- Increase water volume in dense crops
- Product combines active substances with different modes of action and is effective against strains of potato blight that are insensitive to phenylamide fungicides
- To minimise the likelihood of development of resistance to QoI fungicides these products should be used in a planned Resistance Management strategy. See Section 5 for more information

SEE SECTION 3 FOR PRODUCTS ALSO REGISTERED

Pesticide Profiles

Restrictions
- Maximum number of treatments 6 per crop
- Consult processors before using on crops for processing
- Do not apply more than 3 consecutive treatments containing QoI active fungicides

Crop-specific information
- HI 14 d for potatoes

Environmental safety
- Dangerous for the environment
- Very toxic to aquatic organisms
- Dangerous to fish or other aquatic life. Do not contaminate surface waters or ditches with chemical or used container
- LERAP Category B

Hazard classification and safety precautions
Hazard Harmful, Dangerous for the environment
Transport code 9
Packaging group III
UN number 3077
Risk phrases R22a, R50, R53a
Operator protection A, D; U05a, U20b
Environmental protection E13b, E16a, E16b, E34, E38
Storage and disposal D01, D02, D09a, D10b, D12a
Medical advice M03

125   cymoxanil + fludioxonil + metalaxyl-M

A fungicide seed dressing for peas
FRAC mode of action code: 27 + 12 + 4

See also fludioxonil
   metalaxyl-M

Products
   Wakil XL   Syngenta   10:5:17.5% w/w   WS   10562

Uses
- Ascochyta in **combining peas** *(seed treatment)*, **vining peas** *(seed treatment)*
- Damping off in **chard** *(off-label)*, **chives** *(off-label)*, **combining peas** *(seed treatment)*, **herbs (see appendix 6)** *(off-label)*, **leaf spinach** *(off-label)*, **lupins** *(off-label)*, **parsley** *(off-label)*, **poppies for morphine production** *(off-label - seed treatment)*, **red beet** *(off-label - seed treatment)*, **salad brassicas** *(off-label)*, **spinach beet** *(off-label)*, **vining peas** *(seed treatment)*
- Downy mildew in **broad beans** *(off-label - seed treatment)*, **combining peas** *(seed treatment)*, **lupins** *(off-label)*, **poppies for morphine production** *(off-label - seed treatment)*, **spring field beans** *(off-label - seed treatment)*, **vining peas** *(seed treatment)*, **winter field beans** *(off-label - seed treatment)*
- Pythium in **carrots** *(off-label - seed treatment)*, **combining peas** *(seed treatment)*, **lupins** *(off-label)*, **parsnips** *(off-label - seed treatment)*, **poppies for morphine production** *(off-label - seed treatment)*, **vining peas** *(seed treatment)*

Specific Off-Label Approvals (SOLAs)
- **broad beans** *(seed treatment)* 20023932
- **carrots** *(seed treatment)* 20021191
- **chard** 20070791
- **chives** 20070791
- **herbs (see appendix 6)** 20070791
- **leaf spinach** 20070791
- **lupins** 20060614
- **parsley** 20070791
- **parsnips** *(seed treatment)* 20021191
- **poppies for morphine production** *(seed treatment)* 20040683
- **red beet** *(seed treatment)* 20032313

FOR FULL CONDITIONS OF USE ALWAYS READ THE PRODUCT LABEL

- **salad brassicas** 20070791
- **spinach beet** 20070791
- **spring field beans** (seed treatment) 20023932
- **winter field beans** (seed treatment) 20023932

Approval information
- Fludioxonil and metalaxyl-M included in Annex I under EC Directive 91/414

Efficacy guidance
- Apply through continuous flow seed treaters which should be calibrated before use

Restrictions
- Max number of treatments 1 per seed batch
- Ensure moisture content of treated seed satisfactory and store in a dry place
- Check calibration of seed drill with treated seed before drilling and sow as soon as possible after treatment
- Consult before using on crops for processing

Crop-specific information
- Latest use: pre-drilling

Environmental safety
- Harmful to aquatic organisms
- Do not use treated seed as food or feed

Hazard classification and safety precautions
UN number N/C
Risk phrases R52, R53a
Operator protection A, D, H; U02a, U05a, U20b
Environmental protection E03, E15a
Storage and disposal D01, D02, D05, D07, D09a, D11a, D12a
Treated seed S02, S04b, S05, S07

126 cymoxanil + mancozeb

A protectant and systemic fungicide for potato blight control
FRAC mode of action code: 27 + M3

See also mancozeb

Products

1	Agriguard Cymoxanil Plus	AgriGuard	4.5:68% w/w	WP	10893
2	Astral	AgChem Access	4.5:68% w/w	WP	14049
3	Besiege	DuPont	4.5:68% w/w	WP	08086
4	Clayton Krypton	Clayton	4.5:68% w/w	WP	09398
5	Clayton Krypton MZ	Clayton	4.5:68% w/w	WP	14010
6	Curzate M WG	DuPont	4.5:68% w/w	WG	11901
7	Globe	Sipcam	6:70% w/w	WP	10339
8	Matilda	Nufarm UK	4.5:68% w/w	WP	12006
9	Nautile DG	United Phosphorus	5:68% w/w	WG	14692
10	Rhapsody	DuPont	4.5:68% w/w	WG	11958
11	Video DG	United Phosphorus	5:68% w/w	WG	14850

Uses
- Blight in **potatoes**

Approval information
- Mancozeb included in Annex I under EC Directive 91/414

Efficacy guidance
- Apply immediately after blight warning or as soon as local conditions dictate and repeat at 7-14 d intervals until haulm dies down or is burnt off
- Spray interval should not be more than 7-10 d in irrigated crops (see product label). Apply treatment after irrigation
- To minimise the likelihood of development of resistance these products should be used in a planned Resistance Management strategy. See Section 5 for more information

Restrictions
- Maximum number of treatments 6 per crop [7]. Not specified for other products
- Do not apply at less than 7 d intervals
- Do not allow packs to become wet during storage

Crop-specific information
- HI 7 d [7]
- Destroy and remove any haulm that remains after harvest of early varieties to reduce blight pressure on neighbouring maincrop potatoes

Environmental safety
- Dangerous for the environment
- Very toxic to aquatic organisms
- Keep product away from fire or sparks

Hazard classification and safety precautions
Hazard Harmful [3, 4, 6, 10]; Irritant [1, 2, 5, 7-9, 11]; Dangerous for the environment [1-6, 9-11]
Transport code 9
Packaging group III
UN number 3077
Risk phrases R36 [1-6, 8, 10]; R37 [1-6, 8-11]; R38 [1, 5, 8]; R42 [1-4, 6, 10]; R43 [1-11]; R50, R53a [2-6, 8-11]
Operator protection A [1-11]; C [1-8, 10]; H [7, 9, 11]; U02a [7]; U05a, U08 [1-11]; U11 [2-5, 8, 9, 11]; U14 [2, 3, 6, 9-11]; U19a [1-6, 8-11]; U20a [1]; U20b [2-11]
Environmental protection E13b [7]; E13c [2, 3, 6, 8, 10]; E15a [1-7, 9-11]; E38 [2, 3, 6, 9-11]
Storage and disposal D01, D02, D09a [1-11]; D06b [9, 11]; D11a [1-8, 10]; D12a [2, 3, 9, 11]
Medical advice M04a, M04b [4, 5]

127 cymoxanil + propamocarb

A cyanoacetamide oxime fungicide + a carbamate fungicide for potatoes
FRAC mode of action code: 27 + 28

Products
Proxanil	Sipcam	50:400 g/l	SC	13945

Uses
- Blight in *potatoes*

Approval information
- Propamocarb hydrochloride included in Annex I under EC Directive 91/414

Hazard classification and safety precautions
Hazard Irritant
UN number N/C
Risk phrases R43, R52, R53a
Operator protection A, H; U05a, U08, U14, U20b
Environmental protection E15a, E38
Storage and disposal D01, D02, D09a, D10a

128 cypermethrin

A contact and stomach acting pyrethroid insecticide
IRAC mode of action code: 3

Products
1	Curfew	AgriGuard	100 g/l	EC	14289
2	Cypermethrin Lacquer	Killgerm	7 g/l	LA	H3164
3	Forester	Fargro	100 g/l	EW	13164
4	Permasect C	Nufarm UK	100 g/l	EC	13158
5	Syper 100	Goldengrass	100 g/l	EC	14194
6	Toppel 100 EC	United Phosphorus	100 g/l	EC	13704

cypermethrin

Uses

- Aphids in *apples, vining peas* [1, 4, 6]; *beet leaves* (off-label - outdoor and protected crops), *broad beans* (off-label), *cherries, chinese cabbage* (off-label - outdoor and protected crops), *durum wheat, fodder beet* (off-label), *frise* (off-label), *herbs (see appendix 6)* (off-label), *hops, lamb's lettuce* (off-label), *leaf brassicas* (off-label), *lettuce, mallow (althaea spp.)* (off-label), *ornamental plant production, plums, radicchio* (off-label), *scarole* (off-label), *spinach beet* (off-label), *spring linseed* (off-label), *winter linseed* (off-label) [6]; *broccoli, brussels sprouts, cabbages, calabrese, cauliflowers, winter barley, winter oats, winter rye, winter wheat* [1, 5, 6]; *fodder beet, kale, mangels, potatoes, red beet, spring barley, spring field beans, spring oats, spring oilseed rape, spring wheat, sugar beet, winter field beans, winter oilseed rape* [5]; *pears, spring barley* (autumn sown), *spring wheat* (autumn sown) [1, 6]; *spring barley* (Moderate control), *spring wheat* (moderate control), *winter barley* (moderate control), *winter oats* (moderate control), *winter rye* (moderate control), *winter wheat* (moderate control) [4]; *triticale* [5, 6]
- Asparagus beetle in *asparagus* (off-label) [6]
- Barley yellow dwarf virus vectors in *durum wheat* [6]; *spring barley* (autumn sown), *spring wheat* (autumn sown), *triticale, winter barley, winter oats, winter rye, winter wheat* [1, 4, 6]
- Beetles in *cut logs* [3]
- Bladder pod midge in *spring oilseed rape, winter oilseed rape* [1, 6]
- Blossom beetle in *spring oilseed rape, winter oilseed rape* [4]
- Brassica pod midge in *spring oilseed rape* (some coincidental control only), *winter oilseed rape* (some coincidental control only) [5]
- Cabbage seed weevil in *spring oilseed rape, winter oilseed rape* [4, 5]
- Cabbage stem flea beetle in *spring oilseed rape* [5]; *winter oilseed rape* [1, 4-6]
- Capsids in *apples* [1, 6]; *ornamental plant production* [6]; *pears* [1, 4, 6]
- Caterpillars in *apples, pears* [1, 6]; *beet leaves* (off-label - outdoor and protected crops), *cherries, chinese cabbage* (off-label - outdoor and protected crops), *lettuce, ornamental plant production, plums, spinach beet* (off-label - outdoor and protected crops) [6]; *broccoli, brussels sprouts, cabbages, calabrese, cauliflowers* [1, 4-6]; *fodder beet, kale, mangels, potatoes, red beet, sugar beet* [4, 5]
- Codling moth in *apples* [1, 4, 6]
- Cutworms in *carrots* (off-label), *horseradish* (off-label), *lettuce, ornamental plant production, parsley root* (off-label), *parsnips* (off-label) [6]; *fodder beet, mangels, red beet* [4, 5]; *potatoes, sugar beet* [1, 4-6]
- Frit fly in *grassland* [4, 5]; *permanent grassland* [1, 6]; *rotational grass* [1, 4, 6]
- Insect pests in *farm buildings* [2]
- Large pine weevil in *forest* [3]
- Pea and bean weevil in *spring field beans, vining peas, winter field beans* [1, 4-6]
- Pea moth in *vining peas* [1, 4-6]
- Pod midge in *spring oilseed rape, winter oilseed rape* [4]
- Pollen beetle in *spring oilseed rape, winter oilseed rape* [1, 4-6]
- Rape winter stem weevil in *winter oilseed rape* [1, 6]
- Sawflies in *apples* [1, 6]
- Seed weevil in *spring oilseed rape, winter oilseed rape* [1, 6]
- Suckers in *apples* [4]; *pears* [1, 6]
- Thrips in *ornamental plant production* [6]
- Tortrix moths in *apples* [1, 4, 6]; *pears* [4]
- Whitefly in *broccoli, brussels sprouts, cabbages, calabrese, cauliflowers* [1, 6]; *ornamental plant production, protected cucumbers, protected lettuce, protected ornamentals* [6]
- Winter moth in *apples, pears* [4]
- Yellow cereal fly in *durum wheat* [6]; *spring barley* (autumn sown), *spring wheat* (autumn sown), *triticale, winter barley, winter oats, winter rye* [1, 4, 6]; *winter wheat* [1, 4-6]

Specific Off-Label Approvals (SOLAs)

- *asparagus* 20080639 [6]
- *beet leaves* (outdoor and protected crops) 20080638 [6]
- *broad beans* 20080643 [6]
- *carrots* 20080637 [6]
- *chinese cabbage* (outdoor and protected crops) 20080638 [6]
- *fodder beet* 20080640 [6]

SEE SECTION 3 FOR PRODUCTS ALSO REGISTERED

- *frise* 20080642 [6]
- *herbs (see appendix 6)* 20080642 [6]
- *horseradish* 20080637 [6]
- *lamb's lettuce* 20080642 [6]
- *leaf brassicas* 20080642 [6]
- *mallow (althaea spp.)* 20080642 [6]
- *parsley root* 20080637 [6]
- *parsnips* 20080637 [6]
- *radicchio* 20080642 [6]
- *scarole* 20080642 [6]
- *spinach beet* (outdoor and protected crops) 20080638 [6], 20080638 [6]
- *spring linseed* 20080641 [6]
- *winter linseed* 20080641 [6]

Approval information
- Cypermethrin included in Annex I under EC Directive 91/414
- Following implementation of Directive 98/82/EC, approval for use of cypermethrin on numerous crops was revoked in 1999
- Accepted by BBPA for use on malting barley and hops

Efficacy guidance
- Products combine rapid action, good persistence, and high activity on Lepidoptera.
- As effect is mainly via contact, good coverage is essential for effective action. Spray volume should be increased on dense crops [4, 6]
- A repeat spray after 10-14 d is needed for some pests of outdoor crops, several sprays at shorter intervals for whitefly and other glasshouse pests [4, 6]
- Rates and timing of sprays vary with crop and pest. See label for details [3, 4, 6]
- Where aphids in hops, pear suckers or glasshouse whitefly resistant to cypermethrin occur control is unlikely to be satisfactory [4, 6]
- Apply lacquer undiluted in a 5 cm wide band at strategic places. Ensure surfaces are clean and dry before application [2]
- Lacquer normally remains active for 12 mth [2]
- Temporary loss of activity for about 2 d will occur if lacquer surface is washed, until cypermethrin has migrated to the surface layer [2]

Restrictions
- Maximum number of treatments varies with crop and product. See label or approval notice for details [3, 4, 6]
- Apply by knapsack or hand-held sprayer [3]

Crop-specific information
- HI vining peas 7 d; other crops 0 d
- Test spray sample of new or unusual ornamentals or trees before committing whole batches [3, 6]
- Post-planting forestry applications should be made before damage is seen or at the onset of damage during the first 2 yrs after transplanting [3]

Environmental safety
- Dangerous for the environment [3, 4, 6]
- Very toxic to aquatic organisms
- High risk to bees. Do not apply to crops in flower, or to those in which bees are actively foraging, except as directed. Do not apply when flowering weeds are present
- Flammable [2, 4, 6]
- Do not spray cereals after 31 Mar within 6 m of the edge of the growing crop [3, 4, 6]
- LERAP A [4, 6]
- Broadcast air-assisted LERAP [1, 4-6] (18 m); LERAP Category A [1, 3-6]; LERAP Category B [5]

Hazard classification and safety precautions
Hazard Harmful [1-5]; Irritant [6]; Flammable [1, 2, 4-6]; Dangerous for the environment [1, 3-6]
Transport code 3 [1, 5, 6]; 9 [3]
Packaging group III [1, 3, 5, 6]
UN number 1993 [1, 6]; 3082 [3]; 3295 [5]
Risk phrases R20 [1, 2, 6]; R21, R52 [2]; R22a, R50 [1, 3-6]; R22b, R37, R66 [1, 4-6]; R38, R43 [3]; R41 [1, 6]; R53a [1-6]; R67 [4, 5]

FOR FULL CONDITIONS OF USE ALWAYS READ THE PRODUCT LABEL

cyproconazole

Operator protection A [1-6]; C [1, 2, 4-6]; H, M [3]; P [2]; U02a, U10 [3-5]; U04a [2-5]; U05a [1-6]; U08 [1, 6]; U09a, U15 [2]; U11 [4, 5]; U14 [3]; U16b [2, 3]; U19a [1, 3-6]; U20a [5]; U20b [1, 4, 6]
Environmental protection E02a [2] (48 h); E13a [1, 6]; E15a [2-5]; E16a [5]; E16c [1, 4, 6]; E16d, E34 [1, 3-6]; E17b [1, 4-6] (18 m); E38 [3]
Consumer protection C08, C09, C11, C12 [2]
Storage and disposal D01, D02, D09a [1-6]; D05, D10b [1, 3-6]; D12a [1, 3, 4, 6]; D12b [2]
Medical advice M03 [3]; M04a [2, 4, 5]; M05b [1, 4-6]

129 cyproconazole

A contact and systemic triazole fungicide
FRAC mode of action code: 3

*See also azoxystrobin + cyproconazole
chlorothalonil + cyproconazole
chlorothalonil + cyproconazole + propiconazole
cyproconazole + picoxystrobin*

Products
    Centaur   Bayer CropScience   200 g/l   EC   13852

Uses
- Brown rust in **spring barley**, **spring rye**, **winter barley**, **winter rye**, **winter wheat**
- Chocolate spot in **spring field beans**, **winter field beans**
- Crown rust in **spring oats**, **winter oats**
- Ear diseases in **winter wheat**
- Eyespot in **winter barley** *(useful reduction)*, **winter wheat** *(useful reduction)*
- Foliar disease control in **borage for oilseed production** *(off-label)*, **canary flower (echium spp.)** *(off-label)*, **durum wheat** *(off-label)*, **fodder beet** *(off-label)*, **grass seed crops** *(off-label)*, **linseed** *(off-label)*, **mustard** *(off-label)*, **red beet** *(off-label)*, **triticale** *(off-label)*
- Glume blotch in **winter wheat**
- Light leaf spot in **winter oilseed rape**
- Net blotch in **spring barley** *(useful reduction)*, **winter barley** *(useful reduction)*
- Phoma leaf spot in **winter oilseed rape**
- Powdery mildew in **spring barley**, **spring oats**, **spring rye**, **sugar beet**, **winter barley**, **winter oats**, **winter rye**, **winter wheat**
- Ramularia leaf spots in **sugar beet**
- Rhynchosporium in **spring barley**, **winter barley**
- Rust in **spring field beans**, **winter field beans**
- Septoria leaf blotch in **winter wheat**
- Yellow rust in **spring barley**, **winter barley**, **winter wheat**

Specific Off-Label Approvals (SOLAs)
- *borage for oilseed production* 20081099
- *canary flower (echium spp.)* 20081099
- *durum wheat* 20081101
- *fodder beet* 20081100
- *grass seed crops* 20081101
- *linseed* 20081099
- *mustard* 20081099
- *red beet* 20081100
- *triticale* 20081101

Approval information
- Accepted by BBPA for use on malting barley
- Cyproconazole has not been included in Annex 1 but the date of withdrawal has yet to be set. Current approvals expire in 2013

Efficacy guidance
- Apply at start of disease development or as preventive treatment and repeat as necessary
- Most effective time of treatment varies with disease and use of tank mixes may be desirable. See label for details

SEE SECTION 3 FOR PRODUCTS ALSO REGISTERED

Pesticide Profiles

- Product alone gives useful reduction of cereal eyespot. Where high infections probable a tank mix with prochloraz recommended
- On oilseed rape a two spray autumn/spring programme recommended for high risk situations and on susceptible varieties
- Cyproconazole is a DMI fungicide. Resistance to some DMI fungicides has been identified in Septoria leaf blotch which may seriously affect performance of some products. For further advice contact a specialist advisor and visit the Fungicide Resistance Action Group (FRAG)-UK website

Restrictions
- Maximum total dose equivalent to 1-3 full dose treatments depending on crop treated. See labels for details

Crop-specific information
- HI 6 wk for field beans; 14 d for sugar beet, red beet, leeks
- Latest use: completion of flowering for rye, wheat; emergence of ear complete (GS 59) for barley, oats; before lowest pods more than 2 cm long (GS 5,1) for winter oilseed rape
- Application to winter wheat in spring between the start of stem elongation and the third node detectable stage (GS 30-33) may cause straw shortening, but does not cause loss of yield

Environmental safety
- Dangerous for the environment
- Toxic to aquatic organisms
- Dangerous to fish or other aquatic life. Do not contaminate surface waters or ditches with chemical or used container

Hazard classification and safety precautions
Hazard Harmful, Dangerous for the environment
Transport code 9
Packaging group III
UN number 3082
Risk phrases R36, R37, R43, R51, R53a, R63
Operator protection A, C, H; U05a, U08, U20a
Environmental protection E15a, E38
Storage and disposal D01, D02, D09a, D10a, D12a
Medical advice M03

130 cyproconazole + cyprodinil

A broad spectrum fungicide mixture for cereals
FRAC mode of action code: 3 + 9

See also cyprodinil

Products
Radius   Syngenta   5.33:40 % w/w   WG   09387

Uses
- Brown rust in **spring barley, winter barley, winter wheat**
- Eyespot in **spring barley, winter barley, winter wheat**
- Foliar disease control in **durum wheat** *(off-label)*, **grass seed crops** *(off-label)*, **spring rye** *(off-label)*, **triticale** *(off-label)*, **winter rye** *(off-label)*
- Glume blotch in **winter wheat**
- Net blotch in **spring barley, winter barley**
- Powdery mildew in **spring barley, winter barley, winter wheat**
- Rhynchosporium in **spring barley, winter barley**
- Septoria leaf blotch in **winter wheat**
- Yellow rust in **spring barley, winter barley, winter wheat**

Specific Off-Label Approvals (SOLAs)
- *durum wheat* 20060613
- *grass seed crops* 20060613
- *spring rye* 20060613
- *triticale* 20060613
- *winter rye* 20060613

FOR FULL CONDITIONS OF USE ALWAYS READ THE PRODUCT LABEL

cyproconazole + picoxystrobin

Approval information
- Cyprodinil included in Annex I under EC Directive 91/414
- Accepted by BBPA for use on malting barley

Efficacy guidance
- Best results obtained from treatment at early stages of development. Later treatments may be required if disease pressure remains high
- Eyespot is fully controlled by treatment in spring during stem extension. Control may be reduced when very dry conditions follow application
- Cyproconazole is a DMI fungicide. Resistance to some DMI fungicides has been identified in Septoria leaf blotch which may seriously affect performance of some products. For further advice contact a specialist advisor and visit the Fungicide Resistance Action Group (FRAG)-UK website

Restrictions
- Maximum total dose equivalent to two full dose treatments

Crop-specific information
- Latest use: before first spikelet of inflorescence visible (GS 51) for barley; before caryopsis watery ripe (GS 71) for winter wheat

Environmental safety
- Dangerous for the environment
- Very toxic to aquatic organisms
- Risk to certain non-target insects or other arthropods. For advice on risk management and use in Integrated Pest Management (IPM) see directions for use

Hazard classification and safety precautions
Hazard Harmful, Dangerous for the environment
Transport code 9
Packaging group III
UN number 3077
Risk phrases R38, R43, R50, R53a, R63
Operator protection A, H; U05a, U20a
Environmental protection E15a, E22b, E38
Storage and disposal D01, D02, D05, D07, D09a, D12a

131 cyproconazole + picoxystrobin

A broad spectrum triazole + strobilurin fungicide mixture for use in cereals
FRAC mode of action code: 3 + 11

See also cyproconazole

Products
Furlong   DuPont   80:200 g/l   SC   13818

Uses
- Brown rust in *spring barley*, *spring wheat*, *winter barley*, *winter wheat*
- Crown rust in *spring oats*, *winter oats*
- Glume blotch in *spring wheat*, *winter wheat*
- Net blotch in *spring barley* (moderate control), *winter barley* (moderate control)
- Rhynchosporium in *spring barley* (moderate control), *winter barley* (moderate control)
- Septoria leaf blotch in *spring wheat*, *winter wheat*
- Yellow rust in *spring wheat*, *winter wheat*

Approval information
- Picoxystrobin included in Annex I under EC Directive 91/414

Restrictions
- To reduce the risk of resistance developing in target diseases the total number of applications of products containing QoI fungicides made to any cereal crop must not exceed two.

Hazard classification and safety precautions
Hazard Harmful, Dangerous for the environment
Transport code 9
Packaging group III

SEE SECTION 3 FOR PRODUCTS ALSO REGISTERED

UN number 3082
Risk phrases R50, R53a, R63
Operator protection A, H
Environmental protection E15a, E34, E38
Storage and disposal D01, D02, D05, D09a, D12a

132 cyproconazole + propiconazole

A broad-spectrum mixture of conazole fungicides for wheat and barley
FRAC mode of action code: 3 + 3

See also propiconazole

Products

1	Alto Xtra	Syngenta	160:250 g/l	EC	14437
2	Menara	Syngenta	160:250 g/l	EC	14398

Uses
- Brown rust in **spring barley**, **winter barley**, **winter wheat** [1, 2]
- Foliar disease control in **durum wheat** *(off-label)*, **spring rye** *(off-label)*, **triticale** *(off-label)*, **winter rye** *(off-label)* [2]
- Glume blotch in **winter wheat** [1, 2]
- Late ear diseases in **winter wheat** [1, 2]
- Net blotch in **spring barley**, **winter barley** [1, 2]
- Powdery mildew in **spring barley**, **winter barley** [1, 2]; **winter wheat** [1]
- Rhynchosporium in **spring barley**, **winter barley** [1, 2]
- Septoria leaf blotch in **winter wheat** [1, 2]
- Yellow rust in **spring barley**, **winter barley** [1]; **winter wheat** [1, 2]

Specific Off-Label Approvals (SOLAs)
- **durum wheat** 20090685 [2]
- **spring rye** 20090685 [2]
- **triticale** 20090685 [2]
- **winter rye** 20090685 [2]

Approval information
- Propiconazole included in Annex I under EC Directive 91/414
- Accepted by BBPA for use on malting barley

Efficacy guidance
- Treatment should be made at first sign of disease
- Useful reduction of eyespot when applied in spring during stem extension (GS 30-32)
- Cyproconazole and propiconazole are DMI fungicides. Resistance to some DMI fungicides has been identified in Septoria leaf blotch which may seriously affect performance of some products. For further advice contact a specialist advisor and visit the Fungicide Resistance Action Group (FRAG)-UK website

Restrictions
- Maximum total dose equivalent to one full dose treatment for barley; two full dose treatments for winter wheat

Crop-specific information
- Latest use: before beginning of anthesis (GS 60) for barley; before grain watery ripe (GS 71) for wheat
- Application to winter wheat in spring may cause straw shortening but does not cause loss of yield

Environmental safety
- Dangerous for the environment
- Very toxic to aquatic organisms

Hazard classification and safety precautions
Hazard Harmful, Dangerous for the environment
Transport code 9
Packaging group III
UN number 3082

FOR FULL CONDITIONS OF USE ALWAYS READ THE PRODUCT LABEL

Risk phrases R50, R53a, R63
Operator protection A, C, H; U05a, U08, U20b
Environmental protection E15a, E38
Storage and disposal D01, D02, D05, D07, D09a, D10c, D12a

133 cyproconazole + quinoxyfen

A contact and systemic fungicide mixture for cereals
FRAC mode of action code: 3 + 13

See also quinoxyfen

Products

Excelsior	Dow	80:75 g/l	SC	13378

Uses
- Brown rust in *winter wheat*
- Eyespot in *winter wheat* (reduction)
- Foliar disease control in *grass seed crops* (off-label)
- Glume blotch in *winter wheat*
- Powdery mildew in *winter wheat*
- Septoria leaf blotch in *winter wheat*
- Yellow rust in *winter wheat*

Specific Off-Label Approvals (SOLAs)
- *grass seed crops* 20090887

Approval information
- Quinoxyfen included in Annex I under Directive 91/414
- Accepted by BBPA for use on malting barley

Efficacy guidance
- Best results are achieved when disease is first detected and seen to be active. A second treatment may be needed under prolonged disease pressure
- Treatment for Septoria control should be made as soon as possible after weather that favours infection to prevent spread to the second and flag leaves
- Useful reduction of eyespot when applied between GS 30-34 to control other diseases
- Systemic activity may be reduced under conditions of severe drought stress

Restrictions
- Maximum number of treatments equivalent to a total of two full doses
- Application to winter wheat in spring may cause straw shortening but does not cause loss of yield

Crop-specific information
- Latest use: before ear emergence for wheat

Environmental safety
- Dangerous for the environment
- Very toxic to aquatic organisms
- LERAP Category B

Hazard classification and safety precautions
Hazard Harmful, Dangerous for the environment
Transport code 9
Packaging group III
UN number 3082
Risk phrases R43, R50, R53a, R63
Operator protection A, C, H; U05a, U20b
Environmental protection E15a, E16a, E16b, E38
Storage and disposal D01, D02, D09a, D10b, D12a

SEE SECTION 3 FOR PRODUCTS ALSO REGISTERED

134 cyproconazole + trifloxystrobin

A conazole and strobilurin fungicide mixture for wheat and barley
FRAC mode of action code: 3 + 11

See also trifloxystrobin

Products

1	Escolta	Bayer CropScience	160:375 g/l	SC	13923
2	Sphere	Bayer CropScience	80:187.5 g/l	EC	11429

Uses
- Brown rust in **spring barley, winter barley, winter wheat** [2]
- Cercospora leaf spot in **sugar beet** [1]
- Foliar disease control in **durum wheat** *(off-label)*, **grass seed crops** *(off-label)*, **spring rye** *(off-label)*, **triticale** *(off-label)*, **winter rye** *(off-label)* [2]
- Glume blotch in **winter wheat** [2]
- Net blotch in **spring barley, winter barley** [2]
- Powdery mildew in **spring barley, winter barley, winter wheat** [2]; **sugar beet** [1]
- Ramularia leaf spots in **sugar beet** [1]
- Rhynchosporium in **spring barley, winter barley** [2]
- Rust in **sugar beet** [1]
- Septoria leaf blotch in **winter wheat** [2]
- Yellow rust in **spring barley, winter barley, winter wheat** [2]

Specific Off-Label Approvals (SOLAs)
- *durum wheat* 20060603 [2]
- *grass seed crops* 20060603 [2]
- *spring rye* 20060603 [2]
- *triticale* 20060603 [2]
- *winter rye* 20060603 [2]

Approval information
- Trifloxystrobin included in Annex I under EC Directive 91/414
- Accepted by BBPA for use on malting barley

Efficacy guidance
- Best results obtained from treatment at early stages of disease development. Further treatment may be needed if disease attack is prolonged
- Cyproconazole is a DMI fungicide. Resistance to some DMI fungicides has been identified in Septoria leaf blotch which may seriously affect performance of some products. For further advice contact a specialist advisor and visit the Fungicide Resistance Action Group (FRAG)-UK website
- Trifloxystrobin is a member of the QoI cross resistance group. Product should be used preventatively and not relied on for its curative potential
- Use product as part of an Integrated Crop Management strategy incorporating other methods of control, including where appropriate other fungicides with a different mode of action. Do not apply more than two foliar applications of QoI containing products to any cereal crop
- There is a significant risk of widespread resistance occurring in *Septoria tritici* populations in UK. Failure to follow resistance management action may result in reduced levels of disease control
- Strains of barley powdery mildew resistant to QoIs are common in the UK

Restrictions
- Maximum total dose per crop equivalent to two full dose treatments

Crop-specific information
- HI 35 d

Environmental safety
- Dangerous for the environment
- Very toxic to aquatic organisms

Hazard classification and safety precautions
Hazard Harmful, Dangerous for the environment
Transport code 9
Packaging group III

FOR FULL CONDITIONS OF USE ALWAYS READ THE PRODUCT LABEL

cyprodinil

UN number 3082
Risk phrases R36 [2]; R50, R53a, R63 [1, 2]
Operator protection A [1, 2]; C, H [2]; U02a, U20a [1]; U05a, U09a, U19a [1, 2]; U20b [2]
Environmental protection E15a [2]; E15b [1]; E38 [1, 2]
Storage and disposal D01, D02, D09a, D10b, D12a [1, 2]; D05 [1]
Medical advice M03 [1]

135 cyprodinil

An anilinopyrimidine systemic broad spectrum fungicide for cereals
FRAC mode of action code: 9

See also cyproconazole + cyprodinil

Products

1	Kayak	Syngenta	300 g/l	EC	14847
2	Unix	Syngenta	75% w/w	WG	14846

Uses
- Eyespot in **spring barley**, **winter barley** [1, 2]; **spring wheat**, **winter wheat** [2]
- Net blotch in **spring barley**, **winter barley** [1, 2]
- Powdery mildew in **spring barley**, **winter barley** [1, 2]; **spring wheat** *(moderate control)*, **winter wheat** *(moderate control)* [2]
- Rhynchosporium in **spring barley**, **winter barley** [1]; **spring barley** *(moderate control)*, **winter barley** *(moderate control)* [2]

Approval information
- Cyprodinil included in Annex I under EC Directive 91/414
- Accepted by BBPA for use on malting barley

Efficacy guidance
- Best results obtained from treatment at early stages of disease development
- For best control of eyespot spray before or during the period of stem extension in spring. Control may be reduced if very dry conditions follow treatment

Restrictions
- Maximum number of treatments 2 per crop on wheat and barley [1]

Crop-specific information
- Latest use: before first spikelet of inflorescence just visible stage for barley, before early milk stage for wheat [2]; up to and including first awns visible stage (GS49) in barley [1]

Environmental safety
- Dangerous for the environment
- Very toxic to aquatic organisms
- LERAP Category B

Hazard classification and safety precautions
Hazard Irritant [1]; Dangerous for the environment [1, 2]
Transport code 9
Packaging group III
UN number 3077 [2]; 3082 [1]
Risk phrases R38, R43 [1]; R50, R53a [1, 2]
Operator protection A, H; U05a, U20a [1, 2]; U09a [1]
Environmental protection E15a [2]; E15b, E34 [1]; E16a, E38 [1, 2]
Storage and disposal D01, D02, D05, D07, D09a, D12a [1, 2]; D10c [1]

SEE SECTION 3 FOR PRODUCTS ALSO REGISTERED

136 cyprodinil + fludioxonil

A broad-spectrum fungicide mixture for legumes and strawberries
FRAC mode of action code: 9 + 12

See also fludioxonil

Products

Switch   Syngenta   37.5:25 % w/w   WG   13185

Uses
- Alternaria in **apples**, **beetroot** *(off-label)*, **carrots** *(off-label)*, **crab apples**, **ginger** *(off-label)*, **ginseng root** *(off-label)*, **horseradish** *(off-label)*, **liquorice** *(off-label)*, **parsley root** *(off-label)*, **parsnips** *(off-label)*, **pears**, **quinces**, **salsify** *(off-label)*, **tumeric** *(off-label)*, **valerian root** *(off-label)*
- Ascochyta in **mange-tout peas**, **sugar snap peas**, **vining peas**
- Black spot in **protected strawberries** *(qualified minor use)*, **strawberries** *(qualified minor use)*
- Blossom wilt in **apricots** *(off-label)*, **cherries** *(off-label)*, **peaches** *(off-label)*, **plums** *(off-label)*
- Botrytis in **asparagus** *(off-label)*, **broad beans** *(moderate control)*, **chicory** *(off-label)*, **cress** *(off-label)*, **forest nurseries**, **frise** *(off-label)*, **grapevines** *(off-label)*, **green beans** *(moderate control)*, **lamb's lettuce** *(off-label)*, **lettuce** *(off-label)*, **lupins** *(off-label)*, **mange-tout peas** *(moderate control)*, **onions** *(off-label)*, **ornamental plant production**, **protected aubergines** *(off-label)*, **protected cress** *(off-label)*, **protected cucumbers** *(off-label)*, **protected forest nurseries**, **protected frise** *(off-label)*, **protected lamb's lettuce** *(off-label)*, **protected lettuce** *(off-label)*, **protected ornamentals**, **protected peppers** *(off-label)*, **protected radicchio** *(off-label)*, **protected rocket** *(off-label)*, **protected scarole** *(off-label)*, **protected strawberries**, **protected tomatoes** *(off-label)*, **radicchio** *(off-label)*, **rocket** *(off-label)*, **runner beans** *(moderate control)*, **salad onions** *(off-label)*, **scarole** *(off-label)*, **strawberries**, **sugar snap peas** *(moderate control)*, **vining peas** *(moderate control)*
- Botrytis fruit rot in **apples**, **blackberries**, **crab apples**, **pears**, **quinces**, **raspberries**
- Botrytis root rot in **bilberries** *(qualified minor use recommendation)*, **blackcurrants** *(qualified minor use recommendation)*, **blueberries** *(qualified minor use recommendation)*, **cranberries** *(qualified minor use recommendation)*, **gooseberries** *(qualified minor use recommendation)*, **redcurrants** *(qualified minor use recommendation)*, **whitecurrants** *(qualified minor use recommendation)*
- Fusarium in **protected cucumbers** *(off-label)*
- Fusarium diseases in **apples**, **crab apples**, **pears**, **quinces**
- Gloeosporium rot in **apples**, **crab apples**, **pears**, **quinces**
- Mycosphaerella in **mange-tout peas**, **protected cucumbers** *(off-label)*, **sugar snap peas**, **vining peas**
- Penicillium rot in **apples**, **crab apples**, **pears**, **quinces**
- Sclerotinia in **beetroot** *(off-label)*, **broad beans**, **carrots** *(off-label)*, **ginger** *(off-label)*, **ginseng root** *(off-label)*, **green beans**, **horseradish** *(off-label)*, **liquorice** *(off-label)*, **mange-tout peas**, **parsley root** *(off-label)*, **parsnips** *(off-label)*, **runner beans**, **salsify** *(off-label)*, **sugar snap peas**, **tumeric** *(off-label)*, **valerian root** *(off-label)*, **vining peas**
- Stemphylium in **asparagus** *(off-label)*

Specific Off-Label Approvals (SOLAs)
- *apricots* 20070735
- *asparagus* 20070235
- *beetroot* 20092649
- *carrots* 20092649
- *cherries* 20070735
- *chicory* 20100655
- *cress* 20081931
- *frise* 20081931
- *ginger* 20092649
- *ginseng root* 20092649
- *grapevines* 20070236
- *horseradish* 20092649
- *lamb's lettuce* 20081931
- *lettuce* 20081931

FOR FULL CONDITIONS OF USE ALWAYS READ THE PRODUCT LABEL

- **liquorice** 20092649
- **lupins** 20070237
- **onions** 20100655
- **parsley root** 20092649
- **parsnips** 20092649
- **peaches** 20070735
- **plums** 20070735
- **protected aubergines** 20070238
- **protected cress** 20081931
- **protected cucumbers** 20070234
- **protected frise** 20081931
- **protected lamb's lettuce** 20081931
- **protected lettuce** 20081931
- **protected peppers** 20070238
- **protected radicchio** 20081931
- **protected rocket** 20081931
- **protected scarole** 20081931
- **protected tomatoes** 20071500
- **radicchio** 20081931
- **rocket** 20081931
- **salad onions** 20070239
- **salsify** 20092649
- **scarole** 20081931
- **tumeric** 20092649
- **valerian root** 20092649

Approval information
- Cyprodinil and fludioxonil included in Annex I under EC Directive 91/414

Efficacy guidance
- Best results obtained from application at the earliest stage of disease development or as a protective treatment following disease risk assessment
- Subsequent treatments may follow after a minimum of 10 d if disease pressure remains high
- To minimise the likelihood of development of resistance product should be used in a planned Resistance Management strategy. See Section 5 for more information

Restrictions
- Maximum number of treatments 2 per crop
- Consult processor before treating any crop for processing

Crop-specific information
- HI broad beans, green beans, mange-tout peas, runner beans, sugar snap peas, vining peas 14 d; protected strawberries, strawberries 3 d
- First signs of disease infection likely to be seen from early flowering for all crops
- Ensure peas are free from any stress before treatment. If necessary check wax with a crystal violet test

Environmental safety
- Dangerous for the environment
- Very toxic to aquatic organisms
- LERAP Category B

Hazard classification and safety precautions
    Hazard Dangerous for the environment
    Transport code 9
    Packaging group III
    UN number 3077
    Risk phrases R50, R53a
    Operator protection A, H; U05a, U20a
    Environmental protection E15b, E16a, E16b, E38
    Storage and disposal D01, D02, D05, D07, D09a, D10c, D12a

SEE SECTION 3 FOR PRODUCTS ALSO REGISTERED

137   cyprodinil + isopyrazam

A fungicide mixture for disease control in barley
FRAC mode of action code: 7 + 9

See also cyprodinil
isopyrzam

Products
    Bontima   Syngenta   187.5:62.5 g/l   EC   14899

Uses
- Brown rust in **spring barley, winter barley**
- Net blotch in **spring barley, winter barley**
- Powdery mildew in **spring barley, winter barley**
- Ramularia leaf spots in **spring barley, winter barley**
- Rhynchosporium in **spring barley, winter barley**

Approval information
- Isopyrazam has not yet been included in Annex 1 under EC Directive 91/414
- Accepted by BBPA for use on malting barley

Restrictions
- Isopyrazam is an SDH respiration inhibitor; Do not apply more than two foliar applications of products containing an SDH inhibitor to any cereal crop

Crop-specific information
- There are no restrictions on succeeding crops in a normal rotation

Environmental safety
- LERAP Category B

Hazard classification and safety precautions
  Hazard Harmful, Dangerous for the environment
  Transport code 9
  Packaging group III
  UN number 3082
  Risk phrases R20, R38, R40, R50, R53a, R63
  Operator protection A, C, H; U05a, U09a, U11, U20b
  Environmental protection E15b, E16a, E16b, E38
  Storage and disposal D01, D02, D05, D09a, D10c, D11a, D12a

138   cyprodinil + picoxystrobin

A broad spectrum fungicide mixture for cereals
FRAC mode of action code: 9 + 11

See also picoxystrobin

Products
    Acanto Prima   DuPont   30.8% w/w   WG   14971

Uses
- Brown rust in **spring barley, spring wheat, winter barley, winter wheat**
- Eyespot in **spring barley, spring wheat** (moderate control), **winter barley, winter wheat** (moderate control)
- Glume blotch in **spring wheat, winter wheat**
- Net blotch in **spring barley, winter barley**
- Powdery mildew in **spring barley, winter barley**
- Rhynchosporium in **spring barley, winter barley**
- Septoria leaf blotch in **spring wheat, winter wheat**
- Yellow rust in **spring wheat, winter wheat**

FOR FULL CONDITIONS OF USE ALWAYS READ THE PRODUCT LABEL

Approval information
- Cyprodinil and picoxystrobin included in Annex I under EC Directive 91/414
- Accepted by BBPA for use on malting barley

Efficacy guidance
- Best results obtained from use as a protectant treatment or in the earliest stages of disease development. Further applications may be needed if disease attack is prolonged
- Picoxystrobin is a member of the QoI cross resistance group. Product should be used preventatively and not relied on for its curative potential
- Use product as part of an Integrated Crop Management strategy incorporating other methods of control, including where appropriate other fungicides with a different mode of action. Do not apply more than two foliar applications of QoI containing products to any cereal crop
- There is a significant risk of widespread resistance occurring in *Septoria tritici* populations in UK. Failure to follow resistance management action may result in reduced levels of disease control
- In wheat product must always be used in mixture with another product, recommended for control of the same target disease, that contains a fungicide from a different cross resistance group and is applied at a dose that will give robust control
- Strains of barley powdery mildew resistant to QoIs are common in the UK

Restrictions
- Maximum number of treatments 2 per crop

Crop-specific information
- Latest use: before early milk stage (GS 73) for wheat; before ear just visible (GS 51) for barley

Environmental safety
- Dangerous for the environment
- Very toxic to aquatic organisms
- LERAP Category B

Hazard classification and safety precautions
Hazard Dangerous for the environment
Transport code 9
Packaging group III
UN number 3077
Risk phrases R50, R53a
Operator protection A, C, H; U05a, U09a, U11, U14, U15, U20a
Environmental protection E15a, E16a, E34, E38
Storage and disposal D01, D02, D09a, D10c, D12a

139   2,4-D

A translocated phenoxycarboxylic acid herbicide for cereals, grass and amenity use
HRAC mode of action code: O

See also amitrole + 2,4-D + diuron
clopyralid + 2,4-D + MCPA

Products
1	Depitox	Nufarm UK	500 g/l	SL	13258
2	Dioweed 50	United Phosphorus	500 g/l	SL	13197
3	Headland Staff 500	Headland	500 g/l	SL	13196
4	Herboxone	Headland	500 g/l	SL	13958
5	HY-D Super	Agrichem	500 g/l	SL	13198
6	Zip	AgriChem BV	500 g/l	SL	14806

Uses
- Annual dicotyledons in **amenity grassland, managed amenity turf, spring barley, spring wheat, winter barley, winter rye, winter wheat** [1-6]; **apple orchards, farm forestry** *(off-label)*, **forest nurseries** *(off-label)*, **game cover** *(off-label)*, **ornamental plant production** *(off-label)*, **pear orchards, spring rye** [1]; **grassland** [1, 2, 6]; **miscanthus** *(off-label)* [5]; **permanent grassland** [3-5]; **spring oats** [6]; **undersown barley, undersown oats, undersown rye, undersown wheat** [1, 5, 6]; **undersown spring cereals, undersown winter cereals** [2]; **winter oats** [1-5]

SEE SECTION 3 FOR PRODUCTS ALSO REGISTERED

- Aquatic weeds in **enclosed waters, land immediately adjacent to aquatic areas, open waters** [1]
- Perennial dicotyledons in **amenity grassland, managed amenity turf, spring barley, spring wheat, winter barley, winter rye, winter wheat** [1, 3-6]; **apple orchards, pear orchards, spring rye** [1]; **grassland** [1, 6]; **permanent grassland** [3-5]; **spring oats** [6]; **undersown barley, undersown oats, undersown rye, undersown wheat** [1, 5, 6]; **winter oats** [1, 3-5]

Specific Off-Label Approvals (SOLAs)
- *farm forestry* 20082843 [1]
- *forest nurseries* 20082843 [1]
- *game cover* 20082843 [1]
- *miscanthus* 20082914 [5]
- *ornamental plant production* 20082843 [1]

Approval information
- 2,4-D included in Annex I under EC Directive 91/414
- Approved for aquatic weed control [1]. See Section 5 for information on use of herbicides in or near water
- Accepted by BBPA for use on malting barley

Efficacy guidance
- Best results achieved by spraying weeds in seedling to young plant stage when growing actively in a strongly competing crop
- Most effective stage for spraying perennials varies with species. See label for details
- Spray aquatic weeds when in active growth between May and Sep

Restrictions
- Maximum number of treatments normally 1 per crop and in forestry, 1 per yr in grassland and 2 or 3 per yr in amenity turf. Check individual labels
- Do not use on newly sown leys containing clover
- Do not spray grass seed crops after ear emergence
- Do not spray within 6 mth of laying turf or sowing fine grass
- Do not dump surplus herbicide in water or ditch bottoms
- Do not plant conifers until at least 1 mth after treatment
- Do not spray crops stressed by cold weather or drought or if frost expected
- Do not roll or harrow within 7 d before or after spraying
- Do not spray if rain falling or imminent
- Do not mow or roll turf or amenity grass 4 d before or after spraying. The first 4 mowings after treatment must be composted for at least 6 mth before use
- Do not mow grassland or graze for at least 10 d after spraying

Crop-specific information
- Latest use: before 1st node detectable in cereals; end Aug for conifer plantations; before established grassland 25 cm high
- Spray winter cereals in spring when leaf-sheath erect but before first node detectable (GS 31), spring cereals from 5-leaf stage to before first node detectable (GS 15-31)
- Cereals undersown with grass and/or clover, but not with lucerne, may be treated
- Selective treatment of resistant conifers can be made in Aug when growth ceased and plants hardened off, spray must be directed if applied earlier. See label for details

Following crops guidance
- Do not use shortly before or after sowing any crop
- Do not plant succeeding crops within 3 mth [3]
- Do not direct drill brassicas or grass/clover mixtures within 3 wk of application

Environmental safety
- Dangerous for the environment
- Very toxic to aquatic organisms [3]
- Dangerous to aquatic higher plants. Do not contaminate surface waters or ditches with chemical or used container
- 2,4-D is active at low concentrations. Take extreme care to avoid drift onto neighbouring crops, especially beet crops, brassicas, most market garden crops including lettuce and tomatoes under glass, pears and vines

FOR FULL CONDITIONS OF USE ALWAYS READ THE PRODUCT LABEL

- May be used to control aquatic weeds in presence of fish if used in strict accordance with directions for waterweed control and precautions needed for aquatic use [1]
- Keep livestock out of treated areas for at least 2 wk following treatment and until poisonous weeds, such as ragwort, have died down and become unpalatable
- Water containing the herbicide must not be used for irrigation purposes within 3 wk of treatment or until the concentration in water is below 0.05 ppm

Hazard classification and safety precautions
Hazard Harmful, Dangerous for the environment [2-6]; Irritant [1]
Transport code 9
Packaging group III
UN number 3082
Risk phrases R22a [2-6]; R37 [6]; R38, R52 [1]; R41, R53a [1-6]; R43 [2, 5, 6]; R50 [3, 4, 6]; R51 [2, 5]
Operator protection A, C, H [1-6]; M [1]; U05a, U11 [1-6]; U08, U20b [1, 3-6]; U09a, U20a [2]; U14, U15 [5, 6]
Environmental protection E07a, E34 [1-6]; E15a [1, 2]; E15b, E38 [2-6]
Storage and disposal D01, D02, D05 [1-6]; D09a [1, 3-6]; D10a [3-6]; D10b [1]; D10c [2]; D12a [2-6]
Medical advice M03 [1-6]; M05a [2, 5, 6]

140 2,4-D + dicamba

A translocated herbicide for use on turf
HRAC mode of action code: O + O

See also dicamba

Products
1	Magneto	Nufarm UK	344:120 g/l	SL	12339
2	New Estermone	Vitax	200:35 g/l	EC	13792
3	Scotts Feed and Weed	Scotts	0.8:0.12% w/w	GR	13076
4	Thrust	Nufarm UK	344:120 g/l	SL	12230

Uses
- Annual and perennial weeds in *managed amenity turf* [2]
- Annual dicotyledons in *amenity grassland*, *permanent grassland* [1, 4]; *managed amenity turf* [3]
- Perennial dicotyledons in *amenity grassland*, *permanent grassland* [1, 4]; *managed amenity turf* [3]

Approval information
- 2,4-D and dicamba included in Annex I under EC Directive 91/414
- Accepted by BBPA for use on malting barley

Efficacy guidance
- Best results achieved by application when weeds growing actively in spring or early summer (later with irrigation and feeding)
- More resistant weeds may need repeat treatment after 3 wk
- Improved control of some weeds can be obtained by use of specifed oil adjuvant [1, 4]
- Do not use during drought conditions

Restrictions
- Maximum number of treatments 2 per yr on amenity grass and established grassland [1, 4]
- Do not treat newly sown or turfed areas or grass less than 1 yr old
- Do not mow less than 3 d before treatment and at least 3-4 d afterwards [3]
- Do not treat grass crops intended for seed production [1, 4]
- Do not treat grass suffering from drought, disease or other adverse factors [1, 4]
- Do not roll or harrow for 7 d before or after treatment [1, 4]
- Do not apply when grassland is flowering [1, 4]
- Do not re-seed for 6-8 wk after application [3]
- Avoid spray drift onto cultivated crops or ornamentals
- Do not graze grass for at least seven days after spraying [1, 4]

SEE SECTION 3 FOR PRODUCTS ALSO REGISTERED

Pesticide Profiles

- Do not mow or roll four days before or after application. The first four mowings after treatment must be composted for at least six months before use [1, 4]

Crop-specific information
- Latest use: before grass 25 cm high for amenity grass, established grassland [1, 4]
- The first four mowings after treatment must be composted for at least 6 mth before use

Environmental safety
- Dangerous for the environment [1, 4]
- Toxic to aquatic organisms
- Keep livestock out of treated areas for at least two weeks following treatment and until poisonous weeds, such as ragwort, have died down and become unpalatable [1, 4]

Hazard classification and safety precautions
  Hazard Harmful, Dangerous for the environment [1, 2, 4]
  UN number N/C [2, 3]
  Risk phrases R22a, R43, R51 [1, 2, 4]; R22b, R36, R66 [2]; R41 [1, 4]; R52 [3]; R53a [1-4]
  Operator protection A, C, H [1-4]; D [3]; M [1, 2, 4]; U05a [1, 4]; U08, U14, U15 [2]; U11 [1, 2, 4]; U20b [2, 3]
  Environmental protection E07a, E15a, E38 [1, 2, 4]; E15b [3]; E34 [1, 4]
  Storage and disposal D01 [1, 3, 4]; D02 [1, 4]; D05, D10a, D12a [1, 2, 4]; D09a [1-4]; D11a [3]
  Medical advice M03 [1, 4]; M05b [2]

141 2,4-D + dicamba + dichlorprop-P

A translocated hormone mixture for the control of docks in grassland
HRAC mode of action code: O + O + O

See also dicamba
dichlorprop-P

Products
  Legolas   Nufarm UK   300:120:400 g/l   SL   14439

Uses
- Creeping thistle in *amenity grassland, grassland*
- Dandelions in *amenity grassland, grassland*
- Docks in *amenity grassland, grassland*
- Sowthistle in *amenity grassland, grassland*

Approval information
- 2,4-D, dicamba and dichlorprop-P included in Annex I under EC Directive 91/414
- Accepted by BBPA for use on malting barley

Efficacy guidance
- Make applications when weed growth is vigorous due to warm, moist conditions

Restrictions
- Treatment will severely damage clover. Do not use where clover is an important component of the sward
- Do not mow or roll four days before or after application. The first four mowings after treatment must be composted for at least six months before use.
- Do not treat newly established grass less than one year old.
- Do not treat grass crops intended for seed production

Crop-specific information
- One application per year on established agricultural grassland, two applications per year on amenity grassland

Following crops guidance
- Do not re-seed within 3 months of treatment
- Avoid drift on to suceptible crops such as tomatoes, vegetables, sugar beet, glasshouse crops, fruit and ornamentals

Environmental safety
- Very toxic to aquatic organisms

FOR FULL CONDITIONS OF USE ALWAYS READ THE PRODUCT LABEL

2,4-D + dicamba + fluroxypyr

- Keep livestock out of treated areas for at least 2 weeks following treatment and until poisonous weeds such as ragwort have died and become unpaletable

Hazard classification and safety precautions
Hazard Harmful, Dangerous for the environment
Transport code 9
Packaging group III
UN number 3082
Risk phrases R20, R21, R22a, R41, R43, R50, R53a
Operator protection A, C, H; U05a, U11, U14
Environmental protection E07a, E15a, E34, E38
Storage and disposal D01, D02, D05, D09a, D10a, D12a
Medical advice M03

142 2,4-D + dicamba + fluroxypyr

A translocated and contact herbicide mixture for amenity turf
HRAC mode of action code: O + O + O

See also dicamba
     fluroxypyr

Products
1	Barclay Holster XL	Barclay	285:52.5:105 g/l	EC	13596
2	Mascot Crossbar	Rigby Taylor	285:52.5:105 g/l	EC	14453
3	Swiftsure	Pan Amenity	285:52.5:105 g/l	EC	14468

Uses
- Annual dicotyledons in **managed amenity turf** [1-3]
- Buttercups in **managed amenity turf** [1]
- Clover in **managed amenity turf** [1]
- Daisies in **managed amenity turf** [1]
- Dandelions in **managed amenity turf** [1]
- Yarrow in **managed amenity turf** [1]

Approval information
- 2,4-D, dicamba and fluroxypyr included in Annex I under EC Directive 91/414

Efficacy guidance
- Apply when weeds actively growing (normally between Apr and Sep) and when soil is moist
- Best results obtained from treatment in spring or early summer before weeds begin to flower
- Do not apply if turf is wet or if rainfall expected within 4 h of treatment. Both circumstances will reduce weed control

Restrictions
- Maximum number of treatments 2 per yr
- Do not mow for 3 d before or after treatment
- Avoid overlapping or overdosing, especially on newly sown turf
- Do not spray in drought conditions or if turf under stress from frost, waterlogging, trace element deficiency, pest or disease attack

Crop-specific information
- Latest use: normally Sep for managed amenity turf
- New turf may be treated in spring provided at least 2 mth have elapsed since sowing

Environmental safety
- Dangerous for the environment
- Toxic to aquatic organisms
- Keep livestock out of treated areas for at least two weeks following treatment and until poisonous weeds, such as ragwort, have died down and become unpalatable

Hazard classification and safety precautions
Hazard Harmful
Transport code 9
Packaging group III
UN number 3082

SEE SECTION 3 FOR PRODUCTS ALSO REGISTERED

Pesticide Profiles

Risk phrases R22a, R22b, R36, R38
Operator protection A, C, H, M; U05a, U08, U19a, U20b
Environmental protection E07a, E15a, E34
Storage and disposal D01, D02, D09a, D10a
Medical advice M03

143 2,4-D + dicamba + triclopyr

A translocated herbicide for perennial and woody weed control
HRAC mode of action code: O + O + O

See also dicamba
triclopyr

Products

Broadsword	United Phosphorus	200:85:65 g/l	EC	09140

Uses
- Annual dicotyledons in **land not intended to bear vegetation, permanent grassland, rotational grass**
- Brambles in **farm forestry, forest, natural surfaces not intended to bear vegetation**
- Docks in **land not intended to bear vegetation, permanent grassland, rotational grass**
- Gorse in **farm forestry, forest, natural surfaces not intended to bear vegetation**
- Green cover in **land temporarily removed from production**
- Japanese knotweed in **farm forestry, forest, natural surfaces not intended to bear vegetation**
- Perennial dicotyledons in **farm forestry, forest, land not intended to bear vegetation, natural surfaces not intended to bear vegetation, permanent grassland, rotational grass**
- Plantains in **land not intended to bear vegetation, permanent grassland, rotational grass**
- Rhododendrons in **farm forestry, forest, natural surfaces not intended to bear vegetation**
- Stinging nettle in **permanent grassland, rotational grass**
- Thistles in **land not intended to bear vegetation, permanent grassland, rotational grass**
- Woody weeds in **farm forestry, forest, natural surfaces not intended to bear vegetation**

Approval information
- 2,4-D, dicamba and triclopyr included in Annex I under EC Directive 91/414

Efficacy guidance
- Apply as foliar spray to herbaceous or woody weeds. Timing and growth stage for best results vary with species. See label for details
- Dilute with water for stump treatment and apply after felling up to the start of regrowth. Treat any regrowth with a spray to the growing foliage
- May be applied at 1/3 dilution in weed wipers or 1/8 dilution with ropewick applicators

Restrictions
- Maximum total dose per yr equivalent to two full dose treatments
- Do not use on pasture established less than 1 yr or on grass grown for seed
- Where clover a valued constituent of sward only use as a spot treatment
- Do not graze for 7 d or mow for 14 d after treatment
- Do not direct drill grass, clover or brassicas for at least 6 wk after grassland treatment
- Do not plant trees for 1-3 mth after spraying depending on dose applied. See label
- Avoid spray drift into greenhouses or onto crops or ornamentals. Vapour drift may occur in hot conditions

Crop-specific information
- Latest use: before weed flower buds open in grassland, non-cropped land, land temporarily removed from production; end of summer (spray); after felling up to the start of regrowth (cut stump spray) for farm forestry
- Sprays may be applied in pines, spruce and fir providing drift is avoided. Optimum time is mid-autumn when tree growth ceased but weeds not yet senescent

Environmental safety
- Dangerous for the environment
- Very toxic to aquatic organisms
- Flammable

FOR FULL CONDITIONS OF USE ALWAYS READ THE PRODUCT LABEL

2,4-D + dichlorprop-P + MCPA + mecoprop-P

- Docks and other weeds may become increasingly palatable after treatment and may be preferentially grazed
- Keep livestock out of treated areas for at least two weeks following treatment and until poisonous weeds, such as ragwort, have died down and become unpalatable
- Take extreme care to avoid drift onto neighbouring crops, especially beet crops, brassicas, most market garden crops including lettuce and tomatoes under glass, pears and vines

Hazard classification and safety precautions
Hazard Harmful, Flammable, Dangerous for the environment
Transport code 3
Packaging group III
UN number 3295
Risk phrases R22a, R22b, R36, R37, R38, R50, R53a, R67
Operator protection A, C, H, M; U02a, U04a, U05a, U08, U11, U13, U19a, U20a
Environmental protection E07a, E13b, E34
Consumer protection C01
Storage and disposal D01, D02, D05, D09a, D10b, D12a
Medical advice M03, M05b

144 2,4-D + dichlorprop-P + MCPA + mecoprop-P

A translocated herbicide for use in apple and pear orchards
HRAC mode of action code: O + O + O + O

See also *dichlorprop-P*
MCPA
mecoprop-P

Products
UPL Camppex   United Phosphorus   34.5:66.6:52.8:82.1 g/l SL   11661

Uses
- Annual dicotyledons in **almonds** *(off-label)*, **apple orchards**, **chestnuts** *(off-label)*, **hazel nuts** *(off-label)*, **pear orchards**, **walnuts** *(off-label)*
- Chickweed in **apple orchards**, **pear orchards**
- Cleavers in **apple orchards**, **pear orchards**
- Perennial dicotyledons in **apple orchards**, **pear orchards**

Specific Off-Label Approvals (SOLAs)
- *almonds* 20052244
- *chestnuts* 20052244
- *hazel nuts* 20052244
- *walnuts* 20052244

Approval information
- 2,4-D, dichlorprop-P, MCPA and mecoprop-P included in Annex I under EC Directive 91/414

Efficacy guidance
- Use on emerged weeds in established apple and pear orchards (from 1 yr after planting) as directed application
- Spray when weeds in active growth and at growth stage recommended on label
- Effectiveness may be reduced by rain within 12 h

Restrictions
- Maximum number of treatments 2 per yr
- Applications must be made around, and not directly to, trees. Do not allow drift onto trees
- Do not spray during blossom period. Do not spray to run-off
- Do not roll, harrow or cut grass crops on orchard floor within at least 3 d before or after spraying

Environmental safety
- Harmful to aquatic organisms
- Harmful to fish or other aquatic life. Do not contaminate surface waters or ditches with chemical or used container
- Keep livestock out of treated areas for at least two weeks following treatment and until poisonous weeds, such as ragwort, have died down and become unpalatable

SEE SECTION 3 FOR PRODUCTS ALSO REGISTERED

Pesticide Profiles

- Take extreme care to avoid drift onto neighbouring sensitive crops
- Some pesticides pose a greater threat of contamination of water than others and mecoprop-P is one of these pesticides. Take special care when applying mecoprop-P near water and do not apply if heavy rain is forecast

Hazard classification and safety precautions
Hazard Harmful
UN number N/C
Risk phrases R20, R21, R22a, R43, R52, R53a
Operator protection A, C, H, M; U05a, U08, U20b
Environmental protection E07a, E13c, E34
Storage and disposal D01, D02, D05, D09a, D10c
Medical advice M03, M05a

145   2,4-D + florasulam

A post-emergence herbicide mixture for control of broad leaved weeds in managed amenity turf
HRAC mode of action code: O + B

See also florasulam

Products
Junction   Rigby Taylor   452:6.25 g/l   ME   12493

Uses
- Clover in *managed amenity turf*
- Daisies in *managed amenity turf*
- Dandelions in *managed amenity turf*
- Plantains in *managed amenity turf*
- Sticky mouse-ear in *managed amenity turf*

Approval information
- 2,4-D and florasulam included in Annex I under EC Directive 91/414

Efficacy guidance
- Best results obtained from treatment of actively growing weeds between Mar and Oct when soil is moist
- Do not apply when rain is imminent or during periods of drought unless irrigation is applied

Restrictions
- Maximum number of treatments on managed amenity turf: 1 per yr

Crop-specific information
- Avoid mowing turf 3 d before and after spraying
- Ensure newly sown turf has established before treating. Turf sown in late summer or autumn should not be sprayed until growth is resumed in the following spring

Following crops guidance
- An interval of 4 wk must elapse between application and re-seeding turf

Environmental safety
- Dangerous for the environment
- Toxic to aquatic organisms

Hazard classification and safety precautions
Hazard Harmful, Dangerous for the environment
Transport code 9
Packaging group III
UN number 3082
Risk phrases R22a, R43, R51, R53a
Operator protection A, H; U02a, U05a, U08, U14, U20a
Environmental protection E15a, E34, E38
Storage and disposal D01, D02, D09a, D10b, D12a

FOR FULL CONDITIONS OF USE ALWAYS READ THE PRODUCT LABEL

146 2,4-D + MCPA

A translocated herbicide mixture for cereals and grass
HRAC mode of action code: O + O

See also MCPA

Products

1	Headland Polo	Headland	360:315 g/l	SL	14933
2	Lupo	Nufarm UK	360:315 g/l	SL	14931

Uses
- Annual dicotyledons in **grassland, spring barley, spring wheat, winter barley, winter oats, winter wheat** [1, 2]
- Perennial dicotyledons in **grassland** [2]; **spring barley, spring wheat, winter barley, winter oats, winter wheat** [1, 2]

Approval information
- 2,4-D and MCPA included in Annex I under EC Directive 91/414
- Accepted by BBPA for use on malting barley

Efficacy guidance
- Best results achieved by spraying weeds in seedling to young plant stage when growing actively in a strongly competing crop
- Most effective stage for spraying perennials varies with species. See label for details

Restrictions
- Maximum number of treatments 1 per crop in cereals, 2 per year in grassland and 3 per year in managed amenity turf [1]; 1 per crop in cereals and 1 per year in grassland [2]
- Do not spray if rain falling or imminent
- Do not cut grass or graze for at least 10 d after spraying
- Do not use on newly sown leys containing clover or other legumes
- Do not spray crops stressed by cold weather or drought or if frost expected
- Do not use shortly before or after sowing any crop
- Do not roll or harrow within 7 d before or after spraying

Crop-specific information
- Spray winter cereals in spring when leaf-sheath erect but before first node detectable (GS 31), spring cereals from 5-leaf stage to before first node detectable (GS 15-31)
- Latest use: before first node detectable (GS 31) in cereals;

Environmental safety
- Dangerous for the environment
- Toxic to aquatic organisms
- Keep livestock out of treated areas for at least two weeks following treatment and until poisonous weeds, such as ragwort, have died down and become unpalatable
- 2,4-D and MCPA are active at low concentrations. Take extreme care to avoid drift onto neighbouring crops, especially beet crops, brassicas, most market garden crops including lettuce and tomatoes under glass, pears and vines

Hazard classification and safety precautions
Hazard Harmful, Dangerous for the environment
Transport code 9
Packaging group III
UN number 3082
Risk phrases R20, R21, R22a, R41, R43, R53a [1, 2]; R50 [2]; R51 [1]
Operator protection A, C, H, M; U05a, U08, U11, U14, U20b [1, 2]; U15 [1]
Environmental protection E07a, E15a, E34 [1, 2]; E38 [2]
Storage and disposal D01, D02, D05, D09a, D10a [1, 2]; D12a [2]
Medical advice M03 [1, 2]; M05a [1]

SEE SECTION 3 FOR PRODUCTS ALSO REGISTERED

147 daminozide

A hydrazide plant growth regulator for use in certain ornamentals

Products

1	B-Nine SG	Certis	85% w/w	SG	14435
2	Dazide Enhance	Fine	85% w/w	SG	14433

Uses
- Growth regulation in **azaleas, bedding plants, calibrachoa, chrysanthemums, hortensia, kalanchoes, ornamental specimens, petunia, sunflowers** [2]; **protected ornamentals** [1]

Approval information
- Daminozide included in Annex I under EC Directive 91/414

Efficacy guidance
- Best results obtained by application in late afternoon when glasshouse has cooled down
- Apply a fine spray using compressed air or power sprayers to give good coverage of dry foliage without run off
- Response to treatment differs widely depending on variety, stage of growth and physiological condition of the plant. It is recommended that any new variety is first tested on a small scale to observe if adverse effects occur
- For plants grown in modules or for seedlings prior to pricking out apply at reduced dose on a 7 d regime, starting at cotyledon stage

Restrictions
- Maximum number of treatments 2 per crop
- Apply only to turgid, well watered plants. Do not water for 24 h after spraying
- Do not mix with other spray chemicals unless specifically recommended
- Do not store product in metal containers

Crop-specific information
- Do not use on chrysanthemum variety Fandango
- Evidence of effectiveness on Azaleas and Hydrangeas is limited
- See label for guidance on a range of bedding plant species

Hazard classification and safety precautions
UN number N/C
Operator protection A, H; U02a, U08, U19a, U20c [2]; U05a [1, 2]
Environmental protection E02a [1] (8 days); E15b [1, 2]
Storage and disposal D01, D02, D09a [1, 2]; D10b, D12a [2]
Medical advice M05a [2]

148 dazomet

A methyl isothiocyanate releasing soil fumigant
HRAC mode of action code: Z

Products

Basamid	Certis	97% w/w	GR	11324

Uses
- Nematodes in **field crops** (soil fumigation), **protected crops** (soil fumigation), **vegetables** (soil fumigation)
- Soil insects in **field crops** (soil fumigation), **protected crops** (soil fumigation), **vegetables** (soil fumigation)
- Soil-borne diseases in **field crops** (soil fumigation), **protected crops** (soil fumigation), **vegetables** (soil fumigation)
- Weed seeds in **field crops** (soil fumigation), **protected crops** (soil fumigation), **vegetables** (soil fumigation)

Efficacy guidance
- Dazomet acts by releasing methyl isothiocyanate in contact with moist soil
- Soil sterilization is carried out after harvesting one crop and before planting the next
- The soil must be of fine tilth, free of clods and evenly moist to the depth of sterilization

FOR FULL CONDITIONS OF USE ALWAYS READ THE PRODUCT LABEL

- Soil moisture must not be less than 50% of water-holding capacity or oversaturated. If too dry, water at least 7-14 d before treatment
- In order to obtain short treatment times it is recommended to treat soils when soil temperature is above 7 °C. Treatment should be used outdoors before winter rains make soil too wet to cultivate - usually early Nov
- For club root control treat only in summer when soil temperature above 10 °C
- Where onion white rot is a problem it is unlikely to give effective control where inoculum level high or crop under stress
- Apply granules with suitable applicators, mix into soil immediately to desired depth and seal surface with polythene sheeting, by flooding or heavy rolling. See label for suitable application and incorporation machinery
- With 'planting through' technique polythene seal is left in place to form mulch into which new crop can be planted

Restrictions
- Maximum number of treatments 1 per batch of soil for protected crops or field crop situation
- Do not treat ground where water table may rise into treated layer

Crop-specific information
- Latest use: pre-planting

Following crops guidance
- With 'planting through' technique no gas release cultivations are made and safety test with cress is particularly important. Conduct cress test on soil samples from centre as well as edges of bed. Observe minimum of 30 d from application to cress test in soils at 10 °C or above, at least 50 d in soils below 10 °C. See label for details
- In all other situations 14-28 d after treatment cultivate lightly to allow gas to disperse and conduct cress test after a further 14-28 d (timing depends on soil type and temperature). Do not treat structures containing live plants or any ground within 1 m of live plants

Environmental safety
- Dangerous for the environment
- Very toxic to aquatic organisms

Hazard classification and safety precautions
    Hazard Harmful, Dangerous for the environment
    Transport code 9
    Packaging group III
    UN number 3077
    Risk phrases R22a, R50, R53a
    Operator protection A, M; U04a, U05a, U09a, U19a, U20a
    Environmental protection E15a, E34, E38
    Storage and disposal D01, D02, D07, D09a, D11a, D12a
    Medical advice M03, M05a

149 2,4-DB

A translocated phenoxycarboxylic acid herbicide
HRAC mode of action code: O

Products

1	DB Straight	United Phosphorus	400 g/l	SL	13736
2	Headland Spruce	Headland	400 g/l	SL	15089

Uses
- Annual dicotyledons in **grassland, lucerne, red clover, spring barley, spring oats, spring wheat, white clover, winter barley, winter oats, winter wheat**
- Buttercups in **grassland**
- Fat hen in **grassland**
- Penny cress in **grassland**
- Shepherd's purse in **grassland**

Approval information
- 2,4-DB included in Annex I under EC Directive 91/414
- Accepted by BBPA for use on malting barley

Efficacy guidance
- Best results achieved on young seedling weeds under good growing conditions. Treatment less effective in cold weather and dry soil conditions
- Rain within 12 h may reduce effectiveness

Restrictions
- Do not allow spray drift onto neighbouring crops

Crop-specific information
- Latest use: before first node detectable (GS 31) for undersown cereals; fourth trifoliate leaf for lucerne
- In direct sown lucerne spray when seedlings have reached first trifoliate leaf stage. Optimum time 3-4 trifoliate leaves
- Do not treat any lucerne after fourth trifoliate leaf
- In spring barley and spring oats undersown with lucerne spray from when cereal has 1 leaf unfolded and lucerne has first trifoliate leaf
- In spring wheat undersown with lucerne spray from when cereal has 3 leaves unfolded and lucerne has first trifoliate leaf

Following crops guidance
- Do not sow any crop into soil treated with 2,4-DB for at least 3 months after application.

Environmental safety
- Dangerous for the environment
- Toxic to aquatic organisms
- Keep livestock out of treated areas for at least two weeks following treatment and until poisonous weeds, such as ragwort, have died down and become unpalatable

Hazard classification and safety precautions
Hazard Harmful, Dangerous for the environment
Transport code 9
Packaging group III
UN number 3082
Risk phrases R22a, R41, R43, R51, R53a
Operator protection A, C, H; U05a, U09a, U11, U20a
Environmental protection E07a, E15a, E15b, E34, E38
Storage and disposal D01, D02, D05, D10c, D12a
Medical advice M03, M05a

150   2,4-DB + linuron + MCPA

A translocated herbicide for undersown cereals and grass
HRAC mode of action code: O + C2 + O

See also linuron
MCPA

Products
Alistell	United Phosphorus	220:30:30 g/l	EC	11053

Uses
- Annual dicotyledons in **rotational grass**, **undersown spring cereals** (undersown with clover), **undersown winter cereals** (undersown with clover)

Approval information
- 2,4-DB, linuron and MCPA included in Annex I under EC Directive 91/414
- Accepted by BBPA for use on malting barley

Efficacy guidance
- Best results achieved on young seedling weeds growing actively in warm, moist weather
- May be applied at any time of year provided crop at correct stage and weather suitable
- Avoid spraying if rain falling or imminent

FOR FULL CONDITIONS OF USE ALWAYS READ THE PRODUCT LABEL

2,4-DB + MCPA

Restrictions
- Maximum number of treatments 1 per crop or yr
- Spray winter cereals when fully tillered but before first node detectable (GS 29-30)
- Spray spring wheat from 5-fully expanded leaf stage (GS 15), barley and oats from 2-fully expanded leaves (GS 12)
- Do not spray cereals undersown with lucerne, peas or beans
- Apply to clovers after 1-trifoliate leaf, to grasses after 2-fully expanded leaf stage
- Do not spray in conditions of drought, waterlogging or extremes of temperature
- In frosty weather clover leaf scorch may occur but damage normally outgrown
- Do not use on sand or soils with more than 10% organic matter
- Do not roll or harrow within 7 d before or after spraying
- Avoid drift of spray or vapour onto susceptible crops
- Do not apply by hand-held sprayers

Crop-specific information
- Latest use: before first node detectable (GS 31) for undersown cereals; 2 wk before grazing for grassland

Environmental safety
- Dangerous for the environment
- Toxic to aquatic organisms
- Keep livestock out of treated areas for at least two weeks following treatment and until poisonous weeds, such as ragwort, have died down and become unpalatable
- LERAP Category B

Hazard classification and safety precautions
Hazard Harmful, Dangerous for the environment
Transport code 9
Packaging group III
UN number 3082
Risk phrases R22a, R22b, R40, R41, R51, R53a, R66
Operator protection A, C, H; U05a, U08, U11, U13, U19a, U20b
Environmental protection E07a, E13b, E16a, E34, E38
Storage and disposal D01, D02, D09a, D10c, D12a
Medical advice M03, M05a

151 2,4-DB + MCPA

A translocated herbicide for cereals, clovers and leys
HRAC mode of action code: O + O

See also MCPA

Products

1	Agrichem DB Plus	Agrichem	243:40 g/l	SL	00044
2	Redlegor	United Phosphorus	244:44 g/l	SL	07519

Uses
- Annual dicotyledons in *all cereals* [1]; *rotational grass, undersown spring cereals, undersown winter cereals* [1, 2]; *spring wheat, winter wheat* [2]
- Perennial dicotyledons in *all cereals* [1]; *rotational grass, undersown spring cereals, undersown winter cereals* [1, 2]; *spring wheat, winter wheat* [2]
- Polygonums in *rotational grass, spring wheat, undersown spring cereals, undersown winter cereals, winter wheat* [2]

Approval information
- 2,4-DB and MCPA included in Annex I under EC Directive 91/414
- Accepted by BBPA for use on malting barley
- Approval expiry 31 Oct 2011 [1, 2]

Efficacy guidance
- Best results achieved on young seedling weeds under good growing conditions
- Spray thistles and other perennials when 10-20 cm high provided clover at correct stage
- Effectiveness may be reduced by rain within 12 h, by very cold conditions or drought

SEE SECTION 3 FOR PRODUCTS ALSO REGISTERED

Pesticide Profiles

Restrictions
- Maximum number of treatments 1 per crop
- Do not spray established clover crops or lucerne
- Do not roll or harrow within 7 d before or after spraying
- Do not spray immediately before or after sowing any crop
- Avoid drift onto neighbouring sensitive crops

Crop-specific information
- Latest use: before first node detectable (GS 31) for cereals; 4th trifoliate leaf stage of clover for grass re-seeds
- Apply in spring to winter cereals from leaf sheath erect stage, to spring barley or oats from 2-leaf stage (GS 12), to spring wheat from 5-leaf stage (GS 15)
- Spray clovers as soon as possible after first trifoliate leaf, grasses after 2-3 leaf stage
- Red clover may suffer temporary distortion after treatment

Environmental safety
- Harmful to aquatic organisms
- Harmful to fish or other aquatic life. Do not contaminate surface waters or ditches with chemical or used container
- Keep livestock out of treated areas for at least two weeks following treatment and until poisonous weeds, such as ragwort, have died down and become unpalatable

Hazard classification and safety precautions
  Hazard Harmful
  UN number N/C
  Risk phrases R22a, R41, R52, R53a [1, 2]; R38 [2]
  Operator protection A, C [1]; U05a, U08, U11, U20b [1, 2]; U13, U14, U15 [2]
  Environmental protection E07a, E13c [1, 2]; E19b, E34 [1]
  Storage and disposal D01, D02, D05, D09a, D10c
  Medical advice M03, M05a

152 deltamethrin

A pyrethroid insecticide with contact and residual activity
IRAC mode of action code: 3

Products

1	Bandu	Headland	25 g/l	EC	10994
2	Decis	Bayer CropScience	25 g/l	EC	07172
3	Decis Protech	Bayer CropScience	15 g/l	EW	11502
4	Delta-M 2.5 EC	AgChem Access	25 g/l	EC	12604
5	Grainstore	Pan Agriculture	25 g/l	EC	14932
6	K-Obiol EC 25	Lodi UK	25 g/l	EC	13573
7	K-Obiol ULV 6	Lodi UK	0.69% w/v	UL	13572

Uses
- Aphids in **amenity vegetation**, **apples**, **ornamental plant production**, **protected cucumbers**, **protected peppers**, **protected pot plants**, **protected tomatoes**, **spring barley**, **spring oats**, **spring wheat**, **winter barley**, **winter oats**, **winter wheat** [1-4]; **blackberries** (off-label), **fodder beet** (off-label), **frise** (off-label), **herbs (see appendix 6)** (off-label), **lamb's lettuce** (off-label), **protected aubergines** (off-label), **scarole** (off-label), **spinach** (off-label) [1, 2]; **bulb onions** (off-label), **garlic** (off-label), **leaf brassicas** (off-label), **leeks** (off-label), **protected choi sum** (off-label), **protected cress** (off-label), **protected frise** (off-label), **protected herbs (see appendix 6)** (off-label), **protected lamb's lettuce** (off-label), **protected lettuce** (off-label), **protected ornamentals**, **protected pak choi** (off-label), **protected radicchio** (off-label), **protected scarole** (off-label), **protected spinach** (off-label), **rocket** (off-label), **shallots** (off-label) [3]; **carrots** (off-label), **celery (outdoor)** (off-label), **horseradish** (off-label), **lettuce**, **mallow (althaea spp.)** (off-label), **parsnips** (off-label), **protected celery** (off-label), **protected chinese cabbage** (off-label), **protected rhubarb** (off-label), **protected spring onions** (off-label), **rhubarb** (off-label), **spring onions** (off-label) [1, 3]; **chinese cabbage** (off-label), **durum wheat** (off-label), **spring rye** (off-label), **triticale** (off-label), **winter rye** (off-label) [1-3]; **choi sum** (off-label), **collards** (off-label), **lettuce** (off-label), **pak choi** (off-label), **spring greens** (off-label) [2]; **evening primrose** (off-label), **grass seed crops** (off-label), **marjoram** (off-label), **radicchio** (off-label), **sorrel** (off-label) [2, 3];

FOR FULL CONDITIONS OF USE ALWAYS READ THE PRODUCT LABEL

deltamethrin

lupins (off-label), *parsley root* (off-label), *protected cayenne peppers* (off-label) [1]; *nursery stock* [2, 4]
- Apple sucker in *apples* [1-4]
- Barley yellow dwarf virus vectors in *winter barley*, *winter wheat* [1-4]
- Beet virus yellows vectors in *winter oilseed rape* [1-4]
- Cabbage seed weevil in *mustard*, *spring oilseed rape*, *winter oilseed rape* [1-4]
- Cabbage stem flea beetle in *winter oilseed rape* [1-4]
- Cabbage stem weevil in *mustard*, *spring oilseed rape*, *winter oilseed rape* [1-4]
- Capsids in *amenity vegetation*, *apples*, *ornamental plant production* [1-4]; *nursery stock* [2, 4]
- Caterpillars in *amenity vegetation*, *apples*, *broccoli*, *brussels sprouts*, *cabbages*, *cauliflowers*, *kale*, *ornamental plant production*, *plums*, *protected cucumbers*, *protected peppers*, *protected pot plants*, *protected tomatoes*, *swedes*, *turnips* [1-4]; *black cabbage* (off-label), *borecole* (off-label), *choi sum* (off-label), *collards* (off-label), *mustard*, *orache* (off-label), *pak choi* (off-label), *spring greens* (off-label), *tatsoi* (off-label) [1]; *bulb onions* (off-label), *carrots* (off-label), *celery (outdoor)* (off-label), *chinese cabbage* (off-label), *garlic* (off-label), *horseradish* (off-label), *leeks* (off-label), *lettuce*, *mallow (althaea spp.)* (off-label), *parsnips* (off-label), *protected celery* (off-label), *protected chinese cabbage* (off-label), *protected choi sum* (off-label), *protected cress* (off-label), *protected frise* (off-label), *protected herbs (see appendix 6)* (off-label), *protected lamb's lettuce* (off-label), *protected lettuce* (off-label), *protected ornamentals*, *protected pak choi* (off-label), *protected radicchio* (off-label), *protected rhubarb* (off-label), *protected scarole* (off-label), *protected spinach* (off-label), *protected spring onions* (off-label), *rhubarb* (off-label), *rocket* (off-label), *shallots* (off-label), *spring onions* (off-label) [3]; *evening primrose* (off-label), *grass seed crops* (off-label), *marjoram* (off-label), *radicchio* (off-label), *sorrel* (off-label) [2, 3]; *leaf brassicas* (off-label) [1, 3]; *leaf brassicas* (off-label - baby leaf production), *protected brassica seedlings* (off-label), *protected rocket* (off-label), *rocket* (off-label - baby leaf production) [2]; *nursery stock* [2, 4]
- Codling moth in *apples* [1-4]
- Cutworms in *lettuce* [1, 3]
- Damson-hop aphid in *hops*, *plums* [1-4]
- Flea beetle in *black cabbage* (off-label), *borecole* (off-label), *choi sum* (off-label), *collards* (off-label), *mustard*, *orache* (off-label), *pak choi* (off-label), *spring greens* (off-label), *tatsoi* (off-label) [1]; *broccoli*, *brussels sprouts*, *cabbages*, *cauliflowers*, *kale*, *swedes*, *turnips* [2, 4]; *bulb onions* (off-label), *carrots* (off-label), *celery (outdoor)* (off-label), *chinese cabbage* (off-label), *garlic* (off-label), *horseradish* (off-label), *leeks* (off-label), *lettuce*, *mallow (althaea spp.)* (off-label), *parsnips* (off-label), *protected celery* (off-label), *protected chinese cabbage* (off-label), *protected choi sum* (off-label), *protected pak choi* (off-label), *protected rhubarb* (off-label), *protected spring onions* (off-label), *rhubarb* (off-label), *rocket* (off-label), *shallots* (off-label), *spring onions* (off-label) [3]; *evening primrose* (off-label), *protected frise* (off-label), *radicchio* (off-label), *sorrel* (off-label) [2, 3]; *leaf brassicas* (off-label), *protected scarole* (off-label) [1, 3]; *leaf brassicas* (off-label - baby leaf production), *protected brassica seedlings* (off-label), *protected rocket* (off-label), *rocket* (off-label - baby leaf production) [2]; *protected cress* (off-label), *protected herbs (see appendix 6)* (off-label), *protected lamb's lettuce* (off-label), *protected lettuce* (off-label), *protected radicchio* (off-label), *protected spinach* (off-label) [1-3]; *sugar beet* [1-4]
- Fruit tree tortrix moth in *plums* [1-4]
- Insect pests in *blackberries* (off-label), *frise* (off-label), *herbs (see appendix 6)* (off-label), *lamb's lettuce* (off-label), *lupins* (off-label), *protected aubergines* (off-label), *scarole* (off-label), *spinach* (off-label) [1-3]; *bulb onions* (off-label), *choi sum* (off-label), *collards* (off-label), *garlic* (off-label), *pak choi* (off-label), *shallots* (off-label), *spring greens* (off-label) [2]; *carrots* (off-label), *celery (outdoor)* (off-label), *fodder beet* (off-label), *horseradish* (off-label), *mallow (althaea spp.)* (off-label), *parsley root* (off-label), *parsnips* (off-label), *protected celery* (off-label), *protected chinese cabbage* (off-label), *protected rhubarb* (off-label), *protected spring onions* (off-label), *rhubarb* (off-label), *spring onions* (off-label) [1, 2]; *chinese cabbage* (off-label) [1]; *durum wheat* (off-label), *protected cayenne peppers* (off-label), *spring rye* (off-label), *triticale* (off-label), *winter rye* (off-label) [1, 3]; *grain stores* [5, 6]; *protected herbs (see appendix 6)* (off-label) [3]; *stored grain*, *stored pulses* [5-7]; *stored grain* (off-label) [7]
- Leafhoppers in *bulb onions* (off-label), *carrots* (off-label), *celery (outdoor)* (off-label), *chinese cabbage* (off-label), *garlic* (off-label), *horseradish* (off-label), *leaf brassicas* (off-label), *leeks*

SEE SECTION 3 FOR PRODUCTS ALSO REGISTERED

Pesticide Profiles

(off-label), **lettuce**, **mallow (althaea spp.)** (off-label), **parsnips** (off-label), **protected celery** (off-label), **protected chinese cabbage** (off-label), **protected choi sum** (off-label), **protected pak choi** (off-label), **protected rhubarb** (off-label), **protected spring onions** (off-label), **rhubarb** (off-label), **rocket** (off-label), **shallots** (off-label), **spring onions** (off-label) [3]; **evening primrose** (off-label), **grass seed crops** (off-label), **marjoram** (off-label), **protected frise** (off-label), **radicchio** (off-label), **sorrel** (off-label) [2, 3]; **protected cress** (off-label), **protected herbs (see appendix 6)** (off-label), **protected lamb's lettuce** (off-label), **protected lettuce** (off-label), **protected radicchio** (off-label), **protected spinach** (off-label) [1-3]; **protected scarole** (off-label) [1, 3]
- Mealybugs in **amenity vegetation**, **ornamental plant production**, **protected cucumbers**, **protected peppers**, **protected pot plants**, **protected tomatoes** [1-4]; **nursery stock** [2, 4]; **protected ornamentals** [3]
- Pea and bean weevil in **broad beans**, **peas**, **spring field beans**, **winter field beans** [1-4]
- Pea midge in **peas** [2-4]
- Pea moth in **peas** [1-4]
- Pear sucker in **pears** [1-4]
- Phorid flies in **protected edible fungi** (off-label) [1, 3]; **protected mushrooms** (off-label) [3]
- Pollen beetle in **bulb onions** (off-label), **carrots** (off-label), **celery (outdoor)** (off-label), **chinese cabbage** (off-label), **garlic** (off-label), **horseradish** (off-label), **leaf brassicas** (off-label), **leeks** (off-label), **lettuce**, **mallow (althaea spp.)** (off-label), **parsnips** (off-label), **protected celery** (off-label), **protected chinese cabbage** (off-label), **protected choi sum** (off-label), **protected cress** (off-label), **protected frise** (off-label), **protected herbs (see appendix 6)** (off-label), **protected lamb's lettuce** (off-label), **protected lettuce** (off-label), **protected pak choi** (off-label), **protected radicchio** (off-label), **protected rhubarb** (off-label), **protected scarole** (off-label), **protected spinach** (off-label), **protected spring onions** (off-label), **rhubarb** (off-label), **rocket** (off-label), **shallots** (off-label), **spring onions** (off-label) [3]; **evening primrose** (off-label), **grass seed crops** (off-label), **marjoram** (off-label), **radicchio** (off-label), **sorrel** (off-label) [2, 3]; **mustard**, **spring oilseed rape**, **winter oilseed rape** [1-4]
- Raspberry beetle in **raspberries** [1-4]
- Sawflies in **apples**, **plums** [1-4]
- Scale insects in **amenity vegetation**, **ornamental plant production**, **protected cucumbers**, **protected peppers**, **protected pot plants**, **protected tomatoes** [1-4]; **nursery stock** [2, 4]; **protected ornamentals** [3]
- Sciarid flies in **protected edible fungi** (off-label) [1-3]; **protected mushrooms** (off-label) [2, 3]
- Thrips in **amenity vegetation**, **ornamental plant production** [1-4]; **bulb onions** (off-label), **garlic** (off-label), **shallots** (off-label) [1, 3]; **carrots** (off-label), **celery (outdoor)** (off-label), **chinese cabbage** (off-label), **evening primrose** (off-label), **grass seed crops** (off-label), **horseradish** (off-label), **leaf brassicas** (off-label), **lettuce**, **mallow (althaea spp.)** (off-label), **marjoram** (off-label), **parsnips** (off-label), **protected celery** (off-label), **protected chinese cabbage** (off-label), **protected choi sum** (off-label), **protected cress** (off-label), **protected frise** (off-label), **protected herbs (see appendix 6)** (off-label), **protected lamb's lettuce** (off-label), **protected lettuce** (off-label), **protected ornamentals**, **protected pak choi** (off-label), **protected radicchio** (off-label), **protected rhubarb** (off-label), **protected scarole** (off-label), **protected spinach** (off-label), **protected spring onions** (off-label), **radicchio** (off-label), **rhubarb** (off-label), **rocket** (off-label), **sorrel** (off-label), **spring onions** (off-label) [3]; **leeks** (off-label), **protected courgettes** (off-label), **protected cucumbers**, **protected gherkins** (off-label), **protected peppers**, **protected tomatoes** [1-3]; **nursery stock** [2, 4]; **protected aubergines** (off-label), **protected cayenne peppers** (off-label), **protected ornamentals** (off-label) [2]; **protected non-edible flowers** (off-label) [1]
- Tortrix moths in **apples** [1-4]
- Western flower thrips in **protected aubergines** (off-label), **protected cayenne peppers** (off-label), **protected ornamentals** (off-label) [2]; **protected courgettes** (off-label), **protected cucumbers**, **protected gherkins** (off-label), **protected peppers**, **protected tomatoes** [1-3]; **protected non-edible flowers** (off-label) [1]; **protected ornamentals** [3]
- Whitefly in **amenity vegetation**, **ornamental plant production**, **protected cucumbers**, **protected peppers**, **protected pot plants**, **protected tomatoes** [1-4]; **nursery stock** [2, 4]; **protected ornamentals** [3]

Specific Off-Label Approvals (SOLAs)
- **black cabbage** *20071582* [1]
- **blackberries** *20071605* [1], *20071705* [2], *20071654* [3]

FOR FULL CONDITIONS OF USE ALWAYS READ THE PRODUCT LABEL

deltamethrin

- **borecole** *20071582* [1]
- **bulb onions** *20071613* [1], *20071698* [2], *20071158* [3]
- **carrots** *20071610* [1], *20071696* [2], *20071158* [3]
- **celery (outdoor)** *20071579* [1], *20071697* [2], *20071158* [3]
- **chinese cabbage** *20071606* [1], *20071612* [1], *20071695* [2], *20071158* [3]
- **choi sum** *20071582* [1], *20071704* [2]
- **collards** *20071582* [1], *20082458* [2]
- **durum wheat** *20071603* [1], *20071701* [2], *20071656* [3]
- **evening primrose** *20071694* [2], *20071158* [3]
- **fodder beet** *20071609* [1], *20071702* [2]
- **frise** *20071604* [1], *20071703* [2], *20071655* [3]
- **garlic** *20071613* [1], *20071698* [2], *20071158* [3]
- **grass seed crops** *20071694* [2], *20071158* [3]
- **herbs (see appendix 6)** *20071604* [1], *20071703* [2], *20071655* [3]
- **horseradish** *20071610* [1], *20071696* [2], *20071158* [3]
- **lamb's lettuce** *20071604* [1], *20071703* [2], *20071655* [3]
- **leaf brassicas** *20071582* [1], *(baby leaf production) 20071708* [2], *20071158* [3]
- **leeks** *20071614* [1], *20071699* [2], *20071158* [3]
- **lettuce** *20071695* [2]
- **lupins** *20071583* [1], *20071700* [2], *20071653* [3]
- **mallow (althaea spp.)** *20071610* [1], *20071696* [2], *20071158* [3]
- **marjoram** *20071694* [2], *20071158* [3]
- **orache** *20071582* [1]
- **pak choi** *20071582* [1], *20082458* [2]
- **parsley root** *20071610* [1], *20071696* [2]
- **parsnips** *20071610* [1], *20071696* [2], *20071158* [3]
- **protected aubergines** *20071608* [1], *20071693* [2], *20071707* [2], *20071649* [3]
- **protected brassica seedlings** *20071708* [2]
- **protected cayenne peppers** *20071607* [1], *20071693* [2], *20071657* [3]
- **protected celery** *20071579* [1], *20071697* [2], *20071158* [3]
- **protected chinese cabbage** *20071579* [1], *20071697* [2], *20071158* [3]
- **protected choi sum** *20071158* [3]
- **protected courgettes** *20071611* [1], *20071693* [2], *20071650* [3]
- **protected cress** *20080175* [1], *20071691* [2], *20071709* [2], *20071158* [3]
- **protected edible fungi** *20071581* [1], *20071692* [2], *20071648* [3]
- **protected frise** *20071691* [2], *20071158* [3]
- **protected gherkins** *20071611* [1], *20071693* [2], *20071650* [3]
- **protected herbs (see appendix 6)** *20080175* [1], *20071691* [2], *20071709* [2], *20071158* [3], *20071652* [3]
- **protected lamb's lettuce** *20080175* [1], *20071691* [2], *20071158* [3]
- **protected lettuce** *20080175* [1], *20071691* [2], *20071158* [3]
- **protected mushrooms** *20071692* [2], *20071648* [3]
- **protected non-edible flowers** *20071611* [1]
- **protected ornamentals** *20071693* [2]
- **protected pak choi** *20071158* [3]
- **protected radicchio** *20080175* [1], *20071691* [2], *20071158* [3]
- **protected rhubarb** *20071579* [1], *20071697* [2], *20071158* [3]
- **protected rocket** *20071708* [2]
- **protected scarole** *20080175* [1], *20071158* [3]
- **protected spinach** *20080175* [1], *20071691* [2], *20071158* [3]
- **protected spring onions** *20071579* [1], *20071697* [2], *20071158* [3]
- **radicchio** *20071694* [2], *20071158* [3]
- **rhubarb** *20071579* [1], *20071697* [2], *20071158* [3]
- **rocket** *(baby leaf production) 20071708* [2], *20071158* [3]
- **scarole** *20071604* [1], *20071703* [2], *20071655* [3]
- **shallots** *20071613* [1], *20071698* [2], *20071158* [3]
- **sorrel** *20071694* [2], *20071158* [3]
- **spinach** *20071604* [1], *20071703* [2], *20071655* [3]
- **spring greens** *20071582* [1], *20082458* [2]

SEE SECTION 3 FOR PRODUCTS ALSO REGISTERED

- **spring onions** *20071579* [1], *20071697* [2], *20071158* [3]
- **spring rye** *20071603* [1], *20071701* [2], *20071656* [3]
- **stored grain** *20091011* [7]
- **tatsoi** *20071582* [1]
- **triticale** *20071603* [1], *20071701* [2], *20071656* [3]
- **winter rye** *20071603* [1], *20071701* [2], *20071656* [3]

Approval information
- Deltamethrin included in Annex I under EC Directive 91/414
- Accepted by BBPA for use on malting barley and hops

Efficacy guidance
- A contact and stomach poison with 3-4 wk persistence, particularly effective on caterpillars and sucking insects
- Normally applied at first signs of damage with follow-up treatments where necessary at 10-14 d intervals. Rates, timing and recommended combinations with other pesticides vary with crop and pest. See label for details
- Spray is rainfast within 1 h
- May be applied in frosty weather provided foliage not covered in ice
- Temperatures above 35 °C may reduce effectiveness or persistence

Restrictions
- Maximum number of treatments varies with crop and pest, 4 per crop for wheat and barley, only 1 application between 1 Apr and 31 Aug. See label for other crops
- Do not apply more than 1 aphicide treatment to cereals in summer
- Do not spray crops suffering from drought or other physical stress
- Consult processer before treating crops for processing
- Do not apply to a cereal crop if any product containing a pyrethroid insecticide or dimethoate has been applied to that crop after the start of ear emergence (GS 51)
- Do not spray cereals after 31 Mar in the year of harvest within 6 m of the outside edge of the crop
- Reduced volume spraying must not be used on cereals after 31 Mar in yr of harvest

Crop-specific information
- Latest use: early dough (GS 83) for barley, oats, wheat; before flowering for mustard, oilseed rape; before 31 Mar for grass seed crops [1-3]

Environmental safety
- Dangerous for the environment
- Toxic to aquatic organisms
- Flammable [1, 2]
- Extremely dangerous to fish or other aquatic life. Do not contaminate surface waters or ditches with chemical or used container
- High risk to bees. Do not apply to crops in flower or to those in which bees are actively foraging. Do not apply when flowering weeds are present
- Do not apply in tank-mixture with a triazole-containing fungicide when bees are likely to be actively foraging in the crop
- High risk to non-target insects or other arthropods. Do not spray within 6 m of the field boundary
- Broadcast air-assisted LERAP [2-5] (18 m); LERAP Category A [1-5]

Hazard classification and safety precautions
Hazard Harmful, Flammable [1, 2, 4, 5]; Dangerous for the environment [1-5]
Transport code 3 [1-5]; 9 [6, 7]
Packaging group III
UN number 1993 [1-5]; 3082 [6, 7]
Risk phrases R20, R22a, R22b, R38, R41, R51 [1, 2, 4, 5]; R50 [3]; R53a [1-5]
Operator protection A, C, H [1-7]; D, M [5-7]; U04a, U05a, U19a [1, 2, 4-7]; U08 [1, 2, 4, 5]; U09a, U20c [6, 7]; U10 [1]; U11, U14, U15 [3]; U20b [1-5]
Environmental protection E12a [2-5]; E12c [1]; E12f [1] (cereals, oilseed rape, peas, beans); E12f [2-5] (cereals, oilseed rape, peas, beans - see label for guidance); E15a, E16c, E16d, E38 [1-5]; E15b [6, 7]; E17b [2-5] (18 m); E22a [3]; E34 [1, 2, 4-7]
Storage and disposal D01, D02, D09a, D10c [1-7]; D05 [1, 2, 4, 5]; D12a [2-5]
Medical advice M03 [1, 2, 4-7]; M05b [1, 2, 4, 5]

FOR FULL CONDITIONS OF USE ALWAYS READ THE PRODUCT LABEL

153 desmedipham

A contact phenyl carbamate herbicide available only in mixtures
HRAC mode of action code: C1

154 desmedipham + ethofumesate + lenacil + phenmedipham

A selective contact and residual herbicide mixture for weed control in beet
HRAC mode of action code: C1 + N + C1 + C1

Products
Betanal MaxxPro   Bayer CropScience   47:75:27:60 g/l   OD   15086

Uses
- Annual dicotyledons in **fodder beet**, **mangels**, **sugar beet**
- Annual meadow grass in **fodder beet**, **mangels**, **sugar beet**

Approval information
- Desmedipham, ethofumesate, lenacil and phenmedipham included in Annex I under EC Directive 91/414

Efficacy guidance
- On soils with more than 5% organic matter content, residual activity may be reduced.

Restrictions
- The maximum total dose must not exceed 1.0 kg/ha ethofumesate in any 3 year period.

Following crops guidance
- Beet crops may be sown at any time after the use of Betanal MaxxPro. Any other crop may be sown 3 months after using Betanal MaxxPro. Ploughing (mould board) to a minimum depth of 15 cm should precede preparation of a new seed bed.
- If a crop is suffering from manganese deficiency it may be checked. To avoid crop check, manganese should ideally be applied to the crop first.
- Crops suffering from lime deficiency may also be checked. Growers should ensure that the lime status of the soil is satisfactory before drilling.
- When the temperature is, or is likely to be, above 21 °C (70 °F) on the day of spraying, application should be made after 5 pm otherwise crop check may occur. If crops are subjected to substantial day to night temperature changes shortly before or after spraying, a check may occur from which the crop may not fully recover.

Hazard classification and safety precautions
Hazard Irritant, Dangerous for the environment
Transport code 9
Packaging group III
UN number 3082
Risk phrases R41, R43, R50, R53a
Operator protection A, C, H; U05a, U11, U14, U20b
Environmental protection E15b, E38, E39
Storage and disposal D09a, D10b, D12a
Medical advice M03

155 desmedipham + ethofumesate + metamitron + phenmedipham

A contact and residual herbicide mixture for beet
HRAC mode of action code: C1 + N + C1 + C1

See also ethofumesate
metamitron
phenmedipham

Products
Betanal Quattro   Bayer CropScience   20;100;200;60 g/l   SE   12779

SEE SECTION 3 FOR PRODUCTS ALSO REGISTERED

Uses
- Annual dicotyledons in **sugar beet**
- Annual meadow grass in **sugar beet**

Approval information
- Desmedipham, ethofumesate and phenmedipham included in Annex I under EC Directive 91/414

Restrictions
- Avoid spraying within 6 m of the field boundary to reduce the effects on non-target insects or arthropods.
- The maximum total dose must not exceed 1.0 kg ethofumesate per hectare in any three year period

Crop-specific information
- A check to crop growth can occur if application is made to crops suffering stress from application of other treatments, manganese or lime deficiency, high temperatures or high light intensity, frost, wind or insect attack.

Following crops guidance
- Only sow sugar beet, fodder beet or mangels within 4 months of application. Ploughing to 15 cm followed by thorough cultivation is recommended before planting any crop. Winter cereals can be sown in the same season provided 16 weeks have elapsed since the last application. Any spring crop may be planted in the following year.

Hazard classification and safety precautions
Hazard Dangerous for the environment
Transport code 9
Packaging group III
UN number 3082
Risk phrases R51, R53a
Operator protection A; U08, U20a
Environmental protection E15b, E38, E40
Storage and disposal D05, D09a, D10b
Medical advice M03

156 desmedipham + ethofumesate + phenmedipham

A selective contact and residual herbicide for beet
HRAC mode of action code: C1 + N + C1

See also ethofumesate
phenmedipham

Products

1	Betanal Expert	Bayer CropScience	25:151:75 g/l	EC	14034
2	Betasana Trio SC	United Phosphorus	15:115:75 g/l	SC	14262
3	Trilogy	United Phosphorus	15:115:75 g/l	SC	14464

Uses
- Annual dicotyledons in **fodder beet**, **mangels** [2, 3]; **game cover** *(off-label)*, **ornamental plant production** *(off-label)* [1]; **sugar beet** [1-3]
- Annual meadow grass in **sugar beet** [1]

Specific Off-Label Approvals (SOLAs)
- *game cover* 20082823 [1]
- *ornamental plant production* 20082823 [1]

Approval information
- Desmedipham, ethofumesate and phenmedipham included in Annex I under EC Directive 91/414

Efficacy guidance
- Product recommended for low-volume overall application in a planned spray programme
- Product acts mainly by contact action. A full programme also gives some residual control but this may be reduced on soils with more than 5% organic matter
- Best results achieved from treatments applied at fully expanded cotyledon stage of largest weeds present. Occasional larger weeds will usually be controlled from a full programme of sprays

FOR FULL CONDITIONS OF USE ALWAYS READ THE PRODUCT LABEL

- Where a pre-emergence band spray has been applied treatment must be timed according to size of the untreated weeds between the rows
- Susceptible weeds may not all be killed by the first spray. Repeat applications as each flush of weeds reaches cotyledon size normally necessary for season long control
- Sequential treatments should be applied when the previous one is still showing an effect on the weeds
- Various mixtures with other beet herbicides are recommended. See label for details

Restrictions
- Maximum total dose 4.5 l/ha [1], 7.0 l/ha [2, 3]
- Do not spray crops stressed by nutrient deficiency, wind damage, pest or disease attack, or previous herbicide treatments. Stressed crops treated under conditions of high light intensity may be checked and not recover fully
- If temperature likely to exceed 21 °C spray after 5 pm
- Crystallisation may occur if spray volume exceeds that recommended or spray mixture not used within 2 h, especially if the water temperature is below 5 °C
- Before use, wash out sprayer to remove all traces of previous products, especially hormone and sulfonyl urea weedkillers.
- The maximum total dose must not exceed 1.0 kg ethofumesate per hectare in any three year period

Crop-specific information
- Latest use: before crop meets between rows
- Apply first treatment when majority of crop plants have reached the fully expanded cotyledon stage
- Frost within 7 d of treatment may cause check from which the crop may not recover

Following crops guidance
- Beet crops may be sown at any time after treatment. Any other crop may be sown 3 mth after treatment following mouldboard ploughing to 15 cm minimum

Environmental safety
- Dangerous for the environment
- Toxic to aquatic organisms

Hazard classification and safety precautions
Hazard Irritant [2, 3]; Dangerous for the environment [1-3]
Transport code 9
Packaging group III
UN number 3082
Risk phrases R36, R50 [2, 3]; R51 [1]; R53a [1-3]
Operator protection A, H [1]; U08 [1-3]; U20a [1]
Environmental protection E15a [1]; E15b [2, 3]; E38 [1-3]
Storage and disposal D05, D09a, D12a [1-3]; D10b [1]; D10c [2, 3]
Treated seed S06a [2, 3]
Medical advice M03

157   desmedipham + phenmedipham

A mixture of contact herbicides for use in sugar beet
HRAC mode of action code: C1 + C1

See also phenmedipham

Products
Betanal Maxxim   Bayer CropScience   160;160 g/l   EC   14186

Uses
- Annual dicotyledons in **sugar beet**

Approval information
- Desmedipham and phenmedipham included in Annex I under EC Directive 91/414

SEE SECTION 3 FOR PRODUCTS ALSO REGISTERED

Pesticide Profiles

Efficacy guidance
- Product is recommended for low volume overall spraying in a planned programme involving pre- and/or post-emergence treatments at doses recommended for low volume programmes
- Best results obtained from treatment when earliest germinating weeds have reached cotyledon stage
- Further treatments must be applied as each flush of weeds reaches cotyledon stage but allowing a minimum of 7 d between each spray
- Where a pre-emergence band spray has been applied, the first treatment should be timed according to the size of the weeds in the untreated area between the rows
- Product is absorbed by leaves of emerged weeds which are killed by scorching action in 2-10 d. Apply overall as a fine spray to optimise weed cover and spray retention
- Various tank mixtures and sequences recommended to widen weed spectrum and add residual activity - see label for details

Restrictions
- Maximum total dose equivalent to three full dose treatments
- Do not spray crops stressed by nutrient deficiency, frost, wind damage, pest or disease attack, or previous herbicide treatments. Stressed crops may be checked and not recover fully
- If temperature likely to exceed 21 °C spray after 5 pm
- Before use, wash out sprayer to remove all traces of previous products, especially hormone and sulfonyl urea weedkillers
- Crystallisation may occur if spray volume exceeds that recommended or spray mixture not used within 2 h, especially if the water temperature is below 5 °C
- Product may cause non-reinforced PVC pipes and hoses to soften and swell. Wherever possible, use reinforced PVC or synthetic rubber hoses

Crop-specific information
- Latest use: before crop leaves meet between rows
- Product safe to use on all soil types

Environmental safety
- Dangerous for the environment
- Toxic to aquatic organisms
- Risk to certain non-target insects or other arthropods. Avoid spraying within 6 m of field boundary

Hazard classification and safety precautions
    Hazard Dangerous for the environment
    Transport code 9
    Packaging group III
    UN number 3082
    Risk phrases R50, R53a
    Operator protection A; U05a, U08, U19a, U20a
    Environmental protection E15b, E16b, E38, E39
    Storage and disposal D01, D02, D09a, D10b, D12a

158 dicamba

A translocated benzoic herbicide available only in mixtures
HRAC mode of action code: O

See also 2,4-D + dicamba
          2,4-D + dicamba + dichlorprop-P
          2,4-D + dicamba + fluroxypyr
          2,4-D + dicamba + MCPA + mecoprop-P
          2,4-D + dicamba + triclopyr

FOR FULL CONDITIONS OF USE ALWAYS READ THE PRODUCT LABEL

159 dicamba + dichlorprop-P + ferrous sulphate + MCPA

A herbicide/fertilizer combination for moss and weed control in turf
HRAC mode of action code: O + O + O

See also dichlorprop-P
ferrous sulphate
MCPA

Products

Renovator 2	Scotts	0.03:0.22:16.3: 0.22% w/w	GR	11411

Uses
- Annual dicotyledons in *managed amenity turf*
- Moss in *managed amenity turf*
- Perennial dicotyledons in *managed amenity turf*

Approval information
- Dicamba, dichlorprop-P and MCPA included in Annex I under EC Directive 91/414

Efficacy guidance
- Apply from mid-Apr to mid-Aug when weeds are growing
- For best control of moss scarify vigorously after 2 wk to remove dead moss
- Where regrowth of moss or weeds occurs a repeat treatment may be made after 6 wk
- Avoid treatment of wet grass or during drought. If no rain falls within 48 h water in thoroughly

Restrictions
- Do not treat new turf until established for 6-9 mth
- The first 4 mowings after treatment should not be used to mulch cultivated plants unless composted at least 6 mth
- Avoid walking on treated areas until it has rained or they have been watered
- Do not re-seed or turf within 8 wk of last treatment
- Do not cut grass for at least 3 d before and at least 4 d after treatment
- Do not apply during freezing conditions or when rain imminent

Crop-specific information
- Apply with a suitable calibrated fertilizer distributor

Hazard classification and safety precautions
UN number N/C
Operator protection U09a, U20a
Environmental protection E15a
Storage and disposal D09a, D11a

160 dicamba + dichlorprop-P + MCPA

A translocated herbicide mixture for turf
HRAC mode of action code: O + O + O

See also dichlorprop-P
MCPA

Products

Intrepid 2	Scotts	20.8:167:167 g/l	SL	11594

Uses
- Annual dicotyledons in *managed amenity turf*
- Perennial dicotyledons in *managed amenity turf*

Approval information
- Dicamba, dichlorprop-P and MCPA included in Annex I under EC Directive 91/414
- Accepted by BBPA for use on malting barley

Efficacy guidance
- Apply as directed on label between Apr and Sep when weeds growing actively

SEE SECTION 3 FOR PRODUCTS ALSO REGISTERED

Pesticide Profiles

- Do not mow for at least 3 d before and 3-4 d after treatment so that there is sufficient leaf growth for spray uptake and sufficient time for translocation. See label for details
- If re-growth occurs or new weeds germinate re-treatment recommended

Restrictions
- Do not use during drought unless irrigation is carried out before and after treatment
- Do not apply during freezing conditions or when heavy rain is imminent
- Do not treat new turf until established for about 6 mth after seeding or turfing
- The first 4 mowings after treatment should not be used to mulch cultivated plants unless composted for at least 6 mth
- Do not re-seed turf within 8 wk of last treatment

Crop-specific information
- Latest use: end Sep for managed amenity turf

Environmental safety
- Prevent spray running or drifting onto cultivated plants, including shrubs and trees

Hazard classification and safety precautions
Hazard Harmful
UN number N/C
Risk phrases R22a, R38, R41, R43
Operator protection A, C; U05a, U08, U11, U19a, U20a
Environmental protection E15a, E34
Storage and disposal D01, D02, D05, D09a, D10b
Medical advice M03, M05a

161   dicamba + MCPA + mecoprop-P

A translocated herbicide for cereals, grassland, amenity grass and orchards
HRAC mode of action code: O + O + O

See also MCPA
mecoprop-P

Products

1	Field Marshal	United Phosphorus	18:360:80 g/l	SL	08956
2	Greencrop Triathlon	Greencrop	25:200:200 g/l	SL	10956
3	Headland Relay P	Headland	25:200:200 g/l	SL	08580
4	Headland Relay Turf	Headland Amenity	25:200:200 g/l	SL	08935
5	Headland Transfer	Headland	18:315:50 g/l	SL	11010
6	Headland Trinity	Headland	18:315:50 g/l	SL	10842
7	Hycamba Plus	Agrichem	20.4:125.9:240.2 g/l	SL	10180
8	Hyprone-P	Agrichem	16:101:92 g/l	SL	09125
9	Hysward-P	Agrichem	16:101:92 g/l	SL	09052
10	Mircam Plus	Nufarm UK	19.5:245:43.3 g/l	SL	11525
11	Outrun	Barclay	20.4:125.9:240.2 g/l	SL	12838
12	Pasturol Plus	Headland	25:200:200 g/l	SL	10278
13	Pierce	Nufarm UK	31.2:256.2:237.5 g/l	SL	11924
14	Re-act	Scotts	31:256:237 g/l	SL	12231
15	Super Selective Plus	Rigby Taylor	31.25:256:238 g/l	SL	11928
16	T2 Green	Nufarm UK	31.25: 256: 237 g/l	SL	11925
17	Tribute	Nomix Enviro	18:252:42 g/l	SL	13864
18	Tribute Plus	Nomix Enviro	18:252:42 g/l	SL	13865
19	Tritox	Scotts	12:139:41 g/l	SL	07764
20	UPL Grassland Herbicide	United Phosphorus	25:200:200 g/l	SL	08934

Uses
- Annual dicotyledons in **almonds** *(off-label)*, **chestnuts** *(off-label)*, **hazel nuts** *(off-label)*, **walnuts** *(off-label)* [5, 6]; **almonds** *(off-label - undersown with rotational grass)*, **chestnuts** *(off-label - undersown with rotational grass)*, **durum wheat**, **hazel nuts** *(off-label - undersown with rotational grass)*, **rye** *(off-label)*, **triticale**, **walnuts** *(off-label - undersown with rotational grass)* [10]; **amenity grassland** [2, 6, 7, 9, 11, 12, 14, 16, 18, 20]; **apple orchards**, **durum wheat** *(off-label)*, **pear orchards**, **triticale** *(off-label)* [6, 10]; **grass seed crops** [5, 6, 8, 10, 13]; **grassland**, **green cover on land temporarily removed from production** [18]; **managed amenity turf** [2, 4, 6, 7, 9-12, 14-19]; **permanent grassland** [2, 3, 5, 6, 9, 10, 12, 13, 20]; **rotational grass** [1-3, 5, 6, 8-10, 12,

---

FOR FULL CONDITIONS OF USE ALWAYS READ THE PRODUCT LABEL

dicamba + MCPA + mecoprop-P

13, 20]; **spring barley, spring oats, spring wheat, winter barley, winter oats, winter wheat** [1, 5, 6, 8, 10]; **spring rye, undersown spring cereals, undersown winter cereals, winter rye** [6]; **undersown barley, undersown oats, undersown wheat** [8]; **undersown spring cereals** (grass only), **undersown winter cereals** (grass only) [1, 5]
- Docks in **amenity grassland** [9, 20]; **apple orchards, pear orchards** [6]; **grass seed crops** [6, 8, 13]; **managed amenity turf** [9]; **permanent grassland, rotational grass** [6, 9, 13, 20]
- Perennial dicotyledons in **almonds** (off-label), **chestnuts** (off-label), **hazel nuts** (off-label), **walnuts** (off-label) [5, 6]; **almonds** (off-label - undersown with rotational grass), **chestnuts** (off-label - undersown with rotational grass), **hazel nuts** (off-label - undersown with rotational grass), **triticale, walnuts** (off-label - undersown with rotational grass) [10]; **amenity grassland** [2, 6, 7, 9, 11, 12, 14, 16, 18, 20]; **apple orchards, pear orchards** [6, 10]; **grass seed crops** [5, 6, 8, 10, 13]; **grassland, green cover on land temporarily removed from production** [18]; **managed amenity turf** [2, 4, 6, 7, 9-12, 14-19]; **permanent grassland** [2, 3, 5, 6, 9, 10, 12, 13, 20]; **rotational grass** [1-3, 5, 6, 8-10, 12, 13, 20]; **spring barley, spring oats, spring wheat, winter barley, winter oats, winter wheat** [1, 5, 6, 8, 10]; **spring rye, undersown spring cereals, undersown winter cereals, winter rye** [6]; **undersown barley, undersown oats, undersown wheat** [8]; **undersown spring cereals** (grass only), **undersown winter cereals** (grass only) [1, 5]

Specific Off-Label Approvals (SOLAs)
- **almonds** 20060514 [5], 20060513 [6], (undersown with rotational grass) 20061130 [10]
- **chestnuts** 20060514 [5], 20060513 [6], (undersown with rotational grass) 20061130 [10]
- **durum wheat** 20060526 [6], 20061131 [10]
- **hazel nuts** 20060514 [5], 20060513 [6], (undersown with rotational grass) 20061130 [10]
- **rye** 20061131 [10]
- **triticale** 20060526 [6], 20061131 [10]
- **walnuts** 20060514 [5], 20060513 [6], (undersown with rotational grass) 20061130 [10]

Approval information
- Dicamba, MCPA and mecoprop-P included in Annex I under EC Directive 91/414
- Accepted by BBPA for use on malting barley

Efficacy guidance
- Treatment should be made when weeds growing actively. Weeds hardened by winter weather may be less susceptible
- For best results apply in fine warm weather, preferably when soil is moist. Do not spray if rain expected within 6 h or in drought
- Application of fertilizer 1-2 wk before spraying aids weed control in turf
- Where a second treatment later in the season is needed in amenity situations and on grass allow 4-6 wk between applications to permit sufficient foliage regrowth for uptake

Restrictions
- Maximum number of treatments (including other mecoprop-P products) or maximum total dose varies with crop and product. See label for details. The total amount of mecoprop-P applied in a single yr must not exceed the maximum total dose approved for any single product for the crop/situation
- Do not apply to cereals after the first node is detectable (GS 31), or to grass under stress from drought or cold weather
- Do not spray cereals undersown with clovers or legumes, to be undersown with grass or legumes or grassland where clovers or other legumes are important
- Do not spray leys established less than 18 mth or orchards established less than 3 yr
- Do not roll or harrow within 7 d before or after treatment, or graze for at least 7 d afterwards (longer if poisonous weeds present)
- Do not use on turf or grass in year of establishment. Allow 6-8 wk after treatment before seeding bare patches
- The first mowings after use should not be used for mulching unless composted for 6 mth
- Turf should not be mown for 24 h before or after treatment (3-4 d for closely mown turf)
- Avoid drift onto all broad-leaved plants outside the target area

Crop-specific information
- Latest use: before first node detectable (GS 31) for cereals; 5-6 wk before head emergence for grass seed crops; mid-Oct for established grass
- HI 7-14 d before cutting or grazing for leys, permanent pasture

SEE SECTION 3 FOR PRODUCTS ALSO REGISTERED

- Apply to winter cereals from the leaf sheath erect stage (GS 30), and to spring cereals from the 5 expanded leaf stage (GS 15)
- Spray grass seed crops 4-6 wk before flower heads begin to emerge (timothy 6 wk)
- Turf containing bulbs may be treated once the foliage has died down completely [3, 9]

Environmental safety
- Harmful to aquatic organisms
- Keep livestock out of treated areas for at least 2 wk following treatment and until poisonous weeds, such as ragwort, have died down and become unpalatable
- Harmful to fish or other aquatic life. Do not contaminate surface waters or ditches with chemical or used container
- Some pesticides pose a greater threat of contamination of water than others and mecoprop-P is one of these pesticides. Take special care when applying mecoprop-P near water and do not apply if heavy rain is forecast

Hazard classification and safety precautions
Hazard Harmful [1-7, 10-18]; Irritant [8, 9, 19, 20]
UN number N/C [1, 2, 4, 11, 14, 19, 20]
Risk phrases R20 [3, 5, 6, 12]; R21 [2-6, 10, 12, 17, 18]; R22a [1-7, 10-16]; R36 [2, 3, 8-10, 12, 17-20]; R38 [5-7, 10, 11, 20]; R41 [1, 3-7, 11-16]; R43 [20]; R52 [1, 3-10, 12-20]; R53a [1, 5-10, 13-16, 19, 20]
Operator protection A, C, H, M; U05a [2-20]; U08, U20b [1-12, 17-20]; U11 [1, 3, 5-16, 19, 20]; U15 [3, 12]; U19a [2, 7, 9, 11, 19]
Environmental protection E07a [1-3, 5-19]; E13c [1-9, 11-16, 20]; E15a [1, 10, 13, 14, 16-19]; E23 [17, 18]; E34 [1-6, 10, 12-19]; E38 [3, 12, 17, 18]
Storage and disposal D01, D02, D09a [1-20]; D05 [1-18, 20]; D10a [5, 6, 15, 19]; D10b [2-4, 10, 12-14, 16-18]; D10c [1, 7-9, 11, 20]; D12a [1, 17, 18, 20]
Medical advice M03 [1-6, 10, 12-16, 20]; M05a [1, 6-9, 11, 20]

162   dicamba + mecoprop-P

A translocated post-emergence herbicide for cereals and grassland
HRAC mode of action code: O + O

See also mecoprop-P

Products

1	Di-Farmon R	Headland	42:319 g/l	SL	08472
2	Dockmaster	Nufarm UK	18.7:150 g/l	SL	11651
3	Foundation	Headland	84:600 g/l	SL	11708
4	Headland Saxon	Headland	84:600 g/l	SL	11947
5	High Load Mircam	Nufarm UK	80:600 g/l	SL	11930
6	Hyban-P	Agrichem	18.7:150 g/l	SL	09129
7	Hygrass-P	Agrichem	18.7:150 g/l	SL	09130
8	Mircam	Nufarm UK	18.7:150 g/l	SL	11707
9	Prompt	Headland	84:600 g/l	SL	11948

Uses
- Annual dicotyledons in *canary seed (off-label)* [5]; *durum wheat (off-label)*, *spring rye (off-label)*, *triticale (off-label)*, *winter rye (off-label)* [3, 4, 9]; *permanent grassland* [1, 3-5, 7, 9]; *rotational grass* [5, 7]; *spring barley*, *spring oats*, *spring wheat*, *winter barley*, *winter oats*, *winter wheat* [1-6, 8, 9]; *spring rye*, *triticale*, *undersown rye*, *winter rye* [6]; *undersown barley*, *undersown oats*, *undersown wheat* [5, 6]
- Buttercups in *permanent grassland*, *rotational grass* [2, 6, 8]
- Chickweed in *permanent grassland* [1, 2, 6, 8]; *rotational grass* [2, 5, 6, 8]; *spring barley*, *spring oats*, *spring wheat*, *winter barley*, *winter oats*, *winter wheat* [1-6, 8, 9]; *spring rye*, *triticale*, *undersown rye*, *winter rye* [6]; *undersown barley*, *undersown oats*, *undersown wheat* [5, 6]
- Cleavers in *permanent grassland* [1]; *rotational grass* [5]; *spring barley*, *spring oats*, *spring wheat*, *winter barley*, *winter oats*, *winter wheat* [1-6, 8, 9]; *spring rye*, *triticale*, *undersown rye*, *winter rye* [6]; *undersown barley*, *undersown oats*, *undersown wheat* [5, 6]
- Creeping thistle in *permanent grassland*, *rotational grass* [2, 6, 8]
- Docks in *permanent grassland*, *rotational grass* [2, 6-8]

FOR FULL CONDITIONS OF USE ALWAYS READ THE PRODUCT LABEL

dicamba + mecoprop-P

- Mayweeds in **permanent grassland** [1]; **rotational grass** [5]; **spring barley, spring oats, spring wheat, winter barley, winter oats, winter wheat** [1-6, 8, 9]; **spring rye, triticale, undersown rye, winter rye** [6]; **undersown barley, undersown oats, undersown wheat** [5, 6]
- Perennial dicotyledons in **permanent grassland** [3-5, 7, 9]; **rotational grass** [7]; **spring barley, spring oats, spring wheat, winter barley, winter oats, winter wheat** [2, 6, 8]; **spring rye, triticale, undersown barley, undersown oats, undersown rye, undersown wheat, winter rye** [6]
- Plantains in **spring barley, spring oats, spring rye, spring wheat, triticale, undersown barley, undersown oats, undersown rye, undersown wheat, winter barley, winter oats, winter rye, winter wheat** [6]
- Polygonums in **permanent grassland** [1]; **rotational grass** [5]; **spring barley, spring oats, spring wheat, winter barley, winter oats, winter wheat** [1-6, 8, 9]; **spring rye, triticale, undersown rye, winter rye** [6]; **undersown barley, undersown oats, undersown wheat** [5, 6]
- Scentless mayweed in **permanent grassland, rotational grass** [2, 6, 8]
- Stinging nettle in **permanent grassland, rotational grass** [6]
- Thistles in **permanent grassland, rotational grass** [7]

Specific Off-Label Approvals (SOLAs)
- *canary seed* 20100760 [5]
- *durum wheat* 20052953 [3], 20052951 [4], 20052952 [9]
- *spring rye* 20052953 [3], 20052951 [4], 20052952 [9]
- *triticale* 20052953 [3], 20052951 [4], 20052952 [9]
- *winter rye* 20052953 [3], 20052951 [4], 20052952 [9]

Approval information
- Dicamba and mecoprop-P included in Annex I under EC Directive 91/414
- Accepted by BBPA for use on malting barley

Efficacy guidance
- Best results by application in warm, moist weather when weeds are actively growing

Restrictions
- Maximum number of treatments 1 per crop for cereals and 1 or 2 per yr on grass depending on label. The total amount of mecoprop-P applied in a single yr must not exceed the maximum total dose approved for any single product for the crop/situation
- Do not spray in cold or frosty conditions
- Do not spray if rain expected within 6 h
- Do not treat undersown grass until tillering begins
- Do not spray cereals undersown with clover or legume mixtures
- Do not roll or harrow within 7 d before or after spraying
- Do not treat crops suffering from stress from any cause
- Use product immediately following dilution; do not allow diluted product to stand before use [4]
- Avoid treatment when drift may damage neighbouring susceptible crops

Crop-specific information
- Latest use: before 1st node detectable for cereals; 7 d before cutting or 14 d before grazing grass
- Apply to winter sown crops from 5 expanded leaf stage (GS 15)
- Apply to spring sown cereals from 5 expanded leaf stage but before first node is detectable (GS 15-31)
- Treat grassland just before perennial weeds flower
- Transient crop prostration may occur after spraying but recovery is rapid

Environmental safety
- Dangerous for the environment [1, 3-5, 9]
- Toxic to aquatic organisms
- Harmful to fish or other aquatic life. Do not contaminate surface waters or ditches with chemical or used container
- Keep livestock out of treated areas for at least 2 wk and until foliage of poisonous weeds such as ragwort has died and become unpalatable
- Some pesticides pose a greater threat of contamination of water than others and mecoprop-P is one of these pesticides. Take special care when applying mecoprop-P near water and do not apply if heavy rain is forecast

SEE SECTION 3 FOR PRODUCTS ALSO REGISTERED

Pesticide Profiles

Hazard classification and safety precautions
Hazard Harmful [1-5, 8, 9]; Irritant [6, 7]; Dangerous for the environment [1, 3-5, 9]
UN number N/C
Risk phrases R21, R36 [2, 8]; R22a, R38 [1-5, 8, 9]; R41 [1, 3-7, 9]; R51 [1, 3-5, 9]; R52 [2, 6-8]; R53a [1-9]
Operator protection A, C [1-9]; H, M [1-4, 6-9]; U05a, U11 [1-9]; U08 [1, 3, 4, 6, 9]; U09a [2, 5, 7, 8]; U19a [7]; U20a [1]; U20b [2-9]
Environmental protection E07a [1-4, 6-9]; E13c [2, 5-8]; E15a [1, 3, 4, 9]; E34 [1-9]; E38 [1, 3]
Storage and disposal D01, D02, D05, D09a [1-9]; D10b [1-4, 8, 9]; D10c [6, 7]
Medical advice M03 [1-9]; M05a [6, 7]

163 dichlorprop-P

A translocated phenoxy carboxylic acid herbicide for use in cereals
HRAC mode of action code: O

See also 2,4-D + dicamba + dichlorprop-P
2,4-D + dichlorprop-P + MCPA + mecoprop-P
bentazone + dichlorprop-P
dicamba + dichlorprop-P + ferrous sulphate + MCPA
dicamba + dichlorprop-P + MCPA

Products
  Headland Link   Headland   500 g/l   SL   11091

Uses
- Annual dicotyledons in **permanent grassland, rotational grass, spring barley, spring oats, spring wheat, undersown spring cereals, undersown winter cereals, winter barley, winter oats, winter wheat**
- Charlock in **permanent grassland, rotational grass, spring barley, spring oats, spring wheat, undersown spring cereals, undersown winter cereals, winter barley, winter oats, winter wheat**
- Chickweed in **permanent grassland, rotational grass, spring barley, spring oats, spring wheat, undersown spring cereals, undersown winter cereals, winter barley, winter oats, winter wheat**
- Fat hen in **permanent grassland, rotational grass, spring barley, spring oats, spring wheat, undersown spring cereals, undersown winter cereals, winter barley, winter oats, winter wheat**
- Perennial dicotyledons in **permanent grassland, rotational grass, spring barley, spring oats, spring wheat, undersown spring cereals, undersown winter cereals, winter barley, winter oats, winter wheat**
- Redshank in **permanent grassland, rotational grass, spring barley, spring oats, spring wheat, undersown spring cereals, undersown winter cereals, winter barley, winter oats, winter wheat**

Approval information
- Dichlorprop-P included in Annex I under EC Directive 91/414
- Accepted by BBPA for use on malting barley

Efficacy guidance
- Best results achieved by application to young seedling weeds in good growing conditions in a strongly competing crop
- Where cleavers are to be controlled increase water volume if crop or weed is dense
- Control of redshank is best achieved in warm temperatures even if this means allowing the weed to grow beyond the seedling stage

Restrictions
- Maximum number of treatments 1 per crop
- Do not apply during cold weather, drought, rain or when rain is expected
- Do not roll or harrow within 7 d before or after spraying
- Do not apply immediately before sowing any crop

FOR FULL CONDITIONS OF USE ALWAYS READ THE PRODUCT LABEL

dichlorprop-P + ferrous sulphate + MCPA

Crop-specific information
- Latest use: before second node detectable (GS 32)
- Spray winter cereals in the spring from the leaf sheath erect stage; spray spring cereals from the one leaf stage

Following crops guidance
- Keep livestock out of treated areas until foliage of any poisonous weeds such as ragwort has died and become unpalatable

Environmental safety
- Harmful to fish or other aquatic life. Do not contaminate surface waters or ditches with chemical or used container
- Keep livestock out of treated areas for at least 2 wk and until foliage of any poisonous weeds, such as ragwort, has died and become unpalatable
- Do not spray in windy conditions when drift may cause damage to neighbouring sensitive crops

Hazard classification and safety precautions
    Hazard Harmful, Irritant
    UN number N/C
    Risk phrases R22a, R41, R43
    Operator protection A, C, H; U05a, U08, U20b
    Environmental protection E07a, E13c, E34
    Storage and disposal D01, D02, D05, D09a, D10a
    Medical advice M03

164 dichlorprop-P + ferrous sulphate + MCPA

A herbicide/fertilizer combination for moss and weed control in turf
HRAC mode of action code: O + O

See also ferrous sulphate
    MCPA

Products

1	SHL Granular Feed, Weed & Mosskiller	Sinclair	0.2:10.9:0.3% w/w	GR	10972
2	SHL Turf Feed, Weed & Mosskiller	Sinclair	0.2:10.9:0.3% w/w	DP	10973

Uses
- Annual dicotyledons in *managed amenity turf*
- Buttercups in *managed amenity turf*
- Moss in *managed amenity turf*
- Perennial dicotyledons in *managed amenity turf*

Approval information
- Dichlorprop-P and MCPA included in Annex I under EC Directive 91/414

Efficacy guidance
- Apply between Mar and Sep when grass in active growth and soil moist
- A repeat treatment may be needed after 4-6 wk to control perennial weeds or if moss regrows
- Water in if rainfall does not occur within 48 h [2]

Restrictions
- Do not treat newly sown grass for 6 mth after establishment [1]
- Do not apply during drought or freezing conditions or when rain imminent
- Avoid walking on treated areas until it has rained or turf has been watered
- Do not mow for 3-4 d before or after application
- Do not use first 4 mowings after treatment for mulching. Mowings should be composted for 6 mth before use
- Do not treat areas of fine turf such as golf or bowling greens [1]
- Avoid contact with tarmac surfaces as staining may occur

Environmental safety
- Avoid drift onto nearby plants and borders [1]

SEE SECTION 3 FOR PRODUCTS ALSO REGISTERED

Pesticide Profiles

Hazard classification and safety precautions
UN number N/C
Operator protection U20c
Storage and disposal D01, D09a, D11a

165 dichlorprop-P + ioxynil

A foliar applied translocated and contact herbicide mixture
HRAC mode of action code: O + C3

See also ioxynil

Products

Mextrol DP	Nufarm UK	500:116 g/l	SL	11529

Uses
- Annual dicotyledons in **grass seed crops** *(off-label)*, **spring barley**, **spring wheat**, **triticale**, **winter barley**, **winter oats**, **winter wheat**

Specific Off-Label Approvals (SOLAs)
- *grass seed crops* 20060525

Approval information
- Dichlorprop-P and ioxynil included in Annex I under EC Directive 91/414
- Accepted by BBPA for use on malting barley

Hazard classification and safety precautions
Hazard Harmful
UN number N/C
Risk phrases R22a, R41, R52, R53a, R63
Operator protection A, C, H; U05a, U11, U15
Environmental protection E07a, E12a, E12e, E15a, E34
Storage and disposal D01, D02, D05, D12b
Medical advice M03, M05a

166 dichlorprop-P + MCPA

A translocated herbicide for use in managed amenity turf
HRAC mode of action code: O + O

See also MCPA

Products

1	SHL Granular Feed and Weed	Sinclair	0.2:0.3% w/w	GR	10970
2	SHL Turf Feed and Weed	Sinclair	0.2:0.3% w/w	DP	10963

Uses
- Annual dicotyledons in **managed amenity turf** [1, 2]
- Buttercups in **managed amenity turf** [1, 2]
- Clovers in **managed amenity turf** [2]
- Daisies in **managed amenity turf** [2]
- Dandelions in **managed amenity turf** [2]
- Perennial dicotyledons in **managed amenity turf** [1, 2]

Approval information
- Dichlorprop-P and MCPA included in Annex I under EC Directive 91/414

Efficacy guidance
- Best results achieved by application to young seedling weeds in active growth between Mar and Sep
- Water in if rainfall does not occur within 48 h [2]

Restrictions
- Do not use on newly sown turf for 6 mth after establishment
- Do not cut or mow for 2-3 d before and after treatment

FOR FULL CONDITIONS OF USE ALWAYS READ THE PRODUCT LABEL

dichlorprop-P + MCPA + mecoprop-P

- Do not spray during cold weather, if rain or frost expected, if crop wet or in drought
- The first four mowings should not be used for mulching and should be composted for 6 mth before use
- Avoid close mowing for 3-4 d before or after treatment
- Do not treat areas of fine turf such as golf or bowling greens [1]

Environmental safety
- Avoid drift onto nearby plants and borders [2]
- Avoid contact with tarmac surfaces as staining may occur

Hazard classification and safety precautions
UN number N/C
Operator protection U20c
Storage and disposal D01, D09a, D11a

167   dichlorprop-P + MCPA + mecoprop-P

A translocated herbicide mixture for winter and spring cereals
HRAC mode of action code: O + O + O

See also MCPA
       mecoprop-P

Products
  Hymec Triple   Agrichem   310:160:130 g/l   SL   09949

Uses
- Annual dicotyledons in **durum wheat, grass seed crops** *(off-label)*, **spring barley, spring oats, spring wheat, winter barley, winter oats, winter wheat**
- Chickweed in **durum wheat, spring barley, spring oats, spring wheat, winter barley, winter oats, winter wheat**
- Cleavers in **durum wheat, spring barley, spring oats, spring wheat, winter barley, winter oats, winter wheat**
- Field pansy in **durum wheat, spring barley, spring oats, spring wheat, winter barley, winter oats, winter wheat**
- Mayweeds in **durum wheat, spring barley, spring oats, spring wheat, winter barley, winter oats, winter wheat**
- Poppies in **durum wheat, spring barley, spring oats, spring wheat, winter barley, winter oats, winter wheat**

Specific Off-Label Approvals (SOLAs)
- *grass seed crops* 20060918

Approval information
- Dichlorprop-P, MCPA and mecoprop-P included in Annex I under EC Directive 91/414
- Accepted by BBPA for use on malting barley

Efficacy guidance
- Best results obtained if application is made while majority of weeds are at seedling stage but not if temperatures are too low
- Optimum results achieved by spraying when temperature is above 10 °C. If temperatures are lower delay spraying until growth becomes more active

Restrictions
- Maximum number of treatments 1 per crop
- Do not spray in windy conditions where spray drift may cause damage to neighbouring crops, especially sugar beet, oilseed rape, peas, turnips and most horticultural crops including lettuce and tomatoes under glass

Crop-specific information
- Latest use: before second node detectable (GS 32) for all crops

Environmental safety
- Harmful to aquatic organisms
- Harmful to fish or other aquatic life. Do not contaminate surface waters or ditches with chemical or used container

SEE SECTION 3 FOR PRODUCTS ALSO REGISTERED

Pesticide Profiles

- Some pesticides pose a greater threat of contamination of water than others and mecoprop-P is one of these pesticides. Take special care when applying mecoprop-P near water and do not apply if heavy rain is forecast

Hazard classification and safety precautions
  Hazard Harmful
  UN number N/C
  Risk phrases R22a, R41, R52, R53a
  Operator protection A, C, H, M; U05a, U08, U11, U20b
  Environmental protection E13c, E34
  Storage and disposal D01, D02, D05, D09a, D10c
  Medical advice M03

168 difenacoum

An anticoagulant coumarin rodenticide

Products

1	Difenag	Killgerm	0.005% w/w	RB	H8492
2	Neosorexa Bait Blocks	BASF	0.005% w/w	RB	H6817
3	Neosorexa Gold	BASF	0.005% w/w	RB	H8301
4	Neosorexa Gold Ratpacks	BASF	0.005% w/w	RB	H8302
5	Neosorexa Pasta Bait	BASF	0.005% w/w	RB	H7758
6	Sakarat D (Whole Wheat)	Killgerm	0.005% w/w	RB	H7109
7	Sakarat D Wax Bait	Killgerm	0.005% w/w	RB	H7489
8	Sorexa D	BASF	0.005% w/w	RB	H8242
9	Sorexa Gel	BASF	0.005% w/w	RB	H6790

Uses
- Mice in *farm buildings*, *farmyards* [2-9]; *farm buildings/yards* [1]
- Rats in *farm buildings*, *farmyards* [2-6, 8]; *farm buildings/yards* [1]

Efficacy guidance
- Difenacoum is a chronic poison and rodents need to feed several times before accumulating a lethal dose. Effective against rodents resistant to other commonly used anticoagulants
- Best results achieved by placing baits at points between nesting and feeding places, at entry points, in holes and where droppings are seen
- A minimum of five baiting points normally required for a small infestation; more than 40 for a large infestation
- Inspect bait sites frequently and top up as long as there is evidence of feeding
- Product formulated for application through a skeleton or caulking gun [9]
- Maintain a few baiting points to guard against reinfestation after a successful control campaign

Restrictions
- Only for use by farmers, horticulturists and other professional users
- When working in rodent infested areas wear synthetic rubber/PVC gloves to protect against rodent-borne diseases

Environmental safety
- Harmful to wildlife
- Cover baits by placing in bait boxes, drain pipes or under boards to prevent access by children, animals or birds
- Products contain human taste deterrent

Hazard classification and safety precautions
  UN number N/C [2-5, 8, 9]
  Operator protection A [9]; U13, U20b
  Environmental protection E10b [2-5, 8]; E15a [9]
  Storage and disposal D09a, D11a
  Vertebrate/rodent control products V01a, V03a, V04a [9]; V01b, V04b [2-5, 7, 8]; V02 [2-5, 7-9]; V03b [7]; V04c [2-5, 8]
  Medical advice M03

FOR FULL CONDITIONS OF USE ALWAYS READ THE PRODUCT LABEL

difenoconazole

169 difenoconazole

A diphenyl-ether triazole protectant and curative fungicide
FRAC mode of action code: 3

See also azoxystrobin + difenoconazole

Products

1	Difcor 250 EC	Q-Chem	250 g/l	EC	13917
2	EA Difcon	European Ag	250 g/l	EC	14965
3	Plover	Syngenta	250 g/l	EC	11763
4	Turnstone	AgChem Access	250 g/l	EC	14019

Uses

- Alternaria in **broccoli, brussels sprouts, cabbages, calabrese, cauliflowers, spring oilseed rape, winter oilseed rape** [1-4]
- Blight in **celeriac** *(off-label)* [1]; **celery (outdoor)** *(off-label)*, **rhubarb** *(off-label)* [1, 3]
- Brown rust in **winter wheat** [1-4]
- Celery leaf spot in **celeriac** *(off-label)* [3]
- Disease control in **borage for oilseed production** *(off-label)*, **canary flower (echium spp.)** *(off-label)*, **durum wheat** *(off-label)*, **evening primrose** *(off-label)*, **grass seed crops** *(off-label)*, **honesty** *(off-label)*, **kale** *(off-label)*, **linseed** *(off-label)*, **mustard** *(off-label)*, **spring rye** *(off-label)*, **triticale** *(off-label)*, **winter rye** *(off-label)* [1]; **protected pinks** *(off-label)*, **protected sweet williams** *(off-label)* [1, 3]
- Foliar disease control in **borage** *(off-label)*, **canary flower (echium spp.)** *(off-label)*, **durum wheat** *(off-label)*, **evening primrose** *(off-label)*, **grass seed crops** *(off-label)*, **honesty** *(off-label)*, **linseed** *(off-label)*, **mustard** *(off-label)*, **spring rye** *(off-label)*, **triticale** *(off-label)*, **winter rye** *(off-label)* [3]
- Late blight in **celeriac** *(off-label)* [1]; **celery (outdoor)** *(off-label)*, **rhubarb** *(off-label)* [1, 3]
- Light leaf spot in **spring oilseed rape, winter oilseed rape** [1-4]
- Ring spot in **broccoli, brussels sprouts, cabbages, calabrese, cauliflowers** [1-4]; **chinese cabbage** *(off-label)*, **choi sum** *(off-label)*, **collards** *(off-label)*, **pak choi** *(off-label)* [1, 3]; **kale** *(off-label)*, **protected chinese cabbage** *(off-label)*, **protected choi sum** *(off-label)*, **protected pak choi** *(off-label)*, **protected pinks** *(off-label - ground grown (or pot grown - SOLA 20050159))*, **protected sweet williams** *(off-label - ground grown (or pot grown - 20050159))*, **protected tat soi** *(off-label)* [3]
- Rust in **asparagus** *(off-label)* [1, 3]; **protected hybrid pinks** *(off-label)*, **protected sweet williams** *(off-label)* [1]; **protected pinks** *(off-label - ground grown (or pot grown - SOLA 20050159))*, **protected sweet williams** *(off-label - ground grown (or pot grown - 20050159))* [3]
- Septoria leaf blotch in **winter wheat** [1-4]
- Stem canker in **spring oilseed rape, winter oilseed rape** [1-4]

Specific Off-Label Approvals (SOLAs)

- ***asparagus*** *20080617* [1], *20050158* [3]
- ***borage*** *20060559* [3]
- ***borage for oilseed production*** *20101167* [1]
- ***canary flower (echium spp.)*** *20101167* [1], *20060559* [3]
- ***celeriac*** *20101168* [1], *20060261* [3]
- ***celery (outdoor)*** *20101169* [1], *20091939* [3]
- ***chinese cabbage*** *20101165* [1], *20050558* [3]
- ***choi sum*** *20101165* [1], *20050558* [3]
- ***collards*** *20101165* [1], *20050558* [3]
- ***durum wheat*** *20101164* [1], *20060558* [3]
- ***evening primrose*** *20101167* [1], *20060559* [3]
- ***grass seed crops*** *20101164* [1], *20060558* [3]
- ***honesty*** *20101167* [1], *20060559* [3]
- ***kale*** *20080616* [1], *20050559* [3]
- ***linseed*** *20101167* [1], *20060559* [3]
- ***mustard*** *20101167* [1], *20060559* [3]
- ***pak choi*** *20101165* [1], *20050558* [3]
- ***protected chinese cabbage*** *20101124* [3]
- ***protected choi sum*** *20101124* [3]

SEE SECTION 3 FOR PRODUCTS ALSO REGISTERED

Pesticide Profiles

- *protected hybrid pinks* 20080613 [1]
- *protected pak choi* 20101124 [3]
- *protected pinks* 20080614 [1], *(ground grown (or pot grown - SOLA 20050159))* 20050156 [3], 20050159 [3]
- *protected sweet williams* 20080613 [1], 20080614 [1], *(ground grown (or pot grown - 20050159))* 20050156 [3], 20050159 [3]
- *protected tat soi* 20101124 [3]
- *rhubarb* 20101169 [1], 20091939 [3]
- *spring rye* 20101164 [1], 20060558 [3]
- *triticale* 20101164 [1], 20060558 [3]
- *winter rye* 20101164 [1], 20060558 [3]

Approval information
- Difenoconazole included in Annex I under EC Directive 91/414

Efficacy guidance
- Improved control of established infections on oilseed rape achieved by mixture with carbendazim. See label
- In brassicas a 3-spray programme should be used starting at the first sign of disease and repeated at 14-21 d intervals
- Product is fully rainfast 2 h after application
- For most effective control of Septoria, apply as part of a programme of sprays which includes a suitable flag leaf treatment
- Difenoconazole is a DMI fungicide. Resistance to some DMI fungicides has been identified in Septoria leaf blotch which may seriously affect performance of some products. For further advice contact a specialist advisor and visit the Fungicide Resistance Action Group (FRAG)-UK website

Restrictions
- Maximum number of treatments 3 per crop for brassicas; 2 per crop for oilseed rape; 1 per crop for wheat [1, 2, 4]
- Maximum total dose equivalent to 3 full dose treatments on brassicas; 2 full dose treatments on oilseed rape; 1 full dose treatment on wheat [3]
- Apply to wheat any time from ear fully emerged stage but before early milk-ripe stage (GS 59-73)

Crop-specific information
- Latest use: before grain early milk-ripe stage (GS 73) for cereals; end of flowering for oilseed rape
- HI brassicas 21 d
- Treat oilseed rape in autumn from 4 expanded true leaf stage (GS 1,4). A repeat spray may be made in spring at the beginning of stem extension (GS 2,0) if visible symptoms develop

Environmental safety
- Dangerous for the environment
- Very toxic to aquatic organisms

Hazard classification and safety precautions
Hazard Irritant [1, 2]; Dangerous for the environment [1-4]
Transport code 9
Packaging group III
UN number 3082
Risk phrases R38, R41, R51 [1, 2]; R50 [3, 4]; R53a [1-4]
Operator protection A, C, H; U05a, U09a, U20b [1-4]; U11 [1, 2]
Environmental protection E15a, E38
Storage and disposal D01, D02, D09a, D12a [1-4]; D05, D07, D10c [3, 4]; D10b [1, 2]

170　difenoconazole + fenpropidin

A triazole morpholine fungicide mixture for beet crops
FRAC mode of action code: 3 + 5

See also fenpropidin

Products

Spyrale	Syngenta	375:100 g/l	EC	12566

FOR FULL CONDITIONS OF USE ALWAYS READ THE PRODUCT LABEL

Uses
- Powdery mildew in **fodder beet**, **sugar beet**
- Ramularia leaf spots in **fodder beet**, **sugar beet**
- Rust in **fodder beet**, **sugar beet**

Approval information
- Difenoconazole and fenpropidin included in Annex I under EC Directive 91/414

Efficacy guidance
- For best results apply as a preventative treatment or as soon as first symptoms of disease are seen
- Product gives prolonged protection against re-infection but a second application may be needed where crops remain under heavy disease pressure

Restrictions
- Maximum number of treatments 2 per crop of sugar beet or fodder beet

Crop-specific information
- HI: 28 d for fodder beet, sugar beet

Environmental safety
- Dangerous for the environment
- Very toxic to aquatic organisms

Hazard classification and safety precautions
Hazard Harmful, Dangerous for the environment
Transport code 9
Packaging group III
UN number 3082
Risk phrases R22a, R36, R37, R38, R50, R53a
Operator protection A, C, H; U02a, U04a, U05a, U09a, U10, U11, U14, U15, U20a
Environmental protection E15a, E34, E38
Storage and disposal D01, D02, D05, D07, D09a, D10c, D12a
Medical advice M03

171 difenoconazole + fludioxonil

A triazole + phenylpyrrole seed treatment for use in cereals
FRAC mode of action code: 3 + 12

See also fludioxonil

Products
Celest Extra	Syngenta	25:25 g/l	FS	14050

Uses
- Bunt in **winter wheat**
- Fusarium foot rot and seedling blight in **winter oats**, **winter wheat**
- Microdochium nivale in **winter oats**, **winter wheat**
- Pyrenophora leaf spot in **winter oats**
- Seed-borne diseases in **winter wheat**
- Seedling blight and foot rot in **winter wheat**
- Septoria seedling blight in **winter wheat**
- Snow mould in **winter wheat**
- Stripe smut in **winter rye**

Approval information
- Difenoconazole and fludioxonil included in Annex I under EC Directive 91/414

Efficacy guidance
- [1] is effective against benzimidazole-resistant and benzimidazole-sensitive strains of Microdochium nivale

Crop-specific information
- Under adverse environmental or soil conditions, seed rates should be increased to compensate for a slight drop in germination capacity. Flow rates of seed treated with [1] should be checked before drilling commences.

SEE SECTION 3 FOR PRODUCTS ALSO REGISTERED

Pesticide Profiles

Hazard classification and safety precautions
 Transport code 9
 Packaging group III
 UN number 3082
 Risk phrases R52, R53a
 Operator protection A, H; U05a, U20c
 Environmental protection E15a, E38
 Storage and disposal D01, D02, D09a, D11a, D12a
 Treated seed S01, S02, S03, S04a, S04b, S05, S06a, S07, S08, S09

172 diflubenzuron

A selective, persistent, contact and stomach acting insecticide
IRAC mode of action code: 15

Products
 Dimilin Flo   Certis   480 g/l   SC   11056

Uses
- Browntail moth in **amenity vegetation**, **hedges**, **nursery stock**, **ornamental plant production**
- Bud moth in **apples**, **pears**
- Carnation tortrix moth in **amenity vegetation**, **hedges**, **nursery stock**, **ornamental plant production**
- Caterpillars in **broccoli**, **brussels sprouts**, **cabbages**, **calabrese**, **cauliflowers**, **chives** (off-label), **endives** (off-label), **frise** (off-label), **herbs (see appendix 6)** (off-label), **leaf spinach** (off-label), **lettuce** (off-label), **parsley** (off-label), **radicchio** (off-label), **salad brassicas** (off-label - for baby leaf production), **scarole** (off-label)
- Clouded drab moth in **apples**, **pears**
- Codling moth in **apples**, **pears**
- Fruit tree tortrix moth in **apples**, **pears**
- Houseflies in **livestock houses**, **manure heaps**, **refuse tips**
- Lackey moth in **amenity vegetation**, **hedges**, **nursery stock**, **ornamental plant production**
- Moths in **almonds** (off-label), **bilberries** (off-label), **blackberries** (off-label), **chestnuts** (off-label), **crab apples** (off-label), **cranberries** (off-label), **gooseberries** (off-label), **hazel nuts** (off-label), **quinces** (off-label), **redcurrants** (off-label), **walnuts** (off-label), **whitecurrants** (off-label)
- Oak leaf roller moth in **forest**
- Pear sucker in **pears**
- Phorid flies in **edible fungi** (off-label - other than mushrooms)
- Pine beauty moth in **forest**
- Pine looper in **forest**
- Plum fruit moth in **plums**
- Rust mite in **apples**, **pears**, **plums**
- Sciarid flies in **edible fungi** (off-label - other than mushrooms)
- Small ermine moth in **amenity vegetation**, **hedges**, **nursery stock**, **ornamental plant production**
- Tortrix moths in **plums**
- Winter moth in **amenity vegetation**, **apples**, **blackcurrants**, **forest**, **hedges**, **nursery stock**, **ornamental plant production**, **pears**, **plums**

Specific Off-Label Approvals (SOLAs)
- **almonds** 20060571
- **bilberries** 20060573
- **blackberries** 20060573
- **chestnuts** 20060571
- **chives** 20051321
- **crab apples** 20060572
- **cranberries** 20060573
- **edible fungi** (other than mushrooms) 20060574
- **endives** 20051321
- **frise** 20051321
- **gooseberries** 20060573

FOR FULL CONDITIONS OF USE ALWAYS READ THE PRODUCT LABEL

diflubenzuron

- *hazel nuts* 20060571
- *herbs (see appendix 6)* 20051321
- *leaf spinach* 20051321
- *lettuce* 20051321
- *parsley* 20051321
- *quinces* 20060572
- *radicchio* 20051321
- *redcurrants* 20060573
- *salad brassicas* (for baby leaf production) 20051321
- *scarole* 20051321
- *walnuts* 20060571
- *whitecurrants* 20060573

Approval information
- Approved for aerial application in forestry when average wind velocity does not exceed 18 knots and gusts do not exceed 20 knots [1]. See Section 5 for more information
- Diflubenzuron included in Annex I under EC Directive 91/414

Efficacy guidance
- Most active on young caterpillars and most effective control achieved by spraying as eggs start to hatch
- Dose and timing of spray treatments vary with pest and crop. See label for details
- Addition of wetter recommended for use on brassicas and for pear sucker control in pears

Restrictions
- Maximum number of treatments 3 per yr for apples, pears; 2 per yr for plums, blackcurrants; 2 per crop for brassicas; 1 per yr for forest
- Before treating ornamentals check varietal tolerance on a small sample
- Do not use as a compost drench or incorporated treatment on ornamental crops
- Do not spray protected plants in flower or with flower buds showing colour
- For use only on the food crops specified on the label
- Do not apply directly to livestock/poultry

Crop-specific information
- HI apples, pears, plums, blackcurrants, brassicas 14 d

Environmental safety
- Dangerous for the environment
- Very toxic to aquatic organisms
- Broadcast air-assisted LERAP [1] (20 m when used in orchards; 10 m when used in blackcurrants, forestry or ornamentals); LERAP Category B [1]

Hazard classification and safety precautions
Hazard Dangerous for the environment
Transport code 9
Packaging group III
UN number 3082
Risk phrases R50
Operator protection U20c
Environmental protection E05a, E15a, E16a, E16b, E18, E38 [1]; E17b [1] (20 m when used in orchards; 10 m when used in blackcurrants, forestry or ornamentals)
Storage and disposal D05, D09a, D11a, D12b

SEE SECTION 3 FOR PRODUCTS ALSO REGISTERED

Pesticide Profiles

173 diflufenican

A shoot absorbed pyridinecarboxamide herbicide for winter cereals
HRAC mode of action code: F1

See also bromoxynil + diflufenican + ioxynil
chlorotoluron + diflufenican
clodinafop-propargyl + diflufenican
clopyralid + diflufenican + MCPA

Products

1	Champion	AgriGuard	500 g/l	SC	14599
2	Crater	Makhteshim	500 g/l	SC	13084
3	Difenikan 500	Goldengrass	500 g/l	SC	14600
4	Diflanil 500 SC	Q-Chem	500 g/l	SC	12489
5	Diflufenican GL 500	Globachem	500 g/l	SC	12531
6	Hurricane SC	Makhteshim	500 g/l	SC	12424
7	Overlord	Makhteshim	500 g/l	SC	13521
8	Sempra	AgriChem BV	500 g/l	SC	13525
9	Solo D	Sipcam	500 g/l	SC	14667
10	Twister	ChemSource	500 g/l	SC	14885

Uses

- Annual dicotyledons in **durum wheat** [4, 5]; **grass seed crops** *(off-label)* [2, 4, 6, 7, 9]; **oats** *(off-label)* [9]; **spring barley** [8, 9]; **spring oats** *(off-label)*, **winter oats** *(off-label)* [2-4, 6, 7]; **triticale, winter barley, winter rye, winter wheat** [4, 5, 8, 9]
- Annual grasses in **spring barley, triticale, winter barley, winter rye, winter wheat** [8]
- Chickweed in **durum wheat** [4, 5]; **spring barley** [9]; **triticale, winter barley, winter rye, winter wheat** [4, 5, 9]
- Cleavers in **durum wheat** [2-7, 10]; **spring barley** [1-3, 6, 7, 9, 10]; **spring wheat** [2, 3, 6, 7, 10]; **triticale, winter barley, winter rye, winter wheat** [1-7, 9, 10]
- Field pansy in **durum wheat, spring wheat** [2, 3, 6, 7, 10]; **spring barley, triticale, winter barley, winter rye, winter wheat** [1-3, 6, 7, 10]
- Field speedwell in **durum wheat** [2-7, 10]; **spring barley** [1-3, 6, 7, 9, 10]; **spring wheat** [2, 3, 6, 7, 10]; **triticale, winter barley, winter rye, winter wheat** [1-7, 9, 10]
- Ivy-leaved speedwell in **durum wheat** [2-7, 10]; **spring barley** [1-3, 6, 7, 9, 10]; **spring wheat** [2, 3, 6, 7, 10]; **triticale, winter barley, winter rye, winter wheat** [1-7, 9, 10]
- Mayweeds in **durum wheat, spring wheat** [2, 3, 6, 7, 10]; **spring barley, triticale, winter barley, winter rye, winter wheat** [1-3, 6, 7, 10]
- Poppies in **durum wheat, spring wheat** [2, 3, 6, 7, 10]; **spring barley, triticale, winter barley, winter rye, winter wheat** [1-3, 6, 7, 10]
- Red dead-nettle in **durum wheat, spring wheat** [2, 3, 6, 7, 10]; **spring barley, triticale, winter barley, winter rye, winter wheat** [1-3, 6, 7, 10]
- Volunteer oilseed rape in **durum wheat** [4, 5]; **spring barley** [9]; **triticale, winter barley, winter rye, winter wheat** [4, 5, 9]

Specific Off-Label Approvals (SOLAs)

- *grass seed crops* 20092671 [2], 20081935 [4], 20092507 [6], 20092443 [7], 20102613 [9]
- *oats* 20102613 [9]
- *spring oats* 20082172 [2], 20102394 [3], 20081936 [4], 20081320 [6], 20082170 [7]
- *winter oats* 20082172 [2], 20102394 [3], 20081936 [4], 20081320 [6], 20082170 [7]

Approval information

- Diflufenican included in Annex I under EC Directive 91/414
- Accepted by BBPA for use on malting barley

Efficacy guidance

- Best results achieved from treatment of small actively growing weeds in early autumn or spring
- Good weed control depends on efficient burial of trash or straw before or during seedbed preparation
- Loose or fluffy seedbeds should be rolled before application
- The final seedbed should be moist, fine and firm with clods no bigger than fist size
- Ensure good even spray coverage and increase spray volume for post-emergence treatments where the crop or weed foliage is dense

FOR FULL CONDITIONS OF USE ALWAYS READ THE PRODUCT LABEL

diflufenican + flufenacet

- Activity may be slow under cool conditions and final level of weed control may take some time to appear
- Where cleavers are a particular problem a separate specific herbicide treatment may be required
- Efficacy may be impaired on soils with a Kd factor greater than 6
- Always follow WRAG guidelines for preventing and managing herbicide resistant weeds. See Section 5 for more information

Restrictions
- Maximum number of treatments 1 per crop
- Do not treat broadcast crops [4-6]
- Do not roll treated crops or harrow at any time after treatment
- Do apply to soils with more than 10% organic matter or on Sands, or very stony or gravelly soils
- Do not treat after a period of cold frosty weather

Crop-specific information
- Latest use: before end of tillering (GS 29) for wheat and barley [4-6], pre crop emergence for triticale and winter rye [4-6]
- Treat only named varieties of rye or triticale [4-6]

Following crops guidance
- Labels vary slightly but in general ploughing to 150 mm and thoroughly mixing the soil is recommended before drilling or planting any succeeding crops either after crop failure or after normal harvest
- In the event of crop failure only winter wheat or winter barley may be re-drilled immediately after ploughing. Spring crops of wheat, barley, oilseed rape, peas, field beans, sugar beet [4-6], potatoes, carrots, edible brassicas or onions may be sown provided an interval of 12 wk has elapsed after ploughing
- After normal harvest of a treated crop winter cereals, oilseed rape, field beans, leaf brassicas, sugar beet seed crops and winter onions may be drilled in the following autumn. Other crops listed above for crop failure may be sown in the spring after normal harvest
- Successive treatments with any products containing diflufenican can lead to soil build-up and inversion ploughing to 150 mm must precede sowing any following non-cereal crop. Even where ploughing occurs some crops may be damaged

Environmental safety
- Dangerous for the environment
- Very toxic to aquatic organisms
- LERAP Category B

Hazard classification and safety precautions
Hazard Dangerous for the environment
Transport code 9
Packaging group III
UN number 3082
Risk phrases R50, R53a
Operator protection A [1-10]; C [2-7, 10]; U05a [1-3, 6-10]; U08, U13, U19a [1-3, 6, 7, 10]; U20a [1-3, 6-8, 10]
Environmental protection E13a [8]; E13c, E34, E38 [1-3, 6, 7, 10]; E15a [4, 5, 9]; E16a [1-10]
Storage and disposal D01, D02 [1-3, 6-10]; D05 [1-7, 9, 10]; D09a, D10b [1-10]; D12a [4, 5, 8, 9]

174 diflufenican + flufenacet

A contact and residual herbicide mixture for cereals
HRAC mode of action code: F1 + K3

See also flufenacet

Products

1	Firebird	Bayer CropScience	200:400 g/l	SC	12421
2	Liberator	Bayer CropScience	100:400 g/l	SC	12032
3	Regatta	Bayer CropScience	100:400 g/l	SC	12054

SEE SECTION 3 FOR PRODUCTS ALSO REGISTERED

Pesticide Profiles

Uses
- Annual dicotyledons in **durum wheat** *(off-label)*, **triticale** *(off-label)*, **winter rye** *(off-label)* [2]; **grass seed crops** *(off-label)* [2, 3]
- Annual grasses in **durum wheat** *(off-label)*, **triticale** *(off-label)*, **winter rye** *(off-label)* [2]; **grass seed crops** *(off-label)* [2, 3]
- Annual meadow grass in **winter barley**, **winter wheat** [1-3]
- Blackgrass in **winter barley**, **winter wheat** [1-3]
- Chickweed in **winter barley**, **winter wheat** [1-3]
- Field pansy in **winter barley**, **winter wheat** [1-3]
- Field speedwell in **winter barley**, **winter wheat** [1-3]
- Ivy-leaved speedwell in **winter barley**, **winter wheat** [1-3]
- Mayweeds in **winter barley**, **winter wheat** [1-3]
- Red dead-nettle in **winter barley**, **winter wheat** [2, 3]
- Shepherd's purse in **winter barley**, **winter wheat** [1]
- Volunteer oilseed rape in **winter barley**, **winter wheat** [1]

Specific Off-Label Approvals (SOLAs)
- *durum wheat* 20063650 [2]
- *grass seed crops* 20063650 [2], 20061029 [3]
- *triticale* 20063650 [2]
- *winter rye* 20063650 [2]

Approval information
- Diflufenican and flufenacet included in Annex I under EC Directive 91/414
- Accepted by BBPA for use on malting barley

Efficacy guidance
- Best results obtained when there is moist soil at and after application and rain falls within 7 d
- Residual control may be reduced under prolonged dry conditions
- Activity may be slow under cool conditions and final level of weed control may take some time to appear
- Good weed control depends on burying any trash or straw before or during seedbed preparation
- Established perennial grasses and broad-leaved weeds will not be controlled
- Do not use as a stand-alone treatment for blackgrass control. Always follow WRAG guidelines for preventing and managing herbicide resistant weeds. Section 5 for more information

Restrictions
- Maximum number of treatments 1 per crop
- Do not treat undersown cereals or those to be undersown
- Do not use on waterlogged soils or soils prone to waterlogging
- Do not use on Sands or Very Light soils, or very stony or gravelly soils, or on soils containing more than 10% organic matter
- Do not treat broadcast crops and treat shallow-drilled crops post-emergence only
- Do not incorporate into the soil or disturb the soil after application by rolling or harrowing
- Avoid treating crops under stress from whatever cause and avoid treating during periods of prolonged or severe frosts

Crop-specific information
- Latest use: before 31 Dec in yr of sowing and before 3rd tiller stage (GS 23) for wheat or 4th tiller stage (GS 24) for barley
- For pre-emergence treatments the seed should be covered with a minimum of 32 mm settled soil

Following crops guidance
- In the event of crop failure wheat, barley or potatoes may be sown provided the soil is ploughed to 15 cm, and a minimum of 12 weeks elapse between treatment and sowing spring wheat or spring barley
- After normal harvest wheat, barley or potatoes my be sown without special cultivations. Soil must be ploughed or cultivated to 15 cm before sowing oilseed rape, field beans, peas, sugar beet, carrots, onions or edible brassicae
- Successive treatments with any products containing diflufenican can lead to soil build-up and inversion ploughing to 150 mm must precede sowing any following non-cereal crop. Even where ploughing occurs some crops may be damaged

FOR FULL CONDITIONS OF USE ALWAYS READ THE PRODUCT LABEL

diflufenican + flufenacet + flurtamone

Environmental safety
- Dangerous for the environment
- Very toxic to aquatic organisms
- Risk to non-target insects or other arthropods. Avoid spraying within 6 m of the field boundary to reduce the effects on non-target insects or other arthropods
- LERAP Category B

Hazard classification and safety precautions
Hazard Harmful, Dangerous for the environment
Transport code 9
Packaging group III
UN number 3082
Risk phrases R22a, R43, R48, R50, R53a
Operator protection A, H; U05a, U14
Environmental protection E15a [1, 3]; E16a, E34, E38 [1-3]; E22c [1, 2]
Storage and disposal D01, D02, D09a, D10b, D12a
Medical advice M03, M05a

175   diflufenican + flufenacet + flurtamone

A herbicide mixture for use in wheat and barley
HRAC mode of action code: F1 + K3

Products

1	Movon	Bayer CropScience	90:240:120 g/l	SC	14784
2	Vigon	Bayer CropScience	60:240:120 g/l	SC	14785

Uses
- Annual meadow grass in *winter barley*, *winter wheat* [1, 2]
- Blackgrass in *winter wheat* *(moderately susceptible)* [1, 2]
- Chickweed in *winter barley*, *winter wheat* [1, 2]
- Field pansy in *winter barley*, *winter wheat* [1, 2]
- Field speedwell in *winter barley*, *winter wheat* [1, 2]
- Forget-me-not in *winter wheat* [1, 2]
- Groundsel in *winter wheat* [1, 2]
- Italian ryegrass in *winter wheat* *(moderately susceptible)* [1, 2]
- Ivy-leaved speedwell in *winter wheat* [1, 2]
- Mayweeds in *winter barley* [1]; *winter wheat* [1, 2]

Approval information
- Diflufenican + flufenacet + flurtamone included in Annex 1 under EC Directive 91/414
- Accepted by BBPA for use on malting barley

Environmental safety
- LERAP Category B

Hazard classification and safety precautions
Hazard Harmful, Dangerous for the environment
Transport code 9
Packaging group III
UN number 3082
Risk phrases R22a, R50, R53a, R69
Operator protection A; U05a
Environmental protection E15b, E16a, E22c, E34, E38, E39, E40b
Storage and disposal D01, D02, D09a, D10b, D12a
Medical advice M03, M05a

SEE SECTION 3 FOR PRODUCTS ALSO REGISTERED

Pesticide Profiles

176 diflufenican + flupyrsulfuron-methyl

A pyridinecarboxamide and sulfonylurea herbicide mixture for wheat
HRAC mode of action code: F1 + B

See also flupyrsulfuron-methyl

Products

Absolute　　　　　　　　　DuPont　　　　　　　　　41.7:8.3% w/w　　　　　WG　12558

Uses
- Annual dicotyledons in **winter oats** *(off-label)*, **winter wheat**
- Blackgrass in **winter oats** *(off-label)*, **winter wheat**

Specific Off-Label Approvals (SOLAs)
- **winter oats** *20082624*

Approval information
- Diflufenican and flupyrsulfuron-methyl included in Annex I under EC Directive 91/414

Efficacy guidance
- Best results achieved from applications made in good growing conditions
- Good spray cover of weeds must be obtained
- Best control of blackgrass obtained from application from the 2-leaf stage but tank mixing or sequencing with a partner product is recommended as a resistance management strategy (see below)
- Growth of weeds is inhibited within hours of treatment but visible symptoms may not be apparent for up to 4 wk
- Product has moderate residual life in soil. Under normal moisture conditions susceptible weeds germinating soon after treatment will be controlled
- Product may be used on all soil types except those with more than 10% organic matter, but residual activity and weed control is reduced by high soil temperatures and when soil is very dry and on highly alkaline soils
- Flupyrsulfuron-methyl is a member of the ALS-inhibitor group of herbicides. To avoid the build up of resistance do not use any product containing an ALS-inhibitor herbicide with claims for control of grass weeds more than once on any crop
- Use these products as part of a resistance management strategy that includes cultural methods of control and does not use ALS inhibitors as the sole chemical method of grass weed control

Restrictions
- Maximum number of treatments 1 per crop
- Do not use on wheat undersown with grasses, clover or other legumes, or any other broad-leaved crop
- Do not apply within 7 d of rolling
- Do not treat broadcast crops. Seed should be covered by at least 25 mm of settled soil
- Do not treat any crop suffering from drought, waterlogging, pest or disease attack, nutrient deficiency, or any other stress factors
- Do not use on soils with more than 10% organic matter
- Specific restrictions apply to use in sequence or tank mixture with other sulfonylurea or ALS-inhibiting herbicides. See label for details

Crop-specific information
- Latest use: before 1st node detectable stage (GS 31) for winter wheat
- Under certain conditions slight chlorosis and stunting may occur from which recovery is rapid

Following crops guidance
- Only cereals, oilseed rape or field beans may be sown in the yr of harvest of a treated crop. Ploughing must be carried out before drilling any non-cereal crop
- An interval of at least 20 wk must elapse after treatment before planting wheat, barley, oilseed rape, peas, field beans, sugar beet, potatoes, carrots, brassicas or onions in the following spring, and only after ploughing
- In the event of failure of a treated crop only winter wheat or spring wheat may be sown within 3 mth of treatment, and only after ploughing and cultivating to at least 15 cm. After 3 mth the normal guidance for following crops may be observed

FOR FULL CONDITIONS OF USE ALWAYS READ THE PRODUCT LABEL

- Successive treatments with any products containing diflufenican can lead to soil build-up and inversion ploughing to 150 mm must precede sowing any following non-cereal crop. Even where ploughing occurs some crops may be damaged

Environmental safety
- Dangerous for the environment
- Very toxic to aquatic organisms
- Take extreme care to avoid damage by drift onto broad-leaved plants outside the target area
- Do not drain or flush spraying equipment onto land planted, or intended for planting, with trees or crops other than cereals
- Poor sprayer cleanout practices may result in inadequate removal of chemical deposits which may lead to damage of subsequently treated non-cereal crops. See label for guidance on cleaning
- LERAP Category B

Hazard classification and safety precautions
Hazard Dangerous for the environment
Transport code 9
Packaging group III
UN number 3077
Risk phrases R50, R53a
Operator protection U08, U19a, U20b
Environmental protection E15a, E16a, E38
Storage and disposal D09a, D12a

177   diflufenican + flurtamone

A contact and residual herbicide mixture for cereals
HRAC mode of action code: F1 + F1

See also flurtamone

Products
    Bacara   Bayer CropScience   100:250 g/l   SC   10744

Uses
- Annual dicotyledons in **grass seed crops** *(off-label)*, **spring barley**, **spring rye** *(off-label)*, **triticale** *(off-label)*, **winter barley**, **winter oats** *(off-label)*, **winter rye** *(off-label)*, **winter wheat**
- Annual grasses in **grass seed crops** *(off-label)*, **spring rye** *(off-label)*, **triticale** *(off-label)*, **winter oats** *(off-label)*, **winter rye** *(off-label)*
- Annual meadow grass in **spring barley**, **winter barley**, **winter wheat**
- Blackgrass in **spring barley**, **winter barley**, **winter wheat**
- Loose silky bent in **spring barley**, **winter barley**, **winter wheat**
- Volunteer oilseed rape in **spring barley**, **winter barley**, **winter wheat**

Specific Off-Label Approvals (SOLAs)
- *grass seed crops* 20060813
- *spring rye* 20060813
- *triticale* 20060813
- *winter oats* 20102425
- *winter rye* 20060813

Approval information
- Diflufenican and flurtamone included in Annex I under EC Directive 91/414
- Accepted by BBPA for use on malting barley

Efficacy guidance
- Apply pre-emergence or from when crop has first leaf unfolded before susceptible weeds pass recommended size
- Best results obtained on firm, fine seedbeds with adequate soil moisture present at and after application. Increase water volume where crop or weed foliage is dense
- Good weed control requires ash, trash and burnt straw to be buried during seed bed preparation
- Loose fluffy seedbeds should be rolled before application and the final seed bed should be fine and firm without large clods
- Speed of control depends on weather conditions and activity can be slow under cool conditions

Pesticide Profiles

- Always follow WRAG guidelines for preventing and managing herbicide resistant weeds. See Section 5 for more information

Restrictions
- Maximum number of treatments 1 per crop
- Crops should be drilled to a normal depth of 25 mm and the seed well covered. Do not treat broadcast crops as uncovered seed may be damaged
- Do not treat spring sown cereals, durum wheat, oats, undersown cereals or those to be undersown
- Do not treat frosted crops or when frost is imminent. Severe frost after application, or any other stress, may lead to transient discoloration or scorch
- Do not use on Sands or Very Light soils or those that are very stony or gravelly
- Do not use on waterlogged soils, or on crops subject to temporary waterlogging by heavy rainfall, as there is risk of persistent crop damage which may result in yield loss
- Do not use on soils with more than 10% organic matter
- Do not harrow at any time after application and do not roll autumn treated crops until spring

Crop-specific information
- Latest use: before 2nd node detectable (GS32)
- Take particular care to match spray swaths otherwise crop discoloration and biomass reduction may occur which may lead to yield reduction

Following crops guidance
- In the event of crop failure winter wheat may be redrilled immediately after normal cultivation, and winter barley may be sown after ploughing. Fields must be ploughed to a depth of 15 cm and 20 wk must elapse before sowing spring crops of wheat, barley, oilseed rape, peas, field beans or potatoes
- After normal harvest autumn cereals can be drilled after ploughing. Thorough mixing of the soil must take place before drilling field beans, leaf brassicae or winter oilseed rape. For sugar beet seed crops and winter onions complete inversion of the furrow slice is essential
- Do not broadcast or direct drill oilseed rape or other brassica crops as a following crop on treated land. See label for detailed advice on preparing land for subsequent autumn cropping in the normal rotation
- Successive treatments with any products containing diflufenican can lead to soil build-up and inversion ploughing to 150 mm must precede sowing any following non-cereal crop. Even where ploughing occurs some crops may be damaged

Environmental safety
- Dangerous for the environment
- Very toxic to aquatic organisms
- Extremely dangerous to fish or other aquatic life. Do not contaminate surface waters or ditches with chemical or used container
- LERAP Category B

Hazard classification and safety precautions
Hazard Dangerous for the environment
Transport code 9
Packaging group III
UN number 3082
Risk phrases R50, R53a
Operator protection A, H; U20c
Environmental protection E13a, E16a, E16b, E38
Storage and disposal D05, D09a, D10b, D12a

178 diflufenican + glyphosate

A foliar non-selective herbicide mixture
HRAC mode of action code: F1 + G

See also glyphosate

Products

1	Pistol	Bayer Environ.	40:250 g/l	SC	12173
2	Proshield	Scotts	40:250 g/l	SC	14767

FOR FULL CONDITIONS OF USE ALWAYS READ THE PRODUCT LABEL

diflufenican + iodosulfuron-methyl-sodium + mesosulfuron-methyl

Uses
- Annual and perennial weeds in **hard surfaces** [2]; **natural surfaces not intended to bear vegetation**, **permeable surfaces overlying soil** [1, 2]

Approval information
- Diflufenican and glyphosate included in Annex I under EC Directive 91/414

Efficacy guidance
- Treat when weeds actively growing from Mar to end Sep and before they begin to senesce
- Performance may be reduced if application is made to plants growing under stress, such as drought or water-logging
- Pre-emergence activity is reduced on soils containing more than 10% organic matter or where organic debris has collected
- Perennial weeds such as docks, perennial sowthistle and willowherb are best treated before flowering or setting seed
- Perennial weeds emerging from established rootstocks after treatment will not be controlled
- A rainfree period of at least 6 h (preferably 24 h) should follow spraying for optimum control
- For optimum control do not cultivate or rake after treatment

Restrictions
- Maximum number of treatments 1 per yr
- May only be used on porous surfaces overlying soil. Must not be used if an impermeable membrane lies between the porous surface and the soil, and must not be used on any non-porous man-made surfaces
- Do not add any wetting agent or adjuvant oil
- Do not spray in windy weather

Following crops guidance
- A period of at least 6 mth must be allowed after treatment of sites that are to be cleared and grubbed before sowing and planting. Soil should be ploughed or dug first to ensure thorough mixing and dilution of any herbicide residues

Environmental safety
- Dangerous for the environment
- Very toxic to aquatic organisms
- Avoid drift onto non-target plants
- Heavy rain after application may wash product onto sensitive areas such as newly sown grass or areas about to be planted
- LERAP Category B

Hazard classification and safety precautions
    Hazard Dangerous for the environment
    Transport code 9
    Packaging group III
    UN number 3082
    Risk phrases R50, R53a
    Operator protection A, C; U20b
    Environmental protection E15a, E16a, E16b, E38
    Storage and disposal D01, D02, D09a, D11a, D12a

179   diflufenican + iodosulfuron-methyl-sodium + mesosulfuron-methyl

A contact and residual herbicide mixture containing sulfonyl ureas for winter wheat
HRAC mode of action code: F1 + B + B

See also iodosulfuron-methyl-sodium
          mesosulfuron-methyl

Products
    Othello   Bayer CropScience   50:2.5:7.5 g/l   OD   12695

Uses
- Annual dicotyledons in **winter wheat**

SEE SECTION 3 FOR PRODUCTS ALSO REGISTERED

- Annual meadow grass in *winter wheat*
- Cleavers in *winter wheat*
- Mayweeds in *winter wheat*
- Rough meadow grass in *winter wheat*
- Volunteer oilseed rape in *winter wheat*

Approval information
- Diflufenican, iodosulfuron-methyl-sodium and mesosulfuron-methyl included in Annex I under EC Directive 91/414

Efficacy guidance
- Optimum control obtained when all weeds are emerged at spraying. Activity is primarily via foliar uptake and good spray coverage of the target weeds is essential
- Translocation occurs readily within the target weeds and growth is inhibited within hours of treatment but symptoms may not be apparent for up to 4 wk, depending on weed species, timing of treatment and weather conditions
- Iodosulfuron-methyl and mesosulfuron-methyl are both members of the ALS-inhibitor group of herbicides. To avoid the build up of resistance do not use any product containing an ALS-inhibitor herbicide with claims for control of grass weeds more than once on any crop
- Use this product as part of a Resistance Management Strategy that includes cultural methods of control and does not use ALS inhibitors as the sole chemical method of weed control in successive crops. See Section 5 for more information

Restrictions
- Maximum number of treatments 1 per crop
- Do not use on crops undersown with grasses, clover or other legumes or any other broad-leaved crop
- Do not use where annual grass weeds other than annual meadow grass and rough meadow grass are present
- Do not use as a stand-alone treatment for control of common chickweed or common poppy. Only use mixtures with non ALS-inhibitor herbicides for these weeds
- Do not use as the sole means of weed control in successive crops
- Specific restrictions apply to use in sequence or tank mixture with other sulfonylurea or ALS-inhibiting herbicides. See label for details
- Do not apply to crops under stress from any cause
- Do not apply when rain is imminent or during periods of frosty weather
- Specified adjuvant must be used. See label

Crop-specific information
- Latest use: before 2nd node detectable for winter wheat
- Transitory crop effects may occur, particularly on overlaps and after late season/spring applications. Recovery is normally complete and yield not affected

Following crops guidance
- In the event of crop failure winter or spring wheat may be drilled after normal cultivation and an interval of 6 wk
- Winter wheat, winter barley or winter oilseed rape may be drilled in the autumn following normal harvest of a treated crop. Spring wheat, spring barley, spring oilseed rape or sugar beet may be drilled in the following spring
- Where the product has been applied in sequence with a permitted ALS-inhibitor herbicide (see label) only winter or spring wheat or barley, or sugar beet may be sown as following crops
- Successive treatments with any products containing diflufenican can lead to soil build-up and inversion ploughing to 150 mm must precede sowing any following non-cereal crop. Even where ploughing occurs some crops may be damaged

Environmental safety
- Dangerous for the environment
- Very toxic to aquatic organisms
- Take extreme care to avoid drift outside the target area
- LERAP Category B

Hazard classification and safety precautions
Hazard Irritant, Dangerous for the environment
Transport code 9

FOR FULL CONDITIONS OF USE ALWAYS READ THE PRODUCT LABEL

Packaging group III
UN number 3082
Risk phrases R36, R50, R53a
Operator protection A, C; U05a, U11, U20b
Environmental protection E15b, E16a, E38
Storage and disposal D01, D02, D05, D09a, D10a, D12a

180   diflufenican + mecoprop-P

A mixture of shoot absorbed pyridinecarboxamide herbicide and a translocated phenoxycarboxylic acid herbicide for winter cereals
HRAC mode of action code: F1 + O

See also mecoprop-P

Products

1	Atom	Nufarm UK	33.3:500 g/l	SC	15038
2	Conga	Nufarm UK	33.3:500 g/l	SC	15011
3	Pixie	Nufarm UK	33.3:500 g/l	SC	13884

Uses
- Annual dicotyledons in **winter barley, winter wheat**

Approval information
- Diflufenican and mecoprop-P included in Annex I under EC Directive 91/414
- Accepted by BBPA for use on malting barley

Environmental safety
- LERAP Category B

Hazard classification and safety precautions
Hazard Harmful, Dangerous for the environment
Transport code 9
Packaging group III
UN number 3082
Risk phrases R22a, R41, R43, R50, R53a
Operator protection A, C, H; U05a, U08, U11, U19a, U20b
Environmental protection E15a, E16a, E34, E38
Storage and disposal D01, D02, D09a, D10b, D12a

181   diflufenican + pendimethalin

A mixture of pyridinecarboxamide and dinitroanaline herbicides, see Section 3
HRAC mode of action code: F1 + K1

See also pendimethalin

Products

1	Churchill	Makhteshim	18.75:300 g/l	EC	14139
2	Omaha	Makhteshim	30:300 g/l	EC	14234

Uses
- Annual dicotyledons in **triticale, winter barley, winter rye, winter wheat** [1, 2]
- Annual grasses in **triticale, winter barley, winter rye, winter wheat** [1]
- Annual meadow grass in **triticale, winter barley, winter rye, winter wheat** [2]

Approval information
- Diflufenican and pendimethalin included in Annex I under EC Directive 91/414

Efficacy guidance
- Do not apply pre-emergence to winter crops drilled after 20th November
- Do not treat broadcast crops

Following crops guidance
- Do not drill autumn-sown broad leaved crops following application to the previous crop [1]
- Plough to 150 mm and thoroughly mix the soil before planting any following crop

SEE SECTION 3 FOR PRODUCTS ALSO REGISTERED

Pesticide Profiles

Environmental safety
- LERAP Category B

Hazard classification and safety precautions
Hazard Toxic, Flammable, Dangerous for the environment
Transport code 3
Packaging group III
UN number 1993
Risk phrases R20 [1]; R36, R37, R38 [2]; R43, R50, R53a, R61 [1, 2]
Operator protection A, C, H; U08, U13, U19a, U20b [1, 2]; U14, U15, U19c [2]; U19b [1]
Environmental protection E15a, E16a, E34, E38 [1, 2]; E23 [1]
Storage and disposal D09a, D10b, D12a
Medical advice M04a

182 dimethenamid-p

A chloroacetamide herbicide available only in mixtures
HRAC mode of action code: K3

183 dimethenamid-p + metazachlor

A soil acting herbicide mixture for oilseed rape
HRAC mode of action code: K3 + K3

See also metazachlor

Products

1	Muntjac	BASF	200:200 g/l	EC	14540
2	Springbok	BASF	200:200 g/l	EC	12983

Uses
- Annual dicotyledons in **winter oilseed rape** [1, 2]
- Chickweed in **winter oilseed rape** [1, 2]
- Cleavers in **winter oilseed rape** [1, 2]
- Common storksbill in **winter oilseed rape** [1, 2]
- Crane's-bill in **swedes** *(off-label)*, **turnips** *(off-label)* [2]
- Field speedwell in **winter oilseed rape** [1, 2]
- Poppies in **winter oilseed rape** [1, 2]
- Scented mayweed in **winter oilseed rape** [1, 2]
- Shepherd's purse in **swedes** *(off-label)*, **turnips** *(off-label)* [2]; **winter oilseed rape** [1, 2]

Specific Off-Label Approvals (SOLAs)
- **swedes** *20102133* [2]
- **turnips** *20102133* [2]

Approval information
- Dimethenamid-p and metazachlor included in Annex I under EC Directive 91/414

Efficacy guidance
- Best results obtained from treatments to fine, firm and moist seedbeds
- Apply pre- or post-emergence of the crop and ideally before weed emergence
- Residual weed control may be reduced under prolonged dry conditions
- Weeds germinating from depth may not be controlled

Restrictions
- Maximum total dose on winter oilseed rape equivalent to one full dose treatment
- Do not disturb soil after application
- Do not treat broadcast crops until they have attained two fully expanded cotyledons
- Do not use on Sands, Very Light soils, or soils containing 10% organic matter
- Do not apply when heavy rain forecast or on soils waterlogged or prone to waterlogging
- Do not treat crops suffering from stress from any cause

FOR FULL CONDITIONS OF USE ALWAYS READ THE PRODUCT LABEL

- Do not use pre-emergence when crop seed has started to germinate or if not covered with 15 mm of soil
- A maximum total dose of 1000 g.ai/ha may only be applied every third year to the same field

Crop-specific information
- Latest use: before 7th true leaf for winter oilseed rape
- All varieties of winter oilseed rape may be treated

Following crops guidance
- Any crop may follow a normally harvested treated winter oilseed rape crop. Ploughing is not essential before a following cereal crop but is required for all other crops
- In the event of failure of a treated crop winter wheat (excluding durum) or winter barley may be drilled in the same autumn, and any cereal (excluding durum wheat), spring oilseed rape, peas or field beans may be sown in the following spring. Ploughing to at least 150 mm should precede in all cases

Environmental safety
- Dangerous for the environment
- Very toxic to aquatic organisms
- Take extreme care to avoid spray drift onto non-crop plants outside the target area
- Some pesticides pose a greater threat of contamination of water than others and metazachlor is one of these pesticides. Take special care when applying metazachlor near water and do not apply if heavy rain is forecast
- LERAP Category B

Hazard classification and safety precautions
  Hazard Harmful, Dangerous for the environment
  Transport code 9
  Packaging group III
  UN number 3082
  Risk phrases R22a, R36, R38, R43, R50, R53a
  Operator protection A, C, H; U05a, U08, U20c
  Environmental protection E15a, E16a, E34, E38
  Storage and disposal D01, D02, D09a, D10c, D12a
  Medical advice M03, M05a

184 dimethenamid-p + metazachlor + quinmerac

A soil-acting herbicide mixture for use in oilseed rape
HRAC mode of action code: K3 + K3 + O

See also metazachlor
         quinmerac

Products

1	Elk	BASF	200:200:100	SE	13974
2	Katamaran Turbo	BASF	200:200:100	SE	13999
3	Shadow	BASF	200:200:100	SE	14000

Uses
- Annual dicotyledons in *winter oilseed rape*
- Annual grasses in *winter oilseed rape*
- Cleavers in *winter oilseed rape*

Approval information
- Dimethenamid-p, metazachlor and quinmerac included in Annex I under EC Directive 91/414.

Restrictions
- A maximum total dose of 1000 g.ai/ha may only be applied every third year to the same field

Environmental safety
- Some pesticides pose a greater threat of contamination of water than others and metazachlor is one of these pesticides. Take special care when applying metazachlor near water and do not apply if heavy rain is forecast
- LERAP Category B

SEE SECTION 3 FOR PRODUCTS ALSO REGISTERED

Pesticide Profiles

Hazard classification and safety precautions
   Hazard Irritant, Dangerous for the environment
   Transport code 9
   Packaging group III
   UN number 3082
   Risk phrases R43, R50, R53a
   Operator protection A, H; U05a, U08, U14, U20c
   Environmental protection E07a, E15b, E16a, E34, E38
   Storage and disposal D01, D02, D09a, D10c, D12a
   Medical advice M03, M05a

185   dimethoate

A contact and systemic organophosphorus insecticide and acaricide
IRAC mode of action code: 1B

Products

1	Danadim Progress	Headland	400 g/l	EC	12208
2	Dimethoate 40	BASF	400 g/l	EC	13949

Uses
- Aphids in **fodder beet** (excluding Myzus persicae), **mangels** (excluding Myzus persicae), **ornamental plant production**, **red beet** (excluding Myzus persicae), **sugar beet** (excluding Myzus persicae), **triticale**, **winter rye**, **winter wheat** [1, 2]; **grass seed crops**, **mangel seed crops**, **spring rye**, **spring wheat**, **sugar beet seed crops** (excluding Myzus persicae) [2]
- Cabbage aphid in **calabrese** [2]; **calabrese** (off-label) [1]
- General insect control in **calabrese** (off-label), **celeriac** (off-label), **chinese cabbage** (off-label), **collards** (off-label), **compost** (off-label - for watercress propagation), **kale** (off-label), **kohlrabi** (off-label), **protected broccoli** (off-label), **protected cabbages** (off-label), **protected calabrese** (off-label), **protected cauliflowers** (off-label), **protected celeriac** (off-label), **protected chinese cabbage** (off-label), **protected collards** (off-label), **protected kale** (off-label), **protected kohlrabi** (off-label), **protected roscoff cauliflowers** (off-label), **protected savoy cabbage** (off-label), **protected spinach** (off-label), **roscoff cauliflowers** (off-label), **savoy cabbage** (off-label) [1]
- Insect pests in **calabrese** [2]
- Leaf miner in **fodder beet**, **mangels**, **ornamental plant production**, **red beet**, **sugar beet** [1, 2]
- Red spider mites in **ornamental plant production** [1, 2]
- Wheat bulb fly in **spring wheat** [2]; **winter wheat** [1, 2]

Specific Off-Label Approvals (SOLAs)
- **calabrese** 20050682 [1]
- **celeriac** 20050682 [1]
- **chinese cabbage** 20050682 [1]
- **collards** 20050682 [1]
- **compost** (for watercress propagation) 20050683 [1]
- **kale** 20050682 [1]
- **kohlrabi** 20050682 [1]
- **protected broccoli** 20050682 [1]
- **protected cabbages** 20050682 [1]
- **protected calabrese** 20050682 [1]
- **protected cauliflowers** 20050682 [1]
- **protected celeriac** 20050682 [1]
- **protected chinese cabbage** 20050682 [1]
- **protected collards** 20050682 [1]
- **protected kale** 20050682 [1]
- **protected kohlrabi** 20050682 [1]
- **protected roscoff cauliflowers** 20050682 [1]
- **protected savoy cabbage** 20050682 [1]
- **protected spinach** 20050682 [1]
- **roscoff cauliflowers** 20050682 [1]
- **savoy cabbage** 20050682 [1]

FOR FULL CONDITIONS OF USE ALWAYS READ THE PRODUCT LABEL

dimethoate

Approval information
- Dimethoate included in Annex I under EC Directive 91/414
- In 2006 CRD required that all products containing this active ingredient should carry the following warning in the main area of the container label: "Dimethoate is an anticholinesterase organophosphate. Handle with care"

Efficacy guidance
- Chemical has quick knock-down effect and systemic activity lasts for up to 14 d
- With some crops, products differ in range of pests listed as controlled. Uses section above provides summary. See labels for details
- For most pests apply when pest first seen and repeat 2-3 wk later or as necessary. Timing and number of sprays varies with crop and pest. See labels for details
- Best results achieved when crop growing vigorously. Systemic activity reduced when crops suffering from drought or other stress
- In hot weather apply in early morning or late evening
- Where aphids or spider mites resistant to organophosphorus compounds occur control is unlikely to be satisfactory and repeat treatments may result in lower levels of control

Restrictions
- Contains an anticholinesterase organophosphorus compound. Do not use if under medical advice not to work with such compounds
- Maximum number of treatments 6 per crop for Brussels sprouts, broccoli, cauliflower, calabrese, lettuce; 4 per crop for grass seed crops, cereals; 2 per crop for beet crops, triticale, rye; 1 per crop for mangel and sugar beet seed crops
- On beet crops only one treatment per crop may be made for control of leaf miners (max 84 g a.i./ha) and black bean aphid (max 400 g a.i./ha)
- In beet crops resistant strains of peach-potato aphid (*Myzus persicae*) are common and dimethoate products must not be used to control this pest
- Test for varietal susceptibility on all unusual plants or new cultivars
- Do not tank mix with alkaline materials. See label for recommended tank-mixes
- Consult processor before spraying crops grown for processing
- Must not be applied to cereals if any product containing a pyrethroid insecticide or dimethoate has been sprayed after the start of ear emergence (GS 51)

Crop-specific information
- Latest use: before 30 Jun in yr of harvest for beet crops; flowering just complete stage (GS 30) for wheat, rye, triticale [2]; before 31 Mar in yr of harvest for aerial application to cereals
- HI brassicas 21 d; grass seed crops, lettuce, cereals 14 d; ornamental plant production 7 d

Environmental safety
- Dangerous for the environment
- Harmful to aquatic organisms
- Flammable
- Harmful to game, wild birds and animals
- Harmful to livestock. Keep all livestock out of treated areas for at least 7 d
- Likely to cause adverse effects on beneficial arthropods
- High risk to non-target insects or other arthropods. Do not treat cereals after 1 Apr within 6 m of edge of crop
- High risk to bees. Do not apply to crops in flower or to those in which bees are actively foraging. Do not apply when flowering weeds are present
- Surface residues may also cause bee mortality following spraying
- Dangerous to fish or other aquatic life. Do not contaminate surface waters or ditches with chemical or used container
- Keep in original container, tightly closed, in a safe place, under lock and key
- LERAP Category A

Hazard classification and safety precautions
Hazard Harmful, Flammable
Transport code 6.1
Packaging group III
UN number 3017
Risk phrases R20, R21, R48 [2]; R22a, R43 [1, 2]; R52 [1]

SEE SECTION 3 FOR PRODUCTS ALSO REGISTERED

Pesticide Profiles

Operator protection A, H, M [1, 2]; C, J [2]; U02a, U04a, U19a [1, 2]; U05a, U10, U13, U20a [2]; U08, U14, U15, U20b [1]
Environmental protection E06c [1, 2] (7 d); E10b, E12a, E12e, E13b, E16c, E16d, E22a, E34 [1, 2]; E18 [2]
Storage and disposal D01, D02, D05, D09b [1, 2]; D08, D10c [2]; D12b [1]
Medical advice M01, M03 [1, 2]; M05a [2]

186 dimethomorph

A cinnamic acid fungicide with translaminar activity available only in mixtures
FRAC mode of action code: 40

See also ametoctradin + dimethomorph

187 dimethomorph + mancozeb

A systemic and protectant fungicide for potato blight control
FRAC mode of action code: 40 + M3

See also mancozeb

Products

1	Dunkirke	AgChem Access	7.5:66.7% w/w	WG	13796
2	Invader	BASF	7.5:66.7% w/w	WG	11978

Uses
- Blight in **potatoes** [1, 2]
- Downy mildew in **bulb onions** *(off-label)*, **chinese cabbage** *(off-label)*, **chives** *(off-label)*, **choi sum** *(off-label)*, **cress** *(off-label)*, **frise** *(off-label)*, **herbs (see appendix 6)** *(off-label)*, **lamb's lettuce** *(off-label)*, **lettuce** *(off-label)*, **pak choi** *(off-label)*, **poppies for morphine production** *(off-label)*, **radicchio** *(off-label)*, **salad brassicas** *(off-label - for baby leaf production)*, **salad onions** *(off-label)*, **scarole** *(off-label)*, **shallots** *(off-label)*, **tatsoi** *(off-label)* [2]
- White tip in **leeks** *(off-label)* [2]

Specific Off-Label Approvals (SOLAs)
- *bulb onions* 20090214 [2]
- *chinese cabbage* 20063044 [2]
- *chives* 20063044 [2]
- *choi sum* 20063044 [2]
- *cress* 20063044 [2]
- *frise* 20063044 [2]
- *herbs (see appendix 6)* 20063044 [2]
- *lamb's lettuce* 20063044 [2]
- *leeks* 20073434 [2]
- *lettuce* 20063044 [2]
- *pak choi* 20063044 [2]
- *poppies for morphine production* 20073433 [2]
- *radicchio* 20063044 [2]
- *salad brassicas (for baby leaf production)* 20063044 [2]
- *salad onions* 20062953 [2]
- *scarole* 20063044 [2]
- *shallots* 20090214 [2]
- *tatsoi* 20063044 [2]

Approval information
- Dimethomorph and mancozeb included in Annex I under EC Directive 91/414

Efficacy guidance
- Commence treatment as soon as there is a risk of blight infection
- In the absence of a warning treatment should start before the crop meets along the rows
- Repeat treatments every 7-14 d depending in the degree of infection risk
- Irrigated crops should be regarded as at high risk and treated every 7 d

FOR FULL CONDITIONS OF USE ALWAYS READ THE PRODUCT LABEL

- For best results good spray coverage of the foliage is essential
- To minimise the likelihood of development of resistance these products should be used in a planned Resistance Management strategy. See Section 5 for more information

Restrictions
- Maximum total dose equivalent to eight full dose treatments on potatoes

Crop-specific information
- HI 7 d for all crops

Environmental safety
- Dangerous for the environment
- Very toxic to aquatic organisms
- LERAP Category B

Hazard classification and safety precautions
    Hazard Irritant, Dangerous for the environment
    Transport code 9
    Packaging group III
    UN number 3077
    Risk phrases R36, R37, R38, R43, R50, R53a
    Operator protection A; U02a, U04a, U05a, U08, U13, U19a, U20b
    Environmental protection E16a, E16b, E38
    Storage and disposal D01, D02, D05, D09a, D12a
    Medical advice M05a

188 dimoxystrobin

A protectant strobilurin fungicide for cereals available only in mixtures
FRAC mode of action code: 11

189 dimoxystrobin + epoxiconazole

A protectant and curative fungicide mixture for cereals
FRAC mode of action code: 11 + 3

See also epoxiconazole

Products
    Swing Gold   BASF   133:50 g/l   SC   11658

Uses
- Brown rust in **winter wheat**
- Foliar disease control in **durum wheat** *(off-label)*, **grass seed crops** *(off-label)*, **spring rye** *(off-label)*, **triticale** *(off-label)*, **winter rye** *(off-label)*
- Fusarium ear blight in **winter wheat**
- Glume blotch in **winter wheat**
- Septoria leaf blotch in **winter wheat**
- Tan spot in **winter wheat**

Specific Off-Label Approvals (SOLAs)
- ***durum wheat*** *20062642*
- ***grass seed crops*** *20062642*
- ***spring rye*** *20062642*
- ***triticale*** *20062642*
- ***winter rye*** *20062642*

Approval information
- Dimoxystrobin and epoxiconazole included in Annex I under EC Directive 91/414

Efficacy guidance
- For best results apply from the start of ear emergence
- Dimoxystrobin is a member of the QoI cross resistance group. Product should be used preventatively and not relied on for its curative potential

SEE SECTION 3 FOR PRODUCTS ALSO REGISTERED

Pesticide Profiles

- Use product as part of an Integrated Crop Management strategy incorporating other methods of control, including where appropriate other fungicides with a different mode of action. Do not apply more than two foliar applications of QoI containing products to any cereal crop
- There is a significant risk of widespread resistance occurring in *Septoria tritici* populations in UK. Failure to follow resistance management action may result in reduced levels of disease control
- Epoxiconazole is a DMI fungicide. Resistance to some DMI fungicides has been identified in Septoria leaf blotch which may seriously affect performance of some products. For further advice contact a specialist advisor and visit the Fungicide Resistance Action Group (FRAG)-UK website

Restrictions
- Maximum number of treatments 1 per crop
- Do not apply before the start of ear emergence

Crop-specific information
- Latest use: up to and including flowering (GS 69)

Environmental safety
- Dangerous for the environment
- Very toxic to aquatic organisms
- Avoid drift onto neighbouring crops
- LERAP Category B

Hazard classification and safety precautions
Hazard Harmful, Dangerous for the environment
Transport code 9
Packaging group III
UN number 3082
Risk phrases R20, R22a, R40, R50, R53a, R63
Operator protection A; U05a, U20b
Environmental protection E15a, E16a, E34, E38
Storage and disposal D01, D02, D05, D09a, D10c, D12a
Medical advice M05a

190 diquat

A non-residual bipyridyl contact herbicide and crop desiccant
HRAC mode of action code: D

Products

1	Agriguard Diquat	AgriGuard	200 g/l	SL	13763
2	Barclay D-Quat	Barclay	200 g/l	SL	14833
3	Brogue	AgriGuard	200 g/l	SL	13328
4	Brogue	AgriGuard	200 g/l	SL	14465
5	ChemSource Diquat	ChemSource	200 g/l	SL	14533
6	Clayton Diquat	Clayton	200 g/l	SL	12739
7	Clayton Diquat 200	Clayton	200 g/l	SL	13942
8	Clayton IQ	Clayton	200 g/l	SL	14519
9	Dessicash 200	Sharda	200 g/l	SL	14848
10	Diqua	Sharda	200 g/l	SL	14032
11	Diquash	Agrichem	200 g/l	SL	14849
12	Dragoon	Belcrop	200 g/l	SL	13927
13	Dragoon Gold	Belcrop	200 g/l	SL	14053
14	Dragoon Gold	Belcrop	200 g/l	SL	14973
15	Knoxdoon	AgChem Access	200 g/l	SL	13987
16	Mission	AgriChem BV	200 g/l	SL	13411
17	Quad-Glob 200 SL	Q-Chem	200 g/l	SL	14758
18	Reglone	Syngenta	200 g/l	SL	10534
19	Retro	Syngenta	200 g/L	SL	13841
20	Roquat 20	Q-Chem	200 g/l	SL	13474
21	Standon Googly	Standon	200 g/l	SL	13281
22	Standon Yorker	Standon	200 g/l	SL	14751
23	UPL Diquat	United Phosphorus	200 g/l	SL	14944
24	Woomera	Belcrop	200 g/l	SL	14972

FOR FULL CONDITIONS OF USE ALWAYS READ THE PRODUCT LABEL

diquat

Uses
- Annual and perennial weeds in **all edible crops (outdoor)** *(around only)*, **all non-edible crops (outdoor)** *(around only)*, **combining peas** *(dry harvested only)*, **hops**, **linseed**, **ornamental plant production**, **potatoes** *(weed control or dessication)*, **red clover** *(seed crop)*, **spring barley** *(animal feed only)*, **spring field beans**, **spring oats** *(animal feed only)*, **spring oilseed rape**, **sugar beet**, **white clover** *(seed crop)*, **winter barley** *(animal feed only)*, **winter field beans**, **winter oats** *(animal feed only)*, **winter oilseed rape** [16]; **barley for animal feed**, **oats for animal feed** [23]; **forest nurseries** *(off-label)*, **poppies for morphine production** *(off-label)* [17]
- Annual dicotyledons in **all edible crops (outdoor)** *(around)*, **all non-edible crops (outdoor)** *(around)*, **ornamental plant production**, **sugar beet** [1, 3, 5-15, 17, 18, 20-24]; **fodder beet** [13, 14]; **lucerne** *(off-label)* [18]
- Basal defoliation in **hops** [8]
- Chemical stripping in **hops** [17, 20]
- Chickweed in **all edible crops (outdoor)**, **all non-edible crops (outdoor)**, **ornamental plant production**, **potatoes**, **sugar beet** [19]
- Desiccation in **barley for animal feed**, **oats for animal feed** [23]; **borage for oilseed production** *(off-label)*, **french beans** *(off-label)*, **hemp for oilseed production** *(off-label)*, **herbs** *(see appendix 6)* *(off-label)*, **poppies for morphine production** *(off-label)*, **poppies for oilseed production** *(off-label)*, **sesame** *(off-label)*, **sunflowers** *(off-label)* [17]; **combining peas**, **linseed**, **red clover**, **spring barley**, **spring field beans**, **spring oats**, **spring oilseed rape**, **white clover**, **winter barley**, **winter field beans**, **winter oats**, **winter oilseed rape** [4]; **crambe** *(off-label)*, **hops** *(off-label)* [18]; **lupins** *(off-label)*, **sweet potato** *(off-label)* [17, 18]; **potatoes** [2, 4]
- General weed control in **chives** *(off-label)*, **herbs (see appendix 6)** *(off-label)*, **parsley** *(off-label)*, **poppies for morphine production** *(off-label)* [18]; **potatoes** [1, 3, 5-15, 17, 18, 20-24]
- Grass weeds in **all edible crops (outdoor)** *(around only)*, **all non-edible crops (outdoor)** *(around only)*, **combining peas** *(dry harvested)*, **hops**, **linseed**, **ornamental plant production**, **potatoes** *(weed control and dessication)*, **red clover** *(seed crop)*, **spring barley** *(animal feed)*, **spring field beans**, **spring oats** *(animal feed)*, **spring oilseed rape**, **sugar beet**, **white clover** *(seed crop)*, **winter barley** *(animal feed)*, **winter field beans**, **winter oats** *(animal feed)*, **winter oilseed rape** [16]
- Growth regulation in **hops** [7]
- Harvest management/desiccation in **combining peas**, **linseed**, **red clover**, **spring barley**, **spring field beans**, **spring oats**, **spring oilseed rape**, **white clover**, **winter barley**, **winter field beans**, **winter oats**, **winter oilseed rape** [4]
- Haulm destruction in **potatoes** [1, 3, 5-15, 17, 18, 20-24]
- Perennial dicotyledons in **lucerne** *(off-label)* [18]
- Pre-harvest desiccation in **borage for oilseed production** *(off-label)*, **brown mustard** *(off-label)*, **french beans** *(off-label)*, **hemp for oilseed production** *(off-label)*, **poppies for morphine production** *(off-label)*, **poppies for oilseed production** *(off-label)*, **sesame** *(off-label)*, **sunflowers** *(off-label)*, **white mustard** *(off-label)* [18]; **broad beans** [5-7, 9-11, 15, 18, 21, 23]; **clover seed crops**, **combining peas**, **linseed**, **spring field beans**, **spring oilseed rape**, **winter field beans**, **winter oilseed rape** [1, 3, 5-15, 17, 18, 20, 21, 23]; **laid barley** *(stockfeed only)*, **laid oats** *(stockfeed only)* [1, 3, 5-15, 17, 18, 20, 21]

Specific Off-Label Approvals (SOLAs)
- **borage for oilseed production** *20093348 expires 31 Dec 2011* [17], *20062390 expires 31 Dec 2011* [18]
- **brown mustard** *20062387 expires 31 Dec 2011* [18]
- **chives** *20062388 expires 31 Dec 2011* [18]
- **crambe** *20082092 expires 31 Dec 2011* [18]
- **forest nurseries** *20093350 expires 31 Dec 2011* [17]
- **french beans** *20093346 expires 31 Dec 2011* [17], *20062389 expires 31 Dec 2011* [18]
- **hemp for oilseed production** *20093351 expires 31 Dec 2011* [17], *20062386 expires 31 Dec 2011* [18]
- **herbs (see appendix 6)** *20093345 expires 31 Dec 2011* [17], *20062388 expires 31 Dec 2011* [18]
- **hops** *20102019 expires 31 Dec 2011* [18]
- **lucerne** *20082842 expires 31 Dec 2011* [18]
- **lupins** *20093353 expires 31 Dec 2011* [17], *20073232 expires 31 Dec 2011* [18]
- **parsley** *20062388 expires 31 Dec 2011* [18]

SEE SECTION 3 FOR PRODUCTS ALSO REGISTERED

- **poppies for morphine production** 20093349 expires 31 Dec 2011 [17], 20062385 expires 31 Dec 2011 [18]
- **poppies for oilseed production** 20093347 expires 31 Dec 2011 [17], 20062387 expires 31 Dec 2011 [18]
- **sesame** 20093347 expires 31 Dec 2011 [17], 20062387 expires 31 Dec 2011 [18]
- **sunflowers** 20093347 expires 31 Dec 2011 [17], 20062387 expires 31 Dec 2011 [18]
- **sweet potato** 20093352 expires 31 Dec 2011 [17], 20070734 expires 31 Dec 2011 [18]
- **white mustard** 20062387 expires 31 Dec 2011 [18]

Approval information
- Diquat included in Annex I under EC Directive 91/414
- All approvals for aquatic weed control expired in 2004
- Approval expiry 05 Mar 2011 [1]
- Approval expiry 31 Dec 2011 [3, 5-8, 12, 14, 16-19, 21-24]
- Approval expiry 04 Mar 2011 [20]
- NOT accepted by BBPA for use on malting barley

Efficacy guidance
- Acts rapidly on green parts of plants and rainfast in 15 min
- Best results for potato desiccation achieved by spraying in bright light and low humidity conditions
- For weed control in row crops apply as overall spray before crop emergence or before transplanting or as an inter-row treatment using a sprayer designed to prevent contamination of crop foliage by spray
- Use of authorised non-ionic wetter for improved weed control essential for most uses except potato desiccation

Restrictions
- Maximum number of treatments 1 per crop in most situations - see labels for details
- Do not apply potato haulm destruction treatment when soil dry. Tubers may be damaged if spray applied during or shortly after dry periods. See label for details of maximum allowable soil-moisture deficit and varietal drought resistance scores
- Do not add wetters to desiccant sprays for potatoes
- Consult processor before adding non-ionic or other wetter on peas. Staining may be increased
- Do not use treated straw or haulm as animal feed or bedding within 4 d of spraying
- Do not apply through ULV or mistblower equipment

Crop-specific information
- Latest use: before crop emergence for sugar beet and before transplanting ornamentals. Not specified for other crops where use varies according to situation
- HI zero when used as a crop desiccant but see label for advisory intervals
- For pre-harvest desiccation, apply to potatoes when tubers the desired size and to other crops when mature or approaching maturity. See label for details of timing and of period to be left before harvesting potatoes
- Apply as a chemical stripping treatment in hops when shoots have reached top wire [20]
- Treatment of hops under stress or when leaves are wet with a heavy dew may cause damage [20]
- If potato tubers are to be stored leave 14 d after treatment before lifting

Environmental safety
- Dangerous for the environment
- Very toxic to aquatic organisms
- Keep all livestock out of treated areas and away from treated water for at least 24 h
- Do not feed treated straw or haulm to livestock within 4 days of spraying
- Do not use on crops if the straw is to be used as animal feed/bedding
- Do not dump surplus herbicide in water or ditch bottoms

Hazard classification and safety precautions
Hazard Very toxic [6]; Toxic [1-5, 7-15, 17-24]; Dangerous for the environment [1-24]
Transport code 8
Packaging group III
UN number 1760
Risk phrases R22a [1-15, 17-24]; R23 [1-5, 7-24]; R25b [23]; R26 [6]; R36 [16, 17, 20]; R37, R38, R50, R53a [1-24]; R43 [7, 16, 17, 20]; R48 [1-15, 17, 18, 20-24]; R69 [16, 19]

FOR FULL CONDITIONS OF USE ALWAYS READ THE PRODUCT LABEL

dithianon

Operator protection A [1-24]; C [1, 3, 4, 16, 17, 20]; D [21]; H [2, 4-16, 18, 19, 21-24]; M [1, 3-20, 22-24]; P [19]; U02a, U04a, U05a, U08, U19a, U20b [1-24]; U04c [16]; U10 [1-5, 8-15, 17-24]; U15 [17, 20]
Environmental protection E06c [1-24] (24 h); E08a [1, 3, 4, 17, 20, 22] (4 d); E09 [2, 5]; E09 [6, 7] (4 d); E09 [8-16, 18, 19, 21]; E09 [23] (4 days); E09 [24]; E15a [16]; E15b [1-15, 17, 18, 20-24]; E34 [1-24]; E38 [1-5, 8-24]
Storage and disposal D01, D02, D09a, D12a [1-24]; D05 [1-15, 17, 18, 20-24]; D10a [16]; D10b [6, 7]; D10c [1-5, 8-15, 17-24]
Medical advice M04a

191 dithianon

A protectant and eradicant dicarbonitrile fungicide for scab control
FRAC mode of action code: M9

Products

1	Dartagnan WG	ChemSource	70% w/w	WG	14963
2	Dithianon Flowable	BASF	750 g/l	SC	10219
3	Dithianon WG	BASF	70% w/w	WG	12538

Uses
- Scab in **apples**, **pears** [1-3]; **quinces** *(off-label)* [2]

Specific Off-Label Approvals (SOLAs)
- *quinces* 20061038 [2]

Approval information
- Dithianon has not been included in Annex 1 but the date of withdrawal has yet to be set.

Efficacy guidance
- Apply at bud-burst and repeat every 7-14 d until danger of scab infection ceases
- Application at high rate within 48 h of a Mills period prevents new infection
- Spray programme also reduces summer infection with apple canker

Restrictions
- Maximum number of treatments on apples and pears 8 per crop [2]; 12 per crop [3]
- Do not use on Golden Delicious apples after green cluster
- Do not mix with lime sulphur or highly alkaline products

Crop-specific information
- HI 4 wk for apples, pears

Environmental safety
- Dangerous for the environment
- Very toxic to aquatic organisms

Hazard classification and safety precautions
Hazard Harmful, Dangerous for the environment
Transport code 9
Packaging group III
UN number 3077 [1, 3]; 3082 [2]
Risk phrases R20 [2]; R22a, R50, R53a [1-3]; R41 [1, 3]
Operator protection A [1-3]; D [1, 3]; H [2]; U05a [1-3]; U08, U19a, U20a [2]; U11 [1, 3]
Environmental protection E15a, E34, E38
Storage and disposal D01, D02, D05 [1-3]; D09a, D10b [2]; D12a [1, 3]
Medical advice M03 [2]; M05a [1-3]

192 dithianon + pyraclostrobin

A protectant fungicide mixture for apples and pears
FRAC mode of action code: M9 + 11

See also pyraclostrobin

Products

Maccani	BASF	12:4% w/w	WG	13227

SEE SECTION 3 FOR PRODUCTS ALSO REGISTERED

Pesticide Profiles

Uses
- Powdery mildew in *apples*
- Scab in *apples*, *pears*

Approval information
- Pyraclostrobin included in Annex I under EC Directive 91/414

Efficacy guidance
- Best resullts obtained from treatments applied from bud burst before disease development and repeated at 10 d intervals
- Water volume should be adjusted according to size of tree and leaf area to ensure good coverage
- Pyraclostrobin is a member of the QoI cross resistance group. Product should be used preventatively at the full recommended dose and not relied on for its curative potential
- Where field performance may be adversely affected apply QoI containing fungicides in mixtures in strict alternation with fungicides from a different cross-resistance group

Restrictions
- Maximum number of treatments 4 per crop. However where the total number of fungicide treatments is less than 12, the maximum number of applications of any QoI containing products should be three
- Consult processors before use on crops destined for fruit preservation

Crop-specific information
- HI 35 d for apples, pears
- Some apple varieties may be considered to be susceptible to russetting caused by the dithianon component during fruit development

Environmental safety
- Dangerous for the environment
- Very toxic to aquatic organisms
- Broadcast air-assisted LERAP (40 m)

Hazard classification and safety precautions
Hazard Harmful, Dangerous for the environment
Transport code 9
Packaging group III
UN number 3077
Risk phrases R20, R22a, R36, R50, R53a
Operator protection A, C, D; U05a, U11, U20b
Environmental protection E15b, E34, E38; E17b (40 m)
Storage and disposal D01, D02, D05, D08, D11a, D12a
Medical advice M03, M05a

193 dodine

A protectant and eradicant guanidine fungicide
FRAC mode of action code: M7

Products
1	Clayton Scabius	Clayton	450 g/l	SC	13339
2	Dodifun 400 SC	Belcrop	400 g/l	SC	11657
3	Greencrop Budburst	Greencrop	450 g/l	SC	11042
4	Radspor FL	Truchem	450 g/l	SC	01685

Uses
- Currant leaf spot in *bilberries* (off-label), *blueberries* (off-label), *cranberries* (off-label), *gooseberries* (off-label), *redcurrants* (off-label), *whitecurrants* (off-label) [4]; *blackcurrants* [1-4]
- Scab in *apples*, *pears* [1-4]; *quinces* (off-label) [4]

Specific Off-Label Approvals (SOLAs)
- *bilberries* 20060617 [4]
- *blueberries* 20060617 [4]
- *cranberries* 20060617 [4]
- *gooseberries* 20060617 [4]
- *quinces* 20060616 [4]

FOR FULL CONDITIONS OF USE ALWAYS READ THE PRODUCT LABEL

epoxiconazole

- *redcurrants* 20060617 [4]
- *whitecurrants* 20060617 [4]

Efficacy guidance
- Apply protective spray on apples and pears at bud-burst and at 10-14 d intervals until late Jun to early Jul
- Apply post-infection spray within 36 h of rain responsible for initiating infection. Where scab already present spray prevents production of spores
- On blackcurrants commence spraying at early grape stage and repeat at 2-3 wk intervals, and at least once after picking

Restrictions
- Do not apply in very cold weather (under 5 °C) or under slow drying conditions to pears or dessert apples during bloom or immediately after petal fall
- Do not mix with lime sulphur or tetradifon
- Consult processors before use on crops grown for processing

Crop-specific information
- Latest use: early Jul for culinary apples; pre-blossom for dessert apples and pears

Environmental safety
- Dangerous for the environment
- Very toxic to aquatic organisms

Hazard classification and safety precautions
Hazard Harmful, Dangerous for the environment
Transport code 9
Packaging group III
UN number 3082
Risk phrases R21 [2, 3]; R22a, R36, R38, R43, R50, R53a [1-4]; R37, R41 [1, 4]
Operator protection A, C; U05a, U08, U19a, U20b [1-4]; U11 [1, 4]
Environmental protection E15a, E34 [1-4]; E38 [4]
Storage and disposal D01, D02, D09a, D10b [1-4]; D05 [1-3]; D12a [1]
Medical advice M03 [1, 4]; M04a [2, 3]; M05a [1]

194 epoxiconazole

A systemic, protectant and curative triazole fungicide for use in cereals
FRAC mode of action code: 3

See also boscalid + epoxiconazole
dimoxystrobin + epoxiconazole

Products

1	Agriguard Epoxiconazole	AgriGuard	125 g/l	SC	09407
2	Clayton Oust	Clayton	125 g/l	SC	11588
3	Clayton Oust SC	Clayton	125 g/l	SC	14871
4	Cortez	Makhteshim	125 g/l	SC	13932
5	Crown	AgriGuard	125 g/l	SC	14830
6	EA Epoxi	European Ag	125 g/l	SC	14832
7	Epic	BASF	125 g/l	SC	12136
8	Epoxi	Goldengrass	125 g/l	SC	14292
9	Epoxi 125	Goldengrass	125 g/l	SC	14291
10	Epoxi 125 SC	Goldengrass	125 g/l	SC	14442
11	Opus	BASF	125 g/l	SC	12057
12	Oracle	ChemSource	125 g/l	SC	14470
13	Oropa	AgChem Access	125 g/l	SC	13435
14	Overture	AgriGuard	125 g/l	SC	12587
15	Rubric	Headland	125 g/l	SC	14118
16	Standon Epoxiconazole	Standon	125 g/l	SC	09517
17	Torrent	AgriGuard	125 g/l	SC	14374

Uses
- Brown rust in **spring barley**, **winter barley**, **winter wheat** [1-17]; **spring rye**, **spring wheat**, **triticale**, **winter rye** [2-15, 17]
- Eyespot in **winter barley** *(reduction)*, **winter wheat** *(reduction)* [1-17]

SEE SECTION 3 FOR PRODUCTS ALSO REGISTERED

Pesticide Profiles

- Foliar disease control in ***durum wheat*** *(off-label)* [4, 7, 11]; ***grass seed crops*** *(off-label)* [1, 2, 4, 11, 16]
- Fusarium ear blight in ***spring wheat***, ***winter wheat*** [2-5, 7-14, 17]; ***winter wheat*** *(reduction)* [1, 16]
- Glume blotch in ***spring wheat***, ***triticale*** [2-15, 17]; ***winter wheat*** [1-17]
- Net blotch in ***spring barley***, ***winter barley*** [1-17]
- Powdery mildew in ***durum wheat*** *(off-label)*, ***grass seed crops*** *(off-label)* [15]; ***spring barley***, ***winter barley***, ***winter wheat*** [1-17]; ***spring oats***, ***spring rye***, ***spring wheat***, ***triticale***, ***winter oats***, ***winter rye*** [2-15, 17]
- Rhynchosporium in ***spring barley***, ***winter barley*** [1-17]; ***spring rye***, ***winter rye*** [2-15, 17]
- Septoria leaf blotch in ***spring wheat***, ***triticale*** [2-15, 17]; ***winter wheat*** [1-17]
- Sooty moulds in ***spring wheat*** *(reduction)* [2-15, 17]; ***winter wheat*** *(reduction)* [1-17]
- Yellow rust in ***spring barley***, ***winter barley***, ***winter wheat*** [1-17]; ***spring rye***, ***spring wheat***, ***triticale***, ***winter rye*** [2-15, 17]

Specific Off-Label Approvals (SOLAs)
- ***durum wheat*** *20080981* [4], *20061053* [7], *20061051* [11], *20090236* [15]
- ***grass seed crops*** *20061056* [1], *20061055* [2], *20080981* [4], *20061051* [11], *20090236* [15], *20061050* [16]

Approval information
- Accepted by BBPA for use on malting barley (before ear emergence only)
- Epoxiconazole included in Annex I under EC Directive 91/414

Efficacy guidance
- Apply at the start of foliar disease attack
- Optimum effect against eyespot achieved by spraying between leaf-sheath erect and second node detectable stages (GS 30-32)
- Best control of ear diseases of wheat obtained by treatment during ear emergence
- Mildew control improved by use of tank mixtures. See label for details
- For Septoria spray after third node detectable stage (GS 33) when weather favouring disease development has occurred
- Epoxiconazole is a DMI fungicide. Resistance to some DMI fungicides has been identified in Septoria leaf blotch which may seriously affect performance of some products. For further advice contact a specialist advisor and visit the Fungicide Resistance Action Group (FRAG)-UK website

Restrictions
- Maximum total dose equivalent to two full dose treatments
- Product may cause damage to broad-leaved plant species
- Avoid spray drift onto neighbouring crops

Crop-specific information
- Latest use: up to and including flowering just complete (GS 69) in wheat, rye, triticale; up to and including emergence of ear just complete (GS 59) in barley, oats

Environmental safety
- Dangerous for the environment
- Very toxic to aquatic organisms
- LERAP Category B

Hazard classification and safety precautions
    Hazard Harmful [2-17]; Dangerous for the environment [1-17]
    Transport code 9
    Packaging group III
    UN number 3082
    Risk phrases R38 [2, 5-17]; R40, R50, R62, R63 [2-17]; R51 [1]; R53a [1-17]
    Operator protection A [1-17]; H [5]; U05a [1-17]; U20a [1]; U20b [2-17]
    Environmental protection E15a, E16a [1-17]; E16b [1, 14, 16]; E38 [4-13, 15, 17]
    Storage and disposal D01, D02, D05, D09a [1-17]; D10b [1, 16]; D10c [2-15, 17]; D12a [2-13, 15, 17]
    Medical advice M05a [2-13, 15, 17]; M05b [16]

FOR FULL CONDITIONS OF USE ALWAYS READ THE PRODUCT LABEL

195 epoxiconazole + fenpropimorph

A systemic, protectant and curative fungicide mixture for cereals
FRAC mode of action code: 3 + 5

See also fenpropimorph

Products

1	Eclipse	BASF	84:250 g/l	SE	11731
2	Opus Team	BASF	84:250 g/l	SE	11759

Uses
- Brown rust in **spring barley**, **spring rye**, **spring wheat**, **triticale**, **winter barley**, **winter rye**, **winter wheat**
- Eyespot in **winter barley** *(reduction)*, **winter wheat** *(reduction)*
- Foliar disease control in **durum wheat** *(off-label)*, **grass seed crops** *(off-label)*
- Fusarium ear blight in **spring wheat**, **winter wheat**
- Glume blotch in **spring wheat**, **triticale**, **winter wheat**
- Net blotch in **spring barley**, **winter barley**
- Powdery mildew in **spring barley**, **spring oats**, **spring rye**, **spring wheat**, **triticale**, **winter barley**, **winter oats**, **winter rye**, **winter wheat**
- Rhynchosporium in **spring barley**, **spring rye**, **winter barley**, **winter rye**
- Septoria leaf blotch in **spring wheat**, **triticale**, **winter wheat**
- Sooty moulds in **spring wheat** *(reduction)*, **winter wheat** *(reduction)*
- Yellow rust in **spring barley**, **spring rye**, **spring wheat**, **triticale**, **winter barley**, **winter rye**, **winter wheat**

Specific Off-Label Approvals (SOLAs)
- ***durum wheat*** *20062644* [1], *20062645* [2]
- ***grass seed crops*** *20062644* [1], *20062645* [2]

Approval information
- Epoxiconazole and fenpropimorph included in Annex I under EC Directive 91/414
- Accepted by BBPA for use on malting barley

Efficacy guidance
- Apply at the start of foliar disease attack
- Optimum effect against eyespot achieved by spraying between leaf-sheath erect and second node detectable stages (GS 30-32)
- Best control of ear diseases obtained by treatment during ear emergence
- For Septoria spray after third node detectable stage (GS 33) when weather favouring disease development has occurred
- Epoxiconazole is a DMI fungicide. Resistance to some DMI fungicides has been identified in Septoria leaf blotch which may seriously affect performance of some products. For further advice contact a specialist advisor and visit the Fungicide Resistance Action Group (FRAG)-UK website

Restrictions
- Maximum total dose equivalent to two full dose treatments
- Product may cause damage to broad-leaved plant species

Crop-specific information
- Latest use: up to and including flowering just complete (GS 69) in wheat, rye, triticale; up to and including emergence of ear just complete (GS 59) in barley, oats

Environmental safety
- Dangerous for the environment
- Toxic to aquatic organisms
- Avoid spray drift onto neighbouring crops
- LERAP Category B

Hazard classification and safety precautions
Hazard Harmful, Dangerous for the environment
Transport code 9
Packaging group III
UN number 3082

SEE SECTION 3 FOR PRODUCTS ALSO REGISTERED

Risk phrases R20, R40, R43, R51, R53a, R62, R63
Operator protection A, H; U05a, U20b
Environmental protection E15a, E16a, E38
Storage and disposal D01, D02, D05, D09a, D10c, D12a
Medical advice M05a

196 epoxiconazole + fenpropimorph + kresoxim-methyl

A protectant, systemic and curative fungicide mixture for cereals
FRAC mode of action code: 3 + 5 + 11

See also fenpropimorph
kresoxim-methyl

Products

1	Bullseye	AgriGuard	125:150:125 g/l	SE	12278
2	Mantra	BASF	125:150:125 g/l	SE	11728

Uses
- Brown rust in **spring barley**, **spring rye**, **spring wheat**, **triticale**, **winter barley**, **winter rye**, **winter wheat** [1, 2]
- Eyespot in **spring oats** (reduction), **spring rye** (reduction), **triticale** (reduction), **winter barley** (reduction), **winter oats** (reduction), **winter rye** (reduction), **winter wheat** (reduction) [1, 2]
- Foliar disease control in **durum wheat** (off-label), **grass seed crops** (off-label) [2]
- Fusarium ear blight in **spring wheat** (reduction), **winter wheat** (reduction) [1, 2]
- Glume blotch in **spring wheat**, **triticale**, **winter wheat** [1, 2]
- Net blotch in **spring barley**, **winter barley** [1, 2]
- Powdery mildew in **spring barley**, **spring oats**, **spring rye**, **triticale**, **winter barley**, **winter oats**, **winter rye** [1, 2]
- Rhynchosporium in **spring barley**, **spring rye**, **winter barley**, **winter rye** [1, 2]
- Septoria leaf blotch in **spring wheat**, **triticale**, **winter wheat** [1, 2]
- Sooty moulds in **spring wheat** (reduction), **winter wheat** (reduction) [1, 2]
- Yellow rust in **spring barley**, **spring rye**, **spring wheat**, **triticale**, **winter barley**, **winter rye**, **winter wheat** [1, 2]

Specific Off-Label Approvals (SOLAs)
- *durum wheat* 20061101 [2]
- *grass seed crops* 20061101 [2]

Approval information
- Kresoxim-methyl, epoxiconazole and fenpropimorph included in Annex I under EC Directive 91/414
- Accepted by BBPA for use on malting barley (before ear emergence only)

Efficacy guidance
- For best results spray at the start of foliar disease attack and repeat if infection conditions persist
- Optimum effect against eyespot obtained by treatment between leaf sheaths erect and first node detectable stages (GS 30-32)
- For protection against ear diseases apply during ear emergence
- Epoxiconazole is a DMI fungicide. Resistance to some DMI fungicides has been identified in Septoria leaf blotch which may seriously affect performance of some products. For further advice contact a specialist advisor and visit the Fungicide Resistance Action Group (FRAG)-UK website
- Kresoxim-methyl is a member of the QoI cross resistance group. Product should be used preventatively and not relied on for its curative potential
- Use product as part of an Integrated Crop Management strategy incorporating other methods of control, including where appropriate other fungicides with a different mode of action. Do not apply more than two foliar applications of QoI containing products to any cereal crop
- There is a significant risk of widespread resistance occurring in *Septoria tritici* populations in UK. Failure to follow resistance management action may result in reduced levels of disease control
- Strains of barley powdery mildew resistant to QoIs are common in the UK

Restrictions
- Maximum total dose equivalent to two full dose treatments

FOR FULL CONDITIONS OF USE ALWAYS READ THE PRODUCT LABEL

epoxiconazole + fenpropimorph + metrafenone

Crop-specific information
- Latest use: mid flowering (GS 65) for wheat, rye, triticale; completion of ear emergence (GS 59) for barley, oats

Environmental safety
- Dangerous for the environment
- Very toxic to aquatic organisms
- Avoid spray drift onto neighbouring crops. Product may damage broad-leaved species
- LERAP Category B

Hazard classification and safety precautions
Hazard Harmful, Dangerous for the environment
Transport code 9
Packaging group III
UN number 3082
Risk phrases R40, R50, R53a, R62, R63
Operator protection A, H; U05a, U14, U20b
Environmental protection E15a, E16a, E38
Storage and disposal D01, D02, D05, D09a, D10c, D12a
Medical advice M05a

197　epoxiconazole + fenpropimorph + metrafenone

A broad spectrum fungicide mixture for cereals
FRAC mode of action code: 3 + 5 + U8

See also fenpropimorph
metrafenone

Products

1	Capalo	BASF	62.5:200:75 g/l	SE	13170
2	Stiletto	BASF	62.5:200:75 g/l	SE	14729

Uses
- Brown rust in **spring barley**, **spring rye**, **spring wheat**, **triticale**, **winter barley**, **winter rye**, **winter wheat**
- Crown rust in **spring oats**, **winter oats**
- Fusarium ear blight in **spring wheat** (reduction), **winter wheat** (reduction)
- Glume blotch in **spring wheat**, **triticale**, **winter wheat**
- Net blotch in **spring barley**, **winter barley**
- Powdery mildew in **spring barley**, **spring oats**, **spring rye**, **spring wheat**, **triticale**, **winter barley**, **winter oats**, **winter rye**, **winter wheat**
- Rhynchosporium in **spring barley**, **spring rye**, **winter barley**, **winter rye**
- Septoria leaf blotch in **spring wheat**, **triticale**, **winter wheat**
- Sooty moulds in **spring wheat** (reduction), **winter wheat** (reduction)
- Yellow rust in **spring barley**, **spring rye**, **spring wheat**, **triticale**, **winter barley**, **winter rye**, **winter wheat**

Approval information
- Epoxiconazole, fenpropimorph and metrafenone included in Annex I under EC Directive 91/414

Efficacy guidance
- For best results spray at the start of foliar disease attack and repeat if infection conditions persist
- Optimum effect against eyespot obtained by treatment between leaf sheaths erect and first node detectable stages (GS 30-32)
- For protection against ear diseases apply during ear emergence
- Treatment for control of Septoria diseases should normally be made after third node detectable stage if the disease is present and favourable weather for disease development has occurred
- Epoxiconazole is a DMI fungicide. Resistance to some DMI fungicides has been identified in Septoria leaf blotch which may seriously affect performance of some products. For further advice contact a specialist advisor and visit the Fungicide Resistance Action Group (FRAG)-UK website
- Should be used as part of a resistance management strategy that includes mixtures or sequences effective against mildew and non-chemical methods

SEE SECTION 3 FOR PRODUCTS ALSO REGISTERED

Restrictions
- Maximum number of treatments 2 per crop
- Do not apply more than two applications of any products containing metrafenone to any one crop

Crop-specific information
- Latest use: up to and including beginning of flowering (GS 61) for rye, wheat, triticale; up to and including ear emergence complete (GS 59) for barley, oats

Following crops guidance
- Cereals. oilseed rape, sugar beet, linseed, maize, clover, field beans, peas, turnips, carrots, cauliflowers, onions, lettuce or potatoes may be sown following normal harvest of a treated crop

Environmental safety
- Dangerous for the environment
- Very toxic to aquatic organisms
- Avoid spray drift onto neighbouring crops. Broad-leaved species may be damaged
- LERAP Category B

Hazard classification and safety precautions
Hazard Harmful, Dangerous for the environment
Transport code 9
Packaging group III
UN number 3082
Risk phrases R38, R40, R43, R50, R53a, R62, R63
Operator protection A, H; U05a, U08, U14, U15, U19a, U20b
Environmental protection E16a, E34, E38
Storage and disposal D01, D02, D05, D08, D09a, D10c, D12a
Medical advice M03

198   epoxiconazole + fenpropimorph + pyraclostrobin

A protectant and systemic fungicide mixture for cereals
FRAC mode of action code: 3 + 5 + 11

See also fenpropimorph
pyraclostrobin

Products

Diamant   BASF   43:214:114 g/l   SE   14149

Uses
- Brown rust in *spring barley*, *spring wheat*, *winter barley*, *winter wheat*
- Crown rust in *spring oats*, *winter oats*
- Fusarium ear blight in *spring wheat* (reduction), *winter wheat* (reduction)
- Glume blotch in *spring wheat*, *winter wheat*
- Net blotch in *spring barley*, *winter barley*
- Powdery mildew in *spring barley*, *spring oats*, *winter barley*, *winter oats*
- Rhynchosporium in *spring barley*, *winter barley*
- Septoria leaf blotch in *spring wheat*, *winter wheat*
- Yellow rust in *spring barley*, *spring wheat*, *winter barley*, *winter wheat*

Approval information
- Epoxiconazole, fenpropimorph and pyraclostrobin included in Annex I under EC Directive 91/414
- Accepted by BBPA for use on malting barley (before ear emergence only)

Efficacy guidance
- For best results spray at the start of foliar disease attack and repeat if infection conditions persist
- For protection against ear diseases apply during ear emergence
- Epoxiconazole is a DMI fungicide. Resistance to some DMI fungicides has been identified in Septoria leaf blotch which may seriously affect performance of some products. For further advice contact a specialist advisor and visit the Fungicide Resistance Action Group (FRAG)-UK website
- Pyraclostrobin is a member of the QoI cross resistance group. Product should be used preventatively and not relied on for its curative potential

FOR FULL CONDITIONS OF USE ALWAYS READ THE PRODUCT LABEL

epoxiconazole + kresoxim-methyl

- Use product as part of an Integrated Crop Management strategy incorporating other methods of control, including where appropriate other fungicides with a different mode of action. Do not apply more than two foliar applications of QoI containing products to any cereal crop
- There is a significant risk of widespread resistance occurring in *Septoria tritici* populations in UK. Failure to follow resistance management action may result in reduced levels of disease control
- Strains of barley powdery mildew resistant to QoIs are common in the UK

Restrictions
- Maximum number of treatments 2 per crop

Crop-specific information
- Latest use: up to and including ear emergence just complete (GS 59) for barley, oats; up to and including flowering just complete (GS 69) for wheat

Environmental safety
- Dangerous for the environment
- Very toxic to aquatic organisms
- Dangerous to fish or other aquatic life. Do not contaminate surface waters or ditches with chemical or used container
- Avoid spray drift onto neighbouring crops. Product may damage broad-leaved species
- LERAP Category B

Hazard classification and safety precautions
Hazard Harmful, Dangerous for the environment
Transport code 9
Packaging group III
UN number 3082
Risk phrases R20, R22a, R38, R40, R50, R53a, R63
Operator protection A; U05a, U14, U20b
Environmental protection E16a, E34, E38
Storage and disposal D01, D02, D05, D09a, D10c, D12a
Medical advice M03, M05a

199   epoxiconazole + kresoxim-methyl

A protectant, systemic and curative fungicide mixture for cereals
FRAC mode of action code: 3 + 11

See also kresoxim-methyl

Products

1	Clayton Gantry	Clayton	125:125 g/l	SC	09482
2	Landmark	BASF	125:125 g/l	SC	11730
3	Serial Duo	AgriGuard	125:125	SC	12252
4	Standon Kresoxim-Epoxiconazole	Standon	125:125 g/l	SC	09281

Uses
- Brown rust in **spring barley**, **winter barley**, **winter wheat** [1-4]; **spring rye**, **spring wheat**, **triticale**, **winter rye** [2, 3]
- Eyespot in **spring oats** *(reduction)*, **spring rye** *(reduction)*, **triticale** *(reduction)*, **winter oats** *(reduction)*, **winter rye** *(reduction)* [2, 3]; **winter barley** *(reduction)*, **winter wheat** *(reduction)* [1-4]
- Foliar disease control in **durum wheat** *(off-label)* [2]; **grass seed crops** *(off-label)* [1, 2, 4]
- Fusarium ear blight in **spring wheat** *(reduction)* [2, 3]; **winter wheat** *(reduction)* [1-4]
- Glume blotch in **spring wheat**, **triticale** [2, 3]; **winter wheat** [1-4]
- Net blotch in **spring barley**, **winter barley** [1-4]
- Powdery mildew in **spring barley**, **winter barley** [1-4]; **spring oats**, **spring rye**, **spring wheat**, **triticale**, **winter oats**, **winter rye**, **winter wheat** [2, 3]
- Rhynchosporium in **spring barley**, **winter barley** [1-4]; **spring rye**, **winter rye** [2, 3]
- Septoria leaf blotch in **spring wheat**, **triticale** [2, 3]; **winter wheat** [1-4]
- Sooty moulds in **spring wheat** *(reduction)* [2, 3]; **winter wheat** *(reduction)* [1-4]
- Yellow rust in **spring barley**, **winter barley**, **winter wheat** [1-4]; **spring rye**, **spring wheat**, **triticale**, **winter rye** [2, 3]

SEE SECTION 3 FOR PRODUCTS ALSO REGISTERED

Specific Off-Label Approvals (SOLAs)
- *durum wheat* 20061084 [2]
- *grass seed crops* 20061047 [1], 20061084 [2], 20061085 [4]

Approval information
- Epoxiconazole and kresoxim-methyl included in Annex I under EC Directive 91/414
- Accepted by BBPA for use on malting barley (before ear emergence only)

Efficacy guidance
- For best results spray at the start of foliar disease attack and repeat if infection conditions persist
- Optimum effect against eyespot obtained by treatment between leaf sheaths erect and first node detectable stages (GS 30-32)
- For protection against ear diseases apply during ear emergence
- Epoxiconazole is a DMI fungicide. Resistance to some DMI fungicides has been identified in Septoria leaf blotch which may seriously affect performance of some products. For further advice contact a specialist advisor and visit the Fungicide Resistance Action Group (FRAG)-UK website
- Kresoxim-methyl is a member of the QoI cross resistance group. Product should be used preventatively and not relied on for its curative potential
- Use product as part of an Integrated Crop Management strategy incorporating other methods of control, including where appropriate other fungicides with a different mode of action. Do not apply more than two foliar applications of QoI containing products to any cereal crop
- There is a significant risk of widespread resistance occurring in *Septoria tritici* populations in UK. Failure to follow resistance management action may result in reduced levels of disease control
- Strains of barley powdery mildew resistant to QoIs are common in the UK

Restrictions
- Maximum total dose equivalent to two full dose treatments

Crop-specific information
- Latest use: mid flowering (GS 65) for wheat, rye, triticale; completion of ear emergence (GS 59) for barley, oats

Environmental safety
- Dangerous for the environment
- Very toxic to aquatic organisms
- Avoid spray drift onto neighbouring crops. Product may damage broad-leaved species
- LERAP Category B

Hazard classification and safety precautions
Hazard Harmful, Dangerous for the environment
Transport code 9
Packaging group III
UN number 3082
Risk phrases R40, R50, R53a, R62, R63 [1-4]; R43 [1, 4]
Operator protection A; U05a, U14, U20b
Environmental protection E15a, E16a [1-4]; E16b [2-4]; E38 [2, 3]
Storage and disposal D01, D02, D09a [1-4]; D05, D10b [1, 4]; D08, D10c [2, 3]; D12a [1-3]
Medical advice M05a [2, 3]; M05b [1, 4]

200 epoxiconazole + kresoxim-methyl + pyraclostrobin

A protectant, systemic and curative fungicide mixture for cereals
FRAC mode of action code: 3 + 11 + 11

See also kresoxim-methyl
pyraclostrobin

Products
1	Covershield	BASF	50:67:133 g/l	SE	10900
2	Opponent	BASF	50:67:133 g/l	SE	10877

Uses
- Brown rust in *spring barley*, *spring wheat*, *winter barley*, *winter wheat*
- Crown rust in *spring oats*, *winter oats*

FOR FULL CONDITIONS OF USE ALWAYS READ THE PRODUCT LABEL

epoxiconazole + kresoxim-methyl + pyraclostrobin

- Foliar disease control in **durum wheat** *(off-label)*, **grass seed crops** *(off-label)*, **spring rye** *(off-label)*, **triticale** *(off-label)*, **winter rye** *(off-label)*
- Glume blotch in **spring wheat**, **winter wheat**
- Net blotch in **spring barley**, **winter barley**
- Rhynchosporium in **spring barley**, **winter barley**
- Septoria leaf blotch in **spring wheat**, **winter wheat**
- Yellow rust in **spring barley**, **spring wheat**, **winter barley**, **winter wheat**

Specific Off-Label Approvals (SOLAs)
- *durum wheat* 20063644 [1], 20060539 [2]
- *grass seed crops* 20063644 [1], 20060539 [2]
- *spring rye* 20063644 [1], 20060539 [2]
- *triticale* 20063644 [1], 20060539 [2]
- *winter rye* 20063644 [1], 20060539 [2]

Approval information
- Epoxiconazole, kresoxim-methyl and pyraclostrobin included in Annex I under EC Directive 91/414
- Accepted by BBPA for use on malting barley (before ear emergence only)

Efficacy guidance
- For best results apply at the start of foliar disease attack
- Good reduction of Fusarium ear blight can be obtained from treatment during flowering
- Yield response may be obtained in the absence of visual disease symptoms
- Best results on Septoria leaf blotch when treated in the latent phase
- Epoxiconazole is a DMI fungicide. Resistance to some DMI fungicides has been identified in Septoria leaf blotch which may seriously affect performance of some products. For further advice contact a specialist advisor and visit the Fungicide Resistance Action Group (FRAG)-UK website
- Kresoxim-methyl and pyraclostrobin are members of the QoI cross resistance group. Product should be used preventatively and not relied on for its curative potential
- Use product as part of an Integrated Crop Management strategy incorporating other methods of control, including where appropriate other fungicides with a different mode of action. Do not apply more than two foliar applications of QoI containing products to any cereal crop
- There is a significant risk of widespread resistance occurring in *Septoria tritici* populations in UK. Failure to follow resistance management action may result in reduced levels of disease control

Restrictions
- Maximum total dose equivalent to two full dose treatments

Crop-specific information
- Latest use: before grain watery ripe (GS 71) for wheat; up to and including emergence of ear just complete (GS 59) for barley and oats

Environmental safety
- Dangerous for the environment
- Very toxic to aquatic organisms
- Dangerous to fish or other aquatic life. Do not contaminate surface waters or ditches with chemical or used container
- LERAP Category B

Hazard classification and safety precautions
Hazard Harmful, Dangerous for the environment
Transport code 9
Packaging group III
UN number 3082
Risk phrases R20, R22a, R40, R50, R53a
Operator protection A; U05a, U14, U20b
Environmental protection E13b [2]; E15a [1]; E16a, E16b, E34, E38 [1, 2]
Storage and disposal D01, D02, D09a, D10c, D12a [1, 2]; D05 [2]
Medical advice M05a

SEE SECTION 3 FOR PRODUCTS ALSO REGISTERED

201   epoxiconazole + metconazole

A triazole mixture for use in cereals
FRAC mode of action code: 3 + 3

See also metconazole

Products

1	Brutus	BASF	37.5;27.5 g/l	EC	14353
2	Icarus	BASF	37.5:25.0 g/l	EC	14471

Uses
- Brown rust in *durum wheat, spring barley, spring rye, spring wheat, triticale, winter barley, winter rye, winter wheat* [1, 2]
- Eyespot in *durum wheat* (reduction), *spring rye* (reduction), *triticale* (reduction), *winter barley* (reduction), *winter rye* (reduction), *winter wheat* (reduction) [1, 2]
- Fusarium ear blight in *durum wheat* (good reduction), *spring wheat* (good reduction), *triticale* (good reduction), *winter wheat* (good reduction) [1, 2]
- Glume blotch in *durum wheat, spring wheat, triticale, winter wheat* [1, 2]
- Net blotch in *spring barley, winter barley* [1, 2]
- Powdery mildew in *durum wheat* (moderate control), *spring barley* (moderate control), *spring wheat* (moderate control), *triticale* (moderate control), *winter barley* (moderate control), *winter wheat* (moderate control) [1, 2]; *spring rye* (moderate control), *winter rye* (moderate control) [1]; *spring rye* (moderate reduction), *winter rye* (moderate reduction) [2]
- Ramularia leaf spots in *spring barley* [2]; *spring barley* (moderate control) [1]; *winter barley* (moderate control) [1, 2]
- Rhynchosporium in *spring barley, winter barley* [1, 2]
- Septoria leaf blotch in *durum wheat, spring wheat, triticale, winter wheat* [1, 2]
- Sooty moulds in *durum wheat* (good reduction), *spring wheat* (good reduction), *winter wheat* (good reduction) [1, 2]
- Tan spot in *durum wheat* (moderate control), *spring wheat* (moderate control), *winter wheat* (moderate control) [1, 2]
- Yellow rust in *durum wheat, spring barley, spring rye, spring wheat, triticale, winter barley, winter rye, winter wheat* [1, 2]

Approval information
- Epoxiconazole and Metconazole included in Annex I under EC Directive 91/414

Following crops guidance
- After ttreating a cereal crop, oilseed rape, cereals, sugar beet, linseed, maize, clover, beans, peas, carrots, potatoes, lettuce, cabbage, sunflower, ryegrass or onions may be sown as a following crop. The effect on other crops has not been evaluated.

Environmental safety
- LERAP Category B

Hazard classification and safety precautions
Hazard Harmful, Dangerous for the environment
Transport code 9
Packaging group III
UN number 3082
Risk phrases R40, R43, R51, R53a, R63
Operator protection A, H; U05a, U14, U20b
Environmental protection E15b, E16a, E34
Storage and disposal D01, D02, D09a, D10c
Medical advice M05a

FOR FULL CONDITIONS OF USE ALWAYS READ THE PRODUCT LABEL

202 epoxiconazole + metrafenone

A broad spectrum fungicide mixture for cereals
FRAC mode of action code: 3 + U8

See also metrafenone

Products

1	Ceando	BASF	83:100 g/l	SC	13271
2	Cloister	BASF	83:100 g/l	SC	14358

Uses
- Brown rust in **spring barley, spring rye, spring wheat, triticale, winter barley, winter rye, winter wheat**
- Crown rust in **spring oats, winter oats**
- Fusarium ear blight in **spring wheat** *(reduction)*, **winter wheat** *(reduction)*
- Glume blotch in **spring wheat, triticale, winter wheat**
- Net blotch in **spring barley, winter barley**
- Powdery mildew in **spring barley, spring oats, spring rye, spring wheat, triticale, winter barley, winter oats, winter rye, winter wheat**
- Rhynchosporium in **spring barley, spring rye, winter barley, winter rye**
- Septoria leaf blotch in **spring wheat, triticale, winter wheat**
- Sooty moulds in **spring wheat** *(reduction)*, **winter wheat** *(reduction)*
- Yellow rust in **spring barley, spring rye, spring wheat, triticale, winter barley, winter rye, winter wheat**

Approval information
- Epoxiconazole and metrafenone included in Annex I under EC Directive 91/414

Efficacy guidance
- For best results spray at the start of foliar disease attack and repeat if infection conditions persist
- Optimum effect against eyespot obtained by treatment between leaf sheaths erect and first node detectable stages (GS 30-32)
- For protection against ear diseases apply during ear emergence
- Treatment for control of Septoria diseases should normally be made after third node detectable stage if the disease is present and favourable weather for disease development has occurred
- Epoxiconazole is a DMI fungicide. Resistance to some DMI fungicides has been identified in Septoria leaf blotch which may seriously affect performance of some products. For further advice contact a specialist advisor and visit the Fungicide Resistance Action Group (FRAG)-UK website
- Should be used as part of a resistance management strategy that includes mixtures or sequences effective against mildew and non-chemical methods
- Where mildew well established at time of treatment product should be mixed with fenpropimorph. See label

Restrictions
- Maximum number of treatments 2 per crop
- Do not apply more than two applications of any products containing metrafenone to any one crop

Crop-specific information
- Latest use: up to and including beginning of flowering (GS 61) for rye, wheat, triticale; up to and including ear emergence complete (GS 59) for barley, oats

Following crops guidance
- Cereals. oilseed rape, sugar beet, linseed, maize, clover, field beans, peas, turnips, carrots, cauliflowers, onions, lettuce or potatoes may be sown following normal harvest of a treated crop

Environmental safety
- Dangerous for the environment
- Very toxic to aquatic organisms
- Avoid spray drift onto neighbouring crops. Broad-leaved species may be damaged
- LERAP Category B

Hazard classification and safety precautions
Hazard Harmful, Dangerous for the environment
UN number N/C

SEE SECTION 3 FOR PRODUCTS ALSO REGISTERED

Risk phrases R40, R50, R53a, R62, R63
Operator protection A, H; U05a, U20b
Environmental protection E15a, E16a, E34, E38
Storage and disposal D01, D02, D05, D08, D09a, D10c
Medical advice M03, M05a

203 epoxiconazole + prochloraz

A systemic, protectant and curative triazole fungicide mixture for use in cereals
FRAC mode of action code: 3 + 3

See also prochloraz

Products
    Ennobe   BASF   62.5 + 225 g/l   DC   14340

Uses
- Brown rust in **spring barley**, **spring rye**, **spring wheat**, **triticale**, **winter barley**, **winter rye**, **winter wheat**
- Eyespot in **spring rye**, **triticale**, **winter barley**, **winter rye**, **winter wheat**
- Fusarium ear blight in **spring wheat** *(good reduction)*, **triticale** *(good reduction)*, **winter wheat** *(good reduction)*
- Glume blotch in **spring wheat**, **triticale**, **winter wheat**
- Net blotch in **spring barley**, **winter barley**
- Powdery mildew in **spring barley** *(moderate control)*, **spring rye** *(moderate control)*, **spring wheat** *(moderate control)*, **triticale** *(moderate control)*, **winter barley**, **winter rye** *(moderate control)*, **winter wheat** *(moderate control)*
- Rhynchosporium in **spring barley**, **spring rye**, **winter barley**, **winter rye**
- Septoria leaf blotch in **spring wheat**, **triticale**, **winter wheat**
- Sooty moulds in **spring wheat** *(good reduction)*, **winter wheat** *(good reduction)*
- Yellow rust in **spring barley**, **spring rye**, **spring wheat**, **triticale**, **winter barley**, **winter rye**, **winter wheat**

Approval information
- Epoxiconazole included in Annex I under EC Directive 91/414
- Prochloraz has not been included in Annex 1 but the withdrawal or inclusion date has yet to be set. Most current approvals now set to expire 2021.

Efficacy guidance
- Use in mixture with Comaptibility Agent CA 259 when tank-mixing with other products. Order of mixing is not important.

Environmental safety
- LERAP Category B

Hazard classification and safety precautions
    Hazard Harmful, Dangerous for the environment
    Transport code 9
    Packaging group III
    UN number 3082
    Risk phrases R20, R22a, R36, R38, R40, R50, R53a, R62, R63
    Operator protection A, H; U05a
    Environmental protection E15b, E16a, E38
    Storage and disposal D01, D02, D08, D10c, D12a
    Medical advice M05a

FOR FULL CONDITIONS OF USE ALWAYS READ THE PRODUCT LABEL

204 epoxiconazole + pyraclostrobin

A protectant, systemic and curative fungicide mixture for cereals
FRAC mode of action code: 3 + 11

See also pyraclostrobin

Products

1	Envoy	BASF	62.5:85 g/l	SE	12750
2	Gemstone	BASF	62.5:80	SE	13918
3	Opera	BASF	50:133 g/l	SE	12167

Uses
- Brown rust in **spring barley**, **spring wheat**, **winter barley**, **winter wheat** [1-3]
- Cercospora leaf spot in **sugar beet** [3]
- Crown rust in **spring oats**, **winter oats** [2, 3]; **spring oats** *(qualified minor use)*, **winter oats** *(qualified minor use)* [1]
- Foliar disease control in **durum wheat** *(off-label)*, **fodder beet** *(off-label)*, **grass seed crops** *(off-label)*, **spring rye** *(off-label)*, **triticale** *(off-label)*, **winter rye** *(off-label)* [3]
- Fusarium ear blight in **spring wheat** *(apply during flowering)*, **winter wheat** *(apply during flowering)* [2]; **spring wheat** *(reduction)*, **winter wheat** *(reduction)* [1]
- Glume blotch in **spring wheat**, **winter wheat** [1-3]
- Net blotch in **spring barley**, **winter barley** [1-3]
- Powdery mildew in **sugar beet** [3]
- Ramularia leaf spots in **sugar beet** [3]
- Rhynchosporium in **spring barley**, **winter barley** [1-3]
- Rust in **sugar beet** [3]
- Septoria leaf blotch in **spring wheat**, **winter wheat** [1-3]
- Tan spot in **spring wheat**, **winter wheat** [1, 2]
- Yellow rust in **spring barley**, **spring wheat**, **winter barley**, **winter wheat** [1-3]

Specific Off-Label Approvals (SOLAs)
- *durum wheat* 20062653 [3]
- *fodder beet* 20062654 [3]
- *grass seed crops* 20062653 [3]
- *spring rye* 20062653 [3]
- *triticale* 20062653 [3]
- *winter rye* 20062653 [3]

Approval information
- Epoxiconazole and pyraclostrobin included in Annex I under EC Directive 91/414
- Accepted by BBPA for use on malting barley (before ear emergence only)

Efficacy guidance
- For best results apply at the start of foliar disease attack
- Good reduction of Fusarium ear blight can be obtained from treatment during flowering
- Yield response may be obtained in the absence of visual disease symptoms
- Epoxiconazole is a DMI fungicide. Resistance to some DMI fungicides has been identified in Septoria leaf blotch which may seriously affect performance of some products. For further advice contact a specialist advisor and visit the Fungicide Resistance Action Group (FRAG)-UK website
- Pyraclostrobin is a member of the QoI cross resistance group. Product should be used preventatively and not relied on for its curative potential
- Use product as part of an Integrated Crop Management strategy incorporating other methods of control, including where appropriate other fungicides with a different mode of action. Do not apply more than two foliar applications of QoI containing products to any cereal crop
- There is a significant risk of widespread resistance occurring in *Septoria tritici* populations in UK. Failure to follow resistance management action may result in reduced levels of disease control

Restrictions
- Maximum number of treatments 2 per crop
- Do not use on sugar beet crops grown for seed [3]

SEE SECTION 3 FOR PRODUCTS ALSO REGISTERED

Crop-specific information
- Latest use: before grain watery ripe (GS 71) for wheat; up to and including emergence of ear just complete (GS 59) for barley and oats
- HI 6 wk for sugar beet [3]

Environmental safety
- Dangerous for the environment
- Very toxic to aquatic organisms
- LERAP Category B

Hazard classification and safety precautions
Hazard Harmful, Dangerous for the environment
Transport code 9
Packaging group III
UN number 3082
Risk phrases R20, R22a, R38, R40, R43, R50, R53a [1-3]; R62, R63 [1, 2]
Operator protection A [1-3]; H [2]; U05a, U14 [1-3]; U20b [1, 3]; U20c [2]
Environmental protection E15a, E16a, E34, E38 [1-3]; E16b [3]
Storage and disposal D01, D02, D09a, D10c, D12a [1-3]; D05 [3]; D08 [2]
Medical advice M03, M05a

205 esfenvalerate

A contact and ingested pyrethroid insecticide
IRAC mode of action code: 3

Products

1	Clayton Cajole	Clayton	25 g/l	EC	14995
2	Greencrop Cajole Ultra	Greencrop	25 g/l	EC	12967
3	Standon Hounddog	Standon	25 g/l	EC	13201
4	Sumi-Alpha	Interfarm	25 g/l	EC	14023
5	Sven	Interfarm	25 g/l	EC	14859

Uses
- Aphids in *broccoli, brussels sprouts, cabbages, calabrese, cauliflowers, chinese cabbage, combining peas, edible podded peas, kale, kohlrabi, potatoes, spring field beans, vining peas, winter field beans* [1, 5]; *grass seed crops* (off-label) [4]; *grassland, managed amenity turf, ornamental plant production* [1]; *spring barley, spring wheat, winter barley, winter wheat* [1-5]
- Bibionids in *grass seed crops, grassland, managed amenity turf* [5]
- Caterpillars in *broccoli, brussels sprouts, cabbages, calabrese, cauliflowers, chinese cabbage, kale, kohlrabi* [5]
- Insect pests in *ornamental plant production, protected ornamentals* [5]
- Pea and bean weevil in *combining peas, edible podded peas, spring field beans, vining peas, winter field beans* [5]

Specific Off-Label Approvals (SOLAs)
- *grass seed crops* 20081266 expires 31 Jul 2011 [4]

Approval information
- Esfenvalerate included in Annex I under EC Directive 91/414
- Accepted by BBPA for use on malting barley
- Approval expiry 31 Jul 2011 [1-5]

Efficacy guidance
- For best reduction of spread of barley yellow dwarf virus winter sown crops at high risk (e.g. after grass or in areas with history of BYDV) should be treated when aphids first seen or by mid-Oct. Otherwise treat in late Oct-early Nov
- High risk winter sown crops will need a second treatment
- Spring sown crops should be treated from the 2-3 leaf stage if aphids are found colonising in the crop and a further application may be needed before the first node stage if they reinfest
- Product also recommended between onset of flowering and milky ripe stages (GS 61-73) for control of summer cereal aphids

FOR FULL CONDITIONS OF USE ALWAYS READ THE PRODUCT LABEL

ethofumesate

Restrictions
- Maximum number of treatments 3 per crop of which 2 may be in autumn
- Do not use if another pyrethroid or dimethoate has been applied to crop after start of ear emergence (GS 51)
- Do not leave spray solution standing in spray tank

Crop-specific information
- Latest use: 31 Mar in yr of harvest (winter use dose); early milk stage for barley (summer use); late milk stage for wheat (summer use)

Environmental safety
- Dangerous for the environment
- Very toxic to aquatic organisms
- High risk to non-target insects or other arthropods. Do not spray within 6 m of the field boundary
- Flammable
- Store product in dark away from direct sunlight
- LERAP Category A

Hazard classification and safety precautions
Hazard Harmful, Flammable, Dangerous for the environment
Transport code 3 [1, 4, 5]; 6.1 [2, 3]
Packaging group II [1, 4, 5]; III [2, 3]
UN number 1993 [1, 4, 5]; 3351 [2, 3]
Risk phrases R20, R22a, R41, R43, R50, R53a
Operator protection A, C, H; U04a, U05a, U08, U11, U14, U19a, U20b
Environmental protection E15a, E22a [1, 2, 4, 5]; E15b, E22b [3]; E16c, E16d, E34 [1-5]; E38 [1, 4, 5]
Storage and disposal D01, D02, D09a [1-5]; D05 [2, 3]; D07, D10b [2]; D10c, D12a [1, 3-5]
Medical advice M03 [1-5]; M05b [3]

206 ethofumesate

A benzofuran herbicide for grass weed control in various crops
HRAC mode of action code: N

See also chloridazon + ethofumesate
             desmedipham + ethofumesate + lenacil + phenmedipham
             desmedipham + ethofumesate + metamitron + phenmedipham
             desmedipham + ethofumesate + phenmedipham

Products

1	Agriguard Ethofumesate 200	AgriGuard	200 g/l	EC	14129
2	Alpha Ethofumesate	Makhteshim	500 g/l	SC	13055
3	Barclay Keeper 500 SC	Barclay	500 g/l	SC	14016
4	Catalyst	AgriGuard	500 g/l	SC	13266
5	Ethosat 500	Makhteshim	500 g/l	SC	13050
6	Fumesate 500 SC	Belcrop	500 g/l	SC	14321
7	Kubist Flo	Nufarm UK	500 g/l	SC	12987
8	Nortron Flo	Bayer CropScience	500 g/l	SC	12986
9	Oblix 500	AgriChem BV	500 g/l	SC	12349

Uses
- Annual dicotyledons in **amenity grassland** [2, 3, 5]; **fodder beet, red beet, sugar beet** [1-9]; **forest nurseries** *(off-label)*, **game cover** *(off-label)*, **ornamental specimens** *(off-label)* [8]; **grass seed crops** [1-3, 5]; **managed amenity turf** *(off-label)* [5]; **mangels** [1-5, 7-9]
- Annual grasses in **amenity grassland, grass seed crops** [3]; **fodder beet, red beet, sugar beet** [3, 6, 9]; **forest nurseries** *(off-label)* [5, 8]; **game cover** *(off-label)*, **ornamental specimens** *(off-label)* [8]; **mangels** [3, 9]
- Annual meadow grass in **amenity grassland** [2, 5]; **fodder beet, mangels, red beet, sugar beet** [1, 2, 4, 5, 7, 8]; **grass seed crops** [1, 2, 5]; **managed amenity turf** *(off-label)* [5]
- Blackgrass in **amenity grassland, grass seed crops** [2, 5]; **fodder beet, mangels, red beet, sugar beet** [1, 2, 4, 5, 7, 8]; **managed amenity turf** *(off-label)* [5]
- Fat hen in **poppies for morphine production** *(off-label)* [8]

SEE SECTION 3 FOR PRODUCTS ALSO REGISTERED

- Grass weeds in *poppies for morphine production* (off-label) [8]
- Mayweeds in *poppies for morphine production* (off-label) [8]

Specific Off-Label Approvals (SOLAs)
- *forest nurseries* 20082846 [5], 20082919 [8]
- *game cover* 20082919 [8]
- *managed amenity turf* 20081026 [5]
- *ornamental specimens* 20082919 [8]
- *poppies for morphine production* 20070077 [8]

Approval information
- Ethofumesate included in Annex I under EC Directive 91/414

Efficacy guidance
- Most products may be applied pre- or post-emergence of crop or weeds but some restricted to pre-emergence or post-emergence use only. Check label
- Some products recommended for use only in mixtures. Check label
- Volunteer cereals not well controlled pre-emergence, weed grasses should be sprayed before fully tillered
- Grass crops may be sprayed during rain or when wet. Not recommended in very dry conditions or prolonged frost

Restrictions
- The maximum total dose must not exceed 1.0 kg ethofumesate per hectare in any three year period.
- Do not use on Sands or Heavy soils, Very Light soils containing a high percentage of stones, or soils with more than 5-10% organic matter (percentage varies according to label)
- Do not use on swards reseeded without ploughing
- Clovers will be killed or severely checked
- Do not graze or cut grass for 14 d after, or roll less than 7 d before or after spraying

Crop-specific information
- Latest use: before crops meet across rows for beet crops and mangels; not specified for other crops
- Apply in beet crops in tank mixes with other pre- or post-emergence herbicides. Recommendations vary for different mixtures. See label for details
- Safe timing on beet crops varies with other ingredient of tank mix. See label for details
- In grass crops apply to moist soil as soon as possible after sowing or post-emergence when crop in active growth, normally mid-Oct to mid-Dec. See label for details
- May be used in Italian, hybrid and perennial ryegrass, timothy, cocksfoot, meadow fescue and tall fescue. Apply pre-emergence to autumn-sown leys, post-emergence after 2-3 leaf stage. See label for details

Following crops guidance
- In the event of failure of a treated crop only sugar beet, red beet, fodder beet and mangels may be redrilled within 3 mth of application
- Any crop may be sown 3 mth after application of mixtures in beet crops following ploughing, 5 mth after application in grass crops

Environmental safety
- Dangerous for the environment
- Toxic to aquatic organisms
- Do not empty into drains
- LERAP Category B [1]

Hazard classification and safety precautions
    Hazard Harmful, Flammable [1]; Dangerous for the environment [1-5, 7, 8]
    Transport code 3 [1]; 9 [2-5, 7, 8]
    Packaging group III [1-5, 7, 8]
    UN number 1993 [1]; 3082 [2-5, 7, 8]; N/C [6, 9]
    Risk phrases R20, R21, R22b, R36, R38 [1]; R51 [2-5, 7, 8]; R52 [1, 6, 9]; R53a [1-9]
    Operator protection A, C [1]; H [1, 9]; U05a [1, 3, 9]; U08 [2-5]; U09a [1, 6-8]; U19a [1-8]; U20a [2, 4, 5, 7, 8]; U20b [1, 3, 6]
    Environmental protection E13c [4]; E15a [2, 4, 5, 7]; E15b [1, 3, 6, 9]; E16a [1]; E19b [2, 5]; E23 [9]; E34 [1-3, 5, 9]; E38 [6-8]

FOR FULL CONDITIONS OF USE ALWAYS READ THE PRODUCT LABEL

Storage and disposal D01, D02 [1, 9]; D05 [2, 4-8]; D09a [1-9]; D10a [3, 6, 9]; D10b [1, 2, 4, 5, 7]; D12a [2, 3, 5-9]
Medical advice M03, M05b [1]

207   ethofumesate + metamitron

A contact and residual herbicide mixture for beet crops
HRAC mode of action code: N + C1

See also metamitron

Products

1	Goltix Super	Makhteshim	150:350 g/l	SC	12216
2	Goltix Uno	Makhteshim	150:350 g/l	SC	12602
3	Oblix MT	AgriChem BV	150:350 g/l	SC	14510
4	Torero	Makhteshim	150:350 g/l	SC	11158

Uses
- Annual dicotyledons in **fodder beet**, **mangels**, **sugar beet**
- Annual meadow grass in **fodder beet**, **mangels**, **sugar beet**

Approval information
- Ethofumesate included in Annex I under EC Directive 91/414

Efficacy guidance
- Best results obtained from a series of treatments applied as an overall fine spray commencing when earliest germinating weeds are no larger than fully expanded cotyledon and the majority of the crop at fully expanded cotyledon
- Apply subsequent sprays as each new flush of weeds reaches early cotyledon and continue until weed emergence ceases
- Product may be used on all soil types but residual activity may be reduced on those with more than 5% organic matter

Restrictions
- Maximum total dose equivalent to three full dose treatments
- The maximum total dose must not exceed 1.0 kg ethofumesate per hectare in any three year period

Crop-specific information
- Latest use: before crop leaves meet between rows
- Crop tolerance may be reduced by stress caused by growing conditions, effects of pests, disease or other pesticides, nutrient deficiency etc

Following crops guidance
- Beet crops may be sown at any time after treatment. Any other crop may be sown after mouldboard ploughing to 15 cm and a minimum interval of 3 mth after treatment

Environmental safety
- Dangerous for the environment
- Very toxic to aquatic organisms
- Do not empty into drains

Hazard classification and safety precautions
Hazard Harmful [3]; Dangerous for the environment [1-4]
Transport code 9
Packaging group III
UN number 3082
Risk phrases R22a [3]; R51, R53a [1-4]
Operator protection U05a, U19a, U20a [1-4]; U08, U14, U15 [1, 2, 4]
Environmental protection E13c, E34 [1, 2, 4]; E15a [3]; E19b [1-4]
Storage and disposal D01, D02, D09a, D10b, D12a [1-4]; D05 [1, 2, 4]
Medical advice M05a [3]

SEE SECTION 3 FOR PRODUCTS ALSO REGISTERED

208 ethofumesate + metamitron + phenmedipham

A contact and residual herbicide mixture for sugar beet
HRAC mode of action code: N + C1 + C1

See also metamitron
phenmedipham

Products
Phemo   AgriChem BV   51:153:51 g/l   SE   14193

Uses
- Annual dicotyledons in **fodder beet**, **mangels**, **red beet**, **sugar beet**
- Annual meadow grass in **fodder beet**, **mangels**, **red beet**, **sugar beet**

Approval information
- Ethofumesate and phenmedipham included in Annex I under EC Directive 91/414

Efficacy guidance
- Best results obtained from a series of treatments applied as an overall fine spray commencing when earliest germinating weeds are no larger than fully expanded cotyledon
- Apply subsequent sprays as each new flush of weeds reaches early cotyledon and continue until weed emergence ceases (maximum 3 sprays)
- Product must be applied with Actipron
- Product may follow certain pre-emergence treatments and be used in conjunction with other post-emergence sprays - see label for details
- Where a pre-emergence band spray has been applied, the first treatment should be timed according to the size of the weeds in the untreated area between the rows

Restrictions
- Maximum number of treatments 3 per crop; maximum total dose 6 kg product per ha
- Product may be used on all soil types but residual activity may be reduced on those with more than 5% organic matter
- Crop tolerance may be reduced by stress caused by growing conditions, effects of pests, disease or other pesticides, nutrient deficiency etc
- Only beet crops should be sown within 4 mth of last treatment; winter cereals may be sown after this interval
- Any spring crop may be sown in the year following use
- Mould-board ploughing to 150 mm followed by thorough cultivation recommended before planting any crop.
- The maximum total dose must not exceed 1.0 kg ethofumesate per hectare in any three year period

Crop-specific information
- Latest use: before crop leaves meet between rows

Environmental safety
- Irritating to eyes
- Harmful to fish or other aquatic life. Do not contaminate surface waters or ditches with chemical or used container

Hazard classification and safety precautions
Hazard Dangerous for the environment
Transport code 9
Packaging group III
UN number 3082
Risk phrases R51, R53a
Operator protection A; U05a, U20c
Environmental protection E15a, E34, E38
Storage and disposal D01, D02, D09a, D11a, D11b, D12a

FOR FULL CONDITIONS OF USE ALWAYS READ THE PRODUCT LABEL

209 ethofumesate + phenmedipham

A contact and residual herbicide for use in beet crops
HRAC mode of action code: N + C1

See also phenmedipham

Products

1	Duo 400 SC	AgriGuard	200:200 g/l	SC	14323
2	Fenlander 2	Makhteshim	94:97 g/l	EC	14030
3	Powertwin	Makhteshim	200:200 g/l	SC	14004
4	Teamforce	AgriChem BV	100:80 g/l	EC	14923
5	Twin	Makhteshim	94:97 g/l	EC	13998

Uses
- Annual dicotyledons in *fodder beet, mangels, sugar beet*
- Annual meadow grass in *fodder beet, mangels, sugar beet*
- Blackgrass in *fodder beet, mangels, sugar beet*

Approval information
- Ethofumesate and phenmedipham included in Annex I under EC Directive 91/414

Efficacy guidance
- Best results achieved by repeat applications to cotyledon stage weeds. Larger susceptible weeds not killed by first treatment usually checked and controlled by second application
- Apply on all soil types at 5-10 d intervals
- On soils with more than 5-10% organic matter residual activity may be reduced

Restrictions
- Maximum number of treatments normally 3 per crop or maximum total dose equivalent to three full dose treatments, or less - see labels for details
- Must be used in conjunction with specified adjuvant. See label [3]
- Do not spray wet foliage or if rain imminent
- Spray in evening if daytime temperatures above 21 °C expected
- Avoid or delay treatment if frost expected within 7 days
- Avoid or delay treating crops under stress from wind damage, manganese or lime deficiency, pest or disease attack etc.
- The maximum total dose must not exceed 1.0 kg ethofumesate per hectare in any three year period

Crop-specific information
- Latest use: before crop foliage meets in the rows
- Check from which recovery may not be complete may occur if treatment made during conditions of sharp diurnal temperature fluctuation

Following crops guidance
- Beet crops may be sown at any time after treatment. Any other crop may be sown after mouldboard ploughing to 15 cm and a minimum interval of 3 mth after treatment

Environmental safety
- Dangerous for the environment [2, 3, 5]
- Toxic to aquatic organisms
- Do not empty into drains
- Extra care necessary to avoid drift because product is recommended for use as a fine spray

Hazard classification and safety precautions
Hazard Harmful [1, 2, 4, 5]; Irritant [3]; Dangerous for the environment [1-5]
Transport code 9
Packaging group III
UN number 3082
Risk phrases R20 [2]; R37, R40 [4, 5]; R43, R51, R53a [1-5]
Operator protection A [1-5]; H [1, 3, 4]; U05a, U08, U19a, U20a [1-5]; U14 [1, 3]
Environmental protection E15a, E19b, E34 [1-5]; E38 [2]
Storage and disposal D01, D09a, D10b, D12a [1-5]; D02 [2-5]; D05 [1, 3]
Medical advice M05a

SEE SECTION 3 FOR PRODUCTS ALSO REGISTERED

210 ethoprophos

An organophosphorus nematicide and insecticide
IRAC mode of action code: 1B

Products

Mocap 10G	Bayer CropScience	10% w/w	GR	10003

Uses
- Potato cyst nematode in *potatoes*
- Wireworm in *potatoes*

Approval information
- Ethoprophos included in Annex I under EC Directive 91/414
- In 2006 CRD required that all products containing this active ingredient should carry the following warning in the main area of the container label: "Ethoprofos is an anticholinesterase organophosphate. Handle with care"

Efficacy guidance
- Broadcast shortly before or during final soil preparation with suitable fertilizer spreader and incorporate immediately to 10-15 cm. Deeper incorporation may reduce efficacy. See label for details
- Treatment can be applied on all soil types. Control of pests reduced on organic soils
- Effectiveness dependent on soil moisture. Drought after application may reduce control
- Where pre-planting nematode populations are high, loss of crop vigour and subsequent loss of yield may occur in spite of treatment

Restrictions
- Contains an anticholinesterase organophosphorous compound. Do not use if under medical advice not to work with such compounds
- Maximum number of treatments 1 per crop
- Do not harvest crops for human or animal consumption for at least 8 wk after application
- Vehicles with a closed cab must be used when applying and incorporating
- Do not apply by hand or hand held equipment

Crop-specific information
- Latest use: pre-planting of crop
- Field scale application must only be through positive displacement type specialist granule applicators or dual purpose fertilizer applicators with microgranule setting

Environmental safety
- Dangerous for the environment
- Toxic to aquatic organisms
- Dangerous to livestock. Keep all livestock out of treated areas/away from treated water for at least 13 wk. Bury or remove spillages
- Dangerous to game, wild birds and animals
- Product supplied in returnable/refillable containers. Follow instructions on label

Hazard classification and safety precautions
    Hazard Toxic, Dangerous for the environment
    Transport code 6.1
    Packaging group II
    UN number 2783
    Risk phrases R22a, R23, R43, R51, R53a
    Operator protection A, D, H; U02a, U04a, U05a, U07, U08, U13, U14, U19a, U20a
    Environmental protection E06b (13 wk); E10a, E13b, E34, E36a, E38
    Consumer protection C02a (8 wk)
    Storage and disposal D01, D02, D09a, D12a, D14
    Medical advice M01, M03, M04a

FOR FULL CONDITIONS OF USE ALWAYS READ THE PRODUCT LABEL

ethylene

211 ethylene

A gas used for fruit ripening and potato storage

Products

1	Ethylene	Biofresh	99.9% w/v	LI	14177
2	ethylene	various	-	GA	-

Uses
- Fruit ripening in **fruit crops** *(in store)* [2]
- Sprout suppression in **potatoes** [1]; **potatoes** *(in store)* [2]

Approval information
- Ethylene is included in Annex 1 under EC Directive 91/414

Efficacy guidance
- For use in stored fruit or potatoes after harvest

Restrictions
- Handling and release of ethylene must only be undertaken by operators suitably trained and competent to carry out the work
- Operators must vacate treated areas immediately after ethylene introduction
- Unprotected persons must be excluded from the treated areas until atmospheres have been thoroughly ventilated for 15 minutes minimum before re-entry
- Ambient atmospheric ethylene concentration must not exceed 1000 ppm. Suitable self-contained breathing apparatus must be worn in atmospheres containing ethylene in excess of 1000 ppm
- A minimum 3 d post treatment period is required before removal of treated crop from storage
- Ethylene treatment must only be undertaken in fully enclosed storage areas that are air tight with appropriate air circulation and venting facilities

212 etoxazole

A mite growth inhibitor
IRAC mode of action code: 10B

Products

Borneo	Interfarm	110 g/l	SC	13919

Uses
- Mites in **protected aubergines**, **protected tomatoes**
- Spider mites in **protected ornamentals** *(off-label)*, **protected strawberries** *(off-label)*
- Two-spotted spider mite in **protected ornamentals** *(off-label)*, **protected strawberries** *(off-label)*

Specific Off-Label Approvals (SOLAs)
- *protected ornamentals* 20081216
- *protected strawberries* 20092400

Approval information
- Etoxazole included in Annex I under EC Directive 91/414

Hazard classification and safety precautions
Hazard Dangerous for the environment
Transport code 9
Packaging group III
UN number 3082
Risk phrases R50, R53a
Operator protection A; U05a, U20b
Environmental protection E15a, E34
Storage and disposal D01, D02, D09b, D11a, D12b

SEE SECTION 3 FOR PRODUCTS ALSO REGISTERED

213 famoxadone

A strobilurin fungicide available only in mixtures
FRAC mode of action code: 11

See also cymoxanil + famoxadone

214 famoxadone + flusilazole

A contact, preventative and curative fungicide mixture for cereals and oilseed rape
FRAC mode of action code: 11 + 3

See also flusilazole

Products

1	Caynil	DuPont	100:106.7 g/l	EC	14641
2	Charisma	DuPont	100:106.7 g/l	EC	10415
3	Jenga	DuPont	100:106.7 g/l	EC	14357
4	Medley	DuPont	100:106.7 g/l	EC	10933

Uses
- Brown rust in **spring barley**, **winter barley**, **winter wheat** [2, 4]
- Foliar disease control in **durum wheat** *(off-label)*, **grass seed crops** *(off-label)*, **triticale** *(off-label)* [2, 4]; **spring rye** *(off-label)*, **winter rye** *(off-label)* [2]
- Glume blotch in **winter wheat** [2, 4]
- Light leaf spot in **spring oilseed rape**, **winter oilseed rape** [1-4]
- Net blotch in **spring barley**, **winter barley** [2, 4]
- Phoma in **spring oilseed rape** *(suppression)*, **winter oilseed rape** *(suppression)* [1-4]
- Rhynchosporium in **spring barley**, **winter barley** [2, 4]
- Septoria leaf blotch in **winter wheat** [2, 4]
- Yellow rust in **spring barley**, **winter barley**, **winter wheat** [2, 4]

Specific Off-Label Approvals (SOLAs)
- ***durum wheat*** *20060505* [2], *20060506* [4]
- ***grass seed crops*** *20060505* [2], *20060506* [4]
- ***spring rye*** *20060505* [2]
- ***triticale*** *20060505* [2], *20060506* [4]
- ***winter rye*** *20060505* [2]

Approval information
- Famoxadone included in Annex I under EC Directive 91/414
- Flusilazole has been reinstated in Annex 1 under Directive 91/414 following an appeal
- Accepted by BBPA for use on malting barley

Efficacy guidance
- Best results obtained from application at early stage of disease development before infection spreads to new growth
- Famoxadone is a member of the QoI cross resistance group. Product should be used preventatively and not relied on for its curative potential
- Use product as part of an Integrated Crop Management strategy incorporating other methods of control, including where appropriate other fungicides with a different mode of action. Do not apply more than two foliar applications of QoI containing products to any cereal crop
- There is a significant risk of widespread resistance occurring in *Septoria tritici* populations in UK. Failure to follow resistance management action may result in reduced levels of disease control
- Flusilazole is a DMI fungicide. Resistance to some DMI fungicides has been identified in Septoria leaf blotch which may seriously affect performance of some products. For further advice contact a specialist advisor and visit the Fungicide Resistance Action Group (FRAG)-UK website

Restrictions
- Maximum number of treatments 2 per crop for wheat and barley; 1 per crop for oilseed rape
- Do not apply to crops under stress
- Do not apply during frosty weather

FOR FULL CONDITIONS OF USE ALWAYS READ THE PRODUCT LABEL

Crop-specific information
- Latest use: before flowering (GS 60) for winter wheat; before quarter ear emergence (GS 53) for barley; up to decline of flowering (GS 67) for oilseed rape

Environmental safety
- Dangerous for the environment
- Very toxic to aquatic organisms
- Dangerous to fish or other aquatic life. Do not contaminate surface waters or ditches with chemical or used container
- LERAP Category B

Hazard classification and safety precautions
Hazard Toxic, Dangerous for the environment
Transport code 9
Packaging group III
UN number 3082
Risk phrases R36, R40, R53a [1-4]; R50 [1, 3]; R51 [2, 4]; R61 [1, 2, 4]; R63 [3]
Operator protection A, C [1-4]; D, M [1, 3]; H [1]; P [3]; U05a, U19a, U20b [1-4]; U16a, U23a [3]
Environmental protection E13b, E16a, E16b, E38
Storage and disposal D01, D02, D05, D09a, D10b, D12a
Medical advice M03 [1, 2, 4]; M04a [1-4]

215 fatty acids

A soap concentrate insecticide and acaricide

Products
Savona   Koppert   49% w/w   SL   06057

Uses
- Aphanomyces cochlioides in *choi sum* (off-label)
- Aphids in *asparagus* (off-label), *aubergines* (off-label), *blackberries* (off-label), *blackcurrants* (off-label), *blueberries* (off-label), *broad beans, broccoli* (off-label), *brussels sprouts, brussels sprouts* (off-label), *bulb onions* (off-label), *cabbages, cabbages* (off-label), *calabrese* (off-label), *carrots* (off-label), *cauliflowers* (off-label), *cayenne pepper* (off-label), *celeriac* (off-label), *celery (outdoor)* (off-label), *chicory* (off-label), *chicory root* (off-label), *chinese cabbage* (off-label), *collards* (off-label), *courgettes* (off-label), *cress* (off-label), *cucumbers, endives* (off-label), *fennel* (off-label), *fodder rape* (off-label), *frise* (off-label), *fruit trees, garlic* (off-label), *globe artichoke* (off-label), *gooseberries* (off-label), *herbs (see appendix 6)* (off-label), *horseradish* (off-label), *jerusalem artichokes* (off-label), *kale* (off-label), *kohlrabi* (off-label), *komatsuna* (off-label), *lamb's lettuce* (off-label), *leaf brassicas* (off-label), *leeks* (off-label), *lentils* (off-label), *lettuce, loganberries* (off-label), *lupins* (off-label), *marrows* (off-label), *melons* (off-label), *mooli* (off-label), *orache* (off-label), *pak choi* (off-label), *parsley root* (off-label), *parsnips* (off-label), *peas, peppers, protected herbs (see appendix 6)* (off-label), *protected tomatoes, pumpkins, radicchio* (off-label), *radishes* (off-label), *raspberries* (off-label), *red beet* (off-label), *redcurrants* (off-label), *rhubarb* (off-label), *ribes hybrids* (off-label), *rubus hybrids* (off-label), *runner beans, salad onions* (off-label), *salsify* (off-label), *scarole* (off-label), *seakale* (off-label), *shallots* (off-label), *spinach* (off-label), *spinach beet* (off-label), *squashes* (off-label), *strawberries* (off-label), *swedes* (off-label), *tomatoes (outdoor), turnips* (off-label), *watermelons* (off-label), *whitecurrants* (off-label), *woody ornamentals*
- Mealybugs in *broad beans, brussels sprouts, cabbages, cucumbers, fruit trees, lettuce, peas, peppers, protected tomatoes, pumpkins, runner beans, tomatoes (outdoor), woody ornamentals*
- Scale insects in *broad beans, brussels sprouts, cabbages, cucumbers, fruit trees, lettuce, peas, peppers, protected tomatoes, pumpkins, runner beans, tomatoes (outdoor), woody ornamentals*
- Spider mites in *broad beans, brussels sprouts, cabbages, cucumbers, fruit trees, lettuce, peas, peppers, protected tomatoes, pumpkins, runner beans, tomatoes (outdoor), woody ornamentals*
- Thrips in *asparagus* (off-label), *aubergines* (off-label), *blackberries* (off-label), *blackcurrants* (off-label), *broccoli* (off-label), *brussels sprouts* (off-label), *bulb onions* (off-label), *cabbages*

SEE SECTION 3 FOR PRODUCTS ALSO REGISTERED

Pesticide Profiles

(off-label), **calabrese** (off-label), **carrots** (off-label), **cauliflowers** (off-label), **cayenne pepper** (off-label), **celeriac** (off-label), **celery (outdoor)** (off-label), **chicory** (off-label), **chicory root** (off-label), **chinese cabbage** (off-label), **choi sum** (off-label), **collards** (off-label), **courgettes** (off-label), **cress** (off-label), **endives** (off-label), **fennel** (off-label), **fodder rape** (off-label), **frise** (off-label), **garlic** (off-label), **globe artichoke** (off-label), **gooseberries** (off-label), **herbs (see appendix 6)** (off-label), **horseradish** (off-label), **jerusalem artichokes** (off-label), **kale** (off-label), **kohlrabi** (off-label), **komatsuna** (off-label), **lamb's lettuce** (off-label), **leaf brassicas** (off-label), **leeks** (off-label), **lentils** (off-label), **loganberries** (off-label), **lupins** (off-label), **marrows** (off-label), **melons** (off-label), **mooli** (off-label), **orache** (off-label), **pak choi** (off-label), **parsley root** (off-label), **parsnips** (off-label), **protected herbs (see appendix 6)** (off-label), **radicchio** (off-label), **radishes** (off-label), **raspberries** (off-label), **red beet** (off-label), **redcurrants** (off-label), **rhubarb** (off-label), **ribes hybrids** (off-label), **rubus hybrids** (off-label), **salad onions** (off-label), **salsify** (off-label), **scarole** (off-label), **seakale** (off-label), **shallots** (off-label), **spinach** (off-label), **spinach beet** (off-label), **squashes** (off-label), **swedes** (off-label), **turnips** (off-label), **watermelons** (off-label)
- Whitefly in **broad beans**, **brussels sprouts**, **cabbages**, **cucumbers**, **fruit trees**, **lettuce**, **peas**, **peppers**, **protected tomatoes**, **pumpkins**, **runner beans**, **tomatoes (outdoor)**, **woody ornamentals**

Specific Off-Label Approvals (SOLAs)
- *asparagus* 20100433
- *aubergines* 20100433
- *blackberries* 20100433
- *blackcurrants* 20100433
- *blueberries* 20102044
- *broccoli* 20100433
- *brussels sprouts* 20100433
- *bulb onions* 20100433
- *cabbages* 20100433
- *calabrese* 20100433
- *carrots* 20100433
- *cauliflowers* 20100433
- *cayenne pepper* 20100433
- *celeriac* 20100433
- *celery (outdoor)* 20100433
- *chicory* 20100433
- *chicory root* 20100433
- *chinese cabbage* 20100433
- *choi sum* 20100433
- *collards* 20100433
- *courgettes* 20100433
- *cress* 20100433
- *endives* 20100433
- *fennel* 20100433
- *fodder rape* 20100433
- *frise* 20100433
- *garlic* 20100433
- *globe artichoke* 20100433
- *gooseberries* 20100433
- *herbs (see appendix 6)* 20100433
- *horseradish* 20100433
- *jerusalem artichokes* 20100433
- *kale* 20100433
- *kohlrabi* 20100433
- *komatsuna* 20100433
- *lamb's lettuce* 20100433
- *leaf brassicas* 20100433
- *leeks* 20100433
- *lentils* 20100433
- *loganberries* 20100433
- *lupins* 20100433

FOR FULL CONDITIONS OF USE ALWAYS READ THE PRODUCT LABEL

- *marrows* 20100433
- *melons* 20100433
- *mooli* 20100433
- *orache* 20100433
- *pak choi* 20100433
- *parsley root* 20100433
- *parsnips* 20100433
- *protected herbs (see appendix 6)* 20100433
- *radicchio* 20100433
- *radishes* 20100433
- *raspberries* 20100433
- *red beet* 20100433
- *redcurrants* 20100433
- *rhubarb* 20100433
- *ribes hybrids* 20100433
- *rubus hybrids* 20100433
- *salad onions* 20100433
- *salsify* 20100433
- *scarole* 20100433
- *seakale* 20100433
- *shallots* 20100433
- *spinach* 20100433
- *spinach beet* 20100433
- *squashes* 20100433
- *strawberries* 20102044
- *swedes* 20100433
- *turnips* 20100433
- *watermelons* 20100433
- *whitecurrants* 20102044

Efficacy guidance
- Use only soft or rain water for diluting spray
- Pests must be sprayed directly to achieve any control. Spray all plant parts thoroughly to run off
- For glasshouse use apply when insects first seen and repeat as necessary. For scale insects apply several applications at weekly intervals after egg hatch
- To control whitefly spray when required and use biological control after 12 h

Restrictions
- Do not use on new transplants, newly rooted cuttings or plants under stress
- Do not use on specified susceptible shrubs. See label for details

Crop-specific information
- HI zero

Environmental safety
- Harmful to fish or other aquatic life. Do not contaminate surface waters or ditches with chemical or used container

Hazard classification and safety precautions
Operator protection U20c
Environmental protection E13c
Storage and disposal D05, D09a, D10a

216 fenamidone

A strobilurin fungicide available only in mixtures
FRAC mode of action code: 11

SEE SECTION 3 FOR PRODUCTS ALSO REGISTERED

217 fenamidone + propamocarb hydrochloride

A systemic fungicide mixture for potatoes
FRAC mode of action code: 11 + 28

See also propamocarb hydrochloride

Products
1	Consento	Bayer CropScience	75:375 g/l	SC	11889
2	Prompto	Bayer CropScience	75:375 g/l	SC	14391

Uses
- Blight in **potatoes** [1]
- Late blight in **potatoes** [2]

Approval information
- Fenamidone and propamocarb hydrochloride included in Annex I under EC Directive 91/414

Efficacy guidance
- Apply as soon as there is risk of blight infection or immediately after a blight warning
- Fenamidone is a member of the QoI cross resistance group. To minimise the likelihood of development of resistance these products should be used in a planned Resistance Management strategy. In addition before application consult and adhere to the latest FRAG-UK resistance guidance on application of QoI fungicides to potatoes. See Section 5 for more information

Restrictions
- Maximum number of treatments (including any other QoI containing fungicide) 6 per yr but no more than three should be applied consecutively. Use in alternation with fungicides from a different cross-resistance group
- Observe a 7 d interval between treatments
- Use only as a protective treatment. Do not use when blight has become readily visible (1% leaf area destroyed)

Crop-specific information
- HI 7 d for potatoes
- All varieties of early, main and seed crop potatoes may be sprayed [2]

Environmental safety
- Dangerous for the environment
- Very toxic to aquatic organisms
- Dangerous to fish or other aquatic life. Do not contaminate surface waters or ditches with chemical or used container
- LERAP Category B

Hazard classification and safety precautions
    Hazard Irritant, Dangerous for the environment
    Transport code 9
    Packaging group III
    UN number 3082
    Risk phrases R36, R50, R53a
    Operator protection A, C; U05a, U08, U11 [1, 2]; U19a, U20b, U20c [2]; U20a [1]
    Environmental protection E13b [1]; E15a [2]; E16a, E38 [1, 2]
    Storage and disposal D01, D02, D09a, D10b, D12a [1, 2]; D05 [1]

218 fenbuconazole

A systemic protectant and curative triazole fungicide for top fruit and grapevines
FRAC mode of action code: 3

Products
1	Agrovista Radni	Agrovista	50 g/l	EW	14404
2	Indar 5 EW	Landseer	50 g/l	EW	09518

Uses
- Blossom wilt in **cherries** *(off-label)*, **mirabelles** *(off-label)*, **plums** *(off-label)* [2]

FOR FULL CONDITIONS OF USE ALWAYS READ THE PRODUCT LABEL

fenhexamid

- Brown rot in **cherries** *(off-label)*, **mirabelles** *(off-label)*, **plums** *(off-label)* [2]
- Powdery mildew in **apples** *(reduction)*, **pears** *(reduction)* [1, 2]; **wine grapes** *(off-label)* [2]
- Scab in **apples**, **pears** [1, 2]

Specific Off-Label Approvals (SOLAs)
- **cherries** *20031372* [2]
- **mirabelles** *20031372* [2]
- **plums** *20031372* [2]
- **wine grapes** *20032081* [2]

Approval information
- Fenbuconazole included in Annex I under EC Directive 91/414

Efficacy guidance
- Most effective when used as part of a routine preventative programme from bud burst to onset of petal fall
- After petal fall, tank mix with other protectant fungicides to enhance scab control
- See label for recommended spray intervals. In periods of rapid growth or high disease pressure, a 7 d interval should be used

Restrictions
- Maximum total dose equivalent to ten full doses per yr on apples, pears
- Consult processors before using on pears for processing
- Do not harvest for human or animal consumption for at least 4 wk after last application

Crop-specific information
- HI 28 d for apples, pears; 21 d for grapevines; 3 d for cherries, mirabelles, plums
- Safe to use on all main commercial varieties of apples and pears in UK

Environmental safety
- Dangerous for the environment
- Toxic to aquatic organisms

Hazard classification and safety precautions
Hazard Irritant, Dangerous for the environment
Transport code 3 [2]; 9 [1]
Packaging group III
UN number 1993 [2]; 3082 [1]
Risk phrases R36, R51, R53a [1, 2]; R41 [2]
Operator protection A, C; U05a, U08, U11 [1, 2]; U20a [2]; U20b [1]
Environmental protection E15a, E38
Consumer protection C02a [1] (4 w); C02a [2] (4 wk)
Storage and disposal D01, D02, D11a, D12a [1, 2]; D05 [2]

219 fenhexamid

A protectant hydroxyanilide fungicide for soft fruit and a range of horticultural crops
FRAC mode of action code: 17

Products
1	Agrovista Fenamid	Agrovista	50% w/w	WG	13733
2	Teldor	Bayer CropScience	50% w/w	WG	11229

Uses
- Botrytis in **bilberries** *(off-label)*, **blackberries**, **blackcurrants**, **blueberries** *(off-label)*, **cherries** *(off-label)*, **gooseberries**, **herbs (see appendix 6)** *(off-label)*, **loganberries**, **plums** *(off-label)*, **protected aubergines** *(off-label)*, **protected courgettes** *(off-label)*, **protected cucumbers** *(off-label)*, **protected gherkins** *(off-label)*, **protected peppers** *(off-label)*, **protected squashes** *(off-label)*, **protected tomatoes** *(off-label)*, **raspberries**, **redcurrants**, **rubus hybrids**, **strawberries**, **tomatoes (outdoor)** *(off-label)*, **whitecurrants** [1, 2]; **frise** *(off-label)*, **hops** *(off-label)*, **lamb's lettuce** *(off-label)*, **leaf brassicas** *(off-label - baby leaf production)*, **protected forest nurseries** *(off-label)*, **rocket** *(off-label)*, **scarole** *(off-label)*, **soft fruit** *(off-label)*, **wine grapes** [2]; **outdoor grapes**, **protected chilli peppers** *(off-label)* [1]

Specific Off-Label Approvals (SOLAs)
- **bilberries** *20080485 expires 31 May 2011* [1], *20061290 expires 31 May 2011* [2]

SEE SECTION 3 FOR PRODUCTS ALSO REGISTERED

- **blueberries** 20080485 expires 31 May 2011 [1], 20061290 expires 31 May 2011 [2]
- **cherries** 20080479 expires 31 May 2011 [1], 20031866 expires 31 May 2011 [2]
- **frise** 20082062 expires 31 May 2011 [2]
- **herbs (see appendix 6)** 20082056 expires 31 May 2011 [1], 20082062 expires 31 May 2011 [2]
- **hops** 20082926 expires 31 May 2011 [2]
- **lamb's lettuce** 20082062 expires 31 May 2011 [2]
- **leaf brassicas** (baby leaf production) 20082062 expires 31 May 2011 [2]
- **plums** 20080479 expires 31 May 2011 [1], 20031866 expires 31 May 2011 [2]
- **protected aubergines** 20080482 expires 31 May 2011 [1], 20042087 [2]
- **protected chilli peppers** 20080481 expires 31 May 2011 [1]
- **protected courgettes** 20080480 expires 31 May 2011 [1], 20042085 [2]
- **protected cucumbers** 20080480 expires 31 May 2011 [1], 20042085 [2]
- **protected forest nurseries** 20082926 expires 31 May 2011 [2]
- **protected gherkins** 20080480 expires 31 May 2011 [1], 20042085 [2]
- **protected peppers** 20080481 expires 31 May 2011 [1], 20042086 [2]
- **protected squashes** 20080480 expires 31 May 2011 [1], 20042085 [2]
- **protected tomatoes** 20080482 expires 31 May 2011 [1], 20042087 [2]
- **rocket** 20082062 expires 31 May 2011 [2]
- **scarole** 20082062 expires 31 May 2011 [2]
- **soft fruit** 20082926 expires 31 May 2011 [2]
- **tomatoes (outdoor)** 20080483 expires 31 May 2011 [1], 20042399 expires 31 May 2011 [2]

Approval information
- Fenhexamid included in Annex I under EC Directive 91/414
- Approval expiry 31 May 2011 [1, 2]

Efficacy guidance
- Use as part of a programme of sprays throughout the flowering period to achieve effective control of Botrytis
- To minimise possibility of development of resistance, no more than two sprays of the product may be applied consecutively. Other fungicides from a different chemical group should then be used for at least two consecutive sprays. If only two applications are made on grapevines, only one may include fenhexamid
- Complete spray cover of all flowers and fruitlets throughout the blossom period is essential for successful control of Botrytis
- Spray programmes should normally start at the start of flowering

Restrictions
- Maximum number of treatments 2 per yr on grapevines; 4 per yr on other listed crops but no more than 2 sprays may be applied consecutively

Crop-specific information
- HI 1 d for strawberries, raspberries, loganberries, blackberries, Rubus hybrids; 3 d for cherries; 7 d for blackcurrants, redcurrants, whitecurrants, gooseberries; 21d for outdoor grapes

Environmental safety
- Dangerous for the environment
- Harmful to fish or other aquatic life. Do not contaminate surface waters or ditches with chemical or used container

Hazard classification and safety precautions
    Hazard Dangerous for the environment
    Transport code 9
    Packaging group III
    UN number 3077
    Risk phrases R53a [2]
    Operator protection A, H [1]; U08, U19a, U20b
    Environmental protection E13c, E38 [2]; E15a [1]
    Storage and disposal D05, D09a, D11a [1, 2]; D12a [2]

FOR FULL CONDITIONS OF USE ALWAYS READ THE PRODUCT LABEL

220 fenoxaprop-P-ethyl

An aryloxyphenoxypropionate herbicide for use in wheat
HRAC mode of action code: A

See also diclofop-methyl + fenoxaprop-P-ethyl

Products

1	Cheetah Super	Bayer CropScience	55 g/l	EW	08723
2	Foxtrot EW	Headland	69 g/l	EW	13243
3	Oskar	Headland	69 g/l	EW	13344
4	Triumph	Bayer CropScience	120 g/l	EC	10902
5	Warrant	Bayer CropScience	83 g/l	EC	13806

Uses
- Annual grasses in *grass seed crops* (off-label) [2]
- Awned canary grass in *spring wheat*, *winter wheat* [5]
- Blackgrass in *spring barley*, *winter barley* [2, 3]; *spring wheat*, *winter wheat* [1-5]
- Canary grass in *spring barley*, *winter barley* [2, 3]; *spring wheat*, *winter wheat* [1-4]
- Grass weeds in *grass seed crops* (off-label), *spring rye* (off-label), *triticale* (off-label), *winter rye* (off-label) [1, 4]
- Perennial grasses in *grass seed crops* (off-label) [2]
- Rough meadow grass in *spring barley*, *winter barley* [2, 3]; *spring wheat*, *winter wheat* [1-5]
- Wild oats in *spring barley*, *winter barley* [2, 3]; *spring wheat*, *winter wheat* [1-5]

Specific Off-Label Approvals (SOLAs)
- *grass seed crops* 20080838 [1], 20101437 [2], 20080839 [4]
- *spring rye* 20080838 [1], 20080839 [4]
- *triticale* 20080838 [1], 20080839 [4]
- *winter rye* 20080838 [1], 20080839 [4]

Approval information
- Fenoxaprop-P-ethyl included in Annex I under EC Directive 91/414

Efficacy guidance
- Treat weeds from 2 fully expanded leaves up to flag leaf ligule just visible; for awned canary-grass and rough meadow-grass from 2 leaves to the end of tillering
- A second application may be made in spring where susceptible weeds emerge after an autumn application
- Spray is rainfast 1 h after application
- Dry conditions resulting in moisture stress may reduce effectiveness
- Fenoxaprop-P-ethyl is an ACCase inhibitor herbicide. To avoid the build up of resistance do not apply products containing an ACCase inhibitor herbicide more than twice to any crop. In addition do not use any product containing fenoxaprop-P-ethyl in mixture or sequence with any other product containing the same ingredient
- Use these products as part of a resistance management strategy that includes cultural methods of control and does not use ACCase inhibitors as the sole chemical method of grass weed control
- Applying a second product containing an ACCase inhibitor to a crop will increase the risk of resistance development; only use a second ACCase inhibitor to control different weeds at a different timing
- Always follow WRAG guidelines for preventing and managing herbicide resistant weeds. See Section 5 for more information

Restrictions
- Maximum total dose equivalent to one or two full dose treatments depending on product used
- Do not apply to barley, durum wheat, undersown crops or crops to be undersown
- Do not roll or harrow within 1 wk of spraying
- Do not spray crops under stress, suffering from drought, waterlogging or nutrient deficiency or those grazed or if soil compacted
- Avoid spraying immediately before or after a sudden drop in temperature or a period of warm days/cold nights
- Do not mix with hormone weedkillers

SEE SECTION 3 FOR PRODUCTS ALSO REGISTERED

Pesticide Profiles

Crop-specific information
- Latest use: before flag leaf sheath extending (GS 41)
- Treat from crop emergence to flag leaf fully emerged (GS 41).
- Product may be sprayed in frosty weather provided crop hardened off but do not spray wet foliage or leaves covered with ice
- Broadcast crops should be sprayed post-emergence after plants have developed well-established root system

Environmental safety
- Dangerous for the environment
- Very toxic (toxic [1]) to aquatic organisms

Hazard classification and safety precautions
  Hazard Irritant [2-5]; Dangerous for the environment [1-5]
  Transport code 9
  Packaging group III
  UN number 3082
  Risk phrases R36, R50 [4]; R38 [2-5]; R43 [2, 3]; R51 [1-3, 5]; R53a [1-5]
  Operator protection A, C, H; U05a, U20b [1-5]; U08 [2-5]; U09a [1]; U11 [5]; U14 [2, 3, 5]; U19a [2, 3]
  Environmental protection E13c [1, 4]; E15a [2, 3, 5]; E34 [2, 3]; E38 [1-5]
  Storage and disposal D01, D02, D09a, D10b [1-5]; D05 [2, 3, 5]; D12a [1, 4, 5]

221 fenoxycarb

An insect specific growth regulator for top fruit
IRAC mode of action code: 7B

Products
  Insegar WG   Syngenta   25% w/w   WG   09789

Uses
- Summer-fruit tortrix moth in *apples*, *pears*

Efficacy guidance
- Best results from application at 5th instar stage before pupation. Product prevents transformation from larva to pupa
- Correct timing best identified from pest warnings
- Because of mode of action rapid knock-down of pest is not achieved and larvae continue to feed for a period after treatment
- Adequate water volume necessary to ensure complete coverage of leaves

Restrictions
- Maximum number of treatments 2 per crop
- Consult processors before use

Crop-specific information
- HI 42 d for apples, pears
- Use on all varieties of apples and pears

Environmental safety
- Dangerous for the environment
- Toxic to aquatic organisms
- High risk to bees. Do not apply to crops in flower or to those in which bees are actively foraging. Do not apply when flowering weeds are present
- Risk to certain non-target insects or other arthropods. See directions for use
- Broadcast air-assisted LERAP (8 m)
- Apply to minimise off-target drift to reduce effects on non-target organisms. Some margin of safety to beneficial arthropods is indicated.
- Broadcast air-assisted LERAP (8 m)

Hazard classification and safety precautions
  Hazard Dangerous for the environment
  Transport code 9
  Packaging group III

FOR FULL CONDITIONS OF USE ALWAYS READ THE PRODUCT LABEL

UN number 3077
Risk phrases R51, R53a
Operator protection A, C, H; U05a, U20a, U23a
Environmental protection E12a, E12e, E15a, E22b, E38; E17b (8 m)
Storage and disposal D01, D02, D07, D09a, D11a, D12a

222 fenpropidin

A systemic, curative and protective piperidine (morpholine) fungicide
FRAC mode of action code: 5

See also difenoconazole + fenpropidin

Products

1	Alpha Fenpropidin 750 EC	Makhteshim	750 g/l	EC	12843
2	Instinct	Headland	750 g/l	EC	14512
3	Tern	Syngenta	750 g/l	EC	08660

Uses
- Brown rust in **spring barley, winter barley** [1-3]
- Foliar disease control in **durum wheat** *(off-label)* [1, 3]; **spring rye** *(off-label)*, **triticale** *(off-label)*, **winter rye** *(off-label)* [3]
- Powdery mildew in **spring barley, spring wheat, winter barley, winter wheat** [1-3]
- Rhynchosporium in **spring barley** *(moderate control)*, **winter barley** *(moderate control)* [1-3]
- Yellow rust in **spring barley** *(moderate control)*, **spring wheat** *(moderate control)*, **winter barley** *(moderate control)*, **winter wheat** *(moderate control)* [1-3]

Specific Off-Label Approvals (SOLAs)
- ***durum wheat*** *20070459* [1], *20060533* [3]
- ***spring rye*** *20060533* [3]
- ***triticale*** *20060533* [3]
- ***winter rye*** *20060533* [3]

Approval information
- Accepted by BBPA for use on malting barley
- Fenpropidin included in Annex I under EC Directive 91/414

Efficacy guidance
- Best results obtained when applied at early stage of disease development. See label for details of recommended timing alone and in mixtures
- Disease control enhanced by vapour-phase activity. Control can persist for 4-6 wk
- Alternate with triazole fungicides to discourage build-up of resistance

Restrictions
- Maximum number of treatments 3 per crop (up to 2 in yr of harvest) for winter crops; 2 per crop for spring crops
- Treated crops must not be harvested for human or animal consumption for at least 5 wk after the last application

Crop-specific information
- Latest use: up to and including ear emergence complete (GS 59).
- HI 5 wk

Environmental safety
- Dangerous for the environment
- Very toxic to aquatic organisms

Hazard classification and safety precautions
Hazard Harmful, Dangerous for the environment
Transport code 6.1 [1]; 9 [2, 3]
Packaging group III
UN number 2902 [1]; 3082 [2, 3]
Risk phrases R20, R21 [1]; R22a [1, 3]; R37 [3]; R38 [2]; R41 [1, 2]; R50, R53a [1-3]
Operator protection A, C, H; U02a, U05a [1-3]; U04a, U10 [1, 3]; U09a [2]; U11, U19a, U20b [1, 2]; U20a [3]

SEE SECTION 3 FOR PRODUCTS ALSO REGISTERED

Environmental protection E15a, E34 [2, 3]; E19b [1]; E38 [1-3]
Consumer protection C02a [2, 3] (5 wk)
Storage and disposal D01, D02, D05, D09a, D12a [1-3]; D10c [2]
Medical advice M03

223   fenpropidin + tebuconazole

A broad spectrum fungicide mixture for cereals
FRAC mode of action code: 5 + 3

See also tebuconazole

Products
Eros   Makhteshim   150:100 g/l   EC   14112

Uses
- Ear diseases in *spring wheat, winter wheat*
- Glume blotch in *spring wheat, winter wheat*
- Septoria leaf blotch in *spring wheat, winter wheat*

Approval information
- Fenpropidin included in Annex I under EC Directive 91/414
- Accepted by BBPA for use on malting barley

Efficacy guidance
- Best disease control and yield benefit obtained when applied at early stage of disease development before infection spreads to new growth
- To protect the flag leaf and ear from Septoria diseases apply from flag leaf emergence to ear fully emerged (GS 37-59)
- Applications once foliar symptoms of *Septoria tritici* are already present on upper leaves will be less effective
- Resistance to some DMI fungicides has been identified in Septoria leaf blotch which may seriously affect the performance of some products.

Restrictions
- Maximum total dose 2.5 l/ha per crop
- Occasional slight temporary leaf speckling may occur on wheat but this has not been shown to reduce yield response or disease control
- Do not treat durum wheat
- Do not apply when temperatures are high

Crop-specific information
- Latest use: Before grain milky ripe stage (GS 73) for winter wheat; up to and including ear emergence just complete (GS 59) for barley and spring wheat
- HI 5 wk for crops for human or animal consumption

Environmental safety
- Harmful if swallowed and in contact with skin
- Irritating to skin
- Risk of serious damage to eyes
- Extremely dangerous to fish or other aquatic life. Do not contaminate surface waters or ditches with chemical or used container

Hazard classification and safety precautions
   Hazard Harmful, Dangerous for the environment
   Transport code 6.1
   Packaging group III
   UN number 2902
   Risk phrases R20, R21, R22a, R51, R53a, R63, R69
   Operator protection A, C, H; U02a, U04a, U05a, U11, U20a
   Environmental protection E15a, E34, E38
   Storage and disposal D02, D09a, D12a
   Medical advice M03

FOR FULL CONDITIONS OF USE ALWAYS READ THE PRODUCT LABEL

224 fenpropimorph

A contact and systemic morpholine fungicide
FRAC mode of action code: 5

See also azoxystrobin + fenpropimorph
epoxiconazole + fenpropimorph
epoxiconazole + fenpropimorph + kresoxim-methyl
epoxiconazole + fenpropimorph + metrafenone
epoxiconazole + fenpropimorph + pyraclostrobin

Products

1	Clayton Spigot	Clayton	750 g/l	EC	11560
2	Corbel	BASF	750 g/l	EC	00578
3	Crebol	AgChem Access	750 g/l	EC	13922
4	Raven	AgriGuard	750 g/l	EC	13188
5	Standon Fenpropimorph 750	Standon	750 g/l	EC	08965

Uses

- Alternaria in **carrots** (off-label), **horseradish** (off-label), **parsley root** (off-label), **parsnips** (off-label), **salsify** (off-label) [2]
- Brown rust in **spring barley**, **spring wheat**, **triticale**, **winter barley**, **winter wheat** [1-5]
- Crown rot in **carrots** (off-label), **horseradish** (off-label), **parsley root** (off-label), **parsnips** (off-label), **salsify** (off-label) [2]
- Foliar disease control in **durum wheat** (off-label) [2]; **grass seed crops** (off-label) [1, 5]
- Powdery mildew in **bilberries** (off-label), **blackcurrants** (off-label), **blueberries** (off-label), **carrots** (off-label), **cranberries** (off-label), **dewberries** (off-label), **gooseberries** (off-label), **hops** (off-label), **horseradish** (off-label), **loganberries** (off-label), **parsley root** (off-label), **parsnips** (off-label), **raspberries** (off-label), **redcurrants** (off-label), **salsify** (off-label), **strawberries** (off-label), **tayberries** (off-label), **whitecurrants** (off-label) [2]; **spring barley**, **spring oats**, **spring wheat**, **winter barley**, **winter oats**, **winter rye**, **winter wheat** [1-5]
- Rhynchosporium in **spring barley**, **winter barley** [1-5]
- Rust in **fodder beet** (off-label), **red beet** (off-label), **sugar beet seed crops** (off-label) [2]; **leeks** [1]
- Yellow rust in **spring barley**, **spring wheat**, **triticale**, **winter barley**, **winter wheat** [1-5]

Specific Off-Label Approvals (SOLAs)

- **bilberries** 20040804 [2]
- **blackcurrants** 20040804 [2]
- **blueberries** 20040804 [2]
- **carrots** 20023753 [2]
- **cranberries** 20040804 [2]
- **dewberries** 20040804 [2]
- **durum wheat** 20061460 [2]
- **fodder beet** 20023757 [2]
- **gooseberries** 20040804 [2]
- **grass seed crops** 20061060 [1], 20061057 [5]
- **hops** 20023759 [2]
- **horseradish** 20023753 [2]
- **loganberries** 20040804 [2]
- **parsley root** 20023753 [2]
- **parsnips** 20023753 [2]
- **raspberries** 20040804 [2]
- **red beet** 20023751 [2]
- **redcurrants** 20040804 [2]
- **salsify** 20023753 [2]
- **strawberries** 20040804 [2]
- **sugar beet seed crops** 20023757 [2]
- **tayberries** 20040804 [2]
- **whitecurrants** 20040804 [2]

SEE SECTION 3 FOR PRODUCTS ALSO REGISTERED

Pesticide Profiles

Approval information
- Accepted by BBPA for use on malting barley and hops
- Fenpropimorph included in Annex I under EC Directive 91/414

Efficacy guidance
- On all crops spray at start of disease attack. See labels for recommended tank mixes. Follow-up treatments may be needed if disease pressure remains high
- Product rainfast after 2 h

Restrictions
- Maximum number of treatments 2 per crop for spring cereals; 3 or 4 per crop for winter cereals; 6 per crop for leeks
- Consult processors before using on crops for processing

Crop-specific information
- HI cereals 5 wk
- Scorch may occur on cereals if applied during frosty weather or in high temperatures
- Some leaf spotting may occur on undersown clovers
- Leeks should be treated every 2-3 wk as required, up to 6 applications

Environmental safety
- Dangerous for the environment
- Very toxic to aquatic organisms

Hazard classification and safety precautions
Hazard Harmful [1-4]; Irritant [5]; Dangerous for the environment [1-5]
Transport code 3 [4]; 9 [1-3, 5]
Packaging group III
UN number 1993 [4]; 3082 [1-3, 5]
Risk phrases R22a [1]; R38, R53a [1-5]; R50 [2-4]; R51 [1, 5]; R63 [1-4]
Operator protection A [1-5]; C [1, 5]; U05a, U08, U14, U15, U19a, U20b
Environmental protection E15a [1-5]; E38 [2-4]
Storage and disposal D01, D02, D09a [1-5]; D05 [1, 5]; D08 [2-4]; D10b [5]; D10c, D12a [1-4]
Medical advice M03 [4]; M05a [1-3, 5]

225   fenpropimorph + flusilazole

A broad-spectrum eradicant and protectant fungicide mixture for cereals
FRAC mode of action code: 5 + 3

See also flusilazole

Products

1	Colstar	DuPont	375:160 g/l	EC	12175
2	Pluton	DuPont	375:160 g/l	EC	12200

Uses
- Brown rust in *spring barley*, *winter barley*, *winter wheat* [1, 2]
- Foliar disease control in *durum wheat* (off-label), *grass seed crops* (off-label) [1, 2]; *spring rye* (off-label), *triticale* (off-label), *winter rye* (off-label) [1]
- Glume blotch in *winter wheat* [1, 2]
- Net blotch in *spring barley*, *winter barley* [1, 2]
- Powdery mildew in *spring barley*, *winter barley*, *winter wheat* [1, 2]
- Rhynchosporium in *spring barley*, *winter barley* [1, 2]
- Septoria leaf blotch in *winter wheat* [1, 2]
- Yellow rust in *spring barley*, *winter barley*, *winter wheat* [1, 2]

Specific Off-Label Approvals (SOLAs)
- *durum wheat* 20060977 [1], 20060985 [2]
- *grass seed crops* 20060977 [1], 20060985 [2]
- *spring rye* 20060977 [1]
- *triticale* 20060977 [1]
- *winter rye* 20060977 [1]

FOR FULL CONDITIONS OF USE ALWAYS READ THE PRODUCT LABEL

Approval information
- Fenpropimorph and flusilazole included in Annex I under EC Directive 91/414
- Flusilazole has been reinstated in Annex 1 under Directive 91/414 following an appeal
- Accepted by BBPA for use on malting barley

Efficacy guidance
- Disease control is more effective if treatment made at an early stage of disease development
- Treat winter cereals in spring or early summer before diseases spread to new growth
- Spring barley should be treated when diseases are first evident
- Treatment may be repeated after 3-4 wk if necessary
- Flusilazole is a DMI fungicide. Resistance to some DMI fungicides has been identified in Septoria leaf blotch which may seriously affect performance of some products. For further advice contact a specialist advisor and visit the Fungicide Resistance Action Group (FRAG)-UK website

Restrictions
- Maximum number of treatments 2 per crop of barley or wheat
- Do not apply to crops under stress
- Do not apply during frosty weather

Crop-specific information
- Latest use: before beginning of anthesis (GS 60) for winter wheat; up to and including completion of ear emergence (GS 59) for barley

Environmental safety
- Dangerous for the environment
- Toxic to aquatic organisms
- Dangerous to fish or other aquatic life. Do not contaminate surface waters or ditches with chemical or used container

Hazard classification and safety precautions
Hazard Toxic, Dangerous for the environment
Transport code 9
Packaging group III
UN number 3082
Risk phrases R36, R40, R51, R53a, R61
Operator protection A, C; U05a, U11, U19a, U20b
Environmental protection E13b, E15a, E38
Storage and disposal D01, D02, D05, D09a, D10b, D12a
Medical advice M04a

226 fenpropimorph + kresoxim-methyl

A protectant and systemic fungicide mixture for cereals
FRAC mode of action code: 5 + 11

See also kresoxim-methyl

Products
Ensign   BASF   300:150 g/l   SE   11729

Uses
- Foliar disease control in **durum wheat** (off-label)
- Glume blotch in **spring wheat** (reduction), **triticale** (reduction), **winter wheat** (reduction)
- Powdery mildew in **spring barley**, **spring oats**, **spring rye**, **triticale**, **winter barley**, **winter oats**, **winter rye**
- Rhynchosporium in **spring barley**, **spring rye**, **triticale**, **winter barley**, **winter rye**
- Septoria leaf blotch in **spring wheat** (reduction), **triticale** (reduction), **winter wheat** (reduction)

Specific Off-Label Approvals (SOLAs)
- *durum wheat* 20062643

Approval information
- Fenpropimorph and kresoxim-methyl included in Annex I under EC Directive 91/414
- Accepted by BBPA for use on malting barley

Efficacy guidance
- For best results spray at the start of foliar disease attack and repeat if infection conditions persist
- Kresoxim-methyl is a member of the QoI cross resistance group. Product should be used preventatively and not relied on for its curative potential
- Use product as part of an Integrated Crop Management strategy incorporating other methods of control, including where appropriate other fungicides with a different mode of action. Do not apply more than two foliar applications of QoI containing products to any cereal crop
- There is a significant risk of widespread resistance occurring in *Septoria tritici* populations in UK. Strains of barley powdery mildew resistant to QoIs are common in UK. Failure to follow resistance management action may result in reduced levels of disease control
- Strains of barley powdery mildew resistant to QoIs are common in the UK

Restrictions
- Maximum total dose equivalent to two full dose treatments for all crops

Crop-specific information
- Latest use: completion of ear emergence (GS 59) for barley and oats; completion of flowering (GS 69) for wheat, rye and triticale

Environmental safety
- Dangerous for the environment
- Very toxic to aquatic organisms

Hazard classification and safety precautions
Hazard Harmful, Dangerous for the environment
Transport code 9
Packaging group III
UN number 3082
Risk phrases R40, R50, R53a, R63
Operator protection A, H; U05a, U14, U20b
Environmental protection E15a, E38
Storage and disposal D01, D02, D05, D09a, D10c, D12a
Medical advice M05a

227 fenpropimorph + pyraclostrobin

A protectant and curative fungicide mixture for cereals
FRAC mode of action code: 5 +11

See also pyraclostrobin

Products
Jenton	BASF	375:100 g/l	EC	11898

Uses
- Brown rust in **spring barley**, **spring wheat**, **winter barley**, **winter wheat**
- Crown rust in **spring oats**, **winter oats**
- Drechslera leaf spot in **spring wheat**, **winter wheat**
- Foliar disease control in **grass seed crops** *(off-label)*, **triticale** *(off-label)*
- Glume blotch in **spring wheat**, **winter wheat**
- Net blotch in **spring barley**, **winter barley**
- Powdery mildew in **spring barley**, **spring oats**, **winter barley**, **winter oats**
- Rhynchosporium in **spring barley**, **winter barley**
- Septoria leaf blotch in **spring wheat**, **winter wheat**
- Yellow rust in **spring barley**, **spring wheat**, **winter barley**, **winter wheat**

Specific Off-Label Approvals (SOLAs)
- *grass seed crops* 20062650
- *triticale* 20062650

Approval information
- Fenpropimorph and pyraclostrobin included in Annex I under EC Directive 91/414
- Accepted by BBPA for use on malting barley

FOR FULL CONDITIONS OF USE ALWAYS READ THE PRODUCT LABEL

fenpropimorph + quinoxyfen

Efficacy guidance
- Best results obtained from treatment at the start of foliar disease attack
- Yield response may be obtained in the absence of visual disease
- Pyraclostrobin is a member of the QoI cross resistance group. Product should be used preventatively and not relied on for its curative potential
- Use product as part of an Integrated Crop Management strategy incorporating other methods of control, including where appropriate other fungicides with a different mode of action. Do not apply more than two foliar applications of QoI containing products to any cereal crop
- There is a significant risk of widespread resistance occurring in *Septoria tritici* populations in UK. Strains of barley powdery mildew resistant to QoIs are common in UK. Failure to follow resistance management action may result in reduced levels of disease control

Restrictions
- Maximum number of treatments 2 per crop

Crop-specific information
- Latest use: up to and including emergence of ear just complete (GS 59) for barley

Environmental safety
- Dangerous for the environment
- Very toxic to aquatic organisms
- LERAP Category B

Hazard classification and safety precautions
  Hazard Harmful, Dangerous for the environment
  Transport code 9
  Packaging group III
  UN number 3082
  Risk phrases R20, R22a, R36, R38, R50, R53a, R63
  Operator protection A, C; U05a, U14, U20b
  Environmental protection E15a, E16a, E34, E38
  Storage and disposal D01, D02, D05, D09a, D10c, D12a
  Medical advice M03, M05a

228 fenpropimorph + quinoxyfen

A systemic fungicide mixture for cereals
FRAC mode of action code: 5 + 13

See also quinoxyfen

Products

Orka	Dow	250:66.7 g/l	EW	08879

Uses
- Powdery mildew in **durum wheat**, **grass seed crops** *(off-label)*, **spring barley**, **spring oats**, **spring rye**, **spring wheat**, **triticale**, **winter barley**, **winter oats**, **winter rye**, **winter wheat**

Specific Off-Label Approvals (SOLAs)
- *grass seed crops* 20060979

Approval information
- Fenpropimorph and quinoxyfen included in Annex I under Directive 91/414
- Accepted by BBPA for use on malting barley

Efficacy guidance
- For best results treat at early stage of disease development before infection spreads to new crop growth. Further treatment may be necessary if disease pressure remains high
- For control of established infections and broad spectrum disease control use in tank mixtures. See label
- Product rainfast after 1 h
- Systemic activity may be reduced in severe drought

Restrictions
- Maximum total dose 3.0 l product per ha
- Apply only in the spring from mid-tillering stage (GS 25)

SEE SECTION 3 FOR PRODUCTS ALSO REGISTERED

Pesticide Profiles

Crop-specific information
- Latest use: when first awns visible (GS 49)
- Crop scorch may occur when treatment made in high temperatures

Environmental safety
- Dangerous for the environment
- Very toxic to aquatic organisms
- LERAP Category B

Hazard classification and safety precautions
Hazard Harmful, Dangerous for the environment
Transport code 9
Packaging group III
UN number 3082
Risk phrases R43, R50, R53a, R61
Operator protection A, H; U05a, U14
Environmental protection E15a, E16a, E34, E38
Storage and disposal D01, D02, D12a

229   fenpyroximate

A mitochondrial electron transport inhibitor (METI) acaricide for apples
IRAC mode of action code: 21

Products
Sequel	Certis	51.3 g/l	SC	12657

Uses
- Fruit tree red spider mite in **apples**
- Red spider mites in **plums** *(off-label)*, **protected strawberries** *(off-label)*, **strawberries** *(off-label)*

Specific Off-Label Approvals (SOLAs)
- **plums** *20093060*
- **protected strawberries** *20093061*
- **strawberries** *20093061*

Approval information
- Fenpyroximate included in Annex I under EC Directive 91/414

Efficacy guidance
- Kills motile stages of fruit tree red spider mite. Best results achieved if applied in warm weather
- Total spray cover of trees essential. Use higher volumes for large trees
- Apply when majority of winter eggs have hatched

Restrictions
- Maximum number of treatments 1 per yr or total dose equivalent to one full dose treatment
- Other mitochondrial electron transport inhibitor (METI) acaricides should not be applied to the same crop in the same calendar yr either separately or in mixture
- Do not apply when apple crops or pollinator are in flower
- Consult processor before use on crops for processing

Crop-specific information
- HI 2 wk

Environmental safety
- Dangerous for the environment
- Very toxic to aquatic organisms
- Risk to non-target insects or other arthropods
- Broadcast air-assisted LERAP (40 m)

Hazard classification and safety precautions
Hazard Harmful, Dangerous for the environment
Transport code 9
Packaging group III
UN number 3082
Risk phrases R20, R36, R43, R50, R53a

FOR FULL CONDITIONS OF USE ALWAYS READ THE PRODUCT LABEL

ferric phosphate

Operator protection A, C, H; U05a, U08, U14, U15, U20b
Environmental protection E13b, E22c; E17b (40 m)
Storage and disposal D01, D02, D05, D09a, D10c, D12a

230 ferric phosphate

A molluscicide bait for controlling slugs and snails

Products

1	Ferramol Max	Certis	2.97% w/w	GB	14463
2	Sluggo	Omex	1% w/w	GB	12529
3	Sluxx	Certis	2.97% w/w	GB	14462

Uses
- Slugs in *all edible crops (outdoor), all non-edible crops (outdoor), protected crops* [2]
- Slugs and snails in *all edible crops (outdoor and protected), all non-edible crops (outdoor), amenity vegetation* [1, 3]
- Snails in *all edible crops (outdoor), all non-edible crops (outdoor), protected crops* [2]

Approval information
- Ferric phosphate included in Annex 1 under EC Directive 91/414
- Approval expiry 31 Oct 2011 [1-3]

Efficacy guidance
- Treat as soon as damage first seen preferably in early evening. Repeat as necessary to maintain control
- Best results obtained from moist soaked granules. This will occur naturally on moist soils or in humid conditions
- Ferric phosphate does not cause excessive slime secretion and has no requirement to collect moribund slugs from the soil surface
- Active ingredient is degraded by micro-organisms to beneficial plant nutrients

Restrictions
- Maximum total dose equivalent to four full dose treatments

Crop-specific information
- Latest use not specified for any crop

Hazard classification and safety precautions
  UN number N/C
  Operator protection U05a [2, 3]; U20a [3]; U20b [2]
  Environmental protection E15a [2]; E34 [2, 3]
  Storage and disposal D01, D09a [2, 3]

231 ferrous sulphate

A herbicide/fertilizer combination for moss control in turf

See also dicamba + dichlorprop-P + ferrous sulphate + MCPA
dichlorprop-P + ferrous sulphate + MCPA

Products

1	Elliott's Lawn Sand	Elliott	9.3% w/w	SA	04860
2	Elliott's Mosskiller	Elliott	24.6% w/w	GR	04909
3	Ferromex Mosskiller Concentrate	Omex	16.4% w/w	SL	13180
4	Greenmaster Autumn	Scotts	18.2% w/w	GR	12196
5	Greenmaster Mosskiller	Scotts	27% w/w	GR	12197
6	Greentec Mosskiller	Headland Amenity	9% w/w	GR	12518
7	Greentec Mosskiller Pro	Headland Amenity	16.78% w/w	GR	14645
8	Landscaper Pro Moss Control + Fertilizer	Scotts	21.7% w/w	GR	14317
9	Mascot Micronised Lawn Sand	Rigby Taylor	8.2% w/w	SA	14783
10	SHL Lawn Sand	Sinclair	5.4% w/w	SA	05254
11	Taylors Lawn Sand	Rigby Taylor	4.1% w/w	SA	04451

SEE SECTION 3 FOR PRODUCTS ALSO REGISTERED

Pesticide Profiles

Uses
- Moss in **amenity grassland** [5]; **managed amenity turf** [1-11]

Efficacy guidance
- For best results apply when turf is actively growing and the soil is moist
- Fertilizer component of most products encourages strong root growth and tillering
- Mow 3 d before treatment and do not mow for 3-4 d afterwards
- Water after 2 d if no rain
- Rake out dead moss thoroughly 7-14 d after treatment. Re-treatment may be necessary for heavy infestations

Restrictions
- Maximum number of treatments - see labels
- Do not apply during drought or when heavy rain expected
- Do not apply in frosty weather or when the ground is frozen
- Do not walk on treated areas until well watered
- Do not apply product undiluted [3]

Crop-specific information
- If spilt on paving, concrete, clothes etc brush off immediately to avoid discolouration
- Observe label restrictions for interval before cutting after treatment

Environmental safety
- Harmful to fish or other aquatic life. Do not contaminate surface waters or ditches with chemical or used container [4]

Hazard classification and safety precautions
Hazard Harmful [7]
UN number N/C [1-8, 10, 11]
Risk phrases R22a, R36, R38 [7]
Operator protection A [3]; U20a [1, 2, 9]; U20b [3, 5-7]; U20c [4, 8, 10, 11]
Environmental protection E13c [4, 8]; E15a [1, 2, 5-7, 9, 11]; E15b [3]
Storage and disposal D01 [3, 5-7, 10]; D09a [1-11]; D11a [1, 2, 4-11]; D12a [3]

232 ferrous sulphate + MCPA + mecoprop-P

A translocated herbicide and moss killer mixture
HRAC mode of action code: O + O

See also MCPA
 mecoprop-P

Products
Renovator Pro   Scotts   0.49:0.29:16.3% w/w   GR   12204

Uses
- Annual dicotyledons in **managed amenity turf**
- Moss in **managed amenity turf**
- Perennial dicotyledons in **managed amenity turf**

Approval information
- MCPA and mecoprop-P included in Annex I under EC Directive 91/414

Efficacy guidance
- Apply from Apr to Sep when weeds are growing
- For best results apply when light showers or heavy dews are expected
- Apply with a suitable calibrated fertilizer distributor
- Retreatment may be necessary after 6 wk if weeds or moss persist
- For best control of moss scarify vigorously after 2 wk to remove dead moss
- Where regrowth of moss or weeds occurs a repeat treatment may be made after 6 wk
- Avoid treatment of wet grass or during drought. If no rain falls within 48 h water in thoroughly

Restrictions
- Maximum number of treatments 3 per yr
- Do not treat new turf until established for 6 mth
- The first 4 mowings after treatment should not be used to mulch cultivated plants unless composted at least 6 mth

FOR FULL CONDITIONS OF USE ALWAYS READ THE PRODUCT LABEL

fipronil

- Avoid walking on treated areas until it has rained or they have been watered
- Do not re-seed or turf within 8 wk of last treatment
- Do not cut grass for at least 3 d before and at least 4 d after treatment
- Do not apply during freezing conditions or when rain imminent

Environmental safety
- Keep livestock out of treated areas for up to two weeks following treatment and until poisonous weeds, such as ragwort, have died down and become unpalatable
- Harmful to fish or other aquatic life. Do not contaminate surface waters or ditches with chemical or used container
- Do not empty into drains
- Some pesticides pose a greater threat of contamination of water than others and mecoprop-P is one of these pesticides. Take special care when applying mecoprop-P near water and do not apply if heavy rain is forecast

Hazard classification and safety precautions
UN number N/C
Operator protection U20b
Environmental protection E07a, E13c, E19b
Storage and disposal D01, D09a, D12a
Medical advice M05a

233 fipronil

A phenylpyrazole insecticide for horticulture; approval expired 2009
IRAC mode of action code: 2B

234 flazasulfuron

A sulfonylurea herbicide for non-crop use
HRAC mode of action code: B

Products
1	Chikara Weed Control	Belchim	25% w/w	WG	14189
2	Chikita	ChemSource	25% w/w	WG	14675
3	Paradise	Pan Agriculture	25% w/w	WG	14504

Uses
- Annual and perennial weeds in **natural surfaces not intended to bear vegetation, permeable surfaces overlying soil, railway tracks**

Environmental safety
- LERAP Category B

Hazard classification and safety precautions
Hazard Dangerous for the environment
Transport code 9
Packaging group III
UN number 3077
Risk phrases R50, R53a
Operator protection A, H, M; U02a, U05a, U20b [1-3]; U08 [1, 2]; U09a [3]
Environmental protection E15a, E16a, E16b, E38
Storage and disposal D01, D02, D09a, D11a, D12a

235 flonicamid

A selective feeding blocker aphicide
IRAC mode of action code: 9C

Products
1	Mainman	Belchim	50% w/w	WG	13123
2	Mutiny	ChemSource	50% w/w	WG	14805
3	Teppeki	Belchim	50% w/w	WG	12402

SEE SECTION 3 FOR PRODUCTS ALSO REGISTERED

Pesticide Profiles

Uses
- Aphids in **apples**, **hops** *(off-label)*, **pears** [1]; **ornamental plant production** *(off-label)* [1, 3]; **potatoes**, **winter wheat** [2, 3]
- Tobacco whitefly in **protected ornamentals** *(off-label)* [1]
- Whitefly in **protected ornamentals** *(off-label)* [1]

Specific Off-Label Approvals (SOLAs)
- **hops** *20090927* [1]
- **ornamental plant production** *20082898* [1], *20082906* [3]
- **protected ornamentals** *20101052* [1]

Approval information
- Flonicamid included in Annex 1 under EC Directive 91/414

Efficacy guidance
- Apply when warning systems forecast significant aphid infestations
- Persistence of action is 21 d
- Do not apply more than two consecutive treatments of flonicamid. If further treatment is needed use an insecticide with a different mode of action

Restrictions
- Maximum number of treatments 2 per crop for potatoes, winter wheat
- Must not be applied to winter wheat before 50% ear emerged stage (GS 53)
- Use a maximum of two consecutive applications of [1] in apples and pears. If further treatments are required, use an insecticide with a different mode of action before applying the final application of [1]

Crop-specific information
- HI: potatoes 14 d; winter wheat 28 d

Environmental safety
- Dangerous for the environment
- Harmful to aquatic organisms

Hazard classification and safety precautions
Hazard Dangerous for the environment [2, 3]
UN number N/C
Risk phrases R52, R53a
Operator protection A, H [1-3]; E, M [1]; U05a [1-3]; U13, U14, U15 [1]
Environmental protection E15a
Storage and disposal D01, D02, D12b [1-3]; D03, D08 [1]; D07 [2, 3]

236 florasulam

A triazolopyrimidine herbicide for cereals
HRAC mode of action code: B

See also 2,4-D + florasulam
clopyralid + florasulam + fluroxypyr

Products
1	Barton WG	Dow	25% w/w	WG	13284
2	Boxer	Dow	50 g/l	SC	09819

Uses
- Annual dicotyledons in **farm forestry** *(off-label)*, **forest nurseries** *(off-label)*, **game cover** *(off-label)*, **grass seed crops** *(off-label)*, **ornamental plant production** *(off-label)*, **spring rye** *(off-label)*, **triticale** *(off-label)*, **winter rye** *(off-label)* [2]; **spring barley**, **spring oats**, **spring wheat**, **winter barley**, **winter oats**, **winter wheat** [1, 2]
- Chickweed in **spring barley**, **spring oats**, **spring wheat**, **winter barley**, **winter oats**, **winter wheat** [1, 2]
- Cleavers in **spring barley**, **spring oats**, **spring wheat**, **winter barley**, **winter oats**, **winter wheat** [1, 2]
- Mayweeds in **spring barley**, **spring oats**, **spring wheat**, **winter barley**, **winter oats**, **winter wheat** [1, 2]

FOR FULL CONDITIONS OF USE ALWAYS READ THE PRODUCT LABEL

florasulam

- Volunteer oilseed rape in *spring barley*, *spring oats*, *spring wheat*, *winter barley*, *winter oats*, *winter wheat* [1, 2]

Specific Off-Label Approvals (SOLAs)
- *farm forestry* 20082826 [2]
- *forest nurseries* 20082826 [2]
- *game cover* 20082826 [2]
- *grass seed crops* 20060997 [2]
- *ornamental plant production* 20082826 [2]
- *spring rye* 20060997 [2]
- *triticale* 20060997 [2]
- *winter rye* 20060997 [2]

Approval information
- Florasulam included in Annex I under EC Directive 91/414
- Accepted by BBPA for use on malting barley

Efficacy guidance
- Best results obtained from treatment of small actively growing weeds in good conditions
- Apply in autumn or spring once crop has 3 leaves
- Product is mainly absorbed by leaves of weeds and is effective on all soil types
- Florasulam is a member of the ALS-inhibitor group of herbicides

Restrictions
- Maximum total dose on any crop equivalent to one full dose treatment
- Do not roll or harrow within 7 d before or after application
- Do not spray when crops under stress from cold, drought, pest damage, nutrient deficiency or any other cause
- Specific restrictions apply to use in sequence or tank mixture with other sulfonylurea or ALS-inhibiting herbicides. See label for details

Crop-specific information
- Latest use: up to and including flag leaf just visible stage for all crops

Following crops guidance
- Where the product has been used in mixture with certain named products (see label) only cereals or grass may be sown in the autumn following harvest [2]
- Unless otherwise restricted cereals, oilseed rape, field beans, grass or vegetable brassicas as transplants may be sown as a following crop in the same calendar yr as treatment. Oilseed rape may show some temporary reduction of vigour after a dry summer, but yields are not affected
- In addition to the above, linseed, peas, sugar beet, potatoes, maize, clover (for use in grass/clover mixtures) or carrots may be sown in the calendar yr following treatment
- In the event of failure of a treated crop in spring only spring wheat, spring barley, spring oats, maize or ryegrass may be sown

Environmental safety
- Dangerous for the environment
- Very toxic to aquatic organisms
- See label for detailed instructions on tank cleaning

Hazard classification and safety precautions
Hazard Dangerous for the environment
Transport code 9
Packaging group III
UN number 3077 [1]; 3082 [2]
Risk phrases R50, R53a
Operator protection U05a
Environmental protection E15a [2]; E15b [1]; E34, E38 [1, 2]
Storage and disposal D01, D02, D05, D09a, D10b, D12a

SEE SECTION 3 FOR PRODUCTS ALSO REGISTERED

237 florasulam + fluroxypyr

A post-emergence herbicide mixture for cereals
HRAC mode of action code: B + O

See also fluroxypyr

Products

1	Cabadex	Headland Amenity	2.5:100 g/l	SE	13948
2	GF 184	Dow	2.5:100 g/l	SE	10878
3	Hiker	Dow	1.0:100 g/l	SE	11451
4	Hunter	Dow	2.5:100 g/l	SE	12836
5	Slalom	Dow	2.5:100 g/l	SE	13772
6	Starane Gold	Dow	1.0:100 g/l	SE	10879
7	Starane Vantage	Dow	1.0:100 g/l	SE	10922
8	Starane XL	Dow	2.5:100 g/l	SE	10921
9	Trafalgar	Pan Amenity	2.5:100 g/l	SE	14888

Uses

- Annual and perennial weeds in **amenity grassland, lawns, managed amenity turf** [1]
- Annual dicotyledons in **amenity grassland, lawns, managed amenity turf** [9]; *durum wheat (off-label)*, **spring rye** *(off-label)*, **triticale** *(off-label)*, **winter rye** *(off-label)* [3, 6-8]; **game cover** *(off-label)* [2, 6, 8]; **grass seed crops** *(off-label)* [2, 3, 6-8]; **ornamental plant production** *(off-label)* [8]; **spring barley, spring oats, spring wheat, winter barley, winter oats, winter wheat** [2, 4, 5, 8]
- Buttercups in **amenity grassland, lawns, managed amenity turf** [1]
- Chickweed in **amenity grassland, lawns, managed amenity turf** [9]; **spring barley, spring oats, spring wheat, winter barley, winter oats, winter wheat** [2-8]
- Cleavers in **spring barley, spring oats, spring wheat, winter barley, winter oats, winter wheat** [2-8]
- Clover in **amenity grassland, lawns, managed amenity turf** [1]
- Daisies in **amenity grassland, lawns, managed amenity turf** [1]
- Dandelions in **amenity grassland, lawns, managed amenity turf** [1]
- Mayweeds in **spring barley, spring oats, spring wheat, winter barley, winter oats, winter wheat** [2, 4, 5, 8]
- Plantains in **amenity grassland, lawns, managed amenity turf** [1]

Specific Off-Label Approvals (SOLAs)

- **durum wheat** 20072810 expires 31 Dec 2011 [3], 20072811 expires 31 Dec 2011 [6], 20072812 expires 31 Dec 2011 [7], 20072815 expires 31 Dec 2011 [8]
- **game cover** 20082858 expires 31 Dec 2011 [2], 20082878 expires 31 Dec 2011 [6], 20082904 expires 31 Dec 2011 [8]
- **grass seed crops** 20072809 expires 31 Dec 2011 [2], 20072810 expires 31 Dec 2011 [3], 20072811 expires 31 Dec 2011 [6], 20072812 expires 31 Dec 2011 [7], 20072815 expires 31 Dec 2011 [8]
- **ornamental plant production** 20082904 expires 31 Dec 2011 [8]
- **spring rye** 20072810 expires 31 Dec 2011 [3], 20072811 expires 31 Dec 2011 [6], 20072812 expires 31 Dec 2011 [7], 20072815 expires 31 Dec 2011 [8]
- **triticale** 20072810 expires 31 Dec 2011 [3], 20072811 expires 31 Dec 2011 [6], 20072812 expires 31 Dec 2011 [7], 20072815 expires 31 Dec 2011 [8]
- **winter rye** 20072810 expires 31 Dec 2011 [3], 20072811 expires 31 Dec 2011 [6], 20072812 expires 31 Dec 2011 [7], 20072815 expires 31 Dec 2011 [8]

Approval information

- Florasulam and fluroxypyr included in Annex I under EC Directive 91/414
- Accepted by BBPA for use on malting barley
- Approval expiry 31 Dec 2011 [1-9]

Efficacy guidance

- Best results obtained when weeds are small and growing actively
- Products are mainly absorbed through weed foliage. Cleavers emerging after application will not be controlled
- Florasulam is a member of the ALS-inhibitor group of herbicides

FOR FULL CONDITIONS OF USE ALWAYS READ THE PRODUCT LABEL

florasulam + pyroxsulam

Restrictions
- Maximum total dose equivalent to one full dose treatment
- Do not roll or harrow 7 d before or after application
- Do not spray when crops are under stress from cold, drought, pest damage or nutrient deficiency
- Do not apply through CDA applicators
- Specific restrictions apply to use in sequence or tank mixture with other sulfonylurea or ALS-inhibiting herbicides. See label for details

Crop-specific information
- Latest use: before flag leaf sheath extended (before GS 41) for spring barley and spring wheat; before flag leaf sheath opening (before GS 47) for winter barley and winter wheat; before second node detectable (before GS 32) for winter oats

Following crops guidance
- Cereals, oilseed rape, field beans or grass may follow treated crops in the same yr. Oilseed rape may suffer temporary vigour reduction after a dry summer
- In addition to the above, linseed, peas, sugar beet, potatoes, maize or clover may be sown in the calendar yr following treatment
- In the event of failure of a treated crop in the spring, only spring cereals, maize or ryegrass may be planted

Environmental safety
- Dangerous for the environment
- Toxic to aquatic organisms
- Take extreme care to avoid drift onto non-target crops or plants

Hazard classification and safety precautions
Hazard Irritant, Dangerous for the environment
Transport code 9
Packaging group III
UN number 3082
Risk phrases R36, R38, R53a, R67 [1-9]; R50 [1]; R51 [2-9]
Operator protection A [1-9]; C [1-8]; U05a, U11, U19a [1-9]; U08, U14, U20b [1]
Environmental protection E15a [2-9]; E15b [1]; E34, E38 [1-9]
Storage and disposal D01, D02, D09a, D10c, D12a [1-9]; D05 [9]
Medical advice M05a [9]

238 florasulam + pyroxsulam

A mixture of two triazolopirimidine sulfonamides for winter wheat
HRAC mode of action code: B

See also pyroxsulam

Products
Broadway Star   Dow   1.42:7.08% w/w   WG   14319

Uses
- Charlock in *triticale, winter rye, winter wheat*
- Chickweed in *triticale, winter rye, winter wheat*
- Cleavers in *triticale, winter rye, winter wheat*
- Field pansy in *triticale, winter rye, winter wheat*
- Field speedwell in *triticale, winter rye, winter wheat*
- Geranium species in *triticale, winter rye, winter wheat*
- Ivy-leaved speedwell in *triticale, winter rye, winter wheat*
- Mayweeds in *triticale, winter rye, winter wheat*
- Poppies in *triticale, winter rye, winter wheat*
- Ryegrass in *triticale, winter rye, winter wheat*
- Sterile brome in *triticale, winter rye, winter wheat*
- Volunteer oilseed rape in *triticale, winter rye, winter wheat*
- Wild oats in *triticale, winter rye, winter wheat*

Approval information
- Florasulam and pyroxsulam included in Annex I under EC Directive 91/414

SEE SECTION 3 FOR PRODUCTS ALSO REGISTERED

Pesticide Profiles

Efficacy guidance
- Rainfast within 1 hour of application
- Requires an authorised adjuvant at application and recommended for use in a programme with herbicides employing a different mode of action.

Crop-specific information
- Crop injury may occur if applied in tank mixture with plant growth regulators - allow a minimum interval of 7 days.
- Crop injury may occur if applied in tank mixture with OP insecticides - allow a minimum interval of 14 days

Following crops guidance
- Crop failure before 1st Feb - allow 6 weeks to elapse and then drill spring wheat, spring barley, grass or maize.
- Crop failure after 1st Feb - plough and allow 6 weeks to elapse before drilling grass or maize.

Environmental safety
- Take extreme care to avoid drift on to susceptible crops, non-target plants or waterways
- LERAP Category B

Hazard classification and safety precautions
Hazard Dangerous for the environment
Transport code 9
Packaging group III
UN number 3077
Risk phrases R50, R53a
Operator protection U05a, U14, U15, U20a
Environmental protection E15b, E16a, E34, E38, E39
Storage and disposal D01, D02, D09a, D12a

239 fluazifop-P-butyl

A phenoxypropionic acid grass herbicide for broadleaved crops
HRAC mode of action code: A

Products

1 Clayton Maximus	Clayton	125 g/l	EC	12543
2 Fusilade Max	Syngenta	125 g/l	EC	11519
3 Greencrop Bantry	Greencrop	125 g/l	EC	12737
4 Howitzer	ChemSource	125 g/l	EC	14942

Uses
- Annual grasses in **almonds** (off-label), **apples** (off-label), **apricots** (off-label), **artichokes** (off-label), **asparagus** (off-label), **bilberries** (off-label), **blackberries** (off-label), **blueberries** (off-label), **borage for oilseed production** (off-label), **broad beans** (off-label), **cabbages** (off-label), **canary flower (echium spp.)** (off-label), **cherries** (off-label), **chicory** (off-label), **choi sum** (off-label), **cob nuts** (off-label), **collards** (off-label), **cranberries** (off-label), **evening primrose** (off-label), **figs** (off-label), **fodder beet** (off-label), **garlic** (off-label), **grapevines** (off-label), **haricot beans** (off-label), **hazel nuts** (off-label), **herbs (see appendix 6)** (off-label), **honesty** (off-label), **lucerne** (off-label), **lupins** (off-label), **mallow (althaea spp.)** (off-label), **mustard** (off-label), **navy beans** (off-label), **non-edible flowers** (off-label), **ornamental plant production** (off-label), **pak choi** (off-label), **parsley root** (off-label), **parsnips** (off-label), **peaches** (off-label), **pears** (off-label), **plums** (off-label), **protected non-edible flowers** (off-label), **protected ornamentals** (off-label), **quinces** (off-label), **red beet** (off-label), **redcurrants** (off-label), **rubus hybrids** (off-label), **shallots** (off-label), **spring greens** (off-label), **tic beans** (off-label), **walnuts** (off-label), **whitecurrants** (off-label) [2]; **blackcurrants, bulb onions, carrots, combining peas, farm forestry, fodder beet, gooseberries, hops, linseed, raspberries, spring field beans, spring oilseed rape, spring oilseed rape for industrial use, strawberries, sugar beet, swedes** (stockfeed only), **turnips** (stockfeed only), **vining peas, winter field beans, winter oilseed rape, winter oilseed rape for industrial use** [1-4]; **field margins, flax, flax for industrial use, linseed for industrial use** [2, 4]; **kale** (stockfeed only) [1, 3]
- Barren brome in **field margins** [2, 4]
- Blackgrass in **hops, spring field beans, winter field beans** [1-4]

FOR FULL CONDITIONS OF USE ALWAYS READ THE PRODUCT LABEL

fluazifop-P-butyl

- Green cover in **land temporarily removed from production** [1-4]
- Perennial grasses in **almonds** *(off-label)*, **apples** *(off-label)*, **apricots** *(off-label)*, **artichokes** *(off-label)*, **asparagus** *(off-label)*, **bilberries** *(off-label)*, **blackberries** *(off-label)*, **blueberries** *(off-label)*, **borage for oilseed production** *(off-label)*, **broad beans** *(off-label)*, **cabbages** *(off-label)*, **canary flower (echium spp.)** *(off-label)*, **cherries** *(off-label)*, **chicory** *(off-label)*, **choi sum** *(off-label)*, **cob nuts** *(off-label)*, **collards** *(off-label)*, **cranberries** *(off-label)*, **evening primrose** *(off-label)*, **figs** *(off-label)*, **fodder beet** *(off-label)*, **garlic** *(off-label)*, **grapevines** *(off-label)*, **haricot beans** *(off-label)*, **hazel nuts** *(off-label)*, **herbs (see appendix 6)** *(off-label)*, **honesty** *(off-label)*, **lucerne** *(off-label)*, **lupins** *(off-label)*, **mallow (althaea spp.)** *(off-label)*, **mustard** *(off-label)*, **navy beans** *(off-label)*, **non-edible flowers** *(off-label)*, **ornamental plant production** *(off-label)*, **pak choi** *(off-label)*, **parsley root** *(off-label)*, **parsnips** *(off-label)*, **peaches** *(off-label)*, **pears** *(off-label)*, **plums** *(off-label)*, **protected non-edible flowers** *(off-label)*, **protected ornamentals** *(off-label)*, **quinces** *(off-label)*, **red beet** *(off-label)*, **redcurrants** *(off-label)*, **rubus hybrids** *(off-label)*, **shallots** *(off-label)*, **spring greens** *(off-label)*, **tic beans** *(off-label)*, **walnuts** *(off-label)*, **whitecurrants** *(off-label)* [2]; **blackcurrants, bulb onions, carrots, combining peas, farm forestry, fodder beet, gooseberries, hops, linseed, raspberries, spring field beans, spring oilseed rape, spring oilseed rape for industrial use, strawberries, sugar beet, swedes** *(stockfeed only)*, **turnips** *(stockfeed only)*, **vining peas, winter field beans, winter oilseed rape, winter oilseed rape for industrial use** [1-4]; **flax, flax for industrial use, linseed for industrial use** [2, 4]; **kale** *(stockfeed only)* [1, 3]
- Volunteer cereals in **blackcurrants, bulb onions, carrots, combining peas, farm forestry, fodder beet, gooseberries, hops, linseed, raspberries, spring field beans, spring oilseed rape, spring oilseed rape for industrial use, strawberries, sugar beet, swedes** *(stockfeed only)*, **turnips** *(stockfeed only)*, **vining peas, winter field beans, winter oilseed rape, winter oilseed rape for industrial use** [1-4]; **field margins, flax, flax for industrial use, linseed for industrial use** [2, 4]; **kale** *(stockfeed only)* [1, 3]
- Wild oats in **blackcurrants, bulb onions, carrots, combining peas, farm forestry, fodder beet, gooseberries, hops, linseed, raspberries, spring field beans, spring oilseed rape, spring oilseed rape for industrial use, strawberries, sugar beet, swedes** *(stockfeed only)*, **turnips** *(stockfeed only)*, **vining peas, winter field beans, winter oilseed rape, winter oilseed rape for industrial use** [1-4]; **field margins, flax, flax for industrial use, linseed for industrial use** [2, 4]; **kale** *(stockfeed only)* [1, 3]

Specific Off-Label Approvals (SOLAs)
- *almonds* 20080323 [2]
- *apples* 20080323 [2]
- *apricots* 20080323 [2]
- *artichokes* 20080324 [2]
- *asparagus* 20080325 [2]
- *bilberries* 20063272 [2]
- *blackberries* 20063274 [2]
- *blueberries* 20063272 [2]
- *borage for oilseed production* 20081832 [2]
- *broad beans* 20081777 [2]
- *cabbages* 20080328 [2]
- *canary flower (echium spp.)* 20081832 [2]
- *cherries* 20080323 [2]
- *chicory* 20080326 [2]
- *choi sum* 20081777 [2]
- *cob nuts* 20080323 [2]
- *collards* 20081777 [2]
- *cranberries* 20063272 [2]
- *evening primrose* 20081832 [2]
- *figs* 20080323 [2]
- *fodder beet* 20063270 [2]
- *garlic* 20081777 [2]
- *grapevines* 20080327 [2]
- *haricot beans* 20081777 [2]
- *hazel nuts* 20080323 [2]
- *herbs (see appendix 6)* 20091176 [2]

SEE SECTION 3 FOR PRODUCTS ALSO REGISTERED

Pesticide Profiles

- ***honesty** 20081832* [2]
- ***lucerne** 20080845* [2]
- ***lupins** 20063268* [2]
- ***mallow (althaea spp.)** 20063273* [2]
- ***mustard** 20081832* [2]
- ***navy beans** 20081777* [2]
- ***non-edible flowers** 20081777* [2]
- ***ornamental plant production** 20081777* [2]
- ***pak choi** 20081777* [2]
- ***parsley root** 20063273* [2]
- ***parsnips** 20081777* [2]
- ***peaches** 20080323* [2]
- ***pears** 20080323* [2]
- ***plums** 20080323* [2]
- ***protected non-edible flowers** 20081777* [2]
- ***protected ornamentals** 20081777* [2]
- ***quinces** 20080323* [2]
- ***red beet** 20081777* [2]
- ***redcurrants** 20063272* [2]
- ***rubus hybrids** 20063274* [2]
- ***shallots** 20063271* [2]
- ***spring greens** 20081777* [2]
- ***tic beans** 20081777* [2]
- ***walnuts** 20080323* [2]
- ***whitecurrants** 20063272* [2]

Approval information
- Accepted by BBPA for use on hops

Efficacy guidance
- Best results achieved by application when weed growth active under warm conditions with adequate soil moisture.
- Spray weeds from 2-expanded leaf stage to fully tillered, couch from 4 leaves when majority of shoots have emerged, with a second application if necessary
- Control may be reduced under dry conditions. Do not cultivate for 2 wk after spraying couch
- Annual meadow grass is not controlled
- May also be used to remove grass cover crops
- Fluazifop-P-butyl is an ACCase inhibitor herbicide. To avoid the build up of resistance do not apply products containing an ACCase inhibitor herbicide more than twice to any crop. In addition do not use any product containing fluazifop-P-butyl in mixture or sequence with any other product containing the same ingredient
- Use these products as part of a resistance management strategy that includes cultural methods of control and does not use ACCase inhibitors as the sole chemical method of grass weed control

Restrictions
- Maximum number of treatments 1 per crop or yr for all crops
- Do not sow cereals or grass crops for at least 8 wk after application of high rate or 2 wk after low rate
- Do not apply through CDA sprayer, with hand-held equipment or from air
- Avoid treatment before spring growth has hardened or when buds opening
- Do not treat bush and cane fruit or hops between flowering and harvest
- Consult processors before treating crops intended for processing
- Oilseed rape, linseed and flax for industrial use must not be harvested for human or animal consumption nor grazed
- Do not use for forestry establishment on land not previously under arable cultivation or improved grassland
- Treated vegetation in field margins, land temporarily removed from production etc, must not be grazed or harvested for human or animal consumption and unprotected persons must be kept out of treated areas for at least 24 h

FOR FULL CONDITIONS OF USE ALWAYS READ THE PRODUCT LABEL

fluazinam

Crop-specific information
- Latest use: before 50% ground cover for swedes, turnips; before 5 leaf stage for spring oilseed rape; before flowering for blackcurrants, gooseberries, hops, raspberries, strawberries; before flower buds visible for field beans, peas, linseed, flax, winter oilseed rape; 2 wk before sowing cereals or grass for field margins, land temporarily removed from production
- HI beet crops, kale, carrots 8 wk; onions 4 wk; oilseed rape for industrial use 2 wk
- Apply to sugar and fodder beet from 1-true leaf to 50% ground cover
- Apply to winter oilseed rape from 1-true leaf to established plant stage
- Apply to spring oilseed rape from 1-true leaf but before 5-true leaves
- Apply in fruit crops after harvest. See label for timing details on other crops
- Before using on onions or peas use crystal violet test to check that leaf wax is sufficient

Environmental safety
- Dangerous for the environment
- Very toxic to aquatic organisms

Hazard classification and safety precautions
Hazard Harmful, Dangerous for the environment
Transport code 9
Packaging group III
UN number 3082
Risk phrases R38, R50, R53a, R63
Operator protection A, C, H, M; U05a, U08, U20b
Environmental protection E15a [1-4]; E38 [2, 4]
Storage and disposal D01, D02, D05, D09a, D12a [1-4]; D10b [1]; D10c [2-4]
Medical advice M05b [1]

240 fluazinam

A dinitroaniline fungicide for use in potatoes
FRAC mode of action code: 29

Products

1	Blizzard	AgriGuard	500 g/l	SC	13831
2	Clayton Solstice	Clayton	500 g/l	SC	13943
3	Floozee	AgChem Access	500 g/l	SC	13856
4	Greencrop Solanum	Greencrop	500 g/l	SC	11052
5	Nando 500 SC	Nufarm UK	500 g/l	SC	14372
6	Ohayo	Syngenta	500 g/l	SC	13734
7	Shirlan	Syngenta	500 g/l	SC	10573
8	Tizca	Headland	500 g/l	SC	13877
9	Volley	Makhteshim	500 g/l	SC	13591

Uses
- Blight in *potatoes* [1-9]; *seed potatoes* (off-label) [1]
- Phytophthora root rot in *blackberries* (off-label), *protected blackberries* (off-label), *protected raspberries* (off-label), *protected rubus hybrids* (off-label), *raspberries* (off-label), *rubus hybrids* (off-label) [5, 9]; *container-grown blackberry* (off-label), *container-grown raspberry* (off-label), *container-grown rubus hybrids* (off-label) [9]
- Powdery scab in *seed potatoes* (off-label) [7]
- Root rot in *blackberries* (off-label), *protected blackberries* (off-label), *protected raspberries* (off-label), *protected rubus hybrids* (off-label), *raspberries* (off-label), *rubus hybrids* (off-label) [7, 8]
- Scab in *seed potatoes* (off-label) [3, 5, 8, 9]

Specific Off-Label Approvals (SOLAs)
- *blackberries* 20092990 [5], 20032168 [7], 20080597 [8], 20073759 [9]
- *container-grown blackberry* 20073759 [9]
- *container-grown raspberry* 20073759 [9]
- *container-grown rubus hybrids* 20073759 [9]
- *protected blackberries* 20092990 [5], 20032168 [7], 20080597 [8], 20073759 [9]
- *protected raspberries* 20092990 [5], 20032168 [7], 20080597 [8], 20073759 [9]
- *protected rubus hybrids* 20092990 [5], 20032168 [7], 20080597 [8], 20073759 [9]

SEE SECTION 3 FOR PRODUCTS ALSO REGISTERED

Pesticide Profiles

- **raspberries** *20092990* [5], *20032168* [7], *20080597* [8], *20073759* [9]
- **rubus hybrids** *20092990* [5], *20032168* [7], *20080597* [8], *20073759* [9]
- **seed potatoes** *20081060* [1], *20100375* [3], *20092989* [5], *20060893* [7], *20080598* [8], *20073757* [9]

Approval information
- Fluazinam included in Annex I under EC Directive 91/414

Efficacy guidance
- Commence treatment at the first blight risk warning (before blight enters the crop). Products are rainfast within 1 h
- In the absence of a warning, treatment should start before foliage of adjacent plants meets in the rows
- Spray at 7-14 d intervals (5-10 d intervals [7]) depending on severity of risk (see label)
- Ensure complete coverage of the foliage and stems, increasing volume as haulm growth progresses, in dense crops and if blight risk increases

Restrictions
- Maximum number of treatments 10 per crop on potatoes [1, 4]
- Maximum total dose equivalent to 7.5 full dose treatments [7, 9]
- Do not use with hand-held sprayers

Crop-specific information
- HI 0 - 10 d for potatoes. Check label
- Ensure complete kill of potato haulm before lifting and do not lift crops for storage while there is any green tissue left on the leaves or stem bases

Environmental safety
- Dangerous for the environment
- Very toxic to aquatic organisms
- LERAP Category B [2-9]

Hazard classification and safety precautions
Hazard Harmful [5]; Irritant [1-4, 6-8]; Dangerous for the environment [1-9]
Transport code 9
Packaging group III
UN number 3082
Risk phrases R36 [1-7]; R41 [8]; R43 [1-8]; R50, R53a [1-9]; R63 [5]
Operator protection A, C, H; U02a, U04a, U08 [2-7, 9]; U05a [1-8]; U11 [8]; U14 [2, 3, 5-8]; U15, U20a [2, 3, 5-7]; U20b [4, 8, 9]
Environmental protection E15a, E16a [2-9]; E16b, E34 [2, 3, 5-7, 9]; E38 [1-3, 5-9]
Storage and disposal D01, D02 [1-8]; D05 [2-8]; D09a [2-9]; D10a [9]; D10b [4]; D10c [2, 3, 5-8]; D12a [1-3, 5-9]
Medical advice M03 [2-7, 9]; M05a [1-3, 5-7]

241 fluazinam + metalaxyl-M

A mixture of contact and systemic fungicides for potatoes
FRAC mode of action code: 29 + 4

See also metalaxyl-M

Products

Epok	Belchim	400:200 g/l	EC	11997

Uses
- Blight in **potatoes**

Approval information
- Fluazinam and metalaxyl-M included in Annex I under EC Directive 91/414

Efficacy guidance
- Commence treatment at the first blight risk warning (before blight enters the crop). Products are rainfast within 1 h
- In the absence of a warning, treatment should start before foliage of adjacent plants meets in the rows

FOR FULL CONDITIONS OF USE ALWAYS READ THE PRODUCT LABEL

fludioxonil

- Spray at 7-14 d intervals depending on severity of risk (see label)
- Ensure complete coverage of the foliage and stems, increasing volume as haulm growth progresses, in dense crops and if blight risk increases
- Metalaxyl-M works best on young actively growing foliage and efficacy declines with the onset of senescence. Therefore it is recommended that the product is only used for the first part of the blight control program, which should be completed with a reliable protectant fungicide

Restrictions
- Maximum number of treatments 3 per crop

Crop-specific information
- HI 7d for potatoes

Environmental safety
- Dangerous for the environment
- Very toxic to aquatic organisms
- LERAP Category B

Hazard classification and safety precautions
Hazard Harmful, Dangerous for the environment
Transport code 9
Packaging group III
UN number 3082
Risk phrases R20, R36, R38, R40, R43, R50, R53a
Operator protection A, C, H; U05a, U07, U11, U14, U15, U19a, U20a
Environmental protection E15a, E16a, E16b, E34, E38
Storage and disposal D01, D02, D09a, D10b, D12a
Medical advice M05a

242 fludioxonil

A phenylpyrrole fungicide seed treatment for wheat and barley
FRAC mode of action code: 12

See also chlorothalonil + fludioxonil + propiconazole
cymoxanil + fludioxonil + metalaxyl-M
cyprodinil + fludioxonil
difenoconazole + fludioxonil

Products
Beret Gold   Syngenta   25 g/l   FS   12625

Uses
- Bunt in **spring wheat** (seed treatment), **winter wheat** (seed treatment)
- Covered smut in **spring barley** (seed treatment), **winter barley** (seed treatment)
- Fusarium foot rot and seedling blight in **spring barley** (seed treatment), **spring oats** (seed treatment), **spring wheat** (seed treatment), **winter barley** (seed treatment), **winter oats** (seed treatment), **winter wheat** (seed treatment)
- Leaf stripe in **spring barley** (seed treatment - reduction), **winter barley** (seed treatment - reduction)
- Pyrenophora leaf spot in **spring oats** (seed treatment), **winter oats** (seed treatment)
- Seed-borne diseases in **durum wheat** (off-label), **spring rye** (off-label), **triticale** (off-label), **winter rye** (off-label)
- Septoria seedling blight in **spring wheat** (seed treatment), **winter wheat** (seed treatment)
- Snow mould in **spring barley** (seed treatment), **spring oats** (seed treatment), **spring wheat** (seed treatment), **winter barley** (seed treatment), **winter oats** (seed treatment), **winter wheat** (seed treatment)

Specific Off-Label Approvals (SOLAs)
- *durum wheat* 20071445
- *spring rye* 20071445
- *triticale* 20071445
- *winter rye* 20071445

SEE SECTION 3 FOR PRODUCTS ALSO REGISTERED

Pesticide Profiles

Approval information
- Fludioxonil included in Annex I under EC Directive 91/414
- Accepted by BBPA for use on malting barley

Efficacy guidance
- Apply direct to seed using conventional seed treatment equipment. Continuous flow treaters should be calibrated using product before use
- Effective against benzimidazole-resistant strains of *Microdochium nivale*

Restrictions
- Maximum number of treatments 1 per seed batch
- Do not apply to cracked, split or sprouted seed
- Sow treated seed within 6 mth

Crop-specific information
- Latest use: before drilling
- Product may reduce flow rate of seed through drill. Recalibrate with treated seed before drilling

Environmental safety
- Dangerous for the environment
- Toxic to aquatic organisms
- Do not use treated seed as food or feed
- Treated seed harmful to game and wildlife

Hazard classification and safety precautions
Hazard Dangerous for the environment
Transport code 9
Packaging group III
UN number 3082
Risk phrases R51, R53a
Operator protection A, H; U05a, U20b
Environmental protection E03, E15a, E38
Storage and disposal D01, D02, D05, D09a, D11a, D12a
Treated seed S02, S05, S07

243   fludioxonil + flutriafol

A fungicide seed treatment mixture for cereals
FRAC mode of action code: 12 + 3

See also flutriafol

Products
Beret Multi   Syngenta   25:25 g/l   FS   13017

Uses
- Bunt in **spring wheat** *(seed treatment)*, **winter wheat** *(seed treatment)*
- Covered smut in **spring barley** *(seed treatment)*, **winter barley** *(seed treatment)*
- Fusarium foot rot and seedling blight in **spring barley** *(seed treatment)*, **spring oats** *(seed treatment)*, **spring wheat** *(seed treatment)*, **winter barley** *(seed treatment)*, **winter oats** *(seed treatment)*, **winter wheat** *(seed treatment)*
- Leaf stripe in **spring barley** *(seed treatment)*, **winter barley** *(seed treatment)*
- Loose smut in **spring barley** *(seed treatment)*, **spring wheat** *(seed treatment)*, **winter barley** *(seed treatment)*, **winter wheat** *(seed treatment)*
- Pyrenophora leaf spot in **spring oats** *(seed treatment)*, **winter oats** *(seed treatment)*
- Septoria seedling blight in **spring wheat** *(seed treatment)*, **winter wheat** *(seed treatment)*
- Snow mould in **spring barley** *(seed treatment)*, **spring wheat** *(seed treatment)*, **winter barley** *(seed treatment)*, **winter wheat** *(seed treatment)*

Approval information
- Fludioxonil included in Annex I under EC Directive 91/414

FOR FULL CONDITIONS OF USE ALWAYS READ THE PRODUCT LABEL

fludioxonil + metalaxyl M

Efficacy guidance
- Apply directly to the seed using conventional seed treatment equipment
- Calibrate continuous flow treaters before use
- Effective against benzimidazole-resistant strains of *Microdochium nivale*

Restrictions
- Maximum number of treatments 1 per seed batch
- Do not apply to cracked, split or sprouted seed

Crop-specific information
- Latest use: before drilling for all cereals
- Treated seed may affect the flow rate through drills. Recalibrate with treated seed before drilling

Environmental safety
- Dangerous for the environment
- Toxic to aquatic organisms

Hazard classification and safety precautions
Hazard Dangerous for the environment
Transport code 9
Packaging group III
UN number 3082
Risk phrases R51, R53a
Operator protection A, H; U05a, U20c
Environmental protection E03, E15a, E38
Storage and disposal D01, D02, D05, D09a, D10c, D11a, D12a
Treated seed S02, S05, S07

244   fludioxonil + metalaxyl M

A seed dressing for use in forage maize
FRAC mode of action code: 12 + 4

Products
    Maxim XL   Syngenta   25:9.69 g/l   FS   14136

Uses
- Damping off in *forage maize*
- Fusarium in *forage maize*
- Pythium in *forage maize*

Approval information
- Fludioxonil and metalaxyl-M included in Annex I under EC Directive 91/414

Restrictions
- For advice on resistance management, refer to the latest Fungicide Resistance Action Group (FRAG) guidelines

Hazard classification and safety precautions
UN number N/C
Risk phrases R52, R53a
Operator protection A, H; U05a
Environmental protection E34, E38
Storage and disposal D01, D02, D05, D09a, D10c, D11a, D12a
Treated seed S01, S02, S03, S04a, S05, S07

SEE SECTION 3 FOR PRODUCTS ALSO REGISTERED

245 fludioxonil + metalaxyl-M + thiamethoxam

A seed dressing for use in oilseed rape, fodder rape and mustard
FRAC mode of action code: 12 + 4 + 4A

See also metalaxyl-M
thiamethoxam

Products
Cruiser OSR   Syngenta   8:32.3:280 g/l   FS   14496

Uses
- Alternaria in **fodder rape** *(moderate control)*, **mustard** *(moderate control)*, **spring oilseed rape** *(moderate control)*, **winter oilseed rape** *(moderate control)*
- Cabbage stem flea beetle in **fodder rape** *(reduction of damage)*, **mustard** *(reduction of damage)*, **spring oilseed rape** *(reduction of damage)*, **winter oilseed rape** *(reduction of damage)*
- Downy mildew in **fodder rape**, **mustard**, **spring oilseed rape**, **winter oilseed rape**
- Flea beetle in **fodder rape** *(reduction of damage)*, **mustard** *(reduction of damage)*, **spring oilseed rape** *(reduction of damage)*, **winter oilseed rape** *(reduction of damage)*
- Phoma lingam in **fodder rape**, **mustard**, **spring oilseed rape**, **winter oilseed rape**
- Pythium in **fodder rape** *(reduction)*, **mustard** *(reduction)*, **poppies for morphine production** *(off-label)*, **spring oilseed rape** *(reduction)*, **winter oilseed rape** *(reduction)*

Specific Off-Label Approvals (SOLAs)
- **poppies for morphine production** 20093080

Approval information
- Fludioxonil, metalaxyl-M and thiamethoxam included in Annex I under EC Directive 91/414

Efficacy guidance
- Contains a neonicotinoid insecticide. The first subsequent foliar insecticide spray should contain an insecticide with a different mode of action to combat resistance - consult the IRAC website for details of modes of action.
- Contains metalaxyl-M, a phenylamide fungicide. Do not use fungicide sprays containing metalaxyl-M on crops grown from treated seed

Crop-specific information
- For use on all varieties of oilseed rape, fodder rape and mustard

Hazard classification and safety precautions
Hazard Dangerous for the environment
UN number N/C
Risk phrases R50, R53a
Operator protection A, D, H; U05a, U07, U20b, U20e
Environmental protection E15b, E34, E38
Storage and disposal D01, D02, D09a, D10a, D10d, D11a, D12a, D14
Treated seed S01, S02, S03, S04a, S04b, S05, S06b, S07, S08, S09

246 fludioxonil + tefluthrin

A fungicide and insecticide seed treatment mixture for cereals
FRAC mode of action code: 12

See also tefluthrin

Products
Austral Plus   Syngenta   10:40 g/l   FS   13314

Uses
- Bunt in **spring wheat**, **winter wheat**
- Covered smut in **spring barley**, **winter barley**
- Fusarium foot rot and seedling blight in **spring barley**, **spring oats**, **spring wheat**, **winter barley**, **winter oats**, **winter wheat**
- Leaf stripe in **spring barley** *(partial control)*, **winter barley** *(partial control)*
- Pyrenophora leaf spot in **spring oats**, **winter oats**

flufenacet

- Seed-borne diseases in **triticale seed crop** *(off-label)*
- Septoria seedling blight in **spring wheat, winter wheat**
- Snow mould in **spring barley, spring wheat, winter barley, winter wheat**
- Wheat bulb fly in **spring barley, spring wheat, winter barley, winter wheat**
- Wireworm in **spring barley, spring oats, spring wheat, winter barley, winter oats, winter wheat**

Specific Off-Label Approvals (SOLAs)
- **triticale seed crop** 20080982

Approval information
- Fludioxonil included in Annex I under EC Directive 91/414
- Accepted by BBPA for use on malting barley

Efficacy guidance
- Apply direct to seed using conventional seed treatment equipment. Continuous flow treaters should be calibrated using product before use
- Best results obtained from seed drilled into a firm even seedbed
- Tefluthrin is released into soil after drilling and repels or kills larvae of wheat bulb fly and wireworm attacking below ground. Pest control may be reduced by deep or shallow drilling
- Where egg counts of wheat bulb fly or population counts of wireworm indicate a high risk of severe attack follow up spray treatments may be needed
- Control of leaf stripe in barley may not be sufficient in crops grown for seed certification
- Effective against benzimidazole-resistant strains of *Microdochium nivale*
- Under adverse soil or environmental conditions seed rates should be increased to compensate for possible reduced germination capacity

Restrictions
- Maximum number of treatments 1 per batch
- Store treated seed in cool dry conditions and drill within 3 mth
- Do not use on seed above 16% moisture content or on sprouted, cracked or damaged seed

Crop-specific information
- Latest use: before drilling

Environmental safety
- Dangerous for the environment
- Very toxic to aquatic organisms

Hazard classification and safety precautions
  Hazard Dangerous for the environment
  Transport code 9
  Packaging group III
  UN number 3082
  Risk phrases R50, R53a
  Operator protection A, C, D, H; U02a, U04a, U05a, U07, U08, U14, U20b
  Environmental protection E03, E15b, E34, E38
  Storage and disposal D01, D02, D05, D09b, D10c, D11a, D12a
  Treated seed S02, S04b, S05, S07, S08

247 flufenacet

A broad spectrum oxyacetamide herbicide available only in mixtures
HRAC mode of action code: K3

See also diflufenican + flufenacet
diflufenican + flufenacet + flurtamone

SEE SECTION 3 FOR PRODUCTS ALSO REGISTERED

248 flufenacet + isoxaflutole

A residual herbicide mixture for maize
HRAC mode of action code: K3 + F2

See also isoxaflutole

Products

1	Amethyst	AgChem Access	48:10% w/w	WG	14079
2	Cadou Star	Bayer CropScience	48:10% w/w	WG	13242

Uses
- Annual dicotyledons in *forage maize, grain maize* [1, 2]; *forage maize (under plastic mulches)* (off-label), *game cover* (off-label), *hops* (off-label), *miscanthus* (off-label), *ornamental plant production* (off-label), *sweetcorn* (off-label), *sweetcorn under plastic mulches* (off-label), *top fruit* (off-label) [2]
- Annual grasses in *forage maize (under plastic mulches)* (off-label), *game cover* (off-label), *hops* (off-label), *miscanthus* (off-label), *ornamental plant production* (off-label), *sweetcorn* (off-label), *sweetcorn under plastic mulches* (off-label), *top fruit* (off-label) [2]
- Grass weeds in *forage maize, grain maize* [1, 2]

Specific Off-Label Approvals (SOLAs)
- *forage maize (under plastic mulches)* 20080960 [2]
- *game cover* 20082829 [2]
- *hops* 20082829 [2]
- *miscanthus* 20082829 [2]
- *ornamental plant production* 20082829 [2]
- *sweetcorn* 20080798 [2]
- *sweetcorn under plastic mulches* 20080960 [2]
- *top fruit* 20082829 [2]

Approval information
- Flufenacet and isoxaflutole included in Annex I under EC Directive 91/414

Efficacy guidance
- Best results obtained from applications to a fine, firm seedbed in the presence of some soil moisture
- Efficacy may be reduced on cloddy seedbeds or under prolonged dry conditions
- Established perennial grasses and broad-leaved weeds growing from rootstocks will not be controlled
- Always follow WRAG guidelines for preventing and managing herbicide resistant weeds. See Section 5 for more information

Restrictions
- Maximum number of treatments 1 per crop
- Do not use on soils both above 70% sand and less than 2% organic matter
- Do not use on maize crops intended for seed production

Crop-specific information
- Latest use: before crop emergence
- Ideally treatment should be made within 4 d of sowing, before tha maize seeds have germinated

Following crops guidance
- In the event of failure of a treated crop maize, sweet corn or potatoes may be grown after surface cultivation or ploughing
- After normal harvest of a treated crop oats or barley may be sown after ploughing and wheat, rye, triticale, sugar beet, field beans, peas, soya, sorghum or sunflowers may be sown after surface cultivation or ploughing.

Environmental safety
- Dangerous for the environment
- Very toxic to aquatic organisms
- Take care to avoid drift over other crops. Beet crops and sunflowers are particularly sensitive
- LERAP Category B

FOR FULL CONDITIONS OF USE ALWAYS READ THE PRODUCT LABEL

flufenacet + metribuzin

Hazard classification and safety precautions
Hazard Harmful, Dangerous for the environment
Transport code 9
Packaging group III
UN number 3077
Risk phrases R22a, R36, R43, R48, R50, R53a, R63
Operator protection A, C, H; U05a, U11, U20b
Environmental protection E15b, E16a, E34, E38
Storage and disposal D01, D02, D05, D09a, D10a, D12a
Medical advice M03

249 flufenacet + metribuzin

A herbicide mixture for potatoes
HRAC mode of action code: K3 + C1

See also metribuzin

Products
    Artist   Bayer CropScience   24:17.5% w/w   WP   11239

Uses
- Annual dicotyledons in **bilberries** *(off-label)*, **blackcurrants** *(off-label)*, **blueberries** *(off-label)*, **cranberries** *(off-label)*, **early potatoes**, **gooseberries** *(off-label)*, **maincrop potatoes**, **redcurrants** *(off-label)*, **ribes hybrids** *(off-label)*
- Annual grasses in **bilberries** *(off-label)*, **blackcurrants** *(off-label)*, **blueberries** *(off-label)*, **cranberries** *(off-label)*, **gooseberries** *(off-label)*, **redcurrants** *(off-label)*, **ribes hybrids** *(off-label)*
- Annual meadow grass in **early potatoes**, **maincrop potatoes**

Specific Off-Label Approvals (SOLAs)
- *bilberries 20092641*
- *blackcurrants 20092641*
- *blueberries 20092641*
- *cranberries 20092641*
- *gooseberries 20092641*
- *redcurrants 20092641*
- *ribes hybrids 20092641*

Approval information
- Flufenacet and metribuzin included in Annex I under EC Directive 91/414

Efficacy guidance
- Product acts through root uptake and needs sufficient soil moisture at and shortly after application
- Effectiveness is reduced under dry soil conditions
- Residual activity is reduced on mineral soils with a high organic matter content and on peaty or organic soils
- Ensure application is made evenly to both sides of potato ridges
- Perennial weeds are not controlled

Restrictions
- Maximum total dose equivalent to one full dose treatment
- Potatoes must be sprayed before emergence of crop and weeds
- See label for list of tolerant varieties. Do not treat Maris Piper grown on Sands or Very Light soils
- Do not use on Sands
- On stony or gravelly soils there is risk of crop damage especially if heavy rain falls soon after application

Crop-specific information
- Latest use: before potato crop emergence
- Consult processor before use on crops for processing

Following crops guidance
- Before drilling or planting any succeeding crop soil must be mouldboard ploughed to at least 15 cm as soon as possible after lifting and no later than end Dec

SEE SECTION 3 FOR PRODUCTS ALSO REGISTERED

- In W Cornwall on soils with more than 5% organic matter treated early potatoes may be followed by summer planted brassica crops 14 wk after treatment and after mouldboard ploughing. Elsewhere cereals or winter beans may be grown in the same year if at least 16 wk have elapsed since treatment
- In the yr following treatment any crop may be grown except lettuce or radish, or vegetable brassica crops on silt soils in Lincs

Environmental safety
- Dangerous for the environment
- Very toxic to aquatic organisms
- Take care to avoid spray drift onto neighbouring crops, especially lettuce or brassicas

Hazard classification and safety precautions
Hazard Harmful, Dangerous for the environment
Transport code 9
Packaging group III
UN number 3077
Risk phrases R22a, R43, R48, R50, R53a
Operator protection A, D, H; U05a, U08, U13, U14, U19a, U20b
Environmental protection E15a, E38
Storage and disposal D01, D02, D09a, D11a, D12a
Medical advice M03

250 flufenacet + pendimethalin

A broad spectrum residual and contact herbicide mixture for winter cereals
HRAC mode of action code: K3 + K1

See also pendimethalin

Products

1	Crystal	BASF	60:300 g/l	EC	13914
2	Ice	BASF	60:300 g/l	EC	13930
3	Rock	Goldengrass	60:300 g/l	EC	14928
4	Shooter	BASF	60:300 g/l	EC	14106
5	Trooper	BASF	60:300 g/l	EC	13924

Uses
- Annual dicotyledons in *game cover* (off-label) [1]; *winter barley*, *winter wheat* [1-5]
- Annual grasses in *winter barley*, *winter wheat* [1-5]
- Annual meadow grass in *game cover* (off-label) [1]; *winter barley*, *winter wheat* [1-5]
- Blackgrass in *game cover* (off-label) [1]; *winter barley*, *winter wheat* [1-5]
- Chickweed in *winter barley*, *winter wheat* [1-5]
- Corn marigold in *winter barley*, *winter wheat* [1-5]
- Field speedwell in *winter barley*, *winter wheat* [1-5]
- Ivy-leaved speedwell in *winter barley*, *winter wheat* [1-5]

Specific Off-Label Approvals (SOLAs)
- *game cover* 20090450 [1]

Approval information
- Flufenacet and pendimethalin included in Annex I under EC Directive 91/414
- Accepted by BBPA for use on malting barley

Efficacy guidance
- Best results achieved when applied from pre-emergence of weeds to 2-leaf stage but post emergence treatment is not recommended on clay soils
- Product requires some soil moisture to be activated ideally from rain within 7 d of application. Prolonged dry conditions may reduce residual control
- Product is slow acting and final level of weed control may take some time to appear
- For effective weed control seed bed preparations should ensure even incorporation of any trash, straw and ash to 15 cm
- Efficacy may be reduced on soils with more than 6% organic matter

FOR FULL CONDITIONS OF USE ALWAYS READ THE PRODUCT LABEL

flumioxazine

- Always follow WRAG guidelines for preventing and managing herbicide resistant weeds. See Section 5 for more information

Restrictions
- Maximum total dose equivalent to one full dose treatment
- For pre-emergence treatments seed should be covered with at least 32 mm settled soil. Shallow drilled crops should be treated post-emergence only
- Do not treat undersown crops
- Avoid spraying during periods of prolonged or severe frosts
- Do not use on stony or gravelly soils or those with more than 10% organic matter
- Pre-emergence treatment may only be used on crops drilled before 30 Nov. All crops must be treated before 31 Dec in yr of planting.
- Concentrated or diluted product may stain clothing or skin

Crop-specific information
- Latest use: before third tiller stage (GS 23) and before 31 Dec in yr of planting
- Very wet weather before and after treatment may result in loss of crop vigour and reduced yield, particularly where soils become waterlogged

Following crops guidance
- Any crop may follow a failed or normally harvested treated crop provided ploughing to at least 15 cm is carried out beforehand

Environmental safety
- Dangerous for the environment
- Very toxic to aquatic organisms
- Risk to certain non-target insects or other arthropods
- Some products supplied in small volume returnable packs. Follow instructions for use
- LERAP Category B

Hazard classification and safety precautions
Hazard Harmful, Dangerous for the environment
Transport code 3
Packaging group III
UN number 1993
Risk phrases R22a, R38, R50, R53a [1-5]; R22b, R40 [1-3, 5]
Operator protection A, C, H; U02a, U05a [1-5]; U14, U20b [4]; U20c [1-3, 5]
Environmental protection E15a [1-3, 5]; E15b [4]; E16a, E16b, E22b, E34, E38 [1-5]
Storage and disposal D01, D02, D09a, D12a [1-5]; D05 [3]; D08, D10c [4]; D10a [1-3, 5]
Medical advice M03 [1-5]; M05b [1-3, 5]

251 flumioxazine

A phenyphthalimide herbicide for winter wheat
HRAC mode of action code: E

Products

1	Digital	Interfarm	300 g/l	SC	13561
2	Guillotine	Interfarm	300 g/l	SC	13562
3	Sumimax	Interfarm	300 g/l	SC	13548

Uses
- Annual dicotyledons in **bulb onions** *(off-label)*, **carrots** *(off-label)*, **ornamental plant production** *(off-label)*, **parsnips** *(off-label)*, **vining peas** *(off-label)*, **winter wheat** [1-3]; **farm forestry** *(off-label)*, **forest nurseries** *(off-label)* [1]; **hops** *(off-label)*, **soft fruit** *(off-label)*, **top fruit** *(off-label)* [2, 3]
- Annual grasses in **winter oats** *(off-label)* [1-3]
- Annual meadow grass in **winter wheat** [1-3]
- Blackgrass in **winter oats** *(off-label)* [1-3]
- Chickweed in **winter wheat** [1-3]
- Cleavers in **winter wheat** [1-3]
- Groundsel in **bulb onions** *(off-label)*, **carrots** *(off-label)*, **ornamental plant production** *(off-label)*, **parsnips** *(off-label)*, **vining peas** *(off-label)* [1-3]; **farm forestry** *(off-label)*, **forest nurseries** *(off-label)* [1]; **hops** *(off-label)*, **soft fruit** *(off-label)*, **top fruit** *(off-label)* [2, 3]

SEE SECTION 3 FOR PRODUCTS ALSO REGISTERED

- Loose silky bent in **winter wheat** [1-3]
- Mayweeds in **winter wheat** [1-3]
- Ryegrass in **winter oats** *(off-label)* [1-3]
- Speedwells in **winter wheat** [1-3]
- Volunteer oilseed rape in **bulb onions** *(off-label)*, **carrots** *(off-label)*, **ornamental plant production** *(off-label)*, **parsnips** *(off-label)*, **vining peas** *(off-label)*, **winter wheat** [1-3]; **farm forestry** *(off-label)*, **forest nurseries** *(off-label)* [1]; **hops** *(off-label)*, **soft fruit** *(off-label)*, **top fruit** *(off-label)* [2, 3]
- Volunteer potatoes in **bulb onions** *(off-label)*, **carrots** *(off-label)*, **ornamental plant production** *(off-label)*, **parsnips** *(off-label)*, **vining peas** *(off-label)* [1-3]; **farm forestry** *(off-label)*, **forest nurseries** *(off-label)* [1]; **hops** *(off-label)*, **soft fruit** *(off-label)*, **top fruit** *(off-label)* [2, 3]

Specific Off-Label Approvals (SOLAs)
- **bulb onions** *20091114* [1], *20091116* [2], *20091108* [3]
- **carrots** *20091112* [1], *20091117* [2], *20091111* [3]
- **farm forestry** *20082844* [1]
- **forest nurseries** *20082844* [1]
- **hops** *20082897* [2], *20082881* [3]
- **ornamental plant production** *20082844* [1], *20082897* [2], *20082881* [3]
- **parsnips** *20091112* [1], *20091117* [2], *20091111* [3]
- **soft fruit** *20082897* [2], *20082881* [3]
- **top fruit** *20082897* [2], *20082881* [3]
- **vining peas** *20091113* [1], *20091115* [2], *20091109* [3]
- **winter oats** *20093121* [1], *20092506* [2], *20093122* [3]

Efficacy guidance
- Best results obtained from applications to moist soil early post-emergence of the crop when weeds are germinating up to 1 leaf stage
- Flumioxazin is contact acting and relies on weeds germinating in moist soil and coming into sufficient contact with the herbicide
- Flumioxazin remains in the surface layer of the soil and does not migrate to lower layers
- Contact activity is long lasting during the autumn
- Seed beds should be firm, fine and free from clods. Crops should be drilled to 25 mm and well covered by soil
- Efficacy is reduced on soils with more than 10% organic matter
- Flumioxazin is a member of the protoporphyrin oxidase (PPO) inhibitor herbicides. To avoid the build up of resistance do not use PPO-inhibitor herbicides more than once on any crop
- Use as part of a resistance management strategy that includes cultural methods of control and does not use PPO-inhibitors as the sole chemical method of grass weed control

Restrictions
- Maximum number of treatments 1 per crop for winter wheat
- Only apply to crops that have been hardened by cool weather. Do not treat crops with lush or soft growth
- Do not apply to waterlogged soils or to soils with more than 10% organic matter
- Do not follow an application with another pesticide for 14 d
- Do not treat undersown crops
- Do not treat broadcast crops until they are past the 3 leaf stage
- Do not treat crops under stress for any reason, or during prolonged frosty weather
- Do not roll or harrow for 2 wk before treatment or at any time afterwards

Crop-specific information
- Latest use: before 5th true leaf stage (GS 15) for winter wheat
- Light discolouration of leaf margins can occur after treatment and treatment of soft crops will cause transient leaf bleaching

Following crops guidance
- Any crop may be sown following normal harvest of a treated crop
- In the event of failure of a treated crop spring cereals, spring oilseed rape, sugar beet, maize or potatoes may be redrilled after ploughing

Environmental safety
- Dangerous for the environment
- Very toxic to aquatic organisms

FOR FULL CONDITIONS OF USE ALWAYS READ THE PRODUCT LABEL

fluopicolide

- Take extreme care to avoid drift onto plants outside the target area
- LERAP Category B

Hazard classification and safety precautions
Hazard Toxic, Dangerous for the environment
Transport code 9
Packaging group III
UN number 3082
Risk phrases R50, R53a, R61
Operator protection A, H; U05a, U19a
Environmental protection E15b, E16a, E34, E38
Storage and disposal D01, D02, D05, D07, D09a, D12b
Medical advice M04a

252 fluopicolide

An benzamide fungicide available only in mixtures
FRAC mode of action code: 43

253 fluopicolide + propamocarb hydrochloride

A protectant and systemic fungicide mixture for potato blight
FRAC mode of action code: 43 + 28

See also propamocarb hydrochloride

Products

Infinito	Bayer CropScience	62.5:625 g/l	SC	12644

Uses
- Blight in **potatoes**
- Disease control in **cabbages** *(off-label)*, **cauliflowers** *(off-label)*

Specific Off-Label Approvals (SOLAs)
- **cabbages** *20101096*
- **cauliflowers** *20101096*

Approval information
- Fluopicolide and propamocarb hydrochloride included in Annex I under EC Directive 91/414

Efficacy guidance
- Commence spray programme before infection appears as soon as weather conditions favourable for disease development occur. At latest the first treatment should be made as the foliage meets along the rows
- Repeat treatments at 7-10 day intervals according to disease incidence and weather conditions
- Reduce spray interval if conditions are conducive to the spread of blight
- Spray as soon as possible after irrigation
- Increase water volume in dense crops
- When used from full canopy development to haulm desiccation as part of a full blight protection programme tubers will be protected from late blight after harvest and tuber blight incidence will be reduced
- To reduce the development of resistance product should be used in single or block applications with fungicides from a different cross-resistance group

Restrictions
- Maximum total dose on potatoes equivalent to four full dose treatments
- Do not apply if rainfall or irrigation is imminent. Product is rainfast in 1 h provided spray has dried on leaf
- Do not apply as a curative treatment when blight is present in the crop
- Do not apply more than 3 consecutive treatments of the product

Crop-specific information
- HI 7 d for potatoes
- All varieties of potatoes, including seed crops, may be treated

SEE SECTION 3 FOR PRODUCTS ALSO REGISTERED

Pesticide Profiles

Environmental safety
- Dangerous for the environment
- Toxic to aquatic organisms

Hazard classification and safety precautions
Hazard Dangerous for the environment
Transport code 9
Packaging group III
UN number 3082
Risk phrases R51, R53a
Operator protection A, H; U05a, U08, U11, U19a, U20a
Environmental protection E15a, E38
Storage and disposal D01, D02, D09a, D10b, D12a

254 fluoxastrobin

A protectant stobilurin fungicide available in mixtures
FRAC mode of action code: 11

255 fluoxastrobin + prothioconazole

A strobilurin and triazole fungicide mixture for cereals
FRAC mode of action code: 11 + 3

See also prothioconazole

Products

1	Clayton Edge	Clayton	100:100 g/l	EC	12911
2	Fandango	Bayer CropScience	100:100 g/l	EC	12276
3	Firefly 155	Bayer CropScience	45:110 g/l	EC	14818
4	Kurdi	AgChem Access	100+100 g/l	EC	13531
5	Maestro	Bayer CropScience	100:100 g/l	EC	12307
6	Redigo Twin TXC	Bayer CropScience	112.5;112.5 g/l	FS	14259
7	Unicur	Bayer CropScience	100:100 g/l	EC	14776

Uses
- Botrytis in **bulb onions** *(useful reduction)* [7]
- Botrytis squamosa in **bulb onions** *(control)* [7]
- Brown rust in **spring barley**, **winter barley** [1, 2, 4, 5]; **spring wheat** [2-5]; **winter rye**, **winter wheat** [1-5]
- Bunt in **durum wheat**, **spring oats**, **spring rye**, **spring wheat**, **triticale**, **winter oats**, **winter rye**, **winter wheat** [6]
- Cladosporium in **bulb onions** *(useful reduction)* [7]
- Covered smut in **spring oats**, **winter oats** [6]
- Crown rust in **spring oats**, **winter oats** [2, 4, 5]
- Downy mildew in **bulb onions** *(control)* [7]
- Eyespot in **spring barley** *(reduction)*, **winter barley** *(reduction)* [1, 2, 4, 5]; **spring oats**, **winter oats** [2, 4, 5]; **spring oats** *(Reduction of incidence and severity)*, **winter oats** *(Reduction of incidence and severity)* [3]; **spring wheat** [5]; **spring wheat** *(reduction)* [2-4]; **winter rye** *(reduction)*, **winter wheat** *(reduction)* [1-5]
- Foliar disease control in **forest nurseries** *(off-label)* [2]
- Fusarium root rot in **spring wheat** *(reduction)*, **winter wheat** *(reduction)* [2, 4]
- Glume blotch in **spring wheat** [2-4]; **spring wheat** *(reduction)* [5]; **winter wheat** [1-5]
- Late ear diseases in **spring barley**, **winter barley**, **winter wheat** [1, 2, 4, 5]; **spring wheat** [2, 4, 5]
- Loose smut in **durum wheat**, **spring oats**, **spring rye**, **spring wheat**, **triticale**, **winter oats**, **winter rye**, **winter wheat** [6]
- Net blotch in **spring barley**, **winter barley** [1, 2, 4, 5]
- Powdery mildew in **spring barley**, **winter barley** [1, 2, 4, 5]; **spring oats**, **winter oats** [2, 4, 5]; **spring wheat** [2-5]; **winter rye**, **winter wheat** [1-5]
- Rhynchosporium in **spring barley**, **winter barley** [1, 2, 4, 5]; **winter rye** [1-5]

FOR FULL CONDITIONS OF USE ALWAYS READ THE PRODUCT LABEL

fluoxastrobin + prothioconazole

- Rust in **bulb onions** *(useful reduction)* [7]
- Seedling blight and foot rot in **durum wheat**, **spring oats**, **spring rye**, **spring wheat**, **triticale**, **winter oats**, **winter rye**, **winter wheat** [6]
- Septoria leaf blotch in **spring wheat** [2-5]; **winter wheat** [1-5]
- Sharp eyespot in **spring wheat** *(reduction)*, **winter wheat** [2, 4]
- Sooty moulds in **spring wheat** *(reduction)*, **winter wheat** *(reduction)* [2, 4]
- Take-all in **spring wheat** *(reduction)*, **winter barley** *(reduction)*, **winter wheat** *(reduction)* [2, 4, 5]
- Tan spot in **spring wheat** [2, 4]; **winter wheat** [1, 2, 4, 5]
- Yellow rust in **spring wheat** [2-5]; **winter wheat** [1-5]

Specific Off-Label Approvals (SOLAs)
- **forest nurseries** 20090226 [2]

Approval information
- Fluoxastrobin and prothioconazole included in Annex I under EC Directive 91/414
- Accepted by BBPA for use on malting barley

Efficacy guidance
- Seed treatments must be applied by manufacturer's recommended treatment application equipment
- Treated cereal seed should preferably be drilled in the same season
- Follow-up treatments will be needed later in the season to give protection against air-borne and splash-borne diseases
- Best results on foliar diseases obtained from treatment at early stages of disease development. Further treatment may be needed if disease attack is prolonged [1, 2, 5]
- Foliar applications to established infections of any disease are likely to be less effective [1, 2, 5]
- Best control of cereal ear diseases obtained by treatment during ear emergence [1, 2, 5]
- Fluoxastrobin is a member of the QoI cross resistance group. Foliar product should be used preventatively and not relied on for its curative potential [1, 2, 5]
- Use product as part of an Integrated Crop Management strategy incorporating other methods of control, including where appropriate other fungicides with a different mode of action. Do not apply more than two foliar applications of QoI containing products to any cereal crop
- There is a significant risk of widespread resistance occurring in *Septoria tritici* populations in UK. Failure to follow resistance management action may result in reduced levels of disease control
- Strains of wheat and barley powdery mildew resistant to QoIs are common in the UK. Control of wheat mildew can only be relied on from the triazole component
- Prothioconazole is a DMI fungicide. Resistance to some DMI fungicides has been identified in Septoria leaf blotch which may seriously affect performance of some products. For further advice contact a specialist advisor and visit the Fungicide Resistance Action Group (FRAG)-UK website
- Where specific control of wheat mildew is required this should be achieved through a programme of measures including products recommended for the control of mildew that contain a fungicide from a different cross-resistance group and applied at a dose that will give robust control

Restrictions
- Maximum number of seed treatments one per batch
- Maximum total dose of foliar sprays equivalent to two full dose treatments for the crop [1, 2, 5]
- Seed treatment must be fully re-dispersed and homogeneous before use
- Do not use on seed with more than 16% moisture content, or on sprouted, cracked or skinned seed
- All seed batches should be tested to ensure they are suitable for treatment

Crop-specific information
- Latest use: pre-drilling for seed treatment [6]; before grain milky ripe for spray treatments on wheat and rye; beginning of flowering for barley, oats [1, 2, 5]
- Some transient leaf chlorosis may occur after treatment of wheat or barley but this has not been found to affect yield [1, 2, 5]

Environmental safety
- Dangerous for the environment
- Toxic to aquatic organisms
- Risk to non-target insects or other arthropods. Avoid spraying within 6 m of the field boundary to reduce the effects on non-target insects or other arthropods [1, 2, 5]
- LERAP Category B [1-5, 7]

SEE SECTION 3 FOR PRODUCTS ALSO REGISTERED

Pesticide Profiles

Hazard classification and safety precautions
   Hazard Dangerous for the environment
   Transport code 9
   Packaging group III
   UN number 3082
   Risk phrases R51, R53a
   Operator protection A [1-5, 7]; H [3, 6, 7]; U05a [1-7]; U09a, U20b [3]; U09b, U20a [1, 2, 4, 5, 7]; U20c, U25 [6]
   Environmental protection E15a, E34 [1-7]; E16a [1-5, 7]; E22c [1, 2, 4, 5, 7]; E38 [2-7]
   Storage and disposal D01, D12a [1-7]; D02 [1, 6]; D05, D09a, D10b [1-5, 7]; D09b, D10d, D14 [6]
   Treated seed S01, S02, S03, S04b, S05, S06b, S07, S08 [6]
   Medical advice M03 [1-5, 7]

256   fluoxastrobin + prothioconazole + trifloxystrobin

A triazole and strobilurin fungicide mixture for cereals
FRAC mode of action code: 11 + 3 + 11

See also prothioconazole
trifloxystrobin

Products
   Jaunt   Bayer CropScience   75:150:75 g/l   EC   12350

Uses
- Brown rust in **spring barley, winter barley, winter wheat**
- Eyespot in **spring barley** *(reduction)*, **winter barley** *(reduction)*, **winter wheat** *(reduction)*
- Glume blotch in **winter wheat**
- Late ear diseases in **spring barley, winter barley, winter wheat**
- Net blotch in **spring barley, winter barley**
- Powdery mildew in **spring barley, winter barley, winter wheat**
- Rhynchosporium in **spring barley, winter barley**
- Septoria leaf blotch in **winter wheat**
- Tan spot in **winter wheat**
- Yellow rust in **winter wheat**

Approval information
- Fluoxastrobin, prothioconazole and trifloxystrobin included in Annex I under EC Directive 91/414
- Accepted by BBPA for use on malting barley

Efficacy guidance
- Best results obtained from treatment at early stages of disease development. Further treatment may be needed if disease attack is prolonged
- Applications to established infections of any disease are likely to be less effective
- Best control of cereal ear diseases obtained by treatment during ear emergence
- Fluoxastrobin and trifloxystrobin are members of the QoI cross resistance group. Product should be used preventatively and not relied on for its curative potential
- Use product as part of an Integrated Crop Management strategy incorporating other methods of control, including where appropriate other fungicides with a different mode of action. Do not apply more than two foliar applications of QoI containing products to any cereal crop
- There is a significant risk of widespread resistance occurring in *Septoria tritici* populations in UK. Failure to follow resistance management action may result in reduced levels of disease control
- Strains of wheat and barley powdery mildew resistant to QoIs are common in the UK. Control of wheat mildew can only be relied on from the triazole component
- Where specific control of wheat mildew is required this should be achieved through a programme of measures including products recommended for the control of mildew that contain a fungicide from a different cross-resistance group and applied at a dose that will give robust control
- Prothioconazole is a DMI fungicide. Resistance to some DMI fungicides has been identified in Septoria leaf blotch which may seriously affect performance of some products. For further advice contact a specialist advisor and visit the Fungicide Resistance Action Group (FRAG)-UK website

FOR FULL CONDITIONS OF USE ALWAYS READ THE PRODUCT LABEL

flupyrsulfuron-methyl

Restrictions
- Maximum total dose equivalent to two full dose treatments

Crop-specific information
- Latest use: before grain milky ripe for winter wheat; up to beginning of anthesis (GS 61) for barley

Environmental safety
- Dangerous for the environment
- Very toxic to aquatic organisms
- Risk to non-target insects or other arthropods. Avoid spraying within 6 m of the field boundary to reduce the effects on non-target insects or other arthropods
- LERAP Category B

Hazard classification and safety precautions
Hazard Irritant, Dangerous for the environment
Transport code 9
Packaging group III
UN number 3082
Risk phrases R37, R50, R53a
Operator protection A, C, H; U05a, U09b, U19a, U20b
Environmental protection E15a, E16a, E22c, E34, E38
Storage and disposal D01, D02, D05, D09a, D10b, D12a
Medical advice M03

257 flupyrsulfuron-methyl

A sulfonylurea herbicide for winter wheat and winter barley
HRAC mode of action code: B

See also carfentrazone-ethyl + flupyrsulfuron-methyl
            diflufenican + flupyrsulfuron-methyl

Products
1	Bullion	DuPont	50% w/w	WG	14058
2	Lexus SX	DuPont	50% w/w	WG	12979
3	Oklar SX	DuPont	50% w/w	WG	15037
4	Oriel 50SX	DuPont	50% w/w	WG	14640

Uses
- Annual dicotyledons in *triticale, winter oats, winter rye* [3, 4]; *winter barley* [1, 2]; *winter linseed* (off-label) [2]; *winter wheat* [1-4]
- Blackgrass in *winter barley, winter wheat* [1, 2]; *winter linseed* (off-label) [2]

Specific Off-Label Approvals (SOLAs)
- *winter linseed* 20063714 expires 30 Jun 2011 [2]

Approval information
- Flupyrsulfuron-methyl included in Annex I under EC Directive 91/414
- Approval expiry 30 Jun 2011 [1-4]

Efficacy guidance
- Best results achieved from applications made in good growing conditions
- Good spray cover of weeds must be obtained
- For pre-emergence control of blackgrass application must be made in tank mixture with specified products (see label). Best control of blackgrass post-emergence in wheat obtained from application from the 1-leaf stage
- Winter barley must be treated pre-emergence and used in mixture. See label
- Growth of weeds is inhibited within hours of treatment but visible symptoms may not be apparent for up to 4 wk
- Flupyrsulfuron-methyl has moderate residual life in soil. Under normal moisture conditions susceptible weeds germinating soon after treatment will be controlled
- May be used on all soil types but residual activity and weed control is reduced on highly alkaline soils

- Flupyrsulfuron-methyl is a member of the ALS-inhibitor group of herbicides. To avoid the build up of resistance do not use any product containing an ALS-inhibitor herbicide with claims for control of grass weeds more than once on any crop
- Use these products as part of a resistance management strategy that includes cultural methods of control and does not use ALS inhibitors as the sole chemical method of grass weed control

Restrictions
- Maximum number of treatments 1 per crop
- Do not use on wheat undersown with grasses, clover or other legumes, or any other broad-leaved crop
- Do not apply within 7 d of rolling
- Do not treat any crop suffering from drought, waterlogging, pest or disease attack, nutrient deficiency, or any other stress factors
- Specific restrictions apply to use in sequence or tank mixture with other sulfonylurea or ALS-inhibiting herbicides. See label for details

Crop-specific information
- Latest use: pre-emergence or before 1st node detectable (GS 31) for wheat; pre-emergence for barley
- Under certain conditions chlorosis and stunting may occur, from which recovery is rapid

Following crops guidance
- Only cereals, oilseed rape, field beans, clover or grass may be sown in the yr of harvest of a treated crop
- In the event of crop failure only winter or spring wheat may be sown within 3 mth of treatment. Land should be ploughed and cultivated to 15 cm minimum before resowing. After 3 mth crops may be sown as shown above

Environmental safety
- Dangerous for the environment
- Very toxic to aquatic organisms
- Take extreme care to avoid drift onto broad-leaved plants outside the target area or onto ponds, waterways or ditches, or onto land intended for cropping
- Spraying equipment should not be drained or flushed onto land planted, or to be planted, with trees or crops other than cereals and should be thoroughly cleansed after use - see label for instructions

Hazard classification and safety precautions
Hazard Dangerous for the environment
Transport code 9
Packaging group III
UN number 3077
Risk phrases R50, R53a [1, 2]
Operator protection U08, U19a, U20b
Environmental protection E15a [1-4]; E38 [1, 2]
Storage and disposal D09a, D11a [1-4]; D12a [1, 2]

258 flupyrsulfuron-methyl + thifensulfuron-methyl

A sulfonylurea herbicide mixture for winter wheat and winter oats
HRAC mode of action code: B + B

See also thifensulfuron-methyl

Products
1	Lancer	Headland	10:40% w/w	WG	13031
2	Lexus Millenium	DuPont	10:40% w/w	WG	09206

Uses
- Annual dicotyledons in *winter oats*, *winter wheat*
- Blackgrass in *winter oats*, *winter wheat*

Approval information
- Flupyrsulfuron-methyl and thifensulfuron-methyl included in Annex I under EC Directive 91/414

FOR FULL CONDITIONS OF USE ALWAYS READ THE PRODUCT LABEL

fluquinconazole

Efficacy guidance
- Best results obtained when applied to small actively growing weeds
- Good spray cover of weeds must be obtained
- Increased degradation of active ingredient in high soil temperatures reduces residual activity
- Product has moderate residual life in soil. Under normal moisture conditions susceptible weeds germinating soon after treatment will be controlled
- Product may be used on all soil types but residual activity and weed control is reduced on highly alkaline soils
- Blackgrass should be treated from 1 leaf up to first node stage
- Flupyrsulfuron-methyl and thifensulfuron-methyl are members of the ALS-inhibitor group of herbicides. To avoid the build up of resistance do not use any product containing an ALS-inhibitor herbicide with claims for control of grass weeds more than once on any crop
- Use these products as part of a resistance management strategy that includes cultural methods of control and does not use ALS inhibitors as the sole chemical method of grass weed control

Restrictions
- Maximum number of treatments 1 per crop
- Do not use on wheat undersown with grasses or legumes, or any other broad-leaved crop
- Do not apply within 7 d of rolling
- Do not treat any crop suffering from drought, waterlogging, pest or disease attack, nutrient deficiency, or any other stress factors
- Contact contract agents before use on crops grown for seed
- Specific restrictions apply to use in sequence or tank mixture with other sulfonylurea or ALS-inhibiting herbicides. See label for details

Crop-specific information
- Latest use: before 1st node detectable (GS 31) on wheat; before 31 Dec in yr of sowing on oats [2]
- Slight chlorosis, speckling and stunting may occur in certain conditions. Recovery is rapid and yield not affected

Following crops guidance
- Only cereals, oilseed rape, field beans or grass may be sown in the yr of harvest of a treated crop. Any crop may be sown the following spring
- In the event of crop failure only winter wheat may be sown before normal harvest date. Land should be ploughed and cultivated to 15 cm minimum before resowing

Environmental safety
- Dangerous for the environment
- Very toxic to aquatic organisms
- LERAP Category B

Hazard classification and safety precautions
    Hazard Irritant, Dangerous for the environment
    Transport code 9
    Packaging group III
    UN number 3077
    Risk phrases R43, R50, R53a
    Operator protection A, H; U05a, U08, U14, U19a, U20b
    Environmental protection E15a, E16a, E38
    Storage and disposal D01, D02, D09a, D11a, D12a

259 fluquinconazole

A protectant, eradicant and systematic triazole fungicide for winter cereals
FRAC mode of action code: 3

Products

1	Flamenco	BASF	100 g/l	SC	11699
2	Galmano	Bayer CropScience	167 g/l	FS	11650
3	Jockey Solo	BASF	167 g/l	FS	14281

Uses
- Brown rust in **winter wheat** [1]
- Bunt in **winter wheat** (seed treatment) [2]

SEE SECTION 3 FOR PRODUCTS ALSO REGISTERED

- Foliar disease control in **grass seed crops** *(off-label)*, **triticale** *(off-label)* [1]
- Glume blotch in **winter wheat** [1]
- Powdery mildew in **winter wheat** [1]
- Seed-borne diseases in **durum wheat** *(off-label)*, **triticale** *(off-label)* [2]; **winter barley**, **winter wheat** [3]
- Septoria leaf blotch in **winter wheat** [1]; **winter wheat** *(seed treatment)* [2]
- Take-all in **winter wheat** *(seed treatment - reduction)* [2]
- Yellow rust in **winter wheat** [1]; **winter wheat** *(seed treatment)* [2]

Specific Off-Label Approvals (SOLAs)
- **durum wheat** 20063065 [2]
- **grass seed crops** 20063472 [1]
- **triticale** 20063472 [1], 20063065 [2]

Approval information
- Fluquinconazole not included in Annex 1 but the date of withdrawal has yet to be set.
- Accepted by BBPA on malting barley before ear emergence only

Efficacy guidance
- Best results obtained from spray treatments when disease first becomes active in crop but before infection spreads to younger leaves [1]
- Adequate disease protection throughout the season will usually require a programme of at least two fungicide treatments [1]
- May also be applied as a protectant at end of ear emergence (but no later) if crop still disease free [1]
- With seed treatments ensure good even coverage of seed to obtain reliable disease control [2]
- Seed treatment provides early control of Septoria leaf blotch and yellow rust but may require foliar treatment for later infection [2]
- Fluquinconazole is a DMI fungicide. Resistance to some DMI fungicides has been identified in Septoria leaf blotch which may seriously affect performance of some products. For further advice contact a specialist advisor and visit the Fungicide Resistance Action Group (FRAG)-UK website

Restrictions
- Maximum total spray dose equivalent to two full doses [1]
- Maximum number of seed treatments 1 per batch [2]
- Do not apply spray treatments after the start of anthesis (GS 59) [1]
- Do not mix with any other products [2]
- Do not treat cracked, split or sprouted seed [2]
- Do not use treated seed as food or feed [2]

Crop-specific information
- Latest use: before beginning of anthesis (GS 59) for spray treatment; before drilling for seed treatment
- Ensure good spray coverage and increase volume in dense crops [1]
- Delayed emergence may result when treated seed is drilled into heavy or poorly drained soils which then become wet or waterlogged [2]

Environmental safety
- Dangerous for the environment
- Toxic to aquatic organisms
- Dangerous to fish or other aquatic life. Do not contaminate surface waters or ditches with chemical or used container
- Special PPE requirements and precautions apply where product supplied in returnable packs. Check label

Hazard classification and safety precautions
Hazard Toxic, Dangerous for the environment
Transport code 9
Packaging group III
UN number 3082
Risk phrases R22a, R51, R53a [1-3]; R25b [3]; R36, R43 [1]; R48 [1, 2]
Operator protection A, H [1-3]; C, M [1]; D [2, 3]; U05a [1-3]; U07 [2]; U11, U14, U15 [1]; U20b [2, 3]; U25 [3]
Environmental protection E13b, E36a [2]; E15a [1, 3]; E34, E38 [1-3]

FOR FULL CONDITIONS OF USE ALWAYS READ THE PRODUCT LABEL

Storage and disposal D01, D02, D12a [1-3]; D05 [1, 2]; D08 [2]; D09a, D14 [2, 3]; D10c [1]; D10d [3]
Treated seed S01, S03, S04b, S06a [2]; S02, S05, S07, S08 [2, 3]
Medical advice M03 [1, 2]; M04a [1, 3]

260 fluquinconazole + prochloraz

A broad-spectrum triazole mixture for use as a spray and seed treatment in winter cereals
FRAC mode of action code: 3 + 3

See also prochloraz

Products

1	Epona	BASF	167:31 g/l	FS	12444
2	Foil	BASF	54:174 g/l	SE	11700
3	Galmano Plus	Bayer CropScience	167:31 g/l	FS	11645
4	Jockey	BASF	167:31 g/l	FS	11689

Uses
- Brown rust in **winter barley**, **winter wheat** [2]; **winter wheat** *(seed treatment - moderate control)* [1]
- Bunt in **winter wheat** *(seed treatment)* [1, 3, 4]
- Covered smut in **winter barley** *(seed treatment)* [4]
- Foliar disease control in **durum wheat** *(off-label)*, **grass seed crops** *(off-label)*, **spring rye** *(off-label)*, **triticale** *(off-label)*, **winter rye** *(off-label)* [2]
- Fusarium foot rot and seedling blight in **winter barley** *(seed treatment)* [4]
- Fusarium root rot in **winter wheat** *(seed treatment)* [1, 3, 4]
- Loose smut in **winter barley** *(seed treatment)* [4]
- Powdery mildew in **winter barley**, **winter wheat** *(moderate control)* [2]
- Rhynchosporium in **winter barley** [2]
- Seed-borne diseases in **durum wheat** *(off-label)*, **triticale** *(off-label)* [3]
- Septoria leaf blotch in **winter wheat** [2]; **winter wheat** *(seed treatment)* [3, 4]; **winter wheat** *(seed treatment - moderate control)* [1]
- Take-all in **winter barley** *(seed treatment - reduction)* [4]; **winter wheat** *(seed treatment - reduction)* [1, 3, 4]
- Yellow rust in **winter barley** *(seed treatment)* [4]; **winter wheat** [2]; **winter wheat** *(seed treatment)* [3, 4]; **winter wheat** *(seed treatment - moderate control)* [1]

Specific Off-Label Approvals (SOLAs)
- *durum wheat* 20060982 [2], 20061093 [3]
- *grass seed crops* 20060982 [2]
- *spring rye* 20060982 [2]
- *triticale* 20060982 [2], 20061093 [3]
- *winter rye* 20060982 [2]

Approval information
- Accepted by BBPA for use on malting barley (before ear emergence only)
- Fluquinconazole and prochloraz have not been included in Annex 1 but the withdrawal or inclusion date has yet to be set. Most current approvals set to expire 2021.

Efficacy guidance
- Best results obtained from spray treatment when disease first becomes active in crop but before infection spreads to younger leaves [2]
- When treating seed ensure good even coverage of seed to obtain reliable disease control [1, 3, 4]
- Seed treatment provides early control of *Septoria* leaf spot and rust but may require foliar treatment for later infection [1, 3, 4]
- Fluquinconazole and prochloraz are DMI fungicides. Resistance to some DMI fungicides has been identified in Septoria leaf blotch which may seriously affect performance of some products. For further advice contact a specialist advisor and visit the Fungicide Resistance Action Group (FRAG)-UK website

SEE SECTION 3 FOR PRODUCTS ALSO REGISTERED

Pesticide Profiles

Restrictions
- Maximum number of treatments 1 per seed batch (seed dressing)
- Maximum total dose equivalent to two full doses (spray)
- Do not treat cracked, split or sprouted seed [1, 3, 4]
- Do not use treated seed as food or feed [1, 3, 4]

Crop-specific information
- Latest use: before 1st awns visible (GS 47) for winter barley; before beginning of anthesis (GS 59) for winter wheat; before drilling (seed treatment)
- Ensure good spray coverage and increase volume in dense crops [2]
- Delayed emergence may result when treated seed is drilled into heavy or poorly drained soils which then become wet or waterlogged [1, 3, 4]

Environmental safety
- Dangerous for the environment
- Toxic to aquatic organisms
- Dangerous to fish or other aquatic life. Do not contaminate surface waters or ditches with chemical or used container
- Product supplied in returnable packs for which special PPE requirements and precautions apply. Check label

Hazard classification and safety precautions
Hazard Toxic [1, 3, 4]; Harmful, Irritant [2]; Dangerous for the environment [1-4]
Transport code 9
Packaging group III
UN number 3082
Risk phrases R22a, R48, R51, R53a [1-4]; R36 [2, 3]; R43, R66 [2]
Operator protection A, H [1-4]; C [2]; D [3, 4]; G [1]; U05a [1-4]; U07 [3, 4]; U20b [1, 3, 4]
Environmental protection E13b [3]; E15a [1, 2, 4]; E34, E38 [1-4]; E36a [3, 4]
Storage and disposal D01, D02, D09a, D12a [1-4]; D05 [2, 3]; D08 [1, 3, 4]; D10c [2]; D11a [1, 4]; D14 [3, 4]
Treated seed S01, S03, S04b, S06a [3, 4]; S02, S05, S07, S08 [1, 3, 4]
Medical advice M03 [2, 3]; M04a [1, 4]; M05b [2]

261 fluroxypyr

A post-emergence pyridinecarboxylic acid herbicide
HRAC mode of action code: O

See also 2,4-D + dicamba + fluroxypyr
aminopyralid + fluroxypyr
bromoxynil + fluroxypyr
bromoxynil + fluroxypyr + ioxynil
clopyralid + florasulam + fluroxypyr
clopyralid + fluroxypyr + MCPA
clopyralid + fluroxypyr + triclopyr
florasulam + fluroxypyr

Products

1	Agriguard Fluroxypyr	AgriGuard	200 g/l	EC	12525
2	Agriguard Fluroxypyr	AgriGuard	200 g/l	EC	13335
3	Barclay Hudson	Barclay	180 g/l	EC	13606
4	Barclay Hurler	Barclay	200 g/l	EC	13458
5	Casino	AgriGuard	25% w/w	WG	14352
6	Crescent	AgriGuard	200 g/l	EC	12559
7	Crescent	AgriGuard	200 g/l	EC	13330
8	Flurostar 200	Globachem	200 g/l	EC	12903
9	Fluroxy 200	Goldengrass	200 g/l	EC	14421
10	Fluxyr 200 EC	AgriChem BV	200 g/l	EC	14086
11	Gala	Dow	200 g/l	EC	12019
12	Greencrop Reaper	Greencrop	200 g/l	EC	12261
13	Hatchet	AgriGuard	200 g/l	EC	12524
14	Hatchet Xtra	AgriGuard	200 g/l	EC	13213
15	Klever	Globachem	200 g/l	EC	13295

FOR FULL CONDITIONS OF USE ALWAYS READ THE PRODUCT LABEL

fluroxypyr

Products – continued

16	Minstrel	United Phosphorus	200 g/l	EC	13745
17	Skora	AgChem Access	200 g/l	EC	13422
18	Standon Homerun 2	Standon	200 g/l	EC	13890
19	Starane 2	Dow	200 g/l	EC	12018
20	Tandus	Nufarm UK	200 g/l	EC	13432
21	Tomahawk	Makhteshim	200 g/l	EC	09249

Uses
- Annual and perennial weeds in **durum wheat, forage maize, spring barley, spring oats, spring wheat, triticale, winter barley, winter oats, winter rye, winter wheat** [11, 18]; **grassland** [11]; **permanent grassland, rotational grass, spring rye** [18]
- Annual dicotyledons in **almonds** (off-label), **apple orchards** (off-label), **chestnuts** (off-label), **crab apples** (off-label), **hazel nuts** (off-label), **pears** (off-label), **poppies for morphine production** (off-label), **sweetcorn** (off-label), **walnuts** (off-label) [21]; **bulb onions** (off-label), **farm forestry** (off-label), **forest nurseries** (off-label), **game cover** (off-label), **garlic** (off-label), **leeks** (off-label), **ornamental plant production** (off-label), **shallots** (off-label) [19, 21]; **durum wheat, forage maize, spring barley, spring oats, spring wheat, triticale, winter barley, winter oats, winter rye, winter wheat** [1-10, 12-17, 19-21]; **grassland** [10, 16]; **miscanthus** (off-label) [1, 19, 21]; **newly sown grass leys** [9, 18, 20]; **permanent grassland, rotational grass** [1-9, 12-15, 17, 19-21]; **spring rye** [1-9, 12-17, 19, 21]
- Black bindweed in **bulb onions** (off-label), **garlic** (off-label), **shallots** (off-label) [19]; **durum wheat, spring barley, spring oats, spring wheat, triticale, winter barley, winter oats, winter rye, winter wheat** [1-9, 11-21]; **forage maize** [1-9, 11-19, 21]; **grassland** [11, 16]; **newly sown grass leys** [9, 18]; **permanent grassland, rotational grass** [1-9, 12-15, 17-19, 21]; **spring rye** [1-9, 12-19, 21]
- Chickweed in **bulb onions** (off-label), **garlic** (off-label), **leeks** (off-label), **shallots** (off-label) [19]; **durum wheat, forage maize, spring barley, spring oats, spring wheat, triticale, winter barley, winter oats, winter rye, winter wheat** [1-9, 11-21]; **grassland** [11, 16]; **newly sown grass leys** [9, 18, 20]; **permanent grassland, rotational grass** [1-9, 12-15, 17-21]; **spring rye** [1-9, 12-19, 21]
- Cleavers in **apple orchards** (off-label), **bulb onions** (off-label), **garlic** (off-label), **leeks** (off-label), **pear orchards** (off-label), **poppies for morphine production** (off-label), **shallots** (off-label) [19]; **durum wheat, spring barley, spring oats, spring wheat, triticale, winter barley, winter oats, winter rye, winter wheat** [1-9, 11-21]; **forage maize** [1-9, 11-19, 21]; **grassland** [11, 16]; **millet** (off-label - grown for wild bird seed production) [21]; **newly sown grass leys** [9, 18]; **permanent grassland, rotational grass** [1-9, 12-15, 17-19, 21]; **spring rye** [1-9, 12-19, 21]
- Docks in **apple orchards** (off-label), **pear orchards** (off-label) [19]; **durum wheat, spring barley, spring oats, spring wheat, triticale, winter barley, winter oats, winter rye, winter wheat** [1-9, 11-21]; **forage maize** [1-9, 11-19, 21]; **grassland** [11]; **newly sown grass leys** [9]; **permanent grassland, rotational grass** [1-9, 12-15, 17-19, 21]; **spring rye** [1-9, 12-19, 21]
- Forget-me-not in **bulb onions** (off-label), **garlic** (off-label), **shallots** (off-label) [19]; **durum wheat, spring barley, spring oats, spring wheat, triticale, winter barley, winter oats, winter rye, winter wheat** [1-9, 11-21]; **forage maize** [1-9, 11-19, 21]; **grassland** [11, 16]; **newly sown grass leys** [9, 18]; **permanent grassland, rotational grass** [1-9, 12-15, 17-19, 21]; **spring rye** [1-9, 12-19, 21]
- Hemp-nettle in **durum wheat, spring barley, spring oats, spring wheat, triticale, winter barley, winter oats, winter rye, winter wheat** [1-9, 11-21]; **forage maize** [1-9, 11-19, 21]; **grassland** [11, 16]; **newly sown grass leys** [9, 18]; **permanent grassland, rotational grass** [1-9, 12-15, 17-19, 21]; **spring rye** [1-9, 12-19, 21]
- Poppies in **poppies for morphine production** (off-label) [19]
- Stinging nettle in **apple orchards** (off-label), **pear orchards** (off-label) [19]
- Volunteer potatoes in **bulb onions** (off-label), **poppies for morphine production** (off-label), **sweetcorn** (off-label) [19, 21]; **durum wheat, forage maize, triticale, winter oats, winter rye** [11, 18]; **garlic** (off-label), **leeks** (off-label), **shallots** (off-label) [21]; **grassland** [11]; **permanent grassland, rotational grass, spring rye** [18]; **spring barley, spring oats, spring wheat** [11, 18, 20]; **winter barley, winter wheat** [1-9, 11-21]

Specific Off-Label Approvals (SOLAs)
- **almonds** 20101920 expires 31 Dec 2011 [21]
- **apple orchards** 20040988 expires 31 Dec 2011 [19], 20101920 expires 31 Dec 2011 [21]

SEE SECTION 3 FOR PRODUCTS ALSO REGISTERED

- **bulb onions** 20081404 expires 31 Dec 2011 [19], 20101919 expires 31 Dec 2011 [21]
- **chestnuts** 20101920 expires 31 Dec 2011 [21]
- **crab apples** 20101920 expires 31 Dec 2011 [21]
- **farm forestry** 20082925 expires 31 Dec 2011 [19], 20101910 expires 31 Dec 2011 [21]
- **forest nurseries** 20082925 expires 31 Dec 2011 [19], 20101910 expires 31 Dec 2011 [21]
- **game cover** 20082925 expires 31 Dec 2011 [19], 20082907 expires 30 Dec 2011 [21], 20101910 expires 31 Dec 2011 [21]
- **garlic** 20081404 expires 31 Dec 2011 [19], 20101919 expires 31 Dec 2011 [21]
- **hazel nuts** 20101920 expires 31 Dec 2011 [21]
- **leeks** 20100390 expires 31 Dec 2011 [19], 20101916 expires 31 Dec 2011 [21]
- **millet** (grown for wild bird seed production) 20080936 expires 31 Dec 2011 [21]
- **miscanthus** 20082803 expires 31 Dec 2011 [1], 20082925 expires 31 Dec 2011 [19], 20101910 expires 31 Dec 2011 [21]
- **ornamental plant production** 20082925 expires 31 Dec 2011 [19], 20101910 expires 31 Dec 2011 [21]
- **pear orchards** 20040988 expires 31 Dec 2011 [19]
- **pears** 20101920 expires 31 Dec 2011 [21]
- **poppies for morphine production** 20052059 expires 31 Dec 2011 [19], 20101908 expires 31 Dec 2011 [21]
- **shallots** 20081404 expires 31 Dec 2011 [19], 20101919 expires 31 Dec 2011 [21]
- **sweetcorn** 20051696 expires 31 Dec 2011 [19], 20101907 expires 31 Dec 2011 [21]
- **walnuts** 20101920 expires 31 Dec 2011 [21]

Approval information
- Fluroxypyr included in Annex I under EC Directive 91/414
- Accepted by BBPA for use on malting barley
- Approval expiry 31 Dec 2011 [1, 2, 4-9, 11-15, 17-21]
- Approval expiry 03 Apr 2011 [3]

Efficacy guidance
- Best results achieved under good growing conditions in a strongly competing crop
- A number of tank mixtures with HBN and other herbicides are recommended for use in autumn and spring to extend range of species controlled. See label for details
- Spray is rainfast in 1 h

Restrictions
- Maximum number of treatments 1 per crop or yr or maximum total dose equivalent to one full dose treatment
- Do not apply in any tank-mix on triticale or forage maize
- Do not use on crops undersown with clovers or other legumes
- Do not treat crops suffering stress caused by any factor
- Do not roll or harrow for 7 d before or after treatment
- Do not spray if frost imminent
- Do not apply before 1 Mar in yr of harvest [20]
- Do not use on crops grown for seed production [20]
- Straw from treated crops must not be returned directly to the soil but must be removed and used only for livestock bedding

Crop-specific information
- Latest use: before flag leaf sheath opening (GS 47) for winter wheat and barley; before flag leaf sheath extending (GS 41) for spring wheat and barley; before second node detectable (GS 32) for oats, rye, triticale and durum wheat; before 7 leaves unfolded and before buttress roots appear for maize
- Apply to new leys from 3 expanded leaf stage
- Timing varies in tank mixtures. See label for details
- Crops undersown with grass may be sprayed provided grasses are tillering

Following crops guidance
- Clovers, peas, beans and other legumes must not be sown for 12 mth following treatment at the highest dose

Environmental safety
- Dangerous for the environment
- Very toxic to aquatic organisms

FOR FULL CONDITIONS OF USE ALWAYS READ THE PRODUCT LABEL

fluroxypyr + thifensulfuron-methyl + tribenuron-methyl

- Flammable
- Keep livestock out of treated areas for at least 3 d following treatment and until poisonous weeds, such as ragwort, have died down and become unpalatable
- Wash spray equipment thoroughly with water and detergent immediately after use. Traces of product can damage susceptible plants sprayed later

Hazard classification and safety precautions

Hazard Harmful [1-4, 6-21]; Irritant [5]; Flammable [1-4, 6-19, 21]; Dangerous for the environment [1-21]

Transport code 3 [1-4, 6-9, 11-21]; 9 [5, 10]

Packaging group III

UN number 1933 [19]; 1993 [1-4, 6-9, 11-18, 20, 21]; 3077 [5]; 3082 [10]

Risk phrases R22b, R67 [1-4, 6-21]; R36 [1, 2, 5-8, 11, 13-15]; R37 [1-4, 6-11, 13-19, 21]; R38 [1-4, 6-8, 11, 13-15]; R43 [1-4, 6-8, 11, 13-15, 20]; R50 [2, 5, 7, 8, 14, 15]; R51 [1, 3, 4, 6, 9-13, 16-21]; R53a [1-21]

Operator protection A [1-9, 11, 13-16, 18, 20]; C [1, 2, 5-9, 11, 13-16, 18]; H [1-4, 6-9, 11, 13-16, 18, 20]; U05a [1-8, 11, 13-15]; U08, U19a [1-21]; U11 [1, 2, 5-8, 11, 13-15]; U14 [1-4, 6-8, 11, 13-15, 20]; U20a [11]; U20b [1-10, 12-19, 21]

Environmental protection E06a [5, 7] (3 d); E07a [1, 2, 6, 8-10, 12-21]; E07c [3, 4] (14 d); E07c [11] (3 days); E15a [9, 10, 12, 16-19, 21]; E15b [1-8, 11, 13-15, 20]; E34 [1-8, 11, 13-15, 21]; E38 [2-4, 7-11, 14-21]

Storage and disposal D01 [1-8, 11, 13-15]; D02 [1-9, 11, 13-15]; D05 [1-4, 7-9, 11, 12, 14, 15, 21]; D09a [1-21]; D10a [3, 4]; D10b [1, 2, 5-10, 12-21]; D10c [11]; D12a [1-4, 6-10, 13-21]

Treated seed S06a [20]

Medical advice M03 [1-8, 11, 13-15]; M05b [1-4, 6-21]

262 fluroxypyr + thifensulfuron-methyl + tribenuron-methyl

A foliar applied herbicide mixture for cereals with some residual activity
HRAC mode of action code: O + B + B

See also thifensulfuron-methyl
tribenuron-methyl

Products

GEX 353	DuPont	200 g/l + 50.25% w/w	KK	12755

Uses
- Annual dicotyledons in **spring barley**, **spring wheat**, **winter barley**, **winter wheat**

Approval information
- Fluroxypyr, thifensulfuron-methyl and tribenuron-methyl included in Annex I under EC Directive 91/414
- Accepted by BBPA for use on malting barley

Efficacy guidance
- Best results obtained when weeds small and actively growing. This is particularly important for cleavers
- Ensure good spray cover
- Susceptible species cease growth immediately after application but may take 2 wk to show symptoms
- Weed control may be reduced in very dry conditions
- Rain within 4 hr may reduce effectiveness
- Thifensulfuron-methyl and tribenuron-methyl are members of the ALS-inhibitor group of herbicides and products should be used in a planned Resistance Management strategy. See Section 5 for more information

Restrictions
- Maximum number of treatments 1 per crop for all cereals
- Must only be applied after 1 Feb in yr of harvest and should not be applied more than once to any cereal crop
- Do not use on any crop suffering stress from drought, waterlogging, frost, deficiency, pest or disease attack or any other cause

SEE SECTION 3 FOR PRODUCTS ALSO REGISTERED

- Do not use on crops undersown with grasses, clover or legumes, or any other broad leaved crop
- Specific restrictions apply to use in sequence or tank mixture with other sulfonylurea or ALS-inhibiting herbicides. See label for details
- Do not apply within 7 d of rolling

Crop-specific information
- Latest use: before flag leaf extending (GS 39) for all crops

Following crops guidance
- Only cereals, field beans or oilseed rape may be sown in the same calendar yr as harvest of a treated crop
- In the event of failure of a treated crop only a cereal crop may be sown within 3 mth

Environmental safety
- Dangerous for the environment
- Very toxic to aquatic organisms
- Extremely dangerous to fish or other aquatic life. Do not contaminate surface waters or ditches with chemical or used container
- Keep livestock out of treated areas for at least 3 days and until foliage of any poisonous weeds such as ragwort has died and become unpalatable
- Take extreme care to avoid damage by drift onto broad-leaved plants outside the target area, or onto ponds, waterways or ditches
- Spraying equipment should not be drained or flushed onto land planted, or to be planted, with trees or crops other than cereals and should be thoroughly cleansed after use - see label for instructions
- LERAP Category B

Hazard classification and safety precautions
Hazard Harmful, Flammable, Dangerous for the environment
Transport code 3
Packaging group II
UN number 1993
Risk phrases R22b, R37, R43, R50, R53a, R67
Operator protection A, C, H; U08, U19a, U20b
Environmental protection E07c (3 d); E13a, E16a, E16b, E38
Storage and disposal D05, D09a, D10b, D12a
Medical advice M05b

263 fluroxypyr + triclopyr

A foliar acting herbicide for docks in grassland
HRAC mode of action code: O + O

See also triclopyr

Products

1	Dovekie	AgChem Access	100:100 g/l	EC	14121
2	Doxstar	Dow	100:100 g/l	EC	11063

Uses
- Chickweed in *rotational grass*
- Docks in *permanent grassland, rotational grass*

Approval information
- Fluroxypyr and triclopyr included in Annex I under EC Directive 91/414

Efficacy guidance
- Seedling docks in new leys are controlled up to 50 mm diameter. In established grass apply in spring or autumn or, at lower dose, in spring and autumn on docks up to 200 mm. A second application in the subsequent yr may be needed
- Allow 2-3 wk after cutting or grazing to allow sufficient regrowth of docks to occur before spraying
- Control may be reduced if rain falls within 2 h of application
- To allow maximum translocation to the roots of docks do not cut grass for 28 d after spraying

FOR FULL CONDITIONS OF USE ALWAYS READ THE PRODUCT LABEL

flurtamone

Restrictions
- Maximum total dose equivalent to one full dose treatment
- Do not roll or harrow for 10 d before or 7 d after spraying
- Do not spray in drought, very hot or very cold weather

Crop-specific information
- Latest use: 7 d before grazing or harvest of grass
- Grass less than one yr old may be treated at half dose from the third leaf visible stage
- Clover will be killed or severely checked by treatment

Following crops guidance
- Do not sow kale, turnips, swedes or grass mixtures containing clover by direct drilling or minimum cultivation techniques within 6 wk of application

Environmental safety
- Dangerous for the environment
- Toxic to aquatic organisms
- Keep livestock out of treated areas for at least 7 d following treatment and until poisonous weeds, such as ragwort, have died down and become unpalatable
- Do not allow drift to come into contact with crops, amenity plantings, gardens, ponds, lakes or watercourses
- Wash spray equipment thoroughly with water and detergent immediately after use. Traces of product can damage susceptible plants sprayed later

Hazard classification and safety precautions
Hazard Harmful, Flammable, Dangerous for the environment
Transport code 3
Packaging group III
UN number 1993
Risk phrases R22a, R22b, R37, R38, R43, R51, R53a, R67
Operator protection A, C; U02a, U05a, U08, U14, U19a, U20b
Environmental protection E07a, E15a, E34, E38
Consumer protection C01
Storage and disposal D01, D02, D05, D09a, D10b, D12a
Medical advice M05b

264 flurtamone

A carotenoid synthesis inhibitor available only in mixtures
HRAC mode of action code: F1

See also diflufenican + flufenacet + flurtamone
diflufenican + flurtamone

265 flusilazole

A systemic, protective and curative triazole fungicide for cereals, oilseed rape and sugar beet
FRAC mode of action code: 3

See also carbendazim + flusilazole
chlorothalonil + flusilazole
famoxadone + flusilazole
fenpropimorph + flusilazole

Products

1	Capitan 25	DuPont	250 g/l	EW	13067
2	Genie 25	DuPont	250 g/l	EW	13065
3	Lyric	DuPont	250 g/l	EW	13022
4	Nustar 25	DuPont	250 g/l	EW	13921
5	Punch 25	DuPont	250 g/l	EW	14497
6	Sanction 25	DuPont	250 g/l	EW	13066

Uses
- Brown rust in *spring barley, winter barley, winter wheat* [1-6]

SEE SECTION 3 FOR PRODUCTS ALSO REGISTERED

Pesticide Profiles

- Canker in **brussels sprouts** *(off-label)*, **swedes** *(off-label)*, **turnips** *(off-label)* [1-3, 6]
- Disease control in **fodder beet** *(off-label)* [1]
- Eyespot in **forage maize** *(off-label)* [1-3, 6]; **grain maize** *(off-label)* [1]; **spring barley, winter barley, winter wheat** [1-6]
- Foliar disease control in **brussels sprouts** *(off-label)*, **fodder beet** *(off-label)*, **forage maize** *(off-label)*, **swedes** *(off-label)*, **turnips** *(off-label)* [4]; **grain maize** *(off-label)* [3, 4]
- Glume blotch in **spring barley, winter barley** [1-3, 6]; **winter wheat** [1-4, 6]
- Light leaf spot in **spring oilseed rape, winter oilseed rape** [1-6]
- Mildew in **spring barley** *(moderate control)*, **winter barley** *(moderate control)*, **winter wheat** [5]
- Net blotch in **spring barley, winter barley** [1-6]
- Powdery mildew in **spring barley, winter barley** [1-3, 6]; **spring barley** *(moderate control)*, **winter barley** *(moderate control)* [4]; **sugar beet** [1-6]; **winter wheat** [1-4, 6]
- Rhynchosporium in **spring barley, winter barley** [1-3, 6]; **spring barley** *(moderate control)*, **winter barley** *(moderate control)* [4, 5]
- Rust in **sugar beet** [1-6]
- Septoria leaf blotch in **spring barley, winter barley** [1-3, 6]; **winter wheat** [1-6]
- Yellow rust in **spring barley, winter barley, winter wheat** [1-6]

Specific Off-Label Approvals (SOLAs)
- **brussels sprouts** *20070102* [1], *20070101* [2], *20063998* [3], *20081104* [4], *20070100* [6]
- **fodder beet** *20101760* [1], *20091618* [4]
- **forage maize** *20080197* [1], *20101759* [1], *20080198* [2], *20080178* [3], *20081102* [4], *20080196* [6]
- **grain maize** *20101759* [1], *20081044* [3], *20081102* [4]
- **swedes** *20073284* [1], *20073283* [2], *20073270* [3], *20081103* [4], *20073285* [6]
- **turnips** *20073284* [1], *20073283* [2], *20073270* [3], *20081103* [4], *20073285* [6]

Approval information
- Flusilazole has been reinstated in Annex 1 under Directive 91/414 following an appeal
- Accepted by BBPA for use on malting barley

Efficacy guidance
- Best results obtained from treatment at onset of disease development
- Best control of eyespot in cereals achieved by spraying between leaf-sheath erect and second node detectable stages (GS 30-32)
- Product active against both MBC-sensitive and MBC-resistant strains of eyespot
- Rain occurring within 2 h after application may reduce effectiveness
- See label for recommended tank-mixes to give broader spectrum control
- Flusilazole is a DMI fungicide. Resistance to some DMI fungicides has been identified in Septoria leaf blotch which may seriously affect performance of some products. For further advice contact a specialist advisor and visit the Fungicide Resistance Action Group (FRAG)-UK website

Restrictions
- Maximum number of treatments (including other products containing flusilazole) on cereals and oilseed rape depends on dose and timing - see labels for details. High rate must not be used more than once in any crop
- Maximum total dose on sugar beet equivalent to one full dose
- Do not apply to crops under stress or during frosty weather

Crop-specific information
- Latest use: high dose before 3rd node detectable (GS 33) plus reduced dose before early milk stage (GS 72) on winter wheat and barley; before first flowers open (GS 4,0) on oilseed rape
- HI 7 wk for sugar beet
- Treat oilseed rape in autumn and/or spring at stem extension stage
- On sugar beet apply at an early stage of disease development, usually in early Aug

Environmental safety
- Dangerous for the environment
- Toxic to aquatic organisms
- Harmful to fish or other aquatic life. Do not contaminate surface waters or ditches with chemical or used container

Hazard classification and safety precautions
Hazard Toxic, Dangerous for the environment

FOR FULL CONDITIONS OF USE ALWAYS READ THE PRODUCT LABEL

Transport code 9
Packaging group III
UN number 3082
Risk phrases R22a, R38, R40, R51, R53a, R61
Operator protection A, C [1-6]; H [4]; M [4, 5]; P [5]; U05a, U11, U19a [1-6]; U16a, U23a [4]; U19c [5]; U20a [1-3, 6]; U20b [4, 5]
Environmental protection E13c, E34, E38
Storage and disposal D01, D02, D10b, D12a [1-6]; D05 [1-3, 5, 6]; D09a [4, 5]
Medical advice M04a

266 flutolanil

An carboxamide fungicide for treatment of potato seed tubers
FRAC mode of action code: 7

Products

1 Rhino	Certis	460 g/l	SC	14311
2 Rhino DS	Certis	6% w/w	DS	12763

Uses
- Black scurf in *potatoes* (tuber treatment)
- Stem canker in *potatoes* (tuber treatment)

Approval information
- Flutolanil included in Annex I under EC Directive 91/414

Efficacy guidance
- Apply to clean tubers before chitting, prior to planting, or at planting
- Apply flowable concentrate through canopied, hydraulic or spinning disc equipment (with or without electrostatics) mounted on a rolling conveyor or table [1]
- Flowable concentrate may be diluted with water up to 2.0 l per tonne to improve tuber coverage. Disease in areas not covered by spray will not be controlled [1]
- Dry powder may be applied at hopper filling, at box filling, to the box surface, to chitting trays or via an on-planter applicator [2]

Restrictions
- Maximum number of treatments 1 per batch of seed tubers
- Check with processor before use on crops for processing
- Treated tubers must be planted within 3 mth [2]

Crop-specific information
- Latest use: at planting
- Seed tubers should be of good quality and free from bacterial rots, physical damage or virus infection, and should not be sprouted to such an extent that mechanical damage to the shoots will occur during treatment or planting

Environmental safety
- Harmful to aquatic organisms

Hazard classification and safety precautions
Hazard Irritant
Transport code 9 [1]
Packaging group III [1]
UN number 3082 [1]; N/C [2]
Risk phrases R36, R52, R53a [2]; R43, R51 [1]
Operator protection A, H [1, 2]; C [2]; U04a, U05a, U19a, U20a [1, 2]; U14 [1]
Environmental protection E15a, E34, E38
Storage and disposal D01, D02, D09a, D12a [1, 2]; D05, D10a [1]; D11a [2]
Treated seed S01, S03, S04a, S05 [1, 2]; S02 [2]
Medical advice M03

SEE SECTION 3 FOR PRODUCTS ALSO REGISTERED

267 flutriafol

A broad-spectrum triazole fungicide for cereals
FRAC mode of action code: 3

See also chlorothalonil + flutriafol
fludioxonil + flutriafol

Products

1	Consul	Headland	125 g/l	SC	12976
2	Pointer	Headland	125 g/l	SC	12975

Uses
- Brown rust in *spring barley, winter barley, winter wheat*
- Powdery mildew in *spring barley, winter barley, winter wheat*
- Rhynchosporium in *spring barley, winter barley*
- Septoria leaf blotch in *winter wheat*
- Yellow rust in *spring barley, winter barley, winter wheat*

Approval information
- Accepted by BBPA for use on malting barley

Efficacy guidance
- Best results obtained from treatment in early stages of disease development. See label for detailed guidance on spray timing for specific diseases and the need for repeat treatments
- Tank mix options available on the label to broaden activity spectrum
- Good spray coverage essential for optimum performance
- Flutriafol is a DMI fungicide. Resistance to some DMI fungicides has been identified in Septoria leaf blotch which may seriously affect performance of some products. For further advice contact a specialist advisor and visit the Fungicide Resistance Action Group (FRAG)-UK website

Restrictions
- Maximum number of treatments 2 per crop (including other products containing flutriafol)

Crop-specific information
- Latest use: before early grain milky ripe stage (GS 73)
- Flag leaf tip scorch on wheat caused by stress may be increased by fungicide treatment

Environmental safety
- Harmful to aquatic organisms
- Harmful to fish or other aquatic life. Do not contaminate surface waters or ditches with chemical or used container

Hazard classification and safety precautions
Hazard Harmful
UN number N/C
Risk phrases R43 [1]; R48, R52 [1, 2]
Operator protection A, H; U05a, U09a, U19a, U20b
Environmental protection E13c, E38
Storage and disposal D01, D02, D09a, D10c

268 fosetyl-aluminium

A systemic phosphonic acid fungicide for various horticultural crops
FRAC mode of action code: 33

Products

1	Aliette 80 WG	Certis	80% w/w	WG	13130
2	Fosal	AgriGuard	80% w/w	WG	13384
3	Standon Fullstop	Standon	80% w/w	WG	13571

Uses
- Collar rot in *apples* [1-3]
- Crown rot in *apples* [1-3]; *protected strawberries* (off-label), *strawberries, strawberries (off-label)* [1]

FOR FULL CONDITIONS OF USE ALWAYS READ THE PRODUCT LABEL

fosetyl-aluminium

- Downy mildew in **beet leaves** *(off-label)*, **chard** *(off-label)*, **chives** *(off-label)*, **combining peas** *(off-label - seed treatment)*, **endives** *(off-label)*, **frise** *(off-label)*, **herbs (see appendix 6)** *(off-label)*, **lamb's lettuce** *(off-label)*, **lettuce** *(off-label)*, **lupins** *(off-label)*, **parsley** *(off-label)*, **protected brassica seedlings** *(off-label)*, **protected chives** *(off-label)*, **protected endives** *(off-label)*, **protected frise**, **protected herbs (see appendix 6)** *(off-label)*, **protected lamb's lettuce** *(off-label)*, **protected parsley** *(off-label)*, **protected radicchio** *(off-label)*, **protected salad brassicas** *(off-label - for baby leaf production)*, **protected scarole** *(off-label)*, **protected spinach** *(off-label)*, **radicchio** *(off-label)*, **salad brassicas** *(off-label - for baby leaf production)*, **salad onions** *(off-label)*, **scarole** *(off-label)*, **spinach** *(off-label)*, **spinach beet** *(off-label)*, **tatsoi** *(off-label)*, **vining peas** *(off-label - seed treatment)*, **watercress** *(off-label - during propagation)*, **wine grapes** *(off-label)* [1]; **broad beans**, **hops**, **protected lettuce** [1-3]; **broccoli** *(off-label)*, **brussels sprouts** *(off-label)*, **cabbages** *(off-label)*, **calabrese** *(off-label)*, **cauliflowers** *(off-label)*, **chinese cabbage** *(off-label)*, **choi sum** *(off-label)*, **collards** *(off-label)*, **kale** *(off-label)*, **pak choi** *(off-label)* [1, 3]; **protected broccoli** *(off-label)*, **protected brussels sprouts** *(off-label)*, **protected cabbages** *(off-label)*, **protected calabrese** *(off-label)*, **protected cauliflowers** *(off-label)*, **protected chinese cabbage** *(off-label)*, **protected collards** *(off-label)*, **protected kale** *(off-label)*, **protected pak choi** *(off-label)*, **watercress** *(off-label)* [3]
- Foliar disease control in **amenity grassland** *(off-label)*, **beet leaves** *(off-label)*, **chard** *(off-label)*, **chicory** *(off-label)*, **endives** *(off-label)*, **frise** *(off-label)*, **herbs (see appendix 6)** *(off-label)*, **lamb's lettuce** *(off-label)*, **managed amenity turf** *(off-label)*, **outdoor grapes** *(off-label)*, **protected endives** *(off-label)*, **protected frise** *(off-label)*, **protected lamb's lettuce** *(off-label)*, **protected radicchio** *(off-label)*, **protected salad brassicas** *(off-label)*, **protected scarole** *(off-label)*, **protected spinach** *(off-label)*, **protected strawberries** *(off-label)*, **radicchio** *(off-label)*, **salad brassicas** *(off-label)*, **salad onions** *(off-label)*, **scarole** *(off-label)*, **spinach** *(off-label)*, **spinach beet** *(off-label)* [3]; **strawberries** [2, 3]
- Phytophthora in **chicory** *(off-label - for forcing)*, **watercress** *(off-label - during propagation)* [1]
- Phytophthora root rot in **capillary benches** [1]; **protected pot plants** [1-3]
- Phytophthora stem rot in **capillary benches** [1]; **protected pot plants** [1-3]
- Phytophthora wilt in **ornamental plant production** [1-3]
- Pythium in **managed amenity turf** *(off-label)*, **watercress** *(off-label - during propagation)* [1]
- Red core in **protected strawberries** *(off-label)*, **strawberries** *(off-label)* [1]; **strawberries** [1-3]
- Root rot in **capillary benches** [1]; **protected pot plants** [1-3]
- Seed-borne diseases in **combining peas** *(off-label)*, **lupins** *(off-label)*, **protected herbs (see appendix 6)** *(off-label)*, **vining peas** *(off-label)* [3]; **protected herbs (see appendix 6)** [2, 3]

Specific Off-Label Approvals (SOLAs)
- *amenity grassland* 20080270 [3]
- *beet leaves* 20063520 [1], 20080267 [3]
- *broccoli* 20063524 [1], 20080271 [3]
- *brussels sprouts* 20063524 [1], 20080271 [3]
- *cabbages* 20063524 [1], 20080271 [3]
- *calabrese* 20063524 [1], 20080271 [3]
- *cauliflowers* 20063524 [1], 20080271 [3]
- *chard* 20063520 [1], 20080267 [3]
- *chicory* (for forcing) 20063521 [1], 20080268 [3]
- *chinese cabbage* 20063524 [1], 20080271 [3]
- *chives* 20063522 [1]
- *choi sum* 20082031 [1], 20080271 [3]
- *collards* 20082031 [1], 20080271 [3]
- *combining peas* (seed treatment) 20063518 [1], 20080265 [3]
- *endives* 20063522 [1], 20080269 [3]
- *frise* 20063522 [1], 20080269 [3]
- *herbs (see appendix 6)* 20063520 [1], 20080267 [3]
- *kale* 20063524 [1], 20080271 [3]
- *lamb's lettuce* 20063522 [1], 20080269 [3]
- *lettuce* 20063522 [1]
- *lupins* 20063518 [1], 20080265 [3]
- *managed amenity turf* 20063523 [1], 20080270 [3]
- *outdoor grapes* 20080266 [3]
- *pak choi* 20082031 [1], 20080271 [3]

SEE SECTION 3 FOR PRODUCTS ALSO REGISTERED

Pesticide Profiles

- *parsley* 20063522 [1]
- *protected brassica seedlings* 20063524 [1]
- *protected broccoli* 20080271 [3]
- *protected brussels sprouts* 20080271 [3]
- *protected cabbages* 20080271 [3]
- *protected calabrese* 20080271 [3]
- *protected cauliflowers* 20080271 [3]
- *protected chinese cabbage* 20080271 [3]
- *protected chives* 20063522 [1]
- *protected collards* 20080271 [3]
- *protected endives* 20063522 [1], 20080269 [3]
- *protected frise* 20080269 [3]
- *protected herbs (see appendix 6)* 20063522 [1], 20080269 [3]
- *protected kale* 20080271 [3]
- *protected lamb's lettuce* 20063522 [1], 20080269 [3]
- *protected pak choi* 20080271 [3]
- *protected parsley* 20063522 [1]
- *protected radicchio* 20063522 [1], 20080269 [3]
- *protected salad brassicas* (for baby leaf production) 20063522 [1], 20080269 [3]
- *protected scarole* 20063522 [1], 20080269 [3]
- *protected spinach* 20063522 [1], 20080269 [3]
- *protected strawberries* 20063517 [1], 20080264 [3]
- *radicchio* 20063522 [1], 20080269 [3]
- *salad brassicas* (for baby leaf production) 20063522 [1], 20080269 [3]
- *salad onions* 20063515 [1], 20080263 [3]
- *scarole* 20063522 [1], 20080269 [3]
- *spinach* 20063520 [1], 20080267 [3]
- *spinach beet* 20063520 [1], 20080267 [3]
- *strawberries* 20063517 [1]
- *tatsoi* 20082031 [1]
- *vining peas* (seed treatment) 20063518 [1], 20080265 [3]
- *watercress* (during propagation) 20071260 [1], 20080272 [3]
- *wine grapes* 20063519 [1]

Approval information
- Fosetyl-aluminium included in Annex I under EC Directive 91/414
- Accepted by BBPA for use on hops

Restrictions
- Maximum number of treatments 1 per seed batch for peas; 1 per batch of compost for lettuce; 1 per yr for strawberries (root dip or foliar spray); 1 per yr for apples (bark paste), 2 per yr for apples (foliar spray); 2 per crop for broad beans, peas; 2 per yr for hops (basal spray); 6 per yr for hops (foliar spray); 1 per crop during propagation, 2 per crop after planting out for lettuce
- Check tolerance of ornamental species before large-scale treatment

Crop-specific information
- Latest use: pre-planting for strawberry runners; pre-sowing for protected lettuce
- HI apples 5 mth (bark paste) or 4 wk (spray); broad beans 17 d; hops 14 d
- Spray young orchards for crown rot protection after blossom when first leaves fully open and repeat after 4-6 wk. Apply as paste to bark of apples to control collar rot
- Apply to broad beans when infection appears (usually at flowering) and 14 d later. Consult before treating crops to be processed
- Use on strawberries only between harvest and 31 Dec
- Spray autumn-planted strawberry runners 2-3 wk after planting or use dip treatment at planting. Spray established crops in late summer/early autumn after picking and repeat annually
- Apply to hops as early season basal spray or as foliar spray every 10-14 d from when training is completed
- Use by wet incorporation in blocking compost for protected lettuce only from Sep to Apr and follow all directions carefully to avoid severe crop injury. Crop maturity may be delayed by a few days

FOR FULL CONDITIONS OF USE ALWAYS READ THE PRODUCT LABEL

fosetyl-aluminium + propamocarb hydrochloride

- Apply as drench to rooted cuttings of hardy nursery stock after first potting and repeat mthly. Up to 6 applications may be needed
- See label for details of application to capillary benches and trays of young plants.

Environmental safety
- Harmful to aquatic organisms

Hazard classification and safety precautions
  Hazard Irritant [3]
  UN number N/C
  Risk phrases R36, R52 [3]
  Operator protection A, C, H; U11 [3]; U20c [1-3]
  Environmental protection E15a
  Storage and disposal D01, D02, D12a [3]; D09a, D11a [1-3]

269 fosetyl-aluminium + propamocarb hydrochloride

A systemic and protectant fungicide mixture for use in horticulture
FRAC mode of action code: 33 + 28

See also propamocarb hydrochloride

Products
  Previcur Energy   Bayer CropScience   310:530 g/l   SL   13342

Uses
- Damping off in *chard* (off-label), *herbs (see appendix 6)* (off-label), *leaf brassicas* (off-label - baby leaf production), *ornamental plant production* (off-label), *protected aubergines* (off-label), *protected cabbages* (off-label), *protected cayenne peppers* (off-label), *protected courgettes* (off-label), *protected gherkins* (off-label), *protected herbs (see appendix 6)* (off-label), *protected marrows* (off-label), *protected melons* (off-label), *protected peppers* (off-label), *protected pumpkins* (off-label), *protected salad brassicas* (off-label - baby leaf production), *protected squashes* (off-label), *protected watermelon* (off-label), *spinach beet* (off-label)
- Downy mildew in *chard* (off-label), *frise* (off-label), *herbs (see appendix 6)* (off-label), *leaf brassicas* (off-label - baby leaf production), *protected chicory* (off-label), *protected herbs (see appendix 6)* (off-label), *protected lettuce*, *protected salad brassicas* (off-label - baby leaf production), *scarole* (off-label), *spinach*, *spinach beet* (off-label)
- Pythium in *protected cucumbers*, *protected lettuce*, *protected tomatoes*

Specific Off-Label Approvals (SOLAs)
- *chard 20082648*
- *frise 20082654*
- *herbs (see appendix 6) 20082668*
- *leaf brassicas* (baby leaf production) *20082668*
- *ornamental plant production 20082667*
- *protected aubergines 20082653*
- *protected cabbages 20082650*
- *protected cayenne peppers 20082652*
- *protected chicory 20082649*
- *protected courgettes 20082653*
- *protected gherkins 20082653*
- *protected herbs (see appendix 6) 20082655*
- *protected marrows 20082651*
- *protected melons 20082651*
- *protected peppers 20082652*
- *protected pumpkins 20082651*
- *protected salad brassicas* (baby leaf production) *20082655*
- *protected squashes 20082651*
- *protected watermelon 20082651*
- *scarole 20082654*
- *spinach beet 20082648*

SEE SECTION 3 FOR PRODUCTS ALSO REGISTERED

Approval information
- Fosetyl-aluminium and propamocarb hydrochloride included in Annex I under EC Directive 91/414

Hazard classification and safety precautions
Hazard Irritant
UN number N/C
Risk phrases R43
Operator protection A, H, M; U04a, U05a, U08, U14, U19a, U20b
Environmental protection E15a
Storage and disposal D09a, D11a
Medical advice M03

270 fosthiazate

An organophosphorus contact nematicide for potatoes
IRAC mode of action code: 1B

Products
Nemathorin 10G   Syngenta   10% w/w   FG   11003

Uses
- Nematodes in *hops* (off-label), *ornamental plant production* (off-label), *soft fruit* (off-label)
- Potato cyst nematode in *potatoes*
- Spraing vectors in *potatoes* (reduction)
- Wireworm in *potatoes* (reduction)

Specific Off-Label Approvals (SOLAs)
- *hops* 20082912
- *ornamental plant production* 20082912
- *soft fruit* 20082912

Approval information
- Fosthiazate included in Annex I under EC Directive 91/414
- In 2006 CRD required that all products containing this active ingredient should carry the following warning in the main area of the container label: "Fosthiazate is an anticholinesterase organophosphate. Handle with care"

Efficacy guidance
- Treatment may be in furrow or applied and incorporated in one operation to a uniform depth of 10-15 cm. Deeper incorporation will reduce control
- Application best achieved using equipment such as Horstine Farmery Microband Applicator, Matco or Stocks Micrometer applicators together with a rear mounted powered rotary cultivator
- Granules must not become wet or damp before use

Restrictions
- Contains an anticholinesterase organophosphorus compound. Do not use if under medical advice not to work with such compounds
- Maximum number of treatments 1 per crop
- Product must only be applied using tractor-mounted/drawn direct placement machinery. Do not use air assisted broadcast machinery other than that referenced on the product label
- Do not allow granules to stand overnight in the application hopper
- Do not apply more than once every four years on the same area of land
- Do not use on crops to be harvested less than 17 wk after treatment
- Consult before using on crops intended for processing

Crop-specific information
- Latest use: at planting
- HI 17 wk for potatoes

Environmental safety
- Dangerous for the environment
- Toxic to aquatic organisms
- Dangerous to game, wild birds and animals
- Dangerous to livestock. Keep all livestock out of treated areas for at least 13 wk

FOR FULL CONDITIONS OF USE ALWAYS READ THE PRODUCT LABEL

fuberidazole

- Incorporation to 10-15 cm and ridging up of treated soil must be carried out immediately after application. Powered rotary cultivators are preferred implements for incorporation but discs, power, spring tine or Dutch harrows may be used provided two passes are made at right angles
- To protect birds and wild mammals remove spillages
- Failure completely to bury granules immediately after application is hazardous to wildlife
- To protect groundwater do not apply any product containing fosthiazate more than once every four yr

Hazard classification and safety precautions
   Hazard Harmful, Dangerous for the environment
   Transport code 9
   Packaging group III
   UN number 3077
   Risk phrases R22a, R43, R51, R53a
   Operator protection A, E, G, H, K, M; U02a, U04a, U05a, U09a, U13, U14, U19a, U20a
   Environmental protection E06b (13 wk); E15b, E34, E38
   Storage and disposal D01, D02, D05, D09a, D11a, D14
   Medical advice M01, M03, M05a

271 fuberidazole

A benzimidazole (MBC) fungicide available only in mixtures
FRAC mode of action code: 1

See also bitertanol + fuberidazole
            bitertanol + fuberidazole + imidacloprid

272 fuberidazole + imidacloprid + triadimenol

A broad spectrum systemic fungicide and insecticide seed treatment for winter cereals
FRAC mode of action code: 1 + IRAC 4A + FRAC 3

See also imidacloprid
            triadimenol

Products
   Tripod Plus   Makhteshim   15:117:125 g/l   LS   13168

Uses
- Blue mould in **winter wheat** *(seed treatment)*
- Brown foot rot in **winter barley** *(seed treatment)*
- Brown rust in **winter barley** *(seed treatment)*, **winter wheat** *(seed treatment)*
- Bunt in **winter wheat** *(seed treatment)*
- Covered smut in **winter barley** *(seed treatment)*
- Fusarium foot rot and seedling blight in **winter barley** *(seed treatment - reduction)*, **winter oats** *(seed treatment - reduction)*, **winter wheat** *(seed treatment - reduction)*
- Loose smut in **winter barley** *(seed treatment)*, **winter oats** *(seed treatment)*, **winter wheat** *(seed treatment)*
- Net blotch in **winter barley** *(seed treatment - seed-borne only)*
- Powdery mildew in **winter barley** *(seed treatment)*, **winter oats** *(seed treatment)*, **winter wheat** *(seed treatment)*
- Pyrenophora leaf spot in **winter oats** *(seed treatment)*
- Septoria leaf blotch in **winter wheat** *(seed treatment)*
- Septoria seedling blight in **winter wheat** *(seed treatment - reduction)*
- Snow rot in **winter barley** *(seed treatment - reduction)*
- Virus vectors in **winter barley** *(seed treatment)*, **winter oats** *(seed treatment)*, **winter wheat** *(seed treatment)*
- Wireworm in **winter barley** *(reduction of damage)*, **winter oats** *(reduction of damage)*, **winter wheat** *(reduction of damage)*
- Yellow rust in **winter barley** *(seed treatment)*, **winter wheat** *(seed treatment)*

SEE SECTION 3 FOR PRODUCTS ALSO REGISTERED

Approval information
- Fuberidazole, imidacloprid and triadimenol included in Annex I under EC Directive 91/414
- Accepted by BBPA for use on malting barley

Efficacy guidance
- Apply through recommended seed treatment machinery
- Evenness of seed cover improved by simultaneous application of equal volumes of product and water
- Calibrate drill for treated seed and drill at 4 cm into firm, well prepared seedbed
- Use minimum 125 kg treated seed per ha and increase seed rate as drilling season progresses. Lower drilling rates and/or early drilling affect duration of BYDV protection needed and may require follow-up aphicide treatment
- When aphid activity unusually late, or is heavy and prolonged in areas of high risk, and mild weather predominates, follow-up treatment may be required
- In addition to seed-borne diseases early attacks of various foliar, air-borne diseases are controlled or suppressed. See label for details

Restrictions
- Maximum number of treatments 1 per batch of seed
- Do not use on naked oats
- Do not drill treated winter wheat seed after end of Nov
- Do not handle seed unnecessarily
- Do not use treated seed as food or feed
- Treated seed should not be broadcast, but drilled to a depth of 4 cm in a well prepared seedbed
- Germination tests should be done on all batches of seed to be treated to ensure seed viability and suitability for treatment
- Do not use on seed with moisture content above 16%, on sprouted, cracked or skinned seed or on seed already treated with another seed treatment

Crop-specific information
- Latest use: before drilling
- Store treated seed in cool, dry, well-ventilated store and drill as soon as possible, preferably in season of purchase
- Treatment may accentuate effects of adverse seedbed conditions on crop emergence

Environmental safety
- Dangerous to fish or other aquatic life. Do not contaminate surface waters or ditches with chemical or used container
- Dangerous to birds, game and other wildlife. Treated seed should not be left on the soil surface. Bury spillages
- Seed should be drilled to a depth of 4 cm into a well-prepared seed bed
- If seed is present on the soil surface, or if spills have occurred, the field should be harrowed and rolled if conditions are appropriate to ensure good incorporation

Hazard classification and safety precautions
Transport code 9
Packaging group III
UN number 3082
Risk phrases R52, R53a
Operator protection A, H; U04a, U05a, U07, U13, U14, U20b
Environmental protection E03, E36a, E38
Storage and disposal D01, D02, D05, D09a, D12a
Treated seed S01, S02, S03, S04c, S05, S06a, S07, S08

273 fuberidazole + triadimenol

A broad spectrum systemic fungicide seed treatment for cereals
FRAC mode of action code: 1 + 3

See also triadimenol

Products
Tripod   Makhteshim   22.5:187.5 g/l   FS   13014

FOR FULL CONDITIONS OF USE ALWAYS READ THE PRODUCT LABEL

fuberidazole + triadimenol

Uses
- Blue mould in **spring wheat** *(seed treatment)*, **triticale** *(seed treatment)*, **winter wheat** *(seed treatment)*
- Brown foot rot in **spring barley** *(seed treatment)*, **spring oats** *(seed treatment)*, **spring rye** *(seed treatment)*, **spring wheat** *(seed treatment)*, **triticale** *(seed treatment)*, **winter barley** *(seed treatment)*, **winter oats** *(seed treatment)*, **winter rye** *(seed treatment)*, **winter wheat** *(seed treatment)*
- Bunt in **spring wheat** *(seed treatment)*, **winter wheat** *(seed treatment)*
- Covered smut in **spring barley** *(seed treatment)*, **winter barley** *(seed treatment)*
- Leaf stripe in **spring barley** *(seed treatment)*
- Loose smut in **spring barley** *(seed treatment)*, **spring oats** *(seed treatment)*, **spring wheat** *(seed treatment)*, **winter barley** *(seed treatment)*, **winter oats** *(seed treatment)*, **winter wheat** *(seed treatment)*
- Pyrenophora leaf spot in **spring oats** *(seed treatment)*, **winter oats** *(seed treatment)*

Approval information
- Fuberidazole and triadimenol included in Annex I under EC Directive 91/414
- Accepted by BBPA for use on malting barley

Efficacy guidance
- Apply through recommended seed treatment machinery
- Calibrate drill for treated seed and drill at 2.5-4 cm into firm, well prepared seedbed
- Product will give control of early attacks of foliar air-borne diseases such as powdery mildew, yellow rust, brown rust, crown rust, Septoria. On autumn-sown crops control of rust is prolonged and usually extends over winter, leading to a delayed build-up of the disease in the following spring, often resulting in fewer fungicide sprays being required to maintain control during spring/summer
- Product will often delay and reduce the likelihood and extent of lodging in autumn sown wheat.

Restrictions
- Maximum number of treatments 1 per batch of seed
- Do not use on naked oats
- Do not drill treated winter wheat or rye seed after end of Nov. Seed rate should be increased as drilling season progresses
- Germination tests should be done on all batches of seed to be treated to ensure seed viability and suitability for treatment
- Do not use on seed with moisture content above 16%, on sprouted, cracked or skinned seed or on seed already treated with another seed treatment
- Do not use treated seed as food or feed

Crop-specific information
- Latest use: before drilling
- Treated spring wheat may be drilled in autumn up to end of Nov or from Feb onwards
- Store treated seed in cool, dry, well-ventilated store and drill as soon as possible, preferably in season of purchase
- Treatment may accentuate effects of adverse seedbed conditions on crop emergence

Environmental safety
- Dangerous for the environment
- Harmful to aquatic organisms

Hazard classification and safety precautions
Hazard Dangerous for the environment
Transport code 9
Packaging group III
UN number 3082
Risk phrases R52, R53a
Operator protection A; U20b
Environmental protection E03, E15a, E34
Storage and disposal D05, D09a, D11a
Treated seed S01, S02, S03, S04a, S05, S06a, S07

SEE SECTION 3 FOR PRODUCTS ALSO REGISTERED

274 gibberellins

A plant growth regulator for use in top fruit

Products

1	Berelex	Interfarm	1 g per tablet	TB	14123
2	GIBB 3	Globachem	10% w/w	TB	12290
3	GIBB Plus	Globachem	10 g/l	SL	11875
4	Gistar	Belcrop	10 g/l	SL	11397
5	Novagib	Fine	10 g/l	SL	08954
6	Regulex 10 SG	Interfarm	10% w/w	SG	14116

Uses

- Growth regulation in **crab apples** (off-label), **ornamental crop production (seed)** (off-label), **quinces** (off-label) [6]; **farm forestry** (off-label), **forest** (off-label) [3, 6]; **pears** (off-label) [5, 6]
- Improved germination in **nothofagus** (do not exceed 2.5 g per 5 litres when used as a seed soak) [6]
- Increasing fruit set in **pears** [1, 2]
- Increasing germination in **nothofagus** (qualified minor use) [3, 4]
- Increasing yield in **celery (outdoor)** [1, 2]; **cherries** (off-label), **protected cherries** (off-label), **rhubarb** [1]; **protected rhubarb** [2]
- Reducing fruit russeting in **apples** [3-5]; **pears** (off-label), **quinces** (off-label) [3]
- Russet reduction in **apples** [6]

Specific Off-Label Approvals (SOLAs)

- **cherries** 20091072 [1]
- **crab apples** 20091078 [6]
- **farm forestry** 20071675 [3], 20091076 [6]
- **forest** 20071675 [3], 20091076 [6]
- **ornamental crop production (seed)** 20091077 [6]
- **pears** 20061627 [3], 20012756 [5], 20091078 [6]
- **protected cherries** 20091072 [1]
- **quinces** 20061627 [3], 20091078 [6]

Approval information

- Gibberellins (GA4, GA7) included in Annex 1 under EC Directive 91/414
- Approval expiry 31 Aug 2011 [4]

Efficacy guidance

- For optimum results on apples spray under humid, slow drying conditions and ensure good spray cover [3-5]
- Treat apples and pears immediately after completion of petal fall and repeat as directed on the label [3-5]
- Nothofagus seeds should be stirred into the solution, soaked for 24 hr and then sown immediately [3, 4]
- Fruit set in pears can be improved when blossom is spare, setting is poor or where frost has killed many flowers [2]

Restrictions

- Maximum total dose varies with crop and product. Check label
- Prepared spray solutions are unstable. Do not leave in the sprayer during meal breaks or overnight
- Return bloom may be reduced in yr following treatment
- Consult before treating apple crops grown for processing [3-5]
- Avoid storage at temperatures above 32 °C [2]

Crop-specific information

- HI zero for apples
- Apply to apples at completion of petal fall and repeat 3 or 4 times at 7-10 d intervals. Number of sprays and spray interval depend on weather conditions and dose (see labels) [3, 5]
- Good results achieved on apple varieties Cox's Orange Pippin, Discovery, Golden Delicious and Karmijn. For other cultivars test on a small number of trees [3, 5]
- Split treatment on pears allows second spray to be omitted if pollinating conditions become very favourable [2]
- Pear variety Conference usually responds well to treatment; Doyenne du Comice can be variable [2]

FOR FULL CONDITIONS OF USE ALWAYS READ THE PRODUCT LABEL

glufosinate-ammonium

- On celery the interval between application and harvest should never exceed 4 wk to avoid risk of premature seeding and root splitting [2]
- Treat washed rhubarb crowns on transfer to forcing shed [2]

Hazard classification and safety precautions
UN number N/C
Operator protection U05a [6]; U08 [2]; U09a [1]; U20b [5]; U20c [1-4, 6]
Environmental protection E15a
Storage and disposal D01 [5, 6]; D02 [6]; D03, D05, D07 [5]; D09a [1-6]; D10b [3, 5, 6]; D10c [4]

275   glufosinate-ammonium

A non-selective, non-residual phosphinic acid contact herbicide
HRAC mode of action code: H

Products

1	Finale	Bayer Environ.	120 g/l	SL	10092
2	Harvest	Bayer CropScience	150 g/l	SL	07321
3	Kaspar	Certis	150 g/l	SL	11214
4	Kibosh	AgChem Access	150 g/l	SL	13976

Uses
- Annual and perennial weeds in *all tree nuts, apple orchards, bilberries, blueberries, cane fruit, cherries, cranberries, cultivated land/soil, currants, damsons, fallows, forest, non-crop farm areas, pear orchards, plums, potatoes* (not for seed), *strawberries, stubbles, sugar beet, vegetables, wine grapes* [2]
- Annual dicotyledons in *all tree nuts, apple orchards, cane fruit, cherries, cultivated land/soil, currants, damsons, fallows, forest, headlands* (uncropped), *non-crop farm areas, pear orchards, plums, strawberries, sugar beet, vegetables, ware potatoes, wine grapes* [3, 4]; *nursery stock, woody ornamentals* [1]
- Annual grasses in *all tree nuts, apple orchards, cane fruit, cherries, cultivated land/soil, currants, damsons, fallows, forest, headlands* (uncropped), *non-crop farm areas, pear orchards, plums, strawberries, sugar beet, vegetables, ware potatoes, wine grapes* [3, 4]; *nursery stock, woody ornamentals* [1]
- Green cover in *land temporarily removed from production* [2-4]
- Harvest management/desiccation in *combining peas, ware potatoes* [3, 4]; *combining peas* (not for seed), *potatoes* [2]; *linseed, spring field beans, spring oilseed rape, winter field beans, winter oilseed rape* [2-4]
- Line marking preparation in *managed amenity turf* [1]
- Perennial dicotyledons in *all tree nuts, apple orchards, cane fruit, cherries, currants, damsons, fallows, forest, headlands* (uncropped), *land temporarily removed from production, non-crop farm areas, pear orchards, plums, strawberries, wine grapes* [3, 4]; *nursery stock, woody ornamentals* [1]
- Perennial grasses in *all tree nuts, apple orchards, cane fruit, cherries, currants, damsons, fallows, forest, headlands* (uncropped), *land temporarily removed from production, non-crop farm areas, pear orchards, plums, strawberries, wine grapes* [3, 4]; *nursery stock, woody ornamentals* [1]
- Sward destruction in *permanent grassland* [2-4]

Approval information
- Glufosinate-ammonium included in Annex I under EC Directive 91/414
- Accepted by BBPA for use on malting barley and hops

Efficacy guidance
- Activity quickest under warm, moist conditions. Light rainfall 3-4 h after application will not affect activity. Do not spray wet foliage or if rain likely within 6 h
- For weed control uses treat when weeds growing actively. Deep rooted weeds may require second treatment
- On uncropped headlands apply in May/Jun to prevent weeds invading field
- Glufosinate kills all green tissue but does not harm mature bark

SEE SECTION 3 FOR PRODUCTS ALSO REGISTERED

Restrictions
- Maximum number of treatments 4 per crop for potatoes (including 2 desiccant uses); 2 per crop (including 1 desiccant use) for oilseed rape, dried peas, field beans, linseed, wheat and barley; 3 per yr for fruit and forestry; 2 per yr for strawberries, non-crop land and land temporarily removed from production, round fruit trees, canes or bushes; 1 per crop for sugar beet, vegetables and other crops; 1 per yr for grassland destruction, on cultivated land prior to planting edible crops
- Pre-harvest desiccation sprays should not be used on seed crops of wheat, barley, peas or potatoes but may be used on seed crops of oilseed rape, field beans and linseed [2, 3]
- Do not desiccate potatoes in exceptionally wet weather or in saturated soil. See label for details
- Do not spray potatoes after emergence if grown from small or diseased seed or under very dry conditions
- Application for weed control and line-marking uses in sports turf must be between 1 Mar and 30 Sep [1]
- Do not spray hedge bottoms or on broadcast crops [2]

Crop-specific information
- Latest use: 30 Sep for use on non-crop land, top fruit, soft fruit, cane fruit, bush fruit, forestry; pre-drilling, pre-planting or pre-emergence in sugar beet, vegetables and other crops; before winter dormancy for grassland destruction.
- HI potatoes, oilseed rape, combining peas, field beans, linseed 7 d; wheat, barley 14 d
- Ploughing or other cultivations can follow 4 h after spraying
- Crops can normally be sown/planted immediately after spraying or sprayed post-drilling. On sand, very light or immature peat soils allow at least 3 d before sowing/planting or expected emergence
- For weed control in potatoes apply pre-emergence or up to 10% emergence on earlies and seed crops, up to 40% on maincrop, on plants up to 15 cm high
- In sugar beet and vegetables apply just before crop emergence, using stale seedbed technique
- In top and soft fruit, grapevines and forestry apply between 1 Mar and 30 Sep as directed sprays
- For grass destruction apply before winter dormancy occurs. Heavily grazed fields should show active regrowth. Plough from the day after spraying
- Apply pre-harvest desiccation treatments 10-21 d before harvest (14-21 d for oilseed rape). See label for timing details on individual crops
- For potato haulm desiccation apply to listed varieties (not seed crops) at onset of senescence, 14-21 d before harvest

Environmental safety
- Harmful to fish or other aquatic life. Do not contaminate surface waters or ditches with chemical or used container
- Keep livestock out of treated areas until foliage of any poisonous weeds such as ragwort has died and become unpalatable
- Risk to certain non-target insects or other arthropods [2, 3]
- Treated pea haulm may be fed to livestock from 7 d after spraying, treated grain from 14 d [2]
- Product supplied in small volume returnable container - see label for filling and mixing instructions [2]

Hazard classification and safety precautions
Hazard Harmful [2-4]; Irritant [1]
Transport code 6.1
Packaging group III
UN number 2902
Risk phrases R21, R22a, R41 [2-4]; R36 [1]
Operator protection A [2-4]; B [1]; C, H, M [1-4]; U04a, U11, U13, U14, U15, U19a, U20a [2-4]; U05a, U08 [1-4]; U20b [1]
Environmental protection E07a [1-4]; E13c [1, 3, 4]; E15a, E22c [2]; E22b [3, 4]; E34 [2-4]
Storage and disposal D01, D02, D09a, D10b [1-4]; D05 [1]; D12a [2-4]
Medical advice M03 [2-4]

FOR FULL CONDITIONS OF USE ALWAYS READ THE PRODUCT LABEL

276 glyphosate

A translocated non-residual glycine derivative herbicide
HRAC mode of action code: G

See also diflufenican + glyphosate
         diuron + glyphosate
         flufenacet + glyphosate + metosulam

Products

#	Name	Supplier	Conc.	Form	Reg. No.
1	Amega Duo	Nufarm UK	540 g/l	SL	13358
2	Amega Duo	Nufarm UK	540 g/l	SL	14131
3	Asteroid	Headland	360 g/l	SL	11118
4	Barclay Barbarian	Barclay	360 g/l	SL	12714
5	Barclay Gallup 360	Barclay	360 g/l	SL	14988
6	Barclay Gallup Amenity	Barclay	360 g/l	SL	13250
7	Barclay Gallup Biograde 360	Barclay	360 g/l	SL	12660
8	Barclay Gallup Biograde Amenity	Barclay	360 g/l	SL	12716
9	Barclay Gallup Hi-Aktiv	Barclay	490 g/l	SL	14987
10	Buggy XTG	Sipcam	36% w/w	SG	12951
11	CDA Vanquish Biactive	Bayer Environ.	120 g/l	UL	12586
12	Charger	AgChem Access	360 g/l	SL	14107
13	Charger B	AgChem Access	360 g/l	SL	14269
14	Charger C	AgChem Access	360 g/l	SL	14216
15	Clinic Ace	Nufarm UK	360 g/l	SL	12980
16	Clinic Ace	Nufarm UK	360 g/l	SL	14040
17	Clipper	Makhteshim	360 g/l	SL	14820
18	Credit DST	Nufarm UK	540 g/l	SL	13822
19	Credit DST	Nufarm UK	540 g/l	SL	14066
20	Discman Biograde	Barclay	216 g/l	SL	12856
21	Dow Agrosciences Glyphosate 360	Dow	360 g/l	SL	12720
22	Envision	Headland	450 g/l	SL	10569
23	Etna	AgriChem BV	360 g/l	SL	14674
24	Glyfo TDI	Q-Chem	360 g/l	SL	13940
25	Glyfo TDI	Q-Chem	360 g/l	SL	14743
26	Glyfos	Headland	360 g/l	SL	10995
27	Glyfos Dakar	Headland	68% w/w	SG	13054
28	Glyfos Dakar Pro	Headland Amenity	68% w/w	SG	13147
29	Glyfos Proactive	Nomix Enviro	360 g/l	SL	11976
30	Glyfos Supreme	Headland	450 g/l	SL	12371
31	Glyfosat 36	Goldengrass	360 g/l	SL	14297
32	Glyphogan	Makhteshim	360 g/l	SL	12668
33	Glypho-Rapid 450	Barclay	450 g/l	SL	13882
34	Greenaway Gly-490	Greenaway	490 g/l	SL	12718
35	Hilite	Nomix Enviro	144 g/l	RH	13871
36	Kernel	Headland	480 g/l	SL	10993
37	Manifest	Headland	360 g/l	SL	11041
38	Nomix Conqueror	Nomix Enviro	144 g/l	UL	12370
39	Nufosate Ace	Nufarm UK	360 g/l	SL	13794
40	Nufosate Ace	Nufarm UK	360 g/l	SL	14959
41	Proliance Quattro	Syngenta	360 g/l	SL	13670
42	Pure Glyphosate 360	Pure Amenity	360 g/l	SL	14834
43	Reaper	AgChem Access	360 g/l	SL	14924
44	Roundup Ace	Monsanto	450 g/l	SL	12772
45	Roundup Biactive	Monsanto	360 g/l	SL	10320
46	Roundup Energy	Monsanto	450 g/l	SL	12945
47	Roundup Klik	Monsanto	450 g/l	SL	12866
48	Roundup Max	Monsanto	68% w/w	SG	12952
49	Roundup Metro	Monsanto	360 g/l	SL	13989
50	Roundup Pro Biactive	Monsanto	360 g/l	SL	10330
51	Roundup ProBiactive 450	Monsanto	450 g/l	SL	12778
52	Roundup Ultimate	Monsanto	540 g/l	SL	13081
53	Rustler	ChemSource	360 g/l	SL	14951
54	Samurai	Monsanto	360 g/l	SL	12674
55	Shyfo	Sharda	360 g/l	SL	15040
56	Slingshot	ChemSource	360 g/l	SL	14505
57	Stacato	Sipcam	360 g/l	SL	12675

SEE SECTION 3 FOR PRODUCTS ALSO REGISTERED

Pesticide Profiles

Products – continued

58	Stamen	AgriGuard	360 g/l	SL	14895
59	Statis	AgriGuard	480 g/l	SL	13079
60	Statis 360	AgriGuard	360 g/l	SL	14568
61	Stirrup	Nomix Enviro	144 g/l	RH	13875
62	Symbol	United Phosphorus	360 g/l	SL	14769
63	Tangent	Headland Amenity	450 g/l	SL	11872
64	Tanker	Nufarm UK	540 g/l	SL	15016
65	Touchdown Quattro	Syngenta	360 g/l	SL	10608
66	Typhoon 360	Makhteshim	360 g/l	SL	12932

Uses

- Annual and perennial weeds in **all edible crops (outdoor and protected)** [1, 2, 15, 16, 18, 19, 33, 39, 40, 57, 62, 64]; **all edible crops (outdoor)** [24, 25, 49, 53]; **all edible crops (outdoor)** *(before planting)*, **all non-edible crops (outdoor)** *(before planting)* [4, 5, 7, 12, 13, 17, 23, 27, 30, 31, 42, 43, 55, 56, 60, 65]; **all non-edible crops (outdoor)** [1, 2, 15, 16, 18, 19, 24, 25, 33, 39, 40, 49, 53, 57, 62, 64]; **almonds** *(off-label)*, **apples** *(off-label)*, **apricots** *(off-label)*, **crab apples** *(off-label)*, **kentish cobnuts** *(off-label)*, **peaches** *(off-label)*, **pears** *(off-label)*, **quinces** *(off-label)*, **wine grapes** *(off-label)* [45]; **amenity grassland**, **amenity grassland** *(wiper application)*, **amenity vegetation** *(wiper application)*, **broad-leaved trees**, **conifers**, **farm buildings/yards**, **fencelines**, **land clearance**, **non-crop areas**, **paths and drives**, **walls**, **woody ornamentals** [50]; **amenity vegetation** [1, 2, 4, 7, 11-13, 15, 16, 18, 19, 31, 34, 35, 38-40, 43, 49, 51, 53, 56-58, 61, 64]; **apple orchards**, **pear orchards** [1-5, 7, 9, 10, 12-19, 21-23, 26, 27, 30-34, 36, 37, 39, 40, 42-44, 46, 47, 52-57, 59, 60, 62, 64, 66]; **apples**, **pears** [24, 25, 38, 49]; **asparagus** [4, 5, 7, 12, 13, 23, 31, 42, 43, 49, 53, 55]; **asparagus** *(off-label)*, **bilberries** *(off-label)*, **blackcurrants** *(off-label)*, **blueberries** *(off-label)*, **chestnuts** *(off-label)*, **cob nuts** *(off-label)*, **cranberries** *(off-label)*, **gooseberries** *(off-label)*, **hazel nuts** *(off-label)*, **redcurrants** *(off-label)*, **rhubarb** *(off-label)*, **vaccinium spp.** *(off-label)*, **walnuts** *(off-label)*, **whitecurrants** *(off-label)* [3, 45]; **beetroot** *(off-label)*, **carrots** *(off-label)*, **garlic** *(off-label)*, **horseradish** *(off-label)*, **leeks** *(off-label)*, **onions** *(off-label)*, **parsley root** *(off-label)*, **parsnips** *(off-label)*, **salad onions** *(off-label)*, **salsify** *(off-label)*, **shallots** *(off-label)*, **swedes** *(off-label)*, **turnips** *(off-label)* [44]; **bog myrtle** *(off-label)* [51]; **buckwheat** *(off-label)*, **canary seed** *(off-label)*, **lupins** *(off-label)*, **millet** *(off-label)*, **quinoa** *(off-label)*, **sorghum** *(off-label)* [46]; **buckwheat** *(off-label - grown for wild bird seed production)*, **canary seed** *(off-label - grown for wild bird seed production)*, **hemp** *(off-label - grown for wild bird seed production)*, **millet** *(off-label - grown for wild bird seed production)*, **quinoa** *(off-label - grown for wild bird seed production)*, **sorghum** *(off-label - grown for wild bird seed production)* [44, 47]; **bulb onions**, **vining peas** [1, 2, 15, 16, 18, 19, 39, 40, 49, 53, 55, 57, 64]; **cherries**, **plums** [1-5, 7, 9, 10, 12-19, 21-27, 30-32, 34, 36-40, 42-47, 49, 52-57, 62, 64, 66]; **combining peas**, **spring barley**, **spring field beans**, **spring wheat**, **winter barley**, **winter field beans**, **winter wheat** [1-3, 15, 16, 18, 19, 24, 25, 33, 39, 40, 49, 53, 57, 62, 64, 65]; **cultivated land/soil** [3, 21, 54]; **damsons** [1-5, 7, 9, 10, 12-19, 21, 22, 24-27, 30-32, 34, 36-40, 42-47, 49, 52, 54, 56, 57, 62, 64, 66]; **durum wheat** [1-3, 15, 16, 18, 19, 33, 39, 40, 49, 53, 57, 62, 64]; **enclosed waters**, **open waters** [6, 8, 24, 25, 33, 38]; **farm forestry** [24, 25, 55]; **farm forestry** *(off-label)*, **game cover** *(off-label)*, **hops** *(off-label)*, **miscanthus** *(off-label)*, **soft fruit** *(off-label)*, **top fruit** *(off-label)*, **woad** *(off-label)* [44-48, 52]; **field crops** *(wiper application)*, **non-crop farm areas** [21, 54]; **forest** [1, 2, 4-10, 12-26, 29-40, 42, 43, 49-51, 53, 55-57, 59-61, 63, 64, 66]; **forest nurseries** [23, 53, 55]; **forestry plantations** [1, 2, 15, 16, 18, 19, 39, 40, 57, 64]; **fruit trees** *(off-label)*, **grapevines** *(off-label)* [3]; **grassland** [1, 2, 15, 16, 18, 19, 24, 25, 33, 39, 40, 49, 53, 57, 59, 60, 62, 64]; **green cover on land temporarily removed from production** [1, 2, 15, 16, 18, 19, 33, 35, 39, 40, 49, 53, 57, 62, 64]; **hard surfaces** [4-9, 12, 13, 17, 20, 23-25, 30, 31, 33-35, 38, 41-43, 49-51, 53, 56, 58-61, 63]; **hemp** *(off-label)* [45, 46, 48, 52]; **industrial sites**, **road verges** [29, 50]; **land immediately adjacent to aquatic areas** [6, 8, 9, 24, 25, 33, 34, 38, 51]; **land not intended to bear vegetation** [9-11, 14, 21, 22, 26, 29, 34, 36, 37, 45, 48, 50, 54, 66]; **land prior to cultivation** [30, 63]; **leeks**, **turnips** [1, 2, 15, 16, 18, 19, 24, 25, 33, 39, 40, 49, 53, 55, 57, 60, 64]; **linseed**, **spring oilseed rape**, **winter oilseed rape** [1-3, 15, 16, 18, 19, 24, 25, 33, 39, 40, 49, 57, 62, 64, 65]; **managed amenity turf** *(pre-establishment)* [35, 61]; **managed amenity turf** *(pre-establishment only)* [38]; **mustard** [1, 2, 15, 16, 18, 19, 24, 25, 33, 39, 40, 49, 53, 57, 62, 64, 65]; **natural surfaces not intended to bear vegetation** [1, 2, 4-9, 12, 13, 15-20, 23-25, 30, 31, 33-35, 38-43, 49, 51, 53, 56-64]; **onions** [24, 25, 33]; **ornamental plant production** [38, 61]; **ornamental plant production** *(off-label)* [15]; **ornamental specimens** *(off-label)* [50, 51]; **permanent grassland** [10, 21, 32, 44-47, 54]; **permeable surfaces overlying soil** [1, 2, 4-9, 12, 13, 15-20, 23-25, 30, 31, 33-35, 39-43, 49, 51, 53, 56-64]; **spring oats**, **winter**

FOR FULL CONDITIONS OF USE ALWAYS READ THE PRODUCT LABEL

glyphosate

oats [1-3, 15, 16, 18, 19, 24, 25, 33, 39, 40, 49, 53, 57, 62, 64]; *stubbles* [3-5, 7, 9, 10, 12, 13, 21, 23, 31, 32, 34, 42-48, 52, 54-56, 59, 60, 65]; *sugar beet, swedes* [1, 2, 15, 16, 18, 19, 24, 25, 33, 39, 40, 49, 53, 55, 57, 64]; *water or waterside areas* [20]
- Annual dicotyledons in *bulb onions, sugar beet, swedes* [4, 5, 7, 12, 13, 23, 27, 42-48, 52, 60]; *combining peas* [4, 5, 7, 12, 13, 17, 23, 27, 42, 43, 48, 59, 60]; *cultivated land/soil* [9, 34]; *durum wheat, spring field beans, spring oilseed rape, winter field beans* [4, 5, 7, 12, 13, 17, 23, 27, 42-48, 52, 59, 60]; *leeks* [4, 5, 7, 12, 13, 23, 27, 42-48, 52]; *linseed, mustard* [4, 5, 7, 12, 13, 17, 23, 27, 42-48, 52, 60]; *peas* [45]; *permanent grassland* [44-47, 52]; *spring barley, spring oats, spring wheat, winter barley, winter oats, winter wheat* [4, 5, 7, 12-14, 17, 22, 23, 26, 27, 30, 36, 37, 42-48, 52, 59, 60, 66]; *turnips* [27, 44-48, 52]; *vining peas* [4, 5, 7, 12, 13, 23, 42-44, 46-48, 52]; *winter oilseed rape* [4, 5, 7, 12, 13, 17, 23, 27, 31, 42-48, 52, 55, 56, 59, 60]
- Annual grasses in *bulb onions, durum wheat, leeks, linseed, mustard, spring barley, spring field beans, spring oats, spring oilseed rape, spring wheat, sugar beet, swedes, turnips, winter barley, winter field beans, winter oats, winter oilseed rape, winter wheat* [27, 44-48, 52]; *combining peas* [27, 48]; *cultivated land/soil* [9, 14, 22, 26, 34, 36, 37, 66]; *peas* [45]; *permanent grassland* [44-47, 52]; *vining peas* [44, 46-48, 52]
- Aquatic weeds in *open waters* [9, 34, 51]
- Black bent in *combining peas, spring field beans, spring oilseed rape, stubbles, winter field beans, winter oilseed rape* [9, 14, 22, 26, 30, 34, 36, 37, 66]; *durum wheat, spring barley, spring oats, spring wheat, winter barley, winter oats, winter wheat* [9, 34]; *linseed* [14, 22, 26, 30, 36, 37, 66]
- Bolters in *sugar beet* (wiper application) [9, 21, 27, 34, 44-47, 52, 54]
- Bracken in *amenity vegetation, fencelines, land clearance, non-crop areas, paths and drives* [50]; *farm forestry* [55]; *forest* [4, 5, 7, 9, 10, 12-14, 17, 21-23, 26, 28-32, 34, 36, 37, 42, 43, 50, 51, 55, 56, 59, 60, 63, 66]; *forest nurseries* [23, 55]; *tolerant conifers* [29, 50]
- Brambles in *tolerant conifers* [29, 50]
- Canary grass in *aquatic areas* [27]; *enclosed waters, land immediately adjacent to aquatic areas* [28]
- Chemical thinning in *farm forestry* [55]; *forest* [4, 5, 7, 9, 12-14, 17, 22, 23, 26, 28, 30-32, 34, 36, 37, 42, 43, 50, 51, 55, 56, 59, 60, 63, 66]; *forest* (stump treatment) [10]; *forest nurseries* [23, 55]
- Couch in *combining peas, spring field beans, spring oilseed rape, stubbles, winter field beans, winter oilseed rape* [3, 9, 10, 14, 21, 22, 26, 27, 30, 32, 34, 36, 37, 44-47, 52, 54, 65, 66]; *durum wheat* [3, 9, 27, 34, 44-47, 52]; *forest* [9, 28, 32, 34, 50, 51]; *linseed* [3, 10, 14, 22, 26, 27, 30, 32, 36, 37, 44-47, 52, 65, 66]; *mustard* [10, 32, 44-47, 52, 65]; *permanent grassland* [10]; *spring barley, spring wheat, winter barley, winter wheat* [3, 9, 10, 21, 27, 32, 34, 44-47, 52, 54, 65]; *spring oats, winter oats* [3, 9, 10, 21, 27, 32, 34, 44-47, 52, 54]
- Creeping bent in *aquatic areas* [27]; *combining peas, spring field beans, spring oilseed rape, stubbles, winter field beans, winter oilseed rape* [9, 14, 22, 26, 30, 34, 36, 37, 66]; *durum wheat, spring barley, spring oats, spring wheat, winter barley, winter oats, winter wheat* [9, 34]; *enclosed waters, land immediately adjacent to aquatic areas* [28]; *linseed* [14, 22, 26, 30, 36, 37, 66]
- Desiccation in *buckwheat* (off-label), *canary seed* (off-label), *hemp* (off-label), *lupins* (off-label), *millet* (off-label), *poppies for morphine production* (off-label), *quinoa* (off-label), *sorghum* (off-label) [46]; *buckwheat* (off-label - for wild bird seed production), *canary seed* (off-label - for wild bird seed production), *hemp* (off-label - for wild bird seed production), *millet* (off-label - for wild bird seed production), *quinoa* (off-label - for wild bird seed production), *sorghum* (off-label - for wild bird seed production) [48]
- Destruction of short term leys in *grassland* [35]
- General weed control in *enclosed waters, land immediately adjacent to aquatic areas, open waters* [58]
- Grass weeds in *all edible crops (outdoor and protected), all non-edible crops (outdoor), apple orchards, combining peas, durum wheat, enclosed waters, grassland, green cover on land temporarily removed from production, land immediately adjacent to aquatic areas, leeks, linseed, mustard, onions, open waters, pear orchards, spring barley, spring field beans, spring oats, spring oilseed rape, spring wheat, sugar beet, swedes, turnips, winter barley, winter field beans, winter oats, winter oilseed rape, winter wheat* [33]; *forest, hard surfaces, natural surfaces not intended to bear vegetation, permeable surfaces overlying soil* [20, 33]; *water or waterside areas* [20]

SEE SECTION 3 FOR PRODUCTS ALSO REGISTERED

Pesticide Profiles

- Green cover in **grassland** [23]; **land temporarily removed from production** [3-5, 7, 9, 10, 12-14, 17, 22, 23, 26, 27, 30, 31, 34, 36, 37, 42-48, 52, 55, 56, 59, 60, 65, 66]
- Growth suppression in **enclosed waters, land immediately adjacent to aquatic areas, open waters** [59, 60]
- Harvest management/desiccation in **combining peas, spring field beans, winter field beans** [4, 5, 7, 10, 12, 13, 17, 21, 23, 31, 32, 42-47, 52, 54-56, 59, 60]; **durum wheat** [4, 5, 7, 12, 13, 17, 21, 23, 27, 31, 42-47, 52, 54-56, 59, 60]; **grassland** [55]; **linseed** [4, 5, 7, 9, 10, 12, 13, 17, 21, 23, 31, 32, 34, 42-47, 52, 54-56, 60]; **mustard** [4, 5, 7, 10, 12, 13, 17, 21, 23, 31, 32, 42-47, 52, 54-56, 60, 65]; **spring barley, winter barley** [4, 5, 7, 10, 12-14, 17, 22, 23, 26, 27, 30-32, 36, 37, 42-47, 52, 55, 56, 59, 60, 66]; **spring oats, spring wheat, winter oats, winter wheat** [4, 5, 7, 10, 12-14, 17, 21-23, 26, 27, 30-32, 36, 37, 42-47, 52, 54-56, 59, 60, 66]; **spring oilseed rape, winter oilseed rape** [3-5, 7, 9, 10, 12-14, 17, 21-23, 26, 27, 30-32, 34, 36, 37, 42-47, 52, 54-56, 59, 60, 65, 66]
- Heather in **farm forestry** [55]; **forest** [4, 5, 7, 9, 10, 12-14, 17, 21-23, 26, 28-31, 34, 36, 37, 42, 43, 51, 55, 56, 59, 60, 63, 66]; **forest nurseries** [23, 55]
- Perennial dicotyledons in **combining peas, linseed, spring field beans, spring oilseed rape, winter field beans, winter oilseed rape** [14, 22, 26, 27, 30, 36, 37, 66]; **forest** [63]; **forest** *(wiper application)* [50]; **permanent grassland** *(wiper application)* [44-47, 52]
- Perennial grasses in **aquatic areas** [3, 9, 10, 21, 22, 26, 29, 34, 36, 37, 50, 63]; **combining peas, linseed, spring field beans, stubbles, winter field beans** [27]; **enclosed waters, land immediately adjacent to aquatic areas** [63]; **tolerant conifers** [29, 50]
- Pre-harvest desiccation in **combining peas, linseed, spring barley, spring field beans, spring wheat, winter barley, winter field beans, winter wheat** [65]; **poppies for morphine production** *(off-label)* [45]
- Reeds in **aquatic areas** [3, 9, 10, 21, 22, 26, 27, 29, 34, 36, 37, 45, 50, 63]; **enclosed waters, land immediately adjacent to aquatic areas** [9, 28, 34, 51, 63]
- Re-growth suppression in **forest** [58]
- Rhododendrons in **farm forestry** [55]; **forest** [4, 5, 7, 9, 10, 12-14, 17, 21-23, 26, 28-31, 34, 36, 37, 42, 43, 50, 51, 55, 56, 59, 60, 63, 66]; **forest nurseries** [23, 55]
- Rushes in **aquatic areas** [3, 9, 10, 21, 22, 26, 27, 29, 34, 36, 37, 45, 50, 63]; **enclosed waters, land immediately adjacent to aquatic areas** [9, 28, 34, 51, 63]; **forest** [32]
- Sedges in **aquatic areas** [3, 9, 10, 21, 22, 26, 27, 29, 34, 36, 37, 45, 50, 63]; **enclosed waters, land immediately adjacent to aquatic areas** [9, 28, 34, 51, 63]
- Sucker control in **apple orchards, cherries, damsons, pear orchards, plums** [44-47, 54]
- Sward destruction in **amenity grassland** [50]; **permanent grassland** [3-5, 7, 9, 10, 12-14, 17, 21, 22, 26, 27, 30-32, 34, 36, 37, 42-47, 52, 54, 56, 65, 66]; **rotational grass** [4, 5, 7, 12, 13, 17, 31, 42, 43, 56, 65]
- Total vegetation control in **all edible crops (outdoor)** *(pre-sowing/planting)*, **all non-edible crops (outdoor)** *(pre-sowing/planting)* [44, 46, 47, 51]; **amenity vegetation** [11, 29, 50]; **fencelines, industrial sites, road verges** [29, 50]; **hard surfaces** [28, 29, 50]; **land not intended to bear vegetation** [11, 29, 50, 65]; **natural surfaces not intended to bear vegetation, permeable surfaces overlying soil** [28]
- Volunteer cereals in **bulb onions, leeks, linseed, mustard, spring field beans, sugar beet, swedes, turnips, winter field beans** [27, 44-48, 52]; **combining peas** [27, 48]; **cultivated land/soil** [9, 14, 22, 26, 34, 36, 37, 66]; **durum wheat, spring barley, spring oats, spring oilseed rape, spring wheat, winter barley, winter oats, winter oilseed rape, winter wheat** [44-48, 52]; **peas** [45]; **stubbles** [3, 9, 10, 14, 21, 22, 26, 27, 30, 32, 34, 36, 37, 44-47, 52, 54, 65, 66]; **vining peas** [44, 46-48, 52]
- Volunteer potatoes in **stubbles** [3, 9, 10, 21, 27, 32, 34, 44-47, 52, 54, 65]
- Waterlilies in **aquatic areas** [3, 9, 10, 21, 22, 26, 27, 29, 34, 36, 37, 50, 63]; **enclosed waters** [9, 28, 34, 51]
- Weed beet in **sugar beet** *(wiper application)* [27, 54]
- Wild oats in **durum wheat, spring barley, spring oats, spring wheat, winter barley, winter oats, winter wheat** [21, 54]
- Woody weeds in **farm forestry** [55]; **forest** [4, 5, 7, 9, 10, 12-14, 17, 21-23, 26, 28-32, 34, 36, 37, 42, 43, 50, 51, 55, 56, 59, 60, 63, 66]; **forest nurseries** [23, 55]; **tolerant conifers** [29, 50]

Specific Off-Label Approvals (SOLAs)
- **almonds** *20072045* [45]
- **apples** *20072045* [45]
- **apricots** *20072045* [45]

FOR FULL CONDITIONS OF USE ALWAYS READ THE PRODUCT LABEL

glyphosate

- **asparagus** *20071476* [3], *20062036* [45]
- **beetroot** *20102519* [44]
- **bilberries** *20071475* [3], *20062035* [45]
- **blackcurrants** *20071475* [3], *20062035* [45]
- **blueberries** *20071475* [3], *20062035* [45]
- **bog myrtle** *20082727* [51]
- **buckwheat** *(grown for wild bird seed production) 20072936* [44], *20071839* [46], *(grown for wild bird seed production) 20072934* [47], *(for wild bird seed production) 20070582* [48]
- **canary seed** *(grown for wild bird seed production) 20072936* [44], *20071839* [46], *(grown for wild bird seed production) 20072934* [47], *(for wild bird seed production) 20070582* [48]
- **carrots** *20102519* [44]
- **chestnuts** *20071477* [3], *20062028* [45], *20072045* [45]
- **cob nuts** *20071477* [3], *20072045* [45]
- **crab apples** *20072045* [45]
- **cranberries** *20071475* [3], *20062035* [45]
- **farm forestry** *20082887* [44], *20082869* [45], *20082888* [46], *20082889* [47], *20082891* [48], *20082892* [52]
- **fruit trees** *20102476* [3]
- **game cover** *20082887* [44], *20082869* [45], *20082888* [46], *20082889* [47], *20082891* [48], *20082892* [52]
- **garlic** *20102519* [44]
- **gooseberries** *20071475* [3], *20062035* [45]
- **grapevines** *20071479* [3]
- **hazel nuts** *20071477* [3], *20062028* [45], *20072045* [45]
- **hemp** *(grown for wild bird seed production) 20072937* [44], *20082869* [45], *20071840* [46], *(grown for wild bird seed production) 20072935* [47], *(for wild bird seed production) 20070581* [48], *20070581* [48], *20082892* [52]
- **hops** *20082887* [44], *20082869* [45], *20082888* [46], *20082889* [47], *20082891* [48], *20082892* [52]
- **horseradish** *20102519* [44]
- **kentish cobnuts** *20062028* [45]
- **leeks** *20102519* [44]
- **lupins** *20071279* [46]
- **millet** *(grown for wild bird seed production) 20072936* [44], *20071839* [46], *(grown for wild bird seed production) 20072934* [47], *(for wild bird seed production) 20070582* [48]
- **miscanthus** *20082887* [44], *20082869* [45], *20082888* [46], *20082889* [47], *20082891* [48], *20082892* [52]
- **onions** *20102519* [44]
- **ornamental plant production** *20082840* [15]
- **ornamental specimens** *20082877* [50], *20082890* [51]
- **parsley root** *20102519* [44]
- **parsnips** *20102519* [44]
- **peaches** *20072045* [45]
- **pears** *20072045* [45]
- **poppies for morphine production** *20062030* [45], *20081058* [46]
- **quinces** *20072045* [45]
- **quinoa** *(grown for wild bird seed production) 20072937* [44], *20071840* [46], *(grown for wild bird seed production) 20072935* [47], *(for wild bird seed production) 20070581* [48]
- **redcurrants** *20071475* [3], *20062035* [45]
- **rhubarb** *20071478* [3], *20062029* [45]
- **salad onions** *20102519* [44]
- **salsify** *20102519* [44]
- **shallots** *20102519* [44]
- **soft fruit** *20082887* [44], *20082869* [45], *20082888* [46], *20082889* [47], *20082891* [48], *20082892* [52]
- **sorghum** *(grown for wild bird seed production) 20072936* [44], *20071839* [46], *(grown for wild bird seed production) 20072934* [47], *(for wild bird seed production) 20070582* [48]
- **swedes** *20102519* [44]

SEE SECTION 3 FOR PRODUCTS ALSO REGISTERED

- **top fruit** *20082887* [44], *20082869* [45], *20082888* [46], *20082889* [47], *20082891* [48], *20082892* [52]
- **turnips** *20102519* [44]
- **vaccinium spp.** *20071475* [3], *20062035* [45]
- **walnuts** *20071477* [3], *20062028* [45], *20072045* [45]
- **whitecurrants** *20071475* [3], *20062035* [45]
- **wine grapes** *20062031* [45]
- **woad** *20082887* [44], *20082869* [45], *20082888* [46], *20082889* [47], *20082891* [48], *20082892* [52]

Approval information
- Glyphosate included in Annex I under EC Directive 91/414
- Accepted by BBPA for use on malting barley and hops

Efficacy guidance
- For best results apply to actively growing weeds with enough leaf to absorb chemical
- For most products a rainfree period of at least 6 h (preferably 24 h) should follow spraying
- Adjuvants are obligatory for some products and recommended for some uses with others. See labels
- Mixtures with other pesticides or fertilizers may lead to reduced control.
- Products are formulated as isopropylamine, ammonium, potassium, or trimesium salts of glyphosate and may vary in the details of efficacy claims. See individual product labels
- Some products in ready-to-use formulations for use through hand-held applicators. See label for instructions [11, 35, 61]
- When applying products through rotary atomisers the spray droplet spectra must have a minimum Volume Median Diameter (VMD) of 200 microns
- With wiper application weeds should be at least 10 cm taller than crop
- Annual weed grasses should have at least 5 cm of leaf and annual broad-leaved weeds at least 2 expanded true leaves
- Perennial grass weeds should have 4-5 new leaves and be at least 10 cm long when treated. Perennial broad-leaved weeds should be treated at or near flowering but before onset of senescence
- Volunteer potatoes and polygonums are not controlled by harvest-aid rates
- Bracken must be treated at full frond expansion
- Fruit tree suckers best treated in late spring
- Chemical thinning treatment can be applied as stump spray or stem injection
- In order to allow translocation, do not cultivate before spraying and do not apply other pesticides, lime, fertilizer or farmyard manure within 5 d of treatment
- Recommended intervals after treatment and before cultivation vary. See labels

Restrictions
- Maximum total dose per crop or season normally equivalent to one full dose treatment on field and edible crops and no restriction for non-crop uses. However some older labels indicate a maximum number of treatments. Check for details
- Do not treat cereals grown for seed or undersown crops
- Consult grain merchant before treating crops grown on contract or intended for malting
- Do not use treated straw as a mulch or growing medium for horticultural crops
- For use in nursery stock, shrubberies, orchards, grapevines and tree nuts care must be taken to avoid contact with the trees. Do not use in orchards established less than 2 yr and keep off low-lying branches
- Certain conifers may be sprayed overall in dormant season. See label for details
- Use a tree guard when spraying in established forestry plantations
- Do not spray root suckers in orchards in late summer or autumn
- Do not use under glass or polythene as damage to crops may result
- Do not mix, store or apply in galvanised or unlined mild steel containers or spray tanks
- Do not leave diluted chemical in spray tanks for long periods and make sure that tanks are well vented

Crop-specific information
- Harvest intervals: 4 wk for blackcurrants, blueberries; 14 d for linseed, oilseed rape; 8 d for mustard; 5-7 d for all other edible crops. Check label for exact details

FOR FULL CONDITIONS OF USE ALWAYS READ THE PRODUCT LABEL

glyphosate

- Latest use: for most products 2-14 d before cultivating, drilling or planting a crop in treated land; after harvest (post-leaf fall) but before bud formation in the following season for nuts and most fruit and vegetable crops; before fruit set for grapevines. See labels for details

Following crops guidance
- Decaying remains of plants killed by spraying must be dispersed before direct drilling
- Crops may be drilled 48 h after application. Trees and shrubs may be planted 7 d after application.

Environmental safety
- Products differ in their hazard and environmental safety classification. See labels
- Do not dump surplus herbicide in water or ditch bottoms or empty into drains
- Check label for maximum permitted concentration in treated water
- The Environment Agency or Local River Purification Authority must be consulted before use in or near water
- Take extreme care to avoid drift and possible damage to neighbouring crops or plants
- Treated poisonous plants must be removed before grazing or conserving
- Do not use in covered areas such as greenhouses or under polythene
- For field edge treatment direct spray away from hedge bottoms
- Some products require livestock to be excluded from treated areas and do not permit treated forage to be used for hay, silage or bedding. Check label for details
- LERAP Category B [49, 52]

Hazard classification and safety precautions

Hazard Harmful [13, 21, 31, 54-57, 62]; Irritant [4-6, 10, 12, 14-17, 23-25, 32, 39, 40, 42, 43, 48, 49, 52, 58-61, 66]; Dangerous for the environment [4-6, 12-17, 21, 23-25, 31, 32, 35-40, 42, 43, 48, 49, 52, 54-62, 65, 66]

Transport code 9 [1, 2, 4-6, 15-17, 21, 23, 26, 29-32, 35-37, 39, 40, 42, 48, 49, 52, 54, 57-62, 66]

Packaging group III [1, 2, 4-6, 15-17, 21, 23, 26, 29-32, 35-37, 39, 40, 42, 48, 49, 52, 54, 57-62, 66]

UN number 3077 [48]; 3082 [1, 2, 4-6, 15-17, 21, 23, 26, 29-32, 35-37, 39, 40, 42, 49, 52, 54, 57-62, 66]; N/C [3, 7-14, 18-20, 22, 27, 28, 33, 38, 41, 43-47, 50, 51, 53, 55, 56, 63-65]

Risk phrases R20 [4, 5, 12, 13, 21, 23, 42, 43, 54, 57, 62]; R36 [21, 32, 61]; R38 [49, 52]; R41 [4, 5, 10, 12-17, 23-25, 31, 39, 40, 42, 43, 48, 54-60, 62, 66]; R50 [35, 38, 49, 52]; R51 [4, 5, 12-17, 21, 23-25, 31, 32, 36, 37, 39, 40, 42, 43, 48, 54-62, 66]; R52 [27, 28, 30, 44, 46, 47, 51, 65]; R53a [1, 2, 4, 5, 12-19, 21, 23-25, 27, 28, 30-32, 35-40, 42-44, 46-49, 51, 52, 54-62, 64-66]

Operator protection A [1-66]; C [3-10, 12-17, 21-26, 29-34, 36, 37, 39, 40, 42-48, 50, 51, 53-60, 62, 63, 66]; D [3, 22, 26, 29, 36, 37, 53, 63]; F [32, 54]; H [1-12, 14, 17-20, 22-30, 32-38, 42, 43, 45, 48-54, 58-66]; M [1-14, 17-30, 32-38, 42-54, 58-64, 66]; N [14, 21, 24, 25, 32, 54, 66]; U02a [1, 2, 6-9, 15, 16, 18-21, 24-28, 30-40, 48, 55-62, 64]; U05a [1, 2, 4, 5, 12-14, 17-19, 21, 23-25, 32, 35, 38, 42, 43, 48, 49, 52, 54, 58, 61, 62, 64-66]; U05b [59, 60]; U08 [6-9, 17, 21, 24-28, 30, 32, 34, 36, 37, 48, 59, 60]; U09a [14, 20, 33, 35, 38, 61, 62, 66]; U11 [4, 5, 10, 12-17, 21, 23-25, 27, 28, 31, 32, 39, 40, 42, 43, 48, 54-62, 66]; U14 [24, 25]; U15 [32, 62]; U16b [61]; U19a [6-9, 11, 14, 17, 20, 21, 24-28, 30, 32-38, 48, 58, 61, 62, 66]; U20a [4, 5, 12, 13, 20, 23, 33, 42, 43, 45, 52, 54, 65]; U20b [1-3, 6-9, 11, 14-19, 21, 22, 24-32, 34-41, 44, 46-51, 53, 55-64, 66]

Environmental protection E06a [41] (2 weeks); E06d, E08b [65]; E13c [9, 26, 32, 34, 35, 38]; E15a [3, 11, 21, 22, 29, 30, 32, 36, 37, 45, 48, 50, 52, 53, 58-60, 62, 63]; E15b [1, 2, 4-8, 10, 12, 13, 15, 16, 18-20, 23-25, 27, 28, 31, 33, 35, 38-44, 46, 47, 49, 51, 54-57, 61, 64, 65]; E16a, E16b [49, 52]; E19a [3, 9, 21, 22, 26, 34, 36, 37, 53, 63]; E19b [14, 66]; E34 [11, 18, 19, 30, 32, 64]; E36a [18, 19, 64]; E38 [4, 5, 10, 12, 13, 15, 16, 21, 23-25, 27, 28, 30-32, 39, 40, 42, 43, 48, 49, 52, 54-57, 61, 62, 65]

Storage and disposal D01 [1-8, 10-14, 17-21, 23-30, 32, 33, 35-38, 42-54, 58-66]; D02 [1, 2, 4-8, 10-14, 17-21, 23-25, 27-29, 32, 33, 35, 38, 42-52, 54, 58-62, 64-66]; D05 [1, 2, 4-9, 11-13, 15, 16, 18-20, 23, 29-35, 38-40, 42-44, 46-48, 50-52, 54-62, 64, 65]; D07 [48]; D09a [3-17, 20-63, 65, 66]; D10a [6-8, 11]; D10b [3, 9, 14, 17, 21, 22, 24, 25, 32-34, 36, 53, 58-60, 63, 66]; D10c [1, 2, 4, 5, 12, 13, 15, 16, 18, 19, 23, 26, 29-31, 37, 39-47, 49-52, 54-57, 62, 64, 65]; D11a [10, 20, 27, 28, 35, 38, 48, 61]; D12a [4, 5, 12-16, 21, 23-25, 27, 28, 30-32, 35, 38-40, 42, 43, 48, 49, 52, 54-57, 61, 62, 65, 66]; D12b [36, 37]; D14 [18, 19, 64]

Medical advice M03 [65]; M05a [14-17, 31, 39, 40, 55-57, 66]

SEE SECTION 3 FOR PRODUCTS ALSO REGISTERED

Pesticide Profiles

277 glyphosate + sulfosulfuron

A translocated non-residual glycine derivative herbicide + a sulfonyl urea herbicide
HRAC mode of action code: G + B

See also sulfosulfuron

Products
   Nomix Dual   Nomix Enviro   120:2.22 g/l   RH   13420

Uses
- Annual and perennial weeds in **amenity vegetation**, **natural surfaces not intended to bear vegetation**, **permeable surfaces overlying soil**

Approval information
- Glyphosate and sulfosulfuron included in Annex I under EC Directive 91/414

Hazard classification and safety precautions
Hazard Dangerous for the environment
Transport code 9
Packaging group III
UN number 3082
Risk phrases R50, R53a
Operator protection A, H, M; U02a, U09a, U19a, U20b
Environmental protection E15b
Storage and disposal D01, D05, D09a, D11a

278 guazatine

A guanidine fungicide seed dressing for cereals
FRAC mode of action code: M7

Products
1 Panoctine   Makhteshim   300 g/l   LS   11404
2 Ravine   Makhteshim   300 g/l   LS   12211

Uses
- Brown foot rot in **spring barley** *(seed treatment - reduction)*, **spring oats** *(seed treatment - reduction)*, **winter barley** *(seed treatment - reduction)*, **winter oats** *(seed treatment - reduction)* [1]
- Foot rot in **spring barley** *(seed treatment)*, **winter barley** *(seed treatment)* [1]
- Fusarium foot rot and seedling blight in **spring barley** *(seed treatment - reduction)*, **spring oats** *(seed treatment - reduction)*, **spring wheat** *(seed treatment - reduction)*, **winter barley** *(seed treatment - reduction)*, **winter oats** *(seed treatment - reduction)*, **winter wheat** *(seed treatment - reduction)* [1, 2]
- Leaf stripe in **spring barley** *(seed treatment)*, **winter barley** *(seed treatment)* [1]
- Net blotch in **spring barley** *(seed treatment)*, **winter barley** *(seed treatment)* [1]
- Pyrenophora leaf spot in **spring oats** *(seed treatment)*, **winter oats** *(seed treatment)* [1]
- Septoria seedling blight in **spring wheat** *(seed treatment)*, **winter wheat** *(seed treatment)* [2]; **spring wheat** *(seed treatment - partial control)*, **winter wheat** *(seed treatment - partial control)* [1]

Approval information
- Accepted by BBPA for use on malting barley
- Guazatine not included in Annex 1 and current approvals set to expire 2021.

Efficacy guidance
- Apply with conventional seed treatment machinery

Restrictions
- Maximum number of treatments 1 per batch
- Do not treat grain with moisture content above 16% and do not allow moisture content of treated seed to exceed 16%
- Do not apply to cracked, split or sprouted seed
- Do not use treated seed as food or feed

FOR FULL CONDITIONS OF USE ALWAYS READ THE PRODUCT LABEL

guazatine + imazalil

Crop-specific information
- Latest use: pre-drilling
- After treating, bag seed immediately and keep in dry, draught-free store
- Treatment may lower germination capacity, particularly if seed grown, harvested or stored under adverse conditions

Environmental safety
- Dangerous for the environment
- Very toxic to aquatic organisms
- Harmful to fish or other aquatic life. Do not contaminate surface waters or ditches with chemical or used container

Hazard classification and safety precautions
Hazard Harmful, Dangerous for the environment
Transport code 9
Packaging group III
UN number 3082
Risk phrases R20, R22a, R37, R41, R43, R50, R53a
Operator protection A, C, D, H; U05a, U20b [1, 2]; U09a, U14, U15 [1]
Environmental protection E13c [1]; E15a [2]; E34, E38 [1, 2]
Storage and disposal D01, D02, D05, D09a, D10a [1, 2]; D14 [1]
Treated seed S01, S02, S04b, S05, S06a, S07
Medical advice M03 [1]; M04a [1, 2]

279   guazatine + imazalil

A fungicide seed treatment for barley and oats
FRAC mode of action code: M7 + 3

See also imazalil

Products

1	Panoctine Plus	Makhteshim	300:25 g/l	LS	11757
2	Ravine Plus	Makhteshim	300:25 g/l	LS	11832

Uses
- Brown foot rot in **spring barley** *(seed treatment)*, **winter barley** *(seed treatment)*
- Covered smut in **spring barley** *(seed treatment)*, **winter barley** *(seed treatment)*
- Leaf stripe in **spring barley** *(seed treatment)*, **winter barley** *(seed treatment)*
- Net blotch in **spring barley** *(seed treatment - moderate control)*, **winter barley** *(seed treatment - moderate control)*
- Pyrenophora leaf spot in **spring oats** *(seed treatment)*, **winter oats** *(seed treatment)*
- Seedling blight and foot rot in **spring barley** *(seed treatment - moderate control)*, **winter barley** *(seed treatment - moderate control)*
- Snow mould in **spring barley** *(seed treatment)*, **winter barley** *(seed treatment)*

Approval information
- Imazalil included in Annex I under EC Directive 91/414
- Accepted by BBPA for use on malting barley

Efficacy guidance
- Apply with conventional seed treatment machinery

Restrictions
- Maximum number of treatments 1 per batch
- Do not mix with other formulations
- Do not treat grain with moisture content above 16% and do not allow moisture content of treated seed to exceed 16%
- Do not apply to cracked, split or sprouted seed
- Do not store treated seed for more than 6 mth
- Treatment may lower germination capacity, particularly if seed grown, harvested or stored under adverse conditions

Crop-specific information
- Latest use: pre-drilling

SEE SECTION 3 FOR PRODUCTS ALSO REGISTERED

Pesticide Profiles

Environmental safety
- Dangerous for the environment
- Very toxic to aquatic organisms
- Do not use treated seed as food or feed

Hazard classification and safety precautions
Hazard Toxic, Dangerous for the environment
Transport code 6.1
Packaging group III
UN number 2902
Risk phrases R22a, R23, R37, R41, R43, R50, R53a
Operator protection A, C, D, H; U05a, U10, U11, U14, U19a, U20b
Environmental protection E15a, E34, E38
Storage and disposal D01, D02, D09a, D10b
Treated seed S02, S04b, S05, S07
Medical advice M04a

280 hymexazol

A systemic heteroaromatic fungicide for pelleting sugar beet seed
FRAC mode of action code: 32

Products

| | Tachigaren 70 WP | Summit Agro | 70% w/w | WP | 12568 |

Uses
- Aphanomyces cochlioides in *red beet* (off-label - seed treatment)
- Black leg in *sugar beet* (seed treatment)

Specific Off-Label Approvals (SOLAs)
- *red beet* (seed treatment) 20062835

Efficacy guidance
- Incorporate into pelleted seed using suitable seed pelleting machinery

Restrictions
- Maximum number of treatments 1 per batch of seed
- Do not use treated seed as food or feed

Crop-specific information
- Latest use: before planting sugar beet seed

Environmental safety
- Harmful to aquatic organisms
- Harmful to fish or other aquatic life. Do not contaminate surface waters or ditches with chemical or used container
- Treated seed harmful to game and wildlife

Hazard classification and safety precautions
Hazard Irritant, Highly flammable
Transport code 3
Packaging group III
UN number 1325
Risk phrases R41, R52, R53a
Operator protection A, C, F; U05a, U11, U20b
Environmental protection E03, E13c
Storage and disposal D01, D02, D09a, D11a
Treated seed S01, S02, S03, S04a, S05

FOR FULL CONDITIONS OF USE ALWAYS READ THE PRODUCT LABEL

281 imazalil

A systemic and protectant imidazole fungicide
FRAC mode of action code: 3

See also guazatine + imazalil

Products

1	Fungazil 100 SL	Certis	100 g/l	SL	14999
2	Fungazil 50LS	Certis	50 g/l	LS	14069
3	IMA 100 SL	Q-Chem	100 g/l	LS	13559
4	Imaz 100 SL	AgriChem BV	100 g/l	SL	13322
5	Imaz 200 EC	AgriChem BV	200 g/l	EC	12991
6	Magnate 100 SL	Makhteshim	100 g/l	SL	14660

Uses
- Dry rot in **seed potatoes**, **ware potatoes** [1, 3, 6]
- Fusarium in **potatoes**, **seed potatoes** [4]
- Gangrene in **potatoes** [4]; **seed potatoes** [1, 3, 4, 6]; **ware potatoes** [1, 3, 6]
- Leaf stripe in **spring barley**, **winter barley** [2]
- Net blotch in **spring barley**, **winter barley** [2]
- Powdery mildew in **protected cucumbers** [5]
- Silver scurf in **potatoes** [4]; **seed potatoes** [1, 3, 4, 6]; **ware potatoes** [1, 3, 6]
- Skin spot in **seed potatoes**, **ware potatoes** [1, 3, 6]

Approval information
- Imazalil included in Annex I under EC Directive 91/414
- Approved by BBPA for use on malting barley
- Approval expiry 31 Dec 2011 [1, 2, 4-6]
- Approval expiry 04 Jun 2011 [3]

Efficacy guidance
- For best control of skin and wound diseases of ware potatoes treat as soon as possible after harvest, preferably within 7-10 d, before any wounds have healed [1, 3, 6]
- Do not use against cucumber powdery mildew if previous applications of DMI fungicides have failed to work [5]

Restrictions
- Maximum number of treatments 1 per batch of ware tubers [1, 3, 6]; 2 per batch of seed tubers [1, 3, 6];
- Consult processor before treating potatoes for processing [1, 3, 6]
- Do not treat protected cucumbers in bright sunny conditions [1, 3, 6]

Crop-specific information
- Latest use: during storage and before chitting for seed potatoes [1, 3, 6]
- HI 1 d for potatoes [3, 6];
- Apply to clean soil-free potatoes post-harvest before putting into store, or at first grading. A further treatment may be applied in early spring before planting [1, 3, 6]
- Apply through canopied hydraulic or spinning disc equipment preferably diluted with up to two litres water per tonne of potatoes to obtain maximum skin cover and penetration [1, 3, 6]
- Use on ware potatoes subject to discharges of imazalil from potato washing plants being within emission limits set by the UK monitoring authority [1, 3, 6]

Environmental safety
- Dangerous for the environment
- Toxic to aquatic organisms
- Do not empty into drains [1, 3, 6]
- Personal protective equipment requirements may vary for each pack size. Check label

Hazard classification and safety precautions
Hazard Harmful [5]; Irritant [1-4, 6]; Flammable [2]; Dangerous for the environment [3-6]
Transport code 9 [2-6]
Packaging group III [2-6]
UN number 3082 [2-6]; N/C [1]
Risk phrases R22a, R66, R67 [5]; R36 [2, 6]; R41 [1, 3, 4]; R50 [6]; R51 [3-5]; R52 [1]; R53a [1, 3-6]

SEE SECTION 3 FOR PRODUCTS ALSO REGISTERED

Pesticide Profiles

Operator protection A [1-6]; C [1, 3, 4, 6]; D [6]; H [1-5]; U04a [1, 3, 4, 6]; U05a [1-6]; U09a [2]; U11 [1, 3, 4]; U14 [1, 3, 5]; U15 [5]; U19a [1, 3, 5, 6]; U20a [1, 3, 6]; U20b [2, 4, 5]
Environmental protection E13c [2, 6]; E15a, E19b [1, 3]; E15b [4, 5]; E34 [1, 3-6]; E38 [4]
Consumer protection C01 [4]
Storage and disposal D01, D02, D09a [1-6]; D05 [1-3, 6]; D10a [2, 4, 6]; D10b, D12b [5]; D10c [1, 3]; D12a [1, 3, 4]
Treated seed S01, S03, S05 [1-4, 6]; S02 [1-4]; S04a [1-3, 6]; S04b [4]; S06a [1, 3, 6]; S07 [2]
Medical advice M03 [2, 4-6]

282   imazalil + pencycuron

A fungicide mixture for treatment of seed potatoes
FRAC mode of action code: 3 + 20

See also pencycuron

Products

Monceren IM   Bayer CropScience   0.6:12.5% w/w   DS   11426

Uses
- Black scurf in **potatoes** *(tuber treatment)*
- Silver scurf in **potatoes** *(tuber treatment - reduction)*
- Stem canker in **potatoes** *(tuber treatment - reduction)*

Approval information
- Imazalil included in Annex I under EC Directive 91/414

Efficacy guidance
- Apply to clean seed tubers during planting (see label for suitable method) or sprinkle over tubers in chitting trays before loading into planter. It is essential to obtain an even distribution over tubers
- If seed tubers become damp from light rain distribution of product should not be affected. Tubers in the hopper should be covered if a shower interrupts planting

Restrictions
- Maximum number of treatments 1 per batch of tubers
- Operators must wear suitable respiratory equipment and gloves when handling product and when riding on planter. Wear gloves when handling treated tubers
- Do not use on tubers which have previously been treated with a dry powder seed treatment or hot water
- Do not use treated tubers for human or animal consumption

Crop-specific information
- Latest use: immediately before planting
- May be used on seed tubers previously treated with a liquid fungicide but not before 8 wk have elapsed if this contained imazalil

Environmental safety
- Harmful to aquatic organisms
- Harmful to fish or other aquatic life. Do not contaminate surface waters or ditches with chemical or used container
- Treated seed harmful to game and wildlife

Hazard classification and safety precautions
UN number N/C
Risk phrases R52, R53a
Operator protection A, D; U20b
Environmental protection E03, E13c
Storage and disposal D09a, D11a
Treated seed S01, S02, S03, S04a, S05, S06a

FOR FULL CONDITIONS OF USE ALWAYS READ THE PRODUCT LABEL

283 imazalil + thiabendazole

A fungicide mixture for treatment of seed potatoes; product approval expired 2008
FRAC mode of action code: 3 + 1

See also thiabendazole

284 imazamox

An imidazolinone contact and residual herbicide available only in mixtures
HRAC mode of action code: B

285 imazamox + pendimethalin

A pre-emergence broad-spectrum herbicide mixture for legumes
HRAC mode of action code: B + K1

See also pendimethalin

Products

Nirvana	BASF	16.7:250 g/l	EC	14256

Uses
- Annual dicotyledons in **broad beans** (off-label), **combining peas**, **hops** (off-label), **ornamental plant production** (off-label), **soft fruit** (off-label), **spring field beans**, **top fruit** (off-label), **vining peas**, **winter field beans**

Specific Off-Label Approvals (SOLAs)
- **broad beans** 20092891
- **hops** 20092894
- **ornamental plant production** 20092894
- **soft fruit** 20092894
- **top fruit** 20092894

Approval information
- Imazamox and pendimethalin included in Annex I under EC Directive 91/414

Efficacy guidance
- Best results obtained from applications to fine firm seedbeds in the presence of adequate moisture
- Weed control may be reduced on cloddy seedbeds and on soils with over 6% organic matter
- Residual control may be reduced under prolonged dry conditions

Restrictions
- Maximum number of treatments 1 per crop
- Seed must be drilled to at least 2.5 cm of settled soil
- Do not use on soils containing more than 10% organic matter
- Do not apply to soils that are waterlogged or are prone to waterlogging
- Do not apply if heavy rain is forecast
- Do not soil incorporate the product or disturb the soil after application
- Consult processors before use on crops destined for processing

Crop-specific information
- Latest use: pre-crop emergence for all crops
- Inadequately covered seed may result in cupping of the leaves after application from which recovery is normally complete
- Crop damage may occur on stony or gravelly soils especially if heavy rain follows treatment
- Winter oilseed rape and other brassica crops should not be drilled as the following crop. A minimum interval of 12 months should elapse between application and sowing red beet, sugar beet and spinach.

SEE SECTION 3 FOR PRODUCTS ALSO REGISTERED

Pesticide Profiles

Following crops guidance
- Winter wheat or winter barley may be drilled as a following crop provided 3 mth have elapsed since treatment and the land has been at least cultivated by a non-inversion technique such as discing
- Winter oilseed rape or other brassica crops should not be drilled as following crops
- A minimum of 12 mth must elapse between treatment and sowing red beet, sugar beet or spinach

Environmental safety
- Dangerous for the environment
- Very toxic to aquatic organisms
- LERAP Category B

Hazard classification and safety precautions
    Hazard Irritant, Dangerous for the environment
    Transport code 9
    Packaging group III
    UN number 3082
    Risk phrases R38, R43, R50, R53a
    Operator protection A, H; U05a, U14, U20b
    Environmental protection E15b, E16a, E38
    Storage and disposal D01, D02, D09a, D10c, D12a
    Medical advice M03

286 imazaquin

An imidazolinone herbicide and plant growth regulator available only in mixtures
HRAC mode of action code: B

See also chlormequat + 2-chloroethylphosphonic acid + imazaquin
          chlormequat + imazaquin

287 imidacloprid

A neonicotinoid insecticide for seed, soil, peat or foliar treatment
IRAC mode of action code: 4A

See also beta-cyfluthrin + imidacloprid
          bitertanol + fuberidazole + imidacloprid
          fuberidazole + imidacloprid + triadimenol

Products

1	Admire	Bayer CropScience	70% w/w	WG	11234
2	Couraze	Solufeed	5% w/w	GR	13862
3	Imidasect 5 GR	Fargro	5% w/w	GR	13581
4	Imidasect 5 GR	Fargro	5% w/w	GR	14142
5	Imidasect 5GR	Fargro	5 % w/w	GR	14574
6	Intercept 5GR	Scotts	5% w/w	GR	14091
7	Intercept 70WG	Scotts	70% w/w	WG	08585
8	Merit Turf	Bayer Environ.	0.5% w/w	GR	12415
9	Nuprid 600FS	Nufarm UK	600 g/l	FS	13855
10	Valiant	Pan Amenity	5% w/w	WG	14456

Uses
- Aphids in *amenity vegetation* (off-label) [1]; *bedding plants, herbaceous perennials, ornamental plant production, pot plants* [7]; *fodder beet, sugar beet* [9]; *ornamental plant production* (container grown) [3-6]; *protected ornamentals* [2-6]
- Chafer grubs in *managed amenity turf* [8]
- Damson-hop aphid in *hops* [1]
- Flea beetle in *fodder beet, sugar beet* [9]
- Glasshouse whitefly in *bedding plants, herbaceous perennials, ornamental plant production, pot plants* [7]; *ornamental plant production* (container grown), *protected ornamentals* [3-6]
- Insect pests in *lawns* [10]

FOR FULL CONDITIONS OF USE ALWAYS READ THE PRODUCT LABEL

imidacloprid

- Leaf miner in **fodder beet, sugar beet** [9]
- Leatherjackets in **managed amenity turf** [8]
- Millipedes in **fodder beet, sugar beet** [9]
- Pygmy beetle in **fodder beet, sugar beet** [9]
- Sciarid flies in **bedding plants, herbaceous perennials, ornamental plant production, pot plants** [7]; **ornamental plant production** (container grown), **protected ornamentals** [3-6]
- Springtails in **fodder beet, sugar beet** [9]
- Symphylids in **fodder beet, sugar beet** [9]
- Tobacco whitefly in **bedding plants, herbaceous perennials, ornamental plant production, pot plants** [7]; **ornamental plant production** (container grown), **protected ornamentals** [3-6]
- Vine weevil in **bedding plants, herbaceous perennials, ornamental plant production, pot plants** [7]; **ornamental plant production** (container grown), **protected ornamentals** [3-6]
- Whitefly in **protected ornamentals** [2]

Specific Off-Label Approvals (SOLAs)
- **amenity vegetation** 20080298 [1]

Approval information
- Imidacloprid is included in Annex I under EC Directive 91/414
- Accepted by BBPA for use on hops
- Approval expiry 31 Jul 2011 [3, 4]

Efficacy guidance
- Treated seed should be drilled within the season of purchase [9]
- Base of hop plants should be free of weeds and debris at application [1]
- Hop bines emerging away from the main stock or adjacent to poles may require a separate application [1]
- Uptake and movement within hops requires soil moisture and good growing conditions [1]
- Control may be impaired in plantations greater than 3640 plants/ha [1]
- To minimise likelihood of resistance in hops do not treat all the crop in any one yr [1]
- Best results on managed amenity turf achieved when treatment is followed by sufficient irrigation to move chemical through the thatch to wet the top 2.5 cm of soil [8]
- To avoid disturbing uniformity of treatment on turf do not mow until irrigation or rainfall has occurred [8]
- To minimise likelihood of resistance when using in compost adopt a planned programme to alternate with pesticides of different types or use other measures [3, 7]
- When applied as drench or incorporated as granules in moist compost imidacloprid is readily absorbed and translocated to aerial parts of plant [3, 7]

Restrictions
- Maximum number of treatments 1 per batch of seed [9] or growing medium [3, 7]; 1 per yr for hops and managed amenity turf [1, 8]
- Must not be used in compost that has already been treated with an imidacloprid-containing product, nor should compost be subsequently re-treated with imidacloprid [3, 7]
- Product formulated for use only as a compost incorporation treatment into peat-based growing media using suitable automated equipment [3]
- For use only on container grown ornamentals [3, 7]
- Users should test treat a small number of all new or unaccustomed species or varieties to satisfy themselves of the safety of the treatment before applying to the bulk of the crop [3]
- Product must not be used on crops for human or animal consumption and treated compost must not be re-used for this purpose [3, 7]
- Do not use treated seed as food or feed
- Do not treat saturated or waterlogged turf, or when raining or in windy conditions [8]
- Avoid puddling or run-off of irrigation water after treatment [8]
- Do not use grass clippings from treated turf in situ, or use them for mulch [8]

Crop-specific information
- Latest use: before bines reach 2 m or before end 1st wk Jun for hops; before drilling for sugar beet; before sowing or planting for brassicas, bedding plants, hardy ornamental nursery stock, pot plants
- Apply to sugar beet seed as part of the normal commercial pelleting process using special treatment machinery [9]

SEE SECTION 3 FOR PRODUCTS ALSO REGISTERED

Pesticide Profiles

- Apply to hops as a directed stem base spray before most bines reach a height of 2 m. If necessary treat both sides of the crown at half the normal concentration [1]
- The rate used per hop plant must be adjusted in accordance with the plant population if greater than 3640 per ha. See label for details [1]
- On managed amenity turf application should be made prior to egg-laying of the target pests [8]
- Low levels of leaf scorch on some *Agrostis* turf species if application is made in mixture with fertilisers [8]

Environmental safety
- Harmful to aquatic organisms [1]
- High risk to bees. Do not apply to crops in flower or to those in which bees are actively foraging. Do not apply when flowering weeds are present [1, 7]
- Treated seed harmful to game and wildlife [9]

Hazard classification and safety precautions
Hazard Harmful [1, 7, 9]; Irritant [8]
Transport code 9 [7, 9]
Packaging group III [7, 9]
UN number 3077 [7]; 3082 [9]; N.C [5]; N/C [1-4, 6, 8, 10]
Risk phrases R22a [1, 7, 9]; R36 [8]; R43 [9]; R52 [1]
Operator protection A [1-6, 8-10]; C, H [9]; U04a, U13, U14 [9]; U05a [1, 3-10]; U20b [1-3, 5-7]; U20c [4, 8-10]
Environmental protection E12a, E12e [1, 7]; E15a [1-10]; E34 [1, 2, 7, 9]
Storage and disposal D01 [1-3, 5-9]; D02 [1, 3, 5-9]; D09a, D11a [1-10]; D12a [4, 10]
Medical advice M03 [9]

288 indol-3-ylacetic acid

A plant growth regulator for promoting rooting of cuttings

Products

1	Rhizopon A Powder	Rhizopon	1% w/w	DP	09087
2	Rhizopon A Tablets	Rhizopon	50 mg a.i.	WT	09088

Uses
- Rooting of cuttings in *ornamental plant production*

Efficacy guidance
- Apply by dipping end of prepared cuttings into powder or dissolved tablet solution
- Shake off excess powder and make planting holes to prevent powder stripping off [1]
- Consult manufacturer for details of application by spray or total immersion

Restrictions
- Maximum number of treatments 1 per cutting
- Store product in a cool, dark and dry place
- Use solutions once only. Discard after use [2]
- Use plastic, not metallic, container for solutions [2]

Crop-specific information
- Latest use: before cutting insertion

Hazard classification and safety precautions
UN number N/C
Operator protection U19a [1]; U20a [1, 2]
Environmental protection E15a
Storage and disposal D09a, D11a

289 4-indol-3-ylbutyric acid

A plant growth regulator promoting the rooting of cuttings

Products

1	Chryzoplus Grey 0.8%	Rhizopon	0.8% w/w	DP	09094
2	Chryzopon Rose 0.1%	Rhizopon	0.1% w/w	DP	09092
3	Chryzosan White 0.6%	Rhizopon	0.6% w/w	DP	09093

FOR FULL CONDITIONS OF USE ALWAYS READ THE PRODUCT LABEL

indoxacarb

Products – continued

4	Chryzotek Beige	Rhizopon	0.4% w/w	DP	09081
5	Chryzotop Green	Rhizopon	0.25% w/w	DP	09085
6	Rhizopon AA Powder (0.5%)	Rhizopon	0.5% w/w	DP	09082
7	Rhizopon AA Powder (1%)	Rhizopon	1% w/w	DP	09084
8	Rhizopon AA Powder (2%)	Rhizopon	2% w/w	DP	09083
9	Rhizopon AA Tablets	Rhizopon	50 mg a.i.	WT	09086

Uses
- Rooting of cuttings in *ornamental plant production*

Efficacy guidance
- Dip base of cuttings into powder immediately before planting
- Powders or solutions of different concentration are required for different types of cutting. Lowest concentration for softwood, intermediate for semi-ripe, highest for hardwood
- See label for details of concentration and timing recommended for different species
- Use of planting holes recommended for powder formulations to ensure product is not removed on insertion of cutting. Cuttings should be watered in if necessary

Restrictions
- Maximum number of treatments 1 per situation
- Use of too strong a powder or solution may cause injury to cuttings
- No unused moistened powder should be returned to container

Crop-specific information
- Latest use: before cutting insertion for ornamental specimens

Hazard classification and safety precautions
UN number N/C
Operator protection U19a, U20a
Environmental protection E15a
Storage and disposal D09a, D11a

290   indoxacarb

An oxadiazine insecticide for caterpillar control in a range of crops
IRAC mode of action code: 22A

Products

1	Rumo	DuPont	30% w/w	WG	14883
2	Steward	DuPont	30% w/w	WG	13149

Uses
- Caterpillars in *all edible seed crops grown outdoors* (off-label), *all non-edible seed crops grown outdoors* (off-label), *apples*, *cherries* (off-label), *forest nurseries* (off-label), *grapevines* (off-label), *hops* (off-label), *ornamental plant production* (off-label), *pears*, *protected forest nurseries* (off-label), *protected hops* (off-label), *protected soft fruit* (off-label), *protected top fruit* (off-label), *soft fruit* (off-label), *top fruit* (off-label) [2]; *broccoli*, *brussels sprouts* (off-label), *cabbages*, *cauliflowers*, *protected aubergines*, *protected courgettes*, *protected cucumbers*, *protected marrows*, *protected melons*, *protected ornamentals*, *protected peppers*, *protected pumpkins*, *protected squashes*, *protected tomatoes*, *sweetcorn* (off-label) [1, 2]
- Diamond-back moth in *sweetcorn* (off-label) [2]
- Insect pests in *all edible seed crops grown outdoors* (off-label), *all non-edible seed crops grown outdoors* (off-label), *forest nurseries* (off-label), *hops* (off-label), *ornamental plant production* (off-label), *protected forest nurseries* (off-label), *protected hops* (off-label) [1]
- Pollen beetle in *spring oilseed rape*, *winter oilseed rape* [1]

Specific Off-Label Approvals (SOLAs)
- *all edible seed crops grown outdoors* 20101537 [1], 20082905 [2]
- *all non-edible seed crops grown outdoors* 20101537 [1], 20082905 [2]
- *brussels sprouts* 20101537 [1], 20081485 [2]
- *cherries* 20092300 [2]
- *forest nurseries* 20101538 [1], 20082905 [2]
- *grapevines* 20072289 [2]
- *hops* 20101537 [1], 20082905 [2]

SEE SECTION 3 FOR PRODUCTS ALSO REGISTERED

Pesticide Profiles

- **ornamental plant production** *20101537* [1], *20082905* [2]
- **protected forest nurseries** *20101537* [1], *20082905* [2]
- **protected hops** *20101537* [1], *20082905* [2]
- **protected soft fruit** *20082905* [2]
- **protected top fruit** *20082905* [2]
- **soft fruit** *20082905* [2]
- **sweetcorn** *20101536* [1], *20072288* [2]
- **top fruit** *20082905* [2]

Approval information
- Indoxacarb included in Annex I under EC Directive 91/414

Efficacy guidance
- Best results in brassica and protected crops obtained from treatment when first caterpillars are detected, or when damage first seen, or 7-10 d after trapping first adults in pheromone traps
- In apples and pears apply at egg-hatch
- Subsequent treatments in all crops may be applied at 8-14 d intervals
- Indoxacarb acts by ingestion and contact. Only larval stages are controlled but there is some ovicidal action against some species

Restrictions
- Maximum number of treatments 3 per crop or yr for apples, pears, brassica crops; 6 per yr for protected crops
- Do not apply to any crop suffering from stress from any cause

Crop-specific information
- HI: brassica crops, protected crops 1 d; apples, pears 7 d

Environmental safety
- Dangerous for the environment
- Toxic to aquatic organisms
- In accordance with good agricultural practice apply in early morning or late evening when bees are less active
- Broadcast air-assisted LERAP (18 m)

Hazard classification and safety precautions
    Hazard Harmful, Dangerous for the environment
    Transport code 9
    Packaging group III
    UN number 3077
    Risk phrases R22a, R51, R53a
    Operator protection A, D; U04a, U05a
    Environmental protection E15b, E34, E38; E17b (18 m)
    Storage and disposal D01, D02, D09a, D12a
    Medical advice M03

291   iodosulfuron-methyl-sodium

A post-emergence sulfonylurea herbicide for winter sown cereals
HRAC mode of action code: B

See also amidosulfuron + iodosulfuron-methyl-sodium
          diflufenican + iodosulfuron-methyl-sodium + mesosulfuron-methyl

Products
    Hussar   Bayer CropScience   5% w/w   WG   12364

Uses
- Annual dicotyledons in **durum wheat** *(off-label)*, **grass seed crops** *(off-label)*
- Chickweed in **spring barley, triticale, winter rye, winter wheat**
- Cleavers in **triticale, winter rye, winter wheat**
- Fat hen in **spring barley**
- Field pansy in **spring barley**
- Field speedwell in **spring barley, triticale, winter rye, winter wheat**
- Hemp-nettle in **spring barley**

FOR FULL CONDITIONS OF USE ALWAYS READ THE PRODUCT LABEL

iodosulfuron-methyl-sodium

- Italian ryegrass in **triticale** *(from seed)*, **winter rye** *(from seed)*, **winter wheat** *(from seed)*
- Ivy-leaved speedwell in **triticale**, **winter rye**, **winter wheat**
- Mayweeds in **spring barley**, **triticale**, **winter rye**, **winter wheat**
- Perennial ryegrass in **triticale** *(from seed)*, **winter rye** *(from seed)*, **winter wheat** *(from seed)*
- Red dead-nettle in **spring barley**, **triticale**, **winter rye**, **winter wheat**
- Ryegrass in **durum wheat** *(off-label - from seed)*, **grass seed crops** *(off-label - from seed)*
- Volunteer oilseed rape in **spring barley**

Specific Off-Label Approvals (SOLAs)
- **durum wheat** *(from seed)* 20080912, 20080912
- **grass seed crops** *(from seed)* 20080912, 20080912

Approval information
- Iodosulfuron-methyl-sodium included in Annex I under EC Directive 91/414
- Accepted by BBPA for use on malting barley

Efficacy guidance
- Best results obtained from treatment in warm weather when soil is moist and the weeds are growing actively
- Weeds must be present at application to be controlled
- Weed control is slow especially under cool dry conditions
- Dry conditions resulting in moisture stress may reduce effectiveness
- Occasionally weeds may only be stunted but they will normally have little or no competitive effect on the crop
- Iodosulfuron is a member of the ALS-inhibitor group of herbicides. To avoid the build up of resistance do not use any product containing an ALS-inhibitor herbicide with claims for control of grass weeds more than once on any crop
- Use these products as part of a resistance management strategy that includes cultural methods of control and does not use ALS inhibitors as the sole chemical method of grass weed control

Restrictions
- Maximum number of treatments 1 per crop
- Must only be applied between 1 Feb in yr of harvest and specified latest time of application
- Do not apply to undersown crops, or crops to be undersown
- Do not roll or harrow within 1 wk of spraying
- Do not spray crops under stress from any cause or if the soil is compacted
- Treat broadcast crops after the plants have a well-established root system
- Do not spray if rain imminent or frost expected
- Do not apply in mixture or in sequence with any other ALS inhibitor

Crop-specific information
- Latest use: before third node detectable (GS 33)

Following crops guidance
- No restrictions apply to the sowing of cereal crops or sugar beet in the spring of the yr following treatment

Environmental safety
- Dangerous for the environment
- Very toxic to aquatic organisms
- Dangerous to fish or other aquatic life. Do not contaminate surface waters or ditches with chemical or used container
- Take extreme care to avoid damage by drift onto broad-leaved plants outside the target area or onto ponds, waterways and ditches
- Observe carefully label instructions for sprayer cleaning
- LERAP Category B

Hazard classification and safety precautions
    Hazard Irritant, Dangerous for the environment
    Transport code 9
    Packaging group III
    UN number 3077
    Risk phrases R41, R50, R53a
    Operator protection A, C, H; U05a, U08, U11, U14, U15, U20b

SEE SECTION 3 FOR PRODUCTS ALSO REGISTERED

Pesticide Profiles

Environmental protection E15b, E16a, E16b, E38
Storage and disposal D01, D02, D10a, D12a

292 iodosulfuron-methyl-sodium + mesosulfuron-methyl

A sulfonyl urea herbicide mixture for winter wheat
HRAC mode of action code: B + B

See also mesosulfuron-methyl

Products

1	Atlantis WG	Bayer CropScience	0.6:3.0% w/w	WG	12478
2	Greencrop Doonbeg 2	Greencrop	0.6:3.0 % w/w	WG	13025
3	Hatra	Bayer CropScience	2:10 g/l	OD	14524
4	Horus	Bayer CropScience	2:10 g/l	OD	14541
5	Nautilus	ChemSource	2:10 g/l	OD	14961
6	Nemo	AgChem Access	0.6:3.0% w/w	WG	14140
7	Pacifica	Bayer CropScience	1.0:3.0% w/w	WG	12049

Uses
- Annual dicotyledons in **durum wheat** *(off-label)*, **spring rye** *(off-label)*, **triticale** *(off-label)*, **winter rye** *(off-label)* [1, 7]; **spring wheat** *(off-label)* [1]
- Annual grasses in **durum wheat** *(off-label)*, **spring rye** *(off-label)*, **triticale** *(off-label)*, **winter rye** *(off-label)* [1, 7]; **spring wheat** *(off-label)* [1]
- Annual meadow grass in **winter wheat** [1, 2, 6, 7]; **winter wheat** *(post-em up to GS31)* [3-5]
- Blackgrass in **winter wheat** [1, 2, 6, 7]; **winter wheat** *(post em up to GS39 (EMR resistant up to GS29))* [3-5]
- Chickweed in **winter wheat** [1, 2, 6, 7]; **winter wheat** *(post-em up to 8 true leaves)* [3-5]
- Italian ryegrass in **winter wheat** [1, 2, 6, 7]; **winter wheat** *(post-em up to GS30)* [3-5]
- Mayweeds in **winter wheat** [1, 2, 6, 7]; **winter wheat** *(post-em up to 8 true leaves)* [3-5]
- Perennial ryegrass in **winter wheat** [1, 2, 6, 7]; **winter wheat** *(post em up to GS31)* [3-5]
- Rough meadow grass in **winter wheat** [1, 2, 6, 7]; **winter wheat** *(post em up to GS29)* [3-5]
- Wild oats in **winter wheat** [1, 2, 6, 7]; **winter wheat** *(post em up to GS29)* [3-5]

Specific Off-Label Approvals (SOLAs)
- **durum wheat** *20063529* [1], *20081264* [7]
- **spring rye** *20063529* [1], *20081264* [7]
- **spring wheat** *20082946* [1]
- **triticale** *20063529* [1], *20081264* [7]
- **winter rye** *20063529* [1], *20081264* [7]

Approval information
- Iodosulfuron-methyl-sodium and mesosulfuron-methyl included in Annex I under EC Directive 91/414

Efficacy guidance
- Optimum grass weed control obtained when all grass weeds are emerged at spraying. Activity is primarily via foliar uptake and good spray coverage of the target weeds is essential
- Translocation occurs readily within the target weeds and growth is inhibited within hours of treatment but symptoms may not be apparent for up to 4 wk, depending on weed species, timing of treatment and weather conditions
- Residual activity is important for best results and is optimised by treatment on fine moist seedbeds. Avoid application under very dry conditions
- Residual efficacy may be reduced by high soil temperatures and cloddy seedbeds
- Iodosulfuron-methyl and mesosulfuron-methyl are both members of the ALS-inhibitor group of herbicides. To avoid the build up of resistance do not use any product containing an ALS-inhibitor herbicide with claims for control of grass weeds more than once on any crop
- Use these products as part of a Resistance Management Strategy that includes cultural methods of control and does not use ALS inhibitors as the sole chemical method of grass weed control. See Section 5 for more information

Restrictions
- Maximum number of treatments 1 per crop with a maximum total dose equivalent to one full dose treatment

FOR FULL CONDITIONS OF USE ALWAYS READ THE PRODUCT LABEL

- Do not use on crops undersown with grasses, clover or other legumes or any other broad-leaved crop
- Do not use as a stand-alone treatment for blackgrass, ryegrass or chickweed control
- Do not use as the sole means of weed control in successive crops
- Do not use in mixture or in sequence with any other ALS-inhibitor herbicide except those (if any) specified on the label
- Do not mix with products containing isoproturon, chlorotoluron or linuron and allow at least 28 d after use of a residual urea herbicide before application, and at least 10 d after application before use of a residual urea herbicide [2]
- Do not apply earlier than 1 Feb in the yr of harvest [7]
- Do not apply to crops under stress from any cause
- Do not apply when rain is imminent or during periods of frosty weather
- Specified adjuvant must be used. See label

Crop-specific information
- Latest use: flag leaf just visible (GS 39)
- Winter wheat may be treated from the two-leaf stage of the crop
- Safety to crops grown for seed not established

Following crops guidance
- In the event of crop failure sow only winter wheat in the same cropping season
- Only winter wheat or winter barley (or winter oilseed rape [1]) may be sown in the year of harvest of a treated crop
- Spring wheat, spring barley, sugar beet or spring oilseed rape may be drilled in the following spring. Plough before drilling oilseed rape

Environmental safety
- Dangerous for the environment
- Very toxic to aquatic organisms
- Dangerous to fish or other aquatic life. Do not contaminate surface waters or ditches with chemical or used container
- Take extreme care to avoid drift onto plants outside the target area or on to ponds, waterways or ditches
- LERAP Category B

Hazard classification and safety precautions
Hazard Irritant, Dangerous for the environment
Transport code 9
Packaging group III
UN number 3077 [1, 2, 6, 7]; 3082 [3-5]
Risk phrases R36, R38 [3-5]; R40 [7]; R41 [1, 2, 6, 7]; R50, R53a [1-7]
Operator protection A, C, H; U05a, U11, U20b
Environmental protection E15a [1, 3-7]; E15b [2]; E16a, E38 [1-7]
Storage and disposal D01, D02, D05, D09a, D10a, D12a

293 ioxynil

A contact acting HBN herbicide for use in turf and onions
HRAC mode of action code: C3

See also bromoxynil + diflufenican + ioxynil
bromoxynil + fluroxypyr + ioxynil
bromoxynil + ioxynil
bromoxynil + ioxynil + mecoprop-P
bromoxynil + ioxynil + triasulfuron
dichlorprop-P + ioxynil

Products
  Totril   Bayer CropScience   225 g/l   EC   14633

Uses
- Annual dicotyledons in **bulb onions**, **garlic**, **leeks**, **salad onions**, **shallots**

SEE SECTION 3 FOR PRODUCTS ALSO REGISTERED

Pesticide Profiles

Approval information
- Ioxynil included in Annex I under EC Directive 91/414
- Ioxynil was reviewed in 1995 and approvals for home garden use, and most hand held applications revoked
- Accepted by BBPA for use on malting barley

Efficacy guidance
- Best results on seedling to 4-leaf stage weeds in active growth during mild weather

Restrictions
- Maximum number of treatments up to 4 per crop at split doses on onions; 1 per crop on carrots, garlic, leeks, shallots
- Do not apply by hand-held equipment or at concentrations higher than those recommended

Crop-specific information
- Latest use: pre-emergence for carrots, parsnips
- HI bulb onions, salad onions, shallots, garlic, leeks 14 d
- Apply to sown onion crops as soon as possible after plants have 3 true leaves or to transplanted crops when established

Environmental safety
- Dangerous for the environment
- Very toxic to aquatic organisms
- Flammable
- Harmful to bees. Do not apply to crops in flower or to those in which bees are actively foraging. Do not apply when flowering weeds are present
- Keep livestock out of treated areas for at least 6 wk after treatment and until foliage of any poisonous weeds such as ragwort has died and become unpalatable

Hazard classification and safety precautions
Hazard Harmful, Flammable, Dangerous for the environment
Transport code 3
Packaging group III
UN number 1993
Risk phrases R22a, R22b, R36, R37, R43, R50, R53a, R63, R66, R67
Operator protection A, C; U05a, U08, U19a, U20b, U23a
Environmental protection E06a (6 wk); E12d, E12e, E15a, E34, E38
Storage and disposal D01, D02, D05, D09a, D10b, D12a
Medical advice M03, M05b

294 ipconazole

A triazole fungicide for disease control in cereals
FRAC mode of action code: 3

Products
Rancona 15 ME   Chemtura   15 g/l   MS   14407

Uses
- Seed-borne diseases in *spring barley*, *spring wheat*, *winter barley*, *winter wheat*
- Soil-borne diseases in *spring barley*, *spring wheat*, *winter barley*, *winter wheat*

Approval information
- Annex 1 approval is pending
- Accepted by BBPA for use on malting barley

Hazard classification and safety precautions
UN number N/C
Operator protection A, H; U05a, U09a, U20b
Environmental protection E03, E15b, E34
Storage and disposal D01, D02, D09a, D10a
Treated seed S01, S02, S03, S04d, S05, S07
Medical advice M03

FOR FULL CONDITIONS OF USE ALWAYS READ THE PRODUCT LABEL

iprodione

295 iprodione

A protectant dicarboximide fungicide with some eradicant activity
FRAC mode of action code: 2

See also carbendazim + iprodione

Products

1	Chipco Green	Bayer Environ.	255 g/l	SC	13843
2	Green Turf King	Standon	255 g/l	SC	14348
3	Mascot Rayzor	Rigby Taylor	255 g/l	SC	14903
4	Panto	Pan Amenity	255 g/l	SC	14307
5	Pure Turf	Pure Amenity	255 g/l	SC	14851
6	Rovral AquaFlo	BASF	500 g/l	SC	14206
7	Rovral WG	BASF	75% w/w	WG	13811
8	Superturf	AgriGuard	255 g/l	SC	14495

Uses

- Alternaria in **brassica seed crops** *(pre-storage only)*, **brussels sprouts, cauliflowers, chicory** *(off-label)*, **dwarf beans** *(off-label)*, **french beans** *(off-label)*, **garlic** *(off-label)*, **grapevines** *(off-label)*, **mange-tout peas** *(off-label)*, **protected herbs (see appendix 6)** *(off-label)*, **runner beans** *(off-label)*, **shallots** *(off-label)*, **spring oilseed rape, winter oilseed rape, witloof** *(off-label)* [7]; **broccoli** *(off-label)*, **brussels sprouts** *(off-label)*, **cabbages** *(off-label)*, **calabrese** *(off-label)*, **cauliflowers** *(off-label)*, **chinese cabbage** *(off-label)*, **collards** *(off-label)*, **kale** *(off-label)* [6]
- Botrytis in **bulb onions, cabbages** *(off-label)*, **chicory** *(off-label)*, **dwarf beans** *(off-label)*, **forest nurseries** *(off-label)*, **french beans** *(off-label)*, **garlic** *(off-label)*, **grapevines** *(off-label)*, **hops** *(off-label)*, **lettuce, mange-tout peas** *(off-label)*, **ornamental plant production, pears** *(off-label)*, **protected aubergines** *(off-label)*, **protected forest nurseries** *(off-label)*, **protected herbs (see appendix 6)** *(off-label)*, **protected hops** *(off-label)*, **protected lettuce, protected rhubarb** *(off-label)*, **protected soft fruit** *(off-label)*, **protected strawberries, protected tomatoes, protected top fruit** *(off-label)*, **raspberries, red beet** *(off-label)*, **runner beans** *(off-label)*, **salad onions, shallots** *(off-label)*, **soft fruit** *(off-label)*, **spring oilseed rape, strawberries, top fruit** *(off-label)*, **winter oilseed rape, witloof** *(off-label)* [7]
- Brown patch in **amenity grassland** [1, 5]; **managed amenity turf** [1, 5, 8]
- Collar rot in **bulb onions, garlic** *(off-label)*, **salad onions, shallots** *(off-label)* [7]
- Disease control in **managed amenity turf** [3, 4]
- Dollar spot in **amenity grassland** [1, 5]; **managed amenity turf** [1, 5, 8]
- Fusarium patch in **amenity grassland** [1, 5]; **managed amenity turf** [1, 2, 5, 8]
- Melting out in **amenity grassland** [1, 5]; **managed amenity turf** [1, 5, 8]
- Red thread in **amenity grassland** [1, 5]; **managed amenity turf** [1, 2, 5, 8]
- Seed-borne diseases in **ornamental crop production (seed), seed potatoes** [6]
- Snow mould in **amenity grassland** [1, 5]; **managed amenity turf** [1, 5, 8]

Specific Off-Label Approvals (SOLAs)

- *broccoli* 20090708 [6]
- *brussels sprouts* 20090708 [6]
- *cabbages* 20090708 [6], 20091845 [7]
- *calabrese* 20090708 [6]
- *cauliflowers* 20090708 [6]
- *chicory* 20081759 expires 31 Jul 2011 [7]
- *chinese cabbage* 20090708 [6]
- *collards* 20090708 [6]
- *dwarf beans* 20081760 [7]
- *forest nurseries* 20082902 [7]
- *french beans* 20081760 [7]
- *garlic* 20081768 [7]
- *grapevines* 20081758 [7]
- *hops* 20082902 [7]
- *kale* 20090708 [6]
- *mange-tout peas* 20081760 [7]
- *pears* 20081761 [7]
- *protected aubergines* 20081767 [7]

SEE SECTION 3 FOR PRODUCTS ALSO REGISTERED

Pesticide Profiles

- ***protected forest nurseries*** *20082902* [7]
- ***protected herbs (see appendix 6)*** *20081757* [7]
- ***protected hops*** *20082902* [7]
- ***protected rhubarb*** *20081763* [7]
- ***protected soft fruit*** *20082902* [7]
- ***protected top fruit*** *20082902* [7]
- ***red beet*** *20081762* [7]
- ***runner beans*** *20081760* [7]
- ***shallots*** *20081768* [7]
- ***soft fruit*** *20082902* [7]
- ***top fruit*** *20082902* [7]
- ***witloof*** *20081759 expires 31 Jul 2011* [7]

Approval information
- Iprodione included in Annex I under EC Directive 91/414
- Accepted by BBPA for use on malting barley

Efficacy guidance
- Many diseases require a programme of 2 or more sprays at intervals of 2-4 wk. Recommendations vary with disease and crop - see label for details
- Use as a drench to control cabbage storage diseases. Spray ornamental pot plants and cucumbers to run-off
- Apply turf and amenity grass treatments after mowing to dry grass, free of dew. Do not mow again for at least 24 h [1]

Restrictions
- Maximum number of treatments or maximum total dose equivalent to 1 per batch for seed treatments; 1 per crop on cereals, cabbage (as drench), 2 per crop on field beans and stubble turnips; 3 per crop on brassicas (including seed crops), oilseed rape, protected winter lettuce (Oct-Feb); 4 per crop on strawberries, grapevines, salad onions, cucumbers; 5 per crop on raspberries; 6 per crop on bulb onions, tomatoes, turf; 7 per crop on lettuce (Mar-Sep)
- A minimum of 3 wk must elapse between treatments on leaf brassicas
- Treated brassica seed crops not to be used for human or animal consumption
- See label for pot plants showing good tolerance. Check other species before applying on a large scale
- Do not treat oilseed rape seed that is cracked or broken or of low viability
- Do not treat oats

Crop-specific information
- Latest use: pre-planting for seed treatments; at planting for potatoes; before grain watery ripe (GS 69) for wheat and barley; 21 d before consumption by livestock for stubble turnips
- HI strawberries, protected tomatoes 1 d; outdoor tomatoes 2 d; bulb onions, salad onions, raspberries, lettuce 7 d; brassicas, brassica seed crops, oilseed rape, mustard, field beans, stubble turnips 21 d
- Turf and amenity grass may become temporarily yellowed if frost or hot weather follows treatment
- Personal protective equipment requirements may vary for each pack size. Check label

Environmental safety
- Dangerous for the environment
- Very toxic to aquatic organisms
- See label for guidance on disposal of spent drench liquor
- Treatment harmless to *Encarsia* or *Phytoseiulus* being used for integrated pest control
- Broadcast air-assisted LERAP [7] (18 m); LERAP Category B [2, 7]

Hazard classification and safety precautions
Hazard Harmful, Dangerous for the environment
Transport code 9
Packaging group III
UN number 3077 [7]; 3082 [1-6, 8]
Risk phrases R36 [1, 3-5, 7, 8]; R38 [1, 3-5, 8]; R40 [1, 3-8]; R50, R53a [1-8]
Operator protection A [1-8]; C [1, 3-5, 7, 8]; H [6-8]; P [2]; U04a, U08 [1, 3-5, 8]; U05a [1-8]; U09a [2]; U11 [6, 7]; U16b, U19a, U20b [7]; U20c [1-6, 8]

FOR FULL CONDITIONS OF USE ALWAYS READ THE PRODUCT LABEL

iprodione + thiophanate-methyl

Environmental protection E15a [1, 3-6, 8]; E15b, E16a [2, 7]; E16b [2]; E17b [7] (18 m); E34 [1-8]; E38 [1, 3-8]
Storage and disposal D01, D02, D09a [1-8]; D05, D10b [1, 3-5, 8]; D10a [2]; D11a [7]; D12a [1, 3-8]
Treated seed S01, S02, S03, S04a, S05 [6]
Medical advice M03 [1, 3-6, 8]; M04a [2]; M05a [2, 6, 7]

296 iprodione + thiophanate-methyl

A protectant and systemic fungicide for oilseed rape
FRAC mode of action code: 2 + 1

See also thiophanate-methyl

Products
   Compass   BASF   167:167 g/l   SC   11740

Uses
- Alternaria in **carrots** *(off-label)*, **horseradish** *(off-label)*, **parsnips** *(off-label)*, **spring oilseed rape**, **winter oilseed rape**
- Botrytis in **meadowfoam (limnanthes alba)** *(off-label)*
- Chocolate spot in **spring field beans**, **winter field beans**
- Crown rot in **carrots** *(off-label)*, **horseradish** *(off-label)*, **parsnips** *(off-label)*
- Grey mould in **spring oilseed rape**, **winter oilseed rape**
- Light leaf spot in **spring oilseed rape**, **winter oilseed rape**
- Sclerotinia stem rot in **spring oilseed rape**, **winter oilseed rape**
- Stem canker in **spring oilseed rape**, **winter oilseed rape**

Specific Off-Label Approvals (SOLAs)
- *carrots* 20040525
- *horseradish* 20040525
- *meadowfoam (limnanthes alba)* 20040507
- *parsnips* 20040525

Approval information
- Iprodione and thiophanate-methyl included in Annex I under EC Directive 91/414

Efficacy guidance
- Timing of sprays on oilseed rape varies with disease, see label for details
- For season-long control product should be applied as part of a disease control programme

Restrictions
- Maximum number of treatments (refers to total sprays containing benomyl, carbendazim or thiophanate methyl) 2 per crop for oilseed rape, field beans, meadowfoam; 3 per crop for carrots, horseradish, parsnips

Crop-specific information
- Latest use: before end of flowering for oilseed rape, field beans
- HI meadowfoam, field beans, oilseed rape 3 wk; carrots, horseradish, parsnips 28 d
- Treatment may extend duration of green leaf in winter oilseed rape

Environmental safety
- Dangerous for the environment
- Very toxic to aquatic organisms

Hazard classification and safety precautions
   Hazard Harmful, Dangerous for the environment
   Transport code 9
   Packaging group III
   UN number 3082
   Risk phrases R40, R43, R50, R53a, R68
   Operator protection A, C, H, M; U05a, U20c
   Environmental protection E15a, E38
   Storage and disposal D01, D02, D08, D09a, D10b, D12a
   Medical advice M05a

SEE SECTION 3 FOR PRODUCTS ALSO REGISTERED

Pesticide Profiles

297 isoproturon

A residual urea herbicide for use in cereals; All approvals expired 30 June 2009.
HRAC mode of action code: C2

298 isopyrzam

A broad spectrum fungicide for use in barley available only in mixtures
FRAC mode of action code: 7

See also cyprodinil + isopyrazam

299 isoxaben

A soil-acting benzamide herbicide for use in grass and fruit
HRAC mode of action code: L

Products

1	Agriguard Isoxaben	AgriGuard	125 g/l	SC	11652
2	Flexidor 125	Landseer	125 g/l	SC	10946

Uses
- Annual dicotyledons in **almonds** *(off-label)*, **amenity vegetation**, **apple orchards**, **blackberries**, **blackcurrants**, **cherries**, **chestnuts** *(off-label)*, **forestry transplants**, **gooseberries**, **hazel nuts** *(off-label)*, **hops**, **ornamental plant production**, **pear orchards**, **plums**, **raspberries**, **rubus hybrids** *(off-label)*, **strawberries**, **walnuts** *(off-label)*, **wine grapes** [1, 2]; **bilberries** *(off-label)*, **blueberries** *(off-label)*, **canary grass** *(off-label - grown for game cover)*, **carrots** *(off-label)*, **clovers** *(off-label - grown for game cover)*, **courgettes** *(off-label)*, **cranberries** *(off-label)*, **forage maize** *(off-label - grown for game cover)*, **marrows** *(off-label)*, **millet** *(off-label - grown for game cover)*, **protected carrots** *(off-label - temporary protection)*, **protected courgettes** *(off-label)*, **protected marrows** *(off-label)*, **protected ornamentals** *(off-label)*, **protected parsnips** *(off-label - temporary protection)*, **protected pumpkins** *(off-label)*, **protected squashes** *(off-label)*, **pumpkins** *(off-label)*, **quinoa** *(off-label - grown for game cover)*, **redcurrants** *(off-label)*, **squashes** *(off-label)*, **sunflowers** *(off-label - grown for game cover)*, **whitecurrants** *(off-label)* [2]; **table grapes** [1]
- Cleavers in **asparagus** *(off-label)* [2]
- Fat hen in **asparagus** *(off-label)* [2]
- Groundsel in **asparagus** *(off-label)* [2]
- Knotgrass in **asparagus** *(off-label)* [2]

Specific Off-Label Approvals (SOLAs)
- *almonds* 20061109 [1], 20061111 [2]
- *asparagus* 20050893 [2]
- *bilberries* 20061113 [2]
- *blueberries* 20061113 [2]
- *canary grass* (grown for game cover) 20050896 [2]
- *carrots* 20050895 [2]
- *chestnuts* 20061109 [1], 20061111 [2]
- *clovers* (grown for game cover) 20050896 [2]
- *courgettes* 20050894 [2]
- *cranberries* 20061113 [2]
- *forage maize* (grown for game cover) 20050896 [2]
- *hazel nuts* 20061109 [1], 20061111 [2]
- *marrows* 20050894 [2]
- *millet* (grown for game cover) 20050896 [2]
- *protected carrots* (temporary protection) 20050892 [2]
- *protected courgettes* 20050894 [2]
- *protected marrows* 20050894 [2]
- *protected ornamentals* 20050891 [2]
- *protected parsnips* (temporary protection) 20050892 [2]

FOR FULL CONDITIONS OF USE ALWAYS READ THE PRODUCT LABEL

isoxaben + terbuthylazine

- *protected pumpkins* 20050894 [2]
- *protected squashes* 20050894 [2]
- *pumpkins* 20050894 [2]
- *quinoa* (grown for game cover) 20050896 [2]
- *redcurrants* 20061113 [2]
- *rubus hybrids* 20061110 [1], 20061112 [2]
- *squashes* 20050894 [2]
- *sunflowers* (grown for game cover) 20050896 [2]
- *walnuts* 20061109 [1], 20061111 [2]
- *whitecurrants* 20061113 [2]

Approval information
- Accepted by BBPA for use on hops
- Isoxaben not included in Annex 1.

Efficacy guidance
- When used alone apply pre-weed emergence
- Effectiveness is reduced in dry conditions. Weed seeds germinating at depth are not controlled
- Activity reduced on soils with more than 10% organic matter. Do not use on peaty soils
- Various tank mixtures are recommended for early post-weed emergence treatment (especially for grass weeds). See label for details

Restrictions
- Maximum number of treatments 1 per crop for all edible crops; 2 per yr on amenity vegetation and non-edible crops

Crop-specific information
- Latest use: before 1 Apr in yr of harvest for edible crops

Following crops guidance
- See label for details of crops which may be sown in the event of failure of a treated crop

Environmental safety
- Keep all livestock out of treated areas for at least 50 d

Hazard classification and safety precautions
UN number N/C
Operator protection A, C; U05a, U20b
Environmental protection E06a (50 d); E15a
Storage and disposal D01, D09a, D11a [1, 2]; D05 [2]

300 isoxaben + terbuthylazine

A contact and residual herbicide for use in peas and spring field beans
HRAC mode of action code: L + C1

See also terbuthylazine

Products
Skirmish	Syngenta	75:420 g/l	SC	08444

Uses
- Annual dicotyledons in *broad beans* (off-label), *combining peas*, *lupins* (off-label), *spring field beans*, *vining peas*
- Annual grasses in *broad beans* (off-label)

Specific Off-Label Approvals (SOLAs)
- *broad beans* 20070076
- *lupins* 20060540

Approval information
- Isoxaben and terbuthylazine not included in Annex 1. Approvals will expire 2021.

Efficacy guidance
- May be applied pre- or post-emergence of crop but before second node stage (GS 102)
- Product will give residual control of germinating weeds on mineral soils for up to 8 wk
- Product is slow acting and control may not be evident for 7-10 d or more after spraying

SEE SECTION 3 FOR PRODUCTS ALSO REGISTERED

Pesticide Profiles

- Best results achieved when soil surface damp and with a fine, firm tilth. Do not use on very cloddy or stony soil
- Rain after spraying will normally improve weed control but excessive rainfall, very dry conditions or unusually low soil temperatures may lead to unsatisfactory control

Restrictions
- Maximum total dose equivalent to one full dose treatment on peas and spring field beans
- Do not use on forage peas
- Do not use on soils lighter than Coarse Sandy Loam, on very stony soils or soils with more than 10% organic matter
- Pea seed should be covered by at least 25 mm of soil
- Heavy rain after application may cause some crop damage, especially on light soils. Do not use on soils where surface water is likely to accumulate
- For post-emergence treatment a crystal violet test for cuticle wax is advised

Crop-specific information
- Latest use: before second node stage (GS 102) for peas; pre-emergence for spring field beans; beans that are emerging at application will be severely damaged or killed
- Product may be used on all varieties of spring sown vining and combining peas and spring field beans but pea varieties Vedette and Printana may be damaged from which recovery may not be complete

Environmental safety
- Dangerous for the environment
- Very toxic to aquatic organisms

Hazard classification and safety precautions
Hazard Harmful, Dangerous for the environment
Transport code 9
Packaging group III
UN number 3082
Risk phrases R22a, R50, R53a
Operator protection A; U05a, U20a
Environmental protection E15a, E38
Storage and disposal D01, D02, D05, D07, D09a, D10c, D12a

301 isoxaflutole

An isoxazole herbicide available only in mixtures
HRAC mode of action code: F2

See also flufenacet + isoxaflutole

302 kresoxim-methyl

A protectant strobilurin fungicide for apples
FRAC mode of action code: 11

See also epoxiconazole + fenpropimorph + kresoxim-methyl
epoxiconazole + kresoxim-methyl
epoxiconazole + kresoxim-methyl + pyraclostrobin
fenpropimorph + kresoxim-methyl

Products

1	Kresoxy 50 WG	AgriGuard	50% w/w	WG	12275
2	Stroby WG	BASF	50% w/w	WG	08653

Uses
- Black spot in **protected roses**, **roses** [1, 2]
- Powdery mildew in **all edible seed crops grown outdoors** (off-label), **all non-edible seed crops grown outdoors** (off-label), **gooseberries** (off-label), **hops** (off-label), **protected soft fruit** (off-label), **quinces** (off-label), **redcurrants** (off-label), **soft fruit** (off-label), **table grapes** (off-label),

FOR FULL CONDITIONS OF USE ALWAYS READ THE PRODUCT LABEL

kresoxim-methyl

whitecurrants (off-label), *wine grapes (off-label)* [2]; *apples (reduction)*, *blackcurrants*, *protected roses*, *protected strawberries*, *roses*, *strawberries* [1, 2]
- Scab in *apples* [1, 2]

Specific Off-Label Approvals (SOLAs)
- *all edible seed crops grown outdoors* 20082917 expires 31 Dec 2011 [2]
- *all non-edible seed crops grown outdoors* 20082917 expires 31 Dec 2011 [2]
- *gooseberries* 20060656 expires 31 Dec 2011 [2]
- *hops* 20082917 expires 31 Dec 2011 [2]
- *protected soft fruit* 20082917 expires 31 Dec 2011 [2]
- *quinces* 20060657 expires 31 Dec 2011 [2]
- *redcurrants* 20060656 expires 31 Dec 2011 [2]
- *soft fruit* 20082917 expires 31 Dec 2011 [2]
- *table grapes* 20092711 expires 31 Dec 2011 [2]
- *whitecurrants* 20060656 expires 31 Dec 2011 [2]
- *wine grapes* 20092711 expires 31 Dec 2011 [2]

Approval information
- Kresoxim-methyl included in Annex I under EC Directive 91/414
- Accepted by BBPA on malting barley
- Approved for use in ULV systems
- Approval expiry 31 Dec 2011 [1, 2]

Efficacy guidance
- Activity is protectant. Best results achieved from treatments prior to disease development. See label for timing details on each crop. Treatments should be repeated at 10-14 d intervals but note limitations below
- To minimise the likelihood of development of resistance to strobilurin fungicides these products should be used in a planned Resistance Management strategy. See Section 5 for more information
- Product may be applied in ultra low volumes (ULV) but disease control may be reduced

Restrictions
- Maximum number of treatments 4 per yr on apples; 3 per yr on other crops. See notes in Efficacy about limitations on consecutive treatments
- Consult before using on crops intended for processing

Crop-specific information
- HI 14 d for blackcurrants, protected strawberries, strawberries; 35 d for apples
- On apples do not spray product more than twice consecutively and separate each block of two consecutive treatments with at least two applications from a different cross-resistance group. For all other crops do not apply consecutively and use a maximum of once in every three fungicide sprays
- Product should not be used as final spray of the season on apples

Environmental safety
- Dangerous for the environment
- Very toxic to aquatic organisms
- Harmless to ladybirds and predatory mites
- Harmless to honey bees and may be applied during flowering. Nevertheless local beekeepers should be notified when treatment of orchards in flower is to occur
- Broadcast air-assisted LERAP (5 m)

Hazard classification and safety precautions
Hazard Harmful, Dangerous for the environment
Transport code 9
Packaging group III
UN number 3077
Risk phrases R40, R50, R53a
Operator protection A, H, J [1]; U05a, U20b
Environmental protection E15a, E38 [1, 2]; E16b [1]; E17b [1, 2] (5 m)
Storage and disposal D01, D02, D09a, D10c, D12a [1, 2]; D05 [1]; D08 [2]
Medical advice M05a

SEE SECTION 3 FOR PRODUCTS ALSO REGISTERED

Pesticide Profiles

303　lambda-cyhalothrin

A quick-acting contact and ingested pyrethroid insecticide
IRAC mode of action code: 3

Products

1	Clayton Lanark	Clayton	100 g/l	CS	12942
2	Clayton Sparta	Clayton	50 g/l	EC	13457
3	Hallmark with Zeon Technology	Syngenta	100 g/l	CS	12629
4	Jackpot 2	AgriGuard	100 g/l	CS	12888
5	Karate 2.5WG	Syngenta	2.5% w/w	GR	14060
6	Major	Makhteshim	50 g/l	CS	14527
7	Markate 50	Agrovista	50 g/l	EC	13529
8	Seal Z	AgChem Access	100 g/l	CS	14201
9	Warrior	Syngenta	100 g/l	CS	13857

Uses

- Aphids in **all edible seed crops grown outdoors** *(off-label)*, **all non-edible seed crops grown outdoors** *(off-label)*, **farm forestry** *(off-label)*, **miscanthus** *(off-label)* [3, 6]; **borage for oilseed production** *(off-label)*, **bulb onions** *(off-label)*, **canary flower (echium spp.)** *(off-label)*, **evening primrose** *(off-label)*, **frise** *(off-label)*, **garlic** *(off-label)*, **grass seed crops** *(off-label)*, **honesty** *(off-label)*, **lamb's lettuce** *(off-label)*, **leaf brassicas** *(off-label)*, **leeks** *(off-label)*, **lupins** *(off-label)*, **mustard** *(off-label)*, **outdoor herbs** *(off-label)*, **salad onions** *(off-label)*, **shallots** *(off-label)*, **short rotation coppice willow** *(off-label)*, **spring linseed** *(off-label)*, **spring rye** *(off-label)*, **triticale** *(off-label)*, **winter linseed** *(off-label)*, **winter rye** *(off-label)* [6]; **chestnuts** *(off-label)*, **hazel nuts** *(off-label)*, **walnuts** *(off-label)* [3, 6, 7]; **cob nuts** *(off-label)*, **edible podded peas, spring field beans, spring oilseed rape, sugar beet, winter field beans, winter oilseed rape** [7]; **combining peas, vining peas** [2, 7]; **crambe** *(off-label)*, **forest nurseries** *(off-label)*, **hops** *(off-label)*, **ornamental plant production** *(off-label)*, **protected forest nurseries** *(off-label)*, **protected ornamentals** *(off-label)*, **protected pak choi** *(off-label)*, **protected tat soi** *(off-label)*, **soft fruit** *(off-label)*, **top fruit** *(off-label)*, **willow (short rotation coppice)** *(off-label)* [3]; **durum wheat** [1, 3-5, 7-9]; **fodder beet** *(off-label)* [1, 6]; **potatoes** [1-4, 6-9]; **spring barley, spring oats** [1, 3, 4, 8, 9]; **spring wheat** [1-4, 7-9]; **winter barley, winter oats** [1, 3-5, 8, 9]; **winter wheat** [1-5, 7-9]
- Barley yellow dwarf virus vectors in **durum wheat, winter barley, winter oats, winter wheat** [1, 3-5, 8, 9]; **spring barley, spring wheat** [4, 9]; **spring oats** [9]
- Beet leaf miner in **sugar beet** [1, 3, 4, 6, 8, 9]
- Beet virus yellows vectors in **spring oilseed rape** [1, 3, 4, 8, 9]; **winter oilseed rape** [1, 3-5, 8, 9]
- Beetles in **combining peas, durum wheat, edible podded peas, potatoes, spring field beans, spring oilseed rape, spring wheat, sugar beet, vining peas, winter field beans, winter oilseed rape, winter wheat** [7]
- Brassica pod midge in **spring oilseed rape, winter oilseed rape** [2]
- Cabbage seed weevil in **spring oilseed rape, winter oilseed rape** [1-4, 8]
- Cabbage stem flea beetle in **spring oilseed rape** [1, 3, 4, 6, 8, 9]; **winter oilseed rape** [1, 3-6, 8, 9]
- Cabbage stem weevil in **celeriac** *(off-label)*, **radishes** *(off-label)* [1, 6, 7]
- Capsids in **blackberries** *(off-label)*, **dewberries** *(off-label)*, **raspberries** *(off-label)*, **rubus hybrids** *(off-label)* [1, 3, 6, 7]; **grapevines** *(off-label)* [1, 3, 7]; **outdoor grapes** *(off-label)* [6]
- Carrot fly in **carrots, parsnips** [9]; **carrots** *(off-label)*, **parsnips** *(off-label)* [7]; **celeriac** *(off-label)*, **radishes** *(off-label)* [1, 6, 7]; **celery (outdoor)** *(off-label)*, **horseradish** *(off-label)*, **mallow (althaea spp.)** *(off-label)*, **parsley root** *(off-label)* [1, 3, 6, 7]; **fennel** *(off-label)* [3, 6]
- Caterpillars in **beetroot** *(off-label)* [1, 3, 6]; **broccoli, brussels sprouts, cabbages, calabrese, cauliflowers** [1, 3, 4, 8, 9]; **celery (outdoor)** *(off-label)*, **navy beans** *(off-label)* [1, 3, 6, 7]; **combining peas, durum wheat, edible podded peas, potatoes, red beet** *(off-label)*, **spring field beans, spring oilseed rape, spring wheat, sugar beet, vining peas, winter field beans, winter oilseed rape, winter wheat** [7]; **dwarf beans** *(off-label)* [6, 7]; **french beans** *(off-label)* [1, 3, 7]; **protected pak choi** *(off-label)*, **protected tat soi** *(off-label)* [6]; **runner beans** *(off-label)* [1, 6, 7]; **swedes** *(off-label)*, **turnips** *(off-label)* [3]
- Clay-coloured weevil in **blackberries** *(off-label)*, **dewberries** *(off-label)*, **raspberries** *(off-label)*, **rubus hybrids** *(off-label)* [1, 3, 6, 7]

FOR FULL CONDITIONS OF USE ALWAYS READ THE PRODUCT LABEL

lambda-cyhalothrin

- Cutworms in **beetroot** *(off-label)* [1, 3, 6]; **carrots** [1, 3, 4, 6, 8]; **celeriac** *(off-label)*, **radishes** *(off-label)* [1, 6, 7]; **chicory** *(off-label)*, **fennel** *(off-label)* [3, 6]; **fodder beet** *(off-label)*, **red beet** *(off-label)* [7]; **lettuce** [1, 3, 4, 8]; **outdoor lettuce** [6, 9]; **parsnips** [3, 6, 8]; **potatoes** *(If present at time of application)* [6]; **sugar beet** [1-4, 6, 8, 9]
- Flea beetle in **crambe** *(off-label)*, **protected pak choi** *(off-label)*, **protected tat soi** *(off-label)* [6]; **fodder beet** *(off-label)* [7]; **poppies for morphine production** *(off-label)* [1, 3, 6]; **spring oilseed rape, sugar beet** [1-4, 6, 8, 9]; **winter oilseed rape** [1-6, 8, 9]
- Frit fly in **sweetcorn** *(off-label)* [1, 3, 6, 7]
- Grain aphid in **durum wheat, spring barley, spring oats, spring wheat, winter barley, winter oats, winter wheat** [6, 9]
- Insect pests in **all edible seed crops grown outdoors** *(off-label)*, **all non-edible seed crops grown outdoors** *(off-label)*, **beetroot** *(off-label)*, **crambe** *(off-label)*, **farm forestry** *(off-label)*, **forest nurseries** *(off-label)*, **hops** *(off-label)*, **lupins** *(off-label)*, **miscanthus** *(off-label)*, **ornamental plant production** *(off-label)*, **poppies for morphine production** *(off-label)*, **protected forest nurseries** *(off-label)*, **protected ornamentals** *(off-label)*, **protected pak choi** *(off-label)*, **protected tat soi** *(off-label)*, **soft fruit** *(off-label)*, **top fruit** *(off-label)*, **willow (short rotation coppice)** *(off-label)* [3]; **blackberries** *(off-label)*, **blackcurrants** *(off-label)*, **bulb onions** *(off-label)*, **dewberries** *(off-label)*, **garlic** *(off-label)*, **gooseberries** *(off-label)*, **leeks** *(off-label)*, **raspberries** *(off-label)*, **redcurrants** *(off-label)*, **rubus hybrids** *(off-label)*, **salad onions** *(off-label)*, **shallots** *(off-label)*, **whitecurrants** *(off-label)* [1, 3, 6, 7]; **borage for oilseed production** *(off-label)*, **canary flower (echium spp.)** *(off-label)*, **evening primrose** *(off-label)*, **fodder beet** *(off-label)*, **grass seed crops** *(off-label)*, **herbs (see appendix 6)** *(off-label)*, **honesty** *(off-label)*, **horseradish** *(off-label)*, **mallow (althaea spp.)** *(off-label)*, **mustard** *(off-label)*, **navy beans** *(off-label)*, **parsley root** *(off-label)*, **runner beans** *(off-label)*, **spring linseed** *(off-label)*, **triticale** *(off-label)*, **winter linseed** *(off-label)* [1, 3, 7]; **broad beans** *(off-label)*, **carrots** *(off-label)*, **cob nuts** *(off-label)*, **dwarf beans** *(off-label)*, **parsnips** *(off-label)*, **red beet** *(off-label)*, **short rotation coppice willow** *(off-label)* [7]; **celeriac** *(off-label)*, **celery (outdoor)** *(off-label)*, **radishes** *(off-label)*, **spring rye** *(off-label)*, **winter rye** *(off-label)* [3, 7]; **chestnuts** *(off-label)*, **hazel nuts** *(off-label)*, **walnuts** *(off-label)* [3, 6, 7]; **french beans** *(off-label)*, **sweetcorn** *(off-label)* [1, 7]; **frise** *(off-label)*, **lamb's lettuce** *(off-label)*, **leaf brassicas** *(off-label)*, **scarole** *(off-label)* [1, 3]; **grapevines** *(off-label)*, **rye** *(off-label)* [1]
- Leaf midge in **blackcurrants** *(off-label)*, **gooseberries** *(off-label)*, **redcurrants** *(off-label)*, **whitecurrants** *(off-label)* [1, 3, 6, 7]
- Leaf miner in **fodder beet** *(off-label)* [7]; **sugar beet** [2]
- Mangold fly in **sugar beet** [9]
- Orange blossom midge in **spring wheat, winter wheat** [9]
- Pea and bean weevil in **combining peas, vining peas** [1-3, 6, 8, 9]; **edible podded peas** [1, 3, 6, 8, 9]; **peas** [4]; **spring field beans, winter field beans** [1-4, 6, 8, 9]
- Pea aphid in **combining peas, edible podded peas, vining peas** [1, 3, 6, 8, 9]; **peas** [4]
- Pea midge in **combining peas, edible podded peas, vining peas** [1, 3, 6, 8]; **peas** [4]
- Pea moth in **combining peas, vining peas** [1-3, 6, 8, 9]; **edible podded peas** [1, 3, 6, 8, 9]; **peas** [4]
- Pear midge in **combining peas, edible podded peas, vining peas** [9]
- Pear sucker in **pears** [1-4, 8, 9]; **winter wheat** [2]
- Pod midge in **spring oilseed rape, winter oilseed rape** [1, 3, 4, 6, 8, 9]
- Pollen beetle in **crambe** *(off-label)* [6]; **poppies for morphine production** *(off-label)* [1, 3, 6]; **spring oilseed rape, winter oilseed rape** [1-4, 6, 8, 9]
- Rose-grain aphid in **durum wheat, spring barley, spring oats, spring wheat, winter barley, winter oats, winter wheat** [6]
- Sawflies in **blackcurrants, gooseberries** *(off-label)*, **redcurrants** *(off-label)*, **whitecurrants** *(off-label)* [1, 3, 6, 7]; **short rotation coppice willow** *(off-label)* [6, 7]
- Seed beetle in **broad beans** *(off-label)* [1, 3, 6, 7]
- Seed weevil in **spring oilseed rape, winter oilseed rape** [6, 9]
- Silver Y moth in **beetroot** *(off-label)* [6]; **celery (outdoor)** *(off-label)*, **dwarf beans** *(off-label)*, **navy beans** *(off-label)* [6, 7]; **french beans** *(off-label)* [3, 7]; **red beet** *(off-label)* [7]; **runner beans** *(off-label)* [3, 6, 7]
- Thrips in **all edible seed crops grown outdoors** *(off-label)*, **all non-edible seed crops grown outdoors** *(off-label)*, **farm forestry** *(off-label)*, **miscanthus** *(off-label)* [6]; **broad beans** *(off-label)*, **bulb onions** *(off-label)*, **garlic** *(off-label)*, **leeks** *(off-label)*, **salad onions** *(off-label)*, **shallots** *(off-label)* [1, 3, 6]

SEE SECTION 3 FOR PRODUCTS ALSO REGISTERED

Pesticide Profiles

- Tobacco whitefly in **all edible seed crops grown outdoors** *(off-label)*, **all non-edible seed crops grown outdoors** *(off-label)*, **borage for oilseed production** *(off-label)*, **canary flower (echium spp.)** *(off-label)*, **evening primrose** *(off-label)*, **farm forestry** *(off-label)*, **honesty** *(off-label)*, **miscanthus** *(off-label)*, **mustard** *(off-label)*, **spring linseed** *(off-label)*, **winter linseed** *(off-label)* [6]
- Wasps in **grapevines** *(off-label)* [1, 3, 7]; **outdoor grapes** *(off-label)* [6]
- Weevils in **combining peas, durum wheat, edible podded peas, potatoes, spring field beans, spring oilseed rape, spring wheat, sugar beet, vining peas, winter field beans, winter oilseed rape, winter wheat** [7]
- Wheat-blossom midge in **winter wheat** [1, 3, 8]
- Whitefly in **all edible seed crops grown outdoors** *(off-label)*, **all non-edible seed crops grown outdoors** *(off-label)*, **borage for oilseed production** *(off-label)*, **canary flower (echium spp.)** *(off-label)*, **evening primrose** *(off-label)*, **farm forestry** *(off-label)*, **honesty** *(off-label)*, **miscanthus** *(off-label)*, **mustard** *(off-label)*, **spring linseed** *(off-label)*, **winter linseed** *(off-label)* [6]; **broccoli, brussels sprouts, cabbages, calabrese, cauliflowers** [1, 3, 4, 8, 9]
- Willow aphid in **short rotation coppice willow** *(off-label)* [6, 7]
- Willow beetle in **short rotation coppice willow** *(off-label)* [6, 7]
- Willow sawfly in **short rotation coppice willow** *(off-label)* [6, 7]
- Yellow cereal fly in **winter wheat** [1-4, 6, 8, 9]

Specific Off-Label Approvals (SOLAs)
- **all edible seed crops grown outdoors** 20082944 expires 31 Dec 2011 [3], 20093498 [6]
- **all non-edible seed crops grown outdoors** 20082944 expires 31 Dec 2011 [3], 20093498 [6]
- **beetroot** 20063761 expires 31 Dec 2011 [1], 20060743 expires 31 Dec 2011 [3], 20093488 [6]
- **blackberries** 20063755 expires 31 Dec 2011 [1], 20060728 expires 31 Dec 2011 [3], 20093478 [6], 20073266 expires 31 Dec 2011 [7]
- **blackcurrants** 20063752 expires 31 Dec 2011 [1], 20060727 expires 31 Dec 2011 [3], 20093477 [6], 20073269 expires 31 Dec 2011 [7]
- **borage for oilseed production** 20063748 expires 31 Dec 2011 [1], 20060634 expires 31 Dec 2011 [3], 20093473 [6], 20073258 expires 31 Dec 2011 [7]
- **broad beans** 20063764 expires 31 Dec 2011 [1], 20060753 expires 31 Dec 2011 [3], 20093493 [6], 20073234 expires 31 Dec 2011 [7]
- **bulb onions** 20063756 expires 31 Dec 2011 [1], 20060730 expires 31 Dec 2011 [3], 20093479 [6], 20073256 expires 31 Dec 2011 [7]
- **canary flower (echium spp.)** 20063748 expires 31 Dec 2011 [1], 20060634 expires 31 Dec 2011 [3], 20093473 [6], 20073258 expires 31 Dec 2011 [7]
- **carrots** 20080201 expires 31 Dec 2011 [7]
- **celeriac** 20063757 expires 31 Dec 2011 [1], 20060731 expires 31 Dec 2011 [3], 20093480 [6], 20080204 expires 31 Dec 2011 [7]
- **celery (outdoor)** 20063762 expires 31 Dec 2011 [1], 20060744 expires 31 Dec 2011 [3], 20093490 [6], 20073257 expires 31 Dec 2011 [7]
- **chestnuts** 20060742 expires 31 Dec 2011 [3], 20093487 [6], 20080206 expires 31 Dec 2011 [7]
- **chicory** 20060740 expires 31 Dec 2011 [3], 20093486 [6]
- **cob nuts** 20080206 expires 31 Dec 2011 [7]
- **crambe** 20081046 expires 31 Dec 2011 [3], 20093496 [6]
- **dewberries** 20063755 expires 31 Dec 2011 [1], 20060728 expires 31 Dec 2011 [3], 20093478 [6], 20073266 expires 31 Dec 2011 [7]
- **dwarf beans** 20093485 [6], 20080202 expires 31 Dec 2011 [7]
- **evening primrose** 20063664 expires 31 Dec 2011 [1], 20060634 expires 31 Dec 2011 [3], 20093473 [6], 20073258 expires 31 Dec 2011 [7]
- **farm forestry** 20082944 expires 31 Dec 2011 [3], 20093498 [6]
- **fennel** 20060733 expires 31 Dec 2011 [3], 20093484 [6]
- **fodder beet** 20063665 expires 31 Dec 2011 [1], 20060637 expires 31 Dec 2011 [3], 20093476 [6], 20073233 expires 31 Dec 2011 [7]
- **forest nurseries** 20082944 expires 31 Dec 2011 [3]
- **french beans** 20063758 expires 31 Dec 2011 [1], 20060739 expires 31 Dec 2011 [3], 20080202 expires 31 Dec 2011 [7]
- **frise** 20063751 expires 31 Dec 2011 [1], 20060636 expires 31 Dec 2011 [3], 20093475 [6]
- **garlic** 20063756 expires 31 Dec 2011 [1], 20060730 expires 31 Dec 2011 [3], 20093479 [6], 20073256 expires 31 Dec 2011 [7]

FOR FULL CONDITIONS OF USE ALWAYS READ THE PRODUCT LABEL

lambda-cyhalothrin

- ***gooseberries*** *20063752 expires 31 Dec 2011* [1], *20060727 expires 31 Dec 2011* [3], *20093477* [6], *20073269 expires 31 Dec 2011* [7]
- ***grapevines*** *20063747 expires 31 Dec 2011* [1], *20060266 expires 31 Dec 2011* [3], *20080205 expires 31 Dec 2011* [7]
- ***grass seed crops*** *20063749 expires 31 Dec 2011* [1], *20060624 expires 31 Dec 2011* [3], *20093472* [6], *20073260 expires 31 Dec 2011* [7]
- ***hazel nuts*** *20060742 expires 31 Dec 2011* [3], *20093487* [6], *20080206 expires 31 Dec 2011* [7]
- ***herbs (see appendix 6)*** *20063751 expires 31 Dec 2011* [1], *20060636 expires 31 Dec 2011* [3], *20073259 expires 31 Dec 2011* [7]
- ***honesty*** *20063748 expires 31 Dec 2011* [1], *20060634 expires 31 Dec 2011* [3], *20093473* [6], *20073258 expires 31 Dec 2011* [7]
- ***hops*** *20082944 expires 31 Dec 2011* [3]
- ***horseradish*** *20071307 expires 31 Dec 2011* [1], *20071301 expires 31 Dec 2011* [3], *20093495* [6], *20080201 expires 31 Dec 2011* [7]
- ***lamb's lettuce*** *20063751 expires 31 Dec 2011* [1], *20060636 expires 31 Dec 2011* [3], *20093475* [6]
- ***leaf brassicas*** *20063751 expires 31 Dec 2011* [1], *20060636 expires 31 Dec 2011* [3], *20093475* [6]
- ***leeks*** *20063756 expires 31 Dec 2011* [1], *20060730 expires 31 Dec 2011* [3], *20093479* [6], *20073256 expires 31 Dec 2011* [7]
- ***lupins*** *20060635 expires 31 Dec 2011* [3], *20093474* [6]
- ***mallow (althaea spp.)*** *20071307 expires 31 Dec 2011* [1], *20071301 expires 31 Dec 2011* [3], *20093495* [6], *20080201 expires 31 Dec 2011* [7]
- ***miscanthus*** *20082944 expires 31 Dec 2011* [3], *20093498* [6]
- ***mustard*** *20063664 expires 31 Dec 2011* [1], *20060634 expires 31 Dec 2011* [3], *20093473* [6], *20073258 expires 31 Dec 2011* [7]
- ***navy beans*** *20063758 expires 31 Dec 2011* [1], *20060739 expires 31 Dec 2011* [3], *20093485* [6], *20080202 expires 31 Dec 2011* [7]
- ***ornamental plant production*** *20082944 expires 31 Dec 2011* [3]
- ***outdoor grapes*** *20093471* [6]
- ***outdoor herbs*** *20093475* [6]
- ***parsley root*** *20071307 expires 31 Dec 2011* [1], *20071301 expires 31 Dec 2011* [3], *20093495* [6], *20080201 expires 31 Dec 2011* [7]
- ***parsnips*** *20080201 expires 31 Dec 2011* [7]
- ***poppies for morphine production*** *20063763 expires 31 Dec 2011* [1], *20060749 expires 31 Dec 2011* [3], *20093492* [6]
- ***protected forest nurseries*** *20082944 expires 31 Dec 2011* [3]
- ***protected ornamentals*** *20082944 expires 31 Dec 2011* [3]
- ***protected pak choi*** *20081263 expires 31 Dec 2011* [3], *20093497* [6]
- ***protected tat soi*** *20081263 expires 31 Dec 2011* [3], *20093497* [6]
- ***radishes*** *20063757 expires 31 Dec 2011* [1], *20060731 expires 31 Dec 2011* [3], *20093480* [6], *20080204 expires 31 Dec 2011* [7]
- ***raspberries*** *20063755 expires 31 Dec 2011* [1], *20060728 expires 31 Dec 2011* [3], *20093478* [6], *20073266 expires 31 Dec 2011* [7]
- ***red beet*** *20073254 expires 31 Dec 2011* [7]
- ***redcurrants*** *20063752 expires 31 Dec 2011* [1], *20060727 expires 31 Dec 2011* [3], *20093477* [6], *20073269 expires 31 Dec 2011* [7]
- ***rubus hybrids*** *20063755 expires 31 Dec 2011* [1], *20060728 expires 31 Dec 2011* [3], *20093478* [6], *20073266 expires 31 Dec 2011* [7]
- ***runner beans*** *20063758 expires 31 Dec 2011* [1], *20060739 expires 31 Dec 2011* [3], *20093485* [6], *20080202 expires 31 Dec 2011* [7]
- ***rye*** *20063749 expires 31 Dec 2011* [1]
- ***salad onions*** *20063756 expires 31 Dec 2011* [1], *20060730 expires 31 Dec 2011* [3], *20093479* [6], *20073256 expires 31 Dec 2011* [7]
- ***scarole*** *20063751 expires 31 Dec 2011* [1], *20060636 expires 31 Dec 2011* [3]
- ***shallots*** *20063756 expires 31 Dec 2011* [1], *20060730 expires 31 Dec 2011* [3], *20093479* [6], *20073256 expires 31 Dec 2011* [7]
- ***short rotation coppice willow*** *20093491* [6], *20073268 expires 31 Dec 2011* [7]
- ***soft fruit*** *20082944 expires 31 Dec 2011* [3]

SEE SECTION 3 FOR PRODUCTS ALSO REGISTERED

Pesticide Profiles

- ***spring linseed*** *20063748 expires 31 Dec 2011* [1], *20060634 expires 31 Dec 2011* [3], *20093473* [6], *20073258 expires 31 Dec 2011* [7]
- ***spring rye*** *20060624 expires 31 Dec 2011* [3], *20093472* [6], *20073260 expires 31 Dec 2011* [7]
- ***swedes*** *20101856 expires 31 Dec 2011* [3]
- ***sweetcorn*** *20063760 expires 31 Dec 2011* [1], *20060732 expires 31 Dec 2011* [3], *20093481* [6], *20080203 expires 31 Dec 2011* [7]
- ***top fruit*** *20082944 expires 31 Dec 2011* [3]
- ***triticale*** *20063749 expires 31 Dec 2011* [1], *20060624 expires 31 Dec 2011* [3], *20093472* [6], *20073260 expires 31 Dec 2011* [7]
- ***turnips*** *20101856 expires 31 Dec 2011* [3]
- ***walnuts*** *20060742 expires 31 Dec 2011* [3], *20093487* [6], *20080206 expires 31 Dec 2011* [7]
- ***whitecurrants*** *20063752 expires 31 Dec 2011* [1], *20060727 expires 31 Dec 2011* [3], *20093477* [6], *20073269 expires 31 Dec 2011* [7]
- ***willow (short rotation coppice)*** *20060748 expires 31 Dec 2011* [3]
- ***winter linseed*** *20063748 expires 31 Dec 2011* [1], *20060634 expires 31 Dec 2011* [3], *20093473* [6], *20073258 expires 31 Dec 2011* [7]
- ***winter rye*** *20060624 expires 31 Dec 2011* [3], *20093472* [6], *20073260 expires 31 Dec 2011* [7]

Approval information
- Lambda-cyhalothrin included in Annex I under EC Directive 91/414
- Accepted by BBPA for use on malting barley
- Approval expiry 31 Dec 2011 [1-5, 7, 9]
- Approval expiry 13 Dec 2011 [8]

Efficacy guidance
- Best results normally obtained from treatment when pest attack first seen. See label for detailed recommendations on each crop
- Timing for control of barley yellow dwarf virus vectors depends on specialist assessment of the level of risk in the area
- Repeat applications recommended in some crops where prolonged attack occurs, up to maximum total dose. See label for details
- Where strains of aphids resistant to lambda-cyhalothrin occur control is unlikely to be satisfactory
- Addition of wetter recommended for control of certain pests in brassicas and oilseed rape
- Use of sufficient water volume to ensure thorough crop penetration recommended for optimum results
- Use of drop-legged sprayer gives improved results in crops such as Brussels sprouts

Restrictions
- Maximum number of applications or maximum total dose per crop varies - see labels
- Do not apply to a cereal crop if any product containing a pyrethroid insecticide or dimethoate has been applied to the crop after the start of ear emergence (GS 51)
- Do not spray cereals in the spring/summer (ie after 1 Apr) within 6 m of edge of crop

Crop-specific information
- Latest use before late milk stage on cereals; before end of flowering for winter oilseed rape
- HI 3 d for radishes, red beet; 7 d for lettuce [1, 3]; 14 d for carrots and parsnips; 25 d for peas, field beans; 6 wk for spring oilseed rape; 8 wk for sugar beet

Environmental safety
- Dangerous for the environment
- Very toxic to aquatic organisms
- Flammable [2]
- Risk to certain non-target insects or other arthropods [3]
- To protect non-target arthropods respect an untreated buffer zone of 5 m to non-crop land
- Broadcast air-assisted LERAP [1, 3, 4, 8, 9] (25 m); Broadcast air-assisted LERAP [2] (38 m); LERAP Category A [2, 7, 9]; LERAP Category B [1, 3-6, 8, 9]

Hazard classification and safety precautions
Hazard Harmful, Dangerous for the environment [1-9]; Corrosive [7]; Flammable [2]
Transport code 3 [2]; 8 [7]; 9 [1, 3-6, 8, 9]
Packaging group III
UN number 1760 [7]; 1993 [2]; 3077 [5]; 3082 [1, 3, 4, 6, 8, 9]
Risk phrases R20, R22a, R53a [1-9]; R21, R37, R67 [2]; R22b [2, 6, 7]; R34 [7]; R36, R38 [2, 5]; R43 [1, 3-6, 8, 9]; R50 [1-5, 8, 9]; R51 [6, 7]

FOR FULL CONDITIONS OF USE ALWAYS READ THE PRODUCT LABEL

lambda-cyhalothrin + pirimicarb

Operator protection A, H [1-9]; C [2, 4-7]; J, K, M [4, 5]; L [4]; U02a [1, 3-9]; U04a [7]; U05a [1-5, 7-9]; U08 [1-4, 7-9]; U09a [5, 6]; U11 [5, 7]; U14 [1, 3-5, 8, 9]; U19a [2, 6, 7]; U20a [2, 7]; U20b [1, 3-6, 8, 9]
Environmental protection E12a [7]; E15b [2, 5-7, 9]; E16a [1, 3-6, 8, 9]; E16b [1, 3, 4, 8]; E16c [2, 7]; E16d [7, 9]; E17b [1] (25 m); E17b [2] (38 m); E17b [3, 4, 8, 9] (25 m); E22a [5, 7]; E22b [3, 6-9]; E34 [1-4, 6-9]; E38 [1, 3-9]
Storage and disposal D01, D02 [1-5, 7-9]; D05 [2, 6]; D07 [2]; D09a [1-4, 6-9]; D10b [2, 7]; D10c [1, 3, 4, 6, 8, 9]; D12a [1-6, 8, 9]
Medical advice M03 [1, 3-5, 8, 9]; M04a [7]; M05b [2, 6, 7]

304   lambda-cyhalothrin + pirimicarb

An insecticide mixture combining translaminar, contact, fumigant and stomach activity
IRAC mode of action code: 3 + 1A

See also pirimicarb

Products

1 Clayton Groove	Clayton	5:100 g/l	EC	13246
2 Dovetail	Syngenta	5:100 g/l	EC	12550
3 Mortice	AgriGuard	5:100 g/l	EC	12961

Uses
- Aphids in **borage** *(off-label)*, **canary flower (echium spp.)** *(off-label)*, **canary seed** *(off-label)*, **evening primrose** *(off-label)*, **fodder beet** *(off-label)*, **grass seed crops** *(off-label)*, **honesty** *(off-label)*, **horseradish** *(off-label)*, **lupins** *(off-label)*, **mallow (althaea spp.)** *(off-label)*, **mustard** *(off-label)*, **parsnips** *(off-label)*, **protected chives** *(off-label)*, **protected frise** *(off-label)*, **protected herbs (see appendix 6)** *(off-label)*, **protected parsley** *(off-label)*, **protected radicchio** *(off-label)*, **protected salad brassicas** *(off-label - for baby leaf production)*, **protected scarole** *(off-label)*, **spring linseed** *(off-label)*, **winter linseed** *(off-label)* [2]; **broccoli, brussels sprouts, cabbages, calabrese, carrots, cauliflowers, lettuce, peas, potatoes, spring barley, spring oats, spring rye, spring wheat, sugar beet, triticale, winter barley, winter oats, winter rye, winter wheat** [1-3]; **durum wheat** [2, 3]
- Bean aphid in **spring field beans, winter field beans** [2]
- Beet virus yellows vectors in **spring oilseed rape, winter oilseed rape** [1-3]
- Cabbage seed weevil in **spring oilseed rape, winter oilseed rape** [1-3]
- Cabbage stem flea beetle in **spring oilseed rape, winter oilseed rape** [1-3]
- Caterpillars in **broccoli, brussels sprouts, cabbages, calabrese, cauliflowers** [1-3]
- Cutworms in **carrots, lettuce, potatoes, sugar beet** [1-3]; **fodder beet** *(off-label)*, **horseradish** *(off-label)*, **mallow (althaea spp.)** *(off-label)*, **parsnips** *(off-label)* [2]
- Flea beetle in **fodder beet** *(off-label)* [2]; **spring oilseed rape, sugar beet, winter oilseed rape** [1-3]
- Insect pests in **canary seed** *(off-label)* [2]
- Leaf miner in **fodder beet** *(off-label)* [2]; **sugar beet** [1-3]
- Mealy aphid in **broccoli, brussels sprouts, cabbages, calabrese, cauliflowers, spring oilseed rape, winter oilseed rape** [1-3]
- Pea and bean weevil in **peas, spring field beans, winter field beans** [1-3]
- Pea midge in **peas** [1-3]
- Pea moth in **peas** [1-3]
- Pod midge in **spring oilseed rape, winter oilseed rape** [1-3]
- Pollen beetle in **broccoli, brussels sprouts, cabbages, calabrese, cauliflowers, spring oilseed rape, winter oilseed rape** [1-3]
- Whitefly in **broccoli, brussels sprouts, cabbages, calabrese, cauliflowers** [1-3]

Specific Off-Label Approvals (SOLAs)
- **borage** *20060641* [2]
- **canary flower (echium spp.)** *20060641* [2]
- **canary seed** *20073139* [2]
- **evening primrose** *20060641* [2]
- **fodder beet** *20060642* [2]
- **grass seed crops** *20060640* [2]

SEE SECTION 3 FOR PRODUCTS ALSO REGISTERED

- **honesty** *20060641* [2]
- **horseradish** *20060639* [2]
- **lupins** *20060638* [2]
- **mallow (althaea spp.)** *20060639* [2]
- **mustard** *20060641* [2]
- **parsnips** *20060639* [2]
- **protected chives** *20060643* [2]
- **protected frise** *20060643* [2]
- **protected herbs (see appendix 6)** *20060643* [2]
- **protected parsley** *20060643* [2]
- **protected radicchio** *20060643* [2]
- **protected salad brassicas** *(for baby leaf production)* *20060643* [2]
- **protected scarole** *20060643* [2]
- **spring linseed** *20060641* [2]
- **winter linseed** *20060641* [2]

Approval information
- Lambda-cyhalothrin and pirimicarb included in Annex I under EC Directive 91/414
- Accepted by BBPA for use on malting barley
- In 2006 CRD required that all products containing pirimicarb should carry the following warning in the main area of the container label: "Pirimicarb is an anticholinesterase carbamate. Handle with care"

Efficacy guidance
- Best results obtained from treatment when pest attack first seen or after warning issued
- Repeat applications recommended in some crops where prolonged attacks occur, up to maximum total dose
- Addition of a non-ionic surfactant that is not an organosilicon recommended for certain uses in brassicas and oilseed rape. See label
- Use of drop-leg sprayers improves efficacy in crops such as Brussels sprouts
- Control unlikely to be satisfactory if aphids resistant to lambda-cyhalothrin or pirimicarb present

Restrictions
- Maximum total dose equivalent to two full dose treatments on carrots, lettuce, peas, field beans; three full dose treatments on brassicas, oilseed rape, cereals; four full dose treatments on sugar beet; eight full dose treatments on potatoes
- Must not be applied to cereals if any product containing a pyrethroid insecticide or dimethoate has been sprayed after the start of ear emergence (GS 51)

Crop-specific information
- Latest use: before late milky ripe (GS 77) for cereals; before end of flowering for oilseed rape
- HI sugar beet 8 wk; carrots 14 d; lettuce 7 d; brassicas, potatoes, peas, field beans 3 d

Environmental safety
- Dangerous for the environment
- Very toxic to aquatic organisms
- Do not spray cereals after 1 Apr within 6 m of the edge of the crop
- Harmful to livestock. Keep all livestock out of treated areas for at least 7 d following treatment
- Keep in original container, tightly closed, in a safe place, under lock and key
- LERAP Category A

Hazard classification and safety precautions
    Hazard Harmful, Dangerous for the environment [1-3]; Flammable [3]
    Transport code 9
    Packaging group III
    UN number 3082
    Risk phrases R20, R22a, R38, R50, R53a [1-3]; R22b, R37, R41, R43 [3]
    Operator protection A, C, D, E, H, J, K, M; U02a, U14 [1, 2]; U05a, U08, U19a, U20b [1-3]
    Environmental protection E06a [1, 3] (7 d); E06c [2] (7 d); E15a, E16c, E16d, E34 [1-3]; E22c [1]; E38 [2]
    Storage and disposal D01, D02, D09b [1-3]; D05, D10c, D12a [1, 2]; D11a [3]
    Medical advice M02, M03 [1-3]; M05b [1, 2]

FOR FULL CONDITIONS OF USE ALWAYS READ THE PRODUCT LABEL

305 Lecanicillium lecanii

A fungal parasite of aphids and whitefly

Products

Mycotal   Koppert   16.1% w/w   WP   04782

Uses

- Whitefly in **aubergines, protected beans, protected bilberries** (off-label), **protected blackberries** (off-label), **protected blackcurrants** (off-label), **protected blueberry** (off-label), **protected cayenne peppers** (off-label), **protected chives** (off-label), **protected cranberries** (off-label), **protected cucumbers, protected edible podded peas** (off-label), **protected gooseberries** (off-label), **protected herbs (see appendix 6)** (off-label), **protected lettuce, protected loganberries** (off-label), **protected ornamentals, protected parsley** (off-label), **protected peppers, protected raspberries** (off-label), **protected redcurrants** (off-label), **protected ribes hybrids** (off-label), **protected rubus hybrids** (off-label), **protected salad brassicas** (off-label), **protected strawberries** (off-label), **protected table grapes** (off-label), **protected tomatoes, protected wine grapes** (off-label)

Specific Off-Label Approvals (SOLAs)

- *protected bilberries* 20101247
- *protected blackberries* 20101248
- *protected blackcurrants* 20101247
- *protected blueberry* 20101247
- *protected cayenne peppers* 20070862
- *protected chives* 20070861
- *protected cranberries* 20101247
- *protected edible podded peas* 20070863
- *protected gooseberries* 20101247
- *protected herbs (see appendix 6)* 20070861
- *protected loganberries* 20101248
- *protected parsley* 20070861
- *protected raspberries* 20101248
- *protected redcurrants* 20101247
- *protected ribes hybrids* 20101247
- *protected rubus hybrids* 20101248
- *protected salad brassicas* 20070861
- *protected strawberries* 20101249
- *protected table grapes* 20101247
- *protected wine grapes* 20101247

Approval information

- Verticillium (Lecanicillium) included in Annex I under EC Directive 91/414

Efficacy guidance

- *Verticillium lecanii* is a pathogenic fungus that infects the target pests and destroys them
- Apply spore powder as spray as part of biological control programme keeping the spray liquid well agitated
- Pre-soak the product for 2-4 h before application to rehydrate the spores and assist in dispersion
- Treat before infestations build to high levels and repeat as directed on the label
- Spray during late afternoon and early evening directing spray onto underside of leaves and to growing points
- Best results require minimum 80% relative humidity and 18 °C within the crop canopy
- Product highly infective to many aphid species except the chrysanthemum aphid. Follow specific label directions for this pest

Restrictions

- Never use in tank mixture
- Do not use a fungicide within 3 d of treatment. Pesticides containing captan, chlorothalonil, fenarimol, dichlofluanid, imazalil, maneb, prochloraz, quinomethionate, thiram or tolylfluanid may not be used on the same crop
- Keep in a refrigerated store at 2-6 °C but do not freeze

SEE SECTION 3 FOR PRODUCTS ALSO REGISTERED

Pesticide Profiles

Environmental safety
- Products have negligible effects on commercially available natural predators or parasites but consult manufacturer before using with a particular biological control agent for the first time

Hazard classification and safety precautions
UN number N/C
Operator protection U19a, U20b
Environmental protection E15a
Storage and disposal D09a, D11a

306 lenacil

A residual, soil-acting uracil herbicide for beet crops
HRAC mode of action code: C1

See also desmedipham + ethofumesate + lenacil + phenmedipham

Products

1	Agriguard Lenacil	AgriGuard	80% w/w	WP	10488
2	Fernpath Lenzo Flo	AgriGuard	440 g/l	SC	11919
3	Lenazar Flo	Belcrop	440 g/l	SC	14791
4	Venzar Flowable	DuPont	440 g/l	SC	06907

Uses
- Annual dicotyledons in **blackberries** *(off-label)*, **blackcurrants** *(off-label)*, **blueberries** *(off-label)*, **cranberries** *(off-label)*, **farm woodland** *(off-label)*, **gooseberries** *(off-label)*, **herbs (see appendix 6)** *(off-label)*, **raspberries** *(off-label)*, **redcurrants** *(off-label)*, **ribes hybrids** *(off-label)*, **rubus hybrids** *(off-label)*, **spinach** *(off-label)*, **spinach beet** *(off-label)*, **strawberries** *(off-label)*, **whitecurrants** *(off-label)* [4]; **fodder beet**, **mangels**, **red beet**, **sugar beet** [1-4]
- Annual grasses in **blackcurrants** *(off-label)*, **blueberries** *(off-label)*, **cranberries** *(off-label)*, **gooseberries** *(off-label)*, **herbs (see appendix 6)** *(off-label)*, **redcurrants** *(off-label)*, **ribes hybrids** *(off-label)*, **spinach** *(off-label)*, **spinach beet** *(off-label)*, **whitecurrants** *(off-label)* [4]
- Annual meadow grass in **farm woodland** *(off-label)* [4]; **fodder beet**, **mangels**, **red beet**, **sugar beet** [1-4]

Specific Off-Label Approvals (SOLAs)
- *blackberries* 20093246 [4]
- *blackcurrants* 970704 [4]
- *blueberries* 970704 [4]
- *cranberries* 970704 [4]
- *farm woodland* 971282 [4]
- *gooseberries* 970704 [4]
- *herbs (see appendix 6)* 970703 [4]
- *raspberries* 20093246 [4]
- *redcurrants* 970704 [4]
- *ribes hybrids* 970704 [4]
- *rubus hybrids* 20093246 [4]
- *spinach* 970703 [4]
- *spinach beet* 970703 [4]
- *strawberries* 20093246 [4]
- *whitecurrants* 970704 [4]

Approval information
- Lenacil included in Annex I under EC Directive 91/414

Efficacy guidance
- Best results, especially from pre-emergence treatments, achieved on fine, even, firm and moist soils free from clods. Continuing presence of moisture from rain or irrigation gives improved residual control of later germinating weeds. Effectiveness may be reduced by dry conditions
- On beet crops may be used pre- or post-emergence, alone or in mixture to broaden weed spectrum
- Apply overall or as band spray to beet crops pre-drilling incorporated, pre- or post-emergence

FOR FULL CONDITIONS OF USE ALWAYS READ THE PRODUCT LABEL

lenacil + triflusulfuron-methyl

- All labels have limitations on soil types that may be treated. Residual activity reduced on soils with high OM content

Restrictions
- Maximum number of treatments in beet crops 1 pre-emergence + 3 post-emergence per crop; 2 per yr on established woody ornamentals and roses
- See label for soil type restrictions
- Do not use any other residual herbicide within 3 mth of the initial application to fruit or ornamental crops
- Do not treat crops under stress from drought, low temperatures, nutrient deficiency, pest or disease attack, or waterlogging

Crop-specific information
- Latest use: pre-emergence for red beet, fodder beet, spinach, spinach beet, mangels; before leaves meet over rows when used on these crops post-emergence; 24 h after planting new strawberry runners or before flowering for established strawberry crops, blackcurrants, gooseberries, raspberries
- Heavy rain after application to beet crops may cause damage especially if followed by very hot weather
- Reduction in beet stand may occur where crop emergence or vigour is impaired by soil capping or pest attack
- Strawberry runner beds to be treated should be level without depressions around the roots
- New soft fruit cuttings should be planted at least 15 cm deep and firmed before treatment
- Check varietal tolerance of ornamentals before large scale treatment

Following crops guidance
- Succeeding crops should not be planted or sown for at least 4 mth (6 mth on organic soils) after treatment following ploughing to at least 150 mm.
- Only beet crops, mangels or strawberries may be sown within 4 months of treatment and no further applications of lenacil should be made for at least 4 months [4]

Environmental safety
- Dangerous for the environment
- Very toxic to aquatic organisms

Hazard classification and safety precautions
    Hazard Irritant [1, 2]; Dangerous for the environment [1-4]
    Transport code 9
    Packaging group III
    UN number 3077 [1]; 3082 [2-4]
    Risk phrases R36, R37, R38 [1, 2]; R50, R53a [2-4]
    Operator protection A [1-4]; C [1-3]; U05a, U08, U19a, U20a [1-4]; U14, U15 [1, 2]
    Environmental protection E13c, E38 [2-4]; E15a, E34 [1]
    Storage and disposal D01, D02, D09a [1-4]; D05, D10a, D12a [2-4]; D10b [1]

307 lenacil + triflusulfuron-methyl

A foliar and residual herbicide mixture for sugar beet
HRAC mode of action code: C1 + B

See also triflusulfuron-methyl

Products
    Safari Lite WSB   DuPont   71.4:5.4% w/w   WG   12169

Uses
- Annual dicotyledons in **fodder beet** *(off-label)*, **sugar beet**
- Volunteer oilseed rape in **sugar beet**

Specific Off-Label Approvals (SOLAs)
- **fodder beet** 20101934

Approval information
- Lenacil and triflusulfuron-methyl included in Annex I under EC Directive 91/414

SEE SECTION 3 FOR PRODUCTS ALSO REGISTERED

Efficacy guidance
- Best results obtained when weeds are small and growing actively
- Product recommended for use in a programme of treatments in tank mixture with a suitable herbicide partner to broaden the weed spectrum
- Ensure good spray cover of weeds. Apply when first weeds have emerged provided crop has reached cotyledon stage
- Susceptible plants cease to grow almost immediately after treatment and symptoms can be seen 5-10 d later
- Weed control may be reduced in very dry soil conditions
- Triflusulfuron-methyl is a member of the ALS-inhibitor group of herbicides

Restrictions
- Maximum number of treatments on sugar beet 3 per crop and do not apply more than 4 applications of any product containing triflusulfuron-methyl
- A maximum total dose of 60 g/ha triflusulfuron may only be applied every third year on the same field.
- Do not apply to crops suffering from stress caused by drought, water-logging, low temperatures, pest or disease attack, nutrient deficiency or any other factors affecting crop growth
- Do not use on Sands, stony or gravelly soils or on soils with more than 10% organic matter
- Do not apply when temperature above or likely to exceed 21 °C on day of spraying or under conditions of high light intensity

Crop-specific information
- Latest use: before crop leaves meet between rows for sugar beet

Following crops guidance
- Only cereals may be sown in the same calendar yr as a treated sugar beet crop. Any crop may be sown in the following spring
- In the event of crop failure sow only sugar beet within 4 mth of treatment
- Applications must not be made to sugar beet or fodder beet when the foliage will be fed to livestock.

Environmental safety
- Dangerous for the environment
- Very toxic to aquatic organisms
- Take extreme care to avoid drift onto broad-leaved plants outside the target area or onto surface waters or ditches, or land intended for cropping
- Spraying equipment should not be drained or flushed onto land planted, or to be planted, with trees or crops other than cereals and should be thoroughly cleansed after use - see label for instructions
- LERAP Category B

Hazard classification and safety precautions
Hazard Irritant, Dangerous for the environment
Transport code 9
Packaging group III
UN number 3077
Risk phrases R36, R37, R38, R43, R50, R53a
Operator protection A, C; U05a, U08, U11, U19a, U20b, U22a
Environmental protection E15a, E16a, E38
Storage and disposal D01, D02, D09a, D12a

FOR FULL CONDITIONS OF USE ALWAYS READ THE PRODUCT LABEL

linuron

308 linuron

A contact and residual urea herbicide for various field crops
HRAC mode of action code: C2

See also 2,4-DB + linuron + MCPA
clomazone + linuron

Products

1	Afalon	Makhteshim	450 g/l	SC	14187
2	Alpha Linuron 50SC	Makhteshim	500 g/l	SC	14215
3	Datura	AgriChem BV	500 g/l	SC	14915
4	Linurex 50 SC	Makhteshim	500 g/l	SC	14652
5	Luas	AgriGuard	500 g/l	SC	14359
6	Nightjar	AgChem Access	450 g/l	SC	14656

Uses

- Annual dicotyledons in **asparagus, celeriac, chervil, dill, forest nurseries, lovage, ornamental plant production** [3]; **bulb onions** (off-label), **garlic** (off-label), **leeks** (off-label), **shallots** (off-label) [1, 2, 4, 6]; **carrots, parsley, parsnips, potatoes** [1-6]; **celeriac** (off-label), **french beans** (off-label), **herbs (see appendix 6)** (off-label), **mallow (althaea spp.)** (off-label), **ornamental plant production** (off-label), **parsley root** (off-label), **runner beans** (off-label) [1, 2, 4]; **celery (outdoor)** (off-label) [4]; **combining peas, spring field beans** [1, 6]; **dandelions** (off-label), **dwarf beans** (off-label), **game cover** (off-label), **ginseng root** (off-label), **liquorice** (off-label), **nettle** (off-label), **valerian root** (off-label) [1]; **dwarf beans, french beans, runner beans** [2]; **forest nurseries** (off-label), **herbs for medicinal uses (see appendix 6)** (off-label) [2, 4]
- Annual grasses in **bulb onions** (off-label), **celeriac** (off-label), **garlic** (off-label), **herbs (see appendix 6)** (off-label), **leeks** (off-label), **mallow (althaea spp.)** (off-label), **ornamental plant production** (off-label), **parsley root** (off-label), **shallots** (off-label) [1, 4]; **carrots, celery (outdoor)** (off-label), **forest nurseries** (off-label), **herbs for medicinal uses (see appendix 6)** (off-label), **parsley, parsnips, potatoes** [4]; **dandelions** (off-label), **dwarf beans** (off-label), **game cover** (off-label), **ginseng root** (off-label), **liquorice** (off-label), **nettle** (off-label), **valerian root** (off-label) [1]; **french beans** (off-label), **runner beans** (off-label) [1, 2, 4]
- Annual meadow grass in **bulb onions** (off-label), **celeriac** (off-label), **dwarf beans, forest nurseries** (off-label), **french beans, garlic** (off-label), **herbs (see appendix 6)** (off-label), **herbs for medicinal uses (see appendix 6)** (off-label), **leeks** (off-label), **mallow (althaea spp.)** (off-label), **ornamental plant production** (off-label), **parsley root** (off-label), **runner beans, shallots** (off-label) [2]; **carrots, parsley, parsnips, potatoes** [1, 2, 5]; **combining peas, spring field beans** [1]

Specific Off-Label Approvals (SOLAs)

- **bulb onions** *20090875* [1], *20090865* [2], *20092471* [4], *20100649* [6]
- **celeriac** *20090885* [1], *20090884* [2], *20092469* [4]
- **celery (outdoor)** *20092468* [4]
- **dandelions** *20090861* [1]
- **dwarf beans** *20090859* [1]
- **forest nurseries** *20082805* [2], *20092473* [4]
- **french beans** *20090859* [1], *20090846* [2], *20092475* [4]
- **game cover** *20082802* [1]
- **garlic** *20090875* [1], *20090865* [2], *20092471* [4], *20100649* [6]
- **ginseng root** *20090861* [1]
- **herbs (see appendix 6)** *20090860* [1], *20090847* [2], *20092476* [4]
- **herbs for medicinal uses (see appendix 6)** *20090848* [2], *20092477* [4]
- **leeks** *20090864* [1], *20090850* [2], *20092472* [4], *20101135* [6]
- **liquorice** *20090861* [1]
- **mallow (althaea spp.)** *20090876* [1], *20090866* [2], *20092470* [4]
- **nettle** *20090861* [1]
- **ornamental plant production** *20090877* [1], *20090867* [2], *20092474* [4]
- **parsley root** *20090876* [1], *20090866* [2], *20092470* [4]
- **runner beans** *20090859* [1], *20090846* [2], *20092475* [4]
- **shallots** *20090875* [1], *20090865* [2], *20092471* [4], *20100649* [6]
- **valerian root** *20090861* [1]

SEE SECTION 3 FOR PRODUCTS ALSO REGISTERED

Pesticide Profiles

Approval information
- Linuron included in Annex I under EC Directive 91/414
- Accepted by BBPA for use on malting barley

Efficacy guidance
- Many weeds controlled pre-emergence or post-emergence to 2-3 leaf stage, some (annual meadow grass, mayweed) only susceptible pre-emergence. See label for details
- Best results achieved by application to firm, moist soil of fine tilth
- Little residual effect on soil with more than 10% organic matter

Restrictions
- Maximum total dose equivalent to one full dose treatment
- Do not use on undersown cereals or crops grown on Sands or Very Light soils or soils heavier than Sandy Clay Loam or with more than 10% organic matter
- Do not apply to emerged crops of carrots, parsnips or parsley under stress
- Do not apply by hand-held sprayers

Crop-specific information
- Latest use: pre-emergence for most crops (products differ, see label for details).
- HI onions, garlic 8 wk; celeriac 12 wk; leeks 16 wk
- Apply to potatoes well earthed up to a rounded ridge pre-crop emergence and do not cultivate after spraying
- Apply to carrots at any time after drilling on organic soils, and within 4 d of drilling on other soils. Apply post-emergence as soon as weeds appear but after first rough leaf stage.
- Recommendations for parsnips, parsley and celery vary. See label for details

Following crops guidance
- Potatoes, carrots and parsnips may be planted at any time after application. Lettuce should not be grown within 12 mth of treatment. Transplanted brassicas may be grown from 3 mth after treatment

Environmental safety
- Dangerous for the environment
- Very toxic to aquatic organisms
- LERAP Category B

Hazard classification and safety precautions
Hazard Toxic [1-4, 6]; Harmful [5]; Dangerous for the environment [1-6]
Transport code 9 [1-5]
Packaging group III [1-5]
UN number 3082 [1-5]
Risk phrases R22a, R48, R61 [1, 5, 6]; R25b, R63 [3]; R40, R50, R53a, R62 [1, 3, 5, 6]
Operator protection A, H [1-6]; C [6]; U05a, U13, U14, U15, U19a [1-6]; U08, U19c, U20b [2-4]
Environmental protection E15b, E16a, E38
Storage and disposal D02, D05, D09a, D11a, D12a [1-6]; D08 [1, 5, 6]
Medical advice M04a

309 magnesium phosphide

A phosphine generating compound used to control insect pests in stored commodities

Products

Degesch Plates	Rentokil	56% w/w	GE	07603

Uses
- Insect pests in *stored grain*

Approval information
- Magnesium phosphide included in Annex I under EC Directive 91/414
- Accepted by BBPA for use in stores for malting barley

Efficacy guidance
- Product acts as fumigant by releasing poisonous hydrogen phosphide gas on contact with moisture in the air

FOR FULL CONDITIONS OF USE ALWAYS READ THE PRODUCT LABEL

maleic hydrazide

- Place plates on the floor or wall of the building or on the surface of the commodity. Exposure time varies depending on temperature and pest. See label

Restrictions
- Magnesium phosphide is subject to the Poisons Rules 1982 and the Poisons Act 1972. See Section 5 for more information
- Only to be used by professional operators trained in the use of magnesium phosphide and familiar with the precautionary measures to be observed. See label for full precautions

Environmental safety
- Highly flammable
- Prevent access to buildings under fumigation by livestock, pets and other non-target mammals and birds
- Dangerous to fish or other aquatic life. Do not contaminate surface waters or ditches with chemical or used container
- Keep in original container, tightly closed, in a safe place, under lock and key
- Do not allow plates or their spent residues to come into contact with food other than raw cereal grains
- Remove used plates after treatment. Do not bulk spent plates and residues: spontaneous ignition could result
- Keep livestock out of treated areas

Hazard classification and safety precautions
    Hazard Very toxic, Highly flammable
    Transport code 4.3
    Packaging group I
    UN number 2011
    Risk phrases R21, R26, R28
    Operator protection A, D, H; U01, U05b, U07, U13, U19a, U20a
    Environmental protection E02a (4 h min); E02b, E13b, E34
    Storage and disposal D01, D02, D05, D07, D09b, D11b
    Medical advice M04a

310   maleic hydrazide

A pyridazine plant growth regulator suppressing sprout and bud growth

Products
    Fazor   Dow   60% w/w   GR   13679

Uses
- Annual grasses in **amenity grassland** *(off-label)*, **hops** *(off-label)*, **ornamental specimens** *(off-label)*, **soft fruit** *(off-label)*
- Growth regulation in **bulb onions**, **carrots** *(off-label)*, **garlic** *(off-label)*, **parsnips** *(off-label)*, **potatoes**, **shallots** *(off-label)*

Specific Off-Label Approvals (SOLAs)
- *amenity grassland* 20082790
- *carrots* 20081326
- *garlic* 20072795
- *hops* 20082848
- *ornamental specimens* 20082848
- *parsnips* 20081326
- *shallots* 20072795
- *soft fruit* 20082848

Approval information
- Maleic hydrazide included in Annex I under EC Directive 91/414

Efficacy guidance
- Apply to grass at any time of yr when growth active, best when growth starting in Apr-May and repeated when growth recommences
- Uniform coverage and dry weather necessary for effective results

SEE SECTION 3 FOR PRODUCTS ALSO REGISTERED

Pesticide Profiles

- Accurate timing essential for good results on potatoes but rain or irrigation within 24 h may reduce effectiveness on onions and potatoes
- Mow 2-3 d before and 5-10 d after spraying for best results. Need for mowing reduced for up to 6 wk
- When used for suppression of volunteer potatoes treatment may also give some suppression of sprouting in store but separate treatment will be necessary if sprouting occurs

Restrictions
- Maximum number of treatments 2 per yr on amenity grass, land not intended for cropping and land adjacent to aquatic areas; 1 per crop on onions, potatoes and on or around tree trunks
- Do not apply in drought or when crops are suffering from pest, disease or herbicide damage. Do not treat fine turf or grass seeded less than 8 mth previously
- Do not treat potatoes within 3 wk of applying a haulm desiccant or if temperatures above 26 °C
- Consult processor before use on potato crops for processing

Crop-specific information
- Latest use: 3 wk before haulm destruction for potatoes; before 50% necking for onions
- HI onions 1 wk; potatoes 3 wk
- Apply to onions at 10% necking and not later than 50% necking stage when the tops are still green
- Only treat onions in good condition and properly cured, and do not treat more than 2 wk before maturing. Treated onions may be stored until Mar but must then be removed to avoid browning
- Apply to second early or maincrop potatoes at least 3 wk before haulm destruction
- Only treat potatoes of good keeping quality; not on seed, first earlies or crops grown under polythene

Environmental safety
- Only apply to grass not to be used for grazing
- Do not use treated water for irrigation purposes within 3 wk of treatment or until concentration in water falls below 0.02 ppm
- Maximum permitted concentration in water 2 ppm
- Do not dump surplus product in water or ditch bottoms
- Avoid drift onto nearby vegetables, flowers or other garden plants

Hazard classification and safety precautions
  Transport code 9
  Packaging group III
  UN number 3077
  Operator protection A, H; U08, U20b
  Environmental protection E15a
  Storage and disposal D09a, D11a

311   maleic hydrazide + pelargonic acid

A pyridazine plant growth regulator suppressing sprout and bud growth + a naturally occuring 9 carbon chain acid
HRAC mode of action code: Not classified

Products
  Ultima   Certis   30:186.7 g/l   SL   13347

Uses
- Algae in *amenity vegetation, natural surfaces not intended to bear vegetation, ornamental plant production, permeable surfaces overlying soil*
- Annual and perennial weeds in *amenity vegetation, natural surfaces not intended to bear vegetation, ornamental plant production, permeable surfaces overlying soil*
- Moss in *amenity vegetation, natural surfaces not intended to bear vegetation, ornamental plant production, permeable surfaces overlying soil*

Approval information
- Maleic hydrazide included in Annex I under EC Directive 91/414 but pelargonic acid not yet listed

Hazard classification and safety precautions
  UN number N/C
  Operator protection A, C, H; U05a, U12, U15

FOR FULL CONDITIONS OF USE ALWAYS READ THE PRODUCT LABEL

maltodextrin

Environmental protection E15a
Storage and disposal D01, D02

312   maltodextrin

A polysaccharide use as a food additive and with activity against red spider mites
IRAC mode of action code: Not classified

Products
1	Eradicoat	Certis	598 g/l	SL	13724
2	Majestik	Certis	598 g/l	SL	14831

Uses
- Aphids in *all protected edible crops, all protected non-edible crops*
- Spider mites in *all protected edible crops, all protected non-edible crops*
- Whitefly in *all protected edible crops, all protected non-edible crops*

Hazard classification and safety precautions
Hazard Irritant
UN number N/C [2]
Risk phrases R43
Operator protection A, H; U05a, U09a, U14, U19a, U20b
Storage and disposal D01, D02, D09a, D10b, D12a

313   mancozeb

A protective dithiocarbamate fungicide for potatoes and other crops
FRAC mode of action code: M3

See also ametoctradin + mancozeb
benalaxyl + mancozeb
benthiavalicarb-isopropyl + mancozeb
chlorothalonil + mancozeb
cymoxanil + mancozeb
dimethomorph + mancozeb
fenamidone + mancozeb

Products
1	Dithane 945	Dow	80% w/w	WP	14547
2	Dithane 945	Interfarm	80% w/w	WP	14621
3	Dithane NT Dry Flowable	Interfarm	75% w/w	WG	12565
4	Dithane NT Dry Flowable	Interfarm	75% w/w	WG	14704
5	Karamate Dry Flo Neotec	Landseer	75% w/w	WG	14632
6	Laminator FL	Interfarm	455 g/l	SC	11072
7	Mewati	AgChem Access	80% w/w	WP	13029
8	Micene 80	Sipcam	80% w/w	WP	09112
9	Micene DF	Sipcam	77% w/w	WG	09957
10	Penncozeb 80 WP	United Phosphorus	80% w/w	WP	14718
11	Penncozeb WDG	United Phosphorus	75% w/w	WG	14719
12	Quell Flo	Interfarm	455 g/l	SC	09894
13	Trimanzone	Intracrop	80% w/w	WP	09584

Uses
- Alternaria blight in *carrots, parsnips* [4]
- Blight in *potatoes* [1-4, 6-13]
- Brown rust in *durum wheat* (useful control), *spring wheat* (useful control), *winter wheat* (useful control) [1, 2, 4]; *spring barley, winter barley* [8, 9, 13]; *spring wheat, winter wheat* [3, 6, 7, 9, 12]
- Disease control in *amenity vegetation, bulb onions, carrots, courgettes, ornamental plant production, parsnips, shallots, wine grapes* [5]; *durum wheat* (off-label) [3]; *grass seed crops* (off-label) [3, 8, 9, 12, 13]; *spring rye* (off-label), *winter rye* (off-label) [12]
- Downy mildew in *bulb onions* [10, 11]; *bulb onions* (off-label), *poppies for morphine production* (off-label), *shallots* (off-label) [3, 6, 12]; *lettuce* (off-label) [5]; *winter oilseed rape* [3, 6, 7, 12]

SEE SECTION 3 FOR PRODUCTS ALSO REGISTERED

Pesticide Profiles

- Glume blotch in **spring wheat**, **winter wheat** [9, 13]
- Net blotch in **spring barley** [8, 9, 13]; **winter barley** [6, 8, 9, 13]
- Phytophthora in **leeks** *(off-label)* [13]
- Rhynchosporium in **spring barley** [9]; **winter barley** [6, 9]
- Rust in **bulb onions**, **shallots** [4]
- Scab in **apples** [5, 8-11, 13]; **pears** [5]
- Septoria leaf blotch in **durum wheat** *(reduction)*, **spring wheat** *(reduction)*, **winter wheat** *(reduction)* [1, 2, 4]; **spring wheat**, **winter wheat** [3, 6-13]
- Sooty moulds in **spring barley** [9]; **spring wheat**, **winter wheat** [3, 6, 7, 9, 12]; **winter barley** [6, 9]
- White tip in **leeks** *(off-label)* [13]
- Yellow rust in **spring barley**, **spring wheat**, **winter barley** [9]; **winter wheat** [3, 6, 7, 9, 12]

Specific Off-Label Approvals (SOLAs)
- **bulb onions** *20100027 expires 31 Oct 2011* [3], *20100051 expires 31 Dec 2011* [6], *20100053 expires 31 Dec 2011* [12]
- **durum wheat** *20100023 expires 31 Oct 2011* [3]
- **grass seed crops** *20100023 expires 31 Oct 2011* [3], *20100067 expires 31 Dec 2011* [8], *20100021 expires 31 Oct 2011* [9], *20100052 expires 31 Dec 2011* [12], *20100056 expires 31 Dec 2011* [13]
- **leeks** *20100055 expires 31 Dec 2011* [13]
- **lettuce** *20101082* [5]
- **poppies for morphine production** *20100026 expires 31 Oct 2011* [3], *20100050 expires 31 Dec 2011* [6], *20100054 expires 31 Dec 2011* [12]
- **shallots** *20100027 expires 31 Oct 2011* [3], *20100051 expires 31 Dec 2011* [6], *20100053 expires 31 Dec 2011* [12]
- **spring rye** *20100052 expires 31 Dec 2011* [12]
- **winter rye** *20100052 expires 31 Dec 2011* [12]

Approval information
- Mancozeb included in Annex I under EC Directive 91/414
- Approved for aerial application on potatoes [8, 9, 12]. See Section 5 for more information
- Accepted by BBPA for use on malting barley
- Approval expiry 31 Oct 2011 [3, 9]
- Approval expiry 31 Dec 2011 [6-8, 12, 13]

Efficacy guidance
- Mancozeb is a protectant fungicide and will give moderate control, suppression or reduction of the cereal diseases listed if treated before they are established but in many cases mixture with carbendazim is essential to achieve satisfactory results. See labels for details
- May be recommended for suppression or control of mildew in cereals depending on product and tank mix. See label for details

Restrictions
- Maximum number of treatments varies with crop and product used - check labels for details
- Check labels for minimum interval that must elapse between treatments
- On protected lettuce only 2 post-planting applications of mancozeb or of any combination of products containing EBDC fungicide (mancozeb, maneb, thiram, zineb) either as a spray or a dust are permitted within 2 wk of planting out and none thereafter.
- Avoid treating wet cereal crops or those suffering from drought or other stress
- Keep dry formulations away from fire and sparks
- Use dry formulations immediately. Do not store

Crop-specific information
- Latest use: before early milk stage (GS 73) for cereals; before 6 true leaf stage and before 31 Dec for winter oilseed rape.
- HI potatoes 7 d; outdoor lettuce 14 d; protected lettuce 21 d; apples, blackcurrants 28 d
- Apply to potatoes before haulm meets across rows (usually mid-Jun) or at earlier blight warning, and repeat every 7-14 d depending on conditions and product used (see label)
- May be used on potatoes up to desiccation of haulm
- On oilseed rape apply as soon as disease develops between cotyledon and 5-leaf stage (GS 1,0-1,5)

FOR FULL CONDITIONS OF USE ALWAYS READ THE PRODUCT LABEL

mancozeb + metalaxyl-M

- Apply to cereals from 4-leaf stage to before early milk stage (GS 71). Recommendations vary, see labels for details
- Treat winter oilseed rape before 6 true leaf stage (GS 1,6) and before 31 Dec

Environmental safety
- Dangerous for the environment
- Very toxic to aquatic organisms
- Harmful to fish or other aquatic life. Do not contaminate surface waters or ditches with chemical or used container
- Do not empty into drains
- LERAP Category B [5]

Hazard classification and safety precautions
Hazard Harmful [10, 11]; Irritant [1-9, 12, 13]; Dangerous for the environment [1-5, 7, 10-12]
Transport code 9
Packaging group III
UN number 3077 [1-5, 7-11, 13]; 3082 [6, 12]
Risk phrases R36, R38 [8, 9, 13]; R37 [1-5, 7-9, 11-13]; R42 [11]; R43 [1-7, 10-12]; R50, R53a [1-5, 7, 10-12]; R63 [10, 11]
Operator protection A [1-9, 11-13]; C [8, 9, 13]; D [1-5, 7-9, 11, 13]; H [1, 2, 11]; U05a [1-9, 11-13]; U08 [6, 8, 9, 12, 13]; U13 [8, 9, 13]; U14 [1-5, 7, 10-12]; U19a [8-11, 13]; U20a [8]; U20b [1-7, 9, 12, 13]
Environmental protection E13c [6, 8, 9, 13]; E15a [1-5, 7, 12]; E15b [10, 11]; E16a [5]; E19b [11]; E34 [6, 12]; E38 [1-5, 7, 10-12]
Storage and disposal D01 [1-5, 7-13]; D02 [1-13]; D05 [1-6, 8, 9, 12]; D07 [6]; D09a [1-9, 11-13]; D11a [1-9, 12, 13]; D12a [1-5, 7, 10-12]
Medical advice M05a [10, 11]

314   mancozeb + metalaxyl-M

A systemic and protectant fungicide mixture
FRAC mode of action code: M3 + 4

See also metalaxyl-M

Products

1	Clayton Erase	Clayton	64:4% w/w	WG	13836
2	Fubol Gold WG	Syngenta	64:4% w/w	WG	14605

Uses
- Blight in **potatoes** [1, 2]
- Disease control in **forest nurseries** *(off-label)*, **herbs (see appendix 6)** *(off-label)*, **hops** *(off-label)*, **lettuce** *(off-label)*, **ornamental plant production** *(off-label)*, **protected forest nurseries** *(off-label)*, **protected hops** *(off-label)*, **protected ornamentals** *(off-label)*, **protected soft fruit** *(off-label)*, **soft fruit** *(off-label)* [2]
- Downy mildew in **bulb onions** *(useful control)*, **rhubarb** *(off-label)*, **salad onions** *(off-label)*, **shallots** *(useful control)* [2]
- Phytophthora fruit rot in **apple orchards** *(off-label)* [2]
- White blister in **cabbages** *(off-label)* [2]

Specific Off-Label Approvals (SOLAs)
- *apple orchards* 20101734 [2]
- *cabbages* 20101733 [2]
- *forest nurseries* 20102567 [2]
- *herbs (see appendix 6)* 20101779 [2]
- *hops* 20102567 [2]
- *lettuce* 20101779 [2]
- *ornamental plant production* 20102567 [2]
- *protected forest nurseries* 20102567 [2]
- *protected hops* 20102567 [2]
- *protected ornamentals* 20102567 [2]
- *protected soft fruit* 20102567 [2]
- *rhubarb* 20101735 [2]

SEE SECTION 3 FOR PRODUCTS ALSO REGISTERED

Pesticide Profiles

- **salad onions** 20101736 [2]
- **soft fruit** 20102567 [2]

Approval information
- Mancozeb and metalaxyl-M included in Annex I under EC Directive 91/414
- Approval expiry 31 Dec 2011 [1]

Efficacy guidance
- Commence potato blight programme before risk of infection occurs as crops begin to meet along the rows and repeat every 7-14 d according to blight risk. Do not exceed a 14 d interval between sprays
- If infection risk conditions occur earlier than the above growth stage commence spraying potatoes immediately
- Complete the potato blight programme using a protectant fungicide starting no later than 10 d after the last phenylamide spray. At least 2 such sprays should be applied
- To minimise the likelihood of development of resistance these products should be used in a planned Resistance Management strategy. See Section 5 for more information

Crop-specific information
- Latest use: before end of active potato haulm growth or before end Aug, whichever is earlier
- HI 7 d for potatoes
- After treating early potatoes destroy and remove any remaining haulm after harvest to minimise blight pressure on neighbouring maincrop potatoes

Environmental safety
- Dangerous for the environment
- Very toxic to aquatic organisms
- Do not harvest crops for human consumption for at least 7 d after final application
- LERAP Category B [2]

Hazard classification and safety precautions
Hazard Harmful, Dangerous for the environment
Transport code 9
Packaging group III
UN number 3077
Risk phrases R37, R43, R50, R53a, R63
Operator protection A; U05a, U08, U20b
Environmental protection E15a, E34, E38 [1, 2]; E16a [2]
Consumer protection C02a (7 d)
Storage and disposal D01, D02, D05, D09a, D11a, D12a

315   mancozeb + zoxamide

A protectant fungicide mixture for potatoes
FRAC mode of action code: M3 + 22

See also zoxamide

Products

1	Electis 75WG	Gowan	66.7:8.3% w/w	WG	14195
2	Roxam 75WG	Gowan	66.7:8.3% w/w	WG	14191

Uses
- Blight in **potatoes**

Approval information
- Mancozeb and zoxamide included in Annex I under EC Directive 91/414

Efficacy guidance
- Apply as protectant spray on potatoes immediately risk of blight in district or as crops begin to meet along the rows and repeat every 7-14 d according to blight risk
- Do not use if potato blight present in crop. Products are not curative
- Spray irrigated potato crops as soon as possible after irrigation once the crop leaves are dry

Restrictions
- Maximum number of treatments 10 per crop for potatoes

FOR FULL CONDITIONS OF USE ALWAYS READ THE PRODUCT LABEL

mandipropamid

Crop-specific information
- HI 7 d for potatoes

Environmental safety
- Dangerous for the environment
- Very toxic to aquatic organisms
- Keep away from fire and sparks
- LERAP Category B

Hazard classification and safety precautions
Hazard Irritant, Dangerous for the environment
Transport code 9
Packaging group III
UN number 3077
Risk phrases R37, R43, R50, R53a
Operator protection A, H; U05a, U14, U20b
Environmental protection E15a, E16a, E16b, E38
Storage and disposal D01, D02, D05, D09a, D11a, D12a

316 mandipropamid

A mandelamide fungicide for the control of potato blight
FRAC mode of action code: 40

Products

Revus	Syngenta	250 g/l	SC	13484

Uses
- Blight in *potatoes*
- Disease control in *ornamental plant production* (off-label)

Specific Off-Label Approvals (SOLAs)
- *ornamental plant production* 20082867 expires 29 Jul 2011

Approval information
- Approval expiry 29 Jul 2011 [1]

Efficacy guidance
- Mandipropamid acts preventatively by preventing spore germination and inhibiting mycelial growth during incubation. Apply immediately after blight warning or as soon as local conditions favour disease development but before blight enters the crop
- Spray at 7-10 d intervals reducing the interval as blight risk increases
- Spray programme should include a complete haulm desiccant to prevent tuber infection at harvest
- Eliminate other potential infection sources
- To minimise the likelihood of development of resistance this product should be used in a planned Resistance Management strategy. See Section 5 for more information
- See label for details of tank mixtures that may be used as part of a resistance management strategy

Restrictions
- Maximum number of treatments 4 per crop. Do not apply more than 3 treatments of this, or any other fungicide in the same resistance category, consecutively

Crop-specific information
- HI 3 d for potatoes

Hazard classification and safety precautions
UN number N/C
Risk phrases R52, R53a
Operator protection A, H; U05a, U20b
Environmental protection E15b, E34, E38
Storage and disposal D01, D02, D05, D09a, D10c, D12a
Medical advice M03, M05a

SEE SECTION 3 FOR PRODUCTS ALSO REGISTERED

Pesticide Profiles

317 maneb

A protectant dithiocarbamate fungicide
FRAC mode of action code: M3

See also carbendazim + maneb

Products

1	Trimangol 80	United Phosphorus	80% w/w	WP	06871
2	Trimangol WDG	United Phosphorus	75	WG	06992

Uses
- Blight in **potatoes** [1, 2]
- Brown rust in **spring barley, spring wheat, winter barley, winter wheat** [1, 2]
- Disease control in **grass seed crops** *(off-label)* [2]; **protected tomatoes** [1, 2]
- Glume blotch in **spring wheat, winter wheat** [1, 2]
- Net blotch in **spring barley, winter barley** [1, 2]
- Rhynchosporium in **spring barley, winter barley** [1, 2]
- Scab in **apples** [1, 2]
- Septoria leaf blotch in **spring wheat, winter wheat** [1, 2]
- Sooty moulds in **spring barley, spring wheat, winter barley, winter wheat** [1, 2]
- Yellow rust in **spring barley, spring wheat, winter barley, winter wheat** [1, 2]

Specific Off-Label Approvals (SOLAs)
- *grass seed crops* 20100031 [2]

Approval information
- Maneb included in Annex I under EC Directive 91/414
- Accepted by BBPA on malting barley
- Approved for aerial application [1, 2]. See Section 5 for more information

Efficacy guidance
- On potatoes first application should be made before blight infection occurs, and further applications made to protect new growth
- Maneb is a protectant fungicide and will give moderate control, suppression or reduction of the cereal diseases listed but mixture with carbendazim is essential to achieve satisfactory results
- Best results on cereals obtained from a programme of a preventative treatment before disease established, an early application at about first node stage (GS 31), a late application after flag leaf emergence (GS 37) and during ear emergence before watery ripe stage (GS 71)

Restrictions
- Maximum number of treatments 2 per crop on cereals; not specified on potatoes
- Do not apply if frost or rain expected, if crop wet or suffering from drought or physical or chemical stress
- A minimum of 7 d must elapse between applications to wheat and barley, and 10 d for potatoes

Crop-specific information
- Latest use: before grain milky-ripe (GS 73) for wheat, barley; before flag leaf sheath opening (GS 45) for barley; HI potatoes 7 d
- Apply to potatoes before blight infection occurs, at blight warning or before haulms meet in row and repeat every 10-14 d

Hazard classification and safety precautions
   Hazard Irritant, Dangerous for the environment [1]
   Transport code 9
   Packaging group III
   UN number 3077
   Risk phrases R37, R43, R50, R53a [1]
   Operator protection A, D [1]; U05a, U14, U20b [1]
   Environmental protection E15a, E38 [1]
   Storage and disposal D01, D02, D09a, D11a, D12a [1]

FOR FULL CONDITIONS OF USE ALWAYS READ THE PRODUCT LABEL

318 MCPA

A translocated phenoxycarboxylic acid herbicide for cereals and grassland
HRAC mode of action code: O

See also 2,4-D + dicamba + MCPA + mecoprop-P
2,4-D + dichlorprop-P + MCPA + mecoprop-P
2,4-D + MCPA
2,4-DB + linuron + MCPA
2,4-DB + MCPA
bentazone + MCPA + MCPB
bifenox + MCPA + mecoprop-P
clopyralid + 2,4-D + MCPA
clopyralid + diflufenican + MCPA
clopyralid + fluroxypyr + MCPA
dicamba + dichlorprop-P + ferrous sulphate + MCPA
dicamba + dichlorprop-P + MCPA
dicamba + MCPA + mecoprop-P
dichlorprop-P + ferrous sulphate + MCPA
dichlorprop-P + MCPA
dichlorprop-P + MCPA + mecoprop-P
ferrous sulphate + MCPA + mecoprop-P

Products

1	Agrichem MCPA-50	Agrichem	500 g/l	SL	04097
2	Agritox	Nufarm UK	500 g/l	SL	14894
3	Agritox 50	Nufarm UK	500 g/l	SL	14814
4	Agroxone	Headland	500 g/l	SL	14909
5	Headland Spear	Headland	500 g/l	SL	14910
6	HY-MCPA	Agrichem	500 g/l	SL	14927
7	Larke	Nufarm UK	750 g/l	SL	14914
8	MCPA 25%	Nufarm UK	500 g/l	SL	14893
9	MCPA 50	United Phosphorus	500 g/l	SL	14908
10	Nufarm MCPA 750	Nufarm UK	750 g/l	SL	14892

Uses

- Annual dicotyledons in **grass seed crops** [2-10]; **grassland, spring barley, spring oats, spring wheat, winter barley, winter oats, winter wheat** [1-10]; **listed cereals u/sown with grass only, listed cereals u/sown with red clover** [9]; **spring rye, winter rye** [1-3, 7-10]; **undersown barley, undersown oats, undersown rye, undersown wheat** [2, 3, 6-8, 10]; **undersown barley** *(red clover or grass)*, **undersown wheat** *(red clover or grass)* [4, 5]
- Charlock in **spring barley, spring oats, spring wheat, winter barley, winter oats, winter wheat** [1-10]; **spring rye** [1-3, 7, 8, 10]; **undersown barley** *(red clover or grass)*, **undersown wheat** *(red clover or grass)* [4, 5]; **winter rye** [1-3, 7-10]
- Fat hen in **spring barley, spring oats, spring wheat, winter barley, winter oats, winter wheat** [1-10]; **spring rye** [1-3, 7, 8, 10]; **undersown barley** *(red clover or grass)*, **undersown wheat** *(red clover or grass)* [4, 5]; **winter rye** [1-3, 7-10]
- Hemp-nettle in **spring barley, spring oats, spring wheat, winter barley, winter oats, winter wheat** [1-10]; **spring rye** [1-3, 7, 8, 10]; **undersown barley** *(red clover or grass)*, **undersown wheat** *(red clover or grass)* [4, 5]; **winter rye** [1-3, 7-10]
- Perennial dicotyledons in **grass seed crops, undersown barley** *(red clover or grass)*, **undersown wheat** *(red clover or grass)* [4, 5]; **grassland** [4, 5, 9]; **listed cereals u/sown with grass only, listed cereals u/sown with red clover** [9]; **spring barley, spring oats, spring wheat, winter barley, winter oats, winter wheat** [1-10]; **spring rye, winter rye** [1-3, 7-10]
- Wild radish in **spring barley, spring oats, spring wheat, winter barley, winter oats, winter wheat** [1-10]; **spring rye** [1-3, 7, 8, 10]; **undersown barley** *(red clover or grass)*, **undersown wheat** *(red clover or grass)* [4, 5]; **winter rye** [1-3, 7-10]

Approval information

- MCPA included in Annex I under EC Directive 91/414
- Accepted by BBPA for use on malting barley
- Approval expiry 31 Oct 2011 [1]

SEE SECTION 3 FOR PRODUCTS ALSO REGISTERED

Efficacy guidance
- Best results achieved by application to weeds in seedling to young plant stage under good growing conditions when crop growing actively
- Spray perennial weeds in grassland before flowering. Most susceptible growth stage varies between species. See label for details
- Do not spray during cold weather, drought, if rain or frost expected or if crop wet

Restrictions
- Maximum number of treatments normally 1 per crop or yr except grass (2 per yr) for some products. See label
- Do not treat grass within 3 mth of germination and preferably not in the first yr of a direct sown ley or after reseeding
- Do not use on cereals before undersowing
- Do not roll, harrow or graze for a few days before or after spraying; see label
- Do not use on grassland where clovers are an important part of the sward
- Do not use on any crop suffering from stress or herbicide damage
- Avoid spray drift onto nearby susceptible crops

Crop-specific information
- Latest use: before 1st node detectable (GS 31) for cereals; 4-6 wk before heading for grass seed crops; before crop 15-25 cm high for linseed
- Apply to winter cereals in spring from fully tillered, leaf sheath erect stage to before first node detectable (GS 31)
- Apply to spring barley and wheat from 5-leaves unfolded (GS 15), to oats from 1-leaf unfolded (GS 11) to before first node detectable (GS 31)
- Apply to cereals undersown with grass after grass has 2-3 leaves unfolded
- Recommendations for crops undersown with legumes vary. Red clover may withstand low doses after 2-trifoliate leaf stage, especially if shielded by taller weeds but white clover is more sensitive. See label for details
- Apply to grass seed crops from 2-3 leaf stage to 5 wk before head emergence
- Temporary wilting may occur on linseed but without long term effects

Following crops guidance
- Do not direct drill brassicas or legumes within 6 wk of spraying grassland

Environmental safety
- Harmful to aquatic organisms
- MCPA is active at low concentrations. Take extreme care to avoid drift onto neighbouring crops, especially beet crops, brassicas, most market garden crops including lettuce and tomatoes under glass, pears and vines
- Keep livestock out of treated areas for at least 2 wk and until foliage of poisonous weeds such as ragwort has died and become unpalatable
- LERAP Category B

Hazard classification and safety precautions
Hazard Harmful [1-10]; Dangerous for the environment [4-6]
Transport code 9 [3, 7, 9, 10]
Packaging group III [3, 7, 9, 10]
UN number 3082 [3, 7, 9, 10]; N/C [1, 2, 4-6, 8]
Risk phrases R20, R21 [4, 9]; R22a [4-6, 9]; R37 [5, 6]; R41 [1-3, 5-10]; R50, R53a [4-6]
Operator protection A, C; U05a, U08, U11 [1-10]; U09a [4]; U15 [5]; U20a [9]; U20b [1-8, 10]
Environmental protection E06a [6] (2 wk); E07a [1-5, 7-10]; E15a [1-4, 6-10]; E15b, E38, E40b [5]; E16a, E34 [1-10]
Storage and disposal D01, D02, D05, D09a [1-10]; D10a [1-4, 6-8, 10]; D10b [5]; D10c [9]
Medical advice M03 [1-10]; M05a [6, 9]

FOR FULL CONDITIONS OF USE ALWAYS READ THE PRODUCT LABEL

319 MCPA + MCPB

A translocated herbicide for undersown cereals, grassland and various legumes
HRAC mode of action code: O + O

See also MCPB

Products
Bellmac Plus   United Phosphorus   38:262 g/l   SL   07521

Uses
- Annual dicotyledons in **rotational grass**, **sainfoin**, **triticale** *(off-label)*, **undersown spring cereals**, **undersown winter cereals**
- Perennial dicotyledons in **rotational grass**, **sainfoin**, **undersown spring cereals**, **undersown winter cereals**

Specific Off-Label Approvals (SOLAs)
- *triticale* 20093026 expires 31 Oct 2011

Approval information
- MCPA and MCPB included in Annex I under EC Directive 91/414
- Accepted by BBPA for use on malting barley
- Approval expiry 31 Oct 2011

Efficacy guidance
- Best results achieved by application to weeds in seedling to young plant stage under good growing conditions when crop growing actively. Spray perennials when adequate leaf surface before flowering. Retreatment often needed in following year
- Spray leys and sainfoin before crop provides cover for weeds
- Rain, cold or drought may reduce effectiveness

Restrictions
- Maximum number of treatments 1 per crop or yr
- Do not spray clovers for seed
- Do not spray peas
- Do not roll or harrow for a few days before or after spraying

Crop-specific information
- Latest use: before first node detectable (GS 31) for cereals; before flower buds visible for peas
- Apply to cereals from 2-expanded leaf stage to before jointing (GS 12-30) and, where undersown, after 1-trifoliate leaf stage of clover
- Apply to direct-sown seedling clover after 1-trifoliate leaf stage
- Apply to mature white clover for fodder at any stage. Do not spray red clover after flower stalk has begun to form
- Apply to sainfoin after first trifoliate leaf stage

Environmental safety
- Harmful to aquatic organisms
- Harmful to fish or other aquatic life. Do not contaminate surface waters or ditches with chemical or used container
- Keep livestock out of treated areas until foliage of any poisonous weeds such as ragwort has died and become unpalatable
- Take extreme care to avoid drift onto neighbouring crops, especially beet crops, brassicas, most market garden crops including lettuce and tomatoes under glass, pears and vines

Hazard classification and safety precautions
Hazard Harmful
UN number N/C
Risk phrases R22a, R38, R41, R52, R53a
Operator protection U08, U11, U14, U15, U20b
Environmental protection E07a, E13c, E34
Storage and disposal D05, D09a, D10c
Medical advice M05a

SEE SECTION 3 FOR PRODUCTS ALSO REGISTERED

320 MCPA + mecoprop-P

A translocated selective herbicide for amenity grass
HRAC mode of action code: O + O

See also mecoprop-P

Products

1	Cleanrun Pro	Scotts	0.49:0.29% w/w	GR	12083
2	Greenmaster Extra	Scotts	0.49:0.29 % w/w	GR	11563

Uses
- Annual dicotyledons in *managed amenity turf*
- Perennial dicotyledons in *managed amenity turf*

Approval information
- MCPA and mecoprop-P included in Annex I under EC Directive 91/414

Efficacy guidance
- Apply from Apr to Sep, when weeds growing actively and have large leaf area available for chemical absorption

Restrictions
- Maximum total dose 105 g/sq m per yr. The total amount of mecoprop-P applied in a single yr must not exceed the maximum total dose approved for any single product for use on turf
- Avoid contact with cultivated plants
- Do not use first 4 mowings as compost or mulch unless composted for 6 mth
- Do not treat newly sown or turfed areas for at least 6 mth
- Do not reseed bare patches for 8 wk after treatment
- Do not apply when heavy rain expected or during prolonged drought. Irrigate after 1-2 d unless rain has fallen
- Do not mow within 2-3 d of treatment
- Treat areas planted with bulbs only after the foliage has died down
- Avoid walking on treated areas until it has rained or irrigation has been applied

Crop-specific information
- Granules contain NPK fertilizer to encourage grass growth

Environmental safety
- Take extreme care to avoid drift onto neighbouring crops, especially beet crops, brassicas, most market garden crops including lettuce and tomatoes under glass, pears and vines
- Harmful to fish or other aquatic life. Do not contaminate surface waters or ditches with chemical or used container
- Keep livestock out of treated areas for at least 2 wk and until foliage of any poisonous weeds such as ragwort has died and become unpalatable
- Some pesticides pose a greater threat of contamination of water than others and mecoprop-P is one of these pesticides. Take special care when applying mecoprop-P near water and do not apply if heavy rain is forecast

Hazard classification and safety precautions
UN number N/C
Operator protection A, C, H, M; U20b
Environmental protection E07a, E13c, E19b
Storage and disposal D01, D09a, D12a
Medical advice M05a

FOR FULL CONDITIONS OF USE ALWAYS READ THE PRODUCT LABEL

321 MCPB

A translocated phenoxycarboxylic acid herbicide
HRAC mode of action code: O

See also bentazone + MCPA + MCPB
bentazone + MCPB
MCPA + MCPB

Products

1	Bellmac Straight	United Phosphorus	400 g/l	SL	07522
2	Butoxone	Headland	400 g/l	SL	14406
3	Tropotox	Nufarm UK	400 g/l	SL	14450

Uses
- Annual dicotyledons in **bilberries** *(off-label)*, **blackcurrants**, **blueberries** *(off-label)*, **clover seed crops**, **cranberries** *(off-label)*, **durum wheat** *(off-label)*, **gooseberries** *(off-label)*, **peas**, **redcurrants** *(off-label)*, **rotational grass**, **rye** *(off-label)*, **triticale** *(off-label)*, **undersown spring cereals**, **undersown winter cereals**, **whitecurrants** *(off-label)* [1]; **combining peas**, **vining peas** [2, 3]
- Docks in **combining peas**, **vining peas** [3]
- Perennial dicotyledons in **blackcurrants**, **clover seed crops**, **peas**, **rotational grass**, **undersown spring cereals**, **undersown winter cereals** [1]; **combining peas**, **vining peas** [2]
- Thistles in **combining peas**, **vining peas** [3]

Specific Off-Label Approvals (SOLAs)
- **bilberries** *20093027 expires 31 Oct 2011* [1]
- **blueberries** *20093027 expires 31 Oct 2011* [1]
- **cranberries** *20093027 expires 31 Oct 2011* [1]
- **durum wheat** *20093028 expires 31 Oct 2011* [1]
- **gooseberries** *20093027 expires 31 Oct 2011* [1]
- **redcurrants** *20093027 expires 31 Oct 2011* [1]
- **rye** *20093028 expires 31 Oct 2011* [1]
- **triticale** *20093028 expires 31 Oct 2011* [1]
- **whitecurrants** *20093027 expires 31 Oct 2011* [1]

Approval information
- MCPB included in Annex I under EC Directive 91/414
- Accepted by BBPA for use on malting barley
- Approval expiry 31 Oct 2011 [1]

Efficacy guidance
- Best results achieved by spraying young seedling weeds in good growing conditions
- Best results on perennials by spraying before flowering
- Effectiveness may be reduced by rain within 12 h, by very cold or dry conditions

Restrictions
- Maximum number of treatments 1 per crop or yr.
- Do not roll or harrow for 7-10 d before or after treatment (check label)

Crop-specific information
- Latest use: first node detectable stage (GS 31) for cereals; before flower buds appear in terminal leaf (GS 201) for peas; before flower buds form for clover; before weeds damaged by frost for cane and bush fruit
- Apply to undersown cereals from 2-leaves unfolded to first node detectable (GS 12-31), and after first trifoliate leaf stage of clover
- Red clover seedlings may be temporarily damaged but later growth is normal
- Apply to white clover seed crops in Mar to early Apr, not after mid-May, and allow 3 wk before cutting and closing up for seed
- Apply to peas from 3-6 leaf stage but before flower bud detectable (GS 103-201). Consult PGRO (see Appendix 2) or label for information on susceptibility of cultivars.
- Do not use on leguminous crops not mentioned on the label
- Apply to cane and bush fruit after harvest and after shoot growth ceased but before weeds are damaged by frost, usually in late Aug or Sep; direct spray onto weeds as far as possible

SEE SECTION 3 FOR PRODUCTS ALSO REGISTERED

Pesticide Profiles

Environmental safety
- Harmful to aquatic organisms
- Harmful to fish or other aquatic life. Do not contaminate surface waters or ditches with chemical or used container
- Keep livestock out of treated areas until foliage of any poisonous weeds such as ragwort has died and become unpalatable
- Take extreme care to avoid drift onto neighbouring sensitive crops

Hazard classification and safety precautions
Hazard Harmful [1-3]; Dangerous for the environment [2]
Transport code 9 [3]
Packaging group III [3]
UN number 3082 [3]; N/C [1, 2]
Risk phrases R22a [1-3]; R38, R41 [1, 2]; R51 [3]; R52 [1]; R53a [1, 3]
Operator protection A, C; U05a, U08, U20b [1-3]; U11, U14, U15 [1, 2]; U19a [3]
Environmental protection E07a, E34 [1-3]; E13c [1, 2]; E15a, E38 [3]
Storage and disposal D01, D02, D09a [1-3]; D05 [1, 2]; D10b [2, 3]; D10c [1]
Medical advice M03, M05a

322 mecoprop-P

A translocated phenoxycarboxylic acid herbicide for cereals and grassland
HRAC mode of action code: O

See also 2,4-D + dicamba + MCPA + mecoprop-P
2,4-D + dichlorprop-P + MCPA + mecoprop-P
2,4-D + mecoprop-P
bifenox + MCPA + mecoprop-P
bromoxynil + ioxynil + mecoprop-P
carfentrazone-ethyl + mecoprop-P
dicamba + MCPA + mecoprop-P
dicamba + mecoprop-P
dichlorprop-P + MCPA + mecoprop-P
diflufenican + mecoprop-P
ferrous sulphate + MCPA + mecoprop-P
fluroxypyr + mecoprop-P
MCPA + mecoprop-P

Products
1	Clenecrop Super	Nufarm UK	600 g/l	SL	14628
2	Compitox Plus	Nufarm UK	600 g/l	SL	14390
3	Duplosan KV	Nufarm UK	600 g/l	SL	13971
4	Headland Charge	Headland	600 g/l	SL	14394
5	Isomec	Nufarm UK	600 g/l	SL	14385
6	Optica	Headland	600 g/l	SL	14373

Uses
- Annual dicotyledons in **amenity grassland, durum wheat** *(off-label)*, **spring durum wheat** *(off-label)*, **spring rye** *(off-label)*, **spring triticale** *(off-label)*, **triticale** *(off-label)*, **winter rye** *(off-label)* [1-3, 5]; **grass seed crops, managed amenity turf, spring barley, spring oats, spring wheat, winter barley, winter oats, winter wheat** [1-6]; **permanent grassland, rotational grass** [4, 6]
- Chickweed in **amenity grassland** [1-3, 5]; **grass seed crops, managed amenity turf, spring barley, spring oats, spring wheat, winter barley, winter oats, winter wheat** [1-6]; **permanent grassland, rotational grass** [4, 6]
- Cleavers in **amenity grassland** [1-3, 5]; **grass seed crops, managed amenity turf, spring barley, spring oats, spring wheat, winter barley, winter oats, winter wheat** [1-6]; **permanent grassland, rotational grass** [4, 6]
- Perennial dicotyledons in **amenity grassland, grass seed crops, managed amenity turf, spring barley, spring oats, spring wheat, winter barley, winter oats, winter wheat** [1-3, 5]

FOR FULL CONDITIONS OF USE ALWAYS READ THE PRODUCT LABEL

mecoprop-P

Specific Off-Label Approvals (SOLAs)
- ***durum wheat*** *20101042* [1], *20093214* [2], *20093211* [3], *20101043* [5]
- ***spring durum wheat*** *20101042* [1], *20093214* [2], *20093211* [3], *20101043* [5]
- ***spring rye*** *20101042* [1], *20093214* [2], *20093211* [3], *20101043* [5]
- ***spring triticale*** *20101042* [1], *20093214* [2], *20093211* [3], *20101043* [5]
- ***triticale*** *20101042* [1], *20093214* [2], *20093211* [3], *20101043* [5]
- ***winter rye*** *20101042* [1], *20093214* [2], *20093211* [3], *20101043* [5]

Approval information
- Mecoprop-P included in Annex I under EC Directive 91/414
- Accepted by BBPA for use on malting barley

Efficacy guidance
- Best results achieved by application to seedling weeds which have not been frost hardened, when soil warm and moist and expected to remain so for several days

Restrictions
- Maximum number of treatments normally 1 per crop for spring cereals and 1 per yr for newly sown grass; 2 per crop or yr for winter cereals and grass crops. Check labels for details
- The total amount of mecoprop-P applied in a single yr must not exceed the maximum total dose approved for any single product for the crop/situation
- Do not spray cereals undersown with clovers or legumes or to be undersown with legumes or grasses
- Do not spray grass seed crops within 5 wk of seed head emergence
- Do not spray crops suffering from herbicide damage or physical stress
- Do not spray during cold weather, periods of drought, if rain or frost expected or if crop wet
- Do not roll or harrow for 7 d before or after treatment

Crop-specific information
- Latest use: generally before 1st node detectable (GS 31) for spring cereals and before 3rd node detectable (GS 33) for winter cereals, but individual labels vary; 5 wk before emergence of seed head for grass seed crops
- Spray winter cereals from 1 leaf stage in autumn up to and including first node detectable in spring (GS 10-31) or up to second node detectable (GS 32) if necessary. Apply to spring cereals from first fully expanded leaf stage (GS 11) but before first node detectable (GS 31)
- Spray cereals undersown with grass after grass starts to tiller
- Spray newly sown grass leys when grasses have at least 3 fully expanded leaves and have begun to tiller. Any clovers will be damaged

Environmental safety
- Harmful to aquatic organisms
- Harmful to fish or other aquatic life. Do not contaminate surface waters or ditches with chemical or used container
- Keep livestock out of treated areas for at least 2 wk and until foliage of any poisonous weeds, such as ragwort, has died and become unpalatable
- Take extreme care to avoid drift onto neighbouring crops, especially beet crops, brassicas, most market garden crops including lettuce and tomatoes under glass, pears and vines
- Some pesticides pose a greater threat of contamination of water than others and mecoprop-P is one of these pesticides. Take special care when applying mecoprop-P near water and do not apply if heavy rain is forecast

Hazard classification and safety precautions
Hazard Harmful, Dangerous for the environment
Transport code 9
Packaging group III
UN number 3082
Risk phrases R22a, R38, R41, R51, R53a [1-6]; R52 [4, 6]
Operator protection A, C [1-6]; H [1-3, 5]; P [4, 6]; U05a, U08, U11, U20b [1-6]; U15 [4, 6]
Environmental protection E07a, E13c [4, 6]; E15b [1-3, 5]; E34, E38 [1-6]
Storage and disposal D01, D02 [1-6]; D05, D10b [4, 6]; D09a, D10c [1-3, 5]
Medical advice M03

SEE SECTION 3 FOR PRODUCTS ALSO REGISTERED

Pesticide Profiles

323 mepanipyrim

An anilinopyrimidine fungicide for use in horticulture
FRAC mode of action code: 9

Products
    Frupica SC   Certis   450 g/l   SC   12067

Uses
- Botrytis in **courgettes** *(off-label)*, **forest nurseries** *(off-label)*, **protected strawberries**, **strawberries**

Specific Off-Label Approvals (SOLAs)
- *courgettes* 20093235
- *forest nurseries* 20082853

Approval information
- Mepanipyrim included in Annex I under Directive 91/414

Efficacy guidance
- Product is protectant and should be applied as a preventative spray when conditions favourable for Botrytis development occur
- To maintain Botrytis control use as part of a programme with other fungicides that control the disease
- To minimise the possibility of development of resistance adopt resistance management procedures by using products from different chemical groups as part of a mixed spray programme

Restrictions
- Maximum number of treatments 2 per crop (including other anilinopyrimidine products)
- Consult processor before use on crops for processing
- Use spray mixture immediately after preparation

Crop-specific information
- HI 3 d

Environmental safety
- Dangerous for the environment
- Very toxic to aquatic organisms
- LERAP Category B

Hazard classification and safety precautions
  Hazard Dangerous for the environment
  Transport code 9
  Packaging group III
  UN number 3082
  Risk phrases R50, R53a
  Operator protection A, H; U20c
  Environmental protection E16a, E16b, E34
  Storage and disposal D01, D02, D05, D10c, D11a, D12b

324 mepiquat chloride

A quaternary ammonium plant growth regulator available only in mixtures

See also 2-chloroethylphosphonic acid + mepiquat chloride
            chlormequat + 2-chloroethylphosphonic acid + mepiquat chloride
            chlormequat + mepiquat chloride

FOR FULL CONDITIONS OF USE ALWAYS READ THE PRODUCT LABEL

325 mepiquat chloride + prohexadione-calcium

A growth regulator mixture for cereals

See also prohexadione-calcium

Products

1	Canopy	BASF	300:50 g/l	SC	13181
2	Medax Top	BASF	300:50 g/l	SC	13282
3	Standon Midget	Standon	300:50	SC	13803

Uses
- Increasing yield in **durum wheat** (off-label) [1]; **triticale** [1, 2]; **winter barley, winter wheat** [1-3]
- Lodging control in **durum wheat** (off-label) [1]; **triticale** [1, 2]; **winter barley, winter wheat** [1-3]

Specific Off-Label Approvals (SOLAs)
- **durum wheat** 20072403 [1]

Approval information
- Mepiquat chloride and prohexadione-calcium included in Annex I under EC Directive 91/414
- Accepted by BBPA for use on malting barley

Efficacy guidance
- Best results obtained from treatments applied to healthy crops from the beginning of stem extension

Restrictions
- Maximum total dose equivalent to one full dose treatment on all crops
- Do not apply to any crop suffering from physical stress caused by waterlogging, drought or other conditions
- Do not treat on soils with a substantial moisture deficit
- Consult grain merchant or processor before use on crops for bread making or brewing. Effects on these processes have not been tested

Crop-specific information
- Latest use: before flag leaf fully emerged on wheat and barley

Following crops guidance
- Any crop may follow a normally harvested treated crop. Ploughing is not essential.

Environmental safety
- Harmful to aquatic organisms
- Avoid spray drift onto neighbouring crops

Hazard classification and safety precautions
Hazard Harmful
UN number N/C
Risk phrases R22a, R52, R53a
Operator protection A; U05a, U20b
Environmental protection E15a, E34, E38
Storage and disposal D01, D02, D05, D09a, D10c
Medical advice M05a

326 meptyldinocap

A protectant dinitrophenyl fungicide for powdery mildew control
FRAC mode of action code: 29

Products

Kindred	Landseer	350 g/l	EC	13891

Uses
- Powdery mildew in **apples** (off-label), **crab apples** (off-label), **pears** (off-label), **quinces** (off-label), **table grapes, wine grapes**

Specific Off-Label Approvals (SOLAs)
- **apples** 20092664

SEE SECTION 3 FOR PRODUCTS ALSO REGISTERED

Pesticide Profiles

- *crab apples* 20092664
- *pears* 20092664
- *quinces* 20092664

Environmental safety
- LERAP Category B

Hazard classification and safety precautions
Hazard Harmful, Flammable, Dangerous for the environment
Transport code 3
Packaging group III
UN number 1993
Risk phrases R22a, R36, R38, R43, R50, R53a, R67
Operator protection A, C, H; U04a, U05a, U11, U14, U19a, U20a
Environmental protection E15b, E16a, E34, E38
Storage and disposal D01, D02, D09a, D10a, D12a
Medical advice M05a

327 mesosulfuron-methyl

A sulfonyl urea herbicide for cereals available only in mixtures
HRAC mode of action code: B

*See also diflufenican + iodosulfuron-methyl-sodium + mesosulfuron-methyl
iodosulfuron-methyl-sodium + mesosulfuron-methyl*

328 mesotrione

A foliar applied triketone herbicide for maize
HRAC mode of action code: F2

Products

1	Callisto	Syngenta	100 g/l	SC	12323
2	Greencrop Goldcob	Greencrop	100 g/l	SC	12863
3	Kalypstowe	AgChem Access	100 g/l	SC	13844
4	Marvel	ChemSource	100 g/l	SC	14508

Uses
- Annual dicotyledons in *forage maize* [1-4]; *game cover* (off-label), *linseed* (off-label), *sweetcorn* (off-label) [1]
- Annual grasses in *game cover* (off-label), *sweetcorn* (off-label) [1]
- Annual meadow grass in *grain maize* [1, 3, 4]; *linseed* (off-label) [1]
- Volunteer oilseed rape in *forage maize* [1-4]; *grain maize* [1, 3, 4]; *linseed* (off-label) [1]

Specific Off-Label Approvals (SOLAs)
- *game cover* 20082830 [1]
- *linseed* 20071706 [1]
- *sweetcorn* 20051893 [1]

Approval information
- Mesotrione included in Annex I under EC Directive 91/414

Efficacy guidance
- Best results obtained from treatment of young actively growing weed seedlings in the presence of adequate soil moisture
- Treatment in poor growing conditions or in dry soil may give less reliable control
- Activity is mostly by foliar uptake with some soil uptake
- To minimise the possible development of resistance where continuous maize is grown the product should not be used for more than two consecutive seasons

Restrictions
- Maximum number of treatments 1 per crop of forage maize
- Do not use on seed crops or on sweetcorn varieties
- Do not spray when crop foliage wet or when excessive rainfall is expected to follow application

FOR FULL CONDITIONS OF USE ALWAYS READ THE PRODUCT LABEL

mesotrione + terbuthylazine

- Do not treat crops suffering from stress from cold or drought conditions, or when wide temperature fluctuations are anticipated

Crop-specific information
- Latest use: 8 leaves unfolded stage (GS 18) for forage maize
- Treatment under adverse conditions may cause mild to moderate chlorosis. The effect is transient and does not affect yield

Following crops guidance
- Winter wheat, durum wheat, winter barley or ryegrass may follow a normally harvested treated crop of maize. Oilseed rape may be sown provided it is preceded by deep ploughing to more than 15 cm
- In the spring following application only forage maize, ryegrass, spring wheat or spring barley may be sown
- In the event of crop failure maize may be re-seeded immediately. Some slight crop effects may be seen soon after emergence but these are normally transient

Environmental safety
- Dangerous for the environment
- Very toxic to aquatic organisms
- Take extreme care to avoid drift onto all plants outside the target area
- LERAP Category B

Hazard classification and safety precautions
Hazard Irritant, Dangerous for the environment
UN number N/C
Risk phrases R36, R50, R53a
Operator protection A, C; U05a, U20b [1-4]; U08, U19a [2]; U09a [1, 3, 4]
Environmental protection E15b, E16a, E16b [1-4]; E38 [1, 3, 4]
Storage and disposal D01, D02, D05, D09a, D12a [1-4]; D10b [2]; D10c, D11a [1, 3, 4]
Medical advice M05a [2]

329 mesotrione + terbuthylazine

A foliar and soil acting herbicide mixture for maize
HRAC mode of action code: F2 + C1

See also terbuthylazine

Products

1 Calaris	Syngenta	70:330 g/l	SC	12405
2 Clayton Faize	Clayton	70:330 g/l	SC	13810
3 Destiny	AgChem Access	70:330 g/l	SC	14159
4 Pan Theta	Pan Agriculture	70:330 g/l	SC	14501

Uses
- Annual dicotyledons in **forage maize, grain maize** [1-4]; **sweetcorn** *(off-label)* [1]
- Annual meadow grass in **forage maize, grain maize** [1-4]; **sweetcorn** *(off-label)* [1]

Specific Off-Label Approvals (SOLAs)
- **sweetcorn** 20051892 [1]

Approval information
- Mesotrione included in Annex I under EC Directive 91/414

Efficacy guidance
- Best results obtained from treatment of young actively growing weed seedlings in the presence of adequate soil moisture
- Treatment in poor growing conditions or in dry soil may give less reliable control
- Residual weed control is reduced on soils with more than 10% organic matter
- To minimise the possible development of resistance where continuous maize is grown the product should not be used for more than two consecutive seasons

Restrictions
- Maximum number of treatments 1 per crop of forage maize
- Do not use on seed crops or on sweetcorn varieties

SEE SECTION 3 FOR PRODUCTS ALSO REGISTERED

Pesticide Profiles

- Do not spray when crop foliage wet or when excessive rainfall is expected to follow application
- Do not treat crops suffering from stress from cold or drought conditions, or when wide temperature fluctuations are anticipated
- Do not apply on Sands or Very Light soils

Crop-specific information
- Latest use: 8 leaves unfolded stage (GS 18) for forage maize
- Treatment under adverse conditions may cause mild to moderate chlorosis. The effect is transient and does not affect yield

Following crops guidance
- Winter wheat, durum wheat, winter barley or ryegrass may follow a normally harvested treated crop of maize. Oilseed rape may be sown provided it is preceded by deep ploughing to more than 15 cm
- In the spring following application forage maize, ryegrass, spring wheat or spring barley may be sown
- Spinach, beet crops, peas, beans, lettuce and cabbages must not be sown in the yr following application
- In the event of crop failure maize may be re-seeded immediately. Some slight crop effects may be seen soon after emergence but these are normally transient

Environmental safety
- Dangerous for the environment
- Very toxic to aquatic organisms
- Take extreme care to avoid drift onto all plants outside the target area
- LERAP Category B

Hazard classification and safety precautions
Hazard Harmful, Dangerous for the environment
Transport code 9
Packaging group III
UN number 3082
Risk phrases R22a, R50, R53a
Operator protection A; U05a
Environmental protection E15a, E16a, E16b, E34, E38
Storage and disposal D01, D02, D05, D09a, D10c, D12a

330 metalaxyl-M

A phenylamide systemic fungicide
FRAC mode of action code: 4

See also chlorothalonil + metalaxyl-M
cymoxanil + fludioxonil + metalaxyl-M
fluazinam + metalaxyl-M
fludioxonil + metalaxyl-M + thiamethoxam
mancozeb + metalaxyl-M

Products

1	Apron XL	Syngenta	339.2 g/l	ES	14654
2	Clayton Tine	Clayton	465 g/l	SL	14072
3	SL 567A	Syngenta	465.2 g/l	SL	12380
4	Subdue	Fargro	465 g/l	SL	12503

Uses
- Cavity spot in **carrots**, **parsnips** *(off-label)* [3]; **carrots** *(reduction only)* [2]
- Crown rot in **protected water lilies** *(off-label)*, **water lilies** *(off-label)* [3]
- Damping off in **watercress** *(off-label)* [3]
- Downy mildew in **asparagus** *(off-label)*, **blackberries** *(off-label)*, **grapevines** *(off-label)*, **herbs (see appendix 6)** *(off-label)*, **hops** *(off-label)*, **horseradish** *(off-label)*, **protected cucumbers** *(off-label)*, **protected herbs (see appendix 6)** *(off-label)*, **protected soft fruit** *(off-label)*, **protected spinach** *(off-label)*, **protected spinach beet** *(off-label)*, **protected water lilies** *(off-label)*, **raspberries** *(off-label)*, **rubus hybrids** *(off-label)*, **salad onions** *(off-label)*, **soft fruit**

FOR FULL CONDITIONS OF USE ALWAYS READ THE PRODUCT LABEL

metalaxyl-M

(off-label), **spinach** (off-label), **spinach beet** (off-label), **water lilies** (off-label), **watercress** (off-label) [3]
- Phytophthora root rot in **ornamental plant production, protected ornamentals** [4]
- Pythium in **beetroot** (off-label), **broccoli, brussels sprouts, bulb onions, cabbages, calabrese, cauliflowers, chinese cabbage, herbs (see appendix 6)** (off-label), **kohlrabi, ornamental plant production** (off-label), **radishes** (off-label), **spinach** [1]; **ornamental plant production, protected ornamentals** [4]
- Root malformation disorder in **red beet** (off-label) [3]
- Storage rots in **cabbages** (off-label) [3]
- White blister in **horseradish** (off-label) [3]

Specific Off-Label Approvals (SOLAs)
- **asparagus** 20051502 [3]
- **beetroot** 20102539 [1]
- **blackberries** 20072195 [3]
- **cabbages** 20062117 [3]
- **grapevines** 20051504 [3]
- **herbs (see appendix 6)** 20102539 [1], 20051507 [3]
- **hops** 20051500 [3]
- **horseradish** 20051499 [3], 20061040 [3]
- **ornamental plant production** 20102539 [1]
- **parsnips** 20051508 [3]
- **protected cucumbers** 20051503 [3]
- **protected herbs (see appendix 6)** 20051507 [3]
- **protected soft fruit** 20082937 [3]
- **protected spinach** 20051507 [3]
- **protected spinach beet** 20051507 [3]
- **protected water lilies** 20051501 [3]
- **radishes** 20102539 [1]
- **raspberries** 20072195 [3]
- **red beet** 20051307 [3]
- **rubus hybrids** 20072195 [3]
- **salad onions** 20072194 [3]
- **soft fruit** 20082937 [3]
- **spinach** 20051507 [3]
- **spinach beet** 20051507 [3]
- **water lilies** 20051501 [3]
- **watercress** 20072193 [3]

Approval information
- Metalaxyl-M included in Annex I under EC Directive 91/414
- Accepted by BBPA for use on hops

Efficacy guidance
- Best results achieved when applied to damp soil or potting media
- Treatments to ornamentals should be followed immediately by irrigation to wash any residues from the leaves and allow penetration to the rooting zone [4]
- Efficacy may be reduced in prolonged dry weather [3]
- Results may not be satisfactory on soils with high organic matter content [3]
- Control of cavity spot on carrots overwintered in the ground or lifted in winter may be lower than expected [3]
- Use in an integrated pest management strategy and, where appropriate, alternate with products from different chemical groups
- Product should ideally be used preventatively and the number of phenylamide applications should be limited to 1-2 consecutive treatments
- Always follow FRAG guidelines for preventing and managing fungicide resistance. See Section 5 for more information

Restrictions
- Maximum number of treatments on ornamentals 1 per situation for media treatment. See label for details of drench treatment of protected ornamentals [4]
- Maximum total dose on carrots equivalent to one full dose treatment [3]

SEE SECTION 3 FOR PRODUCTS ALSO REGISTERED

Pesticide Profiles

- Do not use where carrots have been grown on the same site within the previous eight yrs [3]
- Consult before use on crops intended for processing [3]
- Do not re-use potting media from treated plants for subsequent crops [4]
- Disinfect pots thoroughly prior to re-use [4]

Crop-specific information
- Latest use: 6 wk after drilling for carrots [3]
- Because of the large number of species and ornamental cultivars susceptibility should be checked before large scale treatment [4]
- Treatment of *Viburnum* and *Prunus* species not recommended [4]

Environmental safety
- Harmful to aquatic organisms
- Limited evidence suggests that metalaxyl-M is not harmful to soil dwelling predatory mites

Hazard classification and safety precautions
Hazard Harmful
UN number N/C
Risk phrases R22a, R52, R53a [1-4]; R37 [2-4]
Operator protection A [1-4]; C [4]; D [1]; H [1, 4]; U02a, U05a, U20b [1-4]; U04a, U10, U19a [2-4]
Environmental protection E15b, E34, E38
Storage and disposal D01, D02, D05, D09a, D12a [1-4]; D07, D10c [2-4]; D11a [1]
Treated seed S02, S04d, S05, S07, S08, S09 [1]
Medical advice M03 [2-4]; M05a [1]

331 metaldehyde

A molluscicide bait for controlling slugs and snails

Products

1	Allure	Chiltern	3% w/w	PT	12651
2	Appeal	Chiltern	1.5% w/w	PT	12022
3	Attract	Chiltern	1.5% w/w	PT	12023
4	Carakol 3	Makhteshim	3% w/w	PT	14309
5	Carakol Plus	Makhteshim	5% w/w	PT	13980
6	Certis Metaldehyde 3	Certis	3% w/w	PT	14337
7	Certis Red 3	Certis	3% w/w	PT	14061
8	Chiltern Hundreds	Chiltern	3% w/w	PT	10072
9	Cleancrop Hyde 3	Makhteshim	3% w/w	PT	14310
10	Condor 3	Doff Portland	3% w/w	PT	14324
11	Delicia Slug-lentils	Barclay	3% w/w	PT	13613
12	Desire	Chiltern	1.5% w/w	PT	14048
13	Doff Horticultural Slug Killer Blue Mini Pellets	Doff Portland	3% w/w	RB	11463
14	Duro	Makhteshim	3% w/w	PT	14347
15	Enzo	Makhteshim	3% w/w	PT	14306
16	Escar-go 3	Chiltern	3% w/w	PT	14044
17	ESP	De Sangosse	5% w/w	PT	12999
18	Gusto	Makhteshim	5% w/w	PT	13992
19	Gusto 3	Makhteshim	3% w/w	PT	14308
20	Helimax S	De Sangosse	5% w/w	RB	13109
21	Lynx H	De Sangosse	3% w/w	RB	14426
22	Metadisque	Certis	3% w/w	GB	14665
23	Molotov	Chiltern	3% w/w	PT	12650
24	Monza	Makhteshim	5% w/w	PT	13600
25	Navona	Makhteshim	4% w/w	PT	13892
26	Oreto	Makhteshim	3% w/w	PT	14258
27	Osarex W	De Sangosse	3% w/w	RB	14428
28	Pastel M	Clayton	5% w/w	PT	13479
29	Pesta H	De Sangosse	3% w/w	RB	14427
30	Polymet 5	Unicrop	5% w/w	RB	13480
31	Prego	Makhteshim	5% w/w	RB	13582
32	Regel	De Sangosse	5% w/w	RB	13115
33	Slugdown 3	Doff Portland	3% w/w	PT	14630
34	Steadfast 3	Doff Portland	3% w/w	PT	14367
35	Steam	Chiltern	3% w/w	PT	12473
36	Super 3	Unicrop	3% w/w	RB	14370

FOR FULL CONDITIONS OF USE ALWAYS READ THE PRODUCT LABEL

metaldehyde

Products – continued

37	Super 5	Unicrop	5% w/w	RB	12535
38	TDS Major	De Sangosse	4% w/w	CB	13462
39	TDS Metarex Amba	De Sangosse	4% w/w	CB	13461
40	Tempt	Chiltern	3% w/w	PT	14227
41	Trevi	Makhteshim	4% w/w	PT	13983
42	Trojan	Chiltern	3% w/w	PT	14224
43	Trounce	Chiltern	3% w/w	PT	14222

Uses
- Slugs in *all edible crops (outdoor and protected)* [11]; *all edible crops (outdoor)* [2, 3, 8, 13, 17, 18, 20, 23, 28, 30, 31, 35, 38, 39, 42]; *all edible crops (outdoor)* (around), *all non-edible crops (outdoor)* (around) [37]; *all non-edible crops (outdoor)* [2, 3, 8, 11, 13, 17, 18, 20, 23, 28, 30, 31, 35, 38, 39, 42]; *cultivated land/soil* [2, 3, 8, 13, 23, 35, 42]; *natural surfaces not intended to bear vegetation* [11, 20, 28, 30, 37-39]; *ornamental plant production* [18, 31]; *protected crops* [2, 3, 8, 18, 23, 31, 35, 38, 39, 42]
- Slugs and snails in *all edible crops (outdoor and protected)* [1, 12, 16, 24, 40, 43]; *all edible crops (outdoor and protected)* (do not apply to cauliflowers) [6, 7, 22]; *all edible crops (outdoor)* [10, 21, 27, 29, 33, 34, 36]; *all edible crops (outdoor)* (except potatoes and cauliflowers) [5, 25, 41]; *all edible crops (outdoor)* (excluding potato and cauliflower) [4, 9, 14, 15, 19, 26]; *all edible crops except potatoes and cauliflowers* [32]; *all non-edible crops (outdoor)* [1, 4, 5, 10, 12, 14-16, 19, 21, 22, 24-27, 29, 32-34, 36, 40, 41, 43]; *all non-edible crops (outdoor)* (e) [9]; *amenity grassland, managed amenity turf* [6, 7]; *cauliflowers* [1, 21, 27, 29, 32, 40, 43]; *cultivated land/soil, protected crops* [12, 16]; *natural surfaces not intended to bear vegetation* [6, 7, 10, 21, 22, 27, 29, 32-34, 36]; *potatoes* [1, 4, 6, 7, 9, 14, 15, 19, 21, 22, 26, 27, 29, 32, 40, 43]
- Snails in *all edible crops (outdoor), all non-edible crops (outdoor)* [2, 3, 8, 13, 17, 18, 20, 23, 28, 30, 31, 35, 38, 39, 42]; *all edible crops (outdoor)* (around), *all non-edible crops (outdoor)* (around) [37]; *cultivated land/soil* [2, 3, 8, 13, 23, 35, 42]; *natural surfaces not intended to bear vegetation* [20, 28, 30, 37-39]; *ornamental plant production* [18, 31]; *protected crops* [2, 3, 8, 18, 23, 31, 35, 38, 39, 42]

Approval information
- Metaldehyde has not been included in Annex 1 but the date of withdrawal has not yet been published
- Accepted by BBPA for use on malting barley and hops
- Approval expiry 31 Dec 2011 [24, 25, 28]
- Approval expiry 30 Sep 2011 [30, 37]

Efficacy guidance
- Apply pellets by hand, fiddle drill, fertilizer distributor, by air (check label) or in admixture with seed. See labels for rates and timing.
- Best results achieved from an even spread of granules applied during mild, damp weather when slugs and snails most active. May be applied in standing crops
- To establish the need for pellet application on winter wheat or winter oilseed rape, monitor for slug activity. Where bait traps are used, use a foodstuff attractive to slugs e.g. chicken layer's mash
- Varieties of oilseed rape low in glucosinolates can be more acceptable to slugs than "single low" varieties and control may not be as good
- To prevent slug build up apply at end of season to brassicas and other leafy crops
- To reduce tuber damage in potatoes apply twice in Jul and Aug
- For information on slug trapping and damage risk assessment refer to HGCA Topic Sheets No. 84 (winter wheat) and 85 (winter oilseed rape), available from the HGCA website (www.hgca.com)

Restrictions
- Do not apply when rain imminent or water glasshouse crops within 4 d of application
- Take care to avoid lodging of pellets in the foliage when making late applications to edible crops

Crop-specific information
- Put slug traps out before cultivation, when the soil surface is visibly moist and the weather mild (5-25 °C) (see label for guidance)
- For winter wheat, a catch of 4 or more slugs/trap indicates a possible risk, where soil and weather conditions favour slug activity

SEE SECTION 3 FOR PRODUCTS ALSO REGISTERED

Pesticide Profiles

• For winter oilseed rape a catch of 4 or more slugs in standing cereals, or 1 or more in cereal stubble, if other conditions were met, would indicate possible risk of damage

Environmental safety
• Dangerous to game, wild birds and animals
• Some products contain proprietary cat and dog deterrent
• Keep poultry out of treated areas for at least 7 d
• Do not use slug pellets in traps in winter wheat or winter oilseed rape since they are a potential hazard to wildlife and pets
• Some pesticides pose a greater threat of contamination of water than others and metaldehyde is one of these pesticides. Take special care when applying metaldehyde near water and do not apply if heavy rain is forecast

Hazard classification and safety precautions
   UN number N/C [1-9, 12-33, 35-43]
   Operator protection A, H [1-12, 14-43]; J [17, 28, 30, 37]; U05a [1-12, 14-16, 19, 21-29, 34, 38-43]; U15 [18, 31, 32]; U20a [4, 9, 14, 15, 18, 19, 26, 31]; U20b [5-8, 10, 21-25, 27, 29, 30, 34, 41, 42]; U20c [1-3, 12, 13, 16, 17, 20, 28, 32, 33, 35, 36, 38-40, 43]
   Environmental protection E05b [1-9] (7 d); E05b [10] (7 days); E05b [11] (Keep poultry out of treated areas for at least 7 days following application); E05b [12] (7 days); E05b [13-15] (7 d); E05b [16] (7 days); E05b [17-20] (7 d); E05b [21] (7 days); E05b [22-26] (7 d); E05b [27] (7 days); E05b [28] (7 d); E05b [29] (7 days); E05b [30, 31] (7 d); E05b [32] (7 days); E05b [33] (7 d); E05b [34] (7 days); E05b [35-43] (7 d); E06b [12, 16] (7 d); E07a, E34 [10, 21, 27, 29, 34]; E10a [1-5, 8-21, 23-43]; E10c [6, 7, 22]; E15a [1-43]
   Storage and disposal D01 [1-12, 14-29, 31-43]; D02 [1-12, 14-16, 18-29, 31, 32, 34, 38-43]; D05 [6, 7, 22, 28]; D07 [1-3, 6-8, 10, 12, 13, 16, 18, 20-23, 27, 29-40, 42, 43]; D09a [1-10, 12, 14-43]; D11a [1-10, 12-19, 21-37, 40-43]; D11b [20, 38, 39]; D12a [18, 20, 31]; D12b [32]
   Treated seed S04a [10, 11, 17, 18, 20, 21, 27, 29, 31, 32, 34, 37-39]
   Medical advice M04a [6, 7, 22, 28]; M05a [10, 11, 18, 20, 21, 27, 29, 31, 32, 34, 38, 39]

332   metamitron

A contact and residual triazinone herbicide for use in beet crops
HRAC mode of action code: C1

See also chloridazon + chlorpropham + metamitron
          chloridazon + metamitron
          chlorpropham + metamitron
          desmedipham + ethofumesate + metamitron + phenmedipham
          ethofumesate + metamitron
          ethofumesate + metamitron + phenmedipham

Products

	Name	Supplier	Conc.	Form	Reg. No.
1	Bettix 70 WG	United Phosphorus	70% w/w	WG	11154
2	Bettix Flo	United Phosphorus	700 g/l	SC	11959
3	Celmitron 70% WDG	AgriChem BV	70% w/w	WG	14408
4	Clayton Mitrex	Clayton	70% w/w	WG	11921
5	Defiant SC	United Phosphorus	700 g/l	SC	12302
6	Goldbeet	Makhteshim	90% w/w	WG	11538
7	Goltix 90	Makhteshim	90% w/w	WG	11578
8	Goltix Flowable	Makhteshim	700 g/l	SC	12851
9	Goltix WG	Makhteshim	70% w/w	WG	11539
10	Marquise	Makhteshim	70% w/w	WG	08738
11	Mitron 70 WG	Belcrop	70% w/w	WG	11516
12	Mitron 90 WG	Belcrop	90% w/w	WG	10888
13	Mitron SC	Belcrop	700 g/l	SC	11643
14	MM 70 Flo	Nufarm UK	700 g/l	SC	13318
15	Predator WG	AgriGuard	70% w/w	WG	12630
16	Skater	Makhteshim	700 g/l	SC	12857
17	Target SC	AgriChem BV	700 g/l	SC	13307

Uses
• Annual dicotyledons in **asparagus** *(off-label)* [6-8, 10, 16]; **cardoons** *(off-label)*, **celery (outdoor)** *(off-label)*, **rhubarb** *(off-label)*, **strawberries** *(off-label)* [6-9, 16]; **farm forestry** *(off-label)* [6];

FOR FULL CONDITIONS OF USE ALWAYS READ THE PRODUCT LABEL

metamitron

fodder beet, sugar beet [1-17]; *forest* (off-label) [7, 8, 10, 16]; *herbs (see appendix 6)* (off-label), *horseradish* (off-label), *parsnips* (off-label) [6-8, 16]; *mangels* [1-10, 12-17]; *red beet* [1-10, 12-15, 17]
- Annual grasses in *asparagus* (off-label), *forest* (off-label) [10]; *fodder beet, sugar beet* [1-16]; *herbs (see appendix 6)* (off-label) [7]; *horseradish* (off-label), *parsnips* (off-label) [6-8]; *mangels* [1-10, 12-16]; *red beet* [1-10, 12-15]; *strawberries* (off-label) [8]
- Annual meadow grass in *asparagus* (off-label), *herbs (see appendix 6)* (off-label) [6-8, 16]; *cardoons* (off-label), *celery (outdoor)* (off-label), *rhubarb* (off-label) [6-9]; *farm forestry* (off-label) [6]; *fodder beet, sugar beet* [1-17]; *forest* (off-label) [7, 8, 16]; *horseradish* (off-label), *parsnips* (off-label) [6, 7, 16]; *mangels* [1-10, 12-17]; *red beet* [1-10, 12-15, 17]; *strawberries* (off-label) [6, 7, 9, 16]
- Fat hen in *fodder beet, sugar beet* [1-16]; *herbs (see appendix 6)* (off-label) [7]; *horseradish* (off-label), *parsnips* (off-label) [6, 7]; *mangels* [1-10, 12-16]; *red beet* [1-10, 12-15]
- General weed control in *asparagus* (off-label), *forest* (off-label), *horseradish* (off-label), *parsnips* (off-label) [9]
- Groundsel in *chives* (off-label), *parsley* (off-label) [9, 10]; *herbs (see appendix 6)* (off-label) [6, 8-10]
- Polygonums in *cardoons* (off-label), *celery (outdoor)* (off-label), *rhubarb* (off-label) [16]

Specific Off-Label Approvals (SOLAs)
- *asparagus* 20093358 [6], 20081387 [7], 20100065 [8], 20041758 [9], 20030737 [10], 20100066 [16]
- *cardoons* 20093356 [6], 20081389 [7], 20093520 [8], 20093518 [9], 20070844 [16]
- *celery (outdoor)* 20093356 [6], 20081389 [7], 20093520 [8], 20093518 [9], 20070844 [16]
- *chives* 20041756 [9], 20030737 [10]
- *farm forestry* 20093354 [6]
- *forest* 20081386 [7], 20093519 [8], 20041757 [9], 20030737 [10], 20093525 [16]
- *herbs (see appendix 6)* 20093355 [6], 20081385 [7], 20093521 [8], 20041756 [9], 20030737 [10], 20093522 [16]
- *horseradish* 20081039 [6], 20081041 [7], 20070513 [8], 20061637 [9], 20093524 [16]
- *parsley* 20041756 [9], 20030737 [10]
- *parsnips* 20081039 [6], 20081041 [7], 20070513 [8], 20061637 [9], 20093524 [16]
- *rhubarb* 20093356 [6], 20081389 [7], 20093520 [8], 20093518 [9], 20070844 [16]
- *strawberries* 20093357 [6], 20081388 [7], 20070856 [8], 20093517 [9], 20093523 [16]

Approval information
- Approval expiry 31 Aug 2011 [4, 14]

Efficacy guidance
- May be used pre-emergence alone or post-emergence in tank mixture or with an authorised adjuvant oil
- Low dose programme (LDP). Apply a series of low-dose post-weed emergence sprays, including adjuvant oil, timing each treatment according to weed emergence and size. See label for details and for recommended tank mixes and sequential treatments. On mineral soils the LDP should be preceded by pre-drilling or pre-emergence treatment
- Traditional application. Apply either pre-drilling before final cultivation with incorporation to 8-10 cm, or pre-crop emergence at or soon after drilling into firm, moist seedbed to emerged weeds from cotyledon to first true leaf stage
- On emerged weeds at or beyond 2-leaf stage addition of adjuvant oil advised
- For control of wild oats and certain other weeds, tank mixes with other herbicides or sequential treatments are recommended. See label for details

Restrictions
- Maximum total dose equivalent to three full dose treatments for most products. Check label
- Using traditional method post-crop emergence on mineral soils do not apply before first true leaves have reached 1 cm long

Crop-specific information
- Latest use: before crop foliage meets across rows for beet crops
- HI herbs 6 wk
- Crop tolerance may be reduced by stress caused by growing conditions, effects of pests, disease or other pesticides, nutrient deficiency etc

SEE SECTION 3 FOR PRODUCTS ALSO REGISTERED

Pesticide Profiles

Following crops guidance
- Only sugar beet, fodder beet or mangels may be drilled within 4 mth after treatment. Winter cereals may be sown in same season after ploughing, provided 16 wk passed since last treatment

Environmental safety
- Dangerous for the environment
- Very toxic to aquatic organisms
- Dangerous to fish or other aquatic life. Do not contaminate surface waters or ditches with chemical or used container
- Do not empty into drains

Hazard classification and safety precautions
Hazard Harmful [1-3, 5-10, 12-15, 17]; Irritant [4]; Dangerous for the environment [1-17]
Transport code 9
Packaging group III
UN number 3077 [1, 3, 4, 6, 7, 9-12, 15]; 3082 [2, 5, 8, 13, 14, 16, 17]
Risk phrases R22a [1-3, 5-10, 12-15, 17]; R41 [4]; R43 [2, 4, 5, 10]; R50 [1, 2, 4-8, 10-13]; R51 [3, 9, 14-17]; R53a [1-17]
Operator protection A [2, 4, 5, 8-10, 13-15]; C [4]; H [4, 10, 15]; U05a [1, 4, 10, 11, 14, 17]; U08 [10, 11, 14, 17]; U09a [2, 5]; U11; U15 [4]; U14 [2, 4, 5, 10, 11, 14, 17]; U19a [1, 3, 4, 9-11, 14, 17]; U20a [3, 4, 6, 7, 9-12, 14]; U20b [1, 8, 13, 17]; U20c [2, 5, 16]
Environmental protection E13b [1, 2, 5]; E15a [3, 4, 6-17]; E19b [6-8, 12, 13, 15, 16]; E34 [8, 13, 14, 17]; E38 [3, 9, 11, 17]
Storage and disposal D01 [1, 2, 4-8, 10-15, 17]; D02 [1, 2, 4, 5, 8, 10, 11, 13, 14, 17]; D05 [2, 4, 5, 14]; D09a [1-17]; D10c [2, 5, 16]; D11a [1, 3, 4, 6-14, 17]; D12a [1-9, 12, 13, 15-17]; D12b [10, 14]
Medical advice M03 [1, 2, 5, 8, 13, 17]; M04a [12]; M05a [1, 2, 5-8, 13, 15]

333 metam-sodium

A methyl isothiocyanate producing sterilant for glasshouse, nursery and outdoor soils
HRAC mode of action code: Z

Products

1	Discovery	United Phosphorus	510 g/l	SL	10416
2	Metam 510	Taminco	510 g/l	LI	09796
3	Sistan 51	Unicrop	510 g/l	SL	10046

Uses
- Brown rot in *potting soils, soils* [2]
- Millipedes in *potting soils, soils* [2]
- Nematodes in *glasshouse soils, nursery soils, outdoor soils, potting soils* [1, 3]
- Potato cyst nematode in *potting soils, soils* [2]
- Root rot in *potting soils, soils* [2]
- Root-knot nematodes in *potting soils, soils* [2]
- Soil pests in *glasshouse soils, nursery soils, outdoor soils, potting soils* [1, 3]
- Soil-borne diseases in *glasshouse soils, nursery soils, outdoor soils, potting soils* [1, 3]
- Symphylids in *potting soils, soils* [2]
- Weed seeds in *glasshouse soils, nursery soils, outdoor soils* [1, 3]; *potting soils* [1-3]; *soils* [2]
- Wireworm in *potting soils, soils* [2]

Efficacy guidance
- Metam-sodium is a partial soil sterilant and acts by breaking down in contact with soil to release methyl isothiocyanate (MIT)
- Apply to glasshouse soils as a drench, or inject undiluted to 20 cm at 30 cm intervals and seal immediately, or apply to surface and rotavate
- May also be used by mixing into potting soils
- Apply when soil temperatures exceed 7 °C, preferably above 10 °C, between 1 Apr and 31 Oct. Soil must be of fine tilth, free from debris and with 'potting moisture' content. If soil is too dry postpone treatment and water soil

Restrictions
- No plants must be present during treatment
- Crops must not be planted until a cress germination test has been completed satisfactorily

FOR FULL CONDITIONS OF USE ALWAYS READ THE PRODUCT LABEL

metazachlor

- Do not treat glasshouses within 2 m of growing crops. Fumes are damaging to all plants
- Avoid using in equipment incorporating natural rubber parts

Crop-specific information
- Latest use: pre-planting of crop

Following crops guidance
- Do not plant until soil is entirely free of fumes

Environmental safety
- Dangerous for the environment
- Very toxic to aquatic organisms
- Keep unprotected persons, livestock and pets out of treated areas for at least 24 h following treatment
- When diluted breakdown commences almost immediately. Only quantities for immediate use should be made up
- Divert or block drains which could carry solution under untreated glasshouses
- After treatment allow sufficient time (several weeks) for residues to dissipate and aerate soil by forking. Time varies with soil and season. Soils with high clay or organic matter content will retain gas longer than lighter soils

Hazard classification and safety precautions
Hazard Harmful [1, 3]; Corrosive, Dangerous for the environment [1-3]
Transport code 8 [2, 3]; 9 [1]
Packaging group III
UN number 3082 [1]; 3267 [2, 3]
Risk phrases R22a, R34, R43, R50, R53a
Operator protection A, C, H, M; U02a, U08, U11 [1-3]; U04a, U05a, U16a, U19a, U20c [2]; U07, U20b [1, 3]
Environmental protection E02a [2] (24 hours); E15a [1-3]; E34, E38 [2]; E36a [1, 3]
Storage and disposal D01, D02 [1-3]; D09a, D10a, D12a [2]; D12b, D14 [1, 3]
Medical advice M03 [2]; M04a [1-3]

334 metazachlor

A residual anilide herbicide for use in brassicas, nurseries and forestry
HRAC mode of action code: K3

See also clomazone + metazachlor
dimethenamid-p + metazachlor
dimethenamid-p + metazachlor + quinmerac

Products

1	Alpha Metazachlor 50 SC	Makhteshim	500 g/l	SC	10669
2	Butisan S	BASF	500 g/l	SC	11733
3	Clayton Buzz	Clayton	500 g/l	SC	12509
4	Clayton Metazachlor 50 SC	Clayton	500 g/l	SC	11719
5	Fuego 50	Makhteshim	500 g/l	SC	11473
6	Greencrop Monogram	Greencrop	500 g/l	SC	12048
7	Marksman	AgriGuard	500 g/l	SC	12456
8	Mashona	AgChem Access	500 g/l	SC	13006
9	Metaz 500 SC	Goldengrass	500 g/l	SC	14413
10	Mezzanine	AgriGuard	500 g/l	SC	13719
11	Rapsan 500 SC	Globachem	500 g/l	SC	13916
12	Standon Metazachlor 500	Standon	500 g/l	SC	12012
13	Sultan 50 SC	Makhteshim	500 g/l	SC	10418

Uses
- Annual dicotyledons in **amenity vegetation** [3]; **borage** *(off-label)* [1, 4, 6, 12]; **borage for oilseed production** *(off-label)* [2, 13]; **broccoli** [1-4, 6, 8-10, 12, 13]; **brussels sprouts, cabbages, cauliflowers, swedes, turnips** [1-4, 6-13]; **calabrese** [1-4, 7-13]; **canary flower (echium spp.)** *(off-label)* [1, 2, 4, 6, 12, 13]; **chinese cabbage** *(off-label)*, **choi sum** *(off-label)*, **collards** *(off-label)*, **kale** *(off-label)*, **leeks** *(off-label)*, **pak choi** *(off-label)*, **winter linseed** *(off-label)* [1, 13]; **evening primrose** *(off-label)* [1, 2, 4, 9, 12, 13]; **farm forestry** [4]; **farm woodland** [1, 2, 7, 11, 12]; **forest, nursery fruit trees and bushes** [1-4, 6, 7, 10-12]; **honesty** *(off-label)*, **mustard** *(off-label)* [1, 2, 4,

SEE SECTION 3 FOR PRODUCTS ALSO REGISTERED

Pesticide Profiles

6, 9, 12, 13]; *linseed (off-label)* [2]; *ornamental trees* [1-4, 6, 7, 11, 12]; *shrubs* [1-4, 6, 7, 11-13]; *spring linseed (off-label)* [1]; *spring oilseed rape, winter oilseed rape* [1-13]; *transplanted leeks (off-label)* [1, 2, 13]

- Annual grasses in *amenity vegetation* [3]; *borage (off-label)* [1, 4, 6, 12]; *borage for oilseed production (off-label), transplanted leeks (off-label)* [2, 13]; *broccoli, brussels sprouts, cabbages, calabrese, cauliflowers, swedes, turnips* [8]; *canary flower (echium spp.) (off-label), honesty (off-label), mustard (off-label)* [1, 2, 4, 6, 12, 13]; *chinese cabbage (off-label), choi sum (off-label), collards (off-label), kale (off-label), pak choi (off-label), winter linseed (off-label)* [1, 13]; *evening primrose (off-label)* [1, 2, 4, 12, 13]; *farm woodland* [1, 2, 7, 11, 12]; *forest* [1-4, 6, 7, 10-12]; *leeks (off-label)* [13]; *linseed (off-label)* [2]; *ornamental trees* [1-4, 6, 7, 11, 12]; *shrubs* [1-4, 6, 7, 11-13]; *spring linseed (off-label)* [1]; *spring oilseed rape, winter oilseed rape* [5, 8]

- Annual meadow grass in *amenity vegetation* [3]; *borage for oilseed production (off-label), canary flower (echium spp.) (off-label)* [2, 13]; *broccoli, brussels sprouts, cabbages, cauliflowers, spring oilseed rape, swedes, turnips, winter oilseed rape* [1-4, 6, 7, 9-13]; *calabrese* [1-4, 7, 9-13]; *chinese cabbage (off-label), choi sum (off-label), collards (off-label), kale (off-label), leeks (off-label), pak choi (off-label), transplanted leeks (off-label), winter linseed (off-label)* [13]; *evening primrose (off-label), honesty (off-label), mustard (off-label)* [2, 9, 13]; *farm forestry* [4]; *linseed (off-label)* [2]; *nursery fruit trees and bushes* [1-4, 6, 7, 10-12]; *ornamental trees* [1-4, 6, 7, 11, 12]; *shrubs* [1-4, 6, 7, 11-13]

- Blackgrass in *amenity vegetation* [3]; *borage for oilseed production (off-label), canary flower (echium spp.) (off-label), chinese cabbage (off-label), choi sum (off-label), collards (off-label), evening primrose (off-label), honesty (off-label), kale (off-label), leeks (off-label), mustard (off-label), pak choi (off-label), transplanted leeks (off-label), winter linseed (off-label)* [13]; *broccoli* [1-4, 6, 9, 10, 12, 13]; *brussels sprouts, cabbages, cauliflowers, spring oilseed rape, swedes, turnips, winter oilseed rape* [1-4, 6, 7, 9-13]; *calabrese* [1-4, 7, 9-13]; *farm forestry* [4]; *nursery fruit trees and bushes* [1-4, 6, 7, 10-12]; *ornamental trees* [1-4, 6, 7, 11, 12]; *shrubs* [1-4, 6, 7, 11-13]

- Groundsel in *chinese cabbage (off-label), choi sum (off-label), collards (off-label), kale (off-label), pak choi (off-label)* [2]

- Mayweeds in *chinese cabbage (off-label), choi sum (off-label), collards (off-label), kale (off-label), pak choi (off-label)* [2]

- Shepherd's purse in *chinese cabbage (off-label), choi sum (off-label), collards (off-label), kale (off-label), pak choi (off-label)* [2]

Specific Off-Label Approvals (SOLAs)
- *borage* 20060901 [1], 20060902 [4], 20060898 [6], 20060890 [12]
- *borage for oilseed production* 20060906 [2], 20081419 [13]
- *canary flower (echium spp.)* 20060901 [1], 20060906 [2], 20060902 [4], 20060898 [6], 20060890 [12], 20081419 [13]
- *chinese cabbage* 20073306 [1], 20050344 [2], 20081420 [13]
- *choi sum* 20073306 [1], 20050344 [2], 20081420 [13]
- *collards* 20073306 [1], 20050344 [2], 20081420 [13]
- *evening primrose* 20060901 [1], 20060906 [2], 20060902 [4], 20090935 [9], 20060890 [12], 20081419 [13]
- *honesty* 20060901 [1], 20060906 [2], 20060902 [4], 20060898 [6], 20090935 [9], 20060890 [12], 20081419 [13]
- *kale* 20073306 [1], 20050344 [2], 20081420 [13]
- *leeks* 20081415 [1], 20081421 [13]
- *linseed* 20060906 [2]
- *mustard* 20060901 [1], 20060906 [2], 20060902 [4], 20060898 [6], 20090935 [9], 20060890 [12], 20081419 [13]
- *pak choi* 20073306 [1], 20050344 [2], 20081420 [13]
- *spring linseed* 20060901 [1]
- *transplanted leeks* 20081415 [1], 20063698 [2], 20081421 [13]
- *winter linseed* 20060901 [1], 20081419 [13]

Approval information
- Metazachlor is included in Annex I under EC Directive 91/414

FOR FULL CONDITIONS OF USE ALWAYS READ THE PRODUCT LABEL

metazachlor

Efficacy guidance
- Activity is dependent on root uptake. For pre-emergence use apply to firm, moist, clod-free seedbed
- Some weeds (chickweed, mayweed, blackgrass etc) susceptible up to 2- or 4-leaf stage. Moderate control of cleavers achieved provided weeds not emerged and adequate soil moisture present
- Split pre- and post-emergence treatments recommended for certain weeds in winter oilseed rape on light and/or stony soils
- Effectiveness is reduced on soils with more than 10% organic matter
- Always follow WRAG guidelines for preventing and managing herbicide resistant weeds. See Section 5 for more information

Restrictions
- Maximum number of treatments 1 per crop for spring oilseed rape, swedes, turnips and brassicas; 2 per crop for winter oilseed rape (split dose treatment); 3 per yr for ornamentals, nursery stock, nursery fruit trees, forestry and farm forestry
- Do not use on sand, very light or poorly drained soils
- Do not treat protected crops or spray overall on ornamentals with soft foliage
- Do not spray crops suffering from wilting, pest or disease
- Do not spray broadcast crops or if a period of heavy rain forecast
- When used on nursery fruit trees any fruit harvested within 1 yr of treatment must be destroyed
- A maximum total dose of 1000 g.ai/ha may only be applied every third year to the same field

Crop-specific information
- Latest use: pre-emergence for swedes and turnips; before 10 leaf stage for spring oilseed rape; before end of Jan for winter oilseed rape
- HI brassicas 6 wk
- On winter oilseed rape may be applied pre-emergence from drilling until seed chits, post-emergence after fully expanded cotyledon stage (GS 1,0) or by split dose technique depending on soil and weeds. See label for details
- On spring oilseed rape may also be used pre-weed-emergence from cotyledon to 10-leaf stage of crop (GS 1,0-1,10)
- With pre-emergence treatment ensure seed covered by 15 mm of well consolidated soil. Harrow across slits of direct-drilled crops
- Ensure brassica transplants have roots well covered and are well established. Direct drilled brassicas should not be treated before 3 leaf stage
- In ornamentals and hardy nursery stock apply after plants established and hardened off as a directed spray or, on some subjects, as an overall spray. See label for list of tolerant subjects. Do not treat plants in containers

Following crops guidance
- Any crop can follow normally harvested treated winter oilseed rape. See label for details of crops which may be planted after spring treatment and in event of crop failure

Environmental safety
- Dangerous for the environment
- Very toxic to aquatic organisms
- Keep livestock out of treated areas until foliage of any poisonous weeds such as ragwort has died and become unpalatable
- Keep livestock out of treated areas of swede and turnip for at least 5 wk following treatment
- Some pesticides pose a greater threat of contamination of water than others and metazachlor is one of these pesticides. Take special care when applying metazachlor near water and do not apply if heavy rain is forecast

Hazard classification and safety precautions
Hazard Harmful, Dangerous for the environment
Transport code 9
Packaging group III
UN number 3082
Risk phrases R22a, R43, R50 [1-13]; R38 [1-6, 8, 9, 13]; R41 [10, 11]; R53a [2-4, 7-11]
Operator protection A, C, H [1-13]; M [1-8, 10-13]; U05a, U19a [1-13]; U08, U20b [1-10, 12, 13]; U09a, U20a [11]; U14 [1, 2, 5, 8, 9, 13]; U15 [1, 5, 13]

SEE SECTION 3 FOR PRODUCTS ALSO REGISTERED

Pesticide Profiles

Environmental protection E06a [1-4, 6-9] (5 wk for swedes, turnips); E06a [10] (5 weeks for swedes and turnips); E06a [11-13] (5 wk for swedes, turnips); E07a, E15a, E34 [1-13]; E38 [2, 3, 8, 9]
Consumer protection C02a [11] (5 weeks for swedes and turnips)
Storage and disposal D01, D02, D09a [1-13]; D05 [1, 3-7, 9, 11-13]; D10b [4, 6, 12]; D10c [1-3, 5, 7-11, 13]; D12a [2-4, 8, 9]; D12b [1, 5, 13]
Medical advice M03 [1-13]; M05a [2-4, 8, 9]

335 metazachlor + quinmerac

A residual herbicide mixture for oilseed rape
HRAC mode of action code: K3 + O

See also quinmerac

Products

1	Boomerang	BASF	400:100 g/l	SC	14043
2	Naspar TDI	Q-Chem	400:100 g/l	SC	14584
3	Novall	BASF	400:100 g/l	SC	12031
4	Oryx	BASF	333:83 g/l	SC	13527

Uses
- Annual dicotyledons in **borage** *(off-label)*, **canary flower (echium spp.)** *(off-label)*, **evening primrose** *(off-label)*, **honesty** *(off-label)*, **linseed** *(off-label)*, **mustard** *(off-label)* [3]; **winter oilseed rape** [1-4]
- Annual grasses in **borage** *(off-label)*, **canary flower (echium spp.)** *(off-label)*, **evening primrose** *(off-label)*, **honesty** *(off-label)*, **linseed** *(off-label)*, **mustard** *(off-label)* [3]; **winter oilseed rape** [1]
- Annual meadow grass in **winter oilseed rape** [2-4]
- Blackgrass in **winter oilseed rape** [2-4]
- Cleavers in **winter oilseed rape** [1-4]
- Poppies in **winter oilseed rape** [2-4]

Specific Off-Label Approvals (SOLAs)
- *borage* 20060548 [3]
- *canary flower (echium spp.)* 20060548 [3]
- *evening primrose* 20060548 [3]
- *honesty* 20060548 [3]
- *linseed* 20060548 [3]
- *mustard* 20060548 [3]

Approval information
- Metazachlor and quinmerac included in Annex I under EC Directive 91/414.

Efficacy guidance
- Activity is dependent on root uptake. Pre-emergence treatments should be applied to firm moist seedbeds. Applications to dry soil do not become effective until after rain has fallen
- Maximum activity achieved from treatment before weed emergence for some species
- Weed control may be reduced if excessive rain falls shortly after application especially on light soils
- May be used on all soil types except Sands, Very Light Soils, and soils containing more than 10% organic matter. Crop vigour and/or plant stand may be reduced on brashy and stony soils

Restrictions
- Maximum total dose equivalent to one full dose treatment
- Damage may occur in waterlogged conditions. Do not use on poorly drained soils
- Do not treat stressed crops. In frosty conditions transient scorch may occur
- A maximum total dose of 1000 g.ai/ha may only be applied every third year to the same field

Crop-specific information
- Latest use: end Jan in yr of harvest
- To ensure crop safety it is essential that crop seed is well covered with soil to 15 mm. Loose or puffy seedbeds must be consolidated before treatment. Do not use on broadcast crops
- Crop vigour and possibly plant stand may be reduced if excessive rain falls shortly after treatment especially on light soils

FOR FULL CONDITIONS OF USE ALWAYS READ THE PRODUCT LABEL

metconazole

Following crops guidance
- In the event of crop failure after use, wheat or barley may be sown in the autumn after ploughing to 15 cm. Spring cereals or brassicas may be planted after ploughing in the spring

Environmental safety
- Dangerous for the environment
- Very toxic to aquatic organisms
- Keep livestock out of treated areas until foliage of any poisonous weeds such as ragwort has died and become unpalatable
- To reduce risk of movement to water do not apply to dry soil or if heavy rain is forecast. On clay soils create a fine consolidated seedbed
- Some pesticides pose a greater threat of contamination of water than others and metazachlor is one of these pesticides. Take special care when applying metazachlor near water and do not apply if heavy rain is forecast
- LERAP Category B

Hazard classification and safety precautions
Hazard Irritant, Dangerous for the environment
Transport code 9
Packaging group III
UN number 3082
Risk phrases R43, R50, R53a
Operator protection A, C, H; U04a, U05a, U08, U14, U19a, U20b
Environmental protection E07a, E15a, E16a, E38
Storage and disposal D01, D02, D08, D09a, D10c, D12a
Medical advice M05a

336 metconazole

A triazole fungicide for cereals and oilseed rape
FRAC mode of action code: 3

See also epoxiconazole + metconazole

Products

1	Caramba 90	BASF	90 g/l	SL	12655
2	Clayton Tunik	Clayton	60 g/l	SL	10545
3	Gringo*	AgChem Access	60 g/l	SL	14467
4	Juventus	BASF	90 g/l	SL	13340
5	Pan Metconazole	Pan Agriculture	60 g/l	SL	10473
6	Redstar	AgChem Access	90 g/l	SL	14197
7	Sunorg Pro	BASF	90 g/l	SL	11112

Uses
- Alternaria in **spring oilseed rape, winter oilseed rape** [1-7]
- Ascochyta in **combining peas** *(reduction)*, **lupins** *(qualified minor use)*, **vining peas** *(reduction)* [1, 3, 4, 6, 7]
- Botrytis in **combining peas** *(reduction)*, **lupins** *(qualified minor use)*, **vining peas** *(reduction)* [1, 3, 4, 6, 7]; **spring linseed** *(off-label)*, **winter linseed** *(off-label)* [7]
- Brown rust in **spring barley, winter barley, winter wheat** [1-7]
- Disease control in **linseed** *(off-label)* [1]
- Foliar disease control in **grass seed crops** *(off-label)* [2, 5]
- Fusarium ear blight in **winter wheat** *(reduction)* [1-7]
- Light leaf spot in **spring oilseed rape, winter oilseed rape** [1-7]
- Mycosphaerella in **combining peas** *(reduction)*, **vining peas** *(reduction)* [1, 3, 4, 6, 7]
- Net blotch in **spring barley** *(reduction)*, **winter barley** *(reduction)* [1-7]
- Phoma in **spring oilseed rape** *(reduction)*, **winter oilseed rape** *(reduction)* [1-7]
- Powdery mildew in **spring barley** *(moderate control)*, **winter barley, winter wheat** *(moderate control)* [1-7]
- Rhynchosporium in **spring barley, winter barley** [1-7]
- Rust in **broad beans** *(off-label)* [7]; **combining peas, lupins** *(qualified minor use)*, **spring field beans, vining peas, winter field beans** [1, 3, 4, 6, 7]

SEE SECTION 3 FOR PRODUCTS ALSO REGISTERED

- Septoria leaf blotch in **winter wheat** [1-7]
- Yellow rust in **winter wheat** [1-7]

Specific Off-Label Approvals (SOLAs)
- **broad beans** 20063080 [7]
- **grass seed crops** 20061092 [2], 20061104 [5]
- **linseed** 20100675 [1]
- **spring linseed** 20063079 [7]
- **winter linseed** 20063079 [7]

Approval information
- Metconazole included in Annex I under EC Directive 91/414
- Accepted by BBPA for use on malting barley

Efficacy guidance
- Best results from application to healthy, vigorous crops when disease starts to develop
- Good spray cover of the target is essential for best results. Spray volume should be increased to improve spray penetration into dense crops
- Metconazole is a DMI fungicide. Resistance to some DMI fungicides has been identified in Septoria leaf blotch which may seriously affect performance of some products. For further advice contact a specialist advisor and visit the Fungicide Resistance Action Group (FRAG)-UK website

Restrictions
- Maximum total dose equivalent to two full dose treatments on all crops
- Do not apply to oilseed rape crops that are damaged or stressed from previous treatments, adverse weather, nutrient deficiency or pest attack. Spring application may lead to reduction of crop height
- The addition of adjuvants is neither advised nor necessary and can lead to enhanced growth regulatory effects on stressed crops of oilseed rape
- Ensure sprayer is free from residues of previous treatments that may harm the crop, especially oilseed rape. Use of a detergent cleaner is advised before and after use
- Do not apply with pyrethroid insecticides on oilseed rape at flowering [1, 4, 7]

Crop-specific information
- Latest use: up to and including milky ripe stage (GS 71) for cereals; 10% pods at final size for oilseed rape
- HI: 14 d for peas, field beans, lupins
- Spring treatments on oilseed rape can reduce the height of the crop
- Treatment for Septoria leaf spot in wheat should be made before second node detectable stage and when weather favouring development of the disease has occurred. If conditions continue to favour disease development a follow-up treatment may be needed
- Treat mildew infections in cereals before 3% infection on any green leaf. A specific mildewicide will improve control of established infections
- Treat yellow rust on wheat before 1% infection on any leaf or as preventive treatment after GS 39
- For brown rust in cereals spray susceptible varieties before any of top 3 leaves has more than 2% infection
- Peas should be treated at the start of flowering and repeat 3-4 wk later if required
- Field beans and lupins should be treated at petal fall and repeat 3-4 wk later if required

Following crops guidance
- Only cereals, oilseed rape, sugar beet, linseed, maize, clover, beans, peas, carrots, potatoes or onions may be sown as following crops after treatment

Environmental safety
- Dangerous for the environment
- Very toxic to aquatic organisms
- Flammable [2]
- LERAP Category B but avoid treatment close to field boundary, even if permitted by LERAP assessment, to reduce effects on non-target insects or other arthropods [2]
- LERAP Category B [2, 3, 5]

Hazard classification and safety precautions
Hazard Harmful, Flammable [2, 3, 5]; Irritant [1, 4, 6, 7]; Dangerous for the environment [1-7]
Transport code 3 [2]; 9 [1, 3-7]
Packaging group III

FOR FULL CONDITIONS OF USE ALWAYS READ THE PRODUCT LABEL

methiocarb

UN number 1993 [2, 3, 5]; 3082 [1, 4, 6, 7]
Risk phrases R22b, R38, R41, R43, R50 [2, 3, 5]; R36, R51 [1, 4, 6, 7]; R37 [2]; R53a, R63 [1-7]
Operator protection A, C [1-7]; H [2, 3, 5]; U02a, U05a, U20c [1-7]; U11 [2, 3, 5]; U14 [3, 5]; U19a [2]
Environmental protection E15a, E34 [1-7]; E16a [2, 3, 5]; E16b [3, 5]; E38 [1, 3-7]
Storage and disposal D01, D02, D05, D09a, D10a, D12a [1-7]; D06c [1, 4, 6, 7]
Medical advice M05b [2, 3, 5]

337 methiocarb

A stomach acting carbamate molluscicide and insecticide
IRAC mode of action code: 1A

Products

1	Cobra	Interfarm	4% w/w	PT	14561
2	Decoy Wetex	Bayer CropScience	2% w/w	PT	11266
3	Draza Forte	Bayer CropScience	4% w/w	PT	13306
4	Huron	Bayer CropScience	3% w/w	PT	11288
5	Karan	Bayer CropScience	3% w/w	PT	11289
6	Rivet	Bayer CropScience	3% w/w	PT	11300

Uses
- Cutworms in **sugar beet** *(reduction)* [1-6]
- Leatherjackets in **all cereals** *(reduction)*, **potatoes** *(reduction)*, **sugar beet** *(reduction)* [1-6]; **rotational grass** *(seed admixture)* [2, 4-6]
- Millipedes in **sugar beet** *(reduction)* [1-6]
- Slugs in **all cereals**, **all non-edible crops (outdoor)** *(outdoor only)*, **brussels sprouts**, **cabbages**, **cauliflowers**, **forage maize**, **leaf spinach**, **lettuce**, **potatoes**, **spring oilseed rape**, **sugar beet**, **sunflowers**, **winter oilseed rape** [1-6]; **rotational grass** *(seed admixture)* [2, 4-6]; **strawberries** [1, 2, 4-6]
- Slugs and snails in **strawberries** [3]
- Strawberry seed beetle in **strawberries** [1, 2, 4-6]

Approval information
- Methiocarb included in Annex I under EC Directive 91/414
- In 2006 CRD required that all products containing this active ingredient should carry the following warning in the main area of the container label: "Methiocarb is an anticholinesterase carbamate. Handle with care"
- Accepted by BBPA for use on malting barley

Efficacy guidance
- Use as a surface, overall application when pests active (normally mild, damp weather), pre-drilling or post-emergence. May also be used on cereals or ryegrass in admixture with seed at time of drilling
- Also reduces populations of cutworms and millipedes
- Best on potatoes in late Jul to Aug
- Apply to strawberries before strawing down to prevent seed beetles contaminating crop
- See label for details of suitable application equipment

Restrictions
- This product contains an anticholinesterase carbamate compound. Do not use if under medical advice not to work with such compounds
- Maximum number of treatments 3 per crop for potatoes; 2 per crop for cereals, cauliflowers, maize, oilseed rape, sunflowers; 1 per crop for cabbages, lettuce, leaf spinach, sugar beet; 1 per yr for strawberries. Other crops vary according to product - see labels
- Do not treat any protected crops
- Do not allow pellets to lodge in edible crops

Crop-specific information
- Latest use: before first node detectable for cereals, maize; before three visibly extended internodes for oilseed rape, sunflowers
- HI 7 d for spinach, strawberries; 14 d for Brussels sprouts, cabbages, cauliflowers, lettuce; 18 d for potatoes; 6 mth for sugar beet

SEE SECTION 3 FOR PRODUCTS ALSO REGISTERED

Environmental safety
- Dangerous for the environment [3]
- Very toxic to aquatic organisms [3]
- Harmful to aquatic organisms [2, 4-6]
- Dangerous to game, wild birds and animals
- Risk to certain non-target insects or other arthropods
- Avoid surface broadcasting application within 6 m of field boundary to reduce effects on non-target species
- Admixed seed should be drilled and not broadcast, and may not be applied from the air

Hazard classification and safety precautions
Hazard Harmful [1-6]; Dangerous for the environment [1, 3]
Transport code 9
Packaging group III
UN number 3077
Risk phrases R22a, R53a [1-6]; R50 [1, 3]; R52 [2, 4-6]
Operator protection A, H [1-6]; J [2, 4-6]; U05a, U20b
Environmental protection E05b [1-6] (7 d); E10a, E22b, E34 [1-6]; E13b [2, 4-6]
Storage and disposal D01, D02, D09a, D11a
Medical advice M02, M03

338 methoxyfenozide

A moulting accelerating diacylhydrazine insecticide
IRAC mode of action code: 18A

Products
1	Runner	Bayer CropScience	240 g/l	SC	12940
2	Trotter	Agrovista	240 g/l	SC	14236

Uses
- Codling moth in **apples**, **pears** [1, 2]
- Insect pests in **hops** (off-label), **ornamental specimens** (off-label), **soft fruit** (off-label) [1]
- Tortrix moths in **apples**, **pears** [1, 2]
- Winter moth in **apples**, **pears** [1, 2]

Specific Off-Label Approvals (SOLAs)
- **hops** 20082903 [1]
- **ornamental specimens** 20082903 [1]
- **soft fruit** 20082903 [1]

Efficacy guidance
- To achieve best results uniform coverage of the foliage and full spray penetration of the leaf canopy is important, particularly when spraying post-blossom
- For maximum effectiveness on winter moth and tortrix spray pre-blossom when first signs of active larvae are seen, followed by a further spray in June if larvae of the summer generation are present
- For codling moth spray post-blossom to coincide with early to peak egg deposition. Follow-up treatments will normally be needed
- Methoxyfenozide is a moulting accelerating compound (MAC) and may be used in an anti-resistance strategy with other top fruit insecticides (including chitin biosynthesis inhibitors and juvenile hormones) which have a different mode of action
- To reduce further the likelihood of resistance development use at full recommended dose in sufficient water volume to achieve required spray penetration

Restrictions
- Maximum number of treatments 3 per yr but no more than two should be sprayed consecutively

Crop-specific information
- HI 14 d for apples, pears

Environmental safety
- Risk to non-target insects or other arthropods
- Broadcast air-assisted LERAP (5 m)

FOR FULL CONDITIONS OF USE ALWAYS READ THE PRODUCT LABEL

methyl bromide

Hazard classification and safety precautions
  UN number N/C
  Operator protection U20b
  Environmental protection E15b, E16b, E22c; E17b (5 m)
  Storage and disposal D05, D09a, D11a

339   methyl bromide

The EC Ozone Depleting Substances Regulation (2037/2000) has banned the use of methyl bromide within the UK with certain exceptions, including where a 'Critical Use Exemption' (CUE) has been granted. All other uses ceased in December 2005
IRAC mode of action code: 8A

340   1-methylcyclopropene

An inhibitor of ethylene production for use in stored apples

Products
  SmartFresh   Landseer   3.3% w/w   SP   11799

Uses
  • Ethylene inhibition in **apples** *(post-harvest use)*, **pears** *(off-label - post harvest only)*
  • Scald in **apples** *(post-harvest use)*, **pears** *(off-label - post harvest use only)*

Specific Off-Label Approvals (SOLAs)
  • **pears** *(post harvest only) 20102424, (post harvest use only) 20102424*

Efficacy guidance
  • Best results obtained from treatment of fruit in good condition and of proper quality for long-term storage
  • Effects may be reduced in fruit that is in poor condition or ripe prior to storage or harvested late
  • Product acts by releasing vapour into store when mixed with water
  • Apply as soon as possible after harvest
  • Treatment controls superficial scald and maintains fruit firmness and acid content for 3-6 mth in normal air, and 6-9 mth in controlled atmosphere storage
  • Ethylene production recommences after removal from storage

Restrictions
  • Maximum number of treatments 1 per batch of apples
  • Must only be used by suitably trained and competent persons in fumigation operations
  • Consult processors before treatment of fruit destined for processing or cider making
  • Do not apply in mixture with other products
  • Ventilate all areas thoroughly with all refrigeration fans operating at maximum power for at least 15 min before re-entry

Crop-specific information
  • Latest use: 7d after harvest of apples
  • Product tested on Granny Smith, Gala, Jonagold, Bramley and Cox. Consult distributor or supplier before treating other varieties

Environmental safety
  • Unprotected persons must be kept out of treated stores during the 24 h treatment period
  • Prior to application ensure that the store can be properly and promptly sealed

Hazard classification and safety precautions
  UN number N/C
  Operator protection U05a
  Environmental protection E02a (24 h); E15a, E34
  Consumer protection C12
  Storage and disposal D01, D02, D07, D14

SEE SECTION 3 FOR PRODUCTS ALSO REGISTERED

341 metrafenone

A benzophenone protectant and curative fungicide for cereals
FRAC mode of action code: U8

See also epoxiconazole + fenpropimorph + metrafenone
epoxiconazole + metrafenone
fenpropimorph + metrafenone

Products

1	Attenzo	BASF	300 g/l	SC	11917
2	Flexity	BASF	300 g/l	SC	11775
3	Lexi	AgChem Access	300 g/l	SC	14890

Uses
- Disease control/foliar feed in **ornamental plant production** *(off-label)* [2]
- Eyespot in **spring wheat** *(reduction)*, **winter wheat** *(reduction)* [1-3]
- Powdery mildew in **forest nurseries** *(off-label)* [1]; **ornamental plant production** [2]; **spring barley**, **spring oats** *(evidence of mildew control on oats is limited)*, **spring wheat**, **winter barley**, **winter oats** *(evidence of mildew control on oats is limited)*, **winter wheat** [1-3]

Specific Off-Label Approvals (SOLAs)
- *forest nurseries* 20082814 [1]
- *ornamental plant production* 20082850 [2]

Approval information
- Metrafenone included in Annex I under EC Directive 91/414
- Accepted by BBPA for use on malting barley

Efficacy guidance
- Best results obtained from treatment at the start of foliar disease attack
- Activity against mildew in wheat is mainly protectant with moderate curative control in the latent phase; activity is entirely protectant in barley
- Useful reduction of eyespot in wheat is obtained if treatment applied at GS 30-32
- Should be used as part of a resistance management strategy that includes mixtures or sequences effective against mildew and non-chemical methods

Restrictions
- Maximum number of treatments 2 per crop
- Avoid the use of sequential applications of Flexity unless it is used in tank mixture with other products active against powdery mildew employing a different mode of action

Crop-specific information
- Latest use: beginning of flowering (GS 31) for wheat, barley, oats

Following crops guidance
- Cereals, oilseed rape, sugar beet, linseed, maize, clover, field beans, peas, turnips, carrots, cauliflowers, onions, lettuce or potatoes may follow a treated cereal crop

Environmental safety
- Dangerous for the environment
- Toxic to aquatic organisms

Hazard classification and safety precautions
Hazard Irritant, Dangerous for the environment
UN number N/C
Risk phrases R43, R51, R53a
Operator protection A, H; U05a, U20b
Environmental protection E15a, E34
Storage and disposal D01, D02, D05, D09a, D10c
Medical advice M03

FOR FULL CONDITIONS OF USE ALWAYS READ THE PRODUCT LABEL

metribuzin

342 metribuzin

A contact and residual triazinone herbicide for use in potatoes
HRAC mode of action code: C1

See also clomazone + metribuzin
flufenacet + metribuzin
mecoprop-P + metribuzin

Products

1	Buzin WG	Goldengrass	70% w/w	WG	14415
2	Greencrop Majestic	Greencrop	70% w/w	WG	12422
3	Metribuzin WG	Goldengrass	70% w/w	WG	14315
4	Sarabi	AgChem Access	70% w/w	WG	13453
5	Sencorex WG	Interfarm	70% w/w	WG	14747
6	Shotput	Makhteshim	70% w/w	WG	13788

Uses
- Annual and perennial weeds in *asparagus* (off-label), *sweet potato* (off-label) [6]
- Annual dicotyledons in *asparagus* (off-label), *sweet potato* (off-label) [5]; *early potatoes*, *maincrop potatoes* [2-6]; *potatoes* [1]
- Annual grasses in *early potatoes*, *maincrop potatoes* [2-6]; *potatoes* [1]
- Fool's parsley in *carrots* (off-label), *mallow (althaea spp.)* (off-label), *parsnips* (off-label) [5, 6]; *protected carrots* (off-label), *protected parsnips* (off-label) [6]
- Volunteer oilseed rape in *early potatoes*, *maincrop potatoes* [2-6]
- Wild mignonette in *carrots* (off-label), *mallow (althaea spp.)* (off-label), *parsnips* (off-label) [5]

Specific Off-Label Approvals (SOLAs)
- *asparagus* 20093176 [5], 20080588 [6]
- *carrots* 20093173 [5], 20080589 [6]
- *mallow (althaea spp.)* 20093173 [5], 20080589 [6]
- *parsnips* 20093173 [5], 20080589 [6]
- *protected carrots* 20080589 [6]
- *protected parsnips* 20080589 [6]
- *sweet potato* 20093174 [5], 20080590 [6]

Approval information
- Metribuzin included in Annex I under EC Directive 91/414

Efficacy guidance
- Best results achieved on weeds at cotyledon to 1-leaf stage
- May be applied pre- or post-emergence of crop on named maincrop varieties and cv. Marfona; pre-emergence only on named early varieties
- Apply to moist soil with well-rounded ridges and few clods
- Activity reduced by dry conditions and on soils with high organic matter content
- On fen and moss soils pre-planting incorporation to 10-15 cm gives increased activity. Incorporate thoroughly and evenly
- With named maincrop and second early potato varieties on soils with more than 10% organic matter shallow pre- or post-planting incorporation may be used. See label for details
- Effective control using a programme of reduced doses is made possible by using a spray of smaller droplets, thus improving retention

Restrictions
- Maximum total dose equivalent to one full dose treatment on early potato varieties and one and a third full doses on maincrop varieties
- Only certain varieties may be treated. Apply pre-emergence only on named first earlies, pre- or post-emergence on named second earlies. On named maincrop varieties apply pre-emergence (except for certain varieties on Sands or Very Light soils) or post-emergence. See label
- All post-emergence treatments must be carried out before longest shoots reach 15 cm
- Do not cultivate after treatment
- Some recommended varieties may be sensitive to post-emergence treatment if crop under stress

Crop-specific information
- Latest use: pre-crop emergence for named early potato varieties; before most advanced shoots have reached 15 cm for post-emergence treatment of potatoes

SEE SECTION 3 FOR PRODUCTS ALSO REGISTERED

Pesticide Profiles

- On stony or gravelly soils there is risk of crop damage, especially if heavy rain falls soon after application
- When days are hot and sunny delay spraying until evening

Following crops guidance
- Ryegrass, cereals or winter beans may be sown in same season provided at least 16 wk elapsed after treatment and ground ploughed to 15 cm and thoroughly cultivated as soon as possible after harvest and no later than end Dec
- In W Cornwall on soil with more than 5% organic matter early potatoes treated as recommended may be followed by summer planted brassica crops provided the soil has been ploughed, spring rainfall has been normal and at least 14 wk have elapsed since treatment
- Do not grow any vegetable brassicas, lettuces or radishes on land treated the previous yr. Other crops may be sown normally in spring of next yr

Environmental safety
- Dangerous for the environment
- Very toxic to aquatic organisms
- Do not empty into drains
- LERAP Category B

Hazard classification and safety precautions
Hazard Harmful, Dangerous for the environment
Transport code 9
Packaging group III
UN number 3077
Risk phrases R22a, R50, R53a [1-6]; R43 [6]
Operator protection A, H [6]; U05a [2, 6]; U08, U13 [1-6]; U14, U20b [6]; U19a [1, 3-6]; U20a [1-5]
Environmental protection E15a [1, 3-6]; E16a, E16b [1-6]; E19b [6]; E34 [2]; E38 [1, 3-5]
Storage and disposal D01, D02 [2, 6]; D09a, D11a, D12a [1-6]
Medical advice M03 [2]; M05a [6]

343   metsulfuron-methyl

A contact and residual sulfonylurea herbicide used in cereals, linseed and set-aside
HRAC mode of action code: B

See also carfentrazone-ethyl + metsulfuron-methyl
diflufenican + metsulfuron-methyl
flupyrsulfuron-methyl + metsulfuron-methyl
mecoprop-P + metsulfuron-methyl

Products

1	Alias SX	DuPont	20% w/w	SG	13398
2	Asset	AgriGuard	20% w/w	SG	13902
3	Avail XT	AgriGuard	20% w/w	SG	13194
4	Cimarron	Headland	50% w/w	TB	13408
5	Finy	AgriChem BV	20% w/w	WG	12855
6	Firecrest	European Ag	20% w/w	SG	14486
7	Forge	Interfarm	20% w/w	SG	12848
8	Goldron-M	Goldengrass	20% w/w	SG	13466
9	Jubilee SX	DuPont	20% w/w	SG	12203
10	Lorate	DuPont	20% w/w	SG	12743
11	Orca	ChemSource	20% w/w	WG	14970
12	Standon Metso XT	Standon	20% w/w	SG	13880
13	Standon Mexxon Xtra	Standon	20% w/w	SG	13931

Uses
- Annual dicotyledons in **durum wheat** [7]; **farm forestry** *(off-label)*, **forest nurseries** *(off-label)*, **ornamental plant production** *(off-label)* [9]; **game cover** *(off-label)*, **miscanthus** *(off-label)* [9, 10]; **green cover on land temporarily removed from production** [5, 7, 11-13]; **linseed, spring barley, spring oats, spring wheat, triticale, winter barley, winter oats, winter wheat** [1-3, 5-13]
- Chickweed in **durum wheat** [7]; **linseed, spring barley, spring oats, spring wheat, triticale, winter barley, winter oats, winter wheat** [1-3, 6-10]

FOR FULL CONDITIONS OF USE ALWAYS READ THE PRODUCT LABEL

metsulfuron-methyl

- Docks in **grassland** [4]
- Green cover in **land temporarily removed from production** [1-3, 6, 8-10]
- Mayweeds in **durum wheat** [7]; **linseed, spring barley, spring oats, spring wheat, triticale, winter barley, winter oats, winter wheat** [1-3, 6-10]

Specific Off-Label Approvals (SOLAs)
- **farm forestry** *20082859 expires 30 Jun 2011* [9]
- **forest nurseries** *20082859 expires 30 Jun 2011* [9]
- **game cover** *20082859 expires 30 Jun 2011* [9], *20082860 expires 30 Jun 2011* [10]
- **miscanthus** *20082859 expires 30 Jun 2011* [9], *20082860 expires 30 Jun 2011* [10]
- **ornamental plant production** *20082859 expires 30 Jun 2011* [9]

Approval information
- Metsulfuron-methyl included in Annex I under EC Directive 91/414
- Accepted by BBPA for use on malting barley
- Approval expiry 30 Jun 2011 [1-13]

Efficacy guidance
- Best results achieved on small, actively growing weeds up to 6-true leaf stage. Good spray cover is important
- Weed control may be reduced in dry soil conditions
- Commonly used in tank-mixture on wheat and barley with other cereal herbicides to improve control of resistant dicotyledons (cleavers, fumitory, ivy-leaved speedwell), larger weeds and grasses. See label for recommended mixtures
- Metsulfuron-methyl is a member of the ALS-inhibitor group of herbicides and products should be used in a planned Resistance Management strategy. See Section 5 for more information

Restrictions
- Maximum number of treatments 1 per crop or per yr (for set-aside)
- Product must only be used after 1 Feb
- Do not apply within 7 d of rolling
- Do not use on cereal crops undersown with grass or legumes
- Do not use in tank mixture on oats, triticale or linseed. On linseed allow at least 7 d before or after other treatments
- Do not tank mix with chlorpyrifos. Allow at least 14 d before or after chlorpyrifos treatments
- Consult contract agents before use on a cereal crop for seed
- Do not use on any crop suffering stress from drought, waterlogging, frost, deficiency, pest or disease attack or apply within 7 d of rolling
- Specific restrictions apply to use in sequence or tank mixture with other sulfonylurea or ALS-inhibiting herbicides. See label for details
- Spraying equipment should not be drained or flushed onto land planted, or to be planted, with trees or crops other than cereals and should be thoroughly cleansed after use - see label for instructions

Crop-specific information
- Latest use: before flag leaf sheath extending stage for cereals (GS 41); before flower buds visible or crop 30 cm tall for linseed [3, 9]; before 1 Aug in yr of treatment for land not being used for crop production [3, 9]
- Apply after 1 Feb to wheat, oats and triticale from 2-leaf (GS 12), and to barley from 3-leaf (GS 13) until flag-leaf sheath extending (GS 41)
- On linseed allow at least 7 d (10 d if crop is growing poorly or under stress) before or after other treatments
- Use in set-aside when a full green cover is established and made up predominantly of grassland, wheat, barley, oats or triticale. Do not use on seedling grasses

Following crops guidance
- Only cereals, oilseed rape, field beans or grass may be sown in same calendar year after treating cereals with the product alone. Other restrictions apply to tank mixtures. See label for details
- Only cereals should be planted within 16 mth of applying to a linseed crop or set-aside
- In the event of failure of a treated crop sow only wheat within 3 mth after treatment

Environmental safety
- Dangerous for the environment
- Very toxic to aquatic organisms

SEE SECTION 3 FOR PRODUCTS ALSO REGISTERED

Pesticide Profiles

- Take extreme care to avoid damage by drift onto broad-leaved plants outside the target area, onto surface waters or ditches or onto land intended for cropping
- A range of broad leaved species will be fully or partially controlled when used in land temporarily removed from production, hence product may not be suitable where wild flower borders or other forms of conservation headland are being developed
- Before use on land temporarily removed from production as part of grant-aided scheme, ensure compliance with the management rules
- Green cover on land temporarily removed from production must not be grazed by livestock or harvested for human or animal consumption or used for animal bedding [3, 9]
- LERAP Category B [1-3, 5-9, 11-13]

Hazard classification and safety precautions
Hazard Dangerous for the environment [1-3, 5-13]
Transport code 9
Packaging group III
UN number 3077
Risk phrases R50, R53a [1-3, 5-13]
Operator protection U05a [5, 11]; U08 [3]; U09a [12, 13]; U19a [3, 12, 13]; U20b [3, 4, 12, 13]
Environmental protection E07a [4]; E15b [1-13]; E16a [1-3, 5-9, 11-13]; E16b [2]; E23 [5, 11]; E34 [3, 12, 13]; E38 [1-3, 6-10, 12, 13]
Storage and disposal D01 [3, 5, 11-13]; D02, D12b [5, 11]; D09a [2-5, 8, 11-13]; D11a [2, 5, 8, 11]; D12a [1-4, 6-10, 12, 13]; D17, D18 [4]
Medical advice M05a [3, 12, 13]

344   metsulfuron-methyl + thifensulfuron-methyl

A contact residual and translocated sulfonylurea herbicide mixture for use in cereals
HRAC mode of action code: B + B

See also thifensulfuron-methyl

Products

1	Aria	AgriGuard	4:40 % w/w	SG	14771
2	Avro SX	DuPont	2.9:42.9 % w/w	SG	14313
3	Chimera SX	DuPont	2.9:42.9% w/w	SG	13171
4	Concert SX	DuPont	4:40% w/w	SG	12288
5	Finish SX	DuPont	6.7:33.3% w/w	SG	12259
6	Harmony M SX	DuPont	4:40% w/w	SG	12258
7	Pennant	Headland	4:40% w/w	SG	13350
8	Presite SX	DuPont	6.7:33.3% w/w	SG	12291

Uses

- Annual dicotyledons in **durum wheat** *(off-label)* [8]; **miscanthus** *(off-label)* [3, 6]; **spring barley**, **spring wheat**, **winter wheat** [1-8]; **winter barley** [2, 3, 5, 8]; **winter oats** [5, 8]
- Black bindweed in **spring barley**, **spring wheat**, **winter barley**, **winter wheat** [2]
- Charlock in **spring barley**, **spring wheat**, **winter barley**, **winter wheat** [2]
- Chickweed in **durum wheat** *(off-label)* [8]; **spring barley**, **spring wheat**, **winter wheat** [1-8]; **winter barley** [2, 3, 5, 8]; **winter oats** [5, 8]
- Creeping thistle in **spring barley**, **spring wheat**, **winter barley**, **winter wheat** [2, 3]
- Fat hen in **spring barley**, **spring wheat**, **winter barley**, **winter wheat** [2]
- Field pansy in **durum wheat** *(off-label)* [8]; **spring barley**, **spring wheat**, **winter barley**, **winter wheat** [2, 5, 8]; **winter oats** [5, 8]
- Field speedwell in **durum wheat** *(off-label)* [8]; **spring barley**, **spring wheat**, **winter barley**, **winter wheat** [2, 5, 8]; **winter oats** [5, 8]
- Forget-me-not in **spring barley**, **spring wheat**, **winter barley**, **winter wheat** [2]
- Hemp-nettle in **spring barley**, **spring wheat**, **winter barley**, **winter wheat** [2]
- Ivy-leaved speedwell in **durum wheat** *(off-label)* [8]; **spring barley**, **spring wheat**, **winter barley**, **winter oats**, **winter wheat** [5, 8]
- Knotgrass in **durum wheat** *(off-label)* [8]; **spring barley**, **spring wheat**, **winter barley**, **winter wheat** [2, 5, 8]; **winter oats** [5, 8]
- Mayweeds in **durum wheat** *(off-label)* [8]; **spring barley**, **spring wheat**, **winter wheat** [1-8]; **winter barley** [2, 3, 5, 8]; **winter oats** [5, 8]

FOR FULL CONDITIONS OF USE ALWAYS READ THE PRODUCT LABEL

metsulfuron-methyl + thifensulfuron-methyl

- Polygonums in **spring barley**, **spring wheat**, **winter wheat** [1, 3, 4, 6, 7]; **winter barley** [3]
- Poppies in **spring barley**, **spring wheat**, **winter barley**, **winter wheat** [2]
- Red dead-nettle in **spring barley**, **spring wheat**, **winter wheat** [1-4, 6, 7]; **winter barley** [2, 3]
- Redshank in **spring barley**, **spring wheat**, **winter barley**, **winter wheat** [2]
- Shepherd's purse in **spring barley**, **spring wheat**, **winter wheat** [1-4, 6, 7]; **winter barley** [2, 3]

Specific Off-Label Approvals (SOLAs)
- **durum wheat** 20070452 expires 30 Jun 2011 [8]
- **miscanthus** 20082835 expires 30 Jun 2011 [3], 20082916 expires 30 Jun 2011 [6]

Approval information
- Metsulfuron-methyl and thifensulfuron-methyl included in Annex I under EC Directive 91/414
- Accepted by BBPA for use on malting barley
- Approval expiry 30 Jun 2011 [1-8]

Efficacy guidance
- Best results by application to small, actively growing weeds up to 6-true leaf stage
- Ensure good spray cover
- Susceptible weeds stop growing almost immediately but symptoms may not be visible for about 2 wk
- Effectiveness may be reduced by heavy rain or if soil conditions very dry
- Metsulfuron-methyl and thifensulfuron-methyl are members of the ALS-inhibitor group of herbicides and products should be used in a planned Resistance Management strategy. See Section 5 for more information

Restrictions
- Maximum number of treatments 1 per crop
- Products may only be used after 1 Feb
- Do not use on any crop suffering stress from drought, waterlogging, frost, deficiency, pest or disease attack or any other cause
- Do not use on crops undersown with grasses, clover or legumes, or any other broad leaved crop
- Specific restrictions apply to use in sequence or tank mixture with other sulfonylurea or ALS-inhibiting herbicides. See label for details
- Do not apply within 7 d of rolling
- Consult contract agents before use on a cereal crop grown for seed

Crop-specific information
- Latest use: before flag leaf sheath extending (GS 39) for barley and wheat, before 2nd node (GS32) for winter oats

Following crops guidance
- Only cereals, oilseed rape, field beans or grass may be sown in same calendar year after treatment
- Additional constraints apply after use of certain tank mixtures. See label
- In the event of crop failure sow only winter wheat within 3 mth after treatment and after ploughing and cultivating to a depth of at least 15 cm

Environmental safety
- Dangerous for the environment
- Very toxic to aquatic organisms
- Take extreme care to avoid damage by drift onto broad-leaved plants outside the target area, or onto ponds, waterways or ditches
- Spraying equipment should not be drained or flushed onto land planted, or to be planted, with trees or crops other than cereals and should be thoroughly cleansed after use - see label for instructions
- LERAP Category B

Hazard classification and safety precautions
Hazard Dangerous for the environment
Transport code 9
Packaging group III
UN number 3077
Risk phrases R50, R53a
Operator protection U05a [5, 8]; U08, U19a, U20b [1-8]
Environmental protection E15a [5, 8]; E15b [1, 4, 6, 7]; E16a, E38 [1-8]
Storage and disposal D01, D02 [5, 8]; D09a, D12a [1-8]; D11a [1, 4-8]

SEE SECTION 3 FOR PRODUCTS ALSO REGISTERED

Pesticide Profiles

345   metsulfuron-methyl + tribenuron-methyl

A sulfonylurea herbicide mixture for cereals
HRAC mode of action code: B + B

See also tribenuron-methyl

Products

1	Ally Max SX	DuPont	14.3:14.3% w/w	SG	14835
2	BiPlay SX	DuPont	11.1:22.2% w/w	SG	14836
3	Traton SX	DuPont	11.1:22.2% w/w	SG	14837

Uses
- Annual dicotyledons in **durum wheat** *(off-label)*, **rye** *(off-label)*, **spring oats**, **spring rye** *(off-label)* [1]; **spring barley, spring wheat, triticale, winter barley, winter oats, winter wheat** [1-3]
- Chickweed in **spring barley, spring wheat, triticale, winter barley, winter oats, winter wheat** [1-3]; **spring oats** [1]
- Field pansy in **spring barley, spring wheat, triticale, winter barley, winter oats, winter wheat** [1-3]
- Field speedwell in **spring oats** [1]
- Hemp-nettle in **spring barley, spring wheat, triticale, winter barley, winter oats, winter wheat** [1-3]; **spring oats** [1]
- Mayweeds in **spring barley, spring wheat, triticale, winter barley, winter oats, winter wheat** [1-3]; **spring oats** [1]
- Red dead-nettle in **spring barley, spring wheat, triticale, winter barley, winter oats, winter wheat** [1-3]; **spring oats** [1]
- Volunteer oilseed rape in **spring barley, spring oats, spring wheat, triticale, winter barley, winter oats, winter wheat** [1]
- Volunteer sugar beet in **spring barley, spring oats, spring wheat, triticale, winter barley, winter oats, winter wheat** [1]

Specific Off-Label Approvals (SOLAs)
- *durum wheat* 20101080 expires 30 Jun 2011 [1]
- *rye* 20101080 expires 30 Jun 2011 [1]
- *spring rye* 20101080 expires 30 Jun 2011 [1]

Approval information
- Metsulfuron-methyl and tribenuron-methyl included in Annex I under EC Directive 91/414
- Accepted by BBPA for use on malting barley
- Approval expiry 30 Jun 2011 [1-3]

Efficacy guidance
- Best results obtained when applied to small actively growing weeds
- Product acts by foliar and root uptake. Good spray cover essential but performance may be reduced when soil conditions are very dry and residual effects may be reduced by heavy rain
- Weed growth inhibited within hours of treatment and many show marked colour changes as they die back. Full effects may not be apparent for up to 4 wk
- Metsulfuron-methyl and tribenuron-methyl are members of the ALS-inhibitor group of herbicides and products should be used in a planned Resistance Management strategy. See Section 5 for more information

Restrictions
- Maximum number of treatments 1 per crop
- Product must only be used after 1 Feb and after crop has three leaves
- Do not apply to a crop suffering from drought, water-logging, low temperatures, pest or disease attack, nutrient deficiency, soil compaction or any other stress
- Do not use on crops undersown with grasses, clover or other legumes
- Do not apply within 7 d of rolling
- Specific restrictions apply to use in sequence or tank mixture with other sulfonylurea or ALS-inhibiting herbicides. See label for details

Crop-specific information
- Latest use: before flag leaf sheath extending

FOR FULL CONDITIONS OF USE ALWAYS READ THE PRODUCT LABEL

myclobutanil

Following crops guidance
- Only cereals, field beans or oilseed rape may be sown in the same calendar yr as harvest of a treated crop [1-3]
- In the event of failure of a treated crop only winter wheat may sown within 3 mth of treatment, and only after ploughing and cultivation to 15 cm min

Environmental safety
- Dangerous for the environment
- Very toxic to aquatic organisms
- Some non-target crops are highly sensitive. Take extreme care to avoid drift outside the target area, or onto ponds, waterways or ditches
- Spraying equipment should be thoroughly cleaned in accordance with manufacturer's instructions
- LERAP Category B [2, 3]

Hazard classification and safety precautions
Hazard Irritant, Dangerous for the environment
Transport code 9
Packaging group III
UN number 3077
Risk phrases R43, R50, R53a
Operator protection A, H; U08, U19a, U20b
Environmental protection E15a, E38 [1-3]; E16a [2, 3]
Storage and disposal D01, D02, D09a, D11a, D12a

346 myclobutanil

A systemic, protectant and curative triazole fungicide
FRAC mode of action code: 3

Products

1	Aristocrat	AgriGuard	200 g/l	EW	12343
2	Crassus	Landseer	200 g/l	EW	13899
3	Hithane	ChemSource	200 g/l	EW	14852
4	Masalon	Rigby Taylor	45 g/l	EW	12385
5	Systhane 20EW	Landseer	200 g/l	EW	09396

Uses
- American gooseberry mildew in *blackberries* [2]; *blackcurrants* [1, 3, 5]; *gooseberries* [1-3, 5]
- Black spot in *ornamental plant production*, *roses* [1-3, 5]
- Blossom wilt in *cherries* (off-label), *mirabelles* (off-label) [5]
- Brown rot in *apricots* (off-label), *peaches* (off-label) [5]
- Fusarium patch in *managed amenity turf* [4]
- Plum rust in *plums* (off-label) [5]
- Powdery mildew in *apples*, *ornamental plant production*, *pears*, *roses*, *strawberries* [1-3, 5]; *apricots* (off-label), *artichokes* (off-label), *courgettes* (off-label), *crab apples* (off-label), *gherkins* (off-label), *hops* (off-label), *peaches* (off-label), *protected aubergines* (off-label), *protected blackberries* (off-label), *protected courgettes* (off-label), *protected cucumbers* (off-label), *protected gherkins* (off-label), *protected raspberries* (off-label), *protected rubus hybrids* (off-label), *protected tomatoes* (off-label), *quinces* (off-label), *redcurrants* (off-label), *whitecurrants* (off-label), *wine grapes* (off-label) [5]
- Rust in *ornamental plant production*, *roses* [1-3, 5]
- Scab in *apples*, *pears* [1-3, 5]

Specific Off-Label Approvals (SOLAs)
- *apricots* 20070626 [5]
- *artichokes* 20070624 [5]
- *cherries* 991535 [5]
- *courgettes* 20070627 [5]
- *crab apples* 20080504 [5]
- *gherkins* 20070627 [5]
- *hops* 20021412 [5]
- *mirabelles* 991535 [5]
- *peaches* 20070626 [5]

SEE SECTION 3 FOR PRODUCTS ALSO REGISTERED

- *plums* 20012459 [5]
- *protected aubergines* 20070625 [5]
- *protected blackberries* 20051189 [5]
- *protected courgettes* 20070627 [5]
- *protected cucumbers* 20070627 [5]
- *protected gherkins* 20070627 [5]
- *protected raspberries* 20051189 [5]
- *protected rubus hybrids* 20051189 [5]
- *protected tomatoes* 20070625 [5]
- *quinces* 20080504 [5]
- *redcurrants* 20080505 [5]
- *whitecurrants* 20080505 [5]
- *wine grapes* 20040116 [5]

Approval information
- Accepted by BBPA for use on hops

Efficacy guidance
- Best results achieved when used as part of routine preventive spray programme from bud burst to end of flowering in apples and pears and from just before the signs of mildew infection in blackcurrants and gooseberries [5]
- In strawberries commence spraying at, or just prior to, first flower. Post-harvest sprays may be required on mildew-susceptible varieties where mildew is present and likely to be damaging [5]
- Spray at 7-14 d intervals depending on disease pressure and dose applied [5]
- For improved scab control on apples in post-blossom period tank-mix with mancozeb or captan [5]
- Apply alone from mid-Jun for control of secondary mildew on apples and pears
- On roses spray at first signs of disease and repeat every 2 wk. In high risk areas spray when leaves emerge in spring, repeat 1 wk later and then continue normal programme [5]
- Treatment of managed amenity turf may be carried out at any time of year before, or at, the first sign of disease [4]
- Myclobutanil products should be used in conjunction with other fungicides with a different mode of action to reduce the possibility of resistance developing

Restrictions
- Maximum total dose equivalent to ten full dose treatments in apples and pears; six full dose treatments in blackcurrants, gooseberries, strawberries
- Maximum number of treatments on managed amenity turf: two per yr [4]
- Do not mow turf within 24 hr after treatment [4]

Crop-specific information
- HI 28 d (grapevines); 21 d (cherries, mirabelles); 14 d (apples, pears, blackcurrants, gooseberries, hops); 3 d (plums, protected blackberries, protected raspberries, protected Rubus hybrids, strawberries)
- Product may be applied to newly sown turf after the two-leaf stage but in view of the large number of turf grass cultivars a safety test on a small area is recommended before large scale treatment [4]

Environmental safety
- Dangerous for the environment
- Toxic (harmful [4]) to aquatic organisms
- LERAP Category B [4]

Hazard classification and safety precautions
Hazard Harmful, Dangerous for the environment [1-3, 5]
Transport code 9 [1-3, 5]
Packaging group III [1-3, 5]
UN number 3082 [1-3, 5]; N/C [4]
Risk phrases R22b, R51, R63 [1-3, 5]; R52 [4]; R53a [1-5]
Operator protection A, H [1-5]; J [1-3, 5]; U08, U20a [1, 3-5]; U09a, U20b [2]; U15 [1-3, 5]
Environmental protection E15a [1-5]; E16a [4]
Storage and disposal D01, D02 [1-3, 5]; D05, D10c, D12a [1-5]
Medical advice M05b [1-3, 5]

FOR FULL CONDITIONS OF USE ALWAYS READ THE PRODUCT LABEL

347   1-naphthylacetic acid

A plant growth regulator to promote rooting of cuttings

See also 4-indol-3-ylbutyric acid + 1-naphthylacetic acid

Products

1	Rhizopon B Powder (0.1%)	Rhizopon	0.1% w/w	DP	09089
2	Rhizopon B Powder (0.2%)	Rhizopon	0.2% w/w	DP	09090
3	Rhizopon B Tablets	Rhizopon	25 mg a.i.	WT	09091

Uses
- Rooting of cuttings in **ornamental plant production**

Efficacy guidance
- Dip moistened base of cuttings into powder immediately before planting [1, 2]
- See label for details of concentrations recommended for promotion of rooting in cuttings of different species [3]
- Dip prepared cuttings in solution for 4-24 h depending on species [3]

Restrictions
- Maximum number of treatments 1 per cutting

Crop-specific information
- Latest use: before cutting insertion for ornamental specimens

Hazard classification and safety precautions
  UN number N/C
  Operator protection U08, U19a, U20a
  Environmental protection E15a, E34
  Storage and disposal D09a, D11a

348   napropamide

A soil applied alkanamide herbicide for oilseed rape, fruit and woody ornamentals
HRAC mode of action code: K3

Products

1	AC 650	United Phosphorus	450 g/l	SC	11102
2	Devrinol	United Phosphorus	450 g/l	SC	09374
3	Jouster	AgriGuard	450 g/l	SC	12741
4	MAC-Napropamide 450 SC	AgChem Access	450 g/l	SC	13972
5	Nappa	ChemSource	450 g/l	SC	14881

Uses
- Annual dicotyledons in **bilberries** *(off-label)*, **blackberries** *(off-label)*, **blueberries** *(off-label)*, **chinese cabbage** *(off-label)*, **choi sum** *(off-label)*, **collards** *(off-label)*, **evening primrose** *(off-label)*, **farm forestry** *(off-label)*, **forest** *(off-label)*, **honesty** *(off-label)*, **mustard** *(off-label)*, **pak choi** *(off-label)*, **redcurrants** *(off-label)*, **rubus hybrids** *(off-label)*, **whitecurrants** *(off-label)* [2]; **blackcurrants, broccoli, brussels sprouts, cabbages, calabrese, cauliflowers, forest nurseries, gooseberries, kale, ornamental plant production, raspberries, strawberries** [2-5]; **linseed** *(off-label)* [1, 2]; **winter oilseed rape** [1, 3]; **woody ornamentals** [2, 4, 5]
- Annual grasses in **bilberries** *(off-label)*, **blackberries** *(off-label)*, **blueberries** *(off-label)*, **chinese cabbage** *(off-label)*, **choi sum** *(off-label)*, **collards** *(off-label)*, **evening primrose** *(off-label)*, **farm forestry** *(off-label)*, **forest** *(off-label)*, **honesty** *(off-label)*, **mustard** *(off-label)*, **pak choi** *(off-label)*, **redcurrants** *(off-label)*, **rubus hybrids** *(off-label)*, **whitecurrants** *(off-label)* [2]; **blackcurrants, broccoli, brussels sprouts, cabbages, calabrese, cauliflowers, forest nurseries, gooseberries, kale, ornamental plant production, raspberries, strawberries** [2-5]; **linseed** *(off-label)* [1, 2]; **winter oilseed rape** [1, 3]; **woody ornamentals** [2, 4, 5]
- Cleavers in **blackcurrants, broccoli, brussels sprouts, cabbages, calabrese, cauliflowers, forest nurseries, gooseberries, kale, ornamental plant production, raspberries, strawberries** [2-5]; **winter oilseed rape** [3]; **woody ornamentals** [2, 4, 5]
- Groundsel in **blackcurrants, broccoli, brussels sprouts, cabbages, calabrese, cauliflowers, forest nurseries, gooseberries, kale, ornamental plant production, raspberries, strawberries** [2-5]; **winter oilseed rape** [3]; **woody ornamentals** [2, 4, 5]

SEE SECTION 3 FOR PRODUCTS ALSO REGISTERED

Specific Off-Label Approvals (SOLAs)
- **bilberries** *20090403* [2]
- **blackberries** *20090399* [2]
- **blueberries** *20090403* [2]
- **chinese cabbage** *20090406* [2]
- **choi sum** *20090406* [2]
- **collards** *20090406* [2]
- **evening primrose** *20090400* [2]
- **farm forestry** *20090409* [2]
- **forest** *20090409* [2]
- **honesty** *20090400* [2]
- **linseed** *20090397* [1], *20090400* [2]
- **mustard** *20090400* [2]
- **pak choi** *20090406* [2]
- **redcurrants** *20090403* [2]
- **rubus hybrids** *20090399* [2]
- **whitecurrants** *20090403* [2]

Approval information
- Napropamide included in Annex I under EC Directive 91/414

Efficacy guidance
- Best results obtained from treatment pre-emergence of weeds but product may be used in conjunction with contact herbicides for control of emerged weeds. Otherwise remove existing weeds before application
- Weed control may be reduced where spray is mixed too deeply in the soil
- Manure, crop debris or other organic matter may reduce weed control
- Napropamide broken down by sunlight, so application during such conditions not recommended. Most crops recommended for treatment between Nov and end-Feb
- Apply to winter oilseed rape pre-emergence or as pre-drilling treatment in tank-mixture with trifluralin and incorporate within 30 min [1, 3]
- Post-emergence use of specific grass weedkilller recommended where volunteer cereals are a serious problem
- Increase water volume to ensure adequate dampening of compost when treating containerised nursery stock with dense leaf canopy [2]

Restrictions
- Maximum number of treatments 1 per crop or yr
- Do not use on Sands
- Do not use on soils with more than 10% organic matter
- Strawberries, blackcurrants, gooseberries and raspberries must be treated between 1 Nov and end Feb [2, 3]
- Applications to strawberries must only be made where the mature foliage has been previously removed [2, 3]
- Applications to ornamentals and forest nurseries must be made after 1 Nov and before end Apr
- Do not treat ornamentals in containers of less than 1 litre [2]
- Some phytotoxicity seen on yellow and golden varieties of conifers and container grown alpines. On any ornamental variety treat only a small number of plants in first season [2]
- Consult processors before use on any crop for processing

Crop-specific information
- Latest use: before transplanting for brassicas; pre-emergence for winter oilseed rape; before end Feb for strawberries, bush and cane fruit; before end of Apr for field and container grown ornamental trees and shrubs
- Where minimal cultivation used to establish oilseed rape, tank-mixture may be applied directly to stubble and mixed into top 25 mm as part of surface cultivations [1]
- Apply up to 14 d prior to drilling winter oilseed rape [1, 3]
- Apply to strawberries established for at least one season or to maiden crops as long as planted carefully and no roots exposed, between 1 Nov and end Feb. Do not treat runners of poor vigour or with shallow roots, or runner beds [2, 3]

FOR FULL CONDITIONS OF USE ALWAYS READ THE PRODUCT LABEL

- Newly planted ornamentals should have no roots exposed. Do not treat stock of poor vigour or with shallow roots. Treatments made in Mar and Apr must be followed by 25 mm irrigation within 24 h [2]
- Bush and cane fruit must be established for at least 10 mth before spraying and treated between 1 Nov and end Feb [2, 3]

Following crops guidance
- After use in fruit or ornamentals no crop can be drilled within 7 mth of treatment. Leaf, flowerhead, root and fodder brassica crops may be drilled after 7 mth; potatoes, maize, peas or dwarf beans after 9 mth; autumn sown wheat or grass after 18 mth; any crop after 2 yr
- After use in oilseed rape only oilseed rape, swedes, fodder turnips, brassicas or potatoes should be sown within 12 mth of application
- Soil should be mould-board ploughed to a depth of at least 200 mm before drilling or planting any following crop

Environmental safety
- Dangerous for the environment
- Toxic to aquatic organisms

Hazard classification and safety precautions
  Hazard Irritant, Dangerous for the environment
  Transport code 9
  Packaging group III
  UN number 3082
  Risk phrases R36, R38, R51, R53a
  Operator protection A [1-5]; C [2-5]; U05a, U09a, U11, U20b
  Environmental protection E15a [2-5]; E34, E38 [1-5]
  Storage and disposal D01, D02, D05, D09a, D10c [1-5]; D12a [1]; D12b [2-5]

349 nicosulfuron

A sulfonylurea herbicide for maize
HRAC mode of action code: B

Products

1	Agrotech-Nicosulfron 40 SC	AgChem Access	40 g/l	SC	12938
2	Bishop	AgChem Access	40 g/l	SC	14055
3	Clayton Delilah	Clayton	40 g/l	SC	12610
4	Clayton Delilah XL	Clayton	60 g/l	SC	14443
5	Clayton Myth	Clayton	40 g/l	SC	14489
6	Clayton Myth XL	Clayton	60 g/l	SC	14466
7	Greencrop Folklore	Greencrop	40 g/l	SC	12867
8	Nico 4	Goldengrass	40 g/l	SC	14293
9	Samson Extra 6%	Syngenta	60 g/l	OD	13436
10	Standon Frontrunner	Standon	40 g/l	SC	13308

Uses
- Amaranthus in *forage maize*, *grain maize* [4, 6]
- Annual dicotyledons in *forage maize* [1-3, 5, 7-10]; *grain maize*, *sweetcorn* (off-label) [9]
- Annual grasses in *sweetcorn* (off-label) [9]
- Annual meadow grass in *forage maize* [1-8, 10]; *grain maize* [4, 6]
- Black bindweed in *forage maize*, *grain maize* [4, 6]
- Black nightshade in *forage maize*, *grain maize* [4, 6]
- Chickweed in *forage maize* [1-8, 10]; *grain maize* [4, 6]
- Couch in *forage maize* [1-3, 5, 7, 8, 10]
- Fat hen in *forage maize*, *grain maize* [4, 6]
- Grass weeds in *forage maize*, *grain maize*, *sweetcorn* (off-label) [9]
- Italian ryegrass in *forage maize*, *grain maize* [4, 6]
- Knotgrass in *forage maize*, *grain maize* [4, 6]
- Mayweeds in *forage maize* [1-8, 10]; *grain maize* [4, 6]
- Perennial ryegrass in *forage maize*, *grain maize* [4, 6]
- Redshank in *forage maize*, *grain maize* [4, 6]
- Ryegrass in *forage maize* [1-3, 5, 7, 8, 10]

SEE SECTION 3 FOR PRODUCTS ALSO REGISTERED

Pesticide Profiles

- Shepherd's purse in ***forage maize*** [1-8, 10]; ***grain maize*** [4, 6]
- Volunteer oilseed rape in ***forage maize*** [1-3, 5, 7, 8, 10]

Specific Off-Label Approvals (SOLAs)
- ***sweetcorn*** *20080800* [9], *20102068* [9]

Approval information
- Nicosulfuron included in Annex I under EC Directive 91/414

Efficacy guidance
- Product should be applied post-emergence between 2 and 8 crop leaf stage and to emerged weeds from the 2-leaf stage
- Product acts mainly by foliar activity. Ensure good spray cover of the weeds
- Nicosulfuron is a member of the ALS-inhibitor group of herbicides. To avoid the build up of resistance do not use any product containing an ALS-inhibitor herbicide with claims for control of grass weeds more than once on any crop
- Use these products as part of a resistance management strategy that includes cultural methods of control and does not use ALS inhibitors as the sole chemical method of grass weed control

Restrictions
- Maximum number of treatments 1 per crop
- Do not use if an organophosphorus soil insecticide has been used on the same crop
- Do not mix with foliar or liquid fertilisers or specified herbicides. See label for details
- Do not apply in mixture, or sequence, with any other sulfonyl-urea containing product
- Do not treat crops under stress
- Do not apply if rainfall is forecast to occur within 6 h of application

Crop-specific information
- Latest use: up to and including 8 true leaves for maize
- Some transient yellowing may be seen from 1-2 wk after treatment
- Only healthy maize crops growing in good field conditions should be treated

Following crops guidance
- Winter wheat (not undersown) may be sown 4 mth after treatment; maize may be sown in the spring following treatment; all other crops may be sown from the next autumn

Environmental safety
- Dangerous for the environment [7, 10]
- Very toxic to aquatic organisms [7, 10]
- LERAP Category B

Hazard classification and safety precautions
Hazard Harmful [4, 6, 9]; Irritant [1-3, 5, 7, 8, 10]; Dangerous for the environment [1, 4-7, 9, 10]
Transport code 9
Packaging group III
UN number 3082
Risk phrases R20, R36, R43 [4, 6, 9]; R38 [1-3, 5, 7, 8, 10]; R50, R53a [1, 4-7, 9, 10]; R53b [5]
Operator protection A, H [1-10]; C [4, 6, 9]; U02a, U05a, U14, U20b [1-10]; U10 [1, 4-7, 9, 10]; U11, U19a [4, 6, 9]
Environmental protection E15a [2, 3, 5, 7, 8]; E15b [1, 9, 10]; E16a [1-10]; E16b [1-3, 5-10]; E38 [1, 7, 9, 10]
Storage and disposal D01, D02 [1-3, 5, 7-10]; D05 [1, 7, 10]; D09a [2, 3, 5, 8, 10]; D10b [2, 3, 5, 8]; D10c, D12a [1, 7, 9, 10]
Medical advice M05a

350 oxadiazon

A residual and contact oxadiazole herbicide for fruit and ornamentals
HRAC mode of action code: E

Products

1	Clayton Oxen FL	Clayton	250 g/l	EC	12861
2	Ronstar 2G	Certis	2% w/w	GR	12965
3	Ronstar Liquid	Certis	250 g/l	EC	11215
4	Standon Roxx L	Standon	250 g/l	EC	13309

FOR FULL CONDITIONS OF USE ALWAYS READ THE PRODUCT LABEL

oxadiazon

Uses
- Annual dicotyledons in **almonds** *(off-label)*, **bilberries** *(off-label)*, **blackberries** *(off-label)*, **blueberries** *(off-label)*, **chestnuts** *(off-label)*, **cranberries** *(off-label)*, **hazel nuts** *(off-label)*, **redcurrants** *(off-label)*, **rubus hybrids** *(off-label)*, **walnuts** *(off-label)*, **whitecurrants** *(off-label)* [3]; **apple orchards, blackcurrants, gooseberries, hops, pear orchards, raspberries, wine grapes, woody ornamentals** [1, 3, 4]; **ornamental plant production** [2]
- Annual grasses in **almonds** *(off-label)*, **bilberries** *(off-label)*, **blackberries** *(off-label)*, **blueberries** *(off-label)*, **chestnuts** *(off-label)*, **cranberries** *(off-label)*, **hazel nuts** *(off-label)*, **redcurrants** *(off-label)*, **rubus hybrids** *(off-label)*, **walnuts** *(off-label)*, **whitecurrants** *(off-label)* [3]; **apple orchards, blackcurrants, gooseberries, hops, pear orchards, raspberries, wine grapes, woody ornamentals** [1, 3, 4]; **ornamental plant production** [2]
- Bindweeds in **almonds** *(off-label)*, **bilberries** *(off-label)*, **blackberries** *(off-label)*, **blueberries** *(off-label)*, **chestnuts** *(off-label)*, **cranberries** *(off-label)*, **hazel nuts** *(off-label)*, **redcurrants** *(off-label)*, **rubus hybrids** *(off-label)*, **walnuts** *(off-label)*, **whitecurrants** *(off-label)* [3]; **apple orchards, blackcurrants, gooseberries, hops, pear orchards, raspberries, wine grapes, woody ornamentals** [1, 3, 4]
- Cleavers in **almonds** *(off-label)*, **bilberries** *(off-label)*, **blackberries** *(off-label)*, **blueberries** *(off-label)*, **chestnuts** *(off-label)*, **cranberries** *(off-label)*, **hazel nuts** *(off-label)*, **redcurrants** *(off-label)*, **rubus hybrids** *(off-label)*, **walnuts** *(off-label)*, **whitecurrants** *(off-label)* [3]; **apple orchards, blackcurrants, gooseberries, hops, pear orchards, raspberries, wine grapes, woody ornamentals** [1, 3, 4]
- Knotgrass in **almonds** *(off-label)*, **bilberries** *(off-label)*, **blackberries** *(off-label)*, **blueberries** *(off-label)*, **chestnuts** *(off-label)*, **cranberries** *(off-label)*, **hazel nuts** *(off-label)*, **redcurrants** *(off-label)*, **rubus hybrids** *(off-label)*, **walnuts** *(off-label)*, **whitecurrants** *(off-label)* [3]; **apple orchards, blackcurrants, gooseberries, hops, pear orchards, raspberries, wine grapes, woody ornamentals** [1, 3, 4]

Specific Off-Label Approvals (SOLAs)
- **almonds** *20063295* [3]
- **bilberries** *20063296* [3]
- **blackberries** *20063297* [3]
- **blueberries** *20063296* [3]
- **chestnuts** *20063295* [3]
- **cranberries** *20063296* [3]
- **hazel nuts** *20063295* [3]
- **redcurrants** *20063296* [3]
- **rubus hybrids** *20063297* [3]
- **walnuts** *20063295* [3]
- **whitecurrants** *20063296* [3]

Approval information
- Oxadiazon included in Annex I under EC Directive 91/414
- Accepted by BBPA for use on hops

Efficacy guidance
- Apply as a directed spray or as a spot treatment as directed. Avoid direct contact with the crop. See label [1, 3, 4]
- Apply granules as soon after potting as possible and before crop plant is making soft growth [2]
- Best results from spray treatments obtained when weeds are dry at application and for a period afterwards. Later rain or overhead watering is needed for effective results [1, 3, 4]
- Residual activity reduced on soils with more than 10% organic matter and, in these conditions, post-emergence treatment is more effective [1, 3, 4]
- Best results on bindweed when first shoots are 10-15 cm long [1, 3, 4]
- Adjust water volume to achieve good coverage when weed density is high [1, 3, 4]

Restrictions
- Maximum number of treatments 2 per yr for ornamental plant production [2]
- Maximum total dose per yr equivalent to one full dose treatment on edible crops [1, 3, 4]
- See label for list of ornamental species which may be treated with granules. Treat small numbers of other species to check safety. Do not treat Hydrangea, Spiraea or Genista [2]
- Do not cultivate after treatment [1, 3, 4]

SEE SECTION 3 FOR PRODUCTS ALSO REGISTERED

Pesticide Profiles

- Do not treat container stock under glass or use on plants rooted in media with high sand or non-organic content. Do not apply to plants with wet foliage [2]
- Consult processors before treatment of crops intended for processing [1, 3, 4]

Crop-specific information
- Latest use: Jul for apples, grapevines, hops, pears; Jun for raspberries, woody ornamentals
- Apply spray to apples and pears from Jan to Jul, avoiding young growth [1, 3, 4]
- Treat bush fruit from Jan to bud-break, avoiding bushes, grapevines in Feb/Mar before start of new growth or in Jun/Jul avoiding foliage [1, 3, 4]
- Treat hops cropped for at least 2 yr in Feb or in Jun/Jul after deleafing [1, 3, 4]
- Treat woody ornamentals from Jan to Jun, avoiding young growth. Do not spray container stock overall [1, 3, 4]

Following crops guidance
- A period of at least 6 mth must elapse between treatment at half the recommended maximum dose and planting a following crop. 12 mth must elapse after use of the maximum recommended dose. In either case soil must be ploughed to min 15 cm and cultivated before sowing or planting a following crop [1, 3, 4]

Environmental safety
- Flammable [1, 3, 4]
- Dangerous for the environment [1, 3, 4]
- Very toxic to aquatic organisms [1, 3, 4]
- Harmful to aquatic organisms [2]

Hazard classification and safety precautions
Hazard Harmful, Flammable, Dangerous for the environment [1, 3, 4]
Transport code 3 [1, 3, 4]; 9 [2]
Packaging group III
UN number 1993 [1, 3, 4]; 3077 [2]
Risk phrases R22b, R36, R38, R50, R67 [1, 3, 4]; R41 [3]; R52 [2]; R53a [1-4]
Operator protection A, C [1-4]; H [4]; U02a [1, 4]; U05a, U20b [1-4]; U08 [1, 3, 4]; U09a, U19a [2]; U14, U15 [2-4]
Environmental protection E15a [1-4]; E38 [3]
Storage and disposal D01, D02, D09a, D12a [1-4]; D05 [4]; D10a [1, 3, 4]; D11a [2]
Medical advice M05b [1, 3, 4]

351 oxamyl

A soil-applied, systemic carbamate nematicide and insecticide
IRAC mode of action code: 1A

Products

1	Tevday	AgChem Access	10% w/w	GR	13783
2	Vydate 10G	DuPont	10% w/w	GR	02322

Uses
- American serpentine leaf miner in **aubergines** *(off-label)*, **ornamental plant production** *(off-label)*, **protected broad beans** *(off-label)*, **protected ornamentals** *(off-label)*, **protected peppers** *(off-label)*, **protected soya beans** *(off-label)*, **protected tomatoes** *(off-label)* [2]
- Aphids in **fodder beet** *(off-label)* [2]; **potatoes, sugar beet** [1, 2]
- Docking disorder vectors in **fodder beet** *(off-label)* [2]; **sugar beet** [1, 2]
- Eelworm in **bulb onions** *(off-label)*, **garlic** *(off-label)*, **shallots** *(off-label)* [2]
- Free-living nematodes in **potatoes** [1, 2]
- Insect pests in **garlic** *(off-label)* [2]
- Leaf miner in **aubergines** *(off-label)*, **ornamental plant production** *(off-label)*, **protected broad beans** *(off-label)*, **protected ornamentals** *(off-label)*, **protected peppers** *(off-label)*, **protected soya beans** *(off-label)*, **protected tomatoes** *(off-label)* [2]
- Mangold fly in **fodder beet** *(off-label)* [2]; **sugar beet** [1, 2]
- Millipedes in **fodder beet** *(off-label)* [2]; **sugar beet** [1, 2]
- Potato cyst nematode in **potatoes** [1, 2]
- Pygmy beetle in **fodder beet** *(off-label)* [2]; **sugar beet** [1, 2]

FOR FULL CONDITIONS OF USE ALWAYS READ THE PRODUCT LABEL

oxamyl

- South American leaf miner in *aubergines* (off-label), *ornamental plant production* (off-label), *protected broad beans* (off-label), *protected ornamentals* (off-label), *protected peppers* (off-label), *protected soya beans* (off-label), *protected tomatoes* (off-label) [2]
- Spraing vectors in *potatoes* [1, 2]
- Stem nematodes in *bulb onions* (off-label), *garlic* (off-label), *protected garlic* (off-label), *protected onions* (off-label), *shallots* (off-label) [2]; *carrots*, *parsnips* [1, 2]

Specific Off-Label Approvals (SOLAs)
- *aubergines* 930020 [2]
- *bulb onions* 20061890 [2]
- *fodder beet* 20061107 [2]
- *garlic* 20061890 [2], 920163 [2]
- *ornamental plant production* 930020 [2]
- *protected broad beans* 930020 [2]
- *protected garlic* 940925 [2]
- *protected onions* 940925 [2]
- *protected ornamentals* 930020 [2]
- *protected peppers* 930020 [2]
- *protected soya beans* 930020 [2]
- *protected tomatoes* 930020 [2]
- *shallots* 20061890 [2]

Approval information
- Oxamyl included in Annex I under EC Directive 91/414
- In 2006 CRD required that all products containing this active ingredient should carry the following warning in the main area of the container label: "Oxamyl is an anticholinesterase carbamate. Handle with care"

Efficacy guidance
- Apply granules with suitable applicator before drilling or planting. See label for details of recommended machines
- In potatoes incorporate thoroughly to 10 cm and plant within 3-4 d
- In sugar beet apply in seed furrow at drilling

Restrictions
- Oxamyl is subject to the Poisons Rules 1982 and the Poisons Act 1972. See Section 5 for more information
- Contains an anticholinesterase carbamate compound. Do not use if under medical advice not to work with such compounds
- Maximum number of treatments 1 per crop or yr
- Keep in original container, tightly closed, in a safe place, under lock and key
- Wear protective gloves if handling treated compost or soil within 2 wk after treatment
- Allow at least 12 h, followed by at least 1 h ventilation, before entry of unprotected persons into treated glasshouses

Crop-specific information
- Latest use: at drilling/planting for vegetables; before drilling/planting for potatoes and peas.
- HI:potatoes, sugar beet, carrots, parsnips 12 wk

Environmental safety
- Dangerous for the environment
- Toxic to aquatic organisms
- Dangerous to fish or other aquatic life. Do not contaminate surface waters or ditches with chemical or used container
- Dangerous to game, wild birds and animals. Bury spillages

Hazard classification and safety precautions
Hazard Toxic, Dangerous for the environment [2]
Transport code 6.1
Packaging group II
UN number 2757
Risk phrases R23, R25, R51, R53a [2]
Operator protection A, B, H, K, M [2]; C [2] (or D); U02a, U04a, U05a, U07, U09a, U13, U19a, U20a [2]

SEE SECTION 3 FOR PRODUCTS ALSO REGISTERED

Pesticide Profiles

Environmental protection E10a, E13b, E34, E36a, E38 [2]
Storage and disposal D01, D02, D09a, D11a, D14 [2]
Medical advice M04a [2]

352 paclobutrazol

A triazole plant growth regulator for ornamentals and fruit

Products

1	Agrovista Paclo	Agrovista	250 g/l	SC	13424
2	Bonzi	Syngenta Bioline	4 g/l	SC	13623
3	Cultar	Syngenta	250 g/l	SC	10523
4	Pirouette	Fine	4 g/l	SC	13073
5	Pirouette	Fine	4 g/l	SC	13137

Uses
- Control of shoot growth in **apples, pears** [1, 3]
- Growth regulation in **cherries** *(off-label)*, **plums** *(off-label)* [1, 3]; **quinces** *(off-label)* [1]
- Improving colour in **poinsettias** [2, 4, 5]
- Increasing flowering in **azaleas, bedding plants, begonias, kalanchoes, lilies, roses, tulips** [2, 4, 5]
- Increasing fruit set in **apples, pears** [1, 3]; **quinces** *(off-label)* [3]
- Stem shortening in **azaleas, bedding plants, begonias, kalanchoes, lilies, poinsettias, roses, tulips** [2, 4, 5]

Specific Off-Label Approvals (SOLAs)
- **cherries** *20080461* [1], *20031235* [3]
- **plums** *20080461* [1], *20031235* [3]
- **quinces** *20080462* [1], *20062993* [3]

Efficacy guidance
- Chemical is active via both foliage and root uptake. For best results apply in dull weather when relative humidity not high

Restrictions
- Maximum number of treatments 1 per specimen for some species [4]
- Maximum total dose equivalent to 3 full dose treatments for pears and 4 full dose treatments for apples [1, 3]
- Some varietal restrictions apply in top fruit (see label for details) [1, 3]
- Do not use on trees of low vigour or under stress [1, 3]
- Do not use on trees from green cluster to 2 wk after full petal fall [1, 3]
- Do not use in underplanted orchards or those recently interplanted [1, 3]

Crop-specific information
- HI apples, pears 14 d [1, 3]
- Apply to apple and pear trees under good growing conditions as pre-blossom spray (apples only) and post-blossom at 7-14 d intervals [1, 3]
- Timing and dose of orchard treatments vary with species and cultivar. See label [1, 3]

Following crops guidance
- Chemical has residual soil activity which can affect growth of following crops. Withhold treatment from orchards due for grubbing to allow the following intervals between the last treatment and planting the next crop: apples, pears 1 yr; beans, peas, onions 2 yr; stone fruit 3 yr; cereals, grass, oilseed rape, carrots 4 yr; market brassicas 6 yr; potatoes and other crops 7 yr [1, 3]

Environmental safety
- Harmful to aquatic organisms
- Do not use on food crops [4]
- Keep livestock out of treated areas for at least 2 years after treatment [1, 3]

Hazard classification and safety precautions
Hazard Irritant [1, 3]
UN number N/C
Risk phrases R36, R52, R53a [1, 3]
Operator protection A, H, M [1, 3]; U05a, U20a [1-5]; U08 [2, 4, 5]

FOR FULL CONDITIONS OF USE ALWAYS READ THE PRODUCT LABEL

paraffin oil (commodity substance)

Environmental protection E06a [1, 3] (2 yr); E13c [4, 5]; E15a [2-5]; E15b [1]; E34, E38 [1, 3]
Consumer protection C01 [2, 4, 5]; C02a [1, 3] (2 wk)
Storage and disposal D01, D02, D09a [1-5]; D05 [2, 4, 5]; D10b [4, 5]; D10c [1-3]; D12a [1, 3]
Medical advice M03 [1, 3]

353 paraffin oil (commodity substance)

An agent for the control of birds by egg treatment

Products

paraffin oil	various		OL	-

Uses
- Birds in **miscellaneous pest control situations** (egg treatment)

Approval information
- Approval for the use of paraffin oil as a commodity substance was granted on 25 April 1995 by Ministers under regulation 5 of the Control of Pesticides Regulations 1986
- Only to be used where a licence has been approved in accordance with Section 16(1) of the Wildlife and Countryside Act (1981)
- Paraffin oils included in Annex I under EC Directive 91/414

Efficacy guidance
- Egg treatment should be undertaken as soon as clutch is complete
- Eggs should be treated by complete immersion in liquid paraffin

Restrictions
- Use to control eggs of birds covered by licences issued by the Agriculture and Environment Departments under Section 16(1) of the Wildlife and Countryside Act (1981)
- Treat eggs once only

Hazard classification and safety precautions
Operator protection A, C

354 pelargonic acid

A naturally occuring 9 carbon acid terminating in a carboxylic acid only available in mixtures
HRAC mode of action code: Not classified

See also maleic hydrazide + pelargonic acid

355 penconazole

A protectant triazole fungicide with antisporulant activity
FRAC mode of action code: 3

See also captan + penconazole

Products

1	Agrovista Penco	Agrovista	100 g/l	EC	13425
2	Topas	Syngenta	100 g/l	EC	09717
3	Topenco 100 EC	Globachem	100 g/l	EC	11972

Uses
- Powdery mildew in **apples** [1-3]; **bilberries** (off-label), **blackcurrants, blueberries** (off-label), **cranberries** (off-label), **gooseberries** (off-label), **hops, ornamental trees, protected strawberries** (off-label), **redcurrants** (off-label), **strawberries** (off-label), **whitecurrants** (off-label) [1, 2]
- Rust in **roses** [1, 2]
- Scab in **ornamental trees** [1, 2]

Specific Off-Label Approvals (SOLAs)
- **bilberries** *20080477* [1], *20061106* [2]
- **blueberries** *20080477* [1], *20061106* [2]

SEE SECTION 3 FOR PRODUCTS ALSO REGISTERED

- **cranberries** *20080477* [1], *20061106* [2]
- **gooseberries** *20080477* [1], *20061106* [2]
- **protected strawberries** *20080478* [1], *20073651* [2]
- **redcurrants** *20080477* [1], *20061106* [2]
- **strawberries** *20080478* [1], *20073651* [2]
- **whitecurrants** *20080477* [1], *20061106* [2]

Approval information
- Accepted by BBPA for use on hops

Efficacy guidance
- Use as a protectant fungicide by treating at the earliest signs of disease
- Treat crops every 10-14 d (every 7-10 d in warm, humid weather) at first sign of infection or as a protective spray ensuring complete coverage. See label for details of timing
- Increase dose and volume with growth of hops but do not exceed 2000 l/ha. Little or no activity will be seen on established powdery mildew
- Antisporulant activity reduces development of secondary mildew in apples

Restrictions
- Maximum number of treatments 10 per yr for apples, 6 per yr for hops, 4 per yr for blackcurrants
- Check for varietal susceptibility in roses. Some defoliation may occur after repeat applications on Dearest

Crop-specific information
- HI apples, hops 14 d; currants, bilberries, blueberries, cranberries, gooseberries 4 wk

Environmental safety
- Dangerous for the environment
- Toxic to aquatic organisms

Hazard classification and safety precautions
Hazard Irritant, Dangerous for the environment
Transport code 3
Packaging group III
UN number 1915
Risk phrases R36, R51, R53a [1-3]; R38 [3]
Operator protection A, C; U02a, U05a, U20b [1-3]; U08 [1, 2]; U11 [3]
Environmental protection E15a [1-3]; E22c [3]; E38 [1, 2]
Storage and disposal D01, D02, D05, D09a [1-3]; D07 [1, 3]; D10b [3]; D10c, D12a [1, 2]
Medical advice M05b

356 pencycuron

A non-systemic urea fungicide for use on seed potatoes
FRAC mode of action code: 20

See also imazalil + pencycuron

Products

1	Agriguard Pencycuron	AgriGuard	12.5% w/w	DS	10445
2	Monceren DS	Bayer CropScience	12.5% w/w	DS	11292
3	Pan Pencycuron P	Pan Agriculture	12.5% w/w	DP	10494
4	Pency Flowable	Globachem	250 g/l	FS	12637
5	Standon Pencycuron DP	Standon	12.5% w/w	DS	08774

Uses
- Black scurf in **potatoes** (tuber treatment)
- Stem canker in **potatoes** (tuber treatment)

Efficacy guidance
- Provides control of tuber-borne disease and gives some reduction of stem canker
- Seed tubers should be of good quality and vigour, and free from soil deposits when treated
- Dust formulations should be applied immediately before, or during, the planting process by treating seed tubers in chitting trays, in bulk bins immediately before planting or in hopper at planting. Liquid formulations may be applied by misting equipment at any time, into or out of store but treatment over a roller table at the end of grading out, or at planting, is usually most convenient

FOR FULL CONDITIONS OF USE ALWAYS READ THE PRODUCT LABEL

pendimethalin

- Whichever method is chosen, complete and even distribution on the tubers is essential for optimum efficacy
- Apply in accordance with detailed guidelines in manufacturer's literature
- If rain interrupts planting, cover dust treated tubers in hopper

Restrictions
- Maximum number of treatments 1 per batch of seed potatoes
- Treated tubers must be used only as seed and not for human or animal consumption
- Do not use on tubers previously treated with a dry powder seed treatment or hot water
- To prevent rapid absorption through damaged 'eyes' seed tubers should only be treated before the 'eyes' begin to open or immediately before or during planting [4]
- Tubers removed from cold storage must be allowed to attain a temperature of at least 8 °C before treatment [4]
- Use of suitable dust mask is mandatory when applying dust, filling the hopper or riding on planter [1, 2, 5]
- Treated tubers in bulk bins must be allowed to dry before being stored, especially when intended for cold storage [4]
- Some internal staining of boxes used to store treated tubers may occur. Ware tubers should not be stored or supplied to packers in such stained boxes [4]

Crop-specific information
- Latest use: at planting

Environmental safety
- Do not use treated seed as food or feed
- Treated seed harmful to game and wildlife

Hazard classification and safety precautions
Hazard Irritant [4]
UN number N/C
Risk phrases R43 [4]
Operator protection A [1-5]; F [1-3, 5]; U14, U20a [4]; U20b [1-3, 5]
Environmental protection E03, E15a
Storage and disposal D05 [4]; D07 [1, 3, 5]; D09a, D11a [1-5]
Treated seed S01, S02, S03, S04a, S05, S06a

357 pendimethalin

A residual dinitroaniline herbicide for cereals and other crops
HRAC mode of action code: K1

See also bentazone + pendimethalin
chlorotoluron + pendimethalin
diflufenican + pendimethalin
flufenacet + pendimethalin
imazamox + pendimethalin
isoproturon + pendimethalin

Products

1	Alpha Pendimethalin 330 EC	Makhteshim	330 g/l	EC	13815
2	Aquarius	Makhteshim	400 g/l	CS	14712
3	Blazer M	Headland	330 g/l	EC	15084
4	Cinder	Makhteshim	400 g/l	CS	14526
5	Convey 330EC	AgriGuard	330 g/l	EC	14535
6	Fastnet	Sipcam	330 g/l	EC	14068
7	Nighthawk 330 EC	AgChem Access	330 g/l	EC	14083
8	PDM 330 EC	BASF	330 g/l	EC	13406
9	Sherman	Makhteshim	330 g/l	EC	13859
10	Stamp 330 EC	AgChem Access	330 g/l	EC	13995
11	Stomp 400 SC	BASF	400 g/l	SC	13405
12	Stomp Aqua	BASF	455 g/l	CS	14664
13	Trample	AgChem Access	330 g/l	EC	14173

SEE SECTION 3 FOR PRODUCTS ALSO REGISTERED

Pesticide Profiles

Uses

- Annual dicotyledons in **almonds** *(off-label)*, **asparagus** *(off-label)*, **bilberries** *(off-label)*, **blueberries** *(off-label)*, **chestnuts** *(off-label)*, **cranberries** *(off-label)*, **cress** *(off-label)*, **evening primrose** *(off-label)*, **farm forestry** *(off-label)*, **fennel** *(off-label)*, **frise** *(off-label)*, **grass seed crops** *(off-label)*, **hazel nuts** *(off-label)*, **herbs (see appendix 6)** *(off-label)*, **lamb's lettuce** *(off-label)*, **lettuce** *(off-label)*, **lupins** *(off-label)*, **quinces** *(off-label)*, **radicchio** *(off-label)*, **redcurrants** *(off-label)*, **runner beans** *(off-label)*, **scarole** *(off-label)*, **walnuts** *(off-label)*, **whitecurrants** *(off-label)* [1, 4, 9, 11, 12]; **apple orchards**, **pear orchards** [1, 2, 7, 9, 11]; **apples**, **hops** *(off-label)*, **pears** [4, 12]; **blackberries, blackcurrants, broccoli, brussels sprouts, cabbages, calabrese, carrots, cauliflowers, cherries, gooseberries, leeks, loganberries, parsnips, plums, raspberries, rubus hybrids, strawberries** [1, 2, 4, 7, 9, 11, 12]; **broad beans** *(off-label)*, **spring wheat** *(off-label)* [2, 4, 11, 12]; **bulb onions** [2, 4, 9, 11, 12]; **bulb onions** *(off-label)* [1, 2, 12]; **bulb onions** *(pre + post emergence treatment)* [2, 11]; **cabbages** *(off-label)*, **carrots** *(off-label)*, **french beans, horseradish** *(off-label)*, **navy beans, parsnips** *(off-label)* [12]; **celery (outdoor)** *(off-label)*, **miscanthus** *(off-label)*, **protected forest nurseries** *(off-label)*, **rhubarb** *(off-label)*, **top fruit** *(off-label)* [4, 11, 12]; **chervil** *(off-label)*, **coriander** *(off-label)*, **dill** *(off-label)* [1, 11, 12]; **chives** *(off-label)*, **hops, parsley, salad brassicas** *(off-label - for baby leaf production)*, **sweetcorn** *(off-label - under covers)* [11]; **combining peas, forage maize, potatoes, spring barley, sunflowers, triticale, winter barley, winter rye, winter wheat** [1-13]; **daffodils grown for galanthamine production** *(off-label)* [1, 2, 4, 9]; **durum wheat** [2-13]; **dwarf beans** *(off-label)*, **french beans** *(off-label)*, **navy beans** *(off-label)* [4, 11]; **edible podded peas** *(off-label)*, **vining peas** *(off-label)* [1, 2, 4, 9, 12]; **forage maize (under plastic mulches), grain maize, grain maize under plastic mulches** [1-3, 8, 9]; **forest** *(off-label)* [11, 12]; **forest nurseries** *(off-label)*, **ornamental plant production** *(off-label)* [4, 8, 11, 12]; **game cover** *(off-label)* [1, 4, 11, 12]; **garlic** *(off-label)*, **parsley root** *(off-label)*, **salad onions** *(off-label)*, **shallots** *(off-label)* [1, 2, 4, 9, 11, 12]; **leaf brassicas** *(off-label)*, **parsley** *(off-label)* [1, 12]; **leeks** *(off-label)* [2, 12]; **mallow (althaea spp.)** *(off-label)* [1, 4, 9, 12]; **onion sets** *(off-label)* [1, 4, 9]; **protected cress** *(off-label)*, **protected frise** *(off-label)*, **protected herbs (see appendix 6)** *(off-label)*, **protected lamb's lettuce** *(off-label)*, **protected lettuce** *(off-label)*, **protected salad brassicas** *(off-label)*, **protected scarole** *(off-label)* [1]; **salad brassicas** *(off-label)* [4, 9]; **spring field beans** *(off-label)* [1, 4, 6, 8, 9, 11, 12]; **sweetcorn** *(off-label)* [2, 4, 12]; **sweetcorn under plastic mulches** *(off-label)* [1, 2, 4, 12]; **winter field beans** *(off-label)* [1, 4, 6, 8, 9, 12].

- Annual grasses in **almonds** *(off-label)*, **asparagus** *(off-label)*, **bilberries** *(off-label)*, **blueberries** *(off-label)*, **chestnuts** *(off-label)*, **cranberries** *(off-label)*, **cress** *(off-label)*, **daffodils grown for galanthamine production** *(off-label)*, **edible podded peas** *(off-label)*, **evening primrose** *(off-label)*, **farm forestry** *(off-label)*, **frise** *(off-label)*, **garlic** *(off-label)*, **hazel nuts** *(off-label)*, **herbs (see appendix 6)** *(off-label)*, **lamb's lettuce** *(off-label)*, **lettuce** *(off-label)*, **lupins** *(off-label)*, **mallow (althaea spp.)** *(off-label)*, **parsley root** *(off-label)*, **quinces** *(off-label)*, **radicchio** *(off-label)*, **redcurrants** *(off-label)*, **runner beans** *(off-label)*, **salad onions** *(off-label)*, **scarole** *(off-label)*, **shallots** *(off-label)*, **vining peas** *(off-label)*, **walnuts** *(off-label)*, **whitecurrants** *(off-label)* [1, 2, 4, 9]; **apples, blackberries, blackcurrants, broccoli, brussels sprouts, bulb onions, cabbages, calabrese, carrots, cauliflowers, cherries, gooseberries, leeks, loganberries, parsnips, pears, plums, raspberries, strawberries** [4, 12]; **broad beans** *(off-label)*, **dwarf beans** *(off-label)*, **miscanthus** *(off-label)*, **navy beans** *(off-label)*, **protected forest nurseries** *(off-label)*, **top fruit** *(off-label)* [2, 4, 11]; **bulb onions** *(off-label)*, **leaf brassicas** *(off-label)*, **parsley** *(off-label)*, **protected cress** *(off-label)*, **protected frise** *(off-label)*, **protected herbs (see appendix 6)** *(off-label)*, **protected lamb's lettuce** *(off-label)*, **protected lettuce** *(off-label)*, **protected scarole** *(off-label)* [1, 2]; **cabbages** *(off-label)*, **carrots** *(off-label)*, **celery (outdoor)** *(off-label)*, **hops** *(off-label)*, **horseradish** *(off-label)*, **parsnips** *(off-label)*, **rhubarb** *(off-label)*, **sweetcorn** *(off-label)* [2, 4]; **chervil** *(off-label)*, **coriander** *(off-label)*, **dill** *(off-label)* [1, 2, 11]; **combining peas, durum wheat, forage maize, potatoes, spring barley, sunflowers, triticale, winter barley, winter rye, winter wheat** [3-6, 8, 10, 12, 13]; **fennel** *(off-label)*, **grass seed crops** *(off-label)* [1, 2, 4, 9, 12]; **forage maize (under plastic mulches), grain maize, grain maize under plastic mulches** [3, 8, 9]; **forest nurseries** *(off-label)*, **ornamental plant production** *(off-label)* [2, 4, 8, 11]; **french beans** *(off-label)* [4, 11]; **game cover** *(off-label)* [1, 2, 4, 11]; **leeks** *(off-label)*, **protected leaf brassicas** *(off-label)*, **protected radicchio** *(off-label)* [2]; **onion sets** *(off-label)* [1, 4, 9]; **protected salad brassicas** *(off-label)* [1]; **rubus hybrids** [1, 4, 7, 9, 12]; **salad brassicas** *(off-label)* [4, 9]; **spring field beans** *(off-label)*, **winter field beans** *(off-label)* [1, 2, 4, 6-9]; **spring wheat** *(off-label)* [2, 4, 11, 12]; **sweetcorn under plastic mulches** *(off-label)* [1, 2, 4].

FOR FULL CONDITIONS OF USE ALWAYS READ THE PRODUCT LABEL

pendimethalin

- Annual meadow grass in **apple orchards, blackberries, blackcurrants, broccoli, brussels sprouts, cabbages, calabrese, carrots, cauliflowers, cherries, combining peas, forage maize, gooseberries, leeks, loganberries, parsnips, pear orchards, plums, potatoes, raspberries, spring barley, strawberries, sunflowers, triticale, winter barley, winter rye, winter wheat** [1, 2, 7, 9, 11]; **bulb onions** [2, 9, 11]; **cabbages** *(off-label)* [12]; **durum wheat** [2, 7, 9, 11]; **forage maize (under plastic mulches), grain maize, grain maize under plastic mulches** [1, 2]; **hops, parsley** [11]; **rubus hybrids** [2, 11]
- Blackgrass in **apple orchards, blackberries, blackcurrants, broccoli, brussels sprouts, cabbages, calabrese, carrots, cauliflowers, cherries, combining peas, gooseberries, leeks, loganberries, parsnips, pear orchards, plums, raspberries, spring barley, strawberries, sunflowers, triticale, winter barley, winter rye, winter wheat** [1, 2, 7, 9, 11]; **bulb onions** [2, 9, 11]; **durum wheat** [2, 7, 9, 11]; **forage maize (under plastic mulches), grain maize, grain maize under plastic mulches** [1]; **hops, parsley** [11]; **rubus hybrids** [2, 11]
- Cleavers in **winter field beans** *(off-label)* [1, 7, 11]
- Fat hen in **sweetcorn under plastic mulches** *(off-label)* [1]
- Knotgrass in **daffodils grown for galanthamine production** *(off-label)* [1]; **edible podded peas** *(off-label)*, **vining peas** *(off-label)* [11]
- Rough meadow grass in **apple orchards, blackberries, blackcurrants, broccoli, brussels sprouts, cabbages, calabrese, carrots, cauliflowers, cherries, combining peas, forage maize, gooseberries, leeks, loganberries, parsnips, pear orchards, plums, potatoes, raspberries, spring barley, strawberries, sunflowers, triticale, winter barley, winter rye, winter wheat** [1, 2, 7, 9, 11]; **bulb onions** [2, 9, 11]; **durum wheat** [2, 7, 9, 11]; **forage maize (under plastic mulches), grain maize, grain maize under plastic mulches** [1, 2]; **hops, parsley** [11]; **rubus hybrids** [2, 11]
- Speedwells in **daffodils grown for galanthamine production** *(off-label)* [1]
- Volunteer oilseed rape in **apple orchards, blackberries, blackcurrants, broccoli, brussels sprouts, cabbages, calabrese, carrots, cauliflowers, cherries, gooseberries, leeks, loganberries, parsnips, pear orchards, plums, raspberries, rubus hybrids, spring barley, strawberries, sunflowers** [1, 2, 7, 9, 11]; **bulb onions** [2, 9, 11]; **combining peas, forage maize, potatoes, triticale, winter barley, winter rye, winter wheat** [1-3, 5, 7-11, 13]; **daffodils grown for galanthamine production** *(off-label)* [1]; **durum wheat** [2, 3, 5, 7-11, 13]; **forage maize (under plastic mulches), grain maize, grain maize under plastic mulches** [1-3, 8]; **hops, parsley** [11]; **spring field beans** *(off-label)*, **winter field beans** *(off-label)* [8]
- Wild oats in **durum wheat** [2, 7, 9, 11]; **forage maize (under plastic mulches), grain maize, grain maize under plastic mulches** [1]; **triticale, winter barley, winter rye, winter wheat** [1, 2, 7, 9, 11]

Specific Off-Label Approvals (SOLAs)
- **almonds** *20090767* [1], *20093271* [2], *20092184* [4], *20090812* [9], *20071442* [11], *20092916* [12]
- **asparagus** *20090764* [1], *20101003* [2], *20092197* [4], *20090809* [9], *20071431* [11], *20092920* [12]
- **bilberries** *20090770* [1], *20093289* [2], *20092198* [4], *20090817* [9], *20071444* [11], *20092907* [12]
- **blueberries** *20090770* [1], *20093289* [2], *20092198* [4], *20090817* [9], *20071444* [11], *20092907* [12]
- **broad beans** *20093287* [2], *20092199* [4], *20081016* [11], *20092922* [12]
- **bulb onions** *20090765* [1], *20093284* [2], *20092914* [12]
- **cabbages** *20101902* [2], *20101787* [4], *20100650* [12]
- **carrots** *20101134* [2], *20101132* [4], *20093526* [12]
- **celery (outdoor)** *20093294* [2], *20092206* [4], *20071430* [11], *20092924* [12]
- **chervil** *20090772* [1], *20093303* [2], *20081808* [11], *20092913* [12]
- **chestnuts** *20090767* [1], *20093271* [2], *20092184* [4], *20090812* [9], *20071442* [11], *20092916* [12]
- **chives** *20071432* [11]
- **coriander** *20090772* [1], *20093303* [2], *20081808* [11], *20092913* [12]
- **cranberries** *20090770* [1], *20093289* [2], *20092198* [4], *20090817* [9], *20071444* [11], *20092907* [12]
- **cress** *20090772* [1], *20093303* [2], *20092187* [4], *20090819* [9], *20071432* [11], *20092921* [12]
- **daffodils grown for galanthamine production** *20090762* [1], *20093299* [2], *20092189* [4], *20090807* [9]
- **dill** *20090772* [1], *20093303* [2], *20081808* [11], *20092913* [12]

SEE SECTION 3 FOR PRODUCTS ALSO REGISTERED

- **dwarf beans** *20093285* [2], *20092200* [4], *20080053* [11]
- **edible podded peas** *20090768* [1], *20093300* [2], *20092190* [4], *20090813* [9], *20071437* [11], *20092909* [12]
- **evening primrose** *20090759* [1], *20093301* [2], *20092191* [4], *20090804* [9], *20071434* [11], *20092915* [12]
- **farm forestry** *20090761* [1], *20093286* [2], *20092201* [4], *20090806* [9], *20071436* [11], *20092918* [12]
- **fennel** *20090757* [1], *20093302* [2], *20092186* [4], *20090802* [9], *20071430* [11], *20092924* [12]
- **forest** *20071436* [11], *20092918* [12]
- **forest nurseries** *20093298* [2], *20092208* [4], *20082862* [8], *20082923* [11], *20092919* [12]
- **french beans** *20092200* [4], *20080053* [11], *20092908* [12]
- **frise** *20090772* [1], *20093303* [2], *20092187* [4], *20090819* [9], *20071432* [11], *20092921* [12]
- **game cover** *20082806* [1], *20093298* [2], *20092208* [4], *20082923* [11], *20092919* [12]
- **garlic** *20090765* [1], *20093284* [2], *20092203* [4], *20090810* [9], *20071438* [11], *20092914* [12]
- **grass seed crops** *20090766* [1], *20093297* [2], *20092193* [4], *20090811* [9], *20071429* [11], *20092923* [12]
- **hazel nuts** *20090767* [1], *20093271* [2], *20092184* [4], *20090812* [9], *20071442* [11], *20092916* [12]
- **herbs (see appendix 6)** *20090772* [1], *20093303* [2], *20092187* [4], *20090819* [9], *20071432* [11], *20092921* [12]
- **hops** *20093298* [2], *20092208* [4], *20092919* [12]
- **horseradish** *20101134* [2], *20101132* [4], *20093526* [12]
- **lamb's lettuce** *20090772* [1], *20093303* [2], *20092187* [4], *20090819* [9], *20071432* [11], *20092921* [12]
- **leaf brassicas** *20090772* [1], *20093303* [2], *20092921* [12]
- **leeks** *20093284* [2], *20092914* [12]
- **lettuce** *20090772* [1], *20093303* [2], *20092187* [4], *20090819* [9], *20071432* [11], *20092921* [12]
- **lupins** *20090771* [1], *20093295* [2], *20092194* [4], *20090818* [9], *20071437* [11], *20092909* [12]
- **mallow (althaea spp.)** *20090772* [1], *20093303* [2], *20092187* [4], *20090819* [9], *20092921* [12]
- **miscanthus** *20093298* [2], *20092208* [4], *20082923* [11], *20092919* [12]
- **navy beans** *20093285* [2], *20092200* [4], *20080053* [11], *20092908* [12]
- **onion sets** *20090765* [1], *20092203* [4], *20090810* [9]
- **ornamental plant production** *20093298* [2], *20092208* [4], *20082862* [8], *20082923* [11], *20092919* [12]
- **parsley** *20090772* [1], *20093303* [2], *20092913* [12]
- **parsley root** *20090758* [1], *20093290* [2], *20101134* [2], *20092204* [4], *20101132* [4], *20090803* [9], *20071433* [11], *20092911* [12], *20093526* [12]
- **parsnips** *20101134* [2], *20101132* [4], *20093526* [12]
- **protected cress** *20090772* [1], *20093303* [2]
- **protected forest nurseries** *20093298* [2], *20092208* [4], *20082923* [11], *20092919* [12]
- **protected frise** *20090772* [1], *20093303* [2]
- **protected herbs (see appendix 6)** *20090772* [1], *20093303* [2]
- **protected lamb's lettuce** *20090772* [1], *20093303* [2]
- **protected leaf brassicas** *20093303* [2]
- **protected lettuce** *20090772* [1], *20093303* [2]
- **protected radicchio** *20093303* [2]
- **protected salad brassicas** *20090772* [1]
- **protected scarole** *20090772* [1], *20093303* [2]
- **quinces** *20090769* [1], *20093292* [2], *20092205* [4], *20090814* [9], *20071443* [11], *20092917* [12]
- **radicchio** *20090772* [1], *20093303* [2], *20092187* [4], *20090819* [9], *20071432* [11], *20092921* [12]
- **redcurrants** *20090770* [1], *20093289* [2], *20092198* [4], *20090817* [9], *20071444* [11], *20092907* [12]
- **rhubarb** *20093294* [2], *20092206* [4], *20071430* [11], *20092924* [12]
- **runner beans** *20090760* [1], *20093293* [2], *20092195* [4], *20090805* [9], *20071437* [11], *20092909* [12]
- **salad brassicas** *20092187* [4], *20090819* [9], *(for baby leaf production) 20071432* [11]
- **salad onions** *20090765* [1], *20093284* [2], *20092203* [4], *20090810* [9], *20071438* [11], *20092914* [12]
- **scarole** *20090772* [1], *20093303* [2], *20092187* [4], *20090819* [9], *20071432* [11], *20092921* [12]

FOR FULL CONDITIONS OF USE ALWAYS READ THE PRODUCT LABEL

pendimethalin

- **shallots** *20090765* [1], *20093284* [2], *20092203* [4], *20090810* [9], *20071438* [11], *20092914* [12]
- **spring field beans** *20090401* [1], *20093288* [2], *20092202* [4], *20092685* [6], *20092551* [7], *20070876* [8], *20090404* [9], *20071437* [11], *20092909* [12]
- **spring wheat** *20093291* [2], *20092196* [4], *20082947* [11], *20092910* [12]
- **sweetcorn** *20093296* [2], *20092207* [4], *(under covers) 20071435* [11], *20092912* [12]
- **sweetcorn under plastic mulches** *20090083* [1], *20093296* [2], *20092207* [4], *20092912* [12]
- **top fruit** *20093298* [2], *20092208* [4], *20082923* [11], *20092919* [12]
- **vining peas** *20090768* [1], *20093300* [2], *20092190* [4], *20090813* [9], *20071437* [11], *20092909* [12]
- **walnuts** *20090767* [1], *20093271* [2], *20092184* [4], *20090812* [9], *20071442* [11], *20092916* [12]
- **whitecurrants** *20090770* [1], *20093289* [2], *20092198* [4], *20090817* [9], *20071444* [11], *20092907* [12]
- **winter field beans** *20090763* [1], *20093288* [2], *20092202* [4], *20092685* [6], *20092552* [7], *20070876* [8], *20090808* [9], *20071437* [11], *20092909* [12]

Approval information
- Pendimethalin included in Annex I under EC Directive 91/414
- Accepted by BBPA for use on malting barley and hops

Efficacy guidance
- Apply as soon as possible after drilling. Weeds are controlled as they germinate and emerged weeds will not be controlled by use of the product alone
- For effective blackgrass control apply not more than 2 d after final cultivation and before weed seeds germinate
- Best results by application to fine firm, moist, clod-free seedbeds when rain follows treatment. Effectiveness reduced by prolonged dry weather after treatment
- Effectiveness reduced on soils with more than 6% organic matter. Do not use where organic matter exceeds 10%
- Any trash, ash or straw should be incorporated evenly during seedbed preparation
- Do not disturb soil after treatment
- Apply to potatoes as soon as possible after planting and ridging in tank-mix with metribuzin but note that this will restrict the varieties that may be treated
- Always follow WRAG guidelines for preventing and managing herbicide resistant weeds. See Section 5 for more information

Restrictions
- Maximum number of treatments 1 per crop or yr
- Maximum total dose equivalent to one full dose treatment on most crops; 2 full dose treatments on leeks
- May be applied pre-emergence of cereal crops sown before 30 Nov provided seed covered by at least 32 mm soil, or post-emergence to early tillering stage (GS 23)
- Do not undersow treated crops
- Do not use on crops suffering stress due to disease, drought, waterlogging, poor seedbed conditions or chemical treatment or on soils where water may accumulate

Crop-specific information
- Latest use: pre-emergence for spring barley, carrots, lettuce, fodder maize, parsnips, parsley, combining peas, potatoes, onions and leeks; before transplanting for brassicas; before leaf sheaths erect for winter cereals; before bud burst for blackcurrants, gooseberries, cane fruit, hops; before flower trusses emerge for strawberries; 14 d after transplanting for leaf herbs
- Do not use on spring barley after end Mar (mid-Apr in Scotland on some labels) because dry conditions likely. Do not apply to dry seedbeds in spring unless rain imminent
- Apply to combining peas as soon as possible after sowing. Do not spray if plumule less than 13 mm below soil surface
- Apply to potatoes up to 7 d before first shoot emerges
- Apply to drilled crops as soon as possible after drilling but before crop and weed emergence
- Apply in top fruit, bush fruit and hops from autumn to early spring when crop dormant
- In cane fruit apply to weed free soil from autumn to early spring, immediately after planting new crops and after cutting out canes in established crops
- Apply in strawberries from autumn to early spring (not before Oct on newly planted bed). Do not apply pre-planting or during flower initiation period (post-harvest to mid-Sep)

SEE SECTION 3 FOR PRODUCTS ALSO REGISTERED

Pesticide Profiles

- Apply pre-emergence in drilled onions or leeks, not on Sands, Very Light, organic or peaty soils or when heavy rain forecast
- Apply to brassicas after final plant-bed cultivation but before transplanting. Avoid unnecessary soil disturbance after application and take care not to introduce treated soil into the root zone when transplanting. Follow transplanting with specified post-planting treatments - see label
- Do not use on protected crops or in greenhouses

Following crops guidance
- Before ryegrass is drilled after a very dry season plough or cultivate to at least 15 cm. If treated spring crops are to be followed by crops other than cereals, plough or cultivate to at least 15 cm
- In the event of crop failure land must be ploughed or thoroughly cultivated to at least 15 cm. See label for minimum intervals that should elapse between treatment and sowing a range of replacement crops

Environmental safety
- Dangerous for the environment
- Very toxic to aquatic organisms
- Dangerous to fish or other aquatic life. Do not contaminate surface waters or ditches with chemical or used container
- Flammable [1]
- LERAP Category B [1-3, 5, 7-13]

Hazard classification and safety precautions
Hazard Harmful [1, 3, 5-10, 13]; Flammable [1, 7, 9]; Dangerous for the environment [1-13]
Transport code 9
Packaging group III
UN number 3082
Risk phrases R22a, R36, R38 [3, 5, 6, 8, 10, 13]; R22b [1, 7, 9]; R50 [2-6, 8, 10, 11, 13]; R51 [1, 7, 9, 12]; R53a [1-13]; R70 [12]
Operator protection A [2-8, 10-13]; C [3, 5, 7, 8, 10, 13]; H [5, 12]; U02a, U14 [2, 11, 12]; U05a [1-3, 5-13]; U08, U19a [1-5, 7-13]; U11 [3, 5, 6, 8, 10, 13]; U13 [2-5, 8, 10-13]; U20b [2, 3, 5, 8, 10, 11, 13]; U20c [4]
Environmental protection E15a [1, 7, 9]; E15b [2, 3, 5, 8, 10-13]; E16a [1-3, 5, 7-13]; E34 [1, 3, 5, 7-10, 13]; E38 [1-13]; E39 [12]
Storage and disposal D01, D12a [1-13]; D02 [1-3, 5-13]; D05 [1, 7, 9, 12]; D08 [2, 3, 5, 8, 10, 11, 13]; D09a, D10b [1-3, 5, 7-13]
Medical advice M03, M05b [1, 3, 5, 7-10, 13]; M05a [2, 4, 6, 11]

358   pendimethalin + picolinafen

A post-emergence broad-spectrum herbicide mixture for winter cereals
HRAC mode of action code: K1 + F1

See also picolinafen

Products

1	Galivor	BASF	320:14.5 g/l	SC	14986
2	Orient	BASF	330:7.5 g/l	SC	12541
3	PicoMax	BASF	320:16 g/l	SC	13456
4	Picona	BASF	320:16 g/l	SC	13428
5	PicoPro	BASF	320:16 g/l	SC	13454
6	PicoStomp	BASF	320:16 g/l	SC	13455
7	Sienna	BASF	300:7.5 g/l	SC	14991

Uses
- Annual dicotyledons in *forest nurseries* (off-label) [3, 6]; *game cover* (off-label) [4]; *winter barley, winter wheat* [1-7]
- Annual grasses in *forest nurseries* (off-label) [3, 6]; *game cover* (off-label) [4]
- Annual meadow grass in *winter barley, winter wheat* [1-7]
- Chickweed in *winter barley, winter wheat* [1-7]
- Cleavers in *winter barley, winter wheat* [1, 3-7]
- Field speedwell in *winter barley, winter wheat* [1-7]

FOR FULL CONDITIONS OF USE ALWAYS READ THE PRODUCT LABEL

pendimethalin + picolinafen

- Ivy-leaved speedwell in **winter barley**, **winter wheat** [1-7]
- Rough meadow grass in **winter barley**, **winter wheat** [1-7]

Specific Off-Label Approvals (SOLAs)
- **forest nurseries** *20082863* [3], *20082865* [6]
- **game cover** *20082864* [4]

Approval information
- Pendimethalin and picolinafen included in Annex I under EC Directive 91/414

Efficacy guidance
- Best results obtained on crops growing in a fine, firm tilth and when rain falls within 7 d of application
- Loose or cloddy seed beds must be consolidated prior to application otherwise reduced weed control may occur
- Residual weed control may be reduced on soils with more than 6% organic matter or under prolonged dry conditions
- Always follow WRAG guidelines for preventing and managing herbicide resistant weeds. See Section 5 for more information

Restrictions
- Maximum number of treatments 1 per crop
- Do not treat undersown cereals or those to be undersown
- Do not roll emerged crops before treatment nor autumn treated crops until the following spring
- Do not use on stony or gravelly soils, on soils that are waterlogged or prone to waterlogging, or on soils with more than 10% organic matter
- Do not treat crops under stress from any cause
- Do not apply pre-emergence to crops drilled after 30 Nov [2]
- Consult processor before treating crops for processing [2]

Crop-specific information
- Latest use: before pseudo-stem erect stage (GS 30)
- Transient bleaching may occur after treatment but it does not lead to yield loss
- Do not use pre-emergence on crops drilled after 30th November [3]
- For pre-emergence applications, seed should be covered with a minimum of 3.2 cm settled soil [4-6]

Following crops guidance
- In the event of failure of a treated crop plough to at least 15 cm to ensure any residues are evenly dispersed
- In the event of failure of a treated crop allow at least 8 wk from the time of treatment before re-drilling either spring wheat or spring barley. After a normally harvested crop there are no restrictions on following crops apart from rye-grass, as indicated below [3-6]
- In a very dry season, plough or cultivate to at least 15 cm before drilling rye-grass
- A minimum of 5 mth must follow autumn applications before sowing spring barley, spring wheat, maize, peas, potatoes, spring field beans, dwarf beans, turnips, Brussels sprouts, cabbage, calabrese, linseed, carrots, parsnips, parsley, cauliflower [2]
- After spring or summer applications a minimum of 2 mth must follow before sowing spring field beans, dwarf beans, broad beans, peas, turnips, Brussels sprouts, cabbage, calabrese, cauliflower, carrots, parsnips, parsley, linseed and any crop may be sown after 5 mth except red beet, sugar beet and spinach, for which an interval of 12 mth must be allowed [2]

Environmental safety
- Dangerous for the environment
- Very toxic to aquatic organisms
- Product binds strongly to soil minimising likelihood of movement into groundwater
- LERAP Category B

Hazard classification and safety precautions
Hazard Dangerous for the environment
Transport code 9
Packaging group III
UN number 3082
Risk phrases R50, R53a
Operator protection A, H [2, 3, 7]; U02a, U05a, U20c

SEE SECTION 3 FOR PRODUCTS ALSO REGISTERED

Environmental protection E15a [2]; E15b [1, 3-7]; E16a, E34, E38 [1-7]
Storage and disposal D01, D02, D05, D09a, D10c, D12a
Medical advice M03

359 pendimethalin + pyroxsulam

A dinitroaniline + triazolopyrimidine herbicide mixture for use in winter wheat, rye and triticale
HRAC mode of action code: K1 + B

Products
Broadway Sunrise   Dow   314:5.4 g/l   OD   14960

Uses
- Annual dicotyledons in *triticale*, *winter rye*, *winter wheat*
- Blackgrass in *triticale* (MS only), *winter rye* (MS only), *winter wheat* (MS only)
- Ryegrass in *triticale* (from seed), *winter rye* (from seed), *winter wheat* (from seed)
- Sterile brome in *triticale*, *winter rye*, *winter wheat*
- Wild oats in *triticale*, *winter rye*, *winter wheat*

Approval information
- Pendimethalin included in Annex I under EC Directive 91/414 while pyroxsulam is currently pending inclusion.

Efficacy guidance
- To avoid the build-up of resistance, do not apply this or any other product containing an ALS inhibitor with claims for control of grass weeds more than once to any crop.

Following crops guidance
- In the event of crop failure, sow only spring barley, spring wheat or maize at least 5 months after application and cultivate soil to at least 15 cm.

Environmental safety
- LERAP Category B

Hazard classification and safety precautions
Hazard Irritant, Dangerous for the environment
Transport code 9
Packaging group III
UN number 3082
Risk phrases R38, R43, R50, R53a
Operator protection A, H; U05a, U14, U15, U20b
Environmental protection E15b, E16a, E34, E38
Storage and disposal D01, D02, D05, D10c, D12a

360 pendimethalin + terbuthylazine

A residual dinitroaniline and triazine herbicide mixture for weed control in peas, beans, potatoes and forage maize
HRAC mode of action code: K1 + C1

Products
Bullet XL   Makhteshim   64:270 g/l   SE   14469

Uses
- Annual dicotyledons in *forage maize*
- Annual grasses in *forage maize*

Approval information
- Pendimethalin included in Annex 1 under EC Directive 91/414 but terbuthylazine has not yet been included.

Environmental safety
- LERAP Category B

Hazard classification and safety precautions
Hazard Irritant, Dangerous for the environment

FOR FULL CONDITIONS OF USE ALWAYS READ THE PRODUCT LABEL

peroxyacetic acid (commodity substance)

Transport code 9
Packaging group III
UN number 3082
Risk phrases R41, R50, R53a
Operator protection A, C, H; U05a, U11, U19a, U20b
Environmental protection E15b, E16a, E38
Storage and disposal D01, D02, D05, D09a, D11a, D12a
Medical advice M05a

361 peroxyacetic acid (commodity substance)

An inorganic fungicide for use on flower bulbs and potato tubers for which approval expired 31 Dec 2010

362 petroleum oil

An insecticidal and acaricidal hydrocarbon oil

Products
  Croptex Spraying Oil   Certis   710 g/l   EC   -

Uses
- Mealybugs in **bush fruit, cane fruit, fruit trees, hops, nursery stock, protected cucumbers, protected pot plants, protected tomatoes, wine grapes**
- Red spider mites in **bush fruit, cane fruit, fruit trees, hops, nursery stock, protected cucumbers, protected pot plants, protected tomatoes, wine grapes**
- Scale insects in **bush fruit, cane fruit, fruit trees, hops, nursery stock, protected cucumbers, protected pot plants, protected tomatoes, wine grapes**

Approval information
- Product not controlled by Control of Pesticides Regulations because it acts by physical means only

Efficacy guidance
- Spray at 1% (0.5% on tender foliage) to wet plants thoroughly, particularly the underside of leaves, and repeat as necessary
- Petroleum oil acts by blocking pest breathing pores and making leaf surfaces inhospitable to pests seeking to attack or attach to the leaf surface
- On outdoor crops apply when dormant or on plants with a known tolerance
- On plants of unknown sensitivity test first on a small scale
- Mixtures with certain pesticides may damage crop plants. If mixing, spray a few plants to test for tolerance before treating larger areas

Restrictions
- Do not mix with sulphur or iprodione products or use such mixtures within 28 d of treatment
- Do not treat protected crops in bright sunshine unless the glass is well shaded
- Test species tolerance before large scale treatment

Crop-specific information
- HI zero
- Treat grapevines before flowering

Hazard classification and safety precautions
  Hazard Harmful
  Risk phrases R22a, R36, R37, R38
  Operator protection A, C; U05a, U08, U11, U19a
  Environmental protection E34
  Storage and disposal D01, D02, D09a, D10a, D12a
  Medical advice M03

SEE SECTION 3 FOR PRODUCTS ALSO REGISTERED

363 phenmedipham

A contact phenyl carbamate herbicide for beet crops and strawberries
HRAC mode of action code: C1

See also desmedipham + ethofumesate + lenacil + phenmedipham
desmedipham + ethofumesate + metamitron + phenmedipham
desmedipham + ethofumesate + phenmedipham
desmedipham + phenmedipham
ethofumesate + metamitron + phenmedipham
ethofumesate + phenmedipham

Products

1	Alpha Phenmedipham 320 SC	Makhteshim	320 g/l	SC	14070
2	Betamax	AgriGuard	160 g/l	SC	14371
3	Betasana SC	United Phosphorus	160 g/l	SC	14209
4	Corzal	AgriChem BV	157 g/l	SE	14192
5	Dancer Flow	Sipcam	160 g/l	SE	14395
6	Herbasan Flow	Nufarm UK	160 g/l	SE	13894
7	Mandolin Flow	Nufarm UK	160 g/l	SE	13895
8	Shrapnel	Makhteshim	320 g/l	SC	14088

Uses
- Annual dicotyledons in **fodder beet**, **mangels**, **red beet**, **sugar beet** [1-8]; **herbs (see appendix 6)** *(off-label)* [4]; **strawberries** [1, 8]

Specific Off-Label Approvals (SOLAs)
- **herbs (see appendix 6)** *20101731* [4]

Approval information
- Phenmedipham included in Annex I under EC Directive 91/414

Efficacy guidance
- Best results achieved by application to young seedling weeds, preferably cotyledon stage, under good growing conditions when low doses are effective
- If using a low-dose programme 2-3 repeat applications at 7-10 d intervals are recommended on mineral soils, 3-5 applications may be needed on organic soils
- Addition of adjuvant oil may improve effectiveness on some weeds
- Various tank-mixtures with other beet herbicides recommended. See label for details
- Use of certain pre-emergence herbicides is recommended in combination with post-emergence treatment. See label for details

Restrictions
- Maximum number of treatments and maximum total dose varies with crop and product used. See label for details
- At high temperatures (above 21 °C) reduce rate and spray after 5 pm
- Do not apply immediately after frost or if frost expected
- Do not spray wet foliage or if rain imminent
- Do not spray crops stressed by wind damage, nutrient deficiency, pest or disease attack etc. Do not roll or harrow for 7 d before or after treatment
- Do not use on strawberries under cloches or polythene tunnels
- Do not use after 31 Jul in yr of harvest [6, 7]
- Consult processor before use on crops for processing

Crop-specific information
- Latest use: before crop leaves meet between rows for beet crops; before flowering for strawberries
- Apply to beet crops at any stage as low dose/low volume spray or from fully developed cotyledon stage with full rate. Apply to red beet after fully developed cotyledon stage
- Apply to strawberries at any time when weeds in susceptible stage, except in period from start of flowering to picking

Following crops guidance
- Beet crops may follow at any time after a treated crop. 3 mth must elapse from treatment before any other crop is sown and must be preceded by mould-board ploughing to 15 cm

FOR FULL CONDITIONS OF USE ALWAYS READ THE PRODUCT LABEL

d-phenothrin

Environmental safety
- Dangerous for the environment
- Very toxic to aquatic organisms
- Harmful to fish or other aquatic life. Do not contaminate surface waters or ditches with chemical or used container
- LERAP Category B [4, 6, 7]

Hazard classification and safety precautions
  Hazard Irritant [1-3, 8]; Dangerous for the environment [1-8]
  Transport code 9
  Packaging group III
  UN number 3082
  Risk phrases R36, R51 [2, 3]; R38 [1, 8]; R43 [1-3, 8]; R50 [1, 4-8]; R53a [1-8]
  Operator protection A [1-5, 8]; C [2, 3]; H [1-3, 8]; U05a [1, 2, 4, 8]; U08 [1-4, 8]; U11, U20c [3]; U14, U19a [1, 3, 8]; U20a [1, 6-8]; U20b [2, 5]
  Environmental protection E13c [5]; E15a [1, 2, 6-8]; E15b [3, 4]; E16a [4, 6, 7]; E16b [4]; E19b [1, 8]; E34 [1, 4, 8]; E38 [1, 2, 4-8]
  Storage and disposal D01 [1-4, 8]; D02 [2-4]; D05 [1-3, 7, 8]; D08 [5]; D09a, D12a [1-8]; D10b [1, 4-8]; D10c [2, 3]
  Medical advice M05a [1, 8]

364   d-phenothrin

A non-systemic pyrethroid insecticide available only in mixtures
IRAC mode of action code: 3A

365   d-phenothrin + tetramethrin

A pyrethroid insecticide mixture for control of flying insects
IRAC mode of action code: 3 + 3

See also tetramethrin

Products
Killgerm ULV 500   Killgerm   36.8:18.4 g/l   UL   H4647

Uses
- Flies in *agricultural premises*
- Grain storage mite in *grain stores*
- Mosquitoes in *agricultural premises*
- Wasps in *agricultural premises*

Approval information
- Product approved for ULV application. See label for details
- D-phenothrin and tetramethrin are not yet inncluded in Annex 1 under EC Directive 91/414

Efficacy guidance
- Close doors and windows and spray in all directions for 3-5 sec. Keep room closed for at least 10 min

Restrictions
- For use only by professional operators
- Do not use space sprays containing pyrethrins or pyrethroid more than once per week in intensive or controlled environment animal houses in order to avoid development of resistance. If necessary, use a different control method or product

Crop-specific information
- May be used in the presence of poultry and livestock

Environmental safety
- Dangerous for the environment
- Toxic to aquatic organisms
- Do not apply directly to livestock/poultry

SEE SECTION 3 FOR PRODUCTS ALSO REGISTERED

Pesticide Profiles

- Remove exposed milk and collect eggs before application. Protect milk machinery and containers from contamination

Hazard classification and safety precautions
Hazard Harmful, Dangerous for the environment
Transport code 9
Packaging group III
UN number 3082
Risk phrases R22b, R51, R53a
Operator protection A, C, D, E, H; U02b, U09b, U14, U19a, U20a, U20b
Environmental protection E05a, E15a, E38
Consumer protection C06, C07, C08, C09, C11, C12
Storage and disposal D01, D05, D06a, D09a, D10a, D12a
Medical advice M05b

366 physical pest control

Products that work by physical action only

Products

1	SB Plant Invigorator	Fargro	-	SL	-
2	Silico-Sec	Interfarm	>90% w/w	DS	-

Uses
- Aphids in *all edible crops (outdoor and protected), ornamental plant production, protected ornamentals* [1]
- Mealybugs in *all edible crops (outdoor and protected), ornamental plant production, protected ornamentals* [1]
- Mites in *stored grain* [2]
- Powdery mildew in *all edible crops (outdoor and protected), ornamental plant production, protected ornamentals* [1]
- Scale insects in *all edible crops (outdoor and protected), ornamental plant production, protected ornamentals* [1]
- Spider mites in *all edible crops (outdoor and protected), ornamental plant production, protected ornamentals* [1]
- Suckers in *all edible crops (outdoor and protected), ornamental plant production, protected ornamentals* [1]
- Whitefly in *all edible crops (outdoor and protected), ornamental plant production, protected ornamentals* [1]

Approval information
- Products included in this profile are not subject to the Control of Pesticides Regulations/Plant Protection Products Regulations because they act by physical means only

Efficacy guidance
- Products act by physical means following direct contact with spray. They may therefore be used at any time of year on all pest growth stages
- Ensure thorough spray coverage of plant, paying special attention to growing points and the underside of leaves
- Treat as soon as target pests are seen and repeat as often as necessary

Restrictions
- No limit on the number of treatments and no minimum interval between applications
- Before large scale use on a new crop, treat a few plants to check for crop safety
- Do not treat ornamental crops when in flower

Crop-specific information
- HI zero

Environmental safety
- May be used in conjunction with biological control agents. Spray 24 h before they are introduced

FOR FULL CONDITIONS OF USE ALWAYS READ THE PRODUCT LABEL

picloram

Hazard classification and safety precautions
  UN number N/C
  Operator protection F [2]; U10, U14, U15 [1]
  Storage and disposal D01, D02, D05, D06d [1]

367 picloram

A persistent, translocated pyridine carboxylic acid herbicide for non-crop areas
HRAC mode of action code: O

See also 2,4-D + picloram
  clopyralid + picloram

Products

1	Pantheon 2	Pan Agriculture	240 g/l	SC	14052
2	Tordon 22K	Dow	240 g/l	SC	05083

Uses
  • Annual and perennial weeds in *land not intended for cropping* [1]
  • Annual dicotyledons in *land not intended for cropping* [2]
  • Bracken in *land not intended for cropping* [1, 2]
  • Brambles in *land not intended for cropping* [1]
  • Japanese knotweed in *land not intended for cropping* [2]
  • Perennial dicotyledons in *land not intended for cropping* [2]
  • Woody weeds in *land not intended for cropping* [2]

Approval information
  • Picloram included in Annex I under EC Directive 91/414

Efficacy guidance
  • May be applied at any time of year. Best results achieved by application as foliage spray in late winter to early spring
  • For bracken control apply 2-4 wk before frond emergence
  • Clovers are highly sensitive and eliminated at very low doses
  • Persists in soil for up to 2 yr

Restrictions
  • Maximum number of treatments 1 per yr on land not intended for cropping
  • Do not apply around desirable trees or shrubs where roots may absorb chemical
  • Do not apply on slopes where chemical may be leached onto areas of desirable plants

Environmental safety
  • Harmful to fish or other aquatic life. Do not contaminate surface waters or ditches with chemical or used container
  • Keep livestock out of treated areas for at least 2 wk and until foliage of any poisonous weeds such as ragwort has died and become unpalatable

Hazard classification and safety precautions
  Hazard Irritant
  Transport code 9
  Packaging group III
  UN number 3082
  Risk phrases R43
  Operator protection A, H; U05a, U08, U14, U20b
  Environmental protection E07a, E13c
  Storage and disposal D01, D02, D09a [1, 2]; D10a [1]; D10b [2]

SEE SECTION 3 FOR PRODUCTS ALSO REGISTERED

368 picolinafen

A pyridinecarboxamide herbicide for cereals;
HRAC mode of action code: F1

*See also flupyrsulfuron-methyl + picolinafen
           pendimethalin + picolinafen*

Products
　　Vixen　　　　　　　　　　BASF　　　　　　　　　75% w/w　　　　　　　WG　13621

Uses
- Chickweed in **winter barley, winter wheat**
- Cleavers in **winter barley, winter wheat**
- Field pansy in **winter barley, winter wheat**
- Field speedwell in **winter barley, winter wheat**
- Ivy-leaved speedwell in **winter barley, winter wheat**
- Shepherd's purse in **winter barley, winter wheat**

Approval information
- Picolinafen included in Annex I under EC Directive 91/414

Efficacy guidance
- Picolinafen is contact acting. Ensure good spray cover of the target weeds
- Best results obtained from treatment of small actively growing weeds
- Weeds emerging after application will not be controlled
- Always follow WRAG guidelines for preventing and managing herbicide resistant weeds. See Section 5 for more information

Restrictions
- Maximum number of treatments 1 per crop
- Do not disturb soil surface after application
- Do not roll emerged crops prior to application
- Do not treat undersown crops or those to be undersown
- Do not spray during periods of prolonged or severe frost
- Do not apply to crops under stress from any cause

Crop-specific information
- Latest use: before pseudo-stem erect stage (GS 30) for wheat and barley
- Treatment may cause transient bleaching in some crops

Following crops guidance
- No restrictions on following crops after normal harvest of a treated crop
- In the event of failure of a treated crop further crops of wheat or barley may be drilled after a minimum interval of 8 wk from treatment and after ploughing to at least 15 cm

Environmental safety
- Dangerous for the environment
- Very toxic to aquatic organisms
- Take extreme care to avoid drift onto plants outside the target area
- LERAP Category B

Hazard classification and safety precautions
　Hazard Dangerous for the environment
　Transport code 9
　Packaging group III
　UN number 3077
　Risk phrases R50, R53a
　Operator protection A, D; U05a, U08, U13, U19a, U20b
　Environmental protection E15a, E16a, E38
　Storage and disposal D01, D02, D09a, D10b, D12a

FOR FULL CONDITIONS OF USE ALWAYS READ THE PRODUCT LABEL

369 picoxystrobin

A broad spectrum strobilurin fungicide for cereals
FRAC mode of action code: 11

See also chlorothalonil + picoxystrobin
cyproconazole + picoxystrobin
cyprodinil + picoxystrobin

Products

1	Flanker	DuPont	250 g/l	SC	14760
2	Galileo	DuPont	250 g/l	SC	13252

Uses
- Brown rust in **spring barley, spring wheat, winter barley, winter wheat** [1, 2]
- Crown rust in **spring oats, winter oats** [1, 2]
- Eyespot in **spring wheat** *(reduction)*, **winter wheat** *(reduction)* [1, 2]
- Foliar disease control in **ornamental plant production** *(off-label)* [2]
- Glume blotch in **spring wheat, winter wheat** [1, 2]
- Late ear diseases in **spring wheat, winter wheat** [1, 2]
- Net blotch in **spring barley, winter barley** [1, 2]
- Powdery mildew in **spring barley, spring oats, winter barley, winter oats** [1, 2]
- Rhynchosporium in **spring barley, winter barley** [1, 2]
- Sclerotinia in **spring oilseed rape, winter oilseed rape** [1, 2]
- Septoria leaf blotch in **spring wheat, winter wheat** [1, 2]
- Yellow rust in **spring wheat, winter wheat** [1, 2]

Specific Off-Label Approvals (SOLAs)
- **ornamental plant production** 20082855 [2]

Approval information
- Picoxystrobin included in Annex I under EC Directive 91/414
- Accepted by BBPA for use on malting barley

Efficacy guidance
- Best results achieved from protectant treatments made before disease establishes in the crop
- Use of tank mixtures is recommended where disease has become established at the time of treatment
- Persistence of yellow rust control is less than for other foliar diseases but can be improved by use of an appropriate tank mixture
- Eyespot control is not reliable and an appropriate tank mix should be used where crops are at risk
- Picoxystrobin is a member of the QoI cross resistance group. Product should be used preventatively and not relied on for its curative potential
- Use product as part of an Integrated Crop Management strategy incorporating other methods of control, including where appropriate other fungicides with a different mode of action. Do not apply more than two foliar applications of QoI containing products to any cereal crop
- There is a significant risk of widespread resistance occurring in *Septoria tritici* populations in UK. Failure to follow resistance management action may result in reduced levels of disease control
- On cereal crops product must always be used in mixture with another product, recommended for control of the same target disease, that contains a fungicide from a different cross resistance group and is applied at a dose that will give robust control
- Strains of barley powdery mildew resistant to QoIs are common in the UK

Restrictions
- Maximum number of treatments 2 per crop per yr

Crop-specific information
- Latest use: grain watery ripe (GS 71)

Environmental safety
- Dangerous for the environment
- Very toxic to aquatic organisms

Hazard classification and safety precautions
Hazard Dangerous for the environment

SEE SECTION 3 FOR PRODUCTS ALSO REGISTERED

Pesticide Profiles

Transport code 9
Packaging group III
UN number 3082
Risk phrases R50, R53a
Operator protection A; U05a, U09a, U14, U15, U20a
Environmental protection E15a, E38
Storage and disposal D01, D02, D05, D09a, D10c, D12a

370   pinoxaden

A phenylpyrazoline grass weed herbicide for cereals
HRAC mode of action code: A

See also clodinafop-propargyl + pinoxaden

Products

1	Aoraki	AgChem Access	100 g/l	EC	13528
2	Axial	Syngenta	100 g/l	EC	12521
3	Clayton Tonto	Clayton	100 g/l	EC	14436
4	Greencrop Helvick	Greencrop	100 g/l	EC	13052
5	Liberate	ChemSource	45 g/l	EC	14872
6	Quail	ChemSource	100 g/l	EC	14506
7	Rescue	Syngenta	45 g/l	EC	14518

Uses
- Annual grasses in **game cover** *(off-label)* [2]
- Blackgrass in **spring barley, winter barley** [1-3, 6]
- Italian ryegrass in **spring barley, winter barley, winter wheat** [1-4, 6]; **spring wheat** [1, 2, 6]
- Perennial grasses in **game cover** *(off-label)* [2]
- Perennial ryegrass in **amenity grassland** *(off-label)*, **managed amenity turf** *(off-label)* [7]; **spring barley, winter barley, winter wheat** [1-4, 6]; **spring wheat** [1, 2, 6]
- Ryegrass in **amenity grassland** *(reduction only)*, **managed amenity turf** *(reduction only)* [5, 7]
- Wild oats in **spring barley, winter barley, winter wheat** [1-4, 6]; **spring wheat** [1, 2, 6]

Specific Off-Label Approvals (SOLAs)
- *amenity grassland* 20101705 [7]
- *game cover* 20082815 [2]
- *managed amenity turf* 20101705 [7]

Approval information
- Approval expiry 19 Sep 2011 [1, 3, 6]

Efficacy guidance
- Best results obtained from treatment when all grass weeds have emerged. There is no residual activity
- Broad-leaved weeds are not controlled
- Treat before emerged weed competition reduces yield
- Blackgrass in winter and spring barley is also controlled when used as part of an integrated control strategy. Product is not recommended for blackgrass control in winter wheat
- Grass weed control may be reduced if rain falls within 1 hr of application
- Pinoxaden is an ACCase inhibitor herbicide. To avoid the build up of resistance do not apply products containing an ACCase inhibitor herbicide more than twice to any crop. In addition do not use any product containing pinoxaden in mixture or sequence with any other product containing the same ingredient
- Use these products as part of a resistance management strategy that includes cultural methods of control and does not use ACCase inhibitors as the sole chemical method of grass weed control
- Applying a second product containing an ACCase inhibitor to a crop will increase the risk of resistance development; only use a second ACCase inhibitor to control different weeds at a different timing
- Always follow WRAG guidelines for preventing and managing herbicide resistant weeds. See Section 5 for more information

Restrictions
- Maximum number of treatments 1 per crop

FOR FULL CONDITIONS OF USE ALWAYS READ THE PRODUCT LABEL

pirimicarb

- Do not spray crops under stress or suffering from waterlogging, pest attack, disease or frost damage
- Do not spray crops undersown with grass mixtures
- Avoid the use of hormone-containing herbicides in mixture or in sequence. Allow 21 d after, or 7 d before, hormone application
- Product must be used with prescribed adjuvant. See label

Crop-specific information
- Latest use: before flag leaf extending stage (GS 41)
- Spray cereals in autumn, winter or spring from the 2-leaf stage (GS 12)

Following crops guidance
- There are no restrictions on succeeding crops in a normal rotation
- In the event of failure of a treated crop ryegrass, maize, oats or any broad-leaved crop may be planted after a minimum interval of 4 wk from application

Environmental safety
- Dangerous for the environment
- Toxic to aquatic organisms
- Do not allow spray to drift onto neighbouring crops of oats, ryegrass or maize

Hazard classification and safety precautions
  Hazard Harmful [5, 7]; Irritant [1-4, 6]; Dangerous for the environment [1-7]
  Transport code 9
  Packaging group III
  UN number 3082
  Risk phrases R22b, R43 [5, 7]; R36 [1-4, 6]; R38, R51, R53a [1-7]
  Operator protection A, H [1-7]; C [1-4, 6]; U05a, U20b [1-7]; U09a [1-4, 6]
  Environmental protection E15b, E38
  Storage and disposal D01, D02, D05, D09a, D10c, D12a [1-7]; D11a [1-4, 6]
  Medical advice M05b [5, 7]

371   pirimicarb

A carbamate insecticide for aphid control
IRAC mode of action code: 1A

See also lambda-cyhalothrin + pirimicarb

Products

1	Aphox	Syngenta	50% w/w	WG	10515
2	Arena	AgriGuard	50% w/w	WG	14028
3	Clayton Pirimicarb 50	Clayton	50% w/w	WG	12910
4	Phantom	Syngenta	50% w/w	WG	11954
5	Pirimate	European Ag	50% w/w	WG	14539
6	Pirimicarb 50	Goldengrass	50% w/w	WG	14295
7	Reynard	ChemSource	50% w/w	WG	14523
8	Standon Pirimicarb 50	Standon	50% w/w	WG	13290

Uses
- Aphids in **apples, blackcurrants, cherries, chinese cabbage, collards, cucumbers, durum wheat, forest nurseries, gooseberries, ornamental plant production, pears, peppers, protected lettuce, raspberries, redcurrants, spring rye, strawberries, tomatoes (outdoor), triticale, winter rye** [1-4, 6-8]; **aubergines** (off-label), **bilberries** (off-label), **blueberries** (off-label), **cayenne pepper** (off-label), **cranberries** (off-label), **evening primrose** (off-label), **linseed** (off-label), **mallow (althaea spp.)** (off-label), **mustard** (off-label), **protected lamb's lettuce** (off-label), **protected leaf brassicas** (off-label), **salad brassicas** (for baby leaf production) [1]; **blackberries** (off-label), **celeriac** (off-label), **celery (outdoor)** (off-label), **chicory root** (off-label), **choi sum** (off-label), **courgettes** (off-label), **endives** (off-label), **fennel** (off-label), **fodder beet** (off-label), **frise** (off-label), **gherkins** (off-label), **herbs (see appendix 6)** (off-label), **honesty** (off-label), **horseradish** (off-label), **leaf brassicas** (off-label), **marrows** (off-label), **pak choi** (off-label), **parsley** (off-label), **parsley root** (off-label), **plums** (off-label), **poppies for morphine production** (off-label), **protected courgettes** (off-label), **protected endives** (off-label), **protected frise** (off-label), **protected gherkins** (off-label), **protected herbs (see appendix 6)** (off-label),

SEE SECTION 3 FOR PRODUCTS ALSO REGISTERED

Pesticide Profiles

protected horseradish (off-label), *protected parsley* (off-label), *protected radicchio* (off-label), *protected radishes* (off-label), *protected scarole* (off-label), *protected spinach* (off-label), *protected spinach beet* (off-label), *radicchio* (off-label), *radishes* (off-label), *red beet* (off-label), *rhubarb* (off-label), *rubus hybrids* (off-label), *scarole* (off-label), *spinach* (off-label), *spinach beet* (off-label), *sweetcorn* (off-label) [1, 4]; *broad beans, broccoli, brussels sprouts, cabbages, calabrese, carrots, cauliflowers, dwarf beans, forage maize, kale, parsnips, peas, potatoes, spring barley, spring field beans, spring oats, spring oilseed rape, spring wheat, sugar beet, swedes, sweetcorn, turnips, winter barley, winter field beans, winter oats, winter oilseed rape, winter wheat* [1-8]; *chicory* (off-label - for forcing), *lamb's lettuce* (off-label), *leaf brassicas* (off-label - baby leaf production only), *leaf spinach* (off-label), *protected aubergines* (off-label), *protected cayenne peppers* (off-label), *protected leaf spinach* (off-label), *protected parsley root* (off-label), *salad brassicas* (off-label - for baby leaf production), *salsify* (off-label) [4]; *lettuce* [1, 2, 6-8]; *lettuce* (outdoor crops) [3, 4]; *protected carnations, protected chrysanthemums, protected cinerarias, protected cyclamen, protected roses, protected tomatoes* [1, 3, 4, 6-8]; *protected ornamentals* [2]; *runner beans* [1-3, 5-8]
- Leaf curling plum aphid in *plums* (off-label) [1]
- Mealy plum aphid in *plums* (off-label) [1]
- Thrips in *endives* (off-label), *parsley* (off-label), *protected endives* (off-label), *protected parsley* (off-label) [4]

Specific Off-Label Approvals (SOLAs)
- *aubergines* 20102315 [1]
- *bilberries* 20102319 [1]
- *blackberries* 20102316 [1], 20081186 [4], 20102354 [4]
- *blueberries* 20102319 [1]
- *cayenne pepper* 20102315 [1]
- *celeriac* 20102314 [1], 20051747 [4], 20102351 [4]
- *celery (outdoor)* 20102320 [1], 20051744 [4], 20102347 [4]
- *chicory* (for forcing) 20051749 [4]
- *chicory root* 20102313 [1], 20102349 [4]
- *choi sum* 20102317 [1], 20081190 [4], 20102361 [4]
- *courgettes* 20102328 [1], 20051746 [4], 20102350 [4]
- *cranberries* 20102319 [1]
- *endives* 20102312 [1], 20051741 [4], 20102343 [4]
- *evening primrose* 20102325 [1]
- *fennel* 20102314 [1], 20051747 [4], 20102351 [4]
- *fodder beet* 20102323 [1], 20081187 [4], 20102356 [4]
- *frise* 20102321 [1], 20081189 [4], 20102357 [4]
- *gherkins* 20102328 [1], 20051746 [4], 20102350 [4]
- *herbs (see appendix 6)* 20102321 [1], 20081189 [4], 20102350 [4]
- *honesty* 20102325 [1], 20102326 [1], 20051742 [4], 20102345 [4]
- *horseradish* 20102324 [1], 20102328 [1], 20081191 [4], 20102350 [4], 20102358 [4]
- *lamb's lettuce* 20081189 [4], 20102357 [4]
- *leaf brassicas* 20102311 [1], (baby leaf production only) 20051743 [4], 20102346 [4], 20102357 [4]
- *leaf spinach* 20051746 [4]
- *linseed* 20102325 [1]
- *mallow (althaea spp.)* 20102324 [1]
- *marrows* 20102328 [1], 20051746 [4], 20102350 [4]
- *mustard* 20102325 [1]
- *pak choi* 20102317 [1], 20081190 [4], 20102361 [4]
- *parsley* 20102312 [1], 20102343 [4]
- *parsley root* 20102328 [1], 20051746 [4], 20102350 [4]
- *plums* 20102318 [1], 20090903 [4], 20102355 [4]
- *poppies for morphine production* 20102322 [1], 20051748 [4], 20102353 [4]
- *protected aubergines* 20081188 [4], 20102360 [4]
- *protected cayenne peppers* 20081188 [4], 20102360 [4]
- *protected courgettes* 20102327 [1], 20051745 [4], 20102348 [4]
- *protected endives* 20102312 [1], 20051741 [4], 20102343 [4]
- *protected frise* 20102321 [1], 20081189 [4], 20102357 [4]
- *protected gherkins* 20102327 [1], 20051745 [4], 20102348 [4]

FOR FULL CONDITIONS OF USE ALWAYS READ THE PRODUCT LABEL

- *protected herbs (see appendix 6)* 20102321 [1], 20102350 [4]
- *protected horseradish* 20102328 [1], 20051746 [4], 20102350 [4]
- *protected lamb's lettuce* 20102321 [1]
- *protected leaf brassicas* 20102321 [1]
- *protected leaf spinach* 20051746 [4]
- *protected parsley* 20102328 [1], 20051741 [4], 20102343 [4]
- *protected parsley root* 20102350 [4]
- *protected radicchio* 20102321 [1], 20081189 [4], 20102357 [4]
- *protected radishes* 20102328 [1], 20051746 [4], 20102350 [4]
- *protected scarole* 20102321 [1], 20081189 [4], 20102357 [4]
- *protected spinach* 20102328 [1], 20102350 [4]
- *protected spinach beet* 20102328 [1], 20051746 [4], 20102350 [4]
- *radicchio* 20102321 [1], 20081189 [4], 20102357 [4]
- *radishes* 20102328 [1], 20051746 [4], 20102350 [4]
- *red beet* 20102314 [1], 20051747 [4], 20102351 [4]
- *rhubarb* 20102320 [1], 20051744 [4], 20102347 [4]
- *rubus hybrids* 20102316 [1], 20081186 [4], 20102354 [4]
- *salad brassicas* (for baby leaf production) 20081189 [4]
- *salsify* 20051747 [4], 20102351 [4]
- *scarole* 20102321 [1], 20081189 [4], 20102357 [4]
- *spinach* 20102328 [1], 20102350 [4]
- *spinach beet* 20102328 [1], 20051746 [4], 20102350 [4]
- *sweetcorn* 20102328 [1], 20102350 [4]

Approval information
- Pirimicarb included in Annex I under EC Directive 91/414
- Approved for aerial application on cereals [1, 4]. See Section 5 for more information
- Accepted by BBPA for use on malting barley
- In 2006 CRD required that all products containing this active ingredient should carry the following warning in the main area of the container label: "Pirimicarb is an anticholinesterase carbamate. Handle with care"

Efficacy guidance
- Chemical has contact, fumigant and translaminar activity
- Best results achieved under warm, calm conditions when plants not wilting and spray does not dry too rapidly. Little vapour activity at temperatures below 15 °C
- Apply as soon as aphids seen or warning issued and repeat as necessary
- Addition of non-ionic wetter recommended for use on brassicas
- On cucumbers and tomatoes a root drench is preferable to spraying when using predators in an integrated control programme
- Where aphids resistant to pirimicarb occur control is unlikely to be satisfactory

Restrictions
- Contains an anticholinesterase carbamate compound. Do not use if under medical advice not to work with such compounds
- Maximum number of treatments not specified in some cases but normally 2-6 depending on crop
- When treating ornamentals check safety by treating a small number of plants first
- Spray equipment must only be used where the operator's normal working position is within a closed cab on a tractor or self-propelled sprayer when making air-assisted applications to apples or other top fruit

Crop-specific information
- Latest use in accordance with harvest intervals below
- HI oilseed rape, cereals, maize, sweetcorn 14 d (sweetcorn off-label 3 d); grassland, chicory, plums, swedes and turnips 7 d; lettuce 3 d; cucumbers, tomatoes and peppers under glass 2 d; protected courgettes and gherkins 24 h; other edible crops 3 d; flowers and ornamentals zero

Environmental safety
- Dangerous for the environment
- Very toxic to aquatic organisms
- Dangerous to fish or other aquatic life. Do not contaminate surface waters or ditches with chemical or used container

SEE SECTION 3 FOR PRODUCTS ALSO REGISTERED

Pesticide Profiles

- Chemical has little effect on bees, ladybirds and other insects and is suitable for use in integrated control programmes on apples and pears
- Keep all livestock out of treated areas for at least 7 d. Bury or remove spillages

Hazard classification and safety precautions
Hazard Toxic, Dangerous for the environment
Transport code 6.1
Packaging group III
UN number 2757
Risk phrases R20, R25, R36, R50, R53a
Operator protection A, C, D, H, M [1-8]; J [1, 3-8]; U05a, U08, U19a [1-8]; U20a [3, 4]; U20b [1, 2, 5-8]
Environmental protection E06c [1-8] (7 d); E15a [3]; E15b [1, 2, 4-8]; E34 [1-8]; E38 [1, 4-8]
Storage and disposal D01, D02, D09a, D11a [1-8]; D05 [2]; D07 [3]; D12a [1, 3-8]
Medical advice M02, M04a [1, 3-8]; M03 [1, 2, 4-8]; M04b [3]

372 pirimiphos-methyl

A contact, fumigant and translaminar organophosphorus insecticide
IRAC mode of action code: 1B

Products

1	Actellic 50 EC	Syngenta	500 g/l	EC	12726
2	Actellic Smoke Generator No 20	Syngenta	22.5% w/w	FU	10540
3	Flycatcher	AgChem Access	500 g/l	EC	14135
4	Pan PMT	Pan Agriculture	500 g/l	EC	13996

Uses
- Flour beetle in **stored grain** [1, 3, 4]
- Flour moth in **stored grain** [1, 3, 4]
- Grain beetle in **stored grain** [1, 3, 4]
- Grain storage mite in **stored grain** [1, 3, 4]
- Grain storage pests in **grain stores** [2]
- Grain weevil in **stored grain** [1, 3, 4]
- Warehouse moth in **stored grain** [1, 3, 4]

Approval information
- Pirimiphos-methyl included in Annex I under EC Directive 91/414
- In 2006 PSD required that all products containing this active ingredient should carry the following warning in the main area of the container label: "Pirimiphos-methyl is an anticholinesterase organophosphate. Handle with care"
- Accepted by BBPA for use in stores for malting barley

Efficacy guidance
- Chemical acts rapidly and has short persistence in plants, but persists for long periods on inert surfaces
- Best results for protection of stored grain achieved by cleaning store thoroughly before use and employing a combination of pre-harvest and grain/seed treatments [1, 2]
- Best results for admixture treatment obtained when grain stored at 15% moisture or less. Dry and cool moist grain coming into store but then treat as soon as possible, ideally as it is loaded [1]
- Surface admixture can be highly effective on localised surface infestations but should not be relied on for long-term control unless application can be made to the full depth of the infestation [1]
- Where insect pests resistant to pirimiphos-methyl occur control is unlikely to be satisfactory

Restrictions
- Contains an organophosphorus anticholinesterase compound. Do not use if under medical advice not to work with such compounds
- Maximum number of treatments 2 per grain store [1, 2]; 1 per batch of stored grain [1]
- Do not apply surface admixture or complete admixture using hand held equipment [1]
- Surface admixture recommended only where the conditions for store preparation, treatment and storage detailed in the label can be met [1]

FOR FULL CONDITIONS OF USE ALWAYS READ THE PRODUCT LABEL

Crop-specific information
- Latest use: before storing grain [1, 2]
- Disinfect empty grain stores by spraying surfaces and/or fumigation and treat grain by full or surface admixture. Treat well before harvest in late spring or early summer and repeat 6 wk later or just before harvest if heavily infested. See label for details of treatment and suitable application machinery
- Treatment volumes on structural surfaces should be adjusted according to surface porosity. See label for guidelines
- Treat inaccessible areas with smoke generating product used in conjunction with spray treatment of remainder of store
- Predatory mites (*Cheyletus*) found in stored grain feeding on infestations of grain mites may survive treatment and can lead to rejection by buyers. They do not remain for long but grain should be inspected carefully before selling [1, 2]

Environmental safety
- Dangerous for the environment
- Very toxic to aquatic organisms [1]
- Toxic to aquatic organisms [2]
- Highly flammable [2]
- Flammable [1]
- Ventilate fumigated or fogged spaces thoroughly before re-entry
- Unprotected persons must be kept out of fumigated areas within 3 h of ignition and for 4 h after treatment
- Wildlife must be excluded from buildings during treatment [1]
- Keep away from combustible materials [2]

Hazard classification and safety precautions
Hazard Harmful, Dangerous for the environment [1-4]; Highly flammable [2]; Flammable [1, 3, 4]
Transport code 3 [1, 3, 4]; 9 [2]
Packaging group III
UN number 1993 [1, 3, 4]; 3077 [2]
Risk phrases R20, R53a [1-4]; R22a, R22b, R36, R37, R50 [1, 3, 4]; R51 [2]
Operator protection A, C, J, M [1, 3, 4]; D, H [1-4]; U04a, U09a, U16a, U24 [1, 3, 4]; U05a, U19a, U20b [1-4]; U14 [2]
Environmental protection E02a [2] (4 h); E15a, E34, E38 [1-4]
Consumer protection C09 [1, 3, 4]; C12 [1-4]
Storage and disposal D01, D02, D09a, D12a [1-4]; D05, D10c [1, 3, 4]; D07, D11a [2]
Treated seed S06a [1, 3, 4]
Medical advice M01 [1-4]; M03 [2]; M05b [1, 3, 4]

373 potassium bicarbonate (commodity substance)

An inorganic fungicide for use in horticulture

Products
potassium bicarbonate	various	100% w/w	AP	00000

Uses
- Disease control in **all edible crops (outdoor and protected)**, **all non-edible crops (outdoor)**, **all protected non-edible crops**

Approval information
- Approval for the use of potassium bicarbonate as a commodity substance was granted on 26 July 2005 by Ministers under regulation 5 of the Control of Pesticides Regulations 1986
- Potassium bicarbonate is being supported for review in the fourth stage of the EC Review Programme under Directive 91/414. Whether or not it is included in Annex I, the existing commodity chemical approval will have to be revoked because substances listed in Annex I must be approved for marketing and use under the EC regime. If potassium bicarbonate is included in Annex I, products containing it will need to gain approval in the normal way if they carry label claims for pesticidal activity

SEE SECTION 3 FOR PRODUCTS ALSO REGISTERED

Pesticide Profiles

Efficacy guidance
- Adjuvants authorised by PSD may be used in conjunction with potassium bicarbonate
- Crop phytotoxicity can occur

Restrictions
- Food grade potassium bicarbonate must only be used

374 prochloraz

A broad-spectrum protectant and eradicant imidazole fungicide
FRAC mode of action code: 3

See also carbendazim + prochloraz
epoxiconazole + prochloraz
fenpropidin + prochloraz
fenpropidin + prochloraz + tebuconazole
fluquinconazole + prochloraz

Products

1	Cloraz 400	Goldengrass	400 g/l	EC	14657
2	Mirage 40 EC	Makhteshim	400 g/l	EC	06770
3	Panache 40	DAPT	400 g/l	EC	10632
4	Poraz	BASF	450 g/l	EC	11701
5	Prelude 20LF	Agrichem	200 g/l	LS	04371
6	Prospero	Makhteshim	400 g/l	EC	11931
7	Scarlett	AgriGuard	400 g/l	EC	14271
8	Scotts Octave	Scotts	46% w/w	WP	09275
9	Sprint	AgriGuard	400 g/l	EC	14272

Uses
- Alternaria in **spring oilseed rape** [1, 2, 6, 7, 9]; **winter oilseed rape** [1, 2, 4, 6, 7, 9]
- Disease control in **grass seed crops** *(off-label)* [6]; **herbs (see appendix 6)** *(off-label)*, **protected herbs (see appendix 6)** *(off-label)* [8]; **winter linseed** *(off-label)* [2]
- Eyespot in **spring wheat** [1-3, 6, 7, 9]; **winter barley, winter wheat** [1-4, 6, 7, 9]; **winter rye** [1, 2, 4, 6, 7, 9]
- Foliar disease control in **triticale** *(off-label)* [2]
- Fungus diseases in **hardy ornamentals, ornamental plant production, woody ornamentals** [8]
- Glume blotch in **spring wheat** [1, 2, 6, 7, 9]; **winter wheat** [1, 2, 4, 6, 7, 9]
- Grey mould in **spring oilseed rape** [1, 2, 6, 7, 9]; **winter oilseed rape** [1, 2, 4, 6, 7, 9]
- Light leaf spot in **spring oilseed rape** [1-3, 6, 7, 9]; **winter oilseed rape** [1-4, 6, 7, 9]
- Net blotch in **spring barley, winter barley** [1, 2, 4, 6, 7, 9]
- Phoma in **spring oilseed rape** [1-3, 6, 7, 9]; **winter oilseed rape** [1-4, 6, 7, 9]
- Powdery mildew in **spring barley, spring wheat, winter barley, winter wheat** [4]; **winter rye** [1, 2, 4, 6, 7, 9]
- Rhynchosporium in **spring barley, winter barley, winter rye** [1, 2, 4, 6, 7, 9]
- Sclerotinia stem rot in **spring oilseed rape** [1, 2, 6, 7, 9]; **winter oilseed rape** [1, 2, 4, 6, 7, 9]
- Seed-borne diseases in **flax** *(seed treatment)*, **hemp for oilseed production** *(off-label - seed treatment)*, **linseed** *(seed treatment)* [5]
- Septoria leaf blotch in **spring wheat** [1, 2, 6, 7, 9]; **winter rye** [4]; **winter wheat** [1, 2, 4, 6, 7, 9]
- White leaf spot in **spring oilseed rape** [1, 2, 6, 7, 9]; **winter oilseed rape** [1, 2, 4, 6, 7, 9]

Specific Off-Label Approvals (SOLAs)
- *grass seed crops* 20062920 [6]
- *hemp for oilseed production* (seed treatment) 20052562 [5]
- *herbs (see appendix 6)* 20010650 [8]
- *protected herbs (see appendix 6)* 20010650 [8]
- *triticale* 20062686 [2]
- *winter linseed* 20062686 [2], 20062687 [2]

Approval information
- Prochloraz not included in Annex 1 under EC Directive 91/414
- Accepted by BBPA for use on malting barley

FOR FULL CONDITIONS OF USE ALWAYS READ THE PRODUCT LABEL

prochloraz

Efficacy guidance
- Spray cereals at first signs of disease. Protection of winter crops through season usually requires at least 2 treatments. See label for details of rates and timing. Treatment active against strains of eyespot resistant to benzimidazole fungicides [2-4, 6]
- Tank mixes with other fungicides recommended to improve control of rusts in wheat and barley. See label for details [2-4, 6]
- A period of at least 3 h without rain should follow spraying [2-4, 6]
- Can be used through most seed treatment machines if good even seed coverage is obtained. Check drill calibration before drilling treated seed. Use inert seed flow agent supplied by manufacturer [5]
- Use methylated spirits, rather than water, to clean residual material from seed treatment machinery, then carry out final rinse with water and detergent [5]
- Apply as drench against soil diseases, as a spray against aerial diseases, as a dip for cuttings, or as a drench at propagation. Spray applications may be repeated at 10-14 d intervals. Under mist propagation use 7 d intervals [8]
- Prochloraz is a DMI fungicide. Resistance to some DMI fungicides has been identified in Septoria leaf blotch which may seriously affect performance of some products. For further advice contact a specialist advisor and visit the Fungicide Resistance Action Group (FRAG)-UK website

Restrictions
- Maximum number of treatments 1 per batch of seed for flax, linseed [5]; 2 per crop for foliar spray uses
- Do not treat linseed varieties Linda, Bolas, Karen, Laura, Mikael, Norlin, Moonraker, Abbey, Agriace [5]
- Do not use treated seed as food or feed [5]

Crop-specific information
- Latest use: before drilling for seed treatments on flax or linseed [5]; watery ripe to early milk stage (GS 71-73) for cereals
- HI 6 wk for cereals
- Application with other fungicides may cause cereal crop scorch [2-4, 6]

Environmental safety
- Dangerous for the environment
- Very toxic to aquatic organisms
- Treated seed harmful to game and wildlife. Product contains red dye for easy identification of treated seed [5]
- Flammable [3]

Hazard classification and safety precautions
Hazard Harmful [4, 5, 9]; Irritant [1-3, 6-8]; Flammable [3, 9]; Dangerous for the environment [1, 2, 4-9]
Transport code 3 [1]; 9 [2-9]
Packaging group III
UN number 1993 [1]; 3077 [8]; 3082 [2-7, 9]
Risk phrases R22a [4, 5, 9]; R36 [1-9]; R38 [1-4, 6-9]; R43 [3]; R50 [4, 8, 9]; R51 [1, 2, 5-7]; R53a [1, 2, 4-9]; R67 [9]
Operator protection A [1-9]; C [1-5, 7-9]; U05a, U20b [1-9]; U08 [1-7, 9]; U09a, U19a [8]; U11 [1, 2, 5-7, 9]; U14 [1, 2, 4, 6, 7, 9]; U15 [1, 2, 6, 7, 9]
Environmental protection E13b [3]; E15a [1, 2, 4-9]; E34 [4]; E38 [4, 5]
Storage and disposal D01, D09a [1-9]; D02, D10b [1-4, 6-9]; D05 [1-3, 5-7, 9]; D10a [5]; D12a [4, 5, 8]
Treated seed S02, S03, S04b, S05, S06a, S07 [5]
Medical advice M03, M05a [4]; M05b [1, 2, 6, 7, 9]

SEE SECTION 3 FOR PRODUCTS ALSO REGISTERED

375 prochloraz + propiconazole

A broad spectrum fungicide mixture for wheat, barley and oilseed rape
FRAC mode of action code: 3 + 3

See also propiconazole

Products

1	Bumper Excell	Makhteshim	400:90 g/l	EC	11015
2	Bumper P	Makhteshim	400:90 g/l	EC	08548
3	Greencrop Twinstar	Greencrop	400:90 g/l	EC	09516
4	MAC-Prochloraz Plus 490 EC	AgChem Access	400:90 g/l	EC	13616

Uses
- Disease control in *grass seed crops* (off-label) [1]
- Eyespot in *spring barley, spring wheat* [1, 2, 4]; *winter barley, winter wheat* [1-4]
- Light leaf spot in *spring oilseed rape, winter oilseed rape* [1, 2, 4]
- Phoma in *spring oilseed rape, winter oilseed rape* [1, 2, 4]
- Rhynchosporium in *spring barley* [1, 2, 4]; *winter barley* [1-4]
- Septoria leaf blotch in *spring wheat* [1, 2, 4]; *winter wheat* [1-4]

Specific Off-Label Approvals (SOLAs)
- *grass seed crops* 20052289 [1]

Approval information
- Propiconazole included in Annex I under EC Directive 91/414
- Prochloraz has not been included in Annex 1 but the withdrawal date has yet to be set. Most current approvals set to expire 2013.
- Accepted by BBPA for use on malting barley

Efficacy guidance
- Best results obtained from treatment when disease is active but not well established
- Treat wheat normally from flag leaf ligule just visible stage (GS 39) but earlier if there is a high risk of Septoria
- Treat barley from the first node detectable stage (GS 31)
- If disease pressure persists a second application may be necessary
- Prochloraz and propiconazole are DMI fungicides. Resistance to some DMI fungicides has been identified in Septoria leaf blotch which may seriously affect performance of some products. For further advice contact a specialist advisor and visit the Fungicide Resistance Action Group (FRAG)-UK website

Restrictions
- Maximum number of treatments 2 per crop

Crop-specific information
- Latest use: before grain watery ripe (GS 71) for cereals; before most seeds green for winter oilseed rape

Environmental safety
- Dangerous for the environment
- Very toxic to aquatic organisms

Hazard classification and safety precautions
Hazard Irritant, Dangerous for the environment
Transport code 9
Packaging group III
UN number 3082
Risk phrases R36, R53a [1-4]; R50 [3]; R51 [1, 2, 4]
Operator protection A, C; U05a, U08, U20b [1-4]; U11, U15 [1, 2, 4]
Environmental protection E15a
Storage and disposal D01, D02, D05, D09a, D10b
Medical advice M05a [1, 2, 4]

FOR FULL CONDITIONS OF USE ALWAYS READ THE PRODUCT LABEL

376 prochloraz + proquinazid + tebuconazole

A broad spectrum systemic fungicide mixture for cereals
FRAC mode of action code: 3 + U7 + 3

See also proquinazid
tebuconazole

Products
Vareon   DuPont   320:40:160 g/l   EC   14376

Uses
- Brown rust in **spring barley, spring wheat, winter barley, winter wheat**
- Crown rust in **spring oats** *(qualified minor use recommendation)*, **winter oats** *(qualified minor use recommendation)*
- Ear diseases in **spring wheat** *(reduction only)*, **winter wheat** *(reduction only)*
- Eyespot in **spring barley, spring wheat, winter barley, winter wheat**
- Glume blotch in **spring wheat, winter wheat**
- Net blotch in **spring barley** *(moderate control)*, **winter barley** *(moderate control)*
- Powdery mildew in **spring barley, spring oats, spring wheat, winter barley, winter oats, winter wheat**
- Rhynchosporium in **spring barley** *(moderate control)*, **winter barley** *(moderate control)*
- Septoria leaf blotch in **spring wheat** *(moderate control)*, **winter wheat** *(moderate control)*
- Yellow rust in **spring barley, spring wheat, winter barley, winter wheat**

Approval information
- Prochloraz has not been included in Annex 1 but the withdrawal date has yet to be set. Most current approvals set to expire 2013. Proquinazid and tebuconazole included in Annex 1 under EC Directive 91/414

Efficacy guidance
- Resistance to DMI fungicides has been identified in Septoria leaf blotch and may seriously affect the performance of some products. Advice on resistance management is available from the Fungicide Action Grup (FRAG) web site

Environmental safety
- Avoid spraying within 5 m of the field boundary to reduce the effect on non-target insects or other arthropods
- LERAP Category B

Hazard classification and safety precautions
Hazard Harmful, Dangerous for the environment
Transport code 9
Packaging group III
UN number 3082
Risk phrases R22a, R36, R40, R43, R50, R53a, R63
Operator protection A, C, H
Environmental protection E15a, E16a, E22c, E34, E38
Storage and disposal D01, D02, D09a, D12a

377 prochloraz + tebuconazole

A broad spectrum systemic fungicide mixture for cereals, oilseed rape and amenity use
FRAC mode of action code: 3 + 3

See also tebuconazole

Products
1	Agate	Makhteshim	267:133 g/l	EC	13024
2	Agate	Makhteshim	267:133 g/l	EC	14482
3	Astute	Sherriff Amenity	267:133 g/l	EC	14897
4	Lunar	Bayer Environ.	267:132 g/l	EC	12366
5	Monkey	Makhteshim	267:133 g/l	EC	12906
6	Orius P	Makhteshim	267:133 g/l	EC	12880
7	Pan Proteb	Pan Agriculture	267:132 g/l	EC	13427

SEE SECTION 3 FOR PRODUCTS ALSO REGISTERED

Pesticide Profiles

Products – continued

8	Pan Proteb	Pan Agriculture	267:132 g/l	EC	14896
9	System Turf	AgriGuard	267:133 g/l	EC	14538
10	Throttle	Headland Amenity	267:133 g/l	EC	13900

Uses
- Brown rust in **spring barley, spring rye, spring wheat, winter barley, winter rye, winter wheat** [1, 2, 5, 6]
- Eyespot in **spring barley, spring rye, spring wheat, winter barley, winter rye, winter wheat** [1, 2, 5, 6]
- Fusarium patch in **managed amenity turf** [3, 4, 7-10]
- Glume blotch in **spring wheat, winter wheat** [1, 2, 5, 6]
- Late ear diseases in **spring wheat, winter wheat** [1, 2, 5, 6]
- Net blotch in **spring barley, winter barley** [1, 2, 5, 6]
- Powdery mildew in **spring barley, spring rye, spring wheat, winter barley, winter rye, winter wheat** [1, 2, 5, 6]
- Rhynchosporium in **spring barley, spring rye, winter barley, winter rye** [1, 2, 5, 6]
- Sclerotinia stem rot in **spring oilseed rape, winter oilseed rape** [1, 2, 5, 6]
- Septoria leaf blotch in **spring wheat, winter wheat** [1, 2, 5, 6]
- Yellow rust in **spring barley, spring rye, spring wheat, winter barley, winter rye, winter wheat** [1, 2, 5, 6]

Approval information
- Accepted by BBPA for use on malting barley
- Prochloraz has not been included in Annex 1 but the withdrawal date has yet to be set. Most current approvals set to expire 2013. Tebuconazole is included in Annex 1.
- Approval expiry 31 Aug 2011 [4, 7]

Efficacy guidance
- Complete protection of winter cereals will usually require a programme of at least two treatments [1, 5]
- Protection of flag leaf and ear from Septoria diseases requires treatment at flag leaf emergence and repeated when ear fully emerged [1, 5]
- Two treatments separated by a 2 wk gap may be needed for difficult Fusarium patch attacks [4]
- To avoid resistance, do not apply repeated applications of the single product on the same crop against the same disease [10]
- Optimum application timing is normally when disease first seen but varies with main target disease - see labels
- Prochloraz and tebuconazole are DMI fungicides. Resistance to some DMI fungicides has been identified in Septoria leaf blotch which may seriously affect performance of some products. For further advice contact a specialist advisor and visit the Fungicide Resistance Action Group (FRAG)-UK website

Restrictions
- Maximum total dose on cereals equivalent to two full dose treatments
- Maximum number of treatments on managed amenity turf 2 per yr
- Do not apply during drought or to frozen turf [4]

Crop-specific information
- Latest use: before grain milky ripe (GS 73) for cereals [1, 5]; most seeds green for oilseed rape [1, 5]
- HI 6 wk for cereals [1, 5]
- Occasionally transient leaf speckling may occur after treating wheat. Yield responses should not be affected [1]

Environmental safety
- Dangerous for the environment
- Very toxic to aquatic organisms

Hazard classification and safety precautions
Hazard Harmful, Dangerous for the environment
Transport code 6.1 [7]; 9 [1-6, 8-10]
Packaging group III
UN number 2902 [7]; 3082 [1-6, 8-10]

FOR FULL CONDITIONS OF USE ALWAYS READ THE PRODUCT LABEL

prochloraz + thiram

Risk phrases R22a, R36, R43, R50, R53a [1-10]; R63 [4, 7-10]
Operator protection A, C, H [1-10]; M [4, 7-10]; U05a [1-10]; U08, U20b [9, 10]; U09b, U20a [1-3, 5, 6]; U11 [4, 7-10]; U14 [1-8]
Environmental protection E13b [1-3, 5, 6]; E15a [4, 7-10]; E34, E38 [1-10]
Storage and disposal D01, D02, D09a, D12a [1-10]; D05 [1-8]; D10b [4, 7-10]; D10c [1-3, 5, 6]
Medical advice M03 [1-10]; M05a [4, 7, 8]

378   prochloraz + thiram

A seed treatment fungicide mixture for oilseed rape
FRAC mode of action code: 3 + M3

See also thiram

Products
Agrichem Hy-Pro Duet   Agrichem   150:333 g/l   FS   14018

Uses
- Phoma in **winter oilseed rape** *(seed treatment)*

Approval information
- Thiram included in Annex I under EC Directive 91/414
- Prochloraz has not been included in Annex 1 but the withdrawal date has yet to be set. Most current approvals are now set to expire 2021.

Efficacy guidance
- Can be applied through most types of seed treatment machinery
- Good coverage of the seed surface essential for best results
- Product is active against seed-borne disease. It will not control leaf spotting from leaf spores on trash on the seed bed

Restrictions
- Maximum number of treatments 1 per crop
- Seed must be of satisfactory quality and moisture content
- Sow seed as soon as possible after treatment
- Do not store treated seed from one season to the next
- Do not co-apply product with water or mix with any other seed treatment product

Crop-specific information
- Latest use: pre-sowing
- Product will also control soil-borne fungi that cause damping-off of rape seedlings

Following crops guidance
- Any crop may be grown following a treated winter oilseed rape crop

Hazard classification and safety precautions
Hazard Harmful, Dangerous for the environment
Transport code 9
Packaging group III
UN number 3082
Risk phrases R20, R22a, R36, R38, R43, R50, R53a
Operator protection A, C, H; U02a, U08, U14, U15, U20a
Environmental protection E15a, E36b
Storage and disposal D01, D02, D05, D08, D09a, D12b, D13, D15
Treated seed S01, S02, S03, S04d, S05, S08
Medical advice M04a

SEE SECTION 3 FOR PRODUCTS ALSO REGISTERED

Pesticide Profiles

379   prochloraz + triticonazole

A broad spectrum fungicide seed treatment mixture for winter cereals
FRAC mode of action code: 3 + 3

See also triticonazole

Products
   Kinto   BASF   60:20 g/l   FS   12038

Uses
- Bunt in **durum wheat** *(off-label - seed treatment)*, **winter wheat** *(seed treatment)*
- Covered smut in **winter barley** *(seed treatment)*, **winter oats** *(seed treatment)*
- Fusarium foot rot and seedling blight in **durum wheat** *(off-label - seed treatment)*, **triticale** *(seed treatment)*, **winter barley** *(seed treatment)*, **winter oats** *(seed treatment)*, **winter rye** *(seed treatment)*, **winter wheat** *(seed treatment)*
- Leaf stripe in **winter barley** *(seed treatment)*
- Loose smut in **durum wheat** *(off-label - seed treatment)*, **winter barley** *(seed treatment)*, **winter oats** *(seed treatment)*, **winter wheat** *(seed treatment)*
- Septoria seedling blight in **durum wheat** *(off-label - seed treatment)*, **winter wheat** *(seed treatment)*

Specific Off-Label Approvals (SOLAs)
- *durum wheat (seed treatment)* 20061094

Approval information
- Triticonazole included in Annex I under EC Directive 91/414
- Prochloraz has not been included in Annex 1 but the withdrawal date has yet to be set. Most current approvals set to expire 2013.
- Accepted by BBPA for use on malting barley

Efficacy guidance
- Apply undiluted through conventional seed treatment machine
- Calibrate seed drill with treated seed before sowing
- Bag treated seed immediately and keep in a dry, draught free store
- Ensure good even coverage of seed to achieve best results
- Treatment may cause some slight delay in crop emergence

Restrictions
- Maximum number of treatments 1 per batch of seed
- Do not treat seed with moisture content above 16% and do not allow moisture content of treated seed to exceed 16%
- Do not treat cracked, split or sprouted seed
- Do not handle treated seed unnecessarily
- Do not use treated seed as food or feed

Crop-specific information
- Latest use: before drilling wheat, barley, oats, rye or triticale

Environmental safety
- Dangerous for the environment
- Toxic to aquatic organisms

Hazard classification and safety precautions
   Hazard Dangerous for the environment
   Transport code 9
   Packaging group III
   UN number 3082
   Risk phrases R51, R53a
   Operator protection A, H; U05a, U07, U20b
   Environmental protection E15a, E34, E38
   Storage and disposal D01, D02, D08, D09a, D11a, D12a
   Treated seed S01, S02, S03, S04a, S04b, S05, S07

FOR FULL CONDITIONS OF USE ALWAYS READ THE PRODUCT LABEL

380 prohexadione-calcium

A cyclohexanecarboxylate growth regulator for use in apples

See also mepiquat chloride + prohexadione-calcium

Products
Regalis   BASF   10% w/w   WG   12414

Uses
- Control of shoot growth in **apples**
- Growth regulation in **grapevines** (off-label), **hops** (off-label), **ornamental plant production** (off-label), **soft fruit** (off-label)

Specific Off-Label Approvals (SOLAs)
- **grapevines** 20101467 expires 31 Dec 2011
- **hops** 20082866 expires 31 Dec 2011
- **ornamental plant production** 20082866 expires 31 Dec 2011
- **soft fruit** 20082866 expires 31 Dec 2011

Approval information
- Prohexadione-calcium included in Annex I under EC Directive 91/414
- Approval expiry 31 Dec 2011 [1]

Efficacy guidance
- For a standard orchard treat at the start of active growth and again after 3-5 wk depending on growth conditions
- Alternatively follow the first application with four reduced rate applications at two wk intervals

Restrictions
- Maximum total dose equivalent to two full dose treatments
- Do not apply in conjunction with calcium based foliar fertilisers

Crop-specific information
- HI 55 d for apples

Environmental safety
- Harmful to aquatic organisms
- Avoid spray drift onto neighbouring crops and other non-target plants
- Product does not harm natural insect predators when used as directed

Hazard classification and safety precautions
  UN number N/C
  Risk phrases R52, R53a
  Operator protection U05a
  Environmental protection E38
  Storage and disposal D01, D02, D09a, D10c, D12a

381 propamocarb hydrochloride

A translocated protectant carbamate fungicide
FRAC mode of action code: 28

See also chlorothalonil + propamocarb hydrochloride
          fenamidone + propamocarb hydrochloride
          fluopicolide + propamocarb hydrochloride
          fosetyl-aluminium + propamocarb hydrochloride
          mancozeb + propamocarb hydrochloride

Products
1	Filex	Scotts	722 g/l	SL	07631
2	Flash	AgriGuard	722 g/l	SL	11989
3	Pan PCH	Pan Agriculture	722 g/l	SL	12842
4	Proplant	Fargro	722 g/l	SL	13359

Uses
- Blight in **protected tomatoes**, **tomatoes (outdoor)** [2]

SEE SECTION 3 FOR PRODUCTS ALSO REGISTERED

- Botrytis in **cayenne pepper** *(off-label)*, **frise** *(off-label)*, **horseradish** *(off-label)*, **leaf brassicas** *(off-label)*, **protected horseradish** *(off-label)*, **protected radishes** *(off-label)*, **radishes** *(off-label)* [4]; **chives** *(off-label)*, **cress** *(off-label)*, **parsley** *(off-label)*, **protected herbs (see appendix 6)** *(off-label)* [1]; **herbs (see appendix 6)** *(off-label)* [1, 4]; **lettuce, protected lettuce** [3]
- Damping off in **cayenne pepper** *(off-label)*, **frise** *(off-label)*, **herbs (see appendix 6)** *(off-label)*, **horseradish** *(off-label)*, **leaf brassicas** *(off-label)*, **protected cucumbers** *(off-label)*, **protected horseradish** *(off-label)*, **protected radishes** *(off-label)*, **protected tomatoes, protected watercress** *(off-label)*, **radishes** *(off-label)* [4]; **inert substrate cucumbers** *(off-label)*, **inert substrate peppers** *(off-label)*, **inert substrate tomatoes** *(off-label)*, **nft peppers** *(off-label)*, **nft tomatoes** *(off-label)*, **protected tomatoes** *(off-label)* [1]; **protected peppers** *(off-label)* [1, 4]
- Damping off and foot rot in **lamb's lettuce** *(off-label)*, **lettuce, lettuce** *(off-label)*, **protected herbs (see appendix 6)** *(off-label)*, **protected lamb's lettuce** *(off-label)*, **protected lettuce, protected scarole** *(off-label)*, **protected tomatoes, scarole** *(off-label)* [4]
- Downy mildew in **broccoli, brussels sprouts, cabbages, calabrese, cauliflowers, chinese cabbage** [1-4]; **cayenne pepper** *(off-label)*, **frise** *(off-label)*, **horseradish** *(off-label)*, **leaf brassicas** *(off-label)*, **lettuce** *(off-label)*, **protected horseradish** *(off-label)* [4]; **chives** *(off-label)*, **cress** *(off-label)*, **parsley** *(off-label)*, **protected herbs (see appendix 6)** *(off-label)*, **watercress** *(off-label - under protection)* [1]; **herbs (see appendix 6)** *(off-label)*, **protected radishes** *(off-label)*, **radishes** *(off-label)* [1, 4]; **lettuce, protected lettuce** [3, 4]
- Phytophthora in **aubergines, bedding plants, broccoli, brussels sprouts, bulb onions, cabbages, calabrese, cauliflowers, chinese cabbage, cucumbers, leeks, nursery stock, peppers, pot plants** [1-4]; **flower bulbs, tomatoes (outdoor)** [2-4]; **ornamental plant production** [1, 3, 4]; **protected tomatoes** [2, 4]; **rockwool aubergines, rockwool cucumbers, rockwool peppers, rockwool tomatoes** [2]; **tulips** [1, 2]; **watercress** *(off-label - under protection)* [1]
- Phytophthora root rot in **cayenne pepper** *(off-label)*, **frise** *(off-label)*, **herbs (see appendix 6)** *(off-label)*, **horseradish** *(off-label)*, **leaf brassicas** *(off-label)*, **protected cucumbers** *(off-label)*, **protected horseradish** *(off-label)*, **protected peppers** *(off-label)*, **protected radishes** *(off-label)*, **protected tomatoes, protected watercress** *(off-label)*, **radishes** *(off-label)* [4]
- Pythium in **aubergines, bedding plants, broccoli, brussels sprouts, bulb onions, cabbages, calabrese, cauliflowers, chinese cabbage, cucumbers, flower bulbs, leeks, nursery stock, peppers, pot plants, tomatoes (outdoor)** [1-4]; **ornamental plant production** [1, 3, 4]; **protected tomatoes** [2, 4]; **protected tomatoes** *(off-label)*, **watercress** *(off-label - under protection)* [1]; **rockwool aubergines, rockwool cucumbers, rockwool peppers, rockwool tomatoes, tulips** [1, 2]
- White blister in **horseradish** *(off-label)*, **protected horseradish** *(off-label)* [1, 4]; **protected radishes** *(off-label)*, **radishes** *(off-label)* [4]

Specific Off-Label Approvals (SOLAs)
- **cayenne pepper** *20072948* [4]
- **chives** *20040626* [1]
- **cress** *20040626* [1]
- **frise** *20072949* [4]
- **herbs (see appendix 6)** *20040626* [1], *20072947* [4]
- **horseradish** *992036* [1], *20072947* [4]
- **inert substrate cucumbers** *992032* [1]
- **inert substrate peppers** *992032* [1]
- **inert substrate tomatoes** *992032* [1]
- **lamb's lettuce** *20072947* [4]
- **leaf brassicas** *20072949* [4]
- **lettuce** *20072947* [4]
- **nft peppers** *992032* [1]
- **nft tomatoes** *992032* [1]
- **parsley** *20040626* [1]
- **protected cucumbers** *20072945* [4]
- **protected herbs (see appendix 6)** *20040626* [1], *20072947* [4]
- **protected horseradish** *992036* [1], *20072947* [4]
- **protected lamb's lettuce** *20072947* [4]
- **protected peppers** *992032* [1], *20072945* [4]
- **protected radishes** *992030* [1], *20072947* [4]

FOR FULL CONDITIONS OF USE ALWAYS READ THE PRODUCT LABEL

propaquizafop

- **protected scarole** *20072947* [4]
- **protected tomatoes** *992032* [1]
- **protected watercress** *20072947* [4]
- **radishes** *992030* [1], *20072947* [4]
- **scarole** *20072947* [4]
- **watercress** (under protection) *20010439* [1]

Approval information
- Propamocarb hydrochloride included in Annex I under EC Directive 91/414

Efficacy guidance
- Chemical is absorbed through roots and translocated throughout plant
- Incorporate in compost before use or drench moist compost or soil before sowing, pricking out, striking cuttings or potting up
- Drench treatment can be repeated at 3-6 wk intervals
- Concentrated solution is corrosive to all metals other than stainless steel
- May also be applied in trickle irrigation systems
- To prevent root rot in tulip bulbs apply as dip for 20 min but full protection only achieved by soil drench treatment as well

Restrictions
- Maximum number of treatments 4 per crop for cucumbers, tomatoes, peppers, aubergines; 3 per crop for outdoor and protected lettuce; 1 per crop for listed brassicas; 1 compost incorporation and/or 1 drench treatment for leeks, onions; 1 dip and 1 soil drench for flower bulbs
- Do not apply to lettuce crops grown in non-soil systems
- When applied over established seedlings rinse off foliage with water (except lettuce) and do not apply under hot, dry conditions
- On plants of unknown tolerance test first on a small scale
- Do not apply in a recirculating irrigation/drip system
- Store away from seeds and fertilizers

Crop-specific information
- Latest use: before transplanting for brassicas
- HI 14 d for cucumbers, tomatoes, peppers, aubergines; 4 wk for brassicas; 19 wk for leeks, onions

Hazard classification and safety precautions
UN number N/C
Operator protection A, C, H, K, M [1-4]; B [1, 2]; U02a, U19a [3, 4]; U05a [2-4]; U08, U20b [1-4]
Environmental protection E15a [1-4]; E34 [3, 4]
Storage and disposal D01, D02, D05 [2-4]; D07 [2]; D09a [1-4]; D10b [3, 4]; D11a [1, 2]

382 propaquizafop

A phenoxy alkanoic acid foliar acting herbicide for grass weeds in a range of crops
HRAC mode of action code: A

Products

1	Bulldog	Makhteshim	100 g/l	EC	14565
2	Clayton Orleans	Clayton	100 g/l	EC	14452
3	Clayton Orleans	Clayton	100 g/l	EC	14827
4	Cleancrop GYR 2	Makhteshim	100 g/l	EC	14578
5	Falcon	Makhteshim	100 g/l	EC	10585
6	Greencrop Satchmo	Greencrop	100 g/l	EC	10748
7	Longhorn	AgChem Access	100 g/l	EC	12512
8	MAC-Propaquizafop 100 EC	AgChem Access	100 g/l	EC	13978
9	PQF 100	Goldengrass	100 g/l	EC	14414
10	Raptor	Makhteshim	100 g/l	EC	14566
11	Shogun	Makhteshim	100 g/l	EC	14567
12	Standon Propaquizafop	Standon	100 g/l	EC	11536
13	Standon Zing PQF	Standon	100 g/l	EC	13898
14	Standon Zing PQF	Standon	100 g/l	EC	14949
15	Zealot	AgriGuard	100 g/l	EC	12548

SEE SECTION 3 FOR PRODUCTS ALSO REGISTERED

Pesticide Profiles

Uses

- Annual grasses in **beet leaves** *(off-label)*, **herbs (see appendix 6)** *(off-label)*, **horseradish** *(off-label)*, **leeks** *(off-label)*, **red beet** *(off-label)*, **spinach** *(off-label)* [1, 4, 10, 11]; **bog myrtle** *(off-label)*, **celery (outdoor)** *(off-label)*, **honesty** *(off-label)*, **mallow (althaea spp.)** *(off-label)*, **poppies for morphine production** *(off-label)*, **rhubarb** *(off-label)*, **spring linseed** *(off-label)* [1, 4, 5, 10, 11]; **borage for oilseed production** *(off-label)*, **canary flower (echium spp.)** *(off-label)* [1, 4, 5, 10-12]; **bulb onions, carrots, combining peas, early potatoes, farm forestry, fodder beet, linseed, parsnips, spring oilseed rape, sugar beet, swedes, turnips, winter field beans, winter oilseed rape** [1-15]; **cut logs, forest, forest nurseries** [1, 3-5, 8, 10, 11]; **evening primrose** *(off-label)*, **garlic** *(off-label)*, **lupins** *(off-label)*, **mustard** *(off-label)*, **shallots** *(off-label)* [1, 4-6, 10-12]; **maincrop potatoes** [1-3, 5-15]; **potatoes** [4]; **spring field beans** [1-6, 8-15]
- Annual meadow grass in **beet leaves** *(off-label)*, **bog myrtle** *(off-label)*, **herbs (see appendix 6)** *(off-label)*, **honesty** *(off-label)*, **leeks** *(off-label)*, **mallow (althaea spp.)** *(off-label)*, **poppies for morphine production** *(off-label)*, **red beet** *(off-label)*, **spinach** *(off-label)*, **spring linseed** *(off-label)* [5]; **borage for oilseed production** *(off-label)*, **canary flower (echium spp.)** *(off-label)* [5, 12]; **bulb onions, carrots, combining peas, cut logs, early potatoes, farm forestry, fodder beet, forest, forest nurseries, linseed, parsnips, potatoes, spring field beans, spring oilseed rape, sugar beet, swedes, turnips, winter field beans, winter oilseed rape** [4]; **evening primrose** *(off-label)*, **garlic** *(off-label)*, **lupins** *(off-label)*, **mustard** *(off-label)*, **shallots** *(off-label)* [5, 6, 12]
- Couch in **beet leaves** *(off-label)*, **herbs (see appendix 6)** *(off-label)*, **leeks** *(off-label)*, **red beet** *(off-label)*, **spinach** *(off-label)* [5]
- Perennial grasses in **bog myrtle** *(off-label)*, **celery (outdoor)** *(off-label)*, **honesty** *(off-label)*, **mallow (althaea spp.)** *(off-label)*, **poppies for morphine production** *(off-label)*, **rhubarb** *(off-label)*, **spring linseed** *(off-label)* [5]; **borage for oilseed production** *(off-label)*, **canary flower (echium spp.)** *(off-label)* [5, 12]; **bulb onions, carrots, combining peas, early potatoes, farm forestry, fodder beet, linseed, parsnips, spring oilseed rape, sugar beet, swedes, turnips, winter field beans, winter oilseed rape** [1-15]; **cut logs, forest, forest nurseries** [1, 3-5, 8, 10, 11]; **evening primrose** *(off-label)*, **garlic** *(off-label)*, **lupins** *(off-label)*, **mustard** *(off-label)*, **shallots** *(off-label)* [5, 6, 12]; **maincrop potatoes** [1-3, 5-15]; **potatoes** [4]; **red beet** *(off-label)* [1, 4, 10, 11]; **spring field beans** [1-6, 8-15]
- Volunteer cereals in **beet leaves** *(off-label)*, **bog myrtle** *(off-label)*, **celery (outdoor)** *(off-label)*, **herbs (see appendix 6)** *(off-label)*, **honesty** *(off-label)*, **leeks** *(off-label)*, **mallow (althaea spp.)** *(off-label)*, **poppies for morphine production** *(off-label)*, **red beet** *(off-label)*, **rhubarb** *(off-label)*, **spinach** *(off-label)*, **spring linseed** *(off-label)* [5]; **borage for oilseed production** *(off-label)*, **canary flower (echium spp.)** *(off-label)* [5, 12]; **evening primrose** *(off-label)*, **garlic** *(off-label)*, **lupins** *(off-label)*, **mustard** *(off-label)*, **shallots** *(off-label)* [5, 6, 12]
- Wild oats in **celery (outdoor)** *(off-label)*, **rhubarb** *(off-label)* [5]

Specific Off-Label Approvals (SOLAs)
- **beet leaves** *20092728* [1], *20092740* [4], *20080871* [5], *20092707* [10], *20092784* [11]
- **bog myrtle** *20092731* [1], *20092739* [4], *20091295* [5], *20092700* [10], *20092790* [11]
- **borage for oilseed production** *20092724* [1], *20092737* [4], *20080874* [5], *20092704* [10], *20092789* [11], *20080893* [12]
- **canary flower (echium spp.)** *20092724* [1], *20092737* [4], *20080874* [5], *20092704* [10], *20092789* [11], *20080893* [12]
- **celery (outdoor)** *20092723* [1], *20092735* [4], *20080872* [5], *20092703* [10], *20092788* [11]
- **evening primrose** *20092724* [1], *20092737* [4], *20080874* [5], *20080879* [6], *20092704* [10], *20092789* [11], *20080893* [12]
- **garlic** *20092722* [1], *20092732* [4], *20080875* [5], *20080880* [6], *20092705* [10], *20092785* [11], *20080894* [12]
- **herbs (see appendix 6)** *20092728* [1], *20092740* [4], *20080871* [5], *20092707* [10], *20092784* [11]
- **honesty** *20092724* [1], *20092737* [4], *20080874* [5], *20092704* [10], *20092789* [11]
- **horseradish** *20092725* [1], *20092738* [4], *20092702* [10], *20092787* [11]
- **leeks** *20092728* [1], *20092740* [4], *20080871* [5], *20092707* [10], *20092784* [11]
- **lupins** *20092729* [1], *20092736* [4], *20080876* [5], *20080878* [6], *20092699* [10], *20092783* [11], *20080892* [12]
- **mallow (althaea spp.)** *20092721* [1], *20092733* [4], *20080873* [5], *20092706* [10], *20092786* [11]
- **mustard** *20092724* [1], *20092737* [4], *20080874* [5], *20080879* [6], *20092704* [10], *20092789* [11], *20080893* [12]

FOR FULL CONDITIONS OF USE ALWAYS READ THE PRODUCT LABEL

propaquizafop

- **poppies for morphine production** *20092730* [1], *20092734* [4], *20080877* [5], *20092701* [10], *20092791* [11]
- **red beet** *20092728* [1], *20092740* [4], *20080871* [5], *20092707* [10], *20092784* [11]
- **rhubarb** *20092723* [1], *20092735* [4], *20080872* [5], *20092703* [10], *20092788* [11]
- **shallots** *20092722* [1], *20092732* [4], *20080875* [5], *20080880* [6], *20092705* [10], *20092785* [11], *20080894* [12]
- **spinach** *20092728* [1], *20092740* [4], *20080871* [5], *20092707* [10], *20092784* [11]
- **spring linseed** *20092724* [1], *20092737* [4], *20080874* [5], *20092704* [10], *20092789* [11]

Efficacy guidance
- Apply to emerged weeds when they are growing actively with adequate soil moisture
- Activity is slower under cool conditions
- Broad-leaved weeds and any weeds germinating after treatment are not controlled
- Annual meadow grass up to 3 leaves checked at low doses and severely checked at highest dose
- Spray barley cover crops when risk of wind blow has passed and before there is serious competition with the crop
- Various tank mixtures and sequences recommended for broader spectrum weed control in oilseed rape, peas and sugar beet. See label for details
- Severe couch infestations may require a second application at reduced dose when regrowth has 3-4 leaves unfolded
- Products contain surfactants. Tank mixing with adjuvants not required or recommended
- Propaquizafop is an ACCase inhibitor herbicide. To avoid the build up of resistance do not apply products containing an ACCase inhibitor herbicide more than twice to any crop. In addition do not use any product containing propaquizafop in mixture or sequence with any other product containing the same ingredient
- Use these products as part of a resistance management strategy that includes cultural methods of control and does not use ACCase inhibitors as the sole chemical method of grass weed control
- Applying a second product containing an ACCase inhibitor to a crop will increase the risk of resistance development; only use a second ACCase inhibitor to control different weeds at a different timing
- Always follow WRAG guidelines for preventing and managing herbicide resistant weeds. See Section 5 for more information

Restrictions
- Maximum number of treatments 1 or 2 per crop or yr. See label for details
- See label for list of tolerant tree species
- Do not treat seed potatoes

Crop-specific information
- Latest use: before crop flower buds visible for winter oilseed rape, linseed, field beans; before 8 fully expanded leaf stage for spring oilseed rape; before weeds are covered by the crop for potatoes, sugar beet, fodder beet, swedes, turnips; when flower buds visible for peas
- HI onions, carrots, early potatoes, parsnips 4 wk; combining peas, early potatoes 7 wk; sugar beet, fodder beet, maincrop potatoes, swedes, turnips 8 wk; field beans 14 wk
- Application in high temperatures and/or low soil moisture content may cause chlorotic spotting especially on combining peas and field beans
- Overlaps at the highest dose can cause damage from early applications to carrots and parsnips

Following crops guidance
- In the event of a failed treated crop an interval of 2 wk must elapse between the last application and redrilling with winter wheat or winter barley. 4 wk must elapse before sowing oilseed rape, peas or field beans, and 16 wk before sowing ryegrass or oats

Environmental safety
- Dangerous for the environment
- Toxic to aquatic organisms
- Risk to certain non-target insects or other arthropods. See directions for use

Hazard classification and safety precautions
Hazard Harmful [1, 4, 10, 11]; Irritant [2, 3, 5-9, 12-15]; Dangerous for the environment [1-15]
Transport code 9
Packaging group III
UN number 3082
Risk phrases R20, R41 [4]; R36, R38 [1-3, 5-15]; R51, R53a [1-15]

SEE SECTION 3 FOR PRODUCTS ALSO REGISTERED

Pesticide Profiles

Operator protection A [1-15]; C [2-9, 11-15]; H [1, 4, 7, 10, 11]; U02a, U05a [1-3, 5-15]; U08, U19a, U20b [1-15]; U11 [5, 8, 12-14]; U12 [4]; U13 [1-5, 7-11, 15]; U14, U15 [1-5, 7-15]; U23a [6]
Environmental protection E15a, E22b [1-15]; E34 [1-3, 5, 7-11, 15]; E38 [4]
Storage and disposal D01, D02, D05, D09a [1-15]; D10a, D12a [4]; D10b [1-3, 5-15]; D12b, D14 [1-3, 5, 7-11]
Medical advice M05b

383 propiconazole

A systemic, curative and protectant triazole fungicide
FRAC mode of action code: 3

See also azoxystrobin + propiconazole
chlorothalonil + cyproconazole + propiconazole
chlorothalonil + fludioxonil + propiconazole
chlorothalonil + propiconazole
cyproconazole + propiconazole
fenbuconazole + propiconazole
fenpropidin + propiconazole
fenpropidin + propiconazole + tebuconazole
prochloraz + propiconazole
propiconazole + tebuconazole
propiconazole + trifloxystrobin

Products

1	Anode	Makhteshim	250 g/l	EC	14447
2	Banner Maxx	Syngenta	156 g/l	EC	13167
3	Bounty	AgChem Access	250 g/l	EC	14616
4	Bumper 250 EC	Makhteshim	250 g/l	EC	14399
5	Propi 25 EC	Sharda	250 g/l	EC	14939
6	Tonik	AgriGuard	250 g/l	EC	14761

Uses

- Anthracnose in **amenity grassland** *(qualified minor use)*, **managed amenity turf** *(qualified minor use)* [2]
- Brown patch in **amenity grassland** *(qualified minor use)*, **managed amenity turf** *(qualified minor use)* [2]
- Brown rust in **spring barley**, **spring wheat**, **winter barley**, **winter rye**, **winter wheat** [1, 3-6]
- Crown rust in **grass seed crops**, **permanent grassland**, **spring oats**, **winter oats** [1, 3-6]
- Disease control in **honesty** *(off-label)* [4]
- Dollar spot in **amenity grassland**, **managed amenity turf** [2]
- Downy mildew in **garlic** *(off-label)*, **onions** *(off-label)*, **shallots** *(off-label)* [4]
- Drechslera leaf spot in **grass seed crops**, **permanent grassland** [1, 3-6]
- Fusarium patch in **amenity grassland**, **managed amenity turf** [2]
- Glume blotch in **spring wheat**, **winter rye**, **winter wheat** [1, 3-6]
- Light leaf spot in **spring oilseed rape** *(reduction)*, **winter oilseed rape** *(reduction)* [1, 3-6]
- Mildew in **grass seed crops**, **permanent grassland** [1, 3-6]
- Powdery mildew in **spring barley**, **spring oats**, **spring wheat**, **winter barley**, **winter oats**, **winter rye**, **winter wheat** [1, 3-6]
- Ramularia leaf spots in **sugar beet** *(reduction)* [1, 3-6]
- Rhynchosporium in **grass seed crops**, **permanent grassland**, **spring barley**, **winter barley**, **winter rye** [1, 3-6]
- Rust in **ornamental plant production** *(off-label)*, **protected ornamentals** *(off-label)* [4]; **sugar beet** [1, 3-6]
- Septoria leaf blotch in **spring wheat**, **winter rye**, **winter wheat** [1, 3-6]
- Sooty moulds in **spring wheat**, **winter wheat** [1, 3-6]
- Yellow rust in **spring barley**, **spring wheat**, **winter barley**, **winter wheat** [1, 3-6]

Specific Off-Label Approvals (SOLAs)
- *garlic* 20090705 [4]
- *honesty* 20090706 [4]
- *onions* 20090705 [4]

FOR FULL CONDITIONS OF USE ALWAYS READ THE PRODUCT LABEL

- *ornamental plant production* 20090707 [4]
- *protected ornamentals* 20090707 [4]
- *shallots* 20090705 [4]

Approval information
- Propiconazole included in Annex I under EC Directive 91/414
- Accepted by BBPA for use on malting barley

Efficacy guidance
- Best results achieved by applying at early stage of disease development. Recommended spray programmes vary with crop, disease, season, soil type and product.
- For optimum turf quality and control use in conjunction with turf management practices that promote good plant health [2]
- Propiconazole is a DMI fungicide. Resistance to some DMI fungicides has been identified in Septoria leaf blotch which may seriously affect performance of some products. For further advice contact a specialist advisor and visit the Fungicide Resistance Action Group (FRAG)-UK website

Restrictions
- On oilseed rape do not apply during flowering.
- Grass seed crops must be treated in yr of harvest
- A minimum interval of 14 d must elapse between treatments on leeks and 21 d on sugar beet
- Avoid spraying crops or turf under stress, eg during cold weather or periods of frost

Crop-specific information
- Latest use: before 4 pairs of true leaves for honesty
- HI cereals, grass seed crops 35 d; oilseed rape, sugar beet, garlic, onions, shallots, grass for ensiling 28 d
- May be used on all common turf grass species [2]

Environmental safety
- Dangerous for the environment
- Toxic to aquatic organisms
- Risk to non-target insects or other arthropods [2]
- When treating turf with vehicle mounted or drawn hydraulic boom sprayers avoid spraying within 5 m of unmanaged land to reduce effects on non-target arthropods [2]

Hazard classification and safety precautions
Hazard Harmful [2]; Dangerous for the environment [1-6]
Transport code 9
Packaging group III
UN number 3082
Risk phrases R20, R43, R52 [2]; R22a [6]; R51 [1, 3-6]; R53a [1-6]
Operator protection A [1-4, 6]; C [1, 3, 4, 6]; H [2, 6]; U02a, U05a, U19a, U20b [1-6]; U08 [1, 3-6]; U09a, U14 [2]
Environmental protection E15a, E34 [1, 3-6]; E15b, E22c, E38 [2]
Storage and disposal D01, D02, D05, D09a [1-6]; D10b [1, 3-6]; D10c, D12a [2]
Treated seed S06a [1, 3-6]
Medical advice M05a [1, 3-6]

384 propiconazole + tebuconazole

A broad spectrum systemic fungicide mixture for cereals
FRAC mode of action code: 3 + 3

See also propiconazole
tebuconazole

Products
Cogito	Syngenta	250:250 g/l	EC	13847

Uses
- Brown rust in **spring barley, spring wheat, winter barley, winter wheat**
- Glume blotch in **spring wheat, winter wheat**
- Mildew in **spring barley, spring wheat, winter barley, winter wheat**
- Net blotch in **spring barley, winter barley**

SEE SECTION 3 FOR PRODUCTS ALSO REGISTERED

Pesticide Profiles

- Rhynchosporium in **spring barley, winter barley**
- Septoria leaf blotch in **spring wheat, winter wheat**
- Yellow rust in **spring barley, spring wheat, winter barley, winter wheat**

Approval information
- Propiconazole included in Annex I under EC Directive 91/414
- Accepted by BBPA for use on malting barley

Efficacy guidance
- Best disease control and yield benefit obtained when applied at early stage of disease development before infection has spread to new growth
- To protect the flag leaf and ear from Septoria diseases apply from flag leaf emergence to ear fully emerged (GS 37-59)
- Applications once foliar symptoms of *Septoria tritici* are already present on upper leaves will be less effective
- Some strains of *Septoria tritici* have developed resistance to some DMI fungicides and this may lead to poor control

Restrictions
- Maximum total dose 1 l/ha per crop
- Occasional slight temporary leaf speckling may occur on wheat but this has not been shown to reduce yield response or disease control
- Do not treat durum wheat

Crop-specific information
- Before watery ripe stage (GS 71)

Environmental safety
- Irritating to eyes and skin. May cause sensitization by skin contact
- Propiconazole may produce an allergic reaction
- Dangerous to fish or other aquatic life. Do not contaminate surface waters or ditches with chemical or used container
- LERAP Category B

Hazard classification and safety precautions
 Hazard Harmful, Dangerous for the environment
 Transport code 9
 Packaging group III
 UN number 3082
 Risk phrases R43, R50, R53a, R63
 Operator protection A, H; U05a, U09a, U11, U14, U15, U20b
 Environmental protection E15b, E16a, E16b, E38
 Storage and disposal D01, D02, D09a, D10c, D12a

385 propoxycarbazone-sodium

A sulfonylaminocarbonyltriazolinone residual grass weed herbicide for winter wheat
HRAC mode of action code: B

See also iodosulfuron-methyl-sodium + propoxycarbazone-sodium

Products
 Attribut　　　　　　　　　　Interfarm　　　　　　　　70% w/w　　　　　SG　14749

Uses
- Annual grasses in **miscanthus** *(off-label)*
- Blackgrass in **winter wheat**
- Couch in **miscanthus** *(off-label)*, **winter wheat**

Specific Off-Label Approvals (SOLAs)
- *miscanthus* 20093179

Approval information
- Propoxycarbazone-sodium included in Annex I under EC Directive 91/414

FOR FULL CONDITIONS OF USE ALWAYS READ THE PRODUCT LABEL

propyzamide

Efficacy guidance
- Activity by root and foliar absorption but depends on presence of sufficient soil moisture to ensure root uptake by weeds. Control is enhanced by use of an adjuvant to encourage foliar uptake
- Best results obtained from treatments applied when weed grasses are growing actively. Symptoms may not become apparent for 3-4 wk after application
- Ensure good even spray coverage
- To achieve best control of couch and reduction of infestation in subsequent crop treat between 2 true leaves and first node stage of the weed
- Blackgrass should be treated after using specialist blackgrass herbicides with a different mode of action
- Propoxycarbazone-sodium is a member of the ALS-inhibitor group of herbicides. To avoid the build up of resistance do not use any product containing an ALS-inhibitor herbicide with claims for control of grass weeds more than once on any crop
- Use these products as part of a resistance management strategy that includes cultural methods of control and does not use ALS inhibitors as the sole chemical method of grass weed control

Restrictions
- Maximum total dose equivalent to one full dose treatment
- Do not use in a programme with other aceto-lactase synthesis (ALS) inhibitors
- Do not apply when temperature near or below freezing
- Avoid treatment under dry soil conditions

Crop-specific information
- Latest use: third node detectable in wheat (GS 33)

Following crops guidance
- Only winter wheat, field beans or winter barley may be sown in the autumn following spring treatment
- Any crop may be grown in the spring on land treated during the previous calendar yr

Environmental safety
- Dangerous for the environment
- Very toxic to aquatic organisms
- Take extreme care to avoid drift onto adjacent plants or land as this could result in severe damage
- LERAP Category B

Hazard classification and safety precautions
Hazard Dangerous for the environment
Transport code 9
Packaging group III
UN number 3077
Risk phrases R50, R53a
Operator protection A; U05a, U20b
Environmental protection E15b, E16a, E16b, E38
Storage and disposal D01, D02, D09a, D11a, D12a

386 propyzamide

A residual benzamide herbicide for use in a wide range of crops
HRAC mode of action code: K1

See also clopyralid + propyzamide

Products

1	Careca	AgriChem BV	500 g/l	SC	14948
2	Cohort	Makhteshim	400 g/l	SC	15035
3	Conform	AgChem Access	400 g/l	SC	14602
4	Engage	Interfarm	50% w/w	WP	14233
5	Flomide	Interfarm	400 g/l	SC	14223
6	Kerb 50 W	Dow	50% w/w	WP	13715
7	Kerb Flo	Dow	400 g/l	SC	13716
8	Kerb Granules	Barclay	4% w/w	GR	14213
9	Megaflo	Dow	400 g/l	SC	15034
10	Menace 80 EDF	Dow	80% w/w	WG	13714
11	Pizamide	ChemSource	400 g/l	SC	14684

SEE SECTION 3 FOR PRODUCTS ALSO REGISTERED

Pesticide Profiles

Products – continued

12 Pizza 400 SC	Goldengrass	400 g/l	SC	14430
13 Ponder	ChemSource	400 g/l	SC	14693
14 Precis	Dow	400 g/l	SC	14202
15 Propel Flo	Clayton	400 g/l	SC	15044
16 Proper Flo	Globachem	400 g/l	SC	15102
17 Propyzamide 400	European Ag	400 g/l	SC	14685
18 Quaver Flo	Dow	400 g/l	SC	14203
19 Relva	Belcrop	400 g/l	SC	14873
20 Setanta 50 WP	AgriGuard	50% w/w	WP	14494
21 Setanta 50 WP	AgriGuard	50% w/w	WP	14943
22 Setanta Flo	AgriGuard	400 g/l	SC	14493
23 Setanta Flo	AgriGuard	400 g/l	SC	14678
24 Solitaire	AgriGuard	400 g/l	SC	14698
25 Solitaire 50 WP	AgriGuard	50% w/w	WP	14962
26 Standon Santa Fe 50 WP	Standon	50% w/w	WP	14966
27 Standon Santa Fe Flo	Standon	400 g/l	SC	14727
28 Stroller	AgChem Access	400 g/l	SC	14603

Uses

- Annual and perennial weeds in **amenity vegetation** [23, 24]
- Annual dicotyledons in **all edible seed crops grown outdoors** (off-label), **all non-edible seed crops grown outdoors** (off-label), **game cover** (off-label), **hops** (off-label) [6, 7]; **almonds** (off-label), **bilberries** (off-label), **blueberries** (off-label), **broccoli** (off-label), **calabrese** (off-label), **cauliflowers** (off-label), **cherries** (off-label), **chestnuts** (off-label), **chicory root** (off-label), **cob nuts** (off-label), **courgettes** (off-label), **cranberries** (off-label), **cress** (off-label), **endives** (off-label), **frise** (off-label), **hazel nuts** (off-label), **herbs (see appendix 6)** (off-label), **lamb's lettuce** (off-label), **leaf brassicas** (off-label), **marrows** (off-label), **mirabelles** (off-label), **protected endives** (off-label), **protected forest nurseries** (off-label), **protected herbs (see appendix 6)** (off-label), **protected lettuce** (off-label), **pumpkins** (off-label), **quinces** (off-label), **radicchio** (off-label), **salad brassicas** (off-label), **scarole** (off-label), **squashes** (off-label), **table grapes** (off-label), **walnuts** (off-label), **wine grapes** (off-label) [7]; **amenity vegetation** [1-3, 6, 7, 9-13, 17, 18, 26-28]; **apple orchards**, **clover seed crops**, **pear orchards**, **plums** [1-7, 9-14, 17, 18, 20, 22-24, 26-28]; **blackberries**, **blackcurrants**, **gooseberries**, **loganberries**, **lucerne**, **raspberries** (England only), **redcurrants**, **strawberries** [2-7, 9-14, 17, 18, 20, 22-24, 26-28]; **brassica seed crops**, **lettuce** (outdoor crops), **rhubarb** (outdoor) [10]; **farm forestry** [1-3, 6, 7, 9-13, 15, 17, 21, 23-28]; **fodder rape seed crops**, **kale seed crops**, **turnip seed crops** [1-7, 9, 11-14, 16-18, 20, 22-24, 26-28]; **forest** [1-4, 6-9, 11-13, 15-17, 20, 21, 23-28]; **forest nurseries** [1-3, 6, 7, 9-13, 15, 17, 20, 21, 23-28]; **forest nurseries** (off-label) [14]; **hedges** [1-3, 6, 7, 9-13, 17, 21, 23-28]; **lettuce** [3-7, 9, 11-18, 20-28]; **ornamental plant production** [9, 20]; **red clover**, **white clover** [16]; **rhubarb** [1-7, 9, 11-14, 17, 18, 20, 22-24, 26-28]; **roses** [4, 20]; **sugar beet seed crops** [1-7, 9-14, 16-18, 20, 22-24, 26-28]; **top fruit** (off-label) [6]; **trees and shrubs** [4, 8]; **winter field beans** [2-7, 9-18, 20-28]; **winter oilseed rape** [1-7, 9-28]; **woody ornamentals** [4, 6, 8, 10]
- Annual grasses in **all edible seed crops grown outdoors** (off-label), **all non-edible seed crops grown outdoors** (off-label), **game cover** (off-label), **hops** (off-label) [6, 7]; **almonds** (off-label), **bilberries** (off-label), **blueberries** (off-label), **broccoli** (off-label), **calabrese** (off-label), **cauliflowers** (off-label), **cherries** (off-label), **chestnuts** (off-label), **chicory root** (off-label), **cob nuts** (off-label), **courgettes** (off-label), **cranberries** (off-label), **cress** (off-label), **endives** (off-label), **frise** (off-label), **hazel nuts** (off-label), **herbs (see appendix 6)** (off-label), **lamb's lettuce** (off-label), **leaf brassicas** (off-label), **marrows** (off-label), **mirabelles** (off-label), **protected endives** (off-label), **protected forest nurseries** (off-label), **protected herbs (see appendix 6)** (off-label), **protected lettuce** (off-label), **pumpkins** (off-label), **quinces** (off-label), **radicchio** (off-label), **salad brassicas** (off-label), **scarole** (off-label), **squashes** (off-label), **table grapes** (off-label), **walnuts** (off-label), **wine grapes** (off-label) [7]; **amenity vegetation** [1-3, 6, 7, 9-13, 17, 18, 26-28]; **apple orchards**, **clover seed crops**, **pear orchards**, **plums** [1-7, 9-14, 17, 18, 20, 22-24, 26-28]; **blackberries**, **blackcurrants**, **gooseberries**, **loganberries**, **lucerne**, **raspberries** (England only), **redcurrants**, **strawberries** [2-7, 9-14, 17, 18, 20, 22-24, 26-28]; **brassica seed crops**, **lettuce** (outdoor crops), **rhubarb** (outdoor) [10]; **farm forestry** [1-3, 6, 7, 9-13, 15, 17, 21, 23-28]; **fodder rape seed crops**, **kale seed crops**, **turnip seed crops** [1-7, 9, 11-14, 16-18, 20, 22-24, 26-28]; **forest** [1-4, 6-9, 11-13, 15-17, 20, 21, 23-28]; **forest nurseries** [1-3, 6, 7, 9-13, 15, 17, 20, 21, 23-28]; **forest nurseries** (off-label) [14]; **hedges** [1-3, 6, 7, 9-13, 17, 21, 23-28]; **lettuce** [3-7, 9, 11-18, 20-28]; **ornamental plant production** [9, 20]; **red clover**, **white clover** [16]; **rhubarb** [1-

FOR FULL CONDITIONS OF USE ALWAYS READ THE PRODUCT LABEL

propyzamide

7, 9, 11-14, 17, 18, 20, 22-24, 26-28]; *roses* [4, 20]; *sugar beet seed crops* [1-7, 9-14, 16-18, 20, 22-24, 26-28]; *top fruit (off-label)* [6]; *trees and shrubs* [4, 8]; *winter field beans* [2-7, 9-18, 20-28]; *winter oilseed rape* [1-7, 9-28]; *woody ornamentals* [4, 6, 8, 10]
- Horsetails in *forest* [10]
- Perennial grasses in *all edible seed crops grown outdoors (off-label)*, *all non-edible seed crops grown outdoors (off-label)*, *game cover (off-label)*, *hops (off-label)* [6, 7]; *almonds (off-label)*, *bilberries (off-label)*, *blueberries (off-label)*, *broccoli (off-label)*, *calabrese (off-label)*, *cauliflowers (off-label)*, *cherries (off-label)*, *chestnuts (off-label)*, *chicory root (off-label)*, *cob nuts (off-label)*, *courgettes (off-label)*, *cranberries (off-label)*, *cress (off-label)*, *endives (off-label)*, *frise (off-label)*, *hazel nuts (off-label)*, *herbs (see appendix 6) (off-label)*, *lamb's lettuce (off-label)*, *leaf brassicas (off-label)*, *marrows (off-label)*, *mirabelles (off-label)*, *protected endives (off-label)*, *protected forest nurseries (off-label)*, *protected herbs (see appendix 6) (off-label)*, *protected lettuce (off-label)*, *pumpkins (off-label)*, *quinces (off-label)*, *radicchio (off-label)*, *salad brassicas (off-label)*, *scarole (off-label)*, *squashes (off-label)*, *table grapes (off-label)*, *walnuts (off-label)*, *wine grapes (off-label)* [7]; *amenity vegetation* [1-3, 6, 7, 9-13, 17, 18, 26-28]; *apple orchards, pear orchards, plums* [1-7, 9-14, 17, 18, 20, 22-24, 26-28]; *blackberries, blackcurrants, gooseberries, loganberries, raspberries (England only)*, *redcurrants* [2-7, 9-14, 17, 18, 20, 22-24, 26-28]; *clover seed crops, rhubarb* [1-7, 9, 11-14, 17, 18, 20, 22-24, 26-28]; *farm forestry* [1-3, 6, 7, 9-13, 15, 17, 23, 24, 26-28]; *fodder rape seed crops, kale seed crops, sugar beet seed crops, turnip seed crops* [1-7, 9, 11-14, 16-18, 20, 22-24, 26-28]; *forest* [1-4, 6-13, 15-17, 20, 23, 24, 26-28]; *forest nurseries* [1-3, 6, 7, 9-13, 15, 17, 20, 23, 24, 26-28]; *hedges* [1-3, 6, 7, 9-13, 17, 23, 24, 26-28]; *lettuce* [3-7, 9, 11-18, 20, 22-24, 26-28]; *lucerne, strawberries* [2-7, 9, 11-14, 17, 18, 20, 22-24, 26-28]; *ornamental plant production* [9, 20]; *red clover, white clover* [16]; *rhubarb (outdoor)* [10]; *roses* [4, 20]; *top fruit (off-label)* [6]; *trees and shrubs* [4, 8]; *winter field beans* [2-7, 9, 11-18, 20, 22-24, 26-28]; *winter oilseed rape* [1-7, 9, 11-20, 22-24, 26-28]; *woody ornamentals* [4, 6, 8, 10]
- Sedges in *forest* [10]
- Volunteer cereals in *sugar beet seed crops, winter field beans, winter oilseed rape* [10]
- Wild oats in *sugar beet seed crops, winter field beans, winter oilseed rape* [10]

Specific Off-Label Approvals (SOLAs)
- *all edible seed crops grown outdoors* 20082943 [6], 20082942 [7]
- *all non-edible seed crops grown outdoors* 20082943 [6], 20082942 [7]
- *almonds* 20082420 [7]
- *bilberries* 20082419 [7]
- *blueberries* 20082419 [7]
- *broccoli* 20091902 [7]
- *calabrese* 20091902 [7]
- *cauliflowers* 20091902 [7]
- *cherries* 20082418 [7]
- *chestnuts* 20082420 [7]
- *chicory root* 20091530 [7]
- *cob nuts* 20082420 [7]
- *courgettes* 20082416 [7]
- *cranberries* 20082419 [7]
- *cress* 20082410 [7]
- *endives* 20082410 [7]
- *forest nurseries* 20082910 [14]
- *frise* 20082411 [7]
- *game cover* 20082943 [6], 20082942 [7]
- *hazel nuts* 20082420 [7]
- *herbs (see appendix 6)* 20082412 [7]
- *hops* 20082943 [6], 20082414 [7]
- *lamb's lettuce* 20082410 [7]
- *leaf brassicas* 20082410 [7]
- *marrows* 20082416 [7]
- *mirabelles* 20082418 [7]
- *protected endives* 20082415 [7]
- *protected forest nurseries* 20082942 [7]
- *protected herbs (see appendix 6)* 20082412 [7]

SEE SECTION 3 FOR PRODUCTS ALSO REGISTERED

- **protected lettuce** 20082415 [7]
- **pumpkins** 20082416 [7]
- **quinces** 20082413 [7]
- **radicchio** 20082411 [7]
- **salad brassicas** 20082410 [7]
- **scarole** 20082411 [7]
- **squashes** 20082416 [7]
- **table grapes** 20082417 [7]
- **top fruit** 20082943 [6]
- **walnuts** 20082420 [7]
- **wine grapes** 20082417 [7]

Approval information
- Propyzamide included in Annex I under EC Directive 91/414
- Some products may be applied through CDA equipment. See labels for details
- Accepted by BBPA for use on hops

Efficacy guidance
- Active via root uptake. Weeds controlled from germination to young seedling stage, some species (including many grasses) also when established
- Best results achieved by winter application to fine, firm, moist soil. Rain is required after application if soil dry
- Uptake is slow and and may take up to 12 wk
- Excessive organic debris or ploughed-up turf may reduce efficacy
- For heavy couch infestations a repeat application may be needed in following winter
- Always follow WRAG guidelines for preventing and managing herbicide resistant weeds. See Section 5 for more information

Restrictions
- Maximum number of treatments 1 per crop or yr
- Maximum total dose equivalent to one full dose treatment for all crops
- Do not treat protected crops
- Apply to listed edible crops only between 1 Oct and the date specified as the latest time of application (except lettuce)
- Do not apply in windy weather and avoid drift onto non-target crops
- Do not use on soils with more than 10% organic matter except in forestry

Crop-specific information
- Latest use: labels vary but normally before 31 Dec in year before harvest for rhubarb, lucerne, strawberries and winter field beans; before 31 Jan for other crops
- HI: 6 wk for edible crops
- Apply as soon as possible after 3-true leaf stage of oilseed rape (GS 1,3) and seed brassicas, after 4-leaf stage of sugar beet for seed, within 7 d after sowing but before emergence for field beans, after perennial crops established for at least 1 season, strawberries after 1 yr
- Only apply to strawberries on heavy soils. Do not use on matted row crops
- Only apply to field beans on medium and heavy soils
- Only apply to established lucerne not less than 7 d after last cut
- In lettuce lightly incorporate in top 25 mm pre-drilling or irrigate on dry soil
- See label for lists of ornamental and forest species which may be treated

Following crops guidance
- Following an application between 1 Apr and 31 Jul at any dose the following minimum intervals must be observed before sowing the next crop: lettuce 0 wk; broad beans, chicory, clover, field beans, lucerne, radishes, peas 5 wk; brassicas, celery, leeks, oilseed rape, onions, parsley, parsnips 10 wk
- Following an application between 1 Aug and 31 Mar at any dose the following minimum intervals must be observed before sowing the next crop: lettuce 0 wk; broad beans, chicory, clover, field beans, lucerne, radishes, peas 10 wk; brassicas, celery, leeks, oilseed rape, onions, parsley, parsnips 25 wk or after 15 Jun, whichever occurs sooner
- Cereals or grasses or other crops not listed may be sown 30 wk after treatment up to 840 g ai/ha between 1 Aug and 31 Mar or 40 wk after treatment at higher doses at any time and after mouldboard ploughing to at least 15 cm
- A period of at least 9 mth must elapse between applications of propyzamide to the same land

FOR FULL CONDITIONS OF USE ALWAYS READ THE PRODUCT LABEL

proquinazid

Environmental safety
- Dangerous for the environment
- Very toxic to aquatic organisms
- Some pesticides pose a greater threat of contamination of water than others and propyzamide is one of these pesticides. Take special care when applying propyzamide near water and do not apply if heavy rain is forecast

Hazard classification and safety precautions
Hazard Harmful, Dangerous for the environment [1-17, 19-28]
Transport code 9
Packaging group III
UN number 3077 [4, 6, 8, 10, 20, 21, 25, 26]; 3082 [1-3, 5, 7, 9, 11-19, 22-24, 27, 28]
Risk phrases R40, R53a [1-17, 19-28]; R50 [2, 3, 5-7, 9-17, 19, 22-24, 27, 28]; R51 [1, 4, 8, 20, 21, 25, 26]
Operator protection A, H [1-4, 6-17, 19, 21, 23-28]; D [2, 3, 6, 7, 10-13, 17, 21, 25, 26, 28]; M [2, 3, 6, 7, 9-14, 17, 19, 21, 23-28]; U05a [8]; U20c [1-7, 9-17, 19-28]
Environmental protection E15a [1-5, 7-17, 20, 22, 27, 28]; E15b [6, 19, 21, 23-26]; E34 [1, 15, 16, 20-22, 25-27]; E38 [1-17, 19, 22-24, 27, 28]; E39 [19, 23, 24]
Consumer protection C02a [1-7, 9-17, 20-22, 25-28] (6 wk)
Storage and disposal D01, D02 [1, 8]; D05, D11a [1-7, 9-17, 20-22, 25-28]; D07 [1, 15, 16, 22, 27]; D09a [1-17, 20-22, 25-28]; D12a [1-17, 22, 27, 28]
Medical advice M03 [8]

387 proquinazid

A quinazolinone fungicide for powdery mildew control in cereals
FRAC mode of action code: U7

See also prochloraz + proquinazid + tebuconazole

Products

1	Justice	DuPont	200 g/l	EC	12835
2	Talius	DuPont	200 g/l	EC	12752

Uses
- Powdery mildew in ***durum wheat*** *(off-label),* ***spring barley***, ***spring oats***, ***spring rye***, ***spring wheat***, ***triticale***, ***winter barley***, ***winter oats***, ***winter rye***, ***winter wheat*** [1, 2]; ***forest nurseries*** *(off-label),* ***soft fruit*** *(off-label),* ***top fruit*** *(off-label)* [2]

Specific Off-Label Approvals (SOLAs)
- ***durum wheat*** *20090418 expires 29 Jul 2011* [1], *20090419 expires 29 Jul 2011* [2]
- ***forest nurseries*** *20090420 expires 29 Jul 2011* [2]
- ***soft fruit*** *20090420 expires 29 Jul 2011* [2]
- ***top fruit*** *20090420 expires 29 Jul 2011* [2]

Approval information
- Proquinazid included in Annex 1 under EC Directive 91/414
- Accepted by BBPA on malting barley
- Approval expiry 29 Jul 2011 [1, 2]

Efficacy guidance
- Best results obtained from preventive treatment before disease is established in the crop
- Where mildew has already spread to new growth a tank mix with a curative fungicide with an alternative mode of action should be used
- Use as part of an integrated crop management (ICM) strategy incorporating other methods of control or fungicides with different modes of action

Restrictions
- Maximum number of treatments 2 per crop
- Do not apply to any crop suffering from stress from any cause
- Avoid application in either frosty or hot, sunny conditions

Crop-specific information
- Latest use: before beginning of heading (GS 49) for barley, oats, rye, triticale; before full flowering (GS 65) for wheat

SEE SECTION 3 FOR PRODUCTS ALSO REGISTERED

Pesticide Profiles

Environmental safety
- Dangerous for the environment
- Toxic to aquatic organisms
- Dangerous to fish or other aquatic life. Do not contaminate surface waters or ditches with chemical or used container
- LERAP Category B

Hazard classification and safety precautions
Hazard Harmful, Dangerous for the environment
Transport code 9
Packaging group III
UN number 3082
Risk phrases R38, R40, R41, R51, R53a
Operator protection A, C, H; U05a, U11
Environmental protection E13b, E16a, E16b, E34, E38
Storage and disposal D01, D02, D05, D09a, D10b, D11a, D12a
Medical advice M03, M05a

388 prosulfocarb

A thiocarbamate herbicide for grass and broad-leaved weed control in cereals and potatoes
HRAC mode of action code: N

See also clodinafop-propargyl + prosulfocarb

Products

1	Defy	Syngenta	800 g/l	EC	12606
2	Dian	AgChem Access	800 g/l	EC	13567

Uses
- Annual dicotyledons in **bulb onions** (off-label), **celeriac** (off-label), **celery (outdoor)** (off-label), **durum wheat** (off-label), **garlic** (off-label), **leeks** (off-label), **poppies for morphine production** (off-label), **shallots** (off-label), **spring field beans** (off-label), **spring wheat** (off-label), **winter field beans** (off-label), **winter linseed** (off-label) [1]; **potatoes, winter barley, winter wheat** [1, 2]
- Annual grasses in **bulb onions** (off-label), **carrots** (off-label), **celeriac** (off-label), **celery (outdoor)** (off-label), **durum wheat** (off-label), **garlic** (off-label), **herbs (see appendix 6)** (off-label), **horseradish** (off-label), **leeks** (off-label), **parsley root** (off-label), **parsnips** (off-label), **poppies for morphine production** (off-label), **salsify** (off-label), **shallots** (off-label), **spring field beans** (off-label), **spring onions** (off-label), **spring rye** (off-label), **spring wheat** (off-label), **triticale** (off-label), **winter field beans** (off-label), **winter linseed** (off-label), **winter rye** (off-label) [1]
- Annual meadow grass in **poppies for morphine production** (off-label), **spring barley** (off-label), **spring field beans** (off-label), **spring onions** (off-label), **spring wheat** (off-label), **winter linseed** (off-label) [1]; **potatoes, winter barley, winter wheat** [1, 2]
- Black nightshade in **herbs (see appendix 6)** (off-label) [1]
- Chickweed in **potatoes, winter barley, winter wheat** [1, 2]
- Cleavers in **carrots** (off-label), **herbs (see appendix 6)** (off-label), **horseradish** (off-label), **parsley root** (off-label), **parsnips** (off-label), **salsify** (off-label), **spring field beans** (off-label), **spring wheat** (off-label), **winter field beans** (off-label) [1]; **potatoes, winter barley, winter wheat** [1, 2]
- Fat hen in **carrots** (off-label), **horseradish** (off-label), **parsley root** (off-label), **parsnips** (off-label), **salsify** (off-label) [1]
- Field speedwell in **potatoes, winter barley, winter wheat** [1, 2]
- Fumitory in **herbs (see appendix 6)** (off-label), **spring onions** (off-label), **spring wheat** (off-label) [1]
- Ivy-leaved speedwell in **potatoes, winter barley, winter wheat** [1, 2]
- Knotgrass in **carrots** (off-label), **horseradish** (off-label), **parsley root** (off-label), **parsnips** (off-label), **salsify** (off-label) [1]
- Loose silky bent in **potatoes, winter barley, winter wheat** [1, 2]
- Mayweeds in **carrots** (off-label), **horseradish** (off-label), **parsley root** (off-label), **parsnips** (off-label), **salsify** (off-label) [1]
- Rough meadow grass in **potatoes, winter barley, winter wheat** [1, 2]
- Volunteer oilseed rape in **spring onions** (off-label) [1]

FOR FULL CONDITIONS OF USE ALWAYS READ THE PRODUCT LABEL

prosulfocarb

Specific Off-Label Approvals (SOLAs)
- *bulb onions* 20092745 [1]
- *carrots* 20073774 [1]
- *celeriac* 20073777 [1]
- *celery (outdoor)* 20071752 [1]
- *durum wheat* 20073781 [1]
- *garlic* 20092745 [1]
- *herbs (see appendix 6)* 20072073 [1]
- *horseradish* 20073774 [1]
- *leeks* 20092744 [1]
- *parsley root* 20073774 [1]
- *parsnips* 20073774 [1]
- *poppies for morphine production* 20082243 [1]
- *salsify* 20073774 [1]
- *shallots* 20092745 [1]
- *spring barley* 20073778 [1]
- *spring field beans* 20082932 [1]
- *spring onions* 20080085 [1]
- *spring rye* 20073779 [1]
- *spring wheat* 20090237 [1]
- *triticale* 20073779 [1]
- *winter field beans* 20073780 [1]
- *winter linseed* 20092012 [1]
- *winter rye* 20073779 [1]

Approval information
- Prosulfocarb included in Annex I under EC Directive 91/414
- Accepted by BBPA for use on malting barley

Efficacy guidance
- Best results obtained in cereals from treatment of crops in a firm moist seedbed, free from clods
- Pre-emergence use will reduce blackgrass populations but the product should only be used against this weed as part of a management strategy involving sequences with products of alternative modes of action
- Always follow WRAG guidelines for preventing and managing herbicide resistant weeds. See Section 5 for more information

Restrictions
- Maximum number of treatments 1 per crop for winter barley, winter wheat and potatoes
- Do not apply to crops under stress from any cause. Transient yellowing can occur from which recovery is complete
- Winter cereals must be covered by 3 cm of settled soil

Crop-specific information
- Latest use: at emergence (soil rising over emerging potato shoots) for potatoes, up to and including early tillering (GS 21) for winter barley, winter wheat
- When applied pre-emergence to cereals crop emergence may occasionally be slowed down but yield is not affected
- For potatoes, complete ridge formation before application and do not disturb treated soil afterwards
- Peas are severely damaged or killed.

Following crops guidance
- Do not sow field or broad beans within 12 mth of treatment
- In the event of failure of a treated cereal crop, winter wheat, winter barley or transplanted brassicas may be re-sown immediately. seed brassicas can be sown 14 weeks after application provided that the land is ploughed first. In the following spring sunflowers, maize, flax, spring cereals, peas, oilseed rape or soya beans may be sown without ploughing, and carrots, lettuce, onions, sugar beet or potatoes may be sown or planted after ploughing

SEE SECTION 3 FOR PRODUCTS ALSO REGISTERED

Pesticide Profiles

Environmental safety
- Dangerous for the environment
- Very toxic to aquatic organisms
- LERAP Category B

Hazard classification and safety precautions
Hazard Irritant, Dangerous for the environment
Transport code 9
Packaging group III
UN number 3082
Risk phrases R38, R43, R50, R53a
Operator protection A, C, H; U02a, U05a, U08, U14, U15, U20b
Environmental protection E15b, E16a, E16b, E38
Storage and disposal D01, D02, D05, D09a, D10c, D12a
Medical advice M05a

389 prosulfuron

A contact and residual sulfonyl urea herbicide available only in mixture
HRAC mode of action code: B

See also bromoxynil + prosulfuron

390 prothioconazole

A systemic, protectant and curative triazole fungicide
FRAC mode of action code: 3

See also bixafen + prothioconazole
bixafen + prothioconazole + spiroxamine
bixafen + prothioconazole + tebuconazole
clothianidin + prothioconazole
clothianidin + prothioconazole + tebuconazole + triazoxide
fluoxastrobin + prothioconazole
fluoxastrobin + prothioconazole + tebuconazole
fluoxastrobin + prothioconazole + trifloxystrobin

Products

1	Clayton Impress	Clayton	250 g/l	EC	12912
2	Proline 275	Bayer CropScience	275 g/l	EC	14790
3	Prothio	ChemSource	250 g/l	EC	14945
4	Redigo	Bayer CropScience	100 g/l	FS	12085
5	Rudis	Bayer CropScience	480 g/l	SC	14122
6	Standon Win-Win	Standon	250 g/l	EC	12386

Uses
- Alternaria in **cabbages** [5]
- Brown rust in **durum wheat** [3]; **spring barley**, **winter barley**, **winter rye**, **winter wheat** [1-3, 6]; **spring wheat** [2]
- Bunt in **durum wheat** (seed treatment), **spring rye** (seed treatment), **spring wheat** (seed treatment), **triticale** (seed treatment), **winter rye** (seed treatment), **winter wheat** (seed treatment) [4]
- Covered smut in **winter barley** (seed treatment) [4]
- Crown rust in **spring oats**, **winter oats** [2]
- Eyespot in **durum wheat** [3]; **spring barley**, **winter barley**, **winter rye**, **winter wheat** [1-3, 6]; **spring oats**, **spring wheat**, **winter oats** [2]
- Fusarium foot rot and seedling blight in **durum wheat** (seed treatment), **spring oats** (seed treatment), **spring rye** (seed treatment), **spring wheat** (seed treatment), **triticale** (seed treatment), **winter barley** (seed treatment), **winter oats** (seed treatment), **winter rye** (seed treatment), **winter wheat** (seed treatment) [4]
- Glume blotch in **durum wheat** [3]; **spring wheat** [2]; **winter wheat** [1-3, 6]

FOR FULL CONDITIONS OF USE ALWAYS READ THE PRODUCT LABEL

prothioconazole

- Late ear diseases in **durum wheat** [3]; **spring barley, winter barley, winter wheat** [1-3, 6]; **spring wheat** [2]
- Leaf stripe in **winter barley** *(seed treatment)* [4]
- Light leaf spot in **cabbages** [5]
- Loose smut in **durum wheat** *(seed treatment)*, **spring oats** *(seed treatment)*, **spring wheat** *(seed treatment)*, **winter barley** *(seed treatment)*, **winter oats** *(seed treatment)*, **winter wheat** *(seed treatment)* [4]
- Net blotch in **spring barley, winter barley** [1-3, 6]
- Phoma in **cabbages** [5]
- Powdery mildew in **cabbages** [5]; **durum wheat** [3]; **spring barley, winter barley, winter rye, winter wheat** [1-3, 6]; **spring oats, spring wheat, winter oats** [2]
- Purple blotch in **leeks** [5]
- Rhynchosporium in **spring barley, winter barley, winter rye** [1-3, 6]
- Ring spot in **cabbages** [5]
- Rust in **leeks** [5]
- Sclerotinia stem rot in **winter oilseed rape** [1-3, 6]
- Septoria leaf blotch in **durum wheat** [3]; **spring wheat** [2]; **winter wheat** [1-3, 6]
- Tan spot in **durum wheat** [3]; **spring wheat** [2]; **winter wheat** [1-3, 6]
- Yellow rust in **durum wheat** [3]; **spring wheat** [2]; **winter barley, winter wheat** [1-3, 6]

Approval information
- Accepted by BBPA for use on malting barley
- Prothioconazole included in Annex I under EC Directive 91/414
- Approval expiry 31 Jul 2011 [1, 6]

Efficacy guidance
- Seed treatments must be applied by manufacturer's recommended treatment application equipment [4]
- Treated cereal seed should preferably be drilled in the same season [4]
- Follow-up treatments will be needed later in the season to give protection against air-borne and splash-borne diseases [4]
- Best results on cereal foliar diseases obtained from treatment at early stages of disease development. Further treatment may be needed if disease attack is prolonged [2, 6]
- Foliar applications to established infections of any disease are likely to be less effective [2, 6]
- Best control of cereal ear diseases obtained by treatment during ear emergence [2, 6]
- Treat oilseed rape at early to full flower [2, 6]
- Prothioconazole is a DMI fungicide. Resistance to some DMI fungicides has been identified in Septoria leaf blotch which may seriously affect performance of some products. For further advice contact a specialist advisor and visit the Fungicide Resistance Action Group (FRAG)-UK website

Restrictions
- Maximum number of seed treatments one per batch [4]
- Maximum total dose equivalent to one full dose treatment on oilseed rape, two full dose treatments on barley, oats; three full dose treatments on wheat, rye [2, 6]
- Seed treatment must be fully re-dispersed and homogeneous before use [4]
- Do not use on seed with more than 16% moisture content, or on sprouted, cracked or skinned seed [4]
- All seed batches should be tested to ensure they are suitable for treatment [4]
- Treated winter barley seed must be used within the season of treatment; treated winter wheat seed should preferably be drilled in the same season [4]
- Do not make repeated treatments of the product alone to the same crop against pathogens such as powdery mildew. Use tank mixtures or alternate with fungicides having a different mode of action [4]
- Do not treat oilseed rape crops to be used for seed production [2, 6]

Crop-specific information
- Latest use: pre-drilling for seed treatments [4]; before grain milky ripe for foliar sprays on winter rye, winter wheat; beginning of flowering for barley, oats [2, 6]
- HI 56 d [2] or 65 d [6] for winter oilseed rape

Environmental safety
- Dangerous for the environment [2, 6]
- Toxic to aquatic organisms [2, 6]

SEE SECTION 3 FOR PRODUCTS ALSO REGISTERED

Pesticide Profiles

- Harmful to aquatic organisms [4]
- LERAP Category B [1-3, 5, 6]

Hazard classification and safety precautions
  Hazard Irritant [1-4, 6]; Dangerous for the environment [1-3, 5, 6]
  Transport code 9 [1-3, 5, 6]
  Packaging group III [1-3, 5, 6]
  UN number 3082 [1-3, 5, 6]; N/C [4]
  Risk phrases R36 [1-3, 6]; R43, R52 [4]; R51 [1-3, 5, 6]; R53a [1-6]
  Operator protection A, H [1-6]; C [1-3, 6]; U05a, U20b [1-3, 5, 6]; U07, U14, U20a [4]; U09b [1-3, 6]
  Environmental protection E15a [1-4, 6]; E15b [5]; E16a, E38 [1-3, 5, 6]; E34 [1-6]; E36a [4]
  Storage and disposal D01, D02, D12a [1-6]; D05, D09a [1-4, 6]; D10b [1-3, 5, 6]; D14 [4]
  Treated seed S01, S02, S03, S04a, S04b, S05, S06b, S07, S08 [4]
  Medical advice M03 [1-3, 5, 6]

391 prothioconazole + spiroxamine

A broad spectrum fungicide mixture for cereals
FRAC mode of action code: 3 + 5

See also spiroxamine

Products

1	Helix	Bayer CropScience	160:300 g/l	EC	12264
2	Spiral	AgChem Access	160:300 g/l	EC	13993

Uses
- Brown rust in **spring barley, winter barley, winter rye, winter wheat**
- Crown rust in **spring oats, winter oats**
- Eyespot in **spring barley, spring oats, winter barley, winter oats, winter rye, winter wheat**
- Glume blotch in **winter wheat**
- Late ear diseases in **winter rye, winter wheat**
- Net blotch in **spring barley, winter barley**
- Powdery mildew in **spring barley, spring oats, winter barley, winter oats, winter rye, winter wheat**
- Rhynchosporium in **spring barley, winter barley, winter rye**
- Septoria leaf blotch in **winter wheat**
- Tan spot in **winter wheat**
- Yellow rust in **spring barley, winter barley, winter wheat**

Approval information
- Prothioconazole and spiroxamine included in Annex I under EC Directive 91/414
- Accepted by BBPA for use on malting barley
- Approval expiry 31 Jul 2011 [2]

Efficacy guidance
- Best results obtained from treatment at early stages of disease development. Further treatment may be needed if disease attack is prolonged
- Applications to established infections of any disease are likely to be less effective
- Best control of cereal ear diseases obtained by treatment during ear emergence
- Prothioconazole is a DMI fungicide. Resistance to some DMI fungicides has been identified in Septoria leaf blotch which may seriously affect performance of some products. For further advice contact a specialist advisor and visit the Fungicide Resistance Action Group (FRAG)-UK website

Restrictions
- Maximum total dose equivalent to two full dose treatments on barley and oats and three full dose treatments on wheat and rye

Crop-specific information
- Latest use: before grain watery ripe for rye, oats, winter wheat; up to beginning of anthesis for barley

FOR FULL CONDITIONS OF USE ALWAYS READ THE PRODUCT LABEL

prothioconazole + spiroxamine + tebuconazole

Environmental safety
- Dangerous for the environment
- Very toxic to aquatic organisms
- LERAP Category B

Hazard classification and safety precautions
Hazard Harmful, Dangerous for the environment
Transport code 9
Packaging group III
UN number 3082
Risk phrases R20, R22a, R36, R38, R50, R53a
Operator protection A, C, H; U05a, U09b, U11, U19a, U20b
Environmental protection E15a, E16a, E34, E38
Storage and disposal D01, D02, D05, D09a, D10b, D12a
Medical advice M03

392 prothioconazole + spiroxamine + tebuconazole

A broad spectrum fungicide mixture for cereals
FRAC mode of action code: 3 + 5 + 3

See also spiroxamine
tebuconazole

Products

Cello	Bayer CropScience	100:250:100 g/l	EC	13178

Uses
- Brown rust in **spring barley**, **winter barley**, **winter rye**, **winter wheat**
- Crown rust in **spring oats**, **winter oats**
- Eyespot in **spring barley** *(reduction)*, **spring oats**, **winter barley** *(reduction)*, **winter oats**, **winter rye** *(reduction)*, **winter wheat** *(reduction)*
- Glume blotch in **winter wheat**
- Late ear diseases in **spring barley**, **winter barley**, **winter wheat**
- Net blotch in **spring barley**, **winter barley**
- Powdery mildew in **spring barley**, **spring oats**, **winter barley**, **winter oats**, **winter rye**, **winter wheat**
- Rhynchosporium in **spring barley**, **winter barley**, **winter rye**
- Septoria leaf blotch in **winter wheat**
- Yellow rust in **spring barley**, **winter barley**, **winter wheat**

Approval information
- Prothioconazole and spiroxamine included in Annex I under EC Directive 91/414

Efficacy guidance
- Best results obtained from treatment at early stages of disease development. Further treatment may be needed if disease attack is prolonged
- Applications to established infections of any disease are likely to be less effective
- Best control of cereal ear diseases obtained by treatment during ear emergence
- Prothioconazole and tebuconazole are DMI fungicides. Resistance to some DMI fungicides has been identified in Septoria leaf blotch which may seriously affect performance of some products. For further advice contact a specialist advisor and visit the Fungicide Resistance Action Group (FRAG)-UK website

Restrictions
- Maximum total dose equivalent to two full dose treatments

Crop-specific information
- Latest use: before grain milky ripe stage for rye, wheat; up to beginning of anthesis for barley & oats

Pesticide Profiles

Environmental safety
- Dangerous for the environment
- Very toxic to aquatic organisms
- LERAP Category B

Hazard classification and safety precautions
Hazard Harmful, Dangerous for the environment
Transport code 9
Packaging group III
UN number 3082
Risk phrases R20, R36, R37, R38, R43, R50, R53a, R63
Operator protection A, C, H; U05a, U09b, U11, U14, U19a, U20b
Environmental protection E15a, E16a, E34, E38
Storage and disposal D01, D02, D05, D09a, D10b, D12a
Medical advice M03

393 prothioconazole + tebuconazole

A triazole fungicide mixture for cereals
FRAC mode of action code: 3 + 3

See also tebuconazole

Products

1	Corinth	Bayer CropScience	80;160 g/l	EC	14199
2	Kestrel	Bayer CropScience	160+80 g/L	EC	13809
3	Prosaro	Bayer CropScience	125:125 g/l	EC	12263
4	Proteb	AgChem Access	125:125 g/l	EC	13879
5	Standon Mastana	Standon	125:125 g/l	EC	13795

Uses
- Brown rust in **spring barley**, **winter barley**, **winter rye**, **winter wheat** [2-5]; **spring wheat** [3]
- Crown rust in **spring oats**, **winter oats** [2-4]
- Eyespot in **spring barley**, **winter barley**, **winter rye**, **winter wheat** [2, 5]; **spring barley** *(reduction)*, **winter barley** *(reduction)*, **winter rye** *(reduction)*, **winter wheat** *(reduction)* [3, 4]; **spring oats**, **winter oats** [2-4]; **spring wheat** [3]
- Foliar disease control in **grass seed crops** *(off-label)* [3]
- Fusarium ear blight in **spring barley**, **winter barley**, **winter wheat** [2]; **winter wheat** *(moderate control)* [5]
- Glume blotch in **spring wheat** [3]; **winter wheat** [2-5]
- Late ear diseases in **spring barley**, **winter barley**, **winter wheat** [3, 4]; **spring wheat** [3]
- Light leaf spot in **oilseed rape** [2]; **oilseed rape** *(moderate control only)* [1]; **spring oilseed rape**, **winter oilseed rape** [3, 4]; **winter oilseed rape** *(moderate control)* [5]
- Net blotch in **spring barley**, **winter barley** [2-5]; **winter rye** [5]
- Phoma in **spring oilseed rape** [3, 4]; **winter oilseed rape** [3-5]
- Phoma leaf spot in **oilseed rape** [1, 2]
- Powdery mildew in **spring barley**, **spring oats**, **winter barley**, **winter oats**, **winter rye**, **winter wheat** [2-5]; **spring wheat** [3]
- Rhynchosporium in **spring barley**, **winter barley**, **winter rye** [2-5]
- Sclerotinia stem rot in **oilseed rape** [1, 2]; **spring oilseed rape** [3, 4]; **winter oilseed rape** [3-5]
- Septoria leaf blotch in **spring wheat** [3]; **winter wheat** [2-5]
- Sooty moulds in **spring barley**, **winter barley**, **winter wheat** [2]; **winter wheat** *(reduction)* [5]
- Stem canker in **oilseed rape** [1]
- Tan spot in **spring wheat** [3]; **winter wheat** [2-4]
- Yellow rust in **spring barley**, **winter barley**, **winter wheat** [2-5]; **spring wheat** [3]; **winter rye** [5]

Specific Off-Label Approvals (SOLAs)
- *grass seed crops* 20081806 [3]

Approval information
- Prothioconazole included in Annex I under EC Directive 91/414
- Accepted by BBPA for use on malting barley

FOR FULL CONDITIONS OF USE ALWAYS READ THE PRODUCT LABEL

prothioconazole + tebuconazole + triazoxide

Efficacy guidance
- Best results on cereal foliar diseases obtained from treatment at early stages of disease development. Further treatment may be needed if disease attack is prolonged
- On oilseed rape apply a protective treatment in autumn/winter for Phoma followed by a further spray in early spring from the onset of stem elongation, if necessary. For control of Sclerotinia apply at early to full flower
- Applications to established infections of any disease are likely to be less effective
- Best control of cereal ear diseases obtained by treatment during ear emergence
- Prothioconazole and tebuconazole are DMI fungicides. Resistance to some DMI fungicides has been identified in Septoria leaf blotch which may seriously affect performance of some products. For further advice contact a specialist advisor and visit the Fungicide Resistance Action Group (FRAG)-UK website

Restrictions
- Maximum total dose equivalent to two full dose treatments on barley, oats, oilseed rape; three full dose treatments on wheat and rye

Crop-specific information
- Latest use: before grain milky ripe for rye, wheat; beginning of flowering for barley, oats
- HI 56 d for oilseed rape

Environmental safety
- Dangerous for the environment
- Toxic to aquatic organisms
- Risk to non-target insects or other arthropods. Avoid spraying within 6 m of the field boundary to reduce the effects on non-target insects or other arthropods
- LERAP Category B

Hazard classification and safety precautions
    Hazard Harmful [1, 2, 5]; Irritant [3, 4]; Dangerous for the environment [1-5]
    Transport code 9
    Packaging group III
    UN number 3082
    Risk phrases R36 [1, 2]; R38, R43, R51, R53a, R63 [1-5]
    Operator protection A, H [1-5]; C [1, 2]; U05a [1-5]; U09b, U20b [2-5]; U11, U20c [1]; U19a [1, 3-5]
    Environmental protection E15a, E38 [1-4]; E15b [5]; E16a, E34 [1-5]; E22b [1]; E22c [2-5]
    Storage and disposal D01, D02, D09a, D12a [1-5]; D05, D10b [1-4]; D10a [5]
    Medical advice M03

394 prothioconazole + tebuconazole + triazoxide

A broad spectrum fungicide seed treatment for cereals
FRAC mode of action code: 3 + 3 + 35

See also tebuconazole
           triazoxide

Products

Raxil Pro	Bayer CropScience	25:15:10 g/l	FS	12432

Uses
- Covered smut in *spring barley* (seed treatment), *winter barley* (seed treatment)
- Fusarium foot rot and seedling blight in *spring barley* (seed treatment), *winter barley* (seed treatment)
- Leaf stripe in *spring barley* (seed treatment), *winter barley* (seed treatment)
- Loose smut in *spring barley* (seed treatment), *winter barley* (seed treatment)
- Net blotch in *spring barley* (seed treatment), *winter barley* (seed treatment)

Approval information
- Prothioconazole and tebuconazole included in Annex I under EC Directive 91/414. Triazoxide has not yet been included
- Accepted by BBPA for use on malting barley
- Approval expiry 30 May 2011 [1]

SEE SECTION 3 FOR PRODUCTS ALSO REGISTERED

Pesticide Profiles

Efficacy guidance
- Treatments must be applied by manufacturer's recommended treatment application equipment
- Treated seed should preferably be drilled in the same season
- Follow-up treatments will be needed later in the season to give protection against air-borne and splash-borne diseases

Restrictions
- Maximum number of seed treatments one per batch

Crop-specific information
- Latest use for seed treatment: pre-drilling

Environmental safety
- Harmful to aquatic organisms

Hazard classification and safety precautions
    Hazard Irritant
    Transport code 9
    Packaging group III
    UN number 3082
    Risk phrases R37, R43, R52, R53a
    Operator protection A, H; U07, U14, U19a, U20a
    Environmental protection E15a, E34, E36a, E38
    Storage and disposal D05, D09a, D14
    Treated seed S01, S02, S03, S04b, S05, S06b, S07, S08

395 prothioconazole + trifloxystrobin

A triazole and strobilurin fungicide mixture for cereals
FRAC mode of action code: 3 + 11

See also trifloxystrobin

Products

1	Mobius	Bayer CropScience	175:150 g/l	SC	13395
2	Zephyr	Bayer CropScience	175:88 g/l	SC	13174

Uses
- Brown rust in **spring barley**, **winter barley**, **winter wheat** [1, 2]
- Ear diseases in **winter wheat** [1]
- Eyespot in **spring barley** *(reduction)*, **winter barley** *(reduction)* [2]; **spring barley** *(reduction in severity)*, **winter barley** *(reduction in severity)* [1]; **winter wheat** [1, 2]
- Glume blotch in **winter wheat** [1, 2]
- Net blotch in **spring barley**, **winter barley** [1, 2]
- Powdery mildew in **spring barley**, **winter barley**, **winter wheat** [1, 2]
- Rhynchosporium in **spring barley**, **winter barley** [1, 2]
- Septoria leaf blotch in **winter wheat** [1, 2]
- Yellow rust in **spring barley**, **winter barley**, **winter wheat** [1, 2]

Approval information
- Prothioconazole and trifloxystrobin included in Annex I under EC Directive 91/414
- Accepted by BBPA for use on malting barley

Efficacy guidance
- Best results obtained from treatment at early stages of disease development. Further treatment may be needed if disease attack is prolonged
- Applications to established infections of any disease are likely to be less effective
- Best control of cereal ear diseases obtained by treatment during ear emergence
- Prothioconazole is a DMI fungicide. Resistance to some DMI fungicides has been identified in Septoria leaf blotch which may seriously affect performance of some products. For further advice contact a specialist advisor and visit the Fungicide Resistance Action Group (FRAG)-UK website
- Trifloxystrobin is a member of the QoI cross resistance group. Product should be used preventatively and not relied on for its curative potential

FOR FULL CONDITIONS OF USE ALWAYS READ THE PRODUCT LABEL

pymetrozine

- Use product as part of an Integrated Crop Management strategy incorporating other methods of control, including where appropriate other fungicides with a different mode of action. Do not apply more than two foliar applications of QoI containing products to any cereal crop
- There is a significant risk of widespread resistance occurring in *Septoria tritici* populations in UK. Failure to follow resistance management action may result in reduced levels of disease control
- Strains of wheat and barley powdery mildew resistant to QoIs are common in the UK. Control of wheat mildew can only be relied on from the triazole component
- Where specific control of wheat mildew is required this should be achieved through a programme of measures including products recommended for the control of mildew that contain a fungicide from a different cross-resistance group and applied at a dose that will give robust control

Restrictions
- Maximum total dose equivalent to two full dose treatments

Crop-specific information
- Latest use: before grain milky ripe for wheat; beginning of flowering for barley

Environmental safety
- Dangerous for the environment
- Very toxic to aquatic organisms
- LERAP Category B

Hazard classification and safety precautions
Hazard Irritant, Dangerous for the environment
Transport code 9
Packaging group III
UN number 3082
Risk phrases R37, R50, R53a [1, 2]; R70 [1]
Operator protection A, C, H; U05a, U09b, U19a [1, 2]; U20b [2]; U20c [1]
Environmental protection E15a, E16a, E38 [1, 2]; E34 [2]
Storage and disposal D01, D02, D09a, D10c, D12a
Medical advice M03

396   pymetrozine

A novel azomethine insecticide
IRAC mode of action code: 9B

Products
1	Chess WG	Syngenta Bioline	50% w/w	WG	13310
2	Plenum WG	Syngenta	50% w/w	WG	10652

Uses
- Aphids in **bilberries** (off-label), **blackberries** (off-label), **blackcurrants** (off-label), **blueberries** (off-label), **celeriac** (off-label), **celery (outdoor)** (off-label), **chinese cabbage** (off-label), **choi sum** (off-label), **collards** (off-label), **cranberries** (off-label), **forest nurseries** (off-label), **frise** (off-label), **gooseberries** (off-label), **kale** (off-label), **lamb's lettuce** (off-label), **lettuce** (off-label), **loganberries** (off-label), **pak choi** (off-label), **radicchio** (off-label), **raspberries** (off-label), **redcurrants** (off-label), **rubus hybrids** (off-label), **salad brassicas** (off-label - for baby leaf production), **seed potatoes**, **soft fruit** (off-label), **strawberries** (off-label), **sweetcorn** (off-label), **vaccinium spp.** (off-label), **ware potatoes**, **whitecurrants** (off-label) [2]; **brassica seed beds** (off-label), **ornamental plant production**, **protected aubergines** (off-label), **protected blackberries** (off-label), **protected blackcurrants** (off-label), **protected celery** (off-label), **protected chilli peppers** (off-label), **protected chives** (off-label), **protected courgettes** (off-label), **protected cucumbers**, **protected cucumbers** (off-label), **protected endives** (off-label), **protected forest nurseries** (off-label), **protected gherkins** (off-label), **protected gooseberries** (off-label), **protected herbs (see appendix 6)** (off-label), **protected hops** (off-label), **protected lettuce** (off-label), **protected loganberries** (off-label), **protected marrows** (off-label), **protected melons** (off-label), **protected okra** (off-label), **protected ornamentals**, **protected ornamentals** (off-label), **protected parsley** (off-label), **protected peppers** (off-label), **protected pumpkins** (off-label), **protected raspberries** (off-label), **protected redcurrants** (off-label), **protected ribes hybrids** (off-label), **protected rubus hybrids** (off-label), **protected salad brassicas** (off-label - for baby leaf production), **protected soft fruit** (off-label), **protected**

SEE SECTION 3 FOR PRODUCTS ALSO REGISTERED

Pesticide Profiles

- squashes *(off-label)*, **protected strawberries** *(off-label)*, **protected tomatoes** *(off-label)*, **protected top fruit** *(off-label)* [1]
- Cabbage aphid in **chinese cabbage** *(off-label)*, **choi sum** *(off-label)*, **collards** *(off-label)*, **kale** *(off-label)*, **pak choi** *(off-label)* [2]; **protected chinese cabbage** *(off-label)*, **protected choi sum** *(off-label)*, **protected pak choi** *(off-label)* [1]
- Damson-hop aphid in **hops** *(off-label)* [2]
- Glasshouse whitefly in **protected tomatoes** *(off-label)* [1]
- Mealy aphid in **broccoli** *(useful levels of control)*, **brussels sprouts** *(useful levels of control)*, **cabbages** *(useful levels of control)*, **calabrese** *(useful levels of control)*, **cauliflowers** *(useful levels of control)*, **chinese cabbage** *(off-label)*, **choi sum** *(off-label)*, **collards** *(off-label)*, **kale** *(off-label)*, **pak choi** *(off-label)* [2]
- Peach-potato aphid in **broccoli**, **brussels sprouts**, **cabbages**, **calabrese**, **cauliflowers**, **chinese cabbage** *(off-label)*, **choi sum** *(off-label)*, **collards** *(off-label)*, **kale** *(off-label)*, **pak choi** *(off-label)* [2]; **protected aubergines** *(off-label)*, **protected endives** *(off-label)*, **protected peppers** *(off-label)* [1]
- Tobacco whitefly in **protected tomatoes** *(off-label)* [1]
- Whitefly in **protected aubergines** *(off-label)*, **protected chilli peppers** *(off-label)*, **protected courgettes** *(off-label)*, **protected cucumbers** *(off-label)*, **protected forest nurseries** *(off-label)*, **protected gherkins** *(off-label)*, **protected hops** *(off-label)*, **protected melons** *(off-label)*, **protected okra** *(off-label)*, **protected ornamentals** *(off-label)*, **protected peppers** *(off-label)*, **protected pumpkins** *(off-label)*, **protected soft fruit** *(off-label)*, **protected squashes** *(off-label)*, **protected tomatoes** *(off-label)*, **protected top fruit** *(off-label)* [1]

Specific Off-Label Approvals (SOLAs)
- **bilberries** *20061702 expires 31 Oct 2011* [2]
- **blackberries** *20061633 expires 31 Oct 2011* [2]
- **blackcurrants** *20060946 expires 31 Oct 2011* [2]
- **blueberries** *20061702 expires 31 Oct 2011* [2]
- **brassica seed beds** *20070788 expires 31 Oct 2011* [1]
- **celeriac** *20051062 expires 31 Oct 2011* [2]
- **celery (outdoor)** *20051062 expires 31 Oct 2011* [2]
- **chinese cabbage** *20082246 expires 31 Oct 2011* [2]
- **choi sum** *20082246 expires 31 Oct 2011* [2]
- **collards** *20082081 expires 31 Oct 2011* [2]
- **cranberries** *20061702 expires 31 Oct 2011* [2]
- **forest nurseries** *20082920 expires 31 Oct 2011* [2]
- **frise** *20070060 expires 31 Oct 2011* [2]
- **gooseberries** *20061702 expires 31 Oct 2011* [2]
- **hops** *20031423 expires 31 Oct 2011* [2]
- **kale** *20082081 expires 31 Oct 2011* [2]
- **lamb's lettuce** *20070060 expires 31 Oct 2011* [2]
- **lettuce** *20070060 expires 31 Oct 2011* [2]
- **loganberries** *20061702 expires 31 Oct 2011* [2]
- **pak choi** *20082246 expires 31 Oct 2011* [2]
- **protected aubergines** *20070501 expires 31 Oct 2011* [1], *20092024 expires 31 Oct 2011* [1]
- **protected blackberries** *20070498 expires 31 Oct 2011* [1]
- **protected blackcurrants** *20070504 expires 31 Oct 2011* [1]
- **protected celery** *20070503 expires 31 Oct 2011* [1]
- **protected chilli peppers** *20092024 expires 31 Oct 2011* [1]
- **protected chinese cabbage** *20090649 expires 31 Oct 2011* [1]
- **protected chives** *20070502 expires 31 Oct 2011* [1]
- **protected choi sum** *20090649 expires 31 Oct 2011* [1]
- **protected courgettes** *20081012 expires 31 Oct 2011* [1], *20092024 expires 31 Oct 2011* [1]
- **protected cucumbers** *20092024 expires 31 Oct 2011* [1]
- **protected endives** *20092025 expires 31 Oct 2011* [1]
- **protected forest nurseries** *20082834 expires 31 Oct 2011* [1]
- **protected gherkins** *20081012 expires 31 Oct 2011* [1], *20092024 expires 31 Oct 2011* [1]
- **protected gooseberries** *20070504 expires 31 Oct 2011* [1]
- **protected herbs (see appendix 6)** *20070502 expires 31 Oct 2011* [1]
- **protected hops** *20082834 expires 31 Oct 2011* [1]

FOR FULL CONDITIONS OF USE ALWAYS READ THE PRODUCT LABEL

pymetrozine

- *protected lettuce* 20070502 expires 31 Oct 2011 [1]
- *protected loganberries* 20070504 expires 31 Oct 2011 [1]
- *protected marrows* 20081012 expires 31 Oct 2011 [1]
- *protected melons* 20081012 expires 31 Oct 2011 [1], 20092024 expires 31 Oct 2011 [1]
- *protected okra* 20092024 expires 31 Oct 2011 [1]
- *protected ornamentals* 20082834 expires 31 Oct 2011 [1]
- *protected pak choi* 20090649 expires 31 Oct 2011 [1]
- *protected parsley* 20070502 expires 31 Oct 2011 [1]
- *protected peppers* 20070501 expires 31 Oct 2011 [1], 20092024 expires 31 Oct 2011 [1]
- *protected pumpkins* 20081012 expires 31 Oct 2011 [1], 20092024 expires 31 Oct 2011 [1]
- *protected raspberries* 20070498 expires 31 Oct 2011 [1]
- *protected redcurrants* 20070504 expires 31 Oct 2011 [1]
- *protected ribes hybrids* 20070504 expires 31 Oct 2011 [1]
- *protected rubus hybrids* 20070504 expires 31 Oct 2011 [1]
- *protected salad brassicas* (for baby leaf production) 20070502 expires 31 Oct 2011 [1]
- *protected soft fruit* 20082834 expires 31 Oct 2011 [1]
- *protected squashes* 20081012 expires 31 Oct 2011 [1], 20092024 expires 31 Oct 2011 [1]
- *protected strawberries* 20070499 expires 31 Oct 2011 [1]
- *protected tomatoes* 20070501 expires 31 Oct 2011 [1], 20092024 expires 31 Oct 2011 [1]
- *protected top fruit* 20082834 expires 31 Oct 2011 [1]
- *radicchio* 20070060 expires 31 Oct 2011 [2]
- *raspberries* 20061633 expires 31 Oct 2011 [2]
- *redcurrants* 20061702 expires 31 Oct 2011 [2]
- *rubus hybrids* 20061702 expires 31 Oct 2011 [2]
- *salad brassicas* (for baby leaf production) 20070060 expires 31 Oct 2011 [2]
- *soft fruit* 20082920 expires 31 Oct 2011 [2]
- *strawberries* 20060461 expires 31 Oct 2011 [2]
- *sweetcorn* 20041318 expires 31 Oct 2011 [2]
- *vaccinium spp.* 20061702 expires 31 Oct 2011 [2]
- *whitecurrants* 20061702 expires 31 Oct 2011 [2]

Approval information
- Pymetrozine included in Annex I under EC Directive 91/414
- Accepted by BBPA for use on hops
- Approval expiry 31 Oct 2011 [1, 2]

Efficacy guidance
- Pymetrozine moves systemically in the plant and acts by preventing feeding leading to death by starvation in 1-4 d. There is no immediate knockdown
- Aphids controlled include those resistant to organophosphorus and carbamate insecticides
- To prevent development of resistance do not use continuously or as the sole method of control
- Best results achieved by starting spraying as soon as aphids seen in crop and repeating as necessary.
- To limit spread of persistent viruses such as potato leaf roll virus, apply from 90% crop emergence

Restrictions
- Maximum total dose equivalent to two full dose treatments on ware potatoes; three full dose treatments on seed potatoes, leaf spinach [2]; four full dose treatments on cucumbers, ornamentals [1]
- Consult processors before use on potatoes for processing
- Check tolerance of ornamental species before large scale use. See label for list of species known to have been treated without damage. Visible spray deposits may be seen on leaves of some species [1]

Crop-specific information
- HI cucumbers 3 d; potatoes, leaf spinach 7 d

Environmental safety
- High risk to bees (outdoor use only). Do not apply to crops in flower or to those in which bees are actively foraging. Do not apply when flowering weeds are present
- Avoid spraying within 6 m of field boundaries to reduce effects on non-target insects or arthropods. Risk to certain non-target insects and arthropods.

SEE SECTION 3 FOR PRODUCTS ALSO REGISTERED

Pesticide Profiles

Hazard classification and safety precautions
 Hazard Harmful
 UN number N/C
 Risk phrases R40
 Operator protection A [1, 2]; H [2]; U05a, U20c
 Environmental protection E12a, E12e, E15a
 Storage and disposal D01, D02, D09a, D11a

397 pyraclostrobin

A protectant and curative strobilurin fungicide for cereals
FRAC mode of action code: 11

See also boscalid + pyraclostrobin
 dithianon + pyraclostrobin
 epoxiconazole + fenpropimorph + pyraclostrobin
 epoxiconazole + kresoxim-methyl + pyraclostrobin
 epoxiconazole + pyraclostrobin
 fenpropimorph + pyraclostrobin

Products

1	Comet 200	BASF	200 g/l	EC	12639
2	Flyer	BASF	250 g/l	EC	12654
3	Mascot Eland	Rigby Taylor	20% w/w	WG	14549
4	Platoon 250	BASF	250 g/l	EC	12640
5	Tucana	BASF	250 g/l	EC	10899
6	Vanguard	Sherriff Amenity	20% w/w	WG	13838
7	Vivid	BASF	250 g/l	EC	10898

Uses
- Brown rust in **spring barley**, **spring wheat**, **winter barley**, **winter wheat** [1, 2, 4, 5, 7]
- Crown rust in **spring oats**, **winter oats** [1, 2, 4, 5, 7]
- Dollar spot in **managed amenity turf** *(useful reduction)* [3, 6]
- Foliar disease control in **durum wheat** *(off-label)*, **grass seed crops** *(off-label)*, **spring rye** *(off-label)*, **triticale** *(off-label)*, **winter rye** *(off-label)* [5, 7]; **forage maize** *(off-label)* [1]; **ornamental plant production** *(off-label)* [7]
- Fusarium patch in **managed amenity turf** *(moderate control)* [3]; **managed amenity turf** *(moderate control only)* [6]
- Glume blotch in **spring wheat**, **winter wheat** [1, 2, 4, 5, 7]
- Net blotch in **spring barley**, **winter barley** [1, 2, 4, 5, 7]
- Red thread in **managed amenity turf** [3, 6]
- Rhynchosporium in **spring barley** *(moderate)*, **winter barley** *(moderate)* [1, 2, 4, 5, 7]
- Septoria leaf blotch in **spring wheat**, **winter wheat** [1, 2, 4, 5, 7]
- Yellow rust in **spring barley**, **spring wheat**, **winter barley**, **winter wheat** [1, 2, 4, 5, 7]

Specific Off-Label Approvals (SOLAs)
- *durum wheat* 20062652 [5], 20062649 [7]
- *forage maize* 20090928 [1]
- *grass seed crops* 20062652 [5], 20062649 [7]
- *ornamental plant production* 20082884 [7]
- *spring rye* 20062652 [5], 20062649 [7]
- *triticale* 20062652 [5], 20062649 [7]
- *winter rye* 20062652 [5], 20062649 [7]

Approval information
- Pyraclostrobin included in Annex I under EC Directive 91/414
- Accepted by BBPA for use on malting barley (before ear emergence only)

Efficacy guidance
- For best results apply at the start of disease attack on cereals [1, 2, 4, 5, 7]
- For Fusarium Patch treat early as severe damage to turf can occur once the disease is established [3, 6]

FOR FULL CONDITIONS OF USE ALWAYS READ THE PRODUCT LABEL

- Regular turf aeration, appropriate scarification and judicious use of nitrogenous fertiliser will assist the control of Fusarium Patch [3, 6]
- Best results on Septoria glume blotch achieved when used as a protective treatment and against Septoria leaf blotch when treated in the latent phase [1, 2, 4, 5, 7]
- Yield response may be obtained in the absence of visual disease symptoms [1, 2, 4, 5, 7]
- Pyraclostrobin is a member of the QoI cross resistance group. Product should be used preventatively and not relied on for its curative potential
- Use product as part of an Integrated Crop Management strategy incorporating other methods of control, including where appropriate other fungicides with a different mode of action. Do not apply more than two foliar applications of QoI containing products to any cereal crop or to grass
- There is a significant risk of widespread resistance occurring in *Septoria tritici* populations in UK. Failure to follow resistance management action may result in reduced levels of disease control [1, 2, 4, 5, 7]
- On cereal crops product must always be used in mixture with another product, recommended for control of the same target disease, that contains a fungicide from a different cross resistance group and is applied at a dose that will give robust control [1, 2, 4, 5, 7]

Restrictions
- Maximum number of treatments 2 per crop on cereals [1, 2, 4, 5, 7]
- Maximum total dose on turf equivalent to two full dose treatments [3, 6]
- Do not apply during drought conditions or to frozen turf [3, 6]

Crop-specific information
- Latest use: before grain watery ripe (GS 71) for wheat; up to and including emergence of ear just complete (GS 59) for barley and oats [1, 2, 4, 5, 7]
- Avoid applying to turf immediately after cutting or 48 h before mowing [3, 6]

Environmental safety
- Dangerous for the environment
- Very toxic to aquatic organisms
- LERAP Category A [6]; LERAP Category B [1-7]

Hazard classification and safety precautions
Hazard Harmful, Dangerous for the environment
Transport code 6.1 [1, 2, 4, 5, 7]; 9 [3, 6]
Packaging group III
UN number 2902 [1, 2, 4, 5, 7]; 3077 [3, 6]
Risk phrases R20, R50 [1-7]; R22a, R38, R53a [1, 2, 4, 5, 7]
Operator protection A [1-7]; D [3, 6]; U05a, U14, U20b [1, 2, 4, 5, 7]
Environmental protection E15a, E16d [6]; E15b, E16a, E38 [1-7]; E16b [1-5, 7]; E34 [1, 2, 4, 5, 7]
Storage and disposal D01, D02 [1, 2, 4, 5, 7]; D08, D09a, D10c, D12a [1-5]
Medical advice M03, M05a [1, 2, 4, 5, 7]

398 pyrethrins

A non-persistent, contact acting insecticide extracted from Pyrethrum
IRAC mode of action code: 3

Products

1	Dairy Fly Spray	B H & B	0.75 g/l	AL	H5579
2	Killgerm ULV 400	Killgerm	30 g/l	UL	H4838
3	Pyblast	Agropharm	30 g/l	UL	H7485
4	Pyrethrum 5 EC	Agropharm	50 g/l	EC	12685
5	Pyrethrum Spray	Killgerm	3.3 g/l	UL	H4636
6	Spruzit	Certis	4.59 g/l	EC	13438

Uses
- Aphids in *all edible crops (outdoor)*, *all non-edible crops (outdoor)*, *protected crops* [6]; *broccoli*, *brussels sprouts*, *bush fruit*, *cabbages*, *calabrese*, *cane fruit*, *cauliflowers*, *lettuce*, *ornamental plant production*, *protected broccoli*, *protected brussels sprouts*, *protected bush fruit*, *protected cabbages*, *protected calabrese*, *protected cane fruit*, *protected cauliflowers*, *protected cayenne peppers* (off-label), *protected chilli peppers* (off-label), *protected lettuce*, *protected ornamentals*, *protected peppers* (off-label), *protected tomatoes*, *tomatoes (outdoor)* [4]

SEE SECTION 3 FOR PRODUCTS ALSO REGISTERED

Pesticide Profiles

- Caterpillars in *all edible crops (outdoor)*, *all non-edible crops (outdoor)*, *protected crops* [6]; *broccoli*, *brussels sprouts*, *bush fruit*, *cabbages*, *calabrese*, *cane fruit*, *cauliflowers*, *lettuce*, *ornamental plant production*, *protected broccoli*, *protected brussels sprouts*, *protected bush fruit*, *protected cabbages*, *protected calabrese*, *protected cane fruit*, *protected cauliflowers*, *protected lettuce*, *protected ornamentals*, *protected tomatoes*, *tomatoes (outdoor)* [4]
- Flea beetle in *broccoli*, *brussels sprouts*, *cabbages*, *calabrese*, *cauliflowers*, *protected broccoli*, *protected brussels sprouts*, *protected cabbages*, *protected calabrese*, *protected cauliflowers*, *protected tomatoes*, *tomatoes (outdoor)* [4]; *protected crops* [6]
- Insect pests in *dairies*, *farm buildings*, *livestock houses*, *poultry houses* [1-3, 5]; *glasshouses* [5]
- Macrolophus caliginosus in *protected tomatoes*, *protected tomatoes* (off-label) [4]
- Mealybugs in *protected crops* [6]; *protected tomatoes*, *protected tomatoes* (off-label) [4]
- Scale insects in *protected crops* [6]
- Spider mites in *all edible crops (outdoor)*, *all non-edible crops (outdoor)*, *protected crops* [6]
- Thrips in *all edible crops (outdoor)*, *all non-edible crops (outdoor)*, *protected crops* [6]
- Whitefly in *broccoli*, *brussels sprouts*, *cabbages*, *calabrese*, *cauliflowers*, *protected broccoli*, *protected brussels sprouts*, *protected cabbages*, *protected calabrese*, *protected cauliflowers*, *protected tomatoes*, *tomatoes (outdoor)* [4]; *protected crops* [6]

Specific Off-Label Approvals (SOLAs)
- *protected cayenne peppers* 20091005 [4]
- *protected chilli peppers* 20091005 [4]
- *protected peppers* 20091005 [4]
- *protected tomatoes* 20063026 [4]

Approval information
- Pyrethrins have been included in Annex 1 under EC Directive 91/414
- Products formulated for ULV application [1-3]. May be applied through fogging machine or sprayer [1, 3, 5]. See label for details

Efficacy guidance
- For indoor fly control close doors and windows and spray or apply fog as appropriate [1-3]
- For best fly control outdoors spray during early morning or late afternoon and evening when conditions are still [1-3, 5]
- For all uses ensure good spray coverage of the target area or plants by increasing spray volume where necessary
- Best results on outdoor or protected crops achieved from treatment at first signs of pest attack in early morning or evening [6]
- To avoid possibility of development of resistance do not spray more frequently than once per week

Restrictions
- Maximum number of treatments on edible crops 3 per crop or season [4]
- For use only by professional operators [1-3, 5]
- Do not allow spray to contact open food products or food preparing equipment or utensils [1-3, 5]
- Remove exposed milk and collect eggs before application [1-3, 5]
- Do not treat plants [1-3, 5]
- Avoid direct application to open flowers [6]
- Do not mix with, or apply closely before or after, products containing dithianon or tolylfluanid [6]
- Do not use space sprays containing pyrethrins or pyrethroid more than once per week in intensive or controlled environment animal houses in order to avoid development of resistance. If necessary, use a different control method or product [1-3, 5]
- Store away from strong sunlight [4]

Crop-specific information
- HI: 24 h for edible crops [4]
- Some plant species including *Ageratum*, ferns, *Ficus*, *Lantana*, *Poinsettia*, *Petroselium crispum*, and some strawberries may be sensitive especially where more than one application is made. Test before large scale treatment [6]

Environmental safety
- Dangerous for the environment [2-6]
- Very toxic to aquatic organisms [1, 4, 5]; toxic to aquatic organisms [2, 3, 6]

FOR FULL CONDITIONS OF USE ALWAYS READ THE PRODUCT LABEL

pyridate

- High risk to bees. Do not apply to crops in flower or to those in which bees are actively foraging. Do not apply when flowering weeds are present [4]
- Risk to non-target insects or other arthropods [4]
- Do not apply directly to livestock and exclude all persons and animals during treatment [1-3, 5]
- Wash spray equipment thoroughly after use to avoid traces of pyrethrum causing damage to susceptible crops sprayed later [4]
- LERAP Category A [4]

Hazard classification and safety precautions
Hazard Harmful [1, 2, 5]; Irritant [1, 3, 4]; Dangerous for the environment [2-6]
Transport code 9 [1, 3-6]
Packaging group III [1, 3-6]
UN number 3082 [1, 3-6]; N/C [2]
Risk phrases R22a, R36, R38 [1]; R22b [2, 5]; R41 [3, 4]; R50 [4]; R51 [2, 3, 5, 6]; R53a [2-6]
Operator protection A [1-6]; B [1]; C [2, 3, 5]; D [2, 3]; E [1-3]; H [2-6]; U02b, U09b, U14, U20a [2]; U05a [1, 4, 5]; U09a [1, 5]; U11 [3, 4]; U15 [3]; U19a [1-4]; U20b [1-3, 5]; U20c [4]; U20d [6]
Environmental protection E02c [3]; E02d, E05c [5]; E05a [1-3, 5]; E12a, E12e, E16c, E16d, E22c [4]; E13c [1]; E15a [2-5]; E15b [6]; E19b [3, 6]; E38 [2-4, 6]
Consumer protection C04 [1]; C05 [5]; C06 [1, 2, 5]; C07, C09, C11 [1-3, 5]; C08 [1-3]; C10 [1, 3]; C12 [2, 3, 5]
Storage and disposal D01, D09a [1-6]; D02 [1, 3-6]; D05 [2, 4]; D06a, D10a [2]; D11a [1, 4, 6]; D12a [2-6]
Medical advice M05a [3]; M05b [2, 5]

399   pyridate

A contact phenylpyridazine herbicide for cereals, maize and brassicas
HRAC mode of action code: C3

Products
Lentagran WP   Belchim   45% w/w   WP   14162

Uses
- Annual dicotyledons in **asparagus** *(off-label)*, **broccoli** *(off-label)*, **calabrese** *(off-label)*, **cauliflowers** *(off-label)*, **collards** *(off-label)*, **fodder rape** *(off-label)*, **game cover** *(off-label)*, **garlic** *(off-label)*, **kale** *(off-label)*, **leeks** *(off-label)*, **lupins** *(off-label)*, **oilseed rape** *(off-label)*, **shallots** *(off-label)*, **spring greens** *(off-label)*
- Black nightshade in **brussels sprouts**, **bulb onions**, **cabbages**
- Cleavers in **brussels sprouts**, **bulb onions**, **cabbages**
- Fat hen in **brussels sprouts**, **bulb onions**, **cabbages**

Specific Off-Label Approvals (SOLAs)
- *asparagus* 20091039 *expires 31 Dec 2011*
- *broccoli* 20090786 *expires 31 Dec 2011*
- *calabrese* 20090786 *expires 31 Dec 2011*
- *cauliflowers* 20090786 *expires 31 Dec 2011*
- *collards* 20090785 *expires 31 Dec 2011*
- *fodder rape* 20093230 *expires 31 Dec 2011*
- *game cover* 20090788 *expires 31 Dec 2011*
- *garlic* 20092862 *expires 31 Dec 2011*
- *kale* 20090785 *expires 31 Dec 2011*
- *leeks* 20090784 *expires 31 Dec 2011*
- *lupins* 20090787 *expires 31 Dec 2011*
- *oilseed rape* 20093230 *expires 31 Dec 2011*
- *shallots* 20092862 *expires 31 Dec 2011*
- *spring greens* 20090785 *expires 31 Dec 2011*

Approval information
- Pyridate included in Annex I under EC Directive 91/414
- Approval expiry 31 Dec 2011

SEE SECTION 3 FOR PRODUCTS ALSO REGISTERED

Pesticide Profiles

Efficacy guidance
- Best results achieved by application to actively growing weeds at 6-8 leaf stage when temperatures are above 8 °C before crop foliage forms canopy

Restrictions
- Maximum number of treatments 1 per crop
- Do not apply in mixture with or within 14 d of any other product which may result in dewaxing of crop foliage
- Do not use on crops suffering stress from frost, drought, disease or pest attack

Crop-specific information
- Latest use: before 7 leaf stage for maize; before 5 true leaves for onions; before flower buds visible for oilseed rape
- HI 6 wk for Brussels sprouts, cabbages
- Apply to cabbages and Brussels sprouts after 4 fully expanded leaf stage. Allow 2 wk after transplanting before treating

Environmental safety
- Dangerous for the environment
- Toxic to aquatic organisms

Hazard classification and safety precautions
Hazard Irritant, Dangerous for the environment
Transport code 9
Packaging group III
UN number 3077
Risk phrases R43, R51, R53a
Operator protection A, C; U05a, U14, U22a
Environmental protection E15a, E38
Storage and disposal D01, D02, D09a, D10c, D12a

400 pyrimethanil

An anilinopyrimidine fungicide for apples and strawberries
FRAC mode of action code: 9

See also chlorothalonil + pyrimethanil

Products

1	Scala	BASF	400 g/l	SC	11695
2	Waila	ChemSource	400 g/l	SC	14976

Uses
- Botrytis in *aubergines* (off-label), *bilberries* (off-label), *blackberries* (off-label), *blackcurrants* (off-label), *cranberries* (off-label), *gooseberries* (off-label), *protected blackberries* (off-label), *protected herbs (see appendix 6)* (off-label), *protected raspberries* (off-label), *protected tomatoes* (off-label), *quinces* (off-label), *raspberries* (off-label), *redcurrants* (off-label), *tomatoes (outdoor)* (off-label), *whitecurrants* (off-label), *wine grapes* (off-label) [1]; *strawberries* [1, 2]
- Scab in *apples* [1, 2]; *pears* (off-label) [1]

Specific Off-Label Approvals (SOLAs)
- *aubergines* 20040516 [1]
- *bilberries* 20040519 [1]
- *blackberries* 20051737 [1]
- *blackcurrants* 20040519 [1]
- *cranberries* 20040519 [1]
- *gooseberries* 20040519 [1]
- *pears* 20070200 [1]
- *protected blackberries* 20073636 [1]
- *protected herbs (see appendix 6)* 20073635 [1]
- *protected raspberries* 20073636 [1]
- *protected tomatoes* 20040516 [1]
- *quinces* 20061379 [1]
- *raspberries* 20051737 [1]

FOR FULL CONDITIONS OF USE ALWAYS READ THE PRODUCT LABEL

quinmerac

- **redcurrants** *20040519* [1]
- **tomatoes (outdoor)** *20040516* [1]
- **whitecurrants** *20040519* [1]
- **wine grapes** *20040517* [1]

Approval information
- Pyrimethanil included in Annex I under EC Directive 91/414. However a Standing Committee decision in July 2007 to reduce to the limit of determination the MRL set for pyrimenthanil in cane fruit (other than blackberries and raspberries) and the MRL for brassicas grown for baby leaf production led to immediate revocation of SOLAs for these crops

Efficacy guidance
- On apples a programme of sprays will give early season control of scab. Season long control can be achieved by continuing programme with other approved fungicides
- In strawberries product should be used as part of a programme of disease control treatments which should alternate with other materials to prevent or limit development of less sensitive strains of grey mould

Restrictions
- Maximum number of treatments 5 per yr for apples; 4 per yr for protected crops; 3 per yr for grapevines; 2 per yr for cane fruit, bush fruit, strawberries, tomatoes
- Product does not taint apples. Processors should be consulted before use on strawberries

Crop-specific information
- Latest use: before end of flowering for apples
- HI protected cane fruit, strawberries 1 d; aubergines, tomatoes 3 d; protected lettuce 14 d; bush fruit, grapevines 21 d
- All varieties of apples and strawberries may be treated
- Treat apples from bud burst at 10-14 d intervals
- In strawberries start treatments at white bud to give maximum protection of flowers against grey mould and treat every 7-10 d. Product should not be used more than once in a 3 or 4 spray programme

Environmental safety
- Harmful to aquatic organisms
- Product has negligible effect on hoverflies and lacewings. Limited evidence indicates some margin of safety to *Typhlodromus pyri*
- Broadcast air-assisted LERAP (20 m)

Hazard classification and safety precautions
UN number N/C
Risk phrases R52, R53a
Operator protection U08, U20b
Environmental protection E15a; E17b (20 m)
Storage and disposal D01, D02, D05, D08, D09a, D10b, D12a

401   quinmerac

A residual herbicide available only in mixtures
HRAC mode of action code: O

See also chloridazon + quinmerac
dimethenamid-p + metazachlor + quinmerac
metazachlor + quinmerac

SEE SECTION 3 FOR PRODUCTS ALSO REGISTERED

402 quinoxyfen

A systemic protectant quinoline fungicide for cereals
FRAC mode of action code: 13

See also cyproconazole + quinoxyfen
fenpropimorph + quinoxyfen

Products
Fortress   Dow   500 g/l   SC   08279

Uses
- Powdery mildew in **durum wheat**, **forest nurseries** (off-label), **hops** (off-label), **ornamental plant production** (off-label), **protected forest nurseries** (off-label), **protected ornamentals** (off-label), **protected strawberries** (off-label), **spring barley**, **spring oats**, **spring rye**, **spring wheat**, **strawberries** (off-label), **sugar beet**, **triticale**, **winter barley**, **winter oats**, **winter rye**, **winter wheat**

Specific Off-Label Approvals (SOLAs)
- *forest nurseries* 20082852
- *hops* 20061579
- *ornamental plant production* 20082852
- *protected forest nurseries* 20082852
- *protected ornamentals* 20082852
- *protected strawberries* 20041923
- *strawberries* 20041923

Approval information
- Quinoxyfen included in Annex I under Directive 91/414
- Accepted by BBPA for use on malting barley (before ear emergence only)

Efficacy guidance
- For best results treat at early stage of disease development before infection spreads to new crop growth. Further treatment may be necessary if disease pressure remains high
- Product not curative and will not control latent or established disease infections
- For broad spectrum control in cereals use in tank mixtures - see label
- Product rainfast after 1 h
- Systemic activity may be reduced in severe drought
- Use as part of an integrated crop management (ICM) strategy incorporating other methods of control or fungicides with different modes of action

Restrictions
- Maximum total dose equivalent to two full dose treatments in cereals; see labels for split dose recommendation in sugar beet
- On cereals apply only in the spring from mid-tillering stage (GS 25)

Crop-specific information
- Latest use: first awns visible stage (GS 49) for cereals
- HI sugar beet 28 d

Environmental safety
- Dangerous for the environment
- Very toxic to aquatic organisms
- LERAP Category B

Hazard classification and safety precautions
Hazard Irritant, Dangerous for the environment
Transport code 9
Packaging group III
UN number 3082
Risk phrases R43, R50, R53a
Operator protection A, C, H; U05a, U14
Environmental protection E15a, E16a, E16b, E34, E38
Storage and disposal D01, D02, D12a

FOR FULL CONDITIONS OF USE ALWAYS READ THE PRODUCT LABEL

quizalofop-P-ethyl

403 quizalofop-P-ethyl

An aryl phenoxypropionic acid post-emergence herbicide for grass weed control
HRAC mode of action code: A

Products

1	Leopard 5 EC	Makhteshim	50 g/l	EC	13720
2	Pilot Ultra	Nissan	50 g/l	SC	12929
3	Talon	Makhteshim	50 g/l	EC	13835
4	Visor	Makhteshim	50 g/l	EC	14513

Uses
- Annual grasses in **combining peas, fodder beet, mangels, red beet, spring field beans, spring oilseed rape, sugar beet, winter field beans, winter oilseed rape** [1-4]; **linseed, vining peas** [1, 2, 4]; **spring linseed, winter linseed** [3].
- Barren brome in **combining peas, fodder beet, mangels, red beet, spring field beans, spring linseed, spring oilseed rape, sugar beet, winter field beans, winter linseed, winter oilseed rape** [3]
- Blackgrass in **combining peas, fodder beet, mangels, red beet, spring field beans, spring linseed, spring oilseed rape, sugar beet, winter field beans, winter linseed, winter oilseed rape** [3]
- Couch in **combining peas, fodder beet, mangels, red beet, spring field beans, spring oilseed rape, sugar beet, winter field beans, winter oilseed rape** [1-3]; **linseed, vining peas** [1, 2]; **spring linseed, winter linseed** [3]
- Perennial grasses in **combining peas, fodder beet, linseed, mangels, red beet, spring field beans, spring oilseed rape, sugar beet, vining peas, winter field beans, winter oilseed rape** [1, 2, 4]
- Volunteer cereals in **combining peas, fodder beet, mangels, red beet, spring field beans, spring oilseed rape, sugar beet, winter field beans, winter oilseed rape** [1-3]; **linseed, vining peas** [1, 2]; **spring linseed, winter linseed** [3]
- Wild oats in **combining peas, fodder beet, mangels, red beet, spring field beans, spring linseed, spring oilseed rape, sugar beet, winter field beans, winter linseed, winter oilseed rape** [3]

Approval information
- Quizalofop-P has been included in Annex 1 under EC Directive 91/414
- Approval expiry 30 Nov 2011 [1, 3, 4]

Efficacy guidance
- Best results achieved by application to emerged weeds growing actively in warm conditions with adequate soil moisture
- Weed control may be reduced under conditions such as drought that limit uptake and translocation
- Annual meadow-grass is not controlled
- For effective couch control do not hoe beet crops within 21 d after spraying
- At least 2 h without rain should follow application otherwise results may be reduced
- Quizalofop-P-ethyl is an ACCase inhibitor herbicide. To avoid the build up of resistance do not apply products containing an ACCase inhibitor herbicide more than twice to any crop. In addition do not use any product containing quizalofop-P-ethyl in mixture or sequence with any other product containing the same ingredient
- Use these products as part of a resistance management strategy that includes cultural methods of control and does not use ACCase inhibitors as the sole chemical method of grass weed control
- Applying a second product containing an ACCase inhibitor to a crop will increase the risk of resistance development; only use a second ACCase inhibitor to control different weeds at a different timing
- Always follow WRAG guidelines for preventing and managing herbicide resistant weeds. See Section 5 for more information

Restrictions
- Maximum number of treatments 1 on all recommended crops
- Do not spray crops under stress from any cause or in frosty weather
- Consult processor before use on crops recommended for processing

SEE SECTION 3 FOR PRODUCTS ALSO REGISTERED

- An interval of at least 3 d must elapse between treatment and use of another herbicide on beet crops, 14 d on oilseed rape, 21 d on linseed and other recommended crops
- Avoid drift onto neighbouring crops

Crop-specific information
- HI 16 wk for beet crops; 11 wk for oilseed rape, linseed; 8 wk for field beans; 5 wk for peas
- In some situations treatment can cause yellow patches on foliage of peas, especially vining varieties. Symptoms usually rapidly and completely outgrown
- May cause taint in peas

Following crops guidance
- In the event of failure of a treated crop broad-leaved crops may be resown after a minimum interval of 2 wk, and cereals after 2-6 wk depending on dose applied
- Onions, leeks and maize are not recommended to follow a failed treated crop

Environmental safety
- Dangerous for the environment
- Toxic to aquatic organisms
- Flammable

Hazard classification and safety precautions
Hazard Harmful, Flammable [1, 3, 4]; Dangerous for the environment [1-4]
Transport code 3
Packaging group III
UN number 1993
Risk phrases R22b, R37, R38, R41, R67 [1, 3, 4]; R51, R53a [1-4]
Operator protection A, C [1-4]; H [2-4]; U05a, U11, U19a [1, 3, 4]; U09a [2]; U20b [1-4]
Environmental protection E13b [2]; E15a [1, 3, 4]; E34, E38 [1-4]
Storage and disposal D01, D02, D10b [1, 3, 4]; D03 [3]; D05 [1]; D09a, D12a [1-4]; D10c [2]
Medical advice M03, M05b [1, 3, 4]

404 quizalofop-P-tefuryl

An aryloxyphenoxypropionate herbicide for grass weed control
HRAC mode of action code: A

Products

Panarex   Certis   40 g/l   EC   12532

Uses
- Blackgrass in **amenity vegetation** (off-label), **cereal cover crops, combining peas, fodder beet, natural surfaces not intended to bear vegetation** (off-label), **ornamental plant production** (off-label), **permeable surfaces overlying soil** (off-label), **potatoes, spring field beans, spring linseed, spring oilseed rape, sugar beet, winter field beans, winter linseed, winter oilseed rape**
- Couch in **amenity vegetation** (off-label), **cereal cover crops, combining peas, fodder beet, natural surfaces not intended to bear vegetation** (off-label), **ornamental plant production** (off-label), **permeable surfaces overlying soil** (off-label), **potatoes, spring field beans, spring linseed, spring oilseed rape, sugar beet, winter field beans, winter linseed, winter oilseed rape**
- Italian ryegrass in **cereal cover crops, combining peas, fodder beet, potatoes, spring field beans, spring linseed, spring oilseed rape, sugar beet, winter field beans, winter linseed, winter oilseed rape**
- Perennial ryegrass in **amenity vegetation** (off-label), **cereal cover crops, combining peas, fodder beet, natural surfaces not intended to bear vegetation** (off-label), **ornamental plant production** (off-label), **permeable surfaces overlying soil** (off-label), **potatoes, spring field beans, spring linseed, spring oilseed rape, sugar beet, winter field beans, winter linseed, winter oilseed rape**
- Volunteer cereals in **cereal cover crops, combining peas, fodder beet, potatoes, spring field beans, spring linseed, spring oilseed rape, sugar beet, winter field beans, winter linseed, winter oilseed rape**

FOR FULL CONDITIONS OF USE ALWAYS READ THE PRODUCT LABEL

quizalofop-P-tefuryl

- Wild oats in *cereal cover crops, combining peas, fodder beet, potatoes, spring field beans, spring linseed, spring oilseed rape, sugar beet, winter field beans, winter linseed, winter oilseed rape*

Specific Off-Label Approvals (SOLAs)
- *amenity vegetation* 20060364
- *natural surfaces not intended to bear vegetation* 20060364
- *ornamental plant production* 20060364
- *permeable surfaces overlying soil* 20060364

Approval information
- Quizalofop-P has been included in Annex 1 under EC Directive 91/414

Efficacy guidance
- Best results on annual grass weeds when growing actively and treated from 2 leaves to the start of tillering
- Treat cover crops when they have served their purpose and the threat of wind blow has passed
- Best results on couch achieved when the weed is growing actively and commencing new rhizome growth
- Grass weeds germinating after treatment will not be controlled
- Treatment quickly stops growth and visible colour changes to the leaf tips appear after about 7 d. Complete kill takes 3-4 wk under good growing conditions
- Quizalofop-P-tefuryl is an ACCase inhibitor herbicide. To avoid the build up of resistance do not apply products containing an ACCase inhibitor herbicide more than twice to any crop. In addition do not use any product containing quizalofop-P-tefuryl in mixture or sequence with any other product containing the same ingredient
- Use these products as part of a resistance management strategy that includes cultural methods of control and does not use ACCase inhibitors as the sole chemical method of grass weed control
- Applying a second product containing an ACCase inhibitor to a crop will increase the risk of resistance development; only use a second ACCase inhibitor to control different weeds at a different timing
- Always follow WRAG guidelines for preventing and managing herbicide resistant weeds. See Section 5 for more information

Restrictions
- Maximum number of treatments 1 per crop
- Consult processors before use on peas or potatoes for processing
- Do not treat crops and weeds growing under stress from any cause

Crop-specific information
- HI 60 d for all crops
- Treat oilseed rape from the fully expanded cotyledon stage
- Treat linseed, peas and field beans from 2-3 unfolded leaves
- Treat sugar beet from 2 unfolded leaves

Following crops guidance
- In the event of failure of a treated crop any broad-leaved crop may be planted at any time. Cereals may be drilled from 4 wk after treatment

Environmental safety
- Dangerous for the environment
- Very toxic to aquatic organisms
- Risk to certain non-target insects or other arthropods. For advice on risk management and use in Integrated Pest Management (IPM) see directions for use
- Avoid spraying within 6 m of the field boundary to reduce effects on non-target insects and other arthropods

Hazard classification and safety precautions
Hazard Irritant, Dangerous for the environment
Transport code 9
Packaging group III
UN number 3082
Risk phrases R41, R43, R50, R53a
Operator protection A, C, H; U02a, U04a, U05a, U10, U11, U14, U15, U19a, U20b

SEE SECTION 3 FOR PRODUCTS ALSO REGISTERED

Environmental protection E15a, E22b, E38
Storage and disposal D01, D02, D09a, D10b, D12a

405 rimsulfuron

A selective systemic sulfonylurea herbicide
HRAC mode of action code: B

Products

1	Caesar	AgChem Access	25% w/w	SG	13779
2	Clayton Bramble	Clayton	25% w/w	SG	14856
3	Rigid	AgriGuard	25% w/w	SG	12319
4	Tarot	Makhteshim	25% w/w	SG	11896
5	Titus	Makhteshim	25% w/w	SG	11895

Uses
- Annual dicotyledons in *forage maize, potatoes*
- Volunteer oilseed rape in *forage maize, potatoes*

Approval information
- Rimsulfuron included in Annex I under EC Directive 91/414

Efficacy guidance
- Product should be used with a suitable adjuvant or a suitable herbicide tank-mix partner. See label for details
- Product acts by foliar action. Best results obtained from good spray cover of small actively growing weeds. Effectiveness is reduced in very dry conditions
- Weed spectrum can be broadened by tank mixture with other herbicides. See label for details
- Susceptible weeds cease growth immediately and symptoms can be seen 10 d later
- Rimsulfuron is a member of the ALS-inhibitor group of herbicides

Restrictions
- Maximum number of treatments 1 per crop
- Do not treat maize previously treated with organophosphorus insecticides
- Do not apply to potatoes grown for certified seed
- Consult processor before use on crops grown for processing
- Avoid high light intensity (full sunlight) and high temperatures on the day of spraying
- Do not treat during periods of substantial diurnal temperature fluctuation or when frost anticipated
- Do not apply to any crop stressed by drought, water-logging, low temperatures, pest or disease attack, nutrient or lime deficiency
- Do not apply to forage maize treated with organophosphate insecticides
- Do not apply to forage maize undersown with grass or clover

Crop-specific information
- Latest use: before most advanced potato plants are 25 cm high; before 4-collar stage of fodder maize
- All varieties of ware potatoes may be treated, but variety restrictions of any tank-mix partner must be observed
- Only certain named varieties of forage maize may be treated. See label

Following crops guidance
- Only winter wheat should follow a treated crop in the same calendar yr
- Only barley, wheat or maize should be sown in the spring of the yr following treatment
- In the second autumn after treatment any crop except brassicas or oilseed rape may be drilled

Environmental safety
- Dangerous for the environment
- Toxic to aquatic organisms
- Extremely dangerous to fish or other aquatic life. Do not contaminate surface waters or ditches with chemical or used container
- Herbicide is very active. Take particular care to avoid drift onto plants outside the target area
- Spraying equipment should not be drained or flushed onto land planted, or to be planted, with trees or crops other than potatoes or forage maize and should be thoroughly cleansed after use - see label for instructions

FOR FULL CONDITIONS OF USE ALWAYS READ THE PRODUCT LABEL

rotenone

Hazard classification and safety precautions
Hazard Dangerous for the environment [1, 2, 4, 5]
Transport code 9 [1, 2, 4, 5]
Packaging group III [1, 2, 4, 5]
UN number 3077 [1, 2, 4, 5]; N/C [3]
Risk phrases R51, R53a [1, 2, 4, 5]
Operator protection U08, U19a [1-5]; U20a [3]; U20b [1, 2, 4, 5]
Environmental protection E13a, E34 [1-5]; E38 [1, 2, 4, 5]
Storage and disposal D01, D09a, D11a [1-5]; D05 [3]; D12a [1, 2, 4, 5]

406 rotenone

A natural, contact insecticide of low persistence. Rotenone will not be included in Annex I of EC Directive 91/414. All approvals for the storage and use of products will expire 31/10/2011. IRAC mode of action code: 21

407 silthiofam

A thiophene carboxamide fungicide seed dressing for cereals

Products

1	Latitude	Monsanto	125 g/l	FS	10695
2	Meridian	AgChem Access	125 g/l	FS	14676

Uses
- Seed-borne diseases in **durum wheat** (off-label), **spring rye** (off-label), **triticale** (off-label), **winter rye** (off-label) [1]
- Take-all in **spring wheat** (seed treatment), **winter barley** (seed treatment), **winter wheat** (seed treatment) [1, 2]

Specific Off-Label Approvals (SOLAs)
- *durum wheat* 20060973 [1]
- *spring rye* 20060973 [1]
- *triticale* 20060973 [1]
- *winter rye* 20060973 [1]

Approval information
- Silthiofam included in Annex I under EC Directive 91/414
- Accepted by BBPA for use on malting barley

Efficacy guidance
- Apply using approved seed treatment equipment which has been accurately calibrated
- Apply simultaneously (i.e. not in mixture) with a standard seed treatment
- Drill treated seed in the season of purchase. The viability of treated seed and fungicide activity may be reduced by physical storage
- Drill treated seed at 2.5-4 cm into a well prepared firm seedbed
- As precaution against possible development of disease resistance do not treat more than three consecutive susceptible cereal crops in any one rotation

Restrictions
- Maximum number of treatments one per batch of seed
- Do not use on seed with more than 16% moisture content, or on sprouted, cracked or skinned seed
- Test germination of all seed batches before treatment

Crop-specific information
- Latest use: immediately prior to drilling

Hazard classification and safety precautions
UN number N/C
Operator protection A, H; U20b
Environmental protection E03, E15a, E34

SEE SECTION 3 FOR PRODUCTS ALSO REGISTERED

Storage and disposal D05, D09a, D10a
Treated seed S01, S02, S04b, S05, S06a, S07

408 S-metolachlor

A chloroacetamide residual herbicide for use in forage maize and grain maize
HRAC mode of action code: K3

Products

1 Clayton Smelter	Clayton	960 g/l	EC	14937
2 Dual Gold	Syngenta	960 g/l	EC	14649

Uses
- Annual dicotyledons in **french beans** *(off-label)*, **runner beans** *(off-label)* [2]
- Annual grasses in **begonias** *(off-label)*, **chicory root** *(off-label)*, **endives** *(off-label)*, **strawberries** *(off-label)* [2]
- Annual meadow grass in **forage maize**, **grain maize** [1, 2]; **french beans** *(off-label)*, **runner beans** *(off-label)* [2]
- Black nightshade in **chicory root** *(off-label)*, **endives** *(off-label)* [2]
- Chickweed in **forage maize** *(moderately susceptible)*, **grain maize** *(moderately susceptible)* [1, 2]
- Fat hen in **forage maize** *(moderately susceptible)*, **grain maize** *(moderately susceptible)* [1, 2]
- Field speedwell in **forage maize**, **grain maize** [1, 2]
- Mayweeds in **forage maize**, **grain maize** [1, 2]
- Red dead-nettle in **forage maize**, **grain maize** [1, 2]
- Sowthistle in **forage maize**, **grain maize** [1, 2]

Specific Off-Label Approvals (SOLAs)
- *begonias* 20101390 [2]
- *chicory root* 20101391 [2]
- *endives* 20101391 [2]
- *french beans* 20101388 [2]
- *runner beans* 20101388 [2]
- *strawberries* 20101389 [2]

Approval information
- S-metolachlor included in Annex I under EC Directive 91/414

Environmental safety
- LERAP Category B

Hazard classification and safety precautions
Hazard Irritant, Dangerous for the environment
Transport code 9
Packaging group III
UN number 3082
Risk phrases R43, R50, R53a
Operator protection A [1, 2]; C [2]; H [1]; U05a, U09a, U20b
Environmental protection E15b, E16a, E16b, E38
Storage and disposal D01, D02, D09a, D10c, D11a, D12a

409 sodium hypochlorite (commodity substance)

An inorganic horticultural bactericide for use in mushrooms

Products

sodium hypochlorite	various	100% w/v	ZZ	-

Uses
- Bacterial blotch in **mushrooms**

Approval information
- Approval for the use of sodium hypochlorite as a commodity substance was granted on 5 December 1996 by Ministers under regulation 5 of the Control of Pesticides Regulations 1986

FOR FULL CONDITIONS OF USE ALWAYS READ THE PRODUCT LABEL

spinosad

- Sodium hypochlorite is being supported for review in the fourth stage of the EC Review Programme under Directive 91/414. Whether or not it is included in Annex I, the existing commodity chemical approval will have to be revoked because substances listed in Annex I must be approved for marketing and use under the EC regime. If sodium hypochlorite is included in Annex I, products containing it will need to gain approval in the normal way if they carry label claims for pesticidal activity

Restrictions
- Maximum concentration 315 mg/litre of water
- Mixing and loading must only take place in a ventilated area
- Must only be used by suitably trained and competent operators

Crop-specific information
- HI 1 d

Environmental safety
- Harmful to fish or other aquatic life. Do not contaminate surface waters or ditches with chemical or used container

Hazard classification and safety precautions
Operator protection A, C, H
Environmental protection E13c

410 spinosad

A selective insecticide derived from naturally occurring soil fungi (naturalyte)
IRAC mode of action code: 5

Products

1	Conserve	Fargro	120 g/l	SC	12058
2	Tracer	Landseer	480 g/l	SC	12438

Uses
- Cabbage moth in **brussels sprouts, cabbages, cauliflowers** [2]
- Cabbage white butterfly in **brussels sprouts, cabbages, cauliflowers** [2]
- Codling moth in **apples, pears** [2]
- Diamond-back moth in **brussels sprouts, cabbages, cauliflowers** [2]
- Insect pests in **blackberries** *(off-label)*, **celery leaves, cress, frise, herbs (see appendix 6), hops** *(off-label)*, **lamb's lettuce, lettuce, ornamental specimens** *(off-label)*, **protected herbs (see appendix 6)** *(off-label)*, **radicchio, raspberries** *(off-label)*, **scarole, soft fruit** *(off-label)*, **top fruit** *(off-label)* [2]
- Moths in **hops** *(off-label)*, **ornamental specimens** *(off-label)*, **soft fruit** *(off-label)*, **top fruit** *(off-label)* [2]
- Small white butterfly in **brussels sprouts, cabbages, cauliflowers** [2]
- Thrips in **blackberries** *(off-label)*, **bulb onions, leeks, raspberries** *(off-label)*, **salad onions** [2]
- Tortrix moths in **apples, pears** [2]
- Western flower thrips in **protected aubergines, protected cucumbers, protected ornamentals, protected peppers, protected tomatoes** [1]

Specific Off-Label Approvals (SOLAs)
- **blackberries** *20092118* [2]
- **hops** *20082908* [2]
- **ornamental specimens** *20082908* [2]
- **protected herbs (see appendix 6)** *20081290* [2]
- **raspberries** *20092118* [2]
- **soft fruit** *20082908* [2]
- **top fruit** *20082908* [2]

Approval information
- Spinosad included in Annex I under EC Directive 91/414

Efficacy guidance
- Product enters insects by contact from a treated surface or ingestion of treated plant material therefore good spray coverage is essential

SEE SECTION 3 FOR PRODUCTS ALSO REGISTERED

Pesticide Profiles

- Some plants, for example Fuchsia flowers, can provide effective refuges from spray deposits and control of western flower thrips may be reduced [1]
- Apply to protected crops when western flower thrip nymphs or adults are first seen [1]
- Monitor western flower thrip development carefully to see whether further applications are necessary. A 2 spray programme at 5-7 d intervals may be needed when conditions favour rapid pest development [1]
- Treat top fruit and field crops when pests are first seen or at very first signs of crop damage [2]
- Ensure a rain-free period of 12 h after treatment before applying irrigation [2]
- To reduce possibility of development of resistance, adopt resistance management measures. See label and Section 5 for more information

Restrictions
- Maximum number of treatments 4 per crop for brassicas, onions, leeks [2]; 1 pre-blossom and 3 post blossom for apples, pears [2]
- Apply no more than 2 consecutive sprays. Rotate with another insecticide with a different mode of action or use no further treatment after applying the maximum number of treatments
- Establish whether any incoming plants have been treated and apply no more than 3 consecutive sprays. Maximum number of treatments 6 per structure per yr
- Avoid application in bright sunlight or into open flowers [1]

Crop-specific information
- HI 3 d for protected cucumbers, brassicas; 7 d for apples, pears, onions, leeks
- Test for tolerance on a small number of ornamentals or cucumbers before large scale treatment
- Some spotting of african violet flowers may occur

Environmental safety
- Dangerous for the environment
- Very toxic (toxic [1]) to aquatic organisms
- Whenever possible use an Integrated Pest Management system. Spinosad presents low risk to beneficial arthropods
- Product has low impact on many insect and mite predators but is harmful to adults of most parasitic wasps. Most beneficials may be introduced to treated plants when spray deposits are dry but an interval of 2 wk should elapse before introduction of parasitic wasps. See label for details
- Treatment may cause temporary reduction in abundance of insect and mite predators if present at application
- Exposure to direct spray is harmful to bees but dry spray deposits are harmless. Treatment of field crops should not be made in the heat of the day when bees are actively foraging
- Broadcast air-assisted LERAP [2] (40 m); LERAP Category B [2]

Hazard classification and safety precautions
Hazard Dangerous for the environment
Transport code 9
Packaging group III
UN number 3082
Risk phrases R50 [2]; R51 [1]; R53a [1, 2]
Operator protection A, H [2]; U05a [1]; U08, U20b [2]
Environmental protection E15a, E16a [2]; E15b [1]; E17b [2] (40 m); E34, E38 [1, 2]
Storage and disposal D01, D05 [1]; D02, D10b, D12a [1, 2]

411   spirodiclofen

A tetronic acid which inhibits lipid biosynthesis
IRAC mode of action code: 23

Products

Envidor   Bayer CropScience   240 g/l   SC   13947

Uses
- Mussel scale in *apples*, *pears*
- Pear sucker in *pears*
- Red spider mites in *apples*, *pears*
- Rust mite in *apples*, *pears*

FOR FULL CONDITIONS OF USE ALWAYS READ THE PRODUCT LABEL

spiromesifen

- Spider mites in **hops** *(off-label)*, **ornamental plant production** *(off-label)*, **protected ornamentals** *(off-label)*, **protected strawberries** *(off-label)*, **soft fruit** *(off-label)*, **strawberries** *(off-label)*
- Two-spotted spider mite in **apples**, **pears**, **protected strawberries** *(off-label)*, **strawberries** *(off-label)*

Specific Off-Label Approvals (SOLAs)
- *hops* 20093372 expires 29 Jul 2011
- *ornamental plant production* 20093366 expires 29 Jul 2011
- *protected ornamentals* 20093366 expires 29 Jul 2011
- *protected strawberries* 20093371 expires 29 Jul 2011
- *soft fruit* 20093372 expires 29 Jul 2011
- *strawberries* 20093371 expires 29 Jul 2011

Approval information
- Spirodiclofen included in Annex 1 under EC Directive 91/414
- Approval expiry 29 Jul 2011

Restrictions
- Consult processor on crops grown for processing or for cider production

Environmental safety
- Broadcast air-assisted LERAP (18 m)

Hazard classification and safety precautions
Hazard Harmful
UN number N/C
Risk phrases R40, R43, R52, R53a
Operator protection A, H; U05a, U20c
Environmental protection E12c, E12f, E15b, E22c, E34, E38; E17b (18 m)
Storage and disposal D01, D02, D09a, D10c, D12a
Medical advice M05a

412 spiromesifen

A tetronic acid insecticide for protected tomatoes
IRAC mode of action code: 23

Products
Oberon	Certis	240 g/l	SC	11819

Uses
- Red spider mites in **protected ornamentals** *(off-label)*
- Spider mites in **protected aubergines** *(off-label)*, **protected cayenne peppers** *(off-label)*, **protected cucumbers** *(off-label)*, **protected gherkins** *(off-label)*, **protected peppers** *(off-label)*, **protected strawberries** *(off-label)*
- Whitefly in **inert substrate tomatoes**, **nft tomatoes**, **protected aubergines** *(off-label)*, **protected cayenne peppers** *(off-label)*, **protected cucumbers** *(off-label)*, **protected gherkins** *(off-label)*, **protected ornamentals** *(off-label)*, **protected peppers** *(off-label)*, **protected strawberries** *(off-label)*

Specific Off-Label Approvals (SOLAs)
- *protected aubergines* 20050959, 20063645
- *protected cayenne peppers* 20062149
- *protected cucumbers* 20050958
- *protected gherkins* 20050958
- *protected ornamentals* 20041718
- *protected peppers* 20062149
- *protected strawberries* 20050957

Efficacy guidance
- Apply to run off and ensure that all sides of the crop are covered
- Apply when infestation first observed and repeat 7-10 d later if necessary
- Always follow guidelines for preventing and managing insect resistance. See Section 5 for more information

SEE SECTION 3 FOR PRODUCTS ALSO REGISTERED

Pesticide Profiles

Restrictions
- Maximum number of treatments 2 per cropping cycle
- Do not use on any outdoor crops
- Product should be used in rotation with compounds with different modes of action

Crop-specific information
- HI 3 d for tomatoes

Environmental safety
- Dangerous for the environment
- Very toxic to aquatic organisms

Hazard classification and safety precautions
Hazard Irritant, Dangerous for the environment
Transport code 9
Packaging group III
UN number 3082
Risk phrases R43, R50, R53a
Operator protection A, H; U05a, U14, U20b
Environmental protection E15a, E34, E38
Storage and disposal D01, D02, D05, D09a, D11a, D12a

413 spirotetramat

Tetramic acid derivative that acts on lipid synthesis
IRAC mode of action code: 23

Products
 Movento   Bayer CropScience   150 g/l   OD   14446

Uses
- Aphids in **broccoli**, **brussels sprouts**, **cabbages**, **calabrese**, **cauliflowers**, **chard** *(off-label)*, **chinese cabbage** *(off-label)*, **choi sum** *(off-label)*, **cress** *(off-label)*, **endives** *(off-label)*, **frise** *(off-label)*, **hops** *(off-label)*, **lamb's lettuce** *(off-label)*, **leaf brassicas** *(off-label)*, **lettuce**, **mustard** *(off-label)*, **pak choi** *(off-label)*, **protected chard** *(off-label)*, **protected cress** *(off-label)*, **protected endives** *(off-label)*, **protected frise** *(off-label)*, **protected lamb's lettuce** *(off-label)*, **protected leaf brassicas** *(off-label)*, **protected rocket** *(off-label)*, **protected scarole** *(off-label)*, **protected spinach** *(off-label)*, **rocket** *(off-label)*, **scarole** *(off-label)*, **spinach** *(off-label)*, **tatsoi** *(off-label)*
- Damson-hop aphid in **hops** *(off-label)*
- White tip in **chard** *(off-label)*
- Whitefly in **broccoli**, **brussels sprouts**, **cabbages**, **calabrese**, **cauliflowers**, **chinese cabbage** *(off-label)*, **choi sum** *(off-label)*, **cress** *(off-label)*, **endives** *(off-label)*, **frise** *(off-label)*, **lamb's lettuce** *(off-label)*, **leaf brassicas** *(off-label)*, **mustard** *(off-label)*, **pak choi** *(off-label)*, **protected chard** *(off-label)*, **protected cress** *(off-label)*, **protected endives** *(off-label)*, **protected frise** *(off-label)*, **protected lamb's lettuce** *(off-label)*, **protected leaf brassicas** *(off-label)*, **protected rocket** *(off-label)*, **protected scarole** *(off-label)*, **protected spinach** *(off-label)*, **rocket** *(off-label)*, **scarole** *(off-label)*, **spinach** *(off-label)*, **tatsoi** *(off-label)*

Specific Off-Label Approvals (SOLAs)
- *chard* 20102410
- *chinese cabbage* 20101095
- *choi sum* 20101095
- *cress* 20102410
- *endives* 20102410
- *frise* 20102410
- *hops* 20102117
- *lamb's lettuce* 20102410
- *leaf brassicas* 20102410
- *mustard* 20102410
- *pak choi* 20101095
- *protected chard* 20102410
- *protected cress* 20102410
- *protected endives* 20102410

FOR FULL CONDITIONS OF USE ALWAYS READ THE PRODUCT LABEL

- *protected frise* 20102410
- *protected lamb's lettuce* 20102410
- *protected leaf brassicas* 20102410
- *protected rocket* 20102410
- *protected scarole* 20102410
- *protected spinach* 20102410
- *rocket* 20102410
- *scarole* 20102410
- *spinach* 20102410
- *tatsoi* 20101095

Approval information
- Spirotetramat has not yet been added to Annex 1 listing

Hazard classification and safety precautions
Hazard Irritant, Dangerous for the environment
Transport code 9
Packaging group III
UN number 3082
Risk phrases R43, R51, R53a
Operator protection A; U05a, U14, U20b
Environmental protection E15b, E22c, E34, E38
Storage and disposal D02, D09a, D10c, D12a
Medical advice M03

414 spiroxamine

A spiroketal amine fungicide for cereals
FRAC mode of action code: 5

See also bixafen + prothioconazole + spiroxamine
prothioconazole + spiroxamine
prothioconazole + spiroxamine + tebuconazole

Products
Torch   Bayer CropScience   500 g/l   EW   11258

Uses
- Brown rust in **spring barley**, **spring rye**, **spring wheat**, **winter barley**, **winter rye**, **winter wheat**
- Powdery mildew in **spring barley**, **spring rye**, **spring wheat**, **winter barley**, **winter rye**, **winter wheat**
- Rhynchosporium in **spring barley** (reduction), **spring rye** (reduction), **winter barley** (reduction), **winter rye** (reduction)
- Yellow rust in **spring barley**, **spring rye**, **spring wheat**, **winter barley**, **winter rye**, **winter wheat**

Approval information
- Spiroxamine included in Annex I under EC Directive 91/414
- Accepted by BBPA for use on malting barley
- Approval expiry 31 Dec 2011 [1]

Efficacy guidance
- For best results treat at an early stage of disease development before infection spreads to new growth
- To reduce the risk of development of resistance avoid repeat treatments on diseases such as powdery mildew. If necessary tank mix or alternate with other non-morpholine fungicides

Restrictions
- Maximum total dose equivalent to two full dose treatments

Crop-specific information
- Latest use: before caryopsis watery ripe (GS 71) for wheat, rye; ear emergence complete (GS 59) for barley

SEE SECTION 3 FOR PRODUCTS ALSO REGISTERED

Environmental safety
- Dangerous for the environment
- Very toxic to aquatic organisms
- LERAP Category B

Hazard classification and safety precautions
Hazard Harmful, Dangerous for the environment
Risk phrases R20, R22a, R38, R41, R43, R50, R53a
Operator protection A, C, H; U05a, U11, U13, U14, U15, U20a
Environmental protection E13b, E16a, E34, E38
Storage and disposal D01, D02, D05, D09a, D10b, D12a
Medical advice M03

415 spiroxamine + tebuconazole

A broad spectrum systemic fungicide mixture for cereals
FRAC mode of action code: 5 + 3

See also tebuconazole

Products
Sage	Interfarm	250:133 g/l	EW	11303

Uses
- Brown rust in **spring barley**, **spring rye**, **spring wheat**, **winter barley**, **winter rye**, **winter wheat**
- Disease control in **grass seed crops** *(off-label)*
- Fusarium ear blight in **spring wheat**, **winter wheat**
- Glume blotch in **spring wheat**, **winter wheat**
- Net blotch in **spring barley**, **winter barley**
- Powdery mildew in **spring barley**, **spring rye**, **spring wheat**, **winter barley**, **winter rye**, **winter wheat**
- Rhynchosporium in **spring barley**, **winter barley**
- Septoria leaf blotch in **spring wheat**, **winter wheat**
- Sooty moulds in **spring wheat**, **winter wheat**
- Yellow rust in **spring barley**, **spring rye**, **spring wheat**, **winter barley**, **winter rye**, **winter wheat**

Specific Off-Label Approvals (SOLAs)
- *grass seed crops* 20093180

Approval information
- Spiroxamine included in Annex I under EC Directive 91/414
- Accepted by BBPA for use on malting barley

Efficacy guidance
- Best results achieved from treatment at early stage of disease development before infection spreads to new growth
- To protect flag leaf and ear from Septoria apply from flag leaf emergence to ear fully emerged (GS 37-59). Earlier treatment may be necessary where there is high disease risk
- Control of rusts, powdery mildew, leaf blotch and net blotch may require second treatment 2-3 wk later
- Tebuconazole is a DMI fungicide. Resistance to some DMI fungicides has been identified in Septoria leaf blotch which may seriously affect performance of some products. For further advice contact a specialist advisor and visit the Fungicide Resistance Action Group (FRAG)-UK website

Restrictions
- Maximum total dose equivalent to two full dose treatments
- Do not use on durum wheat

Crop-specific information
- Latest use: before caryopsis watery ripe (GS 71) for rye, wheat; ear emergence complete (GS 59) for barley

FOR FULL CONDITIONS OF USE ALWAYS READ THE PRODUCT LABEL

- Some transient leaf speckling may occur on wheat but this has not been shown to reduce yield reponse or disease control

Environmental safety
- Dangerous for the environment
- Very toxic to aquatic organisms
- Dangerous to fish or other aquatic life. Do not contaminate surface waters or ditches with chemical or used container
- LERAP Category B

Hazard classification and safety precautions
Hazard Harmful, Dangerous for the environment
Transport code 9
Packaging group III
UN number 3082
Risk phrases R20, R22a, R38, R41, R43, R50, R53a
Operator protection A, C, H; U05a, U11, U13, U14, U15, U20b
Environmental protection E13b, E16a, E16b, E34, E38
Storage and disposal D01, D02, D05, D09a, D10b, D12a
Medical advice M03

416 sulfosulfuron

A sulfonylurea herbicide for grass and broad-leaved weed control in winter wheat
HRAC mode of action code: B

See also glyphosate + sulfosulfuron

Products
Monitor   Monsanto   80% w/w   WG   12236

Uses
- Annual dicotyledons in **game cover** *(off-label)*, **grass seed crops** *(off-label)*, **spring rye** *(off-label)*, **triticale** *(off-label)*, **winter rye** *(off-label)*
- Brome grasses in **winter wheat** *(moderate control of barren brome)*
- Chickweed in **winter wheat**
- Cleavers in **winter wheat**
- Couch in **winter wheat** *(moderate control only)*
- Grass weeds in **game cover** *(off-label)*, **grass seed crops** *(off-label)*, **spring rye** *(off-label)*, **triticale** *(off-label)*, **winter rye** *(off-label)*
- Loose silky bent in **winter wheat**
- Mayweeds in **winter wheat**
- Onion couch in **winter wheat** *(moderate control only)*

Specific Off-Label Approvals (SOLAs)
- *game cover* 20082861
- *grass seed crops* 20081279
- *spring rye* 20081279
- *triticale* 20081279
- *winter rye* 20081279

Approval information
- Sulfosulfuron included in Annex I under EC Directive 91/414

Efficacy guidance
- For best results treat in early spring when annual weeds are small and growing actively. Avoid treatment when weeds are dormant for any reason
- Best control of onion couch is achieved when the weed has more than two leaves. Effects on bulbils or on growth in the following yr have not been examined
- An extended period of dry weather before or after treatment may result in reduced control
- The addition of a recommended surfactant is essential for full activity
- Specific follow-up treatments may be needed for complete control of some weeds
- Use only where competitively damaging weed populations have emerged otherwise yield may be reduced

SEE SECTION 3 FOR PRODUCTS ALSO REGISTERED

- Sulfosulfuron is a member of the ALS-inhibitor group of herbicides. To avoid the build up of resistance do not use any product containing an ALS-inhibitor herbicide with claims for control of grass weeds more than once on any crop
- Use these products as part of a resistance management strategy that includes cultural methods of control and does not use ALS inhibitors as the sole chemical method of grass weed control

Restrictions
- Maximum total dose equivalent to one treatment at full dose
- Do not treat crops under stress
- Apply only after 1 Feb and from 3 expanded leaf stage
- Do not treat durum wheat or any undersown wheat crop
- Do not use in mixture or in sequence with any other sulfonyl urea herbicide on the same crop

Crop-specific information
- Latest use: flag leaf ligule just visible for winter wheat (GS 39)

Following crops guidance
- In the autumn following treatment winter wheat, winter rye, winter oats, triticale, winter oilseed rape, winter peas or winter field beans may be sown on any soil, and winter barley on soils with less than 60% sand
- In the next spring following application crops of wheat, barley, oats, maize, peas, beans, linseed, oilseed rape, potatoes or grass may be sown
- In the second autumn following application winter linseed may be sown, and winter barley on soils with more than 60% sand
- Sugar beet or any other crop not mentioned above must not be drilled until the second spring following application
- Where winter oilseed rape is to be sown in the autumn following treatment soil cultivation to a minimum of 10 cm is recommended

Environmental safety
- Dangerous for the environment
- Very toxic to aquatic organisms
- Extremely dangerous to fish or other aquatic life. Do not contaminate surface waters or ditches with chemical or used container
- Take extreme care to avoid drift onto broad leaved plants or other crops, or onto ponds, waterways or ditches.
- Follow detailed label instructions for cleaning the sprayer to avoid damage to sensitive crops during subsequent use
- LERAP Category B

Hazard classification and safety precautions
    Hazard Dangerous for the environment
    UN number N/C
    Risk phrases R50, R53a
    Operator protection U20b
    Environmental protection E15a, E16a, E16b, E34, E38
    Storage and disposal D09a, D11a, D12a

417 sulphur

A broad-spectrum inorganic protectant fungicide, foliar feed and acaricide
FRAC mode of action code: M2

Products

1	Headland Sulphur	Headland	800 g/l	SC	12879
2	Kumulus DF	BASF	80% w/w	WG	04707
3	Microthiol Special	United Phosphorus	80% w/w	MG	06268
4	Solfa WG	Nufarm UK	80% w/w	WG	11602
5	Sulphur Flowable	United Phosphorus	800 g/l	SC	07526
6	Thiovit Jet	Syngenta	80% w/w	WG	10928

Uses
- American gooseberry mildew in *gooseberries* [4]

FOR FULL CONDITIONS OF USE ALWAYS READ THE PRODUCT LABEL

sulphur

- Disease control/foliar feed in **bilberries** *(off-label)*, **blueberries** *(off-label)*, **cranberries** *(off-label)*, **redcurrants** *(off-label)*, **whitecurrants** *(off-label)* [2]
- Foliar feed in **all top fruit**, **bush fruit**, **cane fruit**, **edible brassicas**, **managed amenity turf**, **potatoes**, **spring oats**, **sugar beet**, **swedes**, **turnips**, **vegetables**, **winter oats** [3]; **permanent grassland**, **spring oilseed rape** [3, 4, 6]; **spring barley**, **spring wheat**, **winter barley**, **winter wheat** [2, 3, 6]; **winter oilseed rape** [2-4, 6]
- Gall mite in **blackcurrants** [2, 4]
- Powdery mildew in **apples**, **strawberries** [1, 2, 4, 5]; **aubergines** *(off-label)*, **combining peas** *(off-label)*, **protected chilli peppers** *(off-label)*, **protected chives** *(off-label)*, **protected parsley** *(off-label)*, **vining peas** *(off-label)* [6]; **borage for oilseed production** *(off-label)*, **parsnips** *(off-label)*, **protected cucumbers** *(off-label)*, **protected herbs (see appendix 6)** *(off-label)*, **protected peppers** *(off-label)*, **protected tomatoes** *(off-label)* [3, 4, 6]; **fodder beet** *(off-label)* [1, 3, 4, 6]; **gooseberries** [2]; **grapevines** [3]; **grass seed crops** *(off-label)*, **spring barley**, **spring wheat**, **winter barley**, **winter wheat** [1, 3-5]; **hops** [1, 2, 4-6]; **pears** [1, 5]; **protected aubergines** *(off-label)*, **protected cayenne peppers** *(off-label)*, **turnips** [3, 4]; **quinces** *(off-label)* [5]; **spring oats**, **winter oats** [1, 3, 4]; **spring oilseed rape**, **spring rye**, **triticale** *(off-label)*, **winter oilseed rape**, **winter rye** [1]; **sugar beet** [1-6]; **swedes** [1, 3-6]; **turnips** *(off-label)* [1, 4-6]; **wine grapes** [2, 4]
- Scab in **apples** [1, 2, 5]; **pears** [1, 5]; **quinces** *(off-label)* [1]

Specific Off-Label Approvals (SOLAs)
- **aubergines** 20023652 [6]
- **bilberries** 20061042 [2]
- **blueberries** 20061042 [2]
- **borage for oilseed production** 20072027 [3], 20072238 [4], 20061719 [6]
- **combining peas** 20081361 [6]
- **cranberries** 20061042 [2]
- **fodder beet** 20062690 [1], 20072025 [3], 20072240 [4], 20061383 [6]
- **grass seed crops** 20062691 [1], 20061043 [3], 20061044 [4], 20061375 [5]
- **parsnips** 20072026 [3], 20072242 [4], 20023654 [6]
- **protected aubergines** 20072024 [3], 20072239 [4]
- **protected cayenne peppers** 20072024 [3], 20072239 [4]
- **protected chilli peppers** 20081360 [6]
- **protected chives** 20023652 [6]
- **protected cucumbers** 20072024 [3], 20072239 [4], 20023652 [6]
- **protected herbs (see appendix 6)** 20072024 [3], 20072239 [4], 20023652 [6]
- **protected parsley** 20023652 [6]
- **protected peppers** 20072024 [3], 20072239 [4], 20023652 [6]
- **protected tomatoes** 20072024 [3], 20072239 [4], 20023652 [6]
- **quinces** 20062688 [1], 20061374 [5]
- **redcurrants** 20061042 [2]
- **triticale** 20062691 [1]
- **turnips** 20062689 [1], 20072241 [4], 20061373 [5], 20061384 [6]
- **vining peas** 20081361 [6]
- **whitecurrants** 20061042 [2]

Approval information
- Sulphur is included in Annex 1 under EC Directive 91/414
- Accepted by BBPA for use on malting barley and hops (before burr)

Efficacy guidance
- Apply when disease first appears and repeat 2-3 wk later. Details of application rates and timing vary with crop, disease and product. See label for information
- Sulphur acts as foliar feed as well as fungicide and with some crops product labels vary in whether treatment recommended for disease control or growth promotion
- In grassland best results obtained at least 2 wk before cutting for hay or silage, 3 wk before grazing
- Treatment unlikely to be effective if disease already established in crop

SEE SECTION 3 FOR PRODUCTS ALSO REGISTERED

Pesticide Profiles

Restrictions
- Maximum number of treatments normally 2 per crop for grassland, sugar beet, parsnips, swedes, hops, protected herbs; 3 per yr for blackcurrants, gooseberries; 4 per crop on apples, pears but labels vary
- Do not use on sulphur-shy apples (Beauty of Bath, Belle de Boskoop, Cox's Orange Pippin, Lanes Prince Albert, Lord Derby, Newton Wonder, Rival, Stirling Castle) or pears (Doyenne du Comice)
- Do not use on gooseberry cultivars Careless, Early Sulphur, Golden Drop, Leveller, Lord Derby, Roaring Lion, or Yellow Rough
- Do not use on apples or gooseberries when young, under stress or if frost imminent
- Do not use on fruit for processing, on grapevines during flowering or near harvest on grapes for wine-making
- Do not use on hops at or after burr stage
- Do not spray top or soft fruit with oil or within 30 d of an oil-containing spray

Crop-specific information
- Latest use: before burr stage for hops; before end Sep for parsnips, swedes, sugar beet; fruit swell for gooseberries; milky ripe stage for cereals
- HI cutting grass for hay or silage 2 wk; grazing grassland 3 wk

Environmental safety
- Sulphur products are attractive to livestock and must be kept out of their reach
- Do not empty into drains

Hazard classification and safety precautions
Hazard Irritant [4]
UN number N/C
Risk phrases R37 [4]
Operator protection A, H [6]; U02a, U08, U14 [6]; U05a [4, 6]; U19a [4]; U20a [1]; U20b [3, 5]; U20c [2, 4, 6]
Environmental protection E15a [1-6]; E19b [4]; E34 [1]
Storage and disposal D01 [3, 4, 6]; D02 [4, 6]; D05 [5, 6]; D09a [1-6]; D10b [1]; D10c [5]; D11a [2, 3, 6]

418 tau-fluvalinate

A contact pyrethroid insecticide for cereals and oilseed rape
IRAC mode of action code: 3

Products

1	Greencrop Malin	Greencrop	240 g/l	EW	11787
2	Klartan	Makhteshim	240 g/l	EW	11074
3	Mavrik	Makhteshim	240 g/l	EW	10612
4	Revolt	Makhteshim	240 g/l	EW	13383

Uses
- Aphids in **borage for oilseed production** *(off-label)* [3]; **durum wheat** *(off-label)*, **evening primrose** *(off-label)*, **grass seed crops** *(off-label)*, **honesty** *(off-label)*, **mustard** *(off-label)*, **spring linseed** *(off-label)*, **spring rye** *(off-label)*, **triticale** *(off-label)*, **winter rye** *(off-label)* [2, 3]; **spring barley**, **spring oilseed rape**, **spring wheat**, **winter barley**, **winter oilseed rape**, **winter wheat** [1-4]
- Barley yellow dwarf virus vectors in **winter barley**, **winter wheat** [1-4]
- Cabbage stem flea beetle in **winter oilseed rape** [2-4]
- Insect pests in **grass seed crops** *(off-label)*, **mustard** *(off-label)* [1]
- Pollen beetle in **spring oilseed rape**, **winter oilseed rape** [1-4]

Specific Off-Label Approvals (SOLAs)
- **borage for oilseed production** *20061366* [3]
- **durum wheat** *20061369* [2], *20061365* [3]
- **evening primrose** *20061370* [2], *20061366* [3]
- **grass seed crops** *20061368* [1], *20061369* [2], *20061365* [3]
- **honesty** *20061370* [2], *20061366* [3]
- **mustard** *20061367* [1], *20061370* [2], *20061366* [3]

FOR FULL CONDITIONS OF USE ALWAYS READ THE PRODUCT LABEL

tau-fluvalinate

- **spring linseed** *20061370* [2], *20061366* [3]
- **spring rye** *20061369* [2], *20061365* [3]
- **triticale** *20061369* [2], *20061365* [3]
- **winter rye** *20061369* [2], *20061365* [3]

Approval information
- Tau-fluvalinate is not yet listed in Annex 1 under EC Directive 91/414
- Accepted by BBPA for use on malting barley

Efficacy guidance
- For BYDV control on winter cereals follow local warnings or spray high risk crops in mid-Oct and make repeat application in late autumn/early winter if aphid activity persists
- For summer aphid control on cereals spray once when aphids present on two thirds of ears and increasing
- On oilseed rape treat peach potato aphids in autumn in response to local warning and repeat if necessary
- Best control of pollen beetle in oilseed rape obtained from treatment at green to yellow bud stage and repeat if necessary
- Good spray cover of target essential for best results

Restrictions
- Maximum total dose equivalent to two full dose treatments on oilseed rape. See label for dose rates on cereals
- A minimum of 14 d must elapse between applications to cereals

Crop-specific information
- Latest use: before caryopsis watery ripe (GS 71) for barley; before flowering for oilseed rape; before kernel medium milk (GS 75) for wheat

Environmental safety
- Dangerous for the environment
- Very toxic to aquatic organisms
- High risk to non-target insects or other arthropods. Do not spray within 6 m of the field boundary [1-3]
- Avoid spraying oilseed rape within 6 m of field boundary to reduce effects on certain non-target species or other arthropods
- Must not be applied to cereals if any product containing a pyrethroid insecticide or dimethoate has been sprayed after the start of ear emergence (GS 51)
- LERAP Category A

Hazard classification and safety precautions
Hazard Irritant [1]; Dangerous for the environment [1-4]
Transport code 9
Packaging group III
UN number 3082
Risk phrases R36, R38, R51 [1]; R50 [2-4]; R53a [1-4]
Operator protection A, C, H; U05a, U10, U11, U19a, U20a
Environmental protection E15a, E16c, E16d, E22a
Storage and disposal D01, D02, D05, D10c [1-4]; D09a [2-4]; D09b [1]

SEE SECTION 3 FOR PRODUCTS ALSO REGISTERED

419 tebuconazole

A systemic triazole fungicide for cereals and other field crops
FRAC mode of action code: 3

See also bixafen + prothioconazole + tebuconazole
carbendazim + tebuconazole
chlorothalonil + tebuconazole
clothianidin + prothioconazole + tebuconazole + triazoxide
fenpropidin + prochloraz + tebuconazole
fenpropidin + propiconazole + tebuconazole
fenpropidin + tebuconazole
fluoxastrobin + prothioconazole + tebuconazole
imidacloprid + tebuconazole + triazoxide
prochloraz + proquinazid + tebuconazole
prochloraz + tebuconazole
propiconazole + tebuconazole
prothioconazole + spiroxamine + tebuconazole
prothioconazole + tebuconazole
prothioconazole + tebuconazole + triazoxide
spiroxamine + tebuconazole

Products

1	Alpha Tebuconazole 20 EW	Makhteshim	200 g/l	EW	12893
2	Chani	AgChem Access	250 g/l	EW	13325
3	Clayton Tebucon	Clayton	250 g/l	EW	08707
4	Clayton Tebucon EW	Clayton	250 g/l	EW	14824
5	Deacon	Makhteshim	200 g/l	EW	14270
6	EA Tebuzole	European Ag	250 g/l	EW	15000
7	Fezan	Sipcam	250 g/l	EW	13681
8	Folicur	Bayer CropScience	250 g/l	EW	11278
9	Gizmo	Nufarm UK	60 g/l	FS	14150
10	Greencrop Tabloid	Greencrop	250 g/l	EW	11969
11	Icon	AgriGuard	250 g/l	EW	12202
12	Mitre	Makhteshim	200 g/l	EW	12901
13	Orian	ChemSource	200 g/l	EW	14798
14	Orius 20 EW	Makhteshim	200 g/l	EW	12311
15	Raza	ChemSource	250 g/l	EW	14800
16	Riza	Headland	250 g/l	EW	12696
17	Standon Tebuconazole	Standon	250 g/l	EC	09056
18	Tebcon	Goldengrass	250 g/l	EW	14300
19	Tebucon 250	Goldengrass	250 g/l	EW	14789
20	Tebucur 250	Globachem	250 g/l	EW	13975

Uses

- Alternaria in **brussels sprouts** [13, 14]; **cabbages** [1, 2, 4-6, 8, 10-16, 18-20]; **carrots** [1, 2, 4-6, 8, 10, 11, 15, 16, 18-20]; **horseradish** [1, 2, 4, 5, 8, 10, 11, 18, 19]; **horseradish** *(off-label)* [16]; **parsnips** [6, 15, 16, 20]; **spring oilseed rape, winter oilseed rape** [1-6, 8, 10-20]
- Blight in **blackberries** *(off-label)*, **raspberries** *(off-label)*, **rubus hybrids** *(off-label)* [1, 8, 12, 14]
- Botrytis in **daffodils** *(off-label)* [14]; **daffodils** *(off-label - for galanthamine production)* [1, 8, 12]; **daffodils grown for galanthamine production** *(off-label)* [5]; **linseed** *(reduction)* [1-5, 8, 10, 11, 17-19]
- Brown rust in **spring barley, spring rye, spring wheat, winter barley, winter rye, winter wheat** [1-8, 10-20]; **triticale** *(off-label)* [14]
- Cane blight in **blackberries** *(off-label)*, **raspberries** *(off-label)*, **rubus hybrids** *(off-label)* [1, 5, 8, 12, 14]
- Canker in **apples** *(off-label)*, **chestnuts** *(off-label)*, **cob nuts** *(off-label)*, **crab apples** *(off-label)*, **pears** *(off-label)*, **quinces** *(off-label)*, **walnuts** *(off-label)* [1, 5, 8, 12, 14, 16]
- Chocolate spot in **spring field beans, winter field beans** [1-6, 8, 10-20]
- Crown rust in **spring oats, winter oats** [6, 15, 16, 20]; **spring oats** *(reduction)*, **winter oats** *(reduction)* [2, 4, 8, 10, 11, 18, 19]
- Disease control in **borage for oilseed production** *(off-label)*, **canary flower (echium spp.)** *(off-label)*, **evening primrose** *(off-label)*, **grass seed crops** *(off-label)*, **honesty** *(off-label)*,

FOR FULL CONDITIONS OF USE ALWAYS READ THE PRODUCT LABEL

tebuconazole

mustard *(off-label)* [3, 5, 14, 17]; **kohlrabi** *(off-label)*, **mallow (althaea spp.)** *(off-label)*, **parsley root** *(off-label)* [5, 14]; **triticale** *(off-label)* [5]
- Foliar disease control in **borage** *(off-label)*, **grass seed crops** *(off-label)* [1, 8, 12]; **borage for oilseed production** *(off-label)*, **horseradish** *(off-label)* [16]; **canary flower (echium spp.)** *(off-label)*, **evening primrose** *(off-label)*, **honesty** *(off-label)*, **mallow (althaea spp.)** *(off-label)*, **mustard** *(off-label)*, **parsley root** *(off-label)*, **triticale** *(off-label)* [1, 8, 12, 16]
- Fungus diseases in **kohlrabi** *(off-label)* [1, 8, 12]
- Fusarium ear blight in **spring wheat**, **winter wheat** [1, 2, 4-6, 8, 10-16, 18-20]; **triticale** *(off-label)* [14]
- Glume blotch in **spring wheat**, **winter wheat** [1-8, 10-20]; **triticale** *(off-label)* [14]
- Leaf stripe in **spring barley** [9]
- Light leaf spot in **brussels sprouts** [13, 14]; **cabbages**, **spring oilseed rape**, **winter oilseed rape** [1-6, 8, 10-20]
- Lodging control in **spring oilseed rape**, **winter oilseed rape** [1, 2, 4, 5, 8, 10-14, 18, 19]
- Loose smut in **spring barley**, **winter barley** [9]
- Net blotch in **spring barley**, **winter barley** [1-8, 10-20]
- Phoma in **spring oilseed rape**, **winter oilseed rape** [1, 2, 4-6, 8, 10-16, 18-20]
- Powdery mildew in **brussels sprouts** [13, 14]; **cabbages**, **swedes**, **turnips** [1-6, 8, 10-20]; **carrots**, **linseed** [1, 2, 4-6, 8, 10, 11, 15, 16, 18-20]; **chives** *(off-label)* [1, 5, 8, 12, 14, 16]; **horseradish** *(off-label)* [16]; **parsnips** [1, 2, 4-6, 8, 15, 16, 18-20]; **spring barley**, **spring rye**, **spring wheat**, **winter barley**, **winter rye**, **winter wheat** [1-8, 10-20]; **spring oats**, **winter oats** [2, 4, 6-8, 10, 11, 15-20]; **triticale** *(off-label)* [14]
- Rhynchosporium in **spring barley**, **spring rye**, **winter barley**, **winter rye** [1-8, 10-20]
- Ring spot in **broccoli** *(off-label)*, **cauliflowers** *(off-label)*, **chinese cabbage** *(off-label)*, **choi sum** *(off-label)*, **pak choi** *(off-label)* [1, 5, 8, 12, 14, 16]; **brussels sprouts** [13, 14]; **cabbages** [1-6, 8, 10-20]; **calabrese** *(off-label)* [1, 8, 12, 14, 16]; **kohlrabi** *(off-label)* [16]; **spring oilseed rape** *(reduction)*, **winter oilseed rape** *(reduction)* [1, 2, 4, 5, 8, 10-14, 18, 19]
- Rust in **broad beans** *(off-label)*, **chives** *(off-label)*, **dwarf beans** *(off-label)*, **french beans** *(off-label)*, **runner beans** [1, 5, 8, 12, 14, 16]; **leeks** [1, 2, 4-6, 8, 10-16, 18-20]; **spring field beans**, **winter field beans** [1-6, 8, 10-20]
- Sclerotinia in **carrots** [1, 2, 4-6, 8, 10, 11, 15, 16, 18-20]; **horseradish** *(off-label)* [16]; **parsnips** [6, 15, 16, 20]
- Sclerotinia stem rot in **spring oilseed rape**, **winter oilseed rape** [1-6, 8, 10-20]
- Septoria leaf blotch in **spring wheat**, **winter wheat** [1-8, 10-20]; **triticale** *(off-label)* [14]
- Sooty moulds in **spring wheat**, **winter wheat** [1, 2, 4-6, 8, 10-16, 18-20]
- Stem canker in **spring oilseed rape**, **winter oilseed rape** [1, 2, 4-6, 8, 10-16, 18-20]
- White rot in **bulb onions** *(off-label)*, **garlic** *(off-label)*, **salad onions** *(off-label)*, **shallots** *(off-label)* [1, 5, 8, 9, 12, 14, 16]; **onion sets** *(off-label)* [5, 14, 16]
- Yellow rust in **spring barley**, **spring rye**, **spring wheat**, **winter barley**, **winter rye**, **winter wheat** [1-8, 10-20]; **triticale** *(off-label)* [14]

Specific Off-Label Approvals (SOLAs)
- **apples** *20071380* [1], *20090147* [5], *20082159* [8], *20071368* [12], *20071329* [14], *20070554* [16]
- **blackberries** *20091408* [1], *20091407* [5], *20082160* [8], *20091406* [12], *20091399* [14]
- **borage** *20071389* [1], *20062694* [8], *20071359* [12]
- **borage for oilseed production** *20100850 expires 31 Aug 2011* [3], *20090113* [5], *20071337* [14], *20070558* [16], *20100822 expires 31 Aug 2011* [17]
- **broad beans** *20071390* [1], *20090089* [5], *20031878* [8], *20071364* [12], *20071338* [14], *20070549* [16]
- **broccoli** *20071375* [1], *20090091* [5], *20031874* [8], *20071371* [12], *20071326* [14], *20070546* [16]
- **bulb onions** *20071378* [1], *20071384* [1], *20090090* [5], *20031877* [8], *20031879* [8], *20101121* [9], *20071365* [12], *20071369* [12], *20071327* [14], *20070550* [16]
- **calabrese** *20071375* [1], *20031874* [8], *20071371* [12], *20071326* [14], *20070546* [16]
- **canary flower (echium spp.)** *20071389* [1], *20100850 expires 31 Aug 2011* [3], *20090113* [5], *20062694* [8], *20071359* [12], *20071337* [14], *20070558* [16], *20100822 expires 31 Aug 2011* [17]
- **cauliflowers** *20071375* [1], *20090091* [5], *20031874* [8], *20071371* [12], *20071326* [14], *20070546* [16]
- **chestnuts** *20071393* [1], *20090116* [5], *20082158* [8], *20071372* [12], *20071321* [14], *20070555* [16]

SEE SECTION 3 FOR PRODUCTS ALSO REGISTERED

Pesticide Profiles

- **chinese cabbage** *20071375* [1], *20090091* [5], *20031874* [8], *20071371* [12], *20071326* [14], *20070546* [16]
- **chives** *20082083* [1], *20090088* [5], *20082084* [8], *20082085* [12], *20082087* [14], *20082088* [16]
- **choi sum** *20082247* [1], *20090145* [5], *20082249* [8], *20082250* [12], *20082251* [14], *20082252* [16]
- **cob nuts** *20071393* [1], *20090116* [5], *20082158* [8], *20071372* [12], *20071321* [14], *20070555* [16]
- **crab apples** *20071380* [1], *20090147* [5], *20082159* [8], *20071368* [12], *20071329* [14], *20070554* [16]
- **daffodils** *(for galanthamine production) 20071379* [1], *(for galanthamine production) 20041516* [8], *(for galanthamine production) 20071370* [12], *20071328* [14]
- **daffodils grown for galanthamine production** *20090095* [5]
- **dwarf beans** *20071385* [1], *20090094* [5], *20031880* [8], *20071357* [12], *20071320* [14], *20070551* [16]
- **evening primrose** *20071389* [1], *20100850 expires 31 Aug 2011* [3], *20090113* [5], *20062694* [8], *20071359* [12], *20071337* [14], *20070558* [16], *20100822 expires 31 Aug 2011* [17]
- **french beans** *20071385* [1], *20090094* [5], *20031880* [8], *20071357* [12], *20071320* [14], *20070551* [16]
- **garlic** *20071378* [1], *20090090* [5], *20031879* [8], *20101121* [9], *20071369* [12], *20071327* [14], *20070550* [16]
- **grass seed crops** *20071392* [1], *20100849 expires 31 Aug 2011* [3], *20090144* [5], *20081764* [8], *20071362* [12], *20071340* [14], *20100823 expires 31 Aug 2011* [17]
- **honesty** *20071389* [1], *20100850 expires 31 Aug 2011* [3], *20090113* [5], *20062694* [8], *20071359* [12], *20071337* [14], *20070558* [16], *20100822 expires 31 Aug 2011* [17]
- **horseradish** *20070556* [16]
- **kohlrabi** *20071377* [1], *20090093* [5], *20031881* [8], *20071360* [12], *20071334* [14], *20070552* [16]
- **mallow (althaea spp.)** *20071391* [1], *20090143* [5], *20062692* [8], *20071363* [12], *20071339* [14], *20070556* [16]
- **mustard** *20071389* [1], *20100850 expires 31 Aug 2011* [3], *20090113* [5], *20062694* [8], *20071359* [12], *20071337* [14], *20070558* [16], *20100822 expires 31 Aug 2011* [17]
- **onion sets** *20090096* [5], *20071336* [14], *20070548* [16]
- **pak choi** *20082247* [1], *20090145* [5], *20082249* [8], *20082250* [12], *20082251* [14], *20082252* [16]
- **parsley root** *20071391* [1], *20090143* [5], *20062692* [8], *20071363* [12], *20071339* [14], *20070556* [16]
- **pears** *20071380* [1], *20090147* [5], *20082159* [8], *20071368* [12], *20071329* [14], *20070554* [16]
- **quinces** *20071380* [1], *20090147* [5], *20082159* [8], *20071368* [12], *20071329* [14], *20070554* [16]
- **raspberries** *20091408* [1], *20091407* [5], *20082160* [8], *20091406* [12], *20091399* [14]
- **rubus hybrids** *20091408* [1], *20091407* [5], *20082160* [8], *20091406* [12], *20091399* [14]
- **runner beans** *20071385* [1], *20090094* [5], *20031880* [8], *20071357* [12], *20071320* [14], *20070551* [16]
- **salad onions** *20071387* [1], *20090092* [5], *20042408* [8], *20101121* [9], *20071361* [12], *20071322* [14], *20070553* [16]
- **shallots** *20071378* [1], *20090090* [5], *20031879* [8], *20101121* [9], *20071369* [12], *20071327* [14], *20070550* [16]
- **triticale** *20071388* [1], *20090148* [5], *20062693* [8], *20071366* [12], *20071335* [14], *20070557* [16]
- **walnuts** *20071393* [1], *20090116* [5], *20082158* [8], *20071372* [12], *20071321* [14], *20070555* [16]

Approval information
- Tebuconazole is included in Annex 1 under EC Directive 91/414
- Accepted by BBPA for use on malting barley
- Approval expiry 31 Aug 2011 [3, 17]

Efficacy guidance
- For best results apply at an early stage of disease development before infection spreads to new crop growth
- To protect flag leaf and ear from Septoria diseases apply from flag leaf emergence to ear fully emerged (GS 37-59). Earlier application may be necessary where there is a high risk of infection
- Improved control of established cereal mildew can be obtained by tank mixture with fenpropimorph

FOR FULL CONDITIONS OF USE ALWAYS READ THE PRODUCT LABEL

tebuconazole + triadimenol

- For light leaf spot control in oilseed rape apply in autumn/winter with a follow-up spray in spring/summer if required
- For control of most other diseases spray at first signs of infection with a follow-up spray 2-4 wk later if necessary. See label for details
- For disease control in cabbages a 3-spray programme at 21-28 d intervals will give good control
- Tebuconazole is a DMI fungicide. Resistance to some DMI fungicides has been identified in Septoria leaf blotch which may seriously affect performance of some products. For further advice contact a specialist advisor and visit the Fungicide Resistance Action Group (FRAG)-UK website

Restrictions
- Maximum total dose equivalent to 1 full dose treatment on linseed; 2 full dose treatments on wheat, barley, rye, oats, field beans, swedes, turnips, onions; 2.25 full dose treatments on cabbages; 2.5 full dose treatments on oilseed rape; 3 full dose treatments on leeks, parsnips, carrots
- Do not treat durum wheat
- Apply only to listed oat varieties (see label) and do not apply to oats in tank mixture
- Do not apply before swedes and turnips have a root diameter of 2.5 cm, or before heart formation in cabbages, or before button formation in Brussels sprouts
- Consult processor before use on crops for processing

Crop-specific information
- Latest use: before grain milky-ripe for cereals (GS 71); when most seed green-brown mottled for oilseed rape (GS 6,3); before brown capsule for linseed
- HI field beans, swedes, turnips, linseed, winter oats 35 d; market brassicas, carrots, horseradish, parsnips 21 d; leeks 14 d
- Some transient leaf speckling on wheat or leaf reddening/scorch on oats may occur but this has not been shown to reduce yield response to disease control

Environmental safety
- Dangerous for the environment
- Toxic to aquatic organisms

Hazard classification and safety precautions
    Hazard Harmful [2-4, 6, 8-12, 15-20]; Irritant [1, 5, 7, 13, 14]; Dangerous for the environment [2-4, 6, 8, 10-12, 15-20]
    Transport code 9
    Packaging group III
    UN number 3082
    Risk phrases R20 [2-4, 8, 12, 18, 19]; R22a, R41, R51 [2-4, 6, 8, 10-12, 15-20]; R36 [1, 5, 7, 13, 14]; R38 [1, 5-7, 10, 11, 13-17, 20]; R52 [1, 5, 7, 9, 13, 14]; R53a [1-20]; R63 [3, 9, 20]
    Operator protection A, H [1-20]; C [1-8, 10-20]; U05a, U11, U20a [1-8, 10-20]; U14, U15 [1, 5, 7, 13, 14]; U19a [3, 6, 10, 11, 15-17, 20]; U20b [9]
    Environmental protection E03, E15b [9]; E15a [1-8, 10-20]; E34 [1, 2, 4-8, 10-16, 18-20]; E38 [2, 4, 8, 9, 12, 18, 19]
    Storage and disposal D01, D02 [1-8, 10-20]; D05, D10b [2-4, 6, 8, 10-12, 15-20]; D09a [1-20]; D11a [9]; D12a [2, 4, 8, 12, 18, 19]
    Treated seed S01, S02, S03, S04a, S05, S07 [9]
    Medical advice M03 [2-4, 6, 8, 10-12, 15-20]

420   tebuconazole + triadimenol

A broad spectrum systemic fungicide for cereals
FRAC mode of action code: 3 + 3

See also triadimenol

Products

1	Silvacur	Bayer CropScience	250:125 g/l	EC	11309
2	Veto F	Interfarm	225:75 g/l	EC	14748

Uses
- Brown rust in **spring barley, spring rye, spring wheat, winter barley, winter rye, winter wheat** [1, 2]

SEE SECTION 3 FOR PRODUCTS ALSO REGISTERED

- Crown rust in *spring oats*, *winter oats* [1, 2]
- Foliar disease control in *grass seed crops* (off-label) [1]
- Fusarium ear blight in *spring wheat*, *winter wheat* [1, 2]
- Glume blotch in *spring wheat*, *winter wheat* [1, 2]
- Net blotch in *spring barley*, *winter barley* [1, 2]
- Powdery mildew in *spring barley*, *spring oats*, *spring rye*, *spring wheat*, *winter barley*, *winter oats*, *winter rye*, *winter wheat* [1, 2]
- Rhynchosporium in *spring barley*, *winter barley* [1, 2]
- Septoria leaf blotch in *spring wheat*, *winter wheat* [1, 2]
- Sooty moulds in *spring wheat*, *winter wheat* [1, 2]
- Yellow rust in *spring barley*, *spring rye*, *spring wheat*, *winter barley*, *winter rye*, *winter wheat* [1, 2]

Specific Off-Label Approvals (SOLAs)
- *grass seed crops* 20061023 [1]

Approval information
- Tebuconazole and triadimenol included in Annex I under EC Directive 91/414
- Accepted by BBPA for use on malting barley

Efficacy guidance
- For best results apply at an early stage of disease development before infection spreads to new crop growth
- To protect flag leaf and ear from Septoria diseases apply from flag leaf emergence to ear fully emerged (GS 37-59). Earlier application may be necessary where there is a high risk of infection
- For control of rust, powdery mildew, leaf and net blotch apply at first signs of disease with a second application 2-3 wk later if necessary
- Tebuconazole is a DMI fungicide. Resistance to some DMI fungicides has been identified in Septoria leaf blotch which may seriously affect performance of some products. For further advice contact a specialist advisor and visit the Fungicide Resistance Action Group (FRAG)-UK website

Restrictions
- Maximum total dose equivalent to two full dose treatments
- Do not use on durum wheat
- Use only on listed varieties of oats. See label [1]

Crop-specific information
- Latest use: before grain milky-ripe (GS 71)
- Some transient leaf speckling may occur on wheat or leaf reddening/scorch on oats but this has not been shown to reduce yield response or disease control

Environmental safety
- Harmful to aquatic organisms
- Harmful to fish or other aquatic life. Do not contaminate surface waters or ditches with chemical or used container

Hazard classification and safety precautions
    Hazard Harmful [1]
    Transport code 9 [1]
    Packaging group III [1]
    UN number 3082 [1]; N/C [2]
    Risk phrases R20, R36 [1]; R52, R53a [1, 2]
    Operator protection A, C, H; U05a, U09b, U20a
    Environmental protection E13c [1, 2]; E34 [1]
    Storage and disposal D01, D02, D05, D09a [1, 2]; D10b [1]; D10c [2]
    Medical advice M03 [2]

FOR FULL CONDITIONS OF USE ALWAYS READ THE PRODUCT LABEL

421　tebuconazole + trifloxystrobin

A conazole and stobilurin fungicide mixture for wheat and vegetable crops
FRAC mode of action code: 3 + 11

See also trifloxystrobin

Products

1	Clayton Bestow	Clayton	200:100 g/l	SC	14996
2	Dedicate	Bayer Environ.	200:100 g/l	SC	13612
3	Eradicate	Pan Agriculture	200:100 g/l	SC	13986
4	Mascot Fusion	Rigby Taylor	200:100 g/l	EC	14113
5	Nativo 75 WG	Bayer CropScience	50:25% w/w	WG	13057

Uses

- Alternaria in **broccoli, brussels sprouts, cabbages, calabrese, carrots, cauliflowers** [5]
- Anthracnose in **amenity grassland, managed amenity turf** [4]
- Dollar spot in **amenity grassland, managed amenity turf** [1-4]
- Foliar disease control in **horseradish** *(off-label)*, **parsley root** *(off-label)*, **parsnips** *(off-label)*, **salsify** *(off-label)* [5]
- Fusarium in **amenity grassland, managed amenity turf** [1-3]
- Fusarium patch in **amenity grassland, managed amenity turf** [4]
- Light leaf spot in **broccoli, brussels sprouts, cabbages, calabrese, cauliflowers** [5]
- Phoma leaf spot in **broccoli, brussels sprouts, cabbages, calabrese, cauliflowers** [5]
- Powdery mildew in **broccoli, brussels sprouts, cabbages, calabrese, carrots, cauliflowers, wine grapes** *(off-label)* [5]
- Purple blotch in **leeks** [5]
- Red thread in **amenity grassland, managed amenity turf** [1-4]
- Ring spot in **broccoli, brussels sprouts, cabbages, calabrese, cauliflowers** [5]
- Rust in **leeks** [5]
- Sclerotinia rot in **carrots** [5]
- White blister in **broccoli, brussels sprouts, cabbages, calabrese, cauliflowers** [5]
- White tip in **leeks** [5]

Specific Off-Label Approvals (SOLAs)

- **horseradish** *20064163* [5]
- **parsley root** *20064163* [5]
- **parsnips** *20064163* [5]
- **salsify** *20064163* [5]
- **wine grapes** *20090730* [5]

Approval information

- Tebuconazole and trifloxystrobin included in Annex I under EC Directive 91/414

Efficacy guidance

- Best results obtained from treatment at early stages of disease development. Further treatment may be needed if disease attack is prolonged
- Applications to established infections of any disease are likely to be less effective
- Treatment may give some control of *Stemphyllium botryosum* and leaf blotch on leeks [5]
- Tebuconazole is a DMI fungicide. Resistance to some DMI fungicides has been identified in Septoria leaf blotch which may seriously affect performance of some products. For further advice contact a specialist advisor and visit the Fungicide Resistance Action Group (FRAG)-UK website
- Trifloxystrobin is a member of the QoI cross resistance group. Product should be used preventatively and not relied on for its curative potential
- Use product as part of an Integrated Crop Management strategy incorporating other methods of control, including where appropriate other fungicides with a different mode of action. Do not apply more than two foliar applications of QoI containing products to any cereal crop, broccoli, calabrese or cauliflower. Do not apply more than three applications to Brussels sprouts, cabbage, carrots or leeks
- There is a significant risk of widespread resistance occurring in *Septoria tritici* populations in UK. Failure to follow resistance management action may result in reduced levels of disease control
- Strains of wheat powdery mildew resistant to QoIs are common in the UK. Control of wheat mildew can only be relied on from the triazole component

---

SEE SECTION 3 FOR PRODUCTS ALSO REGISTERED

Pesticide Profiles

- Where specific control of wheat mildew is required this should be achieved through a programme of measures including products recommended for the control of mildew that contain a fungicide from a different cross-resistance group and applied at a dose that will give robust control
- Do not apply to grass during drought conditions nor to frozen turf [2]

Restrictions
- Maximum number of treatments 2 per crop for broccoli, calabrese, cauliflowers; 3 per crop for Brussels sprouts, cabbages, carrots, leeks [5]
- In addition to the maximum number of treatments per crop a maximum of 3 applications may be applied to the same ground in one calendar year [5]
- Consult processor before use on vegetable crops for processing [5]

Crop-specific information
- Latest use: before milky ripe stage on winter wheat
- HI 35 d for wheat; 21 d for all other crops
- Performance against leaf spot diseases of brassicas, *Alternaria* leaf blight of carrots, and rust in leeks may be improved by mixing with an approved sticker/wetter [5]

Environmental safety
- Dangerous for the environment
- Very toxic to aquatic organisms

Hazard classification and safety precautions
Hazard Harmful, Dangerous for the environment
Transport code 9
Packaging group III
UN number 3077 [5]; 3082 [1-4]
Risk phrases R37 [1-4]; R50, R53a, R63 [1-5]; R70 [1-3]
Operator protection A [1-5]; H [1, 4, 5]; P [2, 3]; U05a [1-3, 5]; U09a [1-3]; U09b [4, 5]; U19a [1-4]; U20b [1-5]
Environmental protection E15a, E34, E38
Storage and disposal D01, D02, D09a, D10b, D12a [1-5]; D05 [4]
Medical advice M03

422   tebufenpyrad

A pyrazole mitochondrial electron transport inhibitor (METI) aphicide and acaricide
IRAC mode of action code: 21

Products
Masai   BASF   20% w/w   WB   13082

Uses
- Bud mite in **almonds** *(off-label)*, **chestnuts** *(off-label)*, **hazel nuts** *(off-label)*, **walnuts** *(off-label)*
- Damson-hop aphid in **hops**, **plums** *(off-label)*
- Gall mite in **bilberries** *(off-label)*, **blackberries** *(off-label)*, **blackcurrants** *(off-label)*, **blueberries** *(off-label)*, **cranberries** *(off-label)*, **gooseberries** *(off-label)*, **protected rubus hybrids** *(off-label)*, **raspberries** *(off-label)*, **redcurrants** *(off-label)*, **vaccinium spp.** *(off-label)*, **whitecurrants** *(off-label)*
- Red spider mites in **apples**, **blackberries** *(off-label)*, **pears**, **protected rubus hybrids** *(off-label)*, **raspberries** *(off-label)*
- Two-spotted spider mite in **hops**, **protected roses**, **strawberries**

Specific Off-Label Approvals (SOLAs)
- *almonds* 20080133
- *bilberries* 20080131
- *blackberries* 20100204
- *blackcurrants* 20080131
- *blueberries* 20080131
- *chestnuts* 20080133
- *cranberries* 20080131
- *gooseberries* 20080131
- *hazel nuts* 20080133

FOR FULL CONDITIONS OF USE ALWAYS READ THE PRODUCT LABEL

teflubenzuron

- *plums* 20080132
- *protected rubus hybrids* 20100204
- *raspberries* 20100204
- *redcurrants* 20080131
- *vaccinium spp.* 20080131
- *walnuts* 20080133
- *whitecurrants* 20080131

Approval information
- Tebufenpyrad included in Annex 1 under EC Directive 91/414.
- Accepted by BBPA for use on hops

Efficacy guidance
- Acts on eggs (except winter eggs) and all motile stages of spider mites up to adults
- Treat spider mites from 80% egg hatch but before mites become established
- For effective control total spray cover of the crop is required
- Product can be used in a programme to give season-long control of damson-hop aphids coupled with mite control
- Where aphids resistant to tebufenpyrad occur in hops control is unlikely to be satisfactory and repeat treatments may result in lower levels of control. Where possible use different active ingredients in a programme

Restrictions
- Maximum total dose equivalent to one full dose treatment on apples, pears, strawberries; 3 full dose treatments on hops
- Other mitochondrial electron transport inhibitor (METI) acaricides should not be applied to the same crop in the same calendar yr either separately or in mixture
- Do not treat apples before 90% petal fall
- Small-scale testing of rose varieties to establish tolerance recommended before use
- Inner liner of container must not be removed

Crop-specific information
- Latest use: end of burr stage for hops
- HI strawberries 3 d; apples, bilberries, blackcurrants, blueberries, cranberries, gooseberries, pears, redcurrants, whitecurrants 7d; blackberries, plums, raspberries 21 d
- Product has no effect on fruit quality or finish

Environmental safety
- Dangerous for the environment
- Very toxic to aquatic organisms
- High risk to bees. Do not apply to crops in flower or to those in which bees are actively foraging. Do not apply when flowering weeds are present
- Broadcast air-assisted LERAP (18 m); LERAP Category B

Hazard classification and safety precautions
Hazard Harmful, Dangerous for the environment
Transport code 9
Packaging group III
UN number 3077
Risk phrases R20, R22a, R50, R53a
Operator protection A; U02a, U05a, U09a, U13, U14, U20b
Environmental protection E12a, E12e, E15a, E16a, E16b, E38; E17b (18 m)
Storage and disposal D01, D02, D09a, D11a, D12a
Medical advice M05a

423 teflubenzuron

A benzoylurea insecticide for use on ornamentals
IRAC mode of action code: 15

Products
Nemolt	Fargro	150 g/l	SC	10226

SEE SECTION 3 FOR PRODUCTS ALSO REGISTERED

Uses

- Browntail moth in **ornamental plant production**
- Cabbage moth in **broccoli** (off-label), **brussels sprouts** (off-label), **cauliflowers** (off-label), **chinese cabbage** (off-label)
- Cabbage root fly in **broccoli** (off-label), **brussels sprouts** (off-label), **cauliflowers** (off-label), **chinese cabbage** (off-label)
- Cabbage white butterfly in **broccoli** (off-label), **brussels sprouts** (off-label), **cauliflowers** (off-label), **chinese cabbage** (off-label)
- Caterpillars in **amenity vegetation** (off-label), **forest nurseries** (off-label), **ornamental plant production**, **protected aubergines** (off-label), **protected courgettes** (off-label), **protected cucumbers** (off-label), **protected melons** (off-label), **protected ornamentals** (off-label), **protected peppers** (off-label), **protected tomatoes** (off-label)
- Insect pests in **amenity vegetation** (off-label), **forest nurseries** (off-label), **protected aubergines** (off-label), **protected courgettes** (off-label), **protected cucumbers** (off-label), **protected melons** (off-label), **protected ornamentals** (off-label), **protected peppers** (off-label), **protected tomatoes** (off-label)
- Thrips in **protected aubergines** (off-label), **protected courgettes** (off-label), **protected cucumbers** (off-label), **protected melons** (off-label), **protected peppers** (off-label), **protected tomatoes** (off-label)
- Western flower thrips in **amenity vegetation** (off-label), **forest nurseries** (off-label), **protected ornamentals** (off-label)
- Whitefly in **ornamental plant production**, **protected aubergines** (off-label), **protected courgettes** (off-label), **protected cucumbers** (off-label), **protected melons** (off-label), **protected peppers** (off-label), **protected tomatoes** (off-label)

Specific Off-Label Approvals (SOLAs)

- *amenity vegetation* 20102131
- *broccoli* 20102132
- *brussels sprouts* 20102132
- *cauliflowers* 20102132
- *chinese cabbage* 20102132
- *forest nurseries* 20102131
- *protected aubergines* 20102130
- *protected courgettes* 20102130
- *protected cucumbers* 20102130
- *protected melons* 20102130
- *protected ornamentals* 20102131
- *protected peppers* 20102130
- *protected tomatoes* 20102130

Approval information

- Teflubenzuron included in Annex 1 under EC Directive 91/414

Efficacy guidance

- Product acts as larval stomach poison interfering with moulting process leading to cessation of feeding and larval death
- Apply as soon as first stage larvae seen. This will often coincide with the peak of moth flight

Restrictions

- Maximum number of treatments 3 per yr
- Test specific varieties before carrying out extensive treatments

Environmental safety

- Dangerous for the environment
- Very toxic to aquatic organisms
- Limited evidence suggests some margin of safety to *Encarsia formosa*. Effects on other parasites and predators not fully tested

Hazard classification and safety precautions

Hazard Dangerous for the environment
Transport code 9
Packaging group III
UN number 3082

FOR FULL CONDITIONS OF USE ALWAYS READ THE PRODUCT LABEL

tefluthrin

Risk phrases R50, R53a
Operator protection A, C; U05a, U20c
Environmental protection E15a, E38
Storage and disposal D01, D02, D05, D09a, D11a, D12a

424 tefluthrin

A soil acting pyrethroid insecticide seed treatment
IRAC mode of action code: 3

See also fludioxonil + tefluthrin

Products
Force ST   Syngenta   200 g/l   CF   11752

Uses
- Bean seed fly in **bulb onions** *(off-label - seed treatment)*, **chard** *(off-label - seed treatment)*, **chives** *(off-label - seed treatment)*, **green mustard** *(off-label - seed treatment)*, **herbs (see appendix 6)** *(off-label - seed treatment)*, **leaf spinach** *(off-label - seed treatment)*, **leeks** *(off-label - seed treatment)*, **miduna** *(off-label - seed treatment)*, **mizuna** *(off-label - seed treatment)*, **pak choi** *(off-label - seed treatment)*, **parsley** *(off-label - seed treatment)*, **red mustard** *(off-label - seed treatment)*, **salad onions** *(off-label - seed treatment)*, **spinach beet** *(off-label - seed treatment)*, **tatsoi** *(off-label - seed treatment)*
- Carrot fly in **carrots** *(off-label - seed treatment)*, **parsnips** *(off-label - seed treatment)*
- Millipedes in **fodder beet** *(seed treatment)*, **sugar beet** *(seed treatment)*
- Onion fly in **bulb onions** *(off-label - seed treatment)*, **leeks** *(off-label - seed treatment)*, **salad onions** *(off-label - seed treatment)*
- Pygmy beetle in **fodder beet** *(seed treatment)*, **sugar beet** *(seed treatment)*
- Springtails in **fodder beet** *(seed treatment)*, **sugar beet** *(seed treatment)*
- Symphylids in **fodder beet** *(seed treatment)*, **sugar beet** *(seed treatment)*

Specific Off-Label Approvals (SOLAs)
- *bulb onions (seed treatment) 20050546*
- *carrots (seed treatment) 20050547*
- *chard (seed treatment) 20050545*
- *chives (seed treatment) 20050545*
- *green mustard (seed treatment) 20050545*
- *herbs (see appendix 6) (seed treatment) 20050545*
- *leaf spinach (seed treatment) 20050545*
- *leeks (seed treatment) 20050546*
- *miduna (seed treatment) 20050545*
- *mizuna (seed treatment) 20050545*
- *pak choi (seed treatment) 20050545*
- *parsley (seed treatment) 20050545*
- *parsnips (seed treatment) 20050547*
- *red mustard (seed treatment) 20050545*
- *salad onions (seed treatment) 20050546*
- *spinach beet (seed treatment) 20050545*
- *tatsoi (seed treatment) 20050545*

Approval information
- Tefluthrin not yet included in Annex 1 under EC Directive 91/414
- Accepted by BBPA for use on malting barley

Efficacy guidance
- Apply during process of pelleting beet seed. Consult manufacturer for details of specialist equipment required
- Micro-capsule formulation allows slow release to provide a protection zone around treated seed during establishment

Restrictions
- Maximum number of treatments 1 per batch of seed
- Sow treated seed as soon as possible. Do not store treated seed from one drilling season to next

SEE SECTION 3 FOR PRODUCTS ALSO REGISTERED

Pesticide Profiles

- If used in areas where soil erosion by wind or water likely, measures must be taken to prevent this happening
- Can cause a transient tingling or numbing sensation to exposed skin. Avoid skin contact with product, treated seed and dust throughout all operations in the seed treatment plant and at drilling

Crop-specific information
- Latest use: before drilling seed
- Treated seed must be drilled within the season of treatment

Environmental safety
- Dangerous for the environment
- Very toxic to aquatic organisms
- Keep treated seed secure from people, domestic stock/pets and wildlife at all times during storage and use
- Treated seed harmful to game and wild life. Bury spillages
- In the event of seed spillage clean up as much as possible into the related seed sack and bury the remainder completely
- Do not apply treated seed from the air
- Keep livestock out of areas drilled with treated seed for at least 80 d

Hazard classification and safety precautions
Hazard Harmful, Dangerous for the environment
Transport code 9
Packaging group III
UN number 3082
Risk phrases R20, R22b, R43, R50, R53a
Operator protection A, D, E, H; U02a, U04a, U05a, U08, U20b
Environmental protection E06a (80 d); E15a, E34, E38
Storage and disposal D01, D02, D05, D09a, D11a, D12a
Treated seed S02, S03, S04b, S05, S07
Medical advice M05b

425   tepraloxydim

A systemic post-emergence herbicide for control of annual grass weeds
HRAC mode of action code: A

Products

1	Aramo	BASF	50 g/l	EC	10280
2	Clayton Rally	Clayton	50 g/l	EC	13415
3	Omarra	AgChem Access	50 g/l	EC	13693
4	Standon Tonga	Standon	50 g/l	EC	13394
5	Tepra	Goldengrass	50 g/l	EC	14160

Uses
- Annual grasses in *all edible seed crops grown outdoors* (off-label), *all non-edible seed crops grown outdoors* (off-label), *borage* (off-label), *broad beans* (off-label), *canary flower (echium spp.)* (off-label), *chickpeas* (off-label), *edamame beans* (off-label), *evening primrose* (off-label), *forest nurseries* (off-label), *game cover* (off-label), *golf courses* (off-label), *honesty* (off-label), *hops* (off-label), *lentils* (off-label), *lupins* (off-label), *mustard* (off-label), *ornamental plant production* (off-label), *salad onions* (off-label), *swedes* (off-label), *top fruit* (off-label), *turnips* (off-label) [1]; *bulb onions, cabbages, carrots, cauliflowers, combining peas, fodder beet, land temporarily removed from production, leeks, linseed, spring field beans, sugar beet, vining peas, winter field beans, winter oilseed rape* [1-3, 5]; *flax, linseed for industrial use, winter oilseed rape for industrial use* [1, 3]; *garlic* (off-label), *mallow (althaea spp.)* (off-label), *parsnips* (off-label), *red beet* (off-label), *shallots* (off-label) [1, 5]
- Annual meadow grass in *all edible seed crops grown outdoors* (off-label), *all non-edible seed crops grown outdoors* (off-label), *forest nurseries* (off-label), *game cover* (off-label), *golf courses* (off-label), *hops* (off-label), *ornamental plant production* (off-label), *poppies for morphine production* (off-label), *salad onions* (off-label), *swedes* (off-label), *top fruit* (off-label), *turnips* (off-label) [1]; *bulb onions, cabbages, carrots, cauliflowers, combining peas, fodder beet, leeks, linseed, spring field beans, sugar beet, vining peas, winter field beans, winter*

FOR FULL CONDITIONS OF USE ALWAYS READ THE PRODUCT LABEL

tepraloxydim

oilseed rape [4]; **garlic** *(off-label)*, **mallow (althaea spp.)** *(off-label)*, **parsnips** *(off-label)*, **red beet** *(off-label)*, **shallots** *(off-label)* [5]
- Blackgrass in **bulb onions, cabbages, carrots, cauliflowers, combining peas, fodder beet, leeks, linseed, spring field beans, sugar beet, vining peas, winter field beans, winter oilseed rape** [4]
- Green cover in **land temporarily removed from production** [4]
- Perennial grasses in **all edible seed crops grown outdoors** *(off-label)*, **all non-edible seed crops grown outdoors** *(off-label)*, **borage** *(off-label)*, **canary flower (echium spp.)** *(off-label)*, **evening primrose** *(off-label)*, **forest nurseries** *(off-label)*, **game cover** *(off-label)*, **golf courses** *(off-label)*, **honesty** *(off-label)*, **hops** *(off-label)*, **lupins** *(off-label)*, **mustard** *(off-label)*, **ornamental plant production** *(off-label)*, **swedes** *(off-label)*, **top fruit** *(off-label)*, **turnips** *(off-label)* [1]; **bulb onions, cabbages, carrots, cauliflowers, combining peas, fodder beet, land temporarily removed from production, leeks, linseed, spring field beans, sugar beet, vining peas, winter field beans, winter oilseed rape** [1-3, 5]; **flax, linseed for industrial use, winter oilseed rape for industrial use** [1, 3]; **garlic** *(off-label)*, **mallow (althaea spp.)** *(off-label)*, **parsnips** *(off-label)*, **red beet** *(off-label)*, **shallots** *(off-label)* [1, 5]
- Volunteer barley in **bulb onions, cabbages, carrots, cauliflowers, combining peas, fodder beet, leeks, linseed, spring field beans, sugar beet, vining peas, winter field beans, winter oilseed rape** [4]
- Volunteer cereals in **bulb onions, cabbages, carrots, cauliflowers, combining peas, fodder beet, land temporarily removed from production, leeks, linseed, spring field beans, sugar beet, vining peas, winter field beans, winter oilseed rape** [1-3, 5]; **flax, linseed for industrial use, winter oilseed rape for industrial use** [1, 3]
- Volunteer wheat in **bulb onions, cabbages, carrots, cauliflowers, combining peas, fodder beet, leeks, linseed, spring field beans, sugar beet, vining peas, winter field beans, winter oilseed rape** [4]
- Wild oats in **red beet** *(off-label)* [1]

Specific Off-Label Approvals (SOLAs)
- *all edible seed crops grown outdoors* 20082813 [1]
- *all non-edible seed crops grown outdoors* 20082813 [1]
- *borage* 20064019 [1]
- *broad beans* 20101553 [1]
- *canary flower (echium spp.)* 20064019 [1]
- *chickpeas* 20101553 [1]
- *edamame beans* 20101553 [1]
- *evening primrose* 20064019 [1]
- *forest nurseries* 20082813 [1]
- *game cover* 20082813 [1]
- *garlic* 20064021 [1], 20092128 [5]
- *golf courses* 20091527 [1]
- *honesty* 20064019 [1]
- *hops* 20082813 [1]
- *lentils* 20101553 [1]
- *lupins* 20064022 [1]
- *mallow (althaea spp.)* 20064023 [1], 20092129 [5]
- *mustard* 20064019 [1]
- *ornamental plant production* 20082813 [1]
- *parsnips* 20064023 [1], 20092129 [5]
- *poppies for morphine production* 20064026 [1]
- *red beet* 20071844 [1], 20092127 [5]
- *salad onions* 20101146 [1]
- *shallots* 20064021 [1], 20092128 [5]
- *swedes* 20090082 [1]
- *top fruit* 20082813 [1]
- *turnips* 20090082 [1]

Approval information
- Tepraloxydim included in Annex I under EC Directive 91/414

SEE SECTION 3 FOR PRODUCTS ALSO REGISTERED

Pesticide Profiles

Efficacy guidance
- Best results obtained from applications when weeds small and have not begun to compete with crop
- Only emerged weeds are controlled
- Cool conditions slow down activity and very dry conditions reduce activity by interfering with uptake and translocation
- Foliar death of susceptible weeds evident after 3-4 wks in warm conditions
- Reduced doses must not be used on resistant grass weed populations
- Tepraloxydim is an ACCase inhibitor herbicide. To avoid the build up of resistance do not apply products containing an ACCase inhibitor herbicide more than twice to any crop. In addition do not use any product containing tepraloxydim in mixture or sequence with any other product containing the same ingredient
- Use these products as part of a resistance management strategy that includes cultural methods of control and does not use ACCase inhibitors as the sole chemical method of grass weed control
- Applying a second product containing an ACCase inhibitor to a crop will increase the risk of resistance development; only use a second ACCase inhibitor to control different weeds at a different timing
- Always follow WRAG guidelines for preventing and managing herbicide resistant weeds. See Section 5 for more information

Restrictions
- Maximum number of treatments 1 per crop or yr
- Consult processors or contract agents before treatment of crops intended for processing or for seed
- Do not apply to crops damaged or stressed by factors such as previous herbicide treatments, pest or disease attack
- Do not spray if rain or frost expected, or if foliage is wet
- Do not treat oilseed rape with very low vigour and poor yield potential. Overlapping on oilseed rape may cause damage and reduce yields
- On peas a satisfactory crystal violet leaf wax test must be carried out before treatment. Winter varieties may be treated only in the spring
- For sugar beet, fodder beet, linseed and green cover on land temporarily removed from production, applications are prohibited between 1 Nov and 31 Mar
- For field beans, peas, leeks, bulb onions, carrots, cabbages and cauliflowers applications are prohibited between 1 Nov and 1 Mar

Crop-specific information
- Latest use : before end Nov or before crop has 9 true leaves, whichever is first for winter oilseed rape; before flower buds visible for flax [1], linseed; before head formation for cabbages, cauliflowers
- HI 3 wk for carrots; 4 wk for bulb onions, leeks; 5 wk for peas; 8 wk for fodder beet, field beans, sugar beet
- Treatment prohibited between 1 Nov and 31 Mar on golf courses (off-label use), sugar beet, fodder beet, linseed and green cover on land temporarily removed from production
- Green cover must be fully established on land temporarily removed from production before treatment, and treated plants must not be grazed by livestock or harvested for human or animal consumption

Following crops guidance
- In the event of failure of a treated crop, wheat or barley may be drilled after 2 wk following normal seedbed cultivations, maize or Italian rye-grass may be planted after 8 wk following cultivation to 20 cm
- Graminaceous crops other than those mentioned above should not follow a treated crop in the rotation.
- Broad leaved crops may be planted at any time following a failed or normally harvested treated crop

Environmental safety
- Dangerous for the environment
- Toxic to aquatic organisms

Hazard classification and safety precautions
Hazard Harmful, Dangerous for the environment

FOR FULL CONDITIONS OF USE ALWAYS READ THE PRODUCT LABEL

terbuthylazine

Transport code 9
Packaging group III
UN number 3082
Risk phrases R22b, R38, R51, R53a, R63, R66 [1-5]; R40, R62 [2, 4]
Operator protection A; U05a, U08, U20b, U23a [1-5]; U19a [4]
Environmental protection E15a [1-3, 5]; E15b [4]; E38 [1, 3, 5]
Storage and disposal D01, D02, D09a, D10c, D12a [1-5]; D05 [2, 4]; D07 [2]; D08 [1, 3-5]
Medical advice M04a [4]; M05b [1-5]

426 terbuthylazine

A triazine herbicide available only in mixtures
HRAC mode of action code: C1

See also bromoxynil + terbuthylazine
isoxaben + terbuthylazine
mesotrione + terbuthylazine
pendimethalin + terbuthylazine

427 tetraconazole

A systemic, protectant and curative triazole fungicide for cereals; approval expired 2008
FRAC mode of action code: 3

See also chlorothalonil + tetraconazole

428 tetramethrin

A contact acting pyrethroid insecticide
IRAC mode of action code: 3

See also d-phenothrin + tetramethrin

Products
Killgerm Py-Kill W	Killgerm	10.2% w/v	EC	H4632

Uses
- Flies in **agricultural premises**, **livestock houses**

Efficacy guidance
- Dilute in accordance with directions and apply as space or surface spray

Restrictions
- For use only by professional operators
- Maximum number of treatments 1 per wk in agricultural premises and livestock houses
- Do not apply directly on food or livestock
- Remove exposed milk before application. Protect milk machinery and containers from contamination
- Do not use space sprays containing pyrethrins or pyrethroid more than once per wk in intensive or controlled environment animal houses in order to avoid development of resistance. If necessary, use a different control method or product

Environmental safety
- Dangerous for the environment
- Toxic to aquatic organisms

Hazard classification and safety precautions
Hazard Harmful, Flammable, Dangerous for the environment
UN number N/C
Risk phrases R22b, R41, R51, R53a
Operator protection A, H; U02b, U09a, U19a, U20b
Environmental protection E15a

SEE SECTION 3 FOR PRODUCTS ALSO REGISTERED

Pesticide Profiles

Consumer protection C05, C06, C07, C08, C09, C10, C11
Storage and disposal D01, D09a, D12a
Medical advice M05b

429 thiabendazole

A systemic, curative and protectant benzimidazole (MBC) fungicide
FRAC mode of action code: 1

See also imazalil + thiabendazole
metalaxyl + thiabendazole

Products

1	Hykeep	Agrichem	2% w/w	DP	12704
2	Storite Clear Liquid	Frontier	220 g/l	SL	12706
3	Storite Excel	Frontier	500 g/l	SC	12705
4	Tezate 220 SL	AgriChem BV	220 g/l	SL	14007

Uses
- Basal stem rot in **narcissi** *(off-label)* [2, 4]
- Dry rot in **potatoes**, **seed potatoes** [4]; **seed potatoes** *(tuber treatment - post-harvest)* [3]; **ware potatoes** *(tuber treatment - post-harvest)* [1, 3]
- Fusarium basal neck rot in **narcissi** [4]; **narcissi** *(post-lifting or pre-planting dip)*, **narcissi** *(post-lifting spray)* [2]
- Gangrene in **potatoes**, **seed potatoes** [4]; **seed potatoes** *(tuber treatment - post-harvest)* [3]; **ware potatoes** *(tuber treatment - post-harvest)* [1, 3]
- Neck rot in **narcissi** *(off-label)* [2, 4]
- Silver scurf in **potatoes**, **seed potatoes** [4]; **seed potatoes** *(tuber treatment - post-harvest)* [3]; **ware potatoes** *(tuber treatment - post-harvest)* [1, 3]
- Skin spot in **potatoes**, **seed potatoes** [4]; **seed potatoes** *(tuber treatment - post-harvest)* [3]; **ware potatoes** *(tuber treatment - post-harvest)* [1, 3]

Specific Off-Label Approvals (SOLAs)
- **narcissi** *20070924 expires 31 Dec 2011* [2], *20091180 expires 31 Dec 2011* [4]

Approval information
- Thiabendazole included in Annex I under EC Directive 91/414
- Use of thiabendazole products on ware potatoes requires that the discharge of thiabendazole to receiving water from washing plants is kept within emission limits set by the UK monitoring authority
- Approval expiry 31 Dec 2011 [1-4]

Efficacy guidance
- For best results tuber treatments should be applied as soon as possible after lifting and always within 24 hr
- Dust treatments should be applied evenly over the whole tuber surface
- Thiabendazole should only be used on ware potatoes where there is a likely risk of disease during the storage period and in combination with good storage hygiene and maintenance [1, 3]
- Benzimidazole tolerant strains of silver scurf and skin spot are common in UK and tolerant strains of dry rot have been reported. To reduce the chance of such strains increasing benzimidazole based products should not be used more than once in the cropping cycle [1, 3]
- On narcissi treat as the crop goes into store and only if fungicidal treatment is essential. Adopt a resistance management strategy [2]

Restrictions
- Maximum number of treatments 1 per batch for ware or seed potato tuber treatments [1, 3]; 1 per yr for narcissi bulbs [2]
- Treated seed potatoes must not be used for food or feed [3]
- Do not remove treated potatoes from store for sale, processing or consumption for at least 21 d after application [1, 3]
- Do not mix with any other product
- Off-label use as post-lifting cold water dip or pre-planting hot water dip not to be used on narcissus bulbs grown in the Isles of Scilly [2]

FOR FULL CONDITIONS OF USE ALWAYS READ THE PRODUCT LABEL

thiacloprid

Crop-specific information
- Latest use: 21 d before removal from store for sale, processing or consumption for ware potatoes [1, 3]
- Apply to potatoes as soon as possible after harvest using suitable equipment and always within 2 wk of lifting provided the skins are set. See label for details [1, 3]
- Potatoes should only be treated by systems that provide an accurate dose to tubers not carrying excessive quantities of soil [1, 3]
- Use as a post-lifting treatment to reduce basal and neck rot in narcissus bulbs. Ensure bulbs are clean [2]

Environmental safety
- Dangerous for the environment
- Toxic to aquatic organisms

Hazard classification and safety precautions
Hazard Irritant [3]; Dangerous for the environment [1-4]
Transport code 9
Packaging group III
UN number 3077 [1]; 3082 [2-4]
Risk phrases R43 [3]; R51 [2-4]; R52 [1]; R53a [1-4]
Operator protection A, H [1-4]; C, D [2-4]; U05a [2-4]; U14 [3]; U19a [1-4]; U20b [4]; U20c [1-3]
Environmental protection E15b [1-4]; E38 [2-4]
Storage and disposal D01, D09a, D12a [1-4]; D02, D10c [2-4]; D05 [2, 3]; D07, D11a [1]
Treated seed S03, S05 [4]; S04a [2, 4]

430   thiacloprid

A chloronicotinyl insecticide for use in agriculture and horticulture
IRAC mode of action code: 4A

Products

1	Agrovista Reggae	Agrovista	480 g/l	SC	13706
2	Biscaya	Bayer CropScience	240 g/l	OD	15014
3	Calypso	Bayer CropScience	480 g/l	SC	11257
4	Exemptor	Scotts	10% w/w	GR	13122
5	Standon Zero Tolerance	Standon	240 g/l	OD	13546

Uses
- Aphids in *bedding plants, hardy ornamental nursery stock, ornamental plant production, pot plants, protected ornamentals* [4]; *carrots, parsnips, peas, potatoes* [2]; *leaf brassicas* (off-label - baby leaf production), *protected herbs (see appendix 6)* (off-label), *protected lamb's lettuce* (off-label), *protected lettuce* (off-label) [1]; *seed potatoes, ware potatoes* [2, 5]
- Asparagus beetle in *asparagus* (off-label) [3]
- Beetles in *bedding plants, hardy ornamental nursery stock, ornamental plant production, pot plants, protected ornamentals* [4]
- Bud mite in *almonds* (off-label), *chestnuts* (off-label), *cob nuts* (off-label), *hazel nuts* (off-label), *walnuts* (off-label) [1, 3]; *filberts* (off-label) [3]
- Cabbage aphid in *broccoli, brussels sprouts, cabbages, calabrese, cauliflowers* [2]
- Capsids in *protected strawberries* (off-label) [3]; *strawberries* (off-label) [1, 3]
- Common green capsid in *protected blackberries* (off-label), *protected raspberries* (off-label) [3]
- Damson-hop aphid in *plums* (off-label) [1, 3]
- Gall midge in *protected blueberry* (off-label) [1]
- Insect pests in *bilberries* (off-label), *blackberries* (off-label), *blackcurrants* (off-label), *blueberries* (off-label), *courgettes* (off-label), *cranberries* (off-label), *gherkins* (off-label), *gooseberries* (off-label), *hops* (off-label), *marrows* (off-label), *ornamental plant production* (off-label), *pears* (off-label), *protected forest nurseries* (off-label), *raspberries* (off-label), *redcurrants* (off-label), *rubus hybrids* (off-label), *soft fruit* (off-label), *top fruit* (off-label), *whitecurrants* (off-label) [1, 3]; *cherries* (off-label), *mirabelles* (off-label), *protected blueberry* (off-label) [3]; *cherries* (off-label - under temporary protective rain covers), *mirabelles* (off-label - under temporary protective rain covers), *protected blackberries* (off-label), *protected raspberries* (off-label), *protected strawberries* (off-label) [1]

SEE SECTION 3 FOR PRODUCTS ALSO REGISTERED

Pesticide Profiles

- Leaf miner in **protected aubergines** *(off-label)*, **protected courgettes** *(off-label)*, **protected cucumbers** *(off-label)*, **protected ornamentals** *(off-label)*, **protected peppers** *(off-label)*, **protected tomatoes** *(off-label)* [1, 3]
- Mealy aphid in **broccoli, brussels sprouts, cabbages, calabrese, cauliflowers** [2]
- Moths in **protected blueberry** *(off-label)* [1]
- Pea midge in **peas** [2]
- Peach-potato aphid in **leaf brassicas** *(off-label)* [3]; **leaf brassicas** *(off-label - baby leaf production)* [1]; **protected herbs (see appendix 6)** *(off-label)*, **protected lamb's lettuce** *(off-label)*, **protected lettuce** *(off-label)* [1, 3]
- Pear midge in **pears** *(off-label)* [1, 3]
- Pollen beetle in **mustard** [2]; **spring oilseed rape, winter oilseed rape** [2, 5]
- Raspberry beetle in **protected blackberries** *(off-label)*, **protected raspberries** *(off-label)* [3]
- Rosy apple aphid in **apples** [1, 3]
- Tarnished plant bug in **protected blackberries** *(off-label)*, **protected raspberries** *(off-label)* [3]
- Thrips in **protected aubergines** *(off-label)*, **protected courgettes** *(off-label)*, **protected cucumbers** *(off-label)*, **protected ornamentals** *(off-label)*, **protected peppers** *(off-label)*, **protected tomatoes** *(off-label)* [1, 3]
- Vine weevil in **bedding plants, hardy ornamental nursery stock, ornamental plant production, pot plants, protected ornamentals** [4]
- Western flower thrips in **protected aubergines** *(off-label)*, **protected courgettes** *(off-label)*, **protected cucumbers** *(off-label)*, **protected ornamentals** *(off-label)*, **protected peppers** *(off-label)*, **protected tomatoes** *(off-label)* [1, 3]
- Wheat-blossom midge in **spring wheat, winter wheat** [2]
- Whitefly in **bedding plants, hardy ornamental nursery stock, ornamental plant production, pot plants, protected ornamentals** [4]; **protected aubergines** *(off-label)*, **protected courgettes** *(off-label)*, **protected cucumbers** *(off-label)*, **protected ornamentals** *(off-label)*, **protected peppers** *(off-label)*, **protected tomatoes** *(off-label)* [1]

Specific Off-Label Approvals (SOLAs)
- **almonds** *20080471* [1], *20060341* [3]
- **asparagus** *20101802* [3]
- **bilberries** *20080466* [1], *20060335* [3]
- **blackberries** *20080475* [1], *20060336* [3]
- **blackcurrants** *20080466* [1], *20060335* [3]
- **blueberries** *20080466* [1], *20060335* [3]
- **cherries** *(under temporary protective rain covers)* *20080469* [1], *20060338* [3]
- **chestnuts** *20080471* [1], *20060341* [3]
- **cob nuts** *20080471* [1], *20060341* [3]
- **courgettes** *20102032* [1], *20081006* [3]
- **cranberries** *20080466* [1], *20060335* [3]
- **filberts** *20060341* [3]
- **gherkins** *20102032* [1], *20081006* [3]
- **gooseberries** *20080466* [1], *20060335* [3]
- **hazel nuts** *20080471* [1], *20060341* [3]
- **hops** *20102034* [1], *20082831* [3]
- **leaf brassicas** *(baby leaf production)* *20080470* [1], *20060340* [3]
- **marrows** *20102032* [1], *20081006* [3]
- **mirabelles** *(under temporary protective rain covers)* *20080469* [1], *20060338* [3]
- **ornamental plant production** *20102034* [1], *20082831* [3]
- **pears** *20080464* [1], *20060324* [3]
- **plums** *20080468* [1], *20060337* [3]
- **protected aubergines** *20080474* [1], *20063728* [3]
- **protected blackberries** *20080467* [1], *20070534* [3]
- **protected blueberry** *20102035* [1], *20081332* [3]
- **protected courgettes** *20080474* [1], *20063728* [3]
- **protected cucumbers** *20080474* [1], *20063728* [3]
- **protected forest nurseries** *20102034* [1], *20082831* [3]
- **protected herbs (see appendix 6)** *20080472* [1], *20060453* [3]
- **protected lamb's lettuce** *20080472* [1], *20060453* [3]

FOR FULL CONDITIONS OF USE ALWAYS READ THE PRODUCT LABEL

thiacloprid

- ***protected lettuce*** *20080472* [1], *20060453* [3]
- ***protected ornamentals*** *20080474* [1], *20063728* [3]
- ***protected peppers*** *20080474* [1], *20063728* [3]
- ***protected raspberries*** *20080467* [1], *20070534* [3]
- ***protected strawberries*** *20102033* [1], *20060334* [3]
- ***protected tomatoes*** *20080474* [1], *20063728* [3]
- ***raspberries*** *20080475* [1], *20060336* [3]
- ***redcurrants*** *20080466* [1], *20060335* [3]
- ***rubus hybrids*** *20080475* [1], *20060336* [3]
- ***soft fruit*** *20102034* [1], *20082831* [3]
- ***strawberries*** *20080465* [1], *20060333* [3]
- ***top fruit*** *20102034* [1], *20082831* [3]
- ***walnuts*** *20080471* [1], *20060341* [3]
- ***whitecurrants*** *20080466* [1], *20060335* [3]

Approval information
- Thiacloprid included in Annex I under EC Directive 91/414

Efficacy guidance
- Best results in apples obtained by a programme of sprays commencing pre-blossom at the first sign of aphids [3]
- For best pear midge control treat under warm conditions (mid-day) when the adults are flying (temperatures higher than 12 °C). Applications on cool damp days are unlikely to be successful [3]
- Best results in field crops obtained from treatments when target pests reach threshold levels. For wheat blossom midge use pheromone or sticky yellow traps to monitor adult activity in crops at risk [2, 5]
- Treatment of ornamentals and protected ornamentals is by incorporation into peat-based growing media prior to sowing or planting using suitable automated equipment [4]
- Ensure thorough mixing in the compost to achieve maximum control. Top dressing is ineffective [4]
- Unless transplants have been treated with a suitable pesticide prior to planting in treated compost, full protection against target pests is not guaranteed [4]
- Minimise the possibility of the development of resistance by alternating insecticides with different modes of action in the programme
- In dense canopies and on larger trees increase water volume to ensure full coverage

Restrictions
- Maximum number of treatments 2 per yr on apples, seed potatoes; 1 per yr on other field crops and for compost incorporation
- Consult processor before use on crops for processing [2, 5]
- Do not mix treated compost with any other bulky materials such as perlite or bark [4]
- Use treated compost as soon as possible, and within 4 wk of mixing [4]

Crop-specific information
- Latest use: up to and including flowering just complete (GS 69) for wheat; before sowing or planting for treated compost
- HI apples, potatoes 14 d; oilseed rape, mustard 30 d
- Test tolerance of ornamental species before large scale use [4]

Environmental safety
- Dangerous for the environment [2, 4, 5]
- Very toxic or toxic to aquatic organisms
- Risk to certain non-target insects or other arthropods. See directions for use [3]
- Broadcast air-assisted LERAP [1] (18 m); Broadcast air-assisted LERAP [3] (30 m)

Hazard classification and safety precautions
Hazard Harmful [1-5]; Dangerous for the environment [2, 4, 5]
Transport code 6.1 [1, 3]; 9 [4]
Packaging group III [1, 3, 4]
UN number 2902 [1, 3]; 3077 [4]; N/C [2, 5]
Risk phrases R20, R52 [1, 3]; R22a [1-3, 5]; R36, R38, R51 [2, 5]; R40, R53a [1-5]; R43 [1, 3, 4]; R50 [4]
Operator protection A [1-5]; C [2, 5]; D [4]; H [1, 3]; U05a [1-5]; U11, U19a [2, 5]; U14 [1, 3]; U16a [2]; U20b [1-3, 5]; U20c [4]

SEE SECTION 3 FOR PRODUCTS ALSO REGISTERED

Pesticide Profiles

Environmental protection E13c, E22b [3]; E15a, E38 [4]; E15b [1, 2, 5]; E17b [1] (18 m); E17b [3] (30 m); E22c [1]; E34 [1-5]
Storage and disposal D01, D02, D09a [1-5]; D10b [1, 3]; D10c [2, 5]; D11a [4]; D12a [1, 2, 5]
Treated seed S04a [4]
Medical advice M03 [1-5]; M05a [1]; M05b [3]

431 thiamethoxam

A neonicotinoid insecticide for apples, pears, potatoes and beet crops
IRAC mode of action code: 4A

See also fludioxonil + metalaxyl-M + thiamethoxam

Products

1	Actara	Syngenta	25% w/w	WG	13728
2	Centric	Syngenta	25% w/w	WG	13954
3	Cruiser SB	Syngenta	600 g/l	FS	12958

Uses
- Aphids in **apples**, **hops** *(off-label)*, **pears**, **soft fruit** *(off-label)* [2]; **potatoes, potatoes grown for seed** [1]
- Beet leaf miner in **fodder beet** *(seed treatment)*, **sugar beet** *(seed treatment)* [3]
- Flea beetle in **fodder beet** *(seed treatment)*, **sugar beet** *(seed treatment)* [3]
- Foliar disease control in **ornamental plant production** *(off-label)* [1]
- Green aphid in **apples**, **pears** [2]
- Millipedes in **fodder beet** *(seed treatment)*, **sugar beet** *(seed treatment)* [3]
- Pear sucker in **pears** [2]
- Pygmy beetle in **fodder beet** *(seed treatment)*, **sugar beet** *(seed treatment)* [3]
- Rosy apple aphid in **apples**, **pears** [2]
- Springtails in **fodder beet** *(seed treatment)*, **sugar beet** *(seed treatment)* [3]
- Symphylids in **fodder beet** *(seed treatment)*, **sugar beet** *(seed treatment)* [3]
- Western flower thrips in **ornamental plant production** *(off-label)* [2]
- Wireworm in **fodder beet** *(seed treatment - reduction)*, **sugar beet** *(seed treatment - reduction)* [3]
- Woolly aphid in **apples**, **pears** [2]

Specific Off-Label Approvals (SOLAs)
- *hops* 20082832 [2]
- *ornamental plant production* 20082801 [1], 20082230 [2]
- *soft fruit* 20082832 [2]

Approval information
- Thiamethoxam included in Annex I under EC Directive 91/414

Efficacy guidance
- Apply to sugar beet or fodder beet seed as part of the normal commercial pelleting process using special treatment machinery [3]
- Seed drills must be suitable for use with polymer-coated seeds. Standard drill settings should not need to be changed [3]
- Drill treated seed into a firm even seedbed. Poor seedbed quality or seedbed conditions may results in delayed emergence and poor establishment [3]
- Where very high populations of soil pests are present protection may be inadequate to achieve an optimum plant stand [3]
- Use in line with latest IRAG guidelines

Restrictions
- Maximum number of treatments 1 per seed batch [3]
- Avoid deep or shallow drilling which may adversely affect establishment and reduce the level of pest control [3]
- Do not use herbicides containing lenacil pre-emergence on treated crops [3]

Crop-specific information
- Latest use: before drilling for fodder beet, sugar beet [3]

FOR FULL CONDITIONS OF USE ALWAYS READ THE PRODUCT LABEL

Environmental safety
- Dangerous for the environment
- Very toxic to aquatic organisms

Hazard classification and safety precautions
Hazard Dangerous for the environment
UN number N/C
Risk phrases R50 [1, 3]; R51 [2]; R53a [1-3]
Operator protection A, D, H [3]; U05a, U07 [3]; U05b, U20c [1, 2]
Environmental protection E03, E34, E36a [3]; E12c, E12e [1, 2]; E15b, E38 [1-3]
Storage and disposal D01, D02, D09a, D12a [1-3]; D05, D11a, D14 [3]; D10c [1, 2]
Treated seed S02, S04b, S05, S06a, S06b, S08 [3]

432　thifensulfuron-methyl

A translocated sulfonylurea herbicide
HRAC mode of action code: B

See also　carfentrazone-ethyl + thifensulfuron-methyl
　　　　　flupyrsulfuron-methyl + thifensulfuron-methyl
　　　　　fluroxypyr + thifensulfuron-methyl + tribenuron-methyl
　　　　　metsulfuron-methyl + thifensulfuron-methyl

Products
Pinnacle	Headland	50% w/w	SG	12285

Uses
- Docks in **permanent grassland, rotational grass**
- Green cover in **land temporarily removed from production**

Approval information
- Thifensulfuron-methyl included in Annex I under EC Directive 91/414
- Accepted by BBPA for use on malting barley

Efficacy guidance
- Best results achieved from application to small emerged weeds when growing actively. Broad-leaved docks are susceptible during the rosette stage up to onset of stem extension
- Ensure good spray coverage and apply to dry foliage
- Susceptible weeds stop growing almost immediately but symptoms may not be visible for about 2 wk
- Product sold in twin pack with mecoprop-P to provide option for improved control of cleavers
- Only broad-leaved docks (*Rumex obtusifolius*) are controlled; curled docks (*Rumex crispus*) are resistant
- Docks with developing or mature seed heads should be topped and the regrowth treated later
- Established docks with large tap roots may require follow-up treatment
- High populations of docks in grassland will require further treatment in following yr
- Thifensulfuron-methyl is a member of the ALS-inhibitor group of herbicides and products should be used in a planned Resistance Management strategy. See Section 5 for more information

Restrictions
- Maximum number of treatments 1 per crop for cereals or one per yr for grassland and cereals. Must only be applied from 1 Feb in yr of harvest
- Do not treat new leys in year of sowing
- Do not treat where nutrient imbalances, drought, waterlogging, low temperatures, lime deficiency, pest or disease attack have reduced crop or sward vigour
- Do not roll or harrow within 7 d of spraying
- Do not graze grass crops within 7 d of spraying
- Specific restrictions apply to use in sequence or tank mixture with other sulfonylurea or ALS-inhibiting herbicides. See label for details
- Only one application of a sulfonylurea product may be applied per calendar yr to grassland and green cover on land temporarily removed from production

Crop-specific information
- Latest use: before 1 Aug on grass; flag leaf fully emerged stage (GS 39) on cereals

SEE SECTION 3 FOR PRODUCTS ALSO REGISTERED

Pesticide Profiles

- On grass apply 7-10 d before grazing and do not graze for 7 d afterwards
- Product may cause a check to both sward and clover which is usually outgrown

Following crops guidance
- Only grass or cereals may be sown within 4 wk of application to grassland or setaside, or in the event of failure of any treated crop
- No restrictions apply after normal harvest of a treated cereal crop

Environmental safety
- Dangerous for the environment
- Very toxic to aquatic organisms
- Keep livestock out of treated areas for at least 7 d following treatment
- Take extreme care to avoid drift onto broad-leaved plants outside the target area or onto surface waters or ditches, or land intended for cropping
- Spraying equipment should not be drained or flushed onto land planted, or to be planted, with trees or crops other than cereals and should be thoroughly cleansed after use - see label for instructions

Hazard classification and safety precautions
Hazard Dangerous for the environment
Transport code 9
Packaging group III
UN number 3077
Risk phrases R50, R53a
Operator protection U19a, U20b
Environmental protection E06a (7 d); E15a, E38
Storage and disposal D09a, D11a, D12a

433 thifensulfuron-methyl + tribenuron-methyl

A mixture of two sulfonylurea herbicides for cereals
HRAC mode of action code: B + B

See also tribenuron-methyl

Products

1	Calibre SX	DuPont	33.3:16.7% w/w	SG	15032
2	Inka SX	DuPont	25:25% w/w	SG	13601
3	Ratio SX	DuPont	40:10% w/w	SG	12601

Uses
- Annual dicotyledons in **game cover** *(off-label)*, **spring oats, spring rye, triticale, winter oats, winter rye** [2]; **spring barley, spring wheat, winter barley, winter wheat** [1-3]
- Charlock in **spring barley, spring wheat, winter barley, winter wheat** [1-3]; **spring oats, spring rye, triticale, winter oats, winter rye** [2]
- Chickweed in **spring barley, spring wheat, winter barley, winter wheat** [1-3]; **spring oats, spring rye, triticale, winter oats, winter rye** [2]
- Docks in **spring barley, spring oats, spring rye, spring wheat, triticale, winter barley, winter oats, winter rye, winter wheat** [2]
- Mayweeds in **spring barley, spring wheat, winter barley, winter wheat** [1-3]; **spring oats, spring rye, triticale, winter oats, winter rye** [2]

Specific Off-Label Approvals (SOLAs)
- *game cover* 20101408 [2]

Approval information
- Thifensulfuron-methyl and tribenuron-methyl included in Annex I under EC Directive 91/414
- Accepted by BBPA for use on malting barley

Efficacy guidance
- Apply when weeds are small and actively growing
- Apply after 1 Feb [1-3]
- Apply after end of Feb in year of harvest [2]
- Ensure good spray cover of the weeds

FOR FULL CONDITIONS OF USE ALWAYS READ THE PRODUCT LABEL

- Susceptible weeds cease growth almost immediately after application and symptoms become evident 2 wk later
- Effectiveness reduced by rain within 4 h of treatment and in very dry conditions
- Various tank mixtures recommended to broaden weed control spectrum [1]
- Thifensulfuron-methyl and tribenuron-methyl are members of the ALS-inhibitor group of herbicides and products should be used in a planned Resistance Management strategy. See Section 5 for more information

Restrictions
- Maximum number of treatments 1 per crop
- Do not apply to cereals undersown with grass, clover or other legumes, or any other broad-leaved crop
- Do not apply within 7 d of rolling
- Specific restrictions apply to use in sequence or tank mixture with other sulfonylurea or ALS-inhibiting herbicides. See label for details
- Do not apply to any crop suffering from stress
- Consult contract agents before use on crops grown for seed

Crop-specific information
- Latest use: before flag leaf ligule first visible (GS 39) for all crops

Following crops guidance
- Only cereals, field beans or oilseed rape may be sown in the same calendar year as harvest of a treated crop
- In the event of failure of a treated crop sow only a cereal crop within 3 mth of product application and after ploughing and cultivating to at least 15 cm. After 3 mth field beans or oilseed rape may also be sown [2]

Environmental safety
- Dangerous for the environment
- Very toxic to aquatic organisms
- Spraying equipment should not be drained or flushed onto land planted, or to be planted, with trees or crops other than cereals and should be thoroughly cleansed after use - see label for instructions
- Take particular care to avoid damage by drift onto broad-leaved plants outside the target area or onto surface waters or ditches
- LERAP Category B

Hazard classification and safety precautions
    Hazard Irritant [1, 3]; Dangerous for the environment [1-3]
    Transport code 9
    Packaging group III
    UN number 3077
    Risk phrases R43 [1, 3]; R50, R53a [1-3]
    Operator protection A, H [1, 3]; U05a, U08, U20a [1, 3]; U14 [3]; U20c [2]
    Environmental protection E15a [1, 3]; E15b [2]; E16a, E38 [1-3]; E16b [1]
    Storage and disposal D01, D02 [1, 3]; D09a, D11a, D12a [1-3]

434 thiophanate-methyl

A thiophanate fungicide with protectant and curative activity
FRAC mode of action code: 1

See also iprodione + thiophanate-methyl

Products

1	Cercobin WG	Certis	70% w/w	WG	13854
2	Topsin WG	Certis	70% w/w	WG	13988

Uses
- Canker in **apples** (off-label), **crab apples** (off-label), **pears** (off-label), **quinces** (off-label) [1]
- Disease control in **hops** (off-label - do not harvest for human or animal consumption (including idling) within 12 months of treatment.), **soft fruit** (off-label - any fruit harvested within 12 months of treatment must be destroyed.) [1]

SEE SECTION 3 FOR PRODUCTS ALSO REGISTERED

- Fusarium in **durum wheat**, **spring oilseed rape**, **winter oilseed rape** [2]; **leeks** (off-label) [1]; **spring wheat**, **triticale**, **winter wheat** [1, 2]
- Fusarium root rot in **ornamental plant production** (off-label) [1]
- Gloeosporium rot in **apples** (off-label), **crab apples** (off-label), **pears** (off-label), **quinces** (off-label) [1]
- Grey mould in **protected aubergines** (off-label), **protected melons** (off-label), **protected pumpkins** (off-label), **protected squashes** (off-label) [1]
- Leaf curl in **ornamental plant production** (off-label) [1]
- Mycotoxins in **durum wheat** (reduction only), **spring oilseed rape** (reduction only), **winter oilseed rape** (reduction only) [2]; **spring wheat** (reduction only), **triticale** (reduction only), **winter wheat** (reduction only) [1, 2]
- Phoma in **ornamental plant production** (off-label) [1]
- Sclerotinia in **protected aubergines** (off-label), **protected melons** (off-label), **protected pumpkins** (off-label), **protected squashes** (off-label) [1]
- Verticillium wilt in **protected tomatoes** (off-label), **strawberries** (off-label) [1]

Specific Off-Label Approvals (SOLAs)
- **apples** 20081813 expires 28 Feb 2011 [1]
- **crab apples** 20081813 expires 28 Feb 2011 [1]
- **hops** (do not harvest for human or animal consumption (including idling) within 12 months of treatment.) 20082833 [1]
- **leeks** 20081383 expires 28 Feb 2011 [1]
- **ornamental plant production** 20081384 expires 28 Feb 2011 [1]
- **pears** 20081813 expires 28 Feb 2011 [1]
- **protected aubergines** 20081382 expires 28 Feb 2011 [1]
- **protected melons** 20081382 expires 28 Feb 2011 [1]
- **protected pumpkins** 20081382 expires 28 Feb 2011 [1]
- **protected squashes** 20081382 expires 28 Feb 2011 [1]
- **protected tomatoes** 20091969 [1]
- **quinces** 20081813 expires 28 Feb 2011 [1]
- **soft fruit** (any fruit harvested within 12 months of treatment must be destroyed.) 20082833 [1]
- **strawberries** 20081381 expires 28 Feb 2011 [1]

Approval information
- Thiophanate-methyl included in Annex I under EC Directive 91/414

Efficacy guidance
- To avoid the development of resistance, a maximum of 2 applications of any MBC product (thiophanate-methyl or carbendazim) are allowed in any one crop. Avoid using MBC fungicides alone.

Environmental safety
- Dangerous for the environment
- Very toxic to aquatic organisms
- LERAP Category B

Hazard classification and safety precautions
    Transport code 9
    Packaging group III
    UN number 3077
    Operator protection A; U19a [2]
    Environmental protection E15b, E16a, E16b [1, 2]; E38 [2]
    Storage and disposal D01, D02, D12a [2]; D10b [1, 2]
    Medical advice M05a [2]

FOR FULL CONDITIONS OF USE ALWAYS READ THE PRODUCT LABEL

thiram

435 thiram

A protectant dithiocarbamate fungicide
FRAC mode of action code: M3

See also carboxin + thiram
prochloraz + thiram

Products

1	Agrichem Flowable Thiram	Agrichem	600 g/l	FS	10784
2	Thianosan DG	Unicrop	80% w/w	WG	13404
3	Thiraflo	Chemtura	480 g/l	FS	13338
4	Thyram Plus	Agrichem	600 g/l	FS	10785
5	Triptam	Certis	80% w/w	WG	14014

Uses

- Botrytis in **apples**, **ornamental plant production**, **pears**, **protected strawberries**, **strawberries** [5]; **ornamental plant production** (except Hydrangea), **tulips** [2]
- Botrytis fruit rot in **apples** [2]
- Carnation rust in **carnations** [2]
- Chrysanthemum brown rust in **chrysanthemums** [2]
- Damping off in **broad beans** (seed treatment), **combining peas** (seed treatment), **dwarf beans** (seed treatment), **forage maize** (seed treatment), **grass seed** (seed treatment), **runner beans** (seed treatment), **spring field beans** (seed treatment), **spring oilseed rape** (seed treatment), **vining peas** (seed treatment), **winter field beans** (seed treatment), **winter oilseed rape** (seed treatment) [1, 3, 4]; **bulb onions** (seed treatment), **cabbages** (seed treatment), **cauliflowers** (seed treatment), **edible podded peas** (seed treatment), **leeks** (seed treatment), **radishes** (seed treatment), **salad onions** (seed treatment), **turnips** (seed treatment) [1, 4]; **cress** (off-label), **endives** (off-label), **frise** (off-label), **garlic** (off-label - seed treatment), **lamb's lettuce** (off-label), **lettuce** (off-label), **lupins** (off-label - seed treatment), **poppies for morphine production** (off-label - seed treatment), **radicchio** (off-label), **rhubarb** (off-label - seed treatment), **scarole** (off-label), **shallots** (off-label - seed treatment), **soya beans** (seed treatment - qualified minor use)), **swedes** (off-label - seed treatment) [1]; **soya beans** (seed treatment - qualified minor use) [4]
- Gloeosporium in **apples** [2]
- Grey mould in **chrysanthemums**, **freesias**, **protected strawberries**, **strawberries** [2]
- Pythium in **poppies for morphine production** (off-label - seed treatment) [1]
- Scab in **apples**, **pears** [2]
- Seed-borne diseases in **carrots** (seed soak), **celery (outdoor)** (seed soak), **fodder beet** (seed soak), **mangels** (seed soak), **parsley** (seed soak), **red beet** (seed soak), **sugar beet** (seed soak) [1]; **garlic** (off-label - seed treatment), **lupins** (off-label - seed treatment), **shallots** (off-label - seed treatment), **swedes** (off-label - seed treatment) [4]

Specific Off-Label Approvals (SOLAs)

- *cress* 20070792 [1]
- *endives* 20070792 [1]
- *frise* 20070792 [1]
- *garlic* (seed treatment) 20052395 [1], (seed treatment) 20052391 [4]
- *lamb's lettuce* 20070792 [1]
- *lettuce* 20070792 [1]
- *lupins* (seed treatment) 20052394 [1], (seed treatment) 20052389 [4]
- *poppies for morphine production* (seed treatment) 20040681 [1]
- *radicchio* 20070792 [1]
- *rhubarb* (seed treatment) 20052396 [1]
- *scarole* 20070792 [1]
- *shallots* (seed treatment) 20052395 [1], (seed treatment) 20052391 [4]
- *swedes* (seed treatment) 20052397 [1], (seed treatment) 20052390 [4]

Approval information

- Thiram included in Annex I under EC Directive 91/414

Efficacy guidance

- Spray before onset of disease and repeat every 7-14 d. Spray interval varies with crop and disease. See label for details [2]

SEE SECTION 3 FOR PRODUCTS ALSO REGISTERED

Pesticide Profiles

- Seed treatments may be applied through most types of seed treatment machinery, from automated continuous flow machines to smaller batch treating apparatus [1, 3, 4]
- Co-application of 175 ml water per 100 kg of seed likely to improve evenness of seed coverage [1, 3, 4]

Restrictions
- Maximum number of treatments 3 per crop for protected winter lettuce (thiram based products only; 2 per crop if sequence of thiram and other EBDC fungicides used) [2]; 2 per crop for protected summer lettuce [2]; 1 per batch of seed for seed treatments [1, 3, 4]
- Do not apply to hydrangeas [2]
- Notify processor before dusting or spraying crops for processing [2]
- Do not dip roots of forestry transplants [2]
- Do not spray when rain imminent [2]
- Do not treat seed of tomatoes, peppers or aubergines [1, 3, 4]
- Soya bean seed treatments restricted to crops that are to be harvested as a mature pulse crop only [1, 4]
- Treated seed should not be stored from one season to the next [1, 3, 4]

Crop-specific information
- Latest use: pre-drilling for seed treatments and seed soaks [1, 3, 4]; 21 d after planting out or 21 d before harvest, whichever is earlier, for protected winter lettuce; 14 d after planting out or 21 d before harvest, whichever is earlier, for spraying protected summer lettuce [2]
- HI 21 d for protected lettuce; 14 d for outdoor lettuce; 7 d for apples, pears, blackcurrants, raspberries, strawberries, tomatoes [2]
- Seed to be treated should be of satisfactory quality and moisture content [1, 3, 4]
- Follow label instructions for treating small quantities of seed [1, 3, 4]
- For use on tulips, chrysanthemums and carnations add non-ionic wetter [2]

Environmental safety
- Dangerous for the environment [1, 3, 4]
- Very toxic to aquatic organisms [1, 3, 4]
- Do not use treated seed as food or feed [1, 3, 4]
- Treated seed harmful to game and wildlife [1, 3, 4]
- A red dye is available from manufacturer to colour treated seed, but not recommended as part of the seed steep. See label for details [1]

Hazard classification and safety precautions
Hazard Harmful, Dangerous for the environment
Transport code 9
Packaging group III
UN number 3077 [2, 5]; 3082 [1, 3, 4]
Risk phrases R20, R36, R38 [1, 4]; R22a, R48, R50, R53a [1-5]; R43 [1, 2, 4, 5]
Operator protection A [1-5]; P [5]; U02a, U11 [1, 4]; U05a, U20b [1-5]; U08 [1-4]; U09a [5]; U14, U19a [2, 5]
Environmental protection E13b [2, 5]; E15a, E34 [1, 4]; E38 [3]
Storage and disposal D01, D09a [1-5]; D02, D05 [1, 3, 4]; D10a [3]; D10c, D12b [1, 4]; D11a, D12a [2, 5]
Treated seed S01, S02, S03, S04b, S05, S06a, S07 [1, 3, 4]; S04a [1, 4]
Medical advice M03 [3]; M04a [1, 4]

436 tolclofos-methyl

A protectant nitroaniline fungicide for soil-borne diseases
FRAC mode of action code: 14

Products

1	Basilex	Scotts	50% w/w	WP	07494
2	Rizolex	Interfarm	10% w/w	DP	14204
3	Rizolex Flowable	Interfarm	500 g/l	FS	11399

FOR FULL CONDITIONS OF USE ALWAYS READ THE PRODUCT LABEL

tolclofos-methyl

Uses
- Black scurf and stem canker in **potatoes** *(off-label - tuber treatment only to be used with automatic planters)*, **potatoes** *(tuber treatment only to be used with automatic planters)* [3]; **potatoes** *(tuber treatment)* [2]
- Bottom rot in **protected lettuce** [1]
- Damping off in **seedlings of ornamentals** [1]
- Damping off and foot rot in **radishes** *(off-label)* [1]
- Damping off and wirestem in **brussels sprouts**, **cabbages**, **calabrese**, **cauliflowers**, **chinese cabbage** [1]
- Foot rot in **ornamental plant production**, **seedlings of ornamentals** [1]
- Rhizoctonia in **broccoli** *(off-label)*, **leaf brassicas** *(off-label)*, **leaf brassicas** *(off-label - baby leaf production only)*, **protected celery** *(off-label)*, **protected radishes** *(off-label)*, **swedes** *(off-label)*, **swedes** *(without covers)*, **turnips** *(off-label)*, **turnips** *(with fleece or mesh covers)* [1]; **potatoes** *(chitted seed treatment)*, **potatoes** *(off-label - chitted seed treatment)*, **seed potatoes** *(off-label)* [3]
- Root rot in **ornamental plant production**, **seedlings of ornamentals** [1]

Specific Off-Label Approvals (SOLAs)
- **broccoli** *20063527* [1], *20102302* [1]
- **leaf brassicas** *(baby leaf production only)* *20063528* [1], *20102301* [1]
- **potatoes** *(chitted seed treatment) 20041323* [3], *(tuber treatment only to be used with automatic planters) 20041323* [3]
- **protected celery** *20011055* [1], *20102291* [1]
- **protected radishes** *20011054* [1], *20102290* [1]
- **radishes** *20020214* [1], *20102292* [1]
- **seed potatoes** *20102329* [3]
- **swedes** *20023749* [1], *20102293* [1]
- **turnips** *20023749* [1], *20102293* [1]

Approval information
- Tolclofos-methyl included in Annex I under EC Directive 91/414
- In 2006 CRD required that all products containing this active ingredient should carry the following warning in the main area of the container label: "Tolclofos-methyl is an anticholinesterase organophosphate. Handle with care"
- May be applied by misting equipment mounted over roller table. See label for details [2, 3]

Efficacy guidance
- Apply dust to seed potatoes during hopper loading [2]
- Apply flowable formulation to clean tubers with suitable misting equipment over a roller table. Spray as potatoes taken into store (first earlies) or as taken out of store (second earlies, maincrop, crops for seed) pre-chitting [3]
- Do not mix flowable formulation with any other product [3]
- To control Rhizoctonia in vegetables and ornamentals apply as drench before sowing, pricking out or planting [1]
- On established seedlings and pot plants apply as drench and rinse off foliage [1]

Restrictions
- Tolclofos-methyl is an atypical organophosphorus compound which has weak anticholinesterase activity. Do not use if under medical advice not to work with such compounds
- Maximum number of treatments 1 per batch for seed potatoes; 1 per crop for lettuce, brassicas, swedes, turnips; 1 at each stage of growth (ie sowing, pricking out, potting) to a maximum of 3, for ornamentals
- Only to be used with automatic planters [2]
- Not recommended for use on seed potatoes where hot water treatment used or to be used [2]
- Do not apply as overhead drench to vegetables or ornamentals when hot and sunny [1]
- Do not use on heathers [1]
- Must not be used via hand-held equipment
- Application to protected crops must only be made where the operator is outside the structure at the time of treatment

Crop-specific information
- Latest use: at planting of seed potatoes [2, 3]; before transplanting for lettuce, brassicas, ornamental specimens [1]; before 2 true lvs for swedes, turnips [1]

SEE SECTION 3 FOR PRODUCTS ALSO REGISTERED

Pesticide Profiles

Environmental safety
- Dangerous for the environment [1-3]
- Very toxic to aquatic organisms
- Treated tubers to be used as seed only, not for food or feed [2]
- LERAP Category B [1]

Hazard classification and safety precautions
Hazard Irritant, Dangerous for the environment
Transport code 9
Packaging group III
UN number 3077 [1, 2]; 3082 [3]
Risk phrases R36, R37, R38, R51 [1-3]; R53a [1, 3]
Operator protection A, D, H [1-3]; C [3]; E [2, 3]; J, M [2]; U05a, U11 [2]; U16a [3]; U19a, U20b [1-3]
Environmental protection E15a [2, 3]; E16a [1]
Storage and disposal D01, D02, D12a [2]; D08 [2, 3]; D09a [1-3]; D10a [3]; D11a [1, 2]
Medical advice M01 [1]; M04a [2, 3]

437   tralkoxydim

A foliar applied oxime herbicide for grass weed control in cereals.
HRAC mode of action code: A

Products

1	Grasp 400 SC	Nufarm UK	400 g/l	SC	13594
2	Strimma	Makhteshim	250 g/l	SC	12777

Uses
- Awned canary grass in **durum wheat, spring barley, spring wheat, triticale, winter barley, winter rye, winter wheat** [1]
- Blackgrass in **durum wheat, spring barley, spring wheat, triticale, winter barley, winter rye, winter wheat** [1, 2]
- Italian ryegrass in **durum wheat, spring barley, spring wheat, triticale, winter barley, winter rye, winter wheat** [1]
- Perennial ryegrass in **durum wheat, spring barley, spring wheat, triticale, winter barley, winter rye, winter wheat** [1]
- Wild oats in **durum wheat, spring barley, spring wheat, triticale, winter barley, winter rye, winter wheat** [1, 2]
- Yorkshire fog in **durum wheat, spring barley, spring wheat, triticale, winter barley, winter rye, winter wheat** [1]

Approval information
- Accepted by BBPA for use on malting barley
- Tralkoxydim included in Annex I under EC Directive 91/414
- Approval expiry 30 Apr 2011 [2]

Efficacy guidance
- Product leaf-absorbed and translocated rapidly to growing points. Best results achieved after completion of emergence of grass weeds when growing actively in competitive crops under warm humid conditions with adequate soil moisture
- Activity not dependent on soil type or condition. Weeds germinating after application will not be controlled
- Best control of wild oats obtained from 2 leaf to 1st node detectable stage of weeds, and of blackgrass up to 3 tillers
- Tralkoxydim is an ACCase inhibitor herbicide. To avoid the build up of resistance do not apply products containing an ACCase inhibitor herbicide more than twice to any crop. In addition do not use any product containing tralkoxydim in mixture or sequence with any other product containing the same ingredient
- Use these products as part of a resistance management strategy that includes cultural methods of control and does not use ACCase inhibitors as the sole chemical method of grass weed control

FOR FULL CONDITIONS OF USE ALWAYS READ THE PRODUCT LABEL

- Applying a second product containing an ACCase inhibitor to a crop will increase the risk of resistance development; only use a second ACCase inhibitor to control different weeds at a different timing
- Always follow WRAG guidelines for preventing and managing herbicide resistant weeds. See Section 5 for more information

Restrictions
- Maximum number of treatments 1 per crop. See label
- Authorised adjuvant must always be added. See label
- Do not use on oats
- Do not spray undersown crops or crops to be undersown
- Do not spray when foliage wet or covered in ice or crop otherwise under stress
- Do not spray if a protracted period of cold weather forecast
- Do not spray crops under stress from chemical treatment, grazing, pest attack, mineral deficiency or low fertility. Treatment of stressed crops may be followed by transient foliar discolouration
- Do not roll or harrow within 1 wk of spraying
- Restrictions apply to mixtures or sequences with phenoxy hormone or sulfonylurea herbicides. See label

Crop-specific information
- Latest use: before booting (GS 41)
- Apply to winter cereals from 2 leaves unfolded. If necessary winter cereals may be sprayed twice: once in autumn and once in spring
- Apply to spring cereals from end of tillering

Environmental safety
- Dangerous for the environment [2]
- Harmful to aquatic organisms [2]
- Take care to avoid drift onto neighbouring crops, especially oats

Hazard classification and safety precautions
Hazard Harmful, Dangerous for the environment [2]
UN number N/C
Risk phrases R22a, R52, R53a [2]
Operator protection A, H [1, 2]; C [2]; U02a, U04a, U08, U20b [1, 2]; U05a [1]; U14, U15 [2]
Environmental protection E15a, E38 [2]; E15b [1]
Storage and disposal D01, D02, D05, D09a, D10c [1, 2]; D12a [2]

438 tri-allate

A soil-acting thiocarbamate herbicide for grass weed control
HRAC mode of action code: N

Products
Avadex Excel 15G   Gowan   15% w/w   GR 12109

Uses
- Annual grasses in **durum wheat** (off-label), **linseed** (off-label), **spring oilseed rape** (off-label), **winter oilseed rape** (off-label)
- Blackgrass in **broad beans** (off-label), **canary seed** (off-label - for wild bird seed production), **combining peas, durum wheat, fodder beet, lucerne, lupins** (off-label), **mangels, red beet, red clover, sainfoin, spring barley, spring field beans, sugar beet, triticale, vetches, vining peas, white clover, winter barley, winter field beans, winter wheat**
- Meadow grasses in **broad beans** (off-label), **canary seed** (off-label - for wild bird seed production), **combining peas, durum wheat, fodder beet, lucerne, lupins** (off-label), **mangels, red beet, red clover, sainfoin, spring barley, spring field beans, sugar beet, triticale, vetches, vining peas, white clover, winter barley, winter field beans, winter wheat**
- Wild oats in **broad beans** (off-label), **canary seed** (off-label - for wild bird seed production), **combining peas, durum wheat, fodder beet, lucerne, lupins** (off-label), **mangels, red beet, red clover, sainfoin, spring barley, spring field beans, sugar beet, triticale, vetches, vining peas, white clover, winter barley, winter field beans, winter wheat**

SEE SECTION 3 FOR PRODUCTS ALSO REGISTERED

Specific Off-Label Approvals (SOLAs)
- *broad beans* 20063531
- *canary seed* (for wild bird seed production) 20070867
- *durum wheat* 20063532
- *linseed* 20101945
- *lupins* 20063530
- *spring oilseed rape* 20100135
- *winter oilseed rape* 20100135

Approval information
- Tri-allate included in Annex 1 under EC Directive 91/414
- Approved for aerial application on wheat, barley, rye, triticale, field beans, peas, fodder beet, sugar beet, red beet, lucerne, sainfoin, vetches, clover. See Section 5 for more information
- Accepted by BBPA for use on malting barley

Efficacy guidance
- Incorporate or apply to surface pre-emergence (post-emergence application possible in winter cereals up to 2-leaf stage of wild oats)
- Do not use on soils with more than 10% organic matter
- Wild oats controlled up to 2-leaf stage
- If applied to dry soil, rainfall needed for full effectiveness, especially with granules on surface. Do not use if top 5-8 cm bone dry
- Do not apply with spinning disc granule applicator; see label for suitable types
- Do not apply to cloddy seedbeds
- Use sequential treatments to improve control of barren brome and annual dicotyledons (see label for details)

Restrictions
- Maximum number of treatments 1 per crop
- Consolidate loose, puffy seedbeds before drilling to avoid chemical contact with seed
- Drill cereals well below treated layer of soil (see label for safe drilling depths)
- Do not use on direct-drilled crops or undersow grasses into treated crops
- Do not sow oats or grasses within 1 yr of treatment

Crop-specific information
- Latest use: pre-drilling for beet crops; before crop emergence for field beans, spring barley, peas, forage legumes; before first node detectable stage (GS 31) for winter wheat, winter barley, durum wheat, triticale, winter rye

Environmental safety
- Irritating to eyes and skin
- May cause sensitization by skin contact
- Harmful to fish or other aquatic life. Do not contaminate surface waters or ditches with chemical or used container

Hazard classification and safety precautions
Hazard Irritant
Transport code 9
Packaging group III
UN number 3077
Risk phrases R36, R38, R43
Operator protection A, C, H; U02a, U05a, U20a
Environmental protection E13c
Storage and disposal D01, D02, D09a, D11a

FOR FULL CONDITIONS OF USE ALWAYS READ THE PRODUCT LABEL

439 triazoxide

A benzotriazine fungicide available only in mixtures
FRAC mode of action code: 35

See also clothianidin + prothioconazole + tebuconazole + triazoxide
imidacloprid + tebuconazole + triazoxide
prothioconazole + tebuconazole + triazoxide
tebuconazole + triazoxide

440 tribenuron-methyl

A foliar acting sulfonylurea herbicide with some root activity for use in cereals
HRAC mode of action code: B

See also fluroxypyr + thifensulfuron-methyl + tribenuron-methyl
metsulfuron-methyl + tribenuron-methyl
thifensulfuron-methyl + tribenuron-methyl

Products
Thor Nufarm UK 50% w/w WG 14240

Uses
- Annual dicotyledons in **spring barley**, **spring oats**, **spring wheat**, **triticale**, **winter barley**, **winter oats**, **winter rye**, **winter wheat**

Approval information
- Tribenuron-methyl included in Annex I under EC Directive 91/414
- Accepted by BBPA for use on malting barley

Efficacy guidance
- Best control achieved when weeds small and actively growing
- Good spray cover must be achieved since larger weeds often become less susceptible
- Susceptible weeds cease growth almost immediately after treatment and symptoms can be seen in about 2 wk
- Weed control may be reduced when conditions very dry
- Tribenuron-methyl is a member of the ALS-inhibitor group of herbicides and products should be used in a planned Resistance Management strategy. See Section 5 for more information

Restrictions
- Maximum number of treatments 1 per crop
- Specific restrictions apply to use in sequence or tank mixture with other sulfonylurea or ALS-inhibiting herbicides. See label for details
- Do not apply to crops undersown with grass, clover or other broad-leaved crops
- Do not apply to any crop suffering stress from any cause or not actively growing
- Do not apply within 7 d of rolling

Crop-specific information
- Latest use: up to and including flag leaf ligule/collar just visible (GS 39)
- Apply in autumn or in spring from 3 leaf stage of crop

Following crops guidance
- Only cereals, field beans or oilseed rape may be sown in the same calendar yr as harvest of a treated crop
- In the event of crop failure sow only a cereal within 3 mth of application. After 3 mth field beans or oilseed rape may also be sown

Environmental safety
- Dangerous for the environment
- Very toxic to aquatic organisms
- Take extreme care to avoid drift onto broad-leaved plants outside the target area or onto surface waters or ditches, or land intended for cropping

SEE SECTION 3 FOR PRODUCTS ALSO REGISTERED

- Spraying equipment should not be drained or flushed onto land planted, or to be planted, with trees or crops other than cereals and should be thoroughly cleansed after use - see label for instructions

Hazard classification and safety precautions
Hazard Irritant, Dangerous for the environment
Transport code 9
Packaging group III
UN number 3077
Risk phrases R43, R50, R53a, R70
Operator protection A, H; U05a, U08, U20b
Environmental protection E15a, E38
Storage and disposal D01, D02, D09a, D11a, D12a

441 triclopyr

A pyridinecarboxylic acid herbicide for perennial and woody weed control
HRAC mode of action code: O

See also 2,4-D + dicamba + triclopyr
2,4-D + triclopyr
aminopyralid + triclopyr
clopyralid + fluroxypyr + triclopyr
clopyralid + triclopyr
florasulam + triclopyr
fluroxypyr + triclopyr

Products

1	Thrash	Pan Agriculture	480 g/l	EC	15067
2	Timbrel	Dow	480 g/l	EC	05815
3	Woody	AgChem Access	480 g/l	EC	14163

Uses
- Brambles in *forest, land not intended to bear vegetation* [1-3]
- Broom in *forest, land not intended to bear vegetation* [1-3]
- Brush clearance in *industrial sites* [1-3]
- Docks in *forest, land not intended to bear vegetation* [1-3]
- Gorse in *forest, land not intended to bear vegetation* [1-3]
- Perennial dicotyledons in *asparagus* (off-label - directed treatment) [2]; *forest, industrial sites, land not intended to bear vegetation* [1-3]
- Perennial grasses in *asparagus* (off-label - directed treatment) [2]
- Rhododendrons in *forest, land not intended to bear vegetation* [1-3]
- Scrub clearance in *industrial sites* [1-3]
- Stinging nettle in *forest, land not intended to bear vegetation* [1-3]
- Woody weeds in *forest, industrial sites, land not intended to bear vegetation* [1-3]

Specific Off-Label Approvals (SOLAs)
- *asparagus* (directed treatment) 20050530 [2]

Approval information
- Triclopyr included in Annex I under EC Directive 91/414

Efficacy guidance
- Apply in grassland as spot treatment or overall foliage spray when weeds in active growth in spring or summer. Details of dose and timing vary with species. See label
- Apply to woody weeds as summer foliage, winter shoot, basal bark, cut stump or tree injection treatment
- Apply foliage spray in water when leaves fully expanded but not senescent
- Do not spray in drought, in very hot or cold conditions
- Control may be reduced if rain falls within 2 h of application
- Control of rhododendron can be variable. If higher than 1.8 m cut stump treatment recommended. A follow-up shoot treatment may be required
- See label for maximum concentrations when applying in oil, water or via watering can

FOR FULL CONDITIONS OF USE ALWAYS READ THE PRODUCT LABEL

trifloxystrobin

Restrictions
- Maximum number of treatments 2 per yr
- Do not apply to grass leys less than 1 yr old
- Do not drill kale, swedes, turnips, grass or mixtures containing clover within 6 wk of treatment. Allow at least 6 wk before planting trees
- Not to be used on food crops
- Do not apply through hand held rotary atomisers
- Not to be applied in or near water

Crop-specific information
- Latest use: 6 wk before replanting; 7 d before grazing
- Uses on land not intended for cropping include grassland of no agricultural interest such as roadside verges, railway and motorway embankments
- Apply winter shoot, basal bark or cut stump sprays in paraffin or diesel oil. Dose and timing vary with species. See label for details
- Inject undiluted or 1:1 dilution into cuts spaced every 7.5 cm round trunk
- Clover will be killed or severely checked by application in grassland

Environmental safety
- Dangerous for the environment
- Very toxic to aquatic organisms
- Do not allow spray to drift onto agricultural or horticultural crops, amenity plantings, gardens, ponds, lakes or water courses. Vapour drift may occur under hot conditions
- Keep livestock out of treated areas for at least 7 d and until foliage of any poisonous weeds such as buttercups or ragwort has died and become unpalatable
- LERAP Category B

Hazard classification and safety precautions
Hazard Harmful, Dangerous for the environment
Transport code 9
Packaging group III
UN number 3082
Risk phrases R22a, R22b, R38, R43, R50, R53a
Operator protection A, C, H, M; U02a, U05a, U08, U14, U20b
Environmental protection E07a, E15a, E16a, E16b, E23, E34, E38
Consumer protection C01
Storage and disposal D01, D02, D09a, D10b, D12a
Medical advice M05b

442 trifloxystrobin

A protectant strobilurin fungicide for cereals and managed amenity turf
FRAC mode of action code: 11

See also cyproconazole + trifloxystrobin
          fluoxastrobin + prothioconazole + trifloxystrobin
          propiconazole + trifloxystrobin
          prothioconazole + trifloxystrobin
          tebuconazole + trifloxystrobin

Products

1	Martinet	AgChem Access	500 g/l	SC	14614
2	Mascot Defender	Rigby Taylor	500 g/l	SG	14065
3	Micro-Turf	AgriGuard	50% w/w	WG	12988
4	Pan Aquarius	Pan Agriculture	50% w/w	WG	13018
5	Scorpio	Bayer Environ.	50% w/w	WG	12293
6	Swift SC	Bayer CropScience	500 g/l	SC	11227

Uses
- Brown rust in **spring barley**, **winter barley**, **winter wheat** [1, 6]
- Foliar disease control in **durum wheat** *(off-label)*, **grass seed crops** *(off-label)*, **ornamental specimens** *(off-label)*, **spring rye** *(off-label)*, **triticale** *(off-label)*, **winter rye** *(off-label)* [6]
- Fusarium patch in **amenity grassland**, **managed amenity turf** [2-5]

SEE SECTION 3 FOR PRODUCTS ALSO REGISTERED

- Glume blotch in **winter wheat** [1, 6]
- Net blotch in **spring barley**, **winter barley** [1, 6]
- Red thread in **amenity grassland**, **managed amenity turf** [2-5]
- Rhynchosporium in **spring barley**, **winter barley** [1, 6]
- Septoria leaf blotch in **winter wheat** [1, 6]

Specific Off-Label Approvals (SOLAs)
- **durum wheat** 20061287 [6]
- **grass seed crops** 20061287 [6]
- **ornamental specimens** 20082882 [6]
- **spring rye** 20061287 [6]
- **triticale** 20061287 [6]
- **winter rye** 20061287 [6]

Approval information
- Trifloxystrobin included in Annex I under EC Directive 91/414
- Accepted by BBPA for use on malting barley

Efficacy guidance
- Should be used protectively before disease is established in crop. Further treatment may be necessary if disease attack prolonged
- Treat grass after cutting and do not mow for at least 48 h afterwards to allow adequate systemic movement [3-5]
- Trifloxystrobin is a member of the QoI cross resistance group. Product should be used preventatively and not relied on for its curative potential
- Use product as part of an Integrated Crop Management strategy incorporating other methods of control, including where appropriate other fungicides with a different mode of action. Do not apply more than two foliar applications of QoI containing products to any cereal crop
- There is a significant risk of widespread resistance occurring in *Septoria tritici* populations in UK. Failure to follow resistance management action may result in reduced levels of disease control
- On cereal crops product must always be used in mixture with another product, recommended for control of the same target disease, that contains a fungicide from a different cross resistance group and is applied at a dose that will give robust control
- Strains of barley powdery mildew resistant to QoIs are common in the UK

Restrictions
- Maximum number of treatments 2 per crop per yr
- Do not apply to turf during dought conditions or to frozen turf [3-5]

Crop-specific information
- HI barley, wheat 35 d [6]

Environmental safety
- Dangerous for the environment
- Very toxic to aquatic organisms
- LERAP Category B [2-5]

Hazard classification and safety precautions
Hazard Irritant [2-5]; Dangerous for the environment [1-6]
Transport code 9
Packaging group III
UN number 3077 [3-5]; 3082 [1, 2, 6]
Risk phrases R43 [2-5]; R50, R53a [1-6]
Operator protection A, H; U02a, U09a, U19a [1, 6]; U05a, U20b [1-6]; U11, U13, U14, U15 [2-5]
Environmental protection E13b [1, 6]; E15b, E16a, E16b, E34 [2-5]; E38 [1-6]
Consumer protection C02a [1, 6] (35 d)
Storage and disposal D01, D02, D09a, D10c, D12a
Medical advice M03 [2-5]

FOR FULL CONDITIONS OF USE ALWAYS READ THE PRODUCT LABEL

443 triflusulfuron-methyl

A sulfonyl urea herbicide for beet crops
HRAC mode of action code: B

See also lenacil + triflusulfuron-methyl

Products
Debut	DuPont	50% w/w	WG	07804

Uses
- Annual dicotyledons in **chicory** (off-label), **fodder beet**, **red beet** (off-label), **sugar beet**

Specific Off-Label Approvals (SOLAs)
- **chicory** 20101933 expires 31 Dec 2011
- **red beet** 20101932 expires 31 Dec 2011

Approval information
- Triflusulfuron-methyl included in Annex I under EC Directive 91/414

Efficacy guidance
- Product should be used with a recommended adjuvant or a suitable herbicide tank-mix partner - see label for details
- Product acts by foliar action. Best results obtained from good spray cover of small actively growing weeds
- Susceptible weeds cease growth immediately and symptoms can be seen 5-10 d later
- Best results achieved from a programme of up to 4 treatments starting when first weeds have emerged with subsequent applications every 5-14 d when new weed flushes at cotyledon stage
- Weed spectrum can be broadened by tank mixture with other herbicides. See label for details
- Product may be applied overall or via band sprayer
- Triflusulfuron-methyl is a member of the ALS-inhibitor group of herbicides

Restrictions
- Maximum number of treatments 4 per crop
- Do not apply to any crop stressed by drought, water-logging, low temperatures, pest or disease attack, nutrient or lime deficiency
- A maximum total dose of 60 g/ha triflusulfuron may only be applied every third year on the same field.

Crop-specific information
- Latest use: before crop leaves meet between rows
- HI 4 wk for red beet
- All varieties of sugar beet and fodder beet may be treated from early cotyledon stage until the leaves begin to meet between the rows

Following crops guidance
- Only winter cereals should follow a treated crop in the same calendar yr. Any crop may be sown in the next calendar yr
- Applications must not be made to sugar beet or fodder beet when the foliage will be fed to livestock.
- After failure of a treated crop, sow only spring barley, linseed or sugar beet within 4 mth of spraying unless prohibited by tank-mix partner

Environmental safety
- Dangerous for the environment
- Very toxic to aquatic organisms
- Extremely dangerous to fish or other aquatic life. Do not contaminate surface waters or ditches with chemical or used container
- Take extreme care to avoid drift onto broad-leaved plants outside the target area or onto surface waters or ditches, or land intended for cropping
- Spraying equipment should not be drained or flushed onto land planted, or to be planted, with trees or crops other than sugar beet and should be thoroughly cleansed after use - see label for instructions
- LERAP Category B

SEE SECTION 3 FOR PRODUCTS ALSO REGISTERED

Pesticide Profiles

Hazard classification and safety precautions
  Hazard Dangerous for the environment
  Transport code 9
  Packaging group III
  UN number 3077
  Risk phrases R50, R53a
  Operator protection A; U05a, U08, U19a, U20a
  Environmental protection E13a, E15a, E16a, E16b, E38
  Storage and disposal D01, D02, D05, D09a, D11a, D12a

444 trinexapac-ethyl

A novel cyclohexanecarboxylate plant growth regulator for cereals, turf and amenity grassland

Products

1	Clayton Truss	Clayton	250 g/l	EC	13129
2	Cutaway	Syngenta	121 g/l	SL	14445
3	Moddus	Syngenta	250 g/l	EC	08801
4	Pan Tepee	Pan Agriculture	250 g/l	EC	12279
5	Primo Maxx	Syngenta	121 g/l	SL	14780
6	Pure Max	Pure Amenity	121 g/l	SL	14952
7	Shrink	AgChem Access	250 g/l	EC	13762
8	Standon Pygmy	Standon	250 g/l	EC	13858
9	Stunt	AgriGuard	250 g/l	EC	12859
10	Tacet	Goldengrass	250 g/l	EC	14342
11	Temper Turf	AgriGuard	121 g/l	SL	14819
12	Tempo	Syngenta	250 g/l	EC	13799

Uses
  • Growth regulation in **durum wheat**, **ryegrass seed crops**, **spring barley**, **spring oats**, **spring rye**, **triticale**, **winter barley**, **winter oats**, **winter rye**, **winter wheat** [7, 8, 10]
  • Growth retardation in **amenity grassland** [2, 5, 6, 11]; **managed amenity turf** [5, 6, 11]
  • Lodging control in **canary seed** *(off-label - for wild bird seed production)*, **spring wheat** *(off-label - cv A C Barrie only)* [3]; **durum wheat**, **ryegrass seed crops**, **spring barley**, **spring oats**, **spring rye**, **triticale**, **winter barley**, **winter oats**, **winter rye**, **winter wheat** [1, 3, 4, 9, 12]

Specific Off-Label Approvals (SOLAs)
  • **canary seed** *(for wild bird seed production)* 20070869 [3]
  • **spring wheat** *(cv A C Barrie only)* 20042300 [3]

Approval information
  • Trinexapac-ethyl included in Annex I under EC Directive 91/414
  • Accepted by BBPA for use on malting barley

Efficacy guidance
  • Best results on cereals and ryegrass seed crops obtained from treatment from the leaf sheath erect stage [1, 3, 4, 9]
  • Best results on turf achieved from application to actively growing weed free turf grass that is adequately fertilized and watered and is not under stress. Adequate soil moisture is essential [6]
  • Turf should be dry and weed free before application [6]
  • Environmental conditions, management and cultural practices that affect turf growth and vigour will influence effectiveness of treatment [6]
  • Repeat treatments on turf up to the maximum approved dose may be made as soon as growth resumes [6]

Restrictions
  • Maximum total dose equivalent to one full dose on cereals, ryegrass seed crops [1, 3, 4, 9]
  • Maximum number of treatments on turf equivalent to five full dose treatments [6]
  • Do not apply if rain or frost expected or if crop wet. Products are rainfast after 12 h
  • Only use on crops at risk of lodging [1, 3, 4, 9]
  • Do not apply within 12 h of mowing turf [6]
  • Do not treat newly sown turf [6]
  • Not to be used on food crops [6]
  • Do not compost or mulch grass clippings [6]

FOR FULL CONDITIONS OF USE ALWAYS READ THE PRODUCT LABEL

Crop-specific information
- Latest use: before 2nd node detectable (GS 32) for oats, ryegrass seed crops; before 3rd node detectable (GS 33) for durum wheat, spring barley, triticale, rye; before flag leaf sheath extending (GS 41) for winter barley, winter wheat
- On wheat apply as single treatment between leaf sheath erect stage (GS 30) and flag leaf fully emerged (GS 39) [1, 3, 4, 9]
- On barley, rye, triticale and durum wheat apply as single treatment between leaf sheath erect stage (GS 30) and second node detectable (GS 32), or on winter barley at higher dose between flag leaf just visible (GS 37) and flag leaf fully emerged (GS 39) [1, 3, 4, 9]
- On oats and ryegrass seed crops apply between leaf sheath erect stage (GS 30) and first node detectable stage (GS 32) [1, 3, 4, 9]
- Treatment may cause ears of cereals to remain erect through to harvest [1, 3, 4, 9]
- Turf under stress when treated may show signs of damage [6]
- Any weed control in turf must be carried out before application of the growth regulator [6]

Environmental safety
- Dangerous for the environment [1, 3, 4, 9]
- Toxic to aquatic organisms [1, 3, 4, 9]

Hazard classification and safety precautions
Hazard Irritant, Dangerous for the environment [1, 3, 4, 7-12]
Transport code 9 [1, 3, 4, 7-12]
Packaging group III [1, 3, 4, 7-12]
UN number 3082 [1, 3, 4, 7-12]; N/C [2, 5, 6]
Risk phrases R36, R50 [11]; R43, R51 [1, 3, 4, 7-10, 12]; R53a [1, 3, 4, 7-12]
Operator protection A, C [1-12]; H [1, 2, 4-6, 9]; K [2, 5, 6]; U05a [1-12]; U15, U20c [1, 3, 4, 7-10, 12]; U20b [2, 5, 6, 11]
Environmental protection E15a [3, 4, 7-12]; E15b [1]; E34 [9, 11]; E38 [3, 4, 7, 8, 10, 12]
Consumer protection C01 [2, 5, 6]
Storage and disposal D01, D02, D05 [1-12]; D09a [1, 3, 4, 7-12]; D10b [9]; D10c [1-8, 10-12]; D12a [1, 3, 4, 7, 8, 10, 12]

445 triticonazole

A triazole fungicide available only in mixtures
FRAC mode of action code: 3

See also guazatine + triticonazole
imazalil + triticonazole
prochloraz + triticonazole

446 warfarin

A coumarin anti-coagulant rodenticide

Products
1	Grey Squirrel Bait	Killgerm	0.02% w/w	RB	14807
2	Sakarat Ready-to-Use (Cut Wheat Base)	Killgerm	0.025% w/w	RB	H6807
3	Sewarin Extra	Killgerm	0.05% w/w	RB	H6810
4	Sewarin P	Killgerm	0.025% w/w	RB	H6811
5	Warfarin 0.5% Concentrate	B H & B	0.5% w/w	CB	H6815
6	Warfarin Ready Mixed Bait	B H & B	0.025% w/w	RB	H6816

Uses
- Grey squirrels in **agricultural premises**, **forest**, **trees** (nuts) [1]
- Rats in **agricultural premises** [2-6]

Approval information
- Warfarin included in Annex I under EC Directive 91/414

SEE SECTION 3 FOR PRODUCTS ALSO REGISTERED

Pesticide Profiles

Efficacy guidance
- For rodent control place ready-to-use or prepared baits at many points wherever rats active. Out of doors shelter bait from weather [2-6]
- Inspect baits frequently and replace or top up as long as evidence of feeding. Do not underbait [2-6]
- Use grey squirrel baits when bark stripping is evident in farm forestry or forestry or where there is risk of damage to trees grown for their nuts [1]
- Use squirrel bait in specially constructed hoppers and inspect every 2-3 d. Place bait hoppers so as to prevent rainwater entry and replace as necessary [1]

Restrictions
- For use only by local authorities, professional operators providing a pest control service and persons occupying industrial, agricultural or horticultural premises
- For use only between 15 Mar and 15 Aug for tree protection [1]
- Squirrel bait hoppers must be clearly labelled "Poison" or carry the instruction "Tree protection, do not disturb" [1]
- Warfarin baits for grey squirrel control must only be used outdoors in specially constructed hoppers which comply with the Grey Squirrel Order 1973, in Scotland and specified counties of England and Wales. See label [1]
- The use of warfarin to control grey squirrels is illegal unless the provisions of the Grey Squirrels Order 1973 are observed [1]

Environmental safety
- Harmful to wildlife
- Prevent access to baits by children and animals, especially cats, dogs and pigs
- Rodent bodies must be searched for and burned or buried, not placed in refuse bins or rubbish tips. Remains of bait and containers must be removed after treatment and burned or buried
- Bait must not be used where food, feed or water could become contaminated
- If squirrel bait hoppers have obviously been disturbed by badgers or other animals, change the site or lift onto tables or platforms [1]
- Warfarin baits must nor be used outdoors where pine martens are know to occur naturally [1]
- Under the terms of the Wildlife and Countryside Act 1981 warfarin squirrel baits must not be used outdoors in England, Scotland or Wales where red squrrels are know to occur [1]

Hazard classification and safety precautions
UN number N/C [1-4, 6]
Operator protection A [1-4]; C, D, E, H [2-4]; U13 [1-6]; U20a [2-4]; U20b [1, 5, 6]
Environmental protection E10b, E15a [1]
Storage and disposal D09a [1-6]; D10a [5, 6]; D11a [1-4]
Vertebrate/rodent control products V01a, V03a [5, 6]; V01b [1-4]; V02 [1-6]; V03b, V04b [2-4]; V04a [1, 5, 6]
Medical advice M03 [2-4]

447   zeta-cypermethrin

A contact and stomach acting pyrethroid insecticide
IRAC mode of action code: 3

Products

1 Angri	AgChem Access	100 g/l	EW	13730
2 Fury 10 EW	Belchim	100 g/l	EW	12248
3 Minuet EW	Belchim	100 g/l	EW	12304
4 Symphony	AgriGuard	100 g/l	EW	13021

Uses
- Aphids in **broad beans** *(off-label)*, **durum wheat** *(off-label)*, **fodder beet** *(off-label)*, **grass seed crops** *(off-label)*, **lupins** *(off-label)*, **spring rye** *(off-label)*, **triticale** *(off-label)*, **winter rye** *(off-label)* [2]; **spring barley, spring oats, spring wheat, winter barley, winter oats, winter wheat** [1-4]
- Barley yellow dwarf virus vectors in **spring barley, spring wheat, winter barley, winter wheat** [1-4]
- Cabbage seed weevil in **spring oilseed rape, winter oilseed rape** [1-4]
- Cabbage stem flea beetle in **spring oilseed rape, winter oilseed rape** [1-4]

FOR FULL CONDITIONS OF USE ALWAYS READ THE PRODUCT LABEL

zeta-cypermethrin

- Cutworms in **potatoes**, **sugar beet** [1-4]
- Flax flea beetle in **linseed** [1-4]
- Flea beetle in **spring oilseed rape**, **winter oilseed rape** [1-4]
- Insect pests in **broad beans** (off-label), **durum wheat** (off-label), **fodder beet** (off-label), **grass seed crops** (off-label), **lupins** (off-label), **spring rye** (off-label), **triticale** (off-label), **winter rye** (off-label) [2]
- Large flax flea beetle in **linseed** [1-4]
- Pea and bean weevil in **combining peas**, **spring field beans**, **vining peas**, **winter field beans** [1-4]
- Pea aphid in **combining peas**, **vining peas** [1-4]
- Pea moth in **combining peas**, **vining peas** [1-4]
- Pod midge in **spring oilseed rape**, **winter oilseed rape** [1-4]
- Pollen beetle in **spring oilseed rape**, **winter oilseed rape** [1-4]
- Rape winter stem weevil in **spring oilseed rape**, **winter oilseed rape** [1-4]

Specific Off-Label Approvals (SOLAs)
- **broad beans** 20063573 [2]
- **durum wheat** 20063570 [2]
- **fodder beet** 20063572 [2]
- **grass seed crops** 20063570 [2]
- **lupins** 20063571 [2]
- **spring rye** 20063570 [2]
- **triticale** 20063570 [2]
- **winter rye** 20063570 [2]

Approval information
- Zeta-cypermethrin included in Annex I under EC Directive 91/414
- Accepted by BBPA for use on malting barley

Efficacy guidance
- On winter cereals spray when aphids first found in the autumn for BYDV control. A second spray may be required on late drilled crops or in mild conditions
- For summer aphids on cereals spray when treatment threshold reached
- For listed pests in other crops spray when feeding damage first seen or when treatment threshold reached. Under high infestation pressure a second treatment may be necessary
- Best results for pod midge and seed weevil control in oilseed rape obtained from treatment after pod set but before 80% petal fall
- Pea moth treatments should be applied according to ADAS/PGRO warnings or when economic thresholds reached as indicated by pheromone traps
- Treatments for cutworms should be made at egg hatch and repeated no sooner than 10 d later

Restrictions
- Maximum number of treatments 2 per crop [4]
- Maximum total dose equivalent to two full dose treatments on all crops [2, 3]
- Consult processors before use on crops for processing

Crop-specific information
- Latest use: before 4 true leaves for linseed, before end of flowering for oilseed rape, cereals
- HI potatoes, field beans, peas 14 d; sugar beet 60 d

Environmental safety
- Dangerous for the environment
- Very toxic to aquatic organisms
- High risk to non-target insects or other arthropods. Do not spray within 6 m of the field boundary
- LERAP Category A

Hazard classification and safety precautions
Hazard Harmful, Dangerous for the environment
Transport code 6.1
Packaging group III
UN number 3352
Risk phrases R20, R22a, R43, R50, R53a
Operator protection A, C, H; U05a, U08, U14, U15, U19a, U20b
Environmental protection E15b, E16c, E16d, E22a, E34 [1-4]; E38 [1-3]

SEE SECTION 3 FOR PRODUCTS ALSO REGISTERED

Storage and disposal D01, D02, D09a, D10b [1-4]; D05 [4]; D12a [1-3]
Medical advice M05a [1-3]

448 zoxamide

A substituted benzamide fungicide available only in mixtures
FRAC mode of action code: 22

See also mancozeb + zoxamide

SECTION 3
PRODUCTS ALSO REGISTERED

Products also Registered

Products listed in the table below have not been notified for inclusion in Section 2 of this edition of the *Guide*. These products may legally be stored and used in accordance with their label until their approval expires, but they may not still be available for purchase.

Product	Approval holder	MAPP No.	Expiry Date
abamectin			
Acaramik	Rotam Agrochemical	14344	
Mectinide	AgriGuard	13953	
RouteOne Abamectin 18	Albaugh UK	13560	30 Apr 2011
acetamiprid			
Gazelle	Certis	12909	31 Dec 2014
acetic acid			
Natural Weed Spray NO. 1	Punya	12328	
acibenzolar-S-methyl			
Bion	Syngenta	09803	31 Oct 2011
alpha-cypermethrin			
Alert	BASF	13632	
Alpha C 6 ED	Techneat	13611	28 Feb 2015
Alphamex 100 EC	MAC	14071	
Fastac	BASF	13604	
Fastac Dry	BASF	10221	
aluminium ammonium sulphate			
Curb S	Sphere	13564	
Guardsman	Chiltern	05494	
Rezist Liquid	Barrett	14643	
Sphere ASBO	Amenity Land	14884	31 Aug 2014
aluminium phosphide			
Degesch Fumigation Pellets	Rentokil	11436	
Detia Gas Ex-P	Igrox	09802	
Detia Gas-Ex-B	Detia Degesch	06927	
Detia Gas-Ex-T	Igrox	03792	
Luxan Talunex	Luxan	06563	31 Oct 2011
Phostoxin I	Rentokil	05694	
Talunex	Certis	13798	31 May 2011
amidosulfuron			
Amidosulf	Goldengrass	14409	
Barclay Cleave	Barclay	11340	
Landgold Amidosulfuron	Landgold	09021	31 May 2012

EXPIRY DATE IS 31/12/2021 UNLESS OTHERWISE SHOWN

Products also Registered

Product	Approval holder	MAPP No.	Expiry Date
Landgold Amidosulfuron	Teliton	12110	31 Mar 2013
Pursuit	Bayer CropScience	07333	
Squire	Bayer CropScience	08715	
Squire Ultra	Bayer CropScience	13125	
amidosulfuron + iodosulfuron-methyl-sodium			
Sekator	Bayer CropScience	12634	30 Nov 2013
aminopyralid			
Pro-Banish	Dow	14730	29 Jul 2011
aminopyralid + fluroxypyr			
Halcyon	Dow	14709	29 Jul 2011
amitrole			
Aminotriazole Technical	Nufarm UK	11137	
Weedazol Pro	Nufarm UK	11995	
asulam			
Cleancrop Asulam	United Agri	10465	
I T Asulam	I T Agro	10186	
Milentus Asulam	Milentus	12245	
Noble Asulam	Barclay	12939	
azoxystrobin			
5504	Syngenta	12351	31 Dec 2011
Alpha Azoxystrobin	Makhteshim	14400	31 Dec 2011
Azoxystar	Life Scientific	15105	22 Sep 2014
Barclay ZX	Barclay	11336	31 Dec 2011
Clayton Stobik	Clayton	09440	31 Dec 2011
Cleancrop Celeb	United Agri	13294	31 Dec 2011
Jumbo 250 SC	Aako	14900	24 Feb 2013
Kingdom Turf	AgriGuard	13833	31 Dec 2011
Landgold Azzox	Goldengrass	14003	31 Dec 2011
Landgold Strobilurin 250	Teliton	12128	31 Dec 2011
Me2 Azoxystrobin	Me2	09654	31 Dec 2011
Ortiva	Syngenta	10542	31 Dec 2011
Priori	Syngenta	10543	31 Dec 2011
RouteOne Roxybin 25	Albaugh UK	13653	31 Dec 2011
Roxybin 25	RouteOne	13268	31 Dec 2011
Strobiplus 250 EC	Agform	15094	15 Jul 2014
azoxystrobin + chlorothalonil			
Amistar Opti	Syngenta	12515	31 Aug 2011
BTP	Syngenta	13400	31 Aug 2011
Curator	Syngenta	13936	31 Aug 2011
azoxystrobin + cyproconazole			
Cecure	AgriGuard	14889	

EXPIRY DATE IS 31/12/2021 UNLESS OTHERWISE SHOWN

Products also Registered

Product	Approval holder	MAPP No.	Expiry Date
azoxystrobin + fenpropimorph			
RouteOne Roxypro FP	Albaugh UK	13634	
Bacillus thuringiensis			
DiPel DF	Fargro	11184	31 Aug 2012
benalaxyl + mancozeb			
Galben M	Sipcam	07220	31 Dec 2011
Galben M	Belchim	14661	
Intro Plus	Interfarm	11630	31 Dec 2011
Intro Plus	Belchim	14666	
Tairel	Belchim	14659	
Tairel	Sipcam	07767	31 Dec 2011
bentazone			
Basagran	BASF	00188	31 Jul 2011
Bently	Chem-Wise	14210	31 Jul 2011
Euro Benta 480	Euro	14707	31 Jul 2011
IT Bentazone	I T Agro	13132	31 Jul 2011
Landgold Bentazone SL	Teliton	13418	31 Jul 2011
RouteOne Benta 48	Albaugh UK	13662	31 Jul 2011
RouteOne Bentazone 48	Albaugh UK	13664	31 Jul 2011
bentazone + MCPA + MCPB			
Acumen	BASF	00028	31 Oct 2011
bentazone + MCPB			
Pulsar	BASF	04002	31 Oct 2011
bentazone + pendimethalin			
Impuls	BASF	13372	31 Jul 2011
benthiavalicarb-isopropyl + mancozeb			
En-Garde	Certis	12985	31 Dec 2011
En-Garde	Certis	14901	30 Jun 2016
Valbon	Certis	12603	31 Dec 2011
benzoic acid			
Menno Florades	Brinkman	12472	30 Apr 2012
beta-cyfluthrin			
Gandalf	Makhteshim	12865	31 Dec 2013
beta-cyfluthrin + imidacloprid			
Chinook	Bayer CropScience	13696	
bifenazate			
Inter Bifenazate 240 SC	Iticon	14543	30 Nov 2015

EXPIRY DATE IS 31/12/2021 UNLESS OTHERWISE SHOWN

Products also Registered

Product	Approval holder	MAPP No.	Expiry Date
bifenox			
Cleancrop Diode	United Agri	14620	
bifenox + MCPA + mecoprop-P			
Quickfire	Headland Amenity	13053	31 Mar 2014
Quickfire	Makhteshim	14882	
Sirocco	Makhteshim	13051	
bitertanol + fuberidazole			
UK 743	Bayer CropScience	11315	28 Feb 2011
Bordeaux mixture			
Wetcol 3 Copper Fungicide	Law Fertilisers	13369	30 Nov 2011
boscalid			
RouteOne Boscalid 50	Albaugh UK	13660	31 Jul 2011
boscalid + epoxiconazole			
BAR	BASF	14775	
Maitre	Agro Trade	13878	
RouteOne Rocker	Albaugh UK	13642	
Splice	BASF	12315	
Totem	BASF	12692	
Venture	BASF	12316	
boscalid + pyraclostrobin			
RouteOne Bosca 25/12	Albaugh UK	13952	31 Jul 2011
bromoxynil + diflufenican + ioxynil			
Capture	Bayer CropScience	09982	
bromoxynil + terbuthylazine			
Alpha Bromotril PT	Makhteshim	09435	
Cleancrop Amaize	United Agri	11990	
Candida Oleophila Strain 0			
NEXY 1	BioNext	13609	22 Jul 2013
captan			
Akotan 80 WG	Aako	14490	
Alpha Captan 50 WP	Makhteshim	04797	30 Sep 2012
PP Captan 80 WG	Calliope SAS	12435	
PP Captan 83	Calliope SAS	12330	
carbendazim			
AgriGuard Pro-Turf	AgriGuard	11837	31 Dec 2011
Caste Off	Nufarm UK	12832	
Clayton Am-Carb	Clayton	11906	
Cleancrop Curve	United Agri	11774	
Delsene 50 Flo	Nufarm UK	11452	30 Jun 2012

EXPIRY DATE IS 31/12/2021 UNLESS OTHERWISE SHOWN

Product	Approval holder	MAPP No.	Expiry Date
Mascot Systemic	Scotts	09132	
Nuturf Carbendazim	Nufarm UK	11469	
Turf Systemic Fungicide	Barclay	13291	
Turfclear	Scotts	07506	
carbetamide			
Carbetamex	Makhteshim	13045	
carfentrazone-ethyl			
Aurora 40 WG	Belchim	11614	30 Sep 2013
Harrier	Belchim	14164	30 Sep 2013
Platform	Belchim	11615	30 Sep 2013
carfentrazone-ethyl + metsulfuron-methyl			
Ally Express	DuPont	08640	30 Jun 2011
chloridazon			
Pyramin FL	BASF	11628	
chloridazon + ethofumesate			
Gremlin	Sipcam	09468	
chloridazon + metamitron			
Volcan Combi FL	Sipcam	11442	
chlormequat			
3C Chlormequat 750	BASF	13984	
Adjust	Mandops	13148	30 Nov 2011
Agriguard 5C Chlormequat 460	AgriGuard	09851	
Agriguard Chlormequat 700	AgriGuard	09782	
AgriGuard Chlormequat 720	AgriGuard	14120	
AgriGuard Chlormequat 720	AgriGuard	14125	
AgriGuard Chlormequat 750	AgriGuard	14144	
AgriGuard Chlormequat 750	AgriGuard	14180	
Agriguard Chlormequat 760	AgriGuard	10290	30 Nov 2011
Agrovista 3 See	Agrovista	14181	31 Oct 2011
Alpha Chlormequat 460	Makhteshim	04804	31 Jul 2014
Alpha Pentagan	Makhteshim	04794	
Alpha Pentagan Extra	Makhteshim	04796	
Atlas 5C Quintacel	Nufarm UK	11130	
Barclay Holdup	Barclay	06799	
Barclay Holdup	Barclay	14804	
Barclay Holdup 600	Barclay	11373	30 Nov 2011
Barclay Holdup 600	Barclay	08794	30 Nov 2011
Barclay Holdup 640	Barclay	08795	30 Nov 2011
Barclay Holdup 640	Barclay	11374	30 Nov 2011
Barclay Liffey	Barclay	09856	30 Nov 2011
Barclay Liffey	Barclay	11366	30 Nov 2011

EXPIRY DATE IS 31/12/2021 UNLESS OTHERWISE SHOWN

Products also Registered

Product	Approval holder	MAPP No.	Expiry Date
Barclay Lucan	Barclay	11367	30 Nov 2011
Barclay Lucan	Barclay	09855	30 Nov 2011
Barclay Take 5	Barclay	11368	31 Mar 2014
Barclay Take 5	Barclay	08524	30 Nov 2011
Barclay Take 5	Barclay	14799	
Barleyquat B	Mandops	13187	30 Nov 2011
Barleyquat B	Taminco	13965	
BASF 3C Chlormequat 600	BASF	04077	
BASF 3C Chlormequat 720	BASF	06514	
BASF 3C Chlormequat 750	BASF	06878	
Bettaquat B	Mandops	13192	30 Nov 2011
Chlormequat 46	Nufarm UK	11504	
Ciba Chlormequat 460	Ciba Specialty	09525	30 Nov 2011
Ciba Chlormequat 5C 460:320	Ciba Specialty	09527	30 Nov 2011
Ciba Chlormequat 730	Ciba Specialty	09526	30 Nov 2011
Clayton CCC 750	Clayton	07952	
Clayton Manquat	Clayton	09916	30 Nov 2011
Cleancrop Chlormequat 700	United Agri	10143	30 Nov 2011
Cropsafe 5C Chlormequat	Certis	11179	30 Nov 2011
Cropsafe 5C Chlormequat	Certis	07897	30 Nov 2011
Fernpath Tangent	AgriGuard	10341	30 Nov 2011
Hyquat 70	Agrichem	03364	30 Nov 2011
K2	Mandops	13190	30 Nov 2011
Mandops Chlormequat 460	Taminco	13969	
Mandops Chlormequat 700	Mandops	06002	30 Nov 2011
Manipulator	Mandops	13189	30 Nov 2011
Midget	Nufarm UK	12283	
MSS Mirquat	Mirfield	08166	30 Nov 2011
Portman Chlormequat 400	Agform	01523	30 Nov 2011
Portman Chlormequat 460	Agform	02549	30 Nov 2011
Portman Chlormequat 700	Agform	03465	30 Nov 2011
Portman Supaquat	Agform	03466	30 Nov 2011
Selon	Mandops	13826	30 Nov 2011
Stabilan 750	Nufarm UK	09303	
Supaquat 720	Agform	09381	30 Nov 2011
Tricol	Nufarm UK	12682	
Tripart 5C	Tripart	04726	30 Nov 2011
Tripart Brevis	Tripart	03754	30 Nov 2011
Tripart Brevis 2	Tripart	06612	30 Nov 2011
Tripart Chlormequat 460	Tripart	03685	30 Nov 2011
Turpin	Makhteshim	15057	
Uplift	United Phosphorus	07527	30 Nov 2011

chlormequat + 2-chloroethylphosphonic acid

Barclay Banshee XL	Barclay	11339	

EXPIRY DATE IS 31/12/2021 UNLESS OTHERWISE SHOWN

Products also Registered

Product	Approval holder	MAPP No.	Expiry Date
Sypex	BASF	04650	
Terpal C	BASF	07062	

chlormequat + 2-chloroethylphosphonic acid + imazaquin

Satellite	BASF	10395	

2-chloroethylphosphonic acid

Cerone	Bayer CropScience	15087	27 Sep 2014
Ethro 48	RouteOne	13333	29 Feb 2012
Milentus Ethephon	Milentus	12185	
Padawan	Drax	14853	08 Feb 2014
Ripe-On	AgriGuard	15020	08 Feb 2014
RouteOne Ethro 48	Albaugh UK	13639	31 Jul 2012

2-chloroethylphosphonic acid + mepiquat chloride

Barclay Banshee	Barclay	11343	
Clayton Florin	Clayton	14351	28 Feb 2011
CleanCrop Fonic M	United Agri	09553	
Guilder	Nufarm UK	11894	28 Feb 2011
Mepiquat Plus	RouteOne	13297	29 Feb 2012
RouteOne Mepiquat Plus	Albaugh UK	13636	
Terpitz	Me2	09634	

chloropicrin

Chloropicrin Fumigant	Dewco-Lloyd	04216	
Custo-Fume	Custodian	13272	

chlorothalonil

Agriguard Chlorothalonil	AgriGuard	12201	31 Aug 2011
Bravo 500	Syngenta	10518	31 Aug 2011
Cheer 500	Goldengrass	14418	28 Feb 2011
CleanCrop Rio	United Agri	13332	31 Aug 2011
Cropguard 2	Nufarm UK	13193	31 Aug 2011
CTL 500	Goldengrass	14290	31 Aug 2011
Daconil Turf	Scotts	09265	31 Jul 2011
Daconil Turf	Syngenta	13602	31 Aug 2011
Expert-Turf	AgriGuard	12571	31 Aug 2011
Joules	Nufarm UK	13773	31 Aug 2011
Jupital	Syngenta	10528	31 Aug 2011
Jupital	Syngenta	14558	28 Feb 2016
Landgold Chlorothalonil 50	Teliton	13473	28 Feb 2011
Landgold Chlorothalonil 50	Goldengrass	14011	31 Aug 2011
Mainstay	Quadrangle	05625	31 Aug 2011
Pan Trilby	Pan Agriculture	13558	31 Aug 2011
Repulse	Certis	07641	30 Sep 2011
Repulse	Certis	11328	31 Aug 2011
RouteOne Ronil 50	Albaugh UK	13637	31 Aug 2011

EXPIRY DATE IS 31/12/2021 UNLESS OTHERWISE SHOWN

Products also Registered

Product	Approval holder	MAPP No.	Expiry Date
Supreme	AgriGuard	12233	31 Aug 2011
Thalonil 500	Euro	14926	28 Feb 2016
chlorothalonil + cyproconazole			
Bravo Xtra	Syngenta	11824	
Cleancrop Cyprothal	United Agri	09580	
chlorothalonil + cyproconazole + propiconazole			
Apache	Syngenta	14255	
chlorothalonil + flutriafol			
Argon	Headland	12312	
Halo	Headland	11546	
chlorothalonil + metalaxyl-M			
Folio Gold	Syngenta	10704	31 Aug 2011
chlorothalonil + picoxystrobin			
Credo	DuPont	13042	31 Aug 2011
Zimbrail	DuPont	13172	31 Aug 2011
chlorothalonil + propamocarb hydrochloride			
Merlin	Bayer CropScience	07943	
Pan Wizard	Pan Agriculture	11953	
chlorothalonil + propiconazole			
SIP 313	Sipcam	14755	17 Apr 2012
chlorothalonil + tetraconazole			
Eminent Star	Isagro	10447	
chlorotoluron			
Atol	Nufarm UK	11138	31 Aug 2011
Clayton Chloron 500 FL	Clayton	10791	31 Aug 2011
CTU Minrinse	Makhteshim	12458	31 Aug 2011
Headland Tolerate	Headland	10774	31 Aug 2011
Lentipur 700	Nufarm UK	13526	31 Aug 2011
Lentipur CL 500	Nufarm UK	08743	31 Aug 2011
chlorotoluron + diflufenican			
Tremor	Makhteshim	15001	
chlorpropham			
Aceto Chlorpropham 50M	Aceto	14134	07 Aug 2012
Aceto Sprout Nip	Aceto	14156	02 Sep 2012
BL500	Whyte Agrochemicals	14153	07 Aug 2012
Gro-Stop HN	Certis	14146	31 Jan 2015
Gro-Stop Innovator	Certis	14147	31 Jan 2015
Gro-Stop Ready	Certis	15109	31 Jan 2015
Intruder	AgriChem BV	15076	31 Jan 2015

EXPIRY DATE IS 31/12/2021 UNLESS OTHERWISE SHOWN

Products also Registered

Product	Approval holder	MAPP No.	Expiry Date
LS Chlorpropham 300HN	Life Scientific	14957	31 Aug 2014
MSS CIPC 50LF	United Phosphorus	14635	31 Jan 2015
MSS CIPC 50M	Whyte Agrochemicals	14151	07 Aug 2012
Pro-Long	Whyte Agrochemicals	14152	07 Aug 2012
ProStore HN	Belchim	15002	31 Jan 2015
chlorpyrifos			
Akofos 480 EC	Aako	14211	
Alpha Chlorpyrifos 48 EC	Makhteshim	13532	
Clayton Pontoon	Clayton	13219	30 Jun 2011
Lorsban WG	Landseer	11962	31 Dec 2011
Lorsban WG	Dow	10139	31 Dec 2011
Pirisect	AgriGuard	13371	30 Jun 2011
Spannit Granules	Barclay	12950	31 Dec 2011
Suscon Indigo	Fargro	09902	31 Dec 2011
chlorthal-dimethyl			
Dacthal W-75	AMVAC	10289	23 Mar 2011
Dacthal W-75	Certis	11323	23 Mar 2011
RouteOne Chlorthal-D75	Albaugh UK	13649	23 Mar 2011
cinidon-ethyl			
Lotus	BASF	09231	31 Oct 2011
Lotus	Nufarm UK	13744	30 Sep 2012
clodinafop-propargyl			
Agrilens Clodinafop	Agrilens	14331	31 Jul 2012
Landgold Clodinafop	Teliton	13329	31 Jul 2012
RouteOne Roclod 24	RouteOne	13638	31 Jul 2012
clofentezine			
Acaristop 500 SC	Aako	14405	
clomazone			
Clone	AgriGuard	14320	31 Jul 2014
IT Clomazone	Inter-Trade	14695	
Mazone 360	Euro	14898	
clomazone + metazachlor			
Centium Plus	Belchim	14634	
clopyralid			
Cliophar	Agriphar	13360	31 Oct 2012
Cliophar 400	Agriphar	15008	17 Jun 2014
Glopyr 400	Globachem	15009	17 Jun 2014
Vivendi 200	AgriChem BV	15017	30 Apr 2017
clopyralid + 2,4-D + MCPA			
Lonpar	Dow	08686	31 Oct 2012

EXPIRY DATE IS 31/12/2021 UNLESS OTHERWISE SHOWN

Products also Registered

Product	Approval holder	MAPP No.	Expiry Date
clopyralid + florasulam + fluroxypyr			
Bofix FFC	Dow	13152	31 Oct 2012
Bofix FFC	Dow	14179	31 Dec 2011
Galaxy	Dow	13127	31 Oct 2012
clopyralid + fluroxypyr + MCPA			
Interfix	Iticon	14386	31 Oct 2012
clopyralid + fluroxypyr + triclopyr			
Charter	AgriGuard	13908	
Trinity	AgriGuard	12738	28 Feb 2012
clopyralid + picloram			
Euro Pyralid Extra	Euro	14681	
Landgold Piccant	Goldengrass	13770	31 Oct 2012
Prevail	Dow	13205	
clopyralid + propyzamide			
Matrikerb	Dow	10806	31 Oct 2012
clothianidin			
NipsIT Inside	Interfarm	14744	11 Mar 2012
copper ammonium carbonate			
Croptex Fungex	Certis	11049	30 Nov 2011
copper hydroxide			
Spin Out	DuPont	12069	30 Nov 2011
copper silicate			
Socusil Slug Spray	Doff Portland	11817	30 Nov 2011
copper sulphate			
Cuproxat FL	Nufarm UK	13241	
cyazofamid			
Ranman Top	Belchim	14753	07 Dec 2013
Rithfir	DuPont	14816	30 Jun 2013
cyazofamid + cymoxanil			
Ranman Super	Belchim	14024	
cyazofamid + polyalkyleneoxide modheptamethylsiloxane			
Roazafod	RouteOne	13530	29 Feb 2012
RouteOne Roazafod	Albaugh UK	13666	30 Jun 2013
Cydia pomonella GV			
Cyd-X Xtra	Certis	14397	
cymoxanil			
Cleancrop Covert	United Agri	13840	31 Aug 2011

EXPIRY DATE IS 31/12/2021 UNLESS OTHERWISE SHOWN

Products also Registered

Product	Approval holder	MAPP No.	Expiry Date
Curzate 60 WG	Headland	13904	31 Aug 2011
Cymostraight 45	Belchim	14615	
Cymoxanil 45% WG	Globachem	13727	

cymoxanil + mancozeb

Product	Approval holder	MAPP No.	Expiry Date
Blighter	Nufarm UK	14326	30 Nov 2011
Cleancrop Xanilite	United Agri	10050	
Curzate M68	DuPont	08072	
Curzate M68 WSB	DuPont	08073	31 Aug 2011
Me2 Cymoxeb	Me2	09486	
Nautile DG	Nufarm UK	13416	31 Oct 2013
Rhythm	Interfarm	09636	
Solace	Nufarm UK	11936	
Solace WG	Nufarm UK	13729	31 Aug 2011
Standon Cymoxanil Extra	Standon	09442	
Zetanil	Sipcam	11993	

cymoxanil + propamocarb

Product	Approval holder	MAPP No.	Expiry Date
Axidor	Agriphar	15100	

cypermethrin

Product	Approval holder	MAPP No.	Expiry Date
Afrisect 10	Agriphar	13159	
AgriGuard Cypermethrin 250EC	AgriGuard	14076	31 Aug 2011
Cyperkill 10	Agriphar	13157	
Cyperkill 25	Agriphar	13160	31 Aug 2011
Cyperkill 5	Agriphar	13162	31 Aug 2011
Landgold Cyper 100	Goldengrass	14105	
MAC-Cypermethrin 100 EC	MAC	14934	
MCC 25 EC	Agriphar	13161	31 Aug 2011
Sherpa 100 EC	SBM	13570	
Syper 250	Goldengrass	14460	31 Aug 2011

cyproconazole + propiconazole

Product	Approval holder	MAPP No.	Expiry Date
Alto Xtra	Syngenta	13058	31 Mar 2013
Menara	Syngenta	13035	28 Feb 2013

cyprodinil

Product	Approval holder	MAPP No.	Expiry Date
Barclay Amtrak	Barclay	11338	31 Oct 2012
Cleancrop Cyprodinil	United Agri	09668	31 Oct 2012
Kayak	Syngenta	13119	31 Oct 2012
Standon Cyprodinil	Standon	09345	31 Oct 2012
Unix	Syngenta	11512	31 Oct 2012

cyprodinil + picoxystrobin

Product	Approval holder	MAPP No.	Expiry Date
Acanto Prima	DuPont	13041	31 Oct 2012

2,4-D

Product	Approval holder	MAPP No.	Expiry Date
2,4-D Amine 500	Nufarm UK	14360	30 Sep 2012

EXPIRY DATE IS 31/12/2021 UNLESS OTHERWISE SHOWN

Products also Registered

Product	Approval holder	MAPP No.	Expiry Date
Damine	Agriphar	13366	01 Oct 2012
Growell 2,4-D Amine	GroWell	13146	30 Sep 2012
Herboxone	Nufarm UK	14101	30 Apr 2012
Herboxone	Headland	13144	30 Apr 2012
Herboxone 60	Headland	14080	30 Sep 2012
Herboxone 60	Marks	13145	31 Aug 2012
2,4-D + dicamba			
Compo Floranid + Herbicide	Compo	14503	30 Sep 2012
Landscaper Pro Weed Control + Fertilizer	Scotts	14128	
Lawn Builder Plus Weed Control	Scotts	08499	
Lawn Spot Weeder	Bayer CropScience	13116	
2,4-D + dicamba + dichlorprop-P			
NUB-041	Nufarm UK	14287	31 Mar 2013
2,4-D + dicamba + MCPA + mecoprop-P			
Dicophar	Agriphar	13256	
Longbow	Bayer Environ.	14316	
2,4-D + dicamba + triclopyr			
Cleancrop Broadshot	United Agri	11664	
Greengard	SumiAgro Amenity	11715	
Nu-Shot	Nufarm UK	13550	
2,4-D + MCPA			
Agroxone Combi	Marks	11025	31 Oct 2011
Agroxone Combi	Nufarm UK	14094	31 Oct 2011
Agroxone Combi	Nufarm UK	14907	30 Sep 2012
Headland Polo	Headland	10283	31 Oct 2011
Lupo	Nufarm UK	12725	31 Oct 2011
2,4-D + picloram			
Atladox Hi	Nomix Enviro	13867	31 Mar 2013
2,4-D + triclopyr			
Genoxone ZX EC	Agriphar	13131	
daminozide			
B-Nine SG	Certis	12734	31 Aug 2011
B-Nine SG	Chemtura	12698	31 Aug 2011
B-Nine SG	Chemtura	14434	28 Feb 2016
Dazide Enhance	Fine	11943	31 Aug 2011
dazomet			
Assassin	AgriGuard	13737	
Basamid	Kanesho	12895	

EXPIRY DATE IS 31/12/2021 UNLESS OTHERWISE SHOWN

Products also Registered

Product	Approval holder	MAPP No.	Expiry Date
2,4-DB			
Butoxone DB	Marks	13680	30 Sep 2012
Butoxone DB	Nufarm UK	14100	28 Feb 2014
Butoxone DB	Nufarm UK	14840	31 Dec 2013
2,4-DB + MCPA			
Butoxone DB Extra	Marks	12152	31 Oct 2011
Butoxone DB Extra	Nufarm UK	14095	31 Oct 2011
Headland Cedar	Headland	12180	31 Oct 2011
MSS 2,4-DB + MCPA	Mirfield	01392	31 Oct 2011
deltamethrin			
Agriguard Deltamethrin	AgriGuard	10770	
Agrotech Deltamethrin	Agrotech-Trading	12165	
Cleancrop Decathlon	United Agri	12834	
Delta-M 2.5 EC	MAC	14212	
Landgold Deltaland	Teliton	12114	
Milentus Deltamethrin	Milentus	12219	
Pearl Micro	Bayer CropScience	08620	31 Dec 2011
Routeone Deltam 10	Albaugh UK	13615	
desmedipham + ethofumesate + phenmedipham			
D.E.P. 251	Goldengrass	14922	28 Feb 2013
dicamba			
Cadence	Barclay	09578	
Cadence	Syngenta	08796	
dicamba + MCPA + mecoprop-P			
ALS Premier Selective Plus	Amenity Land	08940	
Banlene Super	Certis	14420	
Broadband	Indigrow	15063	
Grassland Herbicide	United Phosphorus	14623	
Mascot Super Selective Plus	Barclay	12839	
Mircam Super	Nufarm UK	11836	
Nomix Turf	Nomix Enviro	14375	
Nomix Turf Plus	Nomix Enviro	14377	
Premier Amenity Selective	Amenity Land	10098	
Trireme	Nufarm UK	11524	
dicamba + mecoprop-P			
Camber	Headland	09901	
Foundation	Syngenta	08475	
Headland Swift	Headland	11945	
Optica Forte	Marks	11913	31 Dec 2012
Optica Forte	Nufarm UK	14334	28 Feb 2014
Optica Forte	Nufarm UK	14845	

EXPIRY DATE IS 31/12/2021 UNLESS OTHERWISE SHOWN

Products also Registered

Product	Approval holder	MAPP No.	Expiry Date
dichlorprop-P			
Optica DP	Marks	11067	30 Sep 2012
Optica DP	Nufarm UK	14097	
dichlorprop-P + ferrous sulphate + MCPA			
Vitagrow Granular Feed, Weed & Mosskiller	Sinclair	10971	31 Aug 2011
dichlorprop-P + MCPA			
Optica Duo	Marks	10343	30 Sep 2012
Optica Duo	Nufarm UK	14098	28 Feb 2014
Optica Duo	Nufarm UK	14844	
dichlorprop-P + MCPA + mecoprop-P			
Optica Trio	Marks	09747	30 Sep 2012
Optica Trio	Nufarm UK	14099	30 Nov 2012
diclofop-methyl + fenoxaprop-P-ethyl			
Corniche	Bayer CropScience	08947	
Tigress Ultra	Bayer CropScience	08946	
difenoconazole			
COM 302 09 F AL	Westland Horticulture	14021	31 Mar 2013
COM 302 8 F EC	Westland Horticulture	14020	31 Mar 2013
Difcor 250 EC	Nufarm UK	12361	28 Feb 2012
Fungus Attack 2 RTU	Westland Horticulture	14454	30 Sep 2013
Landgold Difenoconazole	Landgold	09964	31 Dec 2011
MAC-Difenoconazole 250 EC	MAC	14935	
Plant Rescue Fungus Ready to Use	Westland Horticulture	14646	
diflubenzuron			
Dimilin 25-WP	Chemtura	08902	
Dimilin Flo	Chemtura	08769	
diflufenican			
Alpha Diflufenican 500 SC	Makhteshim	13182	
Dina 50	Globachem	13627	
MAC-Diflufenican 500 SC	MAC	14109	
RouteOne Flucan 50 UK	Albaugh UK	14130	
Viola	Bayer CropScience	13321	
diflufenican + flufenacet			
Firebird	Bayer CropScience	14826	31 Dec 2013
Plumage	Goldengrass	15085	
diflufenican + flurtamone			
Graduate	Bayer CropScience	10776	
diflufenican + glyphosate			
Proshield	Scotts	14767	

EXPIRY DATE IS 31/12/2021 UNLESS OTHERWISE SHOWN

Products also Registered

Product	Approval holder	MAPP No.	Expiry Date
diflufenican + mecoprop-P			
Atom	Nufarm UK	15038	
diflufenican + metsulfuron-methyl			
Pelican Delta	Headland	14312	
dimethoate			
BASF Dimethoate 40	BASF	00199	
Danadim	Headland	11550	
Rogor L40	Headland	15071	
Rogor L40	Interfarm	07611	
Sector	Cheminova	10492	
dimethomorph + mancozeb			
Saracen	BASF	12005	
diphenylamine			
No Scald DPA 31	Decco	14141	30 May 2011
diquat			
A1412A2	Syngenta	13440	31 Dec 2011
BD-200	Barclay	14918	01 Feb 2014
Cleancrop Diquat	United Agri	13348	31 Dec 2011
CleanCrop Flail	United Agri	14870	31 Dec 2011
Diquanet	Hermoo	13413	31 Dec 2011
I.T. Diquat	I T Agro	13557	31 Dec 2011
Inter Diquat	I T Agro	14196	31 Dec 2011
Kalahari	Barclay	14917	01 Feb 2014
Quad	Q-Chem	13396	31 Dec 2013
Quad S	Q-Chem	13578	28 Jun 2011
Quat 200	Euro	15042	31 Dec 2011
Quit	Belchim	14874	31 Dec 2011
Standon Googly	Standon	12995	31 Dec 2011
Tuber Mission	AgriChem BV	14542	31 Dec 2011
Woomera	Hermoo	14009	30 Jun 2014
dithianon			
Barclay Cluster	Barclay	11347	
RouteOne Dith-WG 70	Albaugh UK	13851	
dodeca-8,10-dienyl acetate			
Exosex CM	Exosect	12103	
dodine			
Barclay Dodex	Barclay	11351	
Syllit 400 SC	Chimac-Agriphar	11079	
Syllit 400 SC	Agriphar	13363	

EXPIRY DATE IS 31/12/2021 UNLESS OTHERWISE SHOWN

Products also Registered

Product	Approval holder	MAPP No.	Expiry Date
epoxiconazole			
Agrotech-Epoxiconazole 125 SC	Agrotech-Trading	12382	
Bassoon	BASF	14402	
Epro 125	RouteOne	13375	29 Feb 2012
Fathom	Headland	14267	
Landgold Epoxi	Makhteshim	14062	30 Nov 2012
Landgold Epoxiconazole	Teliton	12117	
Landgold Epoxiconazole	Landgold	09821	31 May 2012
Landgold Epoxiconazole	Goldengrass	14025	30 Nov 2012
Milentus Epoxiconazole	Milentus	12363	
RouteOne Epro 125	Albaugh UK	13640	
Spike SC	Globachem	14185	30 Apr 2011
Starburst	Chem-Wise	13981	
Strand	Headland	14261	
Supo	Makhteshim	13903	
epoxiconazole + fenpropimorph			
Barclay Riverdance	Barclay	11341	30 Apr 2011
Greencrop Galore	Greencrop	09561	30 Apr 2011
Landgold Epoxiconazole FM	Landgold	08806	30 Apr 2011
Optiorb	AgriGuard	13426	
Standon Epoxifen	Standon	08972	30 Apr 2011
Tango Super	BASF	13283	
epoxiconazole + fenpropimorph + kresoxim-methyl			
Allegro Plus	BASF	12218	
Asana	BASF	11934	
BAS 493F	BASF	11748	
Cleancrop Chant	United Agri	11746	
Mastiff	BASF	11747	
Standon Kresoxim Super	Standon	09794	30 Apr 2011
epoxiconazole + fenpropimorph + pyraclostrobin			
Diamant	BASF	11557	
Diamant	BASF	13253	31 Aug 2011
RouteOne Super 3	Albaugh UK	13721	
Tourmaline	Agro Trade	13934	
epoxiconazole + kresoxim-methyl			
Allegro	BASF	12220	
Barclay Avalon	Barclay	11337	
Cleancrop Kresoxazole	United Agri	09698	
Landgold Strobilurin KE	Landgold	09908	31 May 2012
epoxiconazole + pyraclostrobin			
Euro	Me2	11511	
Ibex	BASF	12168	

EXPIRY DATE IS 31/12/2021 UNLESS OTHERWISE SHOWN

Products also Registered

Product	Approval holder	MAPP No.	Expiry Date
esfenvalerate			
Clayton Slalom	Clayton	15054	31 Jul 2011
Clayton Vindicate	Clayton	15055	31 Jul 2011
Sven	Interfarm	14035	30 Apr 2012
ethanol			
Ethy-Gen II	Banana-Rite	13412	31 Aug 2019
Restrain Fuel	Restrain	14520	
ethofumesate			
Barclay Keeper 500 SC	Barclay	13430	28 Feb 2013
Stelga 500 SC	Novastar	14737	28 Feb 2013
ethofumesate + metamitron			
Galahad	Bayer CropScience	10727	
Goltix Plus	Aako	14461	
ethofumesate + metamitron + phenmedipham			
MAUK 540	Makhteshim	11545	
ethofumesate + phenmedipham			
Betosip Combi FL	Sipcam	14403	28 Feb 2013
Magic Tandem	Bayer CropScience	13925	28 Feb 2013
Thunder	Makhteshim	14031	28 Feb 2013
ethylene			
Biofresh Safestore	Biofresh	14579	
famoxadone + flusilazole			
Mandolin	DuPont	14521	
fatty acids			
Finalsan	Certis	13102	
Finalsan Moss Control for Lawns Concentrate	Vitax	13141	
Finalsan Weed Killer RTU	Vitax	13143	
NEU1170 H	Growing Success	12971	
Organic Moss Control for Lawns Concentrate	Growing Success	13140	
Organic Weed Killer RTU	Growing Success	13142	
Safers Insecticidal Soap	Woodstream	07197	
fenamidone + mancozeb			
Sonata	Bayer CropScience	11570	31 Dec 2011
fenazaquin			
Matador 200 SC	Margarita	11058	
fenbuconazole			
Indar 5 EW	Whelehan	09644	

EXPIRY DATE IS 31/12/2021 UNLESS OTHERWISE SHOWN

Products also Registered

Product	Approval holder	MAPP No.	Expiry Date
fenbuconazole + propiconazole			
Graphic	Interfarm	10987	
fenhexamid			
Druid	AgriGuard	13901	31 May 2011
RouteOne Fenhex 50	Albaugh UK	13665	31 May 2011
fenoxaprop-P-ethyl			
Cheetah Energy	Bayer CropScience	13599	
fenpropidin			
Cleancrop Fulmar	United Agri	12033	
Instinct	Headland	12317	
Landgold Fenpropidin 750	Landgold	08973	31 May 2012
Mallard	Syngenta	08662	
Patrol	Syngenta	10531	
fenpropidin + prochloraz + tebuconazole			
Artemis	Makhteshim	14178	
fenpropidin + propiconazole			
Prophet	Syngenta	08433	
fenpropimorph			
Cleancrop Fenpro	United Agri	09885	
Cleancrop Fenpropimorph	United Agri	09445	
Keetak	BASF	06950	
Landgold Fenpropimorph 750	Landgold	10472	31 May 2012
Landgold Fenpropimorph 750	Teliton	12118	31 Mar 2013
Marnoch Phorm	Me2	11087	
Propimorf 750	Goldengrass	14410	
fenpropimorph + kresoxim-methyl			
Cleancrop Duster	United Agri	12306	30 Apr 2011
Greencrop Monsoon	Greencrop	09573	30 Apr 2011
Landgold Strobilurin KF	Landgold	09196	30 Apr 2011
Standon Kresoxim FM	Standon	08922	30 Apr 2011
fenpropimorph + pyraclostrobin			
BAS 528 00f	BASF	11444	
Jemker	BASF	12225	
fenpropimorph + quinoxyfen			
Jackdraw	BASF	13769	
fenpyroximate			
NNI-850 5 SC	Certis	12658	
ferric phosphate			
Ferramol	Certis	12274	31 Oct 2011

EXPIRY DATE IS 31/12/2021 UNLESS OTHERWISE SHOWN

Products also Registered

Product	Approval holder	MAPP No.	Expiry Date
Ferrox	Neudorff	14356	31 Oct 2011
Growing Success Advanced Slug Killer	Growing Success	12378	31 Oct 2011
NEU 1181 M	Neudorff	14355	31 Oct 2011
NEU 1185	Neudorff	14736	31 Oct 2011
ferrous sulphate			
Aitken's Lawn Sand	Aitken	05253	
Maxicrop Moss Killer & Conditioner	Maxicrop	04635	
Moss Control Plus Lawn Fertilizer	Miracle	07912	
No More Moss Lawn Feed	Wolf	10754	
Pentagon Prestige Lawn Sand	Sinclair	10456	31 Aug 2011
Scotts Moss Control	Scotts	12928	
Vitax Microgran 2	Vitax	04541	31 Aug 2011
Vitax Turf Tonic	Vitax	04354	31 Aug 2011
flazasulfuron			
Chikara	Nomix Enviro	13775	31 May 2014
flonicamid			
RouteOne Ski	Albaugh UK	13777	31 May 2014
florasulam			
Primus 25 SC	Dow	10175	30 Sep 2012
RouteOne Florasul 50	Albaugh UK	13668	30 Sep 2012
Spitfire Solo	Dow	14756	30 Sep 2012
florasulam + fluroxypyr			
Image	Dow	13771	31 Dec 2011
Spitfire	Dow	15101	20 Sep 2014
fluazifop-P-butyl			
Clayton Crowe	Clayton	12572	
Clean Crop Clifford	United Agri	13597	
Fluazifop +	RouteOne	13382	29 Feb 2012
Fusilade 250 EW	Syngenta	10525	
PP 007	Syngenta	10533	
RouteOne Fluazifop +	Albaugh UK	13641	
Wizzard	Syngenta	10539	
fluazinam			
Alpha Fluazinam 50SC	Makhteshim	13622	
Barclay Cobbler	Barclay	11348	
Blizzard	AgriGuard	14363	28 Feb 2011
Blizzard	AgriGuard	12854	
Boyano	Hermoo	13849	28 Feb 2011
Boyano	Belchim	14571	
Euro Zinam 500	Euro	14858	

EXPIRY DATE IS 31/12/2021 UNLESS OTHERWISE SHOWN

Products also Registered

Product	Approval holder	MAPP No.	Expiry Date
Float	Headland	14950	
Fluazinam 500	Goldengrass	14411	
FOLY 500 SC	Aako	14920	
Ibiza 500	Belchim	14393	
Ibiza 500	Q-Chem	14002	28 Feb 2013
Landgold Fluazinam	Landgold	08060	31 May 2012
Landgold Fluazinam	Teliton	12119	31 Mar 2013
Legacy	ISK Biosciences	09966	31 Oct 2011
Legacy	Syngenta	13735	
Shirlan Programme	Syngenta	10574	
Standon Fluazinam 500	Standon	08670	
fluazinam + metalaxyl-M			
Fluazmet M	RouteOne	13519	29 Feb 2012
RouteOne Fluazmet M	Albaugh UK	13650	
fludioxonil			
Beret Gold	BASF	12656	
fludioxonil + tefluthrin			
Austral Plus	Bayer CropScience	13376	
flufenacet + isoxaflutole			
RouteOne Oxanet 481	Albaugh UK	13828	30 Sep 2013
fluoxastrobin			
Bayer UK 831	Bayer CropScience	12091	31 Jul 2018
fluoxastrobin + prothioconazole			
Firefly	Bayer CropScience	13692	19 Jan 2014
Redigo Twin	Bayer CropScience	12314	30 Jul 2011
fluoxastrobin + prothioconazole + tebuconazole			
Scenic	Bayer CropScience	12289	31 Aug 2011
flupyrsulfuron-methyl			
DP 459	DuPont	12996	30 Jun 2011
fluquinconazole			
Diablo	BASF	11688	
Jockey F	BASF	11690	30 Nov 2012
Sahara	Bayer CropScience	11905	
Triplex	BASF	12443	
fluquinconazole + prochloraz			
Baron	BASF	11686	
fluroxypyr			
Agrostar	Agro Trade	14925	31 Dec 2011
Agrotech Fluroxypyr	Agrotech-Trading	12294	31 Dec 2011

EXPIRY DATE IS 31/12/2021 UNLESS OTHERWISE SHOWN

Products also Registered

Product	Approval holder	MAPP No.	Expiry Date
Alpha Fluroxypyr 20EC	Makhteshim	13352	31 Dec 2011
Awac	Globachem	13296	31 Dec 2011
Cascade	Chem-Wise	13939	31 Dec 2011
Cerastar	Cera Chem	13183	31 Dec 2011
Cerfix EC	Novastar	15117	06 Oct 2014
Clean Crop Gallifrey 200	Globachem	13813	31 Dec 2011
Floxy	Agriphar	13367	31 Dec 2011
Floxy	Chimac-Agriphar	12984	31 Dec 2011
Flurox 180	Stockton	13959	30 Dec 2011
Flurox 200	Stockton	13938	31 Dec 2011
Forban EC	Novastar	14862	15 Feb 2014
GAL-GONE	Globachem	13821	31 Dec 2011
Hatchet 180EC	AgriGuard	14054	31 Dec 2011
Hatchet 200 EC	AgriGuard	15097	31 Dec 2011
Landgold Fluroxypyr	Teliton	13356	31 Dec 2011
Milentus Fluroxypyr	Milentus	12266	31 Dec 2011
NUB-010	Nufarm UK	13829	31 Dec 2011
Standon Homerun	Standon	13177	31 Dec 2011
Taipan	Cera Chem	14155	31 Dec 2011

flutolanil

NNF-136	Nihon Nohyaku	13120	31 Dec 2012
NNF-136	Nihon	14302	
Rhino	Certis	13101	31 Dec 2012

flutriafol

Impact	Headland	12776	

fosetyl-aluminium

Routeone Fosetyl 80	Albaugh UK	13625	31 Oct 2012

fuberidazole + imidacloprid + triadimenol

Baytan Secur	Bayer CropScience	11253	28 Feb 2011

fuberidazole + triadimenol

Baytan Flowable	Makhteshim	11714	31 Jan 2012

gibberellins

Berelex	Interfarm	08903	31 Aug 2011
Gibb Plus 10 SL	Q-Chem	14968	
Regulex	Sumitomo	10147	31 Aug 2011
Regulex 10 SG	Sumitomo	13323	31 Aug 2011

gliocladium catenulatum

Prestop	Fargro	15103	31 Mar 2015

glufosinate-ammonium

Basta	Bayer CropScience	13820	

EXPIRY DATE IS 31/12/2021 UNLESS OTHERWISE SHOWN

Products also Registered

Product	Approval holder	MAPP No.	Expiry Date
Challenge	Bayer CropScience	07306	
Challenge 60	Fargro	08236	
glyphosate			
Accelerate	Headland	13390	30 Jun 2012
Acomac	CP AGRO	14254	30 Jun 2012
Acrion	Bayer Environ.	12677	30 Jun 2012
Amenity Glyphosate 360	Monsanto	13000	30 Jun 2012
Azural	Monsanto	14361	30 Jun 2012
Azural	Cardel	12666	30 Jun 2012
Barclay Garryowen	Barclay	12715	30 Jun 2012
Cleancrop Hoedown	United Agri	12913	30 Jun 2012
Cleancrop Tungsten	United Agri	13049	30 Jun 2012
Clinic	Nufarm UK	12678	30 Jun 2012
Clipper	Makhteshim	13217	31 Jan 2012
Ecoplug Max	Monsanto	14741	30 Jun 2012
Emrald Gly 360	Emrald	15078	30 Jun 2012
Euro Glyfo 360	Euro	14691	30 Jun 2012
Euro Glyfo 450	Euro	14936	30 Jun 2012
Frontierland Glyphosate	Soyl	14638	31 Oct 2013
Frontsweep	Soyl	14700	30 Jun 2012
Gallup Hi- Aktiv Amenity	Barclay	12898	30 Jun 2012
Glisor 360	Menora	14974	07 Jul 2014
Glister	Sinon EU	12990	30 Jun 2012
Glycel	Excel	15068	30 Jun 2012
Glydate	Nufarm UK	12679	30 Jun 2012
Glyder 360	Agform	14006	30 Jun 2012
Glyfos Gold	Nomix Enviro	10570	30 Jun 2012
Glyfos Monte	Headland	15069	30 Jun 2012
Glymark	Nomix Enviro	12994	31 Jan 2012
Glymark	Nomix Enviro	13870	30 Jun 2012
Glyper	Interfarm	13105	30 Jun 2012
Glyper	SBM	14383	30 Jun 2012
GlyphoFIT 36 SL	GAT Micro	14777	30 Jun 2012
Glyphosate 360	SumiAgro	12680	30 Jun 2012
Glyphosate 360	Monsanto	12669	30 Jun 2012
Glypoo	AgChem Access	13787	30 Jun 2012
Glyweed	Sabero	14444	30 Jun 2012
Habitat	Barclay	12717	30 Jun 2012
HH-001	CP AGRO	14772	30 Jun 2012
Hi-Fosate	Hockley	15062	30 Jun 2012
Hilite	Nomix Enviro	12922	31 Jan 2012
HY-Gly 360	Agrichem	14919	30 Jun 2012
KN 540	Monsanto	12009	30 Jun 2012
Landgold Glyphosate 360	Goldengrass	13977	30 Jun 2012
MON 79351	Monsanto	12860	30 Jun 2012

EXPIRY DATE IS 31/12/2021 UNLESS OTHERWISE SHOWN

Products also Registered

Product	Approval holder	MAPP No.	Expiry Date
MON 79376	Monsanto	12664	30 Jun 2012
MON 79545	Monsanto	12663	30 Jun 2012
MON 79632	Monsanto	12582	28 Feb 2014
MON 79632	Monsanto	14841	30 Jun 2012
Montana	Sapec	14843	30 Jun 2012
Nomix G	Nomix Enviro	12921	31 Jan 2012
Nomix G	Nomix Enviro	13872	30 Jun 2012
Nomix Nova	Nomix Enviro	13873	30 Jun 2012
Nomix Nova	Nomix Enviro	12920	31 Jan 2012
Nomix Revenge	Nomix Enviro	12993	31 Jan 2012
Nomix Revenge	Nomix Enviro	13874	30 Jun 2012
Nufosate	Nufarm UK	12699	30 Jun 2012
Onslaught	Procam	14564	30 Jun 2012
Pontil 360	Mastra	12733	30 Jun 2012
Pontil 360	Barclay	14322	30 Jun 2012
Preline	Linemark	12756	30 Jun 2012
Rhizeup	Clayton	15043	30 Jun 2012
Romany	Greencrop	12681	30 Jun 2012
Rosate 36	Albaugh UK	14459	30 Jun 2012
Rosate 36	Albaugh UK	14104	30 Jun 2012
Rosate 36 SL	Albaugh UK	14866	13 May 2014
Roundup	Monsanto	12645	30 Jun 2012
Roundup Amenity	Monsanto	12672	30 Jun 2012
Roundup Biactive 3G	Monsanto	13409	30 Jun 2012
Roundup Biactive Dry	Monsanto	12646	30 Jun 2012
Roundup Express	Monsanto	12526	30 Jun 2012
Roundup Gold	Monsanto	10975	30 Jun 2012
Roundup Greenscape	Monsanto	12731	30 Jun 2012
Roundup Metro	Monsanto	14842	30 Jun 2012
Roundup Pro-Green	Monsanto	11907	30 Jun 2012
Roundup Provide	Monsanto	12953	30 Jun 2012
Roundup Rail	Monsanto	12671	30 Jun 2012
Roundup Ultimate	Monsanto	12774	30 Jun 2012
Roundup Ultra ST	Monsanto	12732	30 Jun 2012
RouteOne Glyphosate 360	Albaugh UK	14133	30 Jun 2012
RouteOne Glyphosate 360	Albaugh UK	14458	30 Jun 2012
Routeone Rosate 36	Albaugh UK	13889	30 Jun 2012
RouteOne Rosate 360	Albaugh UK	13659	30 Jun 2012
RouteOne Rosate 360	Albaugh UK	13661	30 Jun 2012
Scorpion	Cardel	12673	30 Jun 2012
Silvio	Rotam Agrochemical	14626	30 Jun 2012
Stirrup	Nomix Enviro	12923	31 Jan 2012
Task 360	Albaugh UK	14570	30 Jun 2012
Total 360 WDC	Extreme Green	14958	30 Jun 2012
Trustee Amenity	Barclay	12897	30 Jun 2012

EXPIRY DATE IS 31/12/2021 UNLESS OTHERWISE SHOWN

Products also Registered

Product	Approval holder	MAPP No.	Expiry Date
Trustee Elite	Barclay	12662	31 Jan 2012
Tumbleweed Pro-Active	Scotts	12729	30 Jun 2012
Typhoon 360	Makhteshim	14817	23 May 2014
Vival	Belchim	14550	30 Jun 2012
Wither 36	Amenity Tech	14867	30 Jun 2012
glyphosate + pyraflufen-ethyl			
Hammer	Scotts	15060	31 Oct 2011
Thunderbolt	Nichino	14102	20 Jul 2012
glyphosate + sulfosulfuron			
Nomix Blade	Nomix Enviro	13907	30 Jun 2012
Nomix Duplex	Nomix Enviro	14953	30 Jun 2012
imazalil			
Fungazil 100 SL	BASF	11762	31 Aug 2014
I.T. Imazalil	Inter-Trade	14166	04 Jun 2011
Magnate 100 SL	Makhteshim	11705	30 Sep 2013
Neozil 10 SL	Laboratorios Agrochem	13034	31 Dec 2011
Sphinx	BASF	11764	31 Dec 2011
imazalil + thiabendazole			
Storite Super	Syngenta	13135	31 Dec 2011
imazamox + pendimethalin			
Nirvana	BASF	13220	30 Jun 2013
imidacloprid			
Amuse	Sherriff Amenity	14379	
Gaucho	Bayer CropScience	11281	
Gaucho FS	Bayer CropScience	11282	
Imidasect	Biologic HC	12899	31 Jul 2011
Imidasect 60 FS	Bioscientific	14077	31 Jul 2011
Imidasect 60 FS	Bioscientific	14573	
Imidasect 70 WS	Biologic HC	14572	
Imidasect 70 WS	Biologic HC	13191	31 Jul 2011
Jive	Makhteshim	14514	
Mido 70% WDG	Sharda	14174	
Neptune	Makhteshim	12992	
Picus 600 FS	Headland	14598	30 Sep 2014
Picus 600 FS	Agrichem	15065	
imidacloprid + tebuconazole + triazoxide			
Raxil Secur	Bayer CropScience	11298	30 May 2011
4-indol-3-ylbutyric acid			
Seradix 1	Certis	10422	
Seradix 1	Certis	11330	
Seradix 2	Certis	11331	

EXPIRY DATE IS 31/12/2021 UNLESS OTHERWISE SHOWN

Products also Registered

Product	Approval holder	MAPP No.	Expiry Date
Seradix 2	Certis	10423	
Seradix 3	Certis	10424	
Seradix 3	Certis	11332	
4-indol-3-ylbutyric acid + 1-naphthylacetic acid			
Synergol	Certis	07386	
iodosulfuron-methyl-sodium + mesosulfuron-methyl			
Ocean	Goldengrass	14422	31 Dec 2013
Octavian	Bayer CropScience	14604	31 Dec 2013
RouteOne Seafarer	Albaugh UK	13654	31 Dec 2013
Teliton Ocean	Teliton	13163	31 Mar 2013
iodosulfuron-methyl-sodium + propoxycarbazone-sodium			
Caliban Duo	Headland	14283	31 Mar 2013
ipconazole			
Crusoe	Chemtura	13955	23 Apr 2012
iprodione			
MAC-Iprodione 255 SC	MAC	14781	31 Dec 2013
Mascot Rayzor	Rigby Taylor	14329	31 Dec 2013
Prime-Turf	ChemSource	14617	31 Jul 2014
isoxaben			
Flexidor	Dow	05121	
Flexidor 125	Dow	05104	
Gallery 125	Rigby Taylor	06889	
Knot Out	Vitax	05163	
lambda-cyhalothrin			
Barbarossa	Clayton	13510	03 Apr 2011
Clayton Zen	Clayton	13512	03 Apr 2011
CleanCrop Corsair	United Agri	14124	31 Dec 2011
Dalda 5	Globachem	13688	31 Dec 2011
Euro Lambda 100 CS	Euro	14794	31 Dec 2011
IT Lambda	I T Agro	13451	03 Apr 2011
IT Lambda	Inter-Trade	15119	31 Dec 2011
Judo 5CS	Syngenta	15106	31 Dec 2011
Komodo 10 EC	United Phosphorus	14703	31 Dec 2012
Lambdastar	Biologic HC	15090	31 Aug 2014
MAC-Lambda-Cyhalothrin 50 EC	MAC	14941	28 Jun 2012
Markate 50	Agrovista	13529	31 Dec 2011
Stealth	Syngenta	14551	31 Dec 2011
laminarin			
Vacciplant	Goemar	13260	31 Mar 2015

SECTION 3

EXPIRY DATE IS 31/12/2021 UNLESS OTHERWISE SHOWN

Products also Registered

Product	Approval holder	MAPP No.	Expiry Date
Lecanicillium lecanii			
Vertalec	Koppert	04781	30 Apr 2011
lenacil			
Venzar 80 WP	DuPont	09981	
linuron			
Aredios 45 SC	Novastar	14586	09 Oct 2012
Kaedyn 500	DAPT	14739	31 Dec 2013
Linurex 50SC	Makhteshim	14268	14 Oct 2012
Messidor 45 SC	Novastar	14710	09 Oct 2012
magnesium phosphide			
Detia Gas-Ex-B Forte	Detia Degesch	10661	
maleic hydrazide			
Cleancrop Malahide	United Agri	13629	22 Jul 2012
Fazor	Chemtura	13617	31 Dec 2013
Source II	Chiltern	13618	22 Jul 2012
mancozeb			
Agrizeb	Chimac-Agriphar	10980	31 Dec 2011
Agrizeb	Agriphar	13365	31 Dec 2011
Agrizeb	Agriphar	14770	30 Nov 2013
Beacon	AgriGuard	13368	31 Dec 2011
Cleancrop Mancozeb	United Agri	11193	31 Dec 2011
Cleancrop Mandrake	United Agri	12500	31 Dec 2011
Cleancrop Mandrake	United Agri	12500	31 Dec 2011
Cleancrop Mandrake	United Agri	14792	30 Jun 2016
Dithane 945	Interfarm	12585	31 Jul 2011
Dithane 945	Dow	12545	31 Jul 2011
Dithane Dry Flowable Neotec	Dow	14631	13 Sep 2013
Dithane Dry Flowable Newtec	Dow	12564	30 Sep 2011
Hotspur	AgriGuard	13449	31 Dec 2011
Karamate Dry Flo Newtec	Landseer	12691	31 Oct 2011
Laminator DG	Interfarm	11073	31 Dec 2011
Laminator WP	Interfarm	11071	31 Dec 2011
Manfil 75 WG	Indofil	14768	30 Jun 2016
Manfil 80 WP	Indofil	14766	26 Nov 2013
Manfil 80 WP	Indofil	12941	31 Dec 2011
Manfill 75 WG	Indofil	12419	31 Dec 2011
Manzate 75 WG	United Phosphorus	15052	18 Oct 2013
Manzate 75 WG	Headland	12070	31 Dec 2011
Micene 80	Sipcam	14622	02 Jul 2013
Micene DF	Sipcam	14705	13 Sep 2013
Penncozeb 80 WP	United Phosphorus	11622	31 Dec 2011
Penncozeb 80 WP	United Phosphorus	14580	31 Oct 2011

EXPIRY DATE IS 31/12/2021 UNLESS OTHERWISE SHOWN

Products also Registered

Product	Approval holder	MAPP No.	Expiry Date
Penncozeb WDG	Nufarm UK	12156	31 Dec 2011
SpudGun	AgriGuard	13447	31 Dec 2011
Yankee	AgriGuard	13905	31 Dec 2011
Zebra WDG	Headland	14875	30 Jun 2016
mancozeb + metalaxyl-M			
Fubol Gold WG	Syngenta	10184	31 Jul 2011
MetMan 680	RouteOne	13553	30 Jun 2011
RouteOne MetMan 680	Albaugh UK	13652	30 Jun 2011
mancozeb + propamocarb hydrochloride			
Tattoo	Bayer CropScience	07293	
mancozeb + zoxamide			
Electis 75 WG	Dow	11013	31 Dec 2011
Roxam 75 WG	Dow	11017	31 Dec 2011
Unikat 75 WG	Gowan	11018	31 Dec 2011
Unikat 75WG	Gowan	14200	
MCPA			
Agricorn 500 II	Nufarm UK	09155	31 Oct 2011
Agritox 50	Nufarm UK	07400	31 Oct 2011
Agritox Dry	Nufarm UK	10554	31 Oct 2011
Agroxone	Headland	09947	31 Oct 2011
Agroxone 40	Nufarm UK	14093	31 Oct 2011
Agroxone 40	Marks	10264	31 Oct 2011
Agroxone 75	Marks	09208	31 Oct 2011
Agroxone 75	Nufarm UK	14419	31 Oct 2011
Campbell's MCPA 50	Nufarm UK	00381	31 Oct 2011
Circium II	Nufarm UK	11801	31 Oct 2011
Dow MCPA Amine 50	Dow	14517	31 Oct 2011
Dow MCPA Amine 50	Dow	14911	18 Mar 2014
Go-Low Power	DuPont	13812	30 Apr 2016
Growell MCPA DMA 500	GroWell	12989	31 Oct 2011
Headland Spear	Headland	07115	31 Oct 2011
HY-MCPA	Agrichem	06293	31 Oct 2011
MCPA 25%	Nufarm UK	07998	31 Oct 2011
MCPA 500	Nufarm UK	08655	31 Oct 2011
Nufarm MCPA 750	Nufarm UK	11768	30 Apr 2011
Nufarm MCPA Amine 50	Nufarm UK	12046	31 Oct 2011
POL-MCPA 500 SL	Nufarm UK	14498	31 Oct 2011
POL-MCPA 500 SL	Zaklady	14528	31 Oct 2011
POL-MCPA 500 SL	Zaklady	14912	18 Mar 2014
Tasker 75	Headland	14913	30 Apr 2016
Tasker 75	Headland	10544	31 Oct 2011

EXPIRY DATE IS 31/12/2021 UNLESS OTHERWISE SHOWN

Products also Registered

Product	Approval holder	MAPP No.	Expiry Date
MCPA + MCPB			
Butoxone Plus	Marks	11080	31 Oct 2011
Impetus	Headland	11021	31 Oct 2011
Trifolex-Tra	BASF	10396	31 Oct 2011
Tropotox Plus	Nufarm UK	11142	31 Oct 2011
MCPA + mecoprop-P			
Cleanrun Pro	Scotts	15073	31 May 2014
Greenmaster Extra	Scotts	15075	31 May 2014
No More Weeds Lawn Feed	Wolf	10747	31 Oct 2011
Optica Combi	Marks	10118	31 Oct 2011
Optica Combi	Nufarm UK	14096	31 Oct 2011
MCPB			
Bellmac Straight	United Phosphorus	14448	30 Apr 2016
Butoxone	Headland	10501	31 Oct 2011
Tropotox	Nufarm UK	11141	31 Oct 2011
mecoprop-P			
Clenecorn Super	Nufarm UK	14628	31 May 2014
mecoprop-P + metsulfuron-methyl			
Headland Neptune	Headland	10230	
mesotrione			
RouteOne Trione 10	Albaugh UK	14765	30 Sep 2013
mesotrione + terbuthylazine			
RouteOne Mesot	Albaugh UK	13576	
metalaxyl-M			
Fongarid Gold	Syngenta	12547	30 Sep 2012
metaldehyde			
Agosto	Makhteshim	14171	31 Dec 2011
Allure	Chiltern	11089	
Antares	De Sangosse	12998	
Appeal	Chiltern	12713	
Aristo	De Sangosse	11797	
Ascari	Makhteshim	13991	
Attract	Chiltern	12712	
Blue Prince	Doff Portland	13061	30 Sep 2011
Brits	Doff Portland	11792	
Brits M	De Sangosse	12956	28 Feb 2011
CDP H	De Sangosse	14432	
CDP Minis	De Sangosse	12955	31 Dec 2011
Certis Deal 5	Certis	13488	
Certis Metaldehyde 5	Certis	13490	31 Mar 2013
Certis Metaldehyde 5	Certis	14457	

EXPIRY DATE IS 31/12/2021 UNLESS OTHERWISE SHOWN

Products also Registered

Product	Approval holder	MAPP No.	Expiry Date
Certis Red 5	Certis	14479	
Certis Red 5	Certis	13491	30 Apr 2013
Chiltern Hundreds	Chiltern	12710	
Clartex	De Sangosse	12935	
Clean Crop Jekyll D 5	Doff Portland	13060	30 Sep 2011
Cleancrop Hyde 2	Makhteshim	14042	
Cleancrop Hyde 2	United Agri	13199	
CleanCrop Jekyll 5	United Agri	14117	30 Jun 2013
CleanCrop Jekyll 5	United Agri	14534	
Cleancrop Jekyll D3	United Agri	14369	
Cleancrop Potent	United Agri	14228	
Cleancrop Tremolo	United Agri	14225	
Condor 5	Doff Portland	12484	30 Sep 2011
Duracoat Slug Killer	Doff Portland	13134	30 Sep 2011
Entice	De Sangosse	12933	28 Feb 2011
ESP W	De Sangosse	14429	
Gusto 4	Makhteshim	14165	
Helimax	De Sangosse	13108	
Helimax W	De Sangosse	14424	
Keel Over	De Sangosse	13110	
Keel Over W	De Sangosse	14431	
Limagri GR	SBM	13475	
Luxan 9363 Blue	Luxan	12391	31 Aug 2012
Luxan Red 5	Luxan	12390	31 Aug 2012
Luxan Trigger 5	Luxan	12389	31 Aug 2012
Lynx S	De Sangosse	13276	31 Dec 2011
MetaPads	Frunol Delicia	14658	
Metarex Amba	De Sangosse	13245	
Metarex Green	De Sangosse	13069	
Metarex RG	De Sangosse	13070	
Mifaslug 5	Luxan	12392	31 Aug 2012
Molotov	Chiltern	08295	
Molto	Makhteshim	14170	
Optimol	De Sangosse	12997	31 Dec 2011
Optimol XL	De Sangosse	12937	28 Feb 2011
Pesta	De Sangosse	11796	
Prego 4	Makhteshim	14168	31 Dec 2011
Regel Star	De Sangosse	13071	
Regel W	De Sangosse	14425	
Regiment	Certis	14338	
Sloggy	SBM	13373	
Slug Pellets	De Sangosse	12936	28 Feb 2011
Slugdown	Doff Portland	13063	30 Sep 2011
Slugdown 3	Doff Portland	14366	31 Aug 2013
Slug-Lentils	Frunol Delicia	13613	

SECTION 3

EXPIRY DATE IS 31/12/2021 UNLESS OTHERWISE SHOWN

Products also Registered

Product	Approval holder	MAPP No.	Expiry Date
Spinner	Certis	14127	
Stanza	Makhteshim	14169	
Steadfast	Doff Portland	13062	30 Sep 2011
Super-Flor 6% Metaldehyde Slug Killer Mini Pellets	CMI	11789	
Terminator	Chiltern	14229	
Tinto	Certis	14336	
Top Gun	Certis	14325	
Trigger 3	Certis	14304	
Trigger 5	Certis	14480	
Trigger 5	Certis	13493	30 Apr 2013
Unicrop 6% Mini Slug Pellets	Unicrop	11795	

metamitron

Product	Approval holder	MAPP No.	Expiry Date
Agriguard Metamitron	AgriGuard	09859	31 Aug 2011
Alpha Metamitron	Makhteshim	11081	
Alpha Metamitron 70 SC	Makhteshim	12868	
Barclay Seismic	Barclay	11378	31 Aug 2011
Barclay Seismic	Barclay	11377	31 Aug 2011
Bettix 70 WG	United Phosphorus	11019	31 Aug 2011
Defiant WG	United Phosphorus	12300	
Fernpath Haptol	AgriGuard	11951	31 Aug 2011
Fernpath Haptol Flo	AgriGuard	13549	31 Aug 2011
Goltix Compact	Aako	14863	
Homer	Makhteshim	11544	
IT Metamitron SC	I T Agro	13059	31 Aug 2011
IT Metamitron WG	I T Agro	12467	31 Aug 2011
Landgold Metamitron	Landgold	06287	31 Aug 2011
Landgold Metamitron	Teliton	12123	31 Aug 2011
Lektan	Makhteshim	11543	
Maymat 70 WDG	Global Crop Care	12477	31 Aug 2011
MM 70	Nufarm UK	11582	31 Aug 2011
MM70	Gharda	09490	31 Aug 2011
Standon Metamitron	Standon	07885	31 Aug 2011
Volcan	Sipcam	09295	31 Aug 2011

metam-sodium

Product	Approval holder	MAPP No.	Expiry Date
Fumetham	Chemical Nutrition	10047	31 Dec 2014
Metham Sodium 400	United Phosphorus	08051	31 Dec 2014
Sistan	Unicrop	01957	31 Dec 2014
Sistan 38	Unicrop	08646	31 Dec 2014
Vapam	Willmot Pertwee	09194	31 Dec 2014

metazachlor

Product	Approval holder	MAPP No.	Expiry Date
Agriguard Metazachlor	AgriGuard	10417	31 Jul 2011
Agrotech Metazachlor 500 SC	Agrotech-Trading	12454	
Barclay Metaza	Barclay	11359	31 Jul 2011

EXPIRY DATE IS 31/12/2021 UNLESS OTHERWISE SHOWN

Products also Registered

Product	Approval holder	MAPP No.	Expiry Date
Booty	Me2	10659	31 Jul 2011
Butey	Chem-Wise	14176	
Butisan S	BASF	00357	31 Jul 2011
Clayton Metazachlor	Clayton	09688	31 Jul 2011
Clean Crop MTZ 500	United Agri	09222	31 Jul 2011
Fuego	Makhteshim	11177	
Gharda Bonanza 500 SC	Nufarm UK	13628	31 Jul 2011
Landgold Metazachlor 50	Teliton	12124	31 Jul 2011
Landgold Metazachlor 50	Landgold	09726	31 Jul 2011
Landgold Metazachlor 50 SC	Teliton	12133	31 Mar 2013
Makila 500 SC	Novastar	14492	
Metarap 500 SC	Global Crop Care	12540	31 Jul 2011
Metaz 50	RouteOne	13467	29 Feb 2012
Metaz 500	Goldengrass	14412	31 Jul 2011
Metazachlore GL 500	Globachem	12528	
Noble Metazachlor	Barclay	12892	31 Jul 2011
Rapsan 500 Sc	Nufarm UK	12365	31 Jul 2011
RouteOne Metaz 50	Albaugh UK	13648	
Standon Metazachlor 50	Standon	05581	31 Jul 2011
Triton SC	Nuvaros	14047	31 Jul 2011

metazachlor + quinmerac

Clayton Mazarac	Clayton	11161	31 Jul 2011
Katamaran	BASF	11732	
Rapsan TDI	Q-Chem	13920	31 Jul 2013
Standon Metazachlor-Q	Standon	09676	31 Jul 2011
Syndicate	AgriGuard	12262	

metconazole

Caramba	BASF	10213	
Conzole	AgriGuard	13444	

methiocarb

Draza Wetex	Bayer CropScience	11271	
Exit Wetex	Bayer CropScience	11276	
Lupus	Bayer CropScience	11291	

1-methylcyclopropene

SmartFresh SmartTabs	Landseer	12684	31 Mar 2016

metiram

Polyram DF	BASF	08234	31 Dec 2011

metribuzin

AgriGuard Metribuzin	AgriGuard	14084	31 Jul 2013
Citation 70	United Phosphorus	09370	
Landgold Metribuzin WG	Teliton	12468	31 Mar 2013
Lexone 2	DuPont	13804	30 Apr 2012

EXPIRY DATE IS 31/12/2021 UNLESS OTHERWISE SHOWN

Products also Registered

Product	Approval holder	MAPP No.	Expiry Date
Lexone 2	DuPont	12051	30 Sep 2011
Milentus Metribuzin 70 WG	Milentus	12224	
Promor	Bayer CropScience	12502	
Python	Makhteshim	13802	
Python	Makhteshim	13802	
Rapture	AgriGuard	14536	
Sencorex WG	Bayer CropScience	11304	30 Nov 2013
Shotput	Makhteshim	11960	30 Sep 2011
metsulfuron-methyl			
Accurate	Headland	13224	30 Jun 2011
Alias	DuPont	13397	30 Jun 2011
Ally SX	DuPont	12059	30 Jun 2011
Cleancrop Mondial	United Agri	13353	30 Jun 2011
Landgold Metsulfuron	Teliton	13421	30 Jun 2011
Metro 20	AgriGuard	13496	30 Jun 2011
Metro Solo	AgriGuard	14067	25 Apr 2011
Minx PX	Nufarm UK	13337	30 Jun 2011
Pike	Nufarm UK	12746	30 Jun 2011
Pike PX	Nufarm UK	13249	30 Jun 2011
Revenge 20	Agform	14449	30 Jun 2011
Revenge 20 SG	Agform	13633	30 Jun 2011
RouteOne Romet 20	Albaugh UK	13657	30 Jun 2011
Savvy	Rotam Agrochemical	14266	30 Jun 2011
Savvy Clear	Rotam Agrochemical	14474	30 Jun 2011
Savvy Premium	Rotam Agrochemical	14782	15 Dec 2013
metsulfuron-methyl + thifensulfuron-methyl			
Accurate Extra	Headland	14167	30 Jun 2011
Choir	Nufarm UK	14081	30 Jun 2011
Ergon	Rotam Agrochemical	14382	30 Jun 2011
metsulfuron-methyl + tribenuron-methyl			
Ally Max SX	DuPont	12631	31 Aug 2011
Biplay SX	DuPont	12246	31 Aug 2011
DP 911 WSB	DuPont	09867	31 Aug 2011
Mettle	AgriGuard	14857	31 Aug 2011
Traton SX	DuPont	12270	31 Aug 2011
myclobutanil			
Robut 20	RouteOne	13437	29 Feb 2012
RouteOne Butanil 20	Albaugh UK	14773	
RouteOne Robut 20	Albaugh UK	13646	31 Oct 2012
Systhane 20 EW	Whelehan	09397	31 Mar 2013
Systhane 6 W	Dow	10808	31 Mar 2013

EXPIRY DATE IS 31/12/2021 UNLESS OTHERWISE SHOWN

Products also Registered

Product	Approval holder	MAPP No.	Expiry Date
1-naphthylacetic acid			
Tipoff	Hatrim	12292	
napropamide			
Associate	FMC	13933	
Banweed	United Phosphorus	09376	
Devrinol	United Phosphorus	09375	
Naprop	Globachem	13402	
nicosulfuron			
Euro Nico 40	Euro	14778	
Landgold Nicosulfuron	Teliton	13464	
Landgold Nicosulfuron	Goldengrass	13997	30 Nov 2012
Maizon	AgriGuard	13222	
RouteOne Strong 4	Albaugh UK	13830	
Samson	Syngenta	12141	
oxadiazon			
Festival	Bayer Environ.	14606	
Noble Oxadiazon	Barclay	12889	
paclobutrazol			
Carousel	Fine	13225	31 Aug 2011
penconazole			
RouteOne Pencon 10	Albaugh UK	13651	
pencycuron			
Me2 Penny	Me2	10369	
Monceren Flowable	Bayer CropScience	11425	
Tubercare 12.5 DS	AgriChem BV	12194	31 Dec 2011
pendimethalin			
Akolin 330 EC	Makhteshim	14005	31 Dec 2012
Akolin 330 EC	Aako	14282	31 Dec 2013
Blazer	Headland	14126	31 Dec 2011
Bunker	Makhteshim	13816	31 Dec 2013
Claymore	BASF	13441	31 Dec 2013
CleanCrop Stomp	United Agri	13443	31 Dec 2013
Pendimet 400	Goldengrass	14715	31 Dec 2013
Routeone Penthal 330	Albaugh UK	13768	31 Dec 2013
Shuttle	AgriGuard	14041	31 Jul 2013
Sovereign	BASF	13442	31 Dec 2013
pendimethalin + picolinafen			
Flight	BASF	12534	30 Sep 2012
Peniophora gigantea			
PG Suspension	Forest Research	11772	

EXPIRY DATE IS 31/12/2021 UNLESS OTHERWISE SHOWN

Products also Registered

Product	Approval holder	MAPP No.	Expiry Date
phenmedipham			
Beetup Flo	United Phosphorus	14328	31 Dec 2013
Betanal Flow	Bayer CropScience	13893	28 Feb 2015
Crotale	Sipcam	13909	28 Feb 2015
Galcon SC	Novastar	14607	31 Dec 2013
Pump	Makhteshim	14087	28 Feb 2015
picloram			
RouteOne Loram 24	Albaugh UK	13951	
Tordon 22K	Dow	13869	31 Mar 2013
picolinafen			
AC 900001	BASF	10714	30 Sep 2012
picoxystrobin			
Acanto	DuPont	13043	31 Dec 2013
pinoxaden			
Roaxe 100	RouteOne	13304	29 Feb 2012
RouteOne Roaxe 100	Albaugh UK	13655	30 Jun 2012
Symmetry	AgriGuard	13577	19 Sep 2011
pirimicarb			
Agrotech-Pirimicarb 50 WG	Agrotech-Trading	12269	31 Jul 2012
Cleancrop Miricide	United Agri	11776	31 Jul 2012
Landgold Pirimicarb 50	Goldengrass	14022	30 Nov 2012
Milentus Pirimicarb	Milentus	12268	31 Jul 2012
Piri 50	Euro	14956	31 Jul 2012
RouteOne Primro 50 WG	Albaugh UK	13644	31 Jul 2012
pirimiphos-methyl			
Actellic Smoke Generator No. 10	Syngenta	10448	
Ultrasect 50EC	AgriGuard	15030	
prochloraz			
Alpha Prochloraz 40 EC	Makhteshim	11002	
Barclay Eyetak 40	Barclay	11352	
Mirage 45 SC	Makhteshim	10894	
Octave	BASF	11741	
Piper	Nufarm UK	13793	30 Sep 2011
Prelude 20 LF	BASF	11904	
Sporgon 50 WP	Sylvan	08802	
Sportak 45 EW	BASF	11696	
Sportak 45 HF	BASF	11697	
Warbler	Nufarm UK	13389	30 Sep 2011
prochloraz + propiconazole			
Mambo	Headland Amenity	13387	

EXPIRY DATE IS 31/12/2021 UNLESS OTHERWISE SHOWN

Products also Registered

Product	Approval holder	MAPP No.	Expiry Date
prochloraz + tebuconazole			
Astute	Sherriff Amenity	13825	31 Aug 2011
Grail	Nufarm UK	13385	30 Nov 2011
propamocarb hydrochloride			
Edito	Agriphar	14869	30 Sep 2017
Previcur N	Bayer CropScience	08575	
propaquizafop			
Agil	Makhteshim	11048	31 Jul 2013
Agil	Makhteshim	14562	
Alpha Propaquizafop 100 EC	Makhteshim	13689	31 Jul 2013
Alpha Propaquizafop 100 EC	Makhteshim	14563	
Barclay Rebel	Barclay	11648	
Bulldog	Makhteshim	11723	31 Jul 2013
Cleancrop GYR	Makhteshim	10646	30 Nov 2011
Falcon	Makhteshim	14552	
Landgold PQF 100	Landgold	10763	31 May 2012
Landgold PQF 100	Teliton	12127	31 Mar 2013
Noble Propaquizafop	Barclay	12890	
Osprey	Makhteshim	13731	
Raptor	Makhteshim	11092	31 Jul 2013
RouteOne Quizro 10	Albaugh UK	13643	
Shogun	Makhteshim	10584	30 Jun 2013
propoxycarbazone-sodium			
Attribut	Bayer CropScience	12730	30 Nov 2013
propyzamide			
Artax Flo	Bayer	15051	13 Sep 2013
Barclay Propyz	Barclay	15083	18 Aug 2014
Dennis	Interfarm	15081	31 Mar 2014
Edge 400	Albaugh UK	14683	14 Sep 2013
Emrald Prop 400 FL	Emrald	15079	14 Sep 2013
Emrald Prop 800 WG	Emrald	15077	31 Mar 2014
Flomide 2	Interfarm	15080	14 Sep 2013
KeMiChem - Propyzamide 400 SC	KeMiChem	14706	31 Mar 2014
Kerb Pro Granules	Barclay	14280	05 Oct 2013
MAC-Propyzamide 400 SC	MAC	14786	31 Mar 2014
Propyzamide Flo	Albaugh UK	14732	14 Sep 2013
RouteOne Zamide Flo	Albaugh UK	14559	31 Mar 2014
Verdah 400	DAPT	14680	31 Mar 2014
Verge 400	Albaugh UK	14682	14 Sep 2013
Zamide 80 WG	Albaugh UK	14723	31 Mar 2014
Zamide Flo	Albaugh UK	14679	14 Sep 2013

EXPIRY DATE IS 31/12/2021 UNLESS OTHERWISE SHOWN

Products also Registered

Product	Approval holder	MAPP No.	Expiry Date
prosulfocarb			
A8545G	Syngenta	14796	
Arcade	Syngenta	12641	
Auros	Syngenta	15049	
RouteOne Rosulfocarb 80	Albaugh UK	13635	
Sincere	AgriGuard	14947	
prothioconazole			
Banguy De	PSI	13575	31 Jul 2011
Proline	Bayer CropScience	12084	31 Jul 2018
Prothro 25	RouteOne	13392	31 Jul 2011
RouteOne Prothro 25	Albaugh UK	13658	31 Jul 2011
prothioconazole + spiroxamine			
RouteOne Prothioxamine	Albaugh UK	13656	31 Jul 2011
RouteOne Prothioxamine	Albaugh UK	14904	27 Jan 2014
Pseudomonas chlororaphis MA 342			
Cerall	Chemtura	14546	11 Jun 2013
pyraclostrobin			
BAS 500 06F	BASF	12338	31 May 2014
BASF Insignia	BASF	11900	31 May 2014
Comet	BASF	10875	31 May 2014
Insignia	Vitax	11865	31 May 2014
LEY	BASF	14774	31 May 2014
Platoon	BASF	12325	31 May 2014
pyraflufen-ethyl			
OS159	Ceres	12723	31 Oct 2011
pyrimethanil			
Pyrus 400 SC	Agriphar	13286	31 May 2012
RouteOne Pyrimet 40	Albaugh UK	13574	
Standon Pyrimethanil	Standon	10576	
pyroxsulam			
Avocet	Dow	14829	05 Aug 2012
GF 1274	Dow	14089	31 Jan 2012
quinoxyfen			
Apres	Dow	08881	01 Sep 2014
Clean Crop QFN	United Agri	11966	31 Mar 2013
Erysto	Dow	08697	31 Mar 2013
quizalofop-P-ethyl			
Sceptre	Bayer CropScience	08043	30 Nov 2011

EXPIRY DATE IS 31/12/2021 UNLESS OTHERWISE SHOWN

Products also Registered

Product	Approval holder	MAPP No.	Expiry Date
quizalofop-P-tefuryl			
Rango	Certis	14039	
rimsulfuron			
Agrotech-Rimsulfuron 25 WG	Agrotech-Trading	12427	31 Jul 2012
RouteOne Empera 25	Albaugh UK	13823	31 Jul 2012
Titus	Makhteshim	15050	31 Jan 2017
rotenone			
Devcol Liquid Derris	Nehra	06063	31 Oct 2011
Liquid Derris	Law Fertilisers	01213	31 Oct 2011
Liquid Derris	Law Fertilisers	13370	31 Oct 2011
spiroxamine			
Torch Extra	Bayer CropScience	12140	31 Dec 2011
Zenon	Bayer CropScience	11232	31 Dec 2011
spiroxamine + tebuconazole			
Draco	Bayer CropScience	11267	31 Aug 2011
Thyme	Bayer CropScience	14745	
sulfuryl fluoride			
Profume	Dow	12035	30 Apr 2013
sulphur			
Cosavet DF	Headland	11477	
Headland Venus	Headland	10611	
Venus	Headland	11856	
tau-fluvalinate			
Alpha Tau-Fluvalinate 240 EW	Makhteshim	13605	
tebuconazole			
Agrotech-Tebuconazole 250 EW	Agrotech-Trading	12523	
Aurigen	Syngenta	12043	31 Aug 2011
Barclay Busker	Barclay	11345	31 Aug 2011
Bayer UK226	Bayer CropScience	12412	31 Aug 2011
Bezel	Bayer CropScience	12753	
CHA 1640	Headland	14583	
Euro Tebu 250	Euro	14754	
Flail	Headland	12728	
Manicure	Chem-Wise	13906	
Me2 Tebuconazole	Me2	09751	31 Aug 2011
Noble Tebuconazole	Barclay	12874	31 Aug 2011
Odin	Rotam Agrochemical	13468	
Patriarch	AgriGuard	13979	31 Aug 2011
Raxil	Bayer CropScience	11295	31 Aug 2011
RouteOne Tebbro 25	RouteOne	13647	
Savannah	Rotam Agrochemical	14821	

EXPIRY DATE IS 31/12/2021 UNLESS OTHERWISE SHOWN

Products also Registered

Product	Approval holder	MAPP No.	Expiry Date
Starpro	Rotam Agrochemical	14823	
Toledo	Rotam Agrochemical	14036	
tebuconazole + triadimenol			
Veto F	Bayer CropScience	11317	30 Nov 2012
Veto F	Bayer CropScience	14274	30 Nov 2013
tebuconazole + triazoxide			
Raxil S	Bayer CropScience	11296	30 May 2011
tebuconazole + trifloxystrobin			
Mascot Fission	Rigby Taylor	14114	
Mascot Indicate	Rigby Taylor	14057	
Pandora	Pan Agriculture	14115	
tebufenpyrad			
Masai G	BASF	10224	
tefluthrin			
Force ST	Bayer CropScience	11671	
tepraloxydim			
Assimilate	AgriGuard	14860	31 May 2015
Landgold Tepraloxydim	Teliton	13379	31 May 2015
Landgold Tepraloxydim	Goldengrass	14026	31 Aug 2012
Omara	Chem-Wise	13682	31 May 2015
RouteOne Tepralox 50	Albaugh UK	13667	31 May 2015
tetraconazole			
Domark	Isagro	13048	
Juggler	Sipcam	13077	
thiacloprid			
Biscaya	Bayer CropScience	12471	31 Jul 2014
thiamethoxam			
Cruiser SB	Syngenta	15012	12 Jul 2014
thifensulfuron-methyl			
Harmony SX	DuPont	12181	30 Jun 2012
Prospect SX	DuPont	12212	30 Jun 2012
thifensulfuron-methyl + tribenuron-methyl			
Calibre	DuPont	07795	31 Aug 2011
Calibre SX	DuPont	12241	31 Jan 2012
thiram			
Thianosan DG	Unicrop	13404	
tolclofos-methyl			
Rizolex	Certis	09673	31 Jul 2012

EXPIRY DATE IS 31/12/2021 UNLESS OTHERWISE SHOWN

Products also Registered

Product	Approval holder	MAPP No.	Expiry Date
Rizolex 50 WP	Interfarm	13592	31 Jul 2012
Rizolex 50 WP	Interfarm	14217	31 Jul 2012
Rizolex Flowable	Interfarm	14207	31 Jul 2012
tralkoxydim			
Alpha Tralkoxydim 250 SC	Makhteshim	12884	30 Apr 2011
Greencrop Gweedore	Greencrop	09882	30 Apr 2011
Landgold Tralkoxydim	Teliton	12130	30 Apr 2011
Landgold Tralkoxydim	Landgold	08604	30 Apr 2011
Standon Tralkoxydim	Standon	09579	30 Apr 2011
triadimenol			
Bayfidan	Bayer CropScience	11417	
Hi-Shot	BASF	10397	
Spinnaker	BASF	10398	
tri-allate			
Avadex BW Granular	Gowan	12104	
tribenuron-methyl			
Helmstar 75 WG	Helm	15070	28 Feb 2016
Nuance	Headland	14813	28 Feb 2016
Quantum	DuPont	06270	
Quantum PX	DuPont	10843	31 Aug 2011
Quantum SX	DuPont	12239	
Triad	Headland	12751	
triclopyr			
Altix 240 EC	Agriphar	13362	
Cleancrop Triptic 48 EC	United Agri	11568	
Cleancrop Unival	United Agri	11388	
Garlon 4	Dow	05090	
Nomix Garlon 4	Nomix Enviro	13868	
Nomix Garlon 4	Nomix Enviro	12081	31 Jan 2012
trifloxystrobin			
Flint	Bayer CropScience	11259	30 Sep 2013
Pan Tees	Pan Agriculture	12894	30 Sep 2013
Twist	Bayer CropScience	11230	30 Sep 2013
Twist 500 SC	Bayer CropScience	11231	30 Sep 2013
Zest SC	Bayer CropScience	11233	30 Sep 2013
triflusulfuron-methyl			
Debut WSB	DuPont	07809	
Landgold TFS 50	Landgold	08941	31 May 2012
Standon Triflusulfuron	Standon	09487	
trinexapac-ethyl			
Cleancrop Alatrin	United Agri	11805	31 Oct 2012

SECTION 3

EXPIRY DATE IS 31/12/2021 UNLESS OTHERWISE SHOWN

Products also Registered

Product	Approval holder	MAPP No.	Expiry Date
Duet Max	ChemSource	14802	30 Jun 2014
Groslow	Pan Amenity	13694	30 Apr 2012
Landgold Tacet	Teliton	13036	31 Mar 2012
Peg	Goldengrass	14416	31 Oct 2012
Primo Maxx	Syngenta	13374	30 Apr 2012
RouteOne Rotrinex 25	RouteOne	13645	
Shortcut	Syngenta	13320	31 Oct 2012
Standon Pygmy	Standon	13001	31 Oct 2012

warfarin

Product	Approval holder	MAPP No.	Expiry Date
Grey Squirrel Bait	Killgerm	13020	31 Mar 2012

zeta-cypermethrin

Product	Approval holder	MAPP No.	Expiry Date
RouteOne Zeta 10	Albaugh UK	13608	

EXPIRY DATE IS 31/12/2021 UNLESS OTHERWISE SHOWN

SECTION 4
ADJUVANTS

Adjuvants

Adjuvants are not themselves classed as pesticides and there is considerable misunderstanding over the extent to which they are legally controlled under the Food and Environment Protection Act. An adjuvant is a substance other than water which enhances the effectiveness of a pesticide with which it is mixed. Consent C(i)5 under the Control of Pesticides Regulations allows that an adjuvant can be used with a pesticide only if that adjuvant is authorised and on a list published from time to time in *The Pesticides Register*. An authorised adjuvant has an *adjuvant number* and may have specific requirements about the circumstances in which it may be used.

Adjuvant product labels must be consulted for full details of authorised use, but the table below provides a summary of the label information to indicate the area of use of the adjuvant. Label precautions refer to the keys given in Appendix 4, and may include warnings about products harmful or dangerous to fish. The table includes all adjuvants notified by suppliers as available in 2010.

Product	Supplier	Adj. No.	Type
Abacus	De Sangosse	A0543	vegetable oil
Contains	53.43% esterified rapeseed oil, 20% w/w alcohol ethoxylate and 9.0% w/w natural fatty acids			
Use with	All approved pesticides on all edible crops when used at half their recommended dose or less, and on all non-edible crops up their full recommended dose. Also at a maximum concentration of 0.1% with approved pesticides on listed crops up to specified growth stages			
Protective clothing	A, C, H			
Precautions	R36, U05a, U11, U14, U19a, U20b, E15a, E19b, E34, D01, D02, D05, D10a, D12a, H04			
Activator 90	De Sangosse	A0547	non-ionic surfactant/wetter
Contains	750 g/l alcohol ethoxylates and 150 g/l natural fatty acids			
Use with	All approved pesticides on all edible crops when used at half their recommended dose or less, and on all non-edible crops up their full recommended dose. Also at a maximum concentration of 0.1% with approved pesticides on listed crops up to specified growth stages			
Protective clothing	A, C, H			
Precautions	R36, R38, R53a, U02a, U04a, U05a, U10, U11, U20b, E15a, E19b, D01, D02, D05, D10a, H04			
Adigor	Syngenta	A0522	wetter
Contains	47% w/w methylated rapeseed oil			
Use with	Topik, Axial, Trazos, Amazon and Viscount on cereals in accordance with recommendations on the respective herbicide labels			
Protective clothing	A, C, H			
Precautions	R43, R51, R53a, U02a, U05a, U09a, U20b, E15b, E38, D01, D02, D05, D09a, D10c, D12a, H04, H11			

SECTION 4

Adjuvants

Product	Supplier	Adj. No.	Type
Admix-P	De Sangosse	A0301	wetter
Contains	80% w/w polyalkylene oxide modified heptamethyltrisiloxane and a maximum of 20% w/w allyloxypolyethylene glycol methyl ether		
Use with	A wide range of pesticides applied as corm, tuber, onion and other bulb treatments in seed production and in seed potato treatment		
Protective clothing	A, C, H		
Precautions	R20, R21, R22a, R36, R43, R48, R51, R58, U11, U15, U19a, E15a, E19b, D01, D02, D05, D10a, D12a, H03, H11		
Agropen	Intracrop	A0227	vegetable oil
Contains	95% refined rapeseed oil		
Use with	Recommended rates of approved pesticides up to the growth stage indicated for specified crops, and with half or less than the recommended rate on these crops after the stated growth stages. Also for use with pesticides on grassland at half their recommended rate		
Protective clothing	A, C		
Precautions	U08, U20b, E13c, E37, D09a, D10b		
Amber	Interagro	A0367	vegetable oil
Contains	95% w/w methylated rapeseed oil		
Use with	Sugar beet herbicides, oilseed rape herbicides, cereal graminicides and a wide range of other pesticides that have a label recommendation for use with authorised adjuvant oils on specified crops. See label for details		
Protective clothing	A, C		
Precautions	U05a, U20b, E15a, D01, D02, D05, D09a, D10a		
Anoint	Intracrop	A0236	spreader/vegetable oil/wetter
Contains	95% rapeseed oil		
Use with	Pesticides that have a recommendation for the addition of a wetter/spreader		
Protective clothing	A, C		
Precautions	R36, R38, U05a, U08, U19a, U20b, E13c, E34, D01, D02, D09a, D10a, H04		
Arma	Interagro	A0306	penetrant
Contains	500 g/l alkoxylated fatty amine + 500 g/l polyoxyethylene monolaurate		
Use with	Cereal growth regulators, cereal herbicides, cereal fungicides, oilseed rape fungicides and a wide range of other pesticides on specified crops		
Protective clothing	A, C		
Precautions	R51, R58, U05a, E15a, E34, E37, D01, D02, D05, D09a, D10a, H11		
Asu-Flex	Greenaway	A0677	spreader/sticker/wetter
Contains	10.0% w/w rapeseed oil		
Use with	Asulox, Greencrop Found, I T Asulam, Inter Asulam and Spitfire		
Protective clothing	A, C		
Precautions	U11, U12, U15, U20b, E15a, E34, D02		

Adjuvants

Product	Supplier	Adj. No.	Type
BackRow	Interagro	A0472	mineral oil
Contains	\multicolumn{3}{l}{60% w/w refined paraffinic petroleum oil}		
Use with	\multicolumn{3}{l}{Pre-emergence herbicides. Refer to label or contact supplier for further details}		
Protective clothing	A		
Precautions	\multicolumn{3}{l}{R22b, R38, U02a, U05a, U08, U20b, E15a, D01, D02, D05, D09a, D10a, M05b, H03}		
Ballista	Interagro	A0524	spreader/wetter
Contains	10% w/w alkoxylated triglycerides		
Use with	All approved pesticides and plant growth regulators for use in winter and spring cereals applied at full rate up to and including GS 52, and at half the approved rate thereafter. May be applied with approved pesticides for other specified crops at full rate up to specified growth stages and at half rate thereafter		
Protective clothing	A		
Precautions	U05a, U20b, E15a, E34, D01, D02, D09a		
Bandrift Plus	Ciba Specialty	-	drift retardant
Contains	A non-ionic polyamide dispersed in oil		
Use with	A wide range of pesticides. See label for restrictions on use with wettable powders, suspension concentrates and water dispersible granules, and other usage limitations		
Protective clothing	A, C, M		
Precautions	U05a, U08, U20a, E13c, E34, D01, D02, D09a		
Banka	Interagro	A0245	spreader/wetter
Contains	29.2% w/w alkyl pyrrolidones		
Use with	Potato fungicides and a wide range of other pesticides on specified crops		
Protective clothing	A, C		
Precautions	R38, R41, R52, R58, U02a, U05a, U11, U19a, U20b, E15a, E34, E37, D01, D02, D05, D09a, H04		
Barramundi	Interagro	A0376	mineral oil
Contains	95% w/w mineral oil		
Use with	Sugar beet herbicides, oilseed rape herbicides, cereal graminicides and a wide range of other pesticides that have a label recommendation for use with authorised adjuvant oils on specified crops. Refer to label or contact supplier for further details		
Protective clothing	A		
Precautions	R22b, R38, U02a, U05a, U08, U20b, E15a, D01, D02, D05, D09a, D10a, M05b, H03		

SECTION 4

Adjuvants

Product	Supplier	Adj. No.	Type
Bazuka	Intracrop	A0241	spreader/wetter
Contains	800 g/l polyalkyleneoxide modified heptamethyltrisiloxane		
Use with	All fungicides applied to spring and winter sown wheat, spring and winter sown barley, spring and winter sown oats, triticale, rye and durum wheat where the pesticide label recommends the addition of wetters and spreaders		
Protective clothing	A, C		
Precautions	R21, R22a, R43, U02a, U05a, U08, U19a, U20b, E13c, E34, D01, D02, D09a, D10a, M03, H03, H04		
Binder	Nufarm UK	A0598	spreader/wetter
Contains	30% alkyl ethoxylate		
Use with	Recommended rates of approved pesticides up to the growth stage indicated for specified crops, and with half or less than the recommended rate on these crops after the stated growth stages		
Protective clothing	A		
Precautions	R41, R52, R53a, U02a, U05a, U08, U11, U13, U15, U20b, E34, E37, D01, D09a, M03, H04		
Bio Syl	Intracrop	A0385	spreader/sticker/wetter
Contains	1% w/w polyoxyethylen-alpha-methyl-omega-[3-(1,3,3,3-tetramethyl-3-trimethylsiloxy)-disiloxanyl] propylether and 32.67% w/w ethylene oxide condensate		
Use with	Recommended rates of approved pesticides on a range of specified fruit and vegetable crops and on managed amenity turf. See label for details		
Protective clothing	A, C		
Precautions	R36, R38, U05a, U08, U20c, E15a, D01, D02, D10a, H04		
Bioduo	Intracrop	A0606	wetter
Contains	85% alkyl polyglycol ether and fatty acids		
Use with	A wide range of pesticides used in grassland, agriculture and horticulture and with pesticides used in non-crop situations		
Protective clothing	A, C		
Precautions	R22a, R36, R38, U05a, U08, U20c, E13c, E34, D01, D02, D09a, D10a, M03, H03, H04, H08		
Biofilm	Intracrop	A0634	sticker/wetter
Contains	96% poly-1-p-menthene		
Use with	All approved fungicides and insecticides on edible crops and all approved formulations of glyphosate used pre-harvest on wheat, barley, oilseed rape, stubble, and in non-crop situations and grassland destruction		
Protective clothing	A, C		
Precautions	U19a, U20c, E13c, D09a, D11a		

Adjuvants

Product	Supplier	Adj. No.	Type
BioPower	Bayer CropScience	A0617	wetter
Contains	6.7% w/w 3,6-dioxaeicosylsulphate sodium salt and 20.2% w/w 3,6-dioxaoctadecylsulphate sodium salt		
Use with	Atlantis and all other approved cereal herbicides		
Protective clothing	A, C		
Precautions	R36, R38, U02a, U05a, U08, U13, U19a, U20b, E13c, E34, D01, D02, D05, D09a, D10a, H04		
Biothene	Intracrop	A0633	sticker/wetter
Contains	96% poly-1-p-menthene		
Use with	All approved fungicides and insecticides on edible crops up to 30 d before harvest, and all approved formulations of glyphosate used pre-harvest on wheat, barley, oilseed rape, stubble, and in non-crop situations and grassland destruction. Must not be used in mixture with adjuvant oils or surfactants		
Precautions	U19a, U20c, E13c, D09a, D11a		
Bond	De Sangosse	A0556	extender/sticker/wetter
Contains	900 g/l synthetic latex solution and 101 g/l alcohol alkoxylate		
Use with	All approved potato blight fungicides. Also with all approved pesticides on all edible crops when used at half their recommended dose or less, and on all non-edible crops up their full recommended dose. Also at a maximum concentration of 0.14% with approved pesticides on listed crops up to specified growth stages		
Protective clothing	A, C, H		
Precautions	R36, R38, U11, U14, U16b, U19a, E15a, E19b, D01, D02, D05, D10a, D12a, H04		
Bonser	Nufarm UK	A0715	spreader/wetter
Contains	258 g/l ammonium sulphate, 15.0 g/l modified soya lecithin, 15.0 g/l propionic acid and 4.0 g/l polyoxyethylene (5-7EO) C10-C15 primary alcohol		
Use with	Use with all approved formulations of Glyphosate and all approved pesticides as per label.		
Bonser	Nufarm UK	A0601	spreader/wetter
Contains	258 g/l ammonium sulphate, 15.0 g/l modified soya lecithin, 15.0 g/l propionic acid and 4.0 g/l polyoxyethylene (5-7EO) C10-C15 primary alcohol		
Use with	Credit DST (MAPP 13822) on specified crops, and with all approved pesticides on non-edible crops at up to their full dose, and with all approved pesticides on edible crops at up to half their approved dose		

SECTION 4

Adjuvants

Product	Supplier	Adj. No.	Type
Break-Thru S 240	PP Products	A0431	wetter
Contains	85% w/w polyalkylene oxide modified heptamethyl trisiloxane		
Use with	All approved pesticides which have a recommendation for use with a wetting agent applied to cereals, permanent grassland and rotational grass. Also with all approved pesticides at full rate on non-edible crops and on listed edible crops up to specified growth stages and on all edible crops when used at half or less of their approved full rate		
Protective clothing	A, C		
Precautions	R20, R21, R38, R41, R51, R53a, U11, U16b, U19a, E19b, D01, D02, D09a, D10a, M05a, H03, H11		
Broad-Flex	Greenaway	A0680	spreader/sticker/wetter
Contains	10.0% w/w rapeseed oil		
Use with	Broad Sword or Green Guard		
Protective clothing	A, C		
Precautions	U11, U12, U15, U20b, E15a, E34, D02		
Byo-Flex	Greenaway	A0545	sticker/wetter
Contains	10.0% w/w rapeseed oil		
Use with	GLY 490 (MAPP 12718)		
Protective clothing	A, C		
Precautions	U11, U12, U15, U20b, E34		
C-Cure	Interagro	A0467	mineral oil
Contains	60% w/w refined mineral oil		
Use with	Pre-emergence herbicides		
Protective clothing	A		
Precautions	R22b, R38, U02a, U05a, U08, U20b, E15a, D01, D02, D05, D09a, D10a, M05b, H03		
Ceres Platinum	Interagro	A0445	penetrant/spreader
Contains	500 g/l alkoxylated fatty amine + 500 g/l polyoxyethylene monolaurate		
Use with	Cereal growth regulators, cereal herbicides, cereal fungicides, oilseed rape fungicides and a wide range of other pesticides on specified crops		
Precautions	R51, R58, U05a, E15a, E34, D01, D02, D05, D09a, D10a, H11		
Cerround	Helena	A0510	adjuvant/vegetable oil
Contains	80% w/w methylated soybean oil and 15% w/w polyalkylene oxide modified heptamethyl trisiloxane		
Use with	Approved pesticides on non-edible crops, cereals and stubbles of all edible crops, grassland (destruction), and pesticides approved for use on growing edible crops when used at half recommended dose or less. On specified crops product may be used at a maximum spray concentration of 0.5% with approved pesticides at their full approved rate up to the growth stages shown in the label		
Protective clothing	A, H		
Precautions	R38, R41, U02a, U05a, U08, U11, U19a, U20b, E13c, E34, E37, D01, D02, D05, D09a, M03, H04		

Adjuvants

Product	Supplier	Adj. No.	Type
Codacide Oil	Microcide	A0629	vegetable oil
Contains	95% emulsified vegetable oil		
Use with	All approved pesticides and tank mixes. See label for details		
Protective clothing	A, C		
Precautions	U20b, D09a, D10b		
Companion PCT12	Ciba Specialty	A0482	spreader/sticker/wetter
Contains	25% w/w polyacrylamide		
Use with	Approved herbicides on cereals and any pesticide on non-food crops. See label for restrictions on use with wettable powders, suspension concentrates and water dispersible granules, and other usage limitations		
Precautions	U05a, U08, E13c, D01, D02, D09a		
Compliment	United Agri	A0578	spreader/vegetable oil/wetter
Contains	75% mixed fatty acid esters of rapeseed oil		
Use with	All approved pesticides in non-edible crops, all approved pesticides on edible crops when used at half or less their recommended rate, morpholine or triazine fungicides on cereals, and with all pesticides on listed crops up to specified growth stages		
Protective clothing	A, C		
Precautions	R43, R52, R53a, U14, E38, D01, D05, D07, D08, H04		
Conjoin	AgriGuard	A0676	adjuvant
Contains	47% w/w methylated rapeseed oil		
Use with	Amazon, Symmetry, Topik, Traxos and Viscount		
Protective clothing	A, C, H		
Precautions	R43, R50, R53a, U02a, U05a, U08, U20b, E13c, E15b, D01, D02, D09a, D10c, H04, H11		
Contact Plus	Interagro	A0418	mineral oil
Contains	95% w/w mineral oil		
Use with	Sugar beet herbicides, oilseed rape herbicides, cereal graminicides and a wide range of other pesticides that have a label recommendation for use with authorised adjuvant oils on specified crops. Refer to label or contact supplier for further details		
Protective clothing	A, C		
Precautions	R22b, R38, U02a, U05a, U08, U20b, E15a, D01, D02, D09a, D10a, M05b, H03		
County Mark	Greenaway	A0689	sticker/wetter
Contains	10% w/w rapeseed oil		
Use with	'Greenaway Gly-490' (MAPP 12718) and all approved 490 g/l glyphosate products		
Protective clothing	A, C		
Precautions	U11, U12, U15, U20b, E34		

SECTION 4

Adjuvants

Product	Supplier	Adj. No.	Type
Cropspray 11-E	De Sangosse	A0537	adjuvant/mineral oil
Contains	99% highly refined paraffinic oil		
Use with	All approved pesticides on all edible crops when used at half their recommended dose or less, and on all non-edible crops up to their full recommended dose. Also with listed herbicides on a range of specified crops and with all pesticides on a range of specified crops up to specified growth stages.		
Protective clothing	A, C		
Precautions	R22a, U10, U16b, U19a, E15a, E19b, E34, E37, D01, D02, D05, D10a, D12a, M05b, H03		
Designer	De Sangosse	A0322	drift retardant/extender/sticker/wetter
Contains	8.44% w/w organosilicone wetter and 50% w/w synthetic latex		
Use with	A wide range of fungicides, insecticides and trace elements for cereals and specified agricultural and horticultural crops		
Protective clothing	A, C, H		
Precautions	R36, R38, R52, U02a, U11, U14, U15, U19a, E15a, E37, D01, D02, D05, D09a, D10a, H04		
Desikote Max	Taminco	A0666	anti-transpirant/extender/sticker/wetter
Contains	400 g/l di-1-p-menthene		
Use with	Approved pesticides on all edible crops (except herbicides on peas) and as an anti-transpirant on vegetables before transplanting, evergreens, deciduous trees, shrubs, bushes and on turf		
Protective clothing	A		
Precautions	R38, R50, R53a, U05a, U14, U20b, E15a, E34, E37, D01, D09a, D10b, D12b, M03, H04, H11		
Diagor	AgChem Access	A0671	wetter
Contains	47% w/w methylated rapeseed oil		
Use with	Topik, Axial, Trazos, Amazon and Viscount on cereals in accordance with recommendations on the respective herbicide labels		
Protective clothing	A, C, H		
Precautions	R43, R51, R53a, U02a, U05a, U09a, U20b, E15b, E38, D01, D02, D05, D09a, D10c, D12a, H04, H11		
Drill	De Sangosse	A0544	adjuvant
Contains	63.34% w/w esterified rapeseed oil, 15% w/w alcohol ethoxylate and 7.5% w/w natural fatty acids		
Use with	All approved pesticides on all edible crops when used at half their recommended dose or less, and on all non-edible crops up to their full recommended dose. Also at a maximum concentration of 0.1% with approved pesticides on listed crops up to specified growth stages		
Protective clothing	A, C, H		
Precautions	R36, U05a, U11, U14, U19a, U20b, E15a, E19b, E34, D01, D02, D05, D10a, D12a, H04		

Adjuvants

Product	Supplier	Adj. No.	Type
Eco-flex	Greenaway	A0696	sticker/wetter
Contains	10% w/w refined rapeseed oil		
Use with	Approved formulations of glyphosate, 2,4-D		
Protective clothing	A, C		
Precautions	U11, U12, U15, U20b, E34		
Elan	Intracrop	A0392	spreader/sticker/wetter
Contains	82% w/w polyoxyethylen-alpha-methyl-omega-disiloxanyl] propylether		
Use with	Recommended rates of approved pesticides up to the growth stage indicated for specified crops, and with half or less than the recommended rate on these crops after the stated growth stages. Also for use with recommended rates of herbicides on managed amenity turf		
Protective clothing	A, C, H		
Precautions	R36, R38, U05a, U08, U20c, E13c, E15a, E37, D01, D02, D09a, D10a, H04		
Emerald	Intracrop	A0636	anti-transpirant/extender
Contains	96% di-1-p-menthene		
Use with	Recommended rates of approved pesticides up to the growth stage indicated for specified crops, and with half or less than the recommended rate on these crops after the stated growth stages. Also for use alone on transplants, turf, fruit crops, glasshouse crops and Christmas trees. Must not be used in mixture with adjuvant oils or surfactants		
Precautions	U19a, U20c, E13c, E37, D09a, D11a		
Esterol	Interagro	A0330	vegetable oil
Contains	95% w/w methylated rapeseed oil		
Use with	Plant protection products that have a label recommendation for use with adjuvant oils. Refer to label or contact supplier for further details		
Precautions	U05a, U20b, E15a, D01, D02, D09a, D10a		
Euroagkem Pace	Intracrop	A0562	spreader/wetter
Contains	42% w/w propionic acid		
Use with	All approved formulations of chlormequat, all approved formulations of glyphosate, diquat, fenoxaprop-P-ethyl, tralkoxydim, clodinafop-propargyl, fluazifop-P-butyl, cycloxydim and propaquizafop at half or less than half the approved pesticide rate in edible crops, at full rate in non-edible crops.		
Protective clothing	A, C		
Precautions	U02a, U04a, U05a, U08, U10, U11, U13, U14, U15, U19a, U20b, E15a, D01, D02, D09a, D10a, M05a, H05		
Euroagkem Pen-e-trate	Intracrop	A0564	spreader/wetter
Contains	350 g/l propionic acid		
Use with	All approved pesticides which have a recommendation for use with a wetting agent; pesticides must be used at half approved dose rate or less on edible crops, up to full dose rate on non-edible crops		
Protective clothing	A, C		
Precautions	U05a, U08, U11, U14, U15, U19a, U20c, E13e, E15a, D01, D02, D09a, D10b, M03, H05		

SECTION 4

Adjuvants

Product	Supplier	Adj. No.	Type
Felix	Intracrop	A0178	spreader/wetter
Contains	60% ethylene oxide condensate		
Use with	Mecoprop, 2,4-D in cereals and amenity turf, and a range of grass weedkillers in agriculture. See label for details		
Protective clothing	A, C		
Precautions	R36, R38, U05a, U08, U20a, E15a, D01, D02, D09a, D10a, H04, H08		
Firebrand	Barclay	-	fertiliser/water conditioner
Contains	500 g/l ammonium sulphate		
Use with	glyphosate		
Frigate	Unicrop	A0325	sticker/wetter
Contains	800 g/l tallow amine ethoxylate		
Use with	Roundup formulations		
Protective clothing	A, C		
Precautions	R22a, R23, R41, R51, R58, R67, U05a, U11, U19a, E38, D01, D02, D12a, M04a, H02, H08, H11		
Gateway	United Agri	A0587	extender/sticker/wetter
Contains	73% w/v synthetic latex solution and 8.5% w/v polyether modified trisiloxane		
Use with	All approved pesticides in crops not destined for human or animal consumption, all approved pesticides on edible crops when used at half or less their recommended rate, and with all pesticides on listed crops up to specified growth stages		
Protective clothing	A, C, H		
Precautions	R41, R52, R53a, U11, U15, E37, E38, D01, D05, D07, D08, H04		
Gladiator	De Sangosse	A0542	penetrant/wetter
Contains	210 g/kg alkoxylated tallow amine, 380 g/kg alcohol ethoxylates, 75 g/l natural fatty acids and 210 g/kg polyalkylene glycol		
Use with	All approved pesticides on all edible crops when used at half their recommended dose or less, and on all non-edible crops up their full recommended dose. Also at a maximum concentration of 0.1% with approved pesticides on listed crops up to specified growth stages		
Protective clothing	A, C, H		
Precautions	R36, R38, R41, R52, R58, U05a, U10, U11, U19a, E15a, E19b, E34, E37, D01, D02, D05, D10a, D12a, M03, H04		
Gly-Flex	Greenaway	A0588	spreader/sticker/wetter
Contains	95% refined rapeseed oil		
Use with	GLY-490 (MAPP 12718) or any approved formulations of 490 g/l glyphosate		
Protective clothing	A, C		
Precautions	R36, U11, U12, U15, U20b, E13c, E34, E37, H04		

Adjuvants

Product	Supplier	Adj. No.	Type
Gly-Plus A	Greenaway	A0377	spreader/sticker/wetter
Contains	95% refined rapeseed oil		
Use with	Any approved 360 g/l glyphosate		
Protective clothing	A, C		
Precautions	R36, U05a, U08, U11, U12, U15, U20b, E13c, E34, E37, E38, H04		
Green Gold	Intracrop	A0250	spreader/wetter
Contains	95% refined rapeseed oil		
Use with	All pesticides which have a recommendation for the addition of a wetter/spreader		
Precautions	U08, U20b, E13c, E34, D09a, D10b		
Greencrop Astra	Greencrop	A0417	adjuvant/penetrant/spreader
Contains	500 g/l alkoxylated fatty amine and 500 g/l polyoxylene monolaurate		
Use with	Approved pesticides on a range of specified arable crops, and in forestry, managed amenity turf, stubbles and non-crop areas and situations		
Precautions	U05a, E13c, E34, D01, D02, D05, D09a, D10b		
Greencrop Dolmen	Greencrop	A0419	adjuvant/spreader
Contains	64% w/w polyalkeneoxide modified heptamethyltrisiloxane		
Use with	Approved pesticides on a range of specified arable crops, and in forestry, managed amenity turf, stubbles and non-crop areas and situations		
Protective clothing	A, C, H		
Precautions	R21, R22a, R38, R41, R43, U02a, U05a, U08, U11, E13c, E34, D01, D02, D09a, D10b, M03, H03, H04		
Greencrop Fenit	Greencrop	A0602	wetter
Contains	47% methylated rapeseed oil		
Use with	Greencrop Helvick		
Precautions	R43, R50, R53a, U02a, U05a, U09a, U20b, E15b, D01, D02, D05, D09a, D10c, D12a, H04, H11		
Greencrop Rookie	Greencrop	A0515	wetter
Contains	20.2% w/w 3,6-dioxoctadecylsulphate sodium salt and 6.7% w/w 3,6-dioxaeicosylsulphate sodium salt		
Use with	All approved herbicides in cereals		
Protective clothing	A, C		
Precautions	R36, R38, R52, U02a, U05a, U08, U11, U13, U19a, E15a, D01, D02, D05, D09a, D10b, H04		
Grenadier	Intracrop	A0608	spreader/sticker/wetter
Contains	700 g/l polyoxyethylene (5-8 EO) C10-C15 primary alcohol and 150 g/l fatty acids		
Use with	All approved pesticides on edible and non-edible crops when used at half their approved dose or less.		
Protective clothing	A, C		

SECTION 4

Adjuvants

Product	Supplier	Adj. No.	Type
Grip	De Sangosse	A0558	extender/sticker/wetter

Contains: 900 g/l synthetic latex solution and 101 g/l alcohol alkoxylate

Use with: All approved potato blight fungicides. Also with all approved pesticides on all edible crops when used at half their recommended dose or less, and on all non-edible crops up their full recommended dose. Also at a maximum concentration of 0.14% with approved pesticides on listed crops up to specified growth stages

Protective clothing: A, C, H

Precautions: R36, R38, U11, U14, U16b, U19a, E15a, E19b, D01, D02, D05, D10a, D12a, H04

Product	Supplier	Adj. No.	Type
Grounded	Helena	A0456	mineral oil

Contains: 732 g/l refined paraffinic petroleum oil

Use with: All approved pesticides on edible and non-edible crops when used at half their approved dose or less. Also with approved pesticides on specified crops, up to specified growth stages, at up to their full approved dose

Protective clothing: A, C

Precautions: R53a, U02a, U05a, U08, U20b, E15a, E34, E37, D01, D02, D05, D09a, D10c, H11

Product	Supplier	Adj. No.	Type
Guide	De Sangosse	A0528	acidifier/drift retardant/penetrant

Contains: 35% w/w lecithin, 35% w/w propionic acid and 9.39% w/w alcohol ethoxylate

Use with: All approved pesticides on all edible crops when used at half their recommended dose or less, and on all non-edible crops up their full recommended dose. Also with morpholine fungicides on cereals and with pirimicarb on legumes when used at full dose; with iprodione on brassicas when used at 75% dose, and at a maximum concentration of 0.5% with all pesticides on listed crops at full dose up to specified growth stages

Protective clothing: A, C, H

Precautions: R36, R38, U11, U14, U15, U19a, E15a, E19b, D01, D02, D05, D10a, D12a, H04

Product	Supplier	Adj. No.	Type
Headland Fortune	Headland	A0703	penetrant/spreader/vegetable oil/wetter

Contains: 75% w/w mixed methylated fatty acid esters of seed oil and N-butanol

Use with: Herbicides and fungicides in a wide range of crops. See label for details

Protective clothing: A, C

Precautions: R43, U02a, U05a, U14, U20a, E15a, E34, E38, D01, D02, D05, D09a, D10b, H04

Product	Supplier	Adj. No.	Type
Headland Guard Pro	Headland Amenity	A0423	extender/sticker

Contains: 52% synthetic latex solution

Use with: All approved pesticides, micronutrients and sea-weed based plant growth stimulants in amenity grass, managed amenity turf and amenity vegetation

Precautions: U05a, U08, U19a, U20b, E13b, E34, E37, D01, D02, D05, D10b

Adjuvants

Product	Supplier	Adj. No.	Type
Headland Inflo XL	Headland Amenity	A0329	wetter
Contains	85% w/w polyalkylene oxide modified heptamethyl siloxane		
Use with	Any pesticide for use on amenity grassland or managed amenity turf (except any product applied in or near water) where the use of a wetting, spreading and penetrating surfactant is recommended to improve foliar coverage, and with any approved pesticide on crops not intended for human or animal consumption		
Protective clothing	A, C		
Precautions	R21, R22a, R38, R41, R43, R51, R58, U02a, U05a, U08, U14, U19a, U20b, E15a, E34, E38, D01, D02, D05, D09a, D10b, M03, H03, H11		
Headland Intake	Headland	A0074	penetrant
Contains	450 g/l propionic acid		
Use with	All approved pesticides on any crop not intended for human or animal consumption, and with all approved pesticides on beans, peas, edible podded peas, oilseed rape, linseed, sugar beet, cereals (except triazole fungicides), maize, Brussels sprouts, potatoes, cauliflowers. See label for detailed advice on timing on these crops		
Protective clothing	A, C		
Precautions	R34, U02a, U05a, U10, U11, U14, U15, U19a, E34, D01, D02, D05, D09b, D10b, M04a, H05		
Headland Rhino	Headland	A0328	wetter
Contains	85% w/w polyalkylene oxide modified heptamethyl siloxane		
Use with	Any herbicide, systemic fungicide, systemic insecticide or plant growth regulator (except any product applied in or near water) where the use of a wetting, spreading and penetrating surfactant is recommended to improve foliar coverage		
Protective clothing	A, C		
Precautions	R21, R22a, R38, R41, R43, R51, R58, U02a, U05a, U08, U11, U14, U19a, U20b, E13c, E34, D01, D02, D05, D09a, D10b, M03, H03, H11		
Impala	Intracrop	A0607	spreader/sticker/wetter
Contains	700 g/l polyoxyethylene (5-8 EO) C10-C15 primary alcohol and 150 g/l fatty acids		
Use with	All approved pesticides at half or less than half the approved pesticide rate on edible crops and all approved pesticides on non-edible crops		
Intracrop Agwet	Intracrop	A0616	sticker/wetter
Contains	900 g/l polyoxyethylene (5-8 EO) C10-C15 primary alcohol		
Use with	All approved pesticides at half or less than half the approved pesticide rate on edible crops and all approved pesticides on non-edible crops		
Intracrop Agwet Gtx	Intracrop	A0646	sticker/wetter
Contains	500 g/l polyoxyethylene (5-8 EO) C10-C15 primary alcohol		
Use with	All approved pesticides at half or less than half the approved pesticide rate on edible crops and all approved pesticides on non-edible crops		

SECTION 4

Adjuvants

Product	Supplier	Adj. No.	Type
Intracrop Archer	Intracrop	A0242	spreader/wetter
Contains	800 g/l polyalkyleneoxide modified heptamethyltrisiloxane		
Use with	All fungicides used in cereals where the addition of wetters/spreaders is recommended on the pesticide label		
Protective clothing	A, C		
Precautions	R21, R22a, R43, U02a, U05a, U08, U19a, U20b, E13c, E34, D01, D02, D09a, D10a, M03, H03, H04		
Intracrop BLA	Intracrop	A0453	extender/sticker
Contains	41.5% w/w styrene butadiene co-polymer solution		
Use with	All potato blight fungicides. Also with recommended rates of approved pesticides in certain non-crop situations and up to the growth stage indicated for specified crops, and with half or less than the recommended rate on these crops after the stated growth stages. Also with pesticides in grassland at half or less their recommended rate		
Precautions	U08, U20b, E13c, E34, E37, D01, D05, D09a, D10a		
Intracrop Bla-Tex	Intracrop	A0656	sticker
Contains	22.0% w/w styrene-butadiene co-polymer		
Use with	All approved pesticides at half or less than half the approved pesticide rate on edible crops, all approved pesticides on non edible crops and with all blight fungicides on potatoes		
Intracrop Boost	Intracrop	A0709	spreader/sticker/wetter
Contains	132.67% w/w Polyoxyethylene (5-8 EO) C10-C15 primary alcohol and 1.0% w/w polyoxyethylen-[-alpha-]-methyl-[-omega-]-[3-(1,3,3,3-tetramethyl-3-trimethylsiloxy)-disiloxanyl] propylether;		
Use with	All approved pesticides at half or less than half the approved pesticide rate on edible crops and all approved pesticides on non-edible crops		
Intracrop Cogent	Intracrop	A0707	spreader/sticker/wetter
Contains	32.67% w/w Polyoxyethylene (5-8 EO) C10-C15 primary alcohol and 1.0% w/w polyoxyethylen-[-alpha-]-methyl-[-omega-]-[3-(1,3,3,3-tetramethyl-3-trimethylsiloxy)-disiloxanyl] propylether		
Use with	All approved pesticides at half or less than half the approved pesticide rate on edible crops and all approved pesticides on non-edible crops		
Intracrop Dictate	Intracrop	A0673	adjuvant
Contains	91% w/w methylated rapeseed oil		
Use with	All approved pesticides on non-edible crops; all approved pesticides applied at half or less than half dose on edible crops		
Protective clothing	A, C		
Precautions	U19a, U20b, D05, D09a, D10b		
Intracrop Era	Intracrop	A0664	spreader/sticker/wetter
Contains	27.0% w/w polyoxyethylen-[-alpha-]-methyl-[-omega-]-[3-(1,3,3,3-tetramethyl-3-trimethylsiloxy)-disiloxanyl] propylether		
Use with	All approved pesticides at half or less than half the approved rate		
Protective clothing	A, C		
Precautions	U05a, U08, U20b, E13e, D01, D02, D09a, D10a		

Adjuvants

Product	Supplier	Adj. No.	Type
Intracrop Green Oil	Intracrop	A0262	wetter
Contains	95% refined rapeseed oil		
Use with	Recommended rates of approved pesticides up to the growth stage indicated for specified crops, and with half or less than the recommended rate on these crops after the stated growth stages. Also for use with pesticides on grassland at half their recommended rate		
Precautions	U08, U20b, E13c, E37, D09a, D10b		
Intracrop Impetus	Intracrop	A0647	spreader/wetter
Contains	50.0% w/w polyoxyethylene (5-8 EO) C10-C15 primary alcohol		
Use with			
Intracrop Inca	Intracrop	A0701	adjuvant
Contains	840 g/l methylated rapeseed oil		
Use with	All approved pesticides at half or less than half the approved pesticide rate on edible crops and all approved pesticides on non-edible crops		
Intracrop Incite	Intracrop	A0702	adjuvant
Contains	840 g/l methylated rapeseed oil		
Use with	All approved pesticides at half or less than half the approved pesticide rate on edible crops and all approved pesticides on non-edible crops		
Intracrop Iona	Intracrop	A0583	spreader/wetter
Contains	90.0% w/w polyoxyethylene (3EO) C12-C15 primary alcohol		
Use with	All approved pesticides at half their recommended dose rate on edible crops and all approved pesticides on non-edible crops. For edible crops see label for latest growth stages at application.		
Intracrop Iona Low Foam	Intracrop	A0582	spreader/wetter
Contains	89.8% w/w polyoxyethylene (3EO) C12-C15 primary alcohol		
Use with	All approved pesticides at half their recommended dose rate on edible crops and all approved pesticides on non-edible crops. For edible crops see label for latest growth stages at application.		
Intracrop Mica	Intracrop	A0663	spreader/sticker/wetter
Contains	27.0% w/w polyoxyethylen-[-alpha-]-methyl-[-omega-]-[3-(1,3,3,3-tetramethyl-3-trimethylsiloxy)-disiloxanyl] propylether		
Use with	All approved pesticides at half or less than half the approved rate		
Protective clothing	A, C		
Precautions	U05a, U08, U20c, E13e, D01, D02, D09a, D10a		
Intracrop Micotex	Intracrop	A0688	activator/spreader/wetter
Contains	25.0% w/w polyoxyethylene (12EO))/polyoxypropylene (30EO) block copolymer		
Use with	All approved pesticides on non-edible crops and non crop production uses; All approved pesticides at half or less than half the approved pesticide rate on edible crops		
Protective clothing	A, C		
Precautions	R36, R38, U05b, U08, U20c, E15a, D01, D02, D09a, D10b, H04		

SECTION 4

Adjuvants

Product	Supplier	Adj. No.	Type
Intracrop Neotex	Intracrop	A0460	extender/sticker
Contains	42.5% w/w styrene butadiene co-polymer solution		
Use with	Recommended rates of approved pesticides up to specified growth stages of a wide range of agricultural arable and horticultural crops, and for non-crop uses. Use on edible crops beyond specified growth stages, and in grass, should only be with half recommended rates of the pesticide or less. See label for details of growth stage restrictions. In addition may be used with all potato blight fungicides at their recommended rates of use up to the latest recommended timing of the fungicide		
Precautions	U08, U20b, E13c, E34, E37, D01, D05, D09a, D10a		
Intracrop Novatex	Intracrop	A0566	extender/sticker
Contains	42.5% w/w styrene butadiene co-polymer solution		
Use with	Recommended rates of approved pesticides up to the growth stage indicated for specified crops, and with half or less than the recommended rate on these crops after the stated growth stages. Also for use with recommended rates of pesticides on grassland and specified non-crop situations		
Precautions	U08, U20b, E13c, E34, E37, D01, D09a, D10a		
Intracrop Perm-E8	Intracrop	A0565	spreader/wetter
Contains	42% w/w propionic acid		
Use with	All approved formulations of chlormequat, all approved formulations of glyphosate, diquat, fenoxaprop-P-ethyl, tralkoxydim, clodinafop-propargyl, fluazifop-P-butyl, cycloxydim and propaquizafop at half or less than half the approved pesticide rate in edible crops, at full rate in non-edible crops.		
Protective clothing	A, C		
Precautions	U02a, U04a, U05a, U08, U10, U11, U13, U14, U15, U19a, U20b, D01, D02, D09a, D10a, H05		
Intracrop Predict	Intracrop	A0503	adjuvant/vegetable oil
Contains	91% w/w methylated rapeseed oil		
Use with	All approved pesticides on non-edible crops, and all pesticides approved for use on growing edible crops when used at half recommended dose or less. On specified crops product may be used at a maximum spray concentration of 1% with approved pesticides at their full approved rate up to the growth stages shown in the label		
Precautions	U19a, U20b, D05, D09a, D10b		
Intracrop Protocol	Intracrop	A0687	spreader/wetter
Contains	25.0% w/w polyoxyethylene (12E))/polyoxypropylene (30EO) block copolymer		
Use with	All approved pesticides on all non-edible crops and non crop production; all approved pesticides at half or less than half the approved pesticide rate on all edible crops		
Protective clothing	A, C		
Precautions	R36, R38, U05a, U08, U20c, E15a, D01, D02, D09a, D10b, H04		

Adjuvants

Product	Supplier	Adj. No.	Type
Intracrop Questor	Intracrop	A0495	activator/non-ionic surfactant/spreader
Contains	\multicolumn{3}{l}{75% polyoxyethylene polypropoxypropanol}		
Use with	All approved pesticides on non-edible crops and pesticides used in non-crop production, and all pesticides approved for use on growing edible crops when used at half recommended dose or less. On specified crops product may be used at a maximum spray concentration of 0.3% with approved pesticides at their full approved rate up to the growth stages shown in the label		
Protective clothing	A, C		
Precautions	R36, R38, U04a, U05a, U08, U19a, E13b, D01, D02, D09a, D10b, M03, M05a, H04		
Intracrop Rapeze	Intracrop	A0643	vegetable oil
Contains	842 g/l methylated rapeseed oil		
Use with			
Intracrop Rapide	Intracrop	A0563	penetrant/surfactant
Contains	40% propionic acid		
Use with	A wide range of pesticides and growth regulators, especially chlormequat		
Protective clothing	A, C		
Precautions	R34, R36, R38, U02a, U05a, U08, U19a, U20a, D01, D02, D09a, D10a, H04, H05		
Intracrop Rapide Beta	Intracrop	A0672	wetter
Contains	350 g/l propionic acid and 100 g/l polyoxyethylene (5-8 EO) C10-C15 primary alcohol		
Use with	All approved pesticides which have a recommendation for use with a wetting agent; pesticides must be used at half approved dose rate or less on edible crops, up to full dose rate on non-edible crops		
Protective clothing	A, C		
Precautions	U02a, U04a, U05a, U08, U10, U11, U13, U14, U15, U19a, U20b, E15a, D01, D02, D09a, D10a, M05a, H05		
Intracrop Retainer	Intracrop	A0508	adjuvant/vegetable oil
Contains	60% w/w methylated rapeseed oil		
Use with	All approved pesticides on non-edible crops, and all pesticides approved for use on growing edible crops when used at half recommended dose or less. On specified crops product may be used at a maximum spray concentration of 1% with approved pesticides at their full approved rate up to the growth stages shown in the label		
Precautions	U19a, U20b, D05, D09a, D10b		
Intracrop Retainer NF	Intracrop	A0711	adjuvant
Contains	91.0% w/w methylated rapeseed oil		
Use with	All approved pesticides at half or less than half the approved pesticide rate on edible crops and all approved pesticides on non-edible crops		

SECTION 4

Adjuvants

Product	Supplier	Adj. No.	Type
Intracrop Rigger	Intracrop	A0572	adjuvant/vegetable oil
Contains	\multicolumn{3}{l|}{91.7% w/w methylated rapeseed oil}		
Use with	\multicolumn{3}{l|}{All approved pesticides on non-edible crops, and all pesticides approved for use on growing edible crops when used at half recommended dose or less. On specified crops product may be used at a maximum spray concentration of 1.78% with approved pesticides at their full approved rate up to the growth stages shown in the label}		
Precautions	U19a, U20b, D05, D09a, D10b		
Intracrop Salute	Intracrop	A0683	spreader/sticker/wetter
Contains	27% w/w polyoxyethylen-methyl- [3-(1,3,3,3-tetramethyl-3-trimethylsiloxy) disiloxanyl] propylether		
Use with	All approved pesticides on non-edible crops and non crop production uses; All approved pesticides at half or less than half the approved pesticide rate on edible crops		
Protective clothing	A, C		
Precautions	U05a, U08, U20c, E13e, D01, D02, D09a, D10a		
Intracrop Sapper	Intracrop	A0684	spreader/sticker/wetter
Contains	27% w/w polyoxyethylen-methyl-[3-(1,3,3,3-tetramethyl- 3-trimethylsiloxy) disiloxanyl] propylether		
Use with	All approved pesticides on non-edible crops and non crop production uses; All approved pesticides at half or less than half the approved pesticide rate on edible crops		
Protective clothing	A, C		
Precautions	U05a, U08, U20c, E13c, D01, D02, D09a, D10a		
Intracrop Saturn	Intracrop	A0494	activator/non-ionic surfactant/spreader
Contains	75% polyoxyethylene polypropoxypropanol		
Use with	All approved pesticides on non-edible crops and pesticides used in non-crop production, and all pesticides approved for use on growing edible crops when used at half recommended dose or less. On specified crops product may be used at a maximum spray concentration of 0.3% with approved pesticides at their full approved rate up to the growth stages shown in the label		
Protective clothing	A, C		
Precautions	R36, R38, U04a, U05a, U08, U19a, E13b, D01, D02, D09a, D10b, M03, M05a, H04		
Intracrop Signal XL	Intracrop	A0659	extender/sticker
Contains	62.5% liquid concentrate formulation containing 22% w/w latex solids co-formulated with 20% alcohol ethoxylate		
Use with	All approved pesticides on non-edible crops and non crop production uses; All approved pesticides at half or less than half the approved pesticide rate on edible crops		
Protective clothing	A, C		
Precautions	U05a, U19a, U20b, E13c, E34, D01, D02, D09a, D10b, M04a, M05a		

Adjuvants

Product	Supplier	Adj. No.	Type
Intracrop Spread-n-wet	Intracrop	A0624	spreader/wetter
Contains	900 g/l polyoxyethylene (5-8 EO) C10-C15 primary alcohol		
Use with			
Intracrop Sprinter	Intracrop	A0513	spreader/wetter
Contains	192 g/l primary alcohol ethoxylate		
Use with	Recommended rates of approved pesticides up to the growth stage indicated for specified crops, and with half or less than the recommended rate on these crops after the stated growth stages. Also with herbicides on managed amenity turf at recommended rates, and on grassland at half or less than recommended rates		
Protective clothing	A, C		
Precautions	R36, R38, R41, U05a, U08, U11, U14, U15, U20c, E15a, D01, D02, D09a, D10b, M03, H04		
Intracrop Status	Intracrop	A0506	adjuvant/vegetable oil
Contains	91% w/w methylated rapeseed oil		
Use with	All approved pesticides on non-edible crops, and all pesticides approved for use on growing edible crops when used at half recommended dose or less. On specified crops product may be used at a maximum spray concentration of 1.0% with approved pesticides at their full approved rate up to the growth stages shown in the label		
Precautions	U19a, U20b, D05, D09a, D10b		
Intracrop Stay-Put	Intracrop	A0507	vegetable oil
Contains	91% w/w methylated rapeseed oil		
Use with	Recommended rates of approved pesticides for non-crop uses and with recommended rates of approved pesticides up to the growth stage indicated for specified crops, and with half or less than the recommended rate on these crops after the stated growth stages		
Precautions	U19a, U20b, D05, D09a, D10b		
Intracrop Super Rapeze MSO	Intracrop	A0491	adjuvant/vegetable oil
Contains	91.7% w/w methylated rapeseed oil		
Use with	All approved pesticides on non-edible crops, and all pesticides approved for use on growing edible crops when used at half recommended dose or less. On specified crops product may be used at a maximum spray concentration of 1.78% with approved pesticides at their full approved rate up to the growth stages shown in the label		
Precautions	U19a, U20b, D05, D09a, D10b		
Intracrop Tonto	Intracrop	A0708	spreader/sticker/wetter
Contains	32.67% w/w Polyoxyethylene (5-8 EO) C10-C15 primary alcohol and 1.0% w/w polyoxyethylen-[-alpha-]-methyl-[-omega-]-[3-(1,3,3,3-tetramethyl-3-trimethylsiloxy)-disiloxanyl] propylether		
Use with	All approved pesticides at half or less than half the approved pesticide rate on edible crops and all approved pesticides on non-edible crops		

SECTION 4

Adjuvants

Product	Supplier	Adj. No.	Type
Intracrop Warrior	Intracrop	A0514	spreader/wetter
Contains	192 g/l primary alcohol ethylene		
Use with	Recommended rates of approved pesticides up to the growth stage indicated for specified crops, and with half or less than the recommended rate on these crops after the stated growth stages. Also with herbicides on managed amenity turf at recommended rates, and on grassland at half or less than recommended rates		
Protective clothing	A, C		
Precautions	R36, R38, R41, U05a, U08, U11, U14, U15, U20c, E15a, D01, D02, D09a, D10b, M03, H04		
Intracrop Zenith	Intracrop	A0665	spreader/sticker/wetter
Contains	27.0% w/w polyoxyethylen-[-alpha-]-methyl-[-omega-]-[3-(1,3,3,3-tetramethyl-3-trimethylsiloxy)-disiloxanyl] propylether co-formulated with a betaine latex polymer emulsifier complex		
Use with	All approved pesticides at half or less than half the approved pesticide rate		
Protective clothing	A, C		
Precautions	R36, R38, R41, U05a, U08, U11, U20c, E13e, D01, D02, D09a, D10a, H04		
Jogral	Intracrop	A0226	cationic surfactant
Contains	800 g/l tallow amine ethoxylate		
Use with	Glyphosate		
Protective clothing	A, C		
Precautions	R22a, U08, U19a, U20a, E13b, E34, D01, D02, D06b, D10b, D11a, M03, H03, H08		
Kantor	Interagro	A0623	spreader/wetter
Contains	790 g/l alkoxylated triglycerides		
Use with	All approved pesticides		
Precautions	U05a, U20b, E15b, E19b, E34, D01, D02, D09a, D10a, D12a		
Katalyst	Interagro	A0450	penetrant/water conditioner
Contains	90% w/w alkoxylated fatty amine		
Use with	Glyphosate and a wide range of other pesticides on specified crops. Refer to label or contact supplier for further details		
Protective clothing	A, C		
Precautions	R22a, R36, R38, R50, R58, U02a, U05a, U08, U19a, U20b, E15a, E34, E37, D01, D02, D05, D09a, D10a, M03, H03, H11		

Adjuvants

Product	Supplier	Adj. No.	Type
Kinetic	Helena	A0252	spreader/wetter
Contains	80% w/w polyoxypropylene-polyoxyethylene glycol and 20% w/w polyalkylene oxide modified heptamethyl trisiloxane		
Use with	Approved pesticides on non-edible crops and cereals and stubbles of all edible crops when used at full recommended dose, and with pesticides approved for use on growing edible crops when used at half recommended dose or less. On specified crops product may be used at a maximum spray concentration of 0.2% with approved pesticides at their full approved rate up to the growth stages shown in the label		
Protective clothing	A, H		
Precautions	R38, R41, U02a, U05a, U08, U19a, U20b, E13c, E34, E37, D01, D02, D05, D09a, D10c, M03, H04		
Klipper	Amega	A0260	spreader/wetter
Contains	600 g/l ethylene oxide condensate		
Use with	Approved salt formulations of mecoprop alone or in mixtures with 2,4-D on managed amenity turf		
Protective clothing	A, C		
Precautions	R36, R38, U05a, U08, U20c, E15a, D01, D02, D09a, D10a, H04, H08		
Leaf-Koat	Helena	A0511	spreader/wetter
Contains	80% w/w polyoxypropylene-polyoxyethylene glycol and 20% w/w polyalkylene oxide modified heptamethyl trisiloxane		
Use with	Approved pesticides on non-edible crops and cereals and stubbles of all edible crops when used at full recommended dose, and with pesticides approved for use on growing edible crops when used at half recommended dose or less. On specified crops product may be used at a maximum spray concentration of 0.2% with approved pesticides at their full approved rate up to the growth stages shown in the label		
Protective clothing	A, H		
Precautions	R38, R41, U02a, U05a, U08, U19a, U20b, E13c, E34, E37, D01, D02, D05, D09a, D10c, M03, H04		
Level	United Agri	A0580	extender/sticker
Contains	52% synthetic latex solution		
Use with	All approved pesticides in non-edible crops, all approved pesticides on edible crops when used at half or less their recommended rate, potato blight fungicides and with all pesticides on listed crops up to specified growth stages		
Precautions	E37, D01, D02, D05		

SECTION 4

Adjuvants

Product	Supplier	Adj. No.	Type
Li-700	De Sangosse	A0529	acidifier/drift retardant/penetrant
Contains	35% w/w lecithin, 35% w/w propionic acid and 9.39% w/w alcohol ethoxylate		
Use with	All approved pesticides on all edible crops when used at half their recommended dose or less, and on all non-edible crops up their full recommended dose. Also with morpholine fungicides on cereals and with pirimicarb on legumes when used at full dose; with iprodione on brassicas when used at 75% dose, and at a maximum concentration of 0.5% with all pesticides on listed crops at full dose up to specified growth stages		
Protective clothing	A, C, H		
Precautions	R36, R38, U11, U14, U15, U19a, E15a, E19b, D01, D02, D05, D10a, D12a, H04		
Libsorb	Ciba Specialty	A0438	spreader/wetter
Contains	alkyl alcohol ethoxylate		
Use with	Any spray for which additional wetter is approved and recommended		
Protective clothing	A, C		
Precautions	R36, R38, U05a, U08, U20a, E13c, D01, D02, D09a, D10a, H04		
Logic	Microcide	A0288	vegetable oil
Contains	95% emulsified vegetable oil		
Use with	All approved pesticides on edible and non-edible crops for ground or aerial application		
Protective clothing	A, C		
Precautions	U19a, U20b, D09a, D10b		
Logic Oil	Microcide	A0630	vegetable oil
Contains	95% emulsified vegetable oil		
Use with	All approved pesticides on edible and non-edible crops for ground or aerial application		
Protective clothing	A, C		
Precautions	U19a, U20b, D09a, D10b		
Low Down	Helena	A0459	mineral oil
Contains	732 g/l refined paraffinic petroleum oil		
Use with	All approved pesticides on edible and non-edible crops when used at half their approved dose or less. Also with approved pesticides on specified crops, up to specified growth stages, at up to their full approved dose		
Protective clothing	A, C		
Precautions	R53a, U02a, U05a, U08, U20b, E15a, E34, E37, D01, D02, D05, D09a, D10c, H11		
Mangard	Taminco	A0667	adjuvant/anti-transpirant
Contains	96% w/w di-1-p-menthene		
Use with	Glyphosate and a wide range of other herbicides, fungicides and insecticides. See label for details		
Precautions	U20c, E15a, D09a, D10a		

Adjuvants

Product	Supplier	Adj. No.	Type
Meco-Flex	Greenaway	A0678	spreader/sticker/wetter
Contains	10.0% w/w rapeseed oil		
Use with	Re-Act, Headland Relay Depitox		
Protective clothing	A, C		
Precautions	U11, U12, U15, U20b, E15a, E34, D02		
Mediator Sun	Nufarm UK	A0512	wetter
Contains	655 g/l terpenic alcohols		
Use with	All approved pesticides and plant growth regulators on a wide range of specified crops at full dose up to specified growth stages, and at half dose at later growth stages		
Precautions	R36, R52, R53a, U11, U15, D01, M05a, H04		
Mixture B NF	Amega	A0570	non-ionic surfactant/spreader/wetter
Contains	42.5% polyoxyethylene (3EO) C12-C15 primary alcohol and 38.25% w/w polyoxyethylene (7EO) C12-C15 primary alcohol		
Use with	All approved pesticides on non-edible crops at their full recommended rate. Also with Timbrel (MAPP 05815) and all approved formulations of glyphosate in non-crop situations. Also with a range of specified herbicides when used at less than half their recommended rate on forest and grassland, and with all approved pesticides at full rate on a wide range of specified edible crops up to specified growth stages		
Protective clothing	A, C		
Precautions	R21, R22a, R36, R38, U05a, U08, U20b, E13c, E34, E37, D01, D02, D09a, D10a, M03, H03, H04		
Nettle	Interagro	A0466	mineral oil
Contains	60% w/w mineral oil		
Use with	Cereal graminicides and a wide range of other pesticides that have a label recommendation for use with authorised adjuvant oils on specified crops. Refer to label or contact supplier for further details		
Protective clothing	A		
Precautions	R22b, R38, U02a, U05a, U08, U20b, E15a, D01, D02, D05, D09a, D10a, M05b, H03		
Newman's T-80	De Sangosse	A0192	spreader/wetter
Contains	78% w/w polyoxyethylene tallow amine		
Use with	Glyphosate		
Protective clothing	A, C, H		
Precautions	R22a, R37, R38, R41, R50, R58, R67, U05a, U11, U14, U19a, U20a, E15a, E19b, E34, E37, D01, D02, D05, D10a, D12a, M03, M04a, H03, H08, H11		

SECTION 4

Adjuvants

Product	Supplier	Adj. No.	Type
Nion	Amega	A0415	spreader/wetter
Contains	90% (900 g/l) ethylene oxide condensate		
Use with	Recommended rates of approved pesticides up to specified growth stages of a wide range of agricultural arable and horticultural crops. Use on edible crops beyond specified growth stages should only be with half recommended rates of the pesticide or less. See label for details of growth stage restrictions.		
Protective clothing	A, C		
Precautions	R36, R38, U05a, U08, U19a, U20b, E13c, E37, D01, D02, D09a, D10a, H04		
Nu Film P	Intracrop	A0635	sticker/wetter
Contains	96% poly-1-p-menthene		
Use with	Glyphosate and many other pesticides and growth regulators for which a protectant is recommended. Do not use in mixture with adjuvant oils or surfactants		
Precautions	U19a, U20b, E13c, D09a, D10a		
Nufarm Cropoil	Nufarm UK	A0447	mineral oil
Contains	99% highly refined mineral oil		
Use with	Approved pesticides for use on certain specified crops (cereals, combinable break crops, field and dwarf beans, peas, oilseed rape, brassicas, potatoes, carrots, parsnips, sugar beet, fodder beet, mangels, red beet, maize, sweetcorn, onions, leeks, horticultural crops, forestry, amenity, grassland)		
Protective clothing	A, C		
Precautions	R43, U19a, U20b, E13c, D05, D09a, D10b, H04		
Orlin-one	Intracrop	A0597	spreader/sticker/wetter
Contains	25.0% w/w polyoxyethylene (12EO)/ polyoxypropylene (30EO) block copolymer		
Use with			
Output	Nufarm UK	A0644	mineral oil/surfactant
Contains	60% w/w mineral oil and 40% w/w surfactants		
Use with	Grasp		
Protective clothing	A, C		
Precautions	R36, R38, R52, U05a, U08, U10, U19a, U20b, E15a, E38, D01, D02, D09a, D10b, D12a, H04		
Output	Syngenta	A0429	mineral oil/surfactant
Contains	60% mineral oil and 40% surfactants		
Use with	Grasp		
Protective clothing	A, C		
Precautions	R38, U05a, U08, U19a, U20a, E13c, D01, D02, D09a, D10b, H04		

Adjuvants

Product	Supplier	Adj. No.	Type
Pan Oasis	Pan Agriculture	A0411	vegetable oil
Contains	95% w/w methylated rapeseed oil		
Use with	A wide range of pesticides that have a label recommendation for use with authorised adjuvant oils. Contact distributor for further details		
Protective clothing	A		
Precautions	R36, U02a, U05a, U08, U20b, E15a, D01, D02, D05, D09a, D10a, H04		
Pan Panorama	Pan Agriculture	A0412	mineral oil
Contains	95% w/w mineral oil		
Use with	A wide range of pesticides that have a label recommendation for use with adjuvant oils. Contact distributor for details		
Protective clothing	A		
Precautions	R22b, R38, U02a, U05a, U08, U20b, E13c, D01, D02, D05, D09a, D10a, M05b, H03		
Paramount	United Agri	A0585	wetter
Contains	85% polyalkylene oxide modified heptamethyl siloxane		
Use with	Approved pesticides on cereals and grassland for which a wetter is recommended, all approved pesticides in non-edible crops, all approved pesticides on edible crops when used at half or less their recommended rate, and with all pesticides on listed crops up to specified growth stages		
Protective clothing	A, C		
Precautions	R21, R22a, R38, R41, R51, R53a, U11, U14, U16b, E37, E38, D01, D05, D07, D08, H03, H11		
Phase II	De Sangosse	A0622	vegetable oil
Contains	95% w/w esterified rapeseed oil		
Use with	Pesticides approved for use in sugar beet, oilseed rape, cereals (for grass weed control), and other specified agricultural and horticultural crops		
Protective clothing	A, C, H		
Precautions	U19a, U20b, E15a, E19b, E34, D01, D02, D05, D10a, D12a		
Pik-Flex	Greenaway	A0681	spreader/sticker/wetter
Contains	10.0% w/w rapeseed oil		
Use with	Tordon 22K or Tordon 101		
Protective clothing	A, C		
Precautions	U11, U12, U15, U20b, E15a, E34, D02		
Pin-o-Film	Intracrop	A0637	sticker/wetter
Contains	96% di-1-p-menthene		
Use with	Recommended rates of approved pesticides up to the growth stage indicated for specified crops, and with half or less than the recommended rate on these crops after the stated growth stages. Also for use with pesticides on grassland at half or less than their recommended rates. Must not be used in mixture with adjuvant oils or surfactants		
Precautions	U19a, U20c, E13c, C02a, D09a, D11a		

SECTION 4

Adjuvants

Product	Supplier	Adj. No.	Type
Planet	Intracrop	A0605	non-ionic surfactant/spreader/wetter
Contains	85% alkyl polyglycol ether and fatty acid		
Use with	Any spray for which additional wetter is recommended		
Precautions	R22a, R36, U05a, U08, U19a, U20b, E13c, D01, D09a, D10a, D11a, H03, H04, H08		
Prima	De Sangosse	A0531	mineral oil
Contains	99% highly refined paraffinic oil		
Use with	All pesticides on all crops when used up to 50% of maximum approved dose for that use (mixtures with Roundup must only be used for treatment of stubbles). Also with all approved pesticides at full dose on listed crops (see label). Also with listed herbicides on specified crops		
Protective clothing	A, C		
Precautions	R22a, U10, U16b, U19a, E15a, E19b, E34, E37, D01, D02, D05, D10a, D12a, M05b, H03		
Profit Oil	Microcide	A0631	extender/sticker/wetter
Contains	95% rape/soya oil		
Use with	All approved pesticides		
Protective clothing	A, C		
Precautions	U19a, U20b, D09a, D10b		
Pryz-Flex	Greenaway	A0679	spreader/sticker/wetter
Contains	95% refined rapeseed oil		
Use with	'Greenaway Gly-490' (MAPP 12718) and all approved 490 g/l glyphosate products on permeable surfaces overlying soil or with 'Kerb Flo' (MAPP 13716) in forest situations.		
Protective clothing	A, C		
Precautions	U11, U12, U15, U20b, E34, E37		
Ranger	De Sangosse	A0532	acidifier/drift retardant/penetrant
Contains	35% w/w lecithin, 35% w/w propionic acid and 9.39%w/w alcohol ethoxylate		
Use with	All approved pesticides on all edible crops when used at half their recommended dose or less, and on all non-edible crops up their full recommended dose. Also with morpholine fungicides on cereals and with pirimicarb on legumes when used at full dose; with iprodione on brassicas when used at 75% dose, and at a maximum concentration of 0.5% with all pesticides on listed crops at full dose up to specified growth stages		
Protective clothing	A, C, H		
Precautions	R36, R38, U11, U14, U15, U19a, E15a, E19b, D01, D02, D05, D10a, D12a, H04		
Reward Oil	Microcide	A0632	extender/sticker/wetter
Contains	95% rape/soya oil		
Use with	All approved pesticides		
Protective clothing	A, C		
Precautions	U19a, U20b, D09a, D10b		

Adjuvants

Product	Supplier	Adj. No.	Type
Roller	Agrovista	A0694	spreader/wetter
Contains	832 g/L co-polymer of ethylene oxide and propylene oxide + 204 g/l polyalkylene oxide modified heptamethyl trisiloxane		
Use with	All approved pesticides up to their maximum dose at a max conc of 0.2% vol/vol.		
Protective clothing	A, C		
Precautions	R36, R38, R41, U02a, U08, U20b, E13c, E34, E40c, D01, D02, D05, D09a, D10a, H04		
Ryda	Interagro	A0168	cationic surfactant/wetter
Contains	800 g/l polyethoxylated tallow amine		
Use with	Glyphosate		
Protective clothing	A, C		
Precautions	R22a, R36, R38, R41, R51, R58, U02a, U05a, U08, U13, U19a, U20a, E15a, E34, E37, D01, D02, D05, D09a, D10b, M03, H03, H08, H11		
Saracen	Interagro	A0368	vegetable oil
Contains	95% w/w methylated vegetable oil		
Use with	Cereal graminicides and a wide range of other pesticides that have a recommendation for use with authorised adjuvant oils on specified crops. Refer to label or contact supplier for further details		
Precautions	U05a, U20b, E15a, D01, D02, D09a, D10a		
SAS 90	Intracrop	A0311	spreader/wetter
Contains	100% polyoxyethylen-alpha-methyl-omega-[3-(1,3,3,3-tetramethyl-3-trimethylsiloxy)-disiloxanyl] propylether		
Use with	A wide range of pesticides on specified crops in agriculture and horticulture, and on amenity vegetation and land not intended for cropping		
Precautions	U02a, U08, U19a, U20b, E13c, E34, D09a, D10a		
Scout	De Sangosse	A0533	acidifier/drift retardant/penetrant
Contains	35% w/w lecithin, 35% w/w propionic acid and 9.39%w/w alcohol ethoxylate		
Use with	All approved pesticides on all edible crops when used at half their recommended dose or less, and on all non-edible crops up their full recommended dose. Also with morpholine fungicides on cereals and with pirimicarb on legumes when used at full dose; with iprodione on brassicas when used at 75% dose, and at a maximum concentration of 0.5% with all pesticides on listed crops at full dose up to specified growth stages		
Protective clothing	A, C, H		
Precautions	R36, R38, U11, U14, U15, U19a, E15a, E19b, D01, D02, D05, D10a, D12a, H04		
Siltex	Intracrop	A0398	spreader/sticker/wetter
Contains	27% w/w polyoxyethylen-[-alpha-]-methyl-[-omega-]-[3-(1,3,3,3-tetramethyl-3-trimethylsiloxy)-disiloxanyl] propylether		
Use with	Approved pesticides used at full dose on a range of specified fruit and vegetable crops, and on managed amenity turf		
Protective clothing	A, C		
Precautions	U05a, U08, U20c, E13c, E37, D01, D02, D09a, D10a		

Adjuvants

Product	Supplier	Adj. No.	Type
Silwet L-77	De Sangosse	A0640	drift retardant/spreader/wetter
Contains	80% w/w polyalkylene oxide modified heptamethyltrisiloxane and a maximum of 20% w/w allyloxypolyethylene glycol methyl ether		
Use with	All approved fungicides on winter and spring sown cereals; all approved pesticides applied at 50% or less of their full approved dose. A wide range of other uses. See label or contact supplier for details		
Protective clothing	A, C, H		
Precautions	R20, R21, R22a, R36, R43, R48, R51, R58, U11, U15, U19a, E15a, E19b, D01, D02, D05, D10a, D12a, H03, H11		
Slippa	Interagro	A0206	spreader/wetter
Contains	655 g/l polyalkyleneoxide modified heptamethyltrisiloxane		
Use with	Cereal fungicides and a wide range of other pesticides and trace elements on specified crops		
Protective clothing	A, C, H		
Precautions	R20, R38, R41, R43, R48, R51, R58, U02a, U05a, U08, U11, U19a, U20a, E15a, E34, E37, D01, D02, D05, D09a, D10a, M03, H03, H11		
SM-99	De Sangosse	A0534	mineral oil
Contains	99% highly refined paraffinic oil		
Use with	All pesticides on all crops when used up to 50% of maximum approved dose for that use (mixtures with Roundup must only be used for treatment of stubbles). Also with all approved pesticides at full dose on listed crops (see label). Also with listed herbicides on specified crops		
Protective clothing	A, C		
Precautions	R22a, U10, U16b, U19a, E15a, E19b, E34, E37, D01, D02, D05, D10a, D12a, M05b, H03		
Solar	Intracrop	A0225	activator/non-ionic surfactant/spreader
Contains	75% polypropoxypropanol		
Use with	Foliar applied plant growth regulators		
Protective clothing	A, C		
Precautions	R36, R38, U04a, U05a, U08, U20a, E13c, E34, D01, D02, D09a, D10b, D11a, M03, H04		
Spartan	Interagro	A0375	penetrant/water conditioner
Contains	500 g/l alkoxylated fatty amine + 400 g/l polyoxyethylene monolaurate		
Use with	Cereal growth regulators, cereal herbicides, oilseed rape fungicides and a wide range of other pesticides on specified crops. Refer to label or contact supplier for further details		
Protective clothing	A, C		
Precautions	R22a, R36, R38, R51, R58, U02a, U05a, U19a, U20b, E15a, E34, E37, D01, D02, D05, D09a, D10a, M03, H03, H11		
Sprayfast	Taminco	A0668	extender/sticker/wetter
Contains	334 g/l di-1-p-menthene		
Use with	Glyphosate and other pesticides, growth regulators or nutrients for which a coating agent is approved and recommended		
Precautions	U20b, E13c, D09a, D10a		

Adjuvants

Product	Supplier	Adj. No.	Type
Spray-fix	De Sangosse	A0559	extender/sticker/wetter
Contains	\multicolumn{3}{l}{900 g/l synthetic latex solution and 101 g/l alcohol alkoxylate}		
Use with	\multicolumn{3}{l}{All approved potato blight fungicides. Also with all approved pesticides on all edible crops when used at half their recommended dose or less, and on all non-edible crops up their full recommended dose. Also at a maximum concentration of 0.14% with approved pesticides on listed crops up to specified growth stages}		
Protective clothing	A, C, H		
Precautions	R36, R38, U11, U14, U16b, U19a, E15a, E19b, D01, D02, D05, D10a, D12a, H04		
Spraygard	Taminco	A0669	extender/sticker/wetter
Contains	400 g/l di-1-p-menthene		
Use with	Approved pesticides on all edible crops (except herbicides on peas) and as an anti-transpirant on vegetables before transplanting, evergreens, deciduous trees, shrubs, bushes and on turf		
Protective clothing	A		
Precautions	R38, R50, R53a, U05a, U14, U20b, E15a, E34, E37, D01, D09a, D10b, D12b, M03, H04, H11		
Spraymac	De Sangosse	A0549	acidifier/non-ionic surfactant
Contains	350 g/l propionic acid and 100 g/l alcohol ethoxylate		
Use with	All approved pesticides on all edible crops when used at half their recommended dose or less, and on all non-edible crops up their full recommended dose. Also at a maximum concentration of 0.5% with approved pesticides on listed crops up to specified growth stages		
Protective clothing	A, C, H		
Precautions	R34, U04a, U10, U11, U14, U19a, U20b, E15a, E19b, E34, D01, D02, D05, D10a, D12a, M04a, H05		
Stamina	Interagro	A0202	penetrant
Contains	100% w/w alkoxylated fatty amine		
Use with	Glyphosate and a wide range of other pesticides on specified crops		
Protective clothing	A, C		
Precautions	R22a, R38, R50, R58, U02a, U05a, U20a, E15a, E34, D01, D02, D05, D09a, D10a, M03, H03, H11		
Standon Shiva	Standon	A0519	wetter
Contains	20.2% w/w sodium salt of 3,6-dioxaoctadecylsulphate and 6.7% w/w sodium salt of 3,6-dioxaeicosylsulphate		
Use with	Approved cereal herbicides		
Protective clothing	A, C		
Precautions	R36, R38, R52, U02a, U05a, U08, U13, U19a, E15a, D01, D02, D05, D09a, D10b, H04		

SECTION 4

Adjuvants

Product	Supplier	Adj. No.	Type
Stika	De Sangosse	A0557	extender/sticker/wetter
Contains	450 g/l synthetic latex solution and 101 g/l alcohol alkoxylate		
Use with	All approved potato blight fungicides. Also with all approved pesticides on all edible crops when used at half their recommended dose or less, and on all non-edible crops up their full recommended dose. Also at a maximum concentration of 0.14% with approved pesticides on listed crops up to specified growth stages		
Protective clothing	A, C, H		
Precautions	R36, R38, U11, U14, U16b, U19a, E15a, E19b, D01, D02, D05, D10a, D12a, H04		
Super Nova	Interagro	A0364	penetrant
Contains	500 g/l alkoxylated fatty amine + 500 g/l polyoxyethylene monolaurate		
Use with	Cereal growth regulators, cereal herbicides, cereal fungicides, oilseed rape fungicides and a wide range of other pesticides on specified crops		
Precautions	R51, R58, U05a, E15a, E34, D01, D02, D05, D09a, D10a, H11		
Sward	Amega	A0215	spreader/wetter
Contains	15.2% w/w polyalkylene oxide modified heptomethyltrisiloxane		
Use with	Specified pesticides in amenity situations		
Protective clothing	A, C		
Precautions	R41, R43, U05a, U08, U20c, E13c, D01, D02, D09a, D10a, H04		
Talzene	Intracrop	A0233	wetter
Contains	800 g/l polyoxyethylene tallow amine		
Use with	Herbicides and dessicants up to their latest recommended time of application in agriculture, horticulture, amenity and forestry		
Protective clothing	A, C		
Precautions	R22a, R36, R38, R41, R48, U05a, U08, U11, U13, U19a, U20b, E13c, E34, D01, D02, D09a, D10a, M03, H03		
Teliton Ocean Power	Teliton	A0615	wetter
Contains	20.2% w/w 3,6-dioxoctadecylsulphate sodium salt and 6.7% w/w 3,6-dioxaeicosylsulphate sodium salt		
Use with	All approved herbicides for use on cereals		
Protective clothing	A, C		
Precautions	R36, R38, R52, U02a, U05a, U08, U13, U19a, E15a, D01, D02, D09a, D10a, H04		
TM 1088	Intracrop	A0675	wetter
Contains	750 g/l polyoxyethylene, polypropoxypropanol		
Use with	All approved pesticides on non-edible crops and non crop production uses; All approved pesticides at half or less than half the approved pesticide rate on edible crops		
Protective clothing	A, C		
Precautions	R36, R38, U04a, U05a, U08, U19a, U20b, E13b, E34, D01, D02, D09a, D10a, M03, M05a, H04		

Adjuvants

Product	Supplier	Adj. No.	Type
Toil	Interagro	A0248	vegetable oil
Contains	95% w/w methylated rapeseed oil		
Use with	Sugar beet herbicides, oilseed rape herbicides, cereal graminicides and a wide range of other pesticides that have a label recommendation for use with authorised adjuvant oils on specified crops. Refer to label or contact supplier for further details		
Precautions	U05a, U20b, E15a, D01, D02, D05, D09a, D10a		
Torpedo-II	De Sangosse	A0541	penetrant/wetter
Contains	210 g/kg alkoxylated tallow amine, 380 g/kg alcohol ethoxylates, 75 g/l natural fatty acids and 210 g/kg polyalkylene glycol		
Use with	All approved pesticides on all edible crops when used at half their recommended dose or less, and on all non-edible crops up their full recommended dose. Also at a maximum concentration of 0.1% with approved pesticides on listed crops up to specified growth stages		
Protective clothing	A, C, H		
Precautions	R36, R38, R41, R52, R58, U05a, U10, U11, U19a, E15a, E19b, E34, E37, D01, D02, D05, D10a, D12a, M03, H04		
Transact	United Agri	A0584	penetrant
Contains	40% w/w propionic acid		
Use with	Plant growth regulators in cereals, all approved pesticides in non-edible crops, all approved pesticides on edible crops when used at half or less their recommended rate, and with all pesticides on listed crops up to specified growth stages		
Protective clothing	A, C		
Precautions	R34, U10, U11, U14, U15, U19a, D01, D05, D09b, M04a, H05		
Transcend	Helena	A0333	adjuvant/vegetable oil
Contains	80% w/w methylated soybean oil and 15% w/w polyalkylene oxide modified heptamethyl trisiloxane		
Use with	Approved pesticides on non-edible crops, cereals and stubbles of all edible crops, grassland (destruction), and pesticides approved for use on growing edible crops when used at half recommended dose or less. On specified crops product may be used at a maximum spray concentration of 0.5% with approved pesticides at their full approved rate up to the growth stages shown in the label		
Protective clothing	A, H		
Precautions	R38, R41, U02a, U05a, U08, U11, U19a, U20b, E13c, E34, E37, D01, D02, D05, D09a, M03, H04		
Try-Flex	Greenaway	A0489	sticker/wetter
Contains	95% refined rapeseed oil		
Use with	Gly-480 (MAPP 12718) and Freeway (MAPP 11129) on natural surfaces not intended to bear vegetation, permeable surfaces overlying soil, hard surfaces overlying soil and forest		
Protective clothing	A, C, D, H, M		
Precautions	U08, U20b, E37, C11, D05, D10a		

SECTION 4

Adjuvants

Product	Supplier	Adj. No.	Type
Validate	De Sangosse	A0500	adjuvant/wetter
Contains	50% w/w lecithin, 25% w/w esterified vegetable oil and 25% w/w alcohol ethoxylate		
Use with	Approved pesticides on non-edible crops where the addition of a wetter/spreader or adjuvant oil is recommended on the pesticide label, and pesticides approved for use on growing edible crops when used at half recommended dose or less. On specified crops product may be used at a maximum spray concentration of 0.5% with approved pesticides at their full approved rate up to the growth stages shown in the label		
Protective clothing	A, C, H		
Precautions	R51, R53a, U05a, U11, U14, U19a, U20b, E15a, E19b, E34, D01, D02, D10a, D12a, H11		
Velocity	Agrovista	A0697	adjuvant
Contains	771.5 g/l methylated oil + 105.9 g/l polyalkylene oxide modified heptamethyl trisiloxane		
Use with	All approved pesticides up to their full approved rate at a maximum conc of 0.5% vol/vol.		
Protective clothing	A, C		
Precautions	U02a, U08, U20b, E13c, E34, E40c, D01, D02, D09a, D10a, H04		
Wetcit	Plant Solutions	A0586	penetrant/wetter
Contains	8.15% alcohol ethoxylate		
Use with	All approved pesticides on all edible crops when used at half their recommended dose or less, and on all non-edible crops up their full recommended dose. Also at a maximum concentration of 0.25% with approved pesticides on listed crops up to specified growth stages at full rate, and at half rate thereafter		
Precautions	R22a, R38, R41, U08, U11, U19a, U20b, E13c, E15a, E34, E37, D05, D08, D09a, D10b, H03		
X-Wet	De Sangosse	A0548	non-ionic surfactant/wetter
Contains	750 g/l alcohol ethoxylates and 150 g/l natural fatty acids		
Use with	All approved pesticides on all edible crops when used at half their recommended dose or less, and on all non-edible crops up their full recommended dose. Also at a maximum concentration of 0.1% with approved pesticides on listed crops up to specified growth stages		
Protective clothing	A, C, H		
Precautions	R36, R38, R53a, U02a, U04a, U05a, U10, U11, U20b, E15a, E19b, D01, D02, D05, D10a, H04		
Zarado	De Sangosse	A0516	vegetable oil
Contains	70% w/w esterified rapeseed oil		
Use with	All approved pesticides on edible and non-edible crops when used at half their approved dose or less. Also with approved pesticides on specified crops, up to specified growth stages, at up to their full approved dose		
Protective clothing	A, C, H		
Precautions	R43, R52, U05a, U13, U19a, U20b, E15a, E19b, E34, D01, D02, D05, D10a, M05a, H04		

Adjuvants

Product	Supplier	Adj. No.	Type
Zeal	Interagro	A0685	water conditioner/wetter
Contains	60% w/w alkoxylated fatty amine		
Use with	Trace elements such as calcium, copper, manganese and sulphur and with macronutrients such as phosphates		
Protective clothing	A, C, H		
Precautions	R22a, R36, R38, R51, R53a, U05a, U09a, U11, U14, U15, U19a, E15b, E34, D01, D02, D09a, D10a, M03, H03, H11		
Zigzag	Headland	A0692	extender/sticker/wetter
Contains	36.9% w/v styrene/butadiene co-polymer and 6.375% w/v polyether-modified trisiloxane		
Use with	Approved glyphosate formulations on cereal crops; All approved pesticides applied at half-rate or less; All approved pesticides for use on crops not destined for human or animal consumption; With all approved pesticides on edible crops up to the latest growth stage specified on the label.		
Protective clothing	C, H		
Precautions	U11, U15, E38, D01		
Zinzan	Certis	A0600	extender/spreader/wetter
Contains	70% 1,2 bis (2-ethylhexyloxycarbonyl) ethanesulponate		
Use with	All approved pesticides on non-edible crops, and all pesticides approved for use on growing edible crops when used at half recommended dose or less. On specified edible crops product may be used at a maximum spray concentration of 0.075% with approved pesticides at their full approved rate up to the growth stages shown in the label		
Protective clothing	A, C, P		
Precautions	R38, R41, U11, U14, U20c, D01, D09a, D10a, H04		

SECTION 4

SECTION 5
USEFUL INFORMATION

Pesticide Legislation

Anyone who advertises, sells, supplies, stores or uses a pesticide is bound by legislation, including those who use pesticides in their own homes, gardens or allotments. There are numerous UK statutory controls, but the major legal instruments are outlined below.

Regulation 1107/2009 – The Replacement for EU 91/414

Regulation 1107/2009 entered into force within the EU on 14 December 2009 and is to be applied to new approval applications from 14 June 2011. On that date, all pesticides that hold a current approval under 91/414 will be deemed to be approved under 1107/2009 and the new criteria for approval will be applied only when the active substance comes up for review. The new regulation largely mirrors the previous regulation, but requires additional approval criteria to be met to weed out the more hazardous chemicals. These include assessing the effect on vulnerable groups, taking into account known synergistic and cumulative effects, considering the effect on coastal and estuarine waters, and effects on the behaviour of non-target organisms, biodiversity and the ecosystem. However, these effects will be evaluated only when scientific methods approved by the European Food Safety Authority (EFSA) have been developed.

The additional criteria will require that no substances that are carcinogens, mutagens or toxic to reproduction (CMRs); endocrine disruptors (not to be defined until 2015!); persistent organic pollutants (POPs); chemicals that are persistent, bioaccumulative or toxic (PBTs); or chemicals that are very persistent and very bioaccumulative (vPVBs) will gain approval unless the exposure is negligible. However, note that for EFSA the term 'negligible' has yet to receive a precise definition.

If an active substance passes these hurdles, the first approval for basic substances will be granted for 10 years (15 years for 'low-risk' substances, 7 years for candidates for substitution – see below), and then at each subsequent renewal a further 15 years will be granted. Actives will be divided between five categories – low-risk substances (e.g. pheromones, semiochemicals, microorganisms and natural plant extracts), basic substances, candidates for substitution, safeners/synergists, and co-formulants. Candidates for substitution will be those chemicals that have just scraped past the approval thresholds but are deemed still to carry some degree of hazard. They will be withdrawn when significantly safer alternatives are available. This includes physical methods of control, but any alternative must not have significant economic or practical disadvantages and must minimise the risk of the development of resistance, and the consequences for any 'minor uses' of the original product must also be considered.

The Food and Environment Protection Act 1985 (FEPA) and Control of Pesticides Regulations 1986 (COPR)

FEPA introduced statutory powers to control pesticides with the aims of protecting human beings, creatures and plants, safeguarding the environment, ensuring safe, effective and humane methods of controlling pests, and making pesticide information available to the public. Control of pesticides is achieved by COPR. These Regulations lay down the Approvals required before any pesticide may be sold, stored, supplied, advertised or used, subject to conditions which are contained in a series of Schedules. Schedule 1 relates to advertisement; Schedule 2 to sale, supply and storage; Schedule 3 to use; and Schedule 4 to aerial application of pesticides. The Schedules may be changed at any time, following Parliamentary approval. Details are given on the websites of the Chemicals Regulation Directorate (CRD) and the Health and Safety Executive (HSE).

The controls currently in force include the following.

- Only approved products may be sold, supplied, stored, advertised or used.
- No advertisement may contain any claim for safety beyond that which is permitted in the approved label text.
- Only products specifically approved for the purpose may be applied from the air.

- A recognised Storeman's Certificate of Competence is required by anyone who stores for sale or supply pesticides approved for agricultural use.
- A recognised Certificate of Competence is required by anyone who gives advice when selling or supplying pesticides approved for agricultural use.
- Users of pesticides must comply with the Conditions of Approval relating to use.
- A recognised Certificate of Competence is required for all contractors and persons born after 31 December 1964 applying pesticides approved for agricultural use (unless working under direct supervision of a certificate holder). Proposals are now in place that every user of agricultural pesticides will have to hold a Certificate of Competence, regardless of age or supervision. A suitable transition period will allow this requirement to be enacted.
- Only those adjuvants authorised by CRD may be used.
- Regarding tank-mixes, 'no person shall combine or mix for use two or more pesticides which are anti-cholinesterase compounds unless the approved label of at least one of the pesticide products states that the mixture may be made; and no person shall combine or mix for use two or more pesticides if all the conditions of the Approval relating to this use cannot be complied with'.

Plant Protection Products Regulation

European Council Directive 91/414/EEC, drafted to harmonise national arrangements for the authorisation of plant protection products within the European Community, is currently being upgraded to a full 'Regulation', which means that it becomes mandatory rather than advisory. It first became effective in the UK on 25 July 1993 and the new proposals will come into force in April 2011. The new rules switch from risk management to hazard criteria, and growers will be encouraged to adopt lower-risk practices and products. The aim of the new legislation is to minimise the risk to health and the environment by reducing the levels of harmful pesticides through substitution with safer alternatives, banning or severely limiting the use of the more dangerous pesticides, and encouraging the adoption of best practice for authorised products.

Some of the changes to be introduced include a ban on aerial spraying, and measures to protect water, public spaces and special conservation areas. National Action Plans will be set to establish indicators of risk and impact reduction, and to encourage the development of integrated pest management and alternative techniques to reduce the perceived dependency on pesticides. To simplify pesticide approval, national approvals will be replaced by zonal approvals, but with national veto if a product is deemed to be unsafe to use in a particular country.

Dangerous Preparations Directive (1999/45/EC)

The Dangerous Preparations Directive (DPD) came into force in UK for pesticide and biocidal products on 30 July 2004. Its aim is to achieve a uniform approach across all Member States to the classification, packaging and labelling of most dangerous preparations, including crop protection products. The Directive is implemented in UK under the Chemicals (Hazard Information and Packaging Supply) Regulations 2002, often referred to under the acronym CHIP3. In most cases the Regulations have led to additional hazard symbols, and associated risk and safety phrases relating to environmental and health hazards, appearing on the label. All products affected by the DPD entering the supply chain from the implementation date above must be so labelled. The new environmental hazard classifications and risk phrases are included.

Under the DPD, the labels of all plant protection products now state 'To avoid risks to man and the environment, comply with the instructions for use'. As the instruction applies to all products, it has not been repeated in each fact sheet.

The Review Programme

The review programme of existing active substances will continue under EC Directive 1107/2009. The programme is designed to ensure that all available plant protection products are supported by up-to-date information on safety and efficacy. As this will now be under 1107/2009, the new standards for approval will be applied at the due date and any that are deemed to be hazardous chemicals will be withdrawn. Any that are just within the safety standards yet to be defined will be called 'candidates for substitution' and granted only a 7-year renewal to allow time for significantly safer alternatives to be developed. If these are available at the time of renewal, the approval will be withdrawn. Chemicals deemed to be low risk, such as natural plant extracts, microorganisms, semiochemicals and pheromones, will be renewed for 15 years, while most active substances will be renewed for a further 10 years.

Control of Substances Hazardous to Health Regulations 1988 (COSHH)

The COSHH regulations, which came into force on 1 October 1989, were made under the Health and Safety at Work Act 1974, and are also important as a means of regulating the use of pesticides. The regulations cover virtually all substances hazardous to health, including those pesticides classed as Very toxic, Toxic, Harmful, Irritant or Corrosive, other chemicals used in farming or industry, and substances with occupational exposure limits. They also cover harmful micro-organisms, dusts and any other material, mixture, or compound used at work which can harm people's health.

The original Regulations, together with all subsequent amendments, have been consolidated into a single set of regulations: The Control of Substances Hazardous to Health Regulations 1994 (COSHH 1994).

The basic principle underlying the COSHH regulations is that the risks associated with the use of any substance hazardous to health must be assessed before it is used, and the appropriate measures taken to control the risk. The emphasis is changed from that pertaining under the Poisonous Substances in Agriculture Regulations 1984 (now repealed) – whereby the principal method of ensuring safety was the use of protective clothing – to the prevention or control of exposure to hazardous substances by a combination of measures. In order of preference the measures should be:

(a) substitution with a less hazardous chemical or product

(b) technical or engineering controls (e.g. the use of closed handling systems, etc.)

(c) operational controls (e.g. operators located in cabs fitted with air-filtration systems, etc.)

(d) use of personal protective equipment (PPE), which includes protective clothing.

Consideration must be given as to whether it is necessary to use a pesticide at all in a given situation and, if so, the product posing the least risk to humans, animals and the environment must be selected. Where other measures do not provide adequate control of exposure and the use of PPE is necessary, the items stipulated on the product label must be used as a minimum. It is essential that equipment is properly maintained and the correct procedures adopted. Where necessary, the exposure of workers must be monitored, health checks carried out, and employees instructed and trained in precautionary techniques. Adequate records of all operations involving pesticide application must be made and retained for at least 3 years.

Certificates of Competence – the roles of BASIS and NPTC

COPR, COSHH and other legislation places certain obligations on those who handle and use pesticides. Minimum standards are laid down for the transport, storage and use of pesticides, and the law requires those who act as storekeepers, sellers and advisors to hold recognised Certificates of Competence.

BASIS

BASIS is an independent Registration Scheme for the pesticide industry. It is responsible for organising training courses and examinations to enable such staff to obtain a Certificate of Competence.

In addition, BASIS undertakes annual assessment of pesticide supply stores, enabling distributors, contractors and seedsmen to meet their obligations under the Code of Practice for Suppliers of Pesticides. Further information can be obtained from BASIS.

Useful Information

Certificates of Competence

Storage

- BASIS Certificate of Competence in the Storage and Handling of Crop Protection Products

Sale and supply

- BASIS Certificate in Crop Protection (Agriculture)
- BASIS Certificate in Crop Protection (Commercial Horticulture)
- BASIS Certificate in Crop Protection (Amenity Horticulture)
- BASIS Certificate in Crop Protection (Forestry)
- BASIS Certificate in Crop Protection (Seed Treatment)
- BASIS Certificate in Crop Protection (Seed Sellers)
- BASIS Certificate in Crop Protection (Field Vegetables)
- BASIS Certificate in Crop Protection (Potatoes)
- BASIS Certificate in Crop Protection (Aquatic)
- BASIS Certificate in Crop Protection (Indoor Landscaping)
- BASIS Certificate in Crop Protection (Grassland and Forage Crops)
- BASIS/LEAF ICM Certificate
- BASIS Advanced Certificate
- FACTS (Fertiliser Advisers Certification and Training Scheme) Certificate

NPTC

Certain spray operators also require certificates of competence under the Control of Pesticides Regulations. NPTC's Pesticides Award is a recognised Certificate of Competence under COPR and it is aimed principally at those people who use pesticide products approved for use in agriculture, horticulture (including amenity horticulture) and forestry. All contractors, and anyone else born after 31 December 1964, must possess a certificate if they spray such products unless they are working under the direct supervision of a certificate holder.

Because the required spraying skills vary widely among the uses listed above, candidates are assessed under one (or more) modules that are most appropriate for their professional work. Assessments are carried out by an approved NPTC or Scottish Skills Testing Service Assessor. All candidates must first complete a foundation module (PA1), for which a certificate is not issued, before taking one of the specialist modules. Certificate holders who change their work so that a different specialist module becomes more appropriate may need to obtain a new certificate under the new module.

Holders are required to produce on demand their Certificate of Competence for inspection to any authorised person. Further information can be obtained from NPTC.

Certificates of Competence for spray operators

- PA1 Foundation Module
- PA2 Ground Crop Sprayers – mounted or trailed
- PA3 Broadcast Air Blast Sprayer. Variable Geometry Boom Air Assisted Sprayer
- PA4 Granule Applicator – Mounted or Trailed
- PA5 Boat Mounted Applicators
- PA6 Hand Held Applicators
- PA7 Aerial Application
- PA8 Mixer/Loader
- PA9 Fogging, Misting and Smokes
- PA10 Dipping Bulbs, Corms, Plant Material or Containers
- PA11 Seed Treating Equipment
- PA12 Application of Pesticides to Material as a Continuous or Batch Process

Maximum Residue Levels

A small number of pesticides are liable to leave residues in foodstuffs, even when used correctly. Where residues can occur, statutory limits, known as maximum residue levels (MRLs), have

Pesticide Legislation

been established. MRLs provide a check that products have been used as directed; **they are not safety limits**. However, they do take account of consumer safety because they are set at levels that ensure normal dietary intake of residues presents no risk to health. Wide safety margins are built in, and eating food containing residues above the MRL does not automatically imply a risk to health. Nevertheless, it is an offence to put into circulation any produce where the MRL is exceeded.

The surrounding legislation is complex. MRLs may be specified by several different bodies. The UK has set statutory MRLs since 1988. The European Union intends eventually to introduce MRLs for all pesticide/commodity combinations. These are being introduced initially by a series of priority lists, but will subsequently be covered by the review programme under Directive 91/414. However, in cases where no information is available, EU Directive 2000/42/EC requires many MRLs to be set at the limit of determination (LOD). As a result, certain approvals are being withdrawn where such use would leave residues above the MRL set in the Directive.

MRLs apply to imported as well as home-produced foodstuffs. Details of those that have been set have been published in *The Pesticides (Maximum Residue Levels in Crops, Food and Feeding Stuffs) Regulations 1994*, and successive amendments to these Regulations. These Statutory Instruments are available from The Stationery Office (www.thestationeryoffice.com).

MRLs are set for many chemicals not currently marketed in Britain. Because of this and the ever-changing information on MRLs, direct access is provided to the comprehensive MRL databases on the Chemicals Regulation Directorate website (www.pesticides.gov.uk). This online database sets out in table form the levels specified by UK Regulations, EC Directives, and the Codex Alimentarius for each commodity.

Approval (On-label and Off-label)

Only officially approved pesticides can be marketed and used in the UK. Approvals are granted by UK Government Ministers in response to applications that are supported by satisfactory data on safety, efficacy and, where relevant, humaneness. The Chemicals Regulation Directorate (CRD, www.pesticides.gov.uk) comes under the Health and Safety Executive (HSE, www.hse.gov.uk), and is the UK Government Agency for regulating agricultural pesticides and plant protection products. The HSE currently fulfils the same role for other pesticides, with the two organisations in the process of being merged. The main focus of the regulatory process in both bodies is the protection of human health and the environment.

Statutory Conditions of Use

Approvals are normally granted only in relation to individual products and for specified uses. It is an offence to use non-approved products or to use approved products in a manner that does not comply with the statutory conditions of use, except where the crop or situation is the subject of an off-label extension of use (see below).

Statutory conditions have been laid down for the use of individual products and may include:

- field of use (e.g. agriculture, horticulture etc.)
- crop or situations for which treatment is permitted
- maximum individual dose
- maximum number of treatments or maximum total dose
- maximum area or quantity which may be treated
- latest time of application or harvest interval
- operator protection or training requirements
- environmental protection
- any other specific restrictions relating to particular pesticides.

Products must display these statutory conditions in a boxed area on the label entitled 'Important Information', or words to that effect. At the bottom of the boxed area must be shown a bold text statement: **'Read the label before use. Using this product in a manner that is inconsistent with the label may be an offence. Follow the Code of Practice for Using Plant Protection Products'**. This requirement came into effect in October 2006, and replaced the previous 'Statutory Box'.

Types of Approval

Where there were once three levels of approval (full; provisional; experimental permit), there will now be just two levels (authorisation for use; limited approval for research and development). Provisional approval used to be granted while further data were generated to justify label claims, but under the new regulations, decisions on authorisation will be reached much more quickly, and this 'half-way stage' is deemed to be unnecessary. The official list of approved products, excluding those approved for research and development only, are shown on the websites of CRD and the Health and Safety Executive (HSE).

Withdrawal of Approval

Product approvals may be reviewed, amended, suspended or revoked at any time. Revocation may occur for various reasons, such as commercial withdrawal, or failure by the approval holder to meet data requirements.

From September 2007, where an approval is being revoked for purely administrative, 'housekeeping' reasons, the existing approval will be revoked and in its place will be issued:

- an approval for advertisement, sale and supply by any person for 24 months, and
- an approval for storage and use by any person for 48 months.

Approval (On-label and Off-label)

Where an approval is replaced by a newer approval, such that there are no safety concerns with the previous approval, but the newer updated approval is more appropriate in terms of reflecting the latest regulatory standard, the existing approval will be revoked and in its place will be issued:

- an approval for advertisement, sale and supply by any person for 12 months, and
- an approval for storage and use by any person for 24 months.

Where there is a need for tighter control of the withdrawal of the product from the supply chain, for example failure to meet data submission deadlines, the current timelines will be retained:

- immediate revocation for advertisement, sale or supply for the approval holder
- approval for 6 months for advertisement, sale and supply by 'others', and
- approval for storage and use by any person for 18 months.

Immediate revocation and product withdrawal remain an option where serious concerns are identified, with immediate revocation of all approvals and approval for storage only by anyone for 3 months, to allow for disposal of product.

The expiry date shown in the product fact sheet is the final date of legal use of the product.

Off-label Extension of Use

Products may legally be used in a manner not covered by the printed label in several ways:

- In accordance with a specific off-label approval (SOLA). SOLAs are uses for which individuals or organisations other than the manufacturers have sought approval. The Notices of Approval are published by CRD and are widely available from ADAS or NFU offices. Users of SOLAs must first obtain a copy of the relevant Notice of Approval and comply strictly with the conditions laid down therein. Users of this Guide will find details of extant SOLAs and direct links to the CRD website, where SOLA notices can also be accessed.
- In tank mixture with other approved pesticides in accordance with Consent C(i) made under FEPA. Full details of Consent C(i) are given in Annex A of Guide to Pesticides on the PSD website, but there are two essential requirements for tank mixes. First, all the conditions of approval of all the components of a mixture must be complied with. Second, no person may mix or combine pesticides that are cholinesterase compounds unless allowed by the label of at least one of the pesticides in the mixture.
- In conjunction with authorised adjuvants.
- In reduced spray volume under certain conditions.
- In the use of certain herbicides on specified set-aside areas subject to restrictions, which differ between Scotland and the rest of UK.
- By mutual recognition of a use fully approved in another Member State of the European Union and authorised by CRD.

Although approved, off-label uses are not endorsed by manufacturers and such treatments are made entirely at the risk of the user.

Using Crop Protection Chemicals

Use of Herbicides In or Near Water

Products in this *Guide* approved for use in or near water are listed in Table 5.1. Before use of any product in or near water, the appropriate water regulatory body (Environment Agency/Local Rivers Purification Authority; or in Scotland the Scottish Environment Protection Agency) must be consulted. Guidance and definitions of the situation covered by approved labels are given in the Defra publication *Guidelines for the Use of Herbicides on Weeds in or near Watercourses and Lakes*. Always read the label before use.

Table 5.1 Products approved for use in or near water

Chemical	Product
2,4-D	Depitox
glyphosate	Asteroid, Barclay Gallup 360, Barclay Gallup Amenity, Barclay Gallup Biograde Amenity, Barclay Gallup Hi-Aktiv, Buggy XTG, Discman Biograde, Dow Agrosciences Glyphosate 360, Envision, Glyfo TDI, Glyfos Dakar, Glyfos Dakar Pro, Glyfos Proactive, Glypho-Rapid 450, Greenaway Gly-490, Kernel, Manifest, Pure Glyphosate 360, Roundup Biactive, Roundup Pro Biactive, Roundup ProBiactive 450, Rustler, Tangent

Use of Pesticides in Forestry

Table 5.2 Products in this *Guide* approved for use in forestry

Chemical	Product	Use
2,4-D + dicamba + triclopyr	Broadsword	Herbicide
aluminium ammonium sulphate	Curb Crop Spray Powder, Liquid Curb Crop Spray, Sphere ASBO	Animal deterrent/repellent
asulam	Asulox, Brack-N, Formule 1, Greencrop Frond	Herbicide
chlorpyrifos	Agriguard Chlorpyrifos, Dursban WG, Equity, Govern, Parapet	Acaricide, Insecticide
cycloxydim	Laser	Herbicide
cypermethrin	Forester	Insecticide
diflubenzuron	Dimilin Flo	Insecticide
ferric phosphate	Ferramol Max, Sluxx	Molluscicide
fluazifop-P-butyl	Clayton Maximus, Fusilade Max, Greencrop Bantry, Howitzer	Herbicide
glufosinate-ammonium	Harvest, Kaspar, Kibosh	Herbicide
glyphosate	Amega Duo, Barclay Gallup 360, Barclay Gallup Amenity, Barclay Gallup Biograde Amenity, Barclay Gallup Hi-Aktiv, Buggy XTG, Charger, Charger B, Charger C, Clinic Ace, Credit DST, Discman Biograde, Dow Agrosciences Glyphosate 360, Envision, Etna, Glyfo TDI, Glyfos, Glyfos Dakar Pro, Glyfos Proactive, Glyfos Supreme, Glyfosat 36, Glyphogan, Glypho-Rapid 450, Greenaway Gly-490, Hilite, Kernel, Manifest, Nomix Conqueror, Nufosate Ace, Pure Glyphosate 360, Reaper, Roundup Metro, Roundup Pro Biactive, Roundup ProBiactive 450, Shyfo, Slingshot, Stacato, Stamen, Statis, Statis 360, Stirrup, Tangent, Tanker, Typhoon 360	Herbicide
isoxaben	Agriguard Isoxaben, Flexidor 125	Herbicide

Useful Information

Chemical	Product	Use
metazachlor	Alpha Metazachlor 50 SC, Clayton Buzz, Clayton Metazachlor 50 SC, Greencrop Monogram, Marksman, Mezzanine, Rapsan 500 SC, Standon Metazachlor 500, Sultan 50 SC	Herbicide
napropamide	MAC-Napropamide 450 SC	Herbicide
propaquizafop	Bulldog, Clayton Orleans, Cleancrop GYR 2, Falcon, Greencrop Satchmo, MAC-Propaquizafop 100 EC, PQF 100, Raptor, Shogun, Standon Propaquizafop, Standon Zing PQF, Zealot	Herbicide
propyzamide	Careca, Cohort, Engage, Kerb 50 W, Kerb Granules, Megaflo, Menace 80 EDF, Pizza 400 SC, Propel Flo, Proper Flo, Quaver Flo, Setanta 50 WP, Solitaire 50 WP, Standon Santa Fe 50 WP, Standon Santa Fe Flo	Herbicide
pyrethrins	Spruzit	Insecticide
triclopyr	Thrash, Timbrel, Woody	Herbicide

Pesticides Used as Seed Treatments

Information on the target pests for these products can be found in the relevant pesticide profile in Section 2.

Table 5.3 Products used as seed treatments (including treatments on seed potatoes)

Chemical	Product	Formulation	Crop(s)
beta-cyfluthrin + clothianidin	Modesto	FS	Poppies for morphine production, Winter oilseed rape
	Poncho Beta	FS	Fodder beet, Sugar beet
beta-cyfluthrin + imidacloprid	Chinook Blue	LS	Evening primrose, Honesty, Linseed, Mustard, Spring oilseed rape, Winter oilseed rape
	Chinook Colourless	LS	Evening primrose, Honesty, Linseed, Mustard, Winter oilseed rape
carboxin + thiram	Anchor	FS	Spring barley, Spring oats, Spring rye, Spring wheat, Triticale, Winter barley, Winter oats, Winter rye, Winter wheat
clothianidin	Deter	FS	Durum wheat, Triticale, Winter barley, Winter oats, Winter rye, Winter wheat
clothianidin + prothioconazole	Redigo Deter	FS	Durum wheat, Triticale, Winter barley, Winter oats, Winter rye, Winter wheat
clothianidin + prothioconazole + tebuconazole + triazoxide	Raxil Deter	FS	Winter barley
cymoxanil + fludioxonil + metalaxyl-M	Wakil XL	WS	Broad beans, Carrots, Chard, Chives, Combining peas, Herbs (see appendix 6), Leaf spinach, Lupins, Parsley, Parsnips, Poppies for morphine production, Red beet, Salad brassicas, Spinach beet, Spring field beans, Vining peas, Winter field beans

Useful Information

Chemical	Product	Formulation	Crop(s)
difenoconazole + fludioxonil	Celest Extra	FS	Winter oats, Winter rye, Winter wheat
fludioxonil	Beret Gold	FS	Durum wheat, Spring barley, Spring oats, Spring rye, Spring wheat, Triticale, Winter barley, Winter oats, Winter rye, Winter wheat
fludioxonil + flutriafol	Beret Multi	FS	Spring barley, Spring oats, Spring wheat, Winter barley, Winter oats, Winter wheat
fludioxonil + metalaxyl M	Maxim XL	FS	Forage maize
fludioxonil + metalaxyl-M + thiamethoxam	Cruiser OSR	FS	Fodder rape, Mustard, Poppies for morphine production, Spring oilseed rape, Winter oilseed rape
fludioxonil + tefluthrin	Austral Plus	FS	Spring barley, Spring oats, Spring wheat, Triticale seed crop, Winter barley, Winter oats, Winter wheat
fluoxastrobin + prothioconazole	Redigo Twin TXC	FS	Durum wheat, Spring oats, Spring rye, Spring wheat, Triticale, Winter oats, Winter rye, Winter wheat
fluquinconazole	Galmano	FS	Durum wheat, Triticale, Winter wheat
	Jockey Solo	FS	Winter barley, Winter wheat
fluquinconazole + prochloraz	Epona	FS	Winter wheat
	Galmano Plus	FS	Durum wheat, Triticale, Winter wheat
	Jockey	FS	Winter barley, Winter wheat
flutolanil	Rhino DS	DS	Potatoes

Chemical	Product	Formulation	Crop(s)
fuberidazole + imidacloprid + triadimenol	Tripod Plus	LS	Winter barley, Winter oats, Winter wheat
fuberidazole + triadimenol	Tripod	FS	Spring barley, Spring oats, Spring rye, Spring wheat, Triticale, Winter barley, Winter oats, Winter rye, Winter wheat
guazatine	Panoctine	LS	Spring barley, Spring oats, Spring wheat, Winter barley, Winter oats, Winter wheat
	Ravine	LS	Spring barley, Spring oats, Spring wheat, Winter barley, Winter oats, Winter wheat
guazatine + imazalil	Panoctine Plus	LS	Spring barley, Spring oats, Winter barley, Winter oats
	Ravine Plus	LS	Spring barley, Spring oats, Winter barley, Winter oats
imazalil	Fungazil 50LS	LS	Spring barley, Winter barley
	IMA 100 SL	LS	Seed potatoes, Ware potatoes
imazalil + pencycuron	Monceren IM	DS	Potatoes
imidacloprid	Nuprid 600FS	FS	Fodder beet, Sugar beet
metalaxyl-M	Apron XL	ES	Beetroot, Broccoli, Brussels sprouts, Bulb onions, Cabbages, Calabrese, Cauliflowers, Chinese cabbage, Herbs (see appendix 6), Kohlrabi, Ornamental plant production, Radishes, Spinach
pencycuron	Agriguard Pencycuron	DS	Potatoes
	Monceren DS	DS	Potatoes
	Pency Flowable	FS	Potatoes

Useful Information

Chemical	Product	Formulation	Crop(s)
	Standon Pencycuron DP	DS	Potatoes
physical pest control	Silico-Sec	DS	Stored grain
prochloraz	Prelude 20LF	LS	Flax, Hemp for oilseed production, Linseed
prochloraz + thiram	Agrichem Hy-Pro Duet	FS	Winter oilseed rape
prochloraz + triticonazole	Kinto	FS	Durum wheat, Triticale, Winter barley, Winter oats, Winter rye, Winter wheat
prothioconazole	Redigo	FS	Durum wheat, Spring oats, Spring rye, Spring wheat, Triticale, Winter barley, Winter oats, Winter rye, Winter wheat
prothioconazole + tebuconazole + triazoxide	Raxil Pro	FS	Spring barley, Winter barley
silthiofam	Latitude	FS	Durum wheat, Spring rye, Spring wheat, Triticale, Winter barley, Winter rye, Winter wheat
	Meridian	FS	Spring wheat, Winter barley, Winter wheat
tebuconazole	Gizmo	FS	Bulb onions, Garlic, Salad onions, Shallots, Spring barley, Winter barley
thiamethoxam	Cruiser SB	FS	Fodder beet, Sugar beet

Using Crop Protection Chemicals

Chemical	Product	Formulation	Crop(s)
thiram	Agrichem Flowable Thiram	FS	Broad beans, Bulb onions, Cabbages, Carrots, Cauliflowers, Celery (outdoor), Combining peas, Cress, Dwarf beans, Edible podded peas, Endives, Fodder beet, Forage maize, Frise, Garlic, Grass seed, Lamb's lettuce, Leeks, Lettuce, Lupins, Mangels, Parsley, Poppies for morphine production, Radicchio, Radishes, Red beet, Rhubarb, Runner beans, Salad onions, Scarole, Shallots, Soya beans, Spring field beans, Spring oilseed rape, Sugar beet, Swedes, Turnips, Vining peas, Winter field beans, Winter oilseed rape
	Thiraflo	FS	Broad beans, Combining peas, Dwarf beans, Forage maize, Grass seed, Runner beans, Spring field beans, Spring oilseed rape, Vining peas, Winter field beans, Winter oilseed rape
	Thyram Plus	FS	Broad beans, Bulb onions, Cabbages, Cauliflowers, Combining peas, Dwarf beans, Edible podded peas, Forage maize, Garlic, Grass seed, Leeks, Lupins, Radishes, Runner beans, Salad onions, Shallots, Soya beans, Spring field beans, Spring oilseed rape, Swedes, Turnips, Vining peas, Winter field beans, Winter oilseed rape
tolclofos-methyl	Rizolex Flowable	FS	Potatoes, Seed potatoes

Aerial Application of Pesticides

Products used to be granted specific approval for aerial application, but under the new regulations there will be a general prohibition of aerial application, except for a limited number of derogations such as the treatment of bracken in upland areas, which would be impossible to treat by any other method. These approvals will specify measures for warning residents and bystanders, will be allowed only under certain weather conditions, and will require application equipment fitted with the best available technology to reduce drift into non-target areas. All EU member states will be required to keep records of these uses.

Table 5.4 Products approved for aerial application

Chemical	Product	Crops
2-chloroethylphosphonic acid	Agriguard Cerusite	Winter barley
	Agrotech Ethephon	Winter barley
	Becki	Winter barley
	Cerone	Winter barley
	Ethefon 480	Winter barley
	Pan Stiffen	Winter barley
asulam	Agrotech Asulam	Amenity grassland, Forest, Permanent grassland, Rotational grass
	Asulox	Amenity grassland, Forest, Permanent grassland, Rotational grass
	Brack-N	Amenity grassland, Forest, Permanent grassland
	Greencrop Frond	Amenity grassland, Forest, Permanent grassland
chlormequat	3C Chlormequat 720	Spring oats, Spring rye, Spring wheat, Triticale, Winter barley, Winter oats, Winter rye, Winter wheat
	Agriguard Chlormequat 700	Spring oats, Spring wheat, Winter oats, Winter wheat
	Agriguard Chlormequat 720	Spring oats, Spring wheat, Winter oats, Winter wheat
	Agrovista 3 See 750	Spring oats, Spring wheat, Winter oats, Winter wheat
	Barclay Holdup	Spring oats, Spring wheat, Winter barley, Winter oats, Winter wheat

Using Crop Protection Chemicals

Chemical	Product	Crops
	CCC 720	Spring oats, Spring rye, Spring wheat, Triticale, Winter barley, Winter oats, Winter rye, Winter wheat
	Greencrop Carna	Spring oats, Spring wheat, Winter barley, Winter oats, Winter wheat
	Hive	Winter wheat
	Mirquat	Spring oats, Spring wheat, Winter oats, Winter wheat
	New 5C Cycocel	Spring oats, Spring rye, Spring wheat, Triticale, Winter barley, Winter oats, Winter rye, Winter wheat
	New 5C Quintacel	Spring oats, Spring rye, Spring wheat, Triticale, Winter barley, Winter oats, Winter rye, Winter wheat
	Sigma PCT	Spring wheat, Winter barley, Winter wheat
	Terbine	Winter wheat
chlorothalonil	Busa	Potatoes
	Greencrop Orchid 2	Potatoes
	Greencrop Orchid B	Potatoes
	Juliet	Potatoes
	Sanspor	Potatoes
	Sonar	Potatoes
copper oxychloride	Cuprokylt	Potatoes
diflubenzuron	Dimilin Flo	Forest
dimethoate	Danadim Progress	Sugar beet, Triticale, Winter rye, Winter wheat
	Dimethoate 40	Spring rye, Spring wheat, Sugar beet, Sugar beet seed crops, Triticale, Winter rye, Winter wheat
mancozeb	Dithane NT Dry Flowable	Potatoes
	Laminator FL	Potatoes

Useful Information

Chemical	Product	Crops
	Micene 80	Potatoes
	Micene DF	Potatoes
	Quell Flo	Potatoes
	Trimanzone	Potatoes
metaldehyde	Doff Horticultural Slug Killer Blue Mini Pellets	All edible crops (outdoor), All non-edible crops (outdoor), Cultivated land/soil
	ESP	All edible crops (outdoor), All non-edible crops (outdoor)
	Polymet 5	All edible crops (outdoor), All non-edible crops (outdoor), Natural surfaces not intended to bear vegetation
	Super 5	All edible crops (outdoor), All non-edible crops (outdoor), Natural surfaces not intended to bear vegetation
pirimicarb	Aphox	Durum wheat, Spring barley, Spring oats, Spring rye, Spring wheat, Triticale, Winter barley, Winter oats, Winter rye, Winter wheat
	Arena	Spring barley, Spring oats, Spring wheat, Winter barley, Winter oats, Winter wheat
	Phantom	Durum wheat, Spring barley, Spring oats, Spring rye, Spring wheat, Triticale, Winter barley, Winter oats, Winter rye, Winter wheat
	Pirimate	Spring barley, Spring oats, Spring wheat, Winter barley, Winter oats, Winter wheat
	Pirimicarb 50	Durum wheat, Spring barley, Spring oats, Spring rye, Spring wheat, Triticale, Winter barley, Winter oats, Winter rye, Winter wheat
	Reynard	Durum wheat, Spring barley, Spring oats, Spring rye, Spring wheat, Triticale, Winter barley, Winter oats, Winter rye, Winter wheat

Chemical	Product	Crops
	Standon Pirimicarb 50	Durum wheat, Spring barley, Spring oats, Spring rye, Spring wheat, Triticale, Winter barley, Winter oats, Winter rye, Winter wheat
tri-allate	Avadex Excel 15G	Combining peas, Fodder beet, Lucerne, Red beet, Red clover, Sainfoin, Spring barley, Spring field beans, Sugar beet, Triticale, Vetches, Vining peas, White clover, Winter barley, Winter field beans, Winter wheat

Resistance Management

Pest species are, by definition, adaptable organisms. The development of resistance to some crop protection chemicals is just one example of this adaptability. Repeated use of products with the same mode of action will clearly favour those individuals in the pest population able to tolerate the treatment. This leads to a situation where the tolerant (or resistant) individuals can dominate the population and the product becomes ineffective. In general, the more rapidly the pest species reproduces and the more mobile it is, the faster the emergence of resistant populations, although some weeds seem able to evolve resistance more quickly than would be expected. In the UK, key independent research organisations, chemical manufacturers and other organisations have collaborated to share knowledge and expertise on resistance issues through three action groups. Participants include ADAS, the Chemicals Regulation Directorate, universities, colleges and the Home Grown Cereals Authority. The groups have a common aim of monitoring resistance in UK and devising and publishing management strategies designed to combat it where it occurs.

- **Weed Resistance Action Group (WRAG)**, formed in 1989 (Secretary: Dr Stephen Moss, Rothamsted Research, Harpenden, Herts AL5 2JQ *Tel: 01582 7631330 ext. 2521*).
- **Fungicide Resistance Action Group (FRAG)**, formed in 1995 (Secretary: Mr Oliver Macdonald, Chemicals Regulation Directorate, Mallard House, King's Pool, 3 Peasholme Green, York YO1 2PX *Tel: 01904 455864*).
- **Insecticide Resistance Action Group (IRAG)**, formed in 1997 (Secretary: Dr Stephen Foster, Rothamsted Research, Harpenden, Herts AL5 2JQ *Tel: 01582 763133 ext. 2324*).

The above groups publish detailed advice on resistance management relevant for each sector and, in some cases, specific to a pest problem. This information, together with further details about the function of each group, can be obtained from the Chemicals Regulation Directorate website (www.pesticides.gov.uk).

The speed at which resistance appears depends on the mode of action of the crop protection chemicals, as well as the manner in which they are used. Resistance among insects and fungal diseases has been evident for much longer than weed resistance to herbicides, but examples in all three categories are now widespread and increasing. This has created a need for agreement on the advice given for the use of crop protection chemicals in order to reduce the likelihood of the development of resistance and to avoid the loss of potentially valuable products in the chemical armoury. Mixing or alternating modes of action is one of the guiding principles of resistance management. To assist appropriate product choices, mode of action codes, published by the international resistance action committees (see below), are shown in the respective active ingredient fact sheets. The product label and/or a professional advisor should always be consulted before making decisions. The general guidelines for resistance management are similar for all three problem areas.

Preparation in advance

- Be aware of factors that favour the development of resistance, such as repeated annual use of the same product, and assess the risk.
- Plan ahead and aim to integrate all possible means of control.
- Use cultural measures, such as rotations, stubble hygiene, variety selection and, for fungicides, removal of primary inoculum sources, to reduce reliance on chemical control.
- Monitor crops regularly.
- Keep aware of local resistance problems.
- Monitor effectiveness of actions taken and take professional advice, especially in cases of unexplained poor control.

Using crop protection products

- Optimise product efficacy by using them as directed, at the right time, in good conditions.
- Treat pest problems early.
- Mix or alternate chemicals with different modes of action.
- Avoid repeated applications of very low doses.
- Keep accurate field records.

Using Crop Protection Chemicals

Label guidance depends on the appropriate strategy for the product. Most frequently it consists of a warning of the possibility of poor performance due to resistance, and a restriction on the number of treatments that should be applied in order to minimise the development of resistance. This information is summarised in the profiles in the active ingredient fact sheets, but detailed guidance must always be obtained by reading the label itself before use.

International Action Committees

Resistance to crop protection products is an international problem. Agrochemical industry collaboration on a global scale is via three action committees whose aims are to support a coordinated industry approach to the management of resistance worldwide. In particular they produce lists of crop protection chemicals classified according to their mode of action. These lists and other information can be obtained from the respective websites.

Herbicide Resistance Action Committee – www.hracglobal.com

Fungicide Resistance Action Committee – www.frac.info/frac/index.htm

Insecticide Resistance Action Committee – www.irac-online.org

Poisons and Poisoning

Chemicals Subject to the Poison Law

Certain products are subject to the provisions of the Poisons Act 1972, the Poisons List order 1982 and the Poisons Rules 1982 (copies obtainable from The Stationery Office, www.tso.co.uk). These Rules include general and specific provisions for the storage and sale and supply of listed non-medicine poisons. Full details can be accessed in Annex C of the Guide to Pesticides on the Chemicals Regulation Directorate (CRD) website. The nature of the formulation and the concentration of the active ingredient allow some products to be exempted from the Rules, while others with the same active ingredient are included (see below). The chemicals approved for use in the UK are specified under Parts I and II of the Poisons List as follows.

Part I Poisons (sale restricted to registered retail pharmacists and to registered non-pharmacy businesses provided sales do not take place on retail premises):

- aluminium phosphide
- chloropicrin
- magnesium phosphide.

Part II Poisons (sale restricted to registered retail pharmacists and listed sellers registered with a local authority):

- formaldehyde
- oxamyl (a).

Notes

(a) Granular formulations that do not contain more than 12% w/w of this or a combination of similarly flagged poisons are exempt.

Occupational Exposure Limits

A fundamental requirement of the COSHH Regulations is that exposure of employees to substances hazardous to health should be prevented or adequately controlled. Exposure by inhalation is usually the main hazard, and in order to measure the adequacy of control of exposure by this route various substances have been assigned occupational exposure limits.

There are two types of occupational exposure limits defined under COSHH: Occupational Exposure Standards (OES) and Maximum Exposure Limits (MEL). The key difference is that an OES is set at a level at which there is no indication of risk to health; for a MEL a residual risk may exist and the level takes socio-economic factors into account. In practice, MELs have been most often allocated to carcinogens and to other substances for which no threshold of effect can be identified and for which there is no doubt about the seriousness of the effects of exposure.

OESs and MELs are set on the recommendation of the Advisory Committee on Toxic Substances (ACTS). Full details are published by HSE in *EH 40/95 Occupational Exposure Limits 1995*, ISBN 0 7176 0876 X.

As far as pesticides are concerned, OESs and MELs have been set for relatively few active ingredients. This is because pesticide products usually contain other substances in their formulation, including solvents, which may have their own OES/MEL. In practice inhalation of solvent may be at least, or more, important than that of the active ingredient. These factors are taken into account by the regulators when approving a pesticide product under the Control of Pesticide Regulations. This indicates one of the reasons why a change of pesticide formulation usually necessitates a new approval assessment under COPR.

First Aid Measures

If pesticides are handled in accordance with the required safety precautions, as given on the container label, poisoning should not occur. It is difficult, however, to guard completely against the occasional accidental exposure. Thus, if a person handling, or exposed to, pesticides becomes ill, it is a wise precaution to apply first aid measures appropriate to pesticide poisoning even though the cause of illness may eventually prove to have been quite different. An employer has a legal duty to make adequate first aid provision for employees. Regular pesticide users should consider appointing a trained first aider even if numbers of employees are not large, since there is a specific hazard.

The first essential in a case of suspected poisoning is for the person involved to stop work, to be moved away from any area of possible contamination and for a doctor to be called at once. If no doctor is available the patient should be taken to hospital as quickly as possible. In either event it is most important that the name of the chemical being used should be recorded and preferably the whole product label or leaflet should be shown to the doctor or hospital concerned.

Some pesticides, which are unlikely to cause poisoning in normal use, are extremely toxic if swallowed accidentally or deliberately. In such cases get the patient to hospital as quickly as possible, with all the information you have. Some labels now include Material Safety Data Sheets and these contain valuable information for both the first aider and medical staff.

General Measures

Measures appropriate in all cases of suspected poisoning include the following.

- Remove any protective or other contaminated clothing (taking care to avoid personal contamination).
- Wash any contaminated areas carefully with water or with soap and water if available.
- In cases of eye contamination, flush with plenty of clean water for at least 15 minutes.
- Lay the patient down, keep at rest and under shelter. Cover with one clean blanket or coat, etc. Avoid overheating.
- Monitor level of consciousness, breathing and pulse rate.
- If consciousness is lost, place the casualty in the recovery position (on his/her side with head down and tongue forward to prevent inhalation of vomit).

Reporting of Pesticide Poisoning

Any cases of poisoning by pesticides must be reported without delay to an HM Agricultural Inspector of the Health and Safety Executive. In addition any cases of poisoning by substances named in schedule 2 of The Reporting of Injuries, Diseases and Dangerous Occurrences Regulations 1985, must also be reported to HM Agricultural Inspectorate (this includes organophosphorus chemicals, mercury and some fumigants).

Cases of pesticide poisoning should also be reported to the manufacturer concerned.

Additional Information

General advice on the safe use of pesticides is given in a range of Health and Safety Executive leaflets available from HSE Books. The major agrochemical companies are able to provide authoritative medical advice about their own pesticide products. Useful information is now available from the Material Safety Data Sheet available from the manufacturer or the manufacturer's website.

A useful booklet, *Guidelines for Emergency Measures in Cases of Pesticide Poisoning*, is available from CropLife International.

New arrangements for the provision of information to doctors about poisons and the management of poisonings have been introduced as part of the modernisation of the National Poisons Information Service (NPIS). The Service provides a year-round, 24-hour-a-day service for healthcare staff on the diagnosis, treatment and management of patients who may have been poisoned. The new arrangements are aimed at moving away from the telephone as the first point

Useful Information

of contact for poisons information, to the use by doctors of an online database, supported by a second-tier, consultant-led information service for more complex clinical advice. NPIS no longer provides direct information on poisoning to members of the public. **Anyone suspecting poisoning by pesticides should seek professional medical help immediately via their GP or NHS Direct (www.nhsdirect.nhs.uk) or NHS 24 (www.nhs24.com).**

Environmental Protection

Protection of Bees

Honey bees

Honey bees are a source of income for their owners and important to farmers and growers as pollinators of their crops. It is irresponsible and unnecessary to use pesticides in such a way that may endanger them. Pesticides vary in their toxicity to bees, but those that present a special hazard carry a specific warning in the precautions section of the label. They are indicated in this *Guide* in the hazard classification and safety precautions section of the pesticide profile.

Product labels indicate the necessary environmental precautions to take, but where use of an insecticide on a flowering crop is contemplated the British Beekeepers Association have produced the following guidelines for growers:

- Target insect pests with the most appropriate product.
- Choose a product that will cause minimal harm to beneficial species.
- Follow the manufacturer's instructions carefully.
- Inspect and monitor crops regularly.
- Avoid spraying crops in flower or where bees are actively foraging.
- Keep down flowering weeds.
- Spray late in the day, in still conditions.
- Avoid excessive spray volume and run-off.
- Adjust sprayer pressure to reduce production of fine droplets and drift.
- Give local beekeepers as much warning of your intention as possible.

Wild bees

Wild bees also play an important role. Bumblebees are useful pollinators of spring flowering crops and fruit trees because they forage in cool, dull weather when honey bees are inactive. They play a particularly important part in pollinating field beans, red and white clover, lucerne and borage. Bumblebees nest and overwinter in field margins and woodland edges. Avoidance of direct or indirect spray contamination of these areas, in addition to the creation of hedgerows and field margins, and late cutting or grazing of meadows, all help the survival of these valuable insects.

The Campaign Against Illegal Poisoning of Wildlife

The Campaign Against Illegal Poisoning of Wildlife, aimed at protecting some of Britain's rarest birds of prey and wildlife whilst also safeguarding domestic animals, was launched in March 1991 by the (then) Ministry of Agriculture, Fisheries and Food and the Department of the Environment, Transport and the Regions. The main objective is to deter those who may be considering using pesticides illegally.

The Campaign is supported by a range of organisations associated with animal welfare, nature preservation, field sports and gamekeeping including the RSPB, English Nature, the Countryside Alliance and the Game Conservancy Trust.

The three objectives are:

- To advise farmers, gamekeepers and other land managers on legal ways of controlling pests;
- To advise the public on how to report illegal poisoning incidents and to respect the need for legal alternatives;
- To investigate incidents and prosecute offenders.

Useful Information

A freephone number (0800 321 600) is available to make it easier for the public to report incidents, and numerous leaflets, posters, postcards and stickers have been created to publicise the existence of the Campaign. A video has also been produced to illustrate the many talks, demonstrations and exhibitions.

The Campaign arose from the results of the Wildlife Incident Investigation Scheme for the investigation of possible cases of illegal poisoning. Under this scheme, all reported incidents are considered and thoroughly investigated where appropriate. Enforcement action is taken wherever sufficient evidence of an offence can be obtained and numerous prosecutions have been made since the start of the Campaign.

Further information about the Campaign is available from: Chemicals Regulation Directorate, Research Co-ordination and Environmental Policy Branch, Room 317, Mallard House, 3 Peasholme Green, York YO1 7PX, or from the website: www.caip-uk.info/default.aspx. E-mail: wiis@hse.gsi.gov.uk

Water Quality

Even when diluted, some pesticides are potentially dangerous to fish and other aquatic life. Not only this, but many watercourses and groundwaters are sources of drinking water, and it requires only a tiny amount of contamination to breach the stringent European Union water quality standards. The EEC Drinking Water Directive sets a maximum admissible level in drinking water for any pesticide, regardless of its toxicity, at 1 part in 10,000 million. As little as 250 grams could be enough to cause the daily supply to a city the size of London to exceed the permitted levels (although it would be very unlikely to present a health hazard to consumers).

The protection of groundwater quality is therefore vital. The Food and Environment Protection Act 1985 (FEPA) places a special obligation on users of pesticides to "safeguard the environment and in particular avoid the pollution of water". Under the Water Resources Act 1991 it is an offence to pollute any controlled waters (watercourses or groundwater), either deliberately or accidentally. Protection of controlled waters from pollution is the responsibility of the Environment Agency (in England and Wales) and the Scottish Environmental Protection Agency. Addresses for both organisations can be found under Contacts.

Users of pesticides therefore have a duty to adopt responsible working practices and, unless they are applying herbicides in or near water, to prevent them getting into water. Guidance on how to achieve this is given in the Defra *Code of Good Agricultural Practice for the Protection of Water*.

The duty of care covers not only the way in which a pesticide is sprayed, but also its storage, preparation and disposal of surplus, sprayer washings and the container. Products that are a major hazard to fish, other aquatic life or aquatic higher plants carry one of several specific label precautions in their fact sheet, depending on the assessed hazard level.

Where to get information

For general enquiries telephone 0645 333 111 (Environment Agency) or 01786 457 700 (SEPA). Both Agencies operate a 24-hour emergency hotline for reporting all environmental incidents relating to air, land and water:

0800 80 70 60 (for incidents in England and Wales)

0345 73 72 71 (for incidents in Scotland)

In addition, printed literature concerning the protection of water is available from the Environment Agency and the Crop Protection Association (see Key References).

Protecting surface waters

Surface waters are particularly vulnerable to contamination. One of the best ways of preventing those pesticides that carry the greatest risk to aquatic wildlife from reaching surface waters is to prohibit their application within a boundary adjacent to the water. Such areas are known as no-spray, or buffer, zones. Certain products are restricted in this way and have a legally binding label precaution to make sure the potential exposure of aquatic organisms to pesticides that might harm them is minimised.

Before 1999 the protected zones were measured from the edge of the water. The distances were 2 metres for hand-held or knapsack sprayers, 6 metres for ground crop sprayers, and a variable distance (but often 18 metres) for broadcast air-assisted applications, such as in orchards. The introduction of LERAPs (see below) has changed the method of measuring buffer zones.

'Surface water' includes lakes, ponds, reservoirs, streams, rivers and watercourses (natural or artificial). It also includes temporarily or seasonally dry ditches, which have the potential to carry water at different times of the year. Buffer zone restrictions do not necessarily apply to all products containing the same active ingredient. Those in formulations that are not likely to contaminate surface water through spray drift do not pose the same risk to aquatic life and are not subject to the restrictions.

Local Environmental Risk Assessment for Pesticides (LERAPs)

Local Environment Risk Assessments for Pesticides (LERAPs) were introduced in March 1999, and revised guidelines were issued in January 2002. They give users of most products currently subject to a buffer zone restriction the option of continuing to comply with the existing buffer zone restriction (using the new method of measurement), or carrying out a LERAP and possibly reducing the size of the buffer zone as a result. In either case, there is a new legal obligation for the user to record his decision, including the results of the LERAP.

The scheme has changed the method of measuring the buffer zone. Previously the zone was measured from the edge of the water, but it is now the distance from the top of the bank of the watercourse to the edge of the spray area.

The LERAP provides a mechanism for taking into account other factors that may reduce the risk, such as dose reduction, the use of low drift nozzles, and whether the watercourse is dry or flowing. The previous arrangements applied the same restriction regardless of whether there was actually water present. Now there is a standard zone of 1 m from the top of a dry ditch bank.

Other factors to include in a LERAP that may allow a reduction in the buffer zone are:

- The size of the watercourse, because the wider it is, the greater the dilution factor, and the lower the risk of serious pollution.
- The dose applied. The lower the dose, the less is the risk.
- The application equipment. Sprayers and nozzles are star-rated according to their ability to reduce spray drift fallout. Equipment offering the greatest reductions achieves the highest rating of three stars. The scheme was originally restricted to ground crop sprayers; new, more flexible rules introduced in February 2002 included broadcast air-assisted orchard and hop sprayers.
- Other changes introduced in 2002 allow the reduction of a buffer zone if there is an appropriate living windbreak between the sprayed area and a watercourse.

Not all products that had a label buffer zone restriction are included in the LERAP scheme. The option to reduce the buffer zone does not to apply to organophosphorus or synthetic pyrethroid insecticides. This group are classified as Category A products. All other products that had a label buffer zone restriction are classified as Category B. In addition some products have a 'Broadcast air-assisted LERAP' classification. The wording of the buffer zone label precautions has been amended for all products to take account of the new method of measurement, and whether or not the particular product qualifies for inclusion in the LERAP scheme.

Products in this *Guide* that are in Category A or B, or have a broadcast air-assisted LERAP, are identified with an appropriate icon on the product fact sheet. Updates to the list are published regularly by CRD and details can be obtained from their website at www.pesticides.gov.uk.

Useful Information

The introduction of LERAPs was an important step forward because it demonstrated a willingness to reduce the impact of regulation on users of pesticides by allowing flexibility where local conditions make it safe to do so. This places a legal responsibility on the user to ensure the risk assessment is done either by himself or by the spray operator or by a professional consultant or advisor. It is compulsory to record the LERAP and make it available for inspection by enforcement authorities.

More details of the LERAP arrangements and guidance on how to carry out assessments are published in the Ministry booklets PB5621 *Local Environmental Risk Assessment for Pesticides – Horizontal Boom Sprayers*, and booklet PB6533 *Local Environment Risk Assessment for Pesticides – Broadcast Air-Assisted Sprayers*. Additional booklets, PB2088 *Keeping Pesticides Out of Water*, and PB3160 *Is Your Sprayer Fit for Work* give general practical guidance.

Groundwater Regulations

Groundwater Regulations were introduced in 1999 to complete the implementation in UK of the EU Groundwater Directive (Protection of Groundwater Against Pollution Caused by Certain Dangerous Substances – 80/68/EEC). These Regulations help prevent the pollution of groundwater by controlling discharges or disposal of certain substances, including all pesticides.

Groundwater is defined under the Regulations as any water contained in the ground below the water table. Pesticides must not enter groundwater unless it is deemed by the appropriate Agency to be permanently unsuitable for other uses. The Agricultural Waste Regulations were introduced in May 2006 and apply to the disposal of all farm wastes, including pesticides. With certain exemptions farmers are required to obtain a waste management licence for most waste disposal activities.

As far as pesticides are concerned the new Regulations made little difference to existing controls, except for the disposal of empty containers and the disposal of sprayer washings and rinsings. It remains an offence to dispose of pesticides onto land without official authorisation. Normal use of a pesticide in accordance with product approval does not require authorisation. This includes spraying the washings and rinsings back on the crop provided that, in so doing, the maximum approved dose for that product on that crop is not exceeded. However, those wishing to use a lined biobed for this purpose need to obtain an exemption from the Agency.

In practice the best advice to farmers and growers is to plan to use all diluted spray within the crop and to dispose of all washings via the same route making sure that they stay within the conditions of approval of the product. The enforcing agencies for these Regulations are the Environment Agency (in England and Wales) and the Scottish Environmental Protection Agency. The Agencies can serve notice at any time to modify the conditions of an authorisation where necessary to prevent pollution of groundwater.

Integrated Farm Management (IFM)

Integrated farm management is a method of farming that balances the requirements of running a profitable farming business with the adoption of responsible and sensitive environmental management. It is a whole-farm, long-term strategy that combines the best of modern technology with some basic principles of good farming practice. It is a realistic way forward that addresses the justifiable concerns of the environmental impact of modern farming practices, at the same time as ensuring that the industry remains viable and continues to provide wholesome, affordable food. IFM embraces arable and livestock management. The phrase 'integrated crop management' is sometimes used where a farm is wholly arable.

Pest control is essential in any management system. Much can be done to minimise the incidence and impact of pests, but their presence is almost inevitable. IFM ensures that, where a pest problem needs to be contained, the action taken is the best combination of all available options. Pesticides are one of these options, and form an essential, but by no means exclusive, part of pest control strategy.

Where chemicals are to be used, the choice of product should be made not only with the pest problem in mind, but with an awareness of the environmental and social risks that might accompany its use. The aim should be to use as much as necessary, but as little as possible. A major part of the approval process aims to safeguard the environment, so that any approved product, when used as directed, will not cause long-term harm to wildlife or the environment. This is achieved by specifying on the label detailed rules for the way in which a product may be used.

The skill in implementing an IFM pest control strategy is the decision on how easily these rules may be complied with in the particular situation where use is contemplated. Although many chemical options may be available, some are likely to be more suitable than others. This *Guide* lists, under the heading **Special precautions/Environmental safety**, the key label precautions that need to be considered in this context, although the actual product label must always be read before a product is used.

Campaign for the Farmed Environment

With the demise of set-aside, it was realised that the environmental benefits achieved by this scheme could be lost. Under the threat of legislation to be brought in during 2012, the Campaign for the Farmed Environment (CFE) scheme seeks to retain or even exceed the environmental benefits achieved by set-aside. It has three main themes:

- farming for cleaner water and a healthier soil
- helping farmland birds thrive
- improving the environment for farm wildlife.

The scheme encourages farmers to sign up for (Entry Level Stewardship (ELS) schemes and to target their already extensive knowledge of habitat management to greater effect. If successful, it will help to fend off the proposal that farmers of cultivated land in England will have to adopt environmental management options on up to 6% of their land.

SECTION 6
APPENDICES

Appendix 1
Suppliers of Pesticides and Adjuvants

AgChem Access: AgChemAccess Limited
Pure House,
64-66 Westwick Street
Norwich
Norfolk
NR2 4SZ
UK
Tel: 01603 624413
Fax: 01759 371971
Email: martin@agchemaccess.co.uk
Web: www.agchemaccess.com

Agform: Agform Ltd
Maidstone Heath
Blundell Lane
Burlesdon
Southampton
SO31 1AA
Tel: 023 8040 7831
Fax: 023 8040 7198
Email: info@agform.com

Agrichem: Agrichem (International) Ltd
Industrial Estate
Station Road
Whittlesey
Cambs.
PE7 2EY
Tel: (01733) 204019
Fax: (01733) 204162
Email: alison.williamson@agrichem.co.uk
Web: www.agrichem.co.uk

AgriChem BV: AgriChem BV
Koopvaardijweg 9
4906 CV Oosterhout
Postbus 295
4900 AG Osterhout
Netherlands
Tel: 0031 1624 31931
Fax: 0031 1624 56797
Email: info@agrichem.com
Web: www.agrichem.com

AgriGuard: AgriGuard Ltd
Unit 1
Broomfield Business Park
Malahide
Co. Dublin
Ireland
Tel: (+353) 1 846 2044
Fax: (+353) 1 846 2489
Email: info@agriguard.ie
Web: www.agriguard.ie

Agropharm: Agropharm Limited
Buckingham Place
Church Road
Penn
High Wycombe
Bucks.
HP10 8LN
Tel: (01494) 816575
Fax: (01494) 816578
Email: sales@agropharm.co.uk
Web: www.agropharm.co.uk

Agrovista: Agrovista UK Ltd
Cambridge House
Nottingham Road
Stapleford
Nottingham
NG9 8AB
Tel: (0115) 939 0202
Fax: (0115) 921 8498
Email: enquiries@agrovista.co.uk
Web: www.agrovista.co.uk

Amega: Amega Sciences
Lanchester Way
Royal Oak Industrial Estate
Daventry
Northants.
NN11 5PH
Tel: (01327) 704444
Fax: (01327) 71154
Email: admin@amega-sciences.com
Web: www.amega-sciences.com

Antec Biosentry: Antec Biosentry
DuPont Animal Health Solutions
Windham Road
Chilton Industrial Estate
Sudbury
Suffolk
CO10 2XD
Tel: (01787) 377305
Fax: (01787) 310846
Email: biosecurity@antecint.com
Web: www.antecint.com

B H & B: Battle Hayward & Bower Ltd
Victoria Chemical Works
Crofton Drive
Allenby Road Industrial Estate
Lincoln
LN3 4NP
Tel: (01522) 529206
Fax: (01522) 538960

Barclay: Barclay Chemicals Manufacturing Ltd
 Damastown Way
 Damastown Industrial Park
 Mulhuddart
 Dublin 15
 Ireland
 Tel: (+353) 1 811 2900
 Fax: (+353) 1 822 4678
 Email: margaret@barclay.ie
 Web: www.barclay.ie

Barrier: Barrier BioTech Ltd
 36/37 Haverscroft Industrial Estate
 New Road
 Attleborough
 Norfolk
 NR17 1YE
 Tel: (01953) 456363
 Fax: (01953) 455594
 Email: sales@barrier-biotech.com
 Web: www.barrier-biotech.com

BASF: BASF plc
 Agricultural Divison
 PO Box 4, Earl Road
 Cheadle Hulme
 Cheshire
 SK8 6QG
 Tel: (0845) 602 2553
 Fax: (0161) 485 2229
 Web: www.agricentre.co.uk

Bayer CropScience:
 Bayer CropScience Limited
 230 Cambridge Science Park
 Milton Road
 Cambridge
 CB4 0WB
 Tel: (01223) 226500
 Fax: (01223) 426240
 Web: www.bayercropscience.co.uk

Bayer Environ.: Bayer Environmental Science
 230 Cambridge Science Park
 Milton Road
 Cambridge
 CB4 0WB
 Tel: (01223) 226680
 Fax: (01223) 226635
 Email: claire.hazell@bayercropscience.com
 Web: www.bayer-escience.co.uk

Belchim: Belchim Crop Protection Ltd
 Unit 1b, Fenice Court
 Eaton Socon
 St Neots
 Cambs.
 PE19 8EP
 Tel: (01480) 403333
 Fax: (01480) 403444
 Email: info@belchim.com
 Web: www.belchim.com

Belcrop: Belcrop (UK)
 21 Victoria Rd
 Wargrave
 Berkshire
 RG10 8AD
 UK
 Tel: (0118) 940 4264
 Fax: (0118) 940 4264
 Email: JOHNHUDSON23@aol.com
 Web: www.BELCROP.com

Biofresh: Biofresh Ltd
 INEX Business Centre
 Herschell Building
 Newcastle University Campus
 Newcastle on Tyne
 Tyne & Wear
 NE1 7RU
 UK
 Tel: 0191 243 0879
 Fax: 0191 243 0244
 Email: info@bio-fresh.co.uk

Certis: Certis
 1b Mills Way
 Boscombe Down Business Park
 Amesbury
 Wilts.
 SP4 7RX
 Tel: (01980) 676500
 Fax: (01980) 626555
 Email: certis@certiseurope.co.uk
 Web: www.certiseurope.co.uk

ChemSource: ChemSource Ltd
 5 Jupiter House
 Calleva Park
 Aldermaston
 Reading
 Berkshire
 RG7 4NN
 UK
 Tel: 0845 459 4039
 Email: emma@chemsource.co.uk

Suppliers of Pesticides and Adjuvants

Chemtura: Chemtura Europe Ltd
Kennet House
4 Langley Quay
Slough
Berks.
SL3 6EH
Tel: (01753) 603000
Fax: (01753) 603077
Web: www.chemtura.com

Chiltern: Chiltern Farm Chemicals Ltd
East Mellwaters
Stainmore
Bowes
Barnard Castle
Co. Durham
DL12 9RH
Tel: (01833) 628282
Fax: (01833) 628020
Web: www.chilternfarm.com

Clayton: Clayton Plant Protection Ltd
Bracetown Business Park
Clonee
Co. Meath
Ireland
Tel: (+353) 1 821 0127
Fax: (+353) 81 841 1084
Email: info@cppltd.eu
Web: www.cppltd.eu

DAPT: DAPT Agrochemicals Ltd
14 Monks Walk
Southfleet
Gravesend
Kent
DA13 9NZ
Tel: (01474) 834448
Fax: (01474) 834449
Email: rkjltd@supanet.com

De Sangosse: De Sangosse Ltd
Hillside Mill
Quarry Lane
Swaffham Bulbeck
Cambridge
CB25 0LU
Tel: (01223) 811215
Fax: (01223) 810020
Email: info@desangosse.co.uk
Web: www.desangosse.co.uk

Doff Portland: Doff Portland Ltd
Aerial Way
Hucknall
Nottingham
NG15 6DW
Tel: (0115) 963 2842
Fax: (0115) 963 8657
Email: info@doff.co.uk
Web: www.doff.co.uk

Dow: Dow AgroSciences Ltd
Latchmore Court
Brand Street
Hitchin
Herts.
SG5 1NH
Tel: (01462) 457272
Fax: (01462) 426605
Email: fhihotl@dow.com
Web: www.dowagro.com/uk

DuPont: DuPont (UK) Ltd
Crop Protection Products Department
Wedgwood Way
Stevenage
Herts.
SG1 4QN
Tel: (01438) 734450
Fax: (01438) 734452
Web: www.gbr.ag.dupont.com

Elliott: Thomas Elliott Fertilisers
Selby Place
Stanley Industrial Estate
Skelmersdale
Lancashire
WN8 8EF
Tel: 01522 811981
Fax: 01522 810238
Email:
christinehall@thomaselliot.demon.co.uk

European Ag: European Agrichemicals Ltd
Unit NR2, Hurst House
City Road
Radnage
Bucks
HP14 4DW
UK
Tel: 01494 483728
Email: Radnage

Fargro: Fargro Ltd
Toddington Lane
Littlehampton
Sussex
BN17 7PP
Tel: (01903) 721591
Fax: (01903) 730737
Email: info@fargro.co.uk
Web: www.fargro.co.uk

Fine: Fine Agrochemicals Ltd
Hill End House
Whittington
Worcester
WR5 2RQ
Tel: (01905) 361800
Fax: (01905) 361810
Email: enquire@fine.eu
Web: www.fine.eu

Appendices

Frontier: Frontier Agriculture Ltd
  Fleet Road Industrial Estate
  Holbeach
  Spalding
  Lincs.
  PE12 8LY
  Tel: (01406) 421405
  Email: les_sykes@bankscargill.co.uk
  Web: www.bankscargill.co.uk

Globachem: Globachem NV
  Leeuwerweg 138
  BE-3803 Sint Truiden
  Belgium
  Tel: (+32) 11 785 717
  Fax: (+32) 11 681 565
  Email: globachem@globachem.com
  Web: www.globachem.com

Goldengrass: Goldengrass Ltd
  PO Box 280
  Ware
  SG12 8XY
  Tel: 01920 486253
  Fax: 01920 485123
  Email: contact@goldengrass.uk.com

Gowan: Gowan International
  Carrera
  Valldaura 49
  08195 Sant Cugat
  Spain
  Tel: 0034 617 064 472
  Email: mcomer@gowanco.com

Greenaway: Greenaway Amenity Ltd
  7 Browntoft Lane
  Donington
  Spalding
  Lincs.
  PE11 4TQ
  Tel: (01775) 821031
  Fax: (01775) 821034
  Email: greenawayamenity@aol.com
  Web: www.greenawaycda.com

Greencrop: Greencrop Technology Ltd
  See Clayton Plant Protection Ltd

Headland: Headland Agrochemicals Ltd
  Rectors Lane
  Pentre
  Deeside
  Flintshire
  CH5 2DH
  Tel: (01244) 537370
  Fax: (01244) 532097
  Email: enquiry@headlandgroup.com
  Web: www.headland-ag.co.uk

Headland Amenity:
Headland Amenity Limited
  1010 Cambourne Business Park
  Cambourne
  Cambs.
  CB3 6DP
  Tel: (01223) 597834
  Fax: (01223) 598052
  Email: info@headlandamenity.com
  Web: www.headlandamenity.com

Helena: Helena Chemical Company
  Cambridge House
  Nottingham Road
  Stapleford
  Nottingham
  NG9 8AB
  Tel: (0115) 939 0202
  Fax: (0115) 939 8031

Interagro: Interagro (UK) Ltd
  230 Avenue West,
  Skyline 120,
  Great Notley
  Braintree
  Essex
  CM77 7AA
  Tel: (01376) 552703
  Fax: (01376) 567608
  Email: info@interagro.co.uk
  Web: www.interagro.co.uk

Interfarm: Interfarm (UK) Ltd
  Kinghams's Place
  36 Newgate Street
  Doddington
  Cambs.
  PE15 0SR
  Tel: (01354) 741414
  Fax: (01354) 741004
  Email: technical@interfarm.co.uk
  Web: www.interfarm.co.uk

Intracrop: Intracrop
  Little Hay
  Broadwell
  Lechlade
  Glos.
  GL7 3QS
  Tel: (01367) 860255
  Fax: (01926) 634798
  Email: catherine@a-o-c.co.uk

Suppliers of Pesticides and Adjuvants

Irish Drugs: Irish Drugs Ltd
Burnfoot
Lifford
Co. Donegal
Ireland
Tel: (+353) 74 9368104
Fax: (+353) 74 9368311
Email: jgmcivor@eircom.net
Web: www.idl-home.com

K & S Fumigation:
K & S Fumigation Services Ltd
Asparagus Farm
Court Lodge Road
Appledore
Ashford
Kent
TN26 2DH
Tel: (01233) 758252
Fax: (01233) 758343
Email: david@kstreatments.co.uk

Killgerm: Killgerm Chemicals Ltd
115 Wakefield Road
Flushdyke
Ossett
W. Yorks.
WF5 9AR
Tel: (01924) 268400
Fax: (01924) 264757
Email: info@killgerm.com
Web: www.killgerm.com

Koppert: Koppert (UK) Ltd
Unit 8
53 Hollands Road
Haverhill
Suffolk
CB9 8PJ
Tel: (01440) 704488
Fax: (01440) 704487
Email: info@koppert.co.uk
Web: www.koppert.com

Landseer: Landseer Ltd
Lodge Farm
Goat Hall Lane
Galleywood
Chelmsford
Essex
CM2 8PH
Tel: (01245) 357109
Fax: (01245) 494165
Web: www.lanfruit.co.uk

Law Fertilisers: Law Fertilisers Ltd
Eastwood End
Wimblington
Cambs.
PE15 0QJ
Tel: (01354) 740740
Fax: (01354) 740720
Email: sales@lawfertilisers.co.uk
Web: www.lawfertilisers.co.uk

Lodi UK: Lodi Uk Limited
Building 69
Third Avenue
Pensnett Trading Estate
Kingswinford
West Midlands
DY6 7FD
Tel: 01384 404242
Fax: 01384 404656
Email: phil@lodi-uk.com
Web: www.lodi-uk.com

Makhteshim: Makhteshim-Agan (UK) Ltd
Unit 16
Thatcham Business Village
Colthrop Way
Thatcham
Berks.
RG19 4LW
Tel: (01635) 860555
Fax: (01635) 861555
Email: admin@mauk.co.uk
Web: www.mauk.co.uk

Microcide: Microcide Ltd
Shepherds Grove
Stanton
Bury St. Edmunds
Suffolk
IP31 2AR
Tel: (01359) 251077
Fax: (01359) 251545
Email: microcide@microcide.co.uk
Web: www.microcide.co.uk

Monsanto: Monsanto (UK) Ltd
PO Box 663
Cambourne
Cambridge
CB1 0LD
UK
Tel: (01954) 717575
Fax: (01954) 717579
Email: technical.helpline.uk@monsanto.com
Web: www.monsanto-ag.co.uk

Appendices

Nissan: Nissan Chemical Europe Sarl
Parc d'Affaires de Crecy
2 Rue Claude Chappe
69371 St Didier au Mont d'Or
France
Tel: (+33) 4 376 44020
Fax: (+33) 4 376 46874

Nomix Enviro: Nomix Enviro, A Division of Frontier Agriculture Ltd
The Grain Silos
Weyhill Road
Andover
Hampshire
SP10 3NT
UK
Tel: 01264 388050
Fax: 01522 866176
Email: nomixenviro@frontierag.co.uk
Web: www.nomix.co.uk

Nufarm UK: Nufarm UK Ltd
Wyke Lane
Wyke
West Yorkshire
BD12 9EJ
UK
Tel: 01274 69 1234
Fax: 01274 69 1176
Email: infouk@uk.nufarm.com
Web: www.nufarm.co.uk

Omex: Omex Agriculture Ltd
Bardney Airfield
Tupholme
Lincoln
LN3 5TP
Tel: (01526) 396000
Fax: (01526) 396001
Email: enquire@omex.com
Web: www.omex.co.uk

Pan Agriculture: Pan Agriculture Ltd
8 Cromwell Mews
Station Road
St Ives
Huntingdon
Cambs.
PE27 5HJ
Tel: (01480) 467790
Fax: (01480) 467041
Email: info@panagriculture.co.uk

Pan Amenity: Pan Amenity Ltd
8 Cromwell Mews
Station Road
St Ives
Cambs.
PE27 5HJ
UK
Tel: 01480 467790
Fax: 01480 467041

Plant Solutions: Plant Solutions
Pyports
Downside Bridge Road
Cobham
Surrey
KT11 3EH
Tel: (01932) 576699
Fax: (01932) 868973
Email: sales@plantsolutionsltd.com
Web: www.plantsolutionsltd.com

PP Products: PP Products
Longmynd
Tunstead Road
Hoveton
Norwich
NR12 8QN
Tel: (01603) 784367
Fax: (01603) 784367
Email: mail@ppproducts.net

Pure Amenity : Pure Amenity Limited
Unit 4A, Primrose Hill
Buttercrambe Road
Stamford Bridge
North Yorks
YO41 1AW
Tel: 0845 257 4710
Email: martin@pureamenity.co.uk

Q-Chem: Q-CHEM nv
Leeuwerweg 138
BE-3803
Belgium
Tel: (+32) 11 785 717
Fax: (+32) 11 681 565
Email: q-chem@skynet.be

Rentokil: Rentokil Initial plc
7-8 Foundry Court,
Foundry Lane
Horsham
W. Sussex
RH13 5PY
Tel: (01403) 214122
Fax: (01403) 214101
Email: keira.corrie@rentokil-initial.com
Web: www.rentokil.com

Suppliers of Pesticides and Adjuvants

Rhizopon: Rhizopon UK Ltd
Croda Rosa
12 Bixley Road
Ipswich
Suffolk
IP3 8PL
Tel: (01473) 712666
Fax: (01473) 712666

Rigby Taylor: Rigby Taylor Ltd
Rigby Taylor House
Crown Lane
Horwich
Bolton
Lancs.
BL6 5HP
Tel: (01204) 677777
Fax: (01204) 677715
Email: info@rigbytaylor.com
Web: www.rigbytaylor.com

Scotts: The Scotts Company (UK) Ltd
Paper Mill Lane
Bramford
Ipswich
Suffolk
IP8 4BZ
Tel: (01473) 830492
Fax: (01473) 830814
Web: www.scottsprofessional.co.uk

Sharda: Sharda Worldwide Exports PVT Ltd
Domnic Holm
29th Road
Bandra (West)
Mumbai
400050
India
Tel: 0032 477 707036

Sherriff Amenity: Sherriff Amenity Services
The Pines
Fordham Road
Newmarket
Cambs
CB8 7LG
Tel: (01638) 721888
Fax: (01638) 721815
Web: www.sherriffamenity.com

Sinclair: William Sinclair Horticulture Ltd
Firth Road
Lincoln
LN6 7AH
Tel: (01522) 537561
Fax: (01522) 513609
Web: www.william-sinclair.co.uk

Sipcam: Sipcam UK Ltd
3 The Barn
27 Kneesworth Street
Royston
Herts.
SG8 5AB
Tel: (01763) 212100
Fax: (01763) 212101
Email: paul@sipcamuk.co.uk

Solufeed: Solufeed Ltd
Highground Orchards, Highground Lane
Barnham
Bognor Regis
West Sussex
PO22 0BT
Tel: 01243 554090
Fax: 01243 554568
Email: enquiries@solufeed.com

Sphere: Sphere Laboratories (London) Ltd
c/o Mainswood
Putley Common
Ledbury
Herefordshire
HR8 2RF
Tel: (07974) 732026
Fax: (01531) 670517
Email: homesfg@aol.com

Standon: Standon Chemicals Ltd
48 Grosvenor Square
London
W1K 2HT
Tel: (020) 7493 8648
Fax: (020) 7493 4219

Sumitomo:
Sumitomo Chemical Agro Europe SA
Horatio House
77-85 Fulham Palace Road
London
W6 8JA
Tel: 0208 600 7700
Fax: 0208 600 7717

Summit Agro: Summit Agro Europe Limited
Vintners' Place
68 Upper Thames Street
London
EC4V 3BJ
Tel: (020) 7246 3697
Fax: (020) 7246 3799
Email: summit@summit-agro.com

Appendices

Syngenta:
Syngenta Crop Protection UK Limited
CPC4
Capital Park
Fulbourn
Cambridge
CB21 5XE
Tel: 01223 883400
Fax: (01223) 493700
Web: www.syngenta-crop.co.uk

Syngenta Bioline: Syngenta Bioline
Telstar Nursery
Holland Road
Little Clacton
Essex
CO16 9QG
Tel: (01255) 863200
Fax: (01255) 863206
Email: syngenta.bioline@syngenta.com
Web: www.syngenta-bioline.co.uk

Taminco: Taminco UK Ltd
15 Rampton Drift
Longstanton
Cambridge
CB24 3EH
Tel: 01954 789941
Email: philip.forster@taminco.com
Web: www.taminco.com

Truchem: Truchem Ltd
The Knoll
The Cross
Horsley
Stroud
Gloucestershire
GL6 0PR
Tel: (01453) 833293
Fax: (01453) 833293
Email: nichollstruchem@aol.com

Unicrop: Universal Crop Protection Ltd
Park House
Maidenhead Road
Cookham
Berks.
SL6 9DS
Tel: (01628) 526083
Fax: (01628) 810457
Email: enquiries@unicrop.com

United Agri: United Agri Products Ltd
The Crossways
Alconbury Hill
Huntingdon
PE28 4JH
Tel: (01480) 418000
Fax: (01480) 418010

United Phosphorus: United Phosphorus Ltd
Chadwick House
Birchwood Park
Warrington
Cheshire
WA3 6AE
Tel: (01925) 819999
Fax: (01925) 817425
Email: lpinto@uniphos.com
Web: www.upleurope.com

Vitax: Vitax Ltd
Owen Street
Coalville
Leicester
LE67 3DE
Tel: (01530) 510060
Fax: (01530) 510299
Email: info@vitax.co.uk
Web: www.vitax.co.uk

Whyte Agrochemicals:
Whyte Agrochemicals Ltd
Marlborough House
298 Regents Park Road
Finchley
London
N3 2UA
Tel: (020) 8346 5946
Fax: (020) 8349 4589
Web: www.whytechemicals.co.uk

Appendix 2
Useful Contacts

Agriculture Industries Confederation Ltd
Confederation House
East of England Showground
Peterborough PE2 6XE
Tel: (01733) 385236
Fax: (01733) 385270
Web: www.agindustries.org.uk

BASIS Ltd
Bank Chambers
34 St John Street
Ashbourne
Derbyshire DE6 1GH
Tel: (01335) 343945/346138
Fax: (01335) 346488
Web: www.basis-reg.co.uk

British Beekeepers' Association
National Agricultural Centre
Stoneleigh
Kenilworth
Warwickshire CV8 2LZ
Tel: (024) 7669 6679
Web: www.bbka.org.uk

British Beer & Pub Association
Market Towers
1 Nine Elms Lane
London SW8 5NQ
Tel: (0207) 627 9191
Fax: (0207) 627 9123

British Crop Production Council (BCPC)
7 Omni Business Centre
Omega Park
Alton
Hampshire GU34 2QD
Tel: (01420) 593200
Fax: (01420) 593209
Web: www.bcpc.org

British Pest Control Association
1 Gleneagles House
Vernon Gate, South Street
Derby DE1 1UP
Tel: (01332) 294288
Fax: (01332) 295904
Web: www.bpca.org.uk

Chemicals Regulation Directorate
Mallard House
King's Pool
3 Peasholme Green
York YO1 2PX
Tel: (01904) 640500
Fax: (01904) 455733
Web: www.pesticides.gov.uk

Crop Protection Association Ltd
2 Swan Court
Cygnet Park
Hampton
Peterborough PE7 8GX
Tel: (01733) 355370
Fax: (01733) 355371
Web: www.cropprotection.org.uk

CropLife International
Avenue Louise 143
B-1050 Brussels
Belgium
Tel: (+32) 2 542 0410
Fax: (+32) 2 542 0419
Web: www.croplife.org

Department of Agriculture and Rural Development (Northern Ireland)
Pesticides Section
Dundonald House
Upper Newtownards Road
Belfast BT4 3SB
Tel: (028) 9052 4704
Fax: (028) 9052 4059
Web: www.dardni.gov.uk

Department of Environment, Food and Rural Affairs (Defra)
Nobel House
17 Smith Square
London SW1P 3JR
Tel: (020) 7238 6000
Fax: (020) 7238 6591
Web: www.defra.gov.uk

Environment Agency
Rio House, Waterside Drive
Aztec West
Almondsbury
Bristol BS12 4UD
Tel: (01454) 624400
Fax: (01454) 624409
Web: www.environment-agency.gov.uk

Appendices

European Crop Protection Association (ECPA)
Avenue E van Nieuwenhuyse 6
B-1160 Brussels
Belgium
Tel: (+32) 2 663 1550
Fax: (+32) 2 663 1560
Web: www.ecpa.be

Farmers' Union of Wales
Llys Amaeth
Queen's Square
Aberystwyth
Dyfed SY23 2EA
Tel: (01970) 612755
Fax: (01970)612755
Web: www.fuw.org.uk

Forestry Commission
231 Corstorphine Road
Edinburgh EH12 7AT
Tel: (0131) 334 0303
Fax: (0131) 334 3047
Web: www.forestry.gov.uk

Health and Safety Executive
Information Services
Room 318 Daniel House
Stanley Precinct, Bootle
Merseyside L20 3TW
Tel: (0151) 951 3191
Web: www.hse.gov.uk

Health and Safety Executive
Biocides & Pesticides Assessment Unit
Room 123 Magdalen House, Bootle
Merseyside L20 3QZ
Tel: (0151) 951 3535
Fax: (0151) 951 3317

Health and Safety Executive – Books
PO Box 1999
Sudbury
Suffolk CO10 2WA
Tel: (01787) 881165
Fax: (01787) 313995

Horticulture Development Council
Bradbourne House
East Malling
Kent ME19 6DZ
Tel: (01732) 848383
Fax: (01732) 848498
Web: www.hdc.org.uk

Lantra (previously ATB-LandBase)
National Agricultural Centre
Stoneleigh
Kenilworth
Warwickshire CV8 2LG
Tel: (024) 7669 6996
Fax: (024) 7669 6732
Web: www.lantra.co.uk

National Association of Agricultural Contractors (NAAC)
Samuelson House, Paxton Road
Orton Centre, Peterborough
Cambs PE2 5LT
Tel: (01733) 362920
Fax: (01733) 362921
Web: www.naac.co.uk

National Chemicals Emergency Centre
Building 329H, Harwell
Didcot
Oxon, OX11 0QJ
Tel: (0870) 190 6621
Fax: (0870) 190 6620
Web: http://the-ncec.com/

National Farmers' Union
Agriculture House
Stoneleigh Park
Stoneleigh
Warwickshire CV8 2TZ
Tel: (024) 7685 0662
Fax: (024) 7685 8501
Web: www.nfu.org.uk

National Poisons Information Service (Birmingham Unit)
City Hospital
Birmingham B18 7QH
Tel: (0121) 507 4123
Fax: (0121) 507 5580
Web: www.npis.org

National Turfgrass Council
Hunter's Lodge
Dr Brown's Road
Minchinhampton
Glos. GL6 9BT
Tel: (01453) 883588
Fax: (01453) 731449

NPTC
National Agricultural Centre
Stoneleigh
Kenilworth
Warwickshire CV8 2LG
Tel: (024) 7669 6553
Fax: (020) 7331 7313
Web: www.nptc.org.uk

Useful Contacts

Processors and Growers Research Organisation
The Research Station
Great North Road, Thornhaugh
Peterborough, Cambs. PE8 6HJ
Tel: (01780) 782585
Fax: (01780) 783993
Web: www.pgro.co.uk

Scottish Beekeepers' Association
North Trinity House
114 Trinity Road
Edinburgh EH5 3JZ
Tel: (0131) 552 5341
Web: www.scottishbeekeepers.org.uk

Scottish Environment Protection Agency (SEPA)
Erskine Court
The Castle Business Park
Stirling FK9 4TR
Tel: (01786) 457 700
Fax: (01786) 446 885
Web: www.sepa.org.uk

TSO (The Stationery Office)
Publications Centre
PO Box 276
London SW8 5DT
Tel: (020) 7873 9090 (orders)
(020) 7873 0011 (enquiries)
Fax: (020) 7873 8200
Web: www.thestationeryoffice.com

Ulster Beekeepers' Association
57 Liberty Road
Carrickfergus
Co. Antrim BT38 9DJ
Tel: (01960) 362998
Web: www.ubka.org

Welsh Beekeepers' Association
Trem y Clawdd
Fron Isaf, Chirk
Wrexham
Clwyd LL14 5AH
Tel/Fax: (01691) 773300

Appendix 3
Keys to Crop and Weed Growth Stages

Decimal Code for the Growth Stages of Cereals
Illustrations of these growth stages can be found in the reference indicated below and in some company product manuals.

0 Germination
- 00 Dryseed
- 03 Imbibition complete
- 05 Radicle emerged from caryopsis
- 07 Coleoptile emerged from caryopsis
- 09 Leaf at coleoptile tip

1 Seedling growth
- 10 First leaf through coleoptile
- 11 First leaf unfolded
- 12 2 leaves unfolded
- 13 3 leaves unfolded
- 14 4 leaves unfolded
- 15 5 leaves unfolded
- 16 6 leaves unfolded
- 17 7 leaves unfolded
- 18 8 leaves unfolded
- 19 9 or more leaves unfolded

2 Tillering
- 20 Main shoot only
- 21 Main shoot and 1 tiller
- 22 Main shoot and 2 tillers
- 23 Main shoot and 3 tillers
- 24 Main shoot and 4 tillers
- 25 Main shoot and 5 tillers
- 26 Main shoot and 6 tillers
- 27 Main shoot and 7 tillers
- 28 Main shoot and 8 tillers
- 29 Main shoot and 9 or more tillers

3 Stem elongation
- 30 Ear at 1 cm
- 31 1st node detectable
- 32 2nd node detectable
- 33 3rd node detectable
- 34 4th node detectable
- 35 5th node detectable
- 36 6th node detectable
- 37 Flag leaf just visible
- 39 Flag leaf ligule/collar just visible

4 Booting
- 41 Flag leaf sheath extending
- 43 Boots just visibly swollen
- 45 Boots swollen
- 47 Flag leaf sheath opening
- 49 First awns visible

5 Inflorescence
- 51 First spikelet of inflorescence just visible
- 52 1/4 of inflorescence emerged
- 55 1/2 of inflorescence emerged
- 57 3/4 of inflorescence emerged
- 59 Emergence of inflorescence completed

6 Anthesis
- 60
- 61 } Beginning of anthesis
- 64
- 65 } Anthesis half way
- 68
- 69 } Anthesis complete

7 Milk development
- 71 Caryopsis watery ripe
- 73 Early milk
- 75 Medium milk
- 77 Late milk

8 Dough development
- 83 Early dough
- 85 Soft dough
- 87 Hard dough

9 Ripening
- 91 Caryopsis hard (difficult to divide by thumb-nail)
- 92 Caryopsis hard (can no longer be dented by thumb-nail)
- 93 Caryopsis loosening in daytime

(From Tottman, 1987. *Annals of Applied Biology*, **110**, 441–454)

Keys to Crop and Weed Growth Stages

Stages in Development of Oilseed Rape
Illustrations of these growth stages can be found in the reference indicated below and in some company product manuals.

0 Germination and emergence

1 Leaf production
  1,0 Both cotyledons unfolded and green
  1,1 First true leaf
  1,2 Second true leaf
  1,3 Third true leaf
  1,4 Fourth true leaf
  1,5 Fifth true leaf
  1,10 About tenth true leaf
  1,15 About fifteenth true leaf

2 Stem extension
  2,0 No internodes ('rosette')
  2,5 About five internodes

3 Flower bud development
  3,0 Only leaf buds present
  3,1 Flower buds present but enclosed by leaves
  3,3 Flower buds visible from above ('green bud')
  3,5 Flower buds raised above leaves
  3,6 First flower stalks extending
  3,7 First flower buds yellow ('yellow bud')

4 Flowering
  4,0 First flower opened
  4,1 10% all buds opened
  4,3 30% all buds opened
  4,5 50% all buds opened

5 Pod development
  5,3 30% potential pods
  5,5 50% potential pods
  5,7 70% potential pods
  5,9 All potential pods

6 Seed development
  6,1 Seeds expanding
  6,2 Most seeds translucent but full size
  6,3 Most seeds green
  6,4 Most seeds green-brown mottled
  6,5 Most seeds brown
  6,6 Most seeds dark brown
  6,7 Most seeds black but soft
  6,8 Most seeds black and hard
  6,9 All seeds black and hard

7 Leaf senescence

8 Stem senescence
  8,1 Most stem green
  8,5 Half stem green
  8,9 Little stem green

9 Pod senescence
  9,1 Most pods green
  9,5 Half pods green
  9.9 Few pods green

(From Sylvester-Bradley, 1985. *Aspects of Applied Biology*, **10**, 395–400)

Stages in Development of Peas

Illustrations of these growth stages can be found in the reference indicated below and in some company product manuals.

0 Germination and emergence
- 000 Dry seed
- 001 Imbibed seed
- 002 Radicle apparent
- 003 Plumule and radicle apparent
- 004 Emergence

1 Vegetative stage
- 101 First node (leaf with one pair leaflets, no tendril)
- 102 Second node (leaf with one pair leaflets, simple tendril)
- 103 Third node (leaf with one pair leaflets, complex tendril)
- •
- •
- l0x X nodes (leaf with more than one pair leaflets, complex tendril)
- •
- •
- 10n Last recorded node

2 Reproductive stage (main stem)
- 201 Enclosed buds
- 202 Visible buds
- 203 First open flower
- 204 Pod set (small immature pod)
- 205 Flat pod
- 206 Pod swell (seeds small, immature)
- 207 Podfill
- 208 Pod green, wrinkled
- 209 Pod yellow, wrinkled (seeds rubbery)
- 210 Dry seed

3 Senescence stage
- 301 Desiccant application stage. Lower pods dry and brown, middle yellow, upper green. Overall moisture content of seed less than 45%
- 302 Pre-harvest stage. Lower and middle pods dry and brown, upper yellow. Overall moisture content of seed less than 30%
- 303 Dry harvest stage. All pods dry and brown, seed dry

(From Knott, 1987. *Annals of Applied Biology*, **111**, 233–244)

Stages in Development of Faba Beans

Illustrations of these growth stages can be found in the reference indicated below and in some company product manuals.

0 Germination and emergence
- 000 Dry seed
- 001 Imbibed seed
- 002 Radicle apparent
- 003 Plumule and radicle apparent
- 004 Emergence
- 005 First leaf unfolding
- 006 First leaf unfolded

1 Vegetative stage
- 101 First node
- 102 Second node
- 103 Third node
- •
- •
- l0x X nodes
- •
- •
- 10n N, last recorded node

2 Reproductive stage (main stem)
- 201 Flower buds visible
- 203 First open flowers
- 204 First pod set
- 205 Pods fully formed, green
- 207 Pod fill, pods green
- 209 Seed rubbery, pods pliable, turning black
- 210 Seed dry and hard, pods dry and black

3 Pod senescence
- 301 10% pods dry and black
- •
- •
- 305 50% pods dry and black
- •
- •
- 308 80% pods dry and black, some upper pods green
- 309 90% pods dry and black, most seed dry. Desiccation stage.
- 310 All pods dry and black, seed hard. Pre-harvest (glyphosate application stage)

4 Stem senescence
- 401 10% stem brown/black
- •
- •
- 405 50% stem brown/black
- •
- •
- 409 90% stem brown/black
- 410 All stems brown/black. All pods dry and black, seed hard.

(From Knott, 1990. *Annals of Applied Biology*, **116**, 391–404)

Stages in Development of Potato

Illustrations of these growth stages can be found in the reference indicated below and in some company product manuals.

0 Seed germination and seedling emergence
- 000 Dry seed
- 001 Imbibed seed
- 002 Radicle apparent
- 003 Elongation of hypocotyl
- 004 Seedling emergence
- 005 Cotyledons unfolded

1 Tuber dormancy
- 100 Innate dormancy (no sprout development under favourable conditions)
- 150 Enforced dormancy (sprout development inhibited by environmental conditions)

2 Tuber sprouting
- 200 Dormancy break, sprout development visible
- 21x Sprout with 1 node
- 22x Sprout with 2 nodes
- •
- •
- 29x Sprout with 9 nodes
- 21x(2) Second generation sprout with 1 node
- 22x(2) Second generation sprout with 2 nodes
- •
- •
- 29x(2) Second generation sprout with 9 nodes

Where $x = 1$, sprout <2 mm;
2, 2-5 mm; 3, 5-20 mm;
4, 20-30 mm; 5, 50-100 mm;
6, 100-150 mm long

3 Emergence and shoot expansion
- 300 Main stem emergence
- 301 Node 1
- 302 Node 2
- •
- •
- 319 Node 19

Second order branch
- 321 Node 1
- •
- •

Nth order branch
- 3N1 Node 1
- •
- •
- 3N9 Node 9

4 Flowering

Primary flower
- 400 No flowers
- 410 Appearance of flower bud
- 420 Flower unopen
- 430 Flower open
- 440 Flower closed
- 450 Berry swelling
- 460 Mature berry

Second order flowers
- 410(2) Appearance of flower bud
- 420(2) Flower unopen
- 430(2) Flower open
- 440(2) Flower closed
- 450(2) Berry swelling
- 460(2) Mature berry

5 Tuber development
- 500 No stolons
- 510 Stolon initials
- 520 Stolon elongation
- 530 Tuber initiation
- 540 Tuber bulking (>10 mm diam)
- 550 Skin set
- 560 Stolon development

6 Senescence
- 600 Onset of yellowing
- 650 Half leaves yellow
- 670 Yellowing of stems
- 690 Completely dead

(From Jefferies & Lawson, 1991. *Annals of Applied Biology*, **119**, 387–389)

Stages in Development of Linseed

Illustrations of these growth stages can be found in the reference indicated below and in some company product manuals.

0 Germination and emergence
- 00 Dry seed
- 01 Imbibed seed
- 02 Radicle apparent
- 04 Hypocotyl extending
- 05 Emergence
- 07 Cotyledon unfolding from seed case
- 09 Cotyledons unfolded and fully expanded

1 Vegetative stage (of main stem)
- 10 True leaves visible
- 12 First pair of true leaves fully expanded
- 13 Third pair of true leaves fully expanded
- 1n n leaf fully expanded

2 Basal branching
- 21 One branch
- 22 Two branches
- 23 Three branches
- 2n n branches

3 Flower bud development (on main stem)
- 31 Enclosed bud visible in leaf axils
- 33 Bud extending from axil
- 35 Corymb formed
- 37 Buds enclosed but petals visible
- 39 First flower open

4 Flowering (whole plant)
- 41 10% of flowers open
- 43 30% of flowers open
- 45 50% of flowers open
- 49 End of flowering

5 Capsule formation (whole plant)
- 51 10% of capsules formed
- 53 30% 0f capsules formed
- 55 50% of capsules formed
- 59 End of capsule formation

6 Capsule senescence (on most advanced plant)
- 61 Capsules expanding
- 63 Capsules green and full size
- 65 Capsules turning yellow
- 67 Capsules all yellow brown but soft
- 69 Capsules brown, dry and senesced

7 Stem senescence (whole plant)
- 71 Stems mostly green below panicle
- 73 Most stems 30% brown
- 75 Most stems 50% brown
- 77 Stems 75% brown
- 79 Stems completely brown

8 Stems rotting (retting)
- 81 Outer tissue rotting
- 85 Vascular tissue easily removed
- 89 Stems completely collapsed

9 Seed development (whole plant)
- 91 Seeds expanding
- 92 Seeds white but full size
- 93 Most seeds turning ivory yellow
- 94 Most seeds turning brown
- 95 All seeds brown and hard
- 98 Some seeds shed from capsule
- 99 Most seeds shed from capsule

(From Freer, 1991. *Aspects of Applied Biology*, **28**, 33–40)

Stages in Development of Annual Grass Weeds
Illustrations of these growth stages can be found in the reference indicated below and in some company product manuals.

0 Germination and emergence
- 00 Dry seed
- 01 Start of imbibition
- 03 Imbibition complete
- 05 Radicle emerged from caryopsis
- 07 Coleoptile emerged from caryopsis
- 09 Leaf just at coleoptile tip

1 Seedling growth
- 10 First leaf through coleoptile
- 11 First leaf unfolded
- 12 2 leaves unfolded
- 13 3 leaves unfolded
- 14 4 leaves unfolded
- 15 5 leaves unfolded
- 16 6 leaves unfolded
- 17 7 leaves unfolded
- 18 8 leaves unfolded
- 19 9 or more leaves unfolded

2 Tillering
- 20 Main shoot only
- 21 Main shoot and 1 tiller
- 22 Main shoot and 2 tillers
- 23 Main shoot and 3 tillers
- 24 Main shoot and 4 tillers
- 25 Main shoot and 5 tillers
- 26 Main shoot and 6 tillers
- 27 Main shoot and 7 tillers
- 28 Main shoot and 8 tillers
- 29 Main shoot and 9 or more tillers

3 Stem elongation
- 31 First node detectable
- 32 2nd node detectable
- 33 3rd node detectable
- 34 4th node detectable
- 35 5th node detectable
- 36 6th node detectable
- 37 Flag leaf just visible
- 39 Flag leaf ligule just visible

4 Booting
- 41 Flag leaf sheath extending
- 43 Boots just visibly swollen
- 45 Boots swollen
- 47 Flag leaf sheath opening
- 49 First awns visible

5 Inflorescence emergence
- 51 First spikelet of inflorescence just visible
- 53 1/4 of inflorescence emerged
- 55 1/2 of inflorescence emerged
- 57 3/4 of inflorescence emerged
- 59 Emergence of inflorescence completed

6 Anthesis
- 61 Beginning of anthesis
- 65 Anthesis half-way
- 69 Anthesis complete

(From Lawson & Read, 1992. *Annals of Applied Biology*, **12**, 211–214)

Growth Stages of Annual Broad-leaved Weeds

Preferred Descriptive Phrases
Illustrations of these growth stages can be found in the reference indicated below and in some company product manuals.

Pre-emergence
Early cotyledons
Expanded cotyledons
One expanded true leaf
Two expanded true leaves
Four expanded true leaves
Six expanded true leaves
Plants up to 25 mm across/high

Plants up to 50 mm across/high
Plants up to 100 mm across/high
Plants up to 150 mm across/high
Plants up to 250 mm across/high
Flower buds visible
Plant flowering
Plant senescent

(From Lutman & Tucker, 1987. *Annals of Applied Biology*, **110**, 683–687)

Appendix 4
Key to Hazard Classifications and Safety Precautions

Every product label contains information to warn users of the risks from using the product, together with precautions that must be followed in order to minimise the risks. A hazard classification (if any) and symbol is shown with associated risk phrases, followed by a series of safety precautions. These are represented in the pesticide profiles in Section 2 by code letters and numbers under the heading **Hazard classification and safety precautions**.

The codes are defined below, under the same sub-headings as they appear in the pesticide profiles.

Where a product label specifies the use of personal protective equipment (PPE), the requirements are listed under the sub-heading **Operator protection**, using letter codes to denote the protective items, according to the list below. Often PPE requirements are different for specified operations, e.g. handling the concentrate, cleaning equipment etc., but it is not possible to list them separately. The lists of PPE are therefore an indication of what the user may need to have available to use the product in different ways. **When making a COSHH assessment it is therefore essential that the product label is consulted for information on the particular use that is being assessed.**

Where the generalised wording includes a phrase such as '... for xx days', the specific requirement for each pesticide is shown in brackets after the code.

Hazard

H01	Very toxic
H02	Toxic
H03	Harmful
H04	Irritant
H05	Corrosive
H06	Extremely flammable
H07	Highly flammable
H08	Flammable
H09	Oxidising agent
H10	Explosive
H11	Dangerous for the environment

Risk phrases

R08	Contact with combustible material may cause fire
R09	Explosive when mixed with combustible material
R16	Explosive when mixed with oxidising substances
R20	Harmful by inhalation
R21	Harmful in contact with skin
R22a	Harmful if swallowed
R22b	May cause lung damage if swallowed
R23	Toxic by inhalation
R24	Toxic in contact with skin
R25	Toxic if swallowed
R25b	Toxic; Danger of serious damage to health by prolonged exposure if swallowed
R26	Very toxic by inhalation
R27	Very toxic in contact with skin
R28	Very toxic if swallowed
R34	Causes burns
R35	Causes severe burns
R36	Irritating to eyes

Key to Hazard Classifications and Safety Precautions

R37	Irritating to respiratory system
R38	Irritating to skin
R39	Danger of very serious irreversible effects
R40	Limited evidence of a carcinogenic effect
R41	Risk of serious damage to eyes
R42	May cause sensitization by inhalation
R43	May cause sensitization by skin contact
R45	May cause cancer
R46	May cause heritable genetic damage
R48	Danger of serious damage to health by prolonged exposure
R50	Very toxic to aquatic organisms
R51	Toxic to aquatic organisms
R52	Harmful to aquatic organisms
R53a	May cause long-term adverse effects in the aquatic environment
R53b	Dangerous to aquatic organisms
R54	Toxic to flora
R55	Toxic to fauna
R56	Toxic to soil organisms
R57	Toxic to bees
R58	May cause long-term adverse effects in the environment
R60	May impair fertility
R61	May cause harm to the unborn child
R62	Possible risk of impaired fertility
R63	Possible risk of harm to the unborn child
R64	May cause harm to breast-fed babies
R66	Repeated exposure may cause skin dryness or cracking
R67	Vapours may cause drowsiness and dizziness
R68	Possible risk of irreversible effects
R69	Danger of serious damage to health by prolonged oral exposure
R70	May produce an allergic reaction

Operator protection

A	Suitable protective gloves (the product label should be consulted for any specific requirements about the material of which the gloves should be made)
B	Rubber gauntlet gloves
C	Face-shield
D	Approved respiratory protective equipment
E	Goggles
F	Dust mask
G	Full face-piece respirator
H	Coverall
J	Hood
K	Apron/Rubber apron
L	Waterproof coat
M	Rubber boots
N	Waterproof jacket and trousers
P	Suitable protective clothing
U01	To be used only by operators instructed or trained in the use of chemical/product/type of produce and familiar with the precautionary measures to be observed
U02a	Wash all protective clothing thoroughly after use, especially the inside of gloves
U02b	Avoid excessive contamination of coveralls and launder regularly
U02c	Remove and wash contaminated gloves immediately
U03	Wash splashes off gloves immediately
U04a	Take off immediately all contaminated clothing
U04b	Take off immediately all contaminated clothing and wash underlying skin. Wash clothes before re-use
U04c	Wash clothes before re-use
U05a	When using do not eat, drink or smoke
U05b	When using do not eat, drink, smoke or use naked lights
U06	Handle with care and mix only in a closed container
U07	Open the container only as directed (returnable containers only)
U08	Wash concentrate/dust from skin or eyes immediately

U09a	Wash any contamination/splashes/dust/powder/concentrate from skin or eyes immediately
U09b	Wash any contamination/splashes/dust/powder/concentrate from eyes immediately
U10	After contact with skin or eyes wash immediately with plenty of water
U11	In case of contact with eyes rinse immediately with plenty of water and seek medical advice
U12	In case of contact with skin rinse immediately with plenty of water and seek medical advice
U13	Avoid contact by mouth
U14	Avoid contact with skin
U15	Avoid contact with eyes
U16a	Ensure adequate ventilation in confined spaces
U16b	Use in a well ventilated area
U18	Extinguish all naked flames, including pilot lights, when applying the fumigant/dust/liquid/product
U19a	Do not breathe dust/fog/fumes/gas/smoke/spray mist/vapour. Avoid working in spray mist
U19b	Do not work in confined spaces or enter spaces in which high concentrations of vapour are present. Where this precaution cannot be observed distance breathing or self-contained breathing apparatus must be worn, and the work should be done by trained operators
U19c	In case of insufficient ventilation, wear suitable respiratory equipment
U20a	Wash hands and exposed skin before eating, drinking or smoking and after work
U20b	Wash hands and exposed skin before eating and drinking and after work
U20c	Wash hands before eating and drinking and after work
U20d	Wash hands after use
U20e	Wash hands and exposed skin after cleaning and re-calibrating equipment
U21	Before entering treated crops, cover exposed skin areas, particularly arms and legs
U22a	Do not touch sachet with wet hands or gloves/Do not touch water soluble bag directly
U22b	Protect sachets from rain or water
U23a	Do not apply by knapsack sprayer/hand-held equipment
U23b	Do not apply through hand held rotary atomisers
U23c	Do not apply via tractor mounted horizontal boom sprayers
U24	Do not handle grain unnecessarily
U25	Open the container only as directed
U26	Keep unprotected workers out of treated areas for at least 5 days after treatment.

Environmental protection

E02a	Keep unprotected persons/animals out of treated/fumigation areas for at least xx hours/days
E02b	Prevent access by livestock, pets and other non-target mammals and birds to buildings under fumigation and ventilation
E02c	Vacate treatment areas before application
E02d	Exclude all persons and animals during treatment
E03	Label treated seed with the appropriate precautions, using the printed sacks, labels or bag tags supplied
E05a	Do not apply directly to livestock/poultry
E05b	Keep poultry out of treated areas for at least xx days/weeks
E05c	Do not apply directly to animals
E06a	Keep livestock out of treated areas for at least xx days/weeks after treatment
E06b	Dangerous to livestock. Keep all livestock out of treated areas/away from treated water for at least xx days/weeks. Bury or remove spillages
E06c	Harmful to livestock. Keep all livestock out of treated areas/away from treated water for at least xx days/weeks. Bury or remove spillages
E06d	Exclude livestock from treated fields. Livestock may not graze or be fed treated forage nor may it be used for hay silage or bedding
E07a	Keep livestock out of treated areas for up to two weeks following treatment and until poisonous weeds, such as ragwort, have died down and become unpalatable
E07b	Dangerous to livestock. Keep livestock out of treated areas/away from treated water for at least xx weeks and until foliage of any poisonous weeds, such as ragwort, has died and become unpalatable

Key to Hazard Classifications and Safety Precautions

E07c	Harmful to livestock. Keep livestock out of treated areas/away from treated water for at least xx days/weeks and until foliage of any poisonous weeds such as ragwort has died and become unpalatable
E07d	Keep livestock out of treated areas for up to 4-6 weeks following treatment and until poisonous weeds, such as ragwort, have died down and become unpalatable
E08a	Do not feed treated straw or haulm to livestock within xx days/weeks of spraying
E08b	Do not use on grassland if the crop is to be used as animal feed or bedding
E09	Do not use straw or haulm from treated crops as animal feed or bedding for at least xx days after last application
E10a	Dangerous to game, wild birds and animals
E10b	Harmful to game, wild birds and animals
E10c	Dangerous to game, wild birds and animals. All spillages must be buried or removed
E11	Paraquat can be harmful to hares; spray stubbles early in the day
E12a	High risk to bees
E12b	Extremely dangerous to bees
E12c	Dangerous to bees
E12d	Harmful to bees
E12e	Do not apply to crops in flower or to those in which bees are actively foraging. Do not apply when flowering weeds are present
E12f	Do not apply to crops in flower, or to those in which bees are actively foraging, except as directed on [crop]. Do not apply when flowering weeds are present
E12g	Apply away from bees
E13a	Extremely dangerous to fish or other aquatic life. Do not contaminate surface waters or ditches with chemical or used container
E13b	Dangerous to fish or other aquatic life. Do not contaminate surface waters or ditches with chemical or used container
E13c	Harmful to fish or other aquatic life. Do not contaminate surface waters or ditches with chemical or used container
E13d	Apply away from fish
E13e	Harmful to fish or other aquatic life. The maximum concentration of active ingredient in treated water must not exceed XX ppm or such lower concentration as the appropriate water regulatory body may require.
E14a	Extremely dangerous to aquatic higher plants. Do not contaminate surface waters or ditches with chemical or used container
E14b	Dangerous to aquatic higher plants. Do not contaminate surface waters or ditches with chemical or used container
E15a	Do not contaminate surface waters or ditches with chemical or used container
E15b	Do not contaminate water with product or its container. Do not clean application equipment near surface water. Avoid contamination via drains from farmyards or roads
E15c	To protect groundwater, do not apply to grass leys less than 1 year old
E16a	Do not allow direct spray from horizontal boom sprayers to fall within 5 m of the top of the bank of a static or flowing waterbody, unless a Local Environment Risk Assessment for Pesticides (LERAP) permits a narrower buffer zone, or within 1 m of the top of a ditch which is dry at the time of application. Aim spray away from water
E16b	Do not allow direct spray from hand-held sprayers to fall within 1 m of the top of the bank of a static or flowing waterbody. Aim spray away from water
E16c	Do not allow direct spray from horizontal boom sprayers to fall within 5 m of the top of the bank of a static or flowing waterbody, or within 1m of the top of a ditch which is dry at the time of application. Aim spray away from water. This product is not eligible for buffer zone reduction under the LERAP horizontal boom sprayers scheme.
E16d	Do not allow direct spray from hand-held sprayers to fall within 1 m of the top of the bank of a static or flowing waterbody. Aim spray away from water. This product is not eligible for buffer zone reduction under the LERAP horizontal boom sprayers scheme scheme.
E16e	Do not allow direct spray from horizontal boom sprayers to fall within 5 m of the top of the bank of a static or flowing water body or within 1 m from the top of any ditch which is dry at the time of application. Spray from hand held sprayers must not in any case be allowed to fall within 1 m of the top of the bank of a static or flowing water body. Always direct spray away from water. The LERAP scheme does not extend to adjuvants. This product is therefore not eligible for a reduced buffer zone under the LERAP scheme.
E16f	Do not allow direct spray/granule applications from vehicle mounted/drawn hydraulic sprayers/applicators to fall within 6 m of surface waters or ditches/Do not allow direct

Appendices

	spray/granule applications from hand-held sprayers/applicators to fall within 2 m of surface waters or ditches. Direct spray/applications away from water
E16g	Do not allow direct spray from train sprayers to fall within 5 m of the top of the bank of any static or flowing waterbody.
E17a	Do not allow direct spray from broadcast air-assisted sprayers to fall within xx m of surface waters or ditches. Direct spray away from water
E17b	Do not allow direct spray from broadcast air-assisted sprayers to fall within xx m of the top of the bank of a static or flowing waterbody, unless a Local Environmental Risk Assessment for Pesticides (LERAP) permits a narrower buffer zone, or within 5 m of the top of a ditch which is dry at the time of application. Aim spray away from water
E18	Do not spray from the air within 250 m horizontal distance of surface waters or ditches
E19a	Do not dump surplus herbicide in water or ditch bottoms
E19b	Do not empty into drains
E20	Prevent any surface run-off from entering storm drains
E21	Do not use treated water for irrigation purposes within xx days/weeks of treatment
E22a	High risk to non-target insects or other arthropods. Do not spray within 6 m of the field boundary
E22b	Risk to certain non-target insects or other arthropods. For advice on risk management and use in Integrated Pest Management (IPM) see directions for use
E22c	Risk to non-target insects or other arthropods
E23	Avoid damage by drift onto susceptible crops or water courses
E34	Do not re-use container for any purpose/Do not re-use container for any other purpose
E35	Do not burn this container
E36a	Do not rinse out container (returnable containers only)
E36b	Do not open or rinse out container (returnable containers only)
E37	Do not use with any pesticide which is to be applied in or near water
E38	Use appropriate containment to avoid environmental contamination
E39	Extreme care must be taken to avoid spray drift onto non-crop plants outside the target area
E40	To protect groundwater/soil organisms the maximum total dose of this or other products containing ethofumesate MUST NOT exceed 1.0 kg ethofumesate per hectare in any three year period
E40b	To protect aquatic organisms respect an unsprayed buffer zone to surface waters in line with LERAP requirements
E40c	This product must not be used with any pesticide to be applied in or near water
E40d	To protect non-target arthropods respect an untreated buffer zone of 5m to non crop land
E41	Hay for silage must not be cut from treated crops for at least 21 days after treatment

Consumer protection

C01	Do not use on food crops
C02a	Do not harvest for human or animal consumption for at least xx days/weeks after last application
C02b	Do not remove from store for sale or processing for at least 21 days after application
C02c	Do not remove from store for sale or processing for at least 2 days after application
C04	Do not apply to surfaces on which food/feed is stored, prepared or eaten
C05	Remove/cover all foodstuffs before application
C06	Remove exposed milk before application
C07	Collect eggs before application
C08	Protect food preparing equipment and eating utensils from contamination during application
C09	Cover water storage tanks before application
C10	Protect exposed water/feed/milk machinery/milk containers from contamination
C11	Remove all pets/livestock/fish tanks before treatment/spraying
C12	Ventilate treated areas thoroughly when smoke has cleared/Ventilate treated rooms thoroughly before occupying

Storage and disposal

D01	Keep out of reach of children
D02	Keep away from food, drink and animal feeding-stuffs
D03	Store away from seeds, fertilizers, fungicides and insecticides
D04	Store well away from corms, bulbs, tubers and seeds

Key to Hazard Classifications and Safety Precautions

D05	Protect from frost
D06a	Store away from heat
D06b	Do not store near heat or open flame
D06c	Do not store in direct sunlight
D06d	Do not store above 30/35 °C
D07	Store under cool, dry conditions
D08	Store in a safe, dry, frost-free place designated as an agrochemical store
D09a	Keep in original container, tightly closed, in a safe place
D09b	Keep in original container, tightly closed, in a safe place, under lock and key
D09c	Store unused sachets in a safe place. Do not store half-used sachets
D10a	Wash out container thoroughly and dispose of safely
D10b	Wash out container thoroughly, empty washings into spray tank and dispose of safely
D10c	Rinse container thoroughly by using an integrated pressure rinsing device or manually rinsing three times. Add washings to sprayer at time of filling and dispose of container safely
D10d	Do not rinse out container
D11a	Empty container completely and dispose of safely/Dispose of used generator safely
D11b	Empty container completely and dispose of it in the specified manner
D12a	This material (and its container) must be disposed of in a safe way
D12b	This material and its container must be disposed of as hazardous waste
D13	Treat used container as if it contained pesticide
D14	Return empty container as instructed by supplier (returnable containers only)
D15	Store container in purpose built chemical store until returned to supplier for refilling (returnable containers only)
D16	Do not store below 4 degrees Centigrade
D17	Use immediately on removal of foil
D18	Place the tablets whole into the spray tank - do not break or crumble the tablets

Treated Seed

S01	Do not handle treated seed unnecessarily
S02	Do not use treated seed as food or feed
S03	Keep treated seed secure from people, domestic stock/pets and wildlife at all times during storage and use
S04a	Bury or remove spillages
S04b	Harmful to birds/game and wildlife. Treated seed should not be left on the soil surface. Bury or remove spillages
S04c	Dangerous to birds/game and wildlife. Treated seed should not be left on the soil surface. Bury or remove spillages
S04d	To protect birds/wild animals, treated seed should not be left on the soil surface. Bury or remove spillages
S05	Do not reuse sacks or containers that have been used for treated seed for food or feed
S06a	Wash hands and exposed skin before meals and after work
S06b	Wash hands and exposed skin after cleaning and re-calibrating equipment
S07	Do not apply treated seed from the air
S08	Treated seed should not be broadcast
S09	Label treated seed with the appropriate precautions using printed sacks, labels or bag tags supplied

Vertebrate/Rodent control products

V01a	Prevent access to baits/powder by children, birds and other animals, particularly cats, dogs, pigs and poultry
V01b	Prevent access to bait/gel/dust by children, birds and non-target animals, particularly dogs, cats, pigs, poultry
V02	Do not prepare/use/lay baits/dust/spray where food/feed/water could become contaminated
V03a	Remove all remains of bait, tracking powder or bait containers after use and burn or bury
V03b	Remove all remains of bait and bait containers/exposed dust/after treatment (except where used in sewers) and dispose of safely (e.g. burn/bury). Do not dispose of in refuse sacks or on open rubbish tips.
V04a	Search for and burn or bury all rodent bodies. Do not place in refuse bins or on rubbish tips

Appendices

V04b Search for rodent bodies (except where used in sewers) and dispose of safely (e.g. burn/bury). Do not dispose of in refuse sacks or on open rubbish tips
V04c Dispose of safely any rodent bodies and remains of bait and bait containers that are recovered after treatment (e.g. burn/bury). Do not dispose of in refuse sacks or on open rubbish tips
V05 Use bait containers clearly marked POISON at all surface baiting points

Medical advice

M01 This product contains an anticholinesterase organophosphorus compound. DO NOT USE if under medical advice NOT to work with such compounds
M02 This product contains an anticholinesterase carbamate compound. DO NOT USE if under medical advice NOT to work with such compounds
M03 If you feel unwell, seek medical advice immediately (show the label where possible)
M04a In case of accident or if you feel unwell, seek medical advice immediately (show the label where possible)
M04b In case of accident by inhalation, remove casualty to fresh air and keep at rest
M05a If swallowed, seek medical advice immediately and show this container or label
M05b If swallowed, do not induce vomiting: seek medical advice immediately and show this container or label
M05c If swallowed induce vomiting if not already occurring and take patient to hospital immediately
M06 This product contains an anticholinesterase carbamoyl triazole compound. DO NOT USE if under medical advice NOT to work with such compounds

Appendix 5
Key to Abbreviations and Acronyms

The abbreviations of formulation types in the following list are used in Section 2 (Pesticide Profiles) and are derived from the *Catalogue of Pesticide Formulation Types and International Coding System* (CropLife International Technical Monograph 2, 5th edn, March 2002).

1 Formulation Types

AB	Grain bait
AE	Aerosol generator
AL	Other liquids to be applied undiluted
AP	Any other powder
BB	Block bait
BR	Briquette
CB	Bait concentrate
CC	Capsule suspension in a suspension concentrate
CF	Capsule suspension for seed treatment
CG	Encapsulated granule (controlled release)
CL	Contact liquid or gel (for direct application)
CP	Contact powder (for direct application)
CR	Crystals
CS	Capsule suspension
DC	Dispersible concentrate
DP	Dustable powder
DS	Powder for dry seed treatment
EC	Emulsifiable concentrate
EG	Emulsifiable granule
EO	Water in oil emulsion
ES	Emulsion for seed treatment
EW	Oil in water emulsion
FG	Fine granules
FP	Smoke cartridge
FS	Flowable concentrate for seed treatment
FT	Smoke tablet
FU	Smoke generator
FW	Smoke pellets
GA	Gas
GB	Granular bait
GE	Gas-generating product
GG	Macrogranules
GL	Emulsifiable gel
GP	Flo-dust (for pneumatic application)
GR	Granules
GS	Grease
GW	Water soluble gel
HN	Hot fogging concentrate
KK	Combi-pack (solid/liquid)
KL	Combi-pack (liquid/liquid)
KN	Cold-fogging concentrate
KP	Combi-pack (solid/solid)
LA	Lacquer
LI	Liquid, unspecified
LS	Solution for seed treatment
ME	Microemulsion
MG	Microgranules
MS	Microemulsion for seed treatment
OD	Oil dispersion
OL	Oil miscible liquid
PA	Paste

PC	Gel or paste concentrate
PS	Seed coated with a pesticide
PT	Pellet
RB	Ready-to-use bait
RH	Ready-to-use spray in hand-operated sprayer
SA	Sand
SC	Suspension concentrate (= flowable)
SE	Suspo-emulsion
SG	Water soluble granules
SL	Soluble concentrate
SP	Water soluble powder
SS	Water soluble powder for seed treatment
ST	Water soluble tablet
SU	Ultra low-volume suspension
TB	Tablets
TC	Technical material
TP	Tracking powder
UL	Ultra low-volume liquid
VP	Vapour releasing product
WB	Water soluble bags
WG	Water dispersible granules
WP	Wettable powder
WS	Water dispersible powder for slurry treatment of seed
WT	Water dispersible tablet
XX	Other formulations
ZZ	Not Applicable

2 Other Abbreviations and Acronyms

ACP	Advisory Committee on Pesticides
ACTS	Advisory Committee on Toxic Substances
ADAS	Agricultural Development and Advisory Service
a.i.	active ingredient
AIC	Agriculture Industries Confederation
BBPA	British Beer and Pub Association
CDA	Controlled droplet application
CPA	Crop Protection Association
cm	centimetre(s)
COPR	Control of Pesticides Regulations 1986
COSHH	Control of Substances Hazardous to Health Regulations
CRD	Chemicals Regulation Directorate
d	day(s)
Defra	Department for Environment, Food and Rural Affairs
EA	Environment Agency
EBDC	ethylene-bis-dithiocarbamate fungicide
FEPA	Food and Environment Protection Act 1985
g	gram(s)
GCPF	Global Crop Protection Federation (now CropLife International)
GS	growth stage (unless in formulation column)
h	hour(s)
ha	hectare(s)
HBN	hydroxybenzonitrile herbicide
HI	harvest interval
HSE	Health and Safety Executive
ICM	integrated crop management
IPM	integrated pest management
kg	kilogram(s)
l	litre(s)
LERAP	Local Environmental Risk Assessments for Pesticides
m	metre(s)
MAFF	Ministry of Agriculture, Fisheries and Food (now Defra)
MBC	methyl benzimidazole carbamate fungicide
MEL	maximum exposure limit

Key to Abbreviations and Acronyms

min	minute(s)
mm	millimetre(s)
MRL	maximum residue level
mth	month(s)
NA	Notice of Approval
NFU	National Farmers' Union
OES	Occupational Exposure Standard
OLA	off-label approval
PPE	personal protective equipment
PPPR	Plant Protection Products Regulations
SOLA	specific off-label approval
ULV	ultra-low volume
VI	Voluntary Initiative
w/v	weight/volume
w/w	weight/weight
wk	week(s)
yr	year(s)

3 UN Number details

UN No.	Substance	EAC	APP	Hazards Class	Sub risks	HIN
1692	Strychnine or Strychnine salts	2X		6.1		66
1760	Corrosive Liquid, N.O.S., Packing groups II & III	2X	B	8		88
1830	Sulphuric acid, with more than 51% acid	2P		8		80
1993	Flammable Liquid, N.O.S., Packing group III	•3Y		3		30
2011	Magnesium Phosphide	4W⁽¹⁾		4.3	6.1	
2588	Pesticide, Solid, TOXIC, N.O.S.	2X		6.1		66/60
2757	Carbamate Pesticide, Solid, Toxic	2X		6.1		66/60
2783	Organophosphorus Pesticide, Solid, TOXIC	2X		6.1		66/60
2902	Pesticide, Liquid, TOXIC, N.O.S., Packing Groups I, II & III	2X	B	6.1		66/60
3077	Environmentally hazardous substance, Solid, N.O.S.	2Z		9		90
3082	Environmentally hazardous substance, Solid, N.O.S.	•3Z		9		90
3265	Corrosive Liquid, Acidic, Organic, N.O.S., Packing group 1, II or III	2X	B	8		80/88
3351	Pyrethroid Pesticide, Liquid, Toxic, Flammable, Flash Point 23 °C or more	•3W	A(fl)	6.1	3	663/63
3352	Pyrethroid Pesticide, Liquid, TOXIC, Packing groups I, II & III	2X	B	6.1		66/60

Ref: Dangerous Goods Emergency Action Code List 2009 (TSO).
⁽¹⁾ Not applicable to the carriage of dangerous goods under RID or ADR.

Appendix 6
Definitions

The descriptions used in this *Guide* for the crops or situations in which products are approved for use are those used on the approved product labels. These are now standardised in a Crop Hierarchy published by the Pesticides Safety Directorate in which definitions are given. To assist users of this *Guide* the definitions of some of the terminology where misunderstandings can occur are reproduced below.

Rotational grass: Short-term grass crops grown on land that is likely to be growing different crops in future years (*e.g. short-term intensively managed leys for one to three years that may include clover*)

Permanent grassland: Grazed areas that are intended to be permanent in nature (*e.g. permanent pasture and moorland that can be grazed*).

Ornamental Plant Production: All ornamental plants that are grown for sale or are produced for replanting into their final growing position (*e.g. flowers, house plants, nursery stock, bulbs grown in containers or in the ground*).

Managed Amenity Turf: Areas of frequently mown, intensively managed, turf that is not intended to flower and set seed. It includes areas that may be for intensive public use (*e.g. all types of sports turf*).

Amenity Grassland: Areas of semi-natural or planted grassland subject to minimal management. It includes areas that may be accessed by the public (*e.g. railway and motorway embankments, airfields, and grassland nature reserves*). These areas may be managed for their botanical interest, and the relevant authority should be contacted before using pesticides in such locations.

Amenity Vegetation: Areas of semi-natural or ornamental vegetation, including trees, or bare soil around ornamental plants, or soil intended for ornamental planting. It includes areas to which the public have access. It does NOT include hedgerows around arable fields.

Natural surfaces not intended to bear vegetation: Areas of soil or natural outcroppings of rock that are not intended to bear vegetation, including areas such as sterile strips around fields. It may include areas to which the public have access. It does not include the land between rows of crops.

Hard surfaces: Man-made impermeable surfaces that are not intended to bear vegetation (*e.g. pavements, tennis courts, industrial areas, railway ballast*).

Permeable surfaces overlying soil: Any man-made permeable surface (excluding railway ballast) such as gravel that overlies soil and is not intended to bear vegetation

Green Cover on Land Temporarily Removed from Production: Includes fields covered by natural regeneration or by a planted green cover crop that will not be harvested (*e.g. green cover on setaside*). It does NOT include industrial crops.

Forest Nursery: Areas where young trees are raised outside for subsequent forest planting.

Forest: Groups of trees being grown in their final positions. Covers all woodland grown for whatever objective, including commercial timber production, amenity and recreation, conservation and landscaping, ancient traditional coppice and farm forestry, and trees from natural regeneration, colonisation or coppicing. Also includes restocking of established woodlands and new planting on both improved and unimproved land.

Key to Abbreviations and Acronyms

Farm forestry: Groups of trees established on arable land or improved grassland including those planted for short rotation coppicing. It includes mature hedgerows around arable fields.

Indoors (for rodenticide use): Situations where the bait is placed within a building or other enclosed structure, and where the target is living or feeding predominantly within that building or structure.

Herbs: Reference to Herbs or Protected Herbs when used in Section 2 may include any or all of the following. The particular label or SOLA Notice will indicate which species are included in the approval.

Agastache spp.
Angelica
Applemint
Balm
Basil
Bay
Borage (except when grown for oilseed)
Camomile
Caraway
Catnip
Chervil
Clary
Clary sage
Coriander
Curry plant
Dill
Dragonhead
English chamomile
Fennel
Fenugreek
Feverfew
French lavender
Gingermint
Hyssop
Korean mint
Land cress
Lavandin
Lavender
Lemon balm
Lemon peppermint
Lemon thyme
Lemon verbena
Lovage
Marigold
Marjoram
Mint
Mother of thyme
Nasturtium
Nettle
Oregano
Origanum heracleoticum
Parsley root
Peppermint
Pineapplemint
Rocket
Rosemary
Rue
Sage
Salad burnet
Savory
Sorrel
Spearmint
Spike lavender
Tarragon
Thyme
Thymus camphoratus
Violet
Winter savory
Woodruff

Herbs for Medicinal Uses: Reference to Herbs for Medicinal Uses when used in Section 2 may include any or all of the following. The particular label or SOLA Notice will indicate which species are included in the approval

Black cohosh
Burdock
Dandelion
Echinacea
Ginseng
Goldenseal
Liquorice
Nettle
Valerian

Appendix 7
References

The information given in *The UK Pesticide Guide* provides some of the answers needed to assess health risks, including the hazard classification and the level of operator protection required. However, the *Guide* cannot provide all the details needed for a complete hazard assessment, which must be based on the product label itself and, where necessary, the Health and Safety Data Sheet and other official literature.

Detailed guidance on how to comply with the Regulations is available from several sources.

Pesticides: Code of Practice
Code of Practice for Using Plant Protection Products, 2006 (Defra publication PB 11090)

Known as the 'Green Code', this revised Code of Practice, which came into effect on 15 December 2005, replaced all previous editions of the *Code of Practice for the Safe Use of Pesticides on Farms and Holdings*. It also replaced the voluntary code of practice for the use of pesticides in amenity and industrial areas (the 'Orange Code'). The code explains how plant protection products can be used safely and so meet the legal requirements of the Control of Pesticides Regulations 1986 (COPR) (ISBN 0-110675-10-X), and the Control of Substances Hazardous to Health Regulations 2002 (COSHH) (ISBN 0-110429-19-2). The principal source of information for making a COSHH assessment is the approved product label. In most cases the label provides all the necessary information, but in certain circumstances other sources must be consulted, and these are listed in the Code of Practice.

The code, which applies to England and Wales only, is available in CD-ROM format and electronically from the Chemicals Regulation Directorate and Health and Safety Executive websites. A Welsh language version is available in printed and electronic formats. A Scottish version (approved by the Scottish Parliament) is being produced in printed and electronic formats. Northern Ireland will produce its own version of the code in due course.

Other Codes of Practice
Code of Practice for Suppliers of Pesticides to Agriculture, Horticulture and Forestry (the 'Yellow Code') (Defra Booklet PB 0091)

Code of Good Agricultural Practice for the Protection of Soil (Defra Booklet PB 0617)

Code of Good Agricultural Practice for the Protection of Water (Defra Booklet PB 0587)

Code of Good Agricultural Practice for the Protection of Air (Defra Booklet PB 0618)

Approved Code of Practice for the Control of Substances Hazardous to Health in Fumigation Operations. Health and Safety Commission (ISBN 0-717611-95-7)

Code of Best Practice: Safe use of Sulphuric Acid as an Agricultural Desiccant. National Association of Agricultural Contractors, 2002. (Also available at www.naac.co.uk/?Codes/acidcode.asp)

Safe Use of Pesticides for Non-agricultural Purposes. HSE Approved Code of Practice. HSE L21 (ISBN 0-717624-88-9)

References

Other Guidance and Practical Advice

HSE (by mail order from HSE Books – See Appendix 2)

COSHH – A brief guide to the Regulations 2003 (INDG136)

A Step by Step Guide to COSHH Assessment, 2004 (HSG97) (ISBN 0-717627-85-3)

Defra (from The Stationery Office – see Appendix 2)

Local Environment Risk Assessments for Pesticides (LERAP): Horizontal Boom Sprayers.

Local Environment Risk Assessments for Pesticides (LERAP): Broadcast Air-assisted Sprayers.

Crop Protection Association and the Voluntary Initiative (see Appendix 2)

Every Drop Counts: Keeping Water Clean

Best Practice Guides. A range of leaflets giving guidance on best practice when dealing with pesticides before, during and after application.

H2OK? Best Practice Advice and Decision Trees - July 2009/10

BCPC (British Crop Production Council – see Appendix 2)

The Pesticide Manual (15th edition) (ISBN 978-1-901396-18-8)

The e-Pesticide Manual (version 5.1) (ISBN 978-1-901396-84-3)

The GM Crop Manual (ISBN 978-1-901396-19-5)

The Manual of Biocontrol Agents (4th edition) (ISBN 978-1-901396-17-1)

IdentiPest PC CD-ROM (ISBN 1-901396-05-3) - from January 2011, available via www.plantprotection.co.uk

Garden Detective PC CD-ROM (ISBN 1-901396-32-0)

Small Scale Spraying (ISBN 1-901396-07-X)

Field Scale Spraying (ISBN 1-901396-08-8)

Using Pesticides (ISBN 1-901396-10-X)

Safety Equipment Handbook (ISBN 1-901396-06-1)

The Environment Agency (see Appendix 2)

Best Farming Practices: Profiting from a Good Environment

Use of Herbicides in or Near Water

SECTION 7
INDEX

Index of Proprietary Names of Products

The references are to entry numbers, not to pages. Adjuvant names are referred to as 'Adj' and are listed separately in Section 4

Name	Ref
3C Chlormequat 720	68
Abacus	Adj
Absolute	176
AC 650	348
Acanto Prima	138
Acclaim	77
Acetum	3
Actara	431
Actellic 50 EC	372
Actellic Smoke Generator No 20	372
Activator 90	Adj
A-Cyper 100EC	4
Adagio	83
Adigor	Adj
Adjust	68
Admire	287
Admix-P	Adj
Afalon	308
Agate	377
Agena	91
Agrichem DB Plus	151
Agrichem Flowable Thiram	435
Agrichem Hy-Pro Duet	378
Agrichem MCPA-50	318
Agriguard Cerusite	73
Agriguard Chlormequat 700	68
Agriguard Chlormequat 720	68
Agriguard Chlorothalonil	77
Agriguard Chlorpyrifos	95
Agriguard Cymoxanil Plus	126
Agriguard Diquat	190
Agriguard Epoxiconazole	194
Agriguard Ethofumesate 200	206
Agriguard Fluroxypyr	261
Agriguard Isoxaben	299
Agriguard Lenacil	306
Agriguard Pencycuron	356
Agrimec	1
Agritox	318
Agritox 50	318
Agrotech Asulam	17
Agrotech Ethephon	73
Agrotech-Nicosulfron 40 SC	349
Agrovista 3 See 750	68
Agrovista Fenamid	219
Agrovista Paclo	352
Agrovista Penco	355
Agrovista Radni	218
Agrovista Reggae	430
Agroxone	318
Alias SX	343
Aliette 80 WG	268
Alistell	150
Allure	331
Ally Max SX	345
Alpha Bromolin 225 EC	46
Alpha Captan 80 WDG	52
Alpha Captan 83 WP	52
Alpha Chlorotoluron 500	90
Alpha Ethofumesate	206
Alpha Fenpropidin 750 EC	222
Alpha Linuron 50SC	308
Alpha Metazachlor 50 SC	334
Alpha Pendimethalin 330 EC	357
Alpha Phenmedipham 320 SC	363
Alpha Tebuconazole 20 EW	419
Alto Elite	78
Alto Xtra	132
Amber	Adj
Amega Duo	276
Amethyst	248
Amicron	18
Amistar	18
Amistar Opti	19
Amistar Pro	22
Amistar Top	21
Anchor	58
Angri	447
Anode	383
Antec Durakil 1.5 SC	4
Antec Durakil 6SC	4
Aoraki	370
Aphox	371
Apollo 50 SC	101
Appeal	331
Apron XL	330
Aquarius	357
Aquarius	442
Aramo	425
Arena	371
Aria	344
Aristocrat	346
Arma	Adj
Artist	249
Aspect	22
Aspire	36
Asset	343
Asset Express	62
Asteroid	276
Astral	126
Astute	377
Asu-Flex	Adj
Asulox	17
Atlantis WG	292
Atom	180
Attenzo	341
Attract	331
Attribut	385
Aubrac	18
Aurora 50 WG	59
Austral Plus	246
Auxiliary	100
Avadex Excel 15G	438
Avail XT	343
Aviator 235 Xpro	39
Avro SX	344
Axial	370
Azzox	18
Bacara	177
BackRow	Adj
Ballad	95
Ballista	Adj
Bandu	152
Banka	Adj
Banner Maxx	383
Bantry	239
Barclay Barbarian	276
Barclay D-Quat	190
Barclay Gallup 360	276
Barclay Gallup Amenity	276
Barclay Gallup Biograde 360	276
Barclay Gallup Biograde Amenity	276
Barclay Gallup Hi-Aktiv	276
Barclay Holdup	68
Barclay Holster XL	142
Barclay Hudson	261
Barclay Hurler	261
Barclay Keeper 500 SC	206
Barleyquat B	68
Barramundi	Adj
Barrier H	97
Barton WG	236
Basagran SG	28
Basamid	148
Basilex	436
Becki	73
Belcocel	68

REFERENCES ARE TO ENTRY NUMBERS NOT PAGES

Index

Belfry 18	Bullion 257	Chiltern Hundreds 331
Bellis 44	Bullseye 196	Chimera SX 344
Bellmac Plus 319	Bumper 250 EC 383	Chinook Blue 34
Bellmac Straight 321	Bumper Excell 375	Chinook Colourless 34
Benta 480 SL 28	Bumper P 375	Chipco Green 295
Bentazone 480 28	Busa 77	Chlobber 95
Berelex 274	Butisan S 334	Chlortoluron 500 90
Beret Gold 242	Butoxone 321	Chord 43
Beret Multi 243	Butryflow 46	Chryzoplus Grey 0.8% . . . 289
Besiege 126	Buzin WG 342	Chryzopon Rose 0.1% . . . 289
Betamax 363	Buzz 334	Chryzosan White 0.6% . . . 289
Betanal Expert 156	Byo-Flex . . . . Adj	Chryzotek Beige 289
Betanal Maxxim 157	Cabadex 237	Chryzotop Green 289
Betanal MaxxPro 154	Cadou Star 248	Churchill 181
Betanal Quattro 155	Caesar 405	Chute 90
Betasana SC 363	Cajole Ultra 205	Cimarron 343
Betasana Trio SC 156	Calaris 329	Cinder 357
Bettaquat B 68	Calibre SX 433	Cirrus CS 102
Better DF 64	Callisto 328	Clayton Abba 1
Better Flowable 64	Calypso 430	Clayton Belfry 18
Bettix 70 WG 332	Canopy 325	Clayton Belstone 36
Bettix Flo 332	Capalo 197	Clayton Bestow 421
Binder . . . . Adj	Capitan 25 265	Clayton Bramble 405
Bio Syl . . . . Adj	Capture 47	Clayton Buzz 334
Bioduo . . . . Adj	Carakol 3 331	Clayton Cajole 205
Biofilm . . . . Adj	Carakol Plus 331	Clayton Coldstream 68
BioPower . . . . Adj	Caramba 90 336	Clayton Delilah 349
BiPlay SX 345	carbon dioxide 56	Clayton Delilah XL 349
Biscaya 430	Careca 386	Clayton Diquat 190
Bishop 349	Carfen 59	Clayton Diquat 200 190
BL 500 92	Carna 68	Clayton Edge 255
Blaster 113	Casino 261	Clayton Erase 314
Blazer M 357	Catalyst 206	Clayton Faize 329
Blizzard 240	Caynil 214	Clayton Gantry 199
Blue Rat Bait 45	CCC 720 68	Clayton Groove 304
B-Nine SG 147	C-Cure . . . . Adj	Clayton Impress 390
Bond . . . . Adj	CDA Vanquish Biactive . . . 276	Clayton IQ 190
Bonser . . . . Adj	Ceando 202	Clayton Krypton 126
Bontima 137	Celest Extra 171	Clayton Krypton MZ 126
Bonzi 352	Cello 392	Clayton Lanark 303
Boogie 40	Celmitron 70% WDG 332	Clayton Maximus 239
Boomerang 335	Centaur 129	Clayton Mepiquat 74
Borneo 212	Centium 360 CS 102	Clayton Metazachlor
Bosco WG 42	Centric 431	50 SC 334
Bounty 383	Cercobin WG 434	Clayton Mitrex 332
Boxer 236	Ceres Platinum . . . . Adj	Clayton Myth 349
Brack-N 17	Cerone 73	Clayton Myth XL 349
Bravo 500 77	Cerround . . . . Adj	Clayton Orleans 382
Break-Thru S 240 . . . . . . Adj	Certis Metaldehyde 3 331	Clayton Oust 194
Brigade 80 SC 37	Certis Red 3 331	Clayton Oust SC 194
Broad-Flex . . . . Adj	Champion 173	Clayton Oxen FL 350
Broadsword 143	Chani 419	Clayton Pirimicarb 50 371
Broadway Star 238	Charger 276	Clayton Pontoon 480EC . . . 95
Broadway Sunrise 359	Charger B 276	Clayton Rally 425
Brogue 190	Charger C 276	Clayton Scabius 193
Bromag 45	Charisma 214	Clayton Smelter 408
Bromag Fresh Bait 45	Cheetah Super 220	Clayton Solstice 240
Brutus 201	Chekker 11	Clayton Sparta 303
Buckler 91	ChemSource Diquat 190	Clayton Spigot 224
Budburst 193	Cherokee 79	Clayton Standup 68
Buggy XTG 276	Chess WG 396	Clayton Tebucon 419
Bulldog 382	Chikara Weed Control . . . 234	Clayton Tebucon EW 419
Bullet XL 360	Chikita 234	Clayton Tine 330

REFERENCES ARE TO ENTRY NUMBERS NOT PAGES

Index

Clayton Tonto 370	Cuprokylt 118	Dioweed 50 139
Clayton Truss 444	Cuprokylt FL 118	Dipel DF 25
Clayton Tunik 336	Curator 19	Diqua 190
Cleancrop GYR 2 382	Curb Crop Spray Powder . . 5	Diquash 190
Cleancrop Hyde 3 331	Curfew 128	Discman Biograde 276
Cleanrun Pro 320	Curzate M WG 126	Discovery 333
Clenecrop Super 322	Cutaway 444	Dithane 945 313
Clinic Ace 276	Cyclade 70	Dithane NT Dry Flowable . . 313
Clipper 276	Cyd-X 121	Dithianon Flowable 191
Clodinafop 240 98	Cyflamid 122	Dithianon WG 191
Cloister 202	Cymbal 123	Dockmaster 162
Clomaz 36 CS 102	Cypermethrin Lacquer 128	Dodifun 400 SC 193
Cloraz 400 374	Cyren 95	Doff Horticultural Slug Killer
Cobra 337	Daconil Weatherstik 77	Blue Mini Pellets 331
Codacide Oil . . . . Adj	Dairy Fly Spray 398	Doonbeg 2 292
Cogito 384	Danadim Progress 185	Dovekie 263
Cohort 386	Dancer Flow 363	Dovetail 304
Coldstream 68	Dartagnan WG 191	Dow Agrosciences
Colstar 225	Datura 308	Glyphosate 360 276
Comet 200 397	Dazide Enhance 147	Dow Shield 106
Compass 296	DB Straight 149	Dow Shield 400 106
Compitox Plus 322	Deacon 419	Doxstar 263
Compliment . . . . Adj	Debut 443	Dragoon 190
Concept 102	Decabane 9	Dragoon Gold 190
Concert SX 344	Decis 152	Drat Rat Bait 75
Condor 3 331	Decis Protech 152	Draza Forte 337
Conform 386	Decoy Wetex 337	Drill . . . . Adj
Conga 180	Dedicate 421	Drum 123
Conjoin . . . . Adj	Defender 442	Dual Gold 408
Consento 217	Defiant SC 332	Dunkirke 187
Conserve 410	Defy 388	Duo 400 SC 209
Consul 267	Degesch Fumigation	Duplosan KV 322
Contact Plus . . . . Adj	Tablets 6	Durakil 1.5SC 4
Contans WG 117	Degesch Plates 309	Durakil 6SC 4
Contest 4	Delicia Slug-lentils 331	Duro 331
Contrast 54	Delilah 349	Dursban WG 95
Convey 330EC 357	Delta-M 2.5 EC 152	Dynamec 1
Coragen 63	Depitox 139	EA Cloporam 112
Corbel 224	Designer . . . . Adj	EA Difcon 169
Corinth 393	Desikote Max . . . . Adj	EA Epoxi 194
Cortez 194	Desire 331	EA Tebuzole 419
Corzal 363	Dessicash 200 190	Eagle 10
County Mark . . . . Adj	Destiny 329	Echo 102
Couraze 287	Deter 114	Eclipse 195
Covershield 200	Deuce 43	Eco-flex . . . . Adj
Crassus 346	Devrinol 348	Edge 255
Crater 173	Diagor . . . . Adj	Electis 75WG 315
Crawler 55	Diamant 198	Elk 184
Crebol 224	Dian 388	Elliott's Lawn Sand 231
Credit DST 276	Dicurane 90	Elliott's Mosskiller 231
Credo 85	Dicurane 70SC 90	Endorats 75
Crescent 261	Dicurane 90WDG 90	Endorats Premium Rat
Cropspray 11-E . . . . Adj	Dicurane Surpass 91	Killer 45
Croptex Spraying Oil 362	Dicurane XL 91	Engage 386
Crossfire 480 95	Di-Farmon R 162	Ennobe 203
Crown 194	Difcor 250 EC 169	Ensign 226
Cruiser OSR 245	Difenag 168	Envidor 411
Cruiser SB 431	Difenikan 500 173	Envision 276
Crystal 250	Diflanil 500 SC 173	Envoy 204
CS Azoxy 18	Diflufenican GL 500 173	Enzo 331
CS Chlormequat 68	Digital 251	Epic 194
CTL 500 77	Dimethoate 40 185	Epok 241
Cultar 352	Dimilin Flo 172	Epona 260

REFERENCES ARE TO ENTRY NUMBERS NOT PAGES

Index

Epoxi 194	Force ST 424	Goldbeet 332
Epoxi 125 194	Forefront 13	Goldcob 328
Epoxi 125 SC 194	Forester 128	Goldron-M 343
Equity 95	Forge 343	Goltix 90 332
Eradicate 421	Formule 1 17	Goltix Flowable 332
Eradicoat 312	Fortress 402	Goltix Super 207
Eros 223	Fosal 268	Goltix Uno 207
Escar-go 3 331	Foundation 162	Goltix WG 332
Escolta 134	Fox 36	Googly 190
ESP 331	Foxtrot EW 220	Govern 95
Esteem 107	Frigate . . . . Adj	Grainstore 152
Esterol . . . . Adj	Frond 17	Grasp 400 SC 437
Ethefon 480 73	Frontrunner 349	Grazon 90 113
Ethosat 500 206	Frupica SC 323	Green Gold . . . . Adj
Ethylene 211	Fubol Gold WG 314	Green Turf King 295
Etna 276	Fuego 50 334	Greenaway Gly-490 276
Euroagkem Pace . . . . Adj	Fullstop 268	Greencrop Astra . . . . Adj
Euroagkem Pen-e-trate . . . Adj	Fulmar 42	Greencrop Bantry 239
Excelsior 133	Fumesate 500 SC 206	Greencrop Budburst 193
Exemptor 430	Fungazil 100 SL 281	Greencrop Cajole Ultra . . . 205
Falcon 382	Fungazil 50LS 281	Greencrop Carna 68
Fandango 255	Furlong 131	Greencrop Dolmen . . . . . . Adj
Fargro Chlormequat 68	Fury 10 EW 447	Greencrop Doonbeg 2 . . . 292
Fastnet 357	Fusilade Max 239	Greencrop Fenit . . . . Adj
Fazor 310	Gadwall 102	Greencrop Folklore 349
Feed and Weed 140	Gala 261	Greencrop Frond 17
Felix . . . . Adj	Galaxy 109	Greencrop Goldcob 328
Fenlander 2 209	Galera 112	Greencrop Helvick 370
Fernpath Lenzo Flo 306	Galileo 369	Greencrop Majestic 342
Ferramol Max 230	Galivor 358	Greencrop Malin 418
Ferromex Mosskiller	Gallup 360 276	Greencrop Monogram . . . 334
Concentrate 231	Gallup Amenity 276	Greencrop Orchid 2 77
Fezan 419	Gallup Biograde Amenity . . 276	Greencrop Orchid B 77
Fiddle 102	Gallup Hi-Aktiv 276	Greencrop Reaper 261
Field Marshal 161	Galmano 259	Greencrop Rookie . . . . Adj
Fiesta T 67	Galmano Plus 260	Greencrop Satchmo 382
Filan 42	Gamit 36 CS 102	Greencrop Solanum 240
Filex 381	Gantry 199	Greencrop Tabloid 419
Finale 275	Gateway . . . . Adj	Greencrop Triathlon 161
Finish SX 344	Gazelle SG 2	Greencrop Twinstar 375
Finy 343	Gemstone 204	Greencrop Tycoon 69
Firebird 174	Genie 25 265	Greenmaster Autumn 231
Firebrand . . . . Adj	GEX 353 262	Greenmaster Extra 320
Firecrest 343	GF 184 237	Greenmaster Mosskiller . . . 231
Firefly 155 255	GIBB 3 274	Greenor 110
Flagon 400 EC 46	GIBB Plus 274	Greentec Mosskiller 231
Flamenco 259	Gistar 274	Greentec Mosskiller Pro . . 231
Flanker 369	Gizmo 419	Grey Squirrel Bait 446
Flash 381	Gladiator . . . . Adj	Gringo 336
Flexidor 125 299	Globe 126	Grip . . . . Adj
Flexity 341	Glopyr 200 SL 106	Groove 304
Flomide 386	Gloster 91	Gro-Stop 100 92
Floozee 240	Gly-Flex . . . . Adj	Gro-Stop Fog 92
Floramite 240 SC 35	Glyfo TDI 276	Gro-Stop Solid 92
Flurostar 200 261	Glyfos 276	Grounded . . . . Adj
Fluroxy 200 261	Glyfos Dakar 276	Guide . . . . Adj
Fluxyr 200 EC 261	Glyfos Dakar Pro 276	Guillotine 251
Flycatcher 372	Glyfos Proactive 276	Guru 83
Flyer 397	Glyfos Supreme 276	Gusto 331
Foil 260	Glyfosat 36 276	Gusto 3 331
Folicur 419	Glyphogan 276	Hallmark with Zeon
Folio Gold 84	Glypho-Rapid 450 276	Technology 303
Folklore 349	Gly-Plus A . . . . Adj	Harmony M SX 344

REFERENCES ARE TO ENTRY NUMBERS NOT PAGES

Index

Harness 18	Icon 419	Junction 145
Harpoon 123	IMA 100 SL 281	Justice 387
Harvesan 54	Imaz 100 SL 281	Juventus 336
Harvest 275	Imaz 200 EC 281	K & S Chlorofume 76
Hatchet 261	Imidasect 5 GR 287	K2 68
Hatchet Xtra 261	Imidasect 5GR 287	Kalypstowe 328
Hatra 292	Impact Excel 82	Kantor . . . . *Adj*
Headland Charge 322	Impress 390	Karamate Dry Flo Neotec . . 313
Headland Fortune . . . . *Adj*	Indar 5 EW 218	Karan 337
Headland Guard Pro . . . . . . *Adj*	Infinito 253	Karate 2.5WG 303
Headland Inflo XL . . . . *Adj*	Injun 43	Kaspar 275
Headland Inorganic Liquid	Inka SX 433	Katalyst . . . . *Adj*
Copper 118	Insegar WG 221	Katamaran Turbo 184
Headland Intake . . . . *Adj*	Instinct 222	Kayak 135
Headland Link 163	Instrata 80	Keeper 500 206
Headland Polo 146	Insyst 2	Kerb 50 W 386
Headland Relay P 161	Intercept 5GR 287	Kerb Flo 386
Headland Relay Turf 161	Intercept 70WG 287	Kerb Granules 386
Headland Rhino . . . . *Adj*	Intracrop BLA . . . . *Adj*	Kernel 276
Headland Saxon 162	Intracrop Boost . . . . *Adj*	Kestrel 393
Headland Spear 318	Intracrop Cogent . . . . *Adj*	Kibosh 275
Headland Spruce 149	Intracrop Dictate . . . . *Adj*	Killgerm Py-Kill W 428
Headland Staff 500 139	Intracrop Era . . . . *Adj*	Killgerm ULV 400 398
Headland Sulphur 417	Intracrop Green Oil . . . . . . *Adj*	Killgerm ULV 500 365
Headland Transfer 161	Intracrop Inca . . . . *Adj*	Kindred 326
Headland Trinity 161	Intracrop Incite . . . . *Adj*	Kinetic . . . . *Adj*
Headlite 59	Intracrop Mica . . . . *Adj*	Kinto 379
Headway 23	Intracrop Micotex . . . . *Adj*	Klartan 418
Hekla 91	Intracrop Perm-E8 . . . . *Adj*	Klever 261
Helimax S 331	Intracrop Predict . . . . *Adj*	Klipper . . . . *Adj*
Helix 391	Intracrop Protocol . . . . *Adj*	Knoxdoon 190
Helvick 370	Intracrop Questor . . . . *Adj*	K-Obiol EC 25 152
Herbasan Flow 363	Intracrop Rapide . . . . *Adj*	K-Obiol ULV 6 152
Herboxone 139	Intracrop Rapide Beta . . . . *Adj*	Kresoxy 50 WG 302
Heritage 18	Intracrop Retainer . . . . *Adj*	Krypton 126
Heritage Maxx 18	Intracrop Retainer NF . . . . . *Adj*	Kubist Flo 206
High Load Mircam 162	Intracrop Rigger . . . . *Adj*	Kula 91
Hiker 237	Intracrop Salute . . . . *Adj*	Kumulus DF 417
Hilite 276	Intracrop Sapper . . . . *Adj*	Kurdi 255
Hithane 346	Intracrop Saturn . . . . *Adj*	Laminator FL 313
Hive 68	Intracrop Signal XL . . . . . . *Adj*	Lanark 303
Holdup 68	Intracrop Sprinter . . . . *Adj*	Lancer 258
Holster XL 142	Intracrop Status . . . . *Adj*	Landmark 199
Horus 292	Intracrop Stay-Put . . . . *Adj*	Landscaper Pro Moss Control +
Hounddog 205	Intracrop Super Rapeze	Fertiliser 231
Howitzer 239	MSO . . . . *Adj*	Larke 318
Hudson 261	Intracrop Tonto . . . . *Adj*	Laser 120
Hunter 237	Intracrop Warrior . . . . *Adj*	Latitude 407
Hurler 261	Intracrop Zenith . . . . *Adj*	Lawn Sand 231
Huron 337	Intrepid 2 160	Leaf-Koat . . . . *Adj*
Hurricane SC 173	Invader 187	Legara 112
Hussar 291	Isomec 322	Legolas 141
Hyban-P 162	Jackpot 2 303	Lenazar Flo 306
Hycamba Plus 161	Jaunt 256	Lentagran WP 399
HY-D Super 139	Jenga 214	Lenzo Flo 306
Hygrass-P 162	Jenton 227	Leopard 5 EC 403
Hykeep 429	Jester 49	Level . . . . *Adj*
HY-MCPA 318	Jewel 61	Lexi 341
Hymec Triple 167	Jockey 260	Lexus Class 60
Hyprone-P 161	Jockey Solo 259	Lexus Millenium 258
Hysward-P 161	Jouster 348	Lexus SX 257
Icarus 201	Jubilee SX 343	Li-700 . . . . *Adj*
Ice 250	Juliet 77	Liberate 370

REFERENCES ARE TO ENTRY NUMBERS NOT PAGES

Index

Name	Ref
Liberator	174
Linford	119
Lingo	103
Link	163
Linurex 50 SC	308
Linzone	103
Liquid Curb Crop Spray	5
Logic	Adj
Logic Oil	Adj
Longhorn	382
Lontrel 200	106
Lorate	343
Low Down	Adj
Luas	308
Lunar	377
Lupo	146
Lynx H	331
Lyric	265
Mac-Bentazone 480 SL	28
Maccani	192
MAC-Napropamide 450 SC	348
MAC-Prochloraz Plus 490 EC	375
MAC-Propaquizafop 100 EC	382
Maestro	255
Magician	86
Magnate 100 SL	281
Magneto	140
Magnum	65
Mainman	235
Majestic	342
Majestik	312
Major	303
Major	331
Malin	418
Mandolin Flow	363
Mandops chlormequant	68
Mangard	Adj
Manifest	276
Manipulator	68
Mantra	196
Markate 50	303
Marksman	334
Marquise	332
Martinet	442
Marvel	328
Masai	422
Masalon	346
Mascot Contact	77
Mascot Crossbar	142
Mascot Defender	442
Mascot Eland	397
Mascot Fusion	421
Mascot Micronised Lawn Sand	231
Mascot Rayzor	295
Mascot Systemic	53
Mashona	334
Mastana	393
Matilda	126
Mavrik	418
Maxim XL	244
Maximus	239
MCPA 25%	318
MCPA 50	318
Meco-Flex	Adj
Medax Top	325
Mediator Sun	Adj
Medley	214
Megaflo	386
Menace 80 EDF	386
Menara	132
Menno Florades	31
Meridian	407
Merit Turf	287
Metadisque	331
Metam 510	333
Metarex Amba	331
Metaz 500 SC	334
Meteor	71
Metribuzin WG	342
Metric	105
Metso	343
Metso XT	343
Mewati	313
Mextrol DP	165
Mextrol-Biox	48
Mexxon	343
Mexxon xtra	343
Mezzanine	334
Micene 80	313
Micene DF	313
Microthiol Special	417
Micro-Turf	442
Midas	81
Midget	325
Mileway	13
Minstrel	261
Minuet EW	447
Mirage 40 EC	374
Mircam	162
Mircam Plus	161
Mirquat	68
Mission	190
Mitre	419
Mitrex	332
Mitron 70 WG	332
Mitron 90 WG	332
Mitron SC	332
Mixture B NF	Adj
MM 70 Flo	332
Mobius	395
Mocap 10G	210
Moddus	444
Modesto	33
Molotov	331
Monceren DS	356
Monceren IM	282
Monitor	416
Monkey	377
Monogram	334
Monza	331
Mortice	304
Movento	413
Movon	175
MSS CIPC 50 M	92
MSS Sprout Nip	92
Muntjac	183
Mutiny	235
Mycotal	305
Name TBC	41
Nando 500 SC	240
Nappa	348
Naspar TDI	335
Nativo 75 WG	421
Natural Weed & Moss Spray No 1	3
Naturalis-L	26
Nautile DG	126
Nautilus	292
Navona	331
Nemathorin 10G	270
Nemo	292
Nemolt	423
Neosorexa Bait Blocks	168
Neosorexa Gold	168
Neosorexa Gold Ratpacks	168
Neosorexa Pasta Bait	168
Nettle	Adj
New 5C Cycocel	68
New 5C Quintacel	68
New Estermone	140
Newgold	94
Newman's T-80	Adj
Nico 4	349
Nighthawk 330 EC	357
Nightjar	308
Nimbus CS	104
Nimrod	51
Nion	Adj
Nirvana	285
Nomix Conqueror	276
Nomix Dual	277
Nortron Flo	206
Novagib	274
Novall	335
Nu Film P	Adj
Nufarm Cropoil	Adj
Nufarm MCPA 750	318
Nufosate Ace	276
Nugget	112
Nuprid 600FS	287
Nustar 25	265
Oberon	412
Oblix 500	206
Oblix MT	207
Octave	374
Octolan	78
Ohayo	240
Oklar SX	257
Olympus	19
Omaha	181
Omarra	425
Opera	204
Opponent	200
Optica	322
Option	123

REFERENCES ARE TO ENTRY NUMBERS NOT PAGES

Index

Opus 194	Penncozeb WDG 313	Proplant 381
Opus Team 195	Pentangle 89	Propyzamide 400 386
Oracle 194	Permasect C 128	Prosaro 393
Oram 36	Pesta H 331	Proshield 178
Orca 343	Phantom 371	Prospa 82
Orchid 2 77	Pharaoh 14	Prospero 374
Orchid B 77	Phase II . . . . Adj	Proteb 393
Oreto 331	Phemo 208	Prothio 390
Orian 419	Phostoxin 6	Proxanil 127
Oriel 50SX 257	Piccant 112	Pryz-Flex . . . . Adj
Orient 358	PicoMax 358	Punch 25 265
Orius 20 EW 419	Picona 358	Punch C 54
Orius P 377	PicoPro 358	Pure Glyphosate 360 276
Orka 228	PicoStomp 358	Pure Max 444
Oropa 194	Pierce 161	Pure Turf 295
Oryx 335	Pik-Flex . . . . Adj	Pyblast 398
Osarex W 331	Pilot Ultra 403	Pygmy 444
Oskar 220	Pinnacle 432	Py-Kill W 428
Othello 179	Pirimate 371	Pyramin DF 64
Oust 194	Pirimicarb 50 371	Pyrethrum 5 EC 398
Output . . . . Adj	Pirouette 352	Pyrethrum Spray 398
Outrun 161	Pistol 178	Pyrinex 48 EC 95
Overlord 173	Pixie 180	Quad-Glob 200 SL 190
Overture 194	Pizamide 386	Quail 370
Oxen FL 350	Pizza 400 SC 386	Quaver Flo 386
Oxytril CM 48	Planet . . . . Adj	Quell Flo 313
Pacifica 292	Plant Invigorator 366	Quintacel 68
Paclo 352	Platform S 61	Radius 130
Pan Aquarius 442	Platoon 250 397	Radspor FL 193
Pan Ethephon 73	Plenum WG 396	Rally 425
Pan Glory 61	Plover 169	Rancona 15 ME 294
Pan Magician 86	Pluton 225	Ranger . . . . Adj
Pan Metconazole 336	PMT 372	Ranman Twinpack 119
Pan Oasis . . . . Adj	Pointer 267	Rapsan 500 SC 334
Pan Panorama . . . . Adj	Polo 146	Raptor 382
Pan PCH 381	Polymet 5 331	Ratio SX 433
Pan Pencycuron P 356	Poncho 114	Raven 224
Pan PMT 372	Poncho Beta 33	Ravine 278
Pan Proteb 377	Ponder 386	Ravine Plus 279
Pan Stiffen 73	Poraz 374	Raxil Deter 116
Pan Tepee 444	potassium bicarbonate 373	Raxil Pro 394
Pan Theta 329	Powertwin 209	Raza 419
Panache 40 374	PQF 100 382	Re-act 161
Panama 18	Prairie 87	Reaper 261
Panarex 404	Praxys 109	Reaper 276
Panoctine 278	Precis 386	Redigo 390
Panoctine Plus 279	Predator WG 332	Redigo Deter 115
Pantheon 2 367	Prego 331	Redigo Twin TXC 255
Panto 295	Prelude 20LF 374	Redlegor 151
Paradise 234	Presite SX 344	Redstar 336
Parador 64	Previcur Energy 269	Regalis 380
paraffin oil 353	Prima . . . . Adj	Regatta 174
Paramount . . . . Adj	Primo Maxx 444	Regel 331
Parapet 95	Priori Xtra 20	Reglone 190
Pastel M 331	Profit Oil . . . . Adj	Regulex 10 SG 274
Pastor 111	Proliance Quattro 276	Relay P 161
Pasturol Plus 161	Proline 275 390	Relay Turf 161
PCH 381	Pro-Long 92	Reldan 22 96
PDM 330 EC 357	Prompt 162	Relva 386
Penco 355	Prompto 217	Renegade 112
Pency Flowable 356	Propel Flo 386	Renovator 2 159
Pennant 344	Proper Flo 386	Renovator Pro 232
Penncozeb 80 WP 313	Propi 25 EC 383	Rescue 370

REFERENCES ARE TO ENTRY NUMBERS NOT PAGES

Index

Entry	Ref
Resplend	8
Retro	190
Revolt	418
Revus	316
Reward Oil	Adj
Reynard	371
Rezacur	77
Rezist	5
Rhapsody	126
Rhino	266
Rhino DS	266
Rhizopon A Powder	288
Rhizopon A Tablets	288
Rhizopon AA Powder (0.5%)	289
Rhizopon AA Powder (1%)	289
Rhizopon AA Powder (2%)	289
Rhizopon AA Tablets	289
Rhizopon B Powder (0.1%)	347
Rhizopon B Powder (0.2%)	347
Rhizopon B Tablets	347
Rigid	405
Ringer	53
Riot	55
Rivet	337
Riza	419
Rizolex	436
Rizolex Flowable	436
Rock	250
Roller	Adj
Ronstar 2G	350
Ronstar Liquid	350
Roquat 20	190
Rotor	90
Roundup Ace	276
Roundup Biactive	276
Roundup Energy	276
Roundup Klik	276
Roundup Max	276
Roundup Metro	276
Roundup Pro Biactive	276
Roundup ProBiactive 450	276
Roundup Ultimate	276
Rovral AquaFlo	295
Rovral WG	295
Roxam 75WG	315
Roxx L	350
Rubric	194
Rudis	390
Rumo	290
Runner	338
Rustler	276
Ryda	Adj
Sabine	36
Safari Lite WSB	307
Sage	415
Sakarat Bromabait	45
Sakarat D (Whole Wheat)	168
Sakarat D Wax Bait	168
Sakarat Ready-to-Use (Cut Wheat Base)	446
Samson Extra 6%	349
Samurai	276
SAN 703	78
Sanction 25	265
Sanspor	77
Sarabi	342
Saracen	Adj
SAS 90	Adj
Satchmo	382
Savona	215
Saxon	162
SB Plant Invigorator	366
Scabius	193
Scala	400
Scarlett	374
Scorpio	442
Scotts Feed and Weed	140
Scotts Octave	374
Scout	Adj
Scout	81
Scrum	55
Seal Z	303
Sekator	11
Selon	68
Sempra	173
Sencorex WG	342
Sequel	229
Serenade ASO	24
Serial Duo	199
Setanta 50 WP	386
Setanta Flo	386
Sewarin Extra	446
Sewarin P	446
Shadow	184
Shark	59
Sherman	357
Shield	106
Shinkon	15
Shirlan	240
SHL Granular Feed and Weed	166
SHL Granular Feed, Weed & Mosskiller	164
SHL Lawn Sand	231
SHL Turf Feed and Weed	166
SHL Turf Feed, Weed & Mosskiller	164
Shogun	382
Shooter	250
Shotput	342
Shrapnel	363
Shrink	444
Shyfo	276
Sienna	358
Sigma PCT	68
Signum	44
Silico-Sec	366
Siltex	Adj
Siltra Xpro	39
Silvacur	420
Silwet L-77	Adj
Sipcam C 50	123
Sipcam C50 WG	123
Sistan 51	333
Skater	332
Skirmish	300
Skora	261
Skyway 285 Xpro	41
SL 567A	330
Slalom	237
Slingshot	276
Slippa	Adj
Slugdown 3	331
Sluggo	230
Sluxx	230
SM-99	Adj
SmartFresh	340
sodium hypochlorite	409
Solanum	240
Solar	Adj
Solfa WG	417
Solitaire	386
Solitaire 50 WP	386
Solo D	173
Sonar	77
Sorexa D	168
Sorexa Gel	168
Sparta	303
Spartan	Adj
Spear	318
Spearhead	108
Sphere	134
Sphere ASBO	5
Spigot	224
Spiral	391
Spotlight Plus	59
Sprayfast	Adj
Spray-fix	Adj
Spraygard	Adj
Spraying Oil	362
Spraymac	Adj
Springbok	183
Sprint	374
Spruzit	398
Spyrale	170
Squire Ultra	10
Stabilan 700	68
Stacato	276
Staff 500	139
Stamen	276
Stamina	Adj
Stamp 330 EC	357
Standon Azoxystrobin	18
Standon Epoxiconazole	194
Standon Fenpropimorph 750	224
Standon Frontrunner	349
Standon Fullstop	268
Standon Girder 720	68
Standon Googly	190
Standon Homerun 2	261
Standon Hounddog	205
Standon Imazaquin 5C	71

Index

Standon Kresoxim-Epoxiconazole 199	Swallow 119	Tizca 240
Standon Mastana 393	Sward . . . . *Adj*	TM 1088 . . . . *Adj*
Standon Mepiquat Plus . . . 74	Swift SC 442	Toil . . . . *Adj*
Standon Metazachlor 500 . . 334	Swiftsure 142	Tolugan 700 90
Standon Metso XT 343	Swing Gold 189	Tolurex 90 WDG 90
Standon Mexxon xtra 343	Switch 136	Tomahawk 261
Standon Midget 325	Sword 98	Tomcat 2 45
Standon Pencycuron DP . . . 356	Symbol 276	Tomcat 2 Blox 45
Standon Pirimicarb 50 . . . 371	Symphony 447	Tonga 425
Standon Propaquizafop . . . 382	Synero 13	Tonik 383
Standon Pygmy 444	Syper 100 128	Topas 355
Standon Roxx L 350	System Turf 377	Topenco 100 EC 355
Standon Santa Fe 50 WP . . 386	Systhane 20EW 346	Topik 98
Standon Santa Fe Flo 386	T2 Green 161	Toppel 100 EC 128
Standon Shiva . . . . *Adj*	Tabloid 419	Topsin WG 434
Standon Tebuconazole . . . 419	Tacet 444	Torch 414
Standon Tonga 425	Tachigaren 70 WP 280	Tordon 22K 367
Standon Win-Win 390	Takron 64	Torero 207
Standon Yorker 190	Talius 387	Torpedo-II . . . . *Adj*
Standon Zero Tolerance . . . 430	Talon 403	Torrent 194
Standon Zing PQF 382	Talstar 80 Flo 37	Totril 293
Starane 2 261	Talunex 6	Touchdown Quattro 276
Starane Gold 237	Tandus 261	Tracer 410
Starane Vantage 237	Tangent 276	Tracker 43
Starane XL 237	Tanker 276	Trafalgar 237
Starion Flo 37	Tanos 124	Trample 357
Statis 276	Target SC 332	Transact . . . . *Adj*
Statis 360 276	Tarot 405	Transcend . . . . *Adj*
Steadfast 3 331	Tawny 91	Traton SX 345
Steam 331	Taylors Lawn Sand 231	Traxos 99
Steel 91	TDS Major 331	Trevi 331
Stellox 48	TDS Metarex Amba 331	Triathlon 161
Steward 290	Teamforce 209	Tribute 161
Stiffen 73	Tebcon 419	Tribute Plus 161
Stika . . . . *Adj*	Tebucon 250 419	Trilogy 156
Stiletto 197	Tebucur 250 419	Trimangol 80 317
Stirrup 276	Teldor 219	Trimangol WDG 317
Stomp 400 SC 357	Temper Turf 444	Trimanzone 313
Stomp Aqua 357	Tempest 64	Tripod 273
Storite Clear Liquid 429	Templar 50	Tripod Plus 272
Storite Excel 429	Tempo 444	Triptam 435
Strate 69	Tempt 331	Tritox 161
Strimma 437	Tepee 444	Triumph 220
Stroby WG 302	Teppeki 235	Trojan 331
Stroller 386	Tepra 425	Trooper 250
Stronghold 72	Terbine 68	Tropotox 321
Stunt 444	Tern 222	Trotter 338
Subdue 330	Terpal 74	Trounce 331
Sulphur Flowable 417	Tevday 351	Troy 480 28
Sultan 50 SC 334	Tezate 220 SL 429	Truss 444
Sumi-Alpha 205	Thianosan DG 435	Try-Flex . . . . *Adj*
Sumimax 251	Thiovit Jet 417	Tucana 397
Sunorg Pro 336	Thiraflo 435	Tuli 98
Super 3 331	Thistlex 113	Turnstone 169
Super 5 331	Thor 440	Twin 209
Super Nova . . . . *Adj*	Thrash 441	Twinstar 375
Super Selective Plus 161	Throttle 377	Twister 173
Superturf 295	Thrust 140	Tycoon 69
Supreme 77	Thyram Plus 435	Typhoon 360 276
Suscon Green Soil Insecticide 95	Timbrel 441	Ultima 311
Sven 205	Timpani 89	Unicur 255
	Tine 330	Unix 135
	Titus 405	Upgrade 69

REFERENCES ARE TO ENTRY NUMBERS NOT PAGES

Index

UPL B Zone 28	Vivendi 200 106	Whistle 43
UPL Camppex 144	Vivid 397	Woody 441
UPL Diquat 190	Vixen 368	Woomera 190
UPL Grassland Herbicide . . 161	Volcan Combi 66	X-Wet . . . . *Adj*
Upright 71	Volley 240	Zampro DM 8
Valbon 30	Vydate 10G 351	Zarado . . . . *Adj*
Valiant 287	Waila 400	Zeal . . . . *Adj*
Validate . . . . *Adj*	Wakil XL 125	Zealot 382
Vanguard 397	Walabi 88	Zephyr 395
Vareon 376	Warfarin 0.5%	Zero Tolerance 430
Velocity . . . . *Adj*	Concentrate 446	Zigzag . . . . *Adj*
Venzar Flowable 306	Warfarin Ready Mixed	Zimbrail 85
Veto F 420	Bait 446	Zing PQF 382
Video DG 126	Warrant 220	Zinzan . . . . *Adj*
Vigon 175	Warrior 303	Zip 139
Viscount 98	Weedazol-TL 16	Zone 48 28
Visor 403	Wetcit . . . . *Adj*	

REFERENCES ARE TO ENTRY NUMBERS NOT PAGES

THE UK PESTICIDE GUIDE 2011

RE-ORDERS

☐ Please send me ____ more copies of The UK Pesticide Guide 2011 at £44.50 each

Postage: single copy £3.95; orders up to £150, £5.95; orders over £150, no charge.

        Outside the UK please add £10 per order for delivery.

Name _____ Position _____

Institution _____ Department _____

Address _____

City _____ Region _____ Postcode _____ Country _____

      Tel _____ Fax _____ E-mail _____

EU countries except UK – VAT No: _____

Payment (pre-payment is required):

☐ I enclose a cheque/draft for £_____ payable to BCPC. Please send me a receipt.

☐ I wish to pay by credit card: ☐ Visa ☐ Mastercard ☐ Amex ☐ Switch

Please charge to my card £_____ and send me a receipt. Name of issuing bank _____

Card no. ☐☐☐☐ ☐☐☐☐ ☐☐☐☐ ☐☐☐☐

Expiry date ☐☐/☐☐ Security code ☐☐☐   Switch cards only: Start date ☐☐/☐☐ Issue no. ☐

Signature _____ Date _____

Name and address of cardholder if different from above: _____

BCPC

Please photocopy and return to:
BCPC Publications Sales, 7 Omni Business Centre, Omega Park, Alton, Hants GU34 2QD, UK
Tel: 01420 593 200, Fax: 01420 593 209, Email: publications@bcpc.org, Web: www.bcpc.org

FUTURE EDITIONS

☐ I wish to take out an annual order for ____ copies of each new edition of The UK Pesticide Guide

☐ Please send me advance price details for the 2012 edition of The UK Pesticide Guide when available

Order by phone: 01420 593 200 or online: www.bcpc.org/shop

Bookshop orders to:

CABI www.cabi.org
Customer Services, CAB International, Nosworthy Way,
Wallingford, Oxfordshire OX10 8DE, UK
Tel: 01491 832111, Fax: 01491 829292,
Email: enquiries@cabi.org, Web: www.cabi.org

Bulk discount:
100+ copies	30%
50–99	25%
10–49	15%
List price £44.50	

Thank you for your order